Edexcel A-Level

Pure
Mathematics

Year 2

This CGP Student Book is the definitive guide to every Pure topic from Year 2 of the Edexcel A-Level Mathematics course.

It contains clear study notes, advice, examples, hundreds of practice questions and a realistic practice exam — with fully worked answers at the back.

How to access your free Online Edition

Go to **cgpbooks.co.uk/extras** and enter this code:

3768 1173 4656 5038

Contents

Contents

This book has been produced to be a complete resource for your learning and practice. Throughout, we've focused on three core concepts of A-level maths — mathematical methods, problem solving and modelling.

Each chapter starts with a page that includes Learning Objectives and a Prior Knowledge Check.

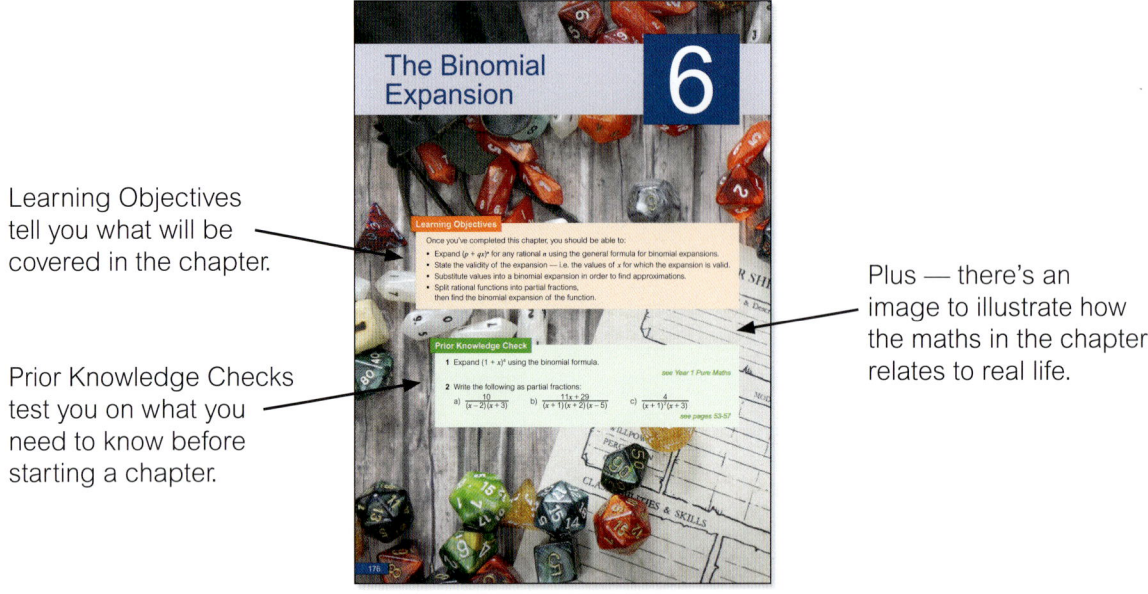

Learning Objectives tell you what will be covered in the chapter.

Prior Knowledge Checks test you on what you need to know before starting a chapter.

Plus — there's an image to illustrate how the maths in the chapter relates to real life.

The main pages have theory, examples and exercises.

Exercises provide lots of practice for every topic, with fully worked answers at the back of the book. Answers to exam-style questions come with a full mark scheme.

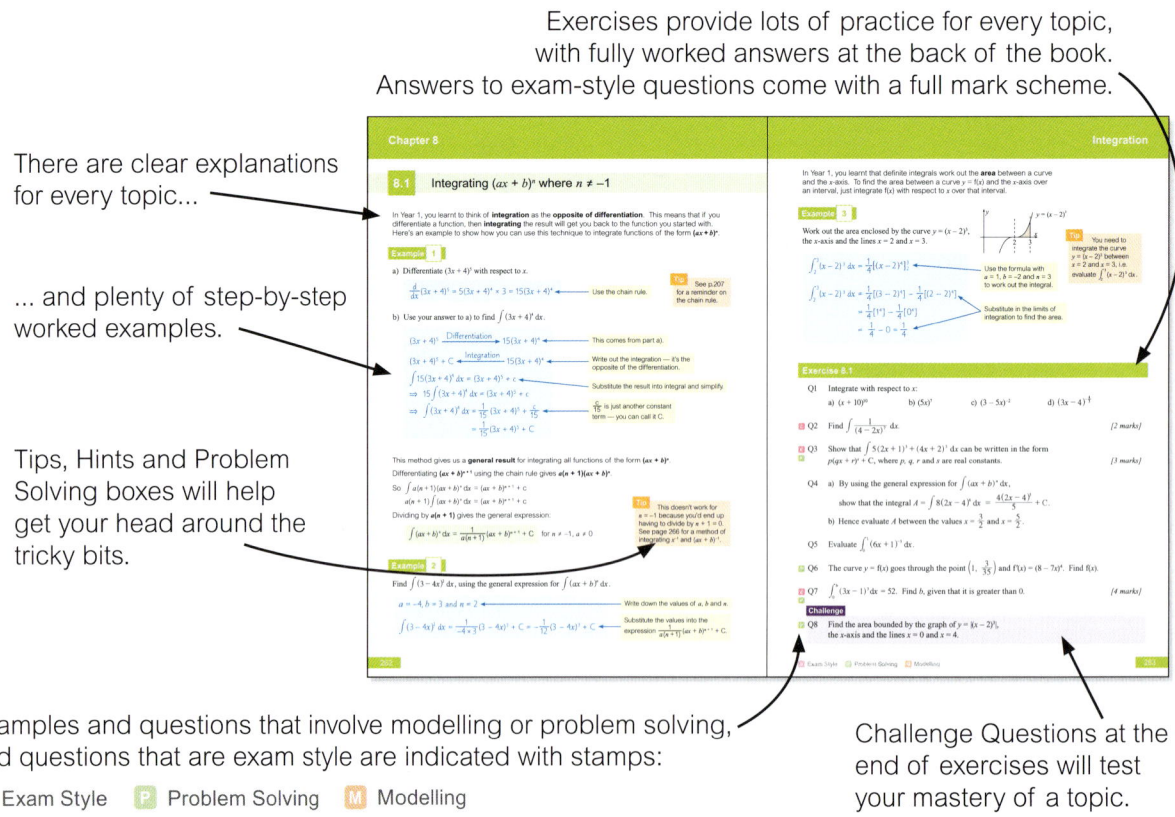

There are clear explanations for every topic...

... and plenty of step-by-step worked examples.

Tips, Hints and Problem Solving boxes will help get your head around the tricky bits.

Examples and questions that involve modelling or problem solving, and questions that are exam style are indicated with stamps:

E Exam Style   **P** Problem Solving   **M** Modelling

The solutions to Problem Solving questions may require knowledge and methods from more than one Chapter of this book.

Challenge Questions at the end of exercises will test your mastery of a topic.

There's a Review Exercise at the end of each chapter and a Practice Paper after the last chapter.

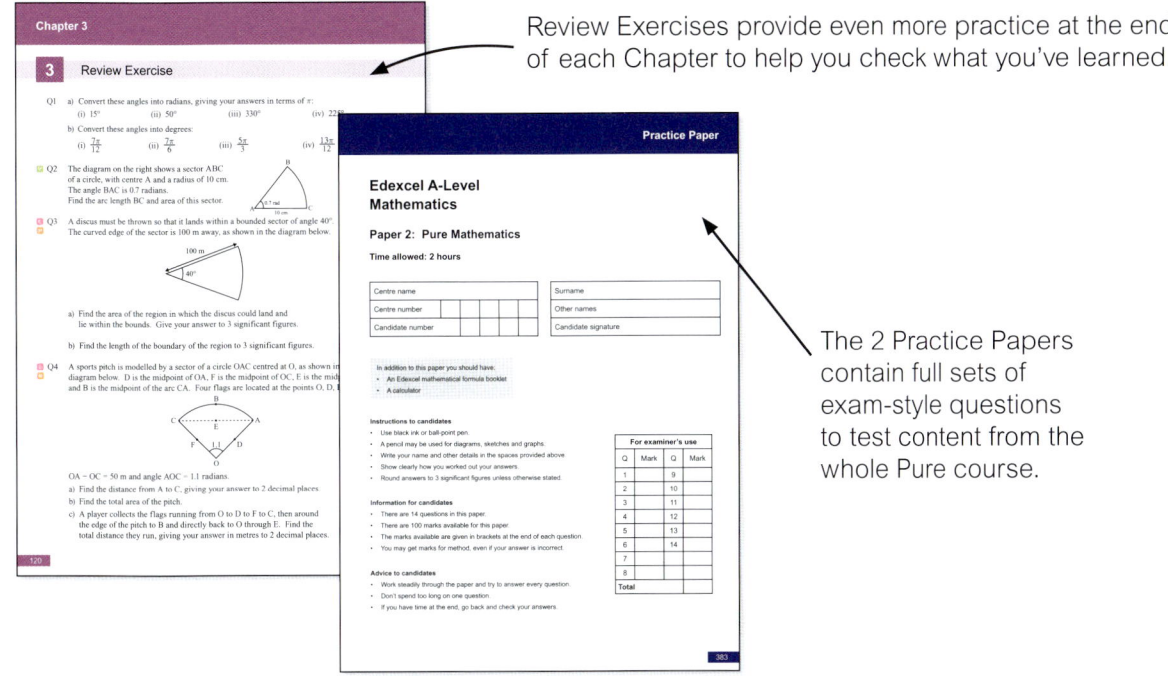

Review Exercises provide even more practice at the end of each Chapter to help you check what you've learned.

The 2 Practice Papers contain full sets of exam-style questions to test content from the whole Pure course.

You can find the Glossary and Formula Sheet at the back of the book.

The Formula Sheet has the relevant formulas you'll get in the exam.

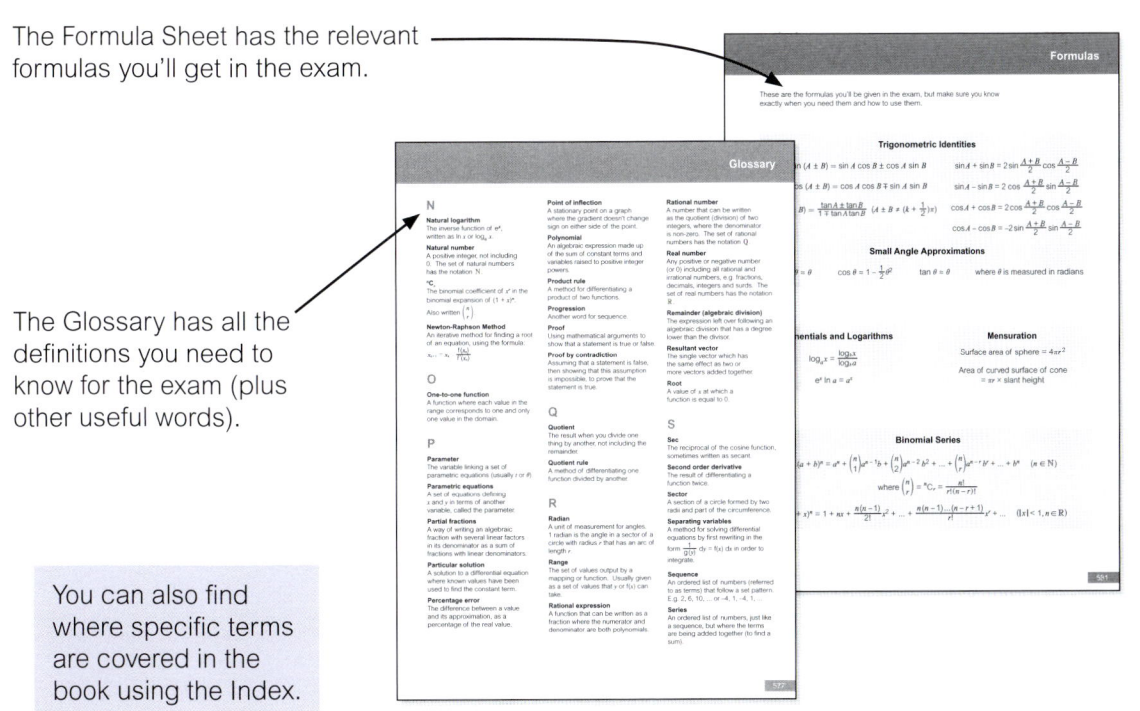

The Glossary has all the definitions you need to know for the exam (plus other useful words).

You can also find where specific terms are covered in the book using the Index.

Published by Coordination Group Publications Ltd

Broughton House, Griffin Street, Broughton-in-Furness, Cumbria, UK, LA20 6HH

www.cgpbooks.co.uk

Text, design, layout and original illustrations
© Coordination Group Publications Ltd (CGP) 2021

Design coordination and cover design by Beckie Doyle and Kirsty Goodall.

Editors:

Michael Bushell, Liam Dyer, Sammy El-Bahrawy, Sarah George, Josie Gilbert
Ruth Greenhalgh, Shaun Harrogate, Rob Hayman, Paul Jordin, Simon Little,
Samuel Mann, Sean McParland, Ali Palin, Rosa Roberts, David Ryan, Ben Train.

Contributors:

Katharine Brown, Margaret Darlington, Paul Freeman, Paul Garrett, Aleksander Goodier,
Dave Harding, Andy Pierson, Rosemary Rogers, Charlotte Young.

Proofreading:

Mona Allen, Rosie Hanson, Glenn Rogers and Charlotte Sheridan.

Photo credits:

Cover image © Tomekbudujedomek/Moment/Getty Images

Page 1 © Bet_Noire/iStock/Getty Images; Page 8 © BrianAJackson/iStock/Getty Images;
Page 66 © Theerawat Payakyut/iStock/Getty Images; Page 127 © frantic00/iStock/Getty Images;
Page 142 © Furtseff/iStock/Getty Images; Page 176 © Firn/iStock/Getty Images; Page 201 ©
ClaudioVentrella/iStock/Getty Images; Page 261 © solarseven/iStock/Getty Images;
Page 331 © joebelanger/iStock/Getty Images; Page 363 © GarryKillian/iStock/Getty Images;

Clipart from Corel®

With thanks to Lottie Edwards for the copyright research.

Printed by Elanders Ltd, Newcastle upon Tyne.

ISBN: 978 1 78908 841 0

Proof

1

Learning Objective

Once you've completed this chapter, you should be able to:

- Use proof by contradiction to show that statements are true.

Prior Knowledge Check

1 In a proof, how might you represent an unknown:
   a) rational number?   b) even number?   c) odd number?   *see Year 1 Pure Maths*

2 Prove that the sum of two consecutive numbers is odd.   *see Year 1 Pure Maths*

3 Prove that when an even number is subtracted from an odd number, the result is always odd.   *see Year 1 Pure Maths*

4 Disprove this statement: "$x^2 > x$ for all real numbers x."   *see Year 1 Pure Maths*

1.1   Types of Proof

Proof in mathematics is all about using logical arguments to show that a statement is true or false. So far, you've learned three main types of proof:

Proof by deduction: using known facts and logic to show that the statement must be true.

Proof by exhaustion: breaking a statement down into two or more cases that cover all possible situations, then showing that the statement is true for all of them.

Disproof by counter-example: giving one example that shows that the statement is not true.

The other method that you need to know is **proof by contradiction**.

1.2   Proof by Contradiction

To prove a statement by **contradiction**, you start by saying "assume the statement **is not true**...". You then show that this would mean that something **impossible** would have to be true, which means that the initial assumption has to be wrong, so the original statement must be true.

Example 1

Prove the following statement: "If x is an integer and x^2 is even, then x must be even."

You can prove this statement by contradiction.

Assume there is an odd number, x, for which x^2 is even.

Assume the statement is not true and that the opposite is actually true.

$x = 2k + 1$, where k is an integer

$x^2 = (2k + 1)^2 = 4k^2 + 4k + 1$

If x is odd, then you can write x as $2k + 1$, where k is an integer, and find x^2 in terms of k.

$4k^2 + 4k = 2(2k^2 + 2k)$ is even because it is 2 × an integer

Rewrite the expression to work out whether x^2 is odd or even.

So $4k^2 + 4k + 1 = 2(2k^2 + 2k) + 1$ must be odd.

So if x^2 is even, then x must be even — a contradiction.
Hence the original statement is true.

State that the original assumption, that there is an odd number x for which x^2 is even, is false.

Example 2

Prove that $\sqrt{2}$ is irrational.

Assume $\sqrt{2}$ is rational so can be written as $\frac{a}{b}$, where a and b are non-zero integers and $\frac{a}{b}$ is a fraction in its simplest form.

Assume that the statement is not true. You can also assume that a and b have no common factors (otherwise they'd cancel down to a different a and b).

$\sqrt{2} = \frac{a}{b} \implies \sqrt{2}b = a \implies 2b^2 = a^2$

Rearrange and square both sides.

$\implies a^2$ is even $\implies a$ is even

This was proved in example 1.

$2b^2 = (2k)^2 = 4k^2 \implies b^2 = 2k^2$

$\implies b^2$ is even $\implies b$ is even

Replace a with $2k$ for some integer k and use the result from example 1 again.

If a and b are both even then $\frac{a}{b}$ is not in its simplest form — a contradiction.

$\implies \sqrt{2} \neq \frac{a}{b}$, so it must be irrational.

You assumed at the start that a and b had no common factors, so you have contradicted your initial assumption. So $\sqrt{2}$ can't be written as $\frac{a}{b}$.

You can use the same method to prove the irrationality of any surd, although you need to prove the statement "*If x^2 is a multiple of a prime number p, then x must also be a multiple of p*", which is a bit trickier than the proof in Example 1.

This next example uses proof by contradiction to prove that there is no 'largest number' of a certain type. You assume that there is one, then you can simply add to it to get a bigger number of the same type, which contradicts your initial assumption. You can use this method in quite a few proofs.

Example 3

Prove by contradiction that there are infinitely many prime numbers.

Assume there are a finite number of primes:
$p_1 = 2, p_2 = 3, p_3 = 5, \ldots, p_{n-1}, p_n$

Assume that the statement is not true, i.e. there are a finite number of primes (say n), and list them all.

$p_1 p_2 p_3 \cdots p_{n-1} p_n$ — call this number, which is a multiple of every prime number, P.

Now multiply all of these primes together to give the number P.

$(P + 1) \div p_1 = (p_1 p_2 p_3 \cdots p_{n-1} p_n + 1) \div p_1$
$= p_2 p_3 \cdots p_{n-1} p_n$ remainder 1
$\implies (P + 1)$ is not divisible by p_1.

Think about $P + 1$ — check if it is divisible by p_1.

Similarly $(P + 1)$ isn't divisible by any of the primes in the list, so either it is prime or it is a product of some primes not in the list.

In fact, dividing $(P + 1)$ by any prime number gives a remainder of 1.

So the list doesn't contain all of the primes — a contradiction. So there must be infinitely many prime numbers.

Tip *Some proofs use $P = p_1 p_2 \cdots p_n + 1$ — but the method is the same.*

Look for a contradiction — you assumed that you had listed all of the primes.

1   Review Exercise

Q1   Tim says, "If f(x) > 0 for all real values of x,
then f'(x) > 0 for all real values of x."
Find a counter-example to show that Tim is wrong.

Q2   Sithuli says, "If a is a real number greater than 1,
then ln a is not a rational number."
Find a counter-example to show that Sithuli is wrong.

Q3   Prove, by contradiction, that:
   a) there is no largest odd integer.
   b) there is no largest multiple of 3.

Q4   Given that x is an integer, use proof by contradiction to prove the following statements.
   a) If x^2 is odd, then x must be odd.
   b) If x^3 is odd, then x must be odd.
   c) If $x^2 + 3$ is odd, then x must be even.

E **Q5**   a) Expand $\left(e^{\frac{1}{2}x} - e^{-\frac{1}{2}x}\right)^2$, simplifying your answer.   *[2 marks]*
   b) Hence or otherwise prove that $e^x + e^{-x} \geq 2$ for all real values of x.   *[2 marks]*

Q6   Prove by contradiction that there is no integer that is both even and odd.

P **Q7**   Prove that, if a and b are integers, then the equation
$3a + 5 = 12b$ has no solutions.

E **Q8**   In this question, a, b and c are integers.
P
Prove by contradiction that if b is odd, then the discriminant
of the quadratic function $ax^2 + bx + c$ cannot be zero.   *[4 marks]*

E **Q9**   Prove that, if x and y are integers, then the equation
P $2x^2 - 6xy = 12y^2 + 1$ has no solutions.   *[3 marks]*

P **Q10**   Prove that, for all real values of x, the following inequality is true.
$$x^2 + x + 4 + \cos x > 1 - x.$$

P **Q11** a) Prove that there are no solutions to the equation
$3^a + b^2 = 6$, where a and b are positive integers.

b) Hence show that the equation $a = \log_3(\sqrt{6} - b) + \log_3(\sqrt{6} + b)$
has no solutions for which a and b are positive integers.

Q12 a) Prove that the product of a non-zero rational number and
an irrational number is always irrational.

b) Disprove the statement that the product of an irrational number
and an irrational number is always irrational.

Q13 Prove that there is no smallest positive rational number.

Q14 Prove that $1 + \sqrt{2}$ is irrational.

E **Q15** Sam attempts to prove that $\sqrt{12}$ is irrational. He writes out the proof below.
P

> Step 1: Suppose there are integers m and n with $n \neq 0$ such that
> $\sqrt{12} = \dfrac{m}{n}$. We can assume that $\dfrac{m}{n}$ is in its simplest form,
> otherwise $\dfrac{m}{n}$ would cancel down to a simpler fraction.
>
> Step 2: Then $12n^2 = m^2$, so 12 is a factor of m^2.
>
> Step 3: Therefore, 12 is a factor of m.
>
> Step 4: This means there must be an integer k such that $m = 12k$,
> so $12n^2 = 144k^2$, and $n^2 = 12k^2$.
>
> Step 5: Since 12 is a factor of n^2, 12 must be a factor of n.
>
> Step 6: This contradicts the assumption that m and n have
> no common factors. Therefore, $\sqrt{12}$ is irrational.

a) At which steps did Sam go wrong? Explain your answer. *[2 marks]*

b) Prove that if x is irrational, then $2x$ is also irrational. *[2 marks]*

c) Sam has already proved that $\sqrt{3}$ is irrational.
Explain how he could use his proof that $\sqrt{3}$ is irrational
to prove that $\sqrt{12}$ is irrational. *[2 marks]*

E **Q16** a) Suppose x is an integer.
Prove by exhaustion that if x^2 is a multiple of 3, then x must be a multiple of 3. *[3 marks]*

b) Hence prove that $\sqrt{3}$ is irrational. *[3 marks]*

E Q17 a) Prove that the sum of a rational number and
an irrational number is always irrational. *[3 marks]*

b) Hence prove that there is no largest irrational number. *[2 marks]*

E Q18 a) Prove that the difference between a rational number
and an irrational number is always irrational. *[4 marks]*

b) Hence or otherwise disprove the statement:
"The sum of an irrational number and an irrational number is always irrational." *[2 marks]*

E Q19 Prove that $\sqrt[3]{5}$ is irrational.

(You may assume that if x^3 is a multiple of 5, then x must be a multiple of 5.) *[3 marks]*

P Q20 Let L be the line with the following equation:

$$y = mx + c, \text{ where } m \text{ and } c \text{ are constants, and } m \text{ is irrational.}$$

Prove by contradiction that there is at most one point on
L with coordinates that are both rational numbers.

E Q21 a) Prove by contradiction that the cube root of 7 is irrational. *[4 marks]*

b) Let x be an irrational number.
Prove by contradiction that the cube root of x is irrational. *[3 marks]*

c) Deduce that $7^{\frac{1}{27}}$ is irrational. *[2 marks]*

E
P Q22 In this question, you may use the fact that e^x is irrational
whenever x is an integer and $x \neq 0$.

Prove that $\ln 5$ is irrational. *[4 marks]*

Challenge

P Q23 a) Prove that every odd number can be written
as the difference of two square numbers.

b) Hence express 197 as the difference of two square numbers.

E
P Q24 By considering $2x^2 + 2y^2 + (x - y)^2$ or otherwise, prove that
$\dfrac{x + y}{\sqrt{x^2 + y^2}} \leq 2$ for all real values of x and y, where $x, y \neq 0$. *[4 marks]*

E
P Q25 Suppose that a, b and c are positive real numbers such that $a^2 + b^2 = c^2$.

a) Prove by contradiction that $a + b > c$. *[4 marks]*

b) Interpret the result of part a) geometrically. *[1 mark]*

1 Chapter Summary

1 There are four types of proof that you need to know about:
- Proof by Deduction — using facts and logic to show a statement is true.
- Proof by Exhaustion — breaking a statement into cases covering all possible situations and proving the statement is true for each case.
- Disproof by Counter-example — give one example where the statement is false.
- Proof by Contradiction — see point 2 below.

2 To prove a statement by contradiction you follow these steps:
- Assume the original statement is not true.
- Write a new statement that says the opposite is true.
- Attempt to prove this new statement is true — expecting to arrive at a contradiction.
- If the new statement is not true, then the original statement must be true.

3 Even numbers can be expressed as $2k$ and odd numbers can be expressed as $2k + 1$, where k is an integer.

4 Rational numbers can be expressed as $\frac{a}{b}$, where a and b are non-zero integers and share no common factors.

5 Common examples of proof by contradiction include:
- Proving that certain numbers, such as $\sqrt{2}$, are irrational.
- Proving that there are infinitely many primes.

Algebra and Functions

Prior Knowledge Check

1. Given $f(x) = 3x^2 - 12$, factorise $f(x - 3)$ completely.   *see Year 1 Pure Maths*

2. Solve the following equations:
   a) $3^{2x-1} = 81$ b) $\log_4 x + \log_4 8 = 4$ *see Year 1 Pure Maths*

3. Sketch the graph of $y = \sin(x + 90°)$ for $0° \le x \le 180°$.   *see Year 1 Pure Maths*

4. Factorise the expression $f(x) = 3x^3 + 2x^2 - 7x + 2$, given that $f\left(\dfrac{1}{3}\right) = 0$.   *see Year 1 Pure Maths*

2.1   Simplifying Algebraic Fractions

Algebraic fractions are a lot like normal fractions — and you can treat them in the same way, whether you're adding, subtracting, multiplying or dividing them. All fractions are much easier to deal with when they're in their **simplest form**, so the first thing to do with algebraic fractions is to simplify them as much as possible.

Look for **common factors** in the numerator and denominator — **factorise** top and bottom and see if there's anything you can **cancel**.

If there's a **fraction** in the numerator or denominator (e.g. $\frac{1}{x}$), **multiply** the whole algebraic fraction (i.e. top and bottom) by the same factor to get rid of it (for $\frac{1}{x}$, you'd multiply through by x).

Example 1

Simplify the following:

a) $\dfrac{x-1}{x^2+3x-4}$

$$\frac{x-1}{x^2+3x-4} = \frac{x-1}{(x-1)(x+4)} = \frac{1}{x+4}$$

Try factorising the denominator first. There's an $(x-1)$ on the top and bottom which will cancel.

b) $\dfrac{3x+6}{x^2-4}$

$$\frac{3x+6}{x^2-4} = \frac{3(x+2)}{(x+2)(x-2)} = \frac{3}{x-2}$$

Both the numerator and the denominator will factorise — the denominator is the difference of two squares. Then cancel $(x+2)$.

c) $\dfrac{x^3-1}{2x^2+5x-7}$

$$\frac{x^3-1}{2x^2+5x-7} = \frac{(x-1)(x^2+x+1)}{(2x+7)(x-1)} = \frac{x^2+x+1}{2x+7}$$

Factorise top and bottom again — the coefficients sum to 0 in each case so $(x-1)$ is a factor. You can then cancel this factor.

d) $\dfrac{2+\frac{1}{2x}}{4x^2+x}$

Tip   Take your time with tricky expressions and work things out in separate steps.

$$\frac{2+\frac{1}{2x}}{4x^2+x} = \frac{2+\frac{1}{2x}}{x(4x+1)} = \frac{\left(2+\frac{1}{2x}\right)\times 2x}{x(4x+1)\times 2x}$$

Start by factorising the denominator. Then multiply the top and bottom by $2x$ to get rid of the fraction in the numerator.

$$= \frac{4x+1}{2x^2(4x+1)} = \frac{1}{2x^2}$$

Simplify by cancelling $(4x+1)$.

Chapter 2

Q1 Simplify the following:

a) $\dfrac{4}{2x+10}$

b) $\dfrac{5x}{x^2+2x}$

c) $\dfrac{6x^2-3x}{3x^2}$

d) $\dfrac{4x^3}{x^3+3x^2}$

Q2 Simplify these algebraic fractions:

a) $\dfrac{3x+6}{x^2+3x+2}$

b) $\dfrac{x^2+3x}{x^2+x-6}$

c) $\dfrac{2x-6}{x^2-9}$

d) $\dfrac{5x^2-20x}{2x^2-5x-12}$

e) $\dfrac{3x^2-7x-6}{2x^2-x-15}$

f) $\dfrac{x^3-2x^2}{x^3-4x}$

P Q3 Give the following in their simplest form:

a) $\dfrac{1+\dfrac{1}{x}}{x+1}$

b) $\dfrac{3+\dfrac{1}{x}}{2+\dfrac{1}{x}}$

c) $\dfrac{1+\dfrac{1}{2x}}{2+\dfrac{1}{x}}$

d) $\dfrac{\dfrac{1}{3x}-1}{3x^2-x}$

e) $\dfrac{2+\dfrac{1}{x}}{6x^2+3x}$

f) $\dfrac{x+4}{x-\dfrac{16}{x}}$

> **Q3c) Problem Solving**
>
> You need to multiply the top and bottom by the same term to get rid of the fractions. Don't just multiply the top by $2x$ and the bottom by x.

P Q4 Give the following in their simplest form:

a) $\dfrac{\dfrac{3x}{x+2}}{\dfrac{x}{x+2}+\dfrac{1}{x+2}}$

b) $\dfrac{2+\dfrac{1}{x+1}}{3+\dfrac{1}{x+1}}$

c) $\dfrac{1-\dfrac{2}{x+3}}{x+2}$

d) $\dfrac{4-\dfrac{1}{x^2}}{2-\dfrac{1}{x}-\dfrac{1}{x^2}}$

e) $\dfrac{\dfrac{4}{x}+\dfrac{x}{4}+2}{\dfrac{4}{x}+1}$

f) $\dfrac{x+\dfrac{13}{2}+\dfrac{3}{x}}{3x-\dfrac{5}{2}-\dfrac{2}{x}}$

> **Q4 Problem Solving**
>
> Multiply by a term that will get rid of all the fractions. E.g. in part d) multiply each term by x^2.

E P Q5 Simplify fully $\dfrac{x^4-16}{x^2+4x+4}$. *[3 marks]*

P Q6 By first factorising the cubic expressions, simplify the following:

a) $\dfrac{x^3-4x^2-19x-14}{x^2-6x-7}$

b) $\dfrac{x^3+6x^2-x-6}{x^3+7x^2+4x-12}$

Challenge

E P Q7 Simplify fully $\dfrac{2^{2x}-5\times2^x+6}{2^{2x}-2^{x+3}+15}$. *[3 marks]*

2.2 Adding and Subtracting Algebraic Fractions

You'll have come across adding and subtracting fractions before, so here's a little reminder of how to do it:

> **Tip** The common denominator should be the lowest common multiple (LCM) of all the denominators.

1. **Find the common denominator**

   Take all the individual 'bits' from the bottom lines and **multiply** them together.
   Only use each bit **once** unless something on the bottom line is raised to a **power**.

2. **Put each fraction over the common denominator**

   Multiply both top and bottom of each fraction by the same factor — whichever factor will turn the denominator into the **common denominator**.

3. **Combine into one fraction**

   Once everything's over the common denominator you can just **add** (or **subtract**) the **numerators**.

Example 1

a) Simplify: $\dfrac{2}{x-1} - \dfrac{3}{3x+2}$

Common denominator: $(x-1)(3x+2)$ ← Multiply the denominators to get the common denominator.

$\dfrac{2 \times (3x+2)}{(x-1) \times (3x+2)} - \dfrac{3 \times (x-1)}{(3x+2) \times (x-1)}$ ← Multiply the top and bottom lines of each fraction by whichever factor changes the denominator into the common denominator.

$= \dfrac{2(3x+2) - 3(x-1)}{(3x+2)(x-1)} = \dfrac{6x+4-3x+3}{(3x+2)(x-1)}$ ← All the denominators are the same — so you can just subtract the numerators and simplify.

$= \dfrac{3x+7}{(3x+2)(x-1)}$

> **Tip** Your final answer needs to be fully simplified to get all the marks in an exam, so always check if there's any more that can be done at the end.

b) Simplify: $\dfrac{2y}{x(x+3)} + \dfrac{1}{y^2(x+3)} - \dfrac{x}{y}$

Common denominator: $xy^2(x+3)$ ← The individual 'bits' here are x, $(x+3)$, y^2 and y. You don't need to multiply by y as it's already a factor of y^2.

$\dfrac{2y \times y^2}{x(x+3) \times y^2} + \dfrac{1 \times x}{y^2(x+3) \times x} - \dfrac{x \times xy(x+3)}{y \times xy(x+3)}$ ← Multiply the top and bottom of each fraction by whichever factor changes the denominator into the common denominator.

$= \dfrac{2y^3 + x - x^2y(x+3)}{xy^2(x+3)}$

$= \dfrac{2y^3 + x - x^3y - 3x^2y}{xy^2(x+3)}$ ← All the denominators are the same — so just add/subtract the numerators and simplify.

Exercise 2.2

Q1 Simplify the following:

   a) $\dfrac{2x}{3} + \dfrac{x}{5}$
      b) $\dfrac{2}{3x} - \dfrac{1}{5x}$
      c) $\dfrac{3}{x^2} + \dfrac{2}{x}$

Q2 Calculate the following, fully simplify your answers:

   a) $\dfrac{x+1}{3} + \dfrac{x+2}{4}$
      b) $\dfrac{2x}{3} + \dfrac{x-1}{7x}$

   c) $\dfrac{3x}{4} - \dfrac{2x-1}{5x}$
      d) $\dfrac{2}{x-1} + \dfrac{3}{x}$

Q3 Write the following as a single fraction:

   a) $\dfrac{3}{x+1} + \dfrac{2}{x+2}$
      b) $\dfrac{4}{x-3} - \dfrac{1}{x+4}$

   c) $\dfrac{6}{x+2} + \dfrac{6}{x-2}$
      d) $\dfrac{3}{x-2} - \dfrac{5}{2x+3}$

   e) $\dfrac{3}{x+2} + \dfrac{x}{x+1}$
      f) $\dfrac{3}{4x-5} - \dfrac{2}{3x-2}$

E P Q4 a) Write $\dfrac{3}{x-1} - \dfrac{2}{x-2}$ as a single fraction in its simplest form.   *[3 marks]*

   b) Deduce that, for $x > 2$, $\dfrac{3}{x-1} - \dfrac{2}{x-2}$ is less than $\dfrac{x}{x^2 - 3x + 2}$.   *[1 mark]*

Q5 Simplify these calculations:

   a) $\dfrac{5x}{(x+1)^2} - \dfrac{3}{x+1}$
      b) $\dfrac{5}{x(x+3)} + \dfrac{3}{x+2}$

   c) $\dfrac{x}{x^2-4} - \dfrac{1}{x+2}$
      d) $\dfrac{3}{x+1} + \dfrac{6}{2x^2+x-1}$

E Q6 Express as a fraction in its simplest form $\dfrac{x+2}{x-3} + \dfrac{1}{2(x^2-9)}$.   *[3 marks]*

Q7 Express the following as single fractions in their simplest form:

   a) $\dfrac{2}{x} + \dfrac{3}{x+1} + \dfrac{4}{x+2}$
      b) $\dfrac{3}{x+4} - \dfrac{2}{x+1} + \dfrac{1}{x-2}$

   c) $2 - \dfrac{3}{x+1} + \dfrac{4}{(x+1)^2}$
      d) $\dfrac{2x^2-x-3}{x^2-1} + \dfrac{1}{x(x-1)}$

E P Q8 $g(x) = \dfrac{1}{x} + \dfrac{2}{x+1} - \dfrac{1}{x+2}$

   a) Express $g(x)$ as a single fraction in its simplest form.   *[4 marks]*

   b) Hence show that $g(x)$ has two real roots.   *[3 marks]*

E P Q9 a) Write $\dfrac{2x^2+5x-2}{x^2+4x+3} + \dfrac{x-1}{x^2-x-2}$ as a single fraction.   *[5 marks]*

   b) Hence write $\dfrac{2x^6+5x^3-2}{x^6+4x^3+3} + \dfrac{x^3-1}{x^6-x^3-2}$ as a single fraction.   *[1 mark]*

2.3   Multiplying and Dividing Algebraic Fractions

Multiplying algebraic fractions

You **multiply** algebraic fractions in exactly the same way that you multiply normal fractions
— multiply the numerators together, then multiply the denominators.
Try to **factorise** and **cancel** any **common factors** before multiplying.

Example   1

a) Simplify $\dfrac{x^3}{2y} \times \dfrac{8y^2}{3}$

$$\dfrac{x^3}{1\,2y} \times \dfrac{8y^2\,4y}{3}$$ ← Cancel all common factors.

$$= \dfrac{x^3 \times 4y}{1 \times 3} = \dfrac{4x^3 y}{3}$$ ← Then multiply the numerators and the denominators.

b) Simplify $\dfrac{x^2 - 2x - 15}{2x + 8} \times \dfrac{x^2 - 16}{x^2 + 3x}$

$$\dfrac{(x+3)(x-5)}{2(x+4)} \times \dfrac{(x+4)(x-4)}{x(x+3)}$$ ← Factorise the expressions in both fractions and cancel.

$$= \dfrac{(x-5)(x-4)}{2x} \left(= \dfrac{x^2 - 9x + 20}{2x} \right)$$ ← Then just multiply as before. (You can multiply out the brackets in the numerator, but you don't have to — both versions are correct.)

Dividing algebraic fractions

To **divide** by an algebraic fraction, you just **multiply** by its reciprocal (as you would for normal fractions).
The reciprocal is 1 ÷ the original thing — for fractions you just turn the fraction upside down.

Example   2

Simplify the following:

a) $\dfrac{8}{5x} \div \dfrac{12}{x^3}$

$$\dfrac{8}{5x} \times \dfrac{x^3}{12}$$ ← Turn the second fraction upside down and change the division to a multiplication.

$$= \dfrac{2\,8}{5x} \times \dfrac{x^3\,x^2}{12\,3}$$ ← Then cancel all common factors.

$$= \dfrac{2 \times x^2}{5 \times 3} = \dfrac{2x^2}{15}$$ ← Now multiply the simplified fractions together.

b) $\dfrac{3x}{5} \div \dfrac{3x^2 - 9x}{20}$

$$\dfrac{3x}{5} \times \dfrac{20}{3x^2 - 9x} = \dfrac{1\,3x}{1\,5} \times \dfrac{20\,4}{3x(x-3)} = \dfrac{4}{x-3}$$ ← Turn the second fraction upside down, factorise any expressions and cancel. Then just do the multiplication as normal.

Exercise 2.3

Q1 Simplify the following:

a) $\dfrac{2x}{3} \times \dfrac{5x}{4}$

b) $\dfrac{6x^3}{7} \times \dfrac{2}{x^2}$

c) $\dfrac{8x^2}{3y^2} \times \dfrac{x^3}{4y}$

d) $\dfrac{8x^4}{3y} \times \dfrac{6y^2}{5x}$

> **Q1 Hint** Remember — cancelling before you multiply will make things a whole lot simpler.

Q2 Work out these divisions, simplifying your answers:

a) $\dfrac{x}{3} \div \dfrac{3}{x}$

b) $\dfrac{4x^3}{3} \div \dfrac{x}{2}$

c) $\dfrac{3}{2x} \div \dfrac{6}{x^3}$

d) $\dfrac{2x^3}{3y} \div \dfrac{4x}{y^2}$

Q3 Write the following as single fractions in their simplest form:

a) $\dfrac{x+2}{4} \times \dfrac{x}{3x+6}$

b) $\dfrac{4x}{5} \div \dfrac{4x^2+8x}{15}$

c) $\dfrac{2}{x^2+4x} \div \dfrac{1}{x+4}$

d) $\dfrac{2x^2+8x}{x^2-2x} \times \dfrac{x-1}{x+4}$

Q4 By factorising the quadratic expressions, simplify the following:

a) $\dfrac{x^2-4}{9} \div \dfrac{x-2}{3}$

b) $\dfrac{2x^2-2}{x} \times \dfrac{5x}{3x-3}$

c) $\dfrac{x^2-1}{3} \div \dfrac{x^2+x}{6}$

d) $\dfrac{2x^2-2x-24}{x^2+7x+12} \div \dfrac{2}{x+4}$

e) $\dfrac{x^2+4x+3}{x^2+5x+6} \times \dfrac{x^2+2x}{x+1}$

f) $\dfrac{x^2+5x+6}{x^2-2x-3} \times \dfrac{3x+3}{x^2+2x}$

g) $\dfrac{x^2-4}{6x-3} \times \dfrac{2x^2+5x-3}{x^2+2x}$

h) $\dfrac{x^2+7x+6}{4x-4} \div \dfrac{x^2+8x+12}{x^2-x}$

E **P** **Q5** a) Simplify $\dfrac{3x}{x^2-4x-21} \times \dfrac{9-x^2}{3x^2+21x}$, giving your answer as a single fraction. *[4 marks]*

b) Hence, or otherwise, simplify

$$\dfrac{3(x-3)}{(x-3)^2-4(x-3)-21} \times \dfrac{9-(x-3)^2}{3(x-3)^2+21(x-3)}.$$ *[2 marks]*

P **Q6** Fully simplify the following calculations:

a) $\dfrac{x^2+4x+4}{x^2-4x+3} \times \dfrac{x^2-2x-3}{2x^2-2x} \times \dfrac{4x-4}{x^2+2x}$

b) $\dfrac{x}{6x+12} \div \dfrac{x^2-x}{x+2} \times \dfrac{3x-3}{x+1}$

> **Q6 Hint** Work from left to right.

c) $\dfrac{x^2+5x}{2x^2+7x+3} \times \dfrac{2x+1}{x^3-x^2} \div \dfrac{x+5}{x^2+x-6}$

d) $\dfrac{3x}{x+2} \div \dfrac{x-2}{x-3} \div \dfrac{x^2-3x}{x^2-4}$

E **P** **Q7** a) Factorise fully t^4-16. *[2 marks]*

b) Hence simplify fully $\dfrac{t^2+8t+12}{t^4-16} \div \dfrac{3t+18}{t^2+4}$. *[3 marks]*

Challenge

E **P** **Q8** Show that $\dfrac{\ln(x)-2}{4x} \div \dfrac{\ln(x^2)-4}{8x^2} = x$. *[3 marks]*

2.4 Algebraic Division

Important terms

Certain terms come up a lot in algebraic division, so make sure you know what they all mean.

> **Polynomial** — an algebraic expression made up of the sum of constant terms and variables raised to **positive integer** powers. For example, $x^3 - 2x + \frac{1}{2}$ is a polynomial, but $x^{-3} - 2x^{\frac{3}{2}}$ is **not** as it has a negative power and a fractional power of x.
>
> **Degree** — the highest power of x in the polynomial. For example, the degree of $4x^5 + 6x^2 - 3x - 1$ is 5.
>
> **Divisor** — this is the thing you're dividing by. For example, if you divide $x^2 + 4x - 3$ by $x + 2$, the divisor is $x + 2$.
>
> **Quotient** — the bit that you get when you divide by the divisor (not including the **remainder** — see below).

Method 1 — using the formula

There's a handy **formula** you can use to do algebraic division.

A polynomial f(x) can be written in the form: $\mathbf{f(x) \equiv q(x)d(x) + r(x)}$

Where q(x) is the quotient, d(x) is the divisor, and r(x) is the remainder.

> **Tip**
> The $\equiv$ symbol means it's an identity.

For example:

This bit is q(x), the quotient. It has a degree of 2.

This bit is r(x), the remainder. It has a degree of 0.

$$x^3 - 2x^2 + 3x + 9 = (x^2 - 3x + 6)(x + 1) + 3.$$

This bit is f(x). It has a degree of 3.

This bit is d(x), the divisor. It has a degree of 1.

In the exam, you'll only have to divide by a **linear factor** (i.e. with a degree of 1). Here's a step-by-step guide to using the formula:

1. First, you have to work out the **degree** of the **quotient**, which will depend on the degree of the polynomial f(x): **deg q(x) = deg f(x) − 1**. The **remainder** will have degree **0**.

2. Write out the division using the formula, but replace q(x) and r(x) with **general polynomials**. For example, a general polynomial of degree 2 is Ax^2 + Bx + C, where A, B and C are constants to be found. A general polynomial of degree 0 is just a constant, e.g. D.

3. The next step is to work out the values of the **constants** (A, B, etc.). You do this by substituting in values for x to make bits disappear, and by **equating coefficients** (i.e. compare the coefficients of each power of x on either side of the identity).

4. It's best to start with the **constant term** and work **backwards** from there.

5. Finally, write out the division again, replacing A, B, C, etc. with the values you've found.

When you're using this method, you might have to use **simultaneous equations** to work out some of the coefficients (have a look back at your Year 1 notes for a reminder of how to do this if you need to). The method looks a bit tricky, but follow through the examples below to see how it works.

Example 1

Divide $x^4 - 3x^3 - 3x^2 + 10x + 5$ by $x - 2$.

$f(x)$ has degree 4, so the quotient $q(x)$ has degree $4 - 1 = 3$. The remainder $r(x)$ has degree 0.

First, work out the degrees of the quotient and remainder.

$x^4 - 3x^3 - 3x^2 + 10x + 5$
$\equiv (Ax^3 + Bx^2 + Cx + D)(x - 2) + E$

Write out the division in the form $f(x) \equiv q(x)d(x) + r(x)$, replacing $q(x)$ and $r(x)$ with general polynomials of degree 3 and 0.

$16 - 3(8) - 3(4) + 10(2) + 5 = 0 + E$
$16 - 24 - 12 + 20 + 5 = E \implies E = 5$

Substitute $x = 2$ into the identity to make the $q(x)d(x)$ bit disappear and give the value of E.

$x^4 - 3x^3 - 3x^2 + 10x + 5$
$\equiv (Ax^3 + Bx^2 + Cx + D)(x - 2) + 5$

So now the identity looks like this.

When $x = 0$, $5 = -2D + 5 \implies D = 0$

Substitute $x = 0$ into the identity to find the value of D.

$x^4 - 3x^3 - 3x^2 + 10x + 5$
$\equiv (Ax^3 + Bx^2 + Cx)(x - 2) + 5$
$\equiv Ax^4 + (B - 2A)x^3 + (C - 2B)x^2 - 2Cx + 5$

Now expand the brackets on the RHS and collect like terms.

$x^4 \equiv Ax^4 \implies A = 1$

$-3x^3 \equiv (B - 2A)x^3 \implies -3 = B - 2(1) \implies B = -1$

$-3x^2 \equiv (C - 2B)x^3 \implies -3 = C - 2(-1) \implies C = -5$

Equate the coefficients to find the values of A, B and C.

$x^4 - 3x^3 - 3x^2 + 10x + 5 \equiv (x^3 - x^2 - 5x)(x - 2) + 5$

So the identity looks like this.

Example 2

Divide $x^3 + 5x^2 - 18x - 18$ by $x - 3$.

$\deg f(x) = 3 \implies \deg q(x)\ 3 - 1 = 2,\ \deg r(x) = 0$

Work out the degrees of the quotient and remainder.

$x^3 + 5x^2 - 18x - 18 \equiv (Ax^2 + Bx + C)(x - 3) + D$

Write out the division in the form $f(x) \equiv q(x)d(x) + r(x)$.

$3^3 + 5(3^2) - 18(3) - 18 \equiv D \implies D = 0$
So $x^3 + 5x^2 - 18x - 18 \equiv (Ax^2 + Bx + C)(x - 3)$

Substitute in $x = 3$ to find D.

When $x = 0$, $-18 = C(-3) \implies C = 6$
So $x^3 + 5x^2 - 18x - 18 \equiv (Ax^2 + Bx + 6)(x - 3)$

Substitute in $x = 0$ to find C.

$x^3 + 5x^2 - 18x - 18 \equiv Ax^3 + (B - 3A)x^2 + (6 - 3B)x - 18$
$x^3 = Ax^3 \implies A = 1$, and $(6 - 3B)x = -18x \implies B = 8$

Expand the brackets.

$x^3 + 5x^2 - 18x - 18 \equiv (x^2 + 8x + 6)(x - 3)$

Equate the coefficients of x^3 and x to find A and B.

Simply stating the identity at the end doesn't always answer the question. If you're asked to divide one thing by another, then you might need to state the **quotient** and the **remainder** which you've worked out using the formula.

So for Example 1 on the previous page: $(x^4 - 3x^3 - 3x^2 + 10x + 5) \div (x - 2) = $ **$x^3 - x^2 - 5x$ remainder 5**.

For Example 2: $(x^3 + 5x^2 - 18x - 18) \div (x - 3) = $ **$x^2 + 8x + 6$** (i.e. remainder 0).

Method 2 — algebraic long division

You can also use **long division** to divide two algebraic expressions (using the same method you'd use for numbers).

> **Tip** You might have come across this method in Year 1.

Example 3

Divide $(2x^3 - 7x^2 - 16x + 11)$ by $(x - 5)$.

$$\begin{array}{r} 2x^2 \\ x-5\overline{)2x^3 - 7x^2 - 16x + 11} \end{array}$$

Start by dividing the first term in the polynomial by the first term of the divisor: $2x^3 \div x = 2x^2$. Write this answer above the polynomial.

$$\begin{array}{r} 2x^2 \\ x-5\overline{)2x^3 - 7x^2 - 16x + 11} \\ 2x^3 - 10x^2 \end{array}$$

Multiply the divisor $(x - 5)$ by this answer $(2x^2)$ to get $2x^3 - 10x^2$. Write it below the division.

$$\begin{array}{r} 2x^2 \\ x-5\overline{)2x^3 - 7x^2 - 16x + 11} \\ -\ (2x^3 - 10x^2) \\ 3x^2 - 16x \end{array}$$

Subtract this from the main expression to get $3x^2$. Bring down the $-16x$ term just to make things clearer for the next subtraction.

$$\begin{array}{r} 2x^2 + 3x \\ x-5\overline{)2x^3 - 7x^2 - 16x + 11} \\ -\ (2x^3 - 10x^2) \\ 3x^2 - 16x \end{array}$$

Now divide the first term of the remaining polynomial $(3x^2)$ by the first term of the divisor (x) to get $3x$ (the second term in the answer).

$$\begin{array}{r} 2x^2 + 3x \\ x-5\overline{)2x^3 - 7x^2 - 16x + 11} \\ -\ (2x^3 - 10x^2) \\ 3x^2 - 16x \\ -\ (3x^2 - 15x) \\ -x + 11 \end{array}$$

Multiply $(x - 5)$ by $3x$ to get $3x^2 - 15x$, then subtract again and bring down the $+11$ term.

$$\begin{array}{r} 2x^2 + 3x - 1 \\ x-5\overline{)2x^3 - 7x^2 - 16x + 11} \\ -\ (2x^3 - 10x^2) \\ 3x^2 - 16x \\ -\ (3x^2 - 15x) \\ -x + 11 \end{array}$$

Divide $-x$ by x to get -1 (the third term in the answer).

$$\begin{array}{r} 2x^2 + 3x - 1 \\ x-5\overline{)2x^3 - 7x^2 - 16x + 11} \\ -\ (2x^3 - 10x^2) \\ 3x^2 - 16x \\ -\ (3x^2 - 15x) \\ -x + 11 \\ -\ (-x + 5) \\ 6 \end{array}$$

Then multiply $(x - 5)$ by -1 to get $-x + 5$. Subtract it from $-x + 11$ to leave 6.

continued on the next page...

So: $(2x^3 - 7x^2 - 16x + 11) \div (x - 5)$
$= 2x^2 + 3x - 1$ remainder 6. ←

> You're left with 6, which has a degree that's less than the degree of the divisor, $(x - 5)$, so it can't be divided. This is the remainder.

Some questions will ask you to give your answer in a certain form, so it's important that you can interpret the result from the long division. Below are two common ways to express the same result shown above.

$$\frac{2x^3 - 7x^2 - 16x + 11}{x - 5} = 2x^2 + 3x - 1 + \frac{6}{x - 5}$$ ←

> This is in the form $Ax^2 + Bx + C + \frac{D}{q(x)}$.

$$2x^3 - 7x^2 - 16x + 11 = (2x^2 + 3x - 1)(x - 5) + 6$$ ←

> Or it could be written in the form $f(x) = q(x)d(x) + r(x)$, like when you use the equating coefficients method.

Exercise 2.4

Q1 Use the formula $f(x) \equiv q(x)d(x) + r(x)$ to divide the following expressions. In each case, state the quotient and remainder.

> **Q1 Hint** If you're told which method to use make sure you show all your working clearly to prove that you know how to use the method.

   a) $(x^3 - 14x^2 + 6x + 11) \div (x + 1)$

   b) $(2x^3 + 5x^2 - 8x - 17) \div (x - 2)$

   c) $(6x^3 + x^2 - 11x - 5) \div (2x + 1)$

E **Q2** Write $3x^4 - 8x^3 - 6x - 4$ in the form $(Ax^3 + Bx^2 + Cx + D)(x - 3) + E$, and hence state the result when $3x^4 - 8x^3 - 6x - 4$ is divided by $x - 3$. *[3 marks]*

Q3 Use long division to divide the following expressions. In each case, state the quotient and remainder. (You will have done some of these before in Q1, but using a different method.)

> **Q3d) Hint** Add in a $0x^2$ term to make sure you don't miss any terms when dividing.

   a) $(x^3 - 14x^2 + 6x + 11) \div (x + 1)$

   b) $(x^3 + 10x^2 + 15x - 13) \div (x + 3)$

   c) $(2x^3 + 5x^2 - 8x - 17) \div (x - 2)$

   d) $(3x^3 - 78x + 9) \div (x + 5)$

   e) $(x^4 - 1) \div (x - 1)$

   f) $(8x^3 - 6x^2 + x + 10) \div (2x - 3)$

In Q4-Q9 you can choose which method to use.

Q4 Divide $10x^3 + 7x^2 - 5x + 21$ by $2x + 1$, stating the quotient and remainder.

Q5 Divide $3x^3 - 8x^2 + 15x - 12$ by $x - 2$, stating the quotient and remainder.

Q6 Divide $16x^4$ by $2x - 3$, stating the quotient and remainder.

E **Q7** Divide $2x^3 - 5x^2 - 21x + 36$ by $2x - 3$, and hence solve $2x^3 - 5x^2 - 21x + 36 = 0$. *[3 marks]*
P

P **Q8** Divide $x^4 + 3x^3 + x^2 + 1$ by $x + 1$, and hence give one solution to $x^4 + 3x^3 + x^2 = -1$.

Q9 Divide $3x^4 + x^3 - 5x^2 - 4x + 4$ by $3x - 2$.

E **Q10** Use algebraic long division to express $f(x) = x^4 + 2x^3 + 1$ in the form
P $f(x) = (x + a)(x^3 + bx^2 + cx + d)$, where a, b, c and d are real integers to be found. *[3 marks]*

E **Q11** Use algebraic long division to find the remainder of
$(2y^4 - 5y^3 + 6y^2 - 11y - 7) \div (2y - 5)$. *[3 marks]*

> **Q13 Problem Solving**
>
> Start off by using the factor theorem to find one solution.

P **Q12** Divide $3x^4 + 7x^3 - 22x^2 - 8x$ by $x - 2$,
and hence solve the equation $3x^4 + 7x^3 - 22x^2 - 8x = 0$.

E **Q13** Using algebraic division, solve the equation
P $2x^4 - 5x^3 - 50x^2 - 85x - 42 = 0$. *[5 marks]*

E **Q14** a) Show that $\dfrac{x^2 + 5}{x + 6} = ax + b + \dfrac{c}{x + 6}$,
P where a, b and c are constants to be found. *[3 marks]*

The graph of the equation $y = \dfrac{x^2 + 5}{x + 6}$ has a vertical asymptote.

b) Using your answer to part a), or otherwise, write down
the equation of the vertical asymptote in the form $x = p$. *[1 mark]*

Challenge

P **Q15** a) Using algebraic long division or otherwise, write $\dfrac{16x^4}{2x + 1}$ in the form
$ax^3 + bx^2 + cx + d + \dfrac{e}{2x + 1}$, where a, b, c, d and e are constants to be found.
b) Hence, or otherwise, write down $\dfrac{32x^5}{2x + 1}$ in the form
$px^4 + qx^3 + rx^2 + sx + t + \dfrac{u}{2x + 1}$, where p, q, r, s, t and u are constants to be found.

E **Q16** In this question, a and b are integers greater than 0.
P a) Using algebraic long division or otherwise, find, in terms of a and b,
the quotient and remainder of $\dfrac{x^2 + 3x + a}{x - b}$. *[3 marks]*

b) Given that $\dfrac{x^2 + 3x + a}{x - b} = Q(x) + \dfrac{13}{x - b}$, find all possible expressions for $Q(x)$. *[2 marks]*

2.5 Mappings and Functions

Mappings

A **mapping** is an operation that takes one number and transforms it into another. For example, 'multiply by 5', 'square root' and 'divide by 7' are all mappings. The set of numbers you start with is called the **domain**, and the set of numbers they become is called the **range**.

Mappings can be drawn as **mapping diagrams**, like the one shown here for 'multiply by 5 and add 1' acting on the domain {–1, 0, 1, 2}:

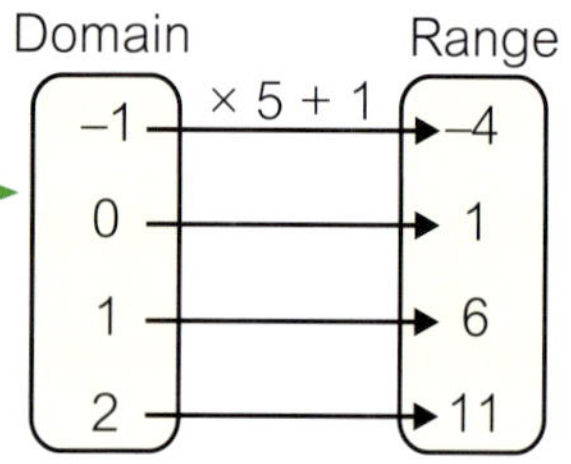

Use the notation {1, 2, ...} for the domain and range if they are a **discrete** list of values. If they can take **any value** above or below a limit, use e.g. $x \geq 0$.

Domain → × 5 + 1 → Range
-1 → -4
0 → 1
1 → 6
2 → 11

The domain and/or range will often be the set of **real numbers**, $\mathbb{R}$. A real number is any positive or negative number (or 0) — including fractions, decimals, integers and surds. If x can take any real value, it's usually written as $x \in \mathbb{R}$ ($\in$ means 'belongs to'). Other sets of numbers include $\mathbb{Z}$, the set of **integers**, and $\mathbb{N}$, the set of **natural numbers** (positive integers, not including 0).

You might have to work out the range of a mapping from the domain you're given. For example, $y = x^2$, $x \in \mathbb{R}$ has the range $y \geq 0$, as the squares of all real numbers are positive or zero.

Functions

Some mappings take every number in the domain to exactly **one** number in the range. These mappings are called **functions**. Functions are written using the following notation:

$$f(x) = 5x + 1 \quad \text{or} \quad f : x \rightarrow 5x + 1$$

You've probably seen at least the $f(x)$ notation before, but you need to be able to understand and use both.

You can substitute values for x into a function to find the value of the function at that point, as shown in Example 1 below.

Example 1

a) Give the value of f(–2) for the function $f(x) = x^2 – 1$.

$f(–2) = (–2)^2 – 1 = 4 – 1 = 3$ ← Just replace each x in the function with –2 and calculate the answer.

b) Find the value of x for which $f(x) = 12$ for the function $f : x \rightarrow 2x – 3$.

$2x – 3 = 12 \Rightarrow 2x = 15 \Rightarrow x = 7.5$ ← Solve this like a normal equation.

Functions can also be given in **several parts** (known as '**piecewise**' functions). Each part of the function will act over a different domain. For example:

$$f(x) = \begin{cases} 2x + 3 & x \le 0 \\ x^2 & x > 0 \end{cases}$$

So f(2) is $2^2 = 4$ (because $x > 0$), but f(−2) is 2(−2) + 3 = −1 (because $x \le 0$).

Tip You'll see different letters used for functions over the next few pages, not just 'f'.

If a mapping takes a number from the domain to **more than one** number in the range (or if it isn't mapped to any number in the range), it's **not** a function.

The mapping shown here is a **function**, because any value of x in the domain maps to only **one value** in the range.

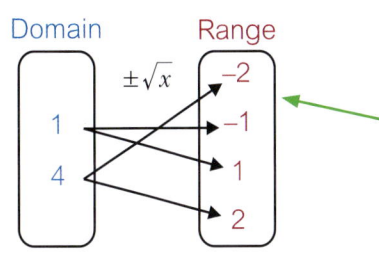

Tip Although each value in the domain only maps to one value in the range, the reverse is not true. This means it's a 'many to one' function — there's more about these on p.27.

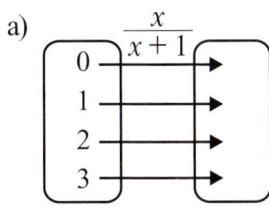

The mapping shown here is **not a function**, because a value of x in the domain can map to **more than one value** in the range.

Exercise 2.5

Q1 Draw a mapping diagram for the map "multiply by 6" acting on the domain {1, 2, 3, 4}.

Q2 $y = x + 4$ is a map with domain $\{x : x \in \mathbb{N}, x \le 7\}$. Draw the mapping diagram.

Q3 Complete the mapping diagrams below:

a)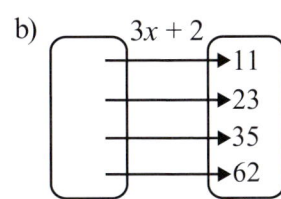

b)

Q4 For the function $g : x \to \dfrac{1}{2x + 1}$, $x > -\dfrac{1}{2}$, evaluate g(0) and g(2).

Q4 Hint 'Evaluate' is just another way of asking you to 'find the value of'.

E **Q5** f defines a function $f : x \to \dfrac{1}{2 + \log_{10} x}$ for the domain $x > 0.01$. Evaluate f(1) and f(100).

[2 marks]

Q6 a) Find the range of the function $h(x) = \sin x$, $0° \le x \le 180°$.

b) Find the range of $j(x) = \cos x$ on the same domain.

E **Q7** The function r is defined by $r(x) = \begin{cases} x^2 + 1, & x \leq 2 \\ 2x^3 + 2, & x > 2 \end{cases}$.

    a) State the range of r. *[1 mark]*

    b) Find all values of x such that $r(x) = 10$. *[3 marks]*

E **P** **M** **Q8** The price £P(x) a customer pays for their smartphone per month when using x GB of mobile data is modelled by:

$$P(x) = \begin{cases} 57, & x \leq 5 \\ 57 + 9.41(x - 5), & x > 5 \end{cases}$$

    a) State the domain and range of P(x). *[2 marks]*

    b) Explain what the value 9.41 represents. *[1 mark]*

    c) A customer paid £87.58 for their smartphone usage in January. How much mobile data did they use during the month of January? *[2 marks]*

Customers are also charged an extra fee for exceeding their allowance of minutes, which for this particular contract is 1000 minutes. For every minute by which a customer exceeds their allowance, they are charged an additional £0.17.

    d) A customer paid £66.58 for their smartphone usage in February and did not use more than 6 GB of mobile data. Find the range of possible values for the number of minutes, m, that the customer used in February. *[4 marks]*

P **Q9** State the largest possible domain and range of each function:

    a) $f(x) = 3^x - 1$ b) $g(x) = (\ln x)^2$

> **Q9 Problem Solving**
>
> To find the largest possible domain, think about what values of x need to be excluded to make the function valid.

Q10 State whether or not each of the mapping diagrams below shows a function, and if not, explain why.

    a) 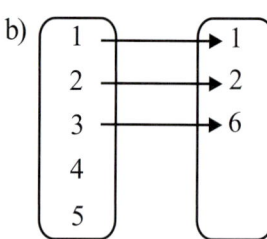   b)   c)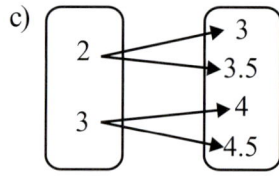

Challenge

E **Q11** The mapping f is defined by $f(x) = \sqrt{e^{2x} - e^x - 2}$, $x \in \mathbb{R}$.

    a) Evaluate f(ln 6), giving an exact answer. *[2 marks]*

    b) Find the largest possible domain over which f is a real-valued function. *[4 marks]*

    c) Find the range of the function f. *[1 mark]*

E **Q12** The function q is defined by $q(x) = ae^{-x} + be^x$, $x \in \mathbb{R}$, where a and b are constants.

    a) Given that $q(\ln 2) = 1$ and $q(\ln 3) = \frac{5}{4}$, find the values of a and b. *[3 marks]*

    b) Hence, explain why the range of q(x) is not q(x) > 0. *[2 marks]*

2.6 Graphs of Functions

Mappings and functions with a **continuous** domain (such as $x \in \mathbb{R}$, i.e. not a discrete set of values) can be drawn as **graphs**. Drawing a graph of, say, $f(x) = x^2$ is exactly the same as drawing a graph of $y = x^2$.

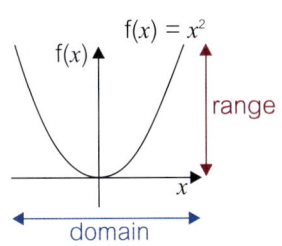

For each value of x in the **domain** (which goes along the horizontal x-axis) you can plot the corresponding value of $f(x)$ in the **range** (up the vertical y-axis):

Drawing graphs can make it easier to **identify functions**, as shown below.

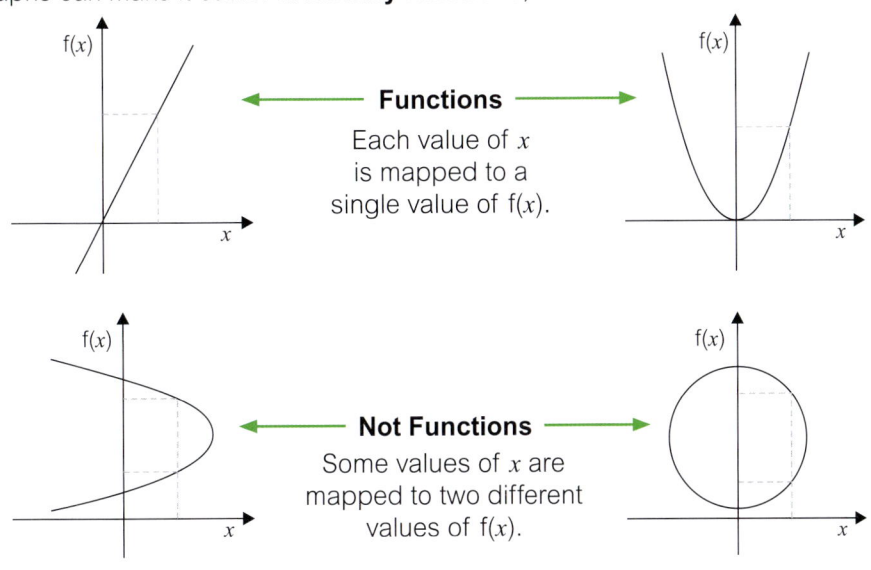

◄——— **Functions** ———►
Each value of x is mapped to a single value of $f(x)$.

◄——— **Not Functions** ———►
Some values of x are mapped to two different values of $f(x)$.

The graph on the right isn't a function for $x \in \mathbb{R}$ because $f(x)$ is not defined for $x < 0$.

This just means that when x is negative there is no real value that $f(x)$ can take.

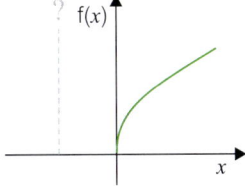

> **Tip** This could be turned into a function by restricting the domain to $x \geq 0$ — see the next page.

Finding ranges and domains using graphs

Sketching a graph can also be really useful when trying to find limits for the domain and range of a function.

Example 1

a) State the range for the function $f(x) = x^2 - 5$, $x \in \mathbb{R}$.

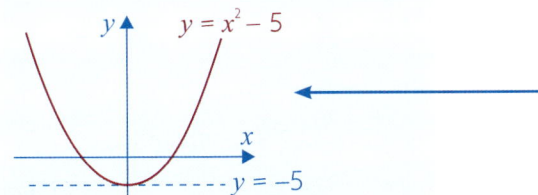

This can be shown clearly by sketching the graph of $y = f(x)$. This is just the graph of $y = x^2$ shifted 5 units down.

The smallest value of $x^2 - 5$ is -5.
So the range is $f(x) \geq -5$.

The smallest possible value of x^2 is 0 when $x = 0$, and there is no largest value.

b) State the domain for $f(x) = \sqrt{(x-4)}$, giving your answer in set notation.

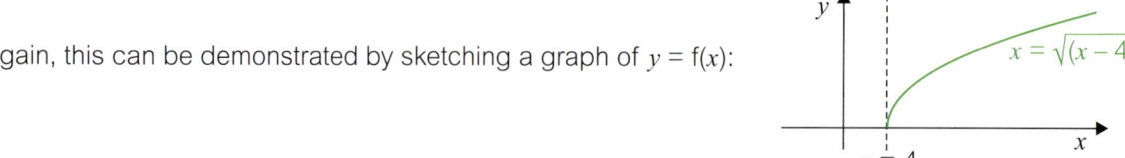

$f(x) = \sqrt{(x-4)}$

There are no **real** solutions for the square root of a negative number.

$\Rightarrow x - 4 \geq 0$

$\Rightarrow x \geq 4$ ← This is the domain. Now just write it in set notation.

$\Rightarrow \{x : x \geq 4\}$

Tip Remember, the : means "such that", so this is the set of x, such that x is greater than or equal to 4. The range for this function, in set notation, is $\{f(x) : f(x) \geq 0\}$.

Again, this can be demonstrated by sketching a graph of $y = f(x)$:

Turning mappings into functions

Some mappings that aren't functions can be turned into functions by **restricting their domain**.

For example, consider the graph of the mapping $y = \dfrac{1}{x-1}$ for $x \in \mathbb{R}$:

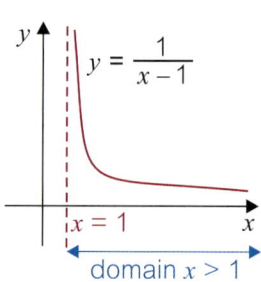

The mapping $y = \dfrac{1}{x-1}$ for $x \in \mathbb{R}$ is **not** a function, because it's not defined at $x = 1$.

(You will often see asymptotes on a graph when a mapping is undefined at a certain value.)

But if you change the domain to $x > 1$, the mapping is now a **function**, as shown:

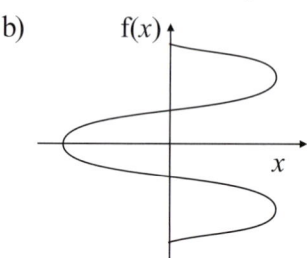

You could also restrict the domain by giving values that x can't be equal to, e.g. $x \in \mathbb{R}, x \neq 1$. In this case the graph would be in two parts like in the first diagram.

Exercise 2.6

Q1 State whether or not each of the graphs below shows a function, and if not, explain why.

a)

b)

Q2　For each of the following functions, sketch the graph of the function for the given domain, marking relevant points on the axes, and state the range.

a)　$f(x) = 3x + 1$　$x \geq -1$

b)　$f(x) = x^2 + 2$　$-3 \leq x \leq 3$

c)　$f(x) = \cos x$　$0° \leq x \leq 360°$

d)　$f(x) = \begin{cases} 5 - x & 0 \leq x < 5 \\ x - 5 & 5 \leq x \leq 10 \end{cases}$

Q3　State the domain and range for the following functions, giving your answers in set notation:

Q3 Hint　Use the given functions to work out the domain from the given range, or the range from the given domain.

a)
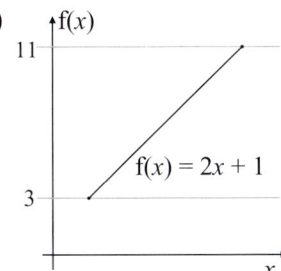

b)
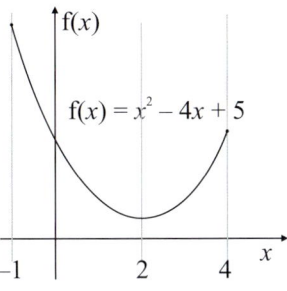

 Q4　The graph below shows the function $f(x) = \dfrac{x + 2}{x + 1}$, defined for the domain $x \geq 0$. State the range.

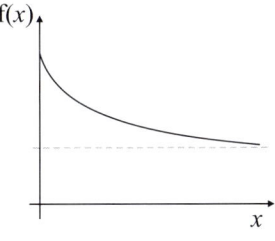

[2 marks]

Q5　The diagram shows the function $f(x) = \dfrac{1}{x - 2}$ drawn over the domain $x > a$. State the value of a.

Q4-5 Hint　Use the functions to work out where the asymptotes lie. The range (or domain) will lie on one side of the asymptote.

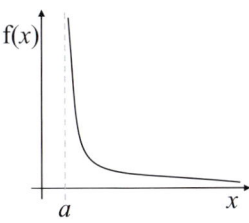

Q6　The diagram shows the function $f(x) = \sqrt{9 - x^2}$ for $x \in \mathbb{R}$, $a \leq x \leq b$. State the values of a and b.

Q6 Problem Solving
$9 - x^2$ cannot be negative, as you can't take the square root of a negative number, so work out the values of x for which $9 - x^2 \geq 0$ and use these as the domain.

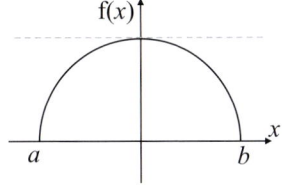

Q7 $h(x) = \sqrt{x+1}$, $x \in \mathbb{R}$. Using set notation, give a restricted domain so that h is a function.

Q8 $k : x \rightarrow \tan x$, $x \in \mathbb{R}$. Give an example of a domain which would make k a function.

E **Q9** $f:x \rightarrow \dfrac{2x+6}{x}$, $x \in \mathbb{R}, x > 0$.

    a) Explain why the domain of f cannot be $\mathbb{R}$.   *[1 mark]*

    b) Sketch the graph of $y = f(x)$.   *[2 marks]*

    c) State the range of f.   *[1 mark]*

P **Q10** $m(x) = \dfrac{1}{x^2 - 4}$. What is the largest continuous domain which would make m(x) a function?

P **Q11** The diagram on the right shows the graph of $y = f(x)$.

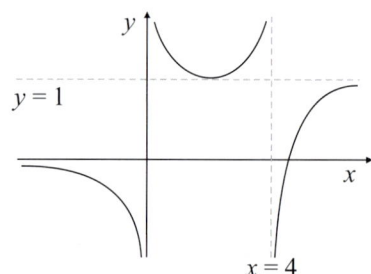

    a) Explain why f is not a function on the domain $x \in \mathbb{R}$.

    b) State the largest possible domain that would make f a function.

E **P** **Q12** The diagram below is the graph of the function $f: \mathbb{R} \rightarrow \mathbb{R}$, $x \geq 0$, $x \neq a, b$, defined by $f(x) = \dfrac{x}{x + 2 - 3\sqrt{x}}$.

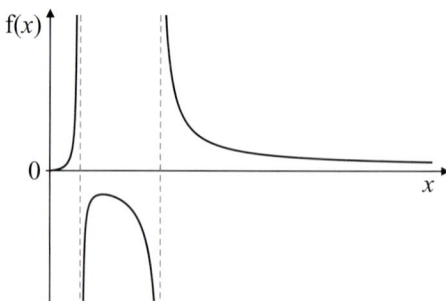

    a) (i) Use algebra to find all values of x such that $x + 2 - 3\sqrt{x} = 0$.   *[3 marks]*

        (ii) Hence state the values of a and b.   *[1 mark]*

    b) Explain why the function is not defined for $x < 0$.   *[1 mark]*

    c) Given that the y-coordinate of the maximum point on the curve is -8, write down the range of the function.   *[2 marks]*

2.7 Types of Function

One-to-one functions

A function is **one-to-one** if each value in the **range** corresponds to **exactly one** value in the **domain**.

Sketching a graph is a good way to help you identify the type of function.

For example, the function $f : x \rightarrow 2x$, $x \in \mathbb{R}$ is one-to-one, as only one value of x in the domain is mapped to each value in the range (the range is also $\mathbb{R}$).

You can see this clearly on a sketch of the function:

Going vertically up or down from any value on the x-axis (the domain) you only ever hit the graph of $f(x) = 2x$ once, so it is a function.

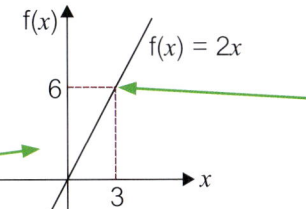

All values of x map to exactly one value of $f(x)$, so it is one-to-one. E.g. Only 3 in the domain is mapped to 6 in the range.

Many-to-one functions

A function is **many-to-one** if some values in the **range** correspond to **more than one (many)** values in the **domain**.

Remember that no element in the domain can map to more than one element in the range, otherwise it wouldn't be a function.

For example, the function $f(x) = x^2$, $x \in \mathbb{R}$ is a many-to-one function, as two elements in the domain map to the same element in the range, as shown:

Going vertically up or down from any value on the x-axis (the domain) you only ever hit the graph of $f(x) = x^2$ once, so it is a function.

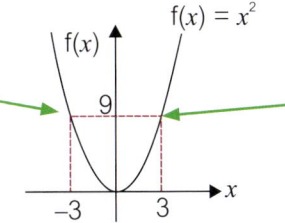

Some values of x map to the same value of $f(x)$ so it is many-to-one. E.g. Both 3 and −3 map to 9.

Exercise 2.7

Q1 State whether each function below is one-to-one or many-to-one.

    a) $f(x) = x^3 \quad x \in \mathbb{R}$

    b) $f : x \rightarrow \sin 2x \quad -180° < x \leq 180°$

    c) $f(x) = \log_{10} x \quad x > 0$

P Q2 State whether each function of the following piecewise functions is one-to-one or many-to-one.

    a) $f(x) = \begin{cases} x + 2 & -2 \leq x < 0 \\ 2 - x & 0 \leq x \leq 2 \end{cases}$
      b) $f(x) = \begin{cases} 2^x & x \geq 0 \\ 1 & x < 0 \end{cases}$

2.8 Composite Functions

- If you have two functions f and g, you can combine them (do one followed by the other) to make a new function. This is called a **composite function**.

- Composite functions are written **fg(x)**. This means 'do **g first**, then **f**'. If it helps, put brackets in until you get used to it, so fg(x) = f(g(x)).

- The **order** is really important — usually fg(x) ≠ gf(x). If you get a composite function that's written f²(x), it means ff(x). This just means you have to do f **twice**.

- Composite functions made up of three or more functions work in exactly the same way — just make sure you get the order right.

Example 1

If f(x) = x − 2 and g(x) = 3x, then find:

a) fg(6)

$$\underset{\text{g(6)}}{6 \longrightarrow \boxed{3 \times 6} \longrightarrow 18} \longrightarrow \underset{\text{f(18)}}{\boxed{18 - 2} \longrightarrow 16} \longleftarrow$$

First substitute 6 into g(x). Then substitute the value that comes out into f(x).

So fg(6) = 16

Tip Comparing the answers to a) and b) you can see that fg(x) ≠ gf(x).

b) gf(6)

$$\underset{\text{f(6)}}{6 \longrightarrow \boxed{6 - 2} \longrightarrow 4} \longrightarrow \underset{\text{g(4)}}{\boxed{3 \times 4} \longrightarrow 12} \longleftarrow$$

This time substitute 6 into f(x) first. Then substitute the value that comes out into g(x).

So gf(6) = 12

c) fg(x)

$$\underset{\text{g(x)}}{x \longrightarrow \boxed{3x} \longrightarrow 3x} \longrightarrow \underset{\text{f(3x)}}{\boxed{(3x) - 2} \longrightarrow 3x - 2} \longleftarrow$$

This time leave everything in terms of x. Do g first, then f.

So fg(x) = 3x − 2

d) gf(x)

$$\underset{\text{f(x)}}{x \longrightarrow \boxed{x - 2} \longrightarrow x - 2} \longrightarrow \underset{\text{g(x − 2)}}{\boxed{3(x - 2)} \longrightarrow 3(x - 2)} \longleftarrow$$

As above, leave everything in terms of x. Do f first, then substitute the expression that comes out into g.

So gf(x) = 3(x − 2) or 3x − 6

The key to composite functions is to work things out in steps. Set out your working for composite functions as shown in the examples below.

Example 2

For the functions $f : x \rightarrow 2x^3, x \in \mathbb{R}$ and $g : x \rightarrow x - 3, x \in \mathbb{R}$, find:

a) $fg(4)$ b) $fg(0)$ c) $gf(0)$ d) $fg(x)$ e) $gf(x)$ f) $f^2(x)$.

a) $fg(4) = f(g(4)) = f(4 - 3) = f(1) = 2 \times 1^3 = 2$

b) $fg(0) = f(g(0)) = f(0 - 3) = f(-3) = 2 \times (-3)^3 = -54$

c) $gf(0) = g(f(0)) = g(2 \times 0^3) = g(0) = 0 - 3 = -3$

d) $fg(x) = f(g(x)) = f(x - 3) = 2(x - 3)^3$

e) $gf(x) = g(f(x)) = g(2x^3) = 2x^3 - 3$

f) $f^2(x) = f(f(x)) = f(2x^3) = 2(2x^3)^3 = 16x^9$

> $f(x) = 2x^3$ and $g(x) = x - 3$. Remember to do the function closest to the brackets first, then apply the outer function to the result.

> **Tip** Don't forget the 2^3 when expanding $(2x^3)^3$ in part f).

Domain and range of composite functions

Two functions with given domains and ranges may form a composite function with a **different** domain and range.

> **Tip** Working out the domains and ranges of composite functions can be tricky — but sketching a graph always helps.

Example 3 P

Give the domain and range of the composite function $fg(x)$, where:
$f(x) = 2x^2 + 1$, domain $x \in \mathbb{R}$, range $f(x) \geq 1$
and $g(x) = \dfrac{1}{x + 3}$, domain $x > -3$, range $g(x) > 0$

$fg(x) = f(g(x))$

$= f\left(\dfrac{1}{x + 3}\right) = 2\left(\dfrac{1}{x + 3}\right)^2 + 1$

> First work out the composite function $fg(x)$ in terms of x.

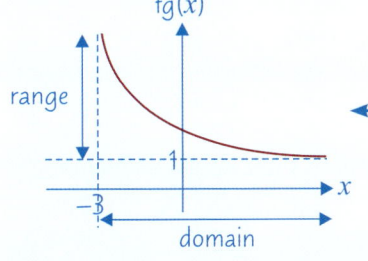

> Next, sketch the graph of the composite function over the domain and range of the original functions.

$\dfrac{1}{x + 3}$ is always greater than 0 for the domain $x > -3$,

so $2\left(\dfrac{1}{x + 3}\right)^2 + 1$ must be greater than $2(0^2) + 1 = 1$.

So $fg(x)$ has a domain of $x > -3$ and a range of $fg(x) > 1$.

> As $g(x)$ is restricted to $x > -3$, the domain of $fg(x)$ is also restricted to $x > -3$.

For a composite function fg(x), the domain can also be found by putting the **range** of **g(x)** into **f(x)**. If the domain of f(x) does not fully include the range of g(x) then the domain of g(x) will have to be **restricted further**. See the example below.

Example 4 P

Find the domain of fg(x), where $f(x) = \sqrt{x}$, $x \geq 0$, and $g(x) = x + 5$, $x \in \mathbb{R}$

$x + 5 \geq 0 \implies x \geq -5$

So the largest possible domain for fg(x) is $x \geq -5$.

The range of g(x) is bigger than the domain of f(x), so the domain of g(x) will need to be restricted. The input into f needs to be ≥ 0, and for fg(x) the input into f is g(x). So the domain is g(x) ≥ 0.

In the example above, if the domain was not restricted, then fg(x) would be undefined in places — e.g. fg(–6) = f(–6 + 5) = f(–1) = $\sqrt{-1}$ (which is undefined).

Exercise 2.8

Q1 $f : x \to x^2$, $x \in \mathbb{R}$ and $g : x \to 2x + 1$, $x \in \mathbb{R}$. Find the values of:

a) fg(3)   b) gf(3)   c) $f^2(5)$ d) $g^2(2)$

Q2 $f(x) = \sin x$ for $\{x : x \in \mathbb{R}\}$ and $g(x) = 2x$ for $\{x : x \in \mathbb{R}\}$. Evaluate fg(90°) and gf(90°).

Q3 $f : x \to \dfrac{3}{x + 2}$, $x > -2$ and $g : x \to 2x$, $x \in \mathbb{R}$.

a) Find the values of gf(1), fg(1) and $f^2(4)$

b) Explain why fg(–1) is undefined.

Q3b) Hint Try to find the value of fg(–1), or consider the domains and ranges of f(x) and g(x).

E Q4 $f(x) = \cos x$, $x \in \mathbb{R}$ and $g(x) = 2x$, $x \in \mathbb{R}$. Find the functions:

a) fg(x)   *[1 mark]*

b) gf(x)   *[1 mark]*

Q5 $f(x) = 2x - 1$, $x \in \mathbb{R}$ and $g(x) = 2^x$, $x \in \mathbb{R}$. Find the functions:

a) fg(x)   b) gf(x)   c) $f^2(x)$

Q6 $f(x) = \dfrac{2}{x - 1}$ for $\{x : x > 1\}$ and $g(x) = x + 4$ for $\{x : x \in \mathbb{R}\}$.

Find the functions fg(x) and gf(x), writing them as single fractions in their simplest forms.

P Q7 $f(x) = \dfrac{x}{1 - x}$ for $\{x : x \in \mathbb{R}, x \neq 1\}$ and $g(x) = x^2$ for $\{x : x \in \mathbb{R}\}$.

Find $f^2(x)$ and gfg(x).

P Q8 $f(x) = x^2$ with domain $x \in \mathbb{R}$, and

$g(x) = 2x - 3$ also with domain $x \in \mathbb{R}$.

a) Find fg(x) and write down its range.

b) Find gf(x) and write down its range.

Q8 Problem Solving Sketching the graphs of the composite functions will help you find the ranges and domains.

P Q9 $f(x) = \dfrac{1}{x}$ and $g(x) = \ln(x + 1)$, both with domain $\{x : x > 0\}$.

       a) Find $gf(x)$ and write down its range and largest possible domain.

       b) Find $fg(x)$ and write down its range and largest possible domain.

P Q10   Given that $f(x) = 3x + 2$, $g(x) = 5x - 1$, and $h(x) = x^2 + 1$ (all with domain $\{x : x \in \mathbb{R}\}$), find $fgh(x)$.

E **P** Q11   a) Hemal claims that for any function f, if $ff(a) = 0$, where a is a constant, then $f(a) = 0$. Show that Hemal is wrong.   *[2 marks]*

       b) Amanda claims that if $ff(x) = gg(x)$ then $f(x) = g(x)$. Show that Amanda is wrong.   *[2 marks]*

E **P** Q12   The functions g and h are defined by

       $g(x) = x^3 - 1$, $h(x) = x^2$, $x \in \mathbb{R}$.

       a) Find $gh(x)$.   *[2 marks]*

       b) Find $hg(x)$.   *[2 marks]*

       c) Do $gh(x)$ and $hg(x)$ have the same range? Explain your answer.   *[2 marks]*

2.9   Solving Composite Function Equations

If you're asked to **solve** an equation such as $fg(x) = 8$, the best way to do it is to work out what $fg(x)$ is, then **rearrange** $fg(x) = 8$ to make x the subject.

Example 1

For the functions $f : x \to \sqrt{x}$ with domain $\{x : x \geq 0\}$ and $g : x \to \dfrac{1}{x - 1}$ with domain $\{x : x > 1\}$, solve the equation $fg(x) = \dfrac{1}{2}$ and state the domain and range of $fg(x)$.

$$fg(x) = f\left(\frac{1}{x-1}\right) = \sqrt{\frac{1}{x-1}} = \frac{1}{\sqrt{x-1}}$$

First, find an expression for $fg(x)$.

$$\Rightarrow \frac{1}{\sqrt{x-1}} = \frac{1}{2} \Rightarrow \sqrt{x-1} = 2$$

Set $fg(x)$ equal to $\frac{1}{2}$ and rearrange to find x.

$$\Rightarrow x - 1 = 4 \Rightarrow x = 5$$

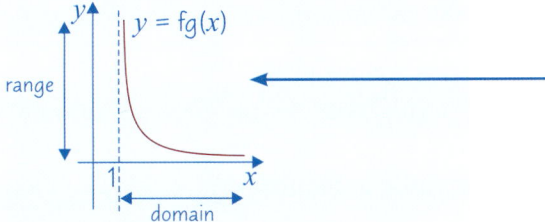

The domain and range can be read from a sketch of the graph of $fg(x)$.

The domain of $fg(x)$ is $\{x : x > 1\}$ and the range is $\{fg(x) : fg(x) > 0\}$.

Tip   Be careful with the domains and ranges of composite functions. Have a look back at pages 29-30 for more on how to find them.

Example 2

For the functions $f : x \rightarrow 2x + 1, x \in \mathbb{R}$ and $g : x \rightarrow x^2, x \in \mathbb{R}$, solve $gf(x) = 16$.

$gf(x) = g(2x + 1) = (2x + 1)^2$ ⟵ Find an expression for $gf(x)$.

$(2x + 1)^2 = 16 \implies 2x + 1 = 4$ or $2x + 1 = -4$ ⟵ Set your expression for $gf(x)$ equal to 16 and solve to find x.

$\implies 2x = 3$ or $2x = -5 \implies x = \dfrac{3}{2}$ or $x = -\dfrac{5}{2}$

Exercise 2.9

Q1 For the following functions, solve the given equations:

a) $f(x) = 2x + 1, x \in \mathbb{R}$ $g(x) = 3x - 4, x \in \mathbb{R}$ $fg(x) = 23$

b) $f(x) = \dfrac{1}{x}$ for $\{x : x \neq 0\}$ $g(x) = 2x + 5$ for $\{x : x \in \mathbb{R}\}$ $gf(x) = 6$

c) $f(x) = x^2$ for $\{x : x \in \mathbb{R}\}$ $g(x) = \dfrac{x}{x - 3}$ for $\{x : x \neq 3\}$ $gf(x) = 4$

d) $f(x) = x^2 + 1, x \in \mathbb{R}$ $g(x) = 3x - 2, x \in \mathbb{R}$ $fg(x) = 50$

e) $f(x) = 2x + 1, x \in \mathbb{R}$ $g(x) = \sqrt{x}, x \geq 0$ $fg(x) = 17$

f) $f(x) = \log_{10} x, x > 0$ $g(x) = 3 - x, x \in \mathbb{R}$ $fg(x) = 0$

g) $f(x) = 2^x$ for $\{x : x \in \mathbb{R}\}$ $g(x) = x^2 + 2x$ for $\{x : x \in \mathbb{R}\}$ $fg(x) = 8$

E **P** **Q2** Given that $f(x) = \dfrac{x}{x + 1}, x \neq -1$ and $g(x) = 2x - 1, x \in \mathbb{R}$, solve $fg(x) = gf(x)$.   *[4 marks]*

P **Q3** $f : x \rightarrow x^2 + b, x \in \mathbb{R}$ and $g : x \rightarrow b - 3x, x \in \mathbb{R}$ (b is a constant).

a) Find fg and gf and give the range of each in terms of b.

b) Given that $gf(2) = -8$, find the value of $fg(2)$.

Challenge

E **P** **Q4** $p(x) = e^{3x} - 3$ and $q(x) = 2\ln(x + 1)$. Both p and q are defined over the domain $x > -1$.

a) Find $pq(x)$, fully simplifying your answer.   *[2 marks]*

b) Solve the equation $pq(x) = 61$.   *[3 marks]*

c) Find the exact value of x for which the graph of $y = pq(x)$ crosses the x-axis.   *[2 marks]*

d) (i) Show that $qq(x) = 2\ln(\ln((x + 1)^2) + 1)$.   *[1 mark]*

     (ii) The function qq is not defined for $x \leq a$, where a is a constant. Find the exact value of a, giving your answer in terms of e.   *[3 marks]*

E **P** **Q5** The functions f and g are defined by $f(x) = \sqrt[3]{2x + 4}$, for all real numbers x, and $g(x) = a + \ln x$, where a is a constant, for all positive real numbers x.

Given that $fg(e^2) = 2$, find a, and hence find the exact solution of $gg(e^x) = 5$.   *[5 marks]*

2.10 Inverse Functions and their Graphs

- An **inverse function** does the **opposite** to the function. So if the function was '+ 1', the inverse would be '− 1', if the function was '× 2', the inverse would be '÷ 2' etc.

- The inverse for a function $f(x)$ is written $\mathbf{f^{-1}(x)}$.

- An inverse function maps an element in the range to an element in the domain — the opposite of a function. This means that only one-to-one functions have inverses, as the inverse of a many-to-one function would be one-to-many, which isn't a function. (See page 27 for more on the different types of functions.)

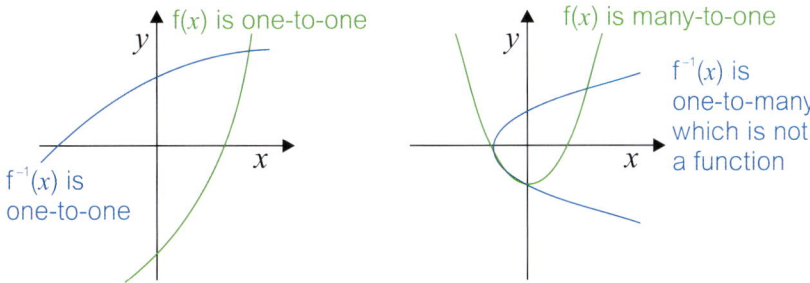

For any inverse $f^{-1}(x)$:

$$f^{-1}f(x) = x = ff^{-1}(x)$$

Doing the function and then the inverse... ...is the same as doing the inverse then doing the function — both just give you x.

> **Tip** $f^{-1}f(x)$ is a composite function (p.28). It just means 'do f then f⁻¹'.

The **domain** of the inverse is the **range** of the function, and the **range** of the inverse is the **domain** of the function.

Example 1

A function $f(x) = x + 7$ has domain $x \geq 0$ and range $f(x) \geq 7$. State whether the function has an inverse. If so, find the inverse, and give its domain and range.

The function $f(x) = x + 7$ is one-to-one, so it does have an inverse. ← To see if it has an inverse check whether it is one-to-one or many-to-one.

The inverse of +7 is −7, so $f^{-1}(x) = x - 7$.

$f^{-1}(x)$ has domain $x \geq 7$ ← The domain is the range of $f(x)$.

and range $f^{-1}(x) \geq 0$ ← The range is the domain of $f(x)$.

For simple functions (e.g. in the example above), it's easy to work out what the inverse is by looking at it. But for more complex functions, you need to **rearrange** the original function to **change the subject**.

Finding the inverse of a function

Here's a general method for finding the inverse of a given function:

1. Replace f(x) with y to get an equation for **y in terms of x**.
2. **Rearrange** the equation to make x the subject.
3. Replace x with $f^{-1}(x)$ and y with x — this is the **inverse function**.
4. **Swap** round the **domain** and **range** of the function.

> **Tip** It's easier to work with y than f(x).

Example 2

Find the inverse of the function $f(x) = \sqrt{2x - 1}$, with domain $x \geq \frac{1}{2}$ and range $f(x) \geq 0$. State the domain and the range of the inverse.

$f(x) = \sqrt{2x - 1}$

$y = \sqrt{2x - 1}$ ← Replace f(x) with y.

$\Rightarrow y^2 = 2x - 1 \Rightarrow x = \dfrac{y^2 + 1}{2}$ ← Rearrange to make x the subject.

$f^{-1}(x) = \dfrac{x^2 + 1}{2}$ ← Replace x with $f^{-1}(x)$ and y with x.

The domain of $f^{-1}(x)$ is $x \geq 0$. ← The domain is the range of f(x).

The range of $f^{-1}(x)$ is $f^{-1}(x) \geq \frac{1}{2}$. ← The range is the domain of f(x).

> **Tip** Breaking it into steps like this means you're less likely to go wrong. It's worth doing it this way even for easier functions.

Example 3

Find the inverse of the function $f(x) = 2^x + 1$ with domain $\{x : x \geq 0\}$, and state its domain and range.

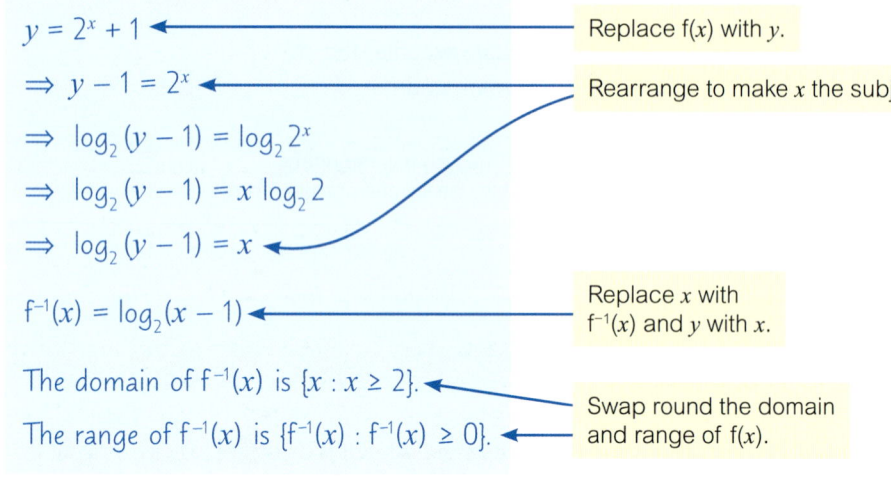

$y = 2^x + 1$ ← Replace f(x) with y.

$\Rightarrow y - 1 = 2^x$ ← Rearrange to make x the subject.

$\Rightarrow \log_2 (y - 1) = \log_2 2^x$

$\Rightarrow \log_2 (y - 1) = x \log_2 2$

$\Rightarrow \log_2 (y - 1) = x$

$f^{-1}(x) = \log_2 (x - 1)$ ← Replace x with $f^{-1}(x)$ and y with x.

The domain of $f^{-1}(x)$ is $\{x : x \geq 2\}$. ← Swap round the domain and range of f(x).

The range of $f^{-1}(x)$ is $\{f^{-1}(x) : f^{-1}(x) \geq 0\}$. ←

> **Problem Solving** If you're not given the domain and / or the range of the original function you'll need to work it out. In this example, x is always at least 0, so f(x) must always be at least $2^0 + 1 = 2$.

Graphs of inverse functions

The inverse of a function is its **reflection** in the line $y = x$.

Sketch the graph of the inverse of the function $f(x) = x^2 - 8$ with domain $x \geq 0$.

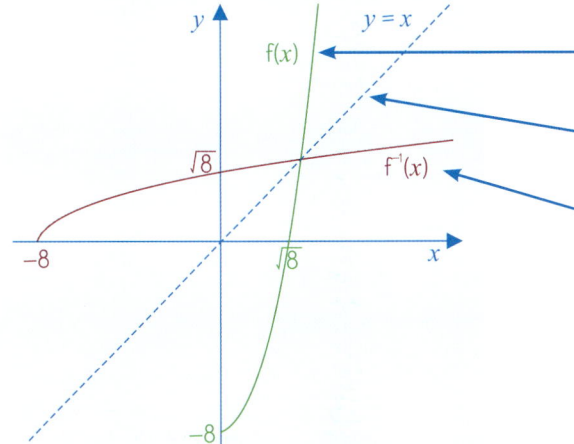

Sketch the graph of $y = f(x)$ over the domain $x \geq 0$. It's the graph of $y = x^2$ shifted 8 units down.

Draw the line $y = x$.

Reflect $f(x)$ in $y = x$ to get $f^{-1}(x)$. The inverse function is $f^{-1}(x) = \sqrt{x + 8}$.

Tip Only sketch the function over the given domain and range. Otherwise you won't be able to see the correct domain and range for the graph of the inverse when you do the reflection.

It's easy to see what the domains and ranges are from the graph in the example above — $f(x)$ has domain $x \geq 0$ and range $f(x) \geq -8$, and $f^{-1}(x)$ has domain $x \geq -8$ and range $f^{-1}(x) \geq 0$.

Exercise 2.10

Q1 Do the functions shown in the diagrams below have inverses? Justify your answers.

a)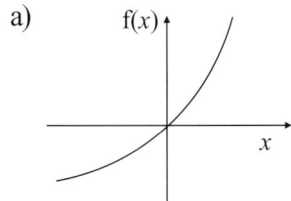

b)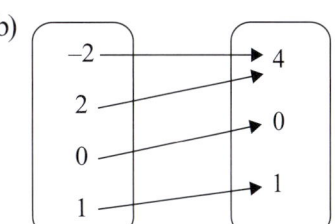

Q2 For the following functions, explain whether or not an inverse f^{-1} exists:

Q2-6 Hint If in doubt, sketch a graph of the function to check domains and ranges.

a) $f(x) = \sin x$, $x \in \mathbb{R}$

b) $f(x) = x^2 + 3$, $x \in \mathbb{R}$

c) $f(x) = (x - 4)^2$ for $\{x : x \geq 4\}$

Q3 Find the inverse of each of the following functions, stating the domain and range:

a) $f(x) = 3x + 4$, $x \in \mathbb{R}$

b) $f(x) = 5(x - 2)$, $x \in \mathbb{R}$

c) $f(x) = \dfrac{1}{x + 2}$, $x > -2$

d) $f(x) = x^2 + 3$ for $\{x : x > 0\}$

Q4 $f(x) = \dfrac{3x}{x+1}$, $x > -1$.

a) Find $f^{-1}(x)$, stating the domain and range.

b) Evaluate $f^{-1}(2)$.

c) Evaluate $f^{-1}\left(\dfrac{1}{2}\right)$.

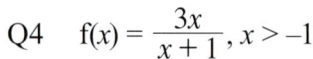

Q4-5 Hint The range of f(x) is quite tricky to find — you might find it helpful to think about what happens to f(x) as $x \to \infty$ and sketch the graph.

Q5 $f(x) = \dfrac{x-4}{x+3}$ for $\{x : x > -3\}$.

a) Find $f^{-1}(x)$, stating the domain and range.

b) Evaluate $f^{-1}(0)$.

c) Evaluate $f^{-1}\left(-\dfrac{2}{5}\right)$.

Q6 Find the domain and range of $f^{-1}(x)$ for the following functions:

a) $f(x) = \log_{10}(x-3)$, $x > 3$

b) $f(x) = 4x - 2$, $1 \le x \le 7$

c) $f(x) = \dfrac{x}{x-2}$ for $\{x : x < 2\}$

d) $f(x) = 3^{x-1}$ for $\{x : x \ge 2\}$

e) $f(x) = \tan x$, $0° \le x < 90°$

f) $f(x) = \ln(x^2)$ for $\{x : 3 \le x \le 4\}$

Q7 Find the inverse $f^{-1}(x)$ for the following functions, giving the domain and range:

a) $f(x) = e^{x+1}$, $x \in \mathbb{R}$

b) $f(x) = x^3$, $x < 0$

c) $f(x) = 2 - \log_2(x)$, $x \ge 1$

d) $f(x) = \dfrac{1}{x-2}$ for $\{x : x \ne 2\}$

e) $f(x) = \dfrac{1}{e^x}$, $x \in \mathbb{R}$

f) $f(x) = \log_{10} e^x$, $x \in \mathbb{R}$

E Q8 $g: x \to x^2$, $x \in \mathbb{R}$, $x < 0$.

a) Find $g^{-1}(x)$. *[2 marks]*

b) State the domain and range of $g^{-1}(x)$. *[2 marks]*

E Q9 $f(x) = 2x + 3$, $x \in \mathbb{R}$. Sketch $y = f(x)$ and $y = f^{-1}(x)$ on the same set of axes, marking the points where the functions cross the axes. *[3 marks]*

Q10 $f(x) = x^2 + 3$, $x > 0$.

a) Sketch the graphs of $f(x)$ and $f^{-1}(x)$ on the same set of axes.

b) State the domain and range of $f^{-1}(x)$.

Q11c) Problem Solving $f(x)$ and $f^{-1}(x)$ are only ever equal on the line $y = x$.

E **P** Q11 $f(x) = x^2 - 5$, $x \ge 0$.

a) Find $f^{-1}(x)$, stating its domain. *[3 marks]*

b) Sketch $y = f(x)$ and $y = f^{-1}(x)$ on the same set of axes. *[2 marks]*

c) Solve $f(x) = f^{-1}(x)$, giving exact answer(s). *[3 marks]*

Q12 $f(x) = \dfrac{1}{x+1}$ for $\{x : x > -1\}$.

a) Sketch the graphs of $f(x)$ and $f^{-1}(x)$ on the same set of axes.

b) Explain how your diagram shows that there is just one solution to the equation $f(x) = f^{-1}(x)$.

P Q13 $f(x) = \dfrac{1}{x-3}$, $x > 3$.

    a) Find $f^{-1}(x)$ and state its domain and range.

    b) Sketch $f(x)$ and $f^{-1}(x)$ on the same set of axes.

    c) How many solutions are there to the equation $f(x) = f^{-1}(x)$?

    d) Solve $f(x) = f^{-1}(x)$.

E
P Q14 The function f is defined for all real numbers x by $f(x) = 5e^{-x} + k$, where k is a constant.

    a) State the range of the function f in terms of k. *[1 mark]*

    b) Find $f^{-1}(x)$ in terms of k. *[3 marks]*

    c) State the domain and range of $f^{-1}(x)$ in terms of k. *[2 marks]*

    d) Given that $y = f^{-1}(x)$ crosses the x-axis at $(8, 0)$, state the value of k. *[2 marks]*

E Q15 $f(x) = \sqrt[3]{1 - 4x}$, $x \in \mathbb{R}$.

    a) Find $f^{-1}(2)$, using the fact that if $f^{-1}(a) = b$ then $f(b) = a$. *[2 marks]*

    b) State the domain and range of $f^{-1}(x)$. *[1 mark]*

    The diagram below shows $y = f(x)$ and $y = x$.

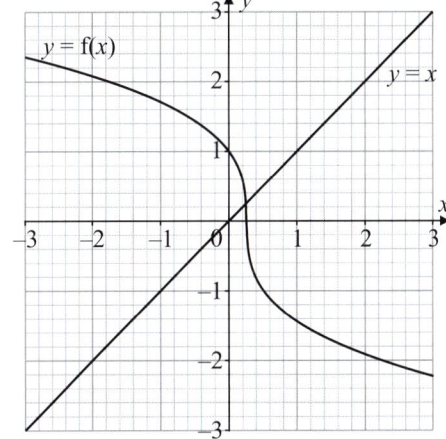

    c) Sketch the graph of $y = f^{-1}(x)$. *[2 marks]*

    d) Deduce the number of real values of x for which $f(x) = f^{-1}(x)$.
       Explain your answer. *[1 mark]*

Challenge

P Q16 $f(x) = \dfrac{3x + 6}{x^2 - 2x - 8}$, $x < 4$.

    a) Find $f^{-1}(x)$ and state its domain and range.

    b) Show that there is one solution to $f^2(x) = f^{-1}(x)$ within the domain of $f^{-1}(x)$.

E Q17 $h(x) = \log_3\left(\dfrac{3^x + 1}{3^x - 1}\right)$, $x > 0$.

    a) Find the range of $h(x)$. Explain your answer. *[2 marks]*

    b) Find $h^{-1}(x)$, stating its domain. *[4 marks]*

E Exam Style   P Problem Solving   M Modelling

2.11 | The Modulus Function

Modulus of a number

The **modulus** of a number is its **size** — it doesn't matter if it's positive or negative. So for a positive number, the modulus is just the same as the number itself, but for a negative number, the modulus is its numerical value without the minus sign. The modulus is sometimes called the **absolute value**.

> The modulus of a number, x, is written $|x|$.
>
> In general terms, for $x \geq 0$, $|x| = x$ and for $x < 0$, $|x| = -x$.

For example, the modulus of 8 is 8, and the modulus of -8 is also 8. This is written $|8| = |-8| = 8$.

Modulus of a function

Functions can have a modulus too — the modulus of a function $f(x)$ is just $f(x)$ but with any negative values that it can take turned positive. Suppose $f(x) = -6$, then $|f(x)| = 6$. In general terms:

> $|f(x)| = f(x)$ when $f(x) \geq 0$ and $|f(x)| = -f(x)$ when $f(x) < 0$.

If the modulus is inside the brackets in the form $f(|x|)$, then you make the x-value positive **before** applying the function. So $f(|-2|) = f(2)$.

Modulus graphs

$y = |f(x)|$

- For the graph of $y = |f(x)|$, any **negative** values of $f(x)$ are made **positive** by **reflecting** them in the **x-axis**.

- This **restricts** the **range** of the modulus function to $|f(x)| \geq 0$ (or some subset within $|f(x)| \geq 0$, e.g. $|f(x)| \geq 1$).

- The easiest way to draw a graph of $y = |f(x)|$ is to initially draw $y = f(x)$, then **reflect** the negative part in the **x-axis**.

$y = f(|x|)$

- For the graph of $y = f(|x|)$, the **negative** x-values produce the same result as the corresponding **positive** x-values. So the graph of $f(x)$ for $x \geq 0$ is **reflected** in the **y-axis** for the negative x-values.

- The range of $f(|x|)$ will be the **same** as the range of $f(x)$ for values of $x \geq 0$.

- To draw a graph of $y = f(|x|)$, first draw the graph of $y = f(x)$ for **positive** values of x, then **reflect** this in the **y-axis** to form the rest of the graph.

For the graph of $y = |f(-x)|$, you reflect the entire graph in the y-axis first and then reflect the negative part in the x-axis.

Example **1**

Given that f(x) = 5x − 5, draw the following graphs and state the range of each.

a) $y = |f(x)|$

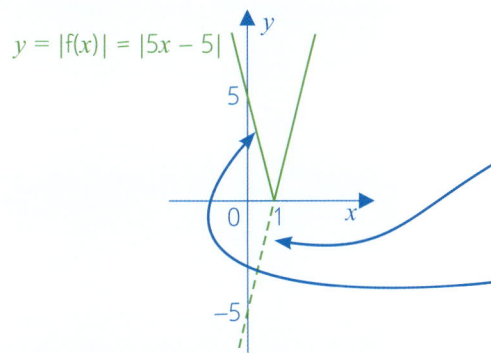

$y = |f(x)| = |5x - 5|$

Draw the graph of $y = 5x - 5$ — where the graph goes below the x-axis (i.e. for negative y-values), draw it as a dashed line.

Reflect the negative (dashed) part of the line in the x-axis. This reflection along with the positive part of the original graph gives the graph of $y = |f(x)|$.

The range of this is $|f(x)| \geq 0$. ← Read the range straight from the graph.

b) $y = f(|x|)$

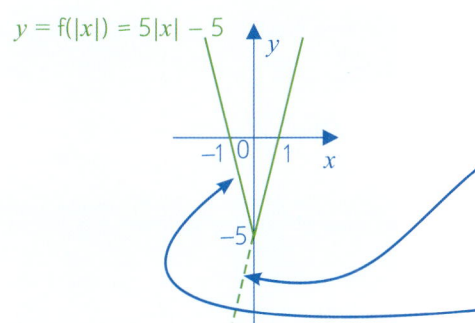

$y = f(|x|) = 5|x| - 5$

Draw the graph of $y = 5x - 5$ again, but this time the dashed part of the line is everything to the left of the y-axis.

For negative x-values, reflect the part of the line where x is positive in the y-axis.

Tip For $y = |f(x)|$ graphs you reflect the dashed line in the x-axis, but for these $y = f(|x|)$ graphs you reflect the solid line in the y-axis.

The range of this is $f(|x|) \geq -5$. ← Read the range straight from the graph.

Example **2**

To the right is the graph of f(x) = x^2 − 4x. Use this to sketch the graph of $y = |f(-x)|$.

$y = f(x) = x^2 - 4x$

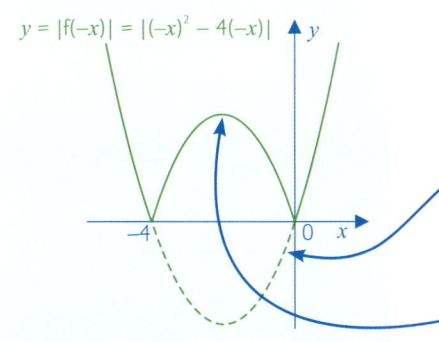

$y = |f(-x)| = |(-x)^2 - 4(-x)|$

Reflect the entire graph of $y = x^2 - 4x$ in the y-axis to give f($-x$) — where the graph goes below the x-axis (i.e. for negative y-values), draw it as a dashed line.

Reflect the negative (dashed) part of the line in the x-axis.

Chapter 2

Exercise 2.11

Q1 Sketch the following graphs, labelling any axis intercepts, and state the range of each:

a) $y = |x + 3|$ b) $y = |5 - x|$ c) $y = |3x - 1|$

d) $y = |x| - 9$ e) $y = 2|x| + 5$ f) $y = 3|x| - 11$

Q2 For each of the following functions, sketch the graph of $y = |f(x)|$, labelling any axis intercepts:

a) $f(x) = 2x + 3$ b) $f(x) = 4 - 3x$ c) $f(x) = -4x$

d) $f(x) = 7 - \frac{1}{2}x$ e) $f(x) = -(x + 2)$ f) $f(x) = -1 - 5x$

Q3 Match up each graph (1-4) with its correct equation (a-d):

a) $y = |x| + 4$

b) $y = |2x - 10|$

c) $y = |x + 1|$

d) $y = |2x| - 2$

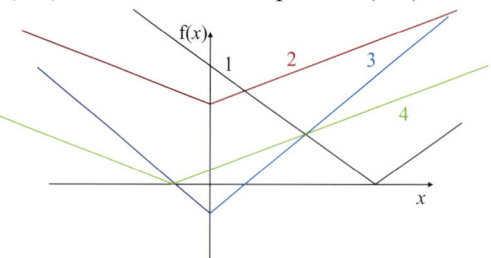

Q3 Hint Try sketching the graphs from the given functions to see what shape they should be, then compare them in terms of where they cross the x and y axes.

Q4 a) For each of the graphs below, sketch the graph of $y = |f(x)|$:

(i)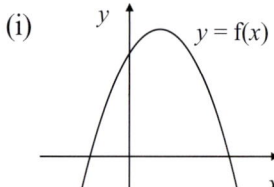

(ii)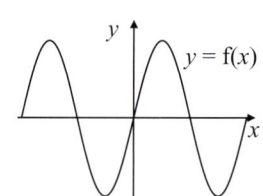

(iii)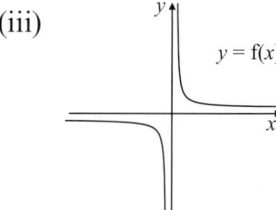

b) For part (i) above, sketch the graph of $y = |f(-x)|$.

E P Q5 Draw the graph of the function $f(x) = \begin{cases} |2x + 4| & x < 0 \\ |x - 4| & x \geq 0 \end{cases}$. Label all axis intercepts. *[4 marks]*

P Q6 For the function $f(x) = 3x - 5$:

a) Draw, on the same axes, the graphs of $y = f(x)$ and $y = |f(x)|$.

b) How many solutions are there to the equation $|3x - 5| = 2$?

Q6b) Problem Solving

Read across from 2 on the y-axis to find the number of values of x for which $|3x - 5| = 2$.

P Q7 For the function $f(x) = 4x + 1$:

a) Draw accurately the graph of $y = |f(-x)|$.

b) Use your graph to solve the equation $|f(-x)| = 3$.

E Q8 The diagram on the right shows the graph of $f(x) = 2 - x$:

a) Let $g(x) = f(|x|)$. Draw accurately the graph of $y = g(x)$. *[1 mark]*

b) Let $h(x) = |g(x)|$. On separate axes, draw the graph of $y = h(x)$. *[1 mark]*

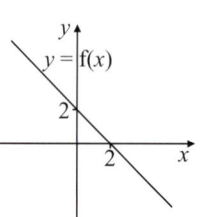

E **P** Q9 The graph of $y = k(x - 2)^2(x - 1)$, where k is a constant, is shown below.

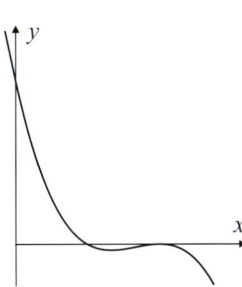

a) Sketch $y = |f(x)|$ on separate axes, showing the coordinates of all points of intersection with the coordinate axes. *[3 marks]*

b) Sketch $y = f(|x|)$ on separate axes, showing the coordinates of all points of intersection with the coordinate axes. *[3 marks]*

c) The equation $f(|x|) = a$ can have at most N solutions, where N varies depending on the value of a.

Write down the greatest possible value of N, giving a reason for your answer. *[2 marks]*

Challenge

E **P** Q10 a) Draw the graph of $y = |3 - x|$ for $0 \leq x \leq 10$. *[2 marks]*

b) The equation $|3 - x| = \frac{1}{2}x + k$, $x \in \mathbb{R}$, has one solution.
Use your graph in part a) to work out the value of k. *[2 marks]*

c) $f(x) = |3 - x|$, $0 \leq x \leq 10$.
Find all possible values of k such that the equation
$f(x) = \frac{1}{2}x + k$ has one real solution. *[2 marks]*

2.12 **Solving Modulus Equations and Inequalities**

You might be asked to substitute a modulus value into an expression to find the possible values that the expression could take.

Example **1**

a) If $|x| = 2$, what are the possible values of $5x - 3$?

$x = 2 \implies 5x - 3 = 5(2) - 3 = 10 - 3 = 7$ ← If $|x| = 2$, then either $x = 2$ or $x = -2$. So substitute these two values into the expression.

$x = -2 \implies 5x - 3 = 5(-2) - 3 = -10 - 3 = -13$

b) Find all of the possible values of $|3x - 4|$ when $|x| = 6$.

$x = 6 \implies |3(6) - 4| = |18 - 4| = |14| = 14$ ← $|x| = 6$ means $x = 6$ or $x = -6$. Substitute each of these values into $|3x - 4|$.

$x = -6 \implies |3(-6) - 4| = |-18 - 4| = |-22| = 22$

|f(x)| = n and |f(x)| = g(x)

The method for solving equations of the form **|f(x)|= n** is shown below.
Solving **|f(x)| = g(x)** is exactly the same — just replace n with g(x).

Step 1: Sketch the functions $y = |f(x)|$ and $y = n$ on the same axes.
The solutions you're trying to find are where they **intersect**.

Step 2: From the graph, work out the ranges of x for which f(x) $\geq$ 0 and f(x) < 0:
e.g. f(x) $\geq$ 0 for $x \leq a$ or $x \geq b$ and f(x) < 0 for $a < x < b$.
These ranges should 'fit together' to cover **all** possible x-values.

Step 3: Use this to write **two new equations**, one true for each range of x:
① f(x) = n   for $x \leq a$ or $x \geq b$
② –f(x) = n   for $a < x < b$

Step 4: Solve each equation and check that any solutions are **valid**.
Get rid of any solutions outside the range of x you have for that equation.

Step 5: Look at the graph and **check** that your solutions look right.

Sketching the graphs will show you where the solutions should be — it's easy to accidentally find solutions of f(x) = g(x) that **don't** satisfy |f(x)| = g(x) if you don't draw the graphs first.

Example 2

Solve $|2x - 4| = 5 - x$.

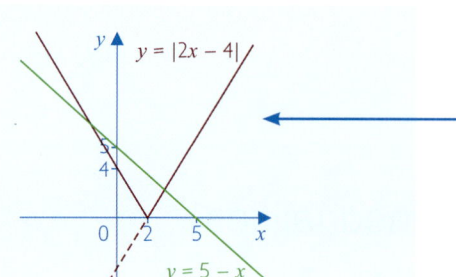

This question is an example of |f(x)| = g(x), where f(x) = 2x – 4 and g(x) = 5 – x.

Sketch $y = |2x - 4|$ and $y = 5 - x$.
To draw $y = |2x - 4|$, start with $y = 2x - 4$ and then reflect the part below the x-axis in the x-axis. The graphs cross twice.

$2x - 4 \geq 0$ when $x \geq 2$
$2x - 4 < 0$ when $x < 2$

Look at where f(x) $\geq$ 0 and where f(x) < 0.

① $2x - 4 = 5 - x$ for $x \geq 2$
② $-(2x - 4) = 5 - x$ for $x < 2$

Form two new equations for f(x) = g(x) and –f(x) = g(x) over the correct domains.

① $3x = 9 \implies x = 3$
(valid because $x \geq 2$)

② $-x = 1 \implies x = -1$
(valid because $x < 2$)

Solve the equations separately.

Tip Make sure that you check whether your solutions are within the valid range for each equation.

It's a good idea to always check your solutions against the graph that you sketched.
In the example above there are two solutions and they're where we'd expect based on the graph.

$|f(x)| = |g(x)|$

When using **graphs** to solve functions of the form $|f(x)| = |g(x)|$ you have to do a bit more work at the start to identify the different areas of the graph. There could be regions where:

- $f(x)$ and $g(x)$ are **both** positive or **both** negative — for solutions in these regions you need to solve the equation **$f(x) = g(x)$**.

- One function is **positive** and the other is **negative** — for solutions in these regions you need to solve the equation **$-f(x) = g(x)$**.

(Solving $-f(x) = -g(x)$ is the same as solving $f(x) = g(x)$, and solving $f(x) = -g(x)$ is the same as solving $-f(x) = g(x)$.)

There is also an **algebraic** method for solving equations of this type:

> If $|a| = |b|$ then $a^2 = b^2$.
> So if $|f(x)| = |g(x)|$ then $[f(x)]^2 = [g(x)]^2$.

> **Tip** The squaring method works for solving $|f(x)| = n$ as well — write $[f(x)]^2 = n^2$, then rearrange and solve.

This is true because squaring gives the **same answer** whether the value is **positive** or **negative**. You'll usually be left with a **quadratic** to solve, but in some cases this might be easier than using a graphical method.

The following example shows how you could use either method to solve the same equation.

Example 3

> **Tip** For this example the algebraic method involves less work than using graphs.

Solve $|x - 2| = |3x + 4|$.

$|x - 2| = |3x + 4|$

$\Rightarrow (x - 2)^2 = (3x + 4)^2$ ⟵ Square both sides of the equation.

$\Rightarrow x^2 - 4x + 4 = 9x^2 + 24x + 16$ ⟵ Expand and rearrange into the form of a general quadratic.

$\Rightarrow 8x^2 + 28x + 12 = 0$

$\Rightarrow 2x^2 + 7x + 3 = 0$

> **Tip** You can use the quadratic formula to solve it if it won't easily factorise.

$\Rightarrow (2x + 1)(x + 3) = 0$ ⟵ Factorise and solve.

$x = -\frac{1}{2}$ and $x = -3$

You can check the solutions in the example above by sketching the graphs:

There are two intersections — one where $f(x)$ is negative but $g(x)$ is positive (shaded green) and the other where $f(x)$ and $g(x)$ are both negative (i.e. where $x < -1\frac{1}{3}$).

These correspond to the two solutions $x = -\frac{1}{2}$ and -3.

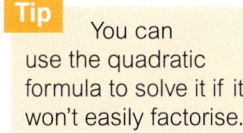

Solving inequalities

Inequalities with a modulus can be a bit nasty to solve. Just remember that $|x| < 5$ means that $x < 5$ and $-x < 5$ (which is the same as $x > -5$). So $|x| < 5 \Rightarrow -5 < x < 5$. Similarly, if $|x| > 5$, you'd end up with $x > 5$ and $x < -5$. In general, for $a > 0$:

$$|x| < a \Rightarrow -a < x < a$$
$$|x| > a \Rightarrow x > a \text{ or } x < -a$$

Using this, you can **rearrange** more complicated inequalities like $|x - a| \leq b$. From the method above, this means that $-b \leq x - a \leq b$, so **adding** a to **each bit** of the inequality gives $a - b \leq x \leq a + b$.

Example 4

Solve $|x - 4| < 7$

$|x - 4| < 7 \Rightarrow -7 < x - 4 < 7$ ◄——— Use the theory above to rewrite the inequality without the modulus.

$-7 + 4 < x < 7 + 4$, so $-3 < x < 11$ ◄——— Now add 4 to each bit to leave x on its own.

For more complicated modulus inequalities, it's often helpful to draw the graph — take a look at this next example:

Example 5

Solve $|2x - 1| > 4 - 0.5x$. Give your answer in set notation.

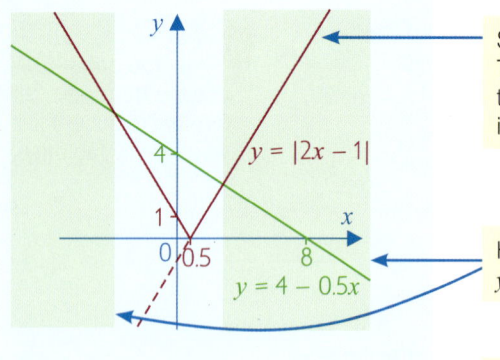

Sketch the graphs of $y = |2x - 1|$ and $y = 4 - 0.5x$. To draw $y = |2x - 1|$, start with $y = 2x - 1$ and then reflect the part below the x-axis in the x-axis. The graphs cross twice.

Highlight the areas where the graph of $y = |2x - 1|$ is above the graph of $y = 4 - 0.5x$.

When $x \geq 0.5$, $2x - 1 \geq 0$:
$2x - 1 = 4 - 0.5x \Rightarrow x = 2$
$2 > 0.5$, so this solution is valid.

When $x < 0.5$, $2x - 1 < 0$:
$-(2x - 1) = 4 - 0.5x \Rightarrow x = -2$
$-2 < 0.5$, so this solution is valid.

Form two equations, for f(x) = g(x) and $-$f(x) = g(x) over the correct domains, then solve. This tells you the bounds of the shaded regions.

$x < -2$ or $x > 2$ ◄——— You can see that there are two regions that satisfy the inequality — the region less than $x = -2$ and the region greater than $x = 2$.

$\{x : x < -2\} \cup \{x : x > 2\}$ ◄——— This is the answer in set notation.

Exercise 2.12

Q1 Solve the following equations:

a) $|x - 2| = 6$ b) $|4x + 2| = 10$ c) $2 - |3x - 4| = 1$

d) $9 - |x + 3| = 0$ e) $|2x + 3| = 1$ f) $|2 - x| = 4$

Q2 If $|x| = 5$, find the possible values of $|3x + 2|$

Q3 If $|x| - 2 = -1$, find the possible values of $|7x - 1|$

Q4 If $|x| = 3$, find the possible values of $|-2x + 1|$

Q5 a) On the same axes sketch the graphs of $|f(x)|$ and $g(x)$, where:
$f(x) = 2x + 3$ and $g(x) = x - 1$

b) Hence find any solutions of the equation $|f(x)| = g(x)$.

Q6 a) On the same axes sketch the graphs of $|f(x)|$ and $|g(x)|$, where:
$f(x) = 5x + 10$ and $g(x) = x + 1$

b) Hence find any solutions of the equation $|f(x)| = |g(x)|$.

> **Q7b)-c) Problem Solving**
>
> Put in your values for C and then rearrange into a more familiar form.

E P M Q7 a) Sketch $y = \left|\frac{1}{4}x - 3\right|$ for $0 \leq x \leq 20$. *[2 marks]*

The cost £C of making an international telephone call lasting t minutes

is modelled by the equation $C = \frac{1}{2}t - 1 + \left|\frac{1}{4}t - 3\right|$, $t \geq 0$.

b) Find the length of a telephone call costing £4.50. *[3 marks]*

c) Find the length of a telephone call costing £36.
Give your answer in minutes and seconds. *[3 marks]*

Q8 Solve, either graphically or otherwise:

a) $|x + 2| = |2x|$ b) $|4x - 1| = |2x + 3|$ c) $|3x - 6| = |10 - 5x|$

P Q9 If $|4x + 1| = 3$, find the possible values of $2|x - 1| + 3$

Q10 Solve the following inequalities:

a) $|x| < 8$ b) $|x| \geq 5$ c) $|2x| > 12$

d) $|4x + 2| \leq 6$ e) $3 \geq |3x - 3|$ f) $6 - 2|x + 4| < 0$

g) $3x + 8 < |x|$ h) $2|x - 4| \geq x$ i) $|x - 3| \geq |2x + 3|$

E Q11 Solve the inequality $x + 6 \leq |3x + 2|$. Give your answer in set notation. *[4 marks]*

E P Q12 a) Solve the equation $|3x + 4| = 1 - x$. *[3 marks]*

b) Hence solve $|3x + 4| \leq 1 - x$, giving your answer in set notation. *[1 mark]*

P Q13 Find the possible values of $|5x + 4|$, given that $|1 + 2x| \leq 3$.

2.13 Transformations of Graphs

The four transformations

The transformations you've met before are translations (a vertical or horizontal shift), stretches (either vertical or horizontal) and reflections in the x- or y-axis. Here's a quick reminder of what each one does.

$y = f(x + a)$

For $a > 0$: $f(x + a)$ is $f(x)$ translated a **left**,

$f(x - a)$ is $f(x)$ translated a **right**.

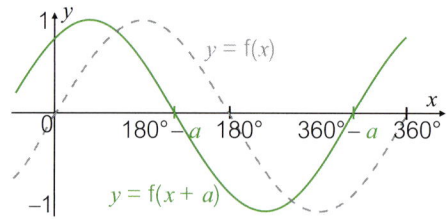

$y = f(x) + a$

For $a > 0$: $f(x) + a$ is $f(x)$ translated a **up**,

$f(x) - a$ is $f(x)$ translated a **down**.

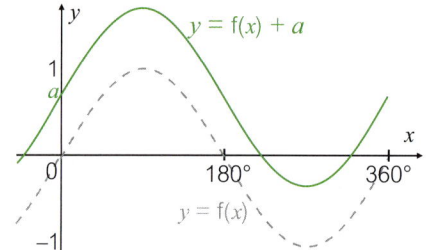

$y = af(x)$

The graph of $af(x)$ is $f(x)$ **stretched** parallel to the **y-axis** (i.e. vertically) by a factor of a.

And if $a < 0$, the graph is also **reflected** in the **x-axis**.

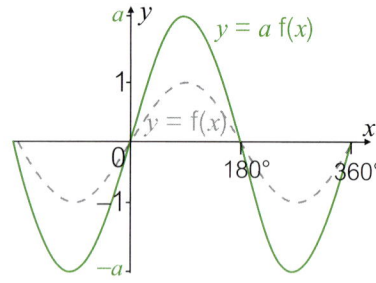

$y = f(ax)$

The graph of $f(ax)$ is $f(x)$ **stretched** parallel to the **x-axis** (i.e. horizontally) by a factor of $\frac{1}{a}$.

And if $a < 0$, the graph is also **reflected** in the **y-axis**.

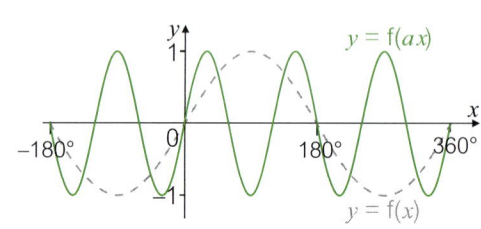

Combinations of transformations

Combinations of transformations can look a bit tricky, but if you take them one step at a time they're not too bad. Don't try and do all the transformations at once — break it up into the separate bits shown on the previous page and draw a graph for each stage.

Example 1

The graph below shows the function $y = f(x)$.

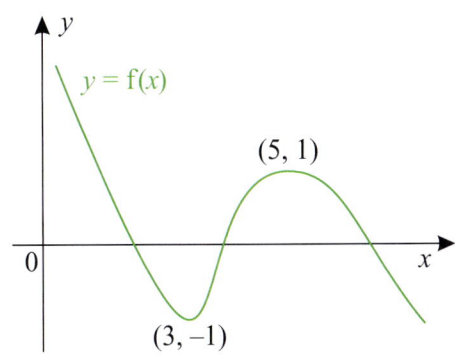

Draw the graph of $y = 3f(x + 2)$, showing the coordinates of the turning points.

Tip Make sure you do the transformations the right way round — you should do the bit in the brackets first.

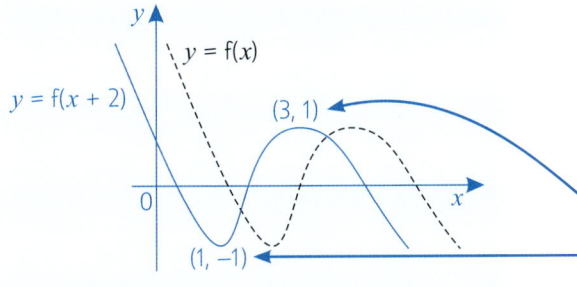

Don't try to do everything at once. First draw the graph of $y = f(x + 2)$ — it is the graph of $f(x)$ translated left by 2 units.

Work out the coordinates of the turning points:
$(5, 1)$ is translated to $(5 - 2, 1) = (3, 1)$
$(3, -1)$ is translated to $(3 - 2, -1) = (1, -1)$

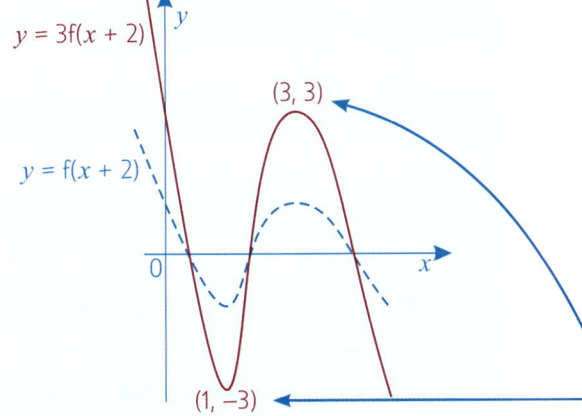

Now use your graph of $y = f(x + 2)$ to draw the graph of $y = 3f(x + 2)$ — it's a stretch in the direction of the y-axis with scale factor 3.

Work out the coordinates of the turning points:
$(3, 1)$ is stretched to $(3, 1 \times 3) = (3, 3)$
$(1, -1)$ is stretched to $(1, -1 \times 3) = (1, -3)$

Example 2

The graph on the right shows the function $f(x) = |x|$.

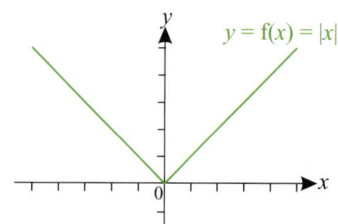

a) Draw the graph of $f(x)$ after a translation by $\begin{pmatrix} -1 \\ 4 \end{pmatrix}$ and give its equation.

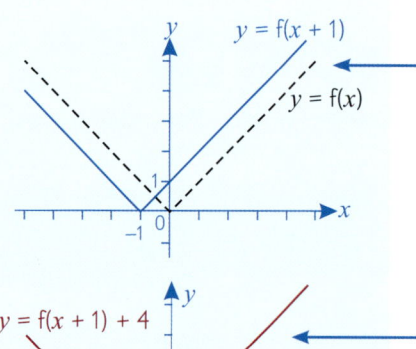

You need to do a horizontal translation left by 1, which translates $f(x)$ into $f(x + 1)$.

Then a vertical translation upwards by 4, which translates $f(x + 1)$ into $f(x + 1) + 4$.

Tip If you're not sure whether your graph is correct, try putting some numbers into the function and checking them against coordinates on the graph.

$y = f(x + 1) + 4$

Write the translated equation in terms of $f(x)$.

$y = |x + 1| + 4$

Replace $f(x)$ with the original function $|x|$.

b) The function $g(x)$ shown on the right is a translation of $f(x)$. Give the translation vector for this transformation, and the equation of $g(x)$.

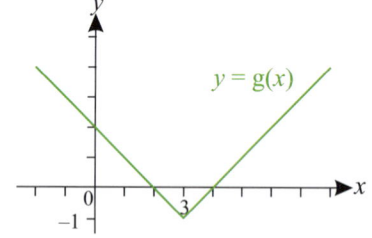

The translation vector is $\begin{pmatrix} 3 \\ -1 \end{pmatrix}$.

The point (0, 0) has moved to (3, −1) so this is a horizontal translation of 3 to the right, and a vertical translation of 1 downwards.

Horizontal translation of 3 to the right $= f(x - 3)$

Vertical translation of 1 downwards $= f(x - 3) - 1$

To find the equation of $g(x)$, use the equations for each translation.

$g(x) = f(x - 3) - 1$

$\Rightarrow g(x) = |x - 3| - 1$

Now replace $f(x)$ with the actual function given above — $|x|$.

Example 3

The graph to the right shows
the function $y = \sin x$, $0° \le x \le 360°$.
Draw the graph of $y = 2 - \sin 2x$, $0° \le x \le 360°$,
giving the coordinates of the turning points.

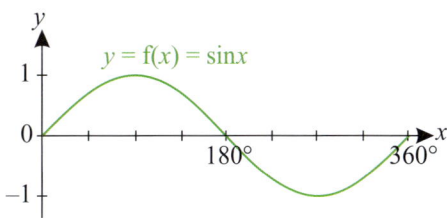

$y = 2 - \sin 2x$
$\Rightarrow y = 2 - f(2x)$
$\Rightarrow y = -f(2x) + 2$ ⟵ Rewrite $y = 2 - \sin 2x$ in terms of $f(x)$.

The transformations are:

A horizontal stretch by a factor of $\frac{1}{2}$, ⟵ Now list the transformations that need to be done.
a vertical stretch by a factor of -1,
and a vertical translation by 2 up.

$f(x)$ turning points: $(90°, 1)$ and $(270°, -1)$. ⟵ Keep track of the turning points with each translation.

$f(2x)$ turning points will be
$(45°, 1)$, $(135°, -1)$, $(225°, 1)$ and $(315°, -1)$. ⟵ $f(2x)$ halves the x-coordinates — so there are an extra two within the domain, each one occurring a further 90° along the x-axis.

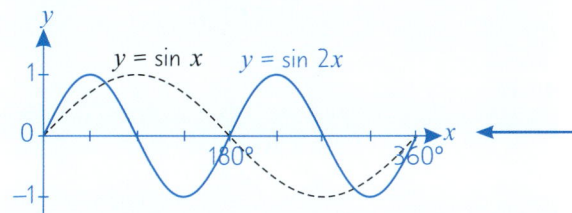

Draw the graph of $y = \sin 2x$, by squashing the graph horizontally by a factor of 2 (i.e. a stretch by a factor of $\frac{1}{2}$).

$-f(2x)$ turning points will be
$(45°, -1)$, $(135°, 1)$, $(225°, -1)$ and $(315°, 1)$. ⟵ $-f(2x)$ flips the turning points, so multiply the y-coordinates by -1.

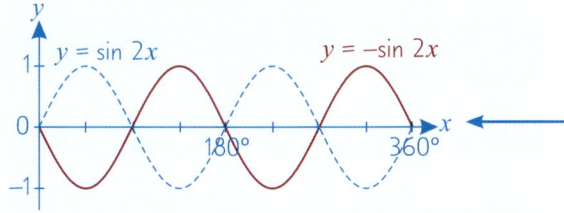

A stretch with a factor of -1 doesn't change the size of the graph — you just have to reflect in the x-axis. So draw $y = -\sin 2x$, by reflecting in the x-axis.

$-f(2x) + 2$ turning points will be
$(45°, 1)$, $(135°, 3)$, $(225°, 1)$ and $(315°, 3)$. ⟵ Add 2 to the y-coordinates of the turning points to give the turning points of $-f(2x) + 2$.

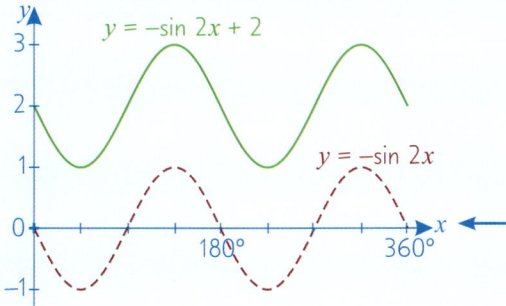

Finally, translate the graph of $y = -\sin 2x$ up by 2 to get the graph of $y = -\sin 2x + 2$ (or $y = 2 - \sin 2x$).

Exercise 2.13

Q1 Given that $f(x) = x^2$, sketch the following graphs on the same axes:

a) $y = f(x)$ b) $y = f(x) + 3$ c) $y = f(x - 2)$ d) $y = f(x + 4) - 1$

In each case, write down the coordinates of the turning point, and in parts b)-d), give the translation as a column vector.

E Q2 The graph of $f(x) = x^3$ is translated to form the graph of $g(x) = f(x - 1) + 4$.

a) Sketch the graphs of $y = f(x)$ and $y = g(x)$.   *[2 marks]*

b) Give this translation as a column vector.   *[1 mark]*

c) What is the equation of $g(x)$?   *[1 mark]*

Q3 Given that $f(x) = |x|$, sketch the following graphs on the same axes:

a) $y = f(x)$ b) $y = f(x) + 2$ c) $y = f(x - 4)$ d) $y = 2f(x + 1)$

In parts b)-d) describe the transformation from $y = f(x)$ in words.

Q4 Let $f(x) = |2x - 6|$. On the same axes sketch the graphs of:

a) $y = f(x)$ b) $y = f(-x)$ c) $y = f(-x) + 2$

Q5 Let $f(x) = \dfrac{1}{x}$. On the same axes sketch the graphs of:

a) $y = f(x)$ b) $y = -f(x)$ c) $y = -f(x) - 3$

Q6 Let $f(x) = e^x$. On the same axes sketch the graphs of:

a) $y = f(x)$ b) $y = f(3x - 2)$ c) $y = f(2x + 1) - 1$

Q7 a) Let $f(x) = \cos x$. Sketch the graph $y = f(x)$ for $0° \leq x \leq 360°$.

b) On the same axes sketch the graph of $y = f(2x)$.

c) On the same axes sketch the graph of $y = 1 + f(2x)$.

d) State the coordinates of the minimum point(s) of $y = \cos 2x + 1$, in the interval $0° \leq x \leq 360°$.

Q8 Complete the following table for the function $f(x) = \sin x$ $(0° \leq x \leq 360°)$.

Transformed function	New equation	Maximum value of transformed function	Minimum value of transformed function
$f(x) + 2$			
$f(x - 90°)$			
$f(3x)$			
$4f(x)$			

Q9 Complete the following table for the function $f(x) = x^3$:

Transformed function	New equation	Coordinates of point of inflection
$f(x) + 1$		
$f(x - 2)$		
$-f(x) - 3$		
$f(-x) + 4$		

E **Q10** $y = \cos x$ is translated by the vector $\begin{pmatrix} 90° \\ 0 \end{pmatrix}$ and stretched by scale factor $\frac{1}{2}$ parallel to the y-axis.

    a) Sketch the new graph for $0 \leq x \leq 360°$.   *[1 mark]*

    b) Write down its equation.   *[1 mark]*

Q11 a) Sketch the graph of $y = f(x)$ where $f(x) = \frac{1}{x}$.

    b) Write down the sequence of transformations needed to map $f(x)$ on to $g(x) = 3 - \frac{1}{x}$.

    c) Sketch the graph of $y = g(x)$.

Q12 Complete the following table:

Original graph	New graph	Sequence of transformations
$y = x^3$	$y = (x - 4)^3 + 5$	
$y = 4^x$	$y = 4^{3x} - 1$	
$y = |x + 1|$	$y = 1 - |2x + 1|$	
$y = \sin x$	$y = -3\sin 2x + 1$	

P **Q13** a) Write $y = 2x^2 - 4x + 6$ in the form $y = a[(x + b)^2 + c]$.

    b) Hence list the sequence of transformations that will map $y = x^2$ on to $y = 2x^2 - 4x + 6$.

    c) Sketch the graph of $y = 2x^2 - 4x + 6$.

    d) Write down the coordinates of the minimum point of the graph.

Q14 Starting with the curve $y = \cos x$, state the sequence of transformations which could be used to sketch the following curves:

    a) $y = 4 \cos 3x$ b) $y = 4 - \cos 2x$ c) $y = 2 \cos(x - 60°)$

Q15 The diagram to the right shows $y = f(x)$ with a minimum point, P, at $(2, -3)$. Copy the diagram and sketch each of the following graphs. In each case state the new coordinates of the point P.

    a) $y = f(x) + 5$

    b) $y = f(x + 4)$

    c) $y = -f(x)$

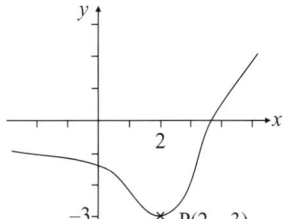

E P Q16 $f(x) = \sin x$, $x \in \mathbb{R}$. The graph of $y = af(bx)$, where a and b are positive constants, satisfies the following properties:

- The graph crosses the x-axis at $x = 0°$ and $x = 540°$ and nowhere in between.
- There is a minimum point at $(-270°, -2)$.

Find the values of a and b. *[2 marks]*

Q17 The diagram shows $y = f(x)$ with a minimum point, Q, at $(-1, -3)$ and a maximum point, P, at $(2, 4)$.

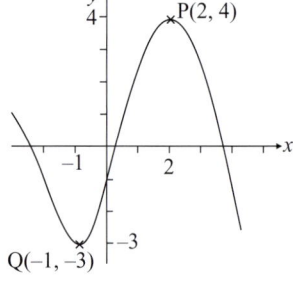

Copy the diagram and sketch each of the following graphs. In each case state the new coordinates of the points P and Q.

a) $y = f(x - 1) + 3$ b) $y = -f(2x)$ c) $y = |f(x + 2)|$

E P Q18 The graph of $y = f(x)$ is shown in the figure below.

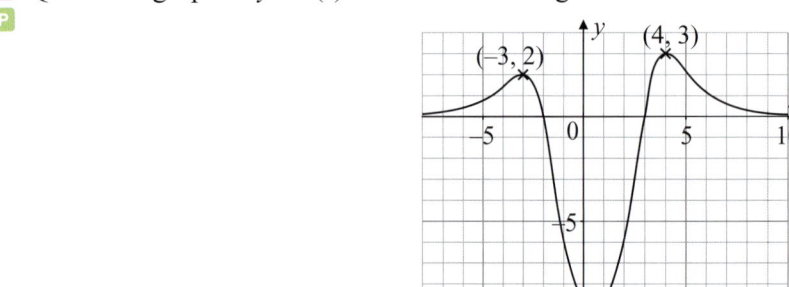

a) Sketch $y = f(|x|)$, labelling all maximum and minimum points. *[2 marks]*

b) Sketch $y = f(|x - 1|)$, labelling all maximum and minimum points. *[2 marks]*

c) Find all possible values of k such that $y = f(|x - k|)$ passes through the origin. *[1 mark]*

d) Find all possible values of k such that $y = f(x) + k$ passes through the origin. *[1 mark]*

E P Q19 $f(x) = 3x^2 + 1$, $x \in \mathbb{R}$.

The curve $y = f(x)$ undergoes the following transformations in the order given:

- It is translated four units to the right.
- It is reflected in the x-axis.
- It is stretched by a scale factor of $\frac{1}{2}$ parallel to the x-axis.
- It is translated two units down.

Find the new equation of the curve in the form $y = ax^2 + bx + c$. *[3 marks]*

Challenge

E P Q20 a) Describe fully the single transformation that transforms the graph of $y = |x + 3|$ into

   (i) the graph of $y = |x - 5|$, *[1 mark]*

   (ii) the graph of $y = |2x + 3|$, *[1 mark]*

b) Describe two different combinations of a translation followed by a stretch that transform the graph of $y = |x + 3|$ into the graph of $y = |2x - 6|$. *[2 marks]*

P Q21 If $f^{-1}(x + 1) = \ln x$, find $f(x)$.

Describe the transformation that maps $y = f(x)$ on to $y = f^{-1}(x + 1)$.

2.14   Partial Fractions — No Repeated Factors

You can split a fraction with **more than one linear factor** in the denominator into **partial fractions**.

This means writing it as a **sum** of two or more **simpler fractions**.

The **denominators** of these simpler fractions will be **factors** of the denominator of the original fraction.

This is useful in lots of areas of maths, such as **integration** (see p.310-311) and **binomial expansions** (p.192-193).

The way that you split the fraction up depends on what the denominator is:

- $\dfrac{7x-7}{(2x+1)(x-3)}$ can be written as partial fractions of the form $\dfrac{A}{(2x+1)} + \dfrac{B}{(x-3)}$.

- $\dfrac{7x-1}{(x-3)(x-1)(x+2)}$ can be written as partial fractions of the form $\dfrac{A}{(x-3)} + \dfrac{B}{(x-1)} + \dfrac{C}{(x+2)}$.

- $\dfrac{21x-2}{9x^2-4}$ can be written as partial fractions of the form $\dfrac{A}{(3x-2)} + \dfrac{B}{(3x+2)}$.

If you're asked to write an algebraic fraction as partial fractions, start by writing the partial fractions out with A, B and C as numerators as shown above. You might have to factorise the denominator first, like in the last example.

The tricky bit is **working out** what A, B and C are. Follow this method:

1. Write out the expression as an identity, e.g. $\dfrac{7x-7}{(2x+1)(x-3)} \equiv \dfrac{A}{(2x+1)} + \dfrac{B}{(x-3)}$.

2. Add the partial fractions together, i.e. write them over a common denominator.

3. Cancel the denominators from both sides (they'll be the same).

4. This will give you an identity for A and B, for example: $7x - 7 \equiv A(x-3) + B(2x+1)$

5. Use the **substitution** method or the **equating coefficients** method:

   - **Substitution** — substitute a suitable number for x to leave you with just one constant on the right-hand side. For example, substituting $x = 3$ into the above equation gives $7(3) - 7 = B(2(3) + 1) \Rightarrow B = 2$.

   - **Equating Coefficients** — equate the constant terms, coefficients of x and coefficients of x^2, then solve the equations simultaneously. For example, expanding the brackets in the above equation and collecting like terms gives $7x - 7 \equiv (A + 2B)x + (-3A + B) \Rightarrow A + 2B = 7$ and $-3A + B = -7$. These two equations can then be solved simultaneously.

Example 1

Express $\dfrac{7x-1}{(x-3)(x-1)(x+2)}$ in partial fractions.

$\dfrac{7x-1}{(x-3)(x-1)(x+2)} \equiv \dfrac{A}{(x-3)} + \dfrac{B}{(x-1)} + \dfrac{C}{(x+2)}$ ← Write it out as an identity.

$\dfrac{A}{(x-3)} + \dfrac{B}{(x-1)} + \dfrac{C}{(x+2)}$

$\equiv \dfrac{A(x-1)(x+2) + B(x-3)(x+2) + C(x-3)(x-1)}{(x-3)(x-1)(x+2)}$

> Add the partial fractions by writing them over a common denominator.

> **Tip** Always check that each term will cancel to produce the original fraction.

$7x-1 \equiv A(x-1)(x+2) + B(x-3)(x+2) + C(x-3)(x-1)$

> Cancel the denominators from both sides of the original identity.

Substituting $x = 3$:
$21 - 1 = A(3-1)(3+2) + 0 + 0$
$\Rightarrow 20 = 10A \Rightarrow A = 2$

Substituting $x = 1$:
$7 - 1 = 0 + B(1-3)(1+2) + 0$
$\Rightarrow 6 = -6B \Rightarrow B = -1$

Substituting $x = -2$:
$-14 - 1 = 0 + 0 + C(-2-3)(-2-1)$
$\Rightarrow -15 = 15C \Rightarrow C = -1$

> Substitute values of x which make one of the expressions in brackets equal zero to get rid of all but one of A, B and C.

$\dfrac{7x-1}{(x-3)(x-1)(x+2)} \equiv \dfrac{2}{(x-3)} - \dfrac{1}{(x-1)} - \dfrac{1}{(x+2)}$

> Finally, replace A, B and C in the original identity.

> **Tip** Don't forget to write out your solution like this once you've done all the working.

Generally it's easier to try the **substitution** method first. In the example above, it's actually quite tricky to solve the simultaneous equations that you get by equating coefficients (see below).

$7x - 1 \equiv A(x - 1)(x + 2) + B(x - 3)(x + 2) + C(x - 3)(x - 1)$
$\equiv A(x^2 + x - 2) + B(x^2 - x - 6) + C(x^2 - 4x + 3)$
$\equiv (A + B + C)x^2 + (A - B - 4C)x + (-2A - 6B + 3C)$

> Compare coefficients in the numerators — expand the brackets and collect like terms.

① $0 = A + B + C$
② $7 = A - B - 4C$
③ $-1 = -2A - 6B + 3C$

> Equating coefficients gives a set of 3 equations in 3 unknowns. This is quite tricky to solve, but you could start by adding all 3 to eliminate A and C.

Solving these equations simultaneously gives
$A = 2$, $B = -1$ and $C = -1$ (as above).

Example 2

Express $\dfrac{3-x}{x^2+x}$ in partial fractions.

$\dfrac{3-x}{x^2+x} \equiv \dfrac{3-x}{x(x+1)}$ — First you need to factorise the denominator.

$\dfrac{3-x}{x(x+1)} \equiv \dfrac{A}{x} + \dfrac{B}{x+1}$ — Now write it as an identity with partial fractions.

$\dfrac{3-x}{x(x+1)} \equiv \dfrac{A(x+1)+Bx}{x(x+1)}$ — Add the partial fractions and cancel the denominators from both sides.

$\Rightarrow 3-x \equiv A(x+1)+Bx$

$\Rightarrow 3-x \equiv A+(A+B)x$

Equating constant terms: $3 = A$ — Here it's easier to equate coefficients because A is the only letter that appears in the constant term.

Equating x coefficients: $-1 = A + B$

$\Rightarrow -1 = 3 + B \Rightarrow B = -4$

$\dfrac{3-x}{x^2+x} \equiv \dfrac{3}{x} - \dfrac{4}{(x+1)}$ — Replace A and B in the identity.

Exercise 2.14

Q1 Express $\dfrac{3x+3}{(x-1)(x-4)}$ in the form $\dfrac{A}{x-1} + \dfrac{B}{x-4}$.

Q2 Express $\dfrac{5x-1}{x(2x+1)}$ in the form $\dfrac{A}{x} + \dfrac{B}{2x+1}$.

Q3 Find the values of the constants A and B in the identity $\dfrac{3x-2}{x^2+x-12} \equiv \dfrac{A}{x+4} + \dfrac{B}{x-3}$.

E Q4 Write $\dfrac{2}{x^2-16}$ in partial fractions. *[3 marks]*

Q5 Factorise x^2-x-6 and hence express $\dfrac{5}{x^2-x-6}$ in partial fractions.

Q6 Write $\dfrac{11x}{2x^2+5x-12}$ in partial fractions.

E P Q7 $f(x) = \dfrac{x+13}{x^2+2x-15}$, $x > 3$

a) Express $f(x)$ in partial fractions. *[3 marks]*

b) Hence, or otherwise, describe the behaviour of $f(x)$ as $x \to \infty$. Explain your answer. *[2 marks]*

Q8 a) Factorise x^3-9x fully.

b) Hence write $\dfrac{12x+18}{x^3-9x}$ in partial fractions.

E **Q9** Write $\dfrac{3x+9}{x^3-36x}$ in the form $\dfrac{A}{x}+\dfrac{B}{x+6}+\dfrac{C}{x-6}$. *[3 marks]*

P **Q10** a) Use the factor theorem to fully factorise x^3-7x-6.

b) Hence write $\dfrac{6x+2}{x^3-7x-6}$ in partial fractions.

P **Q11** Express the following in partial fractions:

a) $\dfrac{6x+4}{(x+4)(x-1)(x+1)}$ b) $\dfrac{15x-27}{x^3-6x^2+3x+10}$ c) $\dfrac{2x+7}{x^3-2x^2-5x+6}$

d) $\dfrac{162}{x^3-81x}$ e) $\dfrac{6-x}{2x^3-7x^2+7x-2}$ f) $\dfrac{x+4}{15x^3-x^2-2x}$

Challenge

P **Q12** Express $\dfrac{x^4}{(x-1)(x-2)}$ in partial fractions.

Q12 Problem Solving

Start with algebraic division.

2.15 Partial Fractions — Repeated Factors

If the denominator of an algebraic fraction has **repeated linear factors** the partial fractions will take a slightly **different form**, as shown in the examples below.

- The **power** of the repeated factor tells you **how many** times that factor should appear in the partial fractions.

  E.g. $\dfrac{7x-3}{(x+1)^2(x-4)}$ is written as $\dfrac{A}{(x+1)}+\dfrac{B}{(x+1)^2}+\dfrac{C}{(x-4)}$

- A factor that's **squared** in the original denominator is in the denominator of **two** of your partial fractions — once squared and once just as it is.

  E.g. $\dfrac{32x-14}{x^2(2x+7)}$ is written as $\dfrac{A}{x}+\dfrac{B}{x^2}+\dfrac{C}{(2x+7)}$

Problem Solving

A factor that's cubed will appear three times — once cubed, once squared and once just as it is.

Example 1

Express $\dfrac{5x+12}{x^2(x-3)}$ in partial fractions.

$\dfrac{5x+12}{x^2(x-3)} \equiv \dfrac{A}{x}+\dfrac{B}{x^2}+\dfrac{C}{(x-3)}$ ← x is a repeated factor so it appears once as x and once as x^2 in the identity.

$\dfrac{5x+12}{x^2(x-3)} \equiv \dfrac{Ax(x-3)+B(x-3)+Cx^2}{x^2(x-3)}$ ← Add the partial fractions.

$5x+12 \equiv Ax(x-3)+B(x-3)+Cx^2$ ← Cancel the denominators from both sides, so that the numerators are equal.

$15+12 = 0+0+C(3^2)$ ← Substitute $x=3$ to get rid of A and B.

$\Rightarrow 27 = 9C \Rightarrow C = 3$

continued on the next page...

$0 + 12 = 0 + B(0 - 3) + 0$ ← Substitute $x = 0$ to get rid of A and C.
$\Rightarrow 12 = -3B \Rightarrow B = -4$

Coefficients of x^2 are: $0 = A + C$ ← There's no value of x you can substitute to get rid of B and C and just leave A,
You know $C = 3$, so: $0 = A + 3 \Rightarrow A = -3$ so equate coefficients of x^2.

$\dfrac{5x + 12}{x^2(x - 3)} \equiv -\dfrac{3}{x} - \dfrac{4}{x^2} + \dfrac{3}{(x - 3)}$ ← Replace A, B and C in the identity.

Example 2

Express $\dfrac{4x + 15}{(x + 2)^2(3x - 1)}$ in partial fractions.

$\dfrac{4x + 15}{(x + 2)^2(3x - 1)} \equiv \dfrac{A}{(x + 2)} + \dfrac{B}{(x + 2)^2} + \dfrac{C}{(3x - 1)}$ ← Write the identity.

$\dfrac{A}{(x + 2)} + \dfrac{B}{(x + 2)^2} + \dfrac{C}{(3x - 1)}$
$\equiv \dfrac{A(x + 2)(3x - 1) + B(3x - 1) + C(x + 2)^2}{(x + 2)^2(3x - 1)}$ ← Add the partial fractions.

$4x + 15 \equiv A(x + 2)(3x - 1) + B(3x - 1) + C(x + 2)^2$ ← Cancel the denominators.

$-8 + 15 = 0 + B(-6 - 1) + 0$ ← Substitute $x = -2$ to get rid of A and C.
$\Rightarrow 7 = -7B \Rightarrow B = -1$

$\dfrac{4}{3} + 15 = 0 + 0 + C\left(\dfrac{1}{3} + 2\right)^2$ ← Substitute $x = \dfrac{1}{3}$ to get rid of A and B.
$\Rightarrow \dfrac{49}{3} = \dfrac{49}{9}C \Rightarrow C = 3$

Equate coefficients of x^2: $0 = 3A + C$ ← There's no value of x you can substitute to get rid of B and C to just leave A, so
You know $C = 3$, so: $-3 = 3A \Rightarrow A = -1$ try equating coefficients instead.

$\dfrac{4x + 15}{(x + 2)^2(3x - 1)} \equiv -\dfrac{1}{(x + 2)} - \dfrac{1}{(x + 2)^2} + \dfrac{3}{(3x - 1)}$ ← Replace A, B and C in the original identity.

There is another method you could have used here to find A — when there's no value of x which will get rid of B and C, you can substitute in any simple value of x (e.g. $x = 0$ or 1) and the values that you have calculated for B and C. From this you can find A.

Exercise 2.15

Q1 Express $\dfrac{3x}{(x+5)^2}$ in the form $\dfrac{A}{(x+5)} + \dfrac{B}{(x+5)^2}$.

Q2 Write $\dfrac{5x+2}{x^2(x+1)}$ in the form $\dfrac{A}{x} + \dfrac{B}{x^2} + \dfrac{C}{(x+1)}$.

Q3 Write the following in partial fractions.

a) $\dfrac{2x-7}{(x-3)^2}$
b) $\dfrac{6x+7}{(2x+3)^2}$
c) $\dfrac{7x}{(x+4)^2(x-3)}$
d) $\dfrac{11x-10}{x(x-5)^2}$

E **Q4** Express $\dfrac{5x+10}{x^3-10x^2+25x}$ in partial fractions. *[4 marks]*

Q5 Express $\dfrac{3x+2}{(x-2)(x^2-4)}$ in partial fractions.

Q6 Express $\dfrac{7x+3}{x^2(3-2x)}$ in partial fractions.

E **Q7** Express $\dfrac{6x^2+3x+6}{(x-2)(x+1)^2}$ in the form $\dfrac{A}{x-2} + \dfrac{B}{x+1} + \dfrac{C}{(x+1)^2}$,
where A, B and C are constants to be found. *[4 marks]*

E **Q8** a) Express $f(x) = \dfrac{(3x-1)(2x-3)}{(x-1)^2(x-3)}$ in partial fractions. *[4 marks]*

b) For $x > 3$, $f(x) > k$, where k is a constant. State the value of k. *[1 mark]*

c) Find the range of $f(x)$ on the domain $1 < x < 3$. Explain your answer. *[2 marks]*

P **Q9** Find the value of c such that $\dfrac{x+17}{(x+1)(x+c)^2} = \dfrac{1}{x+1} - \dfrac{1}{x+c} + \dfrac{5}{(x+c)^2}$

P **Q10** a) Show that $2x^3 + 7x^2 + 4x - 4$ can be factorised as $(x+2)(ax+b)(cx+d)$,
where a, b, c and d are integers to be found.

b) Hence, express $\dfrac{x-13}{2x^3+7x^2+4x-4}$ in partial fractions.

Challenge

Q11 a) Express $\dfrac{2x^2}{(x+1)(2x+1)^2}$ in partial fractions in the form

$\dfrac{A}{x+1} + \dfrac{B}{2x+1} + \dfrac{C}{(2x+1)^2}$, where A, B and C are constants to be found.

The function $g(x) = \dfrac{x^2+3}{(x+a)(2x+1)^2}$ can be written in the form $\dfrac{J}{x+a} + \dfrac{K}{2x+1} + \dfrac{L}{(2x+1)^2}$.

b) Work out the values of J, K and L, giving your answers in terms of a.
Fully simplify your answers.

c) The domain of $g(x)$ is $x \in \mathbb{R}$, $x \neq \pm b$, $b > 0$. What is the value of b? Hence, find a.

2   Review Exercise

Q1   Simplify the following:

a) $\dfrac{4x^2 - 25}{6x - 15}$

b) $\dfrac{2x + 3}{x - 2} \times \dfrac{4x - 8}{2x^2 - 3x - 9}$

c) $\dfrac{x^2 - 3x}{x + 1} \div \dfrac{x}{2}$

E   **Q2**   a)   Use the factor theorem to show that $(x - 1)$ is a factor of $f(x) = x^3 - x^2 + x - 1$.   *[1 mark]*

b)   Hence simplify fully $\dfrac{x^3 - x^2 + x - 1}{x^3 + x}$.   *[2 marks]*

c)   Hence sketch $y = \dfrac{x^3 - x^2 + x - 1}{x^3 + x}$.   *[3 marks]*

E **P**   **Q3**   a)   Simplify fully $f(x) = \dfrac{x^3 - 4x^2 + 3x}{1 - x}$.   *[2 marks]*

b)   Hence, or otherwise, find the equation of the tangent of the curve $y = f(x)$ at the point where $x = 2$.   *[4 marks]*

Q4   Write the following as a single fraction:

a) $\dfrac{x}{2x + 1} + \dfrac{3}{x^2} + \dfrac{1}{x}$

b) $\dfrac{2}{x^2 - 1} - \dfrac{3x}{x - 1} + \dfrac{x}{x + 1}$

c) $\dfrac{2}{(x + 1)^2} - \dfrac{x}{x + 1} + \dfrac{1}{3x}$

E   **Q5** $f(x) = \dfrac{3x - 2}{x^2 + 7x + 12} - \dfrac{2}{x + 4}$.

a)   Write $f(x)$ as a single fraction in its simplest form.   *[3 marks]*

b)   Write down the values of x for which $f(x)$ is undefined.   *[1 mark]*

c)   Solve the equation $f(x) = -2$.   *[4 marks]*

E   **Q6** $f(x) = \dfrac{1}{x} - \dfrac{2x}{x + 1}$

a)   Rewrite $f(x)$ as the quotient $f(x) = \dfrac{g(x)}{h(x)}$, where $g(x)$ and $h(x)$ are both fully factorised quadratic functions to be found.   *[3 marks]*

b)   The graph $y = f(x)$ is plotted. What are the equations of the vertical asymptotes of this graph?   *[2 marks]*

E **P**   **Q7**   a)   Simplify $\dfrac{x^3 - 2x^2 + x}{2x^2} \times \dfrac{4x}{x - 1}$.   *[3 marks]*

b)   Hence show that $\dfrac{\sin^3 x - 2\sin^2 x + \sin x}{2\sin^2 x} \times \dfrac{4\sin x}{\sin x - 1} = \dfrac{1}{4}$ has no real solutions.   *[2 marks]*

E   **Q8**   a)   Simplify fully $\dfrac{y - x}{x^2 - 2xy + y^2} \div \dfrac{2}{x^2 + 2xy + y^2}$.   *[3 marks]*

b)   Hence, or otherwise, explain which values of x and y would result in the calculation giving a negative answer.   *[2 marks]*

Q9   Write $2x^3 + 8x^2 + 7x + 8$ in the form $(Ax^2 + Bx + C)(x + 3) + D$.
      Using your answer, state the result when $2x^3 + 8x^2 + 7x + 8$ is divided by $(x + 3)$.

P Q10   Divide $x^4 + x^3 - 5x^2 - 7x - 2$ by $x + 1$, hence find a solution to $x^4 + x^3 - 5x^2 - 7x = 2$.

E Q11   Given that $r(x) = 2x^3 + 9x^2 - 5x - 39$, solve $r(x) = x + 1$.   *[6 marks]*
P

Q12   For the following mappings, state the range and say whether or not the mapping is a function.
      If not, explain why, and if so, say whether the function is one-to-one or many-to-one.
      a) $f(x) = x^2 - 16, \ x \geq 0$ b) $f : x \to x^2 - 7x + 10, \ x \in \mathbb{R}$ c) $f(x) = \sqrt{x}, \ x \in \mathbb{R}$

P Q13 $f(x) = \dfrac{5}{2x + 1}$ defines a mapping.
      a) Evaluate $f(0)$ and $f(\frac{1}{2})$.
      b) Draw the mapping diagram for the domain $x \in \mathbb{N}, x < 6$ and list the range.
      c) Is the mapping a function for the domain $x \in \mathbb{Z}$? If not, explain why not.
      d) Is the mapping a function for the domain $x \in \mathbb{R}$? If not, explain why not.

Q14   a) Sketch the graph of the function $f(x) = \begin{cases} x^2 - 2 & -2 < x < 2 \\ 2 & \text{otherwise} \end{cases}$
      b) State the range of the function.

E Q15   The share price £$P(t)$ of Company A, which manufactures widgets, is modelled by
P   the function $P(t)$, where t is time in days after 31st August 2020. It is given that:
M
$$P(t) = \begin{cases} 2.4t + 36, & 0 \leq t < 10 \\ 0.25t + 57.5, & 10 \leq t < 50 \\ -3t + 220, & 50 \leq t < 65 \\ 2t - 105, & 65 \leq t < 90 \end{cases}$$

      a) Explain what the value 36 represents in the first equation.   *[1 mark]*
      b) Sketch the graph of $P(t)$ against t.   *[2 marks]*
      c) Give the range of the function P.   *[1 mark]*

      The share price of a different company, Company B, over the same time period
      is modelled by the function $S(t) = \frac{1}{3}t + 30, \ 0 \leq t < 90$.

      d) Give the range of function S.   *[1 mark]*
      e) Find all values of t when the share prices of both companies are equal.   *[3 marks]*
      f) For how many days during the 90-day period does Company B have
          a share price greater than or equal to the share price of Company A?   *[1 mark]*

Q16   For each pair of functions f and g, find fg(2), gf(1) and fg(x).
      a) $f(x) = \dfrac{3}{x}, x > 0$ and $g(x) = 2x + 3, \ x \in \mathbb{R}$
      b) $f(x) = 3x^2, x \geq 0$ and $g(x) = x + 4, \ x \in \mathbb{R}$

Q17 $f(x) = \log_{10} x$ and $g(x) = 10^{x+1}$.

a) Find the values of fg(1), gf(1), $f^2(10)$ and $g^2(-1)$.

b) Explain why $f^2(1)$ is undefined.

Q18 $f(x) = 3x$ and $g(x) = x + 7$, both with domain $x \in \mathbb{R}$.
Find the composite functions fg(x), gf(x) and $g^2(x)$.

E **Q19** a) For the functions

$$f(x) = 2x - 3, x \in \mathbb{R} \quad , \quad g(x) = \frac{1}{x}, x > 0$$

find the following composite functions, along with the range and domain of each.

(i) fg(x)

(ii) gf(x) *[4 marks]*

b) f(x) is combined with a third function, h(x), to form the composite function hf(x), where h(x) and hf(x) are both linear. This is shown in the diagram below.

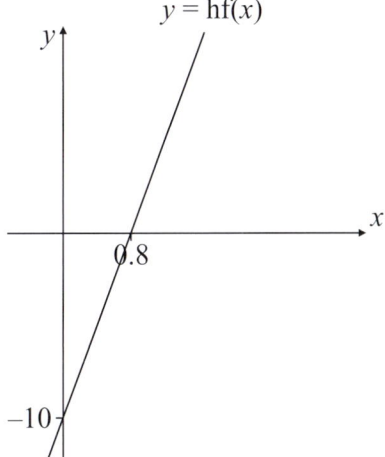

$y = hf(x)$

Use the information given in the diagram to find an expression for h(x). *[4 marks]*

Q20 A one-to-one function f has domain $x \in \mathbb{R}$ and range $f(x) \geq 3$.
Does this function have an inverse? If so, state its domain and range.

E **Q21** Using algebra, find the inverse of the function $f(x) = \sqrt{2x - 4}, x \geq 2$.
State the domain and range of the inverse. *[3 marks]*

Q22 $f(x) = \cos x, 0 \leq x \leq \frac{\pi}{2}$.
Does the inverse function $f^{-1}(x)$ exist? Justify your answer.

P **Q23** The function $f(x) = \frac{x}{x - 1}, x < 1$. Show that $f(x) = f^{-1}(x)$, hence find $f^2(x)$.

Q24 $f(x) = \log_{10}(x + 4), x > -4$. Find $f^{-1}(x)$.

Q25 For the function $f(x) = 2x - 1$, $x \in \mathbb{R}$, sketch the graphs of:

    a) $|f(x)|$ b) $f(|x|)$

Q26 Solve the equation $|3x - 1| = |4 - x|$.

E P M Q27 An assembled tent is modelled to be a triangular prism of length 4 m.
The edges of the triangular faces of the tent are modelled by the equations $y = 3 - 2|x|$
and $y = 0$, where values on the x-axis and y-axis are measured in metres.

    a) Find the volume of the tent in m³.   *[4 marks]*

    b) Assuming that a person requires 2 m² of floor space to sleep in the tent,
       find the maximum number of people that could sleep in the tent.   *[1 mark]*

    c) Explain why the actual number of people with enough room to sleep
       in the tent might be different from the value found in part b).   *[1 mark]*

P Q28 $f(x) = x^2 - 2x - 8$.

    a) On the same axes sketch the graphs of $|f(x)|$ and $f(|x|)$.

    b) Use your graphs to help you solve the equation $f(|x|) = -5$.

E P Q29 a) $f(x) = |2x - 5|$ and $g(x) = 3|x| + 4$.
       On the same set of axes, sketch the following, clearly labelling any axis-intercepts:

      (i) $y = f(x)$

      (ii) $y = g(x)$ *[4 marks]*

    b) Hence, or otherwise, solve the inequality $f(x) > g(x)$.   *[3 marks]*

    c) The inequality $f(x) > g(x) + k$, where k is a real number, has no real solutions.
       Find the smallest possible value of k.   *[1 mark]*

E Q30 In this question, a is a positive constant.

    a) Sketch $y = a + 3|x|$.   *[2 marks]*

    b) Solve the equation $3a - x = a + 3|x|$, giving your answers in terms of a.   *[3 marks]*

    c) Given that $a + 3|x| < 3a - x$ for $-20 < x < 10$, find the value of a.   *[2 marks]*

Q31 The graph on the right shows
the function $y = \cos x$, $0° \leq x \leq 360°$.
Sketch the graph of $y = -\cos 2x + 1$.

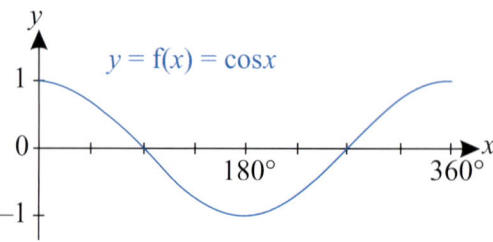

Q32 Sketch the graphs of:

    a) $y = 3x + 2$ b) $y = |3x + 2|$ c) $y = -|3x + 2|$

E Q33 a) Sketch $y = |\tan x|$, for $-180° \le x \le 180°$, indicating clearly on your sketch
 any asymptotes and any points of intersection with the coordinate axes. *[3 marks]*

 b) Solve the equation $|\tan x| = 1$, for $-180° \le x \le 180°$. *[2 marks]*

 c) Sketch $y = 1 - |\tan x|$ for $-180° \le x \le 180°$, indicating clearly on your sketch
 any asymptotes and any point of intersection with the coordinate axes. *[3 marks]*

E Q34 The diagram below shows a sketch of the curve $y = f(x)$, $-12 \le x \le 12$.

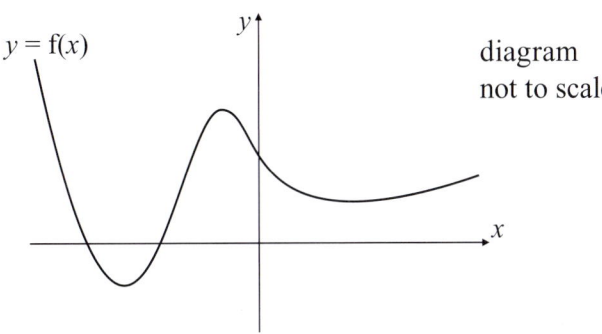

diagram
not to scale

The curve intersects the axes at $x = -9$, $x = -5$ and $y = 4$, and has a minimum at point $(-7, -2)$.

 a) For each of the following equations in the given intervals,
 state all the solutions that can be found with the information
 provided, and state the number of unknown solutions:

 (i) $f(x) = -f(x)$, $-12 \le x \le 12$

 (ii) $f(x) = f(|x|)$, $-12 \le x \le 0$

 (iii) $f(x) = f^{-1}(x)$, $-12 \le x \le 12$ *[6 marks]*

 b) On a separate set of axes, sketch the curve for $y = f(4x) + 2$.
 Clearly label any axis-intercepts and the interval for which it is valid. *[4 marks]*

Q35 Write $\dfrac{2x}{(x-5)(x+5)}$ as partial fractions in the form $\dfrac{A}{(x-5)} + \dfrac{B}{(x+5)}$.

Q36 Find the values of the constants A and B in the identity
 $\dfrac{2-x}{(3x+2)(x+1)} \equiv \dfrac{A}{(3x+2)} + \dfrac{B}{(x+1)}$.

E Q37 Find the values of the constants A and B in the identity
 $\dfrac{x-3}{x^2+3x+2} \equiv \dfrac{A}{(x+1)} + \dfrac{B}{(x+2)}$. *[3 marks]*

Q38 Express the following as partial fractions:

 a) $\dfrac{4}{x^2+x}$

 b) $\dfrac{4x+5}{(x+4)(2x-3)}$

 c) $\dfrac{5x}{x^2+x-6}$

 d) $\dfrac{10x}{(x+3)(2x+4)}$

 e) $\dfrac{6x+10}{(2x+1)(2-3x)}$

 f) $\dfrac{2x+1}{x^2+3x}$

E **Q39** $g(x) = x^3 + x^2 - x - 1$

    a) Given that $(x - 1)$ is a factor of $g(x)$, fully factorise $g(x)$.   *[3 marks]*

    b) Hence, express $\dfrac{x - 5}{x^3 + x^2 - x - 1}$ in partial fractions.   *[4 marks]*

Q40 Show that $\dfrac{2x - 5}{(x - 5)^2}$ can be written in the form $\dfrac{A}{(x - 5)} + \dfrac{B}{(x - 5)^2}$.

Q41 Express the following as partial fractions.

    a) $\dfrac{2x + 2}{(x + 3)^2}$ b) $\dfrac{-18x + 14}{(2x - 1)^2(x + 2)}$ c) $\dfrac{x - 5}{x^3 - x^2}$

P **Q42** Find the value of b such that $\dfrac{bx + 7}{(x + 1)^2(x + 2)} = \dfrac{3}{x + 1} + \dfrac{2}{(x + 1)^2} - \dfrac{3}{x + 2}$.

Challenge

E **P** **M** **Q43** The height, in metres, of a 3-year-old lemon tree is modelled by the function
$H(x) = 2 - (x - 1)^2$, $x \geq 0$, where x is the number of litres of fertiliser used on the tree's soil.
Mature lemon trees (older than 3 years) can be assumed to have stopped growing.

    a) Give one criticism of using this quadratic model for the relationship
    between the height and the amount of fertiliser used.   *[1 mark]*

    b) Using more than n litres of fertiliser will stop the tree from growing at all.
    Find the value of n.   *[3 marks]*

The average mass of lemons, $M(y)$ kg, that a lemon tree of height y metres produces
during one month is modelled by $M(y) = \dfrac{9}{2}(y - 1) + 8$, $y > 0.25$.

    c) Suggest one reason for the restriction $y > 0.25$.   *[1 mark]*

    d) Find the composite function $MH(x)$.   *[2 marks]*

    e) A lemon tree reaches the maximum possible height modelled by $H(x)$. Find the
    mass of lemons produced by this lemon tree in one year once it is mature.   *[3 marks]*

    f) Hence, or otherwise, find the domain and range of $MH(x)$.   *[4 marks]*

E **P** **Q44** The function $g: \mathbb{R} \to \mathbb{R}$ is defined by $g: x \to \begin{cases} 3x + 2, & x \leq 1 \\ 5 - (x - 1)^2, & x > 1 \end{cases}$

    a) Draw the graph of $y = g(x)$ for $-2 \leq x \leq 4$, labelling any axis intercepts.   *[4 marks]*

    b) Explain why $g^{-1}(x)$ is not a function.   *[1 mark]*

    c) Given that the domain of $g^{-1}(x)$ is $x > 0$, find the largest possible range
    of $g^{-1}(x)$, such that $g^{-1}(x)$ is a function.   *[2 marks]*

The outline of a logo in the shape of a shark's fin is modelled by the equation
$y = g(x)$, $A \leq x \leq B$ and the x-axis, where A and B are the roots of g.

    d) Find the area of the logo.   *[5 marks]*

E **Q45** Express $\left(\dfrac{1}{x} - \dfrac{x}{2x - 1}\right) \div \dfrac{x^2 + 2x - 3}{8x - 4}$ in partial fractions.

2   Chapter Summary

1   You can often simplify algebraic fractions by factorising the numerator and/or denominator. Then look for factors on the top that will cancel out with factors on the bottom.

2   To add or subtract algebraic fractions, you need to put both fractions over a common denominator, then you can add or subtract the numerators as normal.

3   To multiply algebraic fractions you just multiply the numerators and multiply the denominators. Dividing something by an algebraic fraction is the same as multiplying it by the reciprocal of the algebraic fraction.

4   There are two methods to divide a polynomial $f(x)$ by a divisor $d(x)$:
- By equating coefficients using the formula $f(x) = q(x)d(x) + r(x)$ where $q(x)$ is the quotient and $r(x)$ is the remainder. See p.15.
- Use algebraic long division — see p.17.

5   A mapping is a rule for transforming one set of numbers (the domain) onto another (the range). Functions transform each domain element to exactly one element in the range. Sketching graphs can help you identify if a mapping is or is not a function.

6   A mapping can be turned into a function by restricting its domain, e.g. $f(x) = \sqrt{x}$ isn't a function for the domain $x \in \mathbb{R}$ but is a function if $x \geq 0$.

7   Functions are one-to-one if each value in the range corresponds to exactly one value in the domain. Functions are many-to-one if some values in the range correspond to many values in the domain.

8   You can combine functions into composite functions by doing one followed by the other. E.g. $gf(x)$ means do f first, then g — order is important so usually $gf(x) \neq fg(x)$.

9   The domain and range of a composite function depends on the domains and ranges of the functions they are made up from. To solve equations involving composite functions you first work out an expression for the composite function and then use that to solve that equation.

10   An inverse function does the opposite of the original function — it transforms an element of the range into an element in the domain. One-to-one functions have inverses but many-to-one functions do not. You can find inverse functions using the method on p.34.

11   The modulus of a number n (usually written $|n|$) is its size, ignoring whether it is positive or negative. For functions $|f(x)| = f(x)$ when $f(x) \geq 0$ and $|f(x)| = -f(x)$ when $f(x) < 0$.

12   To draw $y = |f(x)|$ you can draw $y = f(x)$ and reflect anything below the x-axis in the x-axis. To draw $y = f(|x|)$ you can draw $y = f(x)$ for positive x and reflect in the y-axis to get the rest.

13   You can solve modulus equations of the form $|f(x)| = n$ and $|f(x)| = g(x)$ using the method on p.42. Sketching the modulus graph is often a good place to start.

14   To solve inequalities with a modulus it's usually best to split the modulus into two inequalities — one for positive values and one for negative values.

15   You should know the graph transformations for translations, stretches and reflections — to do combinations of graph transformations, just do them one transformation at a time.

16   You can split up algebraic fractions with more than one linear factor in the denominator into partial fractions (a sum of simpler fractions). The method to do this is given on page 53. It's a bit more complicated when you have repeated linear factors — see p.56.

E Exam Style   **P** Problem Solving   **M** Modelling

Trigonometry

3

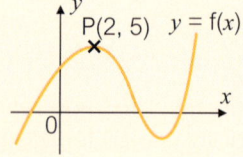

3.1 Radians

A **radian** (rad) is just another unit of measurement for an angle.

1 radian is the angle formed in a **sector** that has an **arc length** that is the same as the **radius**.

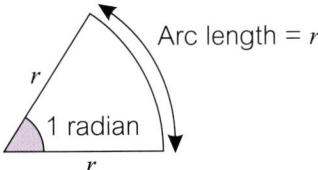

Arc length = r

r

1 radian

r

Tip Radians are sometimes shown by 'rad' after the number, or a 'ᶜ' (e.g. $4\pi^c$), but most of the time angles in radians are given without a symbol.

In other words, if you have a **sector** with an angle of **1 radian**, then the **length** of the **arc** will be exactly the **same length** as the **radius** r.

It's important to know how **radians relate to degrees**:

- 360 degrees (a complete circle) = 2π radians
- 180 degrees = π radians
- 1 radian is about 57 degrees

You also need to know how to **convert** between the two units. The table below shows you how:

Converting angles	
Radians to degrees: Divide by π, multiply by 180.	**Degrees to radians:** Divide by 180, multiply by π.

Here's a table of some of the **common angles** you're going to need, in degrees and radians:

Degrees	0	30	45	60	90	120	180	270	360
Radians	0	$\frac{\pi}{6}$	$\frac{\pi}{4}$	$\frac{\pi}{3}$	$\frac{\pi}{2}$	$\frac{2\pi}{3}$	π	$\frac{3\pi}{2}$	2π

It's a good idea to learn these common angles — they come up a lot.

Example 1

a) Convert $\frac{\pi}{15}$ into degrees.

$$\frac{\pi}{15} \div \pi = \frac{1}{15}$$ — To convert from radians to degrees, divide by π...

$$\frac{1}{15} \times 180 = 12°$$ — ... then multiply by 180.

Tip Notice how the angle was given without a symbol — if this happens, you can just assume it's in radians.

b) Convert 120 degrees into radians.

$$\frac{120}{180} \times \pi = \frac{2\pi}{3}$$

To convert from degrees to radians, divide by 180 and then multiply by π.

Tip This one's in the table on the previous page.

c) Convert 297 degrees into radians.

$$\frac{297}{180} \times \pi = 1.65\pi \text{ or } 5.18 \text{ rad (3 s.f.)}$$

Divide by 180 and then multiply by π.

Exercise 3.1

Q1 Convert the angles below into radians. Give your answers in terms of π.
   a) 180°
   b) 135°
   c) 270°
   d) 70°
   e) 150°
   f) 75°

Q2 Convert the angles below into degrees.
   a) $\frac{\pi}{4}$
   b) $\frac{\pi}{2}$
   c) $\frac{\pi}{3}$
   d) $\frac{5\pi}{2}$
   e) $\frac{3\pi}{4}$
   f) $\frac{7\pi}{3}$

E M Q3 Two projectiles, A and B, are launched from ground level.
Projectile A is launched at an angle of 65° to the horizontal
and projectile B is launched at an angle of $\frac{2\pi}{5}$ radians to the horizontal.
Which projectile has a steeper launch trajectory? Explain your answer. *[1 mark]*

3.2 Arc Length and Sector Area

A **sector** is part of a circle formed by **two radii** and part of the **circumference**. The **arc** of a sector is the **curved** edge of the sector. You can work out the **length** of the **arc**, or the **area** of the sector — as long as you know the **angle** at the **centre** (θ) and the **length** of the **radius** (*r*).

When working out arc length and sector area, you **always** work in radians.

Arc length

For a circle with **radius** *r*, a sector with **angle** θ (measured in **radians**) has **arc length** *s*, given by: $s = r\theta$

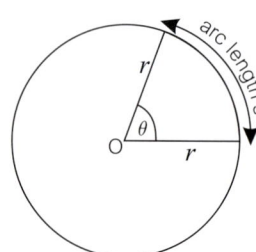

- If you put $\theta = 2\pi$ in this formula (and so make the sector equal to the whole circle), you find that the distance all the way round the outside of the circle is $s = 2\pi r$.

- This is just the normal **circumference** formula.

Sector area

For a circle with **radius r**, a sector with **angle θ** (measured in **radians**) has **area A**, given by:

$$A = \frac{1}{2}r^2\theta$$

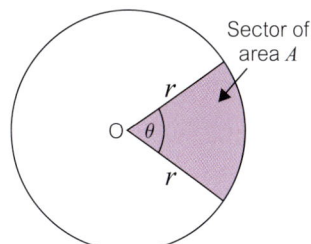

Sector of area A

- Again, if you put $\theta = 2\pi$ in the formula, you find that the area of the whole circle is $A = \frac{1}{2}r^2 \times 2\pi = \pi r^2$.

- This is just the normal '**area of a circle**' formula.

Example 1

Find the exact length L and area A in the diagram to the right.

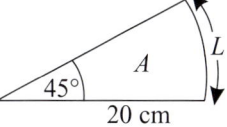

$$45° = \frac{45 \times \pi}{180} = \frac{\pi}{4} \text{ radians} \longleftarrow$$

Convert the angle to radians so you can use the formulas.

$$L = r\theta = 20 \times \frac{\pi}{4} = 5\pi \text{ cm} \longleftarrow$$

$$A = \frac{1}{2}r^2\theta = \frac{1}{2} \times 20^2 \times \frac{\pi}{4} = 50\pi \text{ cm}^2 \longleftarrow$$

Now put everything into each of your formulas.

Tip You could also have done the calculations in degrees:

$$L = \frac{45}{360} \times 2 \times \pi \times 20$$

$$A = \frac{45}{360} \times \pi \times 20^2$$

Example 2

Find the area of the shaded part of the symbol on the right.

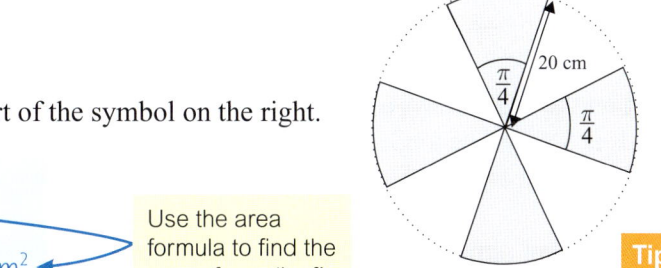

$$A = \frac{1}{2}r^2\theta \longleftarrow$$

$$= \frac{1}{2} \times 20^2 \times \frac{\pi}{4} = 50\pi \text{ cm}^2 \longleftarrow$$

Use the area formula to find the area of one 'leaf'.

$$4 \times 50\pi = 200\pi \text{ cm}^2 \longleftarrow$$

Multiply by 4 to find the total area.

Tip You could have also worked this out in one step using the total angle of all the shaded sectors (π).

Example 3

Find the exact value of θ in the diagram to the right.

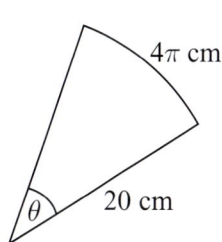

$$s = r\theta \implies 4\pi = 20\theta \longleftarrow$$

Use the formula for the arc length.

$$\implies \theta = \frac{4\pi}{20} = \frac{\pi}{5} \text{ radians} \longleftarrow$$

Rearrange to find the value of θ.

Example **4**

The sector shown has an area of 6π cm^2.
Find the arc length, s.

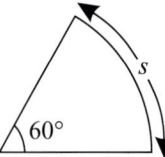

$60° = \dfrac{60 \times \pi}{180} = \dfrac{\pi}{3}$ radians ← First, get the angle in radians.

$A = \dfrac{1}{2}r^2\theta \Rightarrow 6\pi = \dfrac{1}{2} \times r^2 \times \dfrac{\pi}{3}$ ← Use the area formula to work out the radius.

$\Rightarrow 36 = r^2$

$\Rightarrow r = 6$

$s = r\theta = 6 \times \dfrac{\pi}{3} = 2\pi$ cm ← Substitute this value of r into the equation for arc length.

Tip

$\dfrac{\pi}{3}$ is one of the common angles from the table on p.67 — it's definitely worth learning them.

Exercise 3.2

Q1 The diagram below shows a sector OAB. The centre is at O and the radius is 6 cm. The angle AOB is 2 radians. Find the arc length and area of this sector.

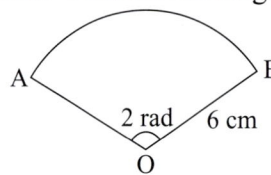

Q2 The diagram on the right shows a sector OAB. The centre is at O and the radius is 8 cm. The angle AOB is 46°. Find the arc length and area of this sector to 1 d.p.

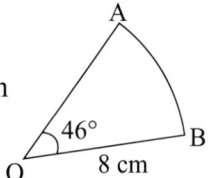

Q3 A sector of a circle of radius 4 cm has an area of 6π cm^2. Find the exact value of the angle θ.

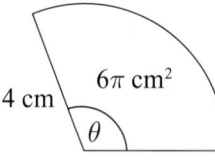

Q4 The diagram on the right shows a sector of a circle with a centre O and radius r cm. The angle AOB shown is θ.

For each of the following values of θ and r, give the arc length and the area of the sector. Where appropriate, give your answers to 3 s.f.

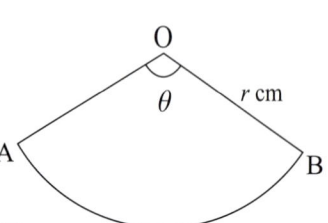

a) $\theta = 1.2$ radians, $r = 5$ cm   b) $\theta = 0.6$ radians, $r = 4$ cm

c) $\theta = 80°$, $r = 9$ cm   d) $\theta = \dfrac{5\pi}{12}$, $r = 4$ cm

E **P** Q5   The diagram below shows a sector ABC of a circle, where the angle BAC is 0.9 radians. Given that the area of the sector is 16.2 cm², find the arc length *s*.

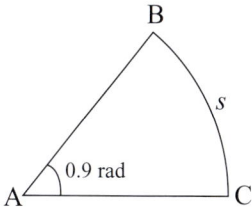

[2 marks]

Q6   A circle C has a radius of length 3 cm with centre O. A sector of this circle is given by angle AOB which is 20°. Find the length of the arc AB and the area of the sector. Give your answer in terms of π.

P Q7   The sector below has an arc length of 7 cm. The angle BAC is 1.4 rad. Find the area of the sector.

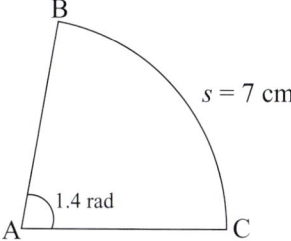

E **P** Q8   The sector ABC below is part of a circle, where the angle BAC is 50°. Given that the area of the sector is 20π cm², find the arc length BC. Give your answer in terms of π.

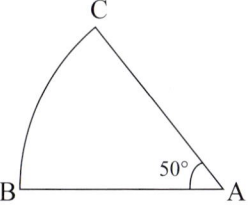

[3 marks]

P Q9   A circle of radius *r* contains a sector of area 80π cm². Given that the arc length of the sector is 16π cm, find the angle of the sector (θ) and the value of *r*, giving your answers to 3 s.f.

E **P** Q10   A sector of a circle has angle θ and radius *r* as shown in the diagram below.

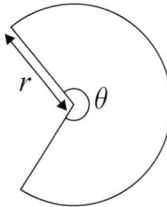

The perimeter of the sector is 22 cm and the area is 28 cm².
a) Find the value of the angle θ, given that $\theta > \pi$.   *[3 marks]*
b) Find the length of the radius, *r* cm.   *[1 mark]*

P Q11 The diagram below shows a semicircle of radius 2 cm, with a smaller sector of radius 1 cm removed. Given that the area of the sector A and the area of B are equal, find the exact value of θ.

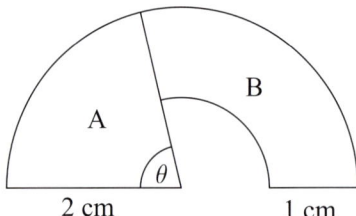

E **P** Q12 PS is an arc of a circle of radius 3 cm centred at O, as shown in the diagram below.

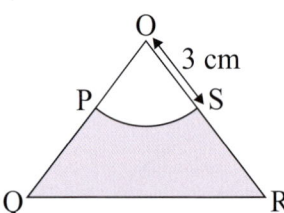

The triangle OQR is isosceles with OQ = OR = 7 cm and QR = 8 cm.

a) Find the angle POS in radians correct to 3 decimal places. *[3 marks]*

b) Find the area of the triangle OQR correct to 3 significant figures. *[2 marks]*

c) Find the percentage of the triangle that is shaded, correct to the nearest 1%. *[3 marks]*

Challenge

E **P** **M** Q13 A logo in the shape of a cat's head is modelled by a circle, centred at the point P with equation $x^2 - 2x + y^2 - 4y = 4$, and two triangles, ABP and CDP, where AB and CD are parallel tangents to the circle at the points A and C respectively. Points B and D both have y-coordinate 7.

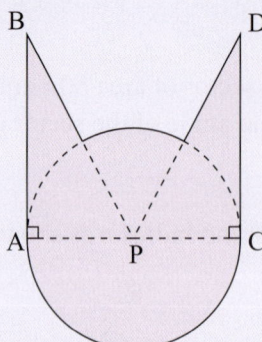

a) Find the coordinates of the centre of the circle and the radius of the circle. *[3 marks]*

b) State the coordinates of the points A and C. *[1 mark]*

c) Find angle APB in radians correct to 3 significant figures. *[2 marks]*

d) Find the area enclosed by the logo correct to 3 significant figures. *[4 marks]*

3.3 Small Angle Approximations

When θ (measured in radians) is very small, you can approximate the value of $\sin \theta$, $\cos \theta$ and $\tan \theta$ using the **small angle approximations**:

$$\sin \theta \approx \theta \qquad\qquad \cos \theta \approx 1 - \frac{1}{2}\theta^2 \qquad\qquad \tan \theta \approx \theta$$

You don't need to know where they come from, but it might help you to understand why they work.

Sin θ, cos θ and tan θ can each be written as an **infinite series** (a sum of terms — see Chapter 5 for more on series and summation notation) known as the **Maclaurin series**:

$$\sin \theta = \theta - \frac{1}{6}\theta^3 + \frac{1}{120}\theta^5 - \ldots = \sum_{n=0}^{\infty}(-1)^n \frac{\theta^{2n+1}}{(2n+1)!}$$

$$\cos \theta = 1 - \frac{1}{2}\theta^2 + \frac{1}{24}\theta^4 - \ldots = \sum_{n=0}^{\infty}(-1)^n \frac{\theta^{2n}}{(2n)!}$$

$$\tan \theta = \theta + \frac{1}{3}\theta^3 + \frac{2}{15}\theta^5 - \ldots$$

> **Tip** The Maclaurin series for tan θ is a bit messy in sum notation, so it's been left out here.

When θ is small ($\theta < 1$), θ^n gets smaller and smaller as n increases, so most of the terms in each series will be tiny (almost 0). Choosing to ignore all of the terms with θ^3 or a higher power leaves you with something close to the actual answer — so $\sin \theta \approx \theta$, $\cos \theta \approx 1 - \frac{1}{2}\theta^2$ and $\tan \theta \approx \theta$.

Note that these approximations only work when $\theta < 1$ — above that, the θ^n terms get bigger instead of smaller, so you can't ignore them.

Remember, θ must be in radians for these to work, not in degrees — you may have to convert before you use the approximations.

Example 1

Give an approximation for cos 0.2.

$\cos \theta \approx 1 - \frac{1}{2}\theta^2$ ← Write down the small angle approximation for cos.

$\cos 0.2 \approx 1 - \frac{1}{2}(0.2)^2$ ← Then substitute in $\theta = 0.2$
$\approx 1 - 0.02 = 0.98$ ← to give the approximation.

The actual value of cos 0.2 is 0.980067 (to 6 d.p.), so this approximation is pretty accurate.

Approximating functions

These approximations are most useful for approximating more **complicated functions**, which could involve sin, cos and tan of **multiples** of θ (when θ is small, you can assume that multiples of θ are also small). Make sure that you apply the approximation to **everything** inside the trig function.

For example: $\tan 4\theta \approx 4\theta$ $\sin \frac{1}{2}\theta \approx \frac{1}{2}\theta$ $\cos 3\theta \approx 1 - \frac{1}{2}(3\theta)^2$

You might have to use more than one formula in a question.

Example 2

Find an approximation for $f(\theta) = 4 \cos \theta \tan 3\theta$ when θ is small.

$f(\theta) \approx 4 \times (1 - \frac{1}{2}\theta^2) \times 3\theta$ ← Replace the functions of cos and tan with their small angle approximations. $\cos \theta \approx 1 - \frac{1}{2}\theta^2$ and $\tan 3\theta \approx 3\theta$.

$= (4 - 2\theta^2) \times 3\theta$

$= 12\theta - 6\theta^3$ (or $6\theta(2 - \theta^2)$) ← Expand and simplify the expression.

Example 3

Show that $\dfrac{2\theta \sin 2\theta}{1 - \cos 5\theta} \approx \dfrac{8}{25}$ when θ is small.

$f(\theta) = \dfrac{2\theta \sin 2\theta}{1 - \cos 5\theta} \approx \dfrac{2\theta(2\theta)}{1 - \left(1 - \frac{1}{2}(5\theta)^2\right)}$ ← Use the small angle approximations for each trig function.

$\sin 2\theta \approx 2\theta$ and $\cos 5\theta \approx 1 - \frac{1}{2}(5\theta)^2$

$= \dfrac{4\theta^2}{\frac{25}{2}\theta^2} = \dfrac{8}{25}$ as required ← Expand and simplify the expression.

Exercise 3.3

Q1   Use the small angle approximations to estimate the following values, then find the actual values on a calculator:

a) $\sin 0.23$ b) $\cos 0.01$ c) $\tan 0.18$

Q2   For the values of θ below, use the small angle approximations to estimate the value of $f(\theta) = \sin \theta + \cos \theta$, then use a calculator to find the actual answer:

a) $\theta = 0.3$ b) $\theta = 0.5$ c) $\theta = 0.25$ d) $\theta = 0.01$

Q3 Find an approximation for the following expressions when θ is small:

a) $\sin\theta\cos\theta$

b) $\theta\tan 5\theta\sin\theta$

c) $\dfrac{\sin 4\theta\cos 3\theta}{2\theta}$

d) $3\tan\theta + \cos 2\theta$

e) $\sin\frac{1}{2}\theta - \cos\theta$

f) $\dfrac{\cos\theta - \cos 2\theta}{1 - (\cos 3\theta + 3\sin\theta\tan\theta)}$

E Q4 Use small angle approximations to find the value of each of the following expressions as $x \to 0$.

a) $\dfrac{\sin x}{x}$ *[1 mark]*

b) $\dfrac{\tan^2 x}{1 - \cos 3x}$ *[2 marks]*

E Q5 Show that $\dfrac{\sin^3 6x + \tan^4 x}{1 - \cos 2x} \approx \frac{1}{2}x^2 + 108x$, for small values of x. *[3 marks]*

E Q6 Use small angle approximations to find an approximation of the negative solution to the equation $\dfrac{1 + \sin\theta}{1 - \cos 2\theta} = 0.62$. *[4 marks]*

E Q7 Show that $\dfrac{\tan kx - \cos kx}{\sin 2kx} \approx \frac{1}{4}kx - \frac{1}{2kx} + \frac{1}{2}$, for small values of x. *[3 marks]*

E P Q8 a) Show that, when θ is a small angle, $\cos 2\theta \approx 1 - 2\theta^2$. *[1 mark]*

b) Hence find an approximation for $\cos^2 10°$, giving your answer to four decimal places. *[2 marks]*

c) Find the percentage error in your approximation in part b) before rounding, giving your answer to three significant figures. *[3 marks]*

P M Q9 A pendulum of length 6 cm follows the arc of a circle. Its straight-line displacement as a vector is given by: $\mathbf{d} = 6\sin\theta\,\mathbf{i} + 6(1 - \cos\theta)\mathbf{j}$.

a) Show that the magnitude of the displacement is $6\sqrt{2(1 - \cos\theta)}$.

b) Show that, when θ is small, the magnitude of the displacement can be approximated by the arc length s.

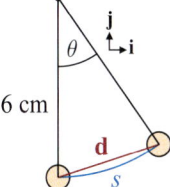

Challenge

E P Q10 In this question, angle $\text{BAC} = \frac{\pi}{6}$, angle $\text{ABC} = \theta$ radians, $\text{BC} = 11$ cm and $\text{AB} = x$ cm, as shown in the diagram below.

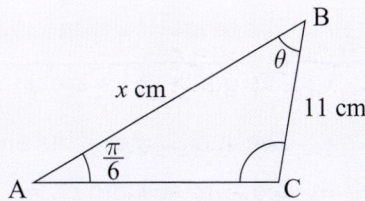

a) Show that the side AC has length $22\sin\theta$ cm. *[2 marks]*

b) Hence show that $\theta^2 \approx \dfrac{(x - 11)^2}{484 - 11x}$ for small values of θ. *[4 marks]*

3.4 Inverse Trig Functions

The inverse trig functions

Arcsin is the inverse of **sin**. You might see it written as arcsine or sin⁻¹.

Arccos is the inverse of **cos**. You might see it written as arccosine or cos⁻¹.

Arctan is the inverse of **tan**. You might see it written as arctangent or tan⁻¹.

The inverse trig functions **reverse** the effect of sin, cos and tan — e.g. $\sin 30° = 0.5$, so $\arcsin 0.5 = 30°$.

You should have buttons for doing arcsin, arccos and arctan on your calculator — they'll probably be labelled sin⁻¹, cos⁻¹ and tan⁻¹.

Graphs of the inverse trig functions

The functions sine, cosine and tangent **aren't one-to-one** mappings (see p.27). This means that more than one value of x gives the same value for $\sin x$, $\cos x$ or $\tan x$. For example: $\cos 0 = \cos 2\pi = \cos 4\pi = 1$, and $\tan 0 = \tan \pi = \tan 2\pi = 0$.

If you want the inverses to be **functions**, you have to **restrict the domains** of the trigonometric functions to make them **one-to-one**.

The graphs of the inverse functions are the **reflections** of the sin, cos and tan graphs in the line $y = x$.

Arcsin

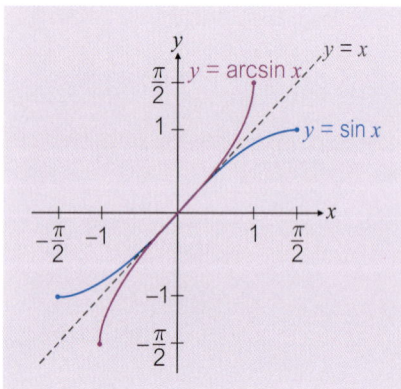

- For arcsin, limit the **domain** of $\sin x$ to $-\dfrac{\pi}{2} \leq x \leq \dfrac{\pi}{2}$ (the range of $\sin x$ is still **−1 ≤ sin x ≤ 1**).

- So the **domain** of arcsin x is **−1 ≤ x ≤ 1**.

- The **range** of arcsin x is $-\dfrac{\pi}{2} \leq \arcsin x \leq \dfrac{\pi}{2}$.

- The graph of **y = arcsin x** goes through the **origin**.

- The coordinates of its **endpoints** are $\left(-1, -\dfrac{\pi}{2}\right)$ and $\left(1, \dfrac{\pi}{2}\right)$.

Arccos

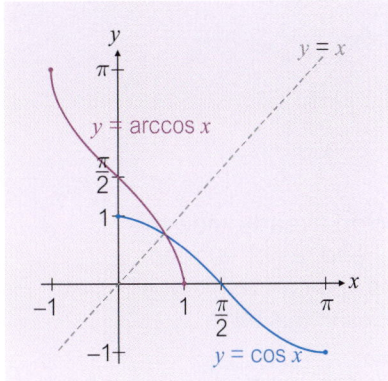

- For arccos, limit the **domain** of cos x to $0 \leq x \leq \pi$ (the range of cos x is still $-1 \leq \cos x \leq 1$.)

- So the **domain** of arccos x is $-1 \leq x \leq 1$.

- The **range** of arccos x is $0 \leq \arccos x \leq \pi$.

- The graph of $y = \arccos x$ crosses the **y-axis** at $\left(0, \frac{\pi}{2}\right)$.

- The coordinates of its **endpoints** are $(-1, \pi)$ and $(1, 0)$.

Arctan

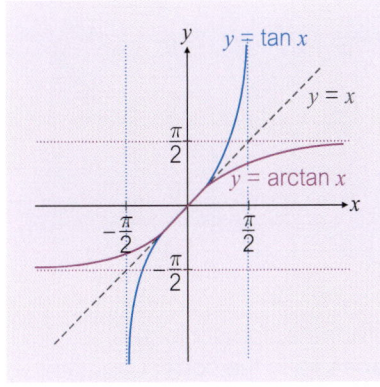

- For arctan, limit the **domain** of tan x to $-\frac{\pi}{2} < x < \frac{\pi}{2}$ (this **doesn't limit** the range of tan x.)

- This means that the **domain** of arctan x **isn't** limited (it's $x \in \mathbb{R}$) — note that there are no endpoints marked on the graph.

- The **range** of arctan x is $-\frac{\pi}{2} < \arctan x < \frac{\pi}{2}$.

- The graph of $y = \arctan x$ goes through the **origin**.

- It has **asymptotes** at $y = \frac{\pi}{2}$ and $y = -\frac{\pi}{2}$.

Learn the **key features** of these inverse trig graphs — you might be asked to **transform** them in different ways. There's more on **transformations of graphs** on pages 46-49.

Evaluating arcsin, arccos and arctan

If a is an angle within the interval $-\frac{\pi}{2} \leq a \leq \frac{\pi}{2}$ (or $-90° \leq a \leq 90°$) such that **sin $a = x$**, then **arcsin $x = a$**. So to evaluate **arcsin x** you need to find the angle a in this interval such that **sin $a = x$**. Using a calculator, this will be the answer you get when you enter "**sin^{-1} x**" (for a given value of x).

Similarly, to find **arccos x**, you need to find the angle a within the interval $0 \leq a \leq \pi$ (or $0° \leq a \leq 180°$) such that **cos $a = x$**, and **arctan x** is the angle a in the interval $-\frac{\pi}{2} < a < \frac{\pi}{2}$ (or $-90° < a < 90°$) such that **tan $a = x$**.

When evaluating inverse trig functions, it's helpful know the **sine**, **cosine** and **tangent** of some **common angles**. Here's a quick recap of the method of drawing triangles — and **SOH CAH TOA.**

Remember: **SOH CAH TOA**...

$$\sin x = \frac{\text{opp}}{\text{hyp}} \qquad \cos x = \frac{\text{adj}}{\text{hyp}} \qquad \tan x = \frac{\text{opp}}{\text{adj}}$$

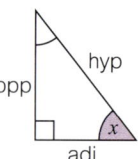

The sin, cos and tan of 30°, 45° and 60° can be found by drawing the following triangles and using **SOH CAH TOA**.

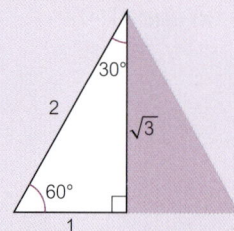

Draw an equilateral triangle with 60° angles and sides of 2 and split it to make a right-angled triangle.

Use Pythagoras to work out the length of the third side.

Draw a right-angled triangle where the edges adjacent to the right angle have length 1.

Use Pythagoras to work out the length of the third side.

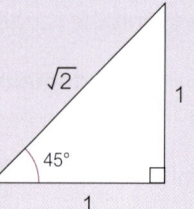

You should already know the sin, cos and tan of 90° and 180°, so you can work out all the values in this table:

$x°$	x (rad)	$\sin x$	$\cos x$	$\tan x$
0	0	0	1	0
30	$\frac{\pi}{6}$	$\frac{1}{2}$	$\frac{\sqrt{3}}{2}$	$\frac{1}{\sqrt{3}}$
45	$\frac{\pi}{4}$	$\frac{1}{\sqrt{2}}$	$\frac{1}{\sqrt{2}}$	1
60	$\frac{\pi}{3}$	$\frac{\sqrt{3}}{2}$	$\frac{1}{2}$	$\sqrt{3}$
90	$\frac{\pi}{2}$	1	0	—
180	π	0	−1	0

Be careful though — the first solution you find might **not** lie within the appropriate domain for the inverse function (see the graphs on pages 76-77). To find a solution that **does** lie in the correct domain, you need to use the **graphs** of the functions, or the **CAST diagram** that was introduced in the first year of this course (you might find it useful to look back at your Year 1 notes on solving trig equations in a given interval). The following examples show how to use these methods.

Example 1

a) Evaluate, without using a calculator, arccos 0.5. Give your answer in degrees.

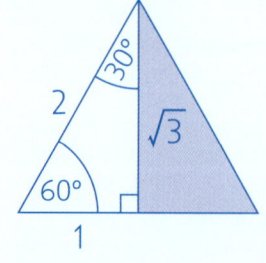

$0.5 = \frac{1}{2}$ which is either

$\sin 30°$ (if using $\frac{\text{opp}}{\text{hyp}}$)

or $\cos 60°$ (if using $\frac{\text{adj}}{\text{hyp}}$).

So take $a = 60°$.

First, work out the angle a for which $\cos a = 0.5$ using a right-angled triangle.

We want the angle for cos as we are looking for the inverse of cos.

For cos, $0 \le \alpha \le \pi$ in radians, which is $0° \le \alpha \le 180°$.

$60°$ lies within this domain, so arccos $0.5 = 60°$.

Check that this answer lies in the appropriate limited domain for cos.

b) Evaluate, without using a calculator, arctan (–1). Give your answer in radians.

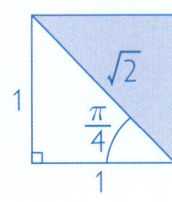

Using this triangle, you can see that $\tan \frac{\pi}{4} = 1$. ◄———— First use the triangles to find the value of a for when $\tan a = 1$.

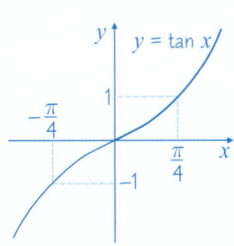

The graph shows that if $\tan \frac{\pi}{4} = 1$, then $\tan -\frac{\pi}{4} = -1$. ◄———— But you need to look at the symmetry of the tan x graph to find the solution for $\tan a = -1$.

The domain of tan is limited to $-\frac{\pi}{2} < a < \frac{\pi}{2}$. $-\frac{\pi}{4}$ is in this interval, so arctan $(-1) = -\frac{\pi}{4}$. ◄———— Check that this answer lies in the appropriate domain for tan.

c) Evaluate $\arcsin\left(-\frac{1}{\sqrt{2}}\right)$ without using your calculator. Give your answer in radians.

$\sin \frac{\pi}{4} = \frac{1}{\sqrt{2}}$ ◄———————————————— Use the triangle from the previous example.

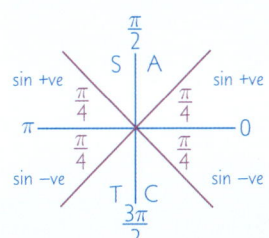

Sin x is negative in the 'T' and 'C' quadrants, which means:

$a = \pi + \frac{\pi}{4} = \frac{5\pi}{4}$, ◄———— To find the angle a such that $\sin a = -\frac{1}{\sqrt{2}}$, look at the CAST diagram.

or $a = 0 - \frac{\pi}{4} = -\frac{\pi}{4}$.

Only the second of these answers lies within the domain $-\frac{\pi}{2} \le a \le \frac{\pi}{2}$, so $\arcsin\left(-\frac{1}{\sqrt{2}}\right) = -\frac{\pi}{4}$. ◄———— Find the answer that lies in the appropriate domain for sin.

Exercise 3.4

Q1 Evaluate the following, giving your answer in radians.

a) arccos 1   b) arcsin $\frac{\sqrt{3}}{2}$ c) arctan $\sqrt{3}$ d) arccos $\frac{1}{\sqrt{2}}$

Q2 a) Sketch the graph of $y = 2 \arccos x$ for $-1 \le x \le 1$.

b) Sketch the graph of $y = \frac{1}{2} \arctan x$ and state the range.

Q3 By drawing the graphs of $y = \frac{x}{2}$ and $y = \cos^{-1} x$, determine the number of real roots of the equation $\cos^{-1} x = \frac{x}{2}$.

Q3 Hint Remember $\cos^{-1} x$ is just another name for arccos x.

Q4 Evaluate the following, giving your answers in radians:

a) $\sin^{-1}(-1)$ b) $\cos^{-1}\left(-\dfrac{\sqrt{3}}{2}\right)$ c) $\tan^{-1}\left(-\dfrac{1}{\sqrt{3}}\right)$ d) $\sin^{-1}\left(-\dfrac{1}{2}\right)$

Q5 Evaluate the following:

a) $\tan\left(\arcsin\dfrac{1}{2}\right)$ b) $\cos^{-1}\left(\cos\dfrac{2\pi}{3}\right)$ c) $\cos\left(\arcsin\dfrac{1}{2}\right)$

P Q6 $f(x) = 1 + \sin 2x$. Find an expression for $f^{-1}(x)$.

E **P** Q7 a) Find the exact values of y that satisfy the equation $8y^2 - 6\pi y + \pi^2 = 0$.   *[2 marks]*

b) Hence solve the equation $8(\arcsin x)^2 - 6\pi\arcsin x + \pi^2 = 0$,
giving exact answers.   *[2 marks]*

P Q8 a) By considering a suitable right-angled triangle,
prove that, for $0 \le x \le 1$, $\arccos x = \arcsin\sqrt{1 - x^2}$.

b) Show that the identity in part a) does not hold
for all real values of x.

> **Q8a) Problem Solving**
>
> Consider a right-angled
> triangle with hypotenuse
> 1 and one side of length x.

P Q9 By considering a suitable right-angled triangle,
prove that $\arctan x = \arcsin\left(\dfrac{x}{\sqrt{x^2 + 1}}\right)$, for $x > 0$.

P Q10 a) By considering a suitable right-angled triangle,
show that $\tan(\arcsin x) = \dfrac{x}{\sqrt{1 - x^2}}$ for $0 \le x \le 1$.

b) Hence find the value of the constant k, given that $\arcsin\dfrac{1}{3} = \arctan k$.

c) Show that the following equation can be written in the form
$(\sin x + a)(\tan x + b) = 0$, where a and b are constants to be found.

$$\sin x\tan x + \dfrac{\sqrt{2}}{12} = \dfrac{1}{3}\tan x + \dfrac{\sqrt{2}}{4}\sin x$$

d) Hence show that the only solution to the equation
$\sin x\tan x + \dfrac{\sqrt{2}}{12} = \dfrac{1}{3}\tan x + \dfrac{\sqrt{2}}{4}\sin x$ has the value $\arcsin\dfrac{1}{3}$.

E **P** Q11 a) Show that 9, 40 and 41 are side lengths in a right-angled triangle.   *[1 mark]*

b) Prove that $\arccos\left(\dfrac{40}{41}\right) + \arctan\left(\dfrac{40}{9}\right) = \dfrac{\pi}{2}$.   *[3 marks]*

E **P** Q12 a) Find the value of $\arcsin\left(\sin\left(\dfrac{2\pi}{3}\right)\right)$.   *[1 mark]*

b) Explain why $\arcsin(\sin x)$ is not equal to x for all real values of x.   *[1 mark]*

c) Sketch the graph $y = \arcsin(\sin x)$ where x is measured in radians
for $-\pi \le x \le \pi$. You must identify any turning points and axis intercepts.   *[3 marks]*

d) Hence or otherwise solve the equation $\arcsin(\sin x) = 0.72$,
where $-\pi \le x \le \pi$. Give your answers to 2 decimal places.   *[2 marks]*

3.5   Graphs of Cosec, Sec and Cot

When you take the **reciprocal** of the three main trig functions, sin, cos and tan, you get three new trig functions — **cosecant** (or **cosec**), **secant** (or **sec**) and **cotangent** (or **cot**).

> **Tip**
> $\tan \theta \equiv \dfrac{\sin \theta}{\cos \theta}$,
> so you can also think of $\cot \theta$ as being $\dfrac{\cos \theta}{\sin \theta}$.

$$\operatorname{cosec} \theta \equiv \frac{1}{\sin \theta} \qquad \sec \theta \equiv \frac{1}{\cos \theta} \qquad \cot \theta \equiv \frac{1}{\tan \theta}$$

Example 1

Write each of the following expressions in terms of sin and cos only:

$$\operatorname{cosec} 20°, \quad \sec \pi, \quad \cot \frac{\pi}{6}$$

$\operatorname{cosec} 20° = \dfrac{1}{\sin 20°}$ ←   Use $\operatorname{cosec} \theta \equiv \dfrac{1}{\sin \theta}$.

$\sec \pi = \dfrac{1}{\cos \pi}$ ←   Use $\sec \theta \equiv \dfrac{1}{\cos \theta}$.

> **Tip**   The trick for remembering which is which is to look at the third letter — co**s**ec (1/sin), se**c** (1/cos) and co**t** (1/tan).

$\cot \dfrac{\pi}{6} = \dfrac{1}{\tan \dfrac{\pi}{6}} = \dfrac{\cos \dfrac{\pi}{6}}{\sin \dfrac{\pi}{6}}$ ←   Use $\cot \theta \equiv \dfrac{1}{\tan \theta} \equiv \dfrac{\cos \theta}{\sin \theta}$.

Graph of cosec

This is the graph of $y = \operatorname{cosec} x$:

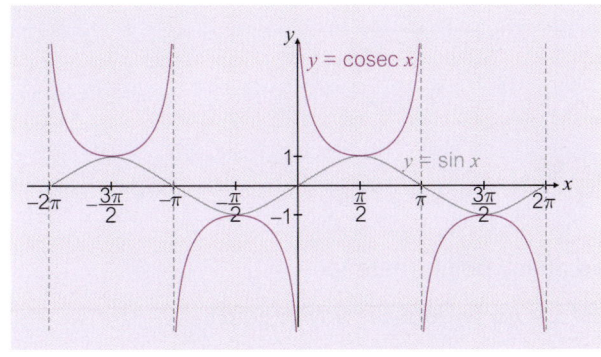

> **Tip**   The x-coordinates of the turning points for $\operatorname{cosec} x$ are the same as for $\sin x$, but a maximum on $\sin x$ becomes a minimum on $\operatorname{cosec} x$, and vice versa.

- Since $\operatorname{cosec} x = \dfrac{1}{\sin x}$, $y = \operatorname{cosec} x$ is **undefined** at any point where $\sin x = 0$.

- So $y = \operatorname{cosec} x$ has **vertical asymptotes** at $x = n\pi$ (where n is any integer).

- The graph $y = \operatorname{cosec} x$ has **minimum** points at $x = ..., -\dfrac{3\pi}{2}, \dfrac{\pi}{2}, \dfrac{5\pi}{2}, ...$ At these points, $y = 1$.

- It has **maximum** points at $x = ..., -\dfrac{\pi}{2}, \dfrac{3\pi}{2}, \dfrac{7\pi}{2}, ...$ At these points, $y = -1$.

Graph of sec

This is the graph of **y = sec** *x*:

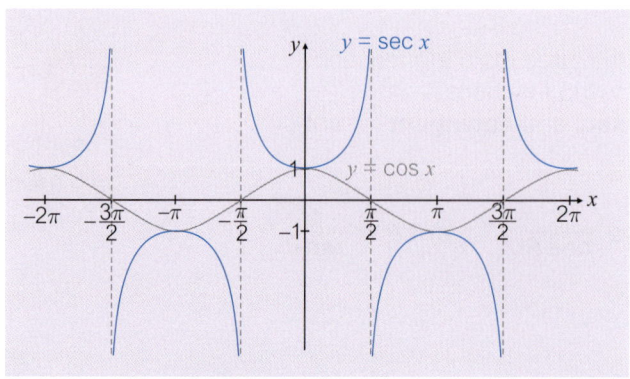

- As sec $x = \frac{1}{\cos x}$, $y = \sec x$ is **undefined** at any point where $\cos x = 0$.

- So $y = \sec x$ has vertical asymptotes at $x = n\pi + \frac{\pi}{2}$ (where n is any integer).

- The graph of $y = \sec x$ has **minimum** points at $x = 0, \pm 2\pi, \pm 4\pi, ...$ (wherever the graph of $y = \cos x$ has a **maximum**). At these points, $y = 1$.

- It has **maximum** points at $x = \pm\pi, \pm 3\pi, ...$ (wherever the graph of $y = \cos x$ has a **minimum**). At these points, $y = -1$.

Graph of cot

This is the graph of **y = cot** *x*:

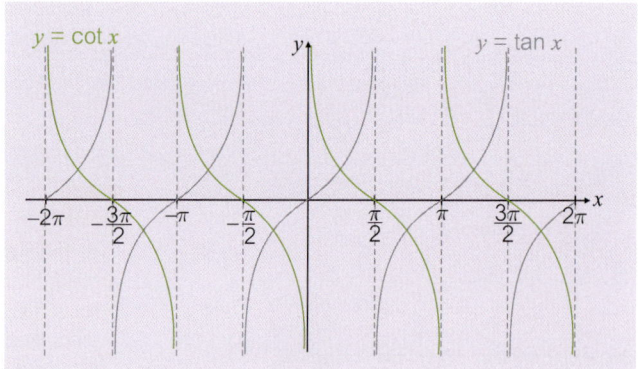

- As cot $x = \frac{1}{\tan x}$, $y = \cot x$ is **undefined** at any point where $\tan x = 0$.

- So $y = \cot x$ has **vertical asymptotes** at $x = n\pi$ (where n is any integer).

- $y = \cot x$ crosses the *x*-axis at every place where the graph of $\tan x$ has an asymptote. This is any point with the coordinates $\left(\left(n\pi + \frac{\pi}{2}\right), 0\right)$.

Just like the graphs of $\sin x$ and $\cos x$, the graphs of $\csc x$ and $\sec x$ have a **period** of **2π radians** — this just means they **repeat themselves** every 2π (or 360°).

The graphs of $\tan x$ and $\cot x$ both have a **period** of π **radians**.

Transformations of cosec, sec and cot

The graphs of the cosec, sec and cot functions can be **transformed** in the same way as other functions.

Example 2

Tip Look back at pages 46-49 for more on transformations of graphs.

a) Sketch the graph of $y = \cot 2x$ over the interval $-\pi \leq x \leq \pi$.

If $f(x) = \cot x$, then $y = f(2x)$ — this transformation is a horizontal stretch by a factor of $\frac{1}{2}$.

Identify the transformation of the graph of $f(x) = \cot x$.

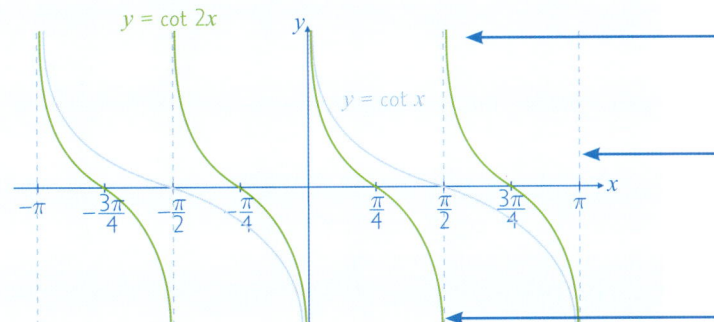

The graph is squashed up in the x-direction by a factor of 2.

The x-coordinates of the asymptotes and the x-intercepts for $y = \cot 2x$ are half of those for $y = \cot x$.

The period of the graph is also halved — $y = \cot 2x$ repeats itself every $\frac{\pi}{2}$ radians.

b) Give the coordinates of the maximum point on the graph of $y = \sec(x - 30°) + 1$, between 0° and 360°.

Let $f(x) = \sec x$, then $y = f(x - 30°) + 1$ — this transformation is a horizontal translation right by 30°, followed by a vertical translation up by 1.

Identify the transformation of the graph of $f(x) = \sec x$.

$f(x) = \sec x$ has a maximum point at $(180°, -1)$.

So $y = f(x - 30°) + 1$ has a maximum point at $(210°, 0)$.

Translate the maximum point of $f(x)$ to $f(x - 30°) + 1$ — the x-coordinate will be increased by 30° and the y-coordinate will be increased by 1.

c) Describe the position of the asymptotes on the graph of $y = \text{cosec}\left(x + \frac{\pi}{3}\right)$.

Let $f(x) = \text{cosec } x$, then $y = f\left(x + \frac{\pi}{3}\right)$ — this transformation is a horizontal translation left by $\frac{\pi}{3}$.

Identify the transformation of the graph of $f(x) = \text{cosec } x$.

The graph of $f(x) = \text{cosec } x$ has asymptotes at $x = n\pi$ (where n is any integer).

So $y = f\left(x + \frac{\pi}{3}\right)$ will have asymptotes at $x = n\pi - \frac{\pi}{3}$ (where n is any integer).

Translate the asymptotes of $f(x)$ to $f\left(x + \frac{\pi}{3}\right)$ — the x-coordinate of each asymptote will be decreased by $\frac{\pi}{3}$.

Exercise 3.5

Q1 a) Sketch the graph of $y = \sec x$ for $-2\pi \leq x \leq 2\pi$.

b) Give the coordinates of the minimum points within this interval.

c) Give the coordinates of the maximum points within this interval.

d) State the range of $y = \sec x$.

> **Q1d) Hint** Think about which values are not included in the range.

Q2 a) Sketch the graph of $y = \operatorname{cosec} x$ for $0 < x < 2\pi$.

b) Give the coordinates of any maximum and minimum points within this interval.

c) State the domain and range of $y = \operatorname{cosec} x$.

Q3 Describe the transformation that maps $y = \sec x$ onto $y = \operatorname{cosec} x$.

> **Q3 Hint** Compare the graphs you've drawn for Q1 and Q2.

E Q4 a) Sketch $y = 1 - 3x$ on the same set of axes as $y = \sec x$ for $-\pi \leq x \leq \pi$. *[3 marks]*

b) Hence state the number of real solutions to the equation $\sec x + 3x = 1$ in the interval $-\pi \leq x \leq \pi$, giving a reason for your answer. *[1 mark]*

c) Find the equation of a line that could be added to the graph of $y = \sec x$ to approximate the real solutions to the equation $\sec x + 8x = 4$. *[1 mark]*

Q5 a) Describe the transformation that maps $y = \cot x$ onto $y = \cot \frac{x}{4}$.

b) What is the period, in degrees, of the graph $y = \cot \frac{x}{4}$?

c) Sketch the graph of $y = \cot \frac{x}{4}$ for $0° \leq x \leq 360°$.

E Q6 a) Sketch the graph of $y = 2 + \sec x$ for $-2\pi \leq x \leq 2\pi$. *[3 marks]*

b) Give the coordinates of any maximum and minimum points within this interval. *[2 marks]*

c) State the domain and range of $y = 2 + \sec x$. *[1 mark]*

Q7 a) Sketch the graph of $y = 2 \operatorname{cosec} 2x$ for $0° \leq x \leq 360°$.

b) Give the coordinates of the minimum points within this interval.

c) Give the coordinates of the maximum points within this interval.

d) For what values of x in this interval is $y = 2 \operatorname{cosec} 2x$ undefined?

Q8 a) Describe the position of the asymptotes on the graph of $y = 2 + 3 \operatorname{cosec} x$.

b) What is the period, in degrees, of the graph $y = 2 + 3 \operatorname{cosec} x$?

c) Sketch the graph of $y = 2 + 3 \operatorname{cosec} x$ for $-180° < x < 180°$.

d) State the range of $y = 2 + 3 \operatorname{cosec} x$.

E Q9 a) Sketch $y = \cot(4x + 1)$ for $10\pi \leq x \leq \frac{21}{2}\pi$, showing any asymptotes.

You need not find the coordinates of the points of intersection with the coordinate axes and you need not state the equations of the asymptotes. *[3 marks]*

b) State the period of $\cot(4x + 1)$. *[1 mark]*

c) Find the number of real solutions of the equation $\cot(4x + 1) = -6$ for $0 \leq x \leq 11\pi$. *[2 marks]*

3.6 Evaluating Cosec, Sec and Cot

To **evaluate** cosec, sec or cot of a number, first evaluate sin, cos or tan, then work out the **reciprocal** of the answer.

Example 1

a) Evaluate $2\sec(-20°) + 5$, giving your answer to 3 significant figures.

$\sec x = \dfrac{1}{\cos x}$, so $2\sec(-20°) + 5 = \dfrac{2}{\cos(-20°)} + 5$ ◄— Write out the expression in terms of sin, cos or tan.

$\dfrac{2}{\cos(-20°)} + 5 = \dfrac{2}{0.93969\ldots} + 5 = 7.13$ to 3 s.f. ◄— Use a calculator to find the answer.

b) Evaluate $\operatorname{cosec}\dfrac{\pi}{4}$ without a calculator. Give your answer in surd form.

$\operatorname{cosec}\dfrac{\pi}{4} = \dfrac{1}{\sin\dfrac{\pi}{4}}$ ◄— Use $\operatorname{cosec} x = \dfrac{1}{\sin x}$.

$\Rightarrow \sin\dfrac{\pi}{4} = \dfrac{1}{\sqrt{2}}$ ◄— Use the right-angled triangle on the left to find $\sin\dfrac{\pi}{4}$.

$\operatorname{cosec}\dfrac{\pi}{4} = \dfrac{1}{\left(\dfrac{1}{\sqrt{2}}\right)} = \sqrt{2}$ ◄— Evaluate $\operatorname{cosec}\dfrac{\pi}{4}$.

c) Find the exact value of $\cot\left(-\dfrac{\pi}{6}\right)$.

$\cot\left(-\dfrac{\pi}{6}\right) = \dfrac{1}{\tan\left(-\dfrac{\pi}{6}\right)}$ ◄— Use $\cot x = \dfrac{1}{\tan x}$.

$\Rightarrow \tan\dfrac{\pi}{6} = \dfrac{1}{\sqrt{3}}$ ◄— Use the right-angled triangle on the left to find $\tan\dfrac{\pi}{6}$.

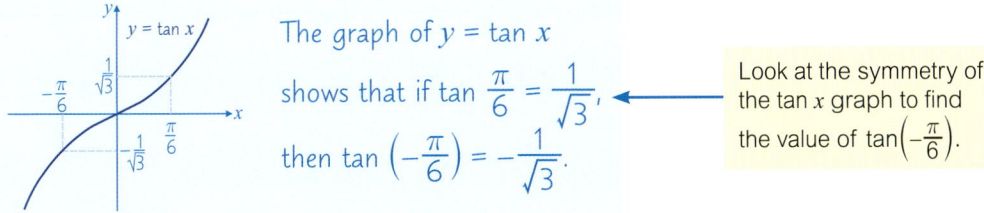

The graph of $y = \tan x$ shows that if $\tan\dfrac{\pi}{6} = \dfrac{1}{\sqrt{3}}$, then $\tan\left(-\dfrac{\pi}{6}\right) = -\dfrac{1}{\sqrt{3}}$. ◄— Look at the symmetry of the tan x graph to find the value of $\tan\left(-\dfrac{\pi}{6}\right)$.

$\cot\left(-\dfrac{\pi}{6}\right) = \dfrac{1}{\left(-\dfrac{1}{\sqrt{3}}\right)} = -\sqrt{3}$ ◄— Evaluate $\cot\left(-\dfrac{\pi}{6}\right)$.

E Exam Style   **P** Problem Solving   **M** Modelling

d) Find cosec 300° without using a calculator.

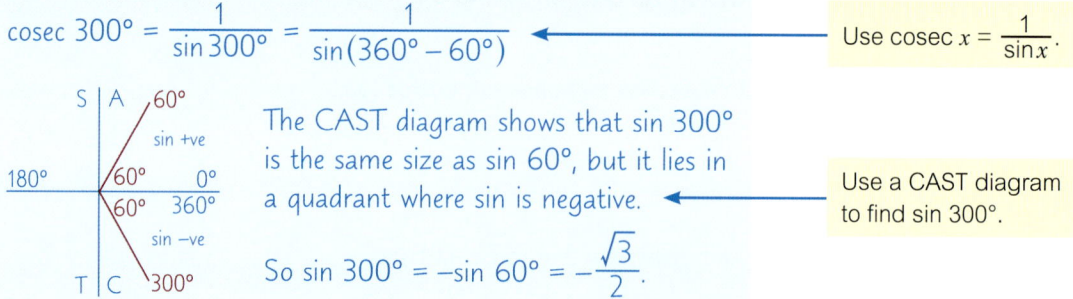

$$\cosec 300° = \frac{1}{\sin 300°} = \frac{1}{\sin(360° - 60°)}$$

Use $\cosec x = \frac{1}{\sin x}$.

The CAST diagram shows that sin 300° is the same size as sin 60°, but it lies in a quadrant where sin is negative.

Use a CAST diagram to find sin 300°.

So $\sin 300° = -\sin 60° = -\frac{\sqrt{3}}{2}$.

So $\cosec 300° = \frac{1}{\left(-\frac{\sqrt{3}}{2}\right)} = -\frac{2}{\sqrt{3}}$

Evaluate cosec 300°.

Exercise 3.6

Q1 Evaluate the following, giving your answers to 2 decimal places:

    a) cosec 80°
    b) sec 75°
    c) cot 30°

    d) sec(–70°)
    e) 3 – cot 250°
    f) 2 cosec 25°

Q2 Evaluate the following, giving your answers to 3 significant figures:

    a) sec 3
    b) cot 0.6
    c) cosec 1.8
    d) sec(–1)

    e) $\cosec \frac{\pi}{8}$
    f) $8 + \cot \frac{\pi}{8}$
    g) $\dfrac{1}{1 + \sec \frac{\pi}{10}}$
    h) $\dfrac{1}{6 + \cot \frac{\pi}{5}}$

Q3 Using the table of common angles on p.78, find the exact values of:

    a) sec 60°
    b) cosec 30°
    c) cot 45°
    d) $\cosec \frac{\pi}{3}$

    e) sec (–180°)
    f) cosec 135°
    g) cot 330°
    h) $\sec \frac{5\pi}{4}$

    i) $\cosec \frac{5\pi}{3}$
    j) $\cosec \frac{2\pi}{3}$
    k) $3 - \cot \frac{3\pi}{4}$
    l) $\dfrac{\sqrt{3}}{\cot \frac{\pi}{6}}$

P Q4 Find, without a calculator, the exact values of:

    a) $\dfrac{1}{1 + \sec 60°}$
    b) $\dfrac{2}{6 + \cot 315°}$
    c) $\dfrac{1}{\sqrt{3} - \sec 30°}$

    d) $1 + \cot 420°$
    e) $\dfrac{2}{7 + \sqrt{3} \cot 150°}$
    f) $\dfrac{2}{1 - \cosec 330°}$

E P Q5 a) Without using a calculator, show that $\left(\dfrac{\tan\left(\frac{\pi}{3}\right) - \cot\left(\frac{\pi}{3}\right)}{\sec\left(\frac{\pi}{6}\right)} \right)^2$ is an integer.   *[3 marks]*

    b) Show that $\cot^2\left(-\frac{\pi}{3}\right) + \cot\left(\frac{5\pi}{3}\right) - \sec\left(\frac{\pi}{6}\right) = p\sqrt{3} + q,$
    where p and q are rational numbers to be found.   *[3 marks]*

Challenge

E P Q6 The graph $y = a \cosec(x + b)° + c$, where a, b and c are constants and $0 < b < 90$, passes through the points $(20, 9)$ and $(260, -3)$, and has a vertical asymptote at $x = 170°$. Find the values of a, b and c.   *[3 marks]*

3.7 Solving Equations Involving Cosec, Sec and Cot

Simplifying expressions

You can use the cosec, sec and cot relationships to **simplify expressions**.
This can make it a lot easier to **solve** trig equations.

Example 1

a) Simplify $\cot^2 x \tan x$.

$$\cot^2 x \tan x = \left(\frac{1}{\tan^2 x}\right) \tan x = \frac{1}{\tan x} = \cot x$$

Use $\cot x = \dfrac{1}{\tan x}$.

b) Show that $\dfrac{\cot x \sec x}{\csc^2 x} \equiv \sin x$.

$$\frac{\cot x \sec x}{\csc^2 x} \equiv \frac{\left(\frac{\cos x}{\sin x}\right)\left(\frac{1}{\cos x}\right)}{\left(\frac{1}{\sin^2 x}\right)} \equiv \frac{\left(\frac{1}{\sin x}\right)}{\left(\frac{1}{\sin^2 x}\right)} \equiv \sin x \text{ as required}$$

Use all three trig relationships and simplify.

c) Write the expression $(\csc x + 1)(\sin x - 1)$ as a single fraction in terms of $\sin x$ only.

$$(\csc x + 1)(\sin x - 1) = \csc x \sin x + \sin x - \csc x - 1$$

Expand the brackets.

$$= \left(\frac{1}{\sin x}\right)\sin x + \sin x - \left(\frac{1}{\sin x}\right) - 1$$

$$= 1 + \sin x - \left(\frac{1}{\sin x}\right) - 1 = \sin x - \frac{1}{\sin x} = \frac{\sin^2 x - 1}{\sin x}$$

Use $\csc x = \dfrac{1}{\sin x}$ and simplify the expression.

Solving equations

You can **solve** equations involving cosec, sec and cot by **rewriting** them in terms of sin, cos or tan and solving as usual (often you'll need to get them all in terms of the same trig function, e.g. all sin).
You may need to use the **CAST diagram** (or a trig graph) to find all the solutions in a given interval.

Example 2

Solve $\sec x = \sqrt{2}$ in the interval $0 \leq x \leq 2\pi$.

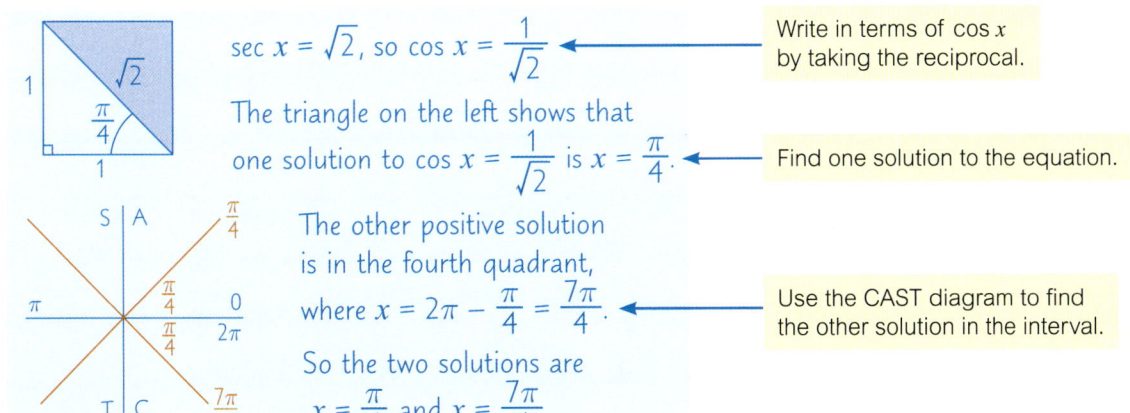

$\sec x = \sqrt{2}$, so $\cos x = \dfrac{1}{\sqrt{2}}$

Write in terms of $\cos x$ by taking the reciprocal.

The triangle on the left shows that one solution to $\cos x = \dfrac{1}{\sqrt{2}}$ is $x = \dfrac{\pi}{4}$.

Find one solution to the equation.

The other positive solution is in the fourth quadrant, where $x = 2\pi - \dfrac{\pi}{4} = \dfrac{7\pi}{4}$.

Use the CAST diagram to find the other solution in the interval.

So the two solutions are
$$x = \frac{\pi}{4} \text{ and } x = \frac{7\pi}{4}.$$

E Exam Style   P Problem Solving   M Modelling

Example 3 **P**

Solve $\cosec^2 x - 3\cosec x + 2 = 0$ in the interval $-180° \le x \le 180°$.

$(\cosec x - 1)(\cosec x - 2) = 0$ ← Factorise the quadratic equation given in cosec x.

$\cosec x - 1 = 0 \Rightarrow \cosec x = 1 \Rightarrow \sin x = 1$
One solution to this is $x = 90°$.

$\cosec x - 2 = 0 \Rightarrow \cosec x = 2 \Rightarrow \sin x = \dfrac{1}{2}$
One solution to this is $x = 30°$.

Solve the equations for x.

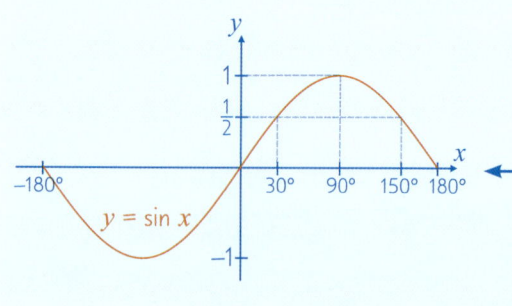

Use the graph of $y = \sin x$ to find other solutions in the interval $-180° \le x \le 180°$.

$\sin x = 1$ only has one solution in this interval.

The symmetry of the graph means there is another solution to $\sin x = \dfrac{1}{2}$ at $x = 180° - 30° = 150°$.

So the solutions are: $x = 30°$, $x = 90°$ and $x = 150°$.

Exercise 3.7

Q1 Simplify the following expressions:

a) $\sec x + \dfrac{1}{\cos x}$

b) $(\cosec^2 x)(\sin^2 x)$

c) $2\cot x + \dfrac{1}{\tan x}$

d) $\dfrac{\sec x}{\cosec x}$

e) $(\cos x)(\cosec x)$

f) $\dfrac{\cosec^2 x}{\cot x}$

g) $5\cot x - \dfrac{\cosec x}{\sec x}$

h) $\dfrac{1}{\sec^2 x} + \dfrac{1}{\cosec^2 x}$

P **Q2** Show that:

a) $\sin x \cot x \equiv \cos x$

b) $\sec x - \cos x \equiv \tan x \sin x$

c) $\dfrac{\sec x}{\cot x} \equiv \sin x \sec^2 x$

d) $\tan x \cosec x \equiv \sec x$

e) $\dfrac{(\tan^2 x)(\cosec x)}{\sin x} \equiv \sec^2 x$

f) $\cosec x (\sin x + \cos x) \equiv 1 + \cot x$

Q3 Solve these equations for $0° \leq x \leq 360°$. Give your answers in degrees to 1 decimal place.

a) $\sec x = 1.9$ b) $\cot x = 2.4$

c) $\operatorname{cosec} x = -2$ d) $\sec x = -1.3$

e) $\cot x = -2.4$ f) $\frac{1}{2} \operatorname{cosec} x = 0.7$

g) $4 \sec 2x = -7$ h) $5 \cot 3x = 4$

Q3 Hint Use a calculator to find one solution, but then look at the graph or the CAST diagram to find any other solutions in the interval.

Q4 Solve these equations for $0 \leq x \leq 2\pi$, giving your answers in radians in terms of π.

a) $\sec x = 2$ b) $\operatorname{cosec} x = -2$

c) $\cot 2x = 1$ d) $\sec 5x = -1$

e) $\operatorname{cosec} 3x = -\sqrt{2}$ f) $\cot 4x = \dfrac{1}{\sqrt{3}}$

E
P
Q5 a) Explain why $\sec x = \operatorname{cosec}\left(x + \dfrac{\pi}{2}\right)$. *[1 mark]*

b) Verify this result by solving $\sec x = 2$ and $\operatorname{cosec}\left(x + \dfrac{\pi}{2}\right) = 2$ for $-\pi \leq x \leq \pi$. *[3 marks]*

E **Q6** Solve the following equations for $-90° < x < 90°$:

a) $\cot^2 x = 3$ *[2 marks]*

b) $\left|\operatorname{cosec} x\right| = \dfrac{2\sqrt{3}}{3}$ *[2 marks]*

c) $\sec \dfrac{1}{2} x = \dfrac{2\sqrt{3}}{3}$ *[2 marks]*

E **Q7** Solve the equation $\cot 2x - 4 = -5$ in the interval $0 \leq x \leq 2\pi$.
Give your answers in terms of π. *[3 marks]*

E **Q8** Solve for $0° \leq x \leq 360°$, $2 \operatorname{cosec} 2x = 3$. Give your answers to 1 decimal place. *[3 marks]*

Q9 Find, for $0 \leq x \leq 2\pi$, all the solutions of the equation $-2 \sec x = 4$.
Give your answers in terms of π.

Q10 Solve $\sqrt{3} \operatorname{cosec} 3x = 2$ for $0 \leq x \leq 2\pi$. Give your answers in terms of π.

P **Q11** Solve the following for $0° \leq x \leq 180°$, giving your answers to 1 d.p. where appropriate:

a) $\sec^2 x - 2\sqrt{2} \sec x + 2 = 0$ b) $2 \cot^2 x + 3 \cot x - 2 = 0$

P **Q12** Solve the equation $(\operatorname{cosec} x - 3)(2 \tan x + 1) = 0$ for $0° \leq x \leq 360°$.
Give your answers to 1 decimal place.

Challenge

E
P
Q13 Solve the following pair of simultaneous equations for the intervals $0 \leq x \leq \pi$ and $0 \leq y \leq \pi$.
$\sin x + \sin y = 1$
$\operatorname{cosec} x + \operatorname{cosec} y = 4$ *[6 marks]*

3.8 Identities Involving Cosec, Sec and Cot

By now you should be familiar with the following **trig identities**:

$$\cos^2\theta + \sin^2\theta \equiv 1 \qquad \tan\theta \equiv \frac{\sin\theta}{\cos\theta}$$

Tip The $\equiv$ sign tells you that this is true for all values of θ, rather than just certain values.

You can use them to produce a couple of other identities:

$$\sec^2\theta \equiv 1 + \tan^2\theta \qquad \text{cosec}^2\theta \equiv 1 + \cot^2\theta$$

You need to know how to **derive** these identities from the ones you already know.

Deriving $\sec^2\theta \equiv 1 + \tan^2\theta$

Start with the identity **$\cos^2\theta + \sin^2\theta \equiv 1$** and divide through by **$\cos^2\theta$**.

$$\cos^2\theta + \sin^2\theta \equiv 1$$

$$\frac{\cos^2\theta}{\cos^2\theta} + \frac{\sin^2\theta}{\cos^2\theta} \equiv \frac{1}{\cos^2\theta}$$

Tip These are examples of proof by deduction, where known facts are used to prove that other relationships are true.

The definition of **$\tan\theta \equiv \dfrac{\sin\theta}{\cos\theta}$**, so replace $\dfrac{\sin^2\theta}{\cos^2\theta}$ with $\tan^2\theta$.

$$1 + \tan^2\theta \equiv \frac{1}{\cos^2\theta}$$

The definition of **$\sec\theta \equiv \dfrac{1}{\cos\theta}$**, so replace $\dfrac{1}{\cos^2\theta}$ with $\sec^2\theta$.

$$1 + \tan^2\theta \equiv \sec^2\theta$$

Rearranging slightly gives... $\qquad \sec^2\theta \equiv 1 + \tan^2\theta$

Deriving $\text{cosec}^2\theta \equiv 1 + \cot^2\theta$

Start again with **$\cos^2\theta + \sin^2\theta \equiv 1$** but this time divide through by **$\sin^2\theta$**.

$$\cos^2\theta + \sin^2\theta \equiv 1$$

$$\frac{\cos^2\theta}{\sin^2\theta} + \frac{\sin^2\theta}{\sin^2\theta} \equiv \frac{1}{\sin^2\theta}$$

The definition of **$\cot\theta \equiv \dfrac{1}{\tan\theta} \equiv \dfrac{\cos\theta}{\sin\theta}$**, so replace $\dfrac{\cos^2\theta}{\sin^2\theta}$ with $\cot^2\theta$.

$$\cot^2\theta + 1 \equiv \frac{1}{\sin^2\theta}$$

The definition of **$\text{cosec}\,\theta = \dfrac{1}{\sin\theta}$**, so replace $\dfrac{1}{\sin^2\theta}$ with $\text{cosec}^2\theta$.

$$\cot^2\theta + 1 \equiv \text{cosec}^2\theta$$

Rearranging slightly gives... $\qquad \text{cosec}^2\theta \equiv 1 + \cot^2\theta$

Using the Identities

You can use identities to get rid of any trig functions that are making an equation difficult to solve.

Example 1

Simplify the expression $3 \tan x + \sec^2 x + 1$.

$3 \tan x + 1 + \tan^2 x + 1$ ← Use $\sec^2 \theta \equiv 1 + \tan^2 \theta$ to swap $\sec^2 x$ for $1 + \tan^2 x$.

$\tan^2 x + 3 \tan x + 2$ ← Now rearrange.

$(\tan x + 1)(\tan x + 2)$ ← Factorise the quadratic in $\tan x$.

Example 2 **P**

Solve the equation $\cot^2 x + 5 = 4 \operatorname{cosec} x$ in the interval $0° \leq x \leq 360°$.

$\operatorname{cosec}^2 x - 1 + 5 = 4 \operatorname{cosec} x$ ← Use $\operatorname{cosec}^2 \theta \equiv 1 + \cot^2 \theta$ to swap $\cot^2 x$ for $\operatorname{cosec}^2 x - 1$.

$\Rightarrow \operatorname{cosec}^2 x + 4 = 4 \operatorname{cosec} x$

$\Rightarrow \operatorname{cosec}^2 x - 4 \operatorname{cosec} x + 4 = 0$

$\Rightarrow (\operatorname{cosec} x - 2)(\operatorname{cosec} x - 2) = 0$ ← Factorise the quadratic in $\operatorname{cosec} x$.

$\Rightarrow (\operatorname{cosec} x - 2) = 0$ ← Solve the quadratic for $\operatorname{cosec} x$.

$\Rightarrow \operatorname{cosec} x = 2$

$\Rightarrow \dfrac{1}{\sin x} = 2$ ← Convert this into $\sin x$ and solve.

$\Rightarrow \sin x = \dfrac{1}{2} \Rightarrow x = 30°$

To find the other values of x, draw a quick sketch of the sine curve.

From the graph, $\sin x$ takes the value of $\dfrac{1}{2}$ twice in the given interval, so the solutions are $x = 30°$ and $x = 180 - 30 = 150°$.

Example 3

Given that $\cot x = \sqrt{8}$, where $0° \leq x \leq 180°$, show how you can use the identity $\operatorname{cosec}^2 \theta \equiv 1 + \cot^2 \theta$ to find the exact value of $\sin x$. Use Pythagoras' theorem to confirm the result.

$\cot x = \sqrt{8} \implies \cot^2 x = 8$

> Use the value for $\cot x$ and the identity $\operatorname{cosec}^2 \theta \equiv 1 + \cot^2 \theta$ to find $\operatorname{cosec} x$.

$\operatorname{cosec}^2 x \equiv 1 + \cot^2 x$

So $\operatorname{cosec}^2 x = 1 + 8 = 9$

$\implies \operatorname{cosec} x = \pm 3$

$\implies \dfrac{1}{\sin x} = \pm 3 \implies \sin x = \pm \dfrac{1}{3}$

> Solve the equation using $\operatorname{cosec} x = \dfrac{1}{\sin x}$.

$0° \leq x \leq 180°$, and $\sin x$ is positive over this interval, so $\sin x = \dfrac{1}{3}$.

$\cot x = \sqrt{8} \implies \tan x = \dfrac{1}{\cot x} = \dfrac{1}{\sqrt{8}}$.

(i.e. the opposite has a length of 1 and the adjacent has a length of $\sqrt{8}$)

> To confirm this using Pythagoras' theorem, start by using $\cot x = \sqrt{8}$ to draw a right-angled triangle.

[Right-angled triangle with vertical side labelled 1, horizontal side labelled $\sqrt{8}$, hypotenuse labelled c, and angle x.]

$c = \sqrt{1^2 + \left(\sqrt{8}\right)^2} = 3$

> Use Pythagoras' theorem to find the length of the hypotenuse.

So $\sin x = \dfrac{\text{opp}}{\text{hyp}} = \dfrac{1}{3}$ as required.

> Use the triangle to find $\sin x$.

Exercise 3.8

Q1 Express $\operatorname{cosec}^2 x + 2 \cot^2 x$ in terms of $\operatorname{cosec} x$ only.

Q2 Simplify the following expression: $\tan^2 x - \dfrac{1}{\cos^2 x}$.

E **Q3** Given that $\tan y = t$, find simplified expressions for the following,
P in terms of t and not involving trigonometric functions:

    a) $\sec y$ *[1 mark]*

    b) $3 \sec^2 y - 4$ *[2 marks]*

    c) $\operatorname{cosec}^2 y$ *[2 marks]*

P **Q4** Given that $x = \sec \theta + \tan \theta$,
show that $x + \dfrac{1}{x} = 2 \sec \theta$.

> **Q4 Problem Solving**
>
> Start by substituting $x = \sec \theta + \tan \theta$ into the left-hand side and writing it as a single fraction.

E **Q5** Given that $x = 5 \sec t$ and $y = 6 \tan t$, show that $\dfrac{36}{25} x^2 - y^2 = 36$.   *[3 marks]*

P **Q6** a) Show that the equation $\tan^2 x = 2 \sec x + 2$
can be written as $\sec^2 x - 2 \sec x - 3 = 0$.

b) Hence solve $\tan^2 x = 2 \sec x + 2$ over the interval $0° \leq x \leq 360°$, giving your answers in degrees to 1 decimal place.

Q6b Problem Solving

Use the result of part a) and solve $\sec^2 x - 2 \sec x - 3 = 0$, as a quadratic in sec x.

P **Q7** a) Show that the equation $2 \csc^2 x = 5 - 5 \cot x$ can be written as $2 \cot^2 x + 5 \cot x - 3 = 0$.

b) Hence solve $2 \csc^2 x = 5 - 5 \cot x$ over the interval $-\pi \leq x \leq \pi$, giving your answers in radians to 2 decimal places.

P **Q8** a) Show that the equation $2 \cot^2 A + 5 \csc A = 10$ can be written as $2 \csc^2 A + 5 \csc A - 12 = 0$.

b) Hence solve $2 \cot^2 A + 5 \csc A = 10$ over the interval $0° \leq x \leq 360°$, giving your answers in degrees to 1 decimal place.

Q9 Problem Solving

First write out the equation in terms of tan x only.

P **Q9** Solve the equation $\sec^2 x + \tan x = 1$
for $0 \leq x \leq 2\pi$, giving exact answers.

P **Q10** a) Given that $\csc^2 \theta + 2 \cot^2 \theta = 2$, find the possible values of $\sin \theta$.

b) Hence solve the equation $\csc^2 \theta + 2 \cot^2 \theta = 2$ in the interval $0° \leq \theta \leq 180°$.

P **Q11** Solve the equation $\sec^2 x = 3 + \tan x$ in the interval $0° \leq x \leq 360°$, giving your answers to 1 d.p.

P **Q12** Solve the equation $\cot^2 x + \csc^2 x = 7$, giving all the solutions in the interval $0 \leq x \leq 2\pi$ in terms of π.

P **Q13** Solve the equation $\tan^2 x + 5 \sec x + 7 = 0$, giving all the solutions in the interval $0 \leq x \leq 2\pi$ to 2 decimal places.

P **Q14** If $\tan \theta = \dfrac{60}{11}$ and $180° \leq \theta \leq 270°$, find the exact value of:

a) $\sin \theta$ b) $\sec \theta$ c) $\csc \theta$

Q14-Q15 Problem Solving

Sketch a right-angled triangle with two sides corresponding to the numerator and denominator to help you see what is going on here.

P **Q15** If $\csc \theta = -\dfrac{17}{15}$ and $180° \leq \theta \leq 270°$, find the exact value of:

a) $\cos \theta$ b) $\sec \theta$ c) $\cot \theta$

E **P** **Q16** Given that $\cos x = \dfrac{1}{6}$, use the identity $\sec^2 \theta = 1 + \tan^2 \theta$
to find the two possible exact values of tan x. *[2 marks]*

E **P** **Q17** Given that $3 \tan x + 1 = \sec^2 x$,

a) Find the possible value(s) of $\tan x$. *[3 marks]*

b) Hence find, to 3 significant figures, the value(s) of x given that $0 < x < \pi$. *[2 marks]*

E **Q18** Given that $\sec^2 x - 6 = \sec x$,

a) Work out the possible values of $\sec x$. *[2 marks]*

b) Use the identity $\sec^2 x \equiv 1 + \tan^2 x$ to find the possible values of $\tan x$. *[2 marks]*

3.9 Proving Other Identities

You can also use identities to prove that two trig expressions are the same, as shown in the examples below. You just need to take one side of the identity and play about with it until you get what's on the other side.

Example 1

a) Show that $\dfrac{\tan^2 x}{\sec x} \equiv \sec x - \cos x$.

$$\dfrac{\tan^2 x}{\sec x} \equiv \dfrac{\sec^2 x - 1}{\sec x}$$

Start with the left-hand side of the identity and replace $\tan^2 x$ with $\sec^2 x - 1$.

$$\equiv \dfrac{\sec^2 x}{\sec x} - \dfrac{1}{\sec x}$$

Rearrange to get the right-hand side of the identity.

$$\equiv \sec x - \cos x$$

b) Prove the identity $\dfrac{\tan^2 x}{\sec x + 1} \equiv \sec x - \cos^2 x - \sin^2 x$.

$$\dfrac{\tan^2 x}{\sec x + 1} \equiv \dfrac{\sec^2 x - 1}{\sec x + 1}$$

Starting with the left-hand side, replace $\tan^2 x$ with $\sec^2 x - 1$.

$$\equiv \dfrac{(\sec x + 1)(\sec x - 1)}{\sec x + 1}$$

Factorise the $\sec^2 x - 1$ — it's the difference of two squares.

$$\equiv \sec x - 1$$

$$\equiv \sec x - (\cos^2 x + \sin^2 x)$$

Cancel and then use $\cos^2 x + \sin^2 x \equiv 1$ to replace the '1'. This gives you the right-hand side.

$$\equiv \sec x - \cos^2 x - \sin^2 x$$

When proving identities, keep checking that you're **getting closer** to the other side of the identity. As well as using the **known identities**, there are lots of little tricks you can use, such as looking for the '**difference of two squares**', and multiplying the top and bottom of a fraction by the **same expression**.

Exercise 3.9

Q1 a) Show that $\sec^2 \theta - \operatorname{cosec}^2 \theta \equiv \tan^2 \theta - \cot^2 \theta$.

b) Hence prove that $(\sec \theta + \operatorname{cosec} \theta)(\sec \theta - \operatorname{cosec} \theta) \equiv (\tan \theta + \cot \theta)(\tan \theta - \cot \theta)$.

Q2 Prove that $\operatorname{cosec} x - \sin x = \cos x \cot x$.

Q3 Prove the identity $(\tan x + \cot x)^2 \equiv \sec^2 x + \operatorname{cosec}^2 x$.

P **Q4** Prove that $\dfrac{1 - \sec x}{\tan x} \equiv \cot x - \operatorname{cosec} x$.

E **Q5** a) Prove the identity $(\cos^2 x - 1)(\cot^2 x - \operatorname{cosec}^2 x) \equiv \sin^2 x$. *[2 marks]*
P
   b) Find the exact value of $(\cos^2 x - 1)(\cot^2 x - \operatorname{cosec}^2 x)$ when $\sqrt{1 + \cot^2 x} = -3$. *[2 marks]*

P **Q6** Prove the identity $\tan^2 x + \cos^2 x \equiv (\sec x + \sin x)(\sec x - \sin x)$.

P **Q7** Prove that $(\sec x + \cot x)(\sec x - \cot x) \equiv \tan^2 x - \operatorname{cosec}^2 x + 2$.

P **Q8** Prove the identity $\sec^2 \theta \operatorname{cosec}^2 \theta \equiv 2 + \cot^2 \theta + \tan^2 \theta$.

P **Q9** Prove that $(\sin x - \operatorname{cosec} x)^3 + \cos^3 x \cot^3 x \equiv 0$.

P **Q10** Prove that $\dfrac{\sec^2 x + 1}{\tan^2 x} \equiv \dfrac{\sec^4 x - 1}{\tan^4 x}$.

> **Q10 Problem Solving**
> Notice that the numerator on the right-hand side of the identity is the difference of two squares.

P **Q11** Prove that $\dfrac{\cos \theta}{\sin \theta} + \dfrac{\sin \theta}{\cos \theta} \equiv \operatorname{cosec} \theta \sec \theta$.

P **Q12** Prove the identity $\dfrac{(\sec x - \tan x)(\tan x + \sec x)}{\operatorname{cosec} x - \cot x} \equiv \cot x + \operatorname{cosec} x$.

P **Q13** Prove that $\dfrac{\cot x}{1 + \operatorname{cosec} x} + \dfrac{1 + \operatorname{cosec} x}{\cot x} \equiv 2 \sec x$.

P **Q14** Prove the identity $\dfrac{\operatorname{cosec} x + 1}{\operatorname{cosec} x - 1} \equiv 2 \sec^2 x + 2 \tan x \sec x - 1$.

E **Q15** a) Prove that $\cot x - \operatorname{cosec} x \sec x + \tan x \equiv 0$. *[3 marks]*
P
   b) Without a calculator, find the value of
   $\cos 13° \cot 13° + \cos 13° \tan 13° - \operatorname{cosec} 13°$. *[2 marks]*

P **Q16** Prove that $8 \sec x + \sec^2 x (4 \cot x - \sin x)^2 \equiv \tan^2 x + 16 \operatorname{cosec}^2 x$.

E **Q17** a) Prove that $\dfrac{\cos 3x}{1 - \cot 3x} + \dfrac{\sin 3x}{1 + \tan 3x} \equiv \dfrac{2 \sin^2 3x \cos 3x}{\sin^2 3x - \cos^2 3x}$. *[4 marks]*
P
   b) Hence or otherwise prove that $\dfrac{\cos 3x}{1 - \cot 3x} + \dfrac{\sin 3x}{1 + \tan 3x} \equiv \dfrac{2 \cos 3x}{2 - \operatorname{cosec}^2 3x}$. *[3 marks]*

E **Q18** a) Prove that $\operatorname{cosec} x + \cot x + \dfrac{1}{\operatorname{cosec} x + \cot x} \equiv 2 \operatorname{cosec} x$. *[3 marks]*
P
   b) Hence or otherwise prove that
   $(\operatorname{cosec} x + \cot x)^2 + \dfrac{1}{(\operatorname{cosec} x + \cot x)^2} \equiv 4 \operatorname{cosec}^2 x - 2$. *[3 marks]*

Challenge

E **Q19** a) Verify that $x^6 - y^6 = (x^2 - y^2)(x^4 + x^2 y^2 + y^4)$. *[1 mark]*
P
   b) Prove that $\operatorname{cosec}^6 x - \cot^6 x \equiv 1 + 3 \operatorname{cosec}^2 x \cot^2 x$. *[4 marks]*
   c) Hence solve the equation $\operatorname{cosec}^6 x - \cot^6 x = 1$ for $0 < x < 2\pi$. *[3 marks]*

E Exam Style **P** Problem Solving **M** Modelling

3.10 The Addition Formulas — Finding Exact Values

The identities shown below are known as the **addition formulas**.

You can use the addition formulas to find the **sin**, **cos** or **tan** of the **sum** or **difference** of two angles, and to 'expand the brackets' in expressions such as $\sin (x + 60°)$ or $\cos \left(n - \frac{\pi}{2}\right)$.

$$\sin (A \pm B) \equiv \sin A \cos B \pm \cos A \sin B$$

$$\cos (A \pm B) \equiv \cos A \cos B \mp \sin A \sin B$$

$$\tan (A \pm B) \equiv \frac{\tan A \pm \tan B}{1 \mp \tan A \tan B}$$

Watch out for the $\pm$ and $\mp$ signs in the formulas — especially for cos and tan.
If you use the sign on the **top** on the **left-hand side** of the identity, you have to use the sign on the **top** on the **right-hand side** too — so $\cos(A + B) = \cos A \cos B - \sin A \sin B$.

Proving the addition formulas

You need to understand the **geometric proof** of these formulas, and although it looks rather complicated, it only uses basic trigonometry.

Step 1:

- Start with a right-angled triangle with a hypotenuse of 1, where one of the angles is $A + B$.

- This is a right-angled triangle, so $\sin (A + B) = \frac{\text{opp}}{\text{hyp}}$ and $\cos (A + B) = \frac{\text{adj}}{\text{hyp}}$, and since the hypotenuse is 1, you can write down the lengths of each side:
  opp $= \sin (A + B)$ and adj $= \cos (A + B)$.

- Add these labels to the diagram as shown.

Step 2:

- If you extend the dotted line, you can form another right-angled triangle with a hypotenuse of 1. This time, the angle is B.

- You can find the lengths of the sides of this triangle in the same way: write out $\sin B = \frac{\text{opp}}{\text{hyp}}$ and $\cos B = \frac{\text{adj}}{\text{hyp}}$, then use the fact that the hypotenuse is 1 to get opp $= \sin B$ and adj $= \cos B$.

- Then add these to the diagram as well.

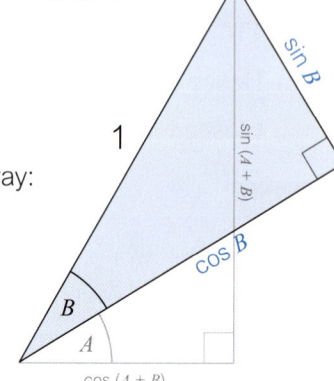

Step 3:

- Now form another right-angled triangle using angle A as shown. This one has a hypotenuse with length $\cos B$.

- So this time, when you rearrange the equations

  $\sin A = \dfrac{\text{opp}}{\text{hyp}}$ and $\cos A = \dfrac{\text{adj}}{\text{hyp}}$, use hyp $= \cos B$,

  which gives opp $= \sin A \cos B$ and adj $= \cos A \cos B$.

- Label these sides, and make sure it's clear which length is $\cos(A + B)$ and which is $\cos A \cos B$.

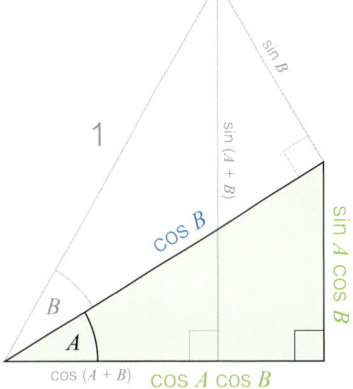

Step 4:

- The last triangle you need to draw has a hypotenuse of $\sin B$ and an angle of A. If you're not sure where this angle comes from, have a look at the diagram.

- So $\sin A = \dfrac{\text{opp}}{\text{hyp}}$ and $\cos A = \dfrac{\text{adj}}{\text{hyp}}$, and since

  hyp $= \sin B$, opp $= \sin A \sin B$ and adj $= \cos A \sin B$.

- Label these sides on the diagram.

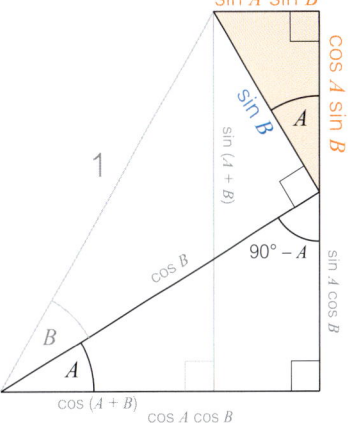

Step 5:

- You've now labelled enough sides to prove the formulas — here's the diagram again, with the labels for $\sin(A + B)$ and $\cos(A + B)$ moved to make things a bit clearer.

- Using the height of the diagram, you can see that:

  $\sin(A + B) = \sin A \cos B + \cos A \sin B$

- The width of the diagram shows you that:

  $\cos A \cos B = \cos(A + B) + \sin A \sin B$

  $\Rightarrow$ **$\cos(A + B) = \cos A \cos B - \sin A \sin B$**

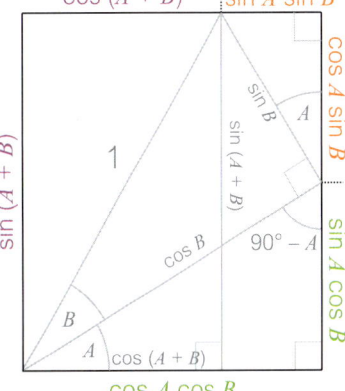

This is the proof of the **sin** and **cos** formulas — see p.101 for a proof of the **tan** formula.

You can work out the subtraction formulas from the addition ones, by using the fact that **sin −B = −sin B** and **cos −B = cos B** — you can see these are true by looking at the graphs of $y = \sin x$ and $y = \cos x$.

$$\sin (A - B) = \sin A \cos (-B) + \cos A \sin (-B) = \sin A \cos B - \cos A \sin B$$

$$\cos (A - B) = \cos A \cos (-B) - \sin A \sin (-B) = \cos A \cos B + \sin A \sin B$$

You can also prove them geometrically using a similar method as for $\sin(A + B)$ and $\cos(A + B)$. You'll end up with the diagram below — work through it yourself to make sure you know where the labels come from.

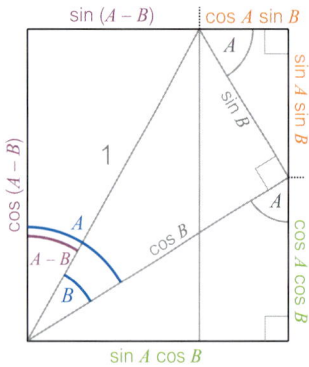

> **Tip** Start with the triangle with angle $A - B$ and hypotenuse 1, and then add on the triangle with angle B below it.

Example 1

a) Find the exact value of $\sin 18° \cos 12° + \cos 18° \sin 12°$.

$$\sin A \cos B + \cos A \sin B \equiv \sin (A + B)$$

$$\sin 18° \cos 12° + \cos 18° \sin 12° = \sin (18° + 12°)$$

Use the sin addition formula, where $A = 18°$ and $B = 12°$.

$$= \sin 30° = \frac{1}{2}$$

b) Write $\dfrac{\tan 5x - \tan 2x}{1 + \tan 5x \tan 2x}$ as a single trigonometric ratio.

$$\frac{\tan 5x - \tan 2x}{1 + \tan 5x \tan 2x} = \tan (5x - 2x) = \tan 3x$$

Use the tan addition formula, with $A = 5x$ and $B = 2x$.

c) Given both x and y are acute, find the exact value of $\cos (x + y)$ if $\sin x = \dfrac{4}{5}$ and $\sin y = \dfrac{15}{17}$.

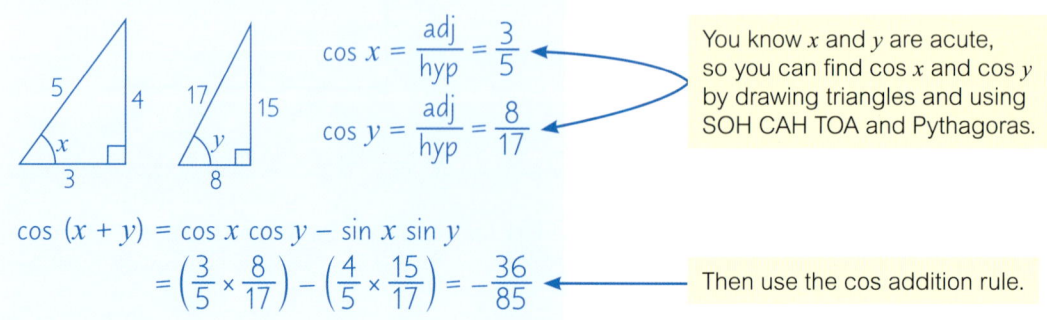

$$\cos x = \frac{\text{adj}}{\text{hyp}} = \frac{3}{5}$$

$$\cos y = \frac{\text{adj}}{\text{hyp}} = \frac{8}{17}$$

You know x and y are acute, so you can find $\cos x$ and $\cos y$ by drawing triangles and using SOH CAH TOA and Pythagoras.

$$\cos (x + y) = \cos x \cos y - \sin x \sin y$$

$$= \left(\frac{3}{5} \times \frac{8}{17}\right) - \left(\frac{4}{5} \times \frac{15}{17}\right) = -\frac{36}{85}$$

Then use the cos addition rule.

You should know the value of sin, cos and tan for **common angles**, in degrees and radians (see p.78). You can use your knowledge of these angles, along with the addition formulas, to find the **exact value** of sin, cos or tan for **other** angles.

Find a **pair** of common angles which **add or subtract** to give the angle you're after, then put them into the addition formula, and work it through.

Example 2

Using the addition formula for tangent, show that $\tan 15° = 2 - \sqrt{3}$.

$\tan 15° = \tan(60° - 45°)$ ← Pick two angles that add or subtract to give 15°. It's easiest to use tan 60° and tan 45°, since neither of them are fractions.

$= \dfrac{\tan 60° - \tan 45°}{1 + \tan 60° \tan 45°}$ ← Put them into the tan addition formula.

$= \dfrac{\sqrt{3} - 1}{1 + (\sqrt{3} \times 1)} = \dfrac{\sqrt{3} - 1}{\sqrt{3} + 1}$ ← Substitute the values for tan 60° ($= \sqrt{3}$) and tan 45° ($= 1$) into the equation.

$= \dfrac{\sqrt{3} - 1}{\sqrt{3} + 1} \times \dfrac{\sqrt{3} - 1}{\sqrt{3} - 1} = \dfrac{3 - 2\sqrt{3} + 1}{3 - \sqrt{3} + \sqrt{3} - 1}$ ← Now rationalise the denominator of the fraction to get rid of the $\sqrt{3}$.

$= \dfrac{4 - 2\sqrt{3}}{2} = 2 - \sqrt{3}$ as required ← Then simplify the expression.

Exercise 3.10

Q1 Use the addition formulas to find the exact values of the following:

a) $\cos 72° \cos 12° + \sin 72° \sin 12°$

b) $\cos 13° \cos 17° - \sin 13° \sin 17°$

c) $\dfrac{\tan 12° + \tan 18°}{1 - \tan 12° \tan 18°}$

d) $\dfrac{\tan 500° - \tan 140°}{1 + \tan 500° \tan 140°}$

e) $\sin 35° \cos 10° + \cos 35° \sin 10°$

f) $\sin 69° \cos 9° - \cos 69° \sin 9°$

Q2 Use the addition formulas to find the exact values of the following:

a) $\sin \dfrac{2\pi}{3} \cos \dfrac{\pi}{2} - \cos \dfrac{2\pi}{3} \sin \dfrac{\pi}{2}$

b) $\cos 4\pi \cos 3\pi + \sin 4\pi \sin 3\pi$

c) $\dfrac{\tan \dfrac{5\pi}{12} + \tan \dfrac{5\pi}{4}}{1 - \tan \dfrac{5\pi}{12} \tan \dfrac{5\pi}{4}}$

d) $\cos \dfrac{\pi}{9} \cos \dfrac{5\pi}{9} - \sin \dfrac{\pi}{9} \sin \dfrac{5\pi}{9}$

Q3 Write the following expressions as a single trigonometric ratio:

a) $\sin 5x \cos 2x - \cos 5x \sin 2x$

b) $\cos 4x \cos 6x - \sin 4x \sin 6x$

c) $\dfrac{\tan 7x + \tan 3x}{1 - \tan 7x \tan 3x}$

d) $5 \sin 2x \cos 3x + 5 \cos 2x \sin 3x$

e) $8 \cos 7x \cos 5x + 8 \sin 7x \sin 5x$

f) $\dfrac{\tan 8x - \tan 5x}{1 + \tan 8x \tan 5x}$

P Q4 $\sin x = \dfrac{5}{13}$ and $\cos y = \dfrac{24}{25}$, where x and y are both acute angles. Calculate the exact value of:

a) $\sin(x - y)$

b) $\cos(x + y)$

c) $\cos(x - y)$

d) $\tan(x + y)$

> **Q4-5 Problem Solving**
>
> You'll need to work out $\cos x$ and $\sin y$ before you can answer parts a)-d). You can use the triangle method or the identity $\cos^2 \theta + \sin^2 \theta \equiv 1$ to work them out.

P Q5 $\sin x = \dfrac{3}{4}$ and $\cos y = \dfrac{3}{\sqrt{10}}$, where x and y are both acute angles. Calculate the exact value of:

a) $\sin(x + y)$

b) $\cos(x - y)$

c) $\operatorname{cosec}(x + y)$

d) $\sec(x - y)$

E Q6 Without using a calculator, work out the exact value of $\sin 28° \cos 32° + \sin 30° \cos 30° + \sin 32° \cos 28°$.

[3 marks]

Q7 Use the addition formulas to find the exact values of:

a) $\arcsin(\sin 17° \cos 5° + \sin 5° \cos 17°)$

b) $\arctan\left(\dfrac{2 \tan\left(\frac{\pi}{7}\right)}{1 - \tan^2\left(\frac{\pi}{7}\right)}\right)$

> **Q7 Problem Solving**
>
> $A = B$ for part b), so you could use the double angle formula instead (see p. 104).

P Q8 Using the addition formula for cos, show that $\cos \dfrac{\pi}{12} = \dfrac{\sqrt{6} + \sqrt{2}}{4}$.

P Q9 Using the addition formula for sin, show that $\sin 75° = \dfrac{\sqrt{6} + \sqrt{2}}{4}$.

P Q10 Using the addition formula for tan, show that $\tan 75° = \dfrac{\sqrt{3} + 1}{\sqrt{3} - 1}$.

E **P** Q11 Given that:

$$\sin \frac{3\pi}{10} = \frac{1}{4}(\sqrt{5} + 1), \quad \sin \frac{\pi}{10} = \frac{1}{4}(\sqrt{5} - 1) \text{ and } \cos \frac{\pi}{5} = \frac{1}{4}(\sqrt{5} + 1)$$

Find the exact value of $\sin \dfrac{\pi}{5} \cos \dfrac{\pi}{10}$.

[3 marks]

E **P** Q12 Given that $\sin x = \dfrac{1}{2}$ and $\cos y = -\dfrac{1}{4}$:

a) Find the two exact values of $\cos x$ in the interval $0 \leq x \leq \pi$. *[2 marks]*

b) Find the two exact values of $\sin y$ in the interval $0 \leq x \leq \pi$. *[2 marks]*

c) Find all four possible exact values of $\cos(x + y)$ in the interval $0 \leq x \leq \pi$. *[2 marks]*

3.11 The Addition Formulas — Equations and Identities

You might be asked to use the addition formulas to **prove an identity**. All you need to do is put the numbers and variables from the left-hand side into the addition formulas and simplify until you get the expression you're after.

Example 1

Prove that $\cos(a + 60°) + \sin(a + 30°) \equiv \cos a$.

$\cos(a + 60°) + \sin(a + 30°)$

$\equiv (\cos a \cos 60° - \sin a \sin 60°) + (\sin a \cos 30° + \cos a \sin 30°)$

$= \dfrac{1}{2}\cos a - \dfrac{\sqrt{3}}{2}\sin a + \dfrac{\sqrt{3}}{2}\sin a + \dfrac{1}{2}\cos a$

$= \dfrac{1}{2}\cos a + \dfrac{1}{2}\cos a = \cos a$ as required

> Put the numbers from the question into the addition formulas.

> Substitute in any sin and cos values that you know — be careful with the + and − signs here.

> Simplify the expression.

Example 2

Use the sine and cosine addition formulas to prove that $\tan(A + B) \equiv \dfrac{\tan A + \tan B}{1 - \tan A \tan B}$.

$\tan(A + B) \equiv \dfrac{\sin(A + B)}{\cos(A + B)}$

$\equiv \dfrac{\sin A \cos B + \cos A \sin B}{\cos A \cos B - \sin A \sin B}$

$\equiv \dfrac{\dfrac{\sin A \cos B}{\cos A \cos B} + \dfrac{\cos A \sin B}{\cos A \cos B}}{\dfrac{\cos A \cos B}{\cos A \cos B} - \dfrac{\sin A \sin B}{\cos A \cos B}}$

$\equiv \dfrac{\dfrac{\sin A}{\cos A} + \dfrac{\sin B}{\cos B}}{1 - \left(\dfrac{\sin A}{\cos A}\right)\left(\dfrac{\sin B}{\cos B}\right)} \equiv \dfrac{\tan A + \tan B}{1 - \tan A \tan B}$

> Start with the identity $\tan \theta \equiv \dfrac{\sin \theta}{\cos \theta}$.

> Replace $\sin(A + B)$ and $\cos(A + B)$ with the addition formulas for each.

> Divide each part of the fraction by $\cos A \cos B$ and cancel where possible.

> **Tip** The first term on the denominator needs to be '1' — divide through by this term to get the 1 in the right place.

Example 3 P

Use the addition formulas to show that $\cos A + \cos B \equiv 2 \cos\left(\dfrac{A+B}{2}\right) \cos\left(\dfrac{A-B}{2}\right)$.

$\cos(x + y) \equiv \cos x \cos y - \sin x \sin y$
$\cos(x - y) \equiv \cos x \cos y + \sin x \sin y$

$\cos(x + y) + \cos(x - y)$

$\equiv \cos x \cos y - \sin x \sin y + \cos x \cos y + \sin x \sin y$

$\equiv 2 \cos x \cos y$

> Start with the cos addition formulas.

> Add the two expressions together and simplify.

continued on the next page...

E Exam Style   P Problem Solving   M Modelling

$A = x + y$ and $B = x - y$ ← Following the cos addition formulas, define A and B in terms of x and y.

$A - B = x + y - (x - y) = 2y$, so $y = \dfrac{A - B}{2}$ ← Subtracting B from A.

$A + B = x + y + (x - y) = 2x$, so $x = \dfrac{A + B}{2}$ ← Adding A and B.

So $\cos A + \cos B \equiv 2 \cos\left(\dfrac{A+B}{2}\right) \cos\left(\dfrac{A-B}{2}\right)$

Tip This is one of the factor formulas. The others are given in the formula booklet — see p.581.

You can also use the addition formulas to **solve** complicated trig equations.

Example 4

Solve $\sin\left(x + \dfrac{\pi}{2}\right) = \sin x$ in the interval $0 \le x \le 2\pi$.

$\sin\left(x + \dfrac{\pi}{2}\right) = \sin x$ ← Replace $\sin\left(x + \dfrac{\pi}{2}\right)$ using the sin addition formula.

$\Rightarrow \sin x \cos\dfrac{\pi}{2} + \cos x \sin\dfrac{\pi}{2} = \sin x$ ← Using the table of common values on page 78, $\cos\dfrac{\pi}{2} = 0$ and $\sin\dfrac{\pi}{2} = 1$.

$\Rightarrow 0 + \cos x = \sin x$

$\dfrac{\cos x}{\cos x} = \dfrac{\sin x}{\cos x}$ ← Divide through by $\cos x$.

$\dfrac{\cancel{\cos x}}{\cancel{\cos x}} = \tan x \Rightarrow \tan x = 1$ ← Replace $\dfrac{\sin x}{\cos x}$ with $\tan x$.

$x = \dfrac{\pi}{4}$ and $\dfrac{5\pi}{4}$ ← Solve for $0 \le x \le 2\pi$.

Exercise 3.11

Q1 Use the sine and cosine addition formulas to prove that $\tan(A - B) \equiv \dfrac{\tan A - \tan B}{1 + \tan A \tan B}$.

Q1 Hint Look at the proof of $\tan(A + B)$ on the previous page.

P Q2 Prove the following identities:

a) $\dfrac{\cos(A - B) - \cos(A + B)}{\cos A \sin B} \equiv 2 \tan A$

b) $\dfrac{1}{2}[\cos(A - B) - \cos(A + B)] \equiv \sin A \sin B$

c) $\sin(x + 90°) \equiv \cos x$

d) $\cos(x + 180°) \equiv -\cos x$

E Q3 Solve $4 \sin\left(x - \dfrac{\pi}{3}\right) = \cos x$ in the interval $-\pi \le x \le \pi$.
Give your answers in radians to 2 decimal places. *[3 marks]*

P Q4 a) Show that $\tan\left(-\frac{\pi}{12}\right) = \sqrt{3} - 2$.

b) Use your answer to a) to solve the equation $\cos x = \cos\left(x + \frac{\pi}{6}\right)$ in the interval $0 \leq x \leq \pi$. Give your answer in terms of π.

Q5 Show that $2\sin(x + 30°) \equiv \sqrt{3}\sin x + \cos x$.

E P Q6 Given that $\tan\left(x + \frac{\pi}{3}\right) = \frac{1}{3}$, show that $\tan x = \frac{6 - 5\sqrt{3}}{3}$. *[4 marks]*

P Q7 a) Show that $\sin(\theta - 45°) \equiv \frac{1}{\sqrt{2}}(\sin\theta - \cos\theta)$.

b) Use your answer to part a) to solve $4\sin(\theta - 45°) = \sqrt{2}\cos\theta$ in the interval $0° \leq \theta \leq 360°$. Give your answers to 3 significant figures.

Q8 Write an expression for $\tan\left(\frac{\pi}{3} - x\right)$ in terms of $\tan x$ only.

Q9 $\tan A = \frac{3}{8}$ and $\tan(A + B) = \frac{1}{4}$. Find the exact value of $\tan B$.

P Q10 Show that $\sin A + \sin B \equiv 2\sin\left(\frac{A + B}{2}\right)\cos\left(\frac{A - B}{2}\right)$.

P Q11 a) Given that $\sin(x + y) = 4\cos(x - y)$, write an expression for $\tan x$ in terms of $\tan y$.

b) Use your answer to a) to solve $\sin\left(x + \frac{\pi}{4}\right) = 4\cos\left(x - \frac{\pi}{4}\right)$ in the interval $0 \leq x \leq 2\pi$.

Q12 Solve the following equations in the given interval. Give your answers to 2 decimal places.

a) $\sqrt{2}\sin(\theta + 45°) = 3\cos\theta$, $0° \leq \theta \leq 360°$

b) $2\cos\left(\theta - \frac{2\pi}{3}\right) - 5\sin\theta = 0$, $0 \leq \theta \leq 2\pi$

c) $\sin(\theta - 30°) - \cos(\theta + 60°) = 0$, $0° \leq \theta \leq 360°$

P Q13 Use the sin addition formula to show that $\sin\left(x + \frac{\pi}{6}\right) \approx \frac{1}{2} + \frac{\sqrt{3}}{2}x - \frac{1}{4}x^2$ when x is small.

E Q14 Given that $\tan(x + 8°) = \cot(x + 80°)$,

a) Show that $\sin(x + 8°)\sin(x + 80°) = \cos(x + 8°)\cos(x + 80°)$. *[2 marks]*

b) Hence find all possible values of x, $0° \leq x \leq 180°$. *[3 marks]*

E P Q15 Prove that $\sec(x + y)(1 - \tan x \tan y) \equiv \sec x \sec y$. *[4 marks]*

P Q16 a) Prove that $\cot(x + y) \equiv \frac{\cot x \cot y - 1}{\cot x + \cot y}$.

b) Prove that $\frac{\cos(x + y)}{\cos x \cos y} \equiv 1 - \tan x \tan y$.

3.12 The Double Angle Formulas — Finding Exact Values

Double angle formulas are a special case of the addition formulas, using $(A + A)$ instead of $(A + B)$. They're called "double angle" formulas because they take an expression with a $2x$ term (a double angle) inside a trig function, and change it into an expression with only single x's inside the trig functions.

You need to know the double angle formulas for **sin**, **cos** and **tan**. Their derivations are given below. You can also prove these **geometrically** in the same way as on pages 96-97 — just replace B with A.

$\sin 2A \equiv 2 \sin A \cos A$

- Start with the **sin addition formula** (see p.96), but replace 'B' with 'A':

$$\sin (A + A) \equiv \sin A \cos A + \cos A \sin A$$

- $\sin (A + A)$ can be written as $\sin 2A$, and so:

$$\sin 2A \equiv \sin A \cos A + \cos A \sin A \equiv 2 \sin A \cos A$$

$\cos 2A \equiv \cos^2 A - \sin^2 A$

- Start with the **cos addition formula**, but again replace 'B' with 'A':

$$\cos (A + A) \equiv \cos A \cos A - \sin A \sin A$$
$$\Rightarrow \cos 2A \equiv \cos^2 A - \sin^2 A$$

- You can then use $\cos^2 A + \sin^2 A \equiv 1$ to get:

$$\cos 2A \equiv \cos^2 A - (1 - \cos^2 A) \text{ and } \cos 2A \equiv (1 - \sin^2 A) - \sin^2 A$$

$$\cos 2A \equiv 2 \cos^2 A - 1 \qquad \cos 2A \equiv 1 - 2 \sin^2 A$$

> **Tip** The double angle formula for cos has three different forms which are all very useful.

$\tan 2A \equiv \dfrac{2 \tan A}{1 - \tan^2 A}$

- Start with the **tan addition formula**, and again replace 'B' with 'A':

$$\tan (A + A) \equiv \frac{\tan A + \tan A}{1 - \tan A \tan A}$$

- Simplifying this gives:

$$\tan 2A \equiv \frac{2 \tan A}{1 - \tan^2 A}$$

Using the double angle formulas

Like the other trig identities covered in this chapter, the double angle formulas are useful when you need to find an **exact value**. You won't usually be told which identity to use, so work on being able to spot the clues — if you're asked for an 'exact value', you should think of your **common angles**. 15° is half of a common angle, so this should get you thinking about double angle formulas.

Example 1

a) Use a double angle formula to work out the exact value of sin 15° cos 15°.

$$\sin 2A \equiv 2 \sin A \cos A$$

Look for an identity that is similar to the expression. The sin double angle formula looks best, but needs rearranging.

$$\Rightarrow \sin A \cos A \equiv \frac{1}{2} \sin 2A$$

$$\sin 15° \cos 15° = \frac{1}{2} \sin 30° = \frac{1}{2} \times \frac{1}{2} = \frac{1}{4}$$

Now put in the numbers from the question.

b) $\sin x = \frac{2}{3}$, where x is acute. Find the exact value of cos $2x$ and sin $2x$.

$$\cos 2A \equiv 1 - 2 \sin^2 A$$

For cos $2x$, use the cos double angle formula in terms of sin and substitute in $\sin x = \frac{2}{3}$.

$$\Rightarrow \cos 2x = 1 - 2\left(\frac{2}{3}\right)^2 = \frac{1}{9}$$

$$\sin 2A \equiv 2 \sin A \cos A$$

For sin $2x$, use the sin double angle formula.

$$\cos x = \frac{adj}{hyp} = \frac{\sqrt{5}}{3}$$

To use the formula, first work out cos x from sin x using the triangle method.

$$\sin 2x = 2 \sin x \cos x$$

$$= 2 \times \frac{2}{3} \times \frac{\sqrt{5}}{3} = \frac{4\sqrt{5}}{9}$$

Now put the values into the sin double angle formula as usual.

The double angle formulas are also handy for **simplifying expressions** in order to solve equations.

Example 2

Write $1 - 2 \sin^2 \left(\frac{3x}{2}\right)$ as a single trigonometric ratio.

$$\cos 2A \equiv 1 - 2 \sin^2 A$$

Look for an identity that is similar to the expression, containing a 'sin²'. The cos double angle formula (in terms of sin) looks best.

$$1 - 2 \sin^2 \left(\frac{3x}{2}\right)$$

Compare the expression with the right-hand side of the identity.

continued on the next page...

So $A = \dfrac{3x}{2} \Rightarrow 2A = 3x$ ←

> Find an expression for $2A$ in terms of x and put this into the identity to get the expression in a single cos term.

$1 - 2\sin^2 \dfrac{3x}{2} \equiv \cos 3x$ ←

Exercise 3.12

Q1 Use the double angle formulas to write down the exact values of:

a) $4\sin \dfrac{\pi}{12} \cos \dfrac{\pi}{12}$

b) $\cos \dfrac{2\pi}{3}$

c) $\dfrac{\sin 120°}{2}$

d) $\dfrac{\tan 15°}{2 - 2\tan^2 15°}$

e) $2\sin^2 15° - 1$

f) $\tan^2 \dfrac{\pi}{3}$

> **Q1 Hint** For some of these there are other ways to find the answer, but if you've been asked to use a certain method then show your working using that method.

Q2 An acute angle x has $\sin x = \dfrac{1}{6}$. Find the exact values of:

a) $\cos 2x$

b) $\sin 2x$

c) $\tan 2x$

d) $\sec 2x$

e) $\operatorname{cosec} 2x$

f) $\cot 2x$

Q3 Angle x has $\sin x = -\dfrac{1}{4}$, and $\pi \leq x \leq \dfrac{3\pi}{2}$. Find the exact values of:

a) $\cos 2x$

b) $\sin 2x$

c) $\tan 2x$

d) $\sec 2x$

e) $\operatorname{cosec} 2x$

f) $\cot 2x$

> **Q3 Hint** Angle x lies in the 3rd quadrant of the CAST diagram, so $\sin x$ and $\cos x$ are negative but $\tan x$ is positive.

E P Q4 Use $\cos 18° = \dfrac{\sqrt{2(5 + \sqrt{5})}}{4}$ to find the exact value of the following in their simplest form:

a) $\cos 36°$ *[3 marks]*

b) $\sec 72°$ *[4 marks]*

Q5 Write the following expressions as a single trigonometric ratio:

a) $\dfrac{\sin 3\theta \cos 3\theta}{3}$

b) $\sin^2 \left(\dfrac{2y}{3}\right) - \cos^2 \left(\dfrac{2y}{3}\right)$

c) $\dfrac{1 - \tan^2 \left(\dfrac{x}{2}\right)}{2\tan \left(\dfrac{x}{2}\right)}$

E Q6 Given that $\cot x = -\dfrac{3}{5}$,

a) use the double angle formula to find the exact value of $\tan 2x$. *[3 marks]*

b) use the addition formula to find the exact value of $\tan 3x$. *[3 marks]*

c) hence find the exact value of $\cot 6x$. *[3 marks]*

Challenge

E P Q7 Express the following fraction as a single trigonometric function

$$\dfrac{3\sin^2(5x) - 3\cos^2(5x)}{\cos^3(5x)\sin(5x) + \sin^3(5x)\cos(5x)}$$

[4 marks]

3.13 The Double Angle Formulas — Equations and Identities

If an equation has a **mixture** of sin x and sin $2x$ terms in it, there's not much that you can do with it in that state. But you can use one of the **double angle formulas** to simplify it, and then solve it.

Example 1 P

Solve the equation $\cos 2x - 5\cos x = 2$ in the interval $0 \le x \le 2\pi$.

$\cos 2A \equiv 2\cos^2 A - 1$

First use the cos double angle formula to get rid of cos $2x$ (this version of the formula will stop you ending up with a mix of sin and cos terms).

$\Rightarrow 2\cos^2 x - 1 - 5\cos x = 2$

$\Rightarrow 2\cos^2 x - 5\cos x - 3 = 0$

Simplify so you have zero on one side.

$\Rightarrow (2\cos x + 1)(\cos x - 3) = 0$

$\Rightarrow (2\cos x + 1) = 0$ or $(\cos x - 3) = 0$

Factorise and solve the quadratic — let $y = \cos x$ and write as a quadratic in y if it helps.

$\cos x - 3 = 0 \Rightarrow \cos x = 3$, which has no solutions, since $-1 \le \cos x \le 1$.

So the only solutions are when

$2\cos x + 1 = 0 \Rightarrow \cos x = -\dfrac{1}{2}$

Find all the solutions in the given interval.

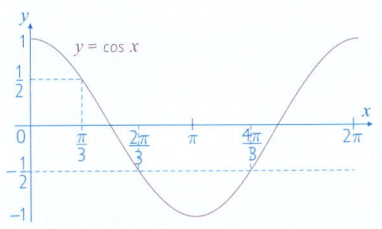

Tip You can either sketch the graph of cos x to find all values of x in the given interval, or use a CAST diagram.

$\cos x = \dfrac{1}{2}$ for $x = \dfrac{\pi}{3}$, so using the symmetry of the graph above, $x = \dfrac{2\pi}{3}$ or $x = \dfrac{4\pi}{3}$.

$\cos x = -\dfrac{1}{2}$ for two values in the interval — once at $\dfrac{2\pi}{3}$ and again at $2\pi - \dfrac{2\pi}{3} = \dfrac{4\pi}{3}$.

The following examples show how the double angle formulas can be used to **prove** other identities.

Example 2 P

Prove that $2\cot\dfrac{x}{2}\left(1 - \cos^2\dfrac{x}{2}\right) \equiv \sin x$.

$\sin^2\theta \equiv 1 - \cos^2\theta \Rightarrow \sin^2\dfrac{x}{2} \equiv 1 - \cos^2\dfrac{x}{2}$

Use the identity $\sin^2\theta + \cos^2\theta \equiv 1$ $\Rightarrow \sin^2\theta \equiv 1 - \cos^2\theta$ to replace $1 - \cos^2\dfrac{x}{2}$ on the left-hand side.

Left-hand side:

$2\cot\dfrac{x}{2}\left(1 - \cos^2\dfrac{x}{2}\right) \equiv 2\cot\dfrac{x}{2}\,\sin^2\dfrac{x}{2}$

continued on the next page...

 Exam Style P Problem Solving M Modelling

$$\cot \theta = \frac{\cos \theta}{\sin \theta} \implies 2\cot \frac{x}{2} \sin^2 \frac{x}{2} \equiv \frac{2\cos \frac{x}{2} \sin^2 \frac{x}{2}}{\sin \frac{x}{2}} \equiv 2\cos \frac{x}{2} \sin \frac{x}{2}$$ ← Simplify the expression.

$$2\cos \frac{x}{2} \sin \frac{x}{2} \equiv \sin x \text{ as required}$$ ← Finally use the sin double angle formula with $A = \frac{x}{2}$.

Clever tricks like splitting up the angle so you can use the addition formulas can really help if you're stuck on a trig identity question.

Example 3 P

Show that $\cos 3\theta \equiv 4\cos^3 \theta - 3\cos \theta$.

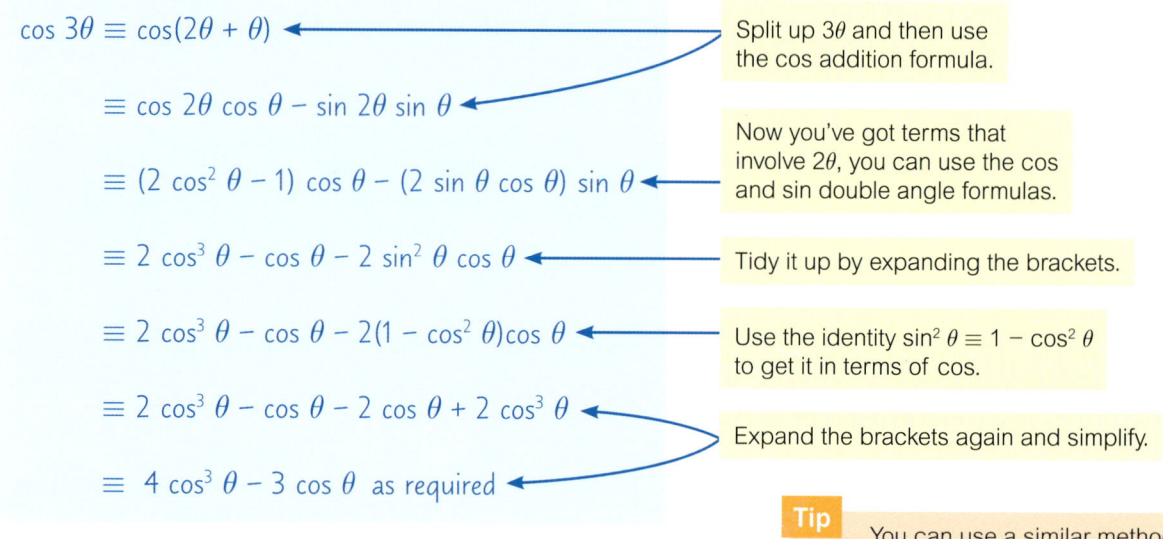

$$\cos 3\theta \equiv \cos(2\theta + \theta)$$ ← Split up 3θ and then use the cos addition formula.

$$\equiv \cos 2\theta \cos \theta - \sin 2\theta \sin \theta$$ ← Now you've got terms that involve 2θ, you can use the cos and sin double angle formulas.

$$\equiv (2\cos^2 \theta - 1)\cos \theta - (2\sin \theta \cos \theta)\sin \theta$$

$$\equiv 2\cos^3 \theta - \cos \theta - 2\sin^2 \theta \cos \theta$$ ← Tidy it up by expanding the brackets.

$$\equiv 2\cos^3 \theta - \cos \theta - 2(1 - \cos^2 \theta)\cos \theta$$ ← Use the identity $\sin^2 \theta \equiv 1 - \cos^2 \theta$ to get it in terms of cos.

$$\equiv 2\cos^3 \theta - \cos \theta - 2\cos \theta + 2\cos^3 \theta$$ ← Expand the brackets again and simplify.

$$\equiv 4\cos^3 \theta - 3\cos \theta \text{ as required}$$

Tip You can use a similar method to show that $\sin 3\theta \equiv 3\sin \theta - 4\sin^3\theta$.

The half angle formulas

The double angle formulas for cos can be **rearranged** to give another three useful identities known as the **half angle formulas**.

The following examples show how you can derive them — you don't need to know these derivations for your exam, but they're good examples of using identities.

Example 4

a) Show that $\cos^2 \left(\frac{\theta}{2}\right) \equiv \frac{1}{2}(1 + \cos \boldsymbol{\theta})$.

$$\cos 2A \equiv 2\cos^2 A - 1$$ ← Start with the double angle formula for cos.

$$\cos \theta \equiv 2\cos^2 \left(\frac{\theta}{2}\right) - 1$$ ← Replace A with $\frac{\theta}{2}$.

$$\cos^2 \left(\frac{\theta}{2}\right) \equiv \frac{1}{2}(1 + \cos \theta)$$ ← Rearrange to get the half angle formula for cos.

b) Show that $\sin^2\left(\frac{\theta}{2}\right) \equiv \frac{1}{2}(1 - \cos\theta)$.

$\cos 2A \equiv 1 - 2\sin^2 A$ — Start with the double angle formula for cos that contains sin.

$\cos\theta \equiv 1 - 2\sin^2\left(\frac{\theta}{2}\right)$ — Again, replace A with $\frac{\theta}{2}$.

$\sin^2\left(\frac{\theta}{2}\right) \equiv \frac{1}{2}(1 - \cos\theta)$ — Rearrange to get the half angle formula for sin.

c) Hence show that $\tan^2\left(\frac{\theta}{2}\right) \equiv \frac{1 - \cos\theta}{1 + \cos\theta}$.

$\tan\left(\frac{\theta}{2}\right) \equiv \dfrac{\sin\left(\frac{\theta}{2}\right)}{\cos\left(\frac{\theta}{2}\right)}$ — Start with the identity $\tan x \equiv \frac{\sin x}{\cos x}$.

$\tan^2\left(\frac{\theta}{2}\right) \equiv \dfrac{\sin^2\left(\frac{\theta}{2}\right)}{\cos^2\left(\frac{\theta}{2}\right)}$ — Square both sides.

$\tan^2\left(\frac{\theta}{2}\right) \equiv \dfrac{\frac{1}{2}(1 - \cos\theta)}{\frac{1}{2}(1 + \cos\theta)}$ — Replace $\sin^2\left(\frac{\theta}{2}\right)$ and $\cos^2\left(\frac{\theta}{2}\right)$ with their half angle formulas from examples a) and b).

$\tan^2\left(\frac{\theta}{2}\right) \equiv \dfrac{1 - \cos\theta}{1 + \cos\theta}$ as required — Simplify to get the half angle formula for tan.

Exercise 3.13

P Q1 Solve the equations below in the interval $0° \leq x \leq 360°$.
Give your answers to 1 decimal place.

a) $4\cos 2x = 14\sin x$

b) $5\cos 2x + 9\cos x = -7$

c) $4\cot 2x + \cot x = 5$

d) $\tan x - 5\sin 2x = 0$

Q1d) Problem Solving

There should be 7 solutions in the given interval, but two of them are easy to miss...

P Q2 Solve the equations below in the interval $0 \leq x \leq 2\pi$.
Give your answers to 3 significant figures.

a) $4\cos 2x - 10\cos x + 1 = 0$

b) $\dfrac{\cos 2x - 3}{2\sin^2 x - 1} = 3$

c) $2\sin x \cos x = 4\cos^2 x - 4\sin^2 x$

d) $2\cos 4x + \sin 2x = -1$

E **P** **Q3** a) Given that $\cos x = a + b \sin^2\left(\frac{x}{2}\right)$, find the values of the constants a and b. *[1 mark]*

b) Hence use the fact that $\sin t \leq t$ for $t \geq 0$ to prove that

$\cos x \geq 1 - \frac{1}{2}x^2$ for all positive real values of x. *[2 marks]*

P **Q4** Solve the equations below in the interval $0 \leq x \leq 2\pi$.
Give your answers in terms of π.

> **Q4b) Problem Solving**
>
> Try writing $\sin x$ as $\sin 2\left(\frac{x}{2}\right)$ and using the sin double angle formula.

a) $\cos 2x + 7 \cos x = -4$ b) $\sin x + \cos \frac{x}{2} = 0$

c) $\sin x - \cos 2x = 0$ d) $\cos x = 7 \cos \frac{x}{2} + 3$

P **Q5** Use the double angle formulas to prove each of the identities below.

a) $\sin 2x \sec^2 x \equiv 2 \tan x$ b) $\dfrac{2}{1 + \cos 2x} \equiv \sec^2 x$

> **Q5d) Problem Solving**
>
> Write the left-hand side in terms of sin and cos first.

c) $\cot x - 2 \cot 2x \equiv \tan x$ d) $\tan 2x + \cot 2x \equiv 2 \operatorname{cosec} 4x$

E **P** **Q6** Prove that $\dfrac{\sin 2x + \cos 2x + 1}{\sin 2x - \cos 2x + 1} \equiv \cot x.$ *[4 marks]*

P **Q7** a) Show that $\dfrac{1 + \cos 2x}{\sin 2x} \equiv \cot x.$

b) Use your answer to part a) to solve $\dfrac{1 + \cos 4\theta}{\sin 4\theta} = 7$
in the interval $0 \leq \theta \leq 360°$. Give your answers to 1 d.p.

P **Q8** a) Show that $\dfrac{\cot^2 x + 1}{\cot^2 x - 1} \equiv \sec 2x.$

b) Use your answer to part a) to solve $\dfrac{\cot^2 2\alpha + 1}{\cot^2 2\alpha - 1} = -3$
in the interval $-\pi \leq \alpha \leq \pi$, giving your answers to 3 s.f.

> **Q9 Problem Solving**
>
> Write $\cot x$ as $\dfrac{1}{\tan x}$, then use $\tan x = \tan 2\left(\frac{x}{2}\right)$.

E **P** **Q9** Solve the equation $\cot x = \tan \frac{x}{2}$ in the interval $0 \leq x \leq 2\pi$,
giving your answer in radians in terms of π. *[3 marks]*

P **Q10** a) Show that $\operatorname{cosec} x - \cot \frac{x}{2} \equiv -\cot x.$

b) Solve $\operatorname{cosec} y = \cot \frac{y}{2} - 2$ in the interval $-\pi \leq y \leq \pi$. Give your answers to 3 s.f.

P **Q11** a) Given that $\sin \theta = \frac{5}{13}$, and that θ is acute, find: (i) $\cos\left(\frac{\theta}{2}\right)$,   (ii) $\sin\left(\frac{\theta}{2}\right)$.

b) Hence find $\tan\left(\frac{\theta}{2}\right)$.

E **P** **Q12** a) Prove that $\tan 4x = \dfrac{4 \tan x (1 - \tan^2 x)}{\tan^4 x - 6 \tan^2 x + 1}.$ *[4 marks]*

b) Hence show that when $\cot x = \dfrac{1}{\sqrt{3}}$, $\tan 4x = \tan x.$ *[3 marks]*

> **Challenge**

E **P** **Q13** a) Prove that $\tan x \sin 4x = 4 \sin^2 x - 8 \sin^4 x.$ *[5 marks]*

b) Given that $\sin x = \dfrac{4}{5}$, find the exact value of $\tan x \sin 4x.$ *[1 mark]*

c) Solve $\cot x \operatorname{cosec} 4x = 2$ on the interval $-\pi \leq x \leq \pi.$ *[5 marks]*

3.14 The R Addition Formulas

If you're solving an equation that contains both sin θ and cos θ terms, e.g. 3 sin θ + 4 cos θ = 1, you need to rewrite it so that it only contains one trig function.

The formulas that you use to do that are known as the **R formulas**:

One set for **sine**: $a \sin \theta \pm b \cos \theta \equiv R \sin (\theta \pm \alpha)$

And one set for **cosine**: $a \cos \theta \pm b \sin \theta \equiv R \cos (\theta \mp \alpha)$

where a, b and R are **positive**, and α is **acute**.

You need to be careful with the + and − signs in the cosine formula. If you have $a \cos \theta$ **+** $b \sin \theta$ then use $R \cos (\theta - \alpha)$.

Using the *R* formulas

Here is the method you'll need to follow to use the R formulas:

- You'll start with an identity like 2 sin x + 5 cos x ≡ R sin (x + α), where R and α need to be found.

- First, **expand** the right-hand side using the **addition formulas** (see p.96):

$$2 \sin x + 5 \cos x \equiv R \sin x \cos \alpha + R \cos x \sin \alpha$$

- **Equate the coefficients** of sin x and cos x. You'll get two equations:
    (1) $R \cos \alpha = 2$ and (2) $R \sin \alpha = 5$

- To find α, **divide** equation (2) by equation (1), (because $\dfrac{R \sin \alpha}{R \cos \alpha} = \tan \alpha$) then take **tan⁻¹** of the result.

- To find R, **square** equations (1) and (2) and **add** them together, then take the **square root** of the answer. This works because:

$$(R \sin \alpha)^2 + (R \cos \alpha)^2 \equiv R^2 (\sin^2 \alpha + \cos^2 \alpha) \equiv R^2$$
$$\text{(using the identity } \sin^2 \alpha + \cos^2 \alpha \equiv 1)$$

This method looks a bit scary, but follow through the next example and it should make more sense.

Example 1

Express $4 \cos x + 5 \sin x$ in the form $R \cos (x \pm \alpha)$.

$a \cos \theta + b \sin \theta \equiv R \cos (\theta - \alpha)$ ← Find the correct R formula to use — the expression is in the form $a \cos \theta + b \sin \theta$

$4 \cos x + 5 \sin x \equiv R \cos (x - \alpha)$ ← and you want the answer in terms of cos.

$4 \cos x + 5 \sin x \equiv R \cos x \cos \alpha + R \sin x \sin \alpha$ ← Expand the right-hand side using the cos addition formula.

① $R \cos \alpha = 4$ and ② $R \sin \alpha = 5$ ← Equate the coefficients of $\sin x$ and $\cos x$ to find α and R.

② ÷ ①: $\tan \alpha = \dfrac{5}{4} \Rightarrow \alpha = 51.3°$ (1 d.p.) ← Divide ② by ① to find α.

①²: $R^2 \cos^2 \alpha = 16$ ②²: $R^2 \sin^2 \alpha = 25$ ← Square ① and ② and add to find R.

$①^2 + ②^2$: $R^2 \cos^2 \alpha + R^2 \sin^2 \alpha = 16 + 25$

$\Rightarrow R^2 (\cos^2 \alpha + \sin^2 \alpha) = 41$

$\Rightarrow R^2 = 41$

$\Rightarrow R = \sqrt{41}$

Tip R is always positive (so take the positive root) and α is always acute (so take the angle in the first quadrant of the CAST diagram when solving).

$4 \cos x + 5 \sin x \equiv \sqrt{41} \cos (x - 51.3°)$ ← Put the values for α and R back into the identity to get the answer.

Example 2

a) Show that $5 \sin x - 5\sqrt{3} \cos x \equiv 10 \sin \left(x - \dfrac{\pi}{3} \right)$.

$a \sin \theta - b \cos \theta \equiv R \sin (\theta - \alpha)$ ← Find the correct R formula to use — the expression is in the form $a \sin \theta - b \cos \theta$

$5 \sin x - 5\sqrt{3} \cos x \equiv R \sin (x - \alpha)$ ← and you want the answer in terms of sin.

$5 \sin x - 5\sqrt{3} \cos x$
$\equiv R \sin x \cos \alpha - R \cos x \sin \alpha$ ← Expand the right-hand side using the sin addition formula.

① $R \cos \alpha = 5$ and ② $R \sin \alpha = 5\sqrt{3}$ ← Equate the coefficients of $\sin x$ and $\cos x$ to find α and R:

② ÷ ①: $\dfrac{R \sin \alpha}{R \cos \alpha} = \tan \alpha = \dfrac{5\sqrt{3}}{5} = \sqrt{3}$ ← Divide ② by ① to find α.

$\Rightarrow \alpha = \dfrac{\pi}{3}$

$①^2 + ②^2$: $R^2 \cos^2 \alpha + R^2 \sin^2 \alpha = 5^2 + (5\sqrt{3})^2$ ← Square ① and ② and add to find R.

$\Rightarrow R^2(1) = 100$

$\Rightarrow R = 10$

$5 \sin x - 5\sqrt{3} \cos x \equiv 10 \sin \left(x - \dfrac{\pi}{3} \right)$ as required ← Then put the values for α and R back into the identity.

b) Hence sketch the graph of $y = 5 \sin x - 5\sqrt{3} \cos x$ in the interval $-\pi \leq x \leq \pi$.

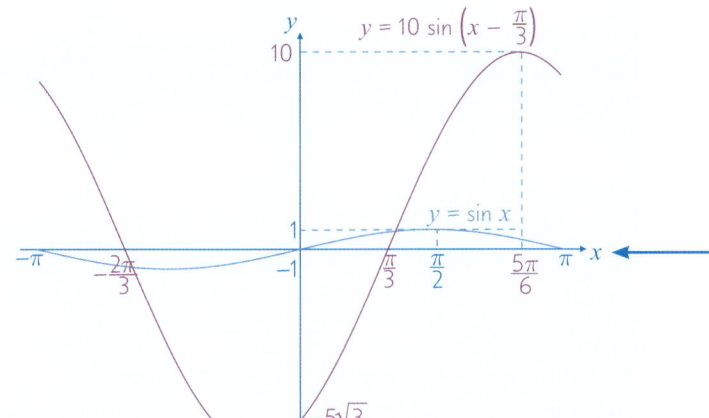

$y = 5 \sin x - 5\sqrt{3} \cos x \Rightarrow y = 10 \sin\left(x - \frac{\pi}{3}\right)$

So it's a transformation of $y = \sin x$
right by $\frac{\pi}{3}$ followed by a stretch
vertically by a scale factor of 10.

> Rewrite the equation using your answer to part a).

> Use the rewritten equation to work out how to transform $y = \sin x$.

> **Tip** Look back at pages 46-49 for a reminder about transformations of graphs.

> Add $\frac{\pi}{3}$ to the x-coordinates of the points on $y = \sin x$ and multiply the y-coordinates by 10.
> E.g. the maximum on $y = \sin x$ at $\left(\frac{\pi}{2}, 1\right)$ gets translated to $\left(\frac{5\pi}{6}, 10\right)$.

Exercise 3.14

Q1 Express $3 \sin x - 2 \cos x$ in the form $R \sin(x - \alpha)$.
Give R in surd form and α in degrees to 1 decimal place.

Q2 Express $6 \cos x - 5 \sin x$ in the form $R \cos(x + \alpha)$.
Give R in surd form and α in degrees to 1 decimal place.

Q3 Express $\sin x + \sqrt{7} \cos x$ in the form $R \sin(x + \alpha)$.
Give R in the form $m\sqrt{2}$ and α in radians to 3 significant figures.

Q4 Write $5 \sin \theta - 6 \cos \theta$ in the form $R \sin(\theta - \alpha)$, where $R > 0$ and $0° \leq \alpha \leq 90°$.

E **Q5** Show that $\sqrt{2} \sin x - \cos x \equiv \sqrt{3} \sin(x - \alpha)$, where $\tan \alpha = \frac{1}{\sqrt{2}}$.   *[4 marks]*

P **Q6** Show that $3 \cos 2x + 5 \sin 2x \equiv \sqrt{34} \cos(2x - \alpha)$, where $\tan \alpha = \frac{5}{3}$.

> **Q6 Problem Solving**
> Treat the $2x$ just the same as if it were an x.

P **Q7** a) Express $\sqrt{3} \sin x + \cos x$ in the form $R \sin(x + \alpha)$.
Give R and α as exact answers and α in radians in terms of π.
 b) Hence sketch the graph of $y = \sqrt{3} \sin x + \cos x$
in the interval $-\pi \leq x \leq \pi$.
 c) State the coordinates of any maximum and minimum points
and intersections with the axes of the graph in b).

> **Q7c) Problem Solving**
> Write down their coordinates on the graph of $y = \sin x$ first, then apply the transformations to them.

E **P** **Q8** Express $\sec x \sin 2x + \dfrac{\cos 2x + 1}{\cos x}$ in the form $R\cos(x - \alpha)$, where $R > 0$
and $0° < \alpha < 90°$, giving the exact value of R and the exact value of α. *[6 marks]*

E **P** **Q9** a) Use the addition formulas to show that
$$2\sin\left(x - \frac{\pi}{6}\right) - 2\cos x = p\sin x + q\cos x,$$
where p and q are constants to be found. *[2 marks]*

b) Express $2\sin\left(x - \frac{\pi}{6}\right) - 2\cos x$ in the form $R\sin(x - \alpha)$,
where $R > 0$ and $0 < \alpha < \frac{\pi}{2}$, giving the exact value of R
and the value of α as an exact multiple of π. *[4 marks]*

c) Hence state the period of $2\sin\left(x - \frac{\pi}{6}\right) - 2\cos x$. *[1 mark]*

P **Q10** a) Express $3\cos(x + 45°) + 2\sin(x - 30°)$ in the form $R\cos(x + \alpha)$,
where $R > 0$ and $0° < \alpha < 90°$, giving the value of R and
the value of α correct to 4 decimal places.

b) Hence express $\dfrac{3}{6\cos(2x + 45°) + 4\sin(2x - 30°)}$ in the form $R\sec(2x + \alpha)$,
where $R > 0$ and $0° < \alpha < 90°$, giving the value of R correct to 1 decimal place
and the value of α correct to 4 decimal places.

3.15 Applying the R Addition Formulas

To solve equations of the form $a \sin \theta + b \cos \theta = c$, it's best to work things out in different **stages**
— first **writing out** the equation in the form of one of the **R formulas**, then **solving** it.

The example below shows you how to do this. Note that you might be asked
for something else too, like the **maximum and minimum values** of the function.

Example 1

a) Solve $2 \sin x - 3 \cos x = 1$ in the interval $0° \leq x \leq 360°$.

$2 \sin x - 3 \cos x \equiv R \sin (x - \alpha)$ ⟵ Before you can solve this equation,
$2 \sin x - 3 \cos x \equiv R \sin x \cos \alpha - R \cos x \sin \alpha$ you need to get $2 \sin x - 3 \cos x$
in the form $R \sin (x - \alpha)$.

① $R \cos \alpha = 2$ and ② $R \sin \alpha = 3$ ⟵ Equate the coefficients.

$\dfrac{R \sin \alpha}{R \cos \alpha} = \tan \alpha = \dfrac{3}{2}$

$\Rightarrow \alpha = \tan^{-1} 1.5 = 56.31°$ (2 d.p.) ⟵ Divide ② by ① to find α.

$R^2 \cos^2 \alpha + R^2 \sin^2 \alpha = 2^2 + 3^2$ ⟵ Square ① and ②
$\Rightarrow R^2 = 13 \Rightarrow R = \sqrt{13}$ and add to find R.

continued on the next page...

So $2 \sin x - 3 \cos x = \sqrt{13} \sin (x - 56.31°)$ ← Then put the values for α and R back into the identity.

$2 \sin x - 3 \cos x = 1$

$\Rightarrow \sqrt{13} \sin (x - 56.31°) = 1$ ← Substitute the result into the original equation and solve.

$\Rightarrow \sin (x - 56.31°) = \dfrac{1}{\sqrt{13}}$

$\Rightarrow x - 56.31° = \sin^{-1}\left(\dfrac{1}{\sqrt{13}}\right) = 16.10°,$ ←

or $x - 56.31° = 180° - 16.10° = 163.90°$ (2 d.p.) ←

Since $0° \leq x \leq 360°$, look for solutions in the interval $-56.31° \leq (x - 56.31°) \leq 303.69°$ (just take 56.31° away from the original interval).

There are no solutions in the range $-56.31° \leq (x - 56.31°) \leq 0°$ since $\sin x$ is negative in the range $-90° \leq x \leq 0°$.

So $x = 16.10 + 56.31 = 72.4°$ (1 d.p.)
or $x = 163.90 + 56.31 = 220.2°$ (1 d.p.)

b) What are the maximum and minimum values of $2 \sin x - 3 \cos x$?

$2 \sin x - 3 \cos x = \sqrt{13} \sin (x - 56.31°)$

So the maximum and minimum values are $\pm\sqrt{13}$. ←

The maximum and minimum values of the sin function are ±1, so the maximum and minimum values of $R \sin (x - \alpha)$ are $\pm R$.

Exercise 3.15

Q1 a) Express $5 \cos \theta - 12 \sin \theta$ in the form $R \cos (\theta + \alpha)$, where $R > 0$ and α is an acute angle in degrees (to 1 d.p.).

b) Hence solve $5 \cos \theta - 12 \sin \theta = 4$ in the interval $0° \leq \theta \leq 360°$.

c) State the maximum and minimum values of $5 \cos \theta - 12 \sin \theta$.

P Q2 a) Express $2 \sin 2\theta + 3 \cos 2\theta$ in the form $R \sin (2\theta + \alpha)$, where $R > 0$ (given in surd form) and $0 < \alpha < \dfrac{\pi}{2}$ (to 3 s.f.).

Q2b) Problem Solving

Take care with the interval here to make sure you get all the correct solutions for θ.

b) Hence solve $2 \sin 2\theta + 3 \cos 2\theta = 1$ in the interval $0 \leq \theta \leq 2\pi$.

Q3 a) Express $3 \sin \theta - 2\sqrt{5} \cos \theta$ in the form $R \sin (\theta - \alpha)$. Give R in surd form and α in degrees to 1 d.p.

b) Hence solve $3 \sin \theta - 2\sqrt{5} \cos \theta = 5$ in the interval $0° \leq \theta \leq 360°$.

c) Find the maximum value of $f(x) = 3 \sin x - 2\sqrt{5} \cos x$ and the smallest positive value of x at which it occurs.

Q4 a) Express $f(x) = 3 \sin x + \cos x$ in the form $R \sin(x + \alpha)$.
where $R > 0$ (given in surd form) and $0° < \alpha < 90°$ (to 1 d.p.).

b) Hence solve the equation $f(x) = 2$ in the interval $0° \leq x \leq 360°$.

c) State the maximum and minimum values of $f(x)$.

P **Q5** a) Express $4 \sin x + \cos x$ in the form $R \sin(x + \alpha)$,
where $R > 0$ (given in surd form) and $0 < \alpha < \frac{\pi}{2}$ (to 3 s.f.).

b) Hence find the greatest value of $(4 \sin x + \cos x)^4$.

c) Solve the equation $4 \sin x + \cos x = 1$ for values of x in the interval $0 \leq x \leq \pi$.

P **Q6** a) Write $f(x) = 8 \cos x + 15 \sin x$ in the form $R \cos(x - \alpha)$, where $R > 0$ and $0 < \alpha < \frac{\pi}{2}$.

b) Solve the equation $f(x) = 5$ in the interval $0 \leq x \leq 2\pi$.

c) Find the minimum value of $g(x) = (8 \cos x + 15 \sin x)^2$
and the smallest positive value of x at which it occurs.

> **Q6c) Problem Solving**
> Think about what happens to the negative values when you square a function.

Q7 The function g is given by $g(x) = 2 \cos x + \sin x$, $x \in \mathbb{R}$.
$g(x)$ can be written as $R \cos(x - \alpha)$, where $R > 0$ and $0° < \alpha < 90°$.

a) Show that $R = \sqrt{5}$, and find the value of α (to 3 s.f.).

b) Hence state the range of $g(x)$.

> **Q7b) Hint** This is just another way of asking for the maximum and minimum values of the function.

E **Q8** Express $3 \sin \theta - \frac{3}{2} \cos \theta$ in the form $R \sin(\theta - \alpha)$,
where $R > 0$ and $0 < \alpha < \frac{\pi}{2}$, and hence solve the equation
$3 \sin \theta - \frac{3}{2} \cos \theta = 3$ for values of θ in the interval $0 \leq \theta \leq 2\pi$. *[6 marks]*

P **Q9** Solve the equation $4 \sin 2\theta + 3 \cos 2\theta = 2$ for values of θ in the interval $0 \leq \theta \leq \pi$.

E **P** **Q10** a) Express $\sin 2x - \cos 2x$ in the form $R \sin(2x - \alpha)$, where $R > 0$ and $0 < \alpha < \frac{\pi}{2}$,
giving the exact value of R and the value of α as an exact multiple of π. *[4 marks]*

b) Solve the equation $\sin 2x - \cos 2x = \frac{\sqrt{3} - 1}{2}$
for values of x in the interval $0 \leq x \leq 2\pi$. *[4 marks]*

c) $\sin 2x - \cos 2x = k$ has no real solutions. State the range of possible values of k. *[2 marks]*

Challenge

E **P** **Q11** a) Express $3 \cos x + \sin x$ in the form $R \cos(x - \alpha)$, where $R > 0$ and $0° < \alpha < 90°$,
giving the exact value of R and the value of α correct to 3 decimal places. *[4 marks]*

b) Given that $f(x) = \frac{1}{(3 \cos x + \sin x)^2}$,

(i) State the least possible value of $f(x)$. *[2 marks]*

(ii) Find the smallest positive value of x for which $f(x)$ is minimised. *[2 marks]*

(iii) Explain why $f(x)$ has no greatest value. *[1 mark]*

c) The graph of $y = f(x)$ has m turning points and n asymptotes for $0° \leq x \leq 3600°$.
Find the values of m and n, explaining your answers. *[3 marks]*

3.16 Modelling with Trig Functions

Trigonometric functions show up a lot in modelling problems. Things that happen in a cycle, like the motion of a child on a swing, or tidal patterns, could be modelled with sin and cos.

When tackling these questions, remember to check whether you should be working in degrees or radians. Make sure that you give your answers to a suitable degree of accuracy (3 s.f. is usually fine).

Example 1

The height of an object bouncing on a spring is modelled by the equation $h = 5 + 2 \sin \left(5t + \frac{\pi}{3}\right)$ where t is the time in seconds and h is the height in cm. Find the first time at which $h = 4$ cm.

$5 + 2 \sin \left(5t + \frac{\pi}{3}\right) = 4$ ← Set $h = 4$ in the equation and solve.

$\Rightarrow 2 \sin \left(5t + \frac{\pi}{3}\right) = -1$

Tip If you see π in a question it means it's in radians.

$\Rightarrow \sin \left(5t + \frac{\pi}{3}\right) = -\frac{1}{2}$

$\Rightarrow \left(5t + \frac{\pi}{3}\right) = \sin^{-1}\left(-\frac{1}{2}\right) = -\frac{\pi}{6}, \frac{7\pi}{6}, \frac{11\pi}{6}, \dots$ etc

$t \geq 0$, which means that $5t + \frac{\pi}{3} \geq \frac{\pi}{3}$. ← Use the fact that time cannot be negative to find the valid solutions.

The first solution to satisfy this inequality is $\frac{7\pi}{6}$:

$5t + \frac{\pi}{3} = \frac{7\pi}{6} \Rightarrow 5t = \frac{5\pi}{6} \Rightarrow t = \frac{\pi}{6}$ ← You want the first time when $h = 4$ cm, so take the smallest solution.

So the first time it has a height of 4 cm is at $t = \frac{\pi}{6} = 0.524$ seconds (3 s.f.)

Problem Solving Be careful when using inverse trig functions to solve equations — you usually get multiple solutions so you need to decide which one(s) to use.

Example 2

Two sound waves are modelled by the functions $f(\theta)$ and $g(\theta)$, where $f(\theta) = 12 \sin \theta$ and $g(\theta) = 4\sqrt{3} \cos \theta$. The two waves combine to produce a new wave given by $h(\theta) = f(\theta) + g(\theta)$.

a) Write the equation for this new wave in the form $R \sin (\theta + \alpha)$, where α is measured in radians, giving your answers as exact values.

$h(\theta) = 12 \sin \theta + 4\sqrt{3} \cos \theta = R \sin (\theta + \alpha)$ ← Set $h(\theta)$ equal to $R \sin (\theta + \alpha)$.

$R \sin (\theta + \alpha) = R \sin \theta \cos \alpha + R \cos \theta \sin \alpha$

$\Rightarrow$ ① $R \cos \alpha = 12$ and ② $R \sin \alpha = 4\sqrt{3}$ ← Expand $R \sin (\theta + \alpha)$ using the sin addition formula and equate the coefficients.

continued on the next page...

$$R = \sqrt{12^2 + \left(4\sqrt{3}\right)^2} = \sqrt{144 + 48} = \sqrt{192} = 8\sqrt{3}$$ ← Square ① and ② and add to find R.

$$\tan \alpha = \frac{4\sqrt{3}}{12} = \frac{\sqrt{3}}{3} = \frac{1}{\sqrt{3}} \Rightarrow \alpha = \frac{\pi}{6}$$ ← Divide ② by ① to find α.

So $h(\theta) = 8\sqrt{3} \sin\left(\theta + \frac{\pi}{6}\right)$ ← Then put the values for α and R back into the equation.

b) The amplitude, a, of a sound wave is equivalent to half the distance between the maximum and minimum values of the wave (as shown). Find the amplitude of $h(\theta)$.

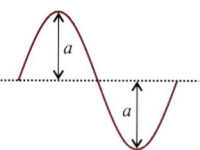

$\sin \theta$ has a maximum of 1 and a minimum of -1. ← Write down the minimum and maximum points of the sin function.

$\sin \theta$ is translated horizontally by $-\frac{\pi}{6}$, and stretched vertically by a scale factor of $8\sqrt{3}$. ← Find the transformation of $\sin \theta$ to $h(\theta)$.

The maximum and minimum values of $h(\theta)$ are $\pm R$ — i.e. $8\sqrt{3}$ and $-8\sqrt{3}$, so its amplitude is $\frac{1}{2}(8\sqrt{3} - (-8\sqrt{3})) = 8\sqrt{3}$. ← Find the amplitude of $h(\theta)$.

Exercise 3.16

M **Q1** A circular plot of land with a radius of 20 m is separated into three gardens. Each garden is a sector of the circle, with angles of 120°, 144° and 96° respectively. Calculate the area and perimeter of each garden (to 3 s.f.).

E **P** **M** **Q2** Adam wants to form a function to model the hours of daylight in his town throughout the year. He knows that the function should be of the form $f(t) = A + B \cos t$, where A and B are positive constants and the time, t, is in radians.

　　a) The daylight hours vary from 7 to 17. Find the values of A and B. *[3 marks]*

　　b) He now wants to adjust the model so that t is measured in months. He rewrites his function as $g(t) = A + B \cos (Ct + D)$. Find the values of C and D, such that the longest day of the year occurs at $t = 0$ and the shortest day of the year occurs at $t = 6$. *[3 marks]*

E **P** **M** **Q3** The speed of an amusement park ride, v ms^{-1}, is modelled by the equation $v(t) = 8 - 8 \cos\left(\frac{\pi t}{5}\right)$, $0 \leq t \leq 50$, where t is the time in seconds after the start of the ride.

　　a) Given that the ride is 50 seconds long, find the number of times during the ride at which the ride instantaneously stops (not counting the start and end times). *[2 marks]*

　　b) (i) Find the maximum speed of the ride. *[1 mark]*

　　　　(ii) Find the first time at which the maximum speed occurs. *[2 marks]*

E **Q4** The temperature in an office, $T\,°C$, is modelled by the equation
P
M $T(t) = 19 + 4\sin\left(t - \dfrac{\pi}{6}\right)$, $t \geq 0$, where t is time in hours after 9 am.

     a) Find the temperature in the office at 1:30 pm,
        giving your answer correct to 3 significant figures. *[1 mark]*

     b) If the temperature in the office falls below 17 °C, the radiator automatically turns on.
        Once the temperature rises above 17 °C, the radiator automatically turns back off.

        (i) Find the time at which the radiator is first turned on. *[3 marks]*

        (ii) Between 9 am and 5 pm on a given day,
            for how long is the radiator turned on? *[3 marks]*

E **Q5** The height of a buoy floating in a harbour, measured in metres, is modelled by the function
M $h(t) = 14 + 5\,(\sin t + \cos t)$, where t is time in hours. By writing h(t) in the form
$14 + R\cos(t - \alpha)$, find the maximum and minimum height of the buoy (to 1 d.p.). *[6 marks]*

P **Q6** Antonia goes on a fairground ride. She is strapped into
M a small spinning disc, which is attached to a large rotating wheel.

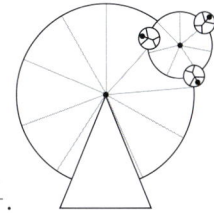

     a) The height, in metres, of the centre of the disc above the ground
        after t seconds is given by $H = 10 + \dfrac{7}{2}(\sin t - \sqrt{3}\,\cos t)$.
        Write this in the form $H = 10 + R\sin(t - \alpha)$, where $R > 0$ and $0 < \alpha \leq \dfrac{\pi}{2}$.

     b) Antonia's height above the ground, h, is given by $h = H - \cos 2(t - \alpha)$.
        Use the double angle formulas to show that $h = A + R\sin(t - \alpha) + B\sin^2(t - \alpha)$,
        and give the values of A and B. It could help to set $(t - \alpha) = x$.

     c) Hence find the first time when Antonia is 13 m above the ground.

E **Q7** The distance of a pendulum, d cm, from a fixed object as it swings, t seconds after it is
P released from rest, is modelled by the equation $d(t) = \cos\left(\dfrac{t}{2}\right) + 5\sin\left(\dfrac{t}{2}\right) + 6$, $t \geq 0$.
M

     a) Write $d(t)$ in the form $A\cos\left(\dfrac{t}{2} - B\right) + C$,
        where A, B and C are constants to be found. *[4 marks]*

     b) Find the exact greatest and least distances of the pendulum from the fixed object. *[2 marks]*

     c) Find the length of time between the first two points
        when the pendulum is 10 cm away from the object. *[4 marks]*

E **Q8** Sarah's diastolic blood pressure, P, is modelled by the equation
P $P = 83 + 6\sin(x° - 30°) + 14\cos(x° + 60°)$, $0 \leq x \leq 720$,
M where x is the number of minutes after 8 am on a given day.

     a) Find Sarah's diastolic blood pressure at 12 noon. *[1 mark]*

     b) Show that $P = 83 + A\sin x° + B\cos x°$, where A and B are constants to be found. *[4 marks]*

     c) Hence show that $P = 83 + R\cos(x° + \alpha)$, where $R > 0$ and $0° < \alpha < 90°$
        are constants to be found. *[4 marks]*

     d) A person is said to have high blood pressure if their diastolic blood pressure is
        greater than 90. Find the earliest time at which Sarah has high blood pressure. *[4 marks]*

     e) Give one criticism of the model. *[1 mark]*

3 Review Exercise

Q1 a) Convert these angles into radians, giving your answers in terms of π:

   (i) 15°   (ii) 50°   (iii) 330°   (iv) 225°

   b) Convert these angles into degrees:

   (i) $\dfrac{7\pi}{12}$ (ii) $\dfrac{7\pi}{6}$ (iii) $\dfrac{5\pi}{3}$ (iv) $\dfrac{13\pi}{12}$

P **Q2** The diagram on the right shows a sector ABC
of a circle, with centre A and a radius of 10 cm.
The angle BAC is 0.7 radians.
Find the arc length BC and area of this sector.

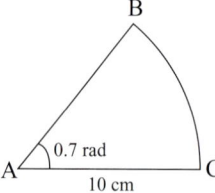

E **Q3** A discus must be thrown so that it lands within a bounded sector of angle 40°.
M The curved edge of the sector is 100 m away, as shown in the diagram below.

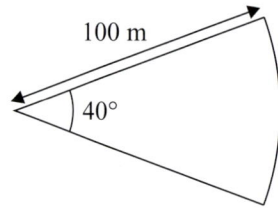

   a) Find the area of the region in which the discus could land and
      lie within the bounds. Give your answer to 3 significant figures.   *[3 marks]*

   b) Find the length of the boundary of the region to 3 significant figures.   *[2 marks]*

E **Q4** A sports pitch is modelled by a sector of a circle OAC centred at O, as shown in the
M diagram below. D is the midpoint of OA, F is the midpoint of OC, E is the midpoint of CA
   and B is the midpoint of the arc CA. Four flags are located at the points O, D, E and F.

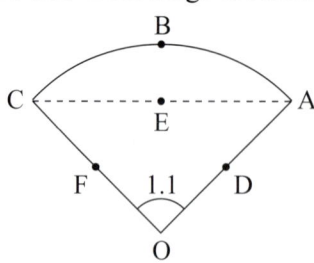

   OA = OC = 50 m and angle AOC = 1.1 radians.

   a) Find the distance from A to C, giving your answer to 2 decimal places.   *[3 marks]*

   b) Find the total area of the pitch.   *[2 marks]*

   c) A player collects the flags running from O to D to F to C, then around
      the edge of the pitch to B and directly back to O through E. Find the
      total distance they run, giving your answer in metres to 2 decimal places.   *[5 marks]*

E **P** Q5 In the diagram below, OAB is a sector of a circle centred at O of radius r, the angle AOB $= \theta$ radians and the angle OBC is a right angle.

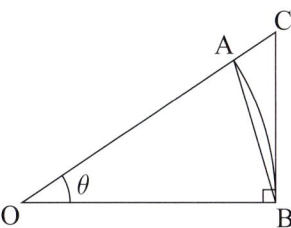

a) In terms of r and θ, give an expression for the area of:

    (i) the triangle OAB,   *[1 mark]*

    (ii) the sector OAB,   *[1 mark]*

    (iii) the triangle OBC.   *[3 marks]*

b) (i) Use your answers in part a) to deduce that $\sin\theta < \theta < \tan\theta$, when $0 < \theta < \frac{\pi}{2}$.   *[2 marks]*

    (ii) Hence deduce that $\cos\theta < \dfrac{\sin\theta}{\theta} < 1$, when $0 < \theta < \frac{\pi}{2}$.   *[2 marks]*

c) Use part b)(ii) to find the value $\dfrac{\sin\theta}{\theta}$ approaches as $\theta \to 0$.

    Explain how this justifies the small angle approximation $\sin\theta \approx \theta$.   *[2 marks]*

Q6 Use the small angle approximations to estimate $\tan\theta - \cos\theta$ for:

    a) $\theta = 0.13$ b) $\theta = 0.07$ c) $\theta = 0.26$

Q7 When θ is small, find an approximation for the expressions below:

    a) $\sin 3\theta \tan 4\theta$ b) $\cos 4\theta + \cos 8\theta$ c) $\dfrac{2\theta^3}{\sin 2\theta \cos\theta}$

E Q8 a) Approximate $\tan 5°$ using an appropriate small angle approximation.   *[2 marks]*

    b) State the percentage error (to 3 s.f.) when using the small angle approximation.   *[2 marks]*

Q9 Using trig values for common angles, evaluate the following in radians, between 0 and $\frac{\pi}{2}$:

    a) $\sin^{-1}\dfrac{1}{\sqrt{2}}$ b) $\cos^{-1}\dfrac{1}{2}$ c) $\tan^{-1}\dfrac{1}{\sqrt{3}}$

Q10 Sketch the graphs of arcsin, arccos and arctan showing their domains and ranges.

E **P** Q11 The graph of $y = \cos^{-1} x$ is shown on the right. Given that A and C are the end points of the graph, state the coordinates of points A, B and C in radians.   *[3 marks]*

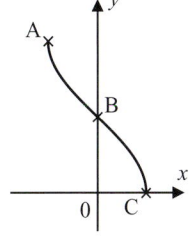

E **Q12** Sketch $y = \frac{1}{3}\arccos(x + 2)$ for $-3 \le x \le 0$.

Q13 The diagram on the right shows the curve

$y = \dfrac{1}{1 + \cos x}$ for $0 \le x \le \dfrac{\pi}{2}$.

a) If $y = f(x)$, show that $f^{-1}(x) = \arccos\left(\dfrac{1}{x} - 1\right)$.

b) State the domain and range of this inverse function.

c) Sketch $y = f^{-1}(x)$ on the same axes as $y = f(x)$.

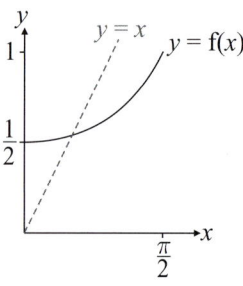

Q14 Given that $f(x) = \sin^{-1} x + \cos^{-1} x + \tan^{-1} x$, find the value, in radians, of:

a) $f(1)$ b) $f(-1)$

Q15 Sketch the graphs of cosecant, secant and cotangent for $-2\pi \le x \le 2\pi$.

Q16 a) Describe the transformation that maps $y = \sec x$ onto $y = \sec 4x$.

b) What is the period, in radians, of the graph $y = \sec 4x$?

c) Sketch the graph of $y = \sec 4x$ for $0 \le x \le \pi$.

d) For what values of x in this interval is $\sec 4x$ undefined?

E **Q17** a) Draw $y = \cot x$ and $y = \tan x$ on the same set of axes for $0 \le x \le 2\pi$, labelling all points of intersection with the coordinate axes and showing clearly any asymptotes.   *[4 marks]*

b) Hence state the number of real solutions to the equation $\cot x = \tan x$ for $0 \le x \le 2\pi$.   *[1 mark]*

c) Find the exact solutions to the equation $\cot x = \tan x$ for $0 \le x \le 2\pi$, giving your answers in terms of π.   *[4 marks]*

E **P** **Q18** a) Write down the range of $f(x) = \sec x$, $0 \le x \le \pi$.   *[1 mark]*

b) Hence write down the domain and range of $f^{-1}(x) = \operatorname{arcsec} x$.   *[2 marks]*

c) Sketch the graph $y = \operatorname{arcsec} x$.   *[3 marks]*

E **P** **Q19** By considering the graph $y = \operatorname{cosec} x$, find the number of real solutions of the equation $\operatorname{cosec}^2 x = 3$ for $0 < x < 2\pi$, explaining your answer.   *[3 marks]*

P **Q20** Find the range of the function $f(x) = \ln(\operatorname{cosec} x)$, $0 < x < \pi$.

Q21 For $\theta = 30°$, find the exact values of:

a) $\operatorname{cosec} \theta$ b) $\sec \theta$ c) $\cot \theta$

E P Q22 a) Find the length of the hypotenuse of the triangle S in the diagram below.

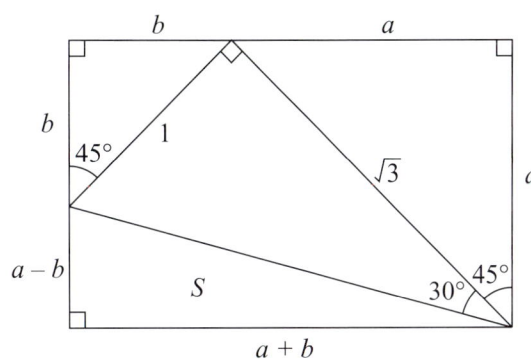

[1 mark]

b) Using the diagram, find the exact values of a and b. [3 marks]

c) By considering the triangle S, find the exact values of the following in their simplest form:

(i) $\sec 75°$ [3 marks]

(ii) $\cot 15°$ [3 marks]

E P Q23 a) Show that $\tan \dfrac{5\pi}{12} = 2 + \sqrt{3}$. [3 marks]

b) Hence, or otherwise, find the exact value of $\cot \dfrac{5\pi}{12}$.
Your answer should be given in the form
$a - \sqrt{b}$, where a and b are constants to be found. [1 mark]

Q24 Given that $x = \cos t + \sec t$ and $y = \cos^3 t + \sec^3 t$, express y in terms of x.

E P Q25 a) Explain why the equation $\sec 3x = \dfrac{1}{2}$ has no real solutions. [1 mark]

b) State the range of possible values of k such that
$\sec^2 3x = k$ has at least one real solution. [1 mark]

c) Solve the equation $\sec^2 3x = 4$ for $-90° \le x \le 90°$. [3 marks]

Q26 Given that $x = \operatorname{cosec} \theta$ and $y = \cot^2 \theta$, show that $y = x^2 - 1$.

P Q27 If $x = \sec \theta$ and $y = 2 \tan \theta$, express y in terms of x only.

E Q28 a) Show that the equation $\operatorname{cosec}^2 x = \dfrac{3 \cot x + 4}{2}$
can be written as $2 \cot^2 x - 3 \cot x - 2 = 0$. [2 marks]

b) Hence solve the equation $\operatorname{cosec}^2 x = \dfrac{3 \cot x + 4}{2}$.
Give all the values of x in the interval $0 \le x \le 2\pi$ in radians to 2 decimal places. [4 marks]

Q29 Solve the following inequalities for $0 \le x \le 2\pi$:

a) $\sec x < -3$ b) $3 - 4 \operatorname{cosec} x < 1$ c) $2 \cot x > 3$

Q30 Given that θ is acute and $\cos\theta = \frac{1}{2}$:

    a) Give the exact value of $\sec\theta$.

    b) Use Pythagoras' theorem to find the value of $\tan\theta$.

    c) Use the identity $\sec^2\theta \equiv 1 + \tan^2\theta$ to find the value of $\tan\theta$ and confirm that it is the same as in part b).

    d) Give the exact value of $\cot\theta$.

    e) Using the identity $\operatorname{cosec}^2\theta \equiv 1 + \cot^2\theta$, give the exact value of $\sin\theta$.

E Q31 Solve $\operatorname{cosec}^2 x = 4 + 2\cot x$ in the interval $0 \le x \le 2\pi$, giving your answers to 3 s.f. *[7 marks]*

P Q32 For each of the following pairs of equations, write y in terms of x.

    a) $x = 2\tan t,\ y = 2\sec t$ b) $x = 3 + \operatorname{cosec} t,\ y = \cot t$

Q33 Express the following as a single trig function:

    a) $\sin 2x \cos 9x + \cos 2x \sin 9x$ b) $3\cos 5x \cos 7x - 3\sin 5x \sin 7x$

    c) $\dfrac{\tan 12x - \tan 8x}{1 + \tan 12x \tan 8x}$ d) $12\sin\dfrac{7x}{2}\cos\dfrac{3x}{2} - 12\cos\dfrac{7x}{2}\sin\dfrac{3x}{2}$

P Q34 Using the addition formula for cos, find the exact value of $\cos\dfrac{7\pi}{12}$.

Q35 $\sin A = \dfrac{4}{5}$, $\sin B = \dfrac{7}{25}$ and both A and B are acute angles. Using the addition formula for sin, find $\sin(A + B)$.

Q36 Use the double angle formula to solve the equation $\sin 2\theta = -\sqrt{3}\sin\theta$ for $0° \le \theta \le 360°$.

Q37 Solve the equations below for $0 \le x \le 2\pi$, giving your answers to 3 s.f. where necessary.

    a) $4\sin x = \sin\dfrac{x}{2}$ b) $\tan\dfrac{x}{2}\tan x = 2$

Q38 Solve the equations below in the interval $0 \le x \le 2\pi$. Give your answers in radians in terms of π.

    a) $2\tan 2x = \tan x$ b) $\sin 6x - \cos 3x = 0$

E Q39 Find the solutions of $2\sin 2\theta - \cos\theta = 0$ in the interval $0 \le \theta \le 2\pi$. Give your answers in radians to 3 s.f. *[5 marks]*

E **P** Q40 a) Show that $\cot\theta + \tan\theta \equiv 2\operatorname{cosec} 2\theta$. *[3 marks]*

    b) Hence, solve $\cot 2\theta + \tan 2\theta = \dfrac{7}{2}$ in the interval $0 \le \theta \le \pi$, giving your answers to 3 s.f. *[5 marks]*

E **P** Q41 Prove that $\tan x\,(\operatorname{cosec} 2x + \cot 2x) \equiv 1$. *[5 marks]*

E Q42 a) Write $3 \cos \theta - 8 \sin \theta$ in the form $R \cos (\theta + \alpha)$, where $R > 0$ and $0° \leq \alpha \leq 90°$. *[4 marks]*

b) Hence solve $3 \cos \theta - 8 \sin \theta = 1.2$ in the interval $0° \leq \theta \leq 360°$. *[2 marks]*

P Q43 The height above the ground, h m, of a rider on a pirate ship at
M a theme park is modelled by the function: $h = 5 \sin t + 3 \cos t + 2$,
where t is the time in seconds. Find the maximum height above
the ground that the rider reaches. Give your answer to 2 decimal places.

E Q44 a) Write $7 \cos x - 11 \sin x$ in the form $R \cos (x + \alpha)$, where $R > 0$ and $0° < \alpha < 90°$. *[3 marks]*

b) Describe the transformations that map the graph of $y = \cos x$
to the graph of $y = 7 \cos x - 11 \sin x$. *[2 marks]*

c) State the maximum value of $y = 7 \cos x - 11 \sin x$
and a value of x for which this maximum occurs. *[2 marks]*

d) Solve $7 \cos x - 11 \sin x = 3$ in the interval $0° \leq x \leq 360°$,
giving your answers to 1 d.p. *[4 marks]*

E Q45 In a beach town, the height of the tide in feet can be modelled by
P
M the function $h(t) = a + b \sin \left(\frac{1}{2}(t - \pi) \right)$, where t is the time in hours.

a) Given that the height of the tide is 9 feet at low tide
and 17 feet at high tide, find the values of a and b. *[3 marks]*

b) How many hours after low tide does the tide first reach 11 feet in this model? *[4 marks]*

Challenge

E Q46 Prove that
P
$$\frac{1}{2} \log \left| \frac{1 + \sin x}{1 - \sin x} \right| = \log | \sec x + \tan x |, \text{ when } \cos x \neq 0,$$

where log denotes a logarithm to any base. *[6 marks]*

P Q47 It is given that the angles x and y satisfy $\sin x = p$ and $\tan y = 2p - 3$,
where p is a constant, and that $\text{cosec}^2 x - \sec^2 y = -1$.

a) Show that $4p^4 - 12p^3 + 9p^2 - 1 = 0$.

b) Show that $2p^2 - 3p - 1$ is a factor of $4p^4 - 12p^3 + 9p^2 - 1$.

c) Hence find all possible values of p.

d) Given that $p = \frac{1}{2}$ and $90° < x < 180°, \ 0° < y < 180°$,

find the exact value of $\sec x + \sec y$.

E Q48 a) Prove that $\sec 3x = \dfrac{1}{4 \cos^3 x - 3 \cos x}$. *[4 marks]*
P
b) Hence solve the equation $\sec 3x - \cos x = 0$ for $0 \leq x \leq 2\pi$. *[5 marks]*

c) (i) Show that $y = \frac{1}{2}$ is a root of the equation $4y^3 - 3y + 1 = 0$. *[1 mark]*

(ii) Given that $\sec 3x = -1$, find the possible exact values of $\cos x$. *[4 marks]*

E Exam Style   **P** Problem Solving   **M** Modelling

3 Chapter Summary

1 A radian (rad) is a unit for measuring angles. One radian is the angle formed in a sector that has an arc length that is the same as the radius.

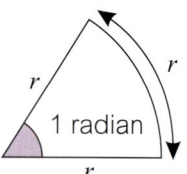

2 You can convert radians to degrees by dividing by π and multiplying by 180; to convert from degrees to radians divide by 180 and then multiply by π.

3 For a circle with radius r and a sector angle θ:

   • Arc Length = $r\theta$ • Sector Area = $\frac{1}{2}r^2\theta$

4 When θ is a very small angle, you can approximate $\sin\theta$, $\cos\theta$ and $\tan\theta$ as follows:

   • $\sin\theta \approx \theta$ • $\cos\theta \approx 1 - \frac{1}{2}\theta^2$ • $\tan\theta \approx \theta$

5 The inverses of sin, cos and tan are arcsin ($\sin^{-1}$), arccos ($\cos^{-1}$) and arctan ($\tan^{-1}$). Their graphs are found by reflecting the graphs of sin, cos and tan in the line $y = x$. If you want arcsin and arccos to be functions you have to restrict their domains so that they are one-to-one — so their domains are $-1 \le x \le 1$.

6 The reciprocals of sin, cos and tan create three new trig functions:

   • $\operatorname{cosec}\theta \equiv \dfrac{1}{\sin\theta}$ • $\sec\theta \equiv \dfrac{1}{\cos\theta}$ • $\cot\theta \equiv \dfrac{1}{\tan\theta}$

7 The graphs of cosec, sec and cot can be transformed in the usual ways.

8 You can evaluate these new trig functions and solve equations containing them. Start by writing them in terms of sin, cos and tan, and then solve using your calculator. You might need to use trig graphs or a CAST diagram to find solutions over a given interval.

9 The following trig identities will help you to solve equations and derive other identities:

   • $\cos^2\theta + \sin^2\theta \equiv 1$ • $\tan\theta = \dfrac{\sin\theta}{\cos\theta}$ • $\sec^2\theta \equiv 1 + \tan^2\theta$ • $\operatorname{cosec}^2\theta \equiv 1 + \cot^2\theta$

10 Here are the trig addition formulas, which can also be used to solve equations:

   • $\sin(A \pm B) \equiv \sin A \cos B \pm \cos A \sin B$
   • $\cos(A \pm B) \equiv \cos A \cos B \mp \sin A \sin B$
   • $\tan(A \pm B) \equiv \dfrac{\tan A \pm \tan B}{1 \mp \tan A \tan B}$

11 Double angle formulas are a special case of the addition formulas (where $A = B$):

   • $\sin(2A) \equiv 2\sin A \cos A$ • $\cos(2A) \equiv \cos^2 A - \sin^2 A$ • $\tan(2A) \equiv \dfrac{2\tan A}{1 - \tan^2 A}$
   • $\cos(2A) \equiv 2\cos^2 A - 1$ • $\cos(2A) \equiv 1 - 2\sin^2 A$

12 Similarly the half angle formulas are defined as follows:

   • $\sin^2\left(\dfrac{\theta}{2}\right) \equiv \dfrac{1}{2}(1 - \cos\theta)$ • $\cos^2\left(\dfrac{\theta}{2}\right) \equiv \dfrac{1}{2}(1 + \cos\theta)$ • $\tan^2\left(\dfrac{\theta}{2}\right) \equiv \dfrac{1 - \cos\theta}{1 + \cos\theta}$

13 The R formulas can be used to solve an equation that contains more than one trig function:

   • $a\sin\theta \pm b\cos\theta \equiv R\sin(\theta \pm \alpha)$ • $a\cos\theta \pm b\sin\theta \equiv R\cos(\theta \mp \alpha)$

Coordinate Geometry in the (x, y) Plane

4

Learning Objectives

Once you've completed this chapter, you should be able to:

- Calculate Cartesian coordinates for a curve given in parametric form.
- Find the coordinates of intersection points between a parametric curve and other lines.
- Convert parametric equations into Cartesian form.

Prior Knowledge Check

1 Find all the solutions to each of the following equations:
   a) $x(x-1)^2 = 0$
   b) $2x^2 - 5x = 3$
   c) $2x^3 - 7x^2 - 5x + 4 = 0$
   ← see Year 1 Pure

2 Find all the solutions in the range $-\pi < \theta \leq \pi$ for the following equations:
   a) $\cos\theta = \dfrac{1}{2}$
   b) $\tan\theta = 1$
   c) $\sin\theta = 0.4$
   ← see Year 1 Pure and page 67

3 Use trig identities to rewrite each of the following as a single trig expression:
   a) $\dfrac{\cot x}{\cos x}$
   b) $\sin^2\left(\dfrac{3x}{4}\right) - \cos^2\left(\dfrac{3x}{4}\right)$
   ← see pages 90-116

4.1 Parametric Equations — Finding Coordinates

Normally, graphs in the (x, y) plane are described using a **Cartesian equation** — a single equation linking x and y. Sometimes, particularly for more complicated graphs, it's easier to have two linked equations, called **parametric equations**.

In parametric equations, x and y are each defined separately in terms of a **third variable**, called a **parameter**. The parameter is usually either t or θ.

Parametric equations are often used to model the path of a moving particle, where its **position** (given by x and y) depends on time, t. You'll use them in kinematics, in the Applied part of this course.

Example 1

Sketch the graph given by the parametric equations $y = t^3 - 1$ and $x = t + 1$.

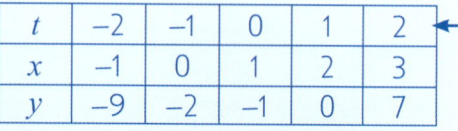

t	−2	−1	0	1	2
x	−1	0	1	2	3
y	−9	−2	−1	0	7

Start by making a table of coordinates. Choose some values for t and calculate x and y at these values. For example:
When $t = -1$, $x = -1 + 1 = 0$ and $y = (-1)^3 - 1 = -2$.
When $t = 1$, $x = 1 + 1 = 2$ and $y = 1^3 - 1 = 0$.

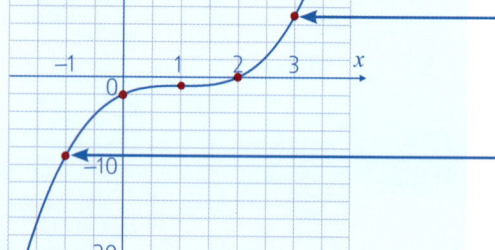

Now plot the Cartesian (x, y) coordinates on a set of axes in the normal way.

Here $t = 2$. So, from the table, this is the point $(3, 7)$.

This point corresponds to $t = -2$, so it has coordinates $x = -1$ and $y = -9$.

Tip You won't necessarily be expected to sketch a curve from its parametric equations in the exam — but finding the Cartesian coordinates like this can make it easier to picture the graph.

Example 2

Sketch the graph given by $x = \cos \theta$ and $y = \sin \theta + 1$.

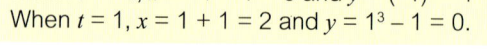

θ	0	$\frac{\pi}{2}$	π	$\frac{3\pi}{2}$	2π
x	1	0	−1	0	1
y	1	2	1	0	1

Make a table as before with values for θ over a suitable interval (such as $0 \le \theta \le 2\pi$). Then calculate the corresponding x and y values.

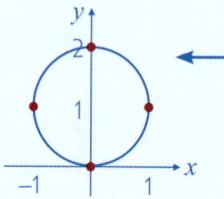

Plot the points — you should spot that they lie on a circle with centre $(0, 1)$ and radius 1. ($x = a + r \cos \theta$, $y = b + r \sin \theta$ are the general parametric equations for a circle with centre (a, b) and radius r.)

You can use the parametric equations to find **x-y values** at a given value of the parameter, and to find the value of the **parameter** for given x- or y-coordinates. There's often a limit on the **domain** of the parameter like the one in Example 3 below — see pages 20-24 for more about domains of functions. For example, the parameter can't take a value that would set the denominator of any fraction to zero.

Example **3**

A flying disc is thrown from the point $(0, 0)$. After t seconds, it has travelled x m horizontally and y m vertically, modelled by the parametric equations $x = t^2 + 2t$ and $y = 6t - t^2$ $(0 \le t \le 6)$.

a) Find the x- and y- values of the position of the disc after 2.5 seconds.

When $t = 2.5$: $x = 2.5^2 + 2 \times 2.5 = 6.25 + 5 = 11.25$
$\qquad\qquad\quad y = 6 \times 2.5 - 2.5^2 = 15 - 6.25 = 8.75$

Just substitute $t = 2.5$ into the equations for x and y.

b) After how many seconds does the disc reach a height of 5 metres?

$5 = 6t - t^2 \Rightarrow t^2 - 6t + 5 = 0$ ← *Find the value of t when $y = 5$.*
$\qquad\qquad \Rightarrow (t - 1)(t - 5) = 0$
$\qquad\qquad \Rightarrow t = 1$ s and $t = 5$ s ← *Factorise the quadratic and solve to give two times when the disc is at 5 m.*

c) What is the value of y when the disc reaches the point $x = 24$ m?

$24 = t^2 + 2t \Rightarrow t^2 + 2t - 24 = 0$ ← *Use the equation for x to find t when $x = 24$.*
$\Rightarrow (t + 6)(t - 4) = 0$
$\Rightarrow t = -6$ or $t = 4$, but $0 \le t \le 6$, so $t = 4$ s ← *Factorise and solve the quadratic.*

$y = 6 \times 4 - 4^2 = 24 - 16 = 8$ m ← *Put $t = 4$ in the other equation to find y.*

Exercise 4.1

Q1 A curve is defined by the parametric equations $x = 3t$, $y = t^2$.
a) Find the coordinates of the point where $t = 5$.
b) Find the value of t at the point where $x = 18$.
c) Find the possible values of x at the point where $y = 36$.

Q2 A curve is defined by the parametric equations $x = 2t - 1$, $y = 4 - t^2$.
a) Find the coordinates of the point where $t = 7$.
b) Find the value of t at the point where $x = 15$.
c) Find the possible values of x at the point where $y = -5$.

Q3 A curve has parametric equations $x = 2 + \sin \theta$, $y = -3 + \cos \theta$.
a) Find the coordinates of the point where $\theta = \dfrac{\pi}{4}$.
b) Find the acute value of θ at the point where $x = \dfrac{4 + \sqrt{3}}{2}$.
c) Find the obtuse value of θ at the point where $y = -\dfrac{7}{2}$.

Q3 Problem Solving
There is only one acute value of θ possible in b), and only one obtuse value of θ possible in c).

Q4 For the curve defined by the parametric equations $x = t^2 + 5t$, $y = 6 + 3t^2$:

a) Find the possible values of t when $y = 33$.

b) Find the possible values of y when $x = 6$.

Q5 Complete the table on the right, and hence sketch the curve represented by the parametric equations:

$x = 5t$, $y = \dfrac{2}{t}$, for $t \neq 0$.

t	-5	-4	-3	-2	-1	1	2	3	4	5
x										
y										

Q6 Sketch the curve represented by the parametric equations $x = 1 + \sin\theta$, $y = 2 + \cos\theta$ for the values $0 \leq \theta \leq 2\pi$.

Use the table below to help you, and give your answers to 2 d.p.

θ	0	$\dfrac{\pi}{4}$	$\dfrac{\pi}{3}$	$\dfrac{\pi}{2}$	$\dfrac{2\pi}{3}$	$\dfrac{3\pi}{4}$	π	$\dfrac{4\pi}{3}$	$\dfrac{3\pi}{2}$	$\dfrac{5\pi}{3}$	2π
x											
y											

Q6 Hint

Make sure your calculator is set to radians.

P Q7 The curve C is defined by the parametric equations $y = 10 - t^2$, $x = 2t^2 - 7t$.

a) Find the value of a if $(a, 1)$ is a point on the curve and $a > 1$.

b) Show that the point $(-6, 4)$ does not lie on curve C.

E P Q8 The curve C has parametric equations $x = \dfrac{1}{t}$ and $y = t(t^2 - a)$, where a is a constant and $t > 0$.

a) Explain why the line with equation $x = 0$ is an asymptote. *[1 mark]*

b) C passes through the point with coordinates $\left(\dfrac{1}{2}, 0\right)$. Find the value of a. *[3 marks]*

c) Describe what happens to the values of x and y for

    (i)   very small values of t, *[1 mark]*

    (ii) very large values of t. *[1 mark]*

d) Sketch the curve C. *[3 marks]*

P M Q9 The orbit of a comet around the Sun is modelled by the parametric equations $x = 3\sin\theta$, $y = 7 + 9\cos\theta$, for $0 \leq \theta \leq 2\pi$, where the point $(0, 0)$ represents the Sun, and where 1 unit on the x- and y-axes represents 1 astronomical unit (AU, 1 AU $\approx$ 150 million km).

How far, in AU, is the comet from the Sun when:

a) $\theta = 0$,   b) $\theta = \dfrac{\pi}{2}$?

Q9b) Problem Solving

You'll need to use Pythagoras' theorem.

P M Q10 The path of a toy plane thrown from a tower is modelled by the parametric equations $x = t^2 + 4t$, $y = 25 - t^2$ for $0 \leq t \leq 5$, where t is the time taken in seconds, and x and y are the horizontal and vertical distances in metres to the plane from the point at ground level at the foot of the tower.

a) How far does the plane travel in the horizontal direction in the first 2 seconds?

b) Sonia is standing 21 m from the base of the tower, in line with the path of the plane. At what height above the ground does the toy plane pass over Sonia's head?

Challenge

E
P
M

Q11 The position of a skateboarder on a half-pipe is modelled by a semi-ellipse with the parametric equations $x = 5\cos(\pi - \theta)$ and $y = 3 - 3\sin(\pi - \theta)$ as shown below.

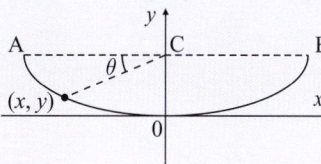

θ is the angle between AC and the line directly to a skateboarder's position, so $0 \leq \theta \leq \pi$. *x* and *y* are measured in metres with (*x*, *y*) indicating a skateboarder's position relative to the centre of the base of the half-pipe.

a) Find the coordinates of the two positions of a skateboarder when they are at the top of either side of the half-pipe (A and B). *[3 marks]*

b) Find the values of θ when a skateboarder is 4 m horizontally from the centre of the half-pipe, giving your answers to three significant figures. *[3 marks]*

A skateboarder jumps from the top of the half-pipe and is currently in the air at (–3, 2). They are currently at angle *p*, measured anticlockwise from the line AC.

c) Find the distance between this point and the point, at the same angle *p*, that they would have been at had they skateboarded directly down the half-pipe. Give your answer to 1 decimal place. *[6 marks]*

4.2 Parametric Equations — Finding Intersections

A lot of parametric equations questions involve identifying points on the curve defined by the equations. You'll often be given the parametric equations of a curve, and asked to find the coordinates of the **points of intersection** of this curve with another line (such as the *x*- or *y*-axis).

Use the information in the question to **solve for *t*** at the intersection point(s). Then **substitute the value(s) of *t*** into the parametric equations to work out the *x* and *y* values (i.e. the **coordinates**) at the intersection point(s). The example below shows how to tackle a question like this.

Example 1

The curve shown has the parametric equations $y = t^3 - t$ and $x = 4t^2 - 1$.

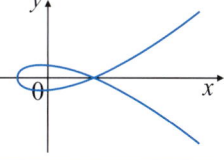

a) Find the coordinates of the points where the graph crosses the *x*-axis.

$0 = t^3 - t$ — On the *x*-axis, $y = 0$. Use the parametric equation for *y* to find the values of *t* where $y = 0$.

$\Rightarrow t(t^2 - 1) = 0$

$\Rightarrow t(t + 1)(t - 1) = 0$

$\Rightarrow t = 0, t = -1, t = 1$ — Factorise and solve the cubic that is formed.

When $t = 0$: $x = 4(0)^2 - 1 = -1$ — Now use those values of *t* to find the *x*-coordinates by putting them into the parametric equation for *x*.

When $t = -1$: $x = 4(-1)^2 - 1 = 3$

When $t = 1$: $x = 4(1)^2 - 1 = 3$

So the graph crosses the *x*-axis at (–1, 0) and (3, 0). — $t = -1$ and $t = 1$ give the same coordinates — that's where the curve crosses over itself.

E Exam Style **P** Problem Solving **M** Modelling

b) Find the coordinates of the points where the graph crosses the y-axis.

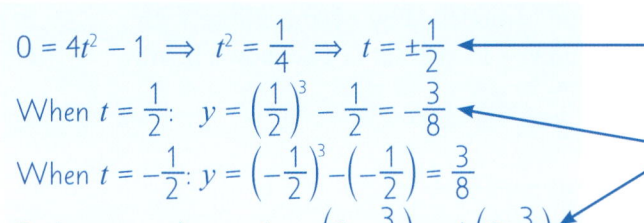

$0 = 4t^2 - 1 \implies t^2 = \dfrac{1}{4} \implies t = \pm\dfrac{1}{2}$ ← On the y-axis, $x = 0$. Use the equation for x this time and solve the quadratic.

When $t = \dfrac{1}{2}$: $y = \left(\dfrac{1}{2}\right)^3 - \dfrac{1}{2} = -\dfrac{3}{8}$ ← Put the values of t into the parametric equation for y to find the y-coordinates.

When $t = -\dfrac{1}{2}$: $y = \left(-\dfrac{1}{2}\right)^3 - \left(-\dfrac{1}{2}\right) = \dfrac{3}{8}$

So it crosses the y-axis at $\left(0, -\dfrac{3}{8}\right)$ and $\left(0, \dfrac{3}{8}\right)$.

Tip The sketch shows there are two points where the graph crosses each axis.

c) Find the coordinates of the points where the graph crosses the line $8y = 3x + 3$.

$8y = 3x + 3 \implies 8(t^3 - t) = 3(4t^2 - 1) + 3$ ← First, substitute the parametric equations into $8y = 3x + 3$.

$8t^3 - 8t = 12t^2 \implies 8t^3 - 12t^2 - 8t = 0$

$\implies t(2t + 1)(t - 2) = 0 \implies t = 0, t = -\dfrac{1}{2}, t = 2$ ← Rearrange and factorise to find the values of t you need.

When $t = 0$: $x = 4(0) - 1 = -1, y = 0^3 - 0 = 0$ ← Put the values of t back into the parametric equations to find the x- and y-coordinates where the graph crosses the line.

When $t = -\dfrac{1}{2}$: $x = 4\left(\dfrac{1}{4}\right) - 1 = 0, y = \left(-\dfrac{1}{2}\right)^3 + \dfrac{1}{2} = \dfrac{3}{8}$

When $t = 2$: $x = 4(4) - 1 = 15, y = 2^3 - 2 = 6$

So it crosses $8y = 3x + 3$ at: $(-1, 0)$, $\left(0, \dfrac{3}{8}\right)$, $(15, 6)$

Tip Check the answers by putting these values back into $8y = 3x + 3$.

Exercise 4.2

Q1 The curve with parametric equations $x = 3 + t, y = -2 + t$ meets the x-axis at the point A and the y-axis at the point B. Find the coordinates of A and B.

Q1 Hint A lies on the x-axis and B lies on the y-axis, so in each case you know one of the coordinates.

Q2 The curve C has parametric equations $x = 2t^2 - 50, y = 3t^3 - 24$.

   a) Find the value of t where the curve meets the x-axis.

   b) Find the values of t where the curve meets the y-axis.

Q3 The curve with parametric equations $x = 64 - t^3, y = \dfrac{1}{t}$, for $t \neq 0$, meets the y-axis at the point P. Find the coordinates of the point P.

Q4 Find the coordinates of the point of intersection, P, of the line $y = x - 3$ and the curve with parametric equations $x = 2t + 1, y = 4t$.

Q4 Hint Replace the x and y in the Cartesian equation with the parametric equations.

Q5 Find the coordinates of the point(s) of intersection of the curve $y = x^2 + 32$ and the curve with parametric equations $x = 2t, y = 6t^2$.

Q6 Find the points of intersection of the circle $x^2 + y^2 = 32$ and the curve with parametric equations $x = t^2, y = 2t$.

P Q7 The curve with parametric equations $x = a(t - 2)$, $y = 2at^2 + 3$ $(a \neq 0)$, meets the y-axis at $(0, 4)$.

a) Find the value of the constant a.

b) Hence determine whether the curve meets the x-axis.

P Q8 A curve has parametric equations $x = \dfrac{2}{t}$, $y = t^2 - 9$, for $t \neq 0$.

a) Find the point(s) at which the curve crosses the x-axis.

b) Does the curve meet the y-axis? Explain your answer.

c) Find the coordinates of the point(s) at which this curve meets the curve $y = \dfrac{10}{x} - 3$.

P Q9 A curve has parametric equations $x = 3 \sin t$, $y = 5 \cos t$
and is defined for the domain $0 \leq t \leq 2\pi$.

a) Determine the coordinates at which this curve meets the x- and y-axes.

b) Find the points where the curve meets the line $y = \left(\dfrac{5\sqrt{3}}{9} \right) x$.

Problem Solving

In Q10, start by setting either the two x-equations or the two y-equations equal to each other.

P M Q10 A simulation models the paths of two ships using parametric equations. The ships are modelled as points, with no width or length. The path taken by the first ship is given by $x_1 = 24 - t$, $y_1 = 10 + 3t$. The path taken by the second ship is given by $x_2 = t + 10$, $y_2 = 12 + 2t - 0.1t^2$. For both sets of equations, $0 \leq t \leq 30$, where t is the time in hours since the start of the simulation, and x and y are measured in miles East and North respectively.

According to the simulation, will the two ships collide?

E P M Q11 Engineers are testing a machine that has been designed to project a ball into the air such that the ball will follow a path programmed into the machine. For safety reasons, the machine should sound an alarm for the duration of time a projected ball remains at a height exceeding the safety limit of 5 m.

To test the alarm the machine is programmed to fire a ball along the path defined by the parametric equations $x = 4t + 5$ and $y = 6t - t^2$, $0 \leq t \leq 6$, where x and y are the horizontal and vertical displacements from the machine, measured in metres, at time t seconds.

Work out how long the alarm should sound for and the horizontal distance the ball will travel during this time. *[5 marks]*

Challenge

E P Q12 The curve C_1 is defined by the parametric equations:
$$x = t - 1 \text{ and } y = t^2 - At + B$$

The curve C_2 is defined by the parametric equations:
$$x = 2u + 1 \text{ and } y = (B - 12)u^2 - 6u + A - 1$$

The curves intersect at the point $(3, 8)$. A and B are integers.

a) Find the values of A and B. *[5 marks]*

b) Given the Cartesian equation of the curve C_1 is $y = x^2 - 5x + 14$, find the other point of intersection of the two curves. *[5 marks]*

c) Prove that the curve C_2 does not intersect the x-axis. *[2 marks]*

4.3 Converting Parametric Equations to Cartesian Equations

Some parametric equations can be **converted** into **Cartesian equations**.
There are two main ways to do this:

- **Rearrange** one of the equations to make the **parameter** the subject, then **substitute** the result into the **other** equation.

- If your equations involve **trig functions**, use **trig identities** to **eliminate** the parameter.

The examples below show how the first method works.

Example 1

Give the Cartesian equations, in the form $y = f(x)$, of the curves represented by the following pairs of parametric equations:

a) $y = t^3 - 1$ and $x = t + 1$,

$x = t + 1 \Rightarrow t = x - 1$

You want the answer in the form $y = f(x)$, so leave y alone for now, and rearrange the equation for x to make t the subject.

$y = t^3 - 1$

$\Rightarrow y = (x - 1)^3 - 1$

Substitute the rearranged equation into the equation for y to eliminate t.

$\Rightarrow y = (x - 1)(x^2 - 2x + 1) - 1$

$\Rightarrow y = x^3 - 2x^2 + x - x^2 + 2x - 1 - 1$

$\Rightarrow y = x^3 - 3x^2 + 3x - 2$

Expand the brackets and simplify.

So the Cartesian equation is
$y = x^3 - 3x^2 + 3x - 2$.

Tip You could use the binomial theorem or Pascal's triangle to find the coefficients in the expansion of $(x - 1)^3$.

b) $y = \dfrac{1}{3t}$ and $x = 2t - 3$, $t \neq 0$.

$x = 2t - 3 \Rightarrow t = \dfrac{x + 3}{2}$

Follow the same method as above, rearrange the equation for x to make t the subject.

So $y = \dfrac{1}{3t} \Rightarrow y = \dfrac{1}{3\left(\dfrac{x + 3}{2}\right)}$

Substitute the rearranged equation into the equation for y to eliminate t.

$\Rightarrow y = \dfrac{1}{\left(\dfrac{3(x + 3)}{2}\right)}$

Now just expand and simplify.

$\Rightarrow y = \dfrac{2}{3x + 9}$

Trigonometric functions

Things get a little trickier when the likes of sin and cos decide to put in an appearance.
For trig functions you need to use **trig identities**.

Example 2

A curve has parametric equations $x = 1 + \sin \theta$, $y = 1 - \cos 2\theta$.
Give the Cartesian equation of the curve in the form $y = f(x)$.

> **Tip** If one of the equations includes $\cos 2\theta$ or $\sin 2\theta$, you'll probably need to use one of the **double angle formulas** from p.104 to get it in terms of $\sin \theta$ or $\cos \theta$.

$y = 1 - \cos 2\theta$
$\quad = 1 - (1 - 2\sin^2 \theta)$
$\quad = 2\sin^2 \theta$

> Instead of making θ the subject, find a way to get both x and y in terms of the same trig function. You can get $\sin \theta$ into the equation for y using the identity $\cos 2\theta \equiv 1 - 2\sin^2 \theta$.

$x = 1 + \sin \theta \implies \sin \theta = x - 1$

> Rearrange the equation for x to make $\sin \theta$ the subject.

$y = 2\sin^2 \theta$
$\implies y = 2(x - 1)^2 = 2x^2 - 4x + 2$

> Replace '$\sin \theta$' in the equation for y with '$x - 1$' to get y in terms of x. Then expand and simplify.

So the equation is $y = 2x^2 - 4x + 2$.

Example 3 P

A curve is defined parametrically by $x = 4\sec \theta$, $y = 4\tan \theta$.
Give the equation of the curve in the form $y^2 = f(x)$, and hence determine whether the curve intersects the line $y + 3x = 4$.

> **Tip** Note that you're asked for the equation in the form $y^2 = f(x)$ not $y = f(x)$.

$x = 4\sec \theta \implies \sec \theta = \dfrac{x}{4}$

$y = 4\tan \theta \implies \tan \theta = \dfrac{y}{4}$

> Start by rearranging each equation to make the trig function the subject.

$\sec^2 \theta \equiv 1 + \tan^2 \theta$

> Find an identity that contains both $\sec \theta$ and $\tan \theta$.

$\left(\dfrac{x}{4}\right)^2 = 1 + \left(\dfrac{y}{4}\right)^2$

> Substitute the trig functions with the x and y terms.

$x^2 = 16 + y^2 \implies y^2 = x^2 - 16$

> Rearrange to get an equation in the form $y^2 = f(x)$.

$y + 3x = 4 \implies y = 4 - 3x$
$\quad\quad\quad \implies (4 - 3x)^2 = x^2 - 16$
$\quad\quad\quad \implies 16 - 24x + 9x^2 = x^2 - 16$
$\quad\quad\quad \implies 8x^2 - 24x + 32 = 0$
$\quad\quad\quad \implies x^2 - 3x + 4 = 0$

> To see whether the curve intersects $y + 3x = 4$, start by rearranging to make y the subject.

> Then substitute for y in the Cartesian equation of the curve. Expand and simplify.

> The curve will intersect the line at the roots of this equation.

There are no real roots to this equation because $b^2 - 4ac < 0$, so the line and the curve do not intersect.

> **Problem Solving** The last bit comes from the quadratic formula. For a quadratic $ax^2 + bx + c = 0$, there are only real solutions when $b^2 - 4ac \geq 0$, because of the $\sqrt{b^2 - 4ac}$ bit in the formula.

Chapter 4

Exercise 4.3

Q1 For each of the following parametrically-defined curves, find the Cartesian equation of the curve in an appropriate form.

a) $x = t + 3$, $y = t^2$

b) $x = 3t$, $y = \dfrac{6}{t}$, $t \neq 0$

c) $x = 2t^3$, $y = t^2$

d) $x = t + 7$, $y = 12 - 2t$

e) $x = t + 4$, $y = t^2 - 9$

f) $x = \dfrac{t+2}{3}$, $y = t^2 - t$

g) $x = t^2 - \dfrac{t}{2}$, $y = 5 - 8t$

h) $x = \sin \theta$, $y = \cos \theta$

i) $x = \sin \theta$, $y = \cos 2\theta$

j) $x = 1 + \sin \theta$, $y = 2 + \cos \theta$

k) $x = \cos \theta$, $y = \cos 2\theta$

l) $x = \cos \theta - 5$, $y = \cos 2\theta$

Q2 By eliminating the parameter θ, express the curve defined by the parametric equations $x = \tan \theta$, $y = \sec \theta$ in the form $y^2 = f(x)$.

Ⓟ Q3 Write the curve $x = 2 \cot \theta$, $y = 3 \csc \theta$ in the form $y^2 = f(x)$.

Q4 A circle is defined by the parametric equations $x = 5 + \sin \theta$, $y = -3 + \cos \theta$.
a) Find the coordinates of the centre of the circle, and the radius of the circle.
b) Write the equation of the curve in Cartesian form.

Q5 A curve has parametric equations $x = \dfrac{1 + 2t}{t}$, $y = \dfrac{3 + t}{t^2}$, $t \neq 0$.
a) Express t in terms of x.
b) Hence show that the Cartesian equation of the curve is: $y = (3x - 5)(x - 2)$.
c) Sketch the curve.

Q6 Express $x = \dfrac{2 - 3t}{1 + t}$, $y = \dfrac{5 - t}{4t + 1}$ ($t \neq -1$, $t \neq -0.25$), in Cartesian form.

Q7 Find the Cartesian equation of the curve defined by the parametric equations $x = 5 \sin^2 \theta$, $y = \cos \theta$. Express your answer in the form $y^2 = f(x)$.

Ⓟ Q8
a) Express $x = a \sin \theta$, $y = b \cos \theta$ in Cartesian form.
b) Use your answer to a) to sketch the curve.
c) What type of curve has the form $x = a \sin \theta$, $y = b \cos \theta$?

Q8b) Problem Solving

Find the x- and y-intercepts first — it will give you an idea of the graph's shape.

Ⓟ Q9 A curve has parametric equations $x = 3t^2$, $y = 2t - 1$.
a) Show that the Cartesian equation of the curve is $x = \dfrac{3}{4}(y + 1)^2$.
b) Hence find the point(s) of intersection of this curve with the line $y = 4x - 3$.

Ⓟ Q10 Find the Cartesian equation of the curve $x = 7t + 2$, $y = \dfrac{5}{t}$, $t \neq 0$, in the form $y = f(x)$ and hence sketch the curve, labelling any asymptotes and points of intersection with the axes clearly.

Q10 Problem Solving

It's easier to sketch if you can transform a standard curve shape — see p.46-49.

Challenge

Ⓔ Q11 A curve with the equation $y = f(x)$ can also be defined parametrically by $x = t + 1$ and $y = 3t^2$.
a) Find a Cartesian equation for the curve in the form $y = f(x)$. *[2 marks]*
b) Sketch the graphs of $y = f(x)$ and $y = f(x - 1) - 3$, labelling any points where the graphs intercept the coordinate axes and the turning point. *[5 marks]*

4 Review Exercise

Q1 A curve is defined by the parametric equations $x = \dfrac{1}{t}$, $y = \dfrac{2}{t^2}$ $(t \neq 0)$.

a) Find the value of t when $x = \dfrac{1}{4}$ and hence find the corresponding y-coordinate.

b) Find the possible values of t when $y = \dfrac{1}{50}$.

E **Q2** The graph with equation $y = \text{f}(x)$ is defined by the following parametric equations,

$$x = t^2 \text{ and } y = \dfrac{1}{2}t^3.$$

a) Complete the table of values for x, y and t given below. *[2 marks]*

t	−3	−2	−1	0	1	2	3
x							
y							

b) Given that $x = 25$, find the possible values of t
and their corresponding y-values. *[2 marks]*

c) The domain of t is $t \in \mathbb{R}$.
Explain why the domain of $\text{f}(x)$ is $x \geq 0$ and find the range of $\text{f}(x)$. *[2 marks]*

E **P** **M** **Q3** A basketball player shoots for a basket by throwing a ball upwards from a height of A m
at time $t = 0$ seconds. Relative to the position of release, the horizontal and vertical
positions of the ball, in m, at time t are modelled by the following parametric equations:

$$x = 4t \text{ and } y = 2.4 + 1.8t - 0.48t^2 \text{ where } t \geq 0.$$

The height of the basket is 3 m above the ground.

a) Write down the value of A.
Briefly explain how you obtained your answer. *[2 marks]*

b) (i) The player scores a basket with the ball
passing through the basket on its way down.
How long is the ball in the air for before the basket is scored? *[2 marks]*

(ii) How far, along the ground, is the basket from
the position the player took their shot from? *[1 mark]*

c) If the basket hadn't been in the way and the ball continued on the same path,
how long would the ball be in the air before it first bounces on the ground? *[2 marks]*

Q4 A curve is defined by the parametric equations $x = t^3 + t^2 - 6t$, $y = 4t - 5$.
Find the coordinates of the points where this curve crosses the y-axis.

Q5 Find the coordinates of the points where the line $y = 10x - 8$ crosses
the curve defined by the parametric equations $x = \dfrac{t+3}{5}$ and $y = t^2 - t$.

Q6 A curve has parametric equations $y = 4 + \dfrac{3}{t}$ and $x = t^2 - 1$ $(t \neq 0)$.
What are the coordinates of the points where this curve crosses:

a) the y-axis,   b) the line $x + 2y = 14$?

P **M** Q7 The movement of a particle is modelled by the parametric
equations $x = -e^t \sin 2t$, $y = e^t \cos 2t$, for $0 \leq t \leq 2\pi$.

a) Find the exact value of x at each point where $y = 0$.

b) If d is the distance of the particle from the origin $(0, 0)$:

    (i) Find an equation for d in terms of t.

    (ii) Find the exact coordinates of the point where $d = e^{\frac{\pi}{8}}$.

E Q8 The diagram below shows the graph of the curve $y = f(x)$ which can
also be defined by the parametric equations $x = e^t$ and $y = 0.1e^{2t}$.

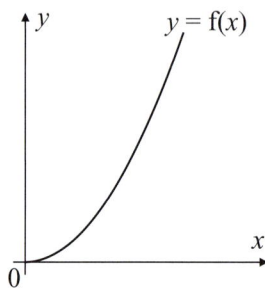

a) (i) Explain why the curve does not intersect the origin.

    (ii) Find the domain and range of $f(x)$.   *[2 marks]*

b) At one point on the curve, the x- and y-coordinates are equal.
Find the coordinates of the point.   *[4 marks]*

c) (i) Make a copy of the diagram above and add a sketch of the graph
defined by the parametric equations $x = 0.1e^{2u}$ and $y = e^u$.

    (ii) Write down the coordinates of any points of intersection
between the two graphs.   *[3 marks]*

Q9 A curve is defined by the parametric equations $y = 2t^2 + t + 4$ and $x = \dfrac{6 - t}{2}$.

a) Find the values of x and y when $t = 0, 1, 2$ and 3.

b) What are the values of t when: (i) $x = -7$ (ii) $y = 19$?

c) Find the Cartesian equation of the curve, in the form $y = f(x)$.

E Q10 A curve has parametric equations:

$$x = \frac{2t - 3}{t} \quad \text{and} \quad y = 2t + 6, \quad 1 < t \leq 6$$

a) Find the points where the curve crosses the coordinate axes.   *[3 marks]*

b) Find the Cartesian equation of the curve in the form $y = f(x)$.   *[3 marks]*

c) Find the range of $f(x)$.   *[2 marks]*

E P Q11 The curve C has parametric equations:

$$x = 2t \quad \text{and} \quad y = t(at + 6), \quad t \in \mathbb{R}$$

a) If the line $y = ax - 4$ is a tangent to the curve, find the possible values for a. *[4 marks]*

b) Sketch the possible curves for C on the same set of axes, showing where they cross the coordinate axes and each other. *[4 marks]*

E P Q12 A curve C has parametric equations:

$$x = 2t^2 + 3t \quad \text{and} \quad y = 2t^2 - 3t, \quad t \in \mathbb{R}$$

a) Show that there are no values of t within the given domain where both x and y take negative values. *[3 marks]*

b) Show that a Cartesian equation for C can be written in the form $0 = x^2 + y^2 + Axy + B(x + y)$ where A and B are constants to be found. *[5 marks]*

Q13 The parametric equations of a curve are $x = 2 \sin \theta$ and $y = \cos^2 \theta + 4$, $-\dfrac{\pi}{2} \leq \theta \leq \dfrac{\pi}{2}$.

a) What are the coordinates of the points where: (i) $\theta = \dfrac{\pi}{4}$ (ii) $\theta = \dfrac{\pi}{6}$?

b) What is the Cartesian equation of the curve?

c) What restrictions are there on the values of x for this curve?

P Q14 The curve C is defined by the parametric equations $x = \dfrac{\sin \theta}{3}$ and $y = 3 + 2 \cos 2\theta$. Find the Cartesian equation of C.

P Q15 The parametric equations describing curve K are $x = 4 - \cos 2\theta$, $y = \sin^4 \theta - \dfrac{1}{2}$, $0 \leq \theta \leq \pi$.

a) Find the coordinates of the point on curve K where $\theta = \dfrac{\pi}{3}$.

b) Explain why curve K does not cross the y-axis.

c) Find the Cartesian equation of curve K, in the form $y = f(x)$.

E P Q16 A curve C has parametric equations:

$$x = 2 \cos t \quad \text{and} \quad y = 5 \cos 3t, \quad -\pi < t < \pi$$

a) Show that a Cartesian equation for C can be written in the form $y = Ax(x^2 - B)$, where A and B are constants to be found, and give the domain of the function. *[6 marks]*

b) Explain why the line $y = 6$ does not cross C. *[1 mark]*

c) The line $y = kx$ crosses C exactly 3 times, where k is a positive constant. Find the range of possible values for k. *[4 marks]*

E P Q17 A circle, C, is described by the parametric equations $x = a + b \cos t$ and $y = 1 - 2a + b \sin t$, $0 \leq t \leq 2\pi$, where a and b are positive constants.

a) Write down, in terms of a and b, the coordinates of the centre of the circle and its radius. *[2 marks]*

b) C passes through the point with coordinates $(-4, 2)$ when $t = \dfrac{3\pi}{4}$. Find the values of a and b. *[4 marks]*

c) Hence write down the Cartesian equation of the circle C. *[1 mark]*

P **Q18** The curve C can be defined by the Cartesian equation $y = f(x)$ or the parametric equations

$x = \dfrac{1}{a}\operatorname{cosec} t$ and $y = \dfrac{1}{a+1}\cot t$, where $0 < t < \pi$. a is a positive integer.

a) (i) Write $\operatorname{cosec}^2 t$ in terms of x and a.

(ii) Write $\cot^2 t$ in terms of y and a.

b) Hence show that a Cartesian equation for C can be written in the form $(a + 1)^2 y^2 = a^2 x^2 - 1$.

c) Given that C intercepts the x-axis when $x = \dfrac{1}{2}$, find the value of a.

d) Find the domain and range of $f(x)$.

E **Q19** An ellipse is defined by the parametric equations $x = 10\sin\theta$ and $y = 6\cos\theta$, $0 \le \theta \le 2\pi$.

a) Find a Cartesian equation of the ellipse. *[4 marks]*

b) Sketch the graph of the ellipse, clearly labelling any intersections with the coordinate axes. *[3 marks]*

c) (i) Write down the equation of a horizontal line that is a tangent to the ellipse. *[1 mark]*

(ii) Write down the equation of a vertical line that is a tangent to the ellipse. *[1 mark]*

Challenge

E
P
M
Q20 A speed skater is practising by skating the path of a figure of eight defined by the parametric equations $x = a\cos t$ and $y = b\sin 2t$, where $t \ge 0$ and a and b are positive constants.
The coordinates of the speed skater at time t seconds are (x, y) where x and y are in metres.

a) After π seconds the speed skater is at the point with x-coordinate -8.
The greatest difference in y-coordinates the speed skater will reach is 6 m.
Find the values of a and b. *[2 marks]*

b) Find all of the times, for $0 \le t \le 2\pi$,
at which the speed skater has a y-coordinate of 0. *[2 marks]*

An obstacle is placed at the point with coordinates $\left(4, -\dfrac{3\sqrt{3}}{2}\right)$.

c) How many times will the speed skater encounter the obstacle in the first twenty seconds of their skate? *[5 marks]*

E
P
M
Q21 The trajectories of some fireworks, launched from the origin, are given by the parametric equations $x = 2t$ and $y = 16t^3 - 84t^2 + 108t$.
x m is the horizontal distance and y m is the vertical distance a firework has travelled from its launch site at time t seconds.

a) Find the time at which a firework returns to the ground. *[3 marks]*

b) Find the Cartesian equation, $y = f(x)$, of the trajectory of a firework and state the domain of $f(x)$. *[3 marks]*

One firework explodes when it reaches its maximum height.

c) Find the maximum height and the time at which the firework explodes.
Give your answers to three significant figures. *[5 marks]*

4   Chapter Summary

1   A Cartesian equation is a single equation linking two variables, usually x and y.
For example, $y = 3x^2 - 5x + 3$ and $x^2 + y^2 = 16$ are both Cartesian equations.

2   Parametric equations define x and y values separately in terms of a third variable (parameter), usually t or θ. For example, $x = 3\sin\theta$ and $y = 1 + 2\cos\theta$ are a set of parametric equations. You can use parametric equations to find x-y values at a given value of the parameter, and vice versa.

3   Parametric equations are useful to model the change in x-y position of an object over time, t, because you can easily calculate the position at a specific time.

4   The domain of the parameter is often restricted to certain values, for example $0 \leq t < 10$ or $0 \leq \theta \leq 2\pi$.

5   You can find points of intersection between a parametric curve and the axes. Just substitute $x = 0$ or $y = 0$ (depending on which intersection you want to find) into the relevant parametric equation, solve for t and then substitute the value(s) of t into the other parametric equation to find the corresponding x or y value(s).

6   You can also find points of intersection between a parametric curve and a line or curve defined by a Cartesian equation. To do this you just substitute both parametric equations into the Cartesian equation and solve for t. Then substitute the value(s) of t back into the parametric equations to find the corresponding x and y values.

7   To convert parametric equations to Cartesian equations:
- Rearrange one of the parametric equations to make the parameter the subject.
- Substitute your rearranged equation into the other parametric equation.

8   If parametric equations involve trig functions (e.g. sin, cos, tan, etc.) then you'll need to use trig identities to convert them into Cartesian equations. Some common trig identities you might find useful are shown below.
- $\sin^2\theta + \cos^2\theta \equiv 1$
- $\cos 2\theta \equiv \cos^2\theta - \sin^2\theta$
- $1 + \tan^2\theta \equiv \sec^2\theta$
- $1 + \cot^2\theta \equiv \operatorname{cosec}^2\theta$

Sequences and Series

5

Learning Objectives

Once you've completed this chapter, you should be able to:

- Understand and use nth term formulas or recurrence relations when working with sequences, including showing that a sequence is increasing, decreasing or periodic.
- Work with arithmetic sequences, including finding an nth term formula and using it to solve problems.
- Find the sum of the first n terms of an arithmetic series and the sum of the first n natural numbers, using sigma notation (Σ).
- Recognise geometric sequences and series.
- Know and use the formulas for the general term and the sum of the first n terms of a geometric sequence or series.
- Recognise convergent geometric series and find their sum to infinity.
- Solve modelling problems involving arithmetic and geometric sequences and series.

Prior Knowledge Check

1 Draw the next two terms in this sequence:

see GCSE Maths

2 What is the nth term of the following sequence?

| 5 | 14 | 23 | 32 | 41 ... |

see GCSE Maths

3 Find the first four terms of a sequence if the nth term is given by:
a) $10 + n$ b) $5 - 3n$
c) $8n - 7$ d) $-6 - 2n$

see GCSE Maths

4 The nth term of a sequence is $4n - 1$.
What are the 5th, 25th and 100th terms in the sequence?

see GCSE Maths

5.1   Sequences

Before we get going with this section, there's some **notation** to learn:

a_n just means the nth **term** of the sequence — e.g. a_4 is the 4th term, and a_{n+1} is the term after a_n.

The idea behind the nth term is that you can use a formula to generate any term in a sequence from its **position**, n, in the sequence.

Example 1

A sequence has n^{th} term $a_n = 4n + 1$.

a) Find the value of a_{10}.

$a_{10} = 4(10) + 1 = 41$ ⟵ Just substitute 10 for n in the nth term formula.

b) A term in the sequence is 33. Find the position of this term.

$33 = 4n + 1$ ⟵ Substitute 33 for a_n in the nth term formula.

$n = (33 - 1) \div 4 = 8$ ⟵ Rearrange to find the value of n — this means that $a_8 = 33$.

Example 2   P

A sequence has the nth term $an^2 + b$, where a and b are constants.

a) If the 3rd term is 7 and the 5th term is 23, find the nth term formula.

For the 3rd term $n = 3$: $a(3^2) + b = 9a + b = 7$ ⟵ Form equations using the information given in the question.

For the 5th term $n = 5$: $a(5^2) + b = 25a + b = 23$ ⟵

$$\begin{array}{r} 25a + b = 23 \\ - \underline{(9a + b = 7)} \\ 16a \quad\;\; = 16 \end{array} \Rightarrow a = 1$$ ⟵ Solve the equations simultaneously to find the value of a or b.

$9(1) + b = 7 \Rightarrow b = -2$ ⟵ Use $a = 1$ to find b using either one of the equations.

So the nth term is $n^2 - 2$

b) Is 35 a term in the sequence?

$n^2 - 2 = 35 \Rightarrow n^2 = 37 \Rightarrow n = \sqrt{37}$ ⟵ Form and solve an equation in n and see if you get a positive whole number.

$\sqrt{37}$ is not an integer, so 35 is not in the sequence.

Increasing and decreasing sequences

There are a few types of sequences that you need to know:

- In an **increasing sequence**, each term is greater than the previous term, so $a_{k+1} > a_k$ for all terms — e.g. the square numbers 1, 4, 9, 16, 25, ...

- In a **decreasing sequence**, each term is less than the previous term, so $a_{k+1} < a_k$ for all terms — e.g. the sequence 16, 13, 10, 7, 4, ...

Example 3

a) Show that the sequence with n^{th} term $7n - 16$ is increasing.

$a_k = 7k - 16$ and $a_{k+1} = 7(k + 1) - 16 = 7k - 9$ ← Use the nth term formula to find expressions for a_k and a_{k+1}.

$7k - 9 > 7k - 16 \Rightarrow -9 > -16$ ← Show that $a_{k+1} > a_k$ for all terms.

This is true, so it is an increasing sequence.

b) Show that the sequence with n^{th} term $\dfrac{1}{2n - 1}$ is decreasing.

$a_k = \dfrac{1}{2k - 1}$ and $a_{k+1} = \dfrac{1}{2(k+1) - 1} = \dfrac{1}{2k + 1}$ ← Find expressions for a_k and a_{k+1}.

$\dfrac{1}{2k+1} < \dfrac{1}{2k-1} \Rightarrow 2k - 1 < 2k + 1 \Rightarrow -1 < 1$ ← Show that $a_{k+1} < a_k$ for all terms. As k is a positive integer, $(2k + 1)$ and $(2k - 1)$ are too. So you can multiply without flipping the inequality sign.

This is true, so it is a decreasing sequence.

You may also see **periodic sequences**, where the terms **repeat** in a cycle. The number of terms in the repeated cycle is known as the **order**. E.g. the sequence 1, 0, 1, 0, 1, 0... is periodic with order 2.

Some sequences aren't increasing, decreasing or periodic. If you can show $a_{k+1} > a_k$ for some values of k and $a_{k+1} < a_k$ for some other values of k, then the sequence is **neither increasing nor decreasing**.

Exercise 5.1

Q1   A sequence has n^{th} term $a_n = 3n - 5$. Find the value of a_{20}.

Q2   Find the 4th term of the sequence with n^{th} term $n(n + 2)$.

Q3   Find the first 5 terms of the sequence with n^{th} term $(n - 1)(n + 1)$.

Q4   The kth term of a sequence is 29. The nth term of this sequence is $4n - 3$. Find the value of k.

Q5   A sequence has the nth term $13 - 6n$. Show that the sequence is decreasing.

Q6   Is the sequence with n^{th} term 3^{-n} increasing or decreasing?

P Q7   A sequence has n^{th} term $= an^2 + b$, where a and b are constants.
         If the 2nd term is 15, and the 5th term is 99, find a and b.

P Q8   A sequence starts 9, 20, 37, ... Its nth term $= en^2 + fn + g$, where e, f and g are constants.
         Find the values of e, f and g.

Q9   The nth term of the sequence is given by $(n - 1)^2$. A term in the sequence is 49.
         Find its position.

Q10 How many terms of the sequence with n^{th} term $15 - 2n$ are positive?

E P Q11 A sequence has n^{th} term $u_n = 4n(n - 3) + 9$.
Prove that all terms in the sequence are positive. *[2 marks]*

P Q12 A sequence has n^{th} term $u_n = p\cos\left(\dfrac{n\pi}{2}\right)$, where $p \neq 0$.

a) (i) Write down, in terms of p, the first five terms in the sequence.

(ii) What are the maximum and minimum values of terms in the sequence?

(iii) Write down the period of the sequence.

b) What is the period of the sequence with:

(i) n^{th} term $u_n = p\cos(n\pi)$, where $p \neq 0$?

(ii) n^{th} term $u_n = p\cos\left(\dfrac{n\pi}{q}\right)$, where $p \neq 0$, and q is a non-zero integer?

Challenge

E P Q13 The nth term of a sequence is given by $u_n = n^2 - 16n + 55$.

a) Find $u_{n+1} - u_n$. *[3 marks]*

b) Find which term is the minimum in the sequence, u_n, and what this value is. *[3 marks]*

c) Given that u_n is finite and is a decreasing sequence, what is
the maximum number of terms in the sequence? Justify your answer. *[1 mark]*

5.2 Recurrence Relations

A **recurrence relation** is another way to describe a sequence.

Recurrence relations tell you how to work out
a term in a sequence from the previous term.

So, using the new notation, what this is saying is that a recurrence
relation describes how to work out a_{k+1} from a_k.

E.g. if each term in the sequence is **2 more** than the previous term:

$a_{k+1} = a_k + 2$ ←⎯⎯ So, if $k = 5$, this says that $a_6 = a_5 + 2$,
that is, the 6th term is equal to the 5th term + 2.

This recurrence relation will be true for many sequences, e.g. 1, 3, 5, 7..., and 4, 6, 8, 10...
So to describe a **particular sequence** you also have to give one term.
E.g. the sequence 1, 3, 5, 7... is described by:

$a_{k+1} = a_k + 2$, $a_1 = 1$ ←⎯⎯ a_1 stands for the 1st term.

Tip You might see
$f(x)$ notation used — e.g.
$f(x + 1) = f(x) + 2$, $f(1) = 1$
would also describe the
sequence on the left.

Most of the sequences you'll see are **infinite** — they go on forever.
But you can have **finite** sequences which stop after a certain number of terms.

Example 1

Describe the following sequence using a recurrence relation: 5, 8, 11, 14, 17, …

$a_{k+1} = a_k + 3$ ← Each term in this sequence equals the one before it, plus 3.

$a_{k+1} = a_k + 3, \ a_1 = 5$ ← The description needs to be more specific, so you've got to give one term in the sequence, as well as the recurrence relation.

Tip You usually give the first value, a_1 — but using $a_2 = 8$ would also describe the sequence.

Note that $a_{k+1} = a_k + 3$ on its own isn't enough to describe 5, 8, 11, 14, 17, ...
For example, it also describes the sequence 1, 4, 7, 10, …

Example 2

A sequence is given by the recurrence relation $a_{k+1} = a_k - 4$, with $a_1 = 20$.
Find the first five terms of this sequence.

$a_{k+1} = a_k - 4$ and $a_1 = 20$
$\Rightarrow a_2 = a_1 - 4 = 20 - 4 = 16$ ← You're given the first term ($a_1 = 20$). So put a_1 into the recurrence relation to find the second term.

$a_3 = a_2 - 4 = 16 - 4 = 12$
$a_4 = a_3 - 4 = 12 - 4 = 8$
$a_5 = a_4 - 4 = 8 - 4 = 4$ ← Repeat this to find the third, fourth and fifth terms. Notice that $a_{k+1} = a_k - 4$ just means each term is 4 less than the previous term.

So the first five terms of the sequence are:
20, 16, 12, 8, 4.

Example 3

A sequence is generated by the recurrence relation $u_{n+1} = 3u_n + k$, with $u_1 = 2$.
a) Find u_3 in terms of k.

$u_{n+1} = 3u_n + k$ and $u_1 = 2$
$\Rightarrow u_2 = 3u_1 + k = 3(2) + k = u_2 = 6 + k$ ← You're given the first term, u_1, so use this to generate the second term, u_2, in terms of k.

$\Rightarrow u_3 = 3u_2 + k = 3(6 + k) + k = 18 + 4k$ ← Use the recurrence relation again to generate the third term, u_3. This time substitute the expression for u_2 in for u_n.

b) Given that $u_4 = 28$, find k.

$u_4 = 3u_3 + k = 3(18 + 4k) + k = 54 + 13k$ ← Form an expression for u_4 using the recurrence relation and the expression for u_3 from part a).

$54 + 13k = 28 \Rightarrow 13k = -26 \Rightarrow k = -2$ ← The question tells you that $u_4 = 28$, so form an equation and solve to find k.

The next example's a bit harder, as there's an n^2 in there. But the method works in exactly the same way — you just end up with a slightly more complicated formula for the recurrence relation.

Example 4

A sequence has the general term $x_n = n^2$.
Describe this sequence using a recurrence relation.

$x_1 = 1^2 = 1$ $x_2 = 2^2 = 4$ ← | Start by finding the first few terms of the sequence (i.e. n = 1, 2, 3, 4...).

$x_3 = 3^2 = 9$ $x_4 = 4^2 = 16$

So the first four terms are: 1, 4, 9, 16 ← | You're at the same point as you were at the start of Example 1. It's trickier to form the recurrence relation here as n in the general term is squared.

$k =$ 1   2   3   4

$x_k =$ 1   4   9   16 | Look at the difference between each term. The difference increases by 2 each time, so the differences form the sequence $2k + 1$.

        +3   +5   +7 ←

    $(2\times1)+1$ $(2\times2)+1$ $(2\times3)+1$

So the recurrence relation is:

$x_{k+1} = x_k + 2k + 1$ ← | To get from term x_k to the next term, x_{k+1}, you take the value of term x_k, and add $2k + 1$ to it. E.g. 3rd term = 2nd term + 2(2) + 1.

$x_{k+1} = x_k + 2k + 1, \; x_1 = 1$ ← | Remember that you need to give one term in the sequence to complete the description.

It's sometimes helpful to draw some little diagrams where n^2 is involved. For the example above these diagrams would help you see that it's $2k + 1$ added each time.

1st   2nd   3rd   4th

Exercise 5.2

Q1 A sequence is defined for $n \geq 1$ by $u_{n+1} = 3u_n$ and $u_1 = 10$.
Find the first 5 terms of the sequence.

Q2 Find the first 4 terms of the sequence in which $u_1 = 2$ and $u_{n+1} = u_n^2$ for $n \geq 1$.

Q3 Find the first 4 terms of the sequence in which $u_1 = 4$ and $u_{n+1} = \dfrac{-1}{u_n}$,
and describe the type of sequence that this forms.

Q4 Describe the following sequence using a recurrence relation: 3, 6, 12, 24, 48, ...

Q5 a) Describe the following sequence using a recurrence relation: 12, 16, 20, 24, 28, ...

     b) The sequence is finite and ends at 100.
       Find the number of terms.

P Q6 Describe the following sequence using a recurrence relation: 7, 4, 7, 4, 7, ...

Q6 Problem Solving You need to find an operation that goes back and forth between 4 and 7. There's more than one possible answer — think about one of these number facts: 7 + 4 = 11 or 7 × 4 = 28.

Q7 In a sequence $u_1 = 4$ and $u_{n+1} = 3u_n - 1$ for $n \geq 1$. Find the value of k if $u_k = 95$.

Q8 In a sequence $x_1 = 9$ and $x_{n+1} = (x_n + 1) \div 2$ for $n \geq 1$.

Find the value of r if $x_r = \dfrac{5}{4}$.

Q9 Find the first 5 terms of the sequence in which $u_1 = 7$ and $u_{n+1} = u_n + n$ for $n \geq 1$.

P Q10 In a sequence $u_1 = 6$, $u_2 = 7$ and $u_3 = 8.5$.
If the recurrence relation is of the form $u_{n+1} = au_n + b$, find the values of the constants a and b.

Q10 Problem Solving Form all the equations you can from the information, and then solve them.

P Q11 A sequence is generated by $u_1 = 8$ and $u_{n+1} = \dfrac{1}{2}u_n$ for $n \geq 1$.
Find the first 5 terms and a formula for u_n in terms of n.

E **P** Q12 A sequence is defined by $u_{n+1} = au_n + 2$, $u_1 = 3$.
  a) Find u_2 and u_3 in terms of the constant a. *[2 marks]*
  b) Given that $u_3 = 87$, find the possible values of a. *[3 marks]*

M Q13 The Fibonacci sequence is defined by the recurrence relation $u_{n+2} = u_{n+1} + u_n$ with $u_1 = u_2 = 1$.
The number of leaves on each branch of a species of tree follows the Fibonacci sequence, where branch n has u_n leaves and the uppermost branch is branch 1.
  a) How many leaves are there on branches 5 and 6 of one of these trees?
  b) A student counts a total of 54 leaves on one tree. How many branches does this tree have?

Challenge

E **P** Q14 A recurrence relation is defined as $u_{n+2} = 2u_{n+1} - u_n$, for $n \geq 1$.
  a) Find the third, fourth and fifth terms of the sequence when $u_1 = 3$ and $u_2 = 6$. *[2 marks]*
  b) (i) Find the third, fourth and fifth terms when $u_1 = a$ and $u_2 = 2a$. *[2 marks]*
     (ii) For the case when $u_2 = 2u_1$, write down a formula for the nth term in terms of n. *[1 mark]*
  c) What would the nth term formula be in the case $u_2 = 3u_1$? *[2 marks]*

P Q15 The nth term of a sequence is given by the recurrence relation $u_{n+1} = u_n^2 - 1$.
  a) Given $u_1 = p$, for some integer p, express u_2 and u_3 in terms of p.
  b) (i) In the case when $p = -1$, find the first four terms in the sequence.
     (ii) Hence describe any properties of the entire sequence when $p = -1$.
  c) Prove that there are only two integer values of p for which this recurrence relation has the same properties you described in part b)(ii).

5.3　Arithmetic Sequences

Here are some examples of **arithmetic sequences**:　5, 7, 9, 11... (add 2 each time);
20, 17, 14, 11... (add –3 each time).

Arithmetic sequences are often referred to as **arithmetic progressions** — it's exactly the same thing.

The formula for the nth term of an arithmetic sequence is:　$u_n = a + (n - 1)d$

where:　a is the **first term** of the sequence.
　　　　d is the amount you add each time — the **common difference**.
　　　　n is the **position** of any term in the sequence.

This box shows you how the formula is derived:

Term	**n**	
1st	1	$u_1 = a$
2nd	2	$u_2 = u_1 + d = a + d$
3rd	3	$u_3 = u_2 + d = (a + d) + d = a + 2d$
4th	4	$u_4 = u_3 + d = (a + 2d) + d = a + 3d$
.	.	.
.	.	.
.	.	.
n^{th}	n	$u_n = a + (n - 1)d$

This is the formula to find the nth term.

> **Tip**　Each term is made up of the previous one, plus d. It's a recurrence relation.

Example 1

For the arithmetic sequence 2, 5, 8, 11, … find u_{20} and the formula for u_n.

$u_n = a + (n - 1)d$
For this sequence, $a = 2$ and $d = 3$.

Write down a (the first term) and d (the common difference).

$u_{20} = 2 + (20 - 1) \times 3 = 2 + 19 \times 3 = 59$
So $u_{20} = 59$.

Put them into the nth term formula with $n = 20$.

$u_n = 2 + (n - 1)3 \implies u_n = 3n - 1$

u_n is the general term, i.e. $a + (n - 1)d$. Substitute in the a and d values and simplify.

You can always check the u_n formula works with a couple of values of n.

$n = 1$ gives $3(1) - 1 = 2$ ✔　　$n = 2$ gives $3(2) - 1 = 5$ ✔

You only actually need to know **two terms** of an arithmetic sequence (and their positions) — then you can work out any other term.

Example 2

The 2nd term of an arithmetic sequence is 21, and the 9th term is –7.
Find the 23rd term of this sequence.

2nd term = 21, so $a + (2 - 1)d = 21 \implies a + d = 21$
9th term = –7, so $a + (9 - 1)d = -7 \implies a + 8d = -7$

Use the formula $u_n = a + (n - 1)d$ to set up equations for the 2nd ($n = 2$) and 9th ($n = 9$) terms.

continued on the next page...

$a + d = 21$ ①   ←   You've now got two simultaneous equations
$a + 8d = -7$ ②   — label them and solve to find a and d.

① − ② : $-7d = 28 \Rightarrow d = -4$ ←   Subtracting one from the other and
rearranging gives you the value of d.

① : $a + d = 21 \Rightarrow a - 4 = 21 \Rightarrow a = 25$ ←   Substitute $d = -4$ back into one of
the equations to find the value of a.

n^{th} term $= a + (n - 1)d = 25 + (n - 1) \times -4$
$\qquad\qquad\qquad = 29 - 4n$ ←

Write the nth term
formula and
use it to find the
23rd term $(n = 23)$.

23rd term $= 29 - 4 \times 23 = -63$ ←

Tip   The first term and
the common difference
have been found along
the way (questions might
ask for these values).

Exercise 5.3

Q1 An arithmetic progression has first term 7 and common difference 5. Find its 10th term.

Q2 Find the nth term for each of the following sequences:
   a) 6, 9, 12, 15, ...   b) 4, 9, 14, 19, ...   c) 12, 8, 4, 0, ...
   d) 1.5, 3.5, 5.5, 7.5 ...   e) 77, 69, 61, 53, ...   f) −2, −2.5, −3, −3.5, ...

Q3 In an arithmetic sequence, the fourth term is 19 and the tenth term is 43.
Find the first term and common difference.

Q4 In an arithmetic progression, $u_1 = -5$ and $u_5 = 19$. Find u_{10}.

Q5 In an arithmetic progression, $u_7 = 8$ and $u_{11} = 10$. Find u_3.

E P Q6 An arithmetic sequence has first four terms 6, 10, 14 and 18.
   a) Write down the formula for the nth term of this sequence.   *[2 marks]*
   b) Prove that every term in the sequence is even.   *[1 mark]*
   c) Find the value of k such that $u_{k+10} = 2u_k + 10$.   *[3 marks]*

P Q7 In an arithmetic sequence, $u_3 = 15$ and $u_7 = 27$. Find the value of k if $u_k = 66$.

E P Q8 An arithmetic sequence has an 8th term of 78 and 12th term of 102.
   a) Find the common difference and the first term.   *[3 marks]*
   b) Which term has the value 150?   *[2 marks]*
   c) Two consecutive terms have a sum of 438.
     Find the two terms and their positions in the sequence.   *[4 marks]*

Challenge

P Q9 In an arithmetic sequence the first three terms are $\ln(x)$, $\ln(x + 8)$, $\ln(x + 48)$.
Find the value of x and the next term in the sequence.

P Q10 An arithmetic progression has exactly 20 terms greater than 0 and smaller than 100.
All the terms are integers. Find two possible values for a if $u_{11} = 47$.

5.4 Arithmetic Series

Here is an arithmetic sequence:
It's an infinite sequence — it goes on forever.

$$5, 8, 11, 14, 17, 20, ...$$

Now suppose you wanted to find the sum of the first 5 terms of this sequence.
You'd write this by replacing the commas with '+' signs like this:

$$5 + 8 + 11 + 14 + 17$$

This is now an **arithmetic series**. It's a finite series with 5 terms. And if you actually added up the numbers you'd find that the **sum** for this series is 55.
So sequences become series when you add up their terms to find sums.

Sum of the first n terms

It would very quickly stop being fun if you had to find the sum of a 100-term series manually. Instead, you can use one of these **two formulas**. S_n represents the **sum of the first n terms**.

$$S_n = \frac{n}{2}[2a + (n-1)d]$$

For this formula, you just need to put in the usual values of **a**, **d** and **n**.

and

$$S_n = \frac{1}{2}n(a + l)$$

Here, **l** represents the **last term**. This formula is a bit easier to use if you know the value of the last term.

> **Tip** You can work out a, d and the nth term for a series, just as you would for a sequence. So in the 5-term series above, $a = 5$, $d = 3$ and n^{th} term = $3n + 2$ (for $1 \leq n \leq 5$). Also, because the series is finite, you can state its last term, which is 17.

It's given in the formula booklet like this, but can also be stated as $S_n = n \times \dfrac{(a + l)}{2}$.

There's a nice little proof for these formulas which you need to know:

1. For any series, you can express S_n as:
   $$S_n = a + (a + d) + (a + 2d) + ... + (a + (n-3)d) + (a + (n-2)d) + (a + (n-1)d)$$

2. Now, if you reverse the order of the terms you can write it as:
   $$S_n = (a + (n-1)d) + (a + (n-2)d) + (a + (n-3)d) + ... + (a + 2d) + (a + d) + a$$

3. Adding the two expressions for S_n gives:
   $$2S_n = (2a + (n-1)d) + (2a + (n-1)d) + (2a + (n-1)d) + ... + (2a + (n-1)d)$$

4. So we've now got the term "$(2a + (n-1)d)$" repeated n times, which is:
   $$2S_n = n \times (2a + (n-1)d) \implies S_n = \frac{n}{2}[2a + (n-1)d]$$

5. So we've derived the first formula. Now to get the second, just replace "$a + (n-1)d$" with l:
   $$S_n = \frac{n}{2}[a + a + (n-1)d], \text{ so } S_n = \frac{1}{2}n[a + l]$$

Now it's time to try out the sum formulas in some worked examples.

Example 1

Find the sum of the series with first term 3, last term 87 and common difference 4.

$$S_n = \frac{1}{2}n(a + l)$$

You're told the last term, so use the S_n formula with l in.

$$a = 3 \text{ and } l = 87$$

You know the first term (a) and the last term (l), but you don't know n yet.

$$a + (n - 1)d = 87$$
$$3 + (n - 1)4 = 87$$
$$4n - 1 = 87$$
$$n = 22$$

Find n by putting what you do know into the 'n^{th} term' formula. $a = 3$ and $d = 4$.

$$S_n = \frac{1}{2}n(a + l)$$
$$S_{22} = \frac{1}{2} \times 22 \times (3 + 87)$$
$$= 11 \times 90 = 990$$

You're now all set to put the values for a, l and n into the S_n formula.

The sum of the series is 990.

This is the sum of the first 22 terms.

You could also have used $S_n = \frac{n}{2}[2a + (n - 1)d]$ in this example: $S_{22} = \frac{22}{2}[2(3) + (22 - 1)(4)] = 990$.

Example 2

This question is about the sequence $-5, -2, 1, 4, 7...$

a) Is 67 a term in the sequence? If it is, give its position.

> **Tip** This question could just have easily been about the series $-5 + -2 + 1 + 4 + 7 + ...$ — the working would have been exactly the same.

$$n^{th} \text{ term} = a + (n - 1)d$$
$$= -5 + (n - 1)3$$
$$= 3n - 8$$

First, find the formula for the nth term of the sequence. $a = -5$ and $d = 3$

$$67 = 3n - 8$$
$$3n = 75 \implies n = 25$$

Put 67 into the formula and see if this gives a whole number for n.

67 is a term in the sequence. It's the 25th term.

The value of n tells you the term's position in the sequence.

b) Find the sum of the first 20 terms.

$$S_n = \frac{n}{2}[2a + (n - 1)d]$$

You aren't given the last term, so use this version of the formula.

$$S_{20} = \frac{20}{2}[2(-5) + (20 - 1)3]$$
$$= 10[-10 + 19 \times 3] = 470$$

The sum of the first 20 terms is 470.

Put $a = -5$, $d = 3$ and $n = 20$ into the formula.

Example 3 **P**

Find the possible numbers of terms in the arithmetic series starting $21 + 18 + 15...$ if the sum of the series is 75.

$a = 21, S_n = 75, d = -3$ ← You need to work out n, the number of terms in the series and you're given the values of a, S_n and d.

$S_n = \frac{n}{2}[2a + (n-1)d]$

$S_n = \frac{n}{2}[2(21) + (n-1) \times -3]$ ← You can use the S_n formula. Put in the values you know and rearrange.

$75 = \frac{n}{2}[42 - 3n + 3]$

$75 = \frac{45n}{2} - \frac{3n^2}{2}$

$-3n^2 + 45n - 150 = 0$ ← Solve this quadratic equation.

$n^2 - 15n + 50 = 0$ ←

$(n - 5)(n - 10) = 0$ ← Divide through by -3 to simplify and then factorise.

$n = 5$ or $n = 10$

There are 5 or 10 terms in the series.

Problem Solving There are two answers because the series goes into negative numbers, so the sum of 75 is reached twice. Look at the first 10 terms of the series to see how:
$21 + 18 + 15 + 12 + 9 + 6 + 3 + 0 + (-3) + (-6)$.

Sigma notation

So far, the letter S has been used for the sum. The Ancient Greeks did a lot of work on this — their capital letter for S is Σ or **sigma**. This is used today, together with the general term, to mean the sum of the series. For example, the following means the sum of the series with n^{th} term $2n + 3$.

Starting with $n = 1$...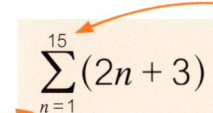

$$\sum_{n=1}^{15}(2n + 3)$$

...and ending with $n = 15$

Tip For all values of n, $\sum_{1}^{n} 1 = n$.

Example 4

Find $\displaystyle\sum_{n=1}^{15}(2n + 3)$.

First term $(n = 1)$ is: $2(1) + 3 = 5$ ←

Last term $(n = 15)$ is: $2(15) + 3 = 33$ ←

Put some values for n into $2n + 3$ to work out a and l for the series.

So $a = 5$ and $l = 33$

$S_n = \frac{1}{2}n(a + l)$ ← You can use the formula containing l to work out the sum of the first 15 terms (i.e. $n = 15$).

$S_{15} = \frac{1}{2} \times 15(5 + 33) = 285$ ←

You get the same answer whichever formula you use. If you wanted to use the other formula here you'd first need to find d by working out that the 2nd term is $2 \times 2 + 3 = 7 \Rightarrow d = 7 - 5 = 2$.

Then $S_n = \frac{n}{2}[2a + (n-1)d] \Rightarrow S_{15} = \frac{15}{2}[2 \times 5 + 14 \times 2] = 285$.

Exercise 5.4

Q1 An arithmetic series has first term 8 and common difference 3.
Find the 10th term and the sum of the first 10 terms.

Q2 In an arithmetic series $u_2 = 16$ and $u_5 = 10$. Find a, d and S_8.

Q3 In an arithmetic series $a = 12$ and $d = 6$. Find u_{100} and S_{100}.

Q4 An arithmetic progression has n^{th} term $8n - 6$. Find the sum of the first 20 terms.

Q5 An arithmetic series has first term -6 and the sum of the first 10 terms is 345.
Find the common difference.

Q6 Find the sum of the arithmetic series with first term 7, last term 79 and common difference 6.

Q7 Find: a) $\displaystyle\sum_{n=1}^{12}(5n - 2)$,   b) $\displaystyle\sum_{n=1}^{9}(20 - 2n)$.

E P Q8 An arithmetic series is given by $\displaystyle\sum_{r=1}^{n}u_r$, where $u_r = 80 - 6r$.

> **Q8 Problem Solving**
>
> This question uses a handy little trick for finding series sums that don't start from $n = 1$.

    a) Find u_6 and u_{10}. *[2 marks]*

    b) (i) Find $\displaystyle\sum_{r=1}^{6}80 - 6r$ and $\displaystyle\sum_{r=1}^{10}80 - 6r$. *[3 marks]*

        (ii) Hence, or otherwise, find $\displaystyle\sum_{r=7}^{10}80 - 6r$. *[1 mark]*

P Q9 In an arithmetic series $a = 3$ and $d = 2$. Find n if $S_n = 960$.

P Q10 Given that $\displaystyle\sum_{n=1}^{k}(5n + 2) = 553$, show that the value of k is 14.

P Q11 An arithmetic sequence begins $x + 11$, $4x + 4$, $9x + 5$, ...
Find the sum of the first 11 terms.

P Q12 An arithmetic progression begins 36, 32, 28, 24, ...
Find the possible values of n if $S_n = 176$.

E Q13 An arithmetic series has 9th term 61 and the sum of the first 12 terms is 522.
    a) Find the first term and the common difference. *[4 marks]*
    b) Find the minimum number of terms needed
       such that their sum exceeds 1000. *[4 marks]*

Challenge

E P Q14 An arithmetic series has first term 5 and common difference d.
The $(k - 4)^{th}$ term is 115. The sum of the first $(k + 4)$ terms is 2000.
Find the possible values of d and k. *[7 marks]*

5.5   Sum of the First n Natural Numbers

The **natural numbers** are the positive whole numbers, i.e. 1, 2, 3, 4...
They form a simple arithmetic progression with $a = 1$ and $d = 1$.

The sum of the first n natural numbers is: $S_n = \frac{1}{2}n(n+1)$

This formula can be derived from the previous sum formulas by putting in values:

> The sum of the first n natural numbers is: $S_n = 1 + 2 + 3 + ... + (n-2) + (n-1) + n$
>
> So $a = 1$, $l = n$ and also $n = n$. $S_n = \frac{1}{2}n(a+l) \longrightarrow S_n = \frac{1}{2}n(n+1)$

You can also derive the formula from first principles —
the proof is almost identical to the one for a general arithmetic series on page 151:

> $S_n = 1 + 2 + 3 + ... + (n-2) + (n-1) + n$ ①
>
> Rewrite ① with the terms reversed:
> $S_n = n + (n-1) + (n-2) + ... + 3 + 2 + 1$ ②
>
> ① + ② gives:
> $2S_n = (n+1) + (n+1) + (n+1) + ... + (n+1) + (n+1) + (n+1)$
> $\Rightarrow 2S_n = n(n+1) \Rightarrow S_n = \frac{1}{2}n(n+1)$

> **Tip**   You could be asked to prove any of these sum formulas, so make sure you know these steps.

Example   1

Find the sum of the first 100 natural numbers.

$S_{100} = \frac{1}{2} \times 100 \times 101 = 5050$ Use the formula
$S_n = \frac{1}{2}n(n+1)$ where $n = 100$.

Sum of the first 100 natural numbers = 5050

Example   2

The sum of the first k natural numbers is 861. Find the value of k.

$\frac{1}{2}k(k+1) = 861$ Form an equation in k using $S_n = \frac{1}{2}n(n+1)$.

$k^2 + k = 1722$ Expand the brackets and rearrange into the general form of a quadratic.

$k^2 + k - 1722 = 0$

$(k + 42)(k - 41) = 0$ We're looking for a whole number for k, so it should factorise. As '$b = 1$', we're looking for two numbers that are 1 apart and multiply to 1722.

$k = -42$ or $k = 41$

We can ignore the negative solution here, so the answer is $k = 41$.   We have k numbers added together, so an answer of $k = -42$ wouldn't make sense.

Exercise 5.5

Q1 Find the sum of the first:

    a) 10 natural numbers,
                    b) 2000 natural numbers.

Q2 Find $\displaystyle\sum_{n=1}^{32} n$.

P Q3 Find $\displaystyle\sum_{n=1}^{10} n$ and $\displaystyle\sum_{n=1}^{20} n$. Hence find $\displaystyle\sum_{n=11}^{20} n$.

Q4 The sum of the first n natural numbers is 66. Find n.

Q5 Find k if $\displaystyle\sum_{n=1}^{k} n = 120$.

P Q6 Find the sum of the series $16 + 17 + 18 + \ldots + 35$.

P Q7 What is the first natural number k for which $\displaystyle\sum_{n=1}^{k} n$ is greater than 1 000 000?

Q8 An arithmetic series S_n is described as the sum of the first n natural numbers.

    a) The sequence, $S_1, S_2, S_3, \ldots$ gives the triangular numbers.
       Draw diagrams for S_3 and S_4 to demonstrate this.

    b) A total of 150 balls are arranged into rows. There are 3 balls in the first row.
       Each row has one more ball than the previous one. How many rows of balls are there?

P Q9 The sum of the first 80 natural numbers is written as $\displaystyle\sum_{n=1}^{80} n$.

    a) Calculate $\displaystyle\sum_{n=1}^{80} n$.

    b) Without doing any calculations, describe in words
       what the result of each of these series gives you:

       (i) $\displaystyle\sum_{n=1}^{80}(n+1)$
           (ii) $\displaystyle\sum_{n=1}^{80}(2n-1)$

E P Q10 The difference between the sum of the first 20 natural numbers and
    the first k natural numbers is 74. Given that $k < 20$, find the value of k.   *[4 marks]*

E P Q11 a) S_n is used to represent the sum of the first n natural numbers. Find:

       (i) $S_3 + S_4$ *[1 mark]*
       (ii) $S_4 + S_5$ *[1 mark]*
    b) Assuming k is a positive integer, prove that $S_k + S_{k+1}$ is a square number.   *[3 marks]*

Challenge

P Q12 The sum of the first p natural numbers is pq. The sum of the first
    q natural numbers is $4p - 3q$. Given that $p > q > 1$, find p and q.

5.6 Geometric Sequences

Remember, with **arithmetic** sequences you get from one term to the next by **adding** a fixed amount each time. They have a **first term** (*a*), and the amount you add to get from one term to the next is called the **common difference** (*d*).

With **geometric sequences**, rather than adding, you get from one term to the next by **multiplying** by a **constant** called the **common ratio** (*r*).

> **Tip** Geometric sequences are also called geometric progressions.

- This is a **geometric sequence** where you find each term by **multiplying** the previous term by 2:

$$\times 2 \quad \times 2 \quad \times 2$$
$$1, \quad 2, \quad 4, \quad 8, \quad 16, \quad 32, \quad 64, \dots$$
$$\times 2 \quad \times 2 \quad \times 2$$

- If the common ratio is **negative**, the signs of the sequence will **alternate**. For this geometric sequence the common ratio is –3.

$$2, \quad -6, \quad 18, \quad -54, \quad 162, \quad -486, \dots$$

- The common ratio might **not** be a **whole number**. Here, it's $\frac{3}{4}$.

$$16, \quad 12, \quad 9, \quad \frac{27}{4}, \quad \frac{81}{16}, \quad \frac{243}{64}, \dots$$

You get each term by multiplying the first term by the common ratio some number of times. In other words, each term is the **first term** multiplied by **some power** of the **common ratio**.

This is how you describe geometric sequences using **algebra**:

The **first term** (u_1) is called '*a*'.

The **common ratio** (the number you multiply by) is called '*r*'.

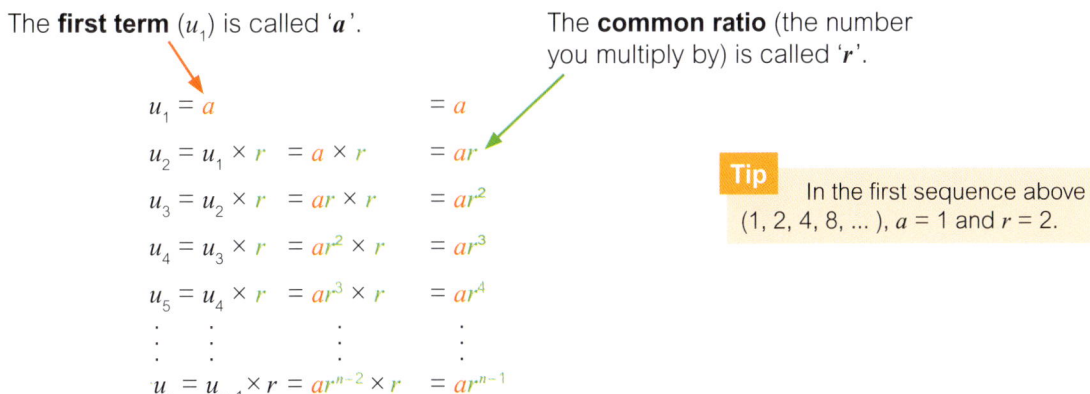

$$u_1 = a \qquad\qquad = a$$
$$u_2 = u_1 \times r \ = a \times r \qquad = ar$$
$$u_3 = u_2 \times r \ = ar \times r \qquad = ar^2$$
$$u_4 = u_3 \times r \ = ar^2 \times r \qquad = ar^3$$
$$u_5 = u_4 \times r \ = ar^3 \times r \qquad = ar^4$$
$$\vdots$$
$$u_n = u_{n-1} \times r = ar^{n-2} \times r \ = ar^{n-1}$$

> **Tip** In the first sequence above (1, 2, 4, 8, ...), $a = 1$ and $r = 2$.

So the **formula** that describes **any term** in a geometric sequence is:

$$u_n = ar^{n-1}$$

The term u_n is called the 'n^{th} term', or 'general term' of the sequence. You need to learn this formula.

If you know the values of a and r, you can substitute them into the **general formula** to find an expression that describes the whole sequence.

Example 1

A chessboard has a 1p piece on the first square, 2p on the second square, 4p on the third, 8p on the fourth and so on until the board is full. Find a formula for the amount of money on each square.

$u_1 = 1, u_2 = 2, u_3 = 4, u_4 = 8, \ldots$ ← Write out the first few terms of the sequence.

$a = u_1 = 1$

$r = \dfrac{u_2}{u_1} = \dfrac{2}{1} = 2$ ← a is the amount on the first square and r is the number you multiply by to get the amount on the next square.

$u_n = ar^{n-1} = 1 \times (2^{n-1}) = 2^{n-1}$ ← Put these into the geometric sequence formula.

Problem Solving

The trick with questions like this is to recognise that you're being asked about a geometric sequence.

You can also use the formula to find the **first term** a, the **common ratio** r or a **particular term** in the sequence, given other information about the sequence.

Example 2

a) Find the 5th term in the geometric sequence 1, 3, 9, ...

$2\text{nd term} = 1\text{st term} \times r \Rightarrow r = \dfrac{2\text{nd term}}{1\text{st term}} = \dfrac{3}{1} = 3$ ← Each term is the previous term multiplied by the common ratio, r, so divide consecutive terms to find r.

$3\text{rd term} = 9$

$4\text{th term} = 3\text{rd term} \times r = 9 \times 3 = 27$

$5\text{th term} = 4\text{th term} \times r = 27 \times 3 = 81$ ← Use the common ratio to find the 4th term then the 5th term.

b) A geometric sequence has first term 2 and common ratio 1.2. Find the 15th term in the sequence to 3 significant figures.

$a = 2$ and $r = 1.2$, so $u_n = ar^{n-1} = 2 \times (1.2)^{n-1}$ ← When you're asked for a higher term, the best method is to find the nth term.

$u_{15} = 2 \times (1.2)^{14} = 2 \times 12.839\ldots = 25.7$ to 3 s.f. ← Then find the 15th term ($n = 15$).

c) A geometric sequence has first term 25 and 10th term 80. Calculate the common ratio. Give your answer to 3 significant figures.

$a = 25$, so $u_n = ar^{n-1} = 25r^{n-1}$ ← Find the nth term in terms of r.

$u_{10} = 80 = 25r^9 \Rightarrow r^9 = \dfrac{80}{25} = 3.2$ ← Form and solve an equation using u_{10}.

$\Rightarrow r = \sqrt[9]{3.2} = 1.137\ldots = 1.14$ to 3 s.f.

d) The 8th term of a geometric sequence is 4374 and the common ratio is 3. What is the first term?

$r = 3$, so $u_n = ar^{n-1} = a(3)^{n-1}$ ← Find the nth term in terms of a.

$u_8 = 4374 = a(3)^7 = 2187a \Rightarrow a = \dfrac{4374}{2187} = 2$ ← Form and solve an equation using u_8.

Exercise 5.6

Q1 Find the common ratio of the geometric progression 3125, 1875, 1125, 675, 405, ...

Q2 Find the seventh term in the geometric progression 2, 3, 4.5, 6.75 ...

Q3 The sixth and seventh terms of a geometric sequence are 2187 and 6561 respectively. What is the first term?

Q4 A geometric sequence is 24, 12, 6, ... What is the 9th term?

Q5 The 14th term of a geometric progression is 9216. The first term is 1.125. Calculate the common ratio.

Q6 Problem Solving

Write yourself an equation or inequality and then solve it using logs. Look back at your Year 1 notes if you need a reminder of how to use logs.

P Q6 The first and second terms of a geometric progression are 1 and 1.1 respectively.
How many terms in this sequence are less than 4?

Q7 A geometric progression has a common ratio of 0.6.
If the first term is 5, what is the difference between the 10th term and the 15th term?
Give your answer to 5 d.p.

P Q8 A geometric sequence has a first term of 25 000 and a common ratio of 0.8.
Which term is the first to be below 1000?

Q9 A geometric sequence is 5, –5, 5, –5, 5, ... Give the common ratio.

Q10 The first three terms of a geometric progression are $\frac{1}{4}$, $\frac{3}{16}$ and $\frac{9}{64}$.
a) Calculate the common ratio.
b) Find the 8th term. Give your answer as a fraction.

Q11 The 7th term of a geometric sequence is 196.608 and the common ratio is 0.8.
What is the first term?

P Q12 A geometric progression begins 2, 6, ...
Which term of the geometric progression equals 1458?

P Q13 3, –2.4, 1.92,... is a geometric progression.
a) What is the common ratio?
b) How many terms are there in the sequence before you reach a term with modulus less than 1?

E Q14 The 8th term of a geometric series is nine times as big as the 6th term.
P Show algebraically that the possible values of the common ratio are ±3. *[3 marks]*

 **E** Exam Style **P** Problem Solving **M** Modelling

E **Q15** A geometric progression, u_n, has first term 6 and common ratio 2.
P Another geometric progression, v_n, has first term 4 and common ratio 2.

Show that $u_i - v_i = 2^i$ for any integer, i. *[3 marks]*

P **Q16** A geometric progression has first term 4 and common ratio $\frac{1}{4}$.

a) Show that the nth term of the sequence can be given by $u_n = 4^{2-n}$.

b) Find the 8th term, both as a fraction and to three significant figures.

c) What is the position of the first term that is less than one billionth (10^{-9})?

E **Q17** A geometric sequence has first term x and common ratio 3.
P
a) Show that $u_n = \frac{1}{3}x(3^n)$. *[2 marks]*

b) What is the first term that is greater than 1 000 000x? *[3 marks]*

Challenge

E **Q18** Three consecutive terms in a geometric sequence are $x - 2$, $2x + 14$ and $8x + 2$.
P Find the possible values of x and the corresponding common ratios. *[5 marks]*

5.7 Geometric Series

A **sequence** becomes a **series** when you **add** the terms to find the **sum**. Geometric series work just like geometric sequences (they have a **first term** and a **common ratio**), but they're written as a **sum of terms** rather than a list.

Geometric sequence: 3, 6, 12, 24, ...

Geometric series: 3 + 6 + 12 + 24 + ...

Sometimes you'll need to find the **sum** of the **first few terms** of a geometric series:

- The sum of the **first n terms** is called S_n.

- S_n can be written in terms of the first term a and the common ratio r:

$$S_n = u_1 + u_2 + u_3 + u_4 + \ldots u_n = a + ar + ar^2 + ar^3 + \ldots + ar^{n-1}$$

- There's a nice formula for finding S_n that doesn't involve loads of adding. You need to know the proof of the formula — luckily it's fairly straightforward:

> **Tip** Geometric series can be infinite sums (i.e. they can go on forever). We're just adding up bits of them for now, but summing a whole series is covered on page 165.

For any geometric sequence:	$S_n = a + ar + ar^2 + \ldots ar^{n-2} + ar^{n-1}$ ①
Multiplying this by r gives:	$rS_n = ar + ar^2 + \ldots ar^{n-2} + ar^{n-1} + ar^n$ ②
Equation ① – Equation ②:	$S_n - rS_n = a - ar^n$
Factorise both sides:	$(1 - r)S_n = a(1 - r^n)$
Divide through by $(1 - r)$:	$S_n = \frac{a(1 - r^n)}{1 - r}$

> **Tip** Subtracting equation ① from equation ② gives:
> $$S_n = \frac{a(r^n - 1)}{r - 1}$$
> which is equivalent.

So the sum of the first n terms of a geometric series is:

$$S_n = \frac{a(1 - r^n)}{1 - r}$$

This formula is given in the formula booklet (but the proof isn't).

Example 1

a) A geometric series has first term 3.5 and common ratio 5.
   Find the sum of the first 6 terms.

$$S_6 = \frac{a(1-r^6)}{1-r}$$ ⟵ You want the sum of the first 6 terms so $n = 6$.

$$= \frac{3.5(1-5^6)}{1-5} = 13\,671$$ ⟵ You're told that $a = 3.5$ and $r = 5$, so just put these values into the formula for S_6.

b) The first two terms in a geometric series are 20, 22.
   To 2 decimal places, the sum of the first k terms of the series is 271.59. Find k.

$$r = \frac{\text{second term}}{\text{first term}} = \frac{22}{20} = 1.1$$ ⟵ Start by finding r — you're told the first two terms (20 and 22), so divide the second by the first.

$$S_k = \frac{a(1-r^k)}{1-r}$$

$$= \frac{20(1-(1.1)^k)}{1-1.1} = -200(1-(1.1)^k)$$ ⟵ $a = 20$, so put a and r into the sum formula to get an expression for S_k.

$$271.59 = -200(1-(1.1)^k)$$ ⟵ Set the expression for S_k equal to the sum given in the question and solve.

$$\Rightarrow -\frac{271.59}{200} - 1 = -(1.1)^k$$

$$\Rightarrow -2.35795 = -(1.1)^k$$

$$\Rightarrow 2.35795 = 1.1^k$$

Tip You're looking for a number of terms so the answer must be a positive integer.

$$\Rightarrow \log(2.35795) = k\log(1.1)$$

$$\Rightarrow k = \frac{\log(2.35795)}{\log(1.1)}$$ ⟵ The answer to this calculation is actually 9.0000102... as the sum given in the question was rounded to 2 d.p.

$$\Rightarrow k = 9$$

Sigma notation

You saw on page 153 that the **sum** of the first n **terms** of a series (S_n) can also be written using **sigma** (Σ) **notation**. For geometric series, sigma notation looks like this:

$$S_n = u_1 + u_2 + u_3 + \ldots + u_n = a + ar + ar^2 + \ldots + ar^{n-1} = \sum_{k=0}^{n-1} ar^k$$

Tip Remember, Σ means sum (it's the Greek letter for S). Here, it's the sum of ar^k from $k = 0$ to $k = n - 1$. Be careful with the limits — it's $n - 1$ on top of the Σ, but the sum is S_n.

So, using the formula from the previous page, the sum of the first n terms can be written:

$$\sum_{k=0}^{n-1} ar^k = \frac{a(1-r^n)}{1-r}$$

Chapter 5

Example 2

$a + ar + ar^2 + \ldots$ is a geometric series, and $\sum_{k=0}^{4} ar^k = 85.2672$.
Given that $r = -1.8$, find the first term a.

$$85.2672 = \sum_{k=0}^{4} ar^k = S_5 = \frac{a(1-r^5)}{1-r}$$ ⟵ Write the sum formula for the first 5 terms ($n = 5$).

$$\Rightarrow 85.2672 = \frac{a(1-(-1.8)^5)}{1-(-1.8)}$$ ⟵ You're told in the question that $r = -1.8$.
Put in that value of r and solve for a.

$$\Rightarrow 85.2672 = a\frac{19.89568}{2.8}$$

$$\Rightarrow 85.2672 = 7.1056a$$

$$\Rightarrow a = \frac{85.2672}{7.1056} = 12$$

Exercise 5.7

Q1 The first term of a geometric sequence is 8 and the common ratio is 1.2.
Find the sum of the first 15 terms.

Q2 A geometric series has first term $a = 25$ and common ratio $r = 0.7$.
Find $\sum_{k=0}^{9} 25(0.7)^k$.

Q3 The sum of the first n terms of a geometric series is 196 605.
The common ratio of the series is 2 and the first term is 3. Find n.

Q4 A geometric progression starts with 4, 5, 6.25.
The first x terms add up to 103.2 to 4 significant figures. Find x.

Q5 The 3rd term of a geometric series is 6 and the 8th term is 192. Find:
   a) the common ratio
   b) the first term
   c) the sum of the first 15 terms

E Q6 A geometric series has first term 2 and common ratio $\frac{1}{2}$.
   a) Show that the sum of the first n terms of
      the series can be written as $S_n = 4 - 2^{2-n}$. *[3 marks]*
   b) (i) Find the 10th term of the series. *[2 marks]*
      (ii) Find the sum of the first 10 terms. *[1 mark]*

P Q7 $m + 10$, m, $2m - 21$, ... is a geometric progression, where m is a positive constant.

     a) Show that $m^2 - m - 210 = 0$.

     b) Hence show that $m = 15$.

> **Q7a) Problem Solving**
>
> The ratio of the first term to the second is the same as the ratio of the second term to the third.

     c) Find the common ratio of this series.

     d) Find the sum of the first 10 terms.

P Q8   The first three terms of a geometric series are 1, x, x^2.
          The sum of these terms is 3 and each term has a different value.

     a) Find x.

     b) Calculate the sum of the first 7 terms.

> **Q9 Problem Solving**
>
> You need to be careful with limits here as it starts from 1 instead of the usual 0.

E
P Q9   A series is defined by $\sum_{k=1}^{n} 3 \times 4^{k-1}$

     a) Find the sum given above in the instances when $n = 4$ and $n = 8$.   *[2 marks]*

     b) Use your answers to part a) to find $\sum_{k=5}^{8} 3 \times 4^{k-1}$.   *[2 marks]*

     c) Find $\sum_{k=7}^{10} 3 \times 4^{k-1}$.   *[3 marks]*

Q10 a, ar, ar^2, ar^3, ... is a geometric progression.

     Given that $a = 7.2$ and $r = 0.38$, find $\sum_{k=0}^{9} ar^k$.

Q11   The sum of the first eight terms of a geometric series is 1.2.

     Find the first term of the series, given that the common ratio is $-\dfrac{1}{3}$.

P Q12 a, $-2a$, $4a$, $-8a$, ... is a geometric sequence.

     Given that $\sum_{k=0}^{12} a(-2)^k = -5735.1$, find a.

E
P Q13   In a geometric series, $S_5 = 781$, $S_6 = 3906$ and $S_7 = 19\,531$.

     a) Find the sixth and seventh terms.   *[2 marks]*

     b) Find the first term and the common ratio.   *[3 marks]*

Challenge

P Q14   The sum of the first nine terms of a geometric series is $19\,682$.
          The sum of the first ten terms of the same series is $59\,048$.
          All terms in the series are positive.

     Given that the difference between the sixth term and the tenth term is $38\,880$, find the first term and the common ratio.

E Exam Style   **P** Problem Solving   **M** Modelling

5.8 Convergent Geometric Series

Convergent sequences

Some geometric sequences **tend towards zero** — in other words, they get closer and closer to zero (but they never actually reach it). For example:

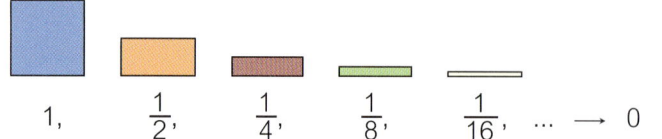

$$1, \quad \frac{1}{2}, \quad \frac{1}{4}, \quad \frac{1}{8}, \quad \frac{1}{16}, \quad ... \longrightarrow 0$$

> **Tip** The arrow here means 'tends to'.

Sequences like this are called **convergent** — the terms **converge** (get closer and closer) to a **limit** (the number they get close to). Geometric sequences either **converge to zero** or **don't converge** at all.

A sequence that doesn't converge is called **divergent**.

A geometric sequence a, ar, ar^2, ar^3, ... will converge to **zero** if each term is **closer** to zero than the one before. This happens when **−1 < r < 1**. You can write this as **|r| < 1**, where **|r|** is the modulus of r (see p.38), so:

$$a, \ ar, \ ar^2, \ ar^3, \ ... \to \mathbf{0} \text{ when } |r| < 1$$

You ignore the sign of r because you can still have a convergent sequence when r is negative. In that case the terms will alternate between > 0 and < 0, but they'll still be getting closer and closer to zero.

Convergent series

When you **sum** a geometric sequence that tends to zero you get a **convergent series**.

Because each term is getting closer and closer to zero, you're **adding smaller and smaller** amounts each time. The sum gets **closer and closer** to a certain number, but never reaches it — this is the **limit** of the series.

For example, the **sum** of the **convergent sequence** $1, \frac{1}{2}, \frac{1}{4}, \frac{1}{8}, \frac{1}{16}, ...$ gets closer and closer to 2:

$$S_1 = 1$$

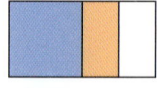
$$S_2 = 1 + \frac{1}{2} = 1\frac{1}{2} = 1.5$$

The more terms you add, the closer you get to the **limit** of the **series**.

$$S_3 = 1 + \frac{1}{2} + \frac{1}{4} = 1\frac{3}{4} = 1.75$$

> **Tip** This is the sum of the sequence above.

$$S_4 = 1 + \frac{1}{2} + \frac{1}{4} + \frac{1}{8} = 1\frac{7}{8} = 1.875$$

$$S_5 = 1 + \frac{1}{2} + \frac{1}{4} + \frac{1}{8} + \frac{1}{16} = 1\frac{15}{16} = 1.9375$$

The terms are getting **closer to 0** at such a rate that the sum will never reach 2.

As you **add more terms**, the sum is **tending to 2**.

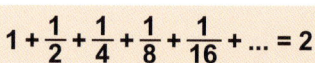
$$1 + \frac{1}{2} + \frac{1}{4} + \frac{1}{8} + \frac{1}{16} + ... = 2$$

So when the **sequence** a, ar, ar^2, ar^3, ... **converges** to zero, the **series** $a + ar + ar^2 + ar^3 + ...$ **converges** to a **limit**.

Like sequences, geometric series converge when $|r| < 1$.

> $a + ar + ar^2 + ar^3 + ...$ with $|r| < 1$ is **convergent**
>
> $a + ar + ar^2 + ar^3 + ...$ with $|r| \geq 1$ is **divergent**

Problem Solving

Not all sequences that tend to zero produce a convergent series — this rule is only true for geometric progressions. E.g. the series $1 + \frac{1}{2} + \frac{1}{3} + \frac{1}{4} + \frac{1}{5} + ...$ diverges.

Example 1

Determine whether or not the following sequences are convergent.

a) 1, 2, 4, 8, 16, ...

$$r = \frac{2^{nd} \text{ term}}{1^{st} \text{ term}} = \frac{2}{1} = 2$$

$|r| = |2| = 2 > 1$, so the series is not convergent.

Both sequences are geometric so all you need to do is work out if r (the common ratio) is between -1 and 1.

b) 81, -27, 9, -3, 1, ...

$$r = \frac{2^{nd} \text{ term}}{1^{st} \text{ term}} = \frac{-27}{81} = -\frac{1}{3}$$

$|r| = \left|-\frac{1}{3}\right| = \frac{1}{3} < 1$, so the series is convergent.

Tip You can usually spot straight away if a geometric series is convergent — the terms will be getting closer and closer to zero. But you still need to check $|r| < 1$ to prove it.

Summing to infinity

When a series is **convergent** you can find its **sum to infinity**. The sum to infinity is called S_∞ — it's the **limit** of S_n as $n \to \infty$. This just means that the sum of the first n terms of the series (S_n) gets closer and closer to S_∞ the more terms you add (the bigger n gets). In other words, it's the **number** that the **series converges to**.

Example 2

If $a = 2$ and $r = \frac{1}{2}$, find the sum to infinity of the geometric series.

$u_1 = 2$ → $S_1 = 2$

$u_2 = 2 \times \frac{1}{2} = 1$ → $S_2 = 2 + 1 = 3$

$u_3 = 1 \times \frac{1}{2} = \frac{1}{2}$ → $S_3 = 2 + 1 + \frac{1}{2} = 3\frac{1}{2}$

$u_4 = \frac{1}{2} \times \frac{1}{2} = \frac{1}{4}$ → $S_4 = 2 + 1 + \frac{1}{2} + \frac{1}{4} = 3\frac{3}{4}$

$u_5 = \frac{1}{4} \times \frac{1}{2} = \frac{1}{8}$ → $S_5 = 2 + 1 + \frac{1}{2} + \frac{1}{4} + \frac{1}{8} = 3\frac{7}{8}$

The sums are getting closer (converging) to 4.
So the sum to infinity is 4.

Work out the nth term and the sum to n terms for each value of n. The nth terms (and so the amounts you add on) get smaller each time.

By the 4th and 5th values you should be able to see that this sequence will never quite reach 4. You're only adding on half of the remaining distance to 4 each time.

You can show series **graphically**. This is the graph for the series in Example 2 — it is getting **closer and closer** to 4, but it'll never actually get there.

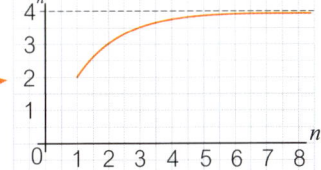

Luckily you don't have to find a list of sums like this to get the sum to infinity — there's a **formula** you use to work out the **sum to infinity** of a geometric series.

- The sum of the **first n terms** of a geometric series is $S_n = \dfrac{a(1-r^n)}{1-r}$.

- If $|r| < 1$ and n is very, very big, then r^n will be very, very **small**, i.e. $r^n \to 0$ as $n \to \infty$.

- This means $(1 - r^n)$ will get really **close to 1**, so $(1 - r^n) \to 1$ as $n \to \infty$.

- Putting this back into the sum formula gives $S_n \to \dfrac{a \times 1}{1-r} = \dfrac{a}{1-r}$ as $n \to \infty$.

- So: $\boxed{S_\infty = \dfrac{a}{1-r}}$

> **Tip**
> The sum of the first n terms formula is on p.160. Both of these formulas are given on the formula sheet.

Example 3

> **Tip**
> This is the same series as Example 2.

a) If $a = 2$ and $r = \dfrac{1}{2}$, find the sum to infinity of the geometric series.

$$S_\infty = \frac{a}{1-r}$$
$$= \frac{2}{1-\frac{1}{2}} = \frac{2}{\frac{1}{2}} = 4$$

$|r| = \left|\dfrac{1}{2}\right| = \dfrac{1}{2} < 1$, so the series converges and you can find its sum to infinity using the formula for S_∞.

b) Find the sum to infinity of the geometric series $8 + 2 + 0.5 + 0.125 + ...$

$$a = 8, \ r = \frac{2^{nd}\text{ term}}{1^{st}\text{ term}} = \frac{2}{8} = 0.25$$

First find the values of a (the first term) and r (the common ratio).

$$S_\infty = \frac{a}{1-r} = \frac{8}{1-0.25} = \frac{32}{3} = 10\frac{2}{3}$$

Again, $|r| < 1$, so the series converges and so you can find the sum to infinity.

Divergent series don't have a sum to infinity.

Because the terms aren't tending to zero, the size of the sum will just keep increasing as you add more terms, so there is **no limit** to the sum.

Exercise 5.8

Q1 State which of these sequences will converge and which will not.

a) 1, 1.1, 1.21, 1.331, ...

b) 0.8, 0.8^2, 0.8^3, ...

c) 1, $\dfrac{1}{4}$, $\dfrac{1}{16}$, $\dfrac{1}{64}$, ...

d) 3, $\dfrac{9}{2}$, $\dfrac{27}{4}$, ...

e) 1, $-\dfrac{1}{2}$, $\dfrac{1}{4}$, $-\dfrac{1}{8}$, $\dfrac{1}{16}$, ...

f) 5, 5, 5, 5, 5, ...

> **Q1 Hint**
> If $r = 1$, the sequence is just the same term repeated, so it diverges. If $r = -1$, the sequence alternates between two terms forever.

Q2 A geometric series is $9 + 8.1 + 7.29 + ...$
Calculate the sum to infinity.

Q3 A geometric series has first term $a = 33$, common ratio $r = 0.25$. Find $\displaystyle\sum_{k=0}^{\infty} ar^k$ for this series.

E **Q4** A geometric series is given as $256 + 64 + 16 + 4 + 1 + ...$

    a) Find the 10th and 12th terms as fractions. *[3 marks]*

    b) Find the sum to infinity as a fraction. *[2 marks]*

Q5 $a, ar, ar^2, ...$ is a geometric sequence. Given that $S_\infty = 2a$, find r.

Q6 The sum to infinity of a geometric progression is 13.5 and the first 3 terms add up to 13.

    a) Find the common ratio r b) Find the first term a.

P **Q7** $a + ar + ar^2 + ...$ is a geometric series. $ra = 3$ and $S_\infty = 12$. Find r and a.

Q8 The sum to infinity of a geometric series is 10 and the first term is 6.

    a) Find the common ratio.   b) What is the 5th term?

Q9 The 2nd term of a geometric progression is -48 and the 5th term is 0.75. Find:

    a) the common ratio   b) the first term   c) the sum to infinity

E **P** **Q10** The first term of a geometric series is q.

    Each subsequent term is 15% less than the previous term.

    a) Explain why this geometric series is convergent. *[2 marks]*

    b) The limit of this series, L, is such that $30\,000 < L < 35\,000$.
        Find the possible values of q. *[2 marks]*

P **Q11** The sum of the terms after the 10th term of a convergent geometric series is less than 1% of the sum to infinity. The first term is positive. Show that the common ratio $|r| < 0.631$.

P **Q12** In a convergent geometric series $S_\infty = \dfrac{9}{8} \times S_4$.

    Find the value of r, given that r is positive and real.

E **P** **Q13** A geometric series with common ratio $2q$, where $q > 0$, has n^{th} term $u_n = p(2q)^n$.

    a) Find the 1st and 10th terms, giving your answers in terms of p and q. *[2 marks]*

    b) State the values of q for which the sum to infinity exists. *[2 marks]*

    c) In the case when $q = \dfrac{1}{8}$ find S_∞ in terms of p. *[2 marks]*

Challenge

E **P** **Q14** The common ratio, r, of a geometric series is $1 - \dfrac{a}{100}$, where a is the first term.

    a) Assuming the series is convergent with limit 100, find:

        (i) the value of the common ratio r, *[2 marks]*

        (ii) the range of possible values for a. *[2 marks]*

    b) Assuming instead that the sequence of terms in the series is periodic, with period 2, find a and state, with a reason, whether the series could converge. *[3 marks]*

E Exam Style   **P** Problem Solving   **M** Modelling

5.9   Modelling Problems

Modelling problems might not say the words sequence, series, arithmetic or geometric in the question — you have to decide what you're dealing with. Look for questions that have a time period (e.g. each year) and describe how values increase or decrease over that time period.

Example 1

Mo is training for a 10 km running race. On the first day he runs 2 km. He schedules his training to increase by 0.5 km each day, so that he runs 2.5 km on the second day and 3 km on the third day and so on. This continues until he reaches the maximum distance of 10 km on the seventeenth day. The distance he runs each day then remains at 10 km until the race. There are 20 days before the race. What is the total distance Mo will run in training?

$a = 2, l = 10, n = 17$ and $d = 0.5$ ← The question describes the sequence 2, 2.5, 3, ... , 10. It's arithmetic because it increases by 0.5 each time. Write down the values of a, l, n and d given in the question.

$S_n = \frac{n}{2}(2a + (n - 1)d)$ ← You're asked for the total distance, so use one of the arithmetic series formulas and put in the numbers.

> **Tip** You could also use the formula: $S_n = \frac{1}{2}n(a + l)$.

$S_{17} = \frac{17}{2}(2 \times 2 + (17 - 1) \times 0.5)$

$= \frac{17}{2}(4 + 8) = 102$ km

$102 + (3 \times 10) = 132$ km ← There are also 3 extra days where he will run 10 km, so add this on.

Example 2

When a baby is born, £3000 is invested in an account with a fixed interest rate of 4% per year.

a) What will the account be worth at the start of the seventh year?

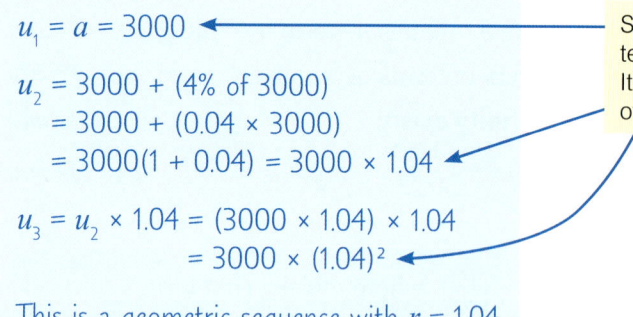

$u_1 = a = 3000$ ← Start by working out the first few terms to see what the sequence is. It'll help to leave the values in terms of the initial value, $a = 3000$.

$u_2 = 3000 + (4\% \text{ of } 3000)$
$= 3000 + (0.04 \times 3000)$
$= 3000(1 + 0.04) = 3000 \times 1.04$

$u_3 = u_2 \times 1.04 = (3000 \times 1.04) \times 1.04$
$= 3000 \times (1.04)^2$

> **Tip** You might recognise that the sequence is geometric with $r = 1.04$ straight away, if you know that a 4% interest rate means you multiply by 1.04.

This is a geometric sequence with $r = 1.04$

$u_n = ar^{n-1} = 3000 \times (1.04)^{n-1}$ ← Write down the nth term of the sequence.

$u_7 = ar^6 = 3000 \times (1.04)^6 = 3795.957...$ ← The 1st term is the value at the start of the first year, so the value at the start of the seventh year is the 7th term ($n = 7$).

So it's £3795.96 (to the nearest penny).

b) After how many full years will the account have doubled in value?

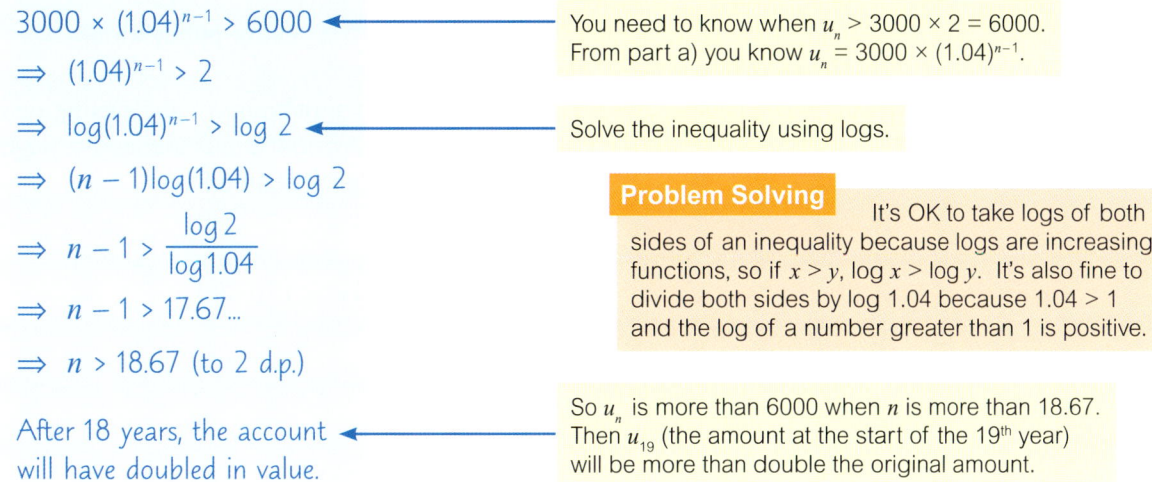

$3000 \times (1.04)^{n-1} > 6000$ ← You need to know when $u_n > 3000 \times 2 = 6000$. From part a) you know $u_n = 3000 \times (1.04)^{n-1}$.

$\Rightarrow (1.04)^{n-1} > 2$

$\Rightarrow \log(1.04)^{n-1} > \log 2$ ← Solve the inequality using logs.

$\Rightarrow (n-1)\log(1.04) > \log 2$

$\Rightarrow n - 1 > \dfrac{\log 2}{\log 1.04}$

Problem Solving It's OK to take logs of both sides of an inequality because logs are increasing functions, so if $x > y$, $\log x > \log y$. It's also fine to divide both sides by $\log 1.04$ because $1.04 > 1$ and the log of a number greater than 1 is positive.

$\Rightarrow n - 1 > 17.67...$

$\Rightarrow n > 18.67$ (to 2 d.p.)

After 18 years, the account ← So u_n is more than 6000 when n is more than 18.67. Then u_{19} (the amount at the start of the 19th year) will have doubled in value.   will be more than double the original amount.

If you have to give an answer in years, make sure you think through which term belongs to which year carefully — it's easy to give the wrong answer even if you've done all the maths right.

Exercise 5.9

M **Q1** Morag starts a new job. In the first week she is paid £60, but this rises by £3 per week, so she earns £63 in the second week and £66 in the third week. How much does she earn in her 12th week?

M **Q2** A collector has 8 china dolls that fit inside each other. The smallest doll is 3 cm high and each doll is 25% taller than the previous one. If he lines them up in height order (shortest to tallest), how tall is the 8th doll?

M **Q3** Mario opens a sandwich shop. On the first day he sells 40 sandwiches. As people hear about the shop, sales increase and on the second day he sells 45 sandwiches. Daily sales rise in an arithmetic sequence. On which day will he sell 80 sandwiches?

P **Q4** A retro cassette player is launched into the market.
M In the first month after launch, the product takes £300 000 of revenue. It takes £270 000 in the second month and £240 000 in the third. If this pattern continues, when would you expect monthly sales to fall below £50 000?

M **Q5** A car depreciates by 15% each year. The value of the car after each year forms a geometric sequence. After 10 years from new the car is valued at £2362. How much was the car when new?

Q5 Hint 'Depreciates' means 'decreases in value'.

M **Q6** A fishing licence cost £120 in 2011. The cost rose 3% each year for the next 5 years.
a) How much was a fishing licence in 2012?
b) Nigel bought a fishing licence every year between 2011 and 2016 (including 2011 and 2016). How much in total did he spend?

Ⓜ Q7 "Cornflake Collector" magazine sells 6000 copies in its first month of publication, 8000 in its second month and 10 000 in its third month.
If this pattern continues, how many copies will it sell in the first year of publication?

Ⓟ Ⓜ Q8 Reina is growing her prize leeks for the Village Show. The bag of compost she's using on her leeks says it will increase their height by 15% every 2 days. After 4 weeks the leeks' height has increased from 5 cm to 25 cm. Has the compost done what it claimed?

Ⓜ Q9 It's predicted that garden gnome value will go up by 2% each year, forming a geometric progression. Jean-Claude has a garden gnome currently valued at £80 000.
If the predicted rate of inflation is correct:
a) What will Jean-Claude's gnome be worth after 1 year?
b) What is the common ratio of the geometric progression?
c) What will Jean-Claude's gnome be worth after 10 years?
d) It will take k years for the value of Jean-Claude's gnome to exceed £120 000. Find k.

Ⓜ Q10 Frazer draws one dot in the first square of his calendar for July, two dots in the second square, and so on up to 31 dots in the last day of the month. How many dots does he draw in total?

Ⓟ Ⓜ Q11 The thickness of a piece of paper is 0.01 cm. The Moon is 384 000 km from the Earth. The piece of paper is on the Earth. Assuming you can fold the piece of paper as many times as you like, how many times would you have to fold it for it to reach the Moon?

> **Q11 Hint**
> Have a go — it really works!*
> *Disclaimer: it gets a bit tricky after 7 folds...

Ⓜ Q12 Allanah is training for an athletics event. Each day she runs 3% farther than the day before. On day 1 she runs 12 miles.
a) How far does she run on day 10?
b) Her training schedule lasts for 20 days.
To the nearest mile, how far does she run altogether?

Ⓟ Ⓜ Q13 Laura puts 1p in her jar on the first day, 2p in on the second day, 3p in on the third day, etc. How many days will it take her to collect over £10?

> **Q14 Hint**
> Call the initial value of her savings a.

Ⓟ Ⓜ Q14 Chardonnay wants to invest her savings for the next 10 years. She wants her investment to double during this time. If interest is added annually, what interest rate does she need?

Challenge

Ⓔ Ⓟ Ⓜ Q15 Water starts leaking from a mains pipe on 4th January. The pipe leaks 2.3 litres of water on the first day, and the amount of water leaking from the pipe increases by 3.2 litres per day. An urgent repair is deemed necessary either when the daily leakage exceeds 80 litres, or when the total amount of leaked water exceeds 1000 litres — whichever comes first.

On what date will an urgent repair be deemed necessary? *[7 marks]*

5   Review Exercise

Q1   Find the 8th term of the sequence $x_n = n^2 - 3$. Is this sequence increasing or decreasing?

E **Q2**   The nth term of a sequence is given by $u_n = an + b$, where a and b are constants.
  a) Find, in terms of a and b, $u_{n+1} - u_n$.   *[2 marks]*
  b) (i) What conditions should be imposed on a and/or b
       to ensure the sequence is increasing? Justify your answer.   *[2 marks]*
     (ii) Write down any conditions on a and/or b
       in order for the sequence to be decreasing.   *[1 mark]*

Q3   In the sequence $u_n = n^2 + 3n + 4$, $u_k = 44$. Find the value of k.

P **Q4**   A sequence has the form $u_n = an^2 + bn$, where a and b are constants.
       If the 3rd term is 18, and the 7th term is 70, find the values of a and b.

Q5   Find the first 5 terms of the sequence in which:
  a) $x_1 = 7$ and $x_{n+1} = x_n + 3$ for $n \geq 1$.   b) $u_1 = 2$ and $u_{n+1} = 6 \div u_n$ for $n \geq 1$.

P **Q6**   Describe a recurrence relation which generates the sequence:
  a) 65 536, 256, 16, 4, 2, ...   b) 40, 38, 34, 28, 20, ...   c) 1, 1, 2, 3, 5, 8, ...

P **Q7**   A sequence is generated by a recurrence relation of the form $u_{n+1} = ku_n + 3$,
       where k is a constant. If $u_1 = 4$ and $u_2 = 11$, find the values of k, u_3 and u_4.

Q8   A sequence is generated by $u_1 = 8$ and $u_{n+1} = 18 - u_n$ for $n \geq 1$. Find the first 5 terms and a
       formula for u_n in terms of n. State whether the sequence is increasing, decreasing or periodic.

E **Q9**   A teacher creates a sequence by using the Fibonacci recurrence
       relation $u_{n+2} = u_{n+1} + u_n$, but she changes the initial two terms (such that $u_1 \neq 1$ and $u_2 \neq 1$).

       Three consecutive terms in the teacher's Fibonacci sequence are such that
       the third of these terms is triple the first, and the difference between
       the first two of these terms is 10. Find these three terms.   *[4 marks]*

E **Q10**   A series has recurrence relation:

$$u_{n+2} = \frac{u_{n+1} + 1}{u_n}, \text{ with } u_1 = 3 \text{ and } u_2 = 5.$$

  a) Find u_6 and u_7.   *[2 marks]*
  b) Find the sum of the first 25 terms.   *[3 marks]*

Q11   Find the common difference in a sequence that starts with –2, ends with 19 and has 29 terms.

E **M** **Q12** A new business models the number of customers they will have each month after the business launches using the sequence $u_n = 20n - 4$, where n represents the month.

    a) How many customers does the model predict the business to have in month 9?   *[1 mark]*

    b) The business becomes profitable the first time the number of customers in a month exceeds 850. According to the model, in which month will the business become profitable?   *[3 marks]*

    c) State a problem with this model for large values of n.   *[1 mark]*

Q13 In an arithmetic series, $u_7 = 8$ and $u_{11} = 10$. Find u_3.

Q14 An arithmetic series has seventh term 36 and tenth term 30. Find the nth term and S_5.

Q15 Find $\sum\limits_{n=1}^{20}(3n - 1)$.

E **P** **Q16** The 6th term of an arithmetic series is p and the 12th term is $4p$.

    a) Write an expression for the common difference, d, in terms of p.   *[3 marks]*

    b) The sum of the first 6 terms is 72 and the sum of the first 12 terms is 288. Find the first 3 terms of the series.   *[4 marks]*

E **P** **Q17** An arithmetic series has first term a, where $a > 0$, and common difference $-b$, where $b > 0$. The fourteenth term is -4 and is the first negative term.

    a) Show that $a + 4 = 13b$.   *[2 marks]*

    b) The sum of the first five terms is seven times the sixth term. Find the values of a and b.   *[5 marks]*

    c) Find the number of terms needed such that the sum of the series is zero.   *[3 marks]*

Q18 Find:

    a) the sum of the first 24 natural numbers,   b) $\sum\limits_{n=13}^{24} n$ c) the value of k if $\sum\limits_{n=1}^{k} n = 630$.

Q19 Write an expression for the nth term of the geometric sequence 3, -9, 27, -81, 243, ...

Q20 For the geometric progression 2, -6, 18, ..., find:

    a) the 10th term,   b) the sum of the first 10 terms.

Q21 A geometric series has first term $a = 7$ and common ratio $r = 0.6$. Find $\sum\limits_{k=0}^{5} 7(0.6)^k$ to 2 d.p.

P **Q22** Show that the sum of the first n terms of a geometric sequence with first term a and common ratio r is: $S_n = \dfrac{a(1 - r^n)}{1 - r}$.

Q23 For the geometric progression 24, 12, 6, ..., find:

    a) the common ratio,   b) the seventh term,

    c) the sum of the first 10 terms,   d) the sum to infinity.

E P Q24 A geometric series has common ratio r, 2nd term 6 and $S_\infty = 8k$.

a) Show that $4kr^2 - 4kr + 3 = 0$. *[3 marks]*

b) Given that there is only one real value of r and $k \neq 0$, find the value of k and the first three terms in the series. *[5 marks]*

E P Q25 The third and sixth terms of a convergent geometric series are $u_3 = \frac{1}{72}$ and $u_6 = \frac{1}{1944}$.

a) Find the common ratio of the series. *[3 marks]*

The first term of the sequence is $u_1 = \frac{1}{8}$.

b) Find the lowest value of n for which the difference between two consecutive terms, u_n and u_{n+1}, is less than 10^{-6}. *[5 marks]*

M Q26 A new shop takes £300 on its first day of business, £315 on its second day and £330 on its third day. If this pattern continues, on which day will the shop first take over £500?

P M Q27 Hussain puts one stone on the wall on his way to school on the first day of term, two stones on the wall on the second day of term and three on the third day of term. The term has 13 weeks of 5 days each. If he continues like this, how many stones will he have put on the wall in total by the holiday?

E M Q28 Samira is training for a triathlon. She plans her training regime as follows:

> Run: Start with 30 minute run and increase by 5 minutes per day.
>
> Cycle: Start with 15 minute cycle then increase by 12% per day.
>
> Swim: Swim for 1 hour each day.

a) How long does Samira exercise for in total on day 3, to the nearest minute? *[3 marks]*

b) On which day does the duration of the cycle and run both individually exceed the duration of the swim for the first time? *[4 marks]*

c) For how long in total does Samira cycle in her second week of training? Give your answer in hours and minutes to the nearest minute. *[3 marks]*

d) Explain one restriction of the models you have used in parts a)-c). *[1 mark]*

Challenge

E P Q29 A geometric series is convergent.
The second term of the series is 2 and the series converges to 9.
Find all possible combinations of values for the first term and common ratio. *[6 marks]*

E P Q30 The 10th term of an arithmetic sequence is double the 4th term, which is non-zero.

a) Show that the first term is three times bigger than the common difference. *[3 marks]*

b) Show that the nth term can be written as $u_n = \frac{1}{3}a(n+2)$. *[2 marks]*

c) Briefly explain why the first term of the sequence cannot be zero. *[1 mark]*

d) k and m are such that $u_m = 2u_k$.
Determine the smallest possible values of k and m. *[2 marks]*

E Exam Style   **P** Problem Solving   **M** Modelling

Challenge

E **Q31** a) Write down the first four terms of the sequence given by $u_n = (-1)^n \times n$.
P State whether this sequence is increasing, decreasing, periodic or none of these. *[2 marks]*

b) Find the first four terms of the sequence given by $v_n = S_n$, where S_n is the sum of the first n terms of the sequence described in part a). *[1 mark]*

c) (i) Write down the value of v_{32}. Justify your answer. *[2 marks]*

(ii) Write down a formula, in terms of n, for v_n, when n is odd. *[1 mark]*

P **Q32** A sequence of concentric circles (circles with the same centre) are sized such that the radius of each circle is 50% larger than the radius of the preceding circle.

a) If the innermost circle has a radius of 2 cm, find the diameter of the tenth circle. Give your answer to the nearest centimetre.

b) (i) Find the position in the sequence of the circle that has an area of 725 cm^2 — this area has been rounded to three significant figures.

(ii) Why is it important to know that the area of 725 cm^2 has been rounded?

c) What is the common ratio of the areas of the circles?

P **Q33** A company building a new sports stadium set up a time-lapse camera so the public can view
M the construction progress online. The camera is switched on at midnight on the first day of construction and runs continuously until the stadium is complete. The nth photo is taken u_n hours after the camera is switched on according to the sequence $u_n = 6n - 2$.

a) By considering the first few terms of the sequence, explain how the camera will take a photo at 4 am, 10 am, 4 pm and 10 pm during each day of construction. Assume the construction is in a country that does not use daylight saving time.

The construction of the stadium is expected to take 5 years to complete.
Upon completion, the photos will be made into a video.
The number of photos used per second of video will follow the sequence $p_n = 2n + 6$,
where p_n is the number of photos in the nth second of the video — this will
create the effect of the construction speeding up as the video progresses.

b) How many seconds long, to the nearest second, will the video be?

E **Q34** In a playground, a series of stepping stones are built such that each stone is 10% higher
P than the previous one. The first stone stands at a height of 10 cm above the level ground.
M
a) Write a formula for the height, in cm, of the nth stepping stone. *[2 marks]*

b) The local health and safety team have decided that the tallest stone in the playground should not be any higher than 0.5 m above the ground. What is the maximum number of stepping stones permitted in the playground? *[3 marks]*

Each stone is made from concrete formed in the shape of a cylinder with a radius of 20 cm, and each stone goes 20 cm under the ground. The concrete company manufacturing the stones needs to know the total volume of concrete required.

c) Given that the playground will have a total of 8 stepping stones, find the total volume of concrete required. Give your answer in cubic metres to three significant figures. *[3 marks]*

5 Chapter Summary

1 a_n is special notation for the nth term of a sequence and a_{n+1} is the term after it.

2 You can use an nth term formula to find the term in any position of a sequence.
E.g. if the nth term formula is $a_n = 4n + 2$, the 100th term is $a_{100} = 4 \times 100 + 2 = 402$.

3 In an increasing sequence each term is greater than the previous term — $a_{k+1} > a_k$.
In a decreasing sequence each term is less than the previous term — $a_{k+1} < a_k$.

4 In periodic sequences, terms repeat — the number of terms in each repeated part is known as the order.

5 A recurrence relation tells you how to work out one term in a sequence from the previous term, e.g. $a_{k+1} = 7a_k + 4$. You'll also need to know one term to get started, e.g. $a_1 = 10$.

6 Arithmetic sequences (or arithmetic progressions) have a common difference between consecutive terms. Their n^{th} term is given by $u_n = a + (n-1)d$, where a is the first term, d is the common difference and n is the position in the sequence.

7 Adding up the terms of an arithmetic sequence gives an arithmetic series.
You can find the sum of the first n terms (S_n) using either $S_n = \frac{n}{2}[2a + (n-1)d]$ or $S_n = \frac{1}{2}n(a+l)$, where a and d are defined as above and l is the last term.

8 You can write series using sigma notation (e.g. $\sum_{n=1}^{10} 5n$). The numbers below and above Σ are the first and last values of n respectively and the bit after is the nth term expression.

9 The sum of the first n natural numbers is $S_n = \frac{1}{2}n(n+1)$
— this is a specific case of the formulas in point 7.

10 Geometric sequences (or geometric progressions) have a common ratio between consecutive terms. Their n^{th} term is given by $u_n = ar^{n-1}$ where a is the first term, r is the common ratio and n is the position in the sequence.

11 Adding up the terms of a geometric sequence gives a geometric series.
You can find the sum of the first n terms (S_n) using $S_n = \sum_{k=0}^{n-1} ar^k = \frac{a(1-r^n)}{1-r}$, where a is the first term and r is the common ratio.

12 Sequences that tend towards a limit are called convergent sequences.
Geometric sequences either converge to 0 or don't converge at all (they're divergent).
A geometric sequence will converge to 0 when the common ratio, r, satisfies $|r| < 1$.

13 When you sum a convergent geometric sequence you get a convergent series.
Geometric series can converge to any limit. This limit is called the sum to infinity, S_∞, and is found using the formula $S_\infty = \frac{a}{1-r}$. Divergent series do not have a limit.

The Binomial Expansion

Learning Objectives

Once you've completed this chapter, you should be able to:

- Expand $(p + qx)^n$ for any rational n using the general formula for binomial expansions.
- State the validity of the expansion — i.e. the values of x for which the expansion is valid.
- Substitute values into a binomial expansion in order to find approximations.
- Split rational functions into partial fractions,
  then find the binomial expansion of the function.

Prior Knowledge Check

1 Expand $(1 + x)^4$ using the binomial formula.

see Year 1 Pure Maths

2 Write the following as partial fractions:

a) $\dfrac{10}{(x-2)(x+3)}$ b) $\dfrac{11x + 29}{(x+1)(x+2)(x-5)}$ c) $\dfrac{4}{(x+1)^2(x+3)}$

see pages 53-57

6.1   Binomial Expansions where n is a Positive Integer

The **binomial expansion** is a way to raise a given expression to any power.
For simpler cases it's basically a fancy way of multiplying out brackets.
You can also use it to approximate more complicated expressions.

In Year 1, you met the formula for the binomial expansion of $(1 + x)^n$:

$$(1 + x)^n = 1 + nx + \frac{n(n-1)}{1 \times 2}x^2 + \dots + \frac{n(n-1)\dots(n-r+1)}{1 \times 2 \times \dots \times r}x^r + \dots$$

From the formula, it looks like the expansion always goes on forever.
But if n **is a positive integer**, the binomial expansion is **finite**.

Example 1

Give the binomial expansion of $(1 + x)^5$.

$(1 + x)^5 = 1 + 5x + \dfrac{5(5-1)}{1 \times 2}x^2 + \dfrac{5(5-1)(5-2)}{1 \times 2 \times 3}x^3$

$\quad + \dfrac{5(5-1)(5-2)(5-3)}{1 \times 2 \times 3 \times 4}x^4$

$\quad + \dfrac{5(5-1)(5-2)(5-3)(5-4)}{1 \times 2 \times 3 \times 4 \times 5}x^5$

$\quad + \dfrac{5(5-1)(5-2)(5-3)(5-4)(5-5)}{1 \times 2 \times 3 \times 4 \times 5 \times 6}x^6 + \dots$

> Use the formula and put in $n = 5$. E.g. the second term is $nx = 5x$, and the numerator in the third term is $n(n-1) = 5(5-1)$.

> You can stop here — the x^6 term is zero, and so are all the remaining terms.

$= 1 + 5x + \dfrac{5 \times 4}{1 \times 2}x^2 + \dfrac{5 \times 4 \times 3}{1 \times 2 \times 3}x^3 + \dfrac{5 \times 4 \times 3 \times 2}{1 \times 2 \times 3 \times 4}x^4$

$\quad + \dfrac{5 \times 4 \times 3 \times 2 \times 1}{1 \times 2 \times 3 \times 4 \times 5}x^5 + \dfrac{5 \times 4 \times 3 \times 2 \times 1 \times 0}{1 \times 2 \times 3 \times 4 \times 5 \times 6}x^6$

$= 1 + 5x + \dfrac{20}{2}x^2 + \dfrac{60}{6}x^3 + \dfrac{120}{24}x^4$

$\quad + \dfrac{120}{120}x^5 + \dfrac{0}{720}x^6$

$= 1 + 5x + 10x^2 + 10x^3 + 5x^4 + x^5$

> **Problem Solving**
>
> The expansion is finite because at some point you introduce an $(n - n)$ (i.e. zero) term in the numerator which then appears in every coefficient from that point on, making them all zero.

The formula still works if the coefficient of x **isn't 1**, i.e. $(1 + ax)^n$ — just **replace** each 'x' in the formula with (ax). The 'a' should be raised to the **same power** as the 'x' in each term, and included in the coefficient when you simplify at the end. The example on the next page shows you how it's done.

Example 2

Give the binomial expansion of $(1 - 3x)^4$.

$(1 - 3x)^4 = 1 + 4(-3x) + \dfrac{4 \times 3}{1 \times 2}(-3x)^2 + \dfrac{4 \times 3 \times 2}{1 \times 2 \times 3}(-3x)^3$

$+ \dfrac{4 \times 3 \times 2 \times 1}{1 \times 2 \times 3 \times 4}(-3x)^4$

$+ \dfrac{4 \times 3 \times 2 \times 1 \times 0}{1 \times 2 \times 3 \times 4 \times 5}(-3x)^5 + \ldots$

$= 1 + 4(-3x) + \dfrac{12}{2}(9x^2)$

$+ \dfrac{4}{1}(-27x^3) + (81x^4) + \dfrac{0}{5}(-243x^5)$

$= 1 - 12x + 54x^2 - 108x^3 + 81x^4$

> **Use the formula** with $n = 4$, but replace every x with $-3x$.
>
> It's useful to cancel down the fractions before you multiply.
>
> Stop here.
>
> Don't forget to square the -3 as well — i.e. $(-3x)^2 = 9x^2$.

> **Tip** Think of this as $(1 + (-3x))^4$ — put the minus into the formula as well as the $3x$.

Validity

Some binomial expansions are **only valid for certain values of** x. When you find a binomial expansion, you usually have to state which values of x the expansion is valid for.

> If n is a **positive integer**, the binomial expansion of $(p + qx)^n$ is valid for **all values of** x.

So far you've only dealt with expansions of $(p + qx)^n$ where p is 1, but there are more complicated examples to come on page 184. There's more on the validity of other expansions on page 180.

Exercise 6.1

Use the formula to expand the following functions in ascending powers of x.

Q1   Expand fully: $(1 + x)^3$

Q2   Expand $(1 + x)^7$ up to and including the term in x^3.

Q3   Expand fully: $(1 - x)^4$

Q4   Give the first 3 terms of $(1 + 3x)^6$.

Q5   Give the first 4 terms of $(1 + 2x)^8$.

Q6   Expand $(1 - 5x)^5$ up to and including the term in x^2.

Q7   Expand fully: $(1 - 4x)^3$

Q8   Expand $(1 + 6x)^6$ up to and including the term in x^3.

E Q9   Expand fully $\left(1 + \dfrac{2}{3}x\right)^4$.

> **Q3 Hint** Watch out for that minus sign — replace 'x' in the formula with $(-x)$.

[2 marks]

Challenge

E **Q10** The coefficient of the x^3 term in the expansion of $(1 + 2kx)^5$ is equal to
P the coefficient of the x^5 term in the expansion of $(1 + kx)^8$.
Find the possible values of k. *[4 marks]*

E **Q11** a) Find the first three terms in the expansion of $(1 + kx)^n$. *[2 marks]*
P b) Given that the coefficient of the x^2 term is 12 times larger than the coefficient
of the x term, that $k \neq 0$, and that $n = 3k$, find the possible values of n. *[3 marks]*

E **Q12** a) Show that the expansion of $(1 + ax)^n - (1 - ax)^n$, where a and n are
P positive integers, will consist of terms involving odd powers of x only. *[2 marks]*
b) Show further that if x is an integer, then the expansion is always even. *[2 marks]*

6.2 Binomial Expansions where n is Negative or a Fraction

n is negative

If n is **negative**, the expansion gets more complicated. You can still use the **formula** in the same way,
but it will produce an **infinite** number of terms (see the example below). You can just write down the
first few terms in the series, but this will only be an **approximation** to the whole expansion.
The question will usually tell you how many terms to give.

This type of expansion can be 'hidden' as a fraction — remember: $\dfrac{1}{(1+x)^a} = (1+x)^{-a}$

Example 1

Find the binomial expansion of $\dfrac{1}{(1+x)^2}$ up to and including the term in x^3.

$\dfrac{1}{(1+x)^2} = (1+x)^{-2}$ ← First, rewrite the expression.

$(1+x)^{-2} = 1 + (-2)x + \dfrac{(-2) \times (-2-1)}{1 \times 2}x^2$ ← Now you can use the formula
for $(1+x)^n$. This time $n = -2$.

$\qquad + \dfrac{(-2) \times (-2-1) \times (-2-2)}{1 \times 2 \times 3}x^3 + \ldots$

Tip Again, you can cancel down before you
multiply — but be careful with those minus signs.

$= 1 + (-2)x + \dfrac{(-2) \times (-3)}{1 \times 2}x^2$

$\qquad + \dfrac{(-2) \times (-3) \times (-4)}{1 \times 2 \times 3}x^3 + \ldots$ ← With a negative n, you'll never get zero
as a coefficient. The question tells you
to stop at the x^3 term, so stop here.

$= 1 + (-2)x + \dfrac{3}{1}x^2 + \dfrac{(-4)}{1}x^3 + \ldots$

$= 1 - 2x + 3x^2 - 4x^3 + \ldots$ ← All the terms after $-4x^3$ have been left out, so
the cubic expression we've ended up with is

$\dfrac{1}{(1+x)^2} \approx 1 - 2x + 3x^2 - 4x^3$ ← an approximation to the original expression.
You can write the answer like this.

n is a fraction

The binomial expansion formula doesn't just work for integer values of *n*.
If *n* is a **fraction**, you'll have to take care multiplying out the fractions in the coefficients,
but otherwise the formula is **exactly the same**.

Remember that **roots** are fractional powers: $\sqrt[a]{1+x} = (1+x)^{\frac{1}{a}}$

Example 2

Find the binomial expansion of $\sqrt[3]{1+2x}$, up to and including the term in x^3.

$$\sqrt[3]{1+2x} = (1+2x)^{\frac{1}{3}}$$

First rewrite the expression as a fractional power.

$$(1+2x)^{\frac{1}{3}} = 1 + \frac{1}{3}(2x) + \frac{\frac{1}{3} \times \left(\frac{1}{3} - 1\right)}{1 \times 2}(2x)^2$$

This time $n = \frac{1}{3}$, and you also need to replace x with $2x$.

$$+ \frac{\frac{1}{3} \times \left(\frac{1}{3} - 1\right) \times \left(\frac{1}{3} - 2\right)}{1 \times 2 \times 3}(2x)^3 + \ldots$$

$$= 1 + \frac{2}{3}x + \frac{\frac{1}{3} \times \left(-\frac{2}{3}\right)}{1 \times 2}(4x^2)$$

Tip   Cancelling down is much trickier with this type of expansion — it's often safer to multiply everything out fully.

$$+ \frac{\frac{1}{3} \times \left(-\frac{2}{3}\right) \times \left(-\frac{5}{3}\right)}{1 \times 2 \times 3}(8x^3) + \ldots$$

$$= 1 + \frac{2}{3}x + \frac{\left(-\frac{2}{9}\right)}{2}(4x^2) + \frac{\left(\frac{10}{27}\right)}{6}(8x^3) + \ldots$$

Tip   Exam questions often ask for the coefficients as simplified fractions.

$$= 1 + \frac{2}{3}x + \left(-\frac{1}{9}\right)(4x^2) + \left(\frac{5}{81}\right)(8x^3) + \ldots$$

$$= 1 + \frac{2}{3}x - \frac{4}{9}x^2 + \frac{40}{81}x^3 - \ldots$$

Validity when *n* is negative or a fraction

Binomial expansions where *n* is negative or a fraction are **not valid** for **all** values of *x*.
The **rule** to work out the validity of an expansion is as follows:

> If *n* is a negative integer or a fraction, the binomial expansion
> of $(p + qx)^n$ is **valid** when $\left|\frac{qx}{p}\right| < 1$, i.e. when $|x| < \left|\frac{p}{q}\right|$.

This just means that the **absolute value** (or **modulus**) of *x* (i.e. ignoring any negative signs)
must be **smaller** than the absolute value of $\frac{p}{q}$ for the expansion to be valid.

You might already know the rules $|ab| = |a||b|$ and $\left|\frac{a}{b}\right| = \frac{|a|}{|b|}$.

If you don't, then get to know them — they're handy for rearranging these limits.

Example 3

State the validity of the expansions given in the previous two examples.

a) Example 1:

$$\frac{1}{(1+x)^2} = (1+x)^{-2} = (p+qx)^n \text{ for } p = 1 \text{ and } q = 1.$$

Put the expression in the form $(p + qx)^n$ to find the values of p and q.

So the expansion $(1+x)^{-2} = 1 - 2x + 3x^2 - 4x^3 + ...$

is valid when $\left|\frac{1x}{1}\right| < 1 \Rightarrow |x| < 1.$

Put p and q into the inequality $\left|\frac{qx}{p}\right| < 1$ and simplify your answer.

b) Example 2:

$$\sqrt[3]{1+2x} = (1+2x)^{\frac{1}{3}} = (p+qx)^n \text{ for } p = 1 \text{ and } q = 2.$$

So the expansion $(1+2x)^{\frac{1}{3}} = 1 + \frac{2}{3}x - \frac{4}{9}x^2 + \frac{40}{81}x^3 - ...$ is

valid when $\left|\frac{2x}{1}\right| < 1$, i.e. $|2x| < 1 \Rightarrow 2|x| < 1 \Rightarrow |x| < \frac{1}{2}$

Tip These examples use the modulus rules on the previous page. For some questions you might also need to use the rule $|-a| = |a|$.

Combinations of expansions

- You can use the binomial expansion formula for more complicated combinations of expansions — e.g. where different brackets raised to different powers are **multiplied together**. Start by dealing with the different expansions **separately**, then **multiply** the expressions together at the end.

- For situations where one bracket is being **divided** by another, change the **sign** of the **power** and **multiply** instead — e.g. $\frac{(1+x)}{(1+2x)^3} = (1+x)(1+2x)^{-3}$.

- You'll usually be asked to give the expansion to a specified number of terms, so the multiplication shouldn't get too complicated as you can ignore terms in higher powers of x.

- For the **combined** expansion to be **valid**, x must be in the valid range for **both** expansions, i.e. where they overlap. In practice this just means sticking to the **narrowest** of the valid ranges for each separate expansion.

Example 4

Write down the first three terms in the expansion of $\frac{(1+2x)^3}{(1-x)^2}$.

State the range of x for which the expansion is valid.

$$\frac{(1+2x)^3}{(1-x)^2} = (1+2x)^3(1-x)^{-2}$$

Re-write as a product of two expansions.

$$(1+2x)^3 = 1 + 3(2x) + \frac{3 \times 2}{1 \times 2}(2x)^2 + \frac{3 \times 2 \times 1}{1 \times 2 \times 3}(2x)^3$$

$$= 1 + 6x + 3(4x^2) + 8x^3 = 1 + 6x + 12x^2 + 8x^3$$

Expand each of these separately using the formula. (You only need to go up to the term in x^2, but it's better to have too many terms than too few at this stage.)

$$(1-x)^{-2} = 1 + (-2)(-x) + \frac{(-2) \times (-3)}{1 \times 2}(-x)^2 + \frac{(-2) \times (-3) \times (-4)}{1 \times 2 \times 3}(-x)^3 + ...$$

$$= 1 + 2x + 3x^2 + 4x^3 + ...$$

continued on the next page...

$(1 + 2x)^3(1 - x)^{-2} = (1 + 6x + 12x^2 + 8x^3)(1 + 2x + 3x^2 + 4x^3 + \ldots)$ ← Multiply the two expansions together.

$= 1(1 + 2x + 3x^2) + 6x(1 + 2x) + 12x^2(1) + \ldots$

$= 1 + 2x + 3x^2 + 6x + 12x^2 + 12x^2 + \ldots$

$= 1 + 8x + 27x^2 + \ldots$

Problem Solving Since you're only asked for the first three terms, ignore any terms with higher powers of x than x^2.

$(1 + 2x)^3$ is valid for all values of x, since n is a positive integer.

$(1 - x)^{-2}$ is valid if $|-x| < 1 \Rightarrow |x| < 1.$

Now find the validity of each expansion.

So the expansion $\dfrac{(1+2x)^3}{(1-x)^2} = 1 + 8x + 27x^2 + \ldots$ is valid if $|x| < 1.$

Use the narrower range of the two separate expansions.

Exercise 6.2

Q1 Use the binomial formula to find the first four terms in the expansion of $(1 + x)^{-4}$.

E **Q2** Find the first three terms in the expansion of $\left(1 - \dfrac{3}{4}x\right)^{-2}$. *[2 marks]*

Q3 a) Find the binomial expansion of $(1 - 6x)^{-3}$, up to and including the term in x^3.
b) For what values of x is this expansion valid?

Q4 Use the binomial formula to find the first three terms in the expansion of:
a) $(1 + 4x)^{\frac{1}{3}}$
b) $(1 + 4x)^{-\frac{1}{2}}$
c) For each of the expansions above, state the range of x for which the expansion is valid.

E **Q5** a) Expand $\left(1 + \dfrac{1}{2}x\right)^{\frac{2}{3}}$ up to and including the term in x^3. *[2 marks]*
b) State the values of x for which the expansion is valid. *[1 mark]*

Q6 Find the binomial expansion of the following functions, up to and including the term in x^3.
a) $\dfrac{1}{(1 - 4x)^2}$ for $|x| < \dfrac{1}{4}$
b) $\sqrt{1 + 6x}$ for $|x| < \dfrac{1}{6}$
c) $\dfrac{1}{\sqrt{1 - 3x}}$ for $|x| < \dfrac{1}{3}$
d) $\sqrt[3]{1 + \dfrac{x}{2}}$ for $|x| < 2$

Q7 a) Find the coefficient of the x^3 term in the expansion of $\dfrac{1}{(1 + 7x)^4}$.
b) For what range of x is this expansion valid?

Q7-8 Hint You don't need to bother doing the full expansion — just work out the coefficient of the term you need.

Q8 a) What is the coefficient of x^5 in the expansion of $\sqrt[4]{1 - 4x}$?
b) For what values of x is the binomial expansion of $\sqrt[4]{1 - 4x}$ valid?

E **P** Q9 The coefficient of the x^2 term in the expansion of $\left(1 - \frac{2}{5}x\right)^n$ is $\frac{1}{40}$.

    a) Find the possible values of n.   *[3 marks]*

    b) Find the values of x for which the expansion is valid.   *[1 mark]*

Q10a) Problem Solving

E **P** Q10 The binomial expansion of $(1 + ax)^n$, where a and n are non-zero real numbers, is valid for $|x| < 5$.

    n lies between 0 and 1, so n is a fraction. Use the corresponding validity rule to find a.

    a) Given that $0 < n < 1$, find the possible values of a.   *[1 mark]*

    b) Given that the coefficient of the x^2 term in the expansion of $(1 + ax)^n$ is the same as the coefficient of the x^2 term in the expansion of $(1 + ax)^{2n}$, find the value of n.   *[4 marks]*

Q11   a) Find the first three terms of the expansion of $(1 - 5x)^{\frac{1}{6}}$.

    b) Hence find the binomial expansion of $(1 + 4x)^4 (1 - 5x)^{\frac{1}{6}}$, up to and including the term in x^2.

    c) State the validity of the expansion in b).

Q12 Expand, up to and including the term in x^2, giving the range for which the expansion is valid:

    a) $(1 + x)^2(1 - 2x)^{-2}$ b) $(1 + 2x)^2(1 + 3x)^{-3}$

    c) $(1 - 7x)(1 + 2x)^{\frac{1}{7}}$ d) $(1 - x)^2(1 + 4x)^{-\frac{1}{3}}$

Q13   a) Find the first three terms of the binomial expansion of $\dfrac{(1 + 3x)^4}{(1 + x)^3}$.

    b) State the range of x for which the expansion is valid.

Challenge

E **P** Q14 The coefficient of the x^2 term in the expansion of $(1 + 2x)^2 (1 + kx)^{-\frac{1}{3}}$ is 2.

    a) Find the value of k.   *[5 marks]*

    b) State the values for which the expansion is valid.   *[1 mark]*

E Q15   a) Find the first three terms of the binomial expansion of $\dfrac{\left(1 - \frac{1}{3}x\right)^{\frac{1}{3}}}{\left(1 + \frac{2}{3}x\right)^{\frac{2}{3}}}$.   *[4 marks]*

    b) State the values of x for which the expansion is valid.   *[1 mark]*

P Q16   a) Write out the expansion of:

      (i) $(1 + ax)^4$ (all terms)   (ii) $(1 - bx)^{-3}$, up to the term in x^2

    b) Hence expand $\dfrac{(1 + ax)^4}{(1 - bx)^3}$, up to and including the term in x^2.

    c) Find the two possible pairs of values of a and b if the first three terms of the expansion of $\dfrac{(1 + ax)^4}{(1 - bx)^3}$ are $1 + x + 24x^2$.

6.3   Expanding $(p + qx)^n$

You've seen over the last few pages that the binomial expansion of $(1 + x)^n$ works for **any n**, and that you can replace the x with other x-terms.

However, the 1 at the start **has to be a 1** before you can expand. If it's **not** a 1, you'll need to **factorise first** before you can use the formula.

This means you have to rewrite the expression as follows:

$$(p + qx)^n = p^n\left(1 + \frac{qx}{p}\right)^n$$

This rearrangement uses the power law $(ab)^k = a^k b^k$
— note that the p outside the brackets is still raised to the power n.

Example 1

Give the binomial expansion of $(3 - x)^4$ and state its validity.

$$3 - x = 3\left(1 - \frac{1}{3}x\right)$$

To use the $(1 + x)^n$ formula, you need the constant term in the brackets to be 1 — you want an expression in the form $c(1 + dx)^n$, where c and d are constants, so factorise the 3 outside the brackets.

$$\Rightarrow (3 - x)^4 = \left[3\left(1 - \frac{1}{3}x\right)\right]^4 = 3^4\left(1 - \frac{1}{3}x\right)^4$$

$$= 81\left(1 - \frac{1}{3}x\right)^4$$

Tip   Don't forget to raise the 3 to the power 4.

$$\left(1 - \frac{1}{3}x\right)^4 = 1 + 4\left(-\frac{1}{3}x\right) + \frac{4 \times 3}{1 \times 2}\left(-\frac{1}{3}x\right)^2$$

$$+ \frac{4 \times 3 \times 2}{1 \times 2 \times 3}\left(-\frac{1}{3}x\right)^3$$

$$+ \frac{4 \times 3 \times 2 \times 1}{1 \times 2 \times 3 \times 4}\left(-\frac{1}{3}x\right)^4$$

Now use the $(1 + x)^n$ formula, with $n = 4$, and $-\frac{1}{3}x$ instead of x.

$$= 1 - \frac{4}{3}x + 6\left(\frac{1}{9}x^2\right) + 4\left(-\frac{1}{27}x^3\right) + \frac{1}{81}x^4$$

$$= 1 - \frac{4x}{3} + \frac{2x^2}{3} - \frac{4x^3}{27} + \frac{x^4}{81}$$

$$(3 - x)^4 = 81\left(1 - \frac{1}{3}x\right)^4$$

Finally, put this back into the original expression.

$$= 81\left(1 - \frac{4x}{3} + \frac{2x^2}{3} - \frac{4x^3}{27} + \frac{x^4}{81}\right)$$

$$= 81 - 108x + 54x^2 - 12x^3 + x^4$$

$n = 4$ is a positive integer, so the expansion is valid for all values of x.

For the example above, you could just pop $a = 3$ and $b = -x$ into the formula for $(a + b)^n$, which will be given to you in the exam.

This formula only works when n is a **natural number** (written $n \in \mathbb{N}$ — these are just positive integers), so you'll need to use the method shown here if the power is negative or a fraction.

Example 2

Give the first 3 terms in the binomial expansion of $(3x + 4)^{\frac{3}{2}}$.

$$3x + 4 = 4 + 3x = 4\left(1 + \frac{3}{4}x\right)$$

Again, you need to factorise before using the formula, taking care to choose the right factor.

$$\Rightarrow (3x + 4)^{\frac{3}{2}} = \left[4\left(1 + \frac{3}{4}x\right)\right]^{\frac{3}{2}}$$

Tip It's tempting to take the 3 outside the brackets because of the order the numbers are written in, but don't be fooled.

$$= 4^{\frac{3}{2}}\left(1 + \frac{3}{4}x\right)^{\frac{3}{2}}$$

$$= 8\left(1 + \frac{3}{4}x\right)^{\frac{3}{2}}$$

$$\left(1 + \frac{3}{4}x\right)^{\frac{3}{2}} = 1 + \frac{3}{2}\left(\frac{3}{4}x\right) + \frac{\frac{3}{2} \times \left(\frac{3}{2} - 1\right)}{1 \times 2}\left(\frac{3}{4}x\right)^2 + \dots$$

Use the $(1 + x)^n$ formula with $n = \frac{3}{2}$, and $\frac{3}{4}x$ instead of x.

$$= 1 + \frac{9}{8}x + \frac{\frac{3}{2} \times \frac{1}{2}}{2}\left(\frac{9}{16}x^2\right) + \dots$$

$$= 1 + \frac{9x}{8} + \frac{27x^2}{128} + \dots$$

Put this back into the original expression.

$$(3x + 4)^{\frac{3}{2}} = 8\left(1 + \frac{3}{4}x\right)^{\frac{3}{2}} = 8\left(1 + \frac{9x}{8} + \frac{27x^2}{128} + \dots\right)$$

$$= 8 + 9x + \frac{27x^2}{16} + \dots$$

Tip This expansion is only valid for $\left|\frac{qx}{p}\right| < 1$, i.e. $\left|\frac{3x}{4}\right| < 1 \Rightarrow |x| < \frac{4}{3}$.

Example 3

Expand $\dfrac{1 + 2x}{(2 - x)^2}$ up to the term in x^3. State the range of x for which the expansion is valid.

$$\frac{1 + 2x}{(2 - x)^2} = (1 + 2x)(2 - x)^{-2}$$

First rearrange and separate the different expansions.

$$2 - x = 2\left(1 - \frac{1}{2}x\right)$$

$$\Rightarrow (2 - x)^{-2} = 2^{-2}\left(1 - \frac{1}{2}x\right)^{-2} = \frac{1}{4}\left(1 - \frac{1}{2}x\right)^{-2}$$

The first bracket doesn't need expanding, so deal with the second bracket as usual, factorising first...

$$\left(1 - \frac{1}{2}x\right)^{-2} = 1 + (-2)\left(-\frac{1}{2}x\right) + \frac{-2 \times -3}{1 \times 2}\left(-\frac{1}{2}x\right)^2$$

... then using the formula with $n = -2$, and $-\frac{1}{2}x$ instead of x.

$$+ \frac{-2 \times -3 \times -4}{1 \times 2 \times 3}\left(-\frac{1}{2}x\right)^3 + \dots$$

Tip Take your time and set everything out in steps, then combine the expansions at the end.

$$= 1 + x + \frac{3x^2}{4} + \frac{x^3}{2} + \dots$$

continued on the next page...

$$(2-x)^{-2} = \frac{1}{4}\left(1-\frac{1}{2}x\right)^{-2} = \frac{1}{4}\left(1 + x + \frac{3x^2}{4} + \frac{x^3}{2} + \dots\right)$$

$$= \frac{1}{4} + \frac{x}{4} + \frac{3x^2}{16} + \frac{x^3}{8} + \dots \qquad \leftarrow \text{Complete the expansion of the second bracket.}$$

$$\frac{1+2x}{(2-x)^2} = (1+2x)(2-x)^{-2} = (1+2x)\left(\frac{1}{4} + \frac{x}{4} + \frac{3x^2}{16} + \frac{x^3}{8} + \dots\right) \qquad \leftarrow \text{Put all this into the original expression and simplify as much as possible.}$$

$$= \frac{1}{4} + \frac{x}{4} + \frac{3x^2}{16} + \frac{x^3}{8} + 2x\left(\frac{1}{4} + \frac{x}{4} + \frac{3x^2}{16}\right) + \dots$$

$$= \frac{1}{4} + \frac{x}{4} + \frac{3x^2}{16} + \frac{x^3}{8} + \frac{x}{2} + \frac{x^2}{2} + \frac{3x^3}{8} + \dots$$

$$= \frac{1}{4} + \frac{3x}{4} + \frac{11x^2}{16} + \frac{x^3}{2} + \dots$$

This is valid only if $\left|\frac{-x}{2}\right| < 1 \implies |x| < 2$

Tip Remember — choose the narrower valid range of x for the combined expansion.

Exercise 6.3

Q1 Find the binomial expansion of the following functions, up to and including the term in x^3:

a) $(2+4x)^3$ b) $(3+4x)^5$ c) $(4+x)^{\frac{1}{2}}$ d) $(8+2x)^{-\frac{1}{3}}$

P Q2 If the x^2 coefficient of the binomial expansion of $(a+5x)^5$ is 2000, what is the value of a?

Q3 a) Find the binomial expansion of $(2-5x)^7$, up to and including the term in x^2.

b) Hence, or otherwise, find the binomial expansion of $(1+6x)^3(2-5x)^7$, up to and including the term in x^2.

E P Q4 The coefficient of the x^2 term in the expansion of $\dfrac{1}{(2+ax)^3}$, where a is a constant, is twice the coefficient of the x^3 term. Find the value of a.   *[4 marks]*

P Q5 a) Find the binomial expansion of $\left(1+\frac{6}{5}x\right)^{-\frac{1}{2}}$, up to and including the term in x^3, stating the range of x for which it is valid.

b) Hence, or otherwise, express $\sqrt{\dfrac{20}{5+6x}}$ in the form $a + bx + cx^2 + dx^3 + \dots$

Q6 $f(x) = \dfrac{1}{\sqrt{5-2x}}$

a) Find the binomial expansion of $f(x)$ in ascending powers of x, up to and including the x^2 term.

b) Hence show that $\dfrac{3+x}{\sqrt{5-2x}} \approx \dfrac{3}{\sqrt{5}} + \dfrac{8x}{5\sqrt{5}} + \dfrac{19x^2}{50\sqrt{5}}$.

Q7   a) Find the binomial expansion of $(9 + 4x)^{-\frac{1}{2}}$, up to and including the term in x^2.

      b) Hence, or otherwise, find the binomial expansion of $\dfrac{(1 + 6x)^4}{\sqrt{9 + 4x}}$, up to and including the term in x^2.

E **P** Q8   A student is attempting to find the first three terms, in ascending powers of x, of the binomial expansion of $(3x + 4)^{-\frac{3}{2}}$. Part of their solution is shown below.

$$(3x + 4)^{-\frac{3}{2}} = (4 + 3x)^{-\frac{3}{2}}$$

$$\Rightarrow (4 + 3x)^{-\frac{3}{2}} = 4\left(1 + \frac{3}{4}x\right)^{-\frac{3}{2}}$$

$$\Rightarrow 4\left(1 + \frac{3}{4}x\right)^{-\frac{3}{2}} \approx 4\left[1 + \left(-\frac{3}{2}\right)\left(\frac{3}{4}x\right) + \frac{\left(-\frac{3}{2}\right) \times \left(-\frac{1}{2}\right)}{1 \times 2}\left(\frac{3}{4}x\right)^2\right]$$

$$\Rightarrow 4\left(1 + \frac{3}{4}x\right)^{-\frac{3}{2}} \approx 4 - \frac{9}{2}x + \frac{27}{32}x^2$$

      a) Identify and explain the two errors in the student's work so far.   *[4 marks]*

      b) Determine the expansion the student should have achieved and the values of x for which it is valid.   *[3 marks]*

Challenge

E **P** Q9   a) Find the binomial expansion of $\left(\sqrt[4]{16 - 80x}\right)^3$, up to and including the term in x^3.   *[4 marks]*

      b) Show that 4 is the smallest positive integer value of x that makes the expansion (up to x^3) an integer.   *[3 marks]*

Q10   Three functions are given below:

$$\mathrm{f}(x) = (4 + x)^{\frac{3}{2}} \qquad \mathrm{g}(x) = \sqrt{4 + 9x} \qquad \mathrm{h}(x) = (3 - x)^3$$

     A fourth function, p(x), is given by $\mathrm{p}(x) = \dfrac{\mathrm{f}(x) \times \mathrm{g}(x)}{\mathrm{h}(x)}$.

      a) Find the first three terms, in ascending powers of x, in the binomial expansion of p(x).

      b) For which values of x is the expansion in part b) valid?

E **P** Q11   In the expansion of $(3 + 2x)^n$, where n is a fraction, the coefficient of the x^5 term is twice as large as the coefficient of the x^3 term.

     Find the possible values of n.   *[6 marks]*

6.4 Using the Binomial Expansion as an Approximation

- When you've done an expansion, you can use it to work out the **value** of the original expression for **given values of** x, by **substituting** those values into the **expansion**.

- For most expansions this will only be an **approximate** answer, because you'll have had to limit the expansion to the first few terms.

- Often you'll have to do some **rearranging** of the expression so that you know what value of x to substitute. For example, $\sqrt[3]{1.3}$ can be written $(1 + 0.3)^{\frac{1}{3}}$, which can be approximated by expanding $(1 + x)^{\frac{1}{3}}$ and substituting $x = 0.3$ into the expansion.

- You also need to check the **validity** of the expansion — the approximation will only work for values of x in the valid range.

Example 1

The binomial expansion of $(1 + 3x)^{-1}$ up to the term in x^3 is:

$$(1 + 3x)^{-1} \approx 1 - 3x + 9x^2 - 27x^3$$

The expansion is valid for $|x| < \frac{1}{3}$.

Use this expansion to approximate $\frac{100}{103}$.
Give your answer to 4 d.p.

$$\frac{100}{103} = \frac{1}{1.03} = \frac{1}{1 + 0.03} = (1 + 0.03)^{-1}$$

First you need to do some clever rearranging to rewrite the value you're given in the form $(1 + 3x)^{-1}$.

$$3x = 0.03 \implies x = 0.01$$

Now work out the value of x:
$(1 + 3x)^{-1} = \frac{100}{103}$ with $x = 0.01$.

$0.01 < \frac{1}{3}$, so the expansion is valid for this value of x.

Check this value of x is in the valid range of the expansion.

$$(1 + 3(0.01))^{-1} \approx 1 - 3(0.01) + 9(0.01^2) - 27(0.01^3)$$
$$= 1 - 0.03 + 0.0009 - 0.000027$$
$$= 1.0009 - 0.030027$$
$$= 0.970873$$

Substitute this value for x into the given expansion and calculate the result.

So $\frac{100}{103} = (1 + 3(0.01))^{-1} \approx 0.9709$ to 4 d.p.

Tip You need to use a "$\approx$" when you give the answer — it's an approximation because you're only using the first few terms of the expansion.

$\frac{100}{103} = 0.97087...$ so this is a pretty good approximation.

In some cases you might have to **rearrange the expansion** first to get it into a form that fits with the given expression.

Example 2

The binomial expansion of $(1 - 5x)^{\frac{1}{2}}$ up to the term in x^2 is $(1 - 5x)^{\frac{1}{2}} \approx 1 - \dfrac{5x}{2} - \dfrac{25x^2}{8}$.

The expansion is valid for $|x| < \dfrac{1}{5}$.

a) Use $x = \dfrac{1}{50}$ in this expansion to show that $\sqrt{10} \approx \dfrac{800}{253}$.

$$\sqrt{\left(1 - 5\left(\frac{1}{50}\right)\right)} \approx 1 - \frac{5}{2}\left(\frac{1}{50}\right) - \frac{25}{8}\left(\frac{1}{50}\right)^2$$

First, substitute $x = \dfrac{1}{50}$ into both sides of the given expansion.

$$\sqrt{\left(1 - \frac{1}{10}\right)} \approx 1 - \frac{1}{20} - \frac{1}{800}$$

$$\sqrt{\frac{9}{10}} \approx \frac{759}{800}$$

Problem Solving If you're not sure how to get the approximation you need from the expansion, try putting the numbers in and see what comes out. It should become clear where the $\sqrt{10}$ is coming from, even if it's not obvious at the start.

$$\sqrt{\frac{9}{10}} = \frac{\sqrt{9}}{\sqrt{10}} = \frac{3}{\sqrt{10}}$$

Now simplify the square root...

So $\dfrac{3}{\sqrt{10}} \approx \dfrac{759}{800}$

... and rearrange to find an estimate for $\sqrt{10}$.

$$3 \times 800 \approx 759\sqrt{10}$$

$$\sqrt{10} \approx \frac{3 \times 800}{759}$$

$$\sqrt{10} \approx \frac{800}{253} \text{ as required.}$$

b) Find the percentage error in your approximation, to 2 s.f.

$$\left|\frac{\text{real value} - \text{estimate}}{\text{real value}}\right| \times 100$$

Work out the percentage error by finding the difference between your estimate and a calculated 'real' value, and give this as a percentage of the real value.

$$= \left|\frac{\sqrt{10} - \frac{800}{253}}{\sqrt{10}}\right| \times 100$$

Tip The modulus sign means you always get a positive answer, whether the estimate is bigger or smaller than the real value. You're only interested in the difference between them.

$$= 0.0070\% \text{ (2 s.f.)}$$

The % error is really small, so the approximation is very close to the real answer.

Exercise 6.4

Q1 a) Find the binomial expansion of $(1 + 6x)^{-1}$, up to and including the term in x^2.

b) What is the validity of the expansion in part a)?

c) Use an appropriate substitution to find an approximation for $\frac{100}{106}$.

d) What is the percentage error of this approximation?
Give your answer correct to 1 significant figure.

> **Q1c) Hint** Always check that the value you've decided to use for x is within the valid range for the expansion.

Q2 a) Use the binomial theorem to expand $(1 + 3x)^{\frac{1}{4}}$ in ascending powers of x, up to and including the term in x^3.

b) For what values of x is this expansion valid?

c) Use this expansion to find an approximate value of $\sqrt[4]{1.9}$ correct to 4 decimal places.

d) Find the percentage error of this approximation, correct to 3 significant figures.

Q3 a) Find the first four terms in the binomial expansion of $(1 - 2x)^{-\frac{1}{2}}$.

b) For what range of x is this expansion valid?

c) Use $x = \frac{1}{10}$ in this expansion to find an approximate value of $\sqrt{5}$.

d) Find the percentage error of this approximation, correct to 2 significant figures.

> **Q3c) Hint** Put the value for x into both sides of the expansion, and rearrange the left-hand side until you get $\sqrt{5}$.

Q4 a) Expand $(2 - 5x)^6$ up to and including the x^2 term.

b) By substituting an appropriate value of x into the expansion in a), find an approximate value for 1.95^6.

c) What is the percentage error of this approximation?
Give your answer to 2 significant figures.

Q5 a) Find the first three terms in the binomial expansion of $\sqrt{3 - 4x}$.

b) For what values of x is this expansion valid?

c) Use $x = \frac{3}{40}$ in this expansion to estimate the value of $\frac{3}{\sqrt{10}}$.
Leave your answer as a fraction.

d) Find the percentage error of this approximation, correct to 1 significant figure.

E Q6 a) Find the first three terms in the expansion of $\left(1 + \frac{1}{4}x\right)^{\frac{2}{3}}$. *[2 marks]*

b) Find the positive value of x that satisfies the equation $\left(1 + \frac{1}{4}x\right)^2 = 1.44$. *[1 mark]*

c) Use your answers to parts a) and b) to find an approximation of $\sqrt[3]{1.44}$ to four decimal places. *[2 marks]*

E P Q7 a) Find the first three terms in the binomial expansion of $(2-x)^{\frac{1}{2}}$. *[3 marks]*

b) Write down the first three terms in the expansion of $(2+x)^{\frac{1}{2}}$. *[1 mark]*

c) Use your answers to parts a) and b) to find the terms up to and including x^2 in the expansion of $(2-x)^{\frac{1}{2}}(2+x)^{\frac{1}{2}}$. *[2 marks]*

d) Use your answer to part c) to find an approximation to $\sqrt{1.98 \times 2.02}$. *[2 marks]*

Q8 a) Expand $\dfrac{(1-5x)}{(1+3x)^{\frac{1}{3}}}$, up to and including the term in x^2.

b) Give the range for which the expansion is valid.

c) Use $x = 0.1$ in this expansion to estimate the value of $\dfrac{1}{2\sqrt[3]{1.3}}$.

d) Find the percentage error of this approximation, correct to 2 significant figures.

E Q9 a) Find the binomial expansion of $\sqrt{5x+4}$, up to and including the term in x^3. *[3 marks]*

b) Find the values for which the expansion in part a) is valid. *[1 mark]*

c) Determine if the expansion in part a) is suitable for estimating the value of $\sqrt{\dfrac{15}{2}}$ or not, fully explaining your decision. *[2 marks]*

E Q10 Functions f(x) and g(x) are given by $f(x) = \dfrac{1}{\sqrt{2+x}}$ and $g(x) = 4 - 3x$.

a) Show that $fg(x) = \dfrac{1}{\sqrt{6-3x}}$. *[1 mark]*

b) Find the binomial expansion of fg(x) up to and including the term in x^2. *[3 marks]*

c) Is the expansion in part b) suitable for estimating the value of $\dfrac{\sqrt{5}}{5}$ or not? Fully justify your answer. *[2 marks]*

Challenge

E P Q11 The angle θ is small and measured in radians.

a) Show that $\sec^6 \theta \approx \left(1 - \dfrac{1}{2}\theta^2\right)^{-6}$. *[2 marks]*

b) Find the first three terms, in ascending powers of θ, in the binomial expansion of $\left(1 - \dfrac{1}{2}\theta^2\right)^{-6}$. *[2 marks]*

c) Use your answer to part b) to estimate the value of $\sec^6 0.2$. *[2 marks]*

E P Q12 The first three terms in the expansion of $(1+ax)^{-\frac{1}{2}}$ are used to estimate the value of $\dfrac{1}{\sqrt{1+\dfrac{a}{2}}}$, where a is a constant. The value of the estimate is 0.875.

Find a and the value being estimated. Fully justify your answer. *[6 marks]*

Chapter 6

6.5 Binomial Expansion and Partial Fractions

You can find the binomial expansion of more complicated rational functions by:

- splitting them into **partial fractions** first,
- expanding **each fraction** using the formula (usually with $n = -1$),
- then **adding** the expansions together.

Example 1

$f(x) = \dfrac{x-1}{(3+x)(1-5x)}$ can be expressed as partial fractions in the form: $\dfrac{A}{(3+x)} + \dfrac{B}{(1-5x)}$.

a) Find the values of A and B, and hence express $f(x)$ as partial fractions.

$$\frac{x-1}{(3+x)(1-5x)} \equiv \frac{A}{(3+x)} + \frac{B}{(1-5x)}$$

Start by writing out the problem as an identity...

$$\Rightarrow \frac{x-1}{(3+x)(1-5x)} \equiv \frac{A(1-5x) + B(3+x)}{(3+x)(1-5x)}$$

$$\Rightarrow \quad x - 1 \equiv A(1-5x) + B(3+x)$$

... and simplify.

Let $x = -3$, then: $-3 - 1 = A(1 - (-15))$

$$\Rightarrow -4 = 16A$$

$$\Rightarrow A = -\frac{1}{4}$$

You can then work out the values of A and B by putting in values of x that make each bracket in turn equal to zero. (This is known as the 'substitution' method.)

Let $x = \frac{1}{5}$, then: $\frac{1}{5} - 1 = B\left(3 + \frac{1}{5}\right)$

$$\Rightarrow -\frac{4}{5} = \frac{16}{5}B$$

$$\Rightarrow B = -\frac{1}{4}$$

Problem Solving

You could also equate the coefficients of x and the constant terms to find A and B.

So $f(x)$ can also be written as:

$$-\frac{1}{4(3+x)} - \frac{1}{4(1-5x)}$$

Put A and B back into the expression to give $f(x)$ as partial fractions.

b) Use your answer to part a) to find the binomial expansion of f(x),
up to and including the term in x^2.

$$f(x) = -\frac{1}{4}(3 + x)^{-1} - \frac{1}{4}(1 - 5x)^{-1}$$

Start by rewriting the partial fractions from a) in $(p + qx)^n$ form.

$$(3 + x)^{-1} = \left(3\left(1 + \frac{1}{3}x\right)\right)^{-1}$$

Now do the two binomial expansions separately.

$$= \frac{1}{3}\left(1 + \frac{1}{3}x\right)^{-1}$$

$$= \frac{1}{3}\left(1 + (-1)\left(\frac{1}{3}x\right) + \frac{(-1)(-2)}{2}\left(\frac{1}{3}x\right)^2 + ...\right)$$

$$= \frac{1}{3}\left(1 - \frac{1}{3}x + \frac{1}{9}x^2 - ...\right)$$

$$= \frac{1}{3} - \frac{1}{9}x + \frac{1}{27}x^2 - ...$$

Tip Questions like these have lots of stages, which means lots of places that you could make a mistake — especially with all the negatives and fractions. Set your working out clearly and don't skip any stages.

$$(1 - 5x)^{-1} = 1 + (-1)(-5x) + \frac{(-1)(-2)}{2}(-5x)^2 + ...$$

$$= 1 + 5x + 25x^2 + ...$$

$$f(x) = -\frac{1}{4}(3 + x)^{-1} - \frac{1}{4}(1 - 5x)^{-1}$$

$$\approx -\frac{1}{4}\left(\frac{1}{3} - \frac{1}{9}x + \frac{1}{27}x^2\right) - \frac{1}{4}(1 + 5x + 25x^2)$$

Finally, put everything together by adding the expansions in the rearranged form of f(x).

$$= -\frac{1}{12} + \frac{x}{36} - \frac{x^2}{108} - \frac{1}{4} - \frac{5x}{4} - \frac{25x^2}{4}$$

$$= -\frac{1}{3} - \frac{11x}{9} - \frac{169x^2}{27}$$

Problem Solving

Remember, the expansion of $(p + qx)^n$ is valid when $\left|\frac{qx}{p}\right| < 1$.

c) Find the range of values of x for which your answer to part b) is valid.

The expansion of $(3 + x)^{-1}$ is valid when:

$$\left|\frac{x}{3}\right| < 1 \implies |x| < 3$$

The two expansions from b) are valid for different values of x.

The expansion of $(1 - 5x)^{-1}$ is valid when:

$$|-5x| < 1 \implies |x| < \frac{1}{5}$$

So the expansion of f(x) is valid for $|x| < \frac{1}{5}$.

The combined expansion of f(x) is valid where these two ranges overlap, i.e. over the narrower of the two ranges. (This is the same as when you combine expansions by multiplying them together, as shown on page 181-182.)

Chapter 6

Exercise 6.5

Q1 a) $\dfrac{5-12x}{(1+6x)(4+3x)} \equiv \dfrac{A}{(1+6x)} + \dfrac{B}{(4+3x)}$. Find A and B.

b) (i) Find the binomial expansion of $(1+6x)^{-1}$, up to and including the term in x^2.

(ii) Find the binomial expansion of $(4+3x)^{-1}$, up to and including the term in x^2.

c) Hence find the binomial expansion of $\dfrac{5-12x}{(1+6x)(4+3x)}$, up to and including the term in x^2.

d) For what values of x is this expansion valid?

Q2 $f(x) = \dfrac{6}{(1-x)(1+x)(1+2x)}$

a) Show that $f(x)$ can be expressed as: $\dfrac{1}{(1-x)} - \dfrac{3}{(1+x)} + \dfrac{8}{(1+2x)}$

b) Give the binomial expansion of $f(x)$ in ascending powers of x, up to and including the term in x^2.

c) Find the percentage error when you use this expansion to estimate $f(0.01)$, giving your answer to 2 significant figures.

Q3 a) Factorise fully $2x^3 + 5x^2 - 3x$.

b) Hence express $\dfrac{5x-6}{2x^3+5x^2-3x}$ as partial fractions.

c) Find the binomial expansion of $\dfrac{5x-6}{2x^3+5x^2-3x}$, up to and including the term in x^2.

> **Q3c) Hint** You'll end up with a term in x^{-1}, which you wouldn't usually get with a binomial expansion — this comes from the partial fractions and you can just leave it as it is.

d) For what values of x is this expansion valid?

Q4 $f(x) = \dfrac{55x+7}{(2x-5)(3x+1)^2}$

a) Express $f(x)$ in the form $\dfrac{A}{(2x-5)} + \dfrac{B}{(3x+1)} + \dfrac{C}{(3x+1)^2}$, where A, B and C are integers to be found.

b) Hence, or otherwise, use the binomial formula to expand $f(x)$ in ascending powers of x, up to and including the term in x^2.

E Q5 a) Show that $\dfrac{q}{(1-4x)(1-2x)}$, where q is a constant, can be written in partial fractions as $\dfrac{2q}{1-4x} - \dfrac{q}{1-2x}$. *[3 marks]*

b) Find the coefficient of the x^3 term in the binomial expansion of:

(i) $(1-4x)^{-1}$ (ii) $(1-2x)^{-1}$ *[2 marks]*

c) (i) Use your answers to parts a) and b) to determine the coefficient of the x^3 term in the binomial expansion of $\dfrac{q}{(1-4x)(1-2x)}$. *[2 marks]*

(ii) Given that the value of the coefficient is 60, find the value of q. *[1 mark]*

E Q6 a) (i) Use the Factor Theorem to show that $(1 + x)$ is a factor of $x^3 + 6x^2 + 11x + 6$.

(ii) Fully factorise $x^3 + 6x^2 + 11x + 6$. *[4 marks]*

b) Express $\dfrac{2}{x^3 + 6x^2 + 11x + 6}$ in partial fractions. *[3 marks]*

c) Find the binomial expansion of $\dfrac{2}{x^3 + 6x^2 + 11x + 6}$ in ascending powers of x, up to and including the term in x^2. *[4 marks]*

d) State the values of x for which the expansion in part c) is valid. *[1 mark]*

E Q7 $f(x) = \dfrac{35x - 33x^2 - 3}{(1 - x)(1 + 2x)(1 - 3x)}$

$f(x)$ can be expressed in the form $\dfrac{A}{1 - x} + \dfrac{B}{1 + 2x} + \dfrac{C}{1 - 3x}$.

a) Find the values of the constants A, B and C . *[3 marks]*

b) Find the binomial expansions of $(1 - x)^{-1}$, $(1 + 2x)^{-1}$ and $(1 - 3x)^{-1}$, giving each in ascending powers of x, up to and including the term in x^2. *[3 marks]*

c) Hence find the binomial expansion of $f(x)$ in ascending powers of x, up to and including the term in x^2 . *[2 marks]*

d) State the range of values for which the expansion in part c) is valid. *[1 mark]*

e) Show how the expansion in part c) indicates $f(x)$ has a root close to $x = 0.1$. *[2 marks]*

Q8a) Problem Solving
Start by using the Factor Theorem to find (at least) one of the brackets.

P Q8 a) Given that $f(0.5) = 0$, write $f(x) = 12x^3 - 8x^2 - x + 1$ in the form $(ax + b)(cx + d)^2$, where a, b, c and d are integers.

b) Use your answer to part a) to express $g(x) = \dfrac{8 - x}{12x^3 - 8x^2 - x + 1}$ as partial fractions.

c) Find the first three terms in the binomial expansion of $\dfrac{8 - x}{12x^3 - 8x^2 - x + 1}$.

d) For what values of x is the expansion in part c) valid?

e) Show that this expansion gives an estimate of $g(0.001)$ correct to 6 decimal places.

Challenge

E P Q9 The coefficient of the x^4 term in the binomial expansion of $\dfrac{p}{(2 - x)(5 - 2x)}$, where p is a constant, is 0.06303. Find the value of p. *[7 marks]*

E P Q10 a) Show that $\dfrac{2kx}{(1 - kx)(1 + kx)}$ can be written in partial fractions as $\dfrac{1}{1 - kx} - \dfrac{1}{1 + kx}$. *[2 marks]*

b) The binomial expansion of $\dfrac{2kx}{(1 - kx)(1 + kx)}$ in ascending powers of x, up to and including the term in x^2, is equal to $4x$. Find the value of k. *[3 marks]*

c) State the values of x for which the expansion is valid. *[1 mark]*

6 Review Exercise

Q1 Give the binomial expansion of:

a) $(1 + 2x)^3$ b) $(1 - x)^5$ c) $(1 - 4x)^4$ d) $\left(1 - \frac{2}{3}x\right)^4$

Q2 For what values of n does the binomial expansion of $(1 + x)^n$ result in a finite expression?

E Q3 The coefficient of the x^3 term in the expansion of $(1 + kx)^8$ is 3584. Find the value of k. *[2 marks]*

E Q4 Find the first three terms in ascending powers of x in the expansion of $\dfrac{3}{(1 - 8x)^3}$. *[2 marks]*

E Q5 Expand $\sqrt[3]{(1 + 2x)^2}$ up to and including the term in x^2. *[2 marks]*

E Q6 The cubic term in the expansion $(1 + ax)^{-4}$ is $-160x^3$. Find the value of a. *[3 marks]*

P Q7 a) If the x^2 coefficient of the binomial expansion of $(1 + ax)^{-2}$ is 48 and a is a positive integer, what is the value of a?

b) If the x^4 coefficient of the binomial expansion of $(1 - ax)^{\frac{1}{3}}$ is $-\dfrac{10}{3}$, and a is positive, what is the value of a?

E P Q8 In the expansion of $(1 + kx)^{-\frac{1}{2}}$ the coefficient of the term in x^4 is $\dfrac{35}{10\,368}$. Find the value of k given that it is positive. *[2 marks]*

E P Q9 When $(1 + ax)^n$ is expanded in ascending powers of x, the first three terms of the expansion are: $1 - 6x + \dfrac{45}{2}x^2$. Find the values of a and n when $n \neq 0$. *[6 marks]*

Q10 If the full binomial expansion of $(c + dx)^n$ is an infinite series, what values of x is the expansion valid for?

Q11 Find the binomial expansion of each of the following, up to and including the term in x^3, stating the range of x for which the expansions are valid.

a) $\dfrac{1}{(1 + x)^5}$ b) $\dfrac{1}{(1 - 3x)^3}$ c) $\sqrt{1 - 5x}$ d) $\dfrac{1}{\sqrt[3]{1 + 2x}}$

E Q12 a) Find the binomial expansion of the following, up to and including the term in x^2.

(i) $(1 - 2x)^{\frac{1}{2}}$ (ii) $(1 - 3x)^{\frac{1}{3}}$ *[3 marks]*

b) Hence:

(i) Find the first three terms in ascending powers of x in the expansion of $(1 - 2x)^{\frac{1}{2}}(1 - 3x)^{\frac{1}{3}}$.

(ii) State the values of x for which the expansion in part b) (i) is valid. *[3 marks]*

E Q13 a) Find the binomial expansion of $\dfrac{4}{(6+2x)^2}$, up to and including the term in x^3. *[3 marks]*

b) Hence, or otherwise, find the binomial expansion of $\dfrac{12x}{(6+2x)^2}$,
up to and including the term in x^3. *[2 marks]*

c) Find the values of x for which both expansions are valid. *[1 mark]*

E Q14 a) Find the expansion of $\left(1 - \dfrac{1}{2}x\right)^{-3}$ up to and including the term in x^2. *[2 marks]*

b) Use your answer to part a) to find an approximation to $\dfrac{1}{(0.95)^3}$. *[2 marks]*

E Q15 a) Find the first three terms in the expansion of $\sqrt{1-3x}$ in ascending powers of x.
State the range of values for which this expansion is valid. *[3 marks]*

b) Hence approximate $\sqrt{0.97}$. *[2 marks]*

E Q16 a) Find the first three terms in ascending
powers of x in the expansion of $(1 - 4x^2)^{\frac{1}{3}}$. *[2 marks]*

b) Find the values of x for which the expansion in part a) is valid. *[1 mark]*

c) Use $x = \dfrac{1}{6}$ and your answer to part a) to find an approximation to $\sqrt[3]{9}$. *[3 marks]*

E
P Q17 a) Find the first three terms in ascending powers of x in the
expansion of $(2 - ax)^{-4}$ where a is a positive integer. *[3 marks]*

b) The expansion found in part a) is used to approximate values.
When $x = 0.1$ the expansion gives the approximation 0.0765625.
Find the value of a. *[2 marks]*

Q18 a) Give the binomial expansions of the following, up to and including the term in x^2.
State the range of x for which each expansion is valid.

(i) $\dfrac{1}{(3+2x)^2}$ (ii) $\sqrt[3]{8-x}$

b) Use your answers to a) to give the binomial expansion of $\dfrac{\sqrt[3]{8-x}}{(3+2x)^2}$, up to and

including the term in x^2. State the range of x for which this expansion is valid.

c) (i) Use the expansion in a) (ii) to find an approximate value of $\sqrt[3]{7}$,
leaving your answer as a fraction.

(ii) Find the percentage error of this approximation, correct to 2 significant figures.

E Q19 a) Find the first three terms in ascending
powers of x in the expansion of $(1 - 4x)^{-\frac{1}{4}}$. *[2 marks]*

b) The approximation in part a) is used to estimate the value of $\dfrac{1}{\sqrt[4]{0.92}}$.

(i) Find an estimate of $\dfrac{1}{\sqrt[4]{0.92}}$ using your answer to part a).

(ii) Find the percentage error in the estimate compared to
the exact value of $\dfrac{1}{\sqrt[4]{0.92}}$ determined by your calculator. *[4 marks]*

E P **Q20** a) Expand $\sqrt{\dfrac{1+x}{1-x}}$ up to and including the term x^3.
State the value of x for which the expansion is valid. *[7 marks]*

b) Hence, by using an appropriate value of x,
find an approximate fractional value for $\sqrt{3}$. *[2 marks]*

E **Q21** $\dfrac{3x^2+x+2}{(x+1)(x-1)(2x+1)} \equiv \dfrac{A}{(x+1)} + \dfrac{B}{(x-1)} + \dfrac{C}{(2x+1)}$

a) Find the values of A, B and C. *[3 marks]*

b) Hence obtain the series expansion for $\dfrac{3x^2+x+2}{(x+1)(x-1)(2x+1)}$ giving all the terms
up to and including the term in x^2. *[6 marks]*

c) State the values of x for which the expansion is valid. *[1 mark]*

Q22 a) Show that $\dfrac{5-10x}{(1+2x)(2-x)}$ can be expressed as: $\dfrac{4}{(1+2x)} - \dfrac{3}{(2-x)}$.

b) Give the binomial expansion of the expression in a), up to and including the term in x^2.

c) Find the percentage error when you use $x = 0.1$ in this expansion to estimate $\dfrac{4}{1.2 \times 1.9}$,
giving your answer to 2 significant figures.

E **Q23** The function is f(x) given by $f(x) = 3x - x^2 - 2x^3$.

a) Fully factorise f(x). *[2 marks]*

b) Express $\dfrac{15}{f(x)}$ as partial fractions. *[4 marks]*

c) Find the binomial expansion for $\dfrac{15}{f(x)}$, up to and including the term in x^3. *[5 marks]*

d) Determine the values for which the expansion from part c) is valid. *[1 mark]*

E **Q24** The function f(x) is given by $f(x) = \dfrac{4x^2 - 8x + 9}{x(2x-3)^2}$.

a) Express f(x) as partial fractions. *[4 marks]*

b) Find the binomial expansion of $(2x-3)^{-2}$, up to and including the term in x^2. *[3 marks]*

c) Hence show that $f(x) \approx x^{-1} + \dfrac{4}{9} + \dfrac{16}{27}x + \dfrac{16}{27}x^2$. *[2 marks]*

d) Determine the values of x for which the result in part c) is valid. *[1 mark]*

Challenge

E P **Q25** The coefficients of the x^2 terms in the binomial expansions of
$3\sqrt{p+4x}$ and $\dfrac{4}{\sqrt{4-px}}$ are equal. Find the value of the constant p. *[5 marks]*

E P **Q26** The coefficient of the x^2 term in the expansion of $\left(1 - \dfrac{kx}{k+1}\right)^5$,
where k is an integer, is $\dfrac{45}{8}$. Find the value of k. *[3 marks]*

E P Q27 a) Show that $(1 + x)^6 - (1 - x)^6 = 4x(1 + 3x^2)(3 + x^2)$. *[3 marks]*

  b) Hence explain why the graph of $y = (1 + x)^6 - (1 - x)^6$
     intercepts the x-axis only at the origin. *[2 marks]*

P Q28 The coefficient of the x^3 term in the expansion of $(1 - 3x)^n$,
    where n is a positive integer, is -108. Find the value of n.

E P Q29 a) Find the binomial expansion of $(6x + a)^{-3}$, where a is a constant,
      in ascending powers of x, up to and including the term in x^2. *[3 marks]*

  b) Given that the coefficient of the x^2 term in the binomial

     expansion of $\dfrac{(1 + 2x)^2}{(6x + a)^3}$ is $-\dfrac{108}{a^5}$, find the value of a. *[4 marks]*

  c) Determine the values of x for which the expansion referred to in part b) is valid. *[1 mark]*

E P Q30 The binomial expansion of $\sqrt{\dfrac{a + x}{a - x}}$, in ascending powers of x,

    up to and including the term in x^2, is used to approximate values.
    When $x = 0.3$, the expansion gives the approximation 0.745.
    Find the possible values of the constant a. *[6 marks]*

P Q31 $f(x) = \dfrac{1}{(a - 2x)^2}$

  a) Find the binomial expansion of $f(x)$, up to and including the term in x^2, in terms of a.

  b) Give the range for which this expansion is valid in terms of a.

  $g(x) = \dfrac{1}{\sqrt{4 - ax}}$

  c) Find the binomial expansion of $g(x)$, up to and including the term in x^2, in terms of a.

  d) Give the range for which this expansion is valid in terms of a.

  e) Hence find the binomial expansion of $f(x) \times g(x)$, up to and including the term in x.

  f) Write down the coefficient of x in the combined expansion of $f(x) \times g(x)$, given that $a = 2$.

E P Q32 a) Show that the function $f(x) = \dfrac{4x^2 - 6x + 3}{2x^2 - 3x + 1}$ can be written in the form

     $A + \dfrac{B}{1 - x} + \dfrac{C}{1 - 2x}$, where A, B and C are integers to be found. *[4 marks]*

  b) (i) Find the binomial expansion of $(1 - x)^{-1}$,
          up to and including the term in x^2.

     (ii) Find the binomial expansion of $(1 - 2x)^{-1}$,
          up to and including the term in x^2. *[3 marks]*

  c) Hence find the binomial expansion of $f(x)$ in ascending powers of x,
     up to and including the term in x^2. *[2 marks]*

  d) Find the values of x for which this expansion is valid. *[1 mark]*

6 Chapter Summary

1   The binomial expansion of $(1 + x)^n$ is given by the formula:

$$(1 + x)^n = 1 + nx + \frac{n(n-1)}{1 \times 2}x^2 + \ldots + \frac{n(n-1)\ldots(n-r+1)}{1 \times 2 \times \ldots \times r}x^r + \ldots$$

2   The formula gives the expansion of $(1 + ax)^n$ too — just replace each 'x' by 'ax'.

3   The formula can be used for $(p + qx)^n$ if you factorise first: $(p + qx)^n = p^n\left(1 + \frac{qx}{p}\right)^n$.

4   If n is a positive integer, the binomial expansion of $(p + qx)^n$ is finite and valid for all values of x.

5   If n is a negative integer or a fraction, the binomial expansion of $(p + qx)^n$ has an infinite number of terms and is valid only when $\left|\frac{qx}{p}\right| < 1$, i.e. $|x| < \left|\frac{p}{q}\right|$.

6   For combined expansions (i.e. where two or more brackets are raised to powers), just expand the brackets separately, then use the results to recombine.

7   For combined expansions, the expansion is valid only where the individual valid ranges overlap.

8   You can use a binomial expansion to approximate a calculation by substituting in a suitable value for x. This will only work if x is within the valid range of the expansion.

9   You can expand rational functions by splitting them into partial fractions. Just expand each partial fraction, and then add the results together.

Differentiation

Learning Objectives

Once you've completed this chapter, you should be able to:

- Use the second derivative of a function to work out when the graph of that function is concave or convex, and to identify points of inflection.
- Use the chain rule to differentiate functions of functions.
- Differentiate e^x, ln x, a^x and functions of the form $y = e^{f(x)}$, $y = \ln(f(x))$ and $y = a^{f(x)}$.
- Differentiate $\sin kx$, $\cos kx$ and $\tan kx$, including $\sin x$ and $\cos x$ from first principles.
- Use the product and quotient rules for differentiation.
- Differentiate $\operatorname{cosec} x$, $\sec x$ and $\cot x$.
- Form differential equations from situations involving connected rates of change.
- Differentiate functions that are defined parametrically or implicitly and find the equations of tangents and normals to them.

Prior Knowledge Check

1. Differentiate $y = 3x^2 + 4$ from first principles. *see Year 1 Pure Maths*

2. Find the second derivative for $y = x^3 + 2x^2 + x - 6$ and say whether the stationary point at $(-1, -6)$ is a maximum or a minimum. *see Year 1 Pure Maths*

3. Find the equation of the normal to the curve $y = 0.5x^2 - 6x$ at the point $(2, -10)$ in the form $y = mx + c$. *see Year 1 Pure Maths*

4. Solve ln $6x = 5$, giving the exact value of x. *see Year 1 Pure Maths*

5. Evaluate arccos (-1) without a calculator, giving your answer in radians. *see pages 77-79*

6. Write the following expressions as a single trigonometric function:
   a) $\sin 5x \cos 5x$ b) $\sin 4x \cos 3x - \cos 4x \sin 3x$ *see pages 96-109*

7. Find the Cartesian equation, in the form $y = f(x)$, for the curve represented by the parametric equations $y = 2t^2 - 7$ and $x = t - 2$. *see pages 127-141*

7.1 Convex and Concave Curves

Continuous curves of the form $y = f(x)$ can be described as **convex** or **concave**.

Convex curves are ones that curve **downwards**. A straight line joining any two points on a convex curve lies **above** the curve between those points.

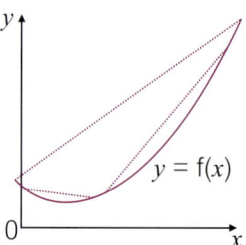

If you join any two points on this curve, the line is above the curve, so $y = f(x)$ is **convex**.

Concave curves are ones that curve **upwards**. A straight line joining any two points on a concave curve lies **below** the curve between those points.

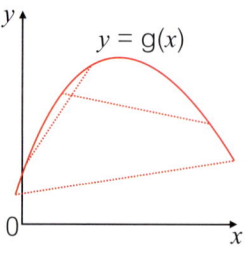

If you join any two points on this curve, the line is below the curve, so $y = g(x)$ is **concave**.

> **Tip** To help you remember: concave is the one that looks like the entrance to a cave.

In Year 1, you learnt how to find **first** and **second derivatives** of a function, and how they relate to the **graph** of that function.

> **Tip** Remember, f'(x) is the first derivative and f''(x) is the second derivative of a function f(x) with respect to x.

$\dfrac{dy}{dx}$ or f'(x) is the **gradient of the graph** $y = f(x)$ — the rate of change of x.

$\dfrac{d^2y}{dx^2}$ or f''(x) is the **rate of change of the gradient** of $y = f(x)$.

Convex curves have an **increasing gradient** and **concave** curves have a **decreasing gradient**. This means you can use the **second derivative** to work out if a curve is convex or concave:

A curve $y = f(x)$ is **convex** if **f''(x) > 0** for all values of x.

A curve $y = f(x)$ is **concave** if **f''(x) < 0** for all values of x.

Most curves aren't **entirely** convex or concave — but the definitions of convex and concave can be applied to **parts** of these curves, so you can divide a curve into concave and convex **sections**.

Example **1**

The graph of $y = x^3 + x^2 - x$ has concave and convex sections.
Find the range of values of x where the graph of $y = x^3 + x^2 - x$ is convex.

$\dfrac{dy}{dx} = 3x^2 + 2x - 1$ ◄——— Find the first derivative.

$\dfrac{d^2y}{dx^2} = 6x + 2$ ◄——— Find the second derivative.

$\dfrac{d^2y}{dx^2} > 0 \Rightarrow 6x + 2 > 0$ ◄——— The graph is convex when the second derivative is positive. Form an inequality and solve.

$\Rightarrow 6x > -2 \Rightarrow x > -\dfrac{1}{3}$

7.2 Points of Inflection

A point where the curve **changes** between concave and convex (i.e. where
$f''(x)$ changes between positive and negative) is called a **point of inflection**.

At a point of inflection, **f''(x) = 0**, but not all points where $f''(x) = 0$ are points of inflection.
You need to look what's happening on **either side** of the point to see if the sign of $f''(x)$
is changing. For example:

- The graph of **$f(x) = x^3 - 6x^2 + 9x + 1$** has
  second derivative $f''(x) = 0$ when $x = 2$.

- When $x < 2$, $f''(x) < 0$ so the curve is **concave**,
  and when $x > 2$, $f''(x) > 0$ so the curve is **convex**.

- So $x = 2$ is a **point of inflection** of the curve.

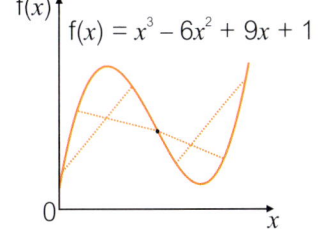

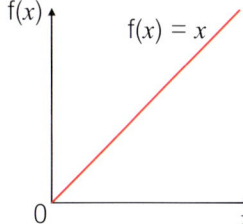

- On the graph of **$f(x) = x$**, $f''(x) = 0$ for **all** values of x.

- So at any particular value of x, $f''(x) = 0$, but it's **not**
  a point of inflection because $f''(x)$ **isn't changing**
  from positive to negative — it's just constant.

- On the graph of **$f(x) = x^4$**, $f''(x) = 12x^2$.

- At the point $(0, 0)$, $f''(x) = 0$, but it's **not** a point of inflection.

- $12x^2$ is positive for **all** non-zero values of x.
  The whole curve is convex, so it doesn't have any points of inflection.

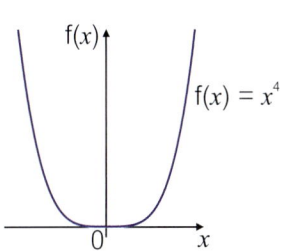

Chapter 7

Example 1

Show that the graph of $y = x^3 - 3x^2 + 3$ has a point of inflection at $x = 1$.

$f(x) = x^3 - 3x^2 + 3 \Rightarrow f'(x) = 3x^2 - 6x$
$\qquad\qquad\qquad\Rightarrow f''(x) = 6x - 6$ ← Start by differentiating twice with respect to x to find $f''(x)$.

At $x = 1$, $f''(x) = 6(1) - 6 = 0$ ← Show that $f''(1) = 0$.

When $x < 1$, $6x < 6 \Rightarrow 6x - 6 < 0 \Rightarrow f''(x) < 0$
When $x > 1$, $6x > 6 \Rightarrow 6x - 6 > 0 \Rightarrow f''(x) > 0$ ← Look at what happens either side of $x = 1$.

The sign of $f''(x)$ changes at $x = 1$, ← At $x = 1$, $f''(x) = 0$ and $f''(x)$ changes sign from negative to positive.
so the graph of $y = x^3 - 3x^2 + 3$ has
a point of inflection at $x = 1$.

Example 2

Find the coordinates of the points of inflection of the graph of $y = 2x^4 + 4x^3 - 72x^2$.

$y = 2x^4 + 4x^3 - 72x^2 \Rightarrow \dfrac{dy}{dx} = 8x^3 + 12x^2 - 144x$

$\qquad\qquad\qquad\qquad\Rightarrow \dfrac{d^2y}{dx^2} = 24x^2 + 24x - 144$ ← Find the second derivative of $y = 2x^4 + 4x^3 - 72x^2$.

$\dfrac{d^2y}{dx^2} = 0 \Rightarrow 24x^2 + 24x - 144 = 0$ ← Now find the points where $\dfrac{d^2y}{dx^2} = 0$.
$\qquad\quad \Rightarrow x^2 + x - 6 = 0$ These could be points of inflection.
$\qquad\quad \Rightarrow (x + 3)(x - 2) = 0$
$\qquad\quad \Rightarrow x = -3$ and $x = 2$

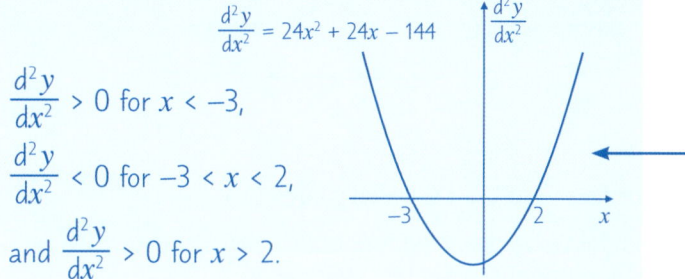

$\dfrac{d^2y}{dx^2} > 0$ for $x < -3$,

$\dfrac{d^2y}{dx^2} < 0$ for $-3 < x < 2$,

and $\dfrac{d^2y}{dx^2} > 0$ for $x > 2$.

← Think about what happens to $\dfrac{d^2y}{dx^2}$ either side of $x = -3$ and $x = 2$. The second derivative is a quadratic, so you can do this easily with a sketch.

$\dfrac{d^2y}{dx^2}$ changes sign at $x = -3$ and $x = 2$,
so both are points of inflection.

Tip If it's not easy to sketch, just work out whether $\dfrac{d^2y}{dx^2}$ is positive or negative either side of each point.

At $x = -3$, $y = 2(-3)^4 + 4(-3)^3 - 72(-3)^2 = -594$ ← The question asks for the coordinates of the points of inflection, so find the y-values.
At $x = 2$, $y = 2(2)^4 + 4(2)^3 - 72(2)^2 = -224$

So the points of inflection are
at $(-3, -594)$ and $(2, -224)$.

7.3   Stationary Points of Inflection

In Year 1, you used the first and second derivatives of a function
to find **maximum** and **minimum points** of its graph.

Stationary points are points where **f′(x) = 0**. A stationary point can be
a **maximum** point, a **minimum** point or a **stationary point of inflection**.
The value of f″(x) can help you work out which type of stationary point it is:

> If f′(x) = 0 and **f″(x) > 0**, it's a **minimum**.
>
> If f′(x) = 0 and **f″(x) < 0**, it's a **maximum**.
>
> If f′(x) = 0 and **f″(x) = 0**, it could be **any one** of the three types
> — so you have to look at f″(x) on either side of the stationary point.

If **f″(x) > 0** on either side of a stationary point, the curve is **convex** near the point, so it's a **minimum**.
If **f″(x) < 0** on either side of a stationary point, the curve is **concave** near the point, so it's a **maximum**.
And if **f″(x) changes** sign either side of stationary point, it's a stationary **point of inflection**.

Example 1

For $f(x) = x^5 - 60x^3$, find the value of x at each stationary point
of the graph of $y = f(x)$, and determine the nature of each one.

$f(x) = x^5 - 60x^3 \implies f'(x) = 5x^4 - 180x^2$ — Find the first derivative.

$\qquad\qquad\qquad\qquad = 5x^2(x^2 - 36)$

$\qquad\qquad\qquad\qquad = 5x^2(x + 6)(x - 6)$ — The stationary points are located where f′(x) = 0. Here you can solve by factorising.

So $f'(x) = 0$ when $x = 0$, $x = -6$ and $x = 6$.

$f'(x) = 5x^4 - 180x^2 \implies f''(x) = 20x^3 - 360x$ — Now find the second derivative.

$f''(-6) = 20(-6)^3 - 360(-6) = -2160$

$f''(-6) < 0$, so the point at $x = -6$ is a maximum. — Use the second derivative to work out the nature of the three stationary points at $x = -6$, $x = 6$ and $x = 0$.

$f''(6) = 20(6)^3 - 360(6) = 2160$

$f''(6) > 0$, so the point at $x = 6$ is a minimum.

$f''(0) = 20(0)^3 - 360(0) = 0$,

so the type of point at $x = 0$ is unknown.

$f''(x) = 20x^3 - 360x$

$f''(-1) = 20(-1)^3 - 360(-1) = -20 + 360 > 0$ — Look at the values of f″(x) close to $x = 0$ (e.g. $x = -1$ and $x = 1$) to determine whether the stationary point at $x = 0$ could be a maximum, a minimum or a point of inflection.

$f''(1) = 20(1)^3 - 360(1) = 20 - 360 < 0$

The sign of $f''(x)$ changes at $x = 0$,

so $x = 0$ is a point of inflection.

Chapter 7

Exercise 7.1-7.3

Q1   Find the values of x for which the following graphs are concave:

a) $y = \frac{1}{6}x^3 - \frac{5}{2}x^2 + \frac{1}{4}x + \frac{1}{9}$

b) $y = 4x^2 - x^4$

> **Problem Solving**
>
> There are only two points where the second derivative is zero in Q2, and the question tells you there are two points of inflection. So you don't actually need to prove they're points of inflection in this case.

Q2   The graph of $y = \frac{3}{2}x^4 - x^2 - 3x$ has two points of inflection. Find the x-coordinates of these two points.

P **Q3**   If $f(x) = \frac{1}{16}x^4 + \frac{3}{4}x^3 - \frac{21}{8}x^2 - 6x + 20$, identify the ranges of values of x for which the graph of $y = f(x)$ is concave and convex.

E **P** **Q4**   Given that $f(x) = x^{\frac{2}{3}}$, $x \in \mathbb{R}$, $x \neq 0$, explain why the graph of $y = f(x)$ is concave.   *[3 marks]*

E **P** **Q5**   Prove that the graph of $y = f(x)$ is convex for $x > 0$, where $f(x) = \left(x + \frac{1}{x}\right)^2$.   *[4 marks]*

Q6   Show that the graph of $y = x^2 - \frac{1}{x}$ has a point of inflection at $(1, 0)$.

E **Q7**   For $f(x) = x^3 + 2x^2 + 3x + 3$, show that:

a) the graph of $y = f(x)$ has one point of inflection,   *[4 marks]*

b) the point of inflection is not a stationary point.   *[2 marks]*

E **Q8**   The curve $y = 5x^3(x - 1)^2$ has stationary points.

a) Find the x-coordinates of the stationary points.   *[4 marks]*

b) Show that there is a point of inflection at $x = 0$.   *[2 marks]*

E **Q9**   A curve has equation $y = -2x^3 + 5x^2 - 4x$.

a) Find $\frac{d^2y}{dx^2}$.   *[2 marks]*

b) Show that the curve has a point of inflection and give its exact coordinates.   *[4 marks]*

c) State whether the point found in part b) is a stationary point of inflection. Explain your answer.   *[1 mark]*

Q10   Find the coordinates of the stationary points of the graph of $y = \frac{1}{10}x^5 - \frac{1}{3}x^3 + \frac{1}{2}x + 4$ and determine their nature.

Challenge

P **Q11**   By finding a suitable example, show that the graph of $y = f(x)$, where $f(x)$ is an increasing function, can also be concave for all positive values of x.

7.4   The Chain Rule

The **chain rule** helps you **differentiate** complicated functions by **splitting them up** into functions that are easier to differentiate. The trick is spotting **how** to split them up, and choosing the right bit to **substitute**.

Once you've worked out how to split up the function, you can differentiate it using this **formula**:

If $y = f(u)$ and $u = g(x)$ then:
$$\frac{dy}{dx} = \frac{dy}{du} \times \frac{du}{dx}$$

> **Tip** It may help you to think of the derivatives as fractions (they're not, but it's a good way to think of it). Then the du's cancel:
> $$\frac{dy}{du} \times \frac{du}{dx} = \frac{dy}{dx}$$
> If you remember this you'll never get the order wrong.

To differentiate a function using the chain rule, just follow these steps:

1. Pick a suitable function of x for 'u' and rewrite y in terms of u.

2. Differentiate u (with respect to x) to get $\dfrac{du}{dx}$...

3. ...and differentiate y (with respect to u) to get $\dfrac{dy}{du}$.

4. Put it all in the formula and write everything in terms of x.

The important part is **choosing** which bit to make into u. The aim is to split it into **two separate functions** that you can **easily differentiate**.

*   If you have a function inside **brackets**, then the part **inside** the brackets is normally u:

    $y = (x + 1)^2$ can be written as $y = u^2$ where $u = x + 1$.

    Now both u^2 and $x + 1$ are **easy to differentiate**, and the chain rule formula does the hard work for you.

*   If there's a **trig function** or **log** involved, it's usually the part **inside** the trig function or log:

    > **Tip** Don't worry if you don't know how to differentiate trig functions yet — it's covered later in this chapter.

    $y = \sin x^2$ can be written as $y = \sin u$ where $u = x^2$

    Again, you end up with 2 functions that are **easy to differentiate**, so you just differentiate each one **separately** then put it all in the **formula**.

Example 1

Find $\dfrac{dy}{dx}$ if $y = (6x - 3)^5$.

$y = (6x - 3)^5$,
so let $y = u^5$ where $u = 6x - 3$ ←

> Replace part of the function with u. The bit inside the brackets is easy to differentiate, so make that u.

$\dfrac{dy}{du} = 5u^4 = 5(6x - 3)^4$ and $\dfrac{du}{dx} = 6$ ←

> Next, differentiate the two parts separately. Substitute $u = 6x - 3$ so they're in terms of x.

$\dfrac{dy}{dx} = \dfrac{dy}{du} \times \dfrac{du}{dx} = 5(6x - 3)^4 \times 6$ ←
$\qquad\qquad\qquad = 30(6x - 3)^4$

> Finally, put everything back into the chain rule formula and simplify.

Now that you can differentiate **functions of functions** using the chain rule, you can find the equation of a **tangent** or **normal** to a curve that has an equation given by a function of a function.

Example 2

Find the equation of the tangent to the curve $y = \dfrac{1}{\sqrt{x^2 + 3x}}$ at $\left(1, \dfrac{1}{2}\right)$.

$y = \dfrac{1}{\sqrt{x^2 + 3x}} = (x^2 + 3x)^{-\frac{1}{2}}$ ←

> This function's a little more complicated than the previous one, so it will help to first rewrite it in terms of powers.

$y = (x^2 + 3x)^{-\frac{1}{2}}$,
so let $y = u^{-\frac{1}{2}}$ where $u = x^2 + 3x$ ←

> Then identify which part to turn into u.

$y = u^{-\frac{1}{2}} \Rightarrow \dfrac{dy}{du} = \left(-\dfrac{1}{2}\right)u^{-\frac{3}{2}} = -\dfrac{1}{2(\sqrt{x^2 + 3x})^3}$

$u = x^2 + 3x \Rightarrow \dfrac{du}{dx} = 2x + 3$ ←

> You now have two parts, so differentiate them separately. Then substitute $u = x^2 + 3x$ so they're both in terms of x.

$\dfrac{dy}{dx} = \dfrac{dy}{du} \times \dfrac{du}{dx} = \left(-\dfrac{1}{2(\sqrt{x^2 + 3x})^3}\right) \times (2x + 3)$ ←
$\qquad\qquad = -\dfrac{2x + 3}{2(\sqrt{x^2 + 3x})^3}$

> Now put it all back into the chain rule formula and simplify.

When $x = 1$, $\dfrac{dy}{dx} = -\dfrac{(2 \times 1) + 3}{2(\sqrt{1^2 + (3 \times 1)})^3} = -\dfrac{5}{16}$ ←

> To find the equation of the tangent, you first need to know the gradient at the point $\left(1, \dfrac{1}{2}\right)$, so put the x-value into your equation for $\dfrac{dy}{dx}$.

$y = mx + c \Rightarrow \dfrac{1}{2} = \left(-\dfrac{5}{16} \times 1\right) + c \Rightarrow c = \dfrac{13}{16}$ ←

> Then use your gradient and the values you're given to find c.

$y = -\dfrac{5}{16}x + \dfrac{13}{16}$ ←

> Now you can write the equation of the tangent at $\left(1, \dfrac{1}{2}\right)$.

Exercise 7.4

Q1 Differentiate with respect to x:

a) $y = (x + 7)^2$

b) $y = (2x - 1)^5$

c) $y = 3(4 - x)^8$

d) $y = (3 - 2x)^7$

e) $y = (x^2 + 3)^5$

f) $y = (5x^2 + 3)^2$

Q2 Find $f'(x)$ for the following:

a) $f(x) = (4x^3 - 9)^8$

b) $f(x) = (6 - 7x^2)^4$

c) $f(x) = (x^2 + 5x + 7)^6$

d) $f(x) = (x + 4)^{-3}$

e) $f(x) = (5 - 3x)^{-2}$

f) $f(x) = \dfrac{1}{(5 - 3x)^4}$

g) $f(x) = (3x^2 + 4)^{\frac{3}{2}}$

h) $f(x) = \dfrac{1}{\sqrt{5 - 3x}}$

i) $f(x) = \dfrac{1}{\sqrt{x^3 + 2x^2}}$

Q3 Find the exact value of $\dfrac{dy}{dx}$ when $x = 1$ for:

a) $y = \dfrac{1}{\sqrt{5x - 3x^2}}$

b) $y = \dfrac{12}{\sqrt[3]{x + 6}}$

Q4 Differentiate $\left(\sqrt{x} + \dfrac{1}{\sqrt{x}}\right)^2$ with respect to x by:

a) Multiplying the brackets out and differentiating term by term.

b) Using the chain rule.

Q5 Find the equation of the tangent to the curve $y = (x - 3)^5$ at $(1, -32)$.

Q6 Find the equation of the normal to the curve $y = \dfrac{1}{4}(x - 7)^4$ when $x = 6$.

> **Q6 Hint**
> The gradient of the normal to a curve is just $-1 \div$ gradient of the tangent (this was covered in Year 1 if you need a refresher).

Q7 Find the value of $\dfrac{dy}{dx}$ when $x = 1$ for $y = (7x^2 - 3)^{-4}$.

Q8 Find $f'(x)$ if $f(x) = \dfrac{7}{\sqrt[3]{3 - 2x}}$.

Q9 Find the equation of the tangent to the curve $y = \sqrt{5x - 1}$ when $x = 2$, in the form $ax + by + c = 0$, $a, b, c \in \mathbb{Z}$.

> **Q9 Hint**
> $a, b, c \in \mathbb{Z}$ just means that a, b and c are integers.

Q10 Find the equation of the normal to the curve $y = \sqrt[3]{3x - 7}$ when $x = 5$.

E **Q11** A curve has the equation $y = \left(\dfrac{1}{x} + 3\right)^4$.

P

a) Find $\dfrac{dy}{dx}$. *[2 marks]*

The line T is the tangent to the curve at the point where $x = -1$.
T intersects the x-axis at A and the y-axis at B. The point O is the origin.

b) Find the area of the triangle OAB. *[4 marks]*

c) Find the coordinates of the minimum point on the curve. *[2 marks]*

Q12 Find the equation of the tangent to the curve $y = (x^4 + x^3 + x^2)^2$ when $x = -1$.

P **Q13** Show that the curve $y = (2x - 3)^7$ has one point of inflection, and find the coordinates of that point.

> **Problem Solving**
>
> You'll need to use the chain rule twice for Q13-14.

P **Q14** Find the ranges of values of x for which the curve

$y = \left(\dfrac{x}{4} - 2\right)^3$ is convex and concave.

Challenge

E **P** **Q15** A curve has equation $y = (ax^2 + bx + c)^{2020}$, where a, b and c are non-zero constants.

    a) Given that $\dfrac{dy}{dx} = (404x - 1010)(ax^2 + bx + c)^{2019}$, find the values of a and b.   *[3 marks]*

    b) Given that the equation of the tangent to the curve at the point where $x = 0$ is $y = 1 - 1010x$, find the value of c.   *[2 marks]*

P **Q16** A function is called 'even' if $f(x) = f(-x)$ for all real values of x, and a function is called 'odd' if $f(x) = -f(-x)$ for all real values of x. Use the chain rule to show that:

    a) the derivative of an even function is odd.

    b) the derivative of an odd function is even.

7.5   Finding $\dfrac{dy}{dx}$ when $x = f(y)$

The principle of the chain rule can also be used where **x is given in terms of y** (i.e. $x = f(y)$). This comes from a little mathematical rearranging, and you'll find it's often quite useful:

$\dfrac{dy}{dx} \times \dfrac{dx}{dy} = \dfrac{dy}{dy} = 1$, so rearranging gives $\dfrac{dy}{dx} = \dfrac{1}{\left(\frac{dx}{dy}\right)}$.

> **Tip** As with the chain rule, treat the derivatives as fractions to make this result easier to follow (they're not actually fractions though).

So to differentiate $x = f(y)$, use: $\boxed{\dfrac{dy}{dx} = \dfrac{1}{\left(\frac{dx}{dy}\right)}}$

Example 1

A curve has the equation $x = y^3 + 2y - 7$. Find $\dfrac{dy}{dx}$ at the point $(-4, 1)$.

$x = y^3 + 2y - 7 \implies \dfrac{dx}{dy} = 3y^2 + 2$ Forget that the x's and y's are in the 'wrong' place and differentiate x with respect to y.

$\dfrac{dy}{dx} = \dfrac{1}{3y^2 + 2}$ Use $\dfrac{dy}{dx} = \dfrac{1}{\left(\frac{dx}{dy}\right)}$ to find $\dfrac{dy}{dx}$.

$\dfrac{dy}{dx} = \dfrac{1}{3(1)^2 + 2} = \dfrac{1}{5} = 0.2$, $y = 1$ at the point $(-4, 1)$, so put this in the equation.

so $\dfrac{dy}{dx} = 0.2$ at the point $(-4, 1)$.

Exercise 7.5

Q1 Find $\frac{dy}{dx}$ for each of the following functions at the given point.

In each case, express $\frac{dy}{dx}$ in terms of y.

a) $x = 3y^2 + 5y + 7$ at $(5, -1)$

b) $x = y^3 - 2y$ at $(-4, -2)$

c) $x = (2y + 1)(y - 2)$ at $(3, -1)$

d) $x = \frac{4 + y^2}{y}$ at $(5, 4)$

P **Q2** Find $\frac{dy}{dx}$ in terms of y if $x = (2y^3 - 5)^3$.

P **Q3** Given that $x = \sqrt{4 + y}$, find $\frac{dy}{dx}$ in terms of x by:

a) finding $\frac{dx}{dy}$ first,

b) rearranging into the form $y = f(x)$.

P **Q4** Given that $x = \frac{1}{\sqrt{2y - 3}}$, find $\frac{dy}{dx}$ in terms of x by:

a) finding $\frac{dx}{dy}$ first,

b) rearranging into the form $y = f(x)$.

E **P** **Q5** A curve drawn on standard x-y axes has equation $x = 3y^2 - 2y + 1$.

a) Find $\frac{dy}{dx}$ in terms of y. *[2 marks]*

b) Find the coordinates of the point on the curve with a vertical tangent. *[2 marks]*

c) Prove that the curve has no stationary points. *[2 marks]*

E **P** **Q6** A curve C drawn on standard x-y axes has equation $x = 2\sqrt{y} - \frac{2}{\sqrt{y}}$.

a) Find the equation of the tangent to C at the point where $y = 4$. *[4 marks]*

b) The tangent to C at $(0, 1)$ intersects the tangent in part a) at point P.
Find the coordinates of P. *[4 marks]*

E **P** **Q7** A curve drawn on standard x-y axes has equation $x = (3y^3 - y - 1)^5$.

a) Find $\frac{dy}{dx}$ in terms of y. *[3 marks]*

b) Find the equation of the normal to the curve at the point $(1, 1)$. *[3 marks]*

The normal to the curve at the point where $x = 0$ is horizontal.

c) Find the coordinates of the point where the normal to the curve at $(1, 1)$
and the tangent to the curve at $x = 0$ intersect. *[2 marks]*

7.6   Differentiating e^x and $e^{f(x)}$

Remember from Year 1 that 'e' is just a number for which the **gradient of e^x is e^x**, which makes it pretty simple to **differentiate**:

$$y = e^x \Rightarrow \frac{dy}{dx} = e^x$$

You can use the chain rule to show another useful relation involving exponentials. If you replace x with $f(x)$, you have a **function of a function**:

$y = e^{f(x)}$, so let $y = e^u$ where $u = f(x)$. So $\frac{dy}{du} = e^u = e^{f(x)}$ and $\frac{du}{dx} = f'(x)$.

Putting it into the chain rule formula you get: $\frac{dy}{dx} = \frac{dy}{du} \times \frac{du}{dx} = e^{f(x)} \times f'(x) = f'(x)e^{f(x)}$

This works because $e^{f(x)}$ is a special case — the 'e' part stays the same when you differentiate, so you only have to worry about the $f(x)$ part. You can just learn the formula:

$$y = e^{f(x)} \Rightarrow \frac{dy}{dx} = f'(x)e^{f(x)}$$

Use this formula to differentiate exponential functions.

Example 1

Find $\frac{dy}{dx}$ if $y = e^{(3x-2)}$.

$y = e^{f(x)}$, where $f(x) = 3x - 2$ ←   Write down $f(x)$.

$f'(x) = 3$ ←   Differentiate $f(x)$.

$y = e^{(3x-2)} \Rightarrow \frac{dy}{dx} = f'(x)e^{f(x)} = 3e^{(3x-2)}$ ←   Now put the right parts back into the formula for $\frac{dy}{dx}$.

Example 2

If $f(x) = e^{x^2} + 2e^x$, find $f'(x)$ when $x = 0$.

$g(x) = 2e^x \Rightarrow g'(x) = 2e^x$ ←   The function is in 2 parts, so differentiate them separately. The second bit's easy so start with that.

$v = e^{x^2}$,
so let $v = e^u$ where $u = x^2$ ←   For the first bit, you could use the formula used on the previous example, but let's use the chain rule here just to show how it works.

$\frac{dv}{du} = e^u$ and $\frac{du}{dx} = 2x$ ←

$\frac{dv}{dx} = \frac{dv}{du} \times \frac{du}{dx} = e^u \times 2x = 2xe^{x^2}$ ←   Both u and v are now easy to differentiate.

$f'(x) = 2xe^{x^2} + 2e^x$ ←   Now put the bits back together.

$f'(0) = (2 \times 0 \times e^{0^2}) + 2e^0$ ←   Finally, work out the value of $f'(x)$ at $x = 0$.
$f'(0) = 0 + 2 = 2$

Example 3

The graph of $y = e^{2x} - 6x^2$ has one point of inflection. Find the exact coordinates of this point.

Let $y_1 = e^{2x}$ and $y_2 = -6x^2$ ◄————— Start by finding $\dfrac{dy}{dx}$. The function is in two parts so differentiate them separately.

$y_1 = e^{2x} = e^{f(x)}$
$f(x) = 2x$ so $f'(x) = 2$
$\Rightarrow \dfrac{dy_1}{dx} = f'(x)e^{f(x)} = 2e^{2x}$ ◄————— For the first bit, use the formula on the previous page.

$y_2 = -6x^2 \Rightarrow \dfrac{dy_2}{dx} = -12x$ ◄————— The second bit is just regular polynomial differentiation.

$\dfrac{dy}{dx} = 2e^{2x} - 12x$ ◄————— Now just put the two parts back together.

$\dfrac{dy}{dx} = 2e^{2x} - 12x = 2e^{f(x)} - 12x$

$\dfrac{d^2y}{dx^2} = 2f'(x)e^{f(x)} - 12$
$\qquad = 2(2 \times e^{2x}) - 12 = 4e^{2x} - 12$ ◄————— Differentiate again to find $\dfrac{d^2y}{dx^2}$.

Tip The question says there's one point of inflection, so this is the point you're looking for — there's no need to prove it's a point of inflection.

Set $\dfrac{d^2y}{dx^2} = 0$ to find the point of inflection.

$4e^{2x} - 12 = 0 \Rightarrow e^{2x} = 3$ ◄—————

$\Rightarrow 2x = \ln 3 \Rightarrow x = \dfrac{1}{2}\ln 3$ ◄—————

$\Rightarrow x = \ln 3^{\frac{1}{2}} = \ln\sqrt{3}$ ◄————— Take logs of both sides to solve for x — remember, $\ln(e^{2x}) = 2x$ and $k\log x = \log x^k$.

$y = e^{2\ln\sqrt{3}} - 6(\ln\sqrt{3})^2$ ◄—————

$\Rightarrow y = e^{\ln 3} - 6(\ln\sqrt{3})^2$

$\Rightarrow y = 3 - 6(\ln\sqrt{3})^2$

Substitute $x = \ln\sqrt{3}$ into $y = e^{2x} - 6x^2$ to find the y-coordinate. You can simplify using $2\ln\sqrt{3} = \ln\sqrt{3}^2 = \ln 3$ and $e^{\ln 3} = 3$.

$(\ln\sqrt{3}, 3 - 6(\ln\sqrt{3})^2)$ ◄————— Write the coordinates for the point of inflection.

Exercise 7.6

Q1 Differentiate with respect to x:
   a) $y = e^{3x}$
   b) $y = e^{2x-5}$
   c) $y = e^{x+7}$
   d) $y = e^{3x+9}$
   e) $y = e^{7-2x}$
   f) $y = e^{x^3}$

Q2 Find $f'(x)$ if:
   a) $f(x) = e^{x^3+3x}$
   b) $f(x) = e^{x^3-3x-5}$
   c) $f(x) = e^{x(2x+1)}$

Q3 Find $f'(x)$ if:
   a) $f(x) = \dfrac{1}{2}(e^x - e^{-x})$
   b) $f(x) = e^{(x+3)(x+4)}$
   c) $f(x) = e^{x^4+3x^2} + 2e^{2x}$

Q4 Find the equation of the tangent to the curve $y = e^{2x}$ at the point $(0, 1)$.

Q5-7 Hint Where the questions ask for exact answers, that means they're likely to include e^a or ln a (where a is a number) rather than working out the actual numbers.

E Q5 Find the exact value of the x-coordinate of the point of the inflection on the curve $y = \frac{1}{2}x^2 - e^{2x-6}$. *[4 marks]*

E Q6 Find the equation of the tangent to the curve $y = e^{2x^2}$ when $x = 1$. Leave the numbers in your answer in exact form. *[4 marks]*

E Q7 If $f(x) = \sqrt{(e^x + e^{2x})}$, find the exact value of $f'(x)$ when $x = 0$. *[4 marks]*

E P Q8 A function is defined as $f(x) = \dfrac{e^x + e^{-x}}{2}$.

    a) Find $f'(x)$ and $f''(x)$. *[3 marks]*

    b) Find the coordinates of the stationary point on the curve $y = f(x)$. *[3 marks]*

    c) Prove that the stationary point found in part b) is a minimum. *[1 mark]*

    d) State the range of the function $f(x)$. *[1 mark]*

P Q9 Show that the curve $y = e^{2x-4} - x$ is convex for all values of x.

Q10 Find the equation of the normal to the curve $y = e^{3x} + 3$ where it cuts the y-axis.

Q11 Find the exact coordinates of any points of inflection on the curve $y = 2e^{2x} - \frac{1}{2}e^{3-4x}$.

E P Q12 A function is defined as $f(x) = e^{-x^2}$.

    a) Find the range of values of x for which $f(x)$ is decreasing. *[3 marks]*

    b) Find the coordinates of the maximum point on the curve $y = f(x)$. *[1 mark]*

    c) Find the range of the function $f(x)$. *[2 marks]*

    d) Find the coordinates of the stationary point on the curve $y = \sqrt{3 - e^{-x^2}}$. *[4 marks]*

Q13 Show that the curve $y = e^{x^3 - 3x - 5}$ has stationary points at $x = \pm 1$.

P Q14 Find the x-coordinate of the stationary point for the curve $y = e^{3x} - 6x$ and determine the nature of this point. Leave the numbers in your answer in exact form.

Problem Solving To determine the nature of the stationary point you need to look at the sign of $\dfrac{d^2y}{dx^2}$ at the point.

7.7 Differentiating ln x and ln (f(x))

The natural logarithm of a function is the logarithm with base e, written as ln x.
Differentiating natural logarithms also uses the chain rule:

- If $y = \ln x$, then $x = e^y$.

- Differentiating gives $\dfrac{dx}{dy} = e^y$, and $\dfrac{dy}{dx} = \dfrac{1}{\left(\dfrac{dx}{dy}\right)} = \dfrac{1}{e^y} = \dfrac{1}{x}$ (since $x = e^y$).

> **Tip** Look back at p.210 for more on $\dfrac{dx}{dy}$.

- This gives the result: $\boxed{y = \ln x \Rightarrow \dfrac{dy}{dx} = \dfrac{1}{x}}$

Example 1

a) Find $\dfrac{dy}{dx}$ if $y = \ln (2x + 3)$.

$y = \ln (2x + 3)$,
so let $y = \ln u$ where $u = 2x + 3$ ← It's a function of a function, so use the chain rule. Start by replacing part of the function with u.

$\Rightarrow \dfrac{dy}{du} = \dfrac{1}{u} = \dfrac{1}{2x+3}$ and $\dfrac{du}{dx} = 2$ ← Next, differentiate the two parts separately.

$\dfrac{dy}{dx} = \dfrac{dy}{du} \times \dfrac{du}{dx} = \dfrac{1}{2x+3} \times 2 = \dfrac{2}{2x+3}$ ← Now put all the parts into the chain rule formula.

b) Find $\dfrac{dy}{dx}$ if $y = \ln (x^2 + 3)$.

$y = \ln u$ and $u = x^2 + 3$, ← Use the chain rule again for this one.

$\dfrac{dy}{du} = \dfrac{1}{u} = \dfrac{1}{x^2+3}$ and $\dfrac{du}{dx} = 2x$ ← Differentiate the two parts separately then put all the parts into the chain rule formula.

$\Rightarrow \dfrac{dy}{dx} = \dfrac{dy}{du} \times \dfrac{du}{dx} = \dfrac{1}{x^2+3} \times 2x = \dfrac{2x}{x^2+3}$

Look at the final answer from those examples. It comes out to $\dfrac{f'(x)}{f(x)}$.

This isn't a coincidence — it will always be the case for $y = \ln (f(x))$, so you can just learn the result:

$\boxed{y = \ln (f(x)) \Rightarrow \dfrac{dy}{dx} = \dfrac{f'(x)}{f(x)}}$

Example 2

Find $g'(x)$ if $g(x) = \ln (x^3 - 4x)$.

$g'(x) = \dfrac{f'(x)}{f(x)}$, $f(x) = x^3 - 4x \Rightarrow f'(x) = 3x^2 - 4$ ← $g(x)$ is in the form $\ln (f(x))$, so use the formula above.

$g(x) = \ln(x^3 - 4x) \Rightarrow g'(x) = \dfrac{f'(x)}{f(x)} = \dfrac{3x^2 - 4}{x^3 - 4x}$ ← Put $f(x)$ and $f'(x)$ back into the formula.

Exercise 7.7

Q1 Differentiate with respect to x:

a) $y = \ln(3x)$ b) $y = \ln(1 + x)$ c) $\ln(3 + 2x)$

d) $y = \ln(1 + 5x)$ e) $y = 4\ln(4x - 2)$ f) $9\ln(3x - 3)$

Q2 Differentiate with respect to x:

a) $y = \ln(1 + x^2)$ b) $y = \ln(2 + x)^2$ c) $\ln(2x - 8)^3$

d) $y = 3\ln x^3$ e) $y = \ln(x^3 + x^2)$ f) $\ln\sqrt{2x^2 - 4}$

Q3 Find $f'(x)$ if:

a) $f(x) = \ln\dfrac{1}{x}$ b) $f(x) = \ln\sqrt{x}$

E Q4 A function is defined as $f(x) = \ln(x + 2) + x^2$.

     a) Find $f''(x)$.   *[2 marks]*

     b) Show that $f(x)$ is convex when $2x^2 + 8x + 7 > 0$.   *[2 marks]*

     c) Hence find the values of x for which $f(x)$ is convex.   *[2 marks]*

P Q5 Find $f'(x)$ if $f(x) = \ln\left((2x + 1)^2\sqrt{x - 4}\right)$.
Give your answer as a single fraction.

> **Problem Solving**
>
> You'll need to rewrite some of these questions as the sum or difference of two logarithms before differentiating them.

P Q6 Find $f'(x)$ if $f(x) = \ln(x - \sqrt{x - 4})$.

P Q7 Find $f'(x)$ if $f(x) = \ln\left(\dfrac{(3x + 1)^2}{\sqrt{2x + 1}}\right)$.

P Q8 Find the equation of the tangent to the curve $y = \ln(3x)^2$:

     a) when $x = -2$ b) when $x = 2$

> **Problem Solving**
>
> For Q8-9, rewrite $\ln(f(x))^k$ as $k\ln(f(x))$ to make differentiation simpler.

P Q9 Find the equation of the normal to the curve $y = \ln(x + 6)^2$:

     a) when $x = -3$ b) when $x = 0$

P Q10 Find any stationary points for the curve $y = \ln(x^3 - 3x^2 + 3x)$.

E P Q11 A curve C has equation $y = \ln(x^4 + 2x^2)$.

     a) Find $\dfrac{dy}{dx}$, simplifying your answer.   *[2 marks]*

The line L is the normal to the curve at the point where $x = 1$.

     b) Find the equation of L, giving your answer in the form $y = ax + b + \ln 3$, where a and b are rational constants to be found.   *[3 marks]*

     c) Find the exact coordinates of the point where L crosses the x-axis.   *[2 marks]*

7.8 Differentiating a^x and $a^{f(x)}$

Differentiating a^x

For any constant a: $\quad \boxed{\dfrac{d}{dx}(a^x) = a^x \ln a}$

Tip The proof for this rule uses implicit differentiation — you can see it on p.249.

Example 1

Differentiate the following:

a) $y = 2^x$

$\quad \dfrac{dy}{dx} = 2^x \ln 2 \longleftarrow$ Use the formula above with $a = 2$.

b) $y = \left(\dfrac{1}{2}\right)^x$

$\quad \dfrac{dy}{dx} = \left(\dfrac{1}{2}\right)^x \ln\left(\dfrac{1}{2}\right) \longleftarrow$ Use the formula above with $a = \dfrac{1}{2}$.

$\quad = 2^{-x} \times (-\ln 2) = -2^{-x} \ln 2 \longleftarrow$ Then use the log laws to tidy it up.

Differentiating $a^{f(x)}$

Use the **chain rule** (see p.207) to differentiate functions of the form $a^{f(x)}$, by differentiating $y = a^u$ and $u = f(x)$ separately, then using the **chain rule formula**: $\quad \dfrac{dy}{dx} = \dfrac{dy}{du} \times \dfrac{du}{dx}$

Example 2

Find the equation of the tangent to the curve $y = 3^{-2x}$ at the point $\left(\dfrac{1}{2}, \dfrac{1}{3}\right)$.

$y = 3^{-2x}$ so let $y = 3^u$ where $u = -2x$ $\longleftarrow$ Rewrite y in terms of u, where u is a function of x.

$\dfrac{du}{dx} = -2$ and $\dfrac{dy}{du} = 3^u \ln 3$ $\longleftarrow$ Differentiate the two equations separately.

$\dfrac{dy}{dx} = \dfrac{dy}{du} \times \dfrac{du}{dx} = 3^u \ln 3 \times -2 = -2(3^{-2x}\ln 3)$ $\longleftarrow$ Put all the parts into the chain rule formula.

At $\left(\dfrac{1}{2}, \dfrac{1}{3}\right)$, $\dfrac{dy}{dx} = -2(3^{-1} \ln 3) = -\dfrac{2}{3}\ln 3$ $\longleftarrow$ Now we can find the gradient of the tangent at $x = \dfrac{1}{2}$.

$\dfrac{1}{3} = (-\dfrac{2}{3} \ln 3)\dfrac{1}{2} + c \Rightarrow c = \dfrac{1}{3} + \dfrac{1}{3} \ln 3$ $\longleftarrow$ The tangent has an equation in the form $y = mx + c$ and meets the curve at $\left(\dfrac{1}{2}, \dfrac{1}{3}\right)$, so you can find c.

$y = -\dfrac{2x}{3} \ln 3 + \dfrac{1}{3} + \dfrac{1}{3} \ln 3$ $\longleftarrow$

or $3y = (1 - 2x)\ln 3 + 1$ $\longleftarrow$ Now write out the equation of the tangent.

Exercise 7.8

Q1 Differentiate the following:

a) $y = 5^x$ b) $y = 3^{2x}$ c) $y = 10^{-x}$ d) $y = p^{qx}$

Q2 A curve has the equation $y = 2^{4x}$.

a) Show that the gradient of the curve is $\dfrac{dy}{dx} = 4(2^{4x} \ln 2)$.

b) Find the equation of the tangent to the curve when $x = 2$.

> **Q2b), Q4 Hint**
> You can leave your answer in terms of ln if you're not asked for a specific degree of accuracy. That way you're giving the exact answer.

P Q3 A curve $y = 2^{px}$ passes through the point (1, 32).

a) Find p.

b) Hence find the gradient of the curve at this point.

P Q4 A curve has the equation $y = p^{x^3}$.

a) Show that the gradient of the curve is $\dfrac{dy}{dx} = 3x^2(p^{x^3} \ln p)$.

b) If the curve passes through the point (2, 6561), find p.

c) Hence find the equation of the tangent to the curve when $x = 1$.

P Q5 The curve $y = 4^{\sqrt{x}}$ passes through the point (25, a).

Show that the equation of the tangent to the curve at (25, a) is $y = 142x - 2520$ (to 3 s.f.).

E P Q6 a) Prove that $f(x) = 3^{2-x}$ is a decreasing function for all real values of x.   *[3 marks]*

b) Prove that the curve $y = f(x)$, where $f(x) = 3^{2-x}$, is convex for all real values of x.   *[2 marks]*

E P Q7 A curve C has the equation $y = 2^{-3x}$. It passes through the point (2, b).

a) Find the gradient $\dfrac{dy}{dx}$ of the curve.   *[2 marks]*

b) Find b and the gradient of the curve at (2, b).   *[2 marks]*

c) Hence show that the equation of the tangent to the curve
at (2, b) is $64y = 1 + 6 \ln 2 - (3 \ln 2)x$.   *[2 marks]*

E P Q8 A curve C has the equation $y = 3^{x^2}$. It passes through the point (2, p).

a) Find the gradient $\dfrac{dy}{dx}$ of the curve.   *[2 marks]*

b) Find p and the gradient of the curve at (2, p).   *[2 marks]*

c) Hence show that the equation of the normal to the curve
at (2, p) is $y = 81 + \dfrac{2 - x}{324 \ln 3}$.   *[3 marks]*

E P Q9 A curve C has the equation $y = 3^x - 2^x$.

a) Find the gradient $\dfrac{dy}{dx}$ of the curve C.   *[2 marks]*

b) Show that the x-coordinate of the stationary point on the curve C is $\log_{\frac{3}{2}}\left(\dfrac{\ln 2}{\ln 3}\right)$.   *[3 marks]*

c) Prove that the stationary point on the curve C is a minimum point.   *[3 marks]*

d) Show that the only point of intersection of C with the coordinate axes is at (0, 0).   *[3 marks]*

e) Sketch the curve C, clearly indicating any stationary points.   *[2 marks]*

7.9   Differentiating Sin, Cos and Tan

For **trigonometric functions** where the angle is measured in **radians** the following rules apply:

$$y = \sin x$$
$$\frac{dy}{dx} = \cos x$$

$$y = \cos x$$
$$\frac{dy}{dx} = -\sin x$$

$$y = \tan x$$
$$\frac{dy}{dx} = \sec^2 x$$

You can prove the rules for sin x and cos x using **differentiation from first principles**.

Recall the diagram from Year 1:

As h gets smaller, the gradient of the line passing through $(x, f(x))$ and $(x + h, f(x + h))$ gets closer to $f'(x)$.

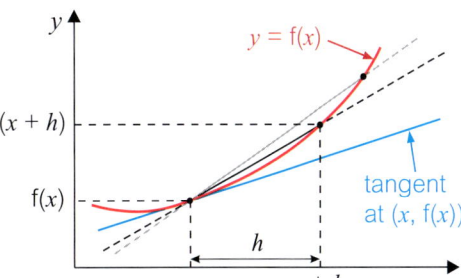

Example 1

Show using differentiation from first principles that if $y = \sin x$, $\dfrac{dy}{dx} = \cos x$.

$$\frac{dy}{dx} = \lim_{h \to 0} \left[\frac{f(x + h) - f(x)}{(x + h) - x} \right]$$

     Write down the formula for differentiating from first principles.

$$= \lim_{h \to 0} \left[\frac{\sin(x + h) - \sin x}{(x + h) - x} \right]$$

     Use the fact that $y = f(x)$ to replace $f(x)$ with sin x.

$$= \lim_{h \to 0} \left[\frac{\sin x \cos h + \cos x \sin h - \sin x}{(x + h) - x} \right]$$

     You can use the sin addition formula (see p.96) to expand sin $(x + h)$ on top of the fraction.

$$= \lim_{h \to 0} \left[\frac{\sin x (\cos h - 1) + \cos x \sin h}{h} \right]$$

     Separate the sin x and cos x terms into two fractions.

$$= \lim_{h \to 0} \left[\frac{\sin x (\cos h - 1)}{h} + \frac{\cos x \sin h}{h} \right]$$

     You're interested in when h gets really small, so you can use the small angle approximations $\sin \theta \approx \theta$ and $\cos \theta \approx 1 - \frac{1}{2}\theta^2$ (see p.73).

$$= \lim_{h \to 0} \left[\frac{\sin x \times \left(-\frac{1}{2}h^2\right)}{h} + \frac{\cos x \times h}{h} \right]$$

$$= \lim_{h \to 0} \left[-\frac{h \sin x}{2} + \cos x \right] = \cos x$$

     As $h \to 0$, the first term $\to 0$, so it disappears.

The equations for differentiating trig functions can be combined with the **chain rule** to differentiate **more complicated** functions.

Example 2

Differentiate the following with respect to x:

a) $y = \cos(2x)$

$y = \cos(2x)$, so let $y = \cos u$ where $u = 2x$ ← Rewrite y in terms of u, where u is a function of x.

$\dfrac{dy}{du} = -\sin u = -\sin(2x)$ and $\dfrac{du}{dx} = 2$ ← Differentiate the parts separately.

$\dfrac{dy}{dx} = \dfrac{dy}{du} \times \dfrac{du}{dx} = -2\sin(2x)$ ← Put the parts back together using the chain rule formula.

b) $y = 4\sin(x^2 + 1)$

$y = 4\sin(x^2 + 1)$, so let $y = 4\sin u$ where $u = x^2 + 1$ ← As before, work out which part needs to be u for the chain rule.

$\dfrac{dy}{du} = 4\cos u = 4\cos(x^2 + 1)$ and $\dfrac{du}{dx} = 2x$ ← Differentiate the parts separately.

$\dfrac{dy}{dx} = \dfrac{dy}{du} \times \dfrac{du}{dx} = 4\cos(x^2 + 1) \times 2x$
$= 8x\cos(x^2 + 1)$ ← Put the parts back together using the chain rule formula.

c) Find $\dfrac{dy}{dx}$ when $x = \tan(3y)$.

$x = \tan u, \; u = 3y$ ← Work out the part that needs to be u — this will be in terms of y in this case.

$\dfrac{dx}{du} = \sec^2 u = \sec^2(3y)$ and $\dfrac{du}{dy} = 3$ ← Differentiate the parts separately.

$\dfrac{dx}{dy} = \dfrac{dx}{du} \times \dfrac{du}{dy} = 3\sec^2(3y)$ ← Find $\dfrac{dx}{dy}$ with the chain rule.

$\dfrac{dy}{dx} = \dfrac{1}{3\sec^2(3y)} = \dfrac{1}{3}\cos^2(3y)$ ← Then use $\dfrac{dy}{dx} = \dfrac{1}{\left(\frac{dx}{dy}\right)}$ to get the final answer.

Once you get the hang of it, you don't need to use the chain rule every time. If it's a **simple function** inside, e.g. $\sin(kx)$, it just differentiates to $k\cos(kx)$. If it's more **complicated** though, like $\sin(x^3)$, it's worth using the **chain rule**.

When differentiating trig functions it's important to know **which part of the function** to turn into u. It's **not** always the part **inside the brackets**.

When you have a trig function multiplied by itself, like $\cos^2 x$, it's often easiest to turn the **trig function itself** into u.

> **Tip** Remember $\cos^2 x$ is another way of writing $(\cos x)^2$.

Example 3

Find $\dfrac{dy}{dx}$ if $y = \sin^3 x$.

> **Tip** Don't get $(\sin x)^3$ confused with $\sin x^3$ — for that, you'd take $u = x^3$, so end up with $3x^2 \cos x^3$ when you differentiate.

$y = \sin^3 x = (\sin x)^3$

so let $y = u^3$ where $u = \sin x$ ← Rewrite y in terms of u, where u is a function of x.

$y = u^3 \implies \dfrac{dy}{du} = 3u^2 = 3\sin^2 x$

and $\dfrac{du}{dx} = \cos x$ ← Differentiate y and u.

$\dfrac{dy}{dx} = \dfrac{dy}{du} \times \dfrac{du}{dx} = 3\sin^2 x \cos x$ ← Then put it all into the chain rule formula.

When you're differentiating trig functions, you'll sometimes be asked to **rearrange** your answer to show it's equal to a **different** trig function. It's worth making sure you're familiar with **trig identities** (see Chapter 3) so you can spot which ones to use and when to use them.

Example 4

For $y = 2\cos^2 x + \sin(2x)$, show that $\dfrac{dy}{dx} = 2(\cos(2x) - \sin(2x))$.

$y = 2\cos^2 x + \sin(2x) \implies y = 2(\cos x)^2 + \sin(2x)$ ← First rewrite the equation to make the chain rule easier to use.

Let $y_1 = 2(\cos x)^2$ and $y_2 = \sin(2x)$ ← You have two terms that are each a function of a function.

$y_1 = 2u^2$ where $u = \cos x$

$\implies \dfrac{dy_1}{du} = 4u = 4\cos x, \dfrac{du}{dx} = -\sin x$ ← Rewrite y_1 in terms of u, where u is a function of x. Then differentiate y_1 and u.

$y_2 = \sin v$ where $v = 2x$

$\implies \dfrac{dy_2}{dv} = \cos v = \cos 2x, \dfrac{dv}{dx} = 2$ ← Rewrite y_2 in terms of v, where v is a function of x. Then differentiate y_2 and v.

$\dfrac{dy}{dx} = [(4\cos x) \times (-\sin x)] + [(\cos(2x)) \times 2]$ ← Put it all back together using the chain rule and simplify.

$= 2\cos(2x) - 4\sin x \cos x$

$\sin(2x) \equiv 2\sin x \cos x \implies 4\sin x \cos x \equiv 2\sin(2x)$ ← From the target answer in the question it looks like you need a $\sin(2x)$ from somewhere, so use the double angle formula for sin (page 104).

$\implies \dfrac{dy}{dx} = 2\cos(2x) - 2\sin(2x)$

$= 2(\cos(2x) - \sin(2x))$ as required.

Exercise 7.9

Q1 Differentiate with respect to x:

a) $y = \sin(3x)$

b) $y = \cos(-2x)$

c) $y = \cos\dfrac{x}{2}$

d) $y = \sin\left(x + \dfrac{\pi}{4}\right)$

e) $y = 6\tan\dfrac{x}{2}$

f) $y = 3\tan(5x)$

> **Q1d) Hint** Don't worry, you don't need to use the sin addition formula — use the chain rule with $u = x + \dfrac{\pi}{4}$.

Q2 Find $f'(x)$ if $f(x) = 3\tan(2x - 1)$.

Q3 Find $f'(x)$ if $f(x) = 3\tan x + \tan(3x)$.

Q4 Find $f'(x)$ if $f(x) = \sin\left(x^2 + \dfrac{\pi}{3}\right)$.

Q5 Find $f'(x)$ if $f(x) = \sin^2 x$.

Q6 Find $f'(x)$ if $f(x) = 2\sin^3 x$.

Q7 a) Find $f'(x)$ if $f(x) = 3\sin x + 2\cos x$.

b) Find the value of x for which $f'(x) = 0$ and $0 \le x \le \dfrac{\pi}{2}$.

Q8 Find $\dfrac{dy}{dx}$ if $y = \dfrac{1}{\cos x}$.

Q9 Use differentiation from first principles to show that $\dfrac{dy}{dx} = -\sin x$ when $y = \cos x$.

P Q10 Differentiate $y = \cos^2 x$ by:

a) Using the chain rule directly.

b) Expressing y in terms of $\cos(2x)$ and differentiating the result.

> **Problem Solving** You'll need the double angle formulas for Q10-11 — see page 104.

P Q11 For $y = 6\cos^2 x - 2\sin(2x)$ show that $\dfrac{dy}{dx} = -6\sin(2x) - 4\cos(2x)$.

Q12 Find the gradient of the curve $y = \sin x$ when $x = \dfrac{\pi}{4}$.

P Q13 Find the equation of the normal to the curve $y = \cos(2x)$ when $x = \dfrac{\pi}{4}$.

E P Q14 A curve has equation $y = \sin\left(\dfrac{1}{x}\right)$ where $x \ne 0$.

a) Find $\dfrac{dy}{dx}$. *[2 marks]*

b) Hence prove that the curve has infinitely many stationary points, justifying your answer. *[2 marks]*

P Q15 For the curve $x = \sin(2y)$:

   a) Find the equation of the tangent at the point $\left(\dfrac{\sqrt{3}}{2}, \dfrac{\pi}{6}\right)$.

   b) Find the equation of the normal at the point $\left(\dfrac{\sqrt{3}}{2}, \dfrac{\pi}{6}\right)$.

P Q16 a) If $y = 2\sin(2x)\cos x$, express y as a difference of two expressions involving $\sin x$ and $\sin^3 x$.

   b) Hence find $\dfrac{dy}{dx}$.

Challenge

E
P Q17 A curve has equation $y = \sin(\sin x)$.

   a) Find $\dfrac{dy}{dx}$. *[2 marks]*

   b) Find the exact coordinates of the stationary points for $0 \leq x \leq 2\pi$ and determine the nature of each one. *[6 marks]*

   c) Hence, or otherwise, explain whether the curve has any points of inflection on the interval $0 \leq x \leq 2\pi$. *[1 mark]*

7.10 ## Differentiating by Using the Chain Rule Twice

Sometimes you'll have to use the chain rule **twice** when you have a function of a function of a function, like $\sin^3(x^2)$.

Example **1**

Find $\dfrac{dy}{dx}$ if $y = \sin^2(2x + 1)$.

$y = \sin^2(2x + 1) = [\sin(2x + 1)]^2$
$\Rightarrow \quad y = u^2, \ u = \sin(2x + 1)$

> Start by setting up the first stage of differentiation with the chain rule, remembering to rewrite the $\sin^2$ part to make differentiating easier.

$\dfrac{dy}{du} = 2u = 2\sin(2x + 1)$

> Finding $\dfrac{dy}{du}$ is easy, so start with that.

$u = \sin(2x + 1)$
so let $u = \sin v$ where $v = 2x + 1$

> To find $\dfrac{du}{dx}$ you need the chain rule again, so just set it up with u in terms of v instead of y in terms of u.

Tip Calling it v just means you don't end up with a load of u's floating around.

$u = \sin v \ \Rightarrow \ \dfrac{du}{dv} = \cos v = \cos(2x + 1)$

$v = 2x + 1 \ \Rightarrow \ \dfrac{dv}{dx} = 2$

> Then go through the usual stages of differentiating u and v separately.

continued on the next page...

$$\frac{du}{dx} = \frac{du}{dv} \times \frac{dv}{dx} = 2\cos(2x+1)$$ ← Use the chain rule formula to find $\frac{du}{dx}$.

$$\frac{dy}{dx} = \frac{dy}{du} \times \frac{du}{dx}$$

$$= [2\sin(2x+1)] \times [2\cos(2x+1)]$$ ← Finally use the chain rule formula again to find $\frac{dy}{dx}$ using the value of $\frac{dy}{du}$ and $\frac{du}{dx}$.

$$= 4\sin(2x+1)\cos(2x+1)$$

Exercise 7.10

Q1 Find $\frac{dy}{dx}$ for the following functions:

a) $y = \sin^2(x+2)$ b) $y = \cos^2(x^2)$ c) $y = \sqrt{\tan(4x)}$

Q2 Find $\frac{dy}{dx}$ if:

a) $y = \sin(\cos(2x))$ b) $y = 2\ln(\cos(3x))$

c) $y = \ln(\tan^2(x))$ d) $y = e^{\tan(2x)}$

Hint Differentiating e was covered on page 212 and differentiating ln was covered on page 215.

Q3 Differentiate the following functions with respect to x:

a) $y = \sin^4(x^2)$ b) $y = e^{\sin^2 x}$

c) $y = \tan^2(3x) + \sin x$ d) $y = e^{2\cos(2x)} + \cos^2(2x)$

E Q4 The function f is defined by $f(x) = e^{1+\sin 2x}$ for $x \in \mathbb{R}$.

a) Find the gradient of the curve $y = f(x)$ at the point where $x = \pi$, giving your answer in terms of e. *[4 marks]*

b) Work out the range for the function f. *[2 marks]*

The graph of $y = f(x)$ has infinitely many stationary points.

c) Find the coordinates of all the stationary points between $x = 0$ and $x = \pi$. Use your answer to part b) to determine their nature. *[4 marks]*

d) Hence, or otherwise, sketch the graph of $y = f(x)$ for $0 \le x \le \pi$. *[3 marks]*

Challenge

E P Q5 The function g is defined by $g(x) = \sin^2(x^2)\cos^2(x^2)$, where $0 < x < \frac{\pi}{2}$.

a) Find $g'(x)$. *[4 marks]*

Problem Solving Use a trig identity to turn this into something you can differentiate.

b) Find the exact x-coordinates of all of the stationary points on the curve $y = g(x)$. *[3 marks]*

c) State the maximum and minimum values of the function g. *[2 marks]*

7.11 Product Rule

To differentiate two functions multiplied together, use the **product rule**:

$$\text{If } y = uv, \text{ then } \frac{dy}{dx} = u\frac{dv}{dx} + v\frac{du}{dx}$$

Where u and v are functions of x, i.e. $u(x)$ and $v(x)$.

Proving this rule is a lot trickier than anything covered in this section (and you won't need to know how to do it in the exam), but it goes like this:

You need the formula for differentiation from first principles: $f'(x) = \lim_{h \to 0} \frac{f(x+h) - f(x)}{h}$

So when you have a product to differentiate, this can be written as: $(fg)'(x) = \lim_{h \to 0} \frac{fg(x+h) - fg(x)}{h} = \lim_{h \to 0} \frac{f(x+h)g(x+h) - f(x)g(x)}{h}$

The numerator of this fraction can be seen as the area of a **rectangle f(x + h) by g(x + h)** minus the area of a **rectangle f(x) by g(x)**.

It can therefore be rewritten as a sum of the areas of the "extra bits" on the diagram:

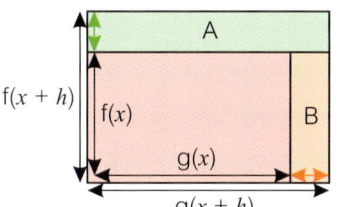

Tip The length of the green arrow is $f(x + h) - f(x)$ and the length of the orange arrow is $g(x + h) - g(x)$.

Area(A) = $g(x + h)[f(x + h) - f(x)]$

Area(B) = $f(x)[g(x + h) - g(x)]$

$(fg)'(x) = \lim_{h \to 0} \left(\frac{g(x+h)[f(x+h) - f(x)] + f(x)[g(x+h) - g(x)]}{h} \right)$

$= \lim_{h \to 0} \left(\frac{g(x+h)[f(x+h) - f(x)]}{h} \right) + \lim_{h \to 0} \left(\frac{f(x)[g(x+h) - g(x)]}{h} \right)$

$= \lim_{h \to 0}(g(x+h)) \lim_{h \to 0} \left(\frac{f(x+h) - f(x)}{h} \right) + \lim_{h \to 0}(f(x)) \lim_{h \to 0} \left(\frac{g(x+h) - g(x)}{h} \right) = g(x)f'(x) + f(x)g'(x)$

As $h \to 0$, $x + h \to x$ so this bit is just $g(x)$.

This is the definition of $f'(x)$.

This bit has no h's in it, so the limit as $h \to 0$ is just $f(x)$.

This is the definition of $g'(x)$.

Example 1

Differentiate $x^3 \tan x$ with respect to x.

$u = x^3$ and $v = \tan x$

The crucial thing is to write down everything in steps. Start by identifying 'u' and 'v'.

$\frac{du}{dx} = 3x^2$ and $\frac{dv}{dx} = \sec^2 x$

Now differentiate these two separately, with respect to x.

Tip See p.219 for how to differentiate $\tan x$.

$\frac{dy}{dx} = u\frac{dv}{dx} + v\frac{du}{dx}$

$= (x^3 \times \sec^2 x) + (\tan x \times 3x^2)$

$= x^3 \sec^2 x + 3x^2 \tan x$

Put all the bits into the formula and simplify.

You might have to differentiate functions using a mixture of the **product rule** and the **chain rule** (as well as the rules for e, ln and trig functions). In a question, you might be told which rules to use, but it's not guaranteed, so make sure you get used to spotting when the different rules are needed.

Example 2

Differentiate $e^{2x}\sqrt{2x-3}$ with respect to x.

$u = e^{2x}$ and $v = \sqrt{2x-3}$ ← It's a product of two functions, so start by identifying 'u' and 'v'.

$\dfrac{du}{dx} = 2e^{2x}$ and $\dfrac{dv}{dx} = \dfrac{1}{\sqrt{2x-3}}$ ← Each of these has been differentiated using the chain rule.

Tip The chain rule bit has been done all in one go here to save time, but if you need to, do it in steps just to make sure nothing goes wrong.

$\dfrac{dy}{dx} = u\dfrac{dv}{dx} + v\dfrac{du}{dx}$

$= \left(e^{2x} \times \dfrac{1}{\sqrt{2x-3}}\right) + \left(\sqrt{2x-3} \times 2e^{2x}\right)$ ← Put it all into the product rule formula.

$\dfrac{dy}{dx} = e^{2x}\left(\dfrac{1}{\sqrt{2x-3}} + 2\left(\sqrt{2x-3}\right)\right)$

$= e^{2x}\left(\dfrac{1 + 2(2x-3)}{\sqrt{2x-3}}\right) = \dfrac{e^{2x}(4x-5)}{\sqrt{2x-3}}$ ← As before, rearrange it and then simplify.

You might also see questions that ask you to rearrange the final answer to 'show that' it's equal to something, or to find the stationary points of a graph.

Example 3

Show that the derivative of $x^2(2x-1)^3$ is $2x(2x-1)^2(5x-1)$.

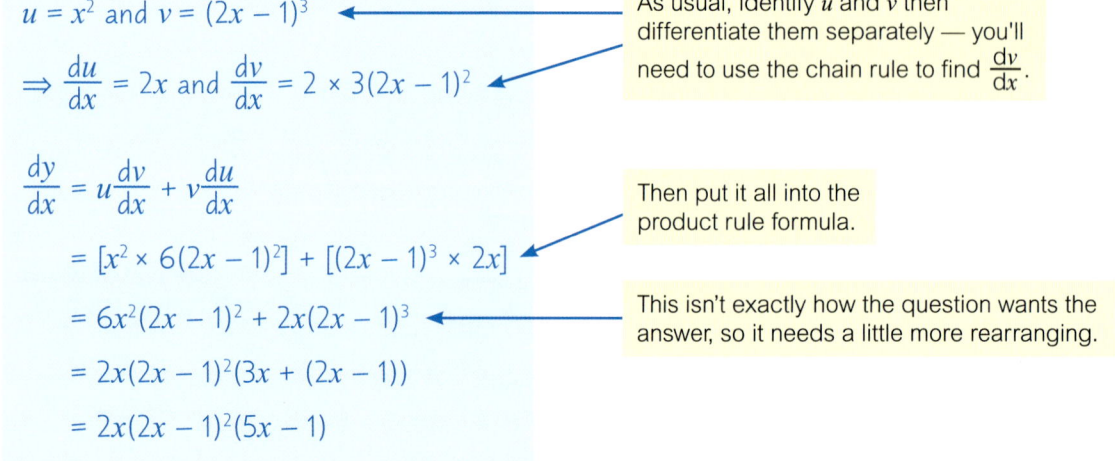

$u = x^2$ and $v = (2x-1)^3$ ← As usual, identify u and v then differentiate them separately — you'll need to use the chain rule to find $\dfrac{dv}{dx}$.

$\Rightarrow \dfrac{du}{dx} = 2x$ and $\dfrac{dv}{dx} = 2 \times 3(2x-1)^2$

$\dfrac{dy}{dx} = u\dfrac{dv}{dx} + v\dfrac{du}{dx}$ ← Then put it all into the product rule formula.

$= [x^2 \times 6(2x-1)^2] + [(2x-1)^3 \times 2x]$

$= 6x^2(2x-1)^2 + 2x(2x-1)^3$ ← This isn't exactly how the question wants the answer, so it needs a little more rearranging.

$= 2x(2x-1)^2(3x + (2x-1))$

$= 2x(2x-1)^2(5x-1)$

Example 4

The graph of $y = x^3 \ln x$ has one point of inflection when $x > 0$.
Find the x-coordinates of this point, leaving your answer as an exact value.

$u = x^3$ and $v = \ln x$

$\Rightarrow \dfrac{du}{dx} = 3x^2$ and $\dfrac{dv}{dx} = \dfrac{1}{x}$

So $\dfrac{dy}{dx} = u\dfrac{dv}{dx} + v\dfrac{du}{dx}$

$= \left(x^3 \times \dfrac{1}{x}\right) + (\ln x \times 3x^2) = x^2 + 3x^2 \ln x$

> You need to find the point where $\dfrac{d^2y}{dx^2} = 0$.

> Start by finding $\dfrac{dy}{dx}$ — identify u and v, differentiate them separately and then use the product rule.

First part: $\dfrac{d}{dx}(x^2) = 2x$

> Now differentiate again to find $\dfrac{d^2y}{dx^2}$ — the first part is easy.

Second part: $\dfrac{d}{dx}(3x^2 \ln x)$

$u = 3x^2$ and $v = \ln x$ $\Rightarrow \dfrac{du}{dx} = 6x$ and $\dfrac{dv}{dx} = \dfrac{1}{x}$

$v\dfrac{du}{dx} + u\dfrac{dv}{dx} = \left(3x^2 \times \dfrac{1}{x}\right) + (\ln x \times 6x)$

$= 3x + 6x \ln x$

> Use the product rule again for the second part.

$\dfrac{d^2y}{dx^2} = 2x + 3x + 6x \ln x = x(5 + 6 \ln x)$

> Put these parts back together to get $\dfrac{d^2y}{dx^2}$.

So $\dfrac{d^2y}{dx^2} = 0 \Rightarrow x = 0$ or $5 + 6\ln x = 0$

$x = 0$ is not a solution as $x > 0$

so $5 + 6\ln x = 0 \Rightarrow \ln x = -\dfrac{5}{6} \Rightarrow x = e^{-\frac{5}{6}}$

> Work out when $\dfrac{d^2y}{dx^2} = 0$.

Exercise 7.11

Q1 Differentiate $y = x(x + 2)$ with respect to x by:
  a) Multiplying the brackets out and differentiating directly.
  b) Using the product rule.

Q2 Differentiate with respect to x:
  a) $y = x^2(x + 6)^3$
  b) $y = x^3(5x + 2)^4$
  c) $y = x^3 e^x$
  d) $y = xe^{4x}$
  e) $y = xe^{x^2}$
  f) $y = e^{2x} \sin x$

Q3 Find f$'(x)$ if:
  a) $f(x) = x^3(x + 3)^{\frac{1}{2}}$
  b) $f(x) = \dfrac{x^2}{\sqrt{x - 7}}$
  c) $f(x) = x^4 \ln x$
  d) $f(x) = 4x \ln x^2$
  e) $f(x) = 2x^3 \cos x$
  f) $f(x) = x^2 \cos(2x)$

Q4 For parts a) and b), multiply out the brackets in your answer and simplify.

    a) Differentiate $y = (x + 1)^2(x^2 - 1)$.

    b) Differentiate $y = (x + 1)^3(x - 1)$.

    c) Your answers to part a) and part b) should be the same.
       Show by rearranging that the expressions for y in parts a) and b) are the same.

P Q5 Find the range of values of x for which
the curve $y = xe^x$ is concave.

> **Problem Solving**
>
> You'll have to differentiate twice in Q5.

P Q6 Find the equation of the tangent to the curve $y = (\sqrt{x + 2})(\sqrt{x + 7})$ at the point $(2, 6)$.
Write your answer in the form $ax + by + c = 0$, where a, b and c are integers.

P Q7 For the curve $y = \dfrac{\sqrt{x - 1}}{\sqrt{x + 4}}$

> **Problem Solving**
>
> To use the product rule in Q7 you'll need to rewrite the function at the bottom as a negative power.

    a) Find the equation of the tangent to the curve when $x = 5$
in the form $ax + by + c = 0$ where a, b and c are integers.

    b) Find the equation of the normal to the curve when $x = 5$
in the form $ax + by + c = 0$ where a, b and c are integers.

Q8 Differentiate $y = e^{x^2\sqrt{x+3}}$.

Q9 Find any stationary points for the curve $y = xe^{x - x^2}$.

Q10 a) Find any stationary points of the curve $y = (x - 2)^2(x + 4)^3$.

    b) By writing the first derivative of $y = (x - 2)^2(x + 4)^3$ in the form
$\dfrac{dy}{dx} = (Ax^2 + Bx + C)(x + D)^n$, find $\dfrac{d^2y}{dx^2}$ and hence identify
the nature of the stationary points of the curve.

Challenge

E **P** Q11 A function is defined such that $f(x) = e^x \sin x$.

    a) Find $f'(x)$ and $f''(x)$. *[4 marks]*

    b) Show that $\dfrac{d^4f}{dx^4} = -4\,f(x)$. *[2 marks]*

    c) Find the 100th derivative of $f(x)$. *[2 marks]*

E **P** Q12 A curve M has equation $y = \sin^3 x$.

    a) Show that $\dfrac{d^2y}{dx^2} = 6\sin x - 9\sin^3 x$. *[5 marks]*

    b) Find the range of values of x in the interval $0 \le x \le \pi$
for which the curve M is convex. *[4 marks]*

    c) Show that the curve M has a point of inflection at $x = n\pi$, where n is an integer. *[3 marks]*

    d) State a deduction you can make about the number of points of inflection. *[1 mark]*

7.12 Quotient Rule

In maths a **quotient** is one thing **divided** by another. As with the product rule, there's a rule that lets you differentiate quotients easily — the **quotient rule**:

$$\text{If } y = \frac{u}{v}, \text{ then } \frac{dy}{dx} = \frac{v\frac{du}{dx} - u\frac{dv}{dx}}{v^2}$$

Where u and v are functions of x, i.e. $u(x)$ and $v(x)$.

> **Tip** The quotient rule is basically just the product rule on $y = uv^{-1}$. It's usually quicker to use the quotient rule though.

There's also a proof for the quotient rule — again you won't need to know it for the exam, but you might find it helpful in understanding how it works.

As before, start with the differentiation from first principles:

$$\frac{d}{dx}f(x) = \lim_{h \to 0} \frac{f(x+h) - f(x)}{h}$$

And so for the quotient $\dfrac{f(x)}{g(x)}$, this becomes:

$$\frac{d}{dx}\left(\frac{f(x)}{g(x)}\right) = \lim_{h \to 0} \frac{\frac{f(x+h)}{g(x+h)} - \frac{f(x)}{g(x)}}{h}$$

Put the top of the fraction over a common denominator $(g(x+h)g(x))$ and multiply this common denominator by the h.

$$\frac{d}{dx}\left(\frac{f(x)}{g(x)}\right) = \lim_{h \to 0} \frac{f(x+h)g(x) - f(x)g(x+h)}{g(x+h)g(x)h}$$

The next stage is to add and subtract $f(x)g(x)$ and then factorise.

$$\frac{d}{dx}\left(\frac{f(x)}{g(x)}\right) = \lim_{h \to 0} \frac{f(x+h)g(x) - f(x)g(x) + f(x)g(x) - f(x)g(x+h)}{g(x+h)g(x)h}$$

$$\frac{d}{dx}\left(\frac{f(x)}{g(x)}\right) = \lim_{h \to 0} \frac{g(x)[f(x+h) - f(x)] - f(x)[g(x+h) - g(x)]}{g(x+h)g(x)h}$$

> **Problem Solving**
> Adding and subtracting the same thing is a classic trick in algebra. It's like adding zero, and can get you from algebraic mess to perfectly formed equations.

You might start to recognise the top row here. To make it a bit clearer, divide both the top and bottom by h, keeping $f(x)$ and $g(x)$ aside.

$$\frac{d}{dx}\left(\frac{f(x)}{g(x)}\right) = \lim_{h \to 0} \frac{g(x)\frac{f(x+h) - f(x)}{h} - f(x)\frac{g(x+h) - g(x)}{h}}{g(x+h)g(x)}$$

These bits are the definitions of $f'(x)$ and $g'(x)$.

As h tends to zero, $g(x+h)$ tends to $g(x)$ so the denominator becomes $g(x)g(x)$, or $(g(x))^2$.

So as $h \to 0$ the equation becomes:

$$\frac{d}{dx}\left(\frac{f(x)}{g(x)}\right) = \frac{g(x)f'(x) - f(x)g'(x)}{(g(x))^2}$$

> **Tip** $\dfrac{f'(x)g(x) - f(x)g'(x)}{(g(x))^2}$ is just another way of writing the quotient rule. It's given like this on the formula sheet.

Example 1

Find $\dfrac{dy}{dx}$ if $y = \dfrac{\sin x}{2x+1}$.

$u = \sin x \implies \dfrac{du}{dx} = \cos x$

and $v = 2x + 1 \implies \dfrac{dv}{dx} = 2$

You can see that y is a quotient in the form of $\frac{u}{v}$. First identify u and v and differentiate them separately.

$\dfrac{dy}{dx} = \dfrac{v\dfrac{du}{dx} - u\dfrac{dv}{dx}}{v^2} = \dfrac{(2x+1)(\cos x) - (\sin x)(2)}{(2x+1)^2}$

Then just put the correct bits into the quotient rule. It's important that you get things in the right order, so concentrate on what's going where.

$\dfrac{dy}{dx} = \dfrac{(2x+1)\cos x - 2\sin x}{(2x+1)^2}$

Now just neaten it up.

Example 2

Find the gradient of the tangent to the curve with equation $y = \dfrac{2x^2 - 1}{3x^2 + 1}$ at the point $(1, 0.25)$.

$u = 2x^2 - 1 \implies \dfrac{du}{dx} = 4x$

and $v = 3x^2 + 1 \implies \dfrac{dv}{dx} = 6x$

'Find the gradient of the tangent' means you have to differentiate. First identify u and v for the quotient rule, and differentiate separately.

$\dfrac{dy}{dx} = \dfrac{v\dfrac{du}{dx} - u\dfrac{dv}{dx}}{v^2}$

Then put everything into the quotient rule.

$= \dfrac{(3x^2 + 1)(4x) - (2x^2 - 1)(6x)}{(3x^2 + 1)^2}$

To make the expression easier to work with, simplify it where possible.

$= \dfrac{2x[2(3x^2 + 1) - 3(2x^2 - 1)]}{(3x^2 + 1)^2}$

$= \dfrac{2x[6x^2 + 2 - 6x^2 + 3]}{(3x^2 + 1)^2} = \dfrac{10x}{(3x^2 + 1)^2}$

Tip You don't necessarily have to simplify in this example as you're not asked for a simplified expression. You could just put $x = 1$ into the first unsimplified expression for the derivative to get the same answer.

$\dfrac{dy}{dx} = \dfrac{10}{(3+1)^2} = 0.625$

Finally, put in $x = 1$ to find the gradient at $(1, 0.25)$.

Example 3

Determine the nature of the stationary point of the curve $y = \dfrac{\ln x}{x^2}$ $(x > 0)$.

$u = \ln x \implies \dfrac{du}{dx} = \dfrac{1}{x}$ and $v = x^2 \implies \dfrac{dv}{dx} = 2x$

First use the quotient rule to find $\dfrac{dy}{dx}$.

$\dfrac{dy}{dx} = \dfrac{(x^2)\left(\dfrac{1}{x}\right) - (\ln x)(2x)}{x^4} = \dfrac{x - 2x\ln x}{x^4} = \dfrac{1 - 2\ln x}{x^3}$

$\dfrac{1 - 2\ln x}{x^3} = 0 \implies \ln x = \dfrac{1}{2} \implies x = e^{\frac{1}{2}}$

The stationary point occurs where $\dfrac{dy}{dx} = 0$ (i.e. zero gradient).

continued on the next page...

$u = 1 - 2\ln x \implies \dfrac{du}{dx} = -\dfrac{2}{x}$ and

$v = x^3 \implies \dfrac{dv}{dx} = 3x^2$

> To find out whether it's a maximum, minimum or point of inflection, differentiate $\dfrac{1-2\ln x}{x^3}$ using the quotient rule to get $\dfrac{d^2y}{dx^2}$.

So $\dfrac{d^2y}{dx^2} = \dfrac{(x^3)\left(-\dfrac{2}{x}\right) - (1 - 2\ln x)(3x^2)}{x^6}$

$= \dfrac{6x^2 \ln x - 5x^2}{x^6} = \dfrac{6\ln x - 5}{x^4}$

> Simplify the expression to make the next calculation easier.

$\dfrac{d^2y}{dx^2} = \dfrac{6\ln e^{\frac{1}{2}} - 5}{\left(e^{\frac{1}{2}}\right)^4} = \dfrac{3-5}{e^2} = -0.27...$

> Now put in the x-value of your stationary point and determine its nature.

$\dfrac{d^2y}{dx^2} < 0$, so it's a maximum stationary point.

As you saw on page 219, the derivative of $\tan x$ is $\sec^2 x$.

Because $\tan x = \dfrac{\sin x}{\cos x}$, you can prove this using the quotient rule.

Example 4

Prove that the derivative of $\tan x$ with respect to x is $\sec^2 x$.

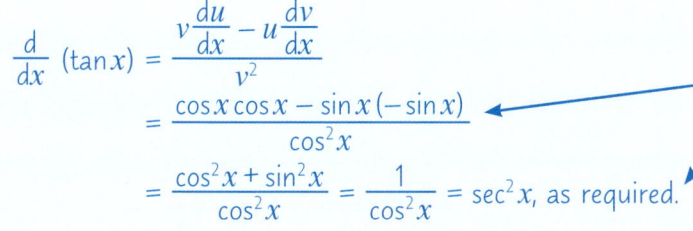

$\tan x = \dfrac{\sin x}{\cos x} = \dfrac{u}{v}$, so $u = \sin x$ and $v = \cos x$

> First write $\tan x$ out as a quotient and set up u and v for the quotient rule.

$\dfrac{du}{dx} = \cos x$ and $\dfrac{dv}{dx} = -\sin x$

$\dfrac{d}{dx}(\tan x) = \dfrac{v\dfrac{du}{dx} - u\dfrac{dv}{dx}}{v^2}$

> Then just put all the right bits into the quotient rule and simplify.

$= \dfrac{\cos x \cos x - \sin x(-\sin x)}{\cos^2 x}$

$= \dfrac{\cos^2 x + \sin^2 x}{\cos^2 x} = \dfrac{1}{\cos^2 x} = \sec^2 x$, as required.

> **Problem Solving**
>
> The identity $\cos^2 x + \sin^2 x \equiv 1$ was used here to simplify.

Exercise 7.12

Q1 Differentiate with respect to x:

a) $y = \dfrac{x+5}{x-3}$

b) $y = \dfrac{(x-7)^4}{(5-x)^3}$

c) $y = \dfrac{e^x}{x^2}$

d) $y = \dfrac{3x}{(x-1)^2}$

e) $y = \dfrac{\ln x^2}{5x}$

f) $y = \dfrac{4x^3}{e^x}$

Q2 Find $f'(x)$ for each of the following functions:

a) $f(x) = \dfrac{x^3}{(x+3)^3}$

b) $f(x) = \dfrac{x^2}{\sqrt{x-7}}$

c) $f(x) = \dfrac{e^{2x}}{e^{2x} + e^{-2x}}$

d) $f(x) = \dfrac{x}{\sin x}$

e) $f(x) = \dfrac{\sin x}{x}$

f) $f(x) = \dfrac{\cos x}{3x}$

Q3 Find f'(x) if $f(x) = \dfrac{x^2}{\tan x}$, giving your answer in terms of cot x and cosec x.

Q4 The graph of $y = \dfrac{5x - 4}{2x^2}$ has one stationary point.

Find the coordinates of this point and show that it is a maximum.

E Q5 A curve has equation $y = \dfrac{4x}{1 + x^2}$ and three points of inflection.

a) Find $\dfrac{d^2y}{dx^2}$, fully simplifying your answer. *[5 marks]*

b) Find the coordinates of the points of inflection on the curve. *[3 marks]*

Q6 Use the quotient rule to find the coordinates of any points of inflection of the graph $y = \dfrac{x}{e^x}$.

Q7 a) Differentiate $y = \dfrac{x}{\cos(2x)}$ with respect to x.

b) Show that $\dfrac{dy}{dx} = 0$ when $x = -\dfrac{1}{2} \cot(2x)$ (you do not need to solve this equation).

> **Q5 & 6 Hint** Remember from p.203 that if (a, b) is a point of inflection: $\dfrac{d^2y}{dx^2} = 0$ when $x = a$, and $\dfrac{d^2y}{dx^2}$ changes sign either side of $x = a$.

Q8 For the curve $y = \dfrac{1}{1 + 4\cos x}$:

a) Find the equation of the tangent to the curve when $x = \dfrac{\pi}{2}$.

b) Find the equation of the normal to the curve when $x = \dfrac{\pi}{2}$.

Q9 For the curve $y = \dfrac{2x}{\cos x}$, find the exact value of $\dfrac{dy}{dx}$ when $x = \dfrac{\pi}{3}$.

E Q10 Show that if $y = \dfrac{x - \sin x}{1 + \cos x}$ then $\dfrac{dy}{dx} = \dfrac{x \sin x}{(1 + \cos x)^2}$. *[3 marks]*

Q11 Find any stationary points on the curve $y = \dfrac{\cos x}{4 - 3\cos x}$ in the range $0 \le x \le 2\pi$.

P Q12 Differentiate $y = e^{\frac{1+x}{1-x}}$ with respect to x.

P Q13 Find the exact set of values of x for which $\dfrac{2 + 3x^2}{3x - 1}$ is increasing.

> **Problem Solving** You've seen increasing functions in Year 1 — a function f(x) is increasing when its gradient f'(x) > 0.

Challenge

E P Q14 A curve has equation $y = \dfrac{\ln x}{x}$.

a) Find $\dfrac{dy}{dx}$, fully simplifying your answer. *[3 marks]*

b) Find the coordinates of the only stationary point on the curve. *[3 marks]*

c) (i) By finding the second derivative, prove that the point found in part b) is a maximum point.

(ii) Hence, without using a calculator, show that $e^2 > 2^e$. *[5 marks]*

7.13 Differentiating Cosec, Sec and Cot

Remember from page 81 the definitions of these trig functions:

$$\operatorname{cosec} x \equiv \frac{1}{\sin x} \qquad \sec x \equiv \frac{1}{\cos x} \qquad \cot x \equiv \frac{1}{\tan x} \equiv \frac{\cos x}{\sin x}$$

Since **cosec**, **sec** and **cot** are just **reciprocals** of **sin**, **cos** and **tan**, the quotient rule can be used to differentiate them.

The following results are in the **formula booklet**, but it will help a lot if you understand where they come from.

$y = \operatorname{cosec} x$	$y = \sec x$	$y = \cot x$
$\dfrac{dy}{dx} = -\operatorname{cosec} x \cot x$	$\dfrac{dy}{dx} = \sec x \tan x$	$\dfrac{dy}{dx} = -\operatorname{cosec}^2 x$

> **Tip** The trig functions beginning with c give negative results when you differentiate them — i.e. cos, cosec and cot.

The formula booklet actually gives these results in the form cosec kx etc — but you can easily get these results by setting $k = 1$.

Example 1

a) Use the quotient rule to differentiate $y = \dfrac{\cos x}{\sin x}$, and hence show that for $y = \cot x$, $\dfrac{dy}{dx} = -\operatorname{cosec}^2 x$.

$u = \cos x$ and $v = \sin x$ ← Start off by identifying u and v.

$\dfrac{du}{dx} = -\sin x$ and $\dfrac{dv}{dx} = \cos x$ ← Differentiate them separately, and use the quotient rule (p.229).

$\dfrac{dy}{dx} = \dfrac{v\dfrac{du}{dx} - u\dfrac{dv}{dx}}{v^2} = \dfrac{(\sin x \times -\sin x) - (\cos x \times \cos x)}{(\sin x)^2}$

$= \dfrac{-\sin^2 x - \cos^2 x}{\sin^2 x} = \dfrac{-(\sin^2 x + \cos^2 x)}{\sin^2 x} = -\dfrac{1}{\sin^2 x}$ ← Simplify using the trig identity $\sin^2 x + \cos^2 x \equiv 1$.

$\tan x \equiv \dfrac{\sin x}{\cos x}$ and $\cot x \equiv \dfrac{1}{\tan x}$, so $y = \dfrac{\cos x}{\sin x} = \cot x$ ← Link this back to the question.

$\dfrac{dy}{dx} = \dfrac{-1}{\sin^2 x} = -\operatorname{cosec}^2 x$ ← You know $\operatorname{cosec} x \equiv \dfrac{1}{\sin x}$.

b) Show that $\dfrac{d}{dx} \operatorname{cosec} x = -\operatorname{cosec} x \cot x$.

$\operatorname{cosec} x = \dfrac{1}{\sin x}$, so ← Rewrite cosec x as $\dfrac{1}{\sin x}$ and use the quotient rule.

$u = 1 \Rightarrow \dfrac{du}{dx} = 0$ and $v = \sin x \Rightarrow \dfrac{dv}{dx} = \cos x$

$\dfrac{dy}{dx} = \dfrac{v\dfrac{du}{dx} - u\dfrac{dv}{dx}}{v^2} = \dfrac{(\sin x \times 0) - (1 \times \cos x)}{\sin^2 x} = -\dfrac{\cos x}{\sin^2 x}$

> **Tip** You could also use the chain rule on $\dfrac{1}{\sin x} = (\sin x)^{-1}$.

continued on the next page...

$$\frac{dy}{dx} = \frac{1}{\sin x} \times \left(-\frac{\cos x}{\sin x}\right) = -\text{cosec}\,x\,\cot x$$

> Simplify using $\cot x \equiv \dfrac{\cos x}{\sin x}$, and $\text{cosec}\,x \equiv \dfrac{1}{\sin x}$.

c) Show that $\dfrac{d}{dx}\sec x = \sec x \tan x$.

$$\sec x \equiv \frac{1}{\cos x}, \text{ so}$$

$$u = 1 \implies \frac{du}{dx} = 0 \text{ and } v = \cos x \implies \frac{dv}{dx} = -\sin x$$

$$\frac{dy}{dx} = \frac{v\dfrac{du}{dx} - u\dfrac{dv}{dx}}{v^2}$$

$$= \frac{(\cos x \times 0) - (1 \times -\sin x)}{\cos^2 x}$$

$$= \frac{\sin x}{\cos^2 x} = \frac{1}{\cos x} \times \frac{\sin x}{\cos x} = \sec x \tan x$$

> $\sec x \equiv \dfrac{1}{\cos x}$, so use the quotient rule.

> **Tip** You could also use the chain rule on $\dfrac{1}{\cos x} = (\cos x)^{-1}$.

> Simplify using $\tan x \equiv \dfrac{\sin x}{\cos x}$, and $\sec x \equiv \dfrac{1}{\cos x}$.

As with other rules covered in this chapter, the rules for $\sec x$, $\text{cosec}\,x$ and $\cot x$ can be used with the **chain**, **product** and **quotient rules** and in combination with all the other functions you've seen so far.

Example 2

a) Find $\dfrac{dy}{dx}$ if $y = \cot\dfrac{x}{2}$.

$$y = \cot u \text{ where } u = \frac{x}{2}$$

$$\implies \frac{dy}{du} = -\text{cosec}^2 u = -\text{cosec}^2 \frac{x}{2} \text{ and } \frac{du}{dx} = \frac{1}{2}$$

$$\text{So } \frac{dy}{dx} = \frac{dy}{du} \times \frac{du}{dx} = -\frac{1}{2}\text{cosec}^2 \frac{x}{2}$$

> y is a function ($\cot$) of a function ($\frac{x}{2}$), so you need the chain rule (see p.207).

> **Tip** Although $\cot x \equiv \dfrac{\cos x}{\sin x}$, you don't need the quotient rule as you know that $\cot x$ differentiates to give $-\text{cosec}^2 x$.

b) Find $\dfrac{dy}{dx}$ if $y = \sec(2x^2)$.

$$y = \sec u \text{ where } u = 2x^2$$

$$\implies \frac{dy}{du} = \sec u \tan u = \sec(2x^2)\tan(2x^2) \text{ and } \frac{du}{dx} = 4x$$

$$\text{So } \frac{dy}{dx} = \frac{dy}{du} \times \frac{du}{dx} = 4x\sec(2x^2)\tan(2x^2)$$

> y is another function ($\sec$) of a function ($2x^2$), so you need the chain rule.

> Differentiate each function separately and then use the chain rule formula.

c) Find $\dfrac{dy}{dx}$ if $y = e^x \cot x$.

$$u = e^x \text{ and } v = \cot x$$

$$\implies \frac{du}{dx} = e^x \text{ and } \frac{dv}{dx} = -\text{cosec}^2 x$$

$$\text{So } \frac{dy}{dx} = u\frac{dv}{dx} + v\frac{du}{dx} = (e^x \times -\text{cosec}^2 x) + (\cot x \times e^x)$$

$$= e^x(\cot x - \text{cosec}^2 x)$$

> y is a product of two functions, so use the product rule (p.225).

> **Tip** If the function inside 'cot' was more difficult, you'd do this in the same way but use the chain rule to work out $\dfrac{dv}{dx}$.

Exercise 7.13

Q1 Differentiate with respect to x:

a) $y = \operatorname{cosec}(2x)$

b) $y = \operatorname{cosec}^2 x$

c) $y = \cot(7x)$

d) $y = \cot^7 x$

e) $y = x^4 \cot x$

f) $y = (x + \sec x)^2$

g) $y = \operatorname{cosec}(x^2 + 5)$

h) $y = e^{3x} \sec x$

i) $y = (2x + \cot x)^3$

Q2 Find f'(x) if:

a) $f(x) = \dfrac{\sec x}{x + 3}$

b) $f(x) = \sec \dfrac{1}{x}$

c) $f(x) = \sec \sqrt{x}$

Q3 The curve A has equation $y = x \cot x$.

a) Find $\dfrac{dy}{dx}$. *[2 marks]*

b) Show that the tangent to curve A at $\left(\dfrac{\pi}{2}, 0\right)$ has the equation $2\pi x + 4y - \pi^2 = 0$. *[3 marks]*

Q4 Find f'(x) if $f(x) = (\sec x + \operatorname{cosec} x)^2$.

Q5 Find f'(x) if $f(x) = \dfrac{1}{x \cot x}$.

Q6 Find f'(x) if $f(x) = e^x \operatorname{cosec} x$.

Q7 Find f'(x) if $f(x) = e^{3x} \cot(4x)$.

Q8 Find f'(x) if $f(x) = e^{-2x} \operatorname{cosec}(4x)$.

Q9 Find f'(x) if $f(x) = \ln(x) \operatorname{cosec} x$.

Q10 Find f'(x) if $f(x) = \sqrt{\sec x}$.

Q11 Find f'(x) if $f(x) = e^{\sec x}$.

Q12 a) Find f'(x) if $f(x) = \ln(\operatorname{cosec} x)$.

b) Show that the function in part a) can be written as $-\ln(\sin x)$ and differentiate it — you should get the same answer.

> **Problem Solving**
>
> In Q12, you'll need to use the log laws from Year 1.

Q13 Find f'(x) if $f(x) = \ln(x + \sec x)$.

Q14 Differentiate $y = \sec(\sqrt{x^2 + 5})$.

Challenge

Q15 a) Prove that the gradient of the curve $y = \dfrac{2 \cot 2x}{x}$ at $x = \dfrac{\pi}{3}$ is $\dfrac{6\sqrt{3} - 16\pi}{\pi^2}$. *[5 marks]*

b) Hence show that the equation of the tangent to the curve $y = \dfrac{2 \cot 2x}{x}$

at the point where $x = \dfrac{\pi}{3}$ is $3\pi^2 y + 6(8\pi - 3\sqrt{3})x = 4\pi(4\pi - 3\sqrt{3})$. *[3 marks]*

7.14 Connected Rates of Change

Some situations have a number of **linked variables**, like length, surface area and volume or distance, speed and acceleration.

If you know the **rate of change** of one of these linked variables, and the equations that connect the variables, you can use the **chain rule** to help you find the rate of change of the other variables.

An equation connecting variables with their rates of change (i.e. with a derivative term) is called a **differential equation**.

When something changes over **time**, the derivative is $\frac{d}{dt}$ of that variable.

Tip There's more on solving differential equations in Chapter 8 (see pages 313-321).

Example 1

If $y = 3e^{5x}$ and $\frac{dx}{dt} = 2$, work out $\frac{dy}{dt}$ when $x = -1$.

$y = 3e^{5x} \implies \frac{dy}{dx} = 15e^{5x}$ — Start off by differentiating the expression for y, with respect to x.

$\frac{dy}{dt} = \frac{dy}{dx} \times \frac{dx}{dt}$ — Write out the chain rule for $\frac{dy}{dt}$.

$\frac{dy}{dt} = 15e^{5x} \times 2 = 30e^{5x}$ — Put in all the things you know to work out $\frac{dy}{dt}$.

When $x = 1$, $\frac{dy}{dt} = 30e^{-5}$ — Now find the value of $\frac{dy}{dt}$.

Example 2 P

y is the surface area of a sphere, and x is its radius.
The rate of change of the radius, $\frac{dx}{dt} = -2$. Find $\frac{dy}{dt}$ when $x = 2.5$.

$y = 4\pi x^2$ — This is trickier because you're not given the expression for y. But you should know (or be able to look up) the surface area of a sphere.

$\frac{dy}{dx} = 8\pi x$ — Now differentiate as normal.

$\frac{dy}{dt} = \frac{dy}{dx} \times \frac{dx}{dt}$
$\quad = 8\pi x \times -2 = -16\pi x$ — Use the chain rule formula to find $\frac{dy}{dt}$.

When $x = 2.5$, $\frac{dy}{dt} = -16\pi \times 2.5$
$\quad\quad\quad\quad = -40\pi$ — Finally, substitute $x = 2.5$ into $\frac{dy}{dt}$.

Tip The formula for the surface area of a sphere is in the formula booklet — given as $4\pi r^2$.

Often you'll see much wordier questions involving related rates of change, like the one in the example below, where you have to do a bit more work to figure out where to start.

Example 3 P M

A scientist is testing how a new material expands when it is gradually heated. The diagram below shows the sample being tested, which is modelled as a triangular prism.

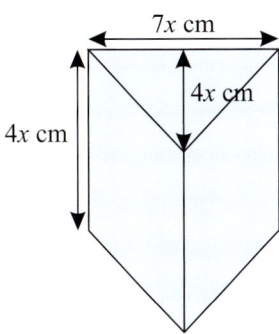

After t minutes, the triangle that forms the base of the prism has base length $7x$ cm and height $4x$ cm, and the length of the prism is also $4x$ cm.

If the sample expands at a constant rate, given by $\frac{dx}{dt} = 0.05$ cm min^{-1}, find an expression in terms of x for $\frac{dV}{dt}$, where V is the volume of the prism.

Area of cross-section $= \frac{1}{2} \times 7x \times 4x$

Length of prism $= 4x$

Use the information you know to write an equation for the volume of the prism in terms for x. Volume = area of cross-section × length.

$V = \left(\frac{1}{2} \times 7x \times 4x\right) \times 4x$
$\Rightarrow V = 56x^3$ cm^3

$\frac{dV}{dx} = 168x^2$

Differentiate this expression for the volume with respect to x.

$\frac{dV}{dt} = \frac{dV}{dx} \times \frac{dx}{dt}$

You know that $\frac{dx}{dt} = 0.05$ cm min^{-1}. So you can use the chain rule to find $\frac{dV}{dt}$.

$\Rightarrow \frac{dV}{dt} = 168x^2 \times 0.05$
$= 8.4x^2$ cm^3 min^{-1}

There are a couple of tricks in this type of question that could catch you out if you're not prepared for them. In the example coming up on the next page, you have to spot that there's a hidden derivative described in words.

You also need to remember the rule: $\frac{dy}{dx} = \frac{1}{\left(\frac{dx}{dy}\right)}$ (see p.210).

Example 4

A giant metal cube from space is cooling after entering the Earth's atmosphere.
As it cools, the surface area of the cube is modelled as decreasing at a constant rate of $0.027 \text{ m}^2 \text{ s}^{-1}$.

If the side length of the cube after t seconds is x m, find $\dfrac{dx}{dt}$ at the point when $x = 15$ m.

The cube has side length x m,

so it has surface area, $A = 6x^2 \Rightarrow \dfrac{dA}{dx} = 12x$

> Start with what you know to work out the surface area (A) of the cube and its rate of change with respect to its side length $\left(\dfrac{dA}{dx}\right)$.

A decreases at a constant rate
of $0.027 \text{ m}^2\text{s}^{-1} \Rightarrow \dfrac{dA}{dt} = -0.027$

> You're given the rate of change in the question. The value is negative because A is decreasing.

$\dfrac{dx}{dt} = \dfrac{dx}{dA} \times \dfrac{dA}{dt} = \dfrac{1}{\left(\dfrac{dA}{dx}\right)} \times \dfrac{dA}{dt}$

> Now use the chain rule to find $\dfrac{dx}{dt}$.

$= \dfrac{1}{12x} \times -0.027 = -\dfrac{0.00225}{x}$

When $x = 15$, $\dfrac{dx}{dt} = -\dfrac{0.00225}{15}$

> This gives the rate that the side length is decreasing by when it is 15 m.

$= -0.00015 \text{ ms}^{-1}$

Exercise 7.14

P **Q1** A cube with sides x cm is cooling and the sides are shrinking by 0.1 cm min^{-1}.
Find an expression for $\dfrac{dV}{dt}$, the rate of change of volume with respect to time.

P **M** **Q2** A cuboid block of sides $2x$ cm by $3x$ cm by $5x$ cm expands when
heated such that x increases at a rate of 0.15 cm °C^{-1}. If the volume
of the cuboid at temperature θ °C is V cm^3, find $\dfrac{dV}{d\theta}$ when $x = 3$.

P **M** **Q3** A snowball of radius r cm is melting. Its radius decreases by
1.6 cm h^{-1}. If the surface area of the snowball at time t hours
is A cm^2, find $\dfrac{dA}{dt}$ when $r = 5.5$ cm. Give your answer to 2 d.p.

> **Problem Solving**
>
> In Q3, model the snowball as a sphere ($A = 4\pi r^2$).

P **M** **Q4** A spherical satellite, radius r m, expands as it enters the atmosphere.
It grows by 2×10^{-2} mm for every 1 °C rise in temperature.
Find an expression $\dfrac{dV}{d\theta}$ for the rate of change of volume with respect to temperature.

E **Q5** Given that $y = \sqrt{16x + 25}$ and $z = 2\ln(y^2 + 1)$,

a) Find:

(i) $\dfrac{dy}{dx}$

(ii) $\dfrac{dz}{dy}$ *[4 marks]*

b) Find the value of $\dfrac{dz}{dx}$ at the instant when $x = 6$. *[2 marks]*

P **M** **Q6** Heat, H, is lost from a closed cylindrical tank of radius r cm and height $3r$ cm at a rate of 2 J cm^{-2} of surface area, A.

Find $\dfrac{dH}{dr}$ when $r = 12.3$. Give your answer to 2 d.p.

P **M** **Q7** A cylindrical polishing block of radius r cm and length H cm is worn down at one circular end at a rate of 0.5 mm h^{-1}. Find an expression for the rate of change of the volume of the block with respect to time.

Problem Solving

In Q7, treat r as a constant as the radius will not change.

E **P** **M** **Q8** A crystal of a salt is shaped like a prism. Its cross section is an equilateral triangle with sides x mm and the height of the crystal is 20 mm. New material is deposited only on the rectangular faces of the prism (i.e. the height does not change), so that x increases at a rate of 0.6 mm per day.

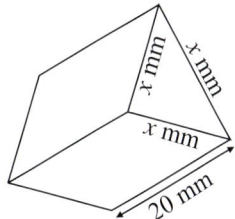

a) Find an expression for the area of the end of the prism in terms of x. *[2 marks]*

b) Find an expression for the rate of change of the volume of the crystal with respect to time. *[3 marks]*

c) Find the rate of change of the volume of the crystal with respect to time when $x = 0.5$. *[2 marks]*

P **M** **Q9** The growth of a population of bacteria in a sample dish is modelled by the equation $D = 1 + 2^{\lambda t}$, where D is the diameter of the colony in mm, t is time in days, and λ is a constant. The number of bacteria in the colony, n, is directly proportional to the diameter. A biologist counts the bacteria in the colony when its diameter is 2 mm and estimates that there are approximately 208 bacteria.

a) Find an expression for the rate of change of n with respect to time.

b) Find the rate of increase in number of bacteria after 1 day if $\lambda = 5$.

E **P** **Q10** The volume of a sphere increases at a rate of 3 cm^3 per second. Find the rate of change of the radius of the sphere with respect to time at the instant when the volume is 100 cm^3. *[4 marks]*

E P M Q11 Water is dripping from a hole in the base of a cylinder of radius
r cm, where the water height is h cm, at a rate of 0.3 cm^3 s^{-1}.

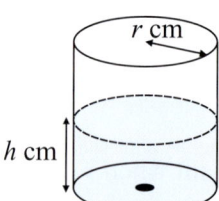

a) Find an expression for $\dfrac{dh}{dt}$, the rate at which the water level falls in the cylinder. *[3 marks]*

b) Hence find the rate of change in the water level, per minute,
in a cylinder of radius 6 cm when the height of water is 4 cm. *[2 marks]*

E P M Q12 The volume, V, of a hemisphere of radius r cm, varies with its temperature, θ, at a rate of
k cm^3 $^\circ$C^{-1}, where k is a constant that depends on the material that the hemisphere is made of.

a) Find an expression for the rate of change of radius with respect to temperature. *[3 marks]*

b) Hence find $\dfrac{dr}{d\theta}$ for a material with $k = 1.5$, when $V = 4$ cm^3. *[2 marks]*

E P M Q13 Water is poured into a container, as shown in the diagram below. When the depth
of water in the container is h cm, the volume, V cm^3, is given by $V = 3h^2(h + 1)$.
The water is poured at a constant rate of 10 cm^3 per second.

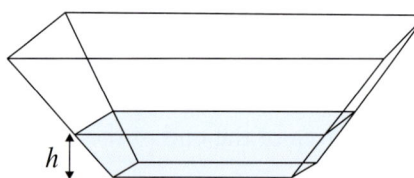

a) Find an expression for $\dfrac{dh}{dt}$ in terms of h. *[3 marks]*

b) Find the rate of change of the depth of the water in the container when $h = 8$ cm. *[2 marks]*

c) Find the volume of water in the container at the instant when the depth of water
in the container is increasing at a rate of 2 cm every 3 seconds. *[3 marks]*

E P M Q14 The kinetic energy, K joules, of a particle of mass m kg travelling
with velocity v ms^{-1} is given by $K = \dfrac{1}{2}mv^2$. Given that the mass
of the moving particle is 5 kg, find the rate of change in the kinetic
energy of the particle at the instant when the acceleration
of the particle is 6 ms^{-2} and the velocity of the particle is 1 ms^{-1}.

[3 marks]

Challenge

P M Q15 Two resistors in a circuit have resistances s_1 and s_2 ohms respectively.
The total resistance in the circuit, r ohms, is given by $\dfrac{1}{r} = \dfrac{1}{s_1} + \dfrac{1}{s_2}$.

The resistance s_1 increases at a rate of 0.3 ohms/min and the resistance s_2
decreases at a rate of 0.1 ohms/min. Find the rate of change of the total
resistance, r, at the instant when $s_1 = 20$ ohms and $s_2 = 30$ ohms.

7.15 Differentiating Parametric Equations

A curve can be defined by **two parametric equations**, often with the parameter t:

$$y = f(t) \text{ and } x = g(t)$$

To find the gradient, $\dfrac{dy}{dx}$, you could convert the equations into **Cartesian** form (see pages 134-135), but this isn't always possible or convenient.

The **chain rule** you met on p.207 can be used to differentiate parametric equations without needing to convert to Cartesian form. It looks like this:

$$\frac{dy}{dx} = \frac{dy}{dt} \div \frac{dx}{dt}$$

Tip On page 207, you saw this given in the form '$\times \dfrac{dt}{dx}$' rather than '$\div \dfrac{dx}{dt}$', but it means the same thing.

So to find $\dfrac{dy}{dx}$ from parametric equations, **differentiate** each equation with respect to the parameter t, then put them into the formula.

Example 1

The curve C is defined by the parametric equations $x = t^2 - 1$ and $y = t^3 - 3t + 4$.

a) Find $\dfrac{dy}{dx}$ in terms of t.

$$x = t^2 - 1 \implies \frac{dx}{dt} = 2t$$

Start by differentiating the two parametric equations with respect to t.

$$y = t^3 - 3t + 4 \implies \frac{dy}{dt} = 3t^2 - 3 = 3(t^2 - 1)$$

$$\frac{dy}{dx} = \frac{dy}{dt} \div \frac{dx}{dt} = \frac{3(t^2 - 1)}{2t}$$

Now use the chain rule to combine them.

b) Find the gradient of C when $t = -2$.

$$\frac{dy}{dx} = \frac{3((-2)^2 - 1)}{2(-2)} = \frac{3(3)}{-4} = -\frac{9}{4}$$

Use the answer to a) to find the gradient for a specific value of t. In this case, when $t = -2$.

E Exam Style   P Problem Solving   M Modelling

c) Find the coordinates of the stationary points.

$$\frac{dy}{dx} = \frac{3\,(t^2 - 1)}{2t} = 0$$ ⟵ 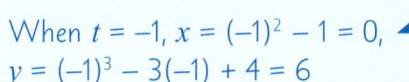 The stationary points occur when $\frac{dy}{dx} = 0$.

$\Rightarrow\ 3(t^2 - 1) = 0 \Rightarrow t^2 = 1 \Rightarrow t = \pm 1$ ⟵ Solve to find the values of t at the stationary points.

When $t = 1$, $x = (1)^2 - 1 = 0$,
$y = (1)^3 - 3(1) + 4 = 2$ ⟵ Now put these values for t into the original parametric equations to find the Cartesian coordinates of the stationary points.

When $t = -1$, $x = (-1)^2 - 1 = 0$,
$y = (-1)^3 - 3(-1) + 4 = 6$

So the stationary points are at
(0, 2) and (0, 6).

Tip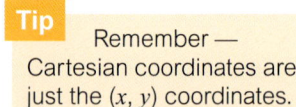
Remember —
Cartesian coordinates are
just the (x, y) coordinates.

Exercise 7.15

Q1 For each curve C, defined by the parametric equations given below, find $\frac{dy}{dx}$ in terms of t.

a) $x = t^2$, $y = t^3 - t$
b) $x = t^3 + t$, $y = 2t^2 + 1$
c) $x = t^4$, $y = t^3 - t^2$
d) $x = \cos t$, $y = 4t - t^2$

Q2 The curve C is defined by the parametric equations $x = t^2$, $y = e^{2t}$.

a) Find $\frac{dy}{dx}$ in terms of t.
b) Find the gradient of C when $t = 1$.

Q3 The curve C is defined by the parametric equations $x = e^{3t}$, $y = 4t^3 - 2t^2$.

a) Find $\frac{dy}{dx}$ in terms of t.
b) Find the gradient of C when $t = 0$.

Q4 The curve C is defined by the parametric equations $x = t^3$, $y = t^2 \cos t$.

a) Find $\frac{dy}{dx}$ in terms of t.
b) Find the gradient of C when $t = \pi$.

Q5 The curve C is defined by the parametric equations $x = t^2 \sin t$, $y = t^3 \sin t + \cos t$.

a) Find $\frac{dy}{dx}$ in terms of t.
b) Find the gradient of C when $t = \pi$.

E Q6 A parametric curve C has equations $x = 3 \cos 2t$, $y = \sin^2 t$.

a) Show that $\frac{dy}{dx}$ is constant, and give its value. *[3 marks]*
b) Find the Cartesian equation of C. *[2 marks]*
c) Sketch the curve C, showing clearly where it intersects the axes. *[2 marks]*

Q7 The curve C is defined by the parametric equations $x = \ln t$, $y = 3t^2 - t^3$.

a) Find $\frac{dy}{dx}$ in terms of t.
b) Evaluate $\frac{dy}{dx}$ when $t = -1$.
c) Find the exact coordinates of the turning point of the curve C.

E **Q8** A curve is defined by the parametric equations $x = t^2$, $y = 3t^3 - 4t$.

    a) Find $\dfrac{dy}{dx}$ for this curve. *[2 marks]*

    b) Find the coordinates of the stationary points of the curve. *[3 marks]*

E **P** **Q9** A parametric curve has equations $x = \dfrac{\cos t}{t}$, $y = \dfrac{\sin t}{t}$, where $0 < t < 4\pi$.

    a) Show that $\dfrac{dy}{dx} = \dfrac{\sin t - t\cos t}{\cos t + t\sin t}$. *[4 marks]*

    b) Find the gradient of the curve at the point where $t = \pi$. *[1 mark]*

    c) State the exact values of x where the curve crosses the x-axis. *[2 marks]*

Challenge

E **P** **Q10** A curve is defined by parametric equations $x = t - \cos t$, $y = \sin t$.

    a) Find $\dfrac{dy}{dx}$ in terms of t. *[2 marks]*

    b) Find $\dfrac{d^2y}{dx^2}$ in terms of t. *[4 marks]*

    c) Explain why the curve does not have any points of inflection. *[1 mark]*

7.16   Finding Tangents and Normals of Parametric Curves

Once you've found the gradient of a parametric curve at a particular point, you can use this to find the **equation** of the **tangent** or **normal** to the curve at that point. You'll have seen this before, but here's a recap of the method:

- The gradient of the **tangent** is the **same** as the gradient of the curve at that point.

- The gradient of the **normal** at that point is $\dfrac{-1}{\text{gradient of tangent}}$.

- Put the values for the gradient, m, and the (x, y) coordinates of the point into $y = mx + c$ to find the equation of the line.

> **Tip** You could also use $y - y_1 = m(x - x_1)$ to get the equation.

Example **1**

The curve C is defined by the following parametric equations: $x = \sin t$, $y = 2t\cos t$.

a) Find the gradient of the curve, and the (x, y) coordinates, when $t = \pi$.

$\dfrac{dx}{dt} = \cos t$, $\dfrac{dy}{dt} = 2\cos t - 2t\sin t$ ←   Find $\dfrac{dx}{dt}$ and $\dfrac{dy}{dt}$.

$\dfrac{dy}{dx} = \dfrac{2\cos t - 2t\sin t}{\cos t} = 2 - 2t\tan t$ ←   Use them to find $\dfrac{dy}{dx}$.

When $t = \pi$, $\dfrac{dy}{dx} = 2 - 2\pi(0) = 2$ ←   Substitute π into $\dfrac{dy}{dx}$ to get the gradient.

When $t = \pi$, $x = 0$, and $y = -2\pi$, ←   Substitute π into the equations for x and y.
so the coordinates are $(0, -2\pi)$.

b) Hence find the equation of the tangent to C when $t = \pi$.

$$y = mx + c$$
$$-2\pi = 2(0) + c$$
$$\Rightarrow c = -2\pi$$

The tangent has an equation in the form $y = mx + c$, so substitute in $m = 2$, $x = 0$ and $y = -2\pi$ and find c.

$$y = 2x - 2\pi \text{ or } y = 2(x - \pi)$$

Put c back into the equation.

c) Find the equation of the normal to C when $t = \pi$.

gradient of normal
$$= -1 \div \text{gradient of tangent} = -\frac{1}{2}$$

Find the gradient of the normal.

$$y = mx + c$$
$$-2\pi = -\frac{1}{2}(0) + c \Rightarrow c = -2\pi$$

Substitute $m = -\frac{1}{2}$, $x = 0$ and $y = -2\pi$ into $y = mx + c$, to find c.

$$y = -\frac{1}{2}x - 2\pi \text{ or } x + 2y + 4\pi = 0$$

Put c back into the equation.

Exercise 7.16

Q1 A curve is defined by the parametric equations $x = t^2$, $y = t^3 - 6t$. Find the equation of the tangent to the curve at $t = 3$, giving your answer in the form $ax + by + c = 0$.

Q2 A curve C is defined parametrically by $x = t^3 - 2t^2$, $y = t^3 - t^2 + 5t$. Find the equation of the tangent at the point $t = -1$.

Q3 A curve C is defined by the parametric equations $x = \sin 2t$, $y = t \cos t + 2 \sin t$. Find the equation of the normal to the curve at $t = \pi$.

Q3-Q5 Hint
You'll need to use the product rule.

Q4 The parametric representation of a curve is given by $x = t \ln t$, $y = t^3 - t^2 + 3$. Find the equation of the tangent to the curve at $t = 1$.

Q5 The path of a particle is given parametrically by $x = \theta \sin 2\theta$, $y = \theta^2 + \theta \cos \theta$. Find the equation of the normal to the particle's path at $\theta = \frac{\pi}{2}$.

P Q6 The motion of a particle is modelled by the parametric equations $x = t^2 - t$, $y = 3t - t^3$.
   a) Find the equation of the tangent to the path of the particle when $t = 2$, giving your answer in a suitable form.
   b) Find the Cartesian coordinates of the point at which the normal to the path at $t = 2$ cuts the x-axis.

E **Q7** A curve has parametric equations $x = 4t^2$, $y = 8t$, $t \in \mathbb{R}$.
P

    a) Show that the equation of the normal at the point (1, 4)
       is given by $x + 2y - 9 = 0$. *[4 marks]*

    b) Find the Cartesian equation of the curve. *[2 marks]*

    c) Find the area enclosed by the curve, the normal to the curve at (1, 4)
       and the x-axis, as shown in the diagram below.

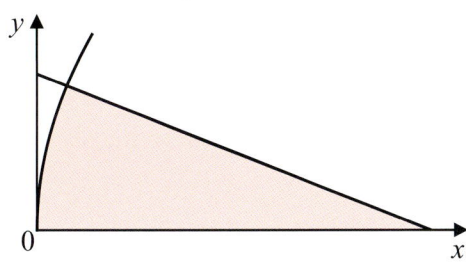

[5 marks]

P **Q8** A particle moves along a path modelled by the
parametric equations $x = \sin 2\theta + 2\cos\theta$, $y = \theta\sin\theta$.

    a) Find the gradient $\dfrac{dy}{dx}$ of the particle's path in terms of θ.

    b) Evaluate $\dfrac{dy}{dx}$ at $\theta = \dfrac{\pi}{2}$ and hence obtain equations
       of the tangent and normal to the path at this point.

E **Q9** A particle moves along a path given parametrically by $x = s^3 \ln s$, $y = s^3 - s^2 \ln s$.
P

    a) Give the value(s) of s at which the path cuts the y-axis. *[2 marks]*

    b) Hence show that the equation of a tangent to the curve when $x = 0$ is $y = 2x + 1$. *[4 marks]*

E **Q10** A curve is defined by the parametric equations $x = 1 - \cos 2t$, $y = \sin t + \cos t$, $0 \le t \le 2\pi$.
P

    a) Show that $\dfrac{dy}{dx} = k(\operatorname{cosec} t - \sec t)$, where k is a constant to be found. *[3 marks]*

    b) Find the equation of the tangent to the curve at the point where $t = \dfrac{\pi}{6}$.
       Give your answer in the form $y = mx + c$. *[4 marks]*

    c) Show that the curve has two stationary points. *[3 marks]*

    d) Find the equations of the tangents to the curve at each of the stationary points. *[2 marks]*

E **Q11** A curve is given parametrically by $x = \theta^2 \sin\theta$, $y = \dfrac{\cos\theta}{\theta^3}$.
P

    a) Show that the gradient of the curve when $\theta = \pi$ is $-\dfrac{3}{\pi^6}$. *[4 marks]*

    b) Hence find the equation of the normal to the curve at this point. *[2 marks]*

E **Q12** Curve C is given by $x = \dfrac{\sin\theta}{\theta^2}$, $y = \theta\cos 2\theta$.
P

    a) Find the gradient of C when $\theta = \dfrac{\pi}{2}$. *[4 marks]*

    b) Hence find the equation of the tangent to the curve at this point. *[2 marks]*

    c) Find the coordinates of the point with $0 < \theta \le \dfrac{\pi}{2}$
       where the curve cuts the x-axis. *[3 marks]*

7.17 Implicit Differentiation

An '**implicit relation**' is the mathematical name for any equation in x and y that's written in the form **f(x, y) = g(x, y)** instead of $y = f(x)$. For example, $y^2 = xy + x + 2$ is implicit.

Some implicit relations are either awkward or impossible to rewrite in the form $y = f(x)$. This can happen, for example, if the equation contains a number of different powers of y, or terms where x is multiplied by y.

> **Tip** $f(x, y)$ and $g(x, y)$ don't actually both have to include x and y — one of them could even be a constant.

This can make implicit relations tricky to differentiate — the solution is **implicit differentiation**.

To find $\dfrac{dy}{dx}$ for an implicit relation between x and y, follow these steps:

Step 1: Differentiate terms in x only (and constant terms) with respect to x, as normal.

Step 2: Use the chain rule to differentiate terms in y only:

$$\frac{d}{dx}f(y) = \frac{d}{dy}f(y) \times \frac{dy}{dx}$$

In practice, this means 'differentiate with respect to y, and put a $\dfrac{dy}{dx}$ on the end'.

> **Tip** $\frac{d}{dx}f(y)$ just means 'the derivative of $f(y)$ with respect to x'.

Step 3: Use the product rule to differentiate terms in both x and y:

$$\frac{d}{dx}u(x)v(y) = u(x)\frac{d}{dx}v(y) + v(y)\frac{d}{dx}u(x)$$

Step 4: Rearrange the resulting equation in x, y and $\dfrac{dy}{dx}$ to make $\dfrac{dy}{dx}$ the subject.

Example 1

Use implicit differentiation to find $\dfrac{dy}{dx}$ for $y^3 + y^2 = e^x + x^3$.

$$\frac{d}{dx}y^3 + \frac{d}{dx}y^2 = \frac{d}{dx}e^x + \frac{d}{dx}x^3$$

You need to differentiate each term of the equation with respect to x. Start by putting '$\frac{d}{dx}$' in front of each term.

$$\frac{d}{dx}y^3 + \frac{d}{dx}y^2 = e^x + 3x^2$$

Differentiate the terms in x only.

$$3y^2\frac{dy}{dx} + 2y\frac{dy}{dx} = e^x + 3x^2$$

Use the chain rule for the terms in y only.

$$(3y^2 + 2y)\frac{dy}{dx} = e^x + 3x^2$$
$$\Rightarrow \frac{dy}{dx} = \frac{e^x + 3x^2}{3y^2 + 2y}$$

There are no terms in both x and y to deal with, so rearrange to make $\dfrac{dy}{dx}$ the subject.

Take your time using the product rule as it's easy to forget terms if you're not careful. Always start by identifying $u(x)$ and $v(y)$, and do it in steps if you need to. Once you're happy with this method you might not need to write all the steps out separately, but you can do if you find it easier.

Example 2

a) Use implicit differentiation to find $\dfrac{dy}{dx}$ for $2x^2y + y^3 = 6x^2 - 15$.

$\dfrac{d}{dx}2x^2y + \dfrac{d}{dx}y^3 = \dfrac{d}{dx}6x^2 - \dfrac{d}{dx}15$ ← Start by putting '$\dfrac{d}{dx}$' in front of each term.

$\dfrac{d}{dx}2x^2y + \dfrac{d}{dx}y^3 = 12x + 0$ ← First, deal with the terms in x and constant terms — in this case that's the two terms on the right-hand side.

$\dfrac{d}{dx}2x^2y + 3y^2\dfrac{dy}{dx} = 12x + 0$ ← Now use the chain rule on the term in y.

$2x^2\dfrac{d}{dx}(y) + y\dfrac{d}{dx}(2x^2) + 3y^2\dfrac{dy}{dx} = 12x + 0$ ← Use the product rule on the term in x and y, where $u(x) = 2x^2$ and $v(y) = y$.

$2x^2\dfrac{dy}{dx} + 4xy + 3y^2\dfrac{dy}{dx} = 12x + 0$

$\dfrac{dy}{dx}$ is just another way of writing $\dfrac{d}{dx}(y)$.

$\dfrac{dy}{dx}(2x^2 + 3y^2) = 12x - 4xy$

$\Rightarrow \dfrac{dy}{dx} = \dfrac{12x - 4xy}{2x^2 + 3y^2}$ ← Finally, rearrange to make $\dfrac{dy}{dx}$ the subject.

b) Find the gradient of the curve $2x^2y + y^3 = 6x^2 - 15$ at the point $(2, 1)$.

$\dfrac{dy}{dx} = \dfrac{12(2) - 4(2)(1)}{2(2)^2 + 3(1)^2} = \dfrac{16}{11}$ ← Just put the values for x and y into $\dfrac{dy}{dx}$.

Exercise 7.17

Q1 Use implicit differentiation to find $\dfrac{dy}{dx}$ for each of these curves:

a) $y + y^3 = x^2 + 4$

b) $x^2 + y^2 = 2x + 2y$

c) $3x^3 - 4y = y^2 + x$

d) $5x - y^2 = x^5 - 6y$

e) $\cos x + \sin y = x^2 + y^3$

f) $x^3y^2 + \cos x = 4xy$

g) $e^x + e^y = x^3 - y$

h) $3xy^2 + 2x^2y = x^3 + 4x$

i) $4x^3y^2 + 3x^2y = 2\sin x - x^4$

Q2 Find the gradient, $\dfrac{dy}{dx}$, for each of these curves given below:

a) $x^3 + 2xy = y^4$ b) $x^2y + y^2 = x^3$ c) $y^3x + y = \sin x$

d) $y \cos x + x \sin y = xy$ e) $e^x + e^y = xy$ f) $\ln x + x^2 = y^3 + y$

g) $e^{2x} + e^{3y} = 3x^2y^2$ h) $x \ln x + y \ln x = x^5 + y^3$

> **Q3a) Hint** Evaluate the left-hand and right-hand sides of the equation separately to show they're the same.

Q3 a) Show that the curve C, defined implicitly by $e^x + 2 \ln y = y^3$, passes through $(0, 1)$.

    b) Find the gradient of the curve at this point.

Q4 A curve is defined implicitly by $x^3 + y^2 - 2xy = 0$.

    a) Find $\dfrac{dy}{dx}$ for this curve.   b) Show that $y = -2 \pm 2\sqrt{3}$ when $x = -2$.

    c) Evaluate the gradient at $(-2, -2 + 2\sqrt{3})$, leaving your answer in surd form.

P Q5 The curve $x^3 - xy = 2y^2$ passes through the points $(1, -1)$ and $(1, a)$.

    a) Find the value of a.

    b) Evaluate the gradient of the curve at each of these points.

P Q6 A curve is defined implicitly by $x^2y + y^2x = xy + 4$.

    a) At which two values of y does the line $x = 1$ cut the curve?

    b) By finding $\dfrac{dy}{dx}$, evaluate the gradient at each of these points.

E **P** Q7 A curve is defined implicitly by $y^2 - 2y - \ln x - \cos(2\pi x) = 0$.

    a) Find $\dfrac{dy}{dx}$ in terms of x and y, simplifying your answer.   *[3 marks]*

    b) Hence, or otherwise, determine the value of y for which the gradient of the curve is undefined for $x \neq 0$.   *[1 mark]*

E **P** Q8 A curve has equation $(x + y)^2 = xy + y^3$.

    a) Show that $\dfrac{dy}{dx} = \dfrac{2x + y}{3y^2 - x - 2y}$.   *[5 marks]*

There are two points on the curve with y-coordinate 1.

    b) Find the x-coordinate at each of these points.   *[2 marks]*

    c) Find the gradient of the curve at each of the points found in part b).   *[2 marks]*

    d) Explain why the curve has a stationary point between the two x-values found in part b).   *[1 mark]*

Challenge

E **P** Q9 A curve has equation $y = x^{\tan x}$.

    a) By taking natural logarithms, find $\dfrac{dy}{dx}$ in terms of x.   *[4 marks]*

    b) Find the exact value of the gradient of the curve at the point $\left(\dfrac{\pi}{4}, \dfrac{\pi}{4}\right)$.   *[2 marks]*

7.18 Applications of Implicit Differentiation

Sometimes you can use implicit differentiation when you might not expect it.
For some equations of the form **y = f(x)**, the easiest way to differentiate them
is to **rearrange** them and use implicit differentiation.

For example, the proof of the rule for **differentiating a^x** from page 217 uses implicit differentiation:

Take **ln** of both sides of the equation:
$$y = a^x \Rightarrow \ln y = \ln a^x$$

Use the **log laws** to rearrange
the right-hand side:
$$\ln y = x \ln a$$

Now use **implicit differentiation**:
$$\frac{d}{dx}(\ln y) = \frac{d}{dx}(x \ln a)$$

Use the **chain rule** to deal with $\frac{d}{dx}(\ln y)$:
$$\frac{1}{y}\frac{dy}{dx} = \ln a$$

> Since a is a constant, $\ln a$ is also a constant.

Rearrange to get $\frac{dy}{dx}$ on its own:
$$\frac{dy}{dx} = y \ln a$$

Use the **original equation** to get rid of y:
$$\frac{dy}{dx} = a^x \ln a$$

This method of rearranging and using implicit differentiation
is also used to differentiate the **inverse trig functions**.

Example 1

Find $\frac{dy}{dx}$ if $y = \arcsin x$ for $-1 \le x \le 1$.

> **Problem Solving**
>
> Arcsin is the inverse of sin, so they cancel out.

$\sin y = \sin(\arcsin x)$
$\Rightarrow \sin y = x$

> Rearrange the equation to get rid of the arcsin by taking sin of both sides.

$\frac{d}{dx}(\sin y) = \frac{d}{dx}(x)$

> Now use implicit differentiation.

$\frac{d}{dx}(\sin y) = 1$

> Differentiate the x-term.

$\cos y \frac{dy}{dx} = 1$

> Use the chain rule on the y-term.

$\frac{dy}{dx} = \frac{1}{\cos y}$

> Rearrange to give an equation for $\frac{dy}{dx}$.

$\frac{dy}{dx} = \frac{1}{\sqrt{\cos^2 y}} = \frac{1}{\sqrt{1 - \sin^2 y}}$

> To get $\frac{1}{\cos y}$ in terms of x, first use the identity $\cos^2 x + \sin^2 x \equiv 1$ to write $\cos y$ in terms of $\sin y$.

$\frac{dy}{dx} = \frac{1}{\sqrt{1 - x^2}}$

> Now, use the equation $\sin y = x$ to get $\frac{dy}{dx}$ in terms of x.

You can use a similar method to differentiate the other inverse trig functions as well — the derivatives are given below:

$$y = \arcsin x$$
$$\frac{dy}{dx} = \frac{1}{\sqrt{1-x^2}}$$

$$y = \arccos x$$
$$\frac{dy}{dx} = -\frac{1}{\sqrt{1-x^2}}$$

$$y = \arctan x$$
$$\frac{dy}{dx} = \frac{1}{1+x^2}$$

Tip You can have a go at finding the derivatives of $\arccos x$ and $\arctan x$ for yourself in the Exercise on p.252-253.

Most implicit differentiation questions aren't that different from any other differentiation questions. Once you've got an expression for the gradient, you'll have to use it to do the sort of things you'd normally expect, like finding **stationary points** of curves and equations of **tangents** and **normals**.

Example 2

Curve A has the equation $x^2 + 2xy - y^2 = 10x + 4y - 21$.

a) Show that when $\frac{dy}{dx} = 0$, $y = 5 - x$.

$$\frac{d}{dx}x^2 + \frac{d}{dx}2xy - \frac{d}{dx}y^2 = \frac{d}{dx}10x + \frac{d}{dx}4y - \frac{d}{dx}21$$

You need to find $\frac{dy}{dx}$ by implicit differentiation so put $\frac{d}{dx}$ in front of each term.

$$\Rightarrow 2x + \frac{d}{dx}2xy - \frac{d}{dx}y^2 = 10 + \frac{d}{dx}4y - 0$$

Differentiate x^2, $10x$ and 21 with respect to x.

$$\Rightarrow 2x + \frac{d}{dx}2xy - 2y\frac{dy}{dx} = 10 + 4\frac{dy}{dx}$$

Use the chain rule to differentiate y^2 and $4y$.

$$\Rightarrow 2x + 2x\frac{dy}{dx} + y\frac{d}{dx}2x - 2y\frac{dy}{dx} = 10 + 4\frac{dy}{dx}$$

Use the product rule to differentiate $2xy$.

$$\Rightarrow 2x + 2x\frac{dy}{dx} + 2y - 2y\frac{dy}{dx} = 10 + 4\frac{dy}{dx}$$

$$\Rightarrow 2x\frac{dy}{dx} - 2y\frac{dy}{dx} - 4\frac{dy}{dx} = 10 - 2x - 2y$$

Rearrange to get $\frac{dy}{dx}$.

$$\Rightarrow (2x - 2y - 4)\frac{dy}{dx} = 10 - 2x - 2y$$

$$\Rightarrow \frac{dy}{dx} = \frac{10 - 2x - 2y}{2x - 2y - 4} = \frac{5 - x - y}{x - y - 2}$$

$$\frac{5 - x - y}{x - y - 2} = 0 \Rightarrow 5 - x - y = 0 \Rightarrow y = 5 - x$$

Find y in terms of x when $\frac{dy}{dx} = 0$.

b) Find the coordinates of the stationary points of A.

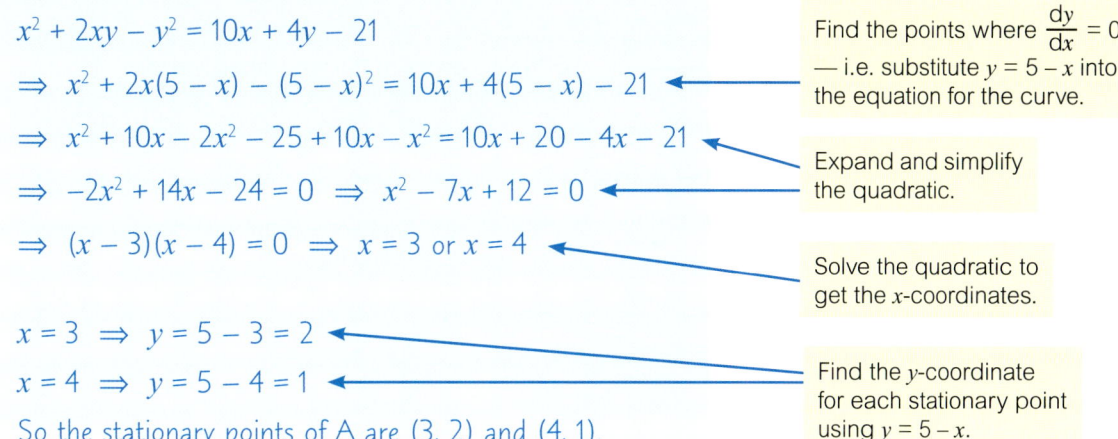

$x^2 + 2xy - y^2 = 10x + 4y - 21$

$\Rightarrow x^2 + 2x(5 - x) - (5 - x)^2 = 10x + 4(5 - x) - 21$ ← Find the points where $\frac{dy}{dx} = 0$ — i.e. substitute $y = 5 - x$ into the equation for the curve.

$\Rightarrow x^2 + 10x - 2x^2 - 25 + 10x - x^2 = 10x + 20 - 4x - 21$ ← Expand and simplify the quadratic.

$\Rightarrow -2x^2 + 14x - 24 = 0 \Rightarrow x^2 - 7x + 12 = 0$

$\Rightarrow (x - 3)(x - 4) = 0 \Rightarrow x = 3$ or $x = 4$ ← Solve the quadratic to get the x-coordinates.

$x = 3 \Rightarrow y = 5 - 3 = 2$

$x = 4 \Rightarrow y = 5 - 4 = 1$ ← Find the y-coordinate for each stationary point using $y = 5 - x$.

So the stationary points of A are (3, 2) and (4, 1).

Example 3

A curve defined implicitly by $\sin x - y \cos x = y^2$
passes through two points (π, a) and (π, b), where a < b.

a) Find the values of a and b.

$\sin \pi - y \cos \pi = y^2 \Rightarrow 0 + y = y^2$ ← Put $x = \pi$ into the equation and solve for y.

$\Rightarrow y^2 - y = 0 \Rightarrow y(y - 1) = 0$

$\Rightarrow y = 0$ and $y = 1$

So a = 0 and b = 1. ← You're looking for the y-coordinate of each point.

b) Find the equations of the tangents to the curve at each of these points.

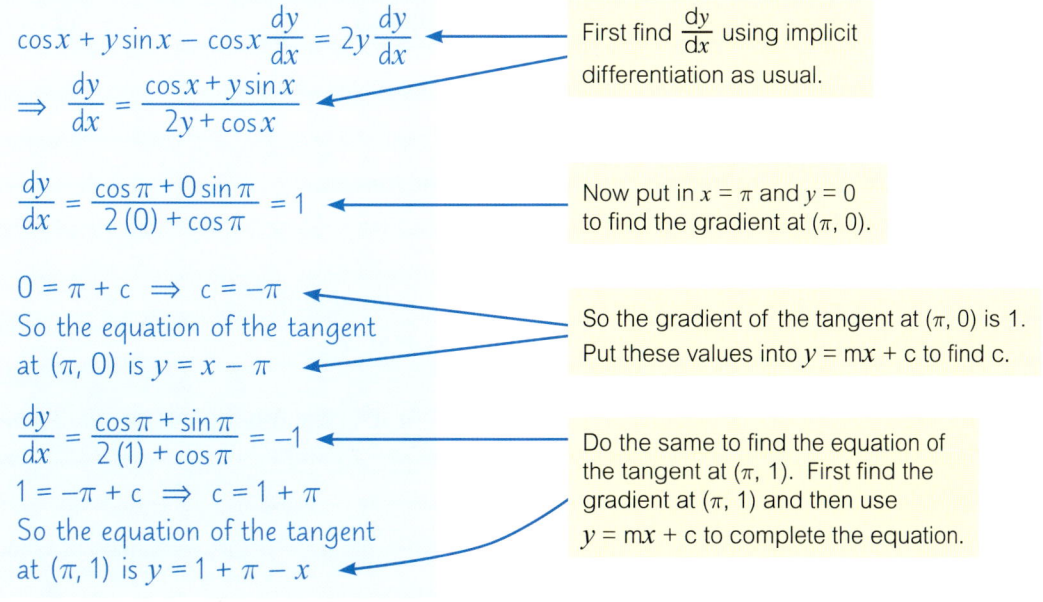

$\cos x + y \sin x - \cos x \frac{dy}{dx} = 2y \frac{dy}{dx}$ ← First find $\frac{dy}{dx}$ using implicit differentiation as usual.

$\Rightarrow \frac{dy}{dx} = \frac{\cos x + y \sin x}{2y + \cos x}$

$\frac{dy}{dx} = \frac{\cos \pi + 0 \sin \pi}{2(0) + \cos \pi} = 1$ ← Now put in $x = \pi$ and $y = 0$ to find the gradient at $(\pi, 0)$.

$0 = \pi + c \Rightarrow c = -\pi$
So the equation of the tangent at $(\pi, 0)$ is $y = x - \pi$ ← So the gradient of the tangent at $(\pi, 0)$ is 1. Put these values into $y = mx + c$ to find c.

$\frac{dy}{dx} = \frac{\cos \pi + \sin \pi}{2(1) + \cos \pi} = -1$ ← Do the same to find the equation of the tangent at $(\pi, 1)$. First find the gradient at $(\pi, 1)$ and then use $y = mx + c$ to complete the equation.

$1 = -\pi + c \Rightarrow c = 1 + \pi$
So the equation of the tangent at $(\pi, 1)$ is $y = 1 + \pi - x$

c) Show that the tangents intersect at the point $\left(\dfrac{1+2\pi}{2}, \dfrac{1}{2}\right)$.

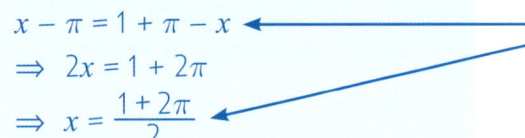

$x - \pi = 1 + \pi - x$ ←——— Set the equations of the tangents equal to one another to find where they intersect.

$\Rightarrow 2x = 1 + 2\pi$

$\Rightarrow x = \dfrac{1 + 2\pi}{2}$ ←———

$y = \left(\dfrac{1+2\pi}{2}\right) - \pi = \dfrac{1}{2} + \pi - \pi = \dfrac{1}{2}$ ←——— Put this value of x into one of the equations to find the corresponding value of y.

So they intersect at $\left(\dfrac{1+2\pi}{2}, \dfrac{1}{2}\right)$.

Exercise 7.18

Q1 A curve is defined implicitly by $x^2 + 2x + 3y - y^2 = 0$.
   a) Find the coordinates of the stationary points (to 2 decimal places).
   b) Show that the curve intersects the y-axis when $y = 0$ and $y = 3$.
      Hence find the equation of the tangent at each of these points.

Q2 A curve is defined implicitly by $x^3 + x^2 + y = y^2$.
   a) Find the coordinates of the stationary points (to 2 decimal places).
   b) Show that the curve intersects the line $x = 2$ when $y = 4$ and $y = -3$.
      Hence find the equation of the tangent at each of these points.

E Q3 A curve is defined implicitly by $x^2y + y^3 = x + 7$.
   a) Calculate the x-coordinates of the points on the curve where $y = 1$
      and hence find the equations of the normals at these points. *[5 marks]*
   b) Find the coordinates of the point where the normals intersect. *[2 marks]*

E Q4 The curve A has equation $x^3 - y^3 = 1$.
   a) Find the gradient of the curve A at the point $(0, -1)$. *[3 marks]*
   b) Find the coordinates of the point of intersection of the tangents
      to the curve A at the points $(0, -1)$ and $(1, 0)$. *[3 marks]*

E P Q5 $e^x + y^2 - xy = 5 - 3y$ is a curve passing through two points $(0, a)$ and $(0, b)$, where a < b.
   a) Find the values of a and b and show that one of these points
      is a stationary point of the curve. *[5 marks]*
   b) Find the equations of the tangent and normal to the curve at the other point. *[4 marks]*

P Q6 Differentiate arccos x with respect to x.

P Q7 If $y = \arctan x$, show that $\dfrac{dy}{dx} = \dfrac{1}{1+x^2}$.

> **Problem Solving**
>
> In Q7, you'll need to use the identity $\sec^2\theta \equiv 1 + \tan^2\theta$.

P **Q8** The curve C is defined by $\ln x + y^2 = x^2 y + 6$.

    a) Show that C passes through $(1, 3)$ and $(1, -2)$.

    b) Find the equations of the normals to the curve at each of these points and explain why these normals cannot intersect.

E **Q9** The ellipse E has equation $(x + 3)^2 + 2(y - 1)^2 = 2$.
P

    a) Find $\dfrac{dy}{dx}$ in terms of x and y. *[3 marks]*

    b) Find the coordinates of the two stationary points on E. *[3 marks]*

P **Q10** A curve is defined implicitly by $e^y + x^2 = y^3 + 4x$. Find the equations of the tangents that touch the curve at $(a, 0)$ and $(b, 0)$. Leave your answer in surd form.

P **Q11** Show that any point on the curve $y \ln x + x^2 = y^2 - y + 1$ which satisfies $y + 2x^2 = 0$ is a stationary point.

P **Q12** If $f(x) = \arccos(x^2)$ for $-1 \le x \le 1$, find the equation of the tangent to the graph of $y = f(x)$ when $x = \dfrac{1}{\sqrt{2}}$ and $0 \le y \le \pi$. Give your answer in the form $y = mx + c$, using exact values for m and c.

Q13 A curve is defined by $e^{2y} + e^x - e^4 = 2xy + 1$.

    a) Find the equation of the tangent to the curve when $y = 0$.

    b) Find the equation of the normal to the curve when $y = 0$.

    c) Show that these two lines intersect when $x = \dfrac{4e^8 + 144}{e^8 + 36}$.

> **Q13 Hint** Remember, $\dfrac{1}{e^x}$ is the same as e^{-x}.

Q14 $y^2 x + 2xy - 3x^3 = x^2 + 2$ passes through two points where $x = 2$. Find the equations of the tangents to the curve at these points, and hence show that they intersect at $\left(-\dfrac{14}{25}, -1\right)$.

E **Q15** The curve B has equation $\sin x = \cos^2 y$.

    a) Verify that the point P with coordinates $\left(\dfrac{\pi}{6}, \dfrac{\pi}{4}\right)$ lies on the curve B. *[1 mark]*

    b) Find the gradient of the curve B at P. *[3 marks]*

    c) Find the equation of the normal to B at P. *[2 marks]*

E **Q16** The curve C is defined implicitly by $\cos y \cos x + \cos y \sin x = \dfrac{1}{2}$.

    a) Find y when $x = \dfrac{\pi}{2}$ and when $x = \pi$, $0 \le y \le \pi$. *[2 marks]*

    b) Find the equations of the tangents at these points. *[4 marks]*

> **Q16 Hint** You won't need your calculator for the trig here but you will need to remember your common angles.

P **Q17** Find the coordinates of the stationary points of the graph $\dfrac{1}{3}y^2 = 6x^3 - 2xy$.

7 Review Exercise

Q1 A curve C has equation $y = 2x^3 - 12x^2 + 18x + 2$.

a) Find the values of for which the curve is:

(i) concave,   (ii) convex.

b) (i) Find the coordinates of the point of inflection.

(ii) Is this a stationary point of inflection? Explain your answer.

E **Q2** a) Find $f'(x)$ given that $f(x) = (3x + 1)^5$.   *[2 marks]*

b) Show that $f(x)$ has one point of inflection
and find the coordinates of that point.   *[3 marks]*

c) Give the range of values for which $f(x)$ is concave.   *[1 mark]*

Q3 A curve C has the equation $y = (x^2 - 1)^3$.

a) Differentiate y with respect to x.

b) Hence find the equation of the normal to the curve C when $x = 2$
in the form $ax + by + c = 0$, where a, b and c are integers.

Q4 Differentiate each of the following with respect to x:

a) $y = \sqrt{x^3 + 2x^2}$

b) $y = e^{5x^2}$

c) $y = \ln(6 - x^2)$

P **Q5** Given that a is a positive constant, $a \neq 1$, prove that $\dfrac{d}{dx} \log_a x = \dfrac{1}{x \ln a}$.

E **Q6** A curve has equation $y = \ln((2x - 3)^4 + 2x - 7)$.

a) Show that $\dfrac{dy}{dx} = \dfrac{g(x)}{(2x - 3)^4 + 2x - 7}$ where $g(x)$ is a function to be found.
You do not need to simplify your answer.   *[2 marks]*

b) Hence find the exact gradient of the curve at the point where $x = 3$,
giving your answer in its simplest form.   *[1 mark]*

Q7 Differentiate the following with respect to x.

a) $3e^{2x+1} - \ln(1 - x^2) + 2x^3$ b) $16x + e^{\sqrt{x}} + \ln(\cos x)$

Q8 a) Find $\dfrac{dy}{dx}$ as a function of x when: (i) $x = 2e^{2y}$ (ii) $x = \ln(2y + 3)$

b) Find $\dfrac{dy}{dx}$ as a function of y when $x = \tan y$.

E P Q9 The curve C has equation $y = (5^x - 1)^2 + 2$.

    a) Find the coordinates of the stationary point on the curve C. *[4 marks]*

    b) Explain briefly why the stationary point found in part a) is a minimum point. *[1 mark]*

    c) As $x \to -\infty$, $\dfrac{dy}{dx} \to k$. Find the value of k, explaining your answer. *[2 marks]*

P Q10 Given that $f(x) = a^x + b^x + c^x$, where a, b and c are positive constants greater than 1.

    a) Prove that the curve $y = f(x)$ does not have any stationary points.

    b) Prove that this result is not true for all positive constants a, b and c.

Q11 Find $f'(x)$ for the following functions:

    a) $f(x) = 2\cos(3x)$ b) $f(x) = \sqrt{\tan x}$

    c) $f(x) = e^{\cos(3x)}$ d) $f(x) = \sin(4x)\tan(x^3)$

E P Q12 a) Prove that $\sin A \sin B = \dfrac{1}{2}(\cos(A-B) - \cos(A+B))$. *[2 marks]*

    b) Deduce that $\cos 10x - \cos 5x = -2\sin\left(\dfrac{15x}{2}\right)\sin\left(\dfrac{5x}{2}\right)$. *[2 marks]*

    c) Hence find $f'(x)$ given that $f(x) = -2\sin\left(\dfrac{15x}{2}\right)\sin\left(\dfrac{5x}{2}\right)$. *[2 marks]*

    d) Find the gradient of the curve $y = -2\sin\left(\dfrac{15x^2}{2}\right)\sin\left(\dfrac{5x^2}{2}\right)$

        at the point where $x = 0$. *[4 marks]*

Q13 Find the value of the gradient for:

    a) $y = e^{2x}(x^2 - 3)$ when $x = 0$ b) $y = (\ln x)(\sin x)$ when $x = 1$

Q14 Find the exact value of $\dfrac{dy}{dx}$ when $x = 1$ if $y = e^{x^2}\sqrt{x+1}$.

E P Q15 Given that $f(x) = a(x-1)(x-2)(x-3)$, prove that

    $\dfrac{f'(x)}{f(x)} = \dfrac{1}{x-1} + \dfrac{1}{x-2} + \dfrac{1}{x-3}$. *[4 marks]*

Q16 Find $\dfrac{dy}{dx}$ if $y = \dfrac{\sqrt{x^2+3}}{\cos(3x)}$.

E Q17 A curve has the equation $y = \dfrac{\cos x^2}{\ln(2x)}$.

    a) Find $\dfrac{dy}{dx}$. *[4 marks]*

    b) Hence calculate the gradient of the curve at $x = 2$.
        Give your answer to 3 significant figures. *[1 mark]*

P Q18 The curve C has equation $y = \dfrac{px - q}{px + q}$, where p and q are positive constants.

    a) Find $\dfrac{dy}{dx}$ in terms of p and q.

    b) Deduce that the curve C has no stationary points.

Q19 Differentiate with respect to x:

    a) $y = \cos x \ln x^2$
                             b) $y = \dfrac{e^{x^2 - x}}{(x + 2)^4}$

Q20 Find the coordinates of the stationary point on the curve $y = \dfrac{e^x}{\sqrt{x}}$.

Q21 Find the equation of the tangent to the curve $y = \dfrac{6x^2 + 3}{4x^2 - 1}$ at the point $(1, 3)$.

E P Q22 A curve has equation $x = \dfrac{y}{y^2 - 1}$.

    a) Find $\dfrac{dy}{dx}$ in terms of y. Fully simplify your answer.   *[4 marks]*

    b) Find the equation of the normal to the curve at the point $(0, 0)$.   *[2 marks]*

    c) The normal to the curve at the point $(0, 0)$ meets the curve again at the points A and B. Find the exact length of the line AB.   *[4 marks]*

Q23 Find the equation of the normal to the curve for $y = 3 \operatorname{cosec} \dfrac{x}{4}$ when $x = \pi$.

Q24 Find $\dfrac{dy}{dx}$ when $x = 0$ for $y = \sec (3x - 2)$.

P Q25 Differentiate with respect to x:

    a) $y = \sqrt{\operatorname{cosec} x}$
                             b) $y = \dfrac{\sec x}{x^2}$

    c) $y = \cot (x^2 + 5)$
                             d) $y = e^{2x} \operatorname{cosec} (5x)$

P M Q26 At the end of its life in the main sequence, a small star like our Sun first expands to a Red Giant and then shrinks to a White Dwarf.

    a) When the star becomes a Red Giant, it expands and cools. The rate of change of radius with respect to temperature is approximately -2500 km K^{-1}. Find an expression for the rate of change of volume (V) with temperature (θ). Model the star as a sphere.

    b) When the star collapses to a White Dwarf, density (ρ) and temperature both increase as the diameter (D) decreases.

       The rate of change of diameter with respect to temperature, $\dfrac{dD}{d\theta}$, is approximately -215 km K^{-1}. Using the expression $V = kD^3$ for the volume of the star, find an expression for the rate of change of density $\dfrac{d\rho}{d\theta}$, if the mass of the star is a constant, m kg ($m = \rho V$).

E **Q27** A cuboid has length x cm, width $2x$ cm and height $3x$ cm.
P
M The temperature (θ) of the cuboid is increased, which causes it to expand.

a) Given that A is the surface area of the cuboid and V is its volume, find:

(i) $\dfrac{dA}{dx}$

(ii) $\dfrac{dV}{dx}$ *[2 marks]*

b) Use your answers from part a) to show that

if $\dfrac{dV}{d\theta} = 3$ cm^{3}K^{-1}, then $\dfrac{dA}{d\theta} = \dfrac{22}{3x}$ cm^{2}K^{-1}.   *[2 marks]*

M **Q28** The path of a particle is described parametrically by $x = t^2 - 6t$, $y = 2t^3 - 6t^2 - 18t$.

a) Find $\dfrac{dy}{dx}$ in terms of t.

b) Hence find any stationary points on the path of the particle.

E **Q29** A curve C defined by $x = t^3 - t^2$, $y = t^3 + 3t^2 - 9t$ has two turning points.

a) Find the coordinates of the turning points.   *[4 marks]*

b) Find the values of t at the points where C cuts the x-axis and hence show
that C passes through the origin. Leave your answers in surd form if necessary. *[2 marks]*

c) Find the equation of the tangent to C when $t = 2$.   *[2 marks]*

E **Q30** A curve is defined by the parametric equations $x = t \ln t$ and $y = 2t^3 - t^2$.

a) Find $\dfrac{dy}{dx}$ in terms of t.   *[3 marks]*

b) Hence find the equation of the tangent at $(0, 1)$.   *[3 marks]*

P **Q31** A curve is given parametrically by $x = 3se^s$, $y = e^{2s} + se^{2s}$

a) Find the equations of the tangents to the curve at $s = 0$ and $s = 2$.

b) Hence find the coordinates of the point of intersection of these tangents,
leaving your answer in terms of e.

P **Q32** The curve $4y + x^2y^2 = 4x$ passes through the two points $(2, a)$ and $(2, b)$, where a > b.

By finding a and b and the gradient $\dfrac{dy}{dx}$ of the curve, show that

the tangents to the curve at $(2, a)$ and $(2, b)$ intersect at $(5, 1)$.

E **Q33** The curve $x^2y + y^2 = x^2 + 1$ passes through the points $(1, -2)$ and $(1, a)$.
P
a) Find the gradient of the curve at $(1, -2)$.   *[4 marks]*

b) Find the value of a and show that $(1, a)$ is a turning point of the curve.   *[2 marks]*

P Q34 The curve $x \ln x + x^2 y = y^2 x - 6x$ passes through two points $(1, a)$ and $(1, b)$, where a > b.

    a) Find a and b.

    b) Use implicit differentiation to find the gradient of the curve at each of these points and hence the equations of the normals passing through the points.

    c) Find the coordinates of the point where the normals intersect.

E Q35 The curve C has the equation $4x^2 - 2y^2 = 7x^2 y$.

    a) Find an expression for the gradient of C using implicit differentiation. *[3 marks]*

    b) Hence find:

        (i)   the gradient of the tangent to C at $(1, -4)$.

        (ii) the equation of the normal to C at $(1, -4)$ in the form $ax + by + c = 0$. *[4 marks]*

E **P** Q36 A curve is defined by $x \cos x + y \sin x = y^3$.

    a) Show that at the stationary points of the curve, $y = x \tan x - 1$. *[5 marks]*

    b) Show that there are three points on the curve which have the x-coordinate $\frac{\pi}{2}$. *[3 marks]*

    c) Find the gradients of the tangents at each of these points and hence show that two of these tangents will never intersect. *[2 marks]*

E **P** Q37 The curve C has equation $x^2 - 2x + xy + y^2 = 20$.

    a) Find $\dfrac{dy}{dx}$ in terms of x and y. *[4 marks]*

    b) Find the coordinates of the point on the curve C with the minimum value of y. *[4 marks]*

    c) The curve C has a vertical tangent at two points, P and Q. Find the exact coordinates of P and Q. *[4 marks]*

P Q38 The function $f(x)$ is defined by $f(x) = \text{arccosec}\, x$, for $x > 1$.

    a) Write down the range of $f(x)$.

    b) Show that $f'(x) = -\dfrac{1}{x\sqrt{x^2 - 1}}$.

    c) Show that $f(x)$ is a decreasing function.

    d) Explain why $f(x)$ does not have any stationary points.

    e) By finding $f''(x)$, show that $f(x)$ is convex on the whole domain.

Challenge

E **P** Q39 a) Prove that the curve with equation $y = ax^3 + bx^2 + cx + d$, where a, b, c and d are constants and $a \neq 0$, $b \neq 0$ has exactly one point of inflection and state the x-coordinate of this point in terms of a, b, c and d. *[4 marks]*

    b) By finding a counter-example, show that not every quartic curve has a point of inflection. *[2 marks]*

E P **Q40** A curve has equation $y = 2^{ax^2+bx+c}$, where a, b and c are constants.
It crosses the y-axis when $y = 4$.

    a) Find the value of c.   *[1 mark]*

    b) Find $\dfrac{dy}{dx}$ in terms of a, b and c.   *[3 marks]*

    c) The curve has a maximum point at $\left(\dfrac{1}{2}, 2^{\frac{9}{4}}\right)$. Find the values of a and b.   *[4 marks]*

E P **Q41** A function f is defined by $f(x) = \dfrac{(2^x + 2^{3x})}{4^x}$.

    a) Find $f'(x)$.   *[3 marks]*

    b) Show that the equation of the normal to $y = f(x)$ at the point
where $x = 1$ is given by $y = -\dfrac{2}{\ln 8}x + \dfrac{5}{2} + \dfrac{2}{\ln 8}$.   *[3 marks]*

E P M **Q42** One end of a rope is attached to the top of a 10 m-tall statue and the other end is
attached to a pulley on the ground with negligible height. The pulley continuously
releases the rope as the statue is being moved away from it in a horizontal direction
at a rate of 1.5 m per second. The rope makes an angle θ with the horizontal, and
the horizontal displacement from the pulley is x m, as shown in the diagram below.

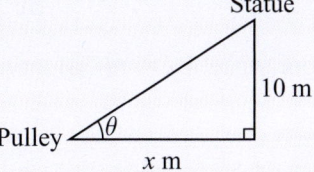

At time T, there is 25 m of rope between the pulley and statue.
Find the rate of change of the angle between the top of the statue
and the horizontal with respect to time at time T.   *[6 marks]*

P **Q43** Use the quotient rule to show that the derivative of $f(x) = \dfrac{\sec x + \operatorname{cosec} x}{\cot x}$
is $f'(x) = \operatorname{cosec} x(\sec^2 x - 1) + \sec x(\sec^2 x + \tan^2 x)$.

E P **Q44** A curve has equation $y = \sec^4 \sqrt{x}$, $x > 0$.

    a) Find $\dfrac{dy}{dx}$, giving your answer in its simplest form.   *[4 marks]*

    b) Find the exact coordinates of the minimum point
on the curve with the smallest positive x-coordinate.   *[3 marks]*

    c) Explain why the distance between the minimum
points increases as x increases.   *[2 marks]*

E Exam Style   P Problem Solving   M Modelling

1. Continuous curves can be convex or concave — convex means it curves downwards and concave means it curves upwards. A straight line joining points on a convex curve will be above the curve, and for concave curves it'll be below the curve.

2. Convex curves have an increasing gradient, $f''(x) > 0$. Concave curves have a decreasing gradient, $f''(x) < 0$. Often curves have both concave and convex parts.

3. The point where a curve changes between convex and concave is a point of inflection. At a point of inflection, $f''(x) = 0$, but not all points where $f''(x) = 0$ are points of inflection.

4. A stationary point occurs when $f'(x) = 0$ — it can be a minimum, a maximum or a point of inflection. It's a minimum if $f''(x) > 0$ and it's a maximum if $f''(x) < 0$. If $f''(x) = 0$, you need to check the sign of $f''(x)$ on either side of the stationary point to determine its nature.

5. The chain rule lets you differentiate functions of functions by splitting them up. If $y = f(u)$ and $u = g(x)$, then $\dfrac{dy}{dx} = \dfrac{dy}{du} \times \dfrac{du}{dx}$.

6. If x is given as a function of y, i.e. $x = f(y)$, you can find $\dfrac{dy}{dx}$ using the formula $\dfrac{dy}{dx} = \dfrac{1}{\left(\dfrac{dx}{dy}\right)}$.

7. e^x differentiates to give e^x. More generally, if $y = e^{f(x)}$ then $\dfrac{dy}{dx} = f'(x)e^{f(x)}$.

8. $\ln x$ differentiates to give $\dfrac{1}{x}$. More generally, if $y = \ln(f(x))$ then $\dfrac{dy}{dx} = \dfrac{f'(x)}{f(x)}$.

9. For any constant a, if $y = a^x$ then $\dfrac{dy}{dx} = a^x \ln a$.

10. The rules for differentiating trig functions measured in radians are:

$y = \sin x \Rightarrow \dfrac{dy}{dx} = \cos x$ $\qquad$ $y = \tan x \Rightarrow \dfrac{dy}{dx} = \sec^2 x$ $\qquad$ $y = \operatorname{cosec} x \Rightarrow \dfrac{dy}{dx} = -\operatorname{cosec} x \cot x$

$y = \cos x \Rightarrow \dfrac{dy}{dx} = -\sin x$ $\qquad$ $y = \cot x \Rightarrow \dfrac{dy}{dx} = -\operatorname{cosec}^2 x$ $\qquad$ $y = \sec x \Rightarrow \dfrac{dy}{dx} = \sec x \tan x$

You can prove the rules for sin and cos using differentiation from first principles.

11. You can differentiate functions of the form $y = uv$, where u and v are each functions of x, using the product rule: $\dfrac{dy}{dx} = u\dfrac{dv}{dx} + v\dfrac{du}{dx}$

12. You can differentiate functions of the form $y = \dfrac{u}{v}$, where u and v are each functions of x, using the quotient rule: $\dfrac{dy}{dx} = \dfrac{v\dfrac{du}{dx} - u\dfrac{dv}{dx}}{v^2}$

13. Connected rates of change occur when you have multiple linked variables — if you know the rate of change of any one of the variables and the equations that connect the variables then you can use the chain rule to find the rates of change of the other variables.

14. You can differentiate parametric curves (e.g. $y = f(t)$ and $x = g(t)$) using the formula $\dfrac{dy}{dx} = \dfrac{dy}{dt} \div \dfrac{dx}{dt}$.

15. You can differentiate equations of the form $f(x, y) = g(x, y)$ using implicit differentiation. Differentiate terms in x only as normal, use the chain rule for terms in y only and use the product rule for terms in both x and y. Rearrange to find $\dfrac{dy}{dx}$ in terms of x and y.

Integration

Learning Objectives

Once you've completed this chapter, you should be able to:

- Integrate functions of the form $(ax + b)^n$, $\dfrac{1}{ax + b}$, e^x, $e^{ax + b}$, and integrate trig functions.
- Integrate functions of the form $\dfrac{f'(x)}{f(x)}$, including multiples of these functions.
- Use the chain rule in reverse to integrate products of the form $\dfrac{du}{dx}f'(u)$ and $f'(x)[f(x)]^n$.
- Use the double angle formulas for sin, cos and tan to simplify difficult trig integrals.
- Use integration to find the area between two curves, or the area between a line and a curve, including finding the area under a curve given its parametric equations.
- Understand and use integration by substitution and integration by parts to integrate functions.
- Use partial fractions to integrate expressions in an appropriate form.
- Formulate and solve differential equations that model real-life situations, interpret the results, identify the limitations of a model and suggest refinements to address them.

Prior Knowledge Check

1 Differentiate, with respect to x, the following expressions:
   a) $7 \cos 3x$ b) $e^{5 - 2x}$ c) $\ln (x^2 + 1)$ *see pages 207-215*

2 Use trig identities to write each of the following as a single trig expression:
   a) $1 + \tan^2 6x$ b) $\cos 4x \sin x + \cos x \sin 4x$ *see pages 90-99*

3 Sketch the curve given by the pair of parametric equations
   $x = t^3 - t$ and $y = t^2$, over the range $-1 \le t \le 1$.   *see Chapter 4*

4 Use the product rule to differentiate the following functions with respect to x:
   a) $y = 4x \tan x$ b) $w = e^{3x} \ln(3x^3 + 3)$ *see pages 225-227*

5 Write $\dfrac{7}{(2x + 1)(x + 4)}$ as partial fractions.   *see pages 53-55*

Chapter 8

8.1 Integrating $(ax + b)^n$ where $n \neq -1$

In Year 1, you learnt to think of **integration** as the **opposite of differentiation**. This means that if you differentiate a function, then **integrating** the result will get you back to the function you started with. Here's an example to show how you can use this technique to integrate functions of the form **$(ax+b)^n$**.

Example 1

a) Differentiate $(3x + 4)^5$ with respect to x.

$$\frac{d}{dx}(3x + 4)^5 = 5(3x + 4)^4 \times 3 = 15(3x + 4)^4$$ ← Use the chain rule.

Tip See p.207 for a reminder on the chain rule.

b) Use your answer to a) to find $\int (3x + 4)^4\, dx$.

$(3x + 4)^5 \xrightarrow{\text{Differentiation}} 15(3x + 4)^4$ ← This comes from part a).

$(3x + 4)^5 + C \xleftarrow{\text{Integration}} 15(3x + 4)^4$ ← Write out the integration — it's the opposite of the differentiation.

$\int 15(3x + 4)^4\, dx = (3x + 4)^5 + c$ ← Substitute the result into the integral and simplify.

$\Rightarrow 15 \int (3x + 4)^4\, dx = (3x + 4)^5 + c$

$\Rightarrow \int (3x + 4)^4\, dx = \frac{1}{15}(3x + 4)^5 + \frac{c}{15}$ ← $\frac{c}{15}$ is just another constant term — you can call it C.

$= \frac{1}{15}(3x + 4)^5 + C$

This method gives us a **general result** for integrating all functions of the form **$(ax + b)^n$**.

Differentiating **$(ax + b)^{n+1}$** using the chain rule gives **$a(n + 1)(ax + b)^n$**.

So $\int a(n+1)(ax + b)^n\, dx = (ax + b)^{n+1} + c$

$a(n + 1) \int (ax + b)^n\, dx = (ax + b)^{n+1} + c$

Dividing by **$a(n + 1)$** gives the general expression:

$$\int (ax + b)^n\, dx = \frac{1}{a(n+1)}(ax + b)^{n+1} + C \quad \text{for } n \neq -1, a \neq 0$$

Tip This doesn't work for $n = -1$ because you'd end up having to divide by $n + 1 = 0$. See page 266 for a method of integrating x^{-1} and $(ax + b)^{-1}$.

Example 2

Find $\int (3 - 4x)^2\, dx$, using the general expression for $\int (ax + b)^n\, dx$.

$a = -4, b = 3$ and $n = 2$ ← Write down the values of a, b and n.

$\int (3 - 4x)^2\, dx = \frac{1}{-4 \times 3}(3 - 4x)^3 + C = -\frac{1}{12}(3 - 4x)^3 + C$ ← Substitute the values into the expression $\frac{1}{a(n+1)}(ax + b)^{n+1} + C$.

In Year 1, you learnt that definite integrals work out the **area** between a curve and the x-axis. To find the area between a curve $y = f(x)$ and the x-axis over an interval, just integrate $f(x)$ with respect to x over that interval.

Example 3

Work out the area enclosed by the curve $y = (x - 2)^3$, the x-axis and the lines $x = 2$ and $x = 3$.

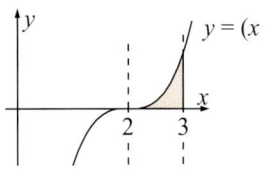

$y = (x - 2)^3$

$$\int_2^3 (x - 2)^3 \, dx = \frac{1}{4}\big[(x - 2)^4\big]_2^3$$

Use the formula with $a = 1$, $b = -2$ and $n = 3$ to work out the integral.

Tip You need to integrate the curve $y = (x - 2)^3$ between $x = 2$ and $x = 3$, i.e. evaluate $\int_2^3 (x - 2)^3 \, dx$.

$$\int_2^3 (x - 2)^3 \, dx = \frac{1}{4}\big[(3 - 2)^4\big] - \frac{1}{4}\big[(2 - 2)^4\big]$$

$$= \frac{1}{4}\big[1^4\big] - \frac{1}{4}\big[0^4\big]$$

$$= \frac{1}{4} - 0 = \frac{1}{4}$$

Substitute in the limits of integration to find the area.

Exercise 8.1

Q1 Integrate with respect to x:

    a) $(x + 10)^{10}$
    b) $(5x)^7$
    c) $(3 - 5x)^{-2}$
    d) $(3x - 4)^{-\frac{4}{3}}$

E **Q2** Find $\displaystyle\int \frac{1}{(4 - 2x)^7} \, dx$.   *[2 marks]*

E **P** **Q3** Show that $\displaystyle\int 5(2x + 1)^3 + (4x + 2)^3 \, dx$ can be written in the form $p(qx + r)^s + C$, where p, q, r and s are real constants.   *[3 marks]*

Q4   a) By using the general expression for $\displaystyle\int (ax + b)^n \, dx$,

        show that the integral $A = \displaystyle\int 8(2x - 4)^4 \, dx = \frac{4(2x - 4)^5}{5} + C$.

    b) Hence evaluate A between the values $x = \frac{3}{2}$ and $x = \frac{5}{2}$.

Q5 Evaluate $\displaystyle\int_0^1 (6x + 1)^{-3} \, dx$.

P **Q6** The curve $y = f(x)$ goes through the point $\left(1, \frac{3}{35}\right)$ and $f'(x) = (8 - 7x)^4$. Find $f(x)$.

E **P** **Q7** $\displaystyle\int_0^b (3x - 1)^3 \, dx = 52$. Find b, given that it is greater than 0.   *[4 marks]*

Challenge

P **Q8** Find the area bounded by the graph of $y = |(x - 2)^3|$, the x-axis and the lines $x = 0$ and $x = 4$.

8.2 Integrating e^x and e^{ax+b}

e^x differentiates to give e^x, so it makes sense that e^x **integrates** to give $\mathbf{e^x + C}$: $\quad \int e^x dx = e^x + C$

Example 1

Integrate the function $6x^2 - 4x + 3e^x$ with respect to x.

$\int 6x^2 - 4x + 3e^x \, dx = \int 6x^2 \, dx - \int 4x \, dx + \int 3e^x \, dx$ ⟵ Integrate each term separately.

$\int 6x^2 \, dx = 2x^3 + c \qquad \int 4x \, dx = 2x^2 + c$ ⟵ Integrate the first two terms as usual.

$\int 3e^x \, dx = 3\int e^x \, dx = 3e^x + c$ ⟵ Use the rule for e^x on the third term.

$\int 6x^2 - 4x + 3e^x \, dx = 2x^3 - 2x^2 + 3e^x + C$ ⟵ Put all the terms together.

On p.262 you saw how to integrate **linear transformations** of x^n by differentiating with the **chain rule** and working backwards. You can do the same with functions of the form $\mathbf{e^{ax+b}}$ where a and b are constants. Start by considering what you get when you **differentiate** functions of the form e^{ax+b} using the chain rule and work backwards.

This method of differentiating $f(ax + b)$ using the chain rule and working backwards to find an integral can be used with **any of the functions** that you know the derivative of.

Example 2

a) Differentiate the function e^{4x-1} with respect to x.

$\frac{d}{dx}(e^{4x-1}) = 4e^{4x-1}$ ⟵ Use the chain rule.

b) Using your answer to a), find the integral $\int e^{4x-1} \, dx$.

$\frac{d}{dx}(e^{4x-1}) = 4e^{4x-1} \Rightarrow \int 4e^{4x-1} \, dx = e^{4x-1} + c$ ⟵ Reverse the process of differentiation to integrate.

$\Rightarrow 4\int e^{4x-1} \, dx = e^{4x-1} + c$

$\Rightarrow \int e^{4x-1} \, dx = \frac{1}{4}e^{4x-1} + C$ ⟵ Take out the factor of 4 and divide by it.

This method gives you a **general rule** for integrating functions of the form $\mathbf{e^{ax+b}}$:

$$\int e^{ax+b} dx = \frac{1}{a}e^{ax+b} + C \qquad \text{for } a \neq 0$$

This means you just need to **divide** by the **coefficient of** x and add a constant of integration — the e^{ax+b} bit **doesn't change**.

Example 3

Integrate the following:

a) e^{7x}

$$\int e^{7x}\, dx = \frac{1}{7}e^{7x} + C$$

e^{7x} is already in the correct form so you just need to divide by 7 (the coefficient of x) when integrating.

b) $e^{\frac{x}{2}}$

$$\int e^{\frac{x}{2}}\, dx = \int e^{\frac{1}{2}x}\, dx = 2e^{\frac{x}{2}} + C$$

The coefficient of x is $\frac{1}{2}$ so you need to divide by $\frac{1}{2}$ (or multiply by 2) to integrate.

c) $2e^{4-3x}$

$$\int 2e^{4-3x}\, dx = -\frac{2}{3}e^{4-3x} + C$$

Multiplying by 2 doesn't change the integration. The coefficient of x is -3, so divide by -3.

Exercise 8.2

Q1 Find the following indefinite integrals:

a) $\displaystyle\int 2e^x\, dx$

b) $\displaystyle\int 4x + 7e^x\, dx$

c) $\displaystyle\int e^{10x}\, dx$

d) $\displaystyle\int e^{-3x} + x\, dx$

e) $\displaystyle\int e^{\frac{7}{2}x}\, dx$

f) $\displaystyle\int e^{4x-2}\, dx$

g) $\displaystyle\int \frac{1}{2}e^{2-\frac{3}{2}x}\, dx$

h) $\displaystyle\int e^{4(\frac{x}{3}+1)}\, dx$

P **Q2** Find the equation of the curve that has the derivative $\dfrac{dy}{dx} = 10e^{-5x-1}$ and passes through the origin.

Q3 Integrate the function e^{8y+5} with respect to y.

E **P** **Q4** f(x) can be written in the form $ae^{bx} + C$.
Given that f$'(x) = e^{\frac{x+10}{5}}$, find the exact values of a and b. *[3 marks]*

E **P** **Q5** Given that f$'(x) = e^{2x+1}$ and f$(-0.5) = 0.5$,
describe the transformation which maps f$'(x)$ onto f(x). *[3 marks]*

Q6 Evaluate the following definite integrals, giving exact answers:

a) $\displaystyle\int_2^3 e^{2x}\, dx$

b) $\displaystyle\int_{-1}^0 12e^{12x+12}\, dx$

c) $\displaystyle\int_{-\frac{\pi}{2}}^{\frac{\pi}{2}} e^{\pi-2x}\, dx$

d) $\displaystyle\int_3^6 \sqrt[6]{e^x} + \frac{1}{\sqrt[3]{e^x}}\, dx$

Q4 d) Hint

Remember:

$$\sqrt[n]{e^x} = e^{\frac{x}{n}}$$

$$\frac{1}{\sqrt[n]{e^x}} = e^{-\frac{x}{n}}$$

E **P** **Q7** Find the exact value of $\int_0^{\ln 2} e^{4x-5}\,dx$.

Give your answer as a fraction in its simplest form. *[4 marks]*

E **P** **Q8** a) Evaluate $\int_{-2}^{2} 20e^{5x+10}\,dx$. *[3 marks]*

b) Describe how your answer to part a) relates to the graph of $y = 20e^{5x+10}$. *[1 mark]*

c) Show that as $k \to \infty$, $\int_{-k}^{k} 20e^{5x+10}\,dx \approx Ae^{Bk}$

and determine the values of A and B. *[3 marks]*

Challenge

P **Q9** The diagram on the right shows the graph of $y = e^{3x+2}$. The point A is the point on the curve that has a gradient of 1.

Find the size of the shaded area.
Give your answer to 2 decimal places.

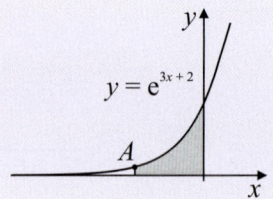

8.3 Integrating $\frac{1}{x}$ and $\frac{1}{ax+b}$

The method for integrating x^n and $(ax + b)^n$ on p.262 doesn't work for $n = -1$. For these functions you need to consider the fact that $\frac{d}{dx}(\ln x) = \frac{1}{x}$, which you should remember from Chapter 7.

Example 1

Let $f(x) = \frac{1}{x}$, $x > 0$. Integrate $f(x)$ with respect to x, given that $\frac{d}{dx}(\ln x) = \frac{1}{x}$.

$$\ln x \xrightarrow{\text{Differentiation}} \frac{1}{x}$$

$$\ln x + C \xleftarrow{\text{Integration}} \frac{1}{x}$$

Write out the integration — it's the opposite of differentiation (don't forget the "+ C").

$$\text{So } \int \frac{1}{x}\,dx = \ln x + C$$

So now we have a general result for integrating $\frac{1}{x}$: $\int \frac{1}{x}\,dx = \ln |x| + C$

Notice that this result uses $|x|$ instead of just x. This is because the function $\ln x$ is not defined for negative values of x. Using the modulus means you'll never end up taking ln of a negative value.

You'll be working with logs all the time when integrating functions of the form $\frac{1}{x}$ — it'll help to remember the log laws:

$$\log(ab) = \log a + \log b$$
$$\log\left(\frac{a}{b}\right) = \log a - \log b$$
$$\log(a^b) = b \log a$$

Example 2

Find the following integrals:

a) $\int \frac{5}{x}\,dx$

$$\int \frac{5}{x}\,dx = 5\int \frac{1}{x}\,dx = 5\ln|x| + C$$ ◄———— Take out a factor of 5 and then integrate $\frac{1}{x}$.

b) $\int_3^9 \frac{1}{3x}\,dx$

$$\int_3^9 \frac{1}{3x}\,dx = \frac{1}{3}\int_3^9 \frac{1}{x}\,dx = \frac{1}{3}\left[\ln|x|\right]_3^9$$ ◄———— Take out a factor of $\frac{1}{3}$ and then integrate $\frac{1}{x}$.

$$= \frac{1}{3}(\ln|9| - \ln|3|) = \frac{1}{3}(\ln(\frac{9}{3})) = \frac{1}{3}\ln 3$$ ◄———— Put in the limits and use the log law $\ln a - \ln b = \ln \frac{a}{b}$ to simplify.

You can integrate **linear transformations** of $\frac{1}{x}$ (i.e. functions of the form $\frac{1}{ax+b}$) by considering the result of differentiating $\ln|ax + b|$.

Example 3

Given that $\frac{d}{dx}(\ln|4x + 2|) = \frac{4}{4x + 2}$, find $\int \frac{1}{4x + 2}\,dx$.

$$\frac{d}{dx}(\ln|4x + 2|) = \frac{4}{4x + 2} \;\Rightarrow\; \int \frac{4}{4x + 2}\,dx = \ln|4x + 2| + c$$ ◄———— Reverse the process of differentiation to integrate.

$$\Rightarrow\; 4\int \frac{1}{4x + 2}\,dx = \ln|4x + 2| + c$$

$$\Rightarrow\; \int \frac{1}{4x + 2}\,dx = \frac{1}{4}\ln|4x + 2| + C$$ ◄———— Take out a factor of 4 from the integral and divide.

This method gives a **general result** for integrating functions of the form $\frac{1}{ax+b}$:

$$\int \frac{1}{ax + b}\,dx = \frac{1}{a}\ln|ax + b| + C$$

Example 4

Find $\int \frac{1}{2x + 5}\,dx$.

$$\int \frac{1}{2x + 5}\,dx = \frac{1}{2}\ln|2x + 5| + C$$ ◄———— Use the general rule with $a = 2$ and $b = 5$.

Chapter 8

Exercise 8.3

Q1 Find the following:

a) $\int \frac{19}{x}\, dx$

b) $\int \frac{1}{7x}\, dx$

c) $\int \frac{1}{7x + 2}\, dx$

d) $\int \frac{4}{1 - 3x}\, dx$

Q2 Integrate $y = \frac{1}{8x} - \frac{20}{x}$ with respect to x.

P **Q3** This is Marianne's attempt at integrating $\frac{3}{3x + 9}$:

$\int \frac{3}{3x + 9}\, dx = \int 3 \times (3x + 9)^{-1}\, dx = \int 3 \times 3(x + 3)^{-1}\, dx = \int 9(x + 3)^{-1}\, dx = 9\,(x + 3)^{-2} + c$

Find her two mistakes and explain what she should have done instead.

E **Q4** a) Express $\frac{9x}{3x + 6}$ in the form $a + \frac{b}{x + 2}$. *[2 marks]*

b) Hence, find $\int \frac{9x}{3x + 6}\, dx$. *[2 marks]*

P **Q5** a) Show that $\int \frac{6}{x} - \frac{3}{x}\, dx = \ln|x^3| + C$.

b) Evaluate $\int_{4}^{5} \frac{6}{x} - \frac{3}{x}\, dx$, giving an exact answer.

> **Q5-6 Problem Solving**
>
> Use the log laws from Year 1 (given on p.266).

P **Q6** Show that $\int_{b}^{a} 15(5 + 3x)^{-1}\, dx = \ln\left|\frac{5 + 3a}{5 + 3b}\right|^5$.

P **Q7** The graph of the curve $y = f(x)$ passes through the point $(1, 2)$. The derivative of $f(x)$ is given by $f'(x) = \frac{4}{10 - 9x}$. Find $f(x)$.

Q8 a) Express the area bounded by the curve $y = \frac{-7}{16 - 2x}$, the x-axis, the y-axis, and the line $x = -3$ as an integral with respect to x.

b) Show that the area is equal to $\ln\left[\left(\frac{8}{11}\right)^{\frac{7}{2}}\right]$.

P **Q9** Given that $\int_{1}^{A} \frac{4}{6x - 5}\, dx = 10$ and $A \geq 1$, find A in terms of e.

E **P** **Q10** Find $\int_{1.5}^{5} \frac{3}{\sqrt{(x + 3)}} \times \frac{(x + 3)^{-0.5}}{6}\, dx$.

Give your answer in the form $\ln a$, where a is a rational number. *[5 marks]*

Challenge

E **P** **Q11** a) Show that $\int \frac{x^2 + 10x + 25}{2x^2 + 13x + 15}\, dx = 0.5 \int 1 + \frac{7}{2x + 3}\, dx$. *[3 marks]*

b) $\int_{0}^{a} \frac{x^2 + 10x + 25}{2x^2 + 13x + 15}\, dx = 0.5a + \ln 128$.

Find a, given that a is greater than 0. *[4 marks]*

8.4 Integrating sin x and cos x

In Chapter 7 you learnt how to differentiate sin$x$ and cos$x$. You should remember that sin$x$ differentiates to cos$x$, and cos$x$ differentiates to –sin$x$.

Working backwards from this, we get:

$$\int \sin x \, dx = -\cos x + C \qquad \qquad \int \cos x \, dx = \sin x + C$$

> **Tip** You won't be given the integrals for sin and cos in your formula booklet, so make sure you remember them.

Example 1

Find the following integrals:

a) $\int 4 \cos x \, dx$

$$\int 4 \cos x \, dx = 4 \int \cos x \, dx = 4 \sin x + C \quad \longleftarrow \quad \text{Use } \int \cos x \, dx = \sin x + C.$$

b) $\int_0^\pi \dfrac{\sin x}{2} + \dfrac{1}{\pi} \, dx$

$$\int_0^\pi \frac{\sin x}{2} + \frac{1}{\pi} \, dx = \int_0^\pi \frac{1}{2}\sin x + \frac{1}{\pi} \, dx$$
$$= \left[\frac{1}{2}(-\cos x) + \frac{1}{\pi}x \right]_0^\pi \quad \longleftarrow \quad \text{Integrate each term separately.}$$
$$= \left[-\frac{1}{2}\cos x + \frac{1}{\pi}x \right]_0^\pi$$
$$= \left[-\frac{1}{2}\cos \pi + \left(\frac{1}{\pi} \times \pi\right) \right] - \left[-\frac{1}{2}\cos 0 + \left(\frac{1}{\pi} \times 0\right) \right]$$
$$= \left[-\frac{1}{2}(-1) + 1 \right] - \left[-\frac{1}{2}(1) + 0 \right]$$
$$= \left[\frac{1}{2} + 1 \right] - \left[-\frac{1}{2} + 0 \right] = 2$$

Put in the limits and simplify to give the exact value.

c) $\int \dfrac{1}{2}(\cos x + 2\sin x) \, dx$

$$\int \frac{1}{2}(\cos x + 2\sin x) \, dx = \int \frac{1}{2}\cos x + \sin x \, dx \quad \longleftarrow \quad \text{Expand the bracket.}$$
$$= \frac{1}{2}\sin x + (-\cos x) + C \quad \longleftarrow \quad \text{Integrate each term separately and simplify.}$$
$$= \frac{1}{2}\sin x - \cos x + C$$

You can integrate **linear transformations** of sin x and cos x of the form sin($ax + b$) and cos($ax + b$).

- **Differentiating** sin($ax + b$) using the chain rule gives: a **cos($ax + b$)**

- So when **integrating** cos($ax + b$), you need to **divide** by a, giving: $\frac{1}{a}$**sin($ax + b$)**

The same can be done when integrating sin ($ax + b$), so we get:

$$\int \sin(ax + b)\, dx = -\frac{1}{a}\cos(ax + b) + C \qquad \int \cos(ax + b)\, dx = \frac{1}{a}\sin(ax + b) + C$$

Example 2

Find $\int \sin(1 - 6x)\, dx$.

$$\int \sin(1 - 6x)\, dx = \frac{1}{-6} \times -\cos(1 - 6x) + C \quad \longleftarrow \quad \text{Use the general formula with } a = -6 \text{ and } b = 1.$$

$$= \frac{1}{6}\cos(1 - 6x) + C$$

Exercise 8.4

Q1 Integrate the following functions with respect to x.

a) $\frac{1}{7}\cos x$

b) $-3 \sin x$

c) $-3 \cos x - 3 \sin x$

d) $\sin 5x$

e) $\cos\left(\frac{x}{7}\right)$

f) $2 \sin(-3x)$

g) $5 \cos\left(3x + \frac{\pi}{5}\right)$

h) $-4 \sin\left(4x - \frac{\pi}{3}\right)$

i) $\cos(4x + 3) + \sin(3 - 4x)$

P Q2 Given that $x = \arcsin y$, find $\int y\, dx$.

Q3 Integrate $\frac{1}{2}\cos 3\theta - \sin \theta$ with respect to θ.

Q4 Evaluate the following definite integrals:

a) $\int_0^{\frac{\pi}{2}} \sin x\, dx$

b) $\int_{\frac{\pi}{6}}^{\frac{\pi}{3}} \sin 3x\, dx$

c) $\int_{-1}^{2} 3 \sin(\pi x + \pi)\, dx$

E P Q5 a) Given that $f'(\alpha) = 4 \sin \alpha + \frac{1}{2}\cos 2\alpha + \alpha$ (where α is in radians) and $f(0) = 0$, find $f(\alpha)$.
 [3 marks]

b) Disprove the following statement: $f(\alpha) \geq 0$ for all $\alpha \in \mathbb{R}$. *[1 mark]*

E P Q6 a) Find $\int a \sin 2x\, dx$ where a is a constant. *[1 mark]*

b) It is given that $g''(x) = a \sin 2x$ and $g(x) = -b \sin bx$ for all values of x, where a and b are constants. Find the values of a and b. *[3 marks]*

Q7 a) Integrate the function $y = 2\pi \cos\left(\frac{\pi x}{2}\right)$ with respect to x between the limits $x = 1$ and $x = 2$.

  b) Given that the function doesn't cross the x-axis between these limits, state whether the area between the curve and the x-axis for $1 \leq x \leq 2$ lies above or below the x-axis, justifying your answer.

E **Q8** Show that $\int_{\frac{\pi}{3}}^{\frac{\pi}{2}} \sin(-x) + \cos(-x)\, dx = \frac{1 - \sqrt{3}}{2}$. *[4 marks]*

Q9 Evaluate the definite integral $\int_{-\frac{\pi}{3}}^{\frac{\pi}{3}} \sin(3x + 2) - \cos x\, dx$.

Q10 Show that the integral of the function $y = 5\cos\left(\frac{x}{6}\right)$ between $x = -2\pi$ and $x = \pi$ is $15(1 + \sqrt{3})$.

8.5 Integrating sec²x

Another trigonometric function which is easy to integrate is the derivative of $\tan x$, **sec² x**. Since $\tan x$ differentiates to $\sec^2 x$, you get:

$$\int \sec^2 x\, dx = \tan x + C$$

> **Tip** This integral is given in the formula booklet — it's given as $\int \sec^2 kx\, dx = \frac{1}{k}\tan kx + C.$

Example 1

Find $\int 2\sec^2 x + 4x\, dx$.

$$\int 2\sec^2 x + 4x\, dx = 2\int \sec^2 x\, dx + \int 4x\, dx = 2\tan x + 2x^2 + C$$

◄ Integrate each term separately.

Unsurprisingly, you can use the chain rule in reverse to integrate functions of the form $\sec^2(ax + b)$:

$$\int \sec^2(ax + b)\, dx = \frac{1}{a}\tan(ax + b) + C$$

Example 2

Find $\int \cos 4x - 2\sin 2x + \sec^2\left(\frac{1}{2}x\right) dx$.

$$\int \cos 4x\, dx = \frac{1}{4}\sin 4x$$

$$\int -2\sin 2x\, dx = -2\left(-\frac{1}{2}\cos 2x\right) = \cos 2x$$

$$\int \sec^2\left(\frac{1}{2}x\right) dx = \frac{1}{\left(\frac{1}{2}\right)}\tan\left(\frac{1}{2}x\right) = 2\tan\left(\frac{1}{2}x\right)$$

Integrate each term separately using the general rules given on p.270 and above.

$$\int \cos 4x - 2\sin 2x + \sec^2\left(\frac{1}{2}x\right) dx$$

$$= \frac{1}{4}\sin 4x + \cos 2x + 2\tan\left(\frac{1}{2}x\right) + C$$

◄ Put them all together — always remember the "+ C".

Exercise 8.5

Q1 Find the following integrals:

a) $\int 2\sec^2 x + 1 \, dx$

b) $\int \sec^2 9x \, dx$

c) $\int 20\sec^2 3y \, dy$

d) $\int \sec^2 \frac{x}{7} \, dx$

e) $\int_0^{\frac{\pi}{3}} -\frac{1}{\cos^2 \theta} \, d\theta$

f) $\int_0^{\frac{\pi}{4}} 3\sec^2(-3x) \, dx$

Q2 Find the value of the integral of the function $y = \sec^2 x$ between the limits $x = \frac{2}{3}\pi$ and $x = \pi$.

E Q3 Find the total area bounded by the graph $y = 2\sec^2 x + \cos x$, the x-axis and the lines $x = -\frac{\pi}{4}$ and $x = \frac{\pi}{4}$. *[3 marks]*

Q4 Integrate $\sec^2(x + \alpha) + \sec^2(3x + \beta)$ with respect to x, where α and β are constants.

Q5 Let A be a constant. Integrate $5A \sec^2\left(\frac{\pi}{3} - 2\theta\right)$ with respect to θ between the limits of $\theta = \frac{\pi}{12}$ and $\theta = \frac{\pi}{6}$.

E Q6 Find $\int_1^2 0.5\sec^2(2\theta + \pi) \, d\theta$. Give your answer to 3 significant figures. *[3 marks]*

Challenge

E
P
Q7 a) Find $\int_{\frac{\pi}{16}}^{\frac{\pi}{12}} (3\sec 4\theta)^2 \, d\theta$. Give your answer in exact form. *[3 marks]*

b) Given that $\int_{\frac{\pi}{16}}^{\frac{\pi}{12}} (3\sec 4\theta)^2 + a\sin(4\theta + \pi) \, d\theta = \frac{9(\sqrt{3} - \sqrt{2})}{4}$,

find the value of the constant a. *[4 marks]*

8.6 Integrating Other Trig Functions

There are some other more complicated trig functions which are really easy to integrate. They are the **derivatives** of the functions **cosec** x, **sec** x and **cot** x.

You may remember these derivatives from Chapter 7, but here's a recap:

$\frac{d}{dx}(\text{cosec } x) = -\text{cosec } x \cot x$

$\frac{d}{dx}(\sec x) = \sec x \tan x$

$\frac{d}{dx}(\cot x) = -\text{cosec}^2 x$

Reversing the differentiation gives the following three integrals. They'll be really useful when integrating complicated trig functions.

$\int \text{cosec } x \cot x \, dx = -\text{cosec } x + C$

$\int \sec x \tan x \, dx = \sec x + C$

$\int \text{cosec}^2 x \, dx = -\cot x + C$

As always, you can integrate **linear transformations** of these functions (i.e. functions of the form **cosec($ax + b$)cot($ax + b$), sec($ax + b$)tan($ax + b$)** and **cosec²($ax + b$)**) by **dividing** by the coefficient of x.

$$\int \operatorname{cosec}(ax + b)\cot(ax + b)\, dx = -\frac{1}{a}\operatorname{cosec}(ax + b) + C$$

$$\int \sec(ax + b)\tan(ax + b)\, dx = \frac{1}{a}\sec(ax + b) + C$$

$$\int \operatorname{cosec}^2(ax + b)\, dx = -\frac{1}{a}\cot(ax + b) + C$$

> **Tip** The $ax + b$ bit has to be the same in each trig function — you couldn't integrate sec x tan $3x$ using these formulas.

Example 1

Find the following:

a) $\int 2\sec x \tan x\, dx$

$$\int 2\sec x \tan x\, dx = 2\int \sec x \tan x\, dx \quad \longleftarrow \quad \text{Take the constant outside the integral.}$$

$$= 2(\sec x + c) \quad \longleftarrow \quad \text{Now just use the rule given on the previous page and simplify.}$$

$$= 2\sec x + C$$

b) $\int_0^\pi \operatorname{cosec}^2\left(\frac{x}{2} - \frac{\pi}{4}\right) dx$

$$\int_0^\pi \operatorname{cosec}^2\left(\frac{x}{2} - \frac{\pi}{4}\right) dx = \left[-\frac{1}{\left(\frac{1}{2}\right)}\cot\left(\frac{x}{2} - \frac{\pi}{4}\right)\right]_0^\pi \quad \longleftarrow \quad \text{Integrate the function, using the rule above where } a = \frac{1}{2} \text{ and } b = \frac{\pi}{4}.$$

$$= -2\left[\cot\left(\frac{x}{2} - \frac{\pi}{4}\right)\right]_0^\pi = -2\left[\frac{1}{\tan\left(\frac{x}{2} - \frac{\pi}{4}\right)}\right]_0^\pi \quad \longleftarrow \quad \text{Simplify by taking out a factor of } -2 \text{ and rewrite cot in terms of tan.}$$

$$= -2\left(\frac{1}{\tan\left(\frac{\pi}{2} - \frac{\pi}{4}\right)} - \frac{1}{\tan\left(0 - \frac{\pi}{4}\right)}\right) \quad \longleftarrow \quad \text{Put in the limits and simplify to evaluate the integral.}$$

$$= -2\left(\frac{1}{\tan\left(\frac{\pi}{4}\right)} - \frac{1}{\tan\left(-\frac{\pi}{4}\right)}\right) = -2\left(\frac{1}{1} - \frac{1}{(-1)}\right) = -4$$

c) $\int 8\operatorname{cosec}(2x + 1)\cot(2x + 1)\, dx$

$$\int 8\operatorname{cosec}(2x + 1)\cot(2x + 1)\, dx$$

$$= 8\int \operatorname{cosec}(2x + 1)\cot(2x + 1)\, dx \quad \longleftarrow \quad \text{Take the constant outside the integral.}$$

$$= 8\left(-\frac{1}{2}\operatorname{cosec}(2x + 1) + c\right) \quad \longleftarrow \quad \text{Integrate the function using the rule above where } a = 2 \text{ and } b = 1, \text{ then simplify.}$$

$$= -4\operatorname{cosec}(2x + 1) + C$$

Example **2**

Find $\int 10 \sec 5x \tan 5x + \frac{1}{2} \csc 3x \cot 3x - \csc^2(6x+1) \, dx$.

$\int 10 \sec 5x \tan 5x \, dx = 10\left(\frac{1}{5} \sec 5x\right) = 2 \sec 5x$

Integrate each term separately — don't forget the minus signs that come from the integration in the second and third terms.

$\int \frac{1}{2} \csc 3x \cot 3x \, dx = \frac{1}{2}\left(-\frac{1}{3} \csc 3x\right) = -\frac{1}{6} \csc 3x$

$\int -\csc^2(6x+1) \, dx = -\left(-\frac{1}{6} \cot(6x+1)\right) = \frac{1}{6} \cot(6x+1)$

$\int 10 \sec 5x \tan 5x + \frac{1}{2} \csc 3x \cot 3x - \csc^2(6x+1) \, dx$

$= 2 \sec 5x - \frac{1}{6} \csc 3x + \frac{1}{6} \cot(6x+1) + C$

Put the terms together and add the constant.

Exercise 8.6

Q1 Find the following integrals:

a) $\int \csc^2 11x \, dx$

b) $\int 5 \sec 10\theta \tan 10\theta \, d\theta$

c) $\int -\csc(x+17)\cot(x+17) \, dx$

d) $\int -3 \csc 3x \cot 3x \, dx$

e) $\int 13 \sec\left(\frac{\pi}{4} - x\right)\tan\left(\frac{\pi}{4} - x\right) dx$

f) $\int 4 \csc^2(5x+3) \, dx$

Q2 Find $\int 10 \csc^2\left(\alpha - \frac{x}{2}\right) - 60 \sec(\alpha - 6x)\tan(\alpha - 6x) \, dx$.

Q3 Integrate the function $6 \sec 2x \tan 2x + 6 \csc 2x \cot 2x$ with respect to x between the limits of $x = \frac{\pi}{12}$ and $x = \frac{\pi}{8}$.

Q4 Find the area of the region bounded by $y = \csc^2(3x)$, the x-axis and the lines $x = \frac{\pi}{12}$ and $x = \frac{\pi}{6}$.

E Q5 Find $\int_0^{\frac{\pi}{3}} \sec 0.5x \tan 0.5x \, dx$, giving your answer as an exact value. *[3 marks]*

E **P** Q6 It is given that $\int_{\frac{7\pi}{3}}^{a} \csc x \cot x \, dx = 0$ and $a \neq \frac{7\pi}{3}$.

a) Show that $\csc a = \frac{2}{\sqrt{3}}$. *[3 marks]*

b) (i) Given that $2\pi < a < 4\pi$, find the value of a.

(ii) Hence find the exact total area of the region bounded by $y = \csc x \cot x$, the x-axis and the lines $x = \frac{7\pi}{3}$ and $x = a$. *[5 marks]*

8.7 Integrating $\frac{f'(x)}{f(x)}$

If you have a fraction that has a function of x as the numerator and a different function of x as the denominator, e.g. $\frac{x-2}{x^3+1}$, you'll probably struggle to integrate it.

However, if you have a fraction where the **numerator** is the **derivative** of the **denominator**, e.g. $\frac{3x^2}{x^3+1}$, it integrates to give ln of the denominator.

In general terms, this is written as:
$$\int \frac{f'(x)}{f(x)} \, dx = \ln|f(x)| + C$$

This rule won't surprise you if you remember differentiating ln (f(x)) using the chain rule (see p.215) — the derivative with respect to x of ln (f(x)) is $\frac{f'(x)}{f(x)}$.

You sometimes see this rule written without the modulus signs, but if it's possible for f(x) to take a negative value, you need the modulus so you don't end up trying to take ln of a negative number.

The hardest bit about integrations like this is recognising that the denominator differentiates to give the numerator — once you've spotted that you can just use the formula.

Example 1

Integrate the following functions with respect to x.

a) $\frac{2x}{x^2+1}$

$\frac{d}{dx}(x^2+1) = 2x$, which is the numerator. ← Differentiate the denominator to see what it gives.

$\int \frac{2x}{x^2+1} \, dx = \ln|x^2+1| + C$ ← The numerator is the derivative of the denominator, so use the formula given above.

Tip You could leave out the modulus sign in part a), as $x^2 + 1 > 0$.

b) $\frac{x(3x-4)}{x^3-2x^2-1}$

$\frac{d}{dx}(x^3-2x^2-1) = 3x^2-4x = x(3x-4)$ ← Differentiate the denominator.

$\int \frac{x(3x-4)}{x^3-2x^2-1} \, dx = \ln|x^3-2x^2-1| + C$ ← $x(3x-4)$ is the numerator, so use the formula given above.

You might see questions where the numerator is a **multiple** of the derivative of the denominator. When this happens, just put the multiple **in front** of the ln.

Example 2

Find:

a) $\int \dfrac{8x^3 - 4}{x^4 - 2x}\, dx$

$\dfrac{d}{dx}(x^4 - 2x) = 4x^3 - 2$ ⟵ —————— Differentiate the denominator.

$8x^3 - 4 = 2(4x^3 - 2)$ ⟵ —————— The numerator is 2 × the derivative of the denominator.

$\int \dfrac{8x^3 - 4}{x^4 - 2x}\, dx = 2\int \dfrac{4x^3 - 2}{x^4 - 2x}\, dx$ ⟵ —— Take out the factor of 2 and then use the formula.

$\qquad\qquad = 2\,\ln|x^4 - 2x| + C$

b) $\int \dfrac{3\sin 3x}{\cos 3x + 2}\, dx$

$\dfrac{d}{dx}(\cos 3x + 2) = -3\sin 3x$ ⟵ —————— Differentiate the denominator.

$\int \dfrac{3\sin 3x}{\cos 3x + 2}\, dx = -\int \dfrac{-3\sin 3x}{\cos 3x + 2}\, dx$ ⟵ —— The numerator is –1 × the derivative of the denominator, so take out a factor of –1.

$\qquad = -\ln|\cos 3x + 2| + C$ ⟵ —— Use the formula.

$\qquad = -\ln|\cos 3x + 2| - \ln k$ ⟵

$\qquad = -\ln|k(\cos 3x + 2)|$ ⟵ —— Simplify by combining it into one logarithm.

C is just a constant, so if we want we can express C as a logarithm — call it **ln k** or **–ln k**, where k is a constant. (Adding in a minus sign makes it easier to simplify in some cases, like in the example above.)

You can use this method to integrate **trig functions** by writing them as fractions:

You can work out the integral of tan$x$ using this method:

$\tan x = \dfrac{\sin x}{\cos x}$, and $\dfrac{d}{dx}(\cos x) = -\sin x$

The numerator is **–1** × the **derivative** of the **denominator**, so:

$\int \tan x\, dx = \int \dfrac{\sin x}{\cos x}\, dx = -\ln|\cos x| + C$

This is a useful result — it's given in the formula booklet in the following form:

$\int \tan kx\, dx = \dfrac{1}{k}\ln|\sec kx| + C$

Tip $-\ln|\cos x|$ is the same as $\ln|\sec x|$ by the laws of logs.

There are some other **trig functions** that you can integrate in the same way.

$$\int \operatorname{cosec} x \, dx = -\ln|\operatorname{cosec} x + \cot x| + C$$
$$\int \sec x \, dx = \ln|\sec x + \tan x| + C$$
$$\int \cot x \, dx = \ln|\sin x| + C$$

> **Tip** The formula booklet actually gives them in the form cosec kx etc. — but you can easily get these results by setting $k = 1$.

As always, if you're integrating a linear transformation of any of these functions of the form f($ax + b$), then divide by a when you integrate.

You can check these results easily by using differentiation — differentiate the right-hand side of the results to get the left-hand sides. Remember that differentiating $\ln|f(x)|$ gives $\frac{f'(x)}{f(x)}$.

Example 3

Find the following integrals:

a) $\int 2 \sec x \, dx$

$$\int 2 \sec x \, dx = 2 \ln|\sec x + \tan x| + C$$ ← Take out the constant of 2 and use the result for sec x above.

b) $\int \frac{\cot x}{5} \, dx$

$$\int \frac{\cot x}{5} \, dx = \frac{1}{5} \ln|\sin x| + C$$ ← Take out the constant of $\frac{1}{5}$ and use the result for cot x above.

c) $\int 2(\operatorname{cosec} x + \sec x) \, dx$

$$\int 2(\operatorname{cosec} x + \sec x) \, dx$$

$$= \int 2 \operatorname{cosec} x + 2 \sec x \, dx$$ ← Expand the brackets.

$$= -2 \ln|\operatorname{cosec} x + \cot x| + 2 \ln|\sec x + \tan x| + C$$ ← Integrate each term separately.

$$= 2 \ln\left|\frac{\sec x + \tan x}{\operatorname{cosec} x + \cot x}\right| + C$$ ← Use log laws to simplify.

d) $\int \frac{1}{2} \operatorname{cosec} 2x \, dx$

The coefficient of x is 2, so you need to divide by 2 when you integrate. ← Work out what happens to the coefficient of x.

$$\int \frac{1}{2} \operatorname{cosec} 2x \, dx$$

$$= -\frac{1}{4} \ln|\operatorname{cosec} 2x + \cot 2x| + C$$ ← Use the result for cosec x — remember to divide by 2.

> **Tip** Check this by differentiating (using the chain rule with $u = \operatorname{cosec} 2x + \cot 2x$).

Chapter 8

Exercise 8.7

Q1 Find the following integrals:

a) $\int \dfrac{4x^3}{x^4 - 1}\, dx$

b) $\int \dfrac{2x - 1}{x^2 - x}\, dx$

c) $\int \dfrac{x^4}{3x^5 + 6}\, dx$

d) $\int \dfrac{12x^3 + 18x^2 - 3}{x^4 + 2x^3 - x}\, dx$

e) $\int \dfrac{8x - 2}{(x - 2)(2x + 3)}\, dx$

f) $\int \dfrac{6 - 4x^2}{2x^3 - 9x + 7}\, dx$

Q2 a) Find $\int \dfrac{9(3x + 2)^2}{(3x + 2)^3}\, dx$, without simplifying the fraction prior to integrating. *[2 marks]*

b) Fully simplify $\dfrac{9(3x + 2)^2}{(3x + 2)^3}$ and integrate the result. *[2 marks]*

c) Verify that the results to part a) and part b) are equivalent. *[1 mark]*

Q3 a) Find the domain of $f(x) = \dfrac{2x + 4}{3x^2 + 12x + 30}$, giving a reason for your answer. *[2 marks]*

b) Find $\displaystyle\int_0^3 \dfrac{2x + 4}{3x^2 + 12x + 30}\, dx$.

Give your answer in the form $a \ln b$, where a and b are constants. *[4 marks]*

c) Does $\displaystyle\int_0^3 \dfrac{2x + 4}{3x^2 + 12x + 30}\, dx$ give the area bounded by the curve

$f(x) = \dfrac{2x + 4}{3x^2 + 12x + 30}$, $x = 0$ and $x = 3$? Explain your answer. *[1 mark]*

Q4 Find the indefinite integrals below:

a) $\int \dfrac{e^x}{e^x + 6}\, dx$

b) $\int \dfrac{2(e^{2x} + 3e^x)}{e^{2x} + 6e^x}\, dx$

c) $\int \dfrac{e^x}{3(e^x + 3)}\, dx$

d) $\int \dfrac{10e^{4x} + 5}{5e^{4x} + 10x}\, dx$

Q5 Find the following integrals:

a) $\int \dfrac{2 \cos 2x}{1 + \sin 2x}\, dx$

b) $\int \dfrac{\sin 3x}{\cos 3x - 1}\, dx$

c) $\int \dfrac{3 \operatorname{cosec} x \cot x + 6x}{\operatorname{cosec} x - x^2 + 4}\, dx$

d) $\int \dfrac{\sec^2 x}{\tan x}\, dx$

e) $\int \dfrac{\sec x \tan x}{\sec x + 5}\, dx$

f) $\int \dfrac{\operatorname{cosec}^2 x}{2 \cot x - 1}\, dx$

Q6 Given that $\int \dfrac{2 f(x)}{3 \tan 4x}\, dx = \ln |\tan 4x| + C$, find $f(x)$. *[2 marks]*

Q7 Show that $\int \dfrac{4 \cos(2x + 7)}{\sin(2x + 7)}\, dx = 2 \ln|k \, \sin(2x + 7)|$.

E P Q8 Show that $\int_{\pi}^{k\pi} \dfrac{16\theta - 12\cos 2\theta}{4\theta^2 - 3\sin 2\theta}\,d\theta = \ln k^4$,

where k is a positive integer greater than 1. *[4 marks]*

P Q9 Prove that:

Problem Solving

In Q9, try multiplying the bit inside the integral by $\dfrac{\sec x + \tan x}{\sec x + \tan x}$ in part a) — there's a similar trick for part b) as well.

a) $\displaystyle\int \sec x\,dx = \ln|\sec x + \tan x| + C$

b) $\displaystyle\int \operatorname{cosec} x\,dx = -\ln|\operatorname{cosec} x + \cot x| + C$

Q10 Find the following integrals:

a) $\displaystyle\int 2\tan x\,dx$

b) $\displaystyle\int \tan 2x\,dx$

c) $\displaystyle\int 4\operatorname{cosec} x\,dx$

d) $\displaystyle\int \cot 3x\,dx$

e) $\displaystyle\int \frac{1}{2}\sec 2x\,dx$

f) $\displaystyle\int 3\operatorname{cosec} 6x\,dx$

Q11 Find $\displaystyle\int \frac{4\sin(3 - 2x)}{3 + \cos(3 - 2x)} + \frac{x^2}{6x^3 - 5}\,dx$.

Q12 Find $\displaystyle\int \frac{\sec^2 x}{2\tan x} - 4\sec 2x\tan 2x + \frac{\operatorname{cosec} 2x\cot 2x - 1}{\operatorname{cosec} 2x + 2x}\,dx$.

E P Q13 The graph of $y = \dfrac{\sin x + \cos x}{\sin x - \cos x}$ is shown below.

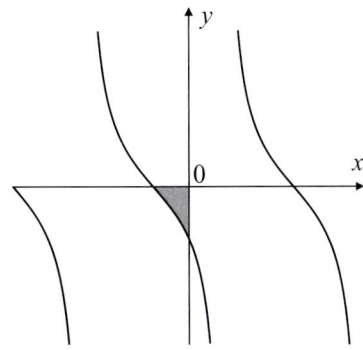

a) Find the exact area bounded by the negative x-axis, the negative y-axis and the curve (shaded on the graph), giving your answer as an exact value. *[5 marks]*

b) Show that the equivalent area for the curve with equation
$y = \dfrac{\sin kx + \cos kx}{\sin kx - \cos kx}$ is inversely proportional to k. *[4 marks]*

Challenge

P Q14 Given that $\displaystyle\int \frac{5\sin 3x + 2\cos x}{g(x)}\,dx = \ln|-5\cos 3x + h(x)| + A$,

where A is a constant and $g(0) = 0$, find $g(x)$.

8.8 Integrating $\frac{du}{dx}f'(u)$

In Chapter 7, you saw the chain rule for differentiating a **function of a function**.

Here it is in the form it was given on page 207: If $y = f(u)$ and $u = g(x)$ then: $\dfrac{dy}{dx} = \dfrac{dy}{du} \times \dfrac{du}{dx}$

Since integration is the opposite of differentiation, you have:

$$y \xrightarrow{\text{Differentiation}} \frac{dy}{du} \times \frac{du}{dx} \qquad y + C \xleftarrow{\text{Integration}} \frac{dy}{du} \times \frac{du}{dx}$$

So $\displaystyle\int \frac{dy}{du} \times \frac{du}{dx}\, dx = y + C$. Then writing $f(u)$ instead of y and $f'(u)$ instead of $\dfrac{dy}{du}$ gives:

$$\int \frac{du}{dx} f'(u)\, dx = f(u) + C$$

> **Tip** To evaluate integrals like this, integrate with respect to u (because $f'(u)$ is $\dfrac{dy}{du}$).

If you're integrating an expression which contains a **function of a function**, $f(u)$, try differentiating the function u. If the **derivative** of u is also part of the expression, you might be able to use the formula above.

This result's quite difficult to grasp — but after a few examples it should make complete sense.

Example 1

a) Differentiate $y = e^{2x^2}$ using the chain rule.

$$\text{Let } u = 2x^2 \text{ and } y = e^u$$

Find a suitable substitution where you can differentiate each equation.

$$\frac{dy}{dx} = \frac{dy}{du} \times \frac{du}{dx} = e^u \times 4x$$
$$= e^{2x^2} \times 4x = 4xe^{2x^2}$$

Differentiate each equation and use the chain rule — remember to replace u and simplify.

b) Find $\displaystyle\int 8xe^{2x^2}\, dx$ using your answer to part a).

$$\int 8xe^{2x^2}\, dx = 2\int 4xe^{2x^2}\, dx$$

Take out a factor of 2 to leave the result obtained in part a). From the formula above, $\dfrac{du}{dx} = 4x$ and $f'(u) = e^u$, where $u = 2x^2$.

$$= 2(e^{2x^2} + c)$$
$$= 2e^{2x^2} + C$$

So replace the integral with $f(u) + C$.

Example 2

Find the following integrals:

a) $\int 6x^5 e^{x^6} \, dx$

Let $u = x^6 \Rightarrow \dfrac{du}{dx} = 6x^5$ ⟵ Identify an expression for u and use it to find $\dfrac{du}{dx}$.

$f'(u) = e^u \Rightarrow f(u) = e^u$ ⟵ Set the rest of the expression equal to $f'(u)$ and find $f(u)$.

$\int 6x^5 e^{x^6} \, dx = \int \dfrac{du}{dx} f'(u) \, dx$ ⟵ Rewrite the integral in terms of $\dfrac{du}{dx}$ and $f'(u)$.

$= f(u) + C = e^u + C = e^{x^6} + C$ ⟵ Use the formula to write down the result.

b) $\int e^{\sin x} \cos x \, dx$

Let $u = \sin x \Rightarrow \dfrac{du}{dx} = \cos x$ ⟵ Identify u and find $\dfrac{du}{dx}$.

$f'(u) = e^u \Rightarrow f(u) = e^u$ ⟵ Set the rest of the expression equal to $f'(u)$ and find $f(u)$.

$\int e^{\sin x} \cos x \, dx = \int \dfrac{du}{dx} f'(u) \, dx$ ⟵ Rewrite the integral and use the

$= f(u) + C = e^u + C = e^{\sin x} + C$ ⟵ formula to write down the result.

Example 3

a) Find $\int x^4 \sin(x^5) \, dx$

Let $u = x^5 \Rightarrow \dfrac{du}{dx} = 5x^4$ ⟵ Identify u and find $\dfrac{du}{dx}$.

$f'(u) = \sin u \Rightarrow f(u) = -\cos u$ ⟵ Set the rest of the expression equal to $f'(u)$ and find $f(u)$.

$\int x^4 \sin(x^5) \, dx = \int \dfrac{1}{5} \dfrac{du}{dx} f'(u) \, dx = \dfrac{1}{5} \int \dfrac{du}{dx} f'(u) \, dx$ ⟵ Write the integral in terms of $\dfrac{du}{dx}$ and $f'(u)$.

$= \dfrac{1}{5}(f(u) + c) = \dfrac{1}{5}(-\cos(u) + c) = -\dfrac{1}{5}\cos(x^5) + C$ ⟵ Now use the formula — remember to multiply by the constant factor.

b) Hence find the exact value of $\int_0^1 x^4 \sin(x^5) \, dx$.

$\int_0^1 x^4 \sin(x^5) \, dx = \left[-\dfrac{1}{5}\cos(x^5) \right]_0^1$ ⟵ Use the result from part a).

$= \left[-\dfrac{1}{5}\cos(1^5) \right] - \left[-\dfrac{1}{5}\cos(0^5) \right]$ ⟵ Put in the limits and simplify.

$= -\dfrac{1}{5}\cos 1 + \dfrac{1}{5} = \dfrac{1}{5}(1 - \cos 1)$

Exercise 8.8

Find the following integrals:

Q1 $\displaystyle\int 2x e^{x^2}\, dx$

Q2 $\displaystyle\int 6x^2 e^{2x^3}\, dx$

Q3 $\displaystyle\int \frac{1}{2\sqrt{x}} e^{\sqrt{x}}\, dx$

Q4 $\displaystyle\int x^3 e^{x^4}\, dx$

Q5 $\displaystyle\int (4x-1) e^{\left(x^2 - \frac{1}{2}x\right)}\, dx$

Q6 $\displaystyle\int 2x \sin(x^2+1)\, dx$

Q7 $\displaystyle\int x^3 \cos(x^4)\, dx$

Q8 $\displaystyle\int x \sec^2(x^2)\, dx$

Q9 $\displaystyle\int e^{\cos x} \sin x\, dx$

Q10 $\displaystyle\int \cos 2x\, e^{\sin 2x}\, dx$

Q11 $\displaystyle\int \sec^2 x\, e^{\tan x}\, dx$

Q12 $\displaystyle\int \sec x \tan x\, e^{\sec x}\, dx$

Q13 $\displaystyle\int 2 \operatorname{cosec}^2 x\, e^{\cot x}\, dx$

Q14 $\displaystyle\int \operatorname{cosec} 3x \cot 3x\, e^{\operatorname{cosec} 3x}\, dx$

E Q15 a) Find $\displaystyle\int (4x+6)\sin(2x^2+6x)\, dx$. *[3 marks]*

b) Hence find $\displaystyle\int 2x\sin(2x^2+6x) + 3\sin(2x^2+6x)\, dx$. *[2 marks]*

E Q16 a) Find $\displaystyle\int (9x^2+18)e^{x^3+6x}\, dx$. *[4 marks]*

b) Given that $f'(x) = (9x^2+18)e^{x^3+6x}$ and $f(1) = e^7$, find $f(x)$. *[2 marks]*

Q17 Evaluate $\displaystyle\int_0^4 (12x+2)\sin(3x^2+x)\, dx$. Give your answer to 3 significant figures.

Q18 a) Find $\displaystyle\int 5\sin(2-5x)e^{\cos(2-5x)}\, dx$

b) Hence evaluate $\displaystyle\int_0^1 5\sin(2-5x)e^{\cos(2-5x)}\, dx$.

Q19 a) Find $\displaystyle\int (8x-1)\cos(4x^2-x)\, e^{\sin(4x^2-x)}\, dx$.

b) Hence evaluate $\displaystyle\int_0^\pi (8x-1)\cos(4x^2-x)\, e^{\sin(4x^2-x)}\, dx$.

E Q20 a) Show that $\dfrac{dy}{dx} = \dfrac{2x\ln x - x}{(\ln x)^2}$ when $y = \dfrac{x^2}{\ln x}$. *[4 marks]*

b) Hence, find $\displaystyle\int \frac{\ln x^{2x} - x}{(\ln x)^2}\sin\left(\frac{x^2}{\ln x}\right) dx$. *[2 marks]*

P Q21 Evaluate the exact value of $\displaystyle\int_0^{\frac{\pi}{6}} \sin x\, e^{3x\sin x} + x\cos x\, e^{3x\sin x}\, dx$.

Challenge

E
P Q22 Given that $\displaystyle\int_0^k (\sin x\cos x + 2\cos x)e^{\sin^2 x} e^{4\sin x} e^3\, dx = 5e^3$ and $k > 0$,

find the smallest possible value of k. Give your answer to 3 significant figures. *[7 marks]*

8.9 Integrating $f'(x) \times [f(x)]^n$

Some products are made up of a **function** and its **derivative**:

This bracket is the **derivative**...

e.g. $3(3x^2 + 4)(x^3 + 4x)^2$

...of this bracket.

If you spot that part of a product is the **derivative** of the other part of it (which is raised to a **power**), you can integrate it using this rule (which is just a special case of the 'reverse chain rule' on p.280):

This function is the **derivative**...

$$\int (n + 1)f'(x)[f(x)]^n \, dx = [f(x)]^{n+1} + C$$

...of this function.

Remember that this result needs you to have a multiple of $n + 1$ (not n) — you can check this by **differentiating** the right-hand side using the **chain rule**.

Watch out for any other multiples too — you might have to **multiply** or **divide** by a **constant**.

This will probably make more sense if you have a look at some examples:

Example 1

Evaluate $\int 12x^3(2x^4 - 5)^2 \, dx$.

$\int (n + 1)f'(x)[f(x)]^n \, dx = [f(x)]^{n+1} + C$ ← Use this formula to integrate.

$f(x) = 2x^4 - 5$, so $f'(x) = 8x^3$. $n = 2$, so $n + 1 = 3$ ← Work out all the terms you need to substitute into the formula and then put them in.

So $\int 3(8x^3)(2x^4 - 5)^2 \, dx = \int 24x^3(2x^4 - 5)^2 \, dx = (2x^4 - 5)^3 + C$

$\Rightarrow \int 12x^3(2x^4 - 5)^2 \, dx = \frac{1}{2}(2x^4 - 5)^3 + C$ ← Divide everything by 2 to match the original integral.

Example 2

Find $\int 8 \cosec^2 x \cot^3 x \, dx$.

$\int (n + 1)f'(x)[f(x)]^n \, dx = [f(x)]^{n+1} + C$ ← Use the formula to integrate.

$f(x) = \cot x$, so $f'(x) = -\cosec^2 x$. $n = 3$, so $n + 1 = 4$. ← Work out all the terms you need to substitute into the formula and then put them in.

So $\int -4 \cosec^2 x \cot^3 x \, dx = \cot^4 x + C$

$\Rightarrow \int 8 \cosec^2 x \cot^3 x \, dx = -2 \cot^4 x + C$ ← Multiply everything by -2 to match the original integral.

Example 3

a) Find $\int (x-2)\sqrt{x^2-4x+5}\ dx$.

$$\int (x-2)\sqrt{x^2-4x+5}\ dx = \int (x-2)(x^2-4x+5)^{\frac{1}{2}}\ dx$$ ← Rewrite the square root as a fractional power.

$$\int (n+1)f'(x)[f(x)]^n\ dx = [f(x)]^{n+1} + C$$ ← Use the formula on the previous page to integrate.

$f(x) = x^2 - 4x + 5$, so $f'(x) = 2x - 4$. $n = \frac{1}{2}$, so $n + 1 = \frac{3}{2}$. ← Work out all the terms you need to substitute into the formula and then put them in.

So $\int \frac{3}{2}(2x-4)(x^2-4x+5)^{\frac{1}{2}}\ dx = (x^2-4x+5)^{\frac{3}{2}} + C$

$\Rightarrow \int 3(x-2)\sqrt{x^2-4x+5}\ dx = (x^2-4x+5)^{\frac{3}{2}} + C$ ← Now just rearrange so that the left-hand side matches the original integral.

$\Rightarrow \int (x-2)\sqrt{x^2-4x+5}\ dx = \frac{1}{3}(x^2-4x+5)^{\frac{3}{2}} + C$

b) Hence solve $\int_2^6 (x-2)\sqrt{x^2-4x+5}\ dx$.

$$\int_2^6 (x-2)\sqrt{x^2-4x+5}\ dx = \left[\frac{1}{3}(x^2-4x+5)^{\frac{3}{2}}\right]_2^6$$ ← Use the result from part a).

$$= \left[\frac{1}{3}(6^2-4(6)+5)^{\frac{3}{2}}\right] - \left[\frac{1}{3}(2^2-4(2)+5)^{\frac{3}{2}}\right]$$ ← Put in the limits and simplify.

$$= \frac{1}{3}\sqrt{17^3} - \frac{1}{3} = \frac{1}{3}(17\sqrt{17} - 1)$$

Example 4

Evaluate $\int \frac{\cos x}{\sin^4 x}\ dx$.

$$\int \frac{\cos x}{\sin^4 x}\ dx = \int \frac{\cos x}{(\sin x)^4}\ dx = \int \cos x(\sin x)^{-4}\ dx$$ ← Write $\frac{1}{\sin^4 x}$ as a negative power.

$$\int (n+1)f'(x)[f(x)]^n\ dx = [f(x)]^{n+1} + C$$ ← Use the formula on the previous page to integrate.

$f(x) = \sin x$, so $f'(x) = \cos x$. $n = -4$, so $n + 1 = -3$. ← Work out all the terms you need to substitute into the formula and then put them in.

So $\int -3\cos x(\sin x)^{-4}\ dx = (\sin x)^{-3} + C$

$\Rightarrow \int \frac{-3\cos x}{\sin^4 x}\ dx = \frac{1}{\sin^3 x} + C$

$\Rightarrow \int \frac{\cos x}{\sin^4 x} = -\frac{1}{3\sin^3 x} + C$ ← Divide everything by −3 to match the original integral.

Exercise 8.9

Q1 Find the following indefinite integrals:

a) $\int 6x(x^2 + 5)^2 \, dx$

b) $\int (2x + 7)(x^2 + 7x)^4 \, dx$

c) $\int (x^3 + 2x)(x^4 + 4x^2)^3 \, dx$

d) $\int \dfrac{2x}{(x^2 - 1)^3} \, dx$

e) $\int \dfrac{6e^{3x}}{(e^{3x} - 5)^2} \, dx$

f) $\int \sin x \cos^5 x \, dx$

g) $\int 2 \sec^2 x \tan^3 x \, dx$

h) $\int 3e^x (e^x + 4)^2 \, dx$

i) $\int 32 (2e^{4x} - 3x)(e^{4x} - 3x^2)^7 \, dx$

j) $\int \dfrac{\cos x}{(2 + \sin x)^4} \, dx$

k) $\int 5 \operatorname{cosec} x \cot x \operatorname{cosec}^4 x \, dx$

l) $\int 2 \operatorname{cosec}^2 x \cot^3 x \, dx$

> **Q1-2 Hint**
> You'll need the derivatives of cosec, sec and cot:
> $\dfrac{d}{dx}(\operatorname{cosec} x) = -\operatorname{cosec} x \cot x$
> $\dfrac{d}{dx}(\sec x) = \sec x \tan x$
> $\dfrac{d}{dx}(\cot x) = -\operatorname{cosec}^2 x$

Q2 Find the following integrals:

a) $\int 6 \tan x \sec^6 x \, dx$

b) $\int \cot x \operatorname{cosec}^3 x \, dx$

[E] [P] Q3 a) Find $\int 4 \sin^3 x \cos x \, dx$. *[3 marks]*

b) Find $\int 4 \cos^3 x \sin x \, dx$. *[3 marks]*

c) Show that $\int 4 \sin^3 x \cos x + 4 \cos^3 x \sin x \, dx = -\cos 2x + C$. *[2 marks]*

Q4 Integrate the following functions with respect to x:

a) $4 \cos x \, e^{\sin x} (e^{\sin x} - 5)^3$

b) $(\sin x \, e^{\cos x} - 4)(e^{\cos x} + 4x)^6$

[E] [P] Q5 a) Find $\int \dfrac{1}{x}(3 + 6x^2)(\ln x + x^2 + 3)^3 \, dx$. *[3 marks]*

b) Given that $f'(x) = \dfrac{1}{x}(3 + 6x^2)(\ln x + x^2 + 3)^2$ and $f(1) = 60$, find $f(x)$. *[2 marks]*

Q6 Integrate:

a) $\int \dfrac{\sec^2 x}{\tan^4 x} \, dx$

b) $\int \cot x \operatorname{cosec} x \sqrt{\operatorname{cosec} x} \, dx$

[E] [P] Q7 a) Find $\int_0^5 (18x + 15)\dfrac{1}{\sqrt{3x^2 + 5x}} \, dx$. *[5 marks]*

b) Hence, or otherwise, show that
$$\int_0^{\frac{\pi}{2}} (18 \sin x + 15) \cos x \dfrac{1}{\sqrt{3 \sin^2 x + 5 \sin x}} \, dx = 12\sqrt{2}.$$ *[3 marks]*

Challenge

[P] Q8 a) Show that $\int e^{\cot 2x} \operatorname{cosec}^2 2x \, e^{3 \cot 2x} \, dx = -\dfrac{1}{8} e^{4 \cot 2x} + C$.

b) Hence solve $\int_{\frac{1}{2}}^1 e^{\cot 2x} \operatorname{cosec}^2 2x \, e^{3 \cot 2x} \, dx$.

8.10 Using Trigonometric Identities in Integration

Integrating using the double angle formulas

If you're given a tricky **trig function** to integrate, you might be able to simplify it using one of the **double angle formulas**. They're especially useful for things like **cos²x**, **sin²x** and **sin x cos x**.

Here are the double angle formulas again:

$$\sin 2x \equiv 2 \sin x \cos x \qquad \cos 2x \equiv \cos^2 x - \sin^2 x \qquad \tan 2x \equiv \frac{2 \tan x}{1 - \tan^2 x}$$

You came across two other ways of writing the double angle formula for cos, which come from using the identity $\cos^2 x + \sin^2 x \equiv 1$:

$$\cos 2x \equiv 2 \cos^2 x - 1 \qquad \cos 2x \equiv 1 - 2 \sin^2 x$$

Once you've rearranged the original function using one of the **double angle formulas**, the function you're left with should be easier to integrate using the rules you've seen in this chapter.

Example 1

Find the following:

a) $\int \sin^2 x \, dx$

Problem Solving
Use one of the cos double angle formulas when you've got a $\cos^2 x$ or a $\sin^2 x$ to integrate.

$$\cos 2x \equiv 1 - 2 \sin^2 x \implies \sin^2 x \equiv \frac{1}{2}(1 - \cos 2x)$$

Rearrange the cos double angle formula.

$$\int \sin^2 x \, dx = \int \frac{1}{2}(1 - \cos 2x) \, dx = \frac{1}{2} \int (1 - \cos 2x) \, dx$$
$$= \frac{1}{2}\left(x - \frac{1}{2} \sin 2x\right) + C = \frac{1}{2}x - \frac{1}{4} \sin 2x + C$$

Rewrite the integration and then integrate the terms separately in the usual way.

b) $\int \cos^2 5x \, dx$

$$\cos 2x \equiv 2 \cos^2 x - 1 \implies \cos^2 x \equiv \frac{1}{2}(\cos 2x + 1)$$

Rearrange the cos double angle formula.

$$\int \cos^2 5x \, dx = \int \frac{1}{2}(\cos 10x + 1) \, dx = \frac{1}{2} \int (\cos 10x + 1) \, dx$$
$$= \frac{1}{2}\left(\frac{1}{10} \sin 10x + x\right) + C = \frac{1}{20} \sin 10x + \frac{1}{2}x + C$$

Rewrite the integration and then integrate in the usual way.

Example 2

Find the following integrals:

a) $\int \sin x \cos x \, dx$

$\sin 2x \equiv 2 \sin x \cos x \implies \sin x \cos x \equiv \frac{1}{2}\sin 2x$ ← Rearrange the sin double angle formula.

$\int \sin x \cos x \, dx = \int \frac{1}{2}\sin 2x \, dx$ ← Rewrite the integration and then integrate in the usual way.

$= \frac{1}{2}\left(-\frac{1}{2}\cos 2x\right) + C = -\frac{1}{4}\cos 2x + C$

b) $\int_0^{\frac{\pi}{4}} \sin 2x \cos 2x \, dx$

$\sin 4x \equiv 2 \sin 2x \cos 2x \implies \sin 2x \cos 2x \equiv \frac{1}{2}\sin 4x$ ← Rearrange the sin double angle formula with x replaced by $2x$.

$\int_0^{\frac{\pi}{4}} \sin 2x \cos 2x \, dx = \int_0^{\frac{\pi}{4}} \frac{1}{2}\sin 4x \, dx = \left[\frac{1}{2}\left(-\frac{1}{4}\cos 4x\right)\right]_0^{\frac{\pi}{4}}$ ← Rewrite the integration and then integrate in the usual way.

$= -\frac{1}{8}[\cos 4x]_0^{\frac{\pi}{4}} = -\frac{1}{8}\left(\left[\cos\frac{4\pi}{4}\right] - [\cos 0]\right)$ ← Substitute in the limits and simplify to evaluate the integral.

$= \frac{1}{8}(\cos 0 - \cos \pi) = \frac{1}{8}(1 - (-1)) = \frac{1}{4}$

c) $\int \dfrac{4\tan\frac{x}{2}}{1 - \tan^2\frac{x}{2}} \, dx$

$\dfrac{4\tan\frac{x}{2}}{1 - \tan^2\frac{x}{2}} = 2\left(\dfrac{2\tan\frac{x}{2}}{1 - \tan^2\frac{x}{2}}\right) = 2\left(\tan\left(2 \times \frac{x}{2}\right)\right) = 2\tan x$ ← Rearrange, then use the double angle formula for tan.

$\int \dfrac{4\tan\frac{x}{2}}{1 - \tan^2\frac{x}{2}} \, dx = \int 2\tan x \, dx = -2\ln|\cos x| + C$ ← Rewrite the integration and then integrate using the rule on p.276.

Integrating using other trigonometric identities

There are a couple of other **identities** you can use to simplify trig functions:

$\sec^2 x \equiv 1 + \tan^2 x$ $\qquad$ $\text{cosec}^2 x \equiv 1 + \cot^2 x$

If you use one of these identities to get rid of a $\cot^2 x$ or a $\tan^2 x$, don't forget the stray 1's flying around — they'll just integrate to x. These identities are really useful if you have to integrate $\tan^2 x$ or $\cot^2 x$, as you already know how to integrate $\sec^2 x$ and $\text{cosec}^2 x$ (see pages 271-273).

$\int \sec^2 x \, dx = \tan x + C$ $\qquad$ $\int \text{cosec}^2 x \, dx = -\cot x + C$

Example 3

a) Find $\int \tan^2 x - 1 \ dx$

$\tan^2 x - 1 = (\sec^2 x - 1) - 1 = \sec^2 x - 2$ ← Rewrite the function in terms of $\sec^2 x$.

$\int \tan^2 x - 1 \ dx = \int \sec^2 x - 2 \ dx = \tan x - 2x + C$ ← Now integrate each term separately.

b) Find $\int \cot^2 3x \ dx$

$\text{cosec}^2 x \equiv 1 + \cot^2 x \Rightarrow \cot^2 3x = \text{cosec}^2 3x - 1$ ← Get the function in terms of $\text{cosec}^2 x$.

$\int \cot^2 3x \ dx = \int \text{cosec}^2 3x - 1 \ dx = -\frac{1}{3} \cot 3x - x + C$ ← Integrate the function.

c) Find $\int \cos^3 x \ dx$

$\cos^3 x = \cos^2 x \cos x = (1 - \sin^2 x) \cos x = \cos x - \cos x (\sin x)^2$ ← Split $\cos^3 x$ into $\cos^2 x$ and $\cos x$ and use identities.

$\int \cos^3 x \ dx = \int \cos x \ dx - \int \cos x \sin^2 x \ dx$ ← Rewrite the integral.

$\int \cos x \ dx = \sin x + c$ ← Integrate $\cos x$ in the usual way.

Let $f(x) = \sin x$, so $f'(x) = \cos x$. $n = 2$, so $n + 1 = 3$.
So $\int 3 \cos x \sin^2 x \ dx = \sin^3 x + c$
$\Rightarrow \int \cos x \sin^2 x \ dx = \frac{1}{3} \sin^3 x + c$

← Integrate $\cos x \sin^2 x$ using the formula on p.283: $\int (n + 1) f'(x) [f(x)]^n \ dx = [f(x)]^{n+1} + C.$

$\int \cos x \ dx - \int \cos x \sin^2 x \ dx = \sin x - \frac{1}{3} \sin^3 x + C$ ← Combine your results to find the whole integral.

d) Evaluate $\int_0^{\frac{\pi}{3}} 6 \sin 3x \cos 3x + \tan^2 \frac{1}{2} x + 1 \ dx$

$6 \sin 3x \cos 3x \equiv 3 \sin 6x$ ← Use the sin double angle formula.

$\tan^2 \frac{1}{2} x + 1 \equiv \sec^2 \frac{1}{2} x$ ← Use the identity for $\tan^2 x$.

$\int_0^{\frac{\pi}{3}} 6 \sin 3x \cos 3x + \tan^2 \frac{1}{2} x + 1 \ dx$ ← Integrate the function.

$= \int_0^{\frac{\pi}{3}} 3 \sin 6x + \sec^2 \frac{1}{2} x \ dx = \left[-\frac{3}{6} \cos 6x + 2 \tan \frac{1}{2} x \right]_0^{\frac{\pi}{3}}$

$= \left[-\frac{1}{2} \cos(2\pi) + 2 \tan\left(\frac{\pi}{6}\right) \right] - \left[-\frac{1}{2} \cos(0) + 2 \tan(0) \right]$ ← Put the limits in and simplify to evaluate the integral.

$= \left[-\frac{1}{2}(1) + 2\left(\frac{1}{\sqrt{3}}\right) \right] - \left[-\frac{1}{2}(1) + 2(0) \right]$

$= -\frac{1}{2} + \frac{2}{\sqrt{3}} + \frac{1}{2} = \frac{2}{\sqrt{3}} = \frac{2\sqrt{3}}{3}$

Exercise 8.10

Q1 Find the following indefinite integrals:

a) $\int \cos^2 x \, dx$

b) $\int 6 \sin x \cos x \, dx$

c) $\int \sin^2 6x \, dx$

d) $\int \dfrac{2 \tan 2x}{1 - \tan^2 2x} \, dx$

e) $\int 2 \sin 4x \cos 4x \, dx$

f) $\int 2 \cos^2 4x \, dx$

g) $\int \cos x \sin x \, dx$

h) $\int \sin 3x \cos 3x \, dx$

i) $\int \dfrac{6 \tan 3x}{1 - \tan^2 3x} \, dx$

j) $\int 5 \sin 2x \cos 2x \, dx$

k) $\int (\sin x + \cos x)^2 \, dx$

l) $\int 4 \sin x \cos x \cos 2x \, dx$

m) $\int (\cos x + \sin x)(\cos x - \sin x) \, dx$

n) $\int \sin^2 x \cot x \, dx$

> **Problem Solving**
>
> In Q1 part l), use the sin double angle formula twice.

Q2 Evaluate the following definite integrals, giving exact answers:

a) $\int_0^{\frac{\pi}{4}} \sin^2 x \, dx$

b) $\int_0^{\pi} \cos^2 2x \, dx$

c) $\int_0^{\pi} \sin \dfrac{x}{2} \cos \dfrac{x}{2} \, dx$

d) $\int_{\frac{\pi}{4}}^{\frac{\pi}{2}} \sin^2 2x \, dx$

e) $\int_0^{\frac{\pi}{4}} \cos 2x \sin 2x \, dx$

f) $\int_{\frac{\pi}{4}}^{\frac{\pi}{2}} \sin^2 x - \cos^2 x \, dx$

E **Q3** $f(\theta) = \sin^2 \theta - 2 \cos^2 \theta$

a) Show that $f(\theta)$ can be written in the form $a - b \cos \alpha$, where a and b are constants to be found and α is given in terms of θ. *[3 marks]*

b) Hence, find $\int \sin^2 \theta - 2 \cos^2 \theta \, d\theta$. *[2 marks]*

P **Q4** Find the exact value of $\int_0^{\frac{\pi}{6}} \dfrac{\tan 7x - \tan 5x}{1 + \tan 7x \tan 5x} \, dx$.

> **Problem Solving**
>
> In Q4, use the tan addition formula and remember that $-\tan x = \tan(-x)$.

Q5 Find the following integrals:

a) $\int \cot^2 x - 4 \, dx$

b) $\int \tan^2 x \, dx$

c) $\int 3 \cot^2 x \, dx$

d) $\int \tan^2 4x \, dx$

E **P** **Q6** Show that $\int_{\frac{\pi}{16}}^{\frac{\pi}{12}} \cot^2 4x \, dx$ can be written in the form $k\left(1 - \dfrac{\pi}{12} - \dfrac{1}{\sqrt{3}}\right)$, where k is a constant to be found. *[4 marks]*

Q7 Find the exact value of $\int_0^{\frac{\pi}{4}} \tan^2 x + \cos^2 x - \sin^2 x \, dx$.

Find the integrals in Q8-11:

Q8 $\displaystyle\int (\sec x + \tan x)^2\, dx$

Q9 $\displaystyle\int (\cot x + \csc x)^2\, dx$

Q10 $\displaystyle\int 4 + \cot^2 3x\, dx$

Q11 $\displaystyle\int \cos^2 4x + \cot^2 4x\, dx$

E P **Q12** a) Show that $\dfrac{\cos 2x}{1 - \cos^2 2x} = \cot 2x \csc 2x$. *[2 marks]*

b) Find $\displaystyle\int \dfrac{\cos 2x}{1 - \cos^2 2x}\, dx$. *[2 marks]*

Q13 Integrate the following functions with respect to x:

a) $\tan^3 x + \tan^5 x$

b) $\cot^5 x + \cot^3 x$

c) $\sin^3 x$

E P **Q14** a) Show that $\dfrac{2\cos^2 2x}{\sin 4x} = \cot 2x$. *[2 marks]*

b) Find $\displaystyle\int_{\frac{\pi}{8}}^{\frac{\pi}{4}} \dfrac{2\cos^2 2x}{\sin 4x}\, dx$, giving your answer in the form $\ln 2^a$, where a is a rational number. *[4 marks]*

P **Q15** Use the identity $\sin A + \sin B \equiv 2\sin\left(\dfrac{A+B}{2}\right)\cos\left(\dfrac{A-B}{2}\right)$ to find $\displaystyle\int 2\sin 4x \cos x\, dx$.

E P **Q16** Find $\displaystyle\int 3\sin x \cos 2x\, dx$. *[5 marks]*

Challenge

E P **Q17** The diagram below shows the graphs of $y = \sin^2 x \cos^2 x$ and $y = p$, where p is a constant.

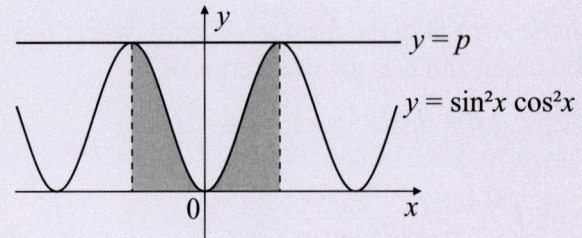

a) (i) Find $\dfrac{dy}{dx}$.

(ii) Find the lowest positive x value for which $\sin^2 x \cos^2 x = p$.

(iii) State the value of p. *[6 marks]*

b) (i) Show that $\sin^2 x \cos^2 x = \dfrac{1 - \cos 4x}{8}$.

(ii) Hence, or otherwise, find the area of the shaded region of the graph. *[6 marks]*

P **Q18** a) Show that $\dfrac{\cos x + \sec x}{1 - \cos x} - \sec x \csc^2 x = \csc^2 x + \cot^2 x + \cot x \csc x$.

b) Hence, find $\displaystyle\int \dfrac{\cos x + \sec x}{1 - \cos x} - \sec x \csc^2 x\, dx$.

8.11   Finding Area using Integration

To find the area between a curve, a line and the x-axis, you'll either have to **add** or **subtract** integrals to find the area you're after — it's always best to **draw a diagram** of the area.

Example 1

Find the area enclosed by the curve $y = x^2$, the line $y = 2 - x$ and the x-axis.

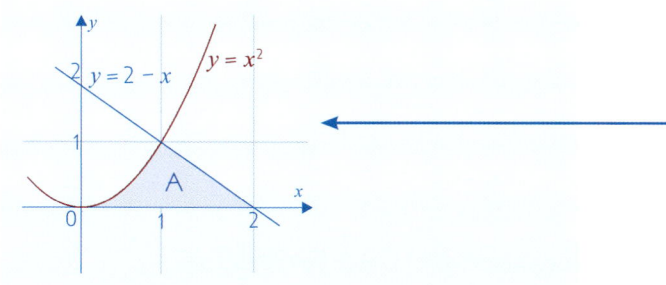

Draw a diagram of the curve and the line. You have to find area A — split it into two smaller bits.

$x^2 = 2 - x \implies x^2 + x - 2 = 0$
$\implies (x - 1)(x + 2) = 0$
So the line and curve meet at $x = 1$
(and at $x = -2$, but this isn't relevant to A).

Find the limits by solving $x^2 = 2 - x$ to find where the line and curve intersect.

$x^2 = 0 \implies x = 0$ and $2 - x = 0 \implies x = 2$

You also need to know the x-intercepts of the line and curve.

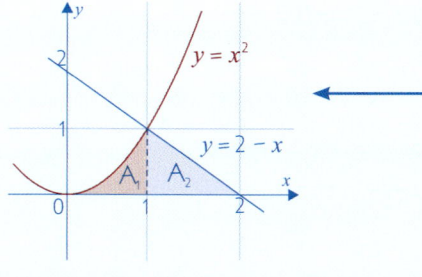

The area A is the area under the red curve between 0 and 1 added to the area under the blue line between 1 and 2.

$\int_0^1 x^2 \, dx = \left[\dfrac{x^3}{3} \right]_0^1 = \dfrac{1}{3} - 0 = \dfrac{1}{3}$

Integrate $y = x^2$ between 0 and 1 to find the area A_1.

$\int_1^2 (2 - x) \, dx = \left[2x - \dfrac{x^2}{2} \right]_1^2$

$= \left(2(2) - \dfrac{2^2}{2} \right) - \left(2(1) - \dfrac{1^2}{2} \right)$

Integrate $y = 2 - x$ between 1 and 2 to find the area A_2.

Tip A_2 is just a triangle with base 1 and height 1, so you could also calculate its area using the formula for the area of a triangle.

$= 2 - \dfrac{3}{2} = \dfrac{1}{2}$

$A = A_1 + A_2 = \dfrac{1}{3} + \dfrac{1}{2} = \dfrac{5}{6}$

Add the areas together to find the area A.

Sometimes you'll need to find the area **enclosed** by the graphs of two functions —
this usually means **subtracting** some area from another. Here is an example with two curves:

Example **2** **P**

The diagram on the right shows the curves
$y = \sin x + 1$ and $y = \cos x + 1$.

Find the area of the shaded grey region.

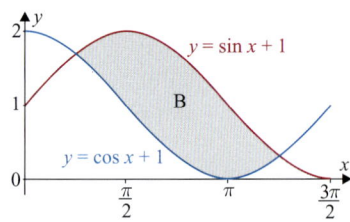

$\sin x + 1 = \cos x + 1 \Rightarrow \sin x = \cos x$ ← Find out where the graphs meet
by solving $\sin x + 1 = \cos x + 1$.

$$\Rightarrow \frac{\sin x}{\cos x} = 1 \Rightarrow \tan x = 1$$

$$\Rightarrow x = \tan^{-1}(1) = \frac{\pi}{4}, \frac{5\pi}{4}$$

They meet at $x = \frac{\pi}{4}$ and $x = \frac{5\pi}{4}$.

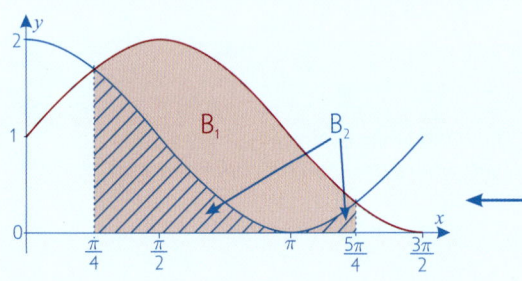

The area of B is the area under
the red curve between $\frac{\pi}{4}$ and $\frac{5\pi}{4}$
(B_1) minus the area under the blue
curve between $\frac{\pi}{4}$ and $\frac{5\pi}{4}$ (B_2).

$$B_1 = \int_{\frac{\pi}{4}}^{\frac{5\pi}{4}} \sin x + 1 \, dx = \left[-\cos x + x \right]_{\frac{\pi}{4}}^{\frac{5\pi}{4}}$$

Integrate $\sin x + 1$ between
$\frac{\pi}{4}$ and $\frac{5\pi}{4}$ to find B_1.

$$= \left(-\cos\left(\frac{5\pi}{4}\right) + \frac{5\pi}{4} \right) - \left(-\cos\left(\frac{\pi}{4}\right) + \frac{\pi}{4} \right)$$

$$= \left(-\left(-\frac{\sqrt{2}}{2}\right) + \frac{5\pi}{4} \right) - \left(-\frac{\sqrt{2}}{2} + \frac{\pi}{4} \right) = \sqrt{2} + \pi$$

Integrate $\cos x + 1$ between
$\frac{\pi}{4}$ and $\frac{5\pi}{4}$ to find B_2.

$$B_2 = \int_{\frac{\pi}{4}}^{\frac{5\pi}{4}} \cos x + 1 \, dx = \left[\sin x + x \right]_{\frac{\pi}{4}}^{\frac{5\pi}{4}}$$

$$= \left(\sin\left(\frac{5\pi}{4}\right) + \frac{5\pi}{4} \right) - \left(\sin\left(\frac{\pi}{4}\right) + \frac{\pi}{4} \right)$$

$$= \left(-\frac{\sqrt{2}}{2} + \frac{5\pi}{4} \right) - \left(\frac{\sqrt{2}}{2} + \frac{\pi}{4} \right) = -\sqrt{2} + \pi$$

Subtract B_2
from B_1 to
find the total
area between
the curves.

Tip You can also work
out questions like this by
integrating the difference
between the functions. In
Example 2, this would give

$$B = \int_{\frac{\pi}{4}}^{\frac{5\pi}{4}} \sin x - \cos x \, dx.$$

You still need to make sure
you're finding the right
area, so drawing a picture
is definitely a good idea.

$$B = B_1 - B_2 = \left(\sqrt{2} + \pi \right) - \left(-\sqrt{2} + \pi \right) = 2\sqrt{2}$$

Sometimes you might need to add **and** subtract integrals to find the right area.
You'll often need to do this when the curve goes **below** the x-axis.
The integrations you need to do should be obvious if you draw a **picture**.

Exercise 8.11

Q1 Find the shaded area in the following diagrams:

a)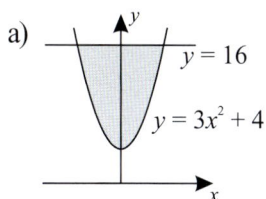

b)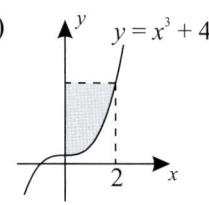

c)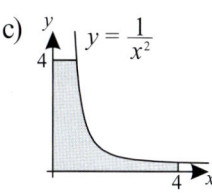

d)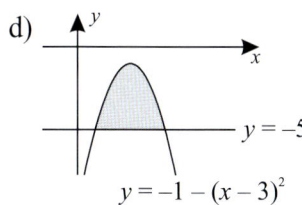

e)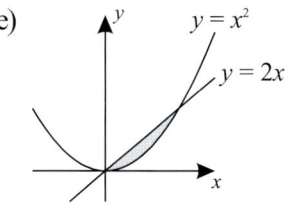

f)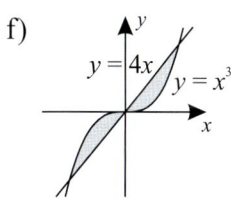

Q2 For each part, find the area enclosed by the curve and line:

a) $y = x^2 + 4$ and $y = x + 4$ b) $y = x^2 + 2x - 3$ and $y = 4x$

> **Q2b) Hint** Consider the bits above and below the x-axis separately.

E **Q3** Find the area bounded by the positive y-axis and the graphs of $y = x^2$ and $y = 4 - x^2$, shown shaded in the diagram on the right.

Give your answer in the from $\frac{a}{b}\sqrt{c}$, where a, b and c are integers.

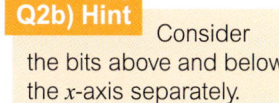

[5 marks]

E **Q4** The diagram on the right shows the graphs of $y = 2 \sin x$ and $y = \tan x$ for $0 \leq x \leq \pi$.

Find the area of the shaded region.
Give your answer to 3 significant figures.

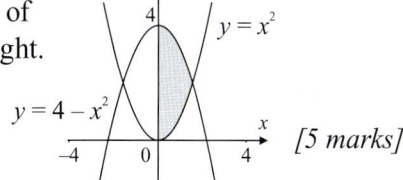

[6 marks]

Challenge

P **Q5** Find the shaded area shown to the right:

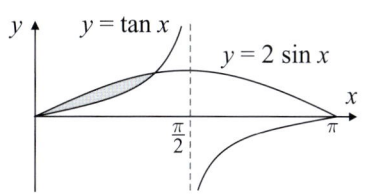

P **Q6** A company has designed a logo based on multiples of the sine curve, shown to the right. Calculate the total area of the grey sections of the logo.

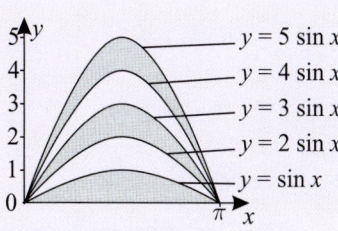

8.12 Parametric Integration

Normally, to find the **area** under a graph, you can do a **simple integration**. But if you've got **parametric equations**, things are more difficult — you can't find $\int y\, dx$ if y isn't written in terms of x.

There's a way to get around this. Suppose your parameter is t. Then:

$$\int y\, dx = \int y\, \frac{dx}{dt}\, dt$$

This comes from the chain rule for differentiation (p.207) — if you think of dx as $\frac{dx}{1}$, then $\frac{dx}{1} = \frac{dx}{dt} \times \frac{dt}{1}$. Both y and $\frac{dx}{dt}$ are written **in terms of** t, so you can **multiply** them together to get an expression you can **integrate with respect to** t.

With a **definite integral**, you need to **alter the limits** as well. So if you have x-values as limits, work out the corresponding values of t before you integrate.

Example **1**

The shaded region marked A on the sketch is bounded by the x-axis, the line $x = 2$, and by the curve with parametric equations $x = t^2 - 2$ and $y = t^2 - 9t + 20$, $t \geq 0$, which crosses the x-axis at $x = 14$.

Find the area of A.

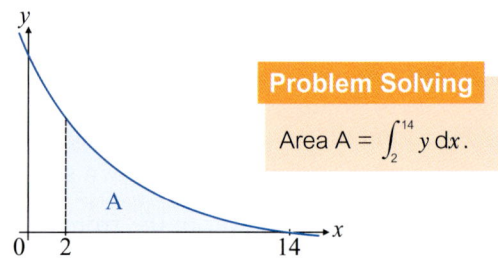

Problem Solving

Area A $= \int_{2}^{14} y\, dx$.

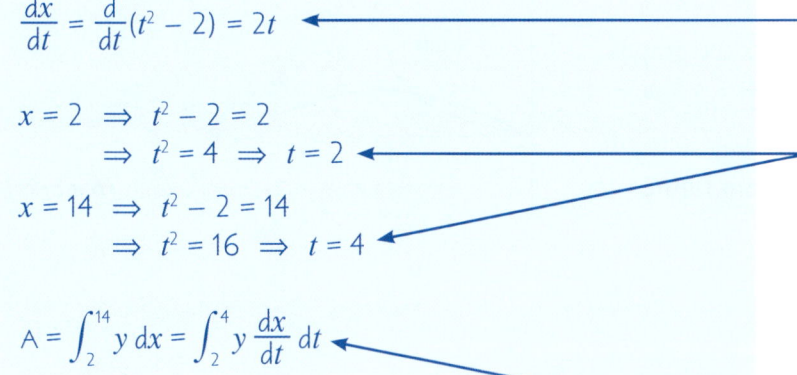

$\dfrac{dx}{dt} = \dfrac{d}{dt}(t^2 - 2) = 2t$ ← Find $\dfrac{dx}{dt}$.

$x = 2 \implies t^2 - 2 = 2$
$\qquad \implies t^2 = 4 \implies t = 2$

$x = 14 \implies t^2 - 2 = 14$
$\qquad \implies t^2 = 16 \implies t = 4$

Sort out the limits — 14 and 2 are the limits for integrating with respect to x, so you need to find the corresponding values of t.

Tip $t \geq 0$, so ignore the negative square roots.

$A = \displaystyle\int_{2}^{14} y\, dx = \int_{2}^{4} y\, \frac{dx}{dt}\, dt$

$\quad = \displaystyle\int_{2}^{4} (t^2 - 9t + 20)(2t)\, dt$

$\quad = \displaystyle\int_{2}^{4} 2t^3 - 18t^2 + 40t\, dt$

Now integrate with respect to t to find the area of A. Remember to change the limits of the integration.

$\quad = \left[\dfrac{1}{2}t^4 - 6t^3 + 20t^2 \right]_{2}^{4}$

$\quad = \left(\dfrac{1}{2}(4)^4 - 6(4)^3 + 20(4)^2 \right) - \left(\dfrac{1}{2}(2)^4 - 6(2)^3 + 20(2)^2 \right)$

$\quad = 64 - 40 = 24$ ← Substitute the limits in and simplify.

Exercise 8.12

Q1 For each of the following curves, find an expression in parametric form
that is equivalent to the indefinite integral $\int y\, dx$.

a) $x = \dfrac{3}{t}, y = 4t^2$

b) $x = \tan 5\theta, y = \sec^2 5\theta$

Q2 For each of the following curves, find an expression equivalent to $\int y\, dx$ and integrate it.

a) $x = (4t - 5)^2, y = t^2 - 3t$

b) $x = t^2 + 3, y = 4t - 1$

Q3 A curve has parametric equations $x = 3t^2, y = \dfrac{5}{t}$, where $t > 0$.
Find an expression for $y\, \dfrac{dx}{dt}$, and hence evaluate $\int_3^{75} y\, dx$.

E **Q4** The curve shown on the right has parametric equations $x = \dfrac{4}{t}$ and
$y = t^2$, where $t > 0$. Use a parametric integral to find the shaded
area between the curve, the x-axis and the lines $x = 1$ and $x = 4$. *[6 marks]*

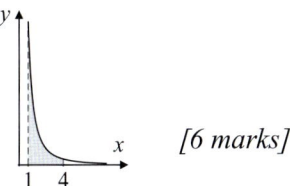

Q5 The curve shown here has parametric equations $x = 4t(t + 1), y = 3t^3$.

a) Find the values t_1 and t_2 that correspond to
$x = 8$ and $x = 120$, given that $t > 0$.

b) Hence find a parametric integral corresponding to $\int_8^{120} y\, dx$,
and evaluate this to find the area A.

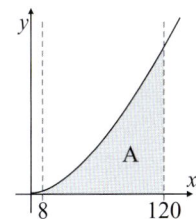

Challenge

E **Q6** The diagram shows the curve with parametric equations
$x = \cos 2t$ and $y = 2 \operatorname{cosec} t$, for $0 \le t < \dfrac{\pi}{2}$.

Find the shaded area enclosed by the curve,
the x-axis and the lines $x = 0$ and $x = \dfrac{1}{2}$. *[7 marks]*

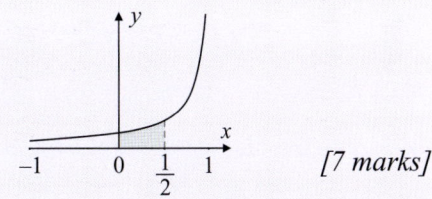

E **P** **Q7** The diagram shows the curve with parametric equations $x = 2t - 6$ and $y = t^2 - 5t + 6$.
The shaded region is bounded by the y-axis, the curve and the normal to the curve at $x = 2$.

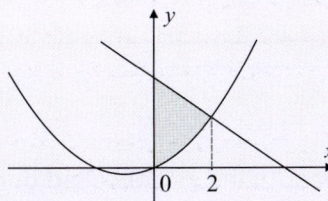

a) Find the area bounded by the curve, the x-axis and the lines $x = 0$ and $x = 2$. *[6 marks]*

b) Hence, find the area of the shaded region. *[6 marks]*

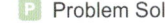

8.13 Integration by Substitution

Integration by substitution is a way of integrating a **function of a function** by simplifying the integral. Like differentiating with the chain rule, to integrate by substitution you have to write part of the function in terms of u, where u is some **function of** x.

Here's the method:

- You'll be given an integral that's made up of **two functions of** x.

- **Substitute** u for one of the functions of x to give a function that's **easier to integrate**.

- Next, find $\dfrac{du}{dx}$, and **rewrite** it so that dx is on its own.

- **Rewrite** the original integral in terms of u and du.

- You should now be left with something that's **easier** to integrate — just **integrate** as normal, then at the last step **replace** u with the **original substitution**.

You won't always be told what substitution to use, so you might have to take an educated guess. Functions inside **brackets**, **denominators of fractions** and **square roots** are often good places to start. There might be **multiple different substitutions** that you could use to get the right answer.

Example 1

Use the substitution $u = x^2 - 2$ to find $\displaystyle\int 4x(x^2 - 2)^4 \, dx$.

$u = x^2 - 2 \Rightarrow \dfrac{du}{dx} = 2x$ ← Differentiate u with respect to x.

$\dfrac{du}{dx} = 2x \Rightarrow du = 2x \, dx \Rightarrow dx = \dfrac{1}{2x} du$ ← Rearrange the equation for $\dfrac{du}{dx}$ to get dx on its own.

Tip $\dfrac{du}{dx}$ isn't really a fraction, but you can treat it like one for this bit.

$\displaystyle\int 4x(x^2 - 2)^4 \, dx = \int 4xu^4 \dfrac{1}{2x} \, du = \int 2u^4 \, du$ ← Substitute what you've got so far back into the original expression and simplify.

$\displaystyle\int 2u^4 \, du = \dfrac{2}{5}u^5 + C$ ← Integrate this simpler expression with respect to u.

$= \dfrac{2}{5}(x^2 - 2)^5 + C$ ← Substitute $u = x^2 - 2$ back in.

That first example worked out nicely, because the x's **cancelled out** when you substituted in the expressions for u and dx. It isn't always quite so straightforward — sometimes you need to get rid of some x's by **rearranging** the equation to get x in terms of u.

Example 2

Find $\int x(3x+2)^3 \, dx$, using the substitution $u = 3x + 2$.

$u = 3x + 2 \implies \dfrac{du}{dx} = 3 \implies dx = \dfrac{1}{3} \, du$ ⟵ Start by finding $\dfrac{du}{dx}$ and then rearrange to get dx on its own.

$\int x(3x+2)^3 \, dx = \int x u^3 \dfrac{1}{3} \, du$ ⟵ If you substitute for u and dx, you end up with an x still in the integral.

$u = 3x + 2 \implies x = \dfrac{u-2}{3}$ ⟵ To get rid of it, rearrange the equation for u.

$\int x(3x+2)^3 \, dx = \int \left(\dfrac{u-2}{3}\right) u^3 \dfrac{1}{3} \, du$ ⟵ Integrate the function in terms of u.

$= \int \dfrac{u^4 - 2u^3}{9} \, du$

$= \dfrac{1}{9}\left(\dfrac{u^5}{5} - \dfrac{u^4}{2}\right) + C$

$= \dfrac{u^5}{45} - \dfrac{u^4}{18} + C$

$= \dfrac{(3x+2)^5}{45} - \dfrac{(3x+2)^4}{18} + C$ ⟵ Substitute $u = 3x + 2$ back in at the end.

Tip Don't forget to add the "+ C" when you integrate. Because it's just a constant, you don't need to do anything with it when you substitute back to get your answer in terms of x.

Some integrations look really tricky, but with a clever substitution they can be made a lot simpler.

Example 3

Find $\int 3x\sqrt{2-x^2} \, dx$, using a suitable substitution.

Let $u = \sqrt{2-x^2}$ ⟵ Choose u to make the integral simpler — $\sqrt{2-x^2}$ looks like the most awkward bit to integrate, so that might be a good choice for the substitution.

$\dfrac{du}{dx} = -\dfrac{x}{\sqrt{2-x^2}} = -\dfrac{x}{u} \implies u \, du = -x \, dx$

$\implies -\dfrac{u}{x} \, du = dx$ Differentiate to find $\dfrac{du}{dx}$.

$\int 3x\sqrt{2-x^2} \, dx = \int 3x \times u \times -\dfrac{u}{x} \, du$ ⟵ Substitute what you've got into the original integral.

$= \int -3u^2 \, du$

$= -u^3 + C$

$= -\left(\sqrt{2-x^2}\right)^3 + C$

Exercise 8.13

Q1 Find the following integrals using the given substitutions:

a) $\int 12(x+3)^5 \, dx, \ u = x+3$

b) $\int (11-x)^4 \, dx, \ u = 11-x$

c) $\int 24x(x^2+4)^3 \, dx, \ u = x^2+4$

d) $\int \sin^5 x \cos x \, dx, \ u = \sin x$

e) $\int x(x-1)^5 \, dx, \ u = x-1$

f) $\int 4x^3(x^2-3)^6 \, dx, \ u = x^2-3$

Q2 Use an appropriate substitution to find:

a) $\int 21(x+2)^6 \, dx$

b) $\int (5x+4)^3 \, dx$

c) $\int x(2x+3)^3 \, dx$

d) $\int 24x(x^2-5)^7 \, dx$

Q3 Use the given substitutions to find the following integrals:

a) $\int 6x\sqrt{x+1} \, dx, \ u = \sqrt{x+1}$

b) $\int \frac{x}{\sqrt{4-x}} \, dx, \ u = \sqrt{4-x}$

c) $\int \frac{15(\ln x)^4}{x} \, dx, \ u = \ln x$

d) $\int \frac{3}{x(\ln(x^2))^3} \, dx, \ u = \ln(x^2)$

Q4 Find the following integrals by substitution:

a) $\int \frac{4x}{\sqrt{(2x-1)}} \, dx$

b) $\int \frac{1}{4-\sqrt{x}} \, dx$

c) $\int \frac{e^{2x}}{1+e^x} \, dx$

E P Q5 The substitution $u = f(x)$ is used to give: $\int 2x\sqrt{3x^2-1} \, dx = \int (3u-1)^{0.5} \, du$.

a) Find $f(x)$ and confirm that the substitution $u = f(x)$ gives the result shown. *[2 marks]*

b) Find $\int 2x\sqrt{3x^2-1} \, dx$ in terms of u. *[1 mark]*

c) Find $\int 2x\sqrt{3x^2-1} \, dx$ in terms of x. *[1 mark]*

E Q6 Find $\int \frac{3^x}{(3^x+2)^3} \, dx$, using $u = 3^x+2$ as a substitution,

giving your answer in the form $-\frac{(f(x))^n}{\ln a} + C$, where a and n are integers. *[5 marks]*

E Q7 Using the substitution $u = x^2-9$, show that $\int x(x^2-9)^{0.5} \, dx = \frac{1}{3}(x^2-9)^{1.5} + C$ *[4 marks]*

E P Q8 a) Show that, if $u = \sqrt{x}$, then $\int \frac{4\sqrt{x}}{x(3-\sqrt{x})} \, dx = \int \frac{8}{3-u} \, du$. *[2 marks]*

b) Hence show that, if $f'(x) = \frac{4\sqrt{x}}{x(3-\sqrt{x})}$ and $f(4) = \ln 2$,

then $f(x) = \ln\left(\frac{2}{(3-\sqrt{x})^8}\right)$. *[4 marks]*

E **Q9** Use a suitable substitution to find $\int \dfrac{e^{4x}}{2 + e^{2x}} \, dx$. *[7 marks]*

P **Q10** Use integration by substitution to prove that $\int (n + 1) f'(x) [f(x)]^n \, dx = [f(x)]^{n+1} + C$.

> **Q10 Tip**
> This is the formula from p.283.

Challenge

E **Q11** a) Using a binomial expansion or otherwise, express $x(x + 1)^3$ in the form $ax^4 + bx^3 + cx^2 + dx$. *[2 marks]*

P

b) Use the substitution $u = x + 1$ to show that $\int x(x + 1)^3 dx = \dfrac{u^4}{20}(4u - 5)) + C$ *[4 marks]*

c) Hence find $\dfrac{(4x - 1)(x + 1)^4}{20}$ as a sum of powers of x. *[4 marks]*

8.14 Integration by Substitution — Definite Integrals

If you're given a **definite integral** to find using a substitution, it's important that you remember to **change the limits** to u. To do this, put the x-limits into the equation for u to find the corresponding values of u.

Doing it this way means you **don't** have to **put x back in** at the last step — just put the values of u into the integration for u.

Example **1**

a) Use the substitution $u = \cos x$ to find $\int_{\frac{\pi}{2}}^{2\pi} -12 \sin x \cos^3 x \, dx$.

> **Problem Solving** You could also solve this using the method on p.283.

$u = \cos x \implies \dfrac{du}{dx} = -\sin x$

$\implies dx = -\dfrac{1}{\sin x} \, du$

> As with indefinite integrals, start by differentiating u, and rearranging to get dx on its own.

$x = \dfrac{\pi}{2} \implies u = \cos \dfrac{\pi}{2} = 0$

$x = 2\pi \implies u = \cos 2\pi = 1$

> Use the substitution to change the limits of the integral from x-values to u-values.

$\int_{\frac{\pi}{2}}^{2\pi} -12 \sin x \cos^3 x \, dx = \int_{0}^{1} -12 \sin x \, u^3 \dfrac{-1}{\sin x} \, du$

> Substitute all that back into the original integral and solve.

$= \int_{0}^{1} 12 u^3 \, du = [3u^4]_{0}^{1}$

$= [3(1)^4] - [3(0)^4]$

$= 3 - 0 = 3$

---

b) Use a suitable substitution to find $\int_2^{\frac{7}{2}} x\sqrt{2x-3}\ dx$.

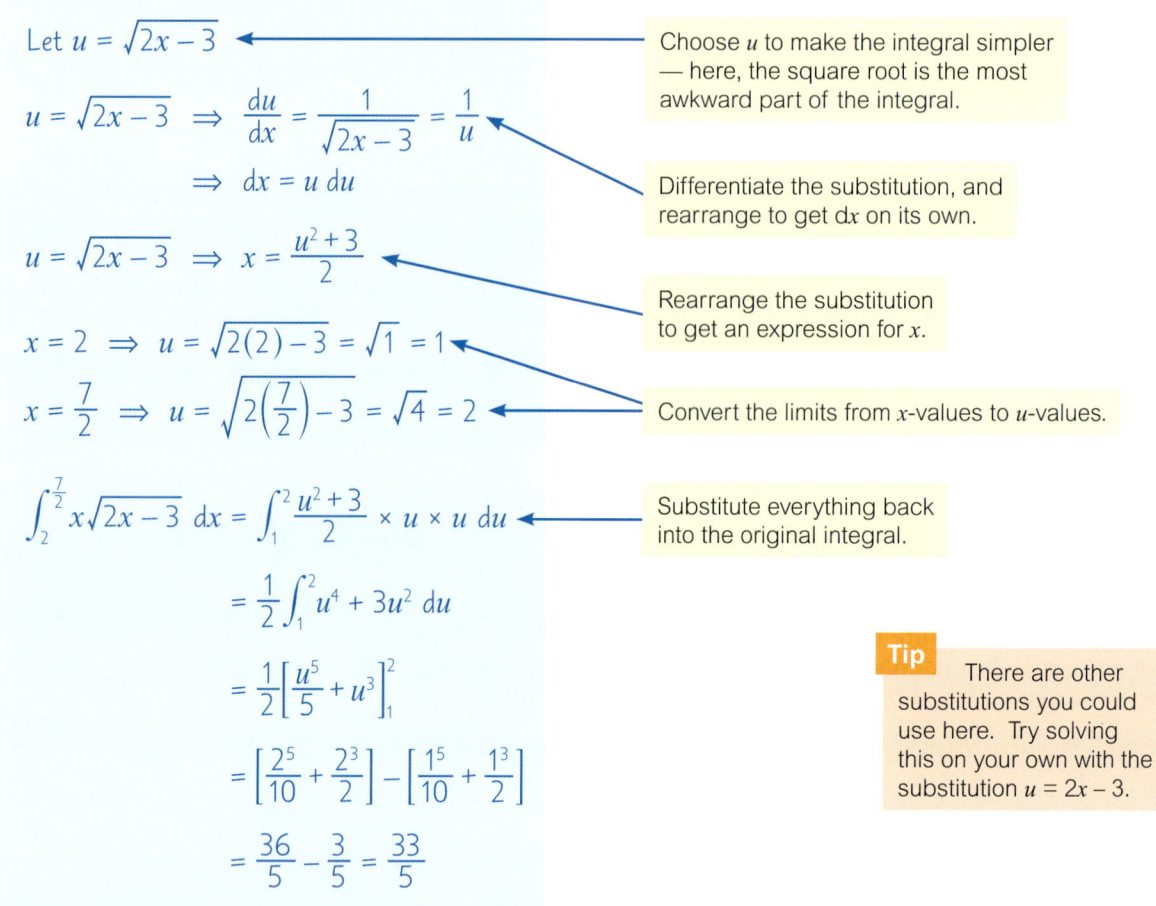

Let $u = \sqrt{2x-3}$ — Choose u to make the integral simpler — here, the square root is the most awkward part of the integral.

$u = \sqrt{2x-3} \implies \dfrac{du}{dx} = \dfrac{1}{\sqrt{2x-3}} = \dfrac{1}{u}$

$\implies dx = u\ du$ — Differentiate the substitution, and rearrange to get dx on its own.

$u = \sqrt{2x-3} \implies x = \dfrac{u^2+3}{2}$ — Rearrange the substitution to get an expression for x.

$x = 2 \implies u = \sqrt{2(2)-3} = \sqrt{1} = 1$

$x = \dfrac{7}{2} \implies u = \sqrt{2\left(\dfrac{7}{2}\right)-3} = \sqrt{4} = 2$ — Convert the limits from x-values to u-values.

$\int_2^{\frac{7}{2}} x\sqrt{2x-3}\ dx = \int_1^2 \dfrac{u^2+3}{2} \times u \times u\ du$ — Substitute everything back into the original integral.

$= \dfrac{1}{2}\int_1^2 u^4 + 3u^2\ du$

$= \dfrac{1}{2}\left[\dfrac{u^5}{5} + u^3\right]_1^2$

$= \left[\dfrac{2^5}{10} + \dfrac{2^3}{2}\right] - \left[\dfrac{1^5}{10} + \dfrac{1^3}{2}\right]$

$= \dfrac{36}{5} - \dfrac{3}{5} = \dfrac{33}{5}$

Tip There are other substitutions you could use here. Try solving this on your own with the substitution $u = 2x - 3$.

Sometimes when you convert the limits of a definite integral, the **upper limit** converts to a **lower number** than the **lower limit** does.

You can either keep the converted limits in the **same places** as the corresponding original limits and carry on as normal or **swap them** so the higher value is the upper limit and put a **minus sign** in front of the whole integral.

Swapping the limits and putting a minus in front of the integral might seem more complicated, but it often **cancels** with another minus, making the whole integration **easier**.

Exercise 8.14

Q1 Find the exact values of the following using the given substitutions:

a) $\int_{\frac{2}{3}}^1 (3x-2)^4\ dx,\ u = 3x - 2$

b) $\int_{-2}^1 2x(x+3)^4\ dx,\ u = x + 3$

c) $\int_0^{\frac{\pi}{6}} 8\sin^3 x \cos x\ dx,\ u = \sin x$

d) $\int_0^3 x\sqrt{x+1}\ dx,\ u = \sqrt{x+1}$

Q2 Use an appropriate substitution to find the exact value of each of the following:

a) $\displaystyle\int_2^{\sqrt{5}} x(x^2 - 3)^4 \, dx$

b) $\displaystyle\int_1^2 x(3x - 4)^3 \, dx$

c) $\displaystyle\int_2^{10} \frac{x}{\sqrt{x-1}} \, dx$

Q3 Integrate the function $y = \dfrac{1}{3 - \sqrt{x}}$ between $x = 1$ and $x = 4$, using a suitable substitution.
Give your answer in the form $a + b \ln 2$, where a and b are integers.

Q4 Find $\displaystyle\int_0^1 2e^x(1 + e^x)^3 \, dx$, using the substitution $u = 1 + e^x$.
Give your answer to 1 decimal place.

Q5 Use integration by substitution to find the integral of the function
$y = \dfrac{x}{\sqrt{3x + 1}}$ between the limits $x = 1$ and $x = 5$.

E Q6 Find $\displaystyle\int_0^1 \frac{x^3}{\sqrt{x^4 + 12}} \, dx$ using a suitable substitution, giving your answer to 3 s.f. *[7 marks]*

Challenge

E **P** Q7 The diagram below shows the graph of $y = \dfrac{5x}{1 + \sqrt{x}}$.

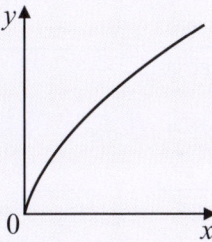

a) Find $\displaystyle\int \frac{5x}{1 + \sqrt{x}} \, dx$, using the substitution $u = 1 + \sqrt{x}$,
giving your answer in terms of u. *[5 marks]*

b) Hence, find, to 3 s.f., the area bounded by the graph of $y = \dfrac{5x}{1 + \sqrt{x}}$,
the y-axis and the lines $y = 2.5$ and $y = 16$. *[7 marks]*

E **P** Q8 a) Use a suitable substitution to show that $\displaystyle\int (2x + 2)^4 \, dx = 0.1(2x + 2)^5 + C$. *[4 marks]*

b) Find the area between the curve $y = (2x + 2)^4$,
the x-axis and the lines $x = -1$ and $x = 0$. *[2 marks]*

c) Without using integration, find the equations of another pair
of vertical lines such that the area between the curve $y = (2x + 2)^4$,
the x-axis and these two vertical lines is equal to the area found in part b). *[2 marks]*

d) Describe the transformation which maps the graph of $f(x) = x^4$
onto the graph of $g(x) = (2x + 2)^4$. *[3 marks]*

e) Hence, find k, such that $\displaystyle\int_{-2}^0 (2x + 2)^4 \, dx = k \int_{-2}^2 x^4 \, dx$. Explain your answer. *[2 marks]*

8.15 Integration by Substitution — Using Trig Identities

As you know by now, there's a vast range of **trig identities** and **formulas** to deal with in A-level maths. This can make for some pretty tricky **integration questions** involving trig functions.

Here are a couple of examples:

Tip If you need a reminder of the trig identities, they're given in Chapter 3.

Example 1

a) Use the substitution $u = \tan x$ to find $\int \dfrac{\sec^4 x}{\sqrt{\tan x}}\, dx$.

$$u = \tan x \implies \frac{du}{dx} = \sec^2 x \implies dx = \frac{1}{\sec^2 x}\, du$$

Work out what the substitutions will be — start by finding dx.

$$\sec^2 x \equiv 1 + \tan^2 x, \text{ so}$$
$$u = \tan x \implies \sec^2 x \equiv 1 + u^2$$

The substitution for dx leaves $\sec^2 x$ on the numerator, so you need to find $\sec^2 x$ in terms of u.

$$\int \frac{\sec^4 x}{\sqrt{\tan x}}\, dx = \int \frac{(1 + u^2) \times \sec^2 x}{\sqrt{u}} \times \frac{1}{\sec^2 x}\, du$$

Substitute all these bits into the integral.

$$= \int \frac{1}{\sqrt{u}} + \frac{u^2}{\sqrt{u}}\, du = \int u^{-\frac{1}{2}} + u^{\frac{3}{2}}\, du$$

$$= 2u^{\frac{1}{2}} + \frac{2}{5}u^{\frac{5}{2}} + C = 2\sqrt{\tan x} + \frac{2}{5}\sqrt{\tan^5 x} + C$$

b) Calculate $\displaystyle\int_{\frac{1}{2}}^{\frac{\sqrt{3}}{2}} \frac{4}{\sqrt{1 - x^2}}\, dx$ using the substitution $x = \sin\theta$, where $-\frac{\pi}{2} \le \theta \le \frac{\pi}{2}$.

$$x = \sin\theta \implies \frac{dx}{d\theta} = \cos\theta \implies dx = \cos\theta\, d\theta$$

Differentiate x with respect to θ, and use the result to find dx.

$$x = \sin\theta \implies \theta = \sin^{-1} x$$

$$\text{So } x = \frac{\sqrt{3}}{2} \implies \theta = \frac{\pi}{3}$$
$$\text{and } x = \frac{1}{2} \implies \theta = \frac{\pi}{6}$$

Use the substitution to convert the limits from x to θ.

Tip $\sin\theta$ has an inverse because θ is restricted to between $-\frac{\pi}{2}$ and $\frac{\pi}{2}$.

$$\int_{\frac{1}{2}}^{\frac{\sqrt{3}}{2}} \frac{4}{\sqrt{1 - x^2}}\, dx = \int_{\frac{\pi}{6}}^{\frac{\pi}{3}} \frac{4}{\sqrt{1 - \sin^2\theta}} \cos\theta\, d\theta$$

$$= \int_{\frac{\pi}{6}}^{\frac{\pi}{3}} \frac{4\cos\theta}{\sqrt{\cos^2\theta}}\, d\theta$$

Now solve the integral.

$$= \int_{\frac{\pi}{6}}^{\frac{\pi}{3}} 4\, d\theta = [4\theta]_{\frac{\pi}{6}}^{\frac{\pi}{3}} = \frac{4\pi}{3} - \frac{2\pi}{3} = \frac{2\pi}{3}$$

Exercise 8.15

Q1 Find the exact value of $\int_0^1 \dfrac{1}{1+x^2}\,dx$
using the substitution $x = \tan\theta$ where $-\dfrac{\pi}{2} < \theta < \dfrac{\pi}{2}$.

Q1-4 Hint Remember, the phrase 'exact value' is usually a clue that the answer will include a surd or π.

Q2 Find the exact value of $\int_0^{\frac{\pi}{6}} 3\sin x \sin 2x\,dx$
using the substitution $u = \sin x$.

Q3 Use the substitution $x = 2\sin\theta$, where $-\dfrac{\pi}{2} \le \theta \le \dfrac{\pi}{2}$,
to find the exact value of $\int_1^{\sqrt{3}} \dfrac{1}{(4-x^2)^{\frac{3}{2}}}\,dx$.

Q4 Find the exact value of $\int_{\frac{1}{2}}^1 \dfrac{1}{x^2\sqrt{1-x^2}}\,dx$. Use the substitution $x = \cos\theta$, where $0 \le \theta \le \pi$.

Q5 Find $\int 2\tan^3 x\,dx$ using the substitution $u = \sec^2 x$.

Q6 Find the exact value of the following integrals:

a) $\int_{-\pi}^{\frac{\pi}{2}} 3\sin\theta\cos^4\theta\,d\theta$, using $u = \cos\theta$

b) $\int_{\frac{\pi}{4}}^{\frac{\pi}{3}} \sec^4 x \tan x\,dx$, using $u = \sec x$

Q7 Find the exact value of $\int_1^{\sqrt{3}} \dfrac{4x}{\sqrt{1+x^2}}\,dx$, using the substitution $x = \cot\theta$, where $-\dfrac{\pi}{2} \le \theta \le \dfrac{\pi}{2}$.

Challenge

E **P** **Q8** a) Find the largest possible domain and range of the function $f(x) = \dfrac{4}{(64-x^2)^{\frac{1}{2}}}$. *[2 marks]*

  b) Showing any asymptotes and the coordinates of any points where it crosses the axes, sketch the graph of $y = \dfrac{4}{(64-x^2)^{\frac{1}{2}}}$. *[4 marks]*

  c) Use a suitable substitution to find the area bounded by this graph, the x-axis, and the lines $x = 0$ and $x = 4$. *[7 marks]*

E **P** **Q9** a) Express $\cos^2 u$ in the form $\dfrac{a + \cos bu}{c}$. *[1 mark]*

  b) Using the substitution $x = \sin u$ for $-\dfrac{\pi}{2} < u < \dfrac{\pi}{2}$, find $\int \sqrt{1-x^2}\,dx$, giving your answer in terms of u. *[4 marks]*

  c) Show algebraically that the graph of $y = \sqrt{1-x^2}$ represents part of a circle. Give the centre and radius of the circle and specify which part of the circle is represented. *[2 marks]*

  d) Verify that using integration to find the area between the curve and the x-axis gives the same result as using the formula for the area of a circle, $A = \pi r^2$. *[3 marks]*

8.16   Integration by Parts

If you have a **product** to integrate but you can't use any of the methods you've learnt so far, you might be able to use **integration by parts**. The **formula** for integrating by parts is:

$$\int u\frac{\mathrm{d}v}{\mathrm{d}x}\,\mathrm{d}x = uv - \int v\frac{\mathrm{d}u}{\mathrm{d}x}\,\mathrm{d}x$$

where u and v are both functions of x.

Here's the **proof** of this formula.
You **don't need** to know it for the exam, but you might find it useful.

Start with the **product rule**:

- If u and v are both functions of x, then:

$$\frac{\mathrm{d}}{\mathrm{d}x}uv = u\frac{\mathrm{d}v}{\mathrm{d}x} + v\frac{\mathrm{d}u}{\mathrm{d}x}$$

- Integrate both sides of the product rule with respect to x:

$$\int \frac{\mathrm{d}}{\mathrm{d}x}uv\,\mathrm{d}x = \int u\frac{\mathrm{d}v}{\mathrm{d}x}\,\mathrm{d}x + \int v\frac{\mathrm{d}u}{\mathrm{d}x}\,\mathrm{d}x$$

- On the left-hand side, uv is differentiated, then integrated — so you end up back at uv:

$$uv = \int u\frac{\mathrm{d}v}{\mathrm{d}x}\,\mathrm{d}x + \int v\frac{\mathrm{d}u}{\mathrm{d}x}\,\mathrm{d}x$$

- Now just rearrange to get:

$$\int u\frac{\mathrm{d}v}{\mathrm{d}x}\,\mathrm{d}x = uv - \int v\frac{\mathrm{d}u}{\mathrm{d}x}\,\mathrm{d}x$$

Tip Take a look back at p.225 for more about the product rule.

Tip The integration by parts formula is sometimes written $\int uv'\,\mathrm{d}x = uv - \int vu'\,\mathrm{d}x$ — you might find this version easier to use.

The hardest thing about integration by parts is **deciding** which bit of your product should be $\boldsymbol{u}$ and which bit should be $\dfrac{\mathrm{d}v}{\mathrm{d}x}$. There's no set rule for this — you just have to look at both parts, see which one **differentiates** to give something **nice**, then set that one as u.

For example, if you have a product that has a **single** x as one part of it, choose this to be u. It'll differentiate to **1**, which will make **integrating** $v\dfrac{\mathrm{d}u}{\mathrm{d}x}$ very easy.

Example 1

Find $\int 2x\,e^x\,dx$.

If $u = 2x$ and $\dfrac{dv}{dx} = e^x$, then: ←————————— Start by working out what should be u
and what should be $\dfrac{dv}{dx}$ — choose them
$\dfrac{du}{dx} = 2$ and $v = e^x \Rightarrow v\dfrac{du}{dx} = 2e^x$. ←——— so that $v\dfrac{du}{dx}$ is easy to integrate.

Or if $u = e^x$ and $\dfrac{dv}{dx} = 2x$, then:

$\dfrac{du}{dx} = e^x$ and $v = x^2 \Rightarrow v\dfrac{du}{dx} = x^2 e^x$. ←——— $2e^x$ is easier to integrate than $x^2 e^x$,
so choose the first option.

$\int 2x\,e^x\,dx = \int u\dfrac{dv}{dx}\,dx = uv - \int v\dfrac{du}{dx}\,dx$ ←——— Put u, v, $\dfrac{du}{dx}$ and $\dfrac{dv}{dx}$ into the
integration by parts formula.

$\qquad = 2x\,e^x - \int 2e^x\,dx$

$\qquad = 2x\,e^x - 2e^x + C$ ←————————————— Don't forget the constant of integration.

You don't always need to work out both possible versions. In Example 1, e^x won't change whether you integrate it or differentiate it, so you just need to think about whether the integration would be made easier by differentiating $2x$ or by integrating it.

Example 2

Find $\int x^3 \ln x\,dx$.

Let $u = \ln x$ and $\dfrac{dv}{dx} = x^3$ ←————————————— Choose u and $\dfrac{dv}{dx}$.

$u = \ln x \Rightarrow \dfrac{du}{dx} = \dfrac{1}{x}$ and $\dfrac{dv}{dx} = x^3 \Rightarrow v = \dfrac{x^4}{4}$

$\int x^3 \ln x\,dx = \ln x \times \dfrac{x^4}{4} - \int \dfrac{x^4}{4} \times \dfrac{1}{x}\,dx$ ←——— Put u, v, $\dfrac{du}{dx}$ and $\dfrac{dv}{dx}$ into the
integration by parts formula.

$\qquad = \dfrac{x^4 \ln x}{4} - \dfrac{1}{4}\int x^3\,dx$

$\qquad = \dfrac{x^4 \ln x}{4} - \dfrac{x^4}{16} + C$

If you have a product that has $\ln x$ as one of its factors, let $u = \ln x$ — it's easy to differentiate but quite tricky to integrate, as shown on the next page.

Until now, you haven't been able to integrate **ln x**, but **integration by parts** gives you a way to get around this. The trick is to write ln x as $(1 \times \ln x)$.

You can write ln x as (ln $x \times 1$). So let $u = \ln x$ and let $\dfrac{dv}{dx} = 1$.

$$u = \ln x \implies \frac{du}{dx} = \frac{1}{x} \qquad \frac{dv}{dx} = 1 \implies v = x$$

Putting these into the formula gives:

$$\int \ln x \; dx = \int (\ln x \times 1) \; dx = \ln x \times x - \int x \frac{1}{x} \; dx$$

$$= x \ln x - \int 1 \; dx = x \ln x - x + C$$

You can use **integration by parts** on definite integrals too. The only change from the method for indefinite integrals is that you have to **apply the limits** of the integral to the uv bit.

The integration by parts formula for definite integrals can be written like this:

$$\int_a^b u \frac{dv}{dx} \; dx = \left[uv \right]_a^b - \int_a^b v \frac{du}{dx} \; dx$$

Example 3

Find the exact value of $\displaystyle\int_0^{\frac{\pi}{2}} 4x \sin\left(\frac{x}{2}\right) dx$.

Let $u = 4x$ and $\dfrac{dv}{dx} = \sin\left(\dfrac{x}{2}\right)$

$\sin\left(\dfrac{x}{2}\right)$ will give a cos function whether you integrate or differentiate it, so the only way to get a simpler $\int v \dfrac{du}{dx} \; dx$ is to make $u = 4x$.

$u = 4x \implies \dfrac{du}{dx} = 4$ and $\dfrac{dv}{dx} = \sin\left(\dfrac{x}{2}\right) \implies v = -2\cos\left(\dfrac{x}{2}\right)$

$$\int_0^{\frac{\pi}{2}} 4x \sin\left(\frac{x}{2}\right) dx = \left[-8x \cos\left(\frac{x}{2}\right) \right]_0^{\frac{\pi}{2}} - \int_0^{\frac{\pi}{2}} -8 \cos\left(\frac{x}{2}\right) dx$$

Substitute everything into the formula and complete the integration.

$$= -8 \left[x \cos\left(\frac{x}{2}\right) \right]_0^{\frac{\pi}{2}} + 16 \left[\sin\left(\frac{x}{2}\right) \right]_0^{\frac{\pi}{2}}$$

$$= -8 \left[\frac{\pi}{2} \cos\left(\frac{\pi}{4}\right) - 0 \cos(0) \right] + 16 \left[\sin\left(\frac{\pi}{4}\right) - \sin(0) \right]$$

$$= -8 \left[\frac{\pi}{2} \frac{1}{\sqrt{2}} \right] + 16 \left[\frac{1}{\sqrt{2}} \right]$$

$$= -\frac{4\pi}{\sqrt{2}} + \frac{16}{\sqrt{2}}$$

$$= 8\sqrt{2} - 2\pi\sqrt{2}$$

Tip See p.269 for a reminder about integrating trig functions.

Exercise 8.16

Q1 Use integration by parts to find:

a) $\int xe^x\,dx$

b) $\int xe^{-x}\,dx$

c) $\int xe^{-\frac{x}{3}}\,dx$

d) $\int x(e^x + 1)\,dx$

Q2 Use integration by parts to find:

a) $\int_0^\pi x\sin x\,dx$

b) $\int 2x\cos x\,dx$

c) $\int 3x\cos\left(\frac{1}{2}x\right)\,dx$

d) $\int_{-\frac{\pi}{2}}^{\frac{\pi}{2}} 2x(1 - \sin x)\,dx$

Q3 Use integration by parts to find:

a) $\int 2\ln x\,dx$

b) $\int x^4\ln x\,dx$

c) $\int \ln 4x\,dx$

d) $\int \ln x^3\,dx$

E Q4 Use integration by parts to find $\int_1^4 x\ln x\,dx$ to 3 s.f. *[5 marks]*

Q5 Use integration by parts to find:

a) $\int_{-1}^1 20x(x + 1)^3$

b) $\int_0^{1.5} 30x\sqrt{2x + 1}\,dx$

Q6 Use integration by parts to find the exact values of the following:

a) $\int_0^1 12xe^{2x}\,dx$

b) $\int_0^{\frac{\pi}{3}} 18x\sin 3x\,dx$

c) $\int_1^2 \frac{1}{x^2}\ln x\,dx$

Q7 Find:

a) $\int \frac{x}{e^{2x}}\,dx$

b) $\int (x + 1)\sqrt{x + 2}\,dx$

c) $\int \ln(x + 1)\,dx$

E P Q8 a) Integrate $3x\cos\left(x + \frac{2\pi}{3}\right)$ with respect to x. *[4 marks]*

b) Use calculus to find which represents the larger area:

- The region between the x-axis and the curve $y = 2\cos x$, for $x = 0$ to $\frac{\pi}{2}$.
- The region between the x-axis and the curve $y = 3x\cos\left(x + \frac{2\pi}{3}\right)$, for $x = 0$ to $\frac{\pi}{2}$.

[5 marks]

E P Q9 The diagram shows the graph of $y = \dfrac{\ln x}{x^3}$.

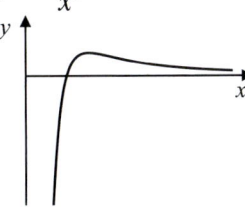

a) Find the gradient function for the curve. *[4 marks]*

b) Give the range of $y = \dfrac{\ln x}{x^3}$. *[2 marks]*

c) Find $\int y\,dx$, giving your answer as an expression in terms of x. *[4 marks]*

d) Hence show that the area between the curve and the x-axis for the x limits 5 to 10 is 409%, to the nearest percentage point, of the area for the x limits 10 to 15. *[5 marks]*

8.17 Repeated Use of Integration by Parts

Sometimes **integration by parts** leaves you with a function for $v\dfrac{du}{dx}$ which is **simpler** than the function you started with, but still **tricky to integrate**.

You might have to carry out integration by parts **again** to find $\int v\dfrac{du}{dx}\,dx$.

Example 1

Find $\int x^2 \sin x\, dx$.

Let $u = x^2$ and let $\dfrac{dv}{dx} = \sin x$ ← Choose u and $\dfrac{dv}{dx}$.

Then $\dfrac{du}{dx} = 2x$ and $v = -\cos x$

$\int x^2 \sin x\, dx = -x^2\cos x - \int -2x \cos x\, dx$ ← Substitute these into the formula.

$\qquad\qquad = -x^2\cos x + \int 2x \cos x\, dx$ ← $2x \cos x$ isn't very easy to integrate, but you can integrate by parts again.

- -

Let $u_1 = 2x$ and let $\dfrac{dv_1}{dx} = \cos x$ ←

Then $\dfrac{du_1}{dx} = 2$ and $v_1 = \sin x$

> **Tip** Using different variable names here (e.g. u_1 and v_1) makes it clear to the examiner that this is a new integration by parts.

$\int 2x \cos x\, dx = 2x \sin x - \int 2 \sin x\, dx$ ←

$\qquad\qquad = 2x \sin x + 2\cos x + C$ ← Substitute into the formula again.

- -

So $\int x^2 \sin x\, dx = -x^2\cos x + \int 2x \cos x\, dx$ ← Now put this result back into what you got earlier.

$\qquad\qquad = -x^2\cos x + 2x \sin x + 2\cos x + C$

Example 2

Use integration by parts to find $\displaystyle\int_2^3 x^2(x-1)^{-4}\, dx$.

Let $u = x^2$ and let $\dfrac{dv}{dx} = (x-1)^{-4}$ ← Choose u and $\dfrac{dv}{dx}$.

Then $\dfrac{du}{dx} = 2x$ and $v = -\dfrac{1}{3}(x-1)^{-3}$

$\displaystyle\int_2^3 x^2(x-1)^{-4}\, dx = \left[-\dfrac{x^2}{3}(x-1)^{-3}\right]_2^3 - \int_2^3 -\dfrac{2x}{3}(x-1)^{-3}\, dx$ ← Put these into the formula.

$\qquad\qquad = \left[-\dfrac{x^2}{3}(x-1)^{-3}\right]_2^3 + \dfrac{2}{3}\int_2^3 x(x-1)^{-3}\, dx$

continued on the next page...

Let $u_1 = x$ and let $\dfrac{dv_1}{dx} = (x-1)^{-3}$ ← This is still tricky to integrate, so use integration by parts again.

Then $\dfrac{du_1}{dx} = 1$ and $v_1 = -\dfrac{1}{2}(x-1)^{-2}$

$\displaystyle\int_2^3 x(x-1)^{-3}\,dx = \left[-\frac{x}{2}(x-1)^{-2}\right]_2^3 - \int_2^3 -\frac{1}{2}(x-1)^{-2}\,dx$ ← Put these into the formula.

$\qquad = \left[-\dfrac{x}{2}(x-1)^{-2}\right]_2^3 - \dfrac{1}{2}\left[(x-1)^{-1}\right]_2^3$

$\qquad = \left[-\dfrac{3}{2}(2)^{-2} + \dfrac{2}{2}(1)^{-2}\right] - \dfrac{1}{2}\left[2^{-1} - 1^{-1}\right]$

$\qquad = \left[-\dfrac{3}{8} + 1\right] - \dfrac{1}{2}\left[\dfrac{1}{2} - 1\right]$

$\qquad = \dfrac{5}{8} + \dfrac{1}{4} = \dfrac{7}{8}$

> **Tip** The formula for integrating $(ax + b)^n$ is used a few times in this example — go back to p.262 if you've forgotten how it works.

- -

$\displaystyle\int_2^3 x^2(x-1)^{-4}\,dx = \left[-\frac{x^2}{3}(x-1)^{-3}\right]_2^3 + \frac{2}{3}\int_2^3 x(x-1)^{-3}\,dx$ ← Now you can evaluate the original integral.

$\qquad = \left[-\dfrac{x^2}{3}(x-1)^{-3}\right]_2^3 + \dfrac{2}{3}\left(\dfrac{7}{8}\right)$

$\qquad = \left[\left(-\dfrac{9}{3}(2)^{-3}\right) - \left(-\dfrac{4}{3}(1)^{-3}\right)\right] + \dfrac{7}{12}$

$\qquad = \left[-\dfrac{9}{24} + \dfrac{4}{3}\right] + \dfrac{7}{12}$

$\qquad = \dfrac{23}{24} + \dfrac{7}{12} = \dfrac{37}{24}$

Exercise 8.17

Q1 Use integration by parts twice to find:

a) $\displaystyle\int x^2 e^x\,dx$

b) $\displaystyle\int x^2 \cos x\,dx$

c) $\displaystyle\int 4x^2 \sin 2x\,dx$

d) $\displaystyle\int 40x^2(2x-1)^4\,dx$

Q2 Find $\displaystyle\int_{-1}^0 x^2(x+1)^4\,dx$ using integration by parts.

P Q3 Use integration by parts to find the area enclosed by the curve $y = x^2 e^{-2x}$, the x-axis and the lines $x = 0$ and $x = 1$.

> **Problem Solving** Since $x^2 \geq 0$ and $e^{-2x} > 0$ for all x, the area will be entirely above the x-axis in Q3.

Challenge

E P Q4 a) Verify that the area bounded by the y-axis, the x-axis from $x = 0$ to $\dfrac{\pi}{2}$, and the curve $y = e^x \cos x$ is equal to $\displaystyle\int_0^{\frac{\pi}{2}} e^x \cos x\,dx$. *[1 mark]*

b) Find $\displaystyle\int e^x \cos x\,dx$. *[7 marks]*

c) Find $\displaystyle\int_0^{\frac{\pi}{2}} e^x \cos x\,dx$ to 3 s.f. *[2 marks]*

8.18 Integration Using Partial Fractions

You can integrate algebraic fractions where the denominator can be written as a product of **linear factors** by splitting them up into **partial fractions**. Each fraction can then be **integrated separately** using the method on pages 266-267.

Example 1

Find $\int \frac{12x+6}{4x^2-9}\,dx$ where $x > 2$.

$$\frac{12x+6}{4x^2-9} \equiv \frac{12x+6}{(2x+3)(2x-3)}$$

The first step is to write the function as partial fractions. Start by factorising the denominator.

$$\frac{12x+6}{(2x+3)(2x-3)} \equiv \frac{A}{2x+3} + \frac{B}{2x-3}$$

Write as an identity with partial fractions.

$$\frac{12x+6}{(2x+3)(2x-3)} \equiv \frac{A(2x-3)+B(2x+3)}{(2x+3)(2x-3)}$$

$$\Rightarrow 12x + 6 \equiv A(2x-3) + B(2x+3)$$

Add the partial fractions and cancel the denominators.

Substituting $x = \frac{3}{2}$ into the identity gives:

$18 + 6 = 0A + 6B \Rightarrow 24 = 6B \Rightarrow B = 4$

Use the substitution method to find A and B.

Substituting $x = -\frac{3}{2}$ into the identity gives:

$-18 + 6 = -6A + 0B \Rightarrow -12 = -6A \Rightarrow A = 2$

Replace A and B in the original identity.

Tip You could also use the 'equating coefficients' method to find A and B.

$$\frac{12x+6}{4x^2-9} \equiv \frac{2}{2x+3} + \frac{4}{2x-3}$$

$$\int \frac{12x+6}{4x^2-9}\,dx = \int \frac{2}{2x+3} + \frac{4}{2x-3}\,dx$$

Integrate each term separately.

$$= 2 \times \frac{1}{2}\ln|2x+3| + 4 \times \frac{1}{2}\ln|2x-3| + C$$

$$= \ln|2x+3| + 2\ln|2x-3| + C$$

$x > 2$ so $2x + 3 > 7$ and $2x - 3 > 1$, so you can remove the modulus signs.

$$= \ln(2x+3) + 2\ln(2x-3) + C$$

$$= \ln(2x+3)(2x-3)^2 + C$$

Tip It's a good idea to tidy up your answers using the log laws.

Example 2

Find the exact value of $\int_3^4 \dfrac{2}{x(x-2)} \, dx$, writing it as a single logarithm.

$\dfrac{2}{x(x-2)} \equiv \dfrac{A}{x} + \dfrac{B}{x-2} \equiv \dfrac{A(x-2)+Bx}{x(x-2)}$ ← Start by writing $\dfrac{2}{x(x-2)}$ as an identity with partial fractions.

$\Rightarrow 2 \equiv A(x-2) + Bx$

Equating constant terms:
$2 = -2A \Rightarrow A = -1$ ← Use the equating coefficients method to find A and B.

Equating x coefficients:
$0 = A + B \Rightarrow 0 = -1 + B \Rightarrow B = 1$

$\dfrac{2}{x(x-2)} \equiv \dfrac{-1}{x} + \dfrac{1}{x-2} \equiv \dfrac{1}{x-2} - \dfrac{1}{x}$ ← Replace A and B in the original identity.

$\int_3^4 \dfrac{2}{x(x-2)} \, dx = \int_3^4 \dfrac{1}{(x-2)} - \dfrac{1}{x} \, dx$

$= \left[\ln|x-2| - \ln|x| \right]_3^4$ ← Integrate each term separately.

$= \left[\ln\left|\dfrac{x-2}{x}\right| \right]_3^4$

$= \ln\left|\dfrac{4-2}{4}\right| - \ln\left|\dfrac{3-2}{3}\right|$

$= \ln\left(\dfrac{1}{2}\right) - \ln\left(\dfrac{1}{3}\right) = \ln\left(\dfrac{3}{2}\right)$ ← Fully simplify your answer by using the law $\log a - \log b = \log\left(\dfrac{a}{b}\right)$.

Exercise 8.18

Q1 Integrate the following functions by writing them as partial fractions:

   a) $\int \dfrac{24(x-1)}{9-4x^2} \, dx$

   b) $\int \dfrac{21x-82}{(x-2)(x-3)(x-4)} \, dx$

Q2 Find $\int_0^1 \dfrac{x}{(x-2)(x-3)} \, dx$ by expressing as partial fractions.
   Give your answer as a single logarithm.

E Q3 a) Express $\dfrac{5x+5}{(x+2)(x-3)}$ as the sum of two fractions, in the form $\dfrac{A}{x+2} + \dfrac{B}{x-3}$. *[3 marks]*

   b) Hence find $\int \dfrac{5x+5}{(x+2)(x-3)} \, dx$. *[2 marks]*

   c) Find the exact value of $\int_{-1}^2 \dfrac{5x+5}{(x+2)(x-3)} \, dx$. *[2 marks]*

Q4 Given that $f(x) = 3x + 5$ and $g(x) = x(x + 10)$, find $\int_1^2 \dfrac{f(x)}{g(x)} \, dx$ to 3 d.p.
   by expressing $\dfrac{f(x)}{g(x)}$ as partial fractions.

Q5 a) Express $\dfrac{6}{2x^2 - 5x + 2}$ in partial fractions.

b) Hence find $\displaystyle\int \dfrac{6}{2x^2 - 5x + 2}\, dx$ where $x > 2$.

c) Evaluate $\displaystyle\int_3^5 \dfrac{6}{2x^2 - 5x + 2}\, dx$, expressing your answer as a single logarithm.

P Q6 Show that $\displaystyle\int_0^{\frac{2}{3}} \dfrac{-(t+3)}{(3t+2)(t+1)}\, dt = 2\ln\dfrac{5}{3} - \dfrac{7}{3}\ln 2$.

E **P** Q7 a) Express $\dfrac{5x + 7}{(x+3)(x-1)}$ as the sum of two fractions with linear denominators. *[3 marks]*

b) Find $\displaystyle\int \dfrac{5x + 7}{(x+3)(x-1)}\, dx$. *[2 marks]*

c) $\displaystyle\int_a^b \dfrac{5x + 7}{(x+3)(x-1)}\, dx \equiv \int_{a+c}^{b+c} \dfrac{5x - 8}{x(x-4)}\, dx$. Find the value of c. *[1 mark]*

E **P** Q8 The diagram shows the graph of $y = \dfrac{-4x - 1}{(x-2)(x+7)}$.

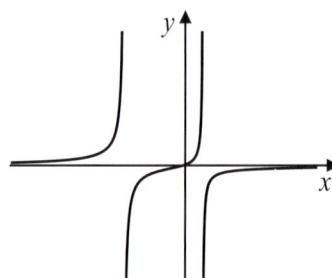

a) Find the equations of any vertical asymptotes of the graph. *[1 mark]*

b) Express $\dfrac{-4x - 1}{(x-2)(x+7)}$ in partial fractions. *[3 marks]*

c) Find $\displaystyle\int \dfrac{-4x - 1}{(x-2)(x+7)}\, dx$. *[2 marks]*

d) Find, to 3 s.f., the area between the graph, the x-axis and the lines $x = 5$ and $x = 10$. *[2 marks]*

P Q9 Use the substitution $u = \sqrt{x}$ to find the exact value of $\displaystyle\int_9^{16} \dfrac{4}{\sqrt{x}(9x - 4)}\, dx$.

Challenge

E **P** Q10 a) $\dfrac{4x + c}{(x+1)(2x-1)}$ can be written in the form $\dfrac{A}{x+1} + \dfrac{B}{2x-1}$.

Express A and B in terms of c. *[3 marks]*

b) The binomial expansion of $\dfrac{4x + c}{(x+1)(2x-1)}$ in ascending powers of x starts: $-1 - 5x...$

Find the values of A, B and c. *[5 marks]*

c) Find $\displaystyle\int \dfrac{4x + c}{(x+1)(2x-1)}\, dx$, for the value of c found in part b). *[2 marks]*

8.19   Differential Equations

A **differential equation** is an equation that includes a **derivative term** such as $\frac{dy}{dx}$ (or $\frac{dP}{dt}$, $\frac{ds}{dt}$, $\frac{dV}{dr}$ etc., depending on the variables), as well as **other variables** (like x **and** y).

Before you even think about **solving** them, you have to be able to **set up** ('**formulate**') differential equations. Differential equations tend to involve a **rate of change** (giving a derivative term) and a **proportion relation**, where the rate of change will be directly or inversely proportional to some function of the variables.

It'll help to think about what the derivative **actually means**. $\frac{dy}{dx}$ is defined as 'the **rate of change** of y with respect to x'. In other words, it tells you how y changes as x changes. One of the variables in a differential equation is often **time**, t, so the question will be about how something **changes over time**.

Example 1

The number of bacteria in a Petri dish, b, is increasing over time, t, at a rate directly proportional to the number of bacteria.
Formulate a differential equation that shows this information.

The number of bacteria (b) increases as time (t) increases — so that's the rate of change of b with respect to t, or $\frac{db}{dt}$.

The question tells you that you need to write a differential equation, so you know there'll be a derivative term — work this out first.

$$\frac{db}{dt} \propto b$$

The rate of change, $\frac{db}{dt}$, is proportional to b.

$$\frac{db}{dt} = kb \text{ for some constant } k, \ k > 0.$$

Rewrite this as an equation, rather than a proportion relation.

Example 2

The volume of interdimensional space jelly, V, in a container is decreasing over time, t, at a rate inversely proportional to the square of its volume.
Show this as a differential equation.

V decreases as t increases, so the derivative term is the rate of change of V with respect to t, or $\frac{dV}{dt}$.

Find the derivative term.

$$\frac{dV}{dt} \propto \frac{1}{V^2}$$

$\frac{dV}{dt}$ is inversely proportional to V^2.

$$\frac{dV}{dt} = -\frac{k}{V^2} \text{ for some constant } k, \ k > 0.$$

The equation needs a minus sign because V is decreasing as t increases.

Chapter 8

Example 3

The rate of cooling of a hot liquid is proportional to the difference
between the temperature of the liquid and the temperature of the room.
Formulate a differential equation to represent this situation.

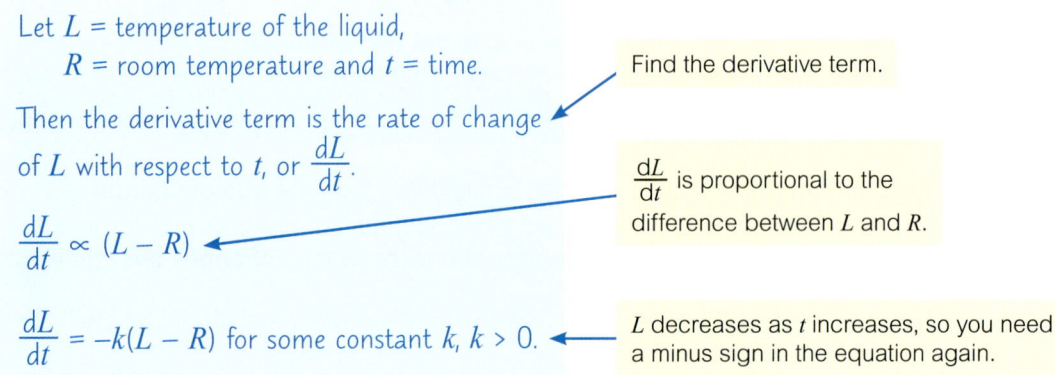

Let L = temperature of the liquid,
R = room temperature and t = time.

Then the derivative term is the rate of change
of L with respect to t, or $\dfrac{dL}{dt}$.

$\dfrac{dL}{dt} \propto (L - R)$

$\dfrac{dL}{dt} = -k(L - R)$ for some constant k, $k > 0$.

Find the derivative term.

$\dfrac{dL}{dt}$ is proportional to the
difference between L and R.

L decreases as t increases, so you need
a minus sign in the equation again.

Exercise 8.19

M **Q1** The number of fleas (N) on a cat is increasing over time, t, at a rate directly proportional
to the number of fleas. Show this as a differential equation.

M **Q2** The value, x, of a house is increasing over time, t, at a rate inversely proportional
to the square of x. Formulate a differential equation to show this.

M **Q3** The rate of depreciation of the amount (£A) a car is worth is directly proportional
to the square root of A. Show this as a differential equation.

M **Q4** The rate of decrease of a population, y, with respect to time is directly proportional to the
difference between y and λ where λ is a constant. Show this as a differential equation.

P
M **Q5** The volume of water which is being poured into a container
is directly proportional to the volume of water (V) in the container.
The container has a hole in it from which water flows out at
a rate of 20 cm³s⁻¹. Formulate a differential equation to show this.

> **Problem Solving**
>
> The rate of change of V
> is the difference between
> the rate water flows in
> and the rate it flows out.

E
M **Q6** A population (p) of meerkats decreases over time at a rate which is directly
proportional to the number of meerkats in the population to the power of 0.8.
Show this information in the form of a differential equation. *[1 mark]*

E
P
M **Q7** A jar of jam is left on the side in a room with temperature 25 °C. The temperature of the
jam (θ °C) starts at below 25 °C and approaches the temperature of the room such that
the rate of change of temperature is directly proportional to the difference between the
temperatures of the jam and the room. Write a differential equation to show this information.
[1 mark]

E **Q8** a) Given that a candle burns at a constant rate and is a prism with the
wick passing down its length, form a differential equation to show
the rate of change of its height (h) in cm with respect to time (t) in hours. *[1 mark]*

P
M
   b) The candle has a constant density. Show that the rate of change
of its mass (m) in kg, with respect to height, is also constant. *[1 mark]*

   c) Using the equations found in parts a) and b), form a third differential equation
to show the rate of change of the mass of the candle with respect to time. *[1 mark]*

8.20 Solving Differential Equations

Solving a differential equation means using it to find an **equation** in terms of the two variables,
without a derivative term. To do this, you need to use **integration**.

The only differential equations containing x and y terms that you'll have to solve in A-Level Maths
are ones with **separable variables** — where x and y can be separated into functions **f(x)** and **g(y)**.
Note that other variables might be used instead — they won't always be x and y.

Solving differential equations

Step 1: Write the differential equation in the form $\dfrac{dy}{dx} = \text{f}(x)\text{g}(y)$.

Step 2: **Rearrange** the equation into the form: $\dfrac{1}{\text{g}(y)}\,dy = \text{f}(x)\,dx$.
To do this, get all the terms containing y on the **left-hand side**,
and all the terms containing x on the **right-hand side** and split up the $\dfrac{dy}{dx}$.

Step 3: Now **integrate both sides**: $\displaystyle\int \dfrac{1}{\text{g}(y)}\,dy = \int \text{f}(x)\,dx$.
Don't forget the **constant of integration** (you only need
one — not one on each side). It might be useful to write
the constant as **ln k** rather than **C** (see p.316).

Step 4: **Rearrange** your answer to get it in a **nice form** —
you might be asked to find it in the form **$y = \text{h}(x)$**.

Step 5: If you're asked for a **general solution**, leave C (or k) in your answer. If they
want a **particular solution**, they'll give you x and y values for a certain point.
All you do is put these values into your equation and use them to **find C** (or k).

Example **1**

Find the general solution of the differential equation $\dfrac{ds}{dt} = -6t^2$.

$\text{f}(t) = -6t^2,\ \text{g}(s) = 1$ ⟵ ———— Step 1 is already done.

$ds = -6t^2\,dt$ ⟵ ———— Step 2 — rearrange the equation.

$\displaystyle\int 1\,ds = \int -6t^2\,dt \implies s = -2t^3 + C$ ⟵— Step 3 — integrate both sides.

Tip Steps 4 and 5
aren't needed here —
the equation doesn't
need rearranging, and
you're only looking for
the general solution.

Example 2

Find the particular solution of $\frac{dy}{dx} = 2y(1 + x)^2$ when $x = -1$ and $y = 4$.

$f(x) = 2(1 + x)^2$ and $g(y) = y$ ←————————— Identify $f(x)$ and $g(y)$.

$\frac{1}{y}\, dy = 2(1 + x)^2\, dx$ ←————————— Separate the variables.

$\int \frac{1}{y}\, dy = \int 2(1 + x)^2\, dx \Rightarrow \ln|y| = \frac{2}{3}(1 + x)^3 + C$ ←——— Integrate both sides.

$\ln 4 = \frac{2}{3}(1 + (-1))^3 + C \Rightarrow \ln 4 = C$ ←——————— Work out the value of C for the given values of x and y.

So $\ln|y| = \frac{2}{3}(1 + x)^3 + \ln 4$

Example 3

Find the general solution of $(x - 2)(2x + 3)\frac{dy}{dx} = xy + 5y$, where $x > 2$.
Give your answer in the form $y = f(x)$.

$\frac{dy}{dx} = \frac{x+5}{(x-2)(2x+3)} \times y \Rightarrow \frac{1}{y}\, dy = \frac{x+5}{(x-2)(2x+3)}\, dx$ ←——— First, separate the variables.

$\frac{x+5}{(x-2)(2x+3)} \equiv \frac{A}{x-2} + \frac{B}{2x+3}$ ←——— Write the right-hand side as partial fractions (see p.310).

$\Rightarrow x + 5 \equiv A(2x + 3) + B(x - 2)$

Solving for A and B gives $A = 1$, $B = -1$, so:

$\frac{1}{y}\, dy = \frac{1}{x-2} - \frac{1}{2x+3}\, dx$

Now you can integrate.

Problem Solving

Since all the other terms are $\ln\,(\text{something})$, it makes sense to use $\ln k$ as the constant of integration, then use the log laws to simplify.

$\int \frac{1}{y}\, dy = \int \frac{1}{x-2} - \frac{1}{2x+3}\, dx$

$\Rightarrow \ln|y| = \ln|x - 2| - \frac{1}{2}\ln|2x + 3| + \ln k$

$\Rightarrow \ln|y| = \ln\left|\frac{k(x-2)}{\sqrt{2x+3}}\right|$

$\Rightarrow y = \frac{k(x-2)}{\sqrt{2x+3}}$ ←——— You know $x > 2$, so $x - 2$ is positive and the modulus can be removed.

Example 4

Find the particular solution to the differential equation $\frac{db}{dt} = 4\sqrt{b}$, given that when $t = 12$, $b = 900$. Give your answer in the form $b = f(t)$.

$\frac{1}{\sqrt{b}}\, db = 4\, dt$ ← Separate the variables.

$\int b^{-\frac{1}{2}}\, db = \int 4\, dt \Rightarrow 2b^{\frac{1}{2}} = 4t + C$ ← Integrate both sides.

$2\sqrt{b} = 4t + C \Rightarrow 2\sqrt{900} = 4(12) + C$
$\Rightarrow 60 = 48 + C$
$\Rightarrow C = 12$

Tip In this case, it's easier to find C for the given values of b and t before you rearrange the equation — rearranging then finding C would give you a quadratic to solve.

$2\sqrt{b} = 4t + 12 \Rightarrow \sqrt{b} = 2t + 6$

$\Rightarrow b = 4t^2 + 24t + 36$ ← Now rearrange to get the form $b = f(t)$.

Exercise 8.20

Q1 Find the general solutions of the following differential equations where $x \geq 0$.
Give your answers in the form $y = f(x)$.

a) $\frac{dy}{dx} = 8x^3$

b) $\frac{dy}{dx} = 5y$

c) $\frac{dy}{dx} = 6x^2y$

d) $\frac{dy}{dx} = \frac{y}{x}$

e) $\frac{dy}{dx} = (y+1)\cos x$

f) $\frac{dy}{dx} = \frac{3xy - 6y}{(x-4)(2x-5)}$

Q1f) Hint You'll need to do some work before you can integrate with respect to x.

Q2 Find the particular solutions of the following differential equations at the given conditions:

a) $\frac{dy}{dx} = -\frac{x}{y}$ $x = 0, y = 2$

b) $\frac{dx}{dt} = \frac{2}{\sqrt{x}}$ $t = 5, x = 9$

c) $\frac{dV}{dt} = 3(V - 1)$ $t = 0, V = 5$

d) $\frac{dy}{dx} = \frac{\tan y}{x}$ $x = 2, y = \frac{\pi}{2}$

e) $\frac{dx}{dt} = 10x(x + 1)$ $t = 0, x = 1$

Q3 a) Find the general solution of the equation $\frac{dx}{d\theta} = \cos^2 x \cot \theta$.
b) Given that $x = \frac{\pi}{4}$ when $\theta = \frac{\pi}{2}$, find a particular solution.
c) Hence find the value of x when $\theta = \frac{\pi}{6}$, for $0 < x < \frac{\pi}{2}$.

E Q4 a) Find the general solution to the equation $2y\frac{dy}{dx} + x^2 = 5x$. *[3 marks]*
b) Find the particular solution to this equation, given that when $x = 6$, $y = 13$. *[1 mark]*

P **Q5** The rate of increase of the variable V at time t satisfies the differential equation $\dfrac{dV}{dt} = a - bV$, where a and b are positive constants.

   a) Show that $V = \dfrac{a}{b} - Ae^{-bt}$, where A is a positive constant.

   b) Given that $V = \dfrac{a}{4b}$ when $t = 0$, find A in terms of a and b.

   c) Find the value V approaches as t gets very large.

E **P** **Q6** a) Find the general solution of the equation $\dfrac{dy}{dx} = xy + x + y + 1$.   *[3 marks]*

   b) When $x = 2$, $y = 0$. Find the particular solution, for values of y greater than -1,

     in the form $y = \dfrac{(e^x)^{f(x)}}{e^a} + b$, where a and b are integer constants.   *[3 marks]*

E **P** **Q7** Given that $\dfrac{dy}{dx} = 3xy^2 + 2y^2$, and that $x = 2$ when $y = 0.25$,

   find the value of y when $x = 4$.   *[5 marks]*

E **Q8** a) $(3y^2 + 6y + 11)\dfrac{dy}{dx} = 3x(y^2 + 2y + 1)$

     Rewrite this differential equation in the form $\displaystyle\int f(y)\,dy = \int f(x)\,dx$.   *[1 mark]*

   b) Find a general solution of the equation.   *[3 marks]*

   c) Given that $x = 1$ when $y = 1$, find the particular solution.   *[1 mark]*

E **P** **Q9** Given that $x = 3$ when $y = 1$ and $e^{y+5-2x}\dfrac{dy}{dx} = 1$, find the value of x when $y = 3$.   *[5 marks]*

E **P** **Q10** a) Find a general solution of the differential equation

     $(x + 1)(x - 2)\dfrac{dy}{dx} = y(-x - 7)$ for $x > 2$ in the form $|y| = f(x)$.   *[7 marks]*

   b) Given that $y = 32$ when $x = 3$, find the particular solution of this equation.   *[2 marks]*

8.21   Applying Differential Equations to Real-Life Problems

- Some questions involve taking **real-life problems** and using differential equations to **model** them.

- **Population** questions come up quite often — the population might be **increasing** or **decreasing**, and you have to find and solve differential equations to show it. In cases like this, one variable will usually be t, **time**.

- You might be given a **starting condition** — e.g. the **initial population**. The important thing to remember is that the starting condition occurs when $t = \mathbf{0}$.

- Once you've solved the differential equation you can use it to **answer questions** about the model. For example, if the equation is for population you might be asked to find the **population** after a certain number of years, or the **number of years** it takes to reach a certain population. Don't forget to relate the answer back to the situation given in the question.

You may also have to identify **limitations** of the model, as well as suggest possible **changes** that would **improve** it. Common things that you should think about are:

- Is there any information **missing** from the model?

- What happens to the model when the variables get really **big/small**?

- Is the model appropriate? Is a **continuous** function used for a **discrete** variable? Does the function allow **negative** values that don't make sense?

- Are there **other factors** that have not been accounted for in the model? Some examples might be **natural immunity** to a disease, **immigration/emigration** of a population or **seasonal variation** in weather.

When suggesting some refinements to the model, you don't have to make up a whole new model — just identify what the changes would be and how you could make them.

Questions like the following examples can seem a bit **overwhelming** at first, but follow things through **step by step** and they shouldn't be too bad.

Example   1

The population of rabbits in a park is decreasing as winter approaches. The rate of decrease is directly proportional to the current number of rabbits (P).

a) Explain why this situation can be modelled by the differential equation $\dfrac{dP}{dt} = -kP$, where t is the time in days and k is a positive constant.

> The model states that the rate of decrease in the rabbit population is proportional to P. ← The rate of change of the rabbit population is $\dfrac{dP}{dt}$.
>
> $\dfrac{dP}{dt} \propto P \Rightarrow \dfrac{dP}{dt} = -kP$ for some $k > 0$, ← Turn this into an equation by introducing a constant of proportionality.
> where the minus sign shows that the population is decreasing.

b) If the initial population is P_0, solve your differential equation to find P in terms of P_0, k and t.

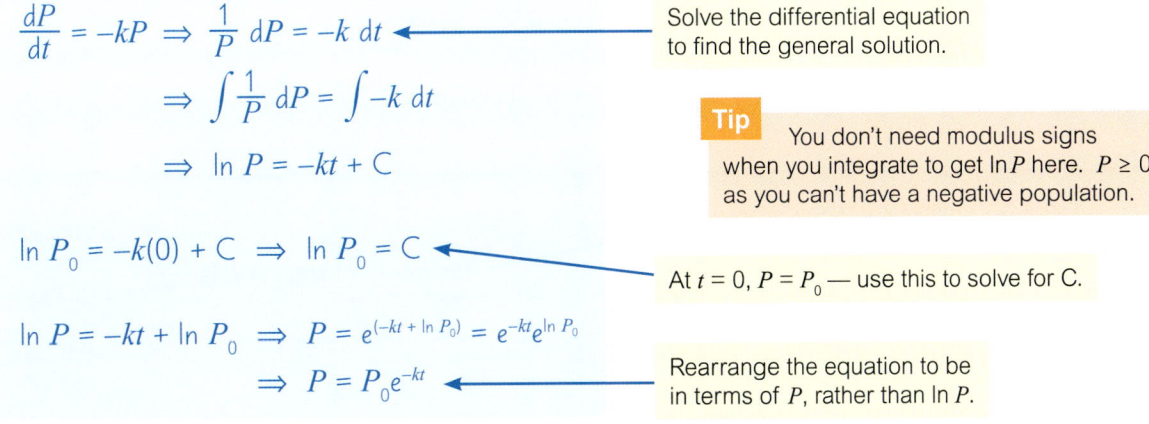

> $\dfrac{dP}{dt} = -kP \Rightarrow \dfrac{1}{P}\, dP = -k\, dt$ ← Solve the differential equation to find the general solution.
>
> $\Rightarrow \displaystyle\int \dfrac{1}{P}\, dP = \int -k\, dt$
>
> $\Rightarrow \ln P = -kt + C$
>
> **Tip** You don't need modulus signs when you integrate to get $\ln P$ here. $P \geq 0$ as you can't have a negative population.
>
> $\ln P_0 = -k(0) + C \Rightarrow \ln P_0 = C$ ← At $t = 0$, $P = P_0$ — use this to solve for C.
>
> $\ln P = -kt + \ln P_0 \Rightarrow P = e^{(-kt + \ln P_0)} = e^{-kt}e^{\ln P_0}$
>
> $\Rightarrow P = P_0 e^{-kt}$ ← Rearrange the equation to be in terms of P, rather than $\ln P$.

c) Given that $k = 0.1$, find the time at which the population of rabbits will have halved, to the nearest day.

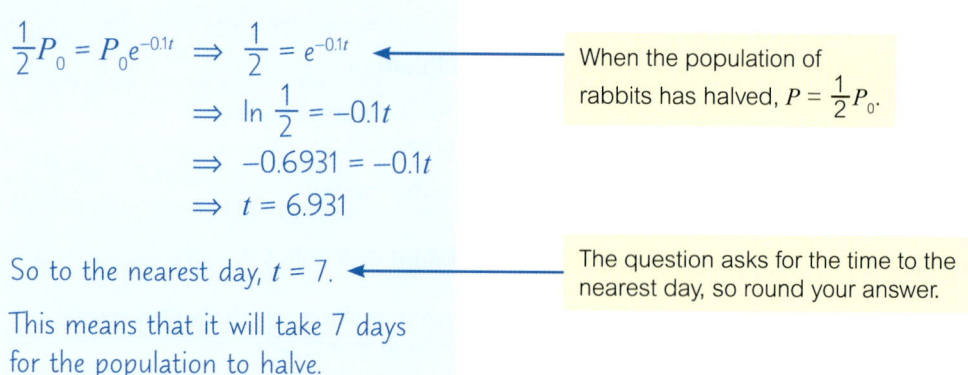

$$\tfrac{1}{2}P_0 = P_0 e^{-0.1t} \implies \tfrac{1}{2} = e^{-0.1t}$$

When the population of rabbits has halved, $P = \tfrac{1}{2}P_0$.

$$\implies \ln \tfrac{1}{2} = -0.1t$$
$$\implies -0.6931 = -0.1t$$
$$\implies t = 6.931$$

So to the nearest day, $t = 7$.

The question asks for the time to the nearest day, so round your answer.

This means that it will take 7 days for the population to halve.

d) Give a limitation of the model and suggest a possible improvement that could be made to address it.

The value of P_0 is not given — the model could be improved by finding a suitable value of P_0.

As t becomes very large, the population becomes increasingly small but never reaches 0 — it may be more realistic to choose a model where the population can reach 0.

Any of these would be suitable answers.

The population of rabbits is a discrete variable while the model is continuous — choosing a function that limits the possible values of P to integers would solve this problem.

The situation being modelled is the approach of winter — the model could give a limit on t to show at which point the model stops being appropriate.

Example 2

Water is leaking from the bottom of a water tank shaped like a vertical cylinder, so that at time t seconds, the depth, D, of water in the tank is decreasing at a rate proportional to $\dfrac{1}{D^2}$.

a) Explain why the depth of water satisfies the differential equation $\dfrac{dD}{dt} = -\dfrac{k}{D^2}$ for some constant $k > 0$.

The rate at which D decreases is inversely proportional to D^2, so:

The rate of change of D is $\dfrac{dD}{dt}$.

$$\dfrac{dD}{dt} \propto \dfrac{1}{D^2} \implies \dfrac{dD}{dt} = -\dfrac{k}{D^2} \text{ for some } k > 0,$$

where the minus sign indicates that the depth of the water is decreasing.

b) Given that D is decreasing at a rate of 2 cm s^{-1} when $D = 40$ cm, find k.

$\dfrac{dD}{dt} = -2$ when $D = 40$

$\dfrac{dD}{dt} = -\dfrac{k}{D^2} \Rightarrow -\dfrac{k}{40^2} = -2 \longleftarrow$ Use the differential equation for D.

$\Rightarrow k = 2 \times 40^2 = 3200$

c) Given that $D = 60$ cm at $t = 0$ s, find a particular solution to the differential equation for D, and hence calculate how long it takes for the tank to empty.

> **Tip** There are a few steps to part c) — first you have to find the general solution, then sub in the values given to find the particular solution, then use this solution to answer the question.

$\dfrac{dD}{dt} = -\dfrac{3200}{D^2} \Rightarrow D^2 \, dD = -3200 \, dt \longleftarrow$

Solve the differential equation, using the value of k from part b), to find the general solution.

$\Rightarrow \displaystyle\int D^2 \, dD = \int -3200 \, dt$

$\Rightarrow \dfrac{1}{3} D^3 = -3200t + C$

$\dfrac{1}{3} 60^3 = -3200(0) + C \Rightarrow C = 72\,000 \longleftarrow$ When $t = 0$, $D = 60$.

So $\dfrac{1}{3} D^3 = 72\,000 - 3200t$

$D = 0 \Rightarrow 72\,000 = 3200t \longleftarrow$ The tank is empty when $D = 0$.

$\Rightarrow t = 22.5$ s

d) Given that the radius of the cylinder is 20 cm, calculate the rate at which the volume of water in the tank is decreasing when $t = 10$ s.

> **Tip** Part d) is a 'connected rates of change' question — see p.236 if you've forgotten how to tackle them.

$V = \pi r^2 h = \pi (20)^2 D = 400\pi D \Rightarrow \dfrac{dV}{dD} = 400\pi \longleftarrow$

Find an expression for the volume of water in the tank and differentiate.

$\dfrac{dV}{dt} = \dfrac{dV}{dD} \times \dfrac{dD}{dt} = 400\pi \times -\dfrac{3200}{D^2} = -\dfrac{1\,280\,000\pi}{D^2} \longleftarrow$

Use the chain rule.

$\dfrac{1}{3} D^3 = 72\,000 - 3200(10) = 40\,000 \longleftarrow$

$\Rightarrow D = \sqrt[3]{3 \times 40\,000} = 49.324\ldots$ cm

Find D when $t = 10$.

$\dfrac{dV}{dt} = \dfrac{-1\,280\,000\pi}{49.324^2} = -1652.87\ldots \text{ cm}^3\text{s}^{-1} \longleftarrow$

Use this value of D to calculate $\dfrac{dV}{dt}$.

So the volume is decreasing at a rate of $1650 \text{ cm}^3\text{s}^{-1}$ (3 s.f.)

Chapter 8

Exercise 8.21

M Q1 A virus spreads so that t hours after infection, the rate of increase of the number of germs (N) in the body of an infected person is directly proportional to the number of germs in the body.

   a) Given that this can be represented by the differential equation $\dfrac{dN}{dt} = kN$, show that the general solution of this equation is $N = Ae^{kt}$, where A and k are positive constants.

   b) Given that a person catching the virus will initially be infected with 200 germs and that this will double to 400 germs in 8 hours, find the number of germs an infected person has after 24 hours.

   c) Give one possible limitation of this model.

M Q2 The rate of depreciation of the value (V) of a car at time t after it is first purchased is directly proportional to V.

   a) If the initial value of the car is V_0, show that $V = V_0 e^{-kt}$, where k is a positive constant.

   b) If the car drops to one half of its initial value in the first year after purchase, how long (to the nearest month) will it take to be worth 5% of its initial value?

M Q3 The population of squirrels is increasing suspiciously quickly.
The rate of increase is directly proportional to the number of squirrels, S.

   a) Formulate a differential equation to model the rate of increase in terms of S, t (time in weeks) and k, a positive constant.

   b) The squirrels need a population of 150 to take over the forest. If, initially, $S = 30$ and $\dfrac{dS}{dt} = 6$, how long (to the nearest week) will it be before they take over?

M Q4 It is thought that the rate of increase of the number of field mice (N) in a given area is directly proportional to N.

   a) Formulate a differential equation for N.

   b) Given that in 4 weeks the number of mice in a particular field has risen from 20 to 30, find the length of time, to the nearest week, before the field is overrun with 1000 mice.

   A biologist believes that the rate of increase of the number field mice is actually directly proportional to the square root of N when natural factors such as predators and disease are taken into account.

   c) Repeat parts a) and b) using this new model.

   d) Suggest another refinement that could be made to improve this model.

P **Q5** A cube has side length x. At time t seconds, the side length is increasing
M at a rate of $\dfrac{1}{x^2(t+1)}$ cms^{-1}.

   a) Show that the volume (V) is increasing at a rate which satisfies the differential equation $\dfrac{dV}{dt} = \dfrac{3}{t+1}$.

   b) Given that the volume of the cube is initially 15 cm³, find the length of time, to 3 s.f., for it to reach a volume of 18 cm³.

E P M Q6 A hot baked potato is placed on a plate. The potato initially has a temperature of 90 °C, and 5 minutes later its temperature is 75 °C. It cools such that the rate of change of temperature is proportional to the difference between the temperature (θ) of the potato and the temperature of the room. The temperature of the room is a constant 20 °C.

a) Find an equation for the temperature, θ °C, in terms of time, t minutes. *[6 marks]*

b) If the temperature of the room had been 30 °C instead and the constant of proportionality and the initial temperature of the potato had been the same, how long would it have taken for the temperature to drop to 75 °C? Give your answer to the nearest whole minute. *[3 marks]*

P M Q7 A local activist is trying to get lots of signatures on his petition, and has just launched a new online campaign. The rate of increase of the number of signatures (y) he's gathered can be represented by the differential equation $\dfrac{dy}{dt} = k(p - y)$, where p is the population of his town and t is the time in days since the new campaign was launched.

a) Find the general solution of this equation.

b) Given that the population of his town is 30 000, he initially has 10 000 signatures on his petition and it takes 5 days for him to reach 12 000 signatures, how long, to the nearest day, will it take for him to reach 25 000 signatures?

c) Draw a graph to show this particular solution.

d) The activist wants 28 000 signatures within 92 days of launching his new campaign. According to the model, will he achieve this?

e) What is a likely limitation of this model and how could it be addressed?

Challenge

E P M Q8 A spherical bath bomb dissolves in water such that its shape always remains spherical. The rate of change of its volume (V) is calculated to be proportional to the square of the radius (r) remaining.

a) Find a differential equation for $\dfrac{dV}{dt}$ in terms of V only. *[3 marks]*

b) The original volume of the bath bomb is 8π cm³. After 200 seconds, its volume is 3.375π cm³. How many seconds would the bath bomb take to dissolve to nothing? *[6 marks]*

P M Q9 A giant balloon is being inflated. Over a period of 2 minutes, air enters the balloon at a rate of $100\ln(t + 100)$ cm³s⁻¹, where t is the time in seconds. The balloon is modelled as a sphere with radius r cm and initially it contains no air.

a) Use the chain rule to find an expression for $\dfrac{dr}{dt}$ in terms of r and t.

b) Show that $\dfrac{\pi r^3}{3} = 25(t + 100)\ln(t + 100) - 25t + C$.

c) Hence find the radius of the balloon after 2 minutes to 3 s.f.

8   Review Exercise

Q1   a) Find $\int \dfrac{1}{\sqrt[3]{(2-11x)}}\,dx$

     b) Show that the area under the curve $y = \dfrac{1}{\sqrt[3]{(2-11x)}}$ from $x = -\dfrac{123}{11}$ to $x = -\dfrac{62}{11}$ is $\dfrac{27}{22}$.

Q2   Find the equation of the curve with derivative $\dfrac{dy}{dx} = (1-7x)^{\frac{1}{2}}$ that goes through the point $(0, 1)$.

Q3   Find the following integrals, giving your answers in terms of e or ln.

     a) $\int 4e^{2x}\,dx$ b) $\int e^{3x-5}\,dx$ c) $\int \dfrac{2}{3x}\,dx$ d) $\int \dfrac{2}{2x+1}\,dx$

Q4   If $\int \dfrac{8}{2-x} - \dfrac{8}{x}\,dx = \ln P + C$, where P is an expression in terms of x and C is a constant, find P.

E **Q5**   a) Determine whether $y = e^x + \dfrac{5}{x}$ is always positive, always negative,
P
       or a mixture of both over the range $2 \leq x \leq 5$.   *[1 mark]*

     b) Find the area bounded by the x-axis, the curve $y = e^x + \dfrac{5}{x}$ and the lines with equations
       $x = 2$ and $x = 5$, giving your answer as an exact value.   *[3 marks]*

E **Q6**   Given that $\int_0^A e^{5x}\,dx = \dfrac{31}{5}$, find the exact value of A.   *[4 marks]*
P

P **Q7**   Find the following integrals (A and B are constants):

     a) $\int \cos(x+A)\,dx$ b) $\int \operatorname{cosec}^2((A+B)t + A + B)\,dt$

E **Q8**   a) Find $\int_{-\pi}^{\pi} \cos 2x\,dx$.   *[3 marks]*

     b) Explain what this result suggests about the area between
       the graph $y = \cos 2x$ and the x-axis between $x = -\pi$ and $x = \pi$.   *[1 mark]*

E **Q9**   Find $\int_{\frac{\pi}{3}}^{\frac{\pi}{2}} \sin(3x - \pi)\,dx$.   *[3 marks]*

Q10   Find the following integrals:

     a) $\int \cos 4x - \sec^2 7x\,dx$ b) $\int 6\sec 3x \tan 3x - \operatorname{cosec}^2 \dfrac{x}{5}\,dx$

E **Q11**   Find $\int \dfrac{4x}{2x^2+1}\,dx$.   *[1 mark]*

E **Q12**   Find $\int \dfrac{x+3}{3x^2+18x-7}\,dx$.   *[2 marks]*

Q13   Find the following integrals:

     a) $\int \dfrac{\cos x}{\sin x}\,dx$ b) $\int \dfrac{20x^4 + 12x^2 - 12}{x^5 + x^3 - 3x}\,dx$

E **Q14** Find the exact value of $\int_2^7 \dfrac{8}{4x-3}\, dx$. *[4 marks]*

Q15 Find the following integrals:

a) $\int 3x^2 e^{x^3}\, dx$ b) $\int 2x \cos(x^2) e^{\sin(x^2)}\, dx$ c) $\int \sec 4x \tan 4x\, e^{\sec 4x}\, dx$

E **Q16** Find $\int 4x \operatorname{cosec}^2(x^2) e^{\cot(x^2)}\, dx$. *[4 marks]*

P **Q17** Use an appropriate trig identity to find $\int \dfrac{2\tan 3x}{1 - \tan^2 3x}\, dx$.

Q18 Find the following integrals:

a) $\int 2\sin^2 x\, dx$ b) $\int \sin 2x \cos 2x\, dx$ c) $\int \tan^2 x + 1\, dx$

E **P** **Q19** Find the integral, with respect to x, of $\dfrac{2\tan 2x}{2 - \sec^2 2x}$. *[3 marks]*

E **Q20** Evaluate $\int_0^{\frac{\pi}{4}} 3\sin 2x \cos 2x\, dx$. *[4 marks]*

Q21 Find the exact shaded area in each of these graphs:

a)

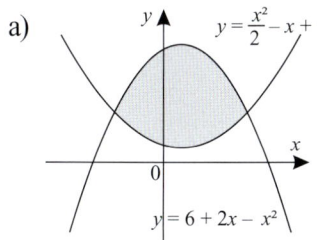

b)

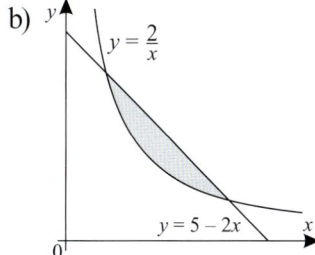

Q22 Find the areas enclosed by the following:

a) The curve $y = \arctan x$, the y-axis and the lines $y = \dfrac{\pi}{6}$ and $y = \dfrac{\pi}{4}$.

b) The curve $y = \dfrac{1}{x^2}$, the y-axis and the lines $y = 4$ and $y = 16$.

E **P** **Q23** a) Find $\int \sqrt{3-x}\, dx$. *[2 marks]*

b) The diagram shows the graph of $y = \sqrt{3-x}$.

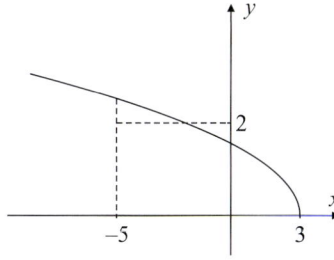

Find the area bounded by the curve, the x-axis, and the lines with equations $y = 2$ and $x = -5$. *[3 marks]*

Q24 The curve on the right has the parametric equations $x = t^2 + 3$, $y = 4t - 1$.

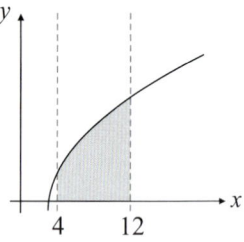

a) Given that $t > 0$, find the values of t when $x = 4$ and $x = 12$.

b) Hence find the shaded area.

E **Q25** The curve below has parametric equations $x = 2t^3$ and $y = \frac{2}{t}$, $t \neq 0$.

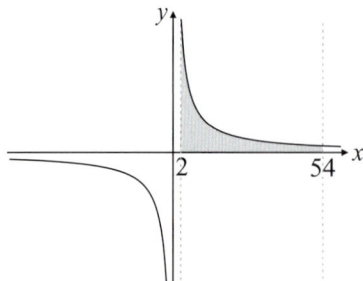

Find the shaded area between the curve, the lines $x = 2$ and $x = 54$ and the x-axis.   *[6 marks]*

Q26 Find the following integrals, using the given substitution in each case.
Where appropriate, give your answers as exact values.

a) $\int 16x(5 - x^2)^5 \, dx$, using $u = 5 - x^2$

b) $\int e^x (e^x + 1)(e^x - 1)^2 \, dx$, using $u = e^x - 1$

c) $\int_2^4 x(x^2 - 4)^3 \, dx$, using $u = x^2 - 4$

d) $\int_3^{11} \frac{2x}{\sqrt{3x - 8}} \, dx$, using $u = \sqrt{3x - 8}$

E **P** **Q27** Use an appropriate substitution to find $\int x(x + 1)^3 \, dx$.   *[3 marks]*

P **Q28** Tucker performs the following integration, using the substitution $u = e^{2x} + 1$:

$$\int \frac{e^{2x}}{e^{2x} + 1} \, dx = \int \frac{u - 1}{u} \, du$$
$$= \int \frac{u}{u} - \frac{1}{u} \, du$$
$$= u \ln|u| - \ln|u| + C = (u - 1)\ln|u| + C$$
$$= e^{2x} \ln|e^{2x} + 1| + C$$

Identify two errors in his working.

Q29 Find $\int_0^{\frac{\pi}{2}} \frac{1}{4} \cos x \sin 2x \, dx$, using the substitution $u = \cos x$.

E **Q30** Use the substitution $x = 3 \tan u$ to find $\int \frac{1}{9 + x^2} \, dx$.   *[6 marks]*

E **P** **Q31** a) $f'(x) = x\sqrt{1 - x^2}$. Use the substitution $x = \cos u$ to find $f(x)$ in terms of both u and the constant of integration, C.   *[5 marks]*

b) Hence find, in terms of C, the maximum value of $f(x)$ and give, in general form, any values of u which would give this maximum.   *[3 marks]*

c) Find the value(s) of x which give the maximum value of $f(x)$.   *[1 mark]*

Q32 Use integration by parts to solve:

a) $\int 3x^2 \ln x \, dx$

b) $\int 4x \cos 4x \, dx$

c) $\int_0^4 e^{\frac{x}{2}} x^2 \, dx$

E **Q33** a) Find $\int x \sin x \, dx$. *[4 marks]*
P

b) Hence find the area between the curve $y = x \sin x$,
the x-axis and the line $x = \frac{\pi}{2}$. *[2 marks]*

c) Using your answer to part b), state the area between
the curve $y = 6x \sin x$, the x-axis and the line $x = \frac{\pi}{2}$. *[1 mark]*

d) State why $\int_{-2\pi}^{\pi} x \sin x \, dx$ is not equal to the area between the graph
and the x-axis bounded by the lines $x = -2\pi$ and $x = \pi$. *[1 mark]*

E **Q34** The diagram shows the graph of $y = f(x)$,
P where $f(x) = (3 - x)\sqrt{x + 1}$.

Find the area between the x-axis
and the part of the graph where $y \geq 0$. *[7 marks]*

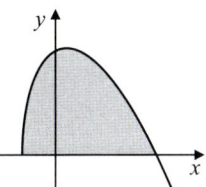

Q35 Use integration by parts twice to find $\int 10x^2 e^{5x} \, dx$.

E **Q36** Evaluate $\int_0^1 2x^2 e^{3x} \, dx$, giving your answer as an exact value in terms of e. *[7 marks]*

E **Q37** The diagram shows the graph of $y = x^2 \cos 4x \, dx$.

a) Find the coordinates of the first point of intersection
between the graph and the x-axis for $x > 0$. *[1 mark]*

b) Hence find the exact area shaded on the diagram. *[7 marks]*

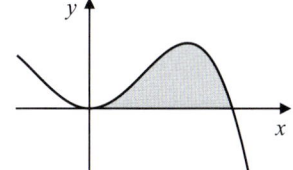

Q38 Given that $\dfrac{3x + 10}{(2x + 3)(x - 4)} \equiv \dfrac{A}{2x + 3} + \dfrac{B}{x - 4}$, find $\int \dfrac{3x + 10}{(2x + 3)(x - 4)} \, dx$.

E **Q39** Find $\int \dfrac{5x + 7}{x^2 + 2x - 3} \, dx$. *[6 marks]*

Q40 Given that $f(x) = \dfrac{13x - 18}{(x - 3)^2 (2x + 1)} \equiv \dfrac{A}{(x - 3)^2} + \dfrac{B}{(x - 3)} + \dfrac{C}{(2x + 1)}$, find $\int_4^9 f(x) \, dx$.

E **Q41** a) Express $\dfrac{10x + 23}{(x + 2)^2 (x + 5)}$ as the sum of partial fractions. *[4 marks]*
P

b) Find $\int \dfrac{10x + 23}{(x + 2)^2 (x + 5)} \, dx$. *[3 marks]*

The diagram shows the graph of $y = \dfrac{10x + 23}{(x + 2)^2 (x + 5)}$.

c) Find the area bounded by the curve,
the x-axis, and the line $x = -4$, shown
shaded in the diagram, to 3 s.f. *[3 marks]*

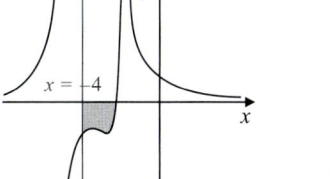

E **Q42** Sally eats sweets on a journey. The mass of sweets she eats per minute (m grams)
M is inversely proportional to the speed (s ms^{-1}) at which she is travelling.
Use a differential equation to show this information. *[1 mark]*

Q43 Find the general solution to the differential equation $\dfrac{dy}{dx} = \dfrac{1}{y}\cos x$.
Give your answer in the form $y^2 = f(x)$.

Q44 Given that $x = 0$ when $t = 1$ and $x = -3$ when $t = 0$, find the particular solution
of the differential equation $\dfrac{dx}{dt} = kte^t$, where k is a constant.

P
M
Q45 The rate of decrease of temperature ($T\,°C$) of a cup of tea with time (t minutes)
satisfies the differential equation $\dfrac{dT}{dt} = -k(T - 21)$, where k is a positive constant.
a) Given that the initial temperature of the tea is 90 °C,
and it cools to 80 °C in 5 minutes, find a particular solution for T.
b) Use this solution to find: (i) the temperature of the tea after 15 minutes,
(ii) the time it takes to drop to 40 °C.
c) Sketch the graph of T against t.

E
P
M
Q46 A population of birds is infected by a disease. The rate of change of the bird population
as a result of the disease is thought to be proportional to the number of birds, P.
Initially, there are 75 birds, and after three weeks, the population falls to 58 birds.
a) Show that the number of birds after t weeks can be modelled by the equation $P = Ae^{-kt}$,
where A and k are constants to be found. *[5 marks]*
b) Find the time it takes, to the nearest week, for the population to fall below 30 birds.
[3 marks]

E
P
M
Q47 From 4 pm each day, the number of people (n) at a theme park is said to decrease at
a rate which is proportional to the number of people in the theme park at that instant.
a) Form a differential equation to model this situation. *[1 mark]*
There are 10 000 people in the park at 4 pm, and after a given period of time,
there are 9000 people.
b) Determine how many people there will be in the park,
according to the model, after this time period has elapsed again. *[6 marks]*

Challenge

E
P
Q48 The graph below shows a shaded region R bounded by the line $y = \dfrac{3}{2}x - 1$,
the curve $y = \dfrac{5x}{1 + x^2}$ and the y-axis.

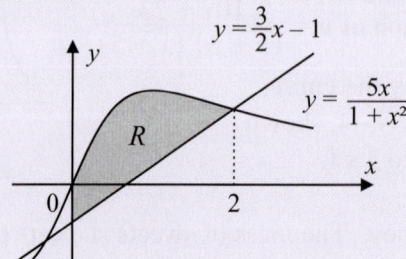

Calculate the exact area of the region R. *[7 marks]*

E **P** **Q49** The graph shows the curve $y = xe^{0.5x^2+x} + e^{0.5x^2+x}$.

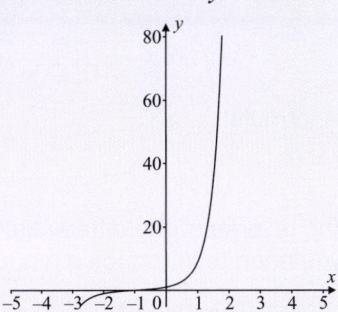

a) Given that $\int_0^a xe^{0.5x^2+x} + e^{0.5x^2+x}\, dx = 2$, where $a > 0$, show that $a^2 + 2a - \ln 9 = 0$ *[5 marks]*

b) Find the value of a, giving your answer correct to 2 decimal places. *[2 marks]*

P **Q50** The graph of the function f(x) is shown in the diagram.
Given that $f'(x) = x^2e^{0.5x}$:

a) Find the equation of f(x).

b) Find $\int f(x)\,dx$.

c) Hence evaluate $\int_0^2 f(x)\,dx$.

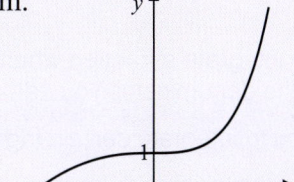

E **P** **Q51** $f'(x) = 60(x^2 + 9)(x + 2)^3$.
Given that f(0) = 2000, use integration by parts twice to find f(1). *[7 marks]*

E **P** **Q52** The diagram shows the graph of $y = \dfrac{19 - 13x}{(x + 2)(x - 1)^2}$
and the line $y = 8x + 16$.

a) Express $\dfrac{19 - 13x}{(x + 2)(x - 1)^2}$ as the sum
of partial fractions. *[4 marks]*

b) Hence find the area between the curve
and the line, shown shaded in the diagram. *[7 marks]*

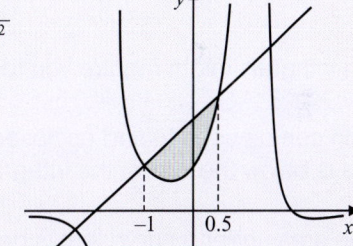

E **P** **M** **Q53** £S represents the total sales made by a company over a time period of t,
measured in years from the beginning of 2010. The sales are thought
to grow at a rate which is proportional to the cube root of the sales to date.

a) Given that the growth is measured as £60 000 per month when the
value of S is £1 000 000, form a differential equation relating S and t. *[2 marks]*

b) Given that at the start of 2010, the value of the sales to date was £64 000,
show that $S = (1600 + 400t)^{\frac{3}{2}}$. *[3 marks]*

c) Jess calculates that the sales by the beginning of 2011 were $(1600 + 400t)^{\frac{3}{2}} \approx$ £89 000,
an increase of £25 000 since the year before, to the nearest £1000.
But when she substitutes $S =$ £64 000 into her equation for $\dfrac{dS}{dt}$, this gives her
a growth rate of £24 000 per year. Why is there a difference in these rates? *[1 mark]*

8 | Chapter Summary

1 You can integrate linear transformations of x^n using this formula:

$$\int (ax+b)^n \, dx = \frac{1}{a(n+1)}(ax+b)^{n+1} + C \text{ for } n \neq -1 \text{ and } a \neq 0.$$

2 Exponentials and logarithms can be integrated using the reverse of the differentiations you saw in Chapter 7, though when you integrate a logarithm, you need to introduce a modulus:

$$\int e^{ax+b} \, dx = \frac{1}{a} e^{ax+b} + C \qquad\qquad \int \frac{1}{ax+b} \, dx = \frac{1}{a} \ln|ax+b| + C$$

3 You can also integrate trig functions using the differentiation results from Chapter 7 in reverse:

$$\int \sin x \, dx = -\cos x + C \qquad \int \cos x \, dx = \sin x + C \qquad \int \sec^2 x \, dx = \tan x + C$$

$$\int \operatorname{cosec} x \cot x \, dx = -\operatorname{cosec} x + C \qquad \int \sec x \tan x \, dx = \sec x + C \qquad \int \operatorname{cosec}^2 x \, dx = -\cot x + C$$

4 If you want to integrate a fraction where the numerator is the derivative of the denominator, you can use the following formula: $\int \frac{f'(x)}{f(x)} \, dx = \ln|f(x)| + C$

This allows you to integrate certain trig functions, such as tan, cosec, sec and cot.

5 If you want to integrate an expression that contains a function of a function, you may be able to use the reverse of the chain rule by writing it in this form:

$\int \frac{du}{dx} f'(u) \, dx = f(u) + C$, where u is a function of x.

A similar result can be used when you have a function raised to a power, multiplied by its derivative: $\int (n+1) f'(x)[f(x)]^n \, dx = [f(x)]^{n+1} + C$

6 Some trig integrals might require you to use any of the identities from Chapter 3.

7 Integration can be used to find enclosed areas, by adding or subtracting one integral from another. If an area is below the x-axis, the integral will come out negative.

8 To find the area under a curve that is defined parametrically with $x(t)$ and $y(t)$, you can use this formula: $\int y \, dx = \int y \frac{dx}{dt} \, dt$

9 Integration by substitution can be used to integrate a function of a function, by substituting u for a function of x and following the method to replace dx with du. If it's a definite integral, you'll also need to change the x-limits into u-limits.

10 Integration by parts is used to integrate a product of two functions.

The formula is: $\int u \frac{dv}{dx} \, dx = uv - \int v \frac{du}{dx} \, dx$

Some integrals may require you to use integration by parts more than once.

11 Some algebraic fractions can be integrated by splitting them into partial fractions.

12 A differential equation is one that includes a derivative term in it.
You can solve them by separating the variables and integrating both sides.
A general solution to a differential equation is one that has the constant of integration, C.
A particular solution can be found by using some given conditions to find the value of C.

Numerical Methods

9

Prior Knowledge Check

1 Sketch the graph of $y = 3x^2 - 5x + 2$, labelling all intersections with the coordinate axes.
see Year 1 Pure Maths

2 Use calculus to find the area between the graph of $y = \frac{1}{4}x^2 + 3x - 1$ and the x-axis between $x = 2$ and $x = 4$.
see Year 1 Pure Maths

3 Differentiate with respect to x:
a) $y = 3x^4 - 2x^2$
b) $y = 2e^{2x-1}$
c) $y = \ln(5x + 4)$
d) $y = \cos(-3x)$
see pages 212-221

9.1   Locating Roots by Changes of Sign

'Solving' or 'finding the roots of' an equation (where $f(x) = 0$) is the same as finding the values of x where the graph crosses the x-axis.

- The graph of the function shows you **how many** roots there are and roughly **where** they are. E.g. the function $f(x) = 3x^2 - x^3 - 2$ (below) has 3 roots, since it crosses the x-axis three times (i.e. there are 3 solutions to the equation $3x^2 - x^3 - 2 = 0$). From the graph you can see there's a root at $x = 1$ and two others near $x = -1$ and $x = 3$.

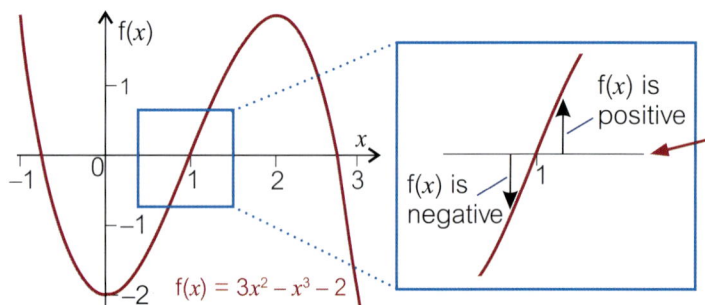

For each root in the graph, $f(x)$ goes from positive to negative or vice versa — **$f(x)$ changes sign as it passes through a root**. So to find if there's a root between two values 'a' and 'b', work out $f(a)$ and $f(b)$. If the signs are different, there's a root somewhere between them.

- Be careful though — this only applies when the **range of x** you're interested in is **continuous** (no break or jump in the line of the graph). Some graphs, like $\tan x$, have an **asymptote** where the line jumps from positive to negative without actually crossing the x-axis, so you might think there's a root when there isn't.

- This method might **fail** to find a particular root. If the function **touches** the x-axis but doesn't cross it, then there won't be a change of sign. An accurate sketch of the graph will help to avoid this.

- Often you'll be given an **approximation** to a root and be asked to show that it's correct to a certain accuracy. To do this, choose the right **upper and lower bounds** and work out if there's a sign change between them.

> The **lower bound** is the **lowest** value a number could have and still be **rounded up** to the correct answer. The **upper bound** is the **upper limit** of the values which will be **rounded down** to the correct answer.

Example   1

Show that one root of the equation $x^3 - x^2 - 9 = 0$ is $x = 2.472$, correct to 3 d.p.

If $x = 2.472$ is a root to 3 d.p., the exact root must lie between the upper and lower bounds of this value — 2.4715 and 2.4725.

Any value in this interval would be rounded to 2.472 to 3 d.p.

| 2.471 | 2.4715 | 2.472 | 2.4725 | 2.473 |

continued on the next page...

$f(2.4715) = 2.4715^3 − 2.4715^2 − 9 = −0.0116...$

$f(2.4725) = 2.4725^3 − 2.4725^2 − 9 = 0.0017...$

> The function $f(x) = x^3 − x^2 − 9$ is continuous, so a root lies in the interval $2.4715 \le x < 2.4725$ if $f(2.4715)$ and $f(2.4725)$ have different signs.

There is a sign change between the upper and lower bounds (and the function is continuous in this interval), so a root must lie between them. Hence a root to 3 d.p. is $x = 2.472$.

> **Tip** The function is continuous because it's just a cubic curve — it has no breaks or jumps.

> Since any value between 2.4715 and 2.4725 would be rounded to 2.472 to 3 d.p. this answer must be correct.

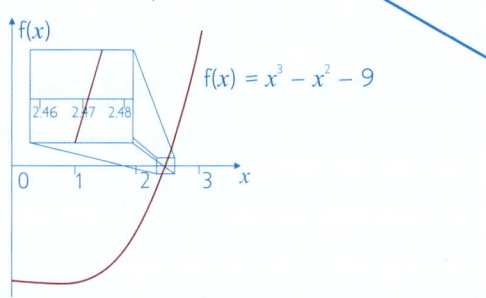

$f(x) = x^3 − x^2 − 9$

Exercise 9.1

Q1 $f(x) = x^3 − 5x + 1$. Show that there is a root of $f(x) = 0$ in the interval $2 < x < 3$.

Q2 $f(x) = \sin 2x − x$ (x is in radians).
Show that there is a root of $f(x) = 0$ between $x = 0.9$ and $x = 1.0$.

Q3 $f(x) = 2x^3 + 5x^2 − 26x + 21$.
Show that the equation $f(x) = 0$ has a root in the interval $[−5.4, −5.3]$.

> **Tip** In interval notation, straight brackets $[a, b]$ means $a \le x \le b$, and curved brackets (a, b) means $a < x < b$.

Q4 $f(x) = x^3 + \ln x − 2$, $x > 0$. Show that there is a root of $f(x) = 0$ in the interval $[1.2, 1.3]$.

Q5 Show that there is a root in the interval:
a) $3 < x < 4$ for $\sin (2x) = 0$ (x in radians) b) $2.1 < x < 2.2$ for $\ln (x − 2) + 2 = 0$
c) $4.3 < x < 4.5$ for $x^3 − 4x^2 = 7$ d) $0 < x < 0.5$ for $e^{2x} + 2e^x = 4$

P Q6 Explain why change of sign methods cannot be used to identify a solution of the equation $f(x) = 0$, where $f(x) = x^2 + 4x + 4$.

M Q7 A bird is observed diving into the sea. Its height above the water after x seconds is modelled by the equation $f(x) = 2x^2 − 8x + 7$, where $f(x)$ is the height in metres. Show that the bird hits the water in the interval $1.2 < x < 1.3$ seconds.

Q8 Show that there are 2 solutions, α and β, to the equation $3x − x^4 + 3 = 0$, such that $1.6 < \alpha < 1.7$ and $−1 < \beta < 0$.

Q9 Show that there are 2 solutions, α and β, to the equation $e^{x-2} - \sqrt{x} = 0$, such that $0.01 < \alpha < 0.02$ and $2.4 < \beta < 2.5$.

Q10 Show that $x = 2.8$ is a solution to the equation $x^3 - 7x - 2 = 0$ to 1 d.p.

Q11 Show that $x = 0.4$ is a solution to the equation $3x^4 + 2x - 1 = 0$ to 1 d.p.

Q12 Show that $x = 0.7$ is a solution to the equation $2x - \dfrac{1}{x} = 0$ to 1 d.p.

E Q13 The equation $2x^4 - 5x^3 - 1 = 0$ has two solutions.
   a) Show that a solution exists in the interval $-1 < x < 0$. *[2 marks]*
   b) By choosing a suitable interval, show that $x = 2.531$ to 3 d.p is a solution. *[3 marks]*

E P Q14 The graph of $f(x) = 5xe^x + x^2 - 6$ crosses the negative x-axis a single point, $(m, 0)$.
   a) Show that $-3 < m < -2$. *[2 marks]*
   b) Show that $m = -2.635$ to 3 d.p. *[3 marks]*

Q15 $f(x) = e^x - x^3 - 5x$. Verify that a root of the equation $f(x) = 0$ is $x = 0.25$ correct to 2 d.p.

P Q16 Show that a solution to the equation $4x - 2x^3 = 15$ lies between -2.3 and -2.2.

P Q17 Show that a solution to the equation $e^x = \dfrac{4}{3x}$ lies between $x = 0.67$ and $x = 0.68$.

P Q18 Show that a solution to the equation $\ln(x + 3) = 5x$ lies between 0.23 and 0.24.

P Q19 Show that a solution to the equation $e^{3x}\sin x = 5$ lies between $x = 0$ and $x = 1$ (x is in radians).

9.2 Sketching Graphs to Find Approximate Roots

Sometimes it's easier to find the number of roots and roughly where they are with a **sketch**.

- In some questions you might be asked to **sketch** the graphs of **two equations** on the same set of axes. Sketching graphs was covered in Year 1 if you need a reminder.

- At the points where they **cross** each other, the two equations are equal. So for $y = x + 3$ and $y = x^2$, at the points of intersection you know that $x + 3 = x^2$, which you can rearrange to get $x^2 - x - 3 = 0$.

- The **number of roots** of this 'combined' equation is the same as the number of **points of intersection** of the original two graphs. The sketch you made will also show roughly **where** the roots are (it's the same x-value for both), so locating them is a bit easier.

> **Tip** Setting the equations equal to each other and then rearranging them to get $f(x) = 0$ gets you to where you were in the previous section.

Example 1

a) On the same set of axes, sketch the graphs $y = \ln x$ and $y = (x - 3)^2$.

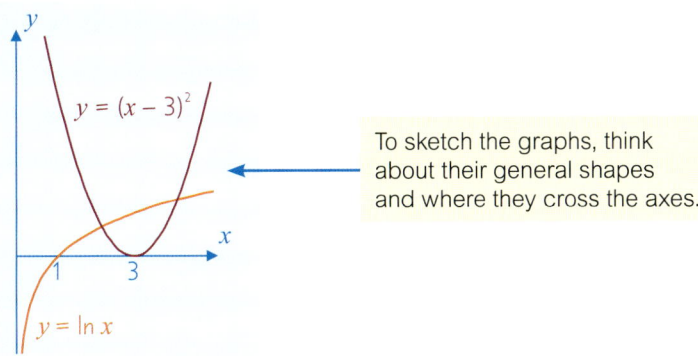

To sketch the graphs, think about their general shapes and where they cross the axes.

Tip Don't worry about trying to make your sketch perfect — the important things are that the graphs are the correct shape and that they cross the axes in the right places.

b) Hence work out the number of roots of the equation $\ln x - (x - 3)^2 = 0$.

The graphs meet where $\ln x = (x - 3)^2 \Rightarrow \ln x - (x - 3)^2 = 0$

So the roots of the equation are where the graphs meet.
The graphs cross twice, so the equation has two roots.

c) Show that there is a root between 2 and 3, and find this root to 1 decimal place.

$\ln 2 - (2 - 3)^2 = -0.306...$
$\ln 3 - (3 - 3)^2 = 1.098...$

A sign change shows there's a root in the interval. Put $x = 2$ and $x = 3$ into the equation and check.

The sign has changed, so there is a root between 2 and 3.

Now use the trial and improvement method to find the root. Test the sign of a point inside the interval to see on which side the root lies. Repeat this until you know the root to the level of accuracy required.

$\ln 2.2 - (2.2 - 3)^2 = 0.148...$
$\Rightarrow$ root is between 2 and 2.2.

On the sketch, the root looks much closer to 2 than 3, so check the sign at $x = 2.2$.

$\ln 2.1 - (2.1 - 3)^2 = -0.068...$
$\Rightarrow$ root is between 2.1 and 2.2.

Check again at $x = 2.1$, as it's halfway between $x = 2$ and $x = 2.2$ and will tell you which is closer to the root.

$\ln 2.15 - (2.15 - 3)^2 = 0.0429...$
$\Rightarrow$ root is between 2.1 and 2.15.

You need the root to 1 d.p., so check if it rounds up to 2.2 or down to 2.1 — try 2.15 as it's the upper bound for 2.1 and lower bound for 2.2.

The answer rounds down to 2.1, so the value of the root is 2.1 to 1 d.p.

Exercise 9.2

Sketch all graphs in this exercise for $-5 < x < 5$ unless otherwise stated.

Q1   a) On the same axes, sketch the graphs of $y = \frac{1}{x}$ and $y = x - 2$.

  b) Using your graphs from part a) write down the number of solutions of the equation $\frac{1}{x} = x - 2$ in this interval.

  c) Show that one solution of the equation $\frac{1}{x} = x - 2$ lies in the interval $2.4 < x < 2.5$.

Q2   a) On the same axes, sketch the graphs of $y = 2x^3 - 7x$ and $y = x^2$.

        b) Using your graphs from part a) write down the number of solutions
           of the equation $2x^3 - x^2 - 7x = 0$ in this interval.

        c) Show that the equation $2x^3 - x^2 - 7x = 0$ has a root between $x = -2$ and $x = -1$.

Q3   a) Sketch the graphs of $y = 2^x - 3$ and $y = \ln x$ on the same axes.

        b) $f(x) = \ln x - 2^x + 3$. Using your graph write down
           the number of roots of the equation $f(x) = 0$.

        c) Show that the equation $f(x) = 0$ has a root between 1.8 and 2.2
           and find this root to 1 decimal place.

E P Q4   a) For $-\pi < x < \pi$, sketch the graphs of $y = \dfrac{5}{x}$ and $y = \tan \dfrac{x}{2}$ on the same axes.   *[3 marks]*

        b) Hence state the number of roots of the equation $f(x) = 0$
           in the interval $-\pi < x < \pi$, where $f(x) = \dfrac{5}{x} - \tan \dfrac{x}{2}$.   *[1 mark]*

        c) Show that the equation $f(x) = 0$ has a root α between $x = 2.2$ and $x = 2.3$.   *[2 marks]*

        d) Given that $\alpha = 2.284$ to three decimal places, write down
           another root of $f(x) = 0$, correct to three decimal places.   *[1 mark]*

Q5   a) Sketch the graphs of $y = \sqrt{x+1}$ and $y = 2x$ on the same axes.

        b) Write down the number of solutions of the equation $\sqrt{x+1} = 2x$.

        c) Show that the equation $\sqrt{x+1} = 2x$ has a solution in the interval $(0.6, 0.7)$.

        d) By rearranging the equation $\sqrt{x+1} = 2x$, use the quadratic formula
           to find the solution of the equation from part c) to 3 s.f.

E Q6   Two functions, f and g, are defined such that $f(x) = e^{-0.5x}$
       and $g(x) = \ln(x + 6)$, both with domain $x > -6$.

        a) On the same axes, sketch the graphs of $y = f(x)$ and $y = g(x)$,
           indicating the exact coordinates of any intersections with the coordinate axes.   *[4 marks]*

        b) State the number of roots of the equation $f(x) = g(x)$.   *[1 mark]*

        c) By considering a change of sign in a suitable interval, verify that
           -0.96 is a root of the equation $f(x) = g(x)$ to 2 significant figures.   *[2 marks]*

Q7   a) Sketch the graphs of $y = e^{2x}$ and $y = 3 - x^2$ on the same axes.

        b) Using your graphs from part a), explain how you know that
           the equation $e^{2x} + x^2 = 3$ has two solutions.

        c) Show that the negative solution of the equation $e^{2x} + x^2 = 3$ lies
           between $x = -2$ and $x = -1$, and find this solution to 1 d.p.

9.3   Using Iteration Formulas

Some equations are too difficult to solve algebraically, so you need to find **approximations** to the roots to a certain level of accuracy. **Iteration** is a numerical method for **solving equations**, like **trial and improvement**:

- You put an approximate value of a root x into an iteration formula, and it gives you a **slightly more accurate** value.

- You then put the **new value** into the iteration formula, and **keep going** until your answers are the same to the level of accuracy needed.

In exam questions you'll usually know a value of x that a root is close to, and then **iteration** does the rest.

Example 1

The iteration formula $x_{n+1} = \sqrt[3]{x_n + 4}$ can be used to find roots of the equation $x^3 - 4 - x = 0$.

a) Starting with $x_0 = 2$, find the values of x_1, x_2 and x_3. Give your answers to 3 decimal places.

$x_0 = 2$, so

$x_1 = \sqrt[3]{x_0 + 4} = \sqrt[3]{2 + 4} = 1.8171...$ Put x_0 in the formula for x_n to get x_1 — the first iteration.

$x_2 = \sqrt[3]{x_1 + 4} = \sqrt[3]{1.8171... + 4} = 1.7984...$ x_1 now gets put back into the formula to find x_2, and then x_2 is used to find x_3.

$x_3 = \sqrt[3]{x_2 + 4} = \sqrt[3]{1.7984... + 4} = 1.7965...$

So $x_1 = 1.817$, $x_2 = 1.798$, $x_3 = 1.797$ to 3 d.p.

Tip   The notation x_n just means the approximation of x at the nth iteration.

b) Hence, find an approximation for a root of $x^3 - 4 - x = 0$ correct to 2 decimal places.

$x_4 = \sqrt[3]{x_3 + 4} = \sqrt[3]{1.7965... + 4} = 1.7963...$

$x_5 = \sqrt[3]{x_4 + 4} = \sqrt[3]{1.7963... + 4} = 1.7963...$

x_4 and x_5 are the same to at least three decimal places, so $x = 1.80$ is a root to 2 d.p.

Continue iterating until your answers are the same when rounded to 2 d.p. — it's best to keep going until the 3rd decimal place stops changing.

It's reasonable to assume that all further iterations will be the same when rounded to 2 d.p., as the iterations seem to be converging towards a root.

Sometimes an iteration formula will just **not find a root**. In these cases, no matter how close to the root you have x_0, the iteration sequence **diverges** — the numbers get further and further apart. The iteration might also **stop working**, like if you have to take the square root of a negative number.

However, you'll nearly always be given a formula that converges to a certain root, otherwise there's not much point in using it. If your formula diverges when it shouldn't, chances are you went wrong somewhere, so go back and double-check every stage.

This doesn't mean you'll never see a diverging formula in an exam, but if you do it will usually be followed by a question like 'what do you notice about the iterations?'. If the iterations seem to bounce up and down, or do something else unexpected, then it probably diverges.

Example 2

The equation $x^3 - x^2 - 9 = 0$ has a root close to $x = 2.5$.
What is the result of using $x_{n+1} = \sqrt{x_n^3 - 9}$ with $x_0 = 2.5$ to find this root?

$x_1 = \sqrt{2.5^3 - 9} = 2.5739...$ ← Start off with x_1.

$x_2 = 2.8376..., x_3 = 3.7214..., x_4 = 6.5221...$ ← Put the value for x_1 back into the formula to find x_2 and onwards.

The results are getting further and further apart with each iteration. So the sequence diverges.

You can use the ANS button on your calculator to speed things up. Enter the starting value and press '=', then type the iteration formula replacing x_n with '**ANS**' — each time you press enter you'll get another iteration. Doing this not only saves time compared to working out the iterations one by one, it also prevents rounding errors and reduces the chance of typing errors.

Exercise 9.3

Q1 a) Show that the equation $x^3 + 3x^2 - 7 = 0$ has a root in the interval $(1, 2)$.

b) Use the iterative formula $x_{n+1} = \sqrt{\dfrac{7 - x_n^3}{3}}$ with $x_0 = 1$
to find values for x_1, x_2, x_3 and x_4 to 3 d.p.

Q2 An intersection of the curves $y = \ln x$ and $y = x - 2$ is at the point $x = \alpha$, where α is 3.1 to 1 d.p.

a) Starting with $x_0 = 3.1$, use the iterative formula $x_{n+1} = 2 + \ln x_n$ to find the first 6 iterations, giving your answers to 4 decimal places.

b) Hence write down an estimate of the value of α correct to 2 decimal places.

Q3 a) Show that the equation $x^4 - 5x + 3 = 0$ has a root between $x = 1.4$ and $x = 1.5$.

b) Use the iterative formula $x_{n+1} = \sqrt[3]{5 - \dfrac{3}{x_n}}$ and $x_0 = 1.4$ to find iterations x_1 to x_6 to 4 d.p.

c) Hence write down an approximation of the root from part a), correct to 2 decimal places.

Q4 a) Show that the function $f(x) = x^2 - 5x - 2$ has a root that lies between $x = 5$ and $x = 6$.

b) The root in part a) can be estimated using the iterative formula $x_{n+1} = \dfrac{2}{x_n} + 5$.
Using a starting value of $x_0 = 5$, find the values of x_1, x_2, x_3 and x_4, giving your answers to 4 significant figures.

E Q5 The solutions of a quartic equation can be estimated using the iterative formula $x_{n+1} = \sqrt[4]{4(1 + x_n)}$.

a) Explain why the formula will fail if $x_0 < -1$. *[1 mark]*

b) Starting at $x_0 = 1.8$, find the values of x_1, x_2 and x_3 to 3 decimal places. *[2 marks]*

c) Hence, find an approximate solution to the equation correct to 2 d.p. *[2 marks]*

Q6 Use the iterative formula $x_{n+1} = 2 - \ln x_n$ with $x_0 = 1.5$ to find iterations x_1 to x_9 to 4 s.f.
Hence, determine whether the formula can be used to estimate a solution to $x = 2 - \ln x$.

E **Q7** a) Show that the equation $e^x - 10x = 0$ has a root in the interval $(3, 4)$. *[2 marks]*

b) Using the iterative formula $x_{n+1} = \ln(10x_n)$ with $x_0 = 3$, find values for x_1, x_2, x_3 and x_4 to 3 d.p. *[2 marks]*

c) Show that $x = 3.577$ is a root of $e^x - 10x = 0$ correct to 3 d.p. *[3 marks]*

d) Determine whether the iteration formula $x_{n+1} = \dfrac{e^{x_n}}{10}$ with $x_0 = 3$ can be used to estimate a root of $e^x - 10x = 0$, justifying your answer. *[2 marks]*

Challenge

Q8 The iterative formula $x_{n+1} = \dfrac{x_n{}^2 - 3x_n}{2} - 5$ is used to try to find approximations to a root of $f(x) = x^2 - 5x - 10$.

a) Find the values of x_1, x_2, x_3 and x_4, starting with $x_0 = -1$, and describe what is happening to the sequence $x_1, x_2, x_3, x_4...$

b) Using the alternative iterative formula $x_{n+1} = \sqrt{5x_n + 10}$ with $x_0 = 6$, find an approximation for a root to the equation $f(x) = 0$ to 3 significant figures. By choosing a suitable interval, verify your answer is correct to this level of accuracy.

9.4 Finding Iteration Formulas

The **iteration formula** is just a **rearrangement** of the equation, leaving a single 'x' on **one side**.

There are often lots of **different ways** to rearrange the equation, so in the exam you might be asked to '**show that**' it can be rearranged in a **certain way**, rather than starting from scratch.

Sometimes a rearrangement of the equation leads to a **divergent** iteration when you come to work out the steps. This is why you probably **won't** be asked to **both** rearrange **and** use a formula to find a root without **prompting**.

Example 1

a) Show that $x^3 - x^2 - 9 = 0$ can be rearranged into $x = \sqrt{\dfrac{9}{x-1}}$. Use this to make an iteration formula and starting at $x_0 = 2.5$, find the values of x_1 and x_2, giving your answers to 2 decimal places.

$x^3 - x^2 - 9 = 0 \implies x^3 - x^2 = 9 \implies x^2(x - 1) = 9$ ← Move the '9' to the RHS, and then factorise the LHS.

$x^2 = \dfrac{9}{x-1}$ ← Get the x^2 on its own by dividing by $(x - 1)$.

$x = \sqrt{\dfrac{9}{x-1}}$ as required ← Take the (positive) square root to get the required formula.

$x_1 = \sqrt{\dfrac{9}{2.5 - 1}} = 2.449... = 2.45$ (2 d.p.) ←

$x_2 = \sqrt{\dfrac{9}{2.449... - 1}} = 2.491... = 2.49$ (2 d.p.) ← Use the iteration formula $x_{n+1} = \sqrt{\dfrac{9}{x_n - 1}}$ to find x_1 and x_2. Remember to use the exact value of x_n for each iteration.

E Exam Style   **P** Problem Solving   **M** Modelling

b) Show that $x^3 - x^2 - 9 = 0$ can also be rearranged into $x = \sqrt{x^3 - 9}$ and use this to make an iteration formula.

$x^2 = x^3 - 9$ ← Start by isolating the x^2 term.

$x = \sqrt{x^3 - 9}$ as required ← Take the positive square root.

$x_{n+1} = \sqrt{x_n^3 - 9}$ ← Replace x with x_{n+1} and x_n to make the iteration formula.

Exercise 9.4

Q1 Show that $x^4 + 7x - 3 = 0$ can be written in the form:

a) $x = \sqrt[4]{3 - 7x}$

b) $x = \dfrac{3 - 5x - x^4}{2}$

c) $x = \dfrac{\sqrt{3 - 7x}}{x}$

Q1 Hint Think about which parts you need to get on their own before starting to rearrange the equation. In part b), for example, turn $7x$ into $5x + 2x$ to get where you want.

Q2 a) Show that the equation $x^3 - 2x^2 - 5 = 0$ can be rewritten as $x = 2 + \dfrac{5}{x^2}$.

b) Use the iterative formula $x_{n+1} = 2 + \dfrac{5}{x_n^2}$ with starting value $x_0 = 2$ to find x_5 to 1 d.p.

c) By choosing a suitable interval, verify that the value found in part b) is a root of the equation $x^3 - 2x^2 - 5 = 0$ correct to that level of accuracy.

P Q3 a) Rearrange the equation $x^2 + 3x - 8 = 0$ into the form $x = \dfrac{a}{x} + b$ where a and b are values to be found.

b) Verify that a root of the equation $x^2 + 3x - 8 = 0$ lies in the interval $(-5, -4)$.

c) (i) Use the iterative formula $x_{n+1} = \dfrac{a}{x_n} + b$ with $x_0 = -5$ to find the values for $x_1, x_2, ..., x_6$, giving your answers to 3 d.p.

(ii) Hence write down an estimate of a root of the equation correct to 1 d.p.

E P Q4 The line $y = 6 - 1.25x$ meets the curve $y = \sin 0.5x$ at a single point, $x = \alpha$.

a) Find an iteration formula of the form $x_{n+1} = a - b \sin 0.5x_n$, where a and b are positive constants, that can be used when trying to find an approximate value of α. *[2 marks]*

b) By letting $x_0 = 4.8$, use the iteration formula from part a) to find the values of x_1 and x_2, giving your answers to 3 d.p. *[2 marks]*

c) Find the smallest value of n for which $x_n = 4.088$ to 3 d.p. *[1 mark]*

P Q5 a) Show that the equation $2^{x-1} = 4\sqrt{x}$ can be written as $x = 2^{2x-6}$.

b) Use the iterative formula $x_{n+1} = 2^{2x_n - 6}$ starting with $x_0 = 1$ to find the values of x_1, x_2, x_3 and x_4, giving your answers to 4 d.p.

c) By choosing a suitable interval, show that the value for x_4 is a correct approximation to 4 d.p. for the root of the equation $2^{x-1} = 4\sqrt{x}$.

Problem Solving In Q5, start by rewriting everything as powers of 2 or x if you're struggling. You're going to need the rules for multiplying powers for this question.

Q6 $f(x) = \ln 2x + x^3$

a) Show that $f(x) = 0$ has a solution in the interval $0.4 < x < 0.5$.

b) Show that $f(x) = 0$ can be rewritten in the form $x = \dfrac{e^{-x^3}}{2}$.

c) Using an iterative formula based on part b) and an appropriate value for x_0, find an approximation of the root of the equation $f(x) = 0$ correct to 2 decimal places.

P Q7 $f(x) = x^2 - 9x - 20$

a) Find an iterative formula for $f(x) = 0$ in the form $x_{n+1} = \sqrt{px_n + q}$ where p and q are constants to be found.

b) By using the formula in part a) with $x_0 = 10$, find x_4 as an approximation to a root of the equation $f(x) = 0$, giving your answer to 3 significant figures.

c) Show that an alternative iterative formula is $x_{n+1} = \dfrac{x_n^2 - 4x_n}{5} - 4$.

d) By using the iterative formula in part c) with starting value $x_0 = 1$, find the value of $x_1, x_2, ..., x_8$.

e) Describe the behaviour of this sequence.

E P M Q8 The vertical heights of two drones flying above horizontal ground are modelled by the equations $h = 9 - 0.1t^2$ and $h = 3^{0.2t}$ for $0 < t \le 8$, where h is the vertical height in metres and t is the time in seconds. The time at which the drones are at the same height, T, is the solution to the equation $f(t) = 0$.

a) Find $f(t)$. *[1 mark]*

b) Find an iteration formula for $f(t) = 0$ in the form $t_{n+1} = \sqrt{g(t_n)}$, where $g(t_n)$ is a function to be found. *[2 marks]*

c) Use this formula with $t_0 = 6.7$ to find t_6 as an approximation to T. Give your answer to 2 significant figures. *[2 marks]*

Challenge

E P M Q9 The value, V (in pounds), of a share in a company is given by the function $V(t) = 4te^{-0.25t} + 0.01t^2$ for $0 \le t \le 14$, where t is the number of days after the company was launched.

a) Find, to the nearest penny, the value of a share two weeks after the company was launched. *[2 marks]*

The value t_{max} for which $V(t)$ is at its maximum satisfies the equation $e^{-0.25t}(4 - t) + 0.02t = 0$.

b) Show that $t = t_{max}$ satisfies the equation $t = \dfrac{4}{1 - 0.02e^{0.25t}}$. *[2 marks]*

c) Using the iteration formula $t_{n+1} = \dfrac{4}{1 - 0.02e^{0.25t_n}}$ with $t_0 = 5$, find an approximation for t_{max} correct to 3 significant figures. *[3 marks]*

d) Estimate the percentage loss in the value of a share from its maximum value to its value two weeks after the company's launch. Give your answer to 2 s.f. *[3 marks]*

9.5   Cobweb and Staircase Diagrams

The instructions below show you how to sketch an iteration diagram.

- First, sketch the graphs of **y = x** and **y = f(x)** (where f(x) is the iterative formula). The point where the two graphs **meet** is the **root** you're aiming for.

- Draw a **vertical line** from the x-value of your starting point (x_0) until it meets the curve **y = f(x)**.

- Now draw a **horizontal line** from this point to the line **y = x**. At this point, the x-value is x_1, the value of your first iteration. This is one **step**.

- Draw another step — a **vertical line** from this point to the curve, and a **horizontal line** joining it to the line $y = x$. Repeat this for each of your iterations.

- If your steps are getting **closer and closer** to the root, the sequence of iterations is **converging**.

- If the steps are moving **further and further away** from the root, the sequence is **diverging**.

The method above produces two different types of diagrams — **cobweb** diagrams and **staircase** diagrams.

Cobweb diagrams

Cobweb diagrams look like they're spiralling in to the root (or away from it). The example below shows a convergent cobweb diagram.

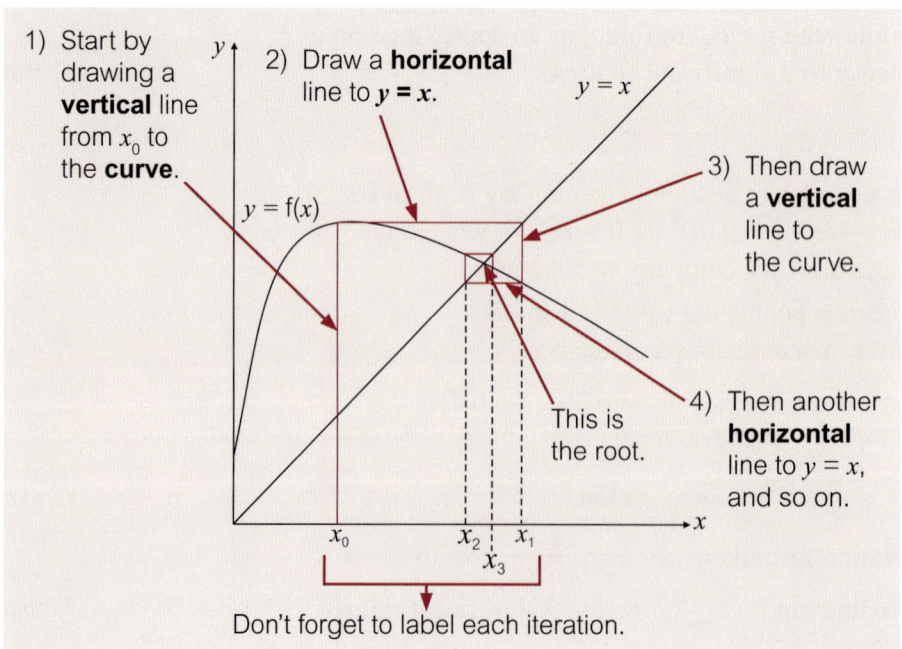

1) Start by drawing a **vertical** line from x_0 to the **curve**.

2) Draw a **horizontal** line to **y = x**.

3) Then draw a **vertical** line to the curve.

4) Then another **horizontal** line to $y = x$, and so on.

This is the root.

$y = x$

$y = f(x)$

Don't forget to label each iteration.

Tip A divergent cobweb diagram would have a similar shape, but each iteration would spiral away from the root rather than towards it.

Staircase diagrams

Staircase diagrams look like a set of steps leading to (or away from) the root.
The examples below show a convergent and a divergent staircase diagram.

This is an example of a **convergent staircase diagram**:

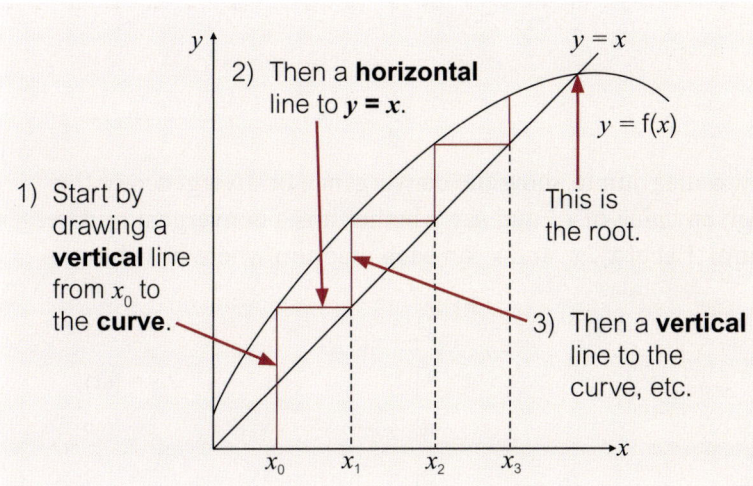

Starting at x_0, the next iterations x_1, x_2 and x_3 are getting **closer** to the point where the two graphs intersect (the root).

This is an example of a **divergent staircase diagram**:

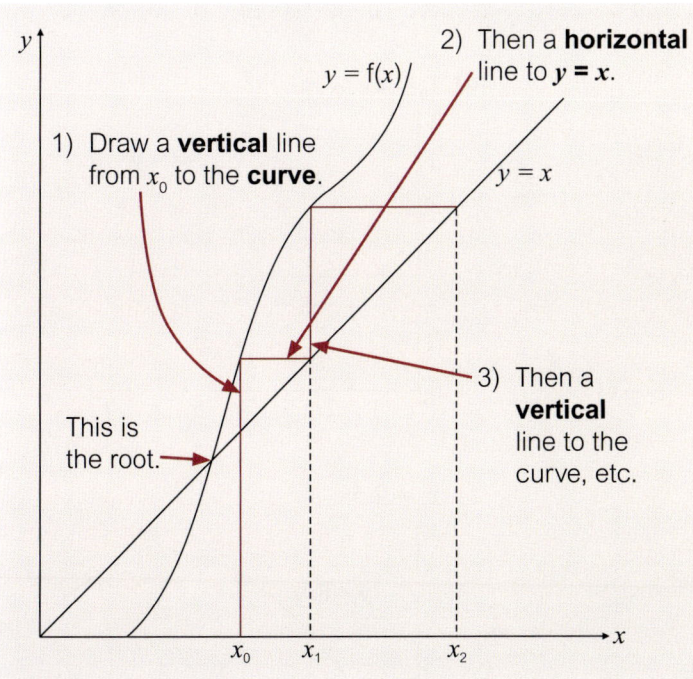

Starting at x_0, the next iterations x_1 and x_2 are getting **further away** from the root.

In general, these diagrams will only converge if your starting value of x_0 is **close enough** to the root, and if the graph of f(x) **isn't too steep**. For a root, a, of a function f(x), the iterations **will converge** if the gradient of f(x) at a is **between −1 and 1** (i.e. |f'(a)| < 1) for a suitable choice of x_0.

Exercise 9.5

Q1 Using the position of x_0 as given on the graph on the right, draw a staircase or cobweb diagram showing how the sequence converges. Label x_1 and x_2 on the diagram.

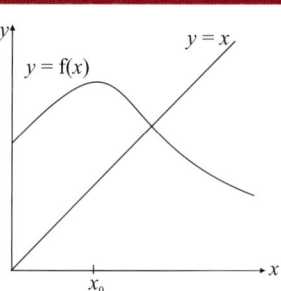

Q2 For each graph below, draw a diagram to show the convergence or divergence of the iterative sequence for the given value of x_0, and say whether it is a convergent or divergent staircase or cobweb diagram. Label x_0, x_1 and x_2 on each diagram where possible.

a) $x_0 = 3.5$

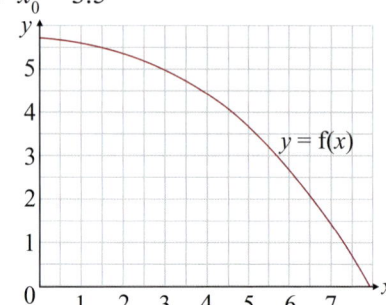

b) $x_0 = 3$

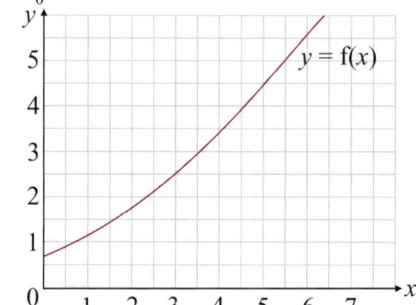

c) $x_0 = 1.75$

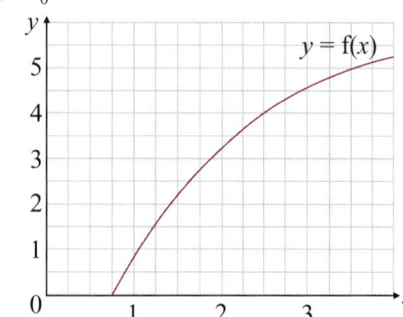

d) $x_0 = 2$

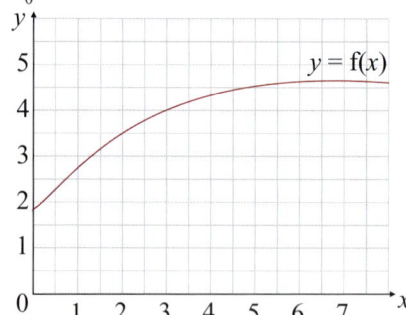

e) $x_0 = 4$

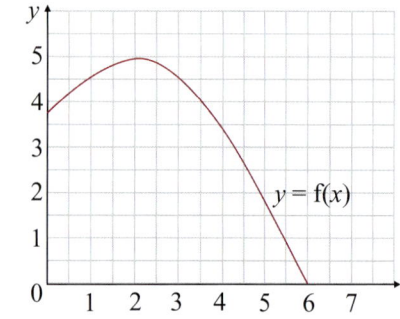

f) $x_0 = 2.5$

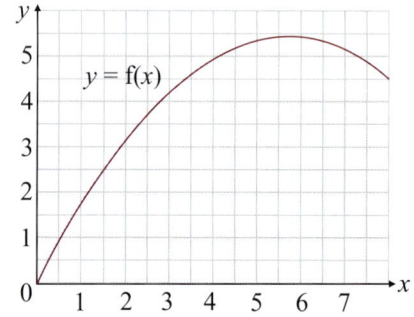

9.6   The Newton-Raphson Method

The **Newton-Raphson method** works by finding the **tangent** to a function at a point x_0, and using its **x-intercept** for the next iteration, x_1. Repeating the process **iteratively** to find x_2, x_3, etc. gets you **closer** to the root, as shown:

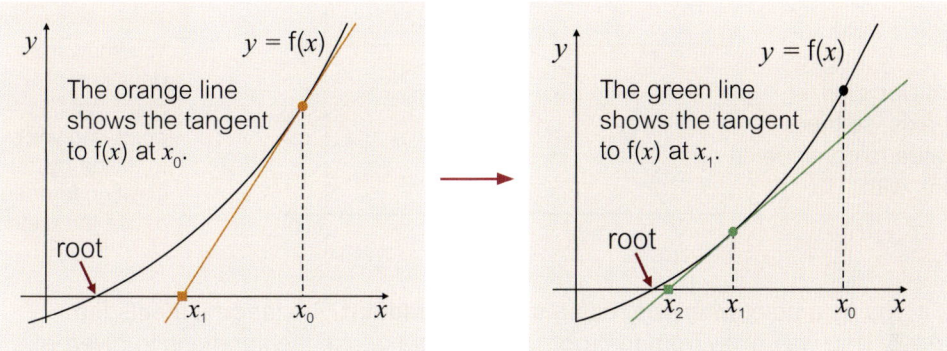

The iteration formula for this method can be derived as follows:

- The tangent to f(x) at the point x_n is a straight line that can be written in the form: $y - y_n = m(x - x_n)$ ← gradient at x_n
$$\Rightarrow y - f(x_n) = f'(x_n)\,(x - x_n)$$

- The next iteration, x_{n+1}, should be at the x-intercept of this tangent line, i.e. $x = x_{n+1}$ when $y = 0$. So substitute $(x_{n+1}, 0)$ into the equation:
$$0 - f(x_n) = f'(x_n)\,(x_{n+1} - x_n) \Rightarrow x_{n+1} - x_n = \frac{-f(x_n)}{f'(x_n)} \Rightarrow \boxed{x_{n+1} = x_n - \frac{f(x_n)}{f'(x_n)}}$$

Example   1

Using the Newton-Raphson method with a starting point of $x_0 = -2$, find x_5 as an approximation for a root of $f(x) = x^3 - 3x + 5$. Give your answer to 5 decimal places.

$f(x) = x^3 - 3x + 5 \Rightarrow f'(x) = 3x^2 - 3$ ←——— Differentiate first.

> **Tip** Make sure you use the ANS button on your calculator when you're doing these iterations, to avoid rounding and typing errors.

$x_{n+1} = x_n - \dfrac{x_n^3 - 3x_n + 5}{3x_n^2 - 3}$ ←——— Substitute f(x_n) and f'(x_n) into the formula for finding x_{n+1}.

$x_1 = -2 - \dfrac{(-2)^3 - 3(-2) + 5}{3(-2)^2 - 3} = -\dfrac{7}{3}$ ←——— Substitute in $x_0 = -2$ to find x_1.

$x_2 = -2.280555...,\quad x_3 = -2.279020...,$
$x_4 = -2.279018...,\quad x_5 = -2.279018...$ ←——— Continue the iterations until you reach x_5.

So a root of f(x) is approximately
$x = -2.27902$ (5 d.p.) ←——— State the approximate root rounded to 5 d.p.

There are times when the Newton-Raphson method can't be used to find a root.

- If the function f(x) cannot be differentiated, then you won't be able to form an iteration formula to use.

- Like other iterative methods, if you choose a start point too far away from the root, the sequence might diverge.

- If the tangent to f(x) is **horizontal** at the point x_n (i.e. if x_n is a **stationary point**), the method will fail. This is shown on the diagram below:

The tangent does not meet the x-axis, so there is no x_{n+1} value to continue the iterations with.

$y = f(x)$

x_n

Problem Solving

You know from the definition of a stationary point that f'(x) = 0 at that point. So you'd have to divide by zero in the iteration formula, which will cause the method to fail.

- Similarly, if you try a point where f(x) has a **shallow gradient**, the tangent meets the x-axis a really long way away from the root, which could cause the iterations to diverge:

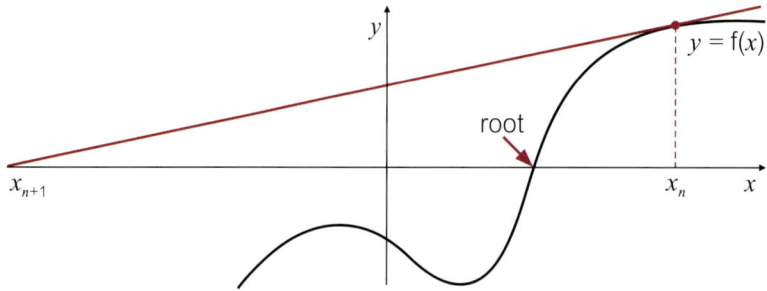

root

x_{n+1}

x_n

$y = f(x)$

Exercise 9.6

Q1 Using the Newton-Raphson method, give an iteration formula for finding the roots of the following:

    a) f(x) = $5x^2 - 6$
    b) g(x) = $e^{3x} - 4x^2 - 1$
    c) h(x) = $\sin x + x^3 - 1$

Q2 Using $x_0 = 2.5$ as a first approximation, apply the Newton-Raphson method to find a second approximation for a root of the equation $x^4 - 2x^3 - 5 = 0$. Give your answer to 3 s.f.

Q3 The function f(x) = $x^2 - 5x - 12$ has a negative root. Using the Newton-Raphson method with $x_0 = -1$, find x_4 as an approximation for this root. Give your answer to 5 s.f.

P Q4 Use the Newton-Raphson method, with the given value of x_0, to find x_4 as an approximate solution to each of the following equations. Give your answers 5 s.f.

    a) $\sin x = x - 1$, using $x_0 = 1$.

    b) $x^2 \ln x = 5$, using $x_0 = 2$.

    c) $e^{-x} = 2 \cos \frac{1}{2}x$, using $x_0 = 1$.

E **P** Q5 The equation $4 \sin 3x + e^{2x} = 0$ has a root α where $-\frac{\pi}{2} < \alpha < -\frac{\pi}{4}$.

Use the Newton-Raphson method starting at $x_0 = -1$ to find:

a) the value of x_1 correct to 5 significant figures,   *[2 marks]*

b) the root α correct to 5 significant figures.   *[2 marks]*

Q6 Show that the Newton-Raphson method will fail to find a root of the equation $2x^3 - 15x^2 + 109 = 0$ if $x_0 = 1$ is used as the starting value.

P Q7 $f(x) = x^3 - 5x^2 - 8x - 3$

a) Use calculus to find the exact coordinates of the maximum turning point of the graph of $y = f(x)$ and to justify that the point is a maximum.

b) With reference to the shape of the graph of $y = f(x)$, explain why there is exactly one real root of the equation $f(x) = 0$.

c) Using $x_0 = 6$ as the first approximation to the root (α), apply the Newton-Raphson method once to obtain a second approximation for α.

d) Show that the Newton-Raphson method will fail to find α if $x_0 = 4$ is taken as the first approximation to α.

e) With reference to the theory of the Newton-Raphson method, explain geometrically why this failure occurs.

Challenge

E **P** **M** Q8 The function $d(t)$ gives the depth relative to sea level of a diver, in metres, as a function of time in minutes. The function is defined such that:

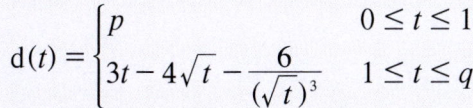

$$d(t) = \begin{cases} p & 0 \le t \le 1 \\ 3t - 4\sqrt{t} - \dfrac{6}{(\sqrt{t})^3} & 1 \le t \le q \end{cases}$$

where p and q are real numbers.

The graph of $y = d(t)$ is shown below.

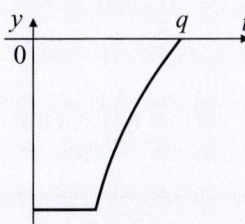

a) Find the value of p.   *[1 mark]*

b) Using $t_0 = 1$ as the starting value, use the Newton-Raphson method once to obtain an approximate value for q.   *[2 marks]*

c) With reference to the graph, explain clearly why the answer to part b) is an underestimate of q.   *[1 mark]*

d) By performing a second iteration of the Newton-Raphson method, find an estimate for the time taken for the diver to ascend to sea level, from the point that he starts ascending. Give your answer in seconds to the nearest second.   *[3 marks]*

9.7   Combining the Iteration Methods

You might see a question that **combines** all (or at least most) of the methods covered so far in this chapter into one long question.

Here you can see an **exam-style question** worked from start to finish, just how they'd want you to do it in the real thing.

Example   **1**

The graph below shows both roots of the continuous function $f(x) = 6x - x^2 + 13$.

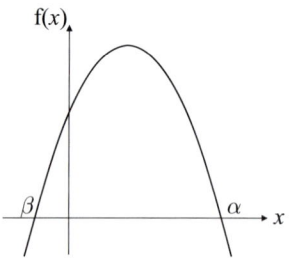

Tip   Remember to show all of your working. That way, even if you get your final answer wrong, you can still get some marks for showing that you understood what the question was asking.

a) Show that the positive root, α, lies in the interval $7 < x < 8$.

$f(7) = (6 \times 7) - 7^2 + 13 = 6$
$f(8) = (6 \times 8) - 8^2 + 13 = -3$

    Check for a sign change between the bounds of the interval.

$f(x)$ is a continuous function and there is a change of sign, so $7 < \alpha < 8$.

b) Show that $6x - x^2 + 13 = 0$ can be rearranged into the formula $x = \sqrt{6x + 13}$.

$6x + 13 = x^2$ Get the x^2 on its own.

$x = \sqrt{6x + 13}$ Now take the (positive) square root.

c) The iteration formula $x_{n+1} = \sqrt{6x_n + 13}$, with $x_0 = 7$, converges to the root α.
Find an approximation for the value of α correct to 1 decimal places.

$x_1 = \sqrt{6 \times 7 + 13} = 7.4161...$ Using $x_{n+1} = \sqrt{6x_n + 13}$ with $x_0 = 7$, find x_1.

$x_2 = \sqrt{6 \times 7.4161... + 13} = 7.5826...$
$x_3 = \sqrt{6 \times 7.5826... + 13} = 7.6482...$ Continue the iterations.
$x_4 = \sqrt{6 \times 7.6482... + 13} = 7.6739...$
$x_5 = \sqrt{6 \times 7.6739... + 13} = 7.6839...$
$x_6 = \sqrt{6 \times 7.6839... + 13} = 7.6879...$
$x_7 = \sqrt{6 \times 7.6879... + 13} = 7.6894...$

Keep going until your answers are the same when rounded to 1 d.p. — it's best to keep going until the 2nd decimal place stops changing.

As x_6 and x_7 are the same to at least 2 d.p., an approximation for the root is $\alpha = 7.7$ to 1 d.p.

d) Sketch a diagram to show the convergence of the sequence for x_1, x_2 and x_3.

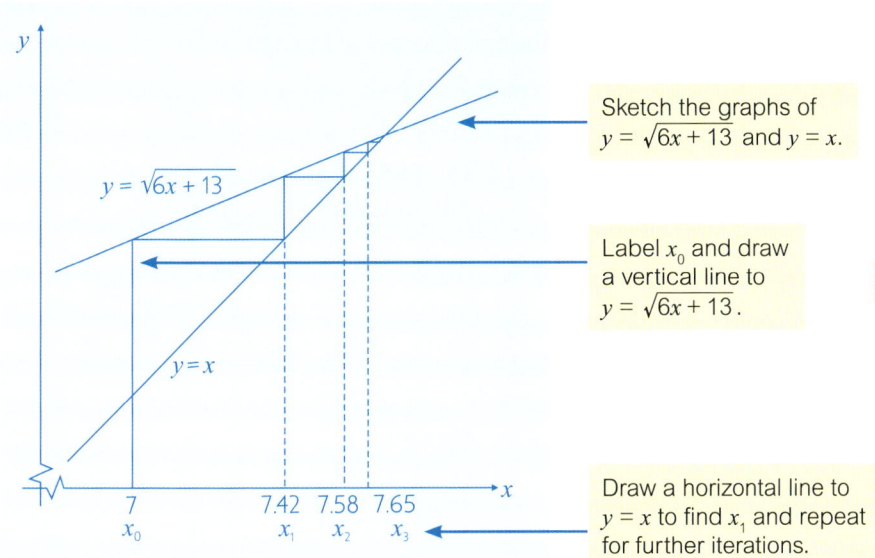

Sketch the graphs of $y = \sqrt{6x + 13}$ and $y = x$.

Tip In an exam question, $y = f(x)$ and $y = x$ would usually be drawn on a graph for you, and the position of x_0 would be marked.

Label x_0 and draw a vertical line to $y = \sqrt{6x + 13}$.

Tip Here the values of x_1, x_2 and x_3 have been rounded to 2 d.p. to make it easier to label them on the graph.

Draw a horizontal line to $y = x$ to find x_1 and repeat for further iterations.

e) Starting with $x_0 = -1$, apply the Newton-Raphson method to find x_4 as an approximation for the negative root, β. Give your answer to 5 significant figures.

$f'(x) = 6 - 2x$ ← Differentiate f(x).

$$x_{n+1} = x_n - \frac{f(x_n)}{f'(x_n)} = x_n - \frac{6x_n - x_n^2 + 13}{6 - 2x_n}$$ ← Put this into the Newton-Raphson formula.

$$x_1 = -1 - \frac{6(-1) - (-1)^2 + 13}{6 - 2(-1)} = -1.75$$ ← Starting with $x_0 = -1$, find x_1.

$x_2 = -1.690789...,\ x_3 = -1.690415...,$
$x_4 = -1.690415...$

← Continue the iterations until you reach x_4.

So β is approximately -1.6904 to 5 s.f.

Exercise 9.7

Q1 Let $f(x) = x^4 + 2x^3 - 4x^2 - 7x + 2$ and $g(x) = x^3 - 4x + 1$.

a) Show that $f(x) = (x + 2)g(x)$. Hence give an integer root of $f(x)$.

b) Show that a root of $g(x)$, α, lies between 1 and 2.

c) Show that $g(x) = 0$ can be rearranged into the formula $x = \sqrt[3]{4x - 1}$.

d) Use the iteration formula $x_{n+1} = \sqrt[3]{4x_n - 1}$ with $x_0 = 2$ to find x_5 as an approximation for α. Give your answer to 4 significant figures.

e) Using the Newton-Raphson method with $x_0 = -2$, show that $\beta = -2.115$ is an approximation for another root of $g(x)$, correct to 4 significant figures.

f) Explain why the Newton-Raphson method for $g(x)$ fails when $x_n = \dfrac{2\sqrt{3}}{3}$.

E **Q2** The positive root of the function $f(x) = x^2 - 7x - 12$
is being found using numerical methods.

a) Show that $f(x) = 0$ can be rearranged into the formula $x = \sqrt{7x + 12}$. *[1 mark]*

b) Use the iteration formula $x_{n+1} = \sqrt{7x_n + 12}$ with $x_0 = 2$
to find x_1 and x_2 to 2 decimal places. *[2 marks]*

c) (i) Complete the table below for $y = \sqrt{7x + 12}$ and plot the graphs of
$y = \sqrt{7x + 12}$ and $y = x$ for $0 \leq x \leq 10$, showing any axes intercepts.

x	0	2	4	6	8	10
y	3.46...	5.10...		7.35...		

> **Tip** Using grid paper will help with the accuracy of your graphs.

(ii) Use your graphs to illustrate the convergence of the
sequence in part b), labelling the x-axis with x_0, x_1 and x_2. *[4 marks]*

d) Show that a second root of $f(x)$ lies between -1 and -2. *[2 marks]*

Challenge

P **Q3** The equation $f(x) = 0$, where $f(x) = 4 - 7x + 10x^2 - x^3 - x^4$, has two real roots.

a) Show that $f(x) = (x + 4)g(x)$ where $g(x) = 1 - 2x + 3x^2 - x^3$.
Hence give an integer root of $f(x)$.

b) Using the Newton-Raphson method with $x_0 = 2$, find x_4 as an approximate
solution to $g(x) = 0$. Give your answer to 4 significant figures.

c) Show that an iteration formula to solve $g(x) = 0$ is $x_{n+1} = \dfrac{1 + 3x_n^2}{2 + x_n^2}$.

d) Using the formula in part c) with $x_0 = 2$, find the lowest value of n for
which x_n matches the answer to part b) to 4 significant figures.

e) Comment on the relative efficiency of the two methods in parts b) and c)
for finding the root of g(x), assuming both converge.

E **P** **M** **Q4** The population of kangaroos, K (in thousands), on an island
is modelled by the function $K(t) = 6 + e^{-t} - 4\sin(0.2t)$ for
$0 \leq t \leq 20$, where t is the number of years after the start of
the year 2000. A sketch of the graph is shown on the right.

a) Find the number of kangaroos predicted
by the model at the start of the year 2000. *[2 marks]*

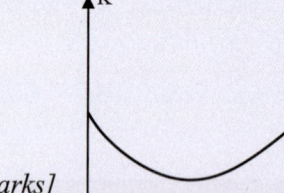

The kangaroo population is at its minimum when $t = t_{min}$.
t_{min} satisfies the equation $t = 5\arccos(-1.25e^{-t})$.

b) Show that t_{min} lies within the interval $(7, 8)$. *[2 marks]*

c) Use the iteration formula $t_{n+1} = 5\arccos(-1.25e^{-t_n})$ starting at $t_0 = 8$
to find t_1 and t_2 to 5 significant figures. *[2 marks]*

d) Hence find, to the nearest whole number, the difference between the
maximum and minimum numbers of kangaroos over the study period. *[2 marks]*

9.8 The Trapezium Rule

When you find yourself with a function which is too difficult to integrate, you can **approximate** the area under the curve using lots of **trapeziums**, which gives an approximate value of the integral.

- The **area** under this curve between a and b can be approximated by the **trapezium** shown.

- It has width $(b - a)$ and parallel sides of length $f(a)$ and $f(b)$.

- The area of the trapezium is an **approximation** of the integral $\int_a^b f(x)\,dx$.

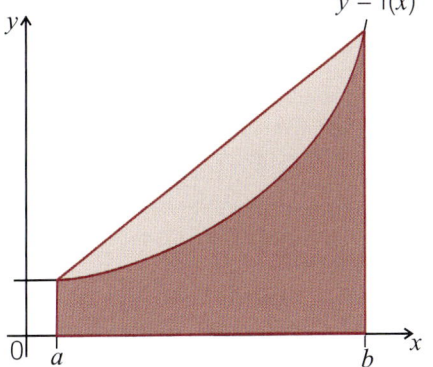

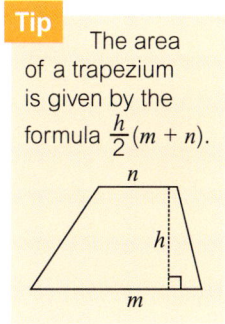

Tip The area of a trapezium is given by the formula $\frac{h}{2}(m + n)$.

It's not a very good approximation, but if you split the area up into **more** trapeziums of equal width, the approximation will get more **accurate** because the **difference** between the trapeziums and the curve will get **smaller**.

The **trapezium rule** for approximating $\int_a^b f(x)\,dx$ works like this:

- n is the **number** of strips i.e. trapeziums.

- h is the **width** of each strip — it's equal to $\dfrac{(b - a)}{n}$.

- The **x-values** go up in steps of h, starting with $x_0 = a$.

- The **y-values** are found by putting the x-values into the equation of the curve — so $y_1 = f(x_1)$. They give the **lengths** of the sides of the trapeziums.

- The **area** of each trapezium is $A = \dfrac{h}{2}(y_r + y_{r+1})$.

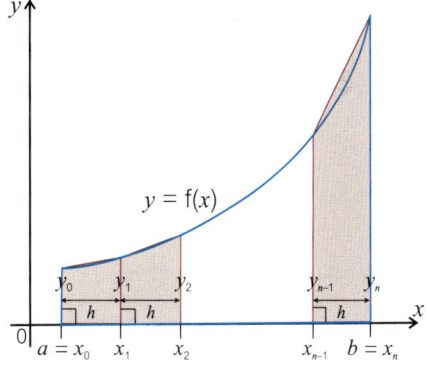

Then an **approximation** for $\int_a^b f(x)\,dx$ is found by **adding** the **areas** of all the trapeziums:

$$\int_a^b f(x)\,dx \approx \frac{h}{2}(y_0 + y_1) + \frac{h}{2}(y_1 + y_2) + \ldots + \frac{h}{2}(y_{n-1} + y_n)$$

$$= \frac{h}{2}\left[y_0 + 2(y_1 + y_2 + \ldots + y_{n-1}) + y_n\right]$$

So the **trapezium rule** says:

$$\int_a^b f(x)\,dx \approx \frac{h}{2}\left[y_0 + 2(y_1 + y_2 + \ldots + y_{n-1}) + y_n\right]$$

Tip The trapezium rule may be given as $\int_a^b y\,dx$, but this is the same since $y = f(x)$.

This may seem like a lot of information, but it's simple if you follow this **step by step** method:

To approximate the integral $\int_a^b f(x)\,dx$:

- Split the interval up into a number of equal sized strips, n. You'll always be told what n is (it could be 4, 5 or even 6).

- Work out the **width** of each strip: $h = \dfrac{(b-a)}{n}$

- Make a **table** of x and y values:

x	$x_0 = a$	$x_1 = a + h$	$x_2 = a + 2h$	...	$x_n = b$
y	$y_0 = f(x_0)$	$y_1 = f(x_1)$	$y_2 = f(x_2)$	...	$y_n = f(x_n)$

- Put all the values into the **trapezium rule**:
$$\int_a^b f(x)\,dx \approx \frac{h}{2}[y_0 + 2(y_1 + y_2 + \dots + y_{n-1}) + y_n]$$

Example 1

Use the trapezium rule with 3 strips to find an approximate value for $\int_0^{1.5} \sqrt{x^2 + 2x}\,dx$.

$n = 3$, $a = 0$ and $b = 1.5$,
so $h = \dfrac{1.5 - 0}{3} = 0.5$ ←——— Work out the width of each strip.

$x_0 = 0$, $x_1 = 0.5$,
$x_2 = 1$ and $x_3 = 1.5$ ←——— Work out the x-values.

x	$y = \sqrt{x^2 + 2x}$
$x_0 = 0$	$y_0 = 0$
$x_1 = 0.5$	$y_1 = \sqrt{1.25} = 1.118\ldots$
$x_2 = 1$	$y_2 = \sqrt{3} = 1.732\ldots$
$x_3 = 1.5$	$y_3 = \sqrt{5.25} = 2.291\ldots$

Calculate the value of y for each of x_0, x_1, x_2, x_3.

$\int_0^{1.5} \sqrt{x^2 + 2x}\,dx$

$\approx \dfrac{h}{2}[y_0 + 2(y_1 + y_2) + y_3]$ ←——— Use the formula to find the approximate value of the integral.

$= \dfrac{0.5}{2}[0 + 2(1.118\ldots + 1.732\ldots) + 2.291\ldots]$

$= \dfrac{1}{4}[7.991\ldots] = 2.00$ (3 s.f.)

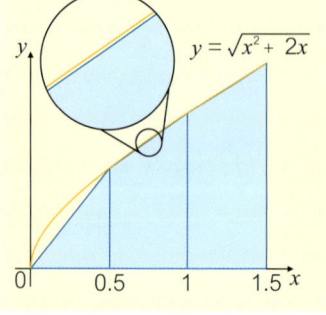

$y = \sqrt{x^2 + 2x}$

Tip There will be always be one more x-value to find (and matching y-value) than the number of strips. E.g., if there are 3 strips, there will be 4 x-values to find.

Tip The graph shows that the estimate here is less than the actual value of the integral — there's a gap between the curve and the top of each strip. Even where it looks like the curve and the top of the trapezium are the same, if you zoom in far enough there will always be a gap.

The **approximation** that the trapezium rule gives will either be an **overestimate** (too big) or an **underestimate** (too small).

This will depend on the **shape** of the graph — a sketch can show whether the tops of the trapeziums lie **above** the curve or stay **below** it.

The estimate is **less** than the real area.

The estimate is **more** than the real area.

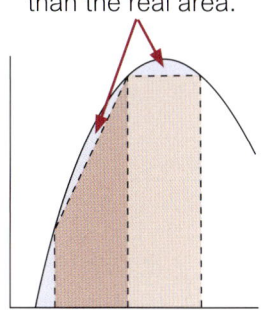

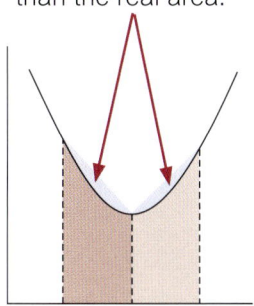

Using **more strips** (i.e. **increasing** n) gives you a **more accurate** approximation.

Example 2

Use the trapezium rule to approximate $\int_0^4 \dfrac{6x^2}{x^3+2}\, dx$ to 3 d.p. using:

a) $n = 2$

$$h = \frac{4-0}{2} = 2$$ ⟵ Work out the width for 2 strips.

$x_0 = 0,\ x_1 = 2$ and $x_2 = 4$ ⟵ Work out the x-values.

x	$y = \dfrac{6x^2}{x^3+2}$
$x_0 = 0$	$y_0 = 0$
$x_1 = 2$	$y_1 = 2.4$
$x_2 = 4$	$y_2 = 1.4545$ (4 d.p.)

⟵ Calculate the corresponding y-values.

$$\int_0^4 \frac{6x^2}{x^3+2}\, dx \approx \frac{h}{2}[y_0 + 2y_1 + y_2]$$
$$= \frac{2}{2}[0 + 2(2.4) + 1.4545]$$ ⟵ Put these values into the formula.
$$= 6.255 \text{ (3 d.p.)}$$

b) $n = 4$

$$h = \frac{4-0}{4} = 1$$ ⟵ Work out the width for 4 strips.

$x_0 = 0,\ x_1 = 1,\ x_2 = 2,\ x_3 = 3$ and $x_4 = 4$ ⟵ Work out the x-values.

continued on the next page...

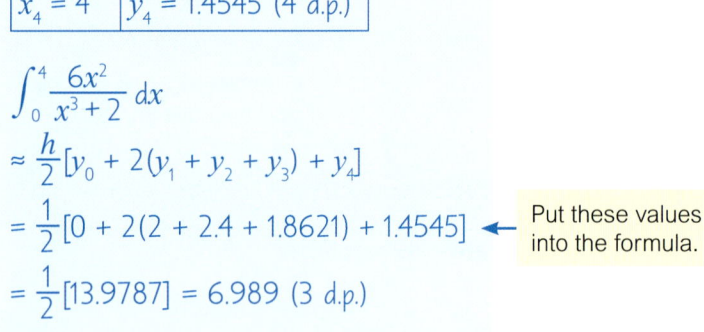

x	$y = \dfrac{6x^2}{x^3 + 2}$
$x_0 = 0$	$y_0 = 0$
$x_1 = 1$	$y_1 = 2$
$x_2 = 2$	$y_2 = 2.4$
$x_3 = 3$	$y_3 = 1.8621$ (4 d.p.)
$x_4 = 4$	$y_4 = 1.4545$ (4 d.p.)

Calculate the corresponding y-values.

$$\int_0^4 \frac{6x^2}{x^3 + 2}\,dx$$

$$\approx \frac{h}{2}[y_0 + 2(y_1 + y_2 + y_3) + y_4]$$

$$= \frac{1}{2}[0 + 2(2 + 2.4 + 1.8621) + 1.4545]$$

Put these values into the formula.

$$= \frac{1}{2}[13.9787] = 6.989 \text{ (3 d.p.)}$$

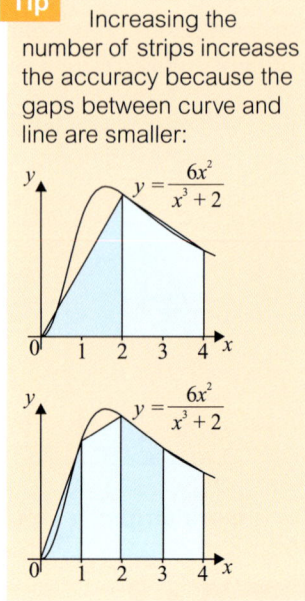

Tip Increasing the number of strips increases the accuracy because the gaps between curve and line are smaller:

$$y = \frac{6x^2}{x^3 + 2}$$

$$y = \frac{6x^2}{x^3 + 2}$$

Upper and lower bounds

You can calculate an **upper** and **lower bound** for the area under a curve using a simplified version of the trapezium rule that uses rectangles:

$$\int_a^b f(x)\,dx \approx h[y_0 + y_1 + y_2 + ... + y_{n-1}]$$

This formula sums the areas of the rectangles which meet f(x) with their **left hand corner**.

In this example, the rectangles are **below** the curve, so they calculate a **lower bound**.

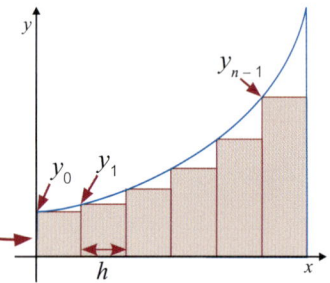

Problem Solving For an increasing function like the one shown here, calculating the bound using the 'left hand corner' will give the lower bound, and using the 'right hand corner' will give the upper bound. But for a decreasing function, it's the other way around. So to calculate a bound of a curve across a turning point (i.e. where the graph switches between increasing and decreasing), do a separate calculation either side of the turning point — one will use the right hand corner, and the other will use the left.

$$\int_a^b f(x)\,dx \approx h[y_1 + y_2 + ... + y_{n-1} + y_n]$$

This formula sums the areas of the rectangles which meet f(x) with their **right hand corner**.

In this example, the rectangles are **above** the curve, so they calculate an **upper bound**.

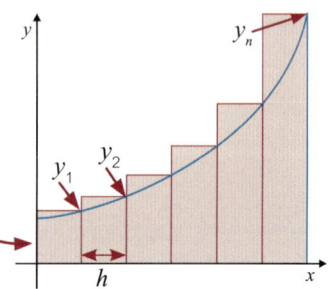

When you approximate the area under a curve using the **trapezium rule**, that value will lie **between** these two bounds.

Integration as the Limit of a Sum

When **differentiating** a function, you're finding the **gradient** of the curve. You saw in Year 1 that you do this by finding the gradient of a straight line over a smaller and smaller interval, until the interval is virtually nothing. So, for a function f(x):

$$f'(x) = \lim_{\delta x \to 0} \left[\frac{f(x + \delta x) - f(x)}{(x + \delta x) - x} \right]$$

Tip On page 219, you saw this with h instead, but it's the same formula. δ is the Greek letter delta, and δx just means "change in x".

Similarly, you can define the **integral** of a curve f(x) between two points a and b using limits:

$$\int_a^b f(x)\, dx = \lim_{\delta x \to 0} \sum_{x=a}^{b} f(x) \times \delta x$$

You saw on the previous page that you can find upper and lower bounds for the area under a curve by adding up the area of rectangles. These are **approximations** of the actual area under the curve.

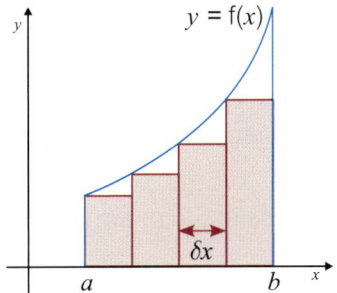

Tip It doesn't matter if you use rectangles above or below the curve — the result will be the same.

- The area of each rectangle is height × width.
  Its height is the y-value, f(x), and its width is just δx.
  So the area of each rectangle is $R = f(x) \times \delta x$.

- Then the **sum** of all the rectangles between a and $b - \delta x$ is:
$$\sum_{x=a}^{b-\delta x} R = \sum_{x=a}^{b-\delta x} f(x)\, \delta x$$

- As δx gets smaller and smaller, this sum of rectangles comes closer and closer to the actual area under the curve. This is how we **define** an integral — as the **limit of a sum** of areas.

- So as the width δx approaches 0, the sum of areas is indistinguishable from the actual area. And as $\delta x \to 0$, $b - \delta x \to b$, so the upper value of the sum can be replaced with b:

$$\lim_{\delta x \to 0} \sum_{x=a}^{b-\delta x} f(x)\, \delta x = \int_a^b f(x)\, dx$$

Problem Solving

You're summing the areas of the rectangles where the left hand corner meets the curve. So, to avoid adding on an extra rectangle beyond the point b, you need to sum only up to the point $b - \delta x$.
If you use rectangles above the curve, then you would sum the 'right hand corner' rectangles from $a + \delta x$ to b.

- This is really just a **change in notation**, so that you don't have to write out the whole limit of a sum each time you want to write an integral.

- When you're writing the **exact** integral, you use **d** instead of δ, and you replace $\sum$ with $\int$, which are both just ways of writing 'S' (for 'sum').

Exercise 9.8

Q1 Use the trapezium rule to find approximations of each of the following integrals. Use the given number of intervals in each case. Give your answers to 3 significant figures.

a) $\int_0^2 \sqrt{x+2} \, dx$, 2 intervals

b) $\int_1^3 2(\ln x)^2 \, dx$, 4 intervals

c) $\int_0^{0.4} e^{x^2} \, dx$, 2 intervals

d) $\int_{-\frac{\pi}{4}}^{\frac{\pi}{4}} 4x \tan x \, dx$, 4 intervals

e) $\int_0^{0.3} \sqrt{e^x + 1} \, dx$, 6 intervals

f) $\int_0^{\pi} \ln(2 + \sin x) \, dx$, 6 intervals

Q2 For Q1 a) and b) above, find an upper and lower bound for the area beneath the curve to 4 s.f.

Q3 Use the trapezium rule with 3 intervals to find an estimate to $\int_0^{\frac{\pi}{2}} \sin^3 \theta \, d\theta$. Give your answer to 3 d.p.

M Q4 The shape of an aeroplane's wing is modelled by the curve $y = \sqrt{\ln x}$, where x and y are measured in metres. Use the trapezium rule with 5 intervals to estimate the area of a cross-section of the wing enclosed by the curve $y = \sqrt{\ln x}$, the x-axis and the lines $x = 2$ and $x = 7$. Give your answer to 3 d.p.

Q5 a) Complete the following table of values to 3 d.p. for $y = e^{\sin x}$.

x	0	$\dfrac{\pi}{8}$	$\dfrac{\pi}{4}$	$\dfrac{3\pi}{8}$	$\dfrac{\pi}{2}$
y	1	1.466			2.718

b) (i) Using the trapezium rule with 2 intervals, estimate $\int_0^{\frac{\pi}{2}} e^{\sin x} \, dx$ to 2 d.p.

   (ii) Repeat the calculation using 4 intervals.

c) Which is the better estimate? Explain your answer.

Q6 The diagram below shows part of the curve $y = \dfrac{3}{\ln x}$.

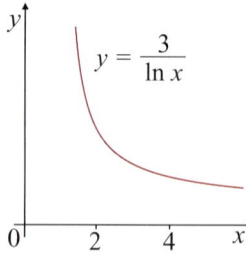

$y = \dfrac{3}{\ln x}$

a) Using the trapezium rule with 4 intervals, find an estimate to 2 d.p. for $\int_2^4 \dfrac{3}{\ln x} \, dx$.

b) Without further calculation, state whether your answer to part a) is an over-estimate or under-estimate of the true area. Explain your answer.

E **P** **Q7** The diagram below shows part of the curve $y = 2^{x-1}$ where $-2 \le x \le 4$.

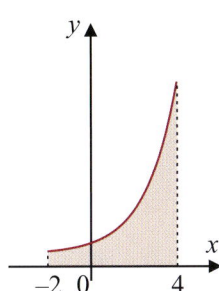

a) Complete the following table of values for $y = 2^{x-1}$.

x	−2	0	2	4
y			2	8

[1 mark]

b) Use the trapezium rule, with all the values of y in the completed table, to find an approximate value for $\int_{-2}^{4} 2^{x-1} \, dx$. *[2 marks]*

c) Without further calculation, explain whether your answer to part b) is an overestimate or underestimate of the true area. *[1 mark]*

d) Explain how the trapezium rule could be used to give a more accurate approximation for the value of the integral in part b). *[1 mark]*

e) Use your answer to part b) to find an approximate value for $\int_{-2}^{4} 10 - 2^{x-1} \, dx$. *[2 marks]*

Challenge

Q8 a) Using the trapezium rule with $h = \dfrac{\pi}{6}$, show that

$\int_{-\frac{\pi}{2}}^{\frac{\pi}{2}} \cos x \, dx$ can be approximated as $\dfrac{\pi(2+\sqrt{3})}{6}$.

b) Without further calculation, state whether this approximation is an under- or over-estimation. Explain your answer.

E **P** **Q9** $I = \displaystyle\int_{0}^{\frac{\pi}{3}} \tan^2 x \, dx$

a) Use the trapezium rule:
(i) with 2 intervals to find an approximate value for I to 4 decimal places. *[3 marks]*
(ii) with 4 intervals to find another approximate value for I to 4 decimal places. *[3 marks]*

b) Show, using integration, that the exact value of I is $\sqrt{3} - \dfrac{\pi}{3}$. *[4 marks]*

c) Hence find the reduction in the percentage error, to 3 significant figures, when four intervals are used in the trapezium rule compared to two intervals. *[2 marks]*

9 Review Exercise

Q1 Alex states that: "If f(a) and f(b) have different signs then there is exactly one root of f(x) = 0 in the interval [a, b]."
a) Give two reasons why Alex's statement may be incorrect.

Bailey states that: "If f(a) and f(b) have the same sign and f(x) is continuous then there are no roots of f(x) = 0 in the interval [a, b]."
b) Explain why Bailey's statement is incorrect using the counterexample f(x) = $x^2 - 6x + 9$.

Q2 By selecting an appropriate interval, show that, to 1 d.p., $x = 1.2$ is a root of the equation $x^3 + x - 3 = 0$.

Q3 a) On the same axes, sketch the graphs of $y = \ln x$ and $y = \dfrac{2}{x}$.
Hence find the number of roots of the equation $\ln x - \dfrac{2}{x} = 0$.
b) Show that there is a root of the equation $\ln x - \dfrac{2}{x} = 0$ between $x = 2$ and $x = 3$.

Q4 a) Sketch the graphs of $y = \dfrac{1}{x+1}$ and $y = x - 2$ on the same axes.
b) Show that f(x) = $\dfrac{1}{x+1} - x + 2$ has a root between $x = -1.4$ and -1.3.
c) Show that the equation f(x) = 0 can be written in the form $x^2 - x - 3 = 0$.

E Q5 a) On the same axes, sketch the graphs of $y = x^2 - 4x + 4$ and $y = |0.5x + 1|$, labelling any points of intersection with the coordinate axes. *[3 marks]*
b) Hence state the number of roots of the equation $x^2 - 4x + 4 - |0.5x + 1| = 0$. *[1 mark]*
c) Verify that a root of this equation lies within the interval [3, 4]. *[2 marks]*

Q6 Solutions to the equation $x^2 - \ln x - 4 = 0$ can be estimated using the formula $x_{n+1} = \sqrt{\ln x_n + 4}$. Starting at $x_0 = 2$, find the values of x_1, x_2, x_3 and x_4 to 3 d.p.

Q7 a) Show that the equation $x^x = 3$ has a root between $x = 1.5$ and $x = 2$.
b) Using the iterative formula $x_{n+1} = 3^{\frac{1}{x_n}}$, with $x_0 = 2$, find x_4 as an approximation for this root. Give your answer to 1 decimal place.

E Q8 The equation $x^2 - 7x - 9 = 0$ has a root in the interval (8, 9).
a) Use the iteration formula $x_{n+1} = \sqrt{7x_n + 9}$ starting at $x_0 = 9$ to find x_{10} as an approximation for the root, giving your answer to 3 decimal places. *[2 marks]*
b) By choosing a suitable interval, verify the accuracy of your answer to part b). *[2 marks]*

An alternative iteration formula is $x_{n+1} = \dfrac{x_n^2 - 9}{7}$.
c) Use this alternative formula starting at $x_0 = -9$ to find values of x_1, x_2 and x_3 to 3 decimal places. *[2 marks]*
d) State, with a reason, whether continued iterations of the alternative formula will find a root. *[1 mark]*

Q9 a) Show that a solution to the equation $2x - 5 \cos x = 0$ lies in the interval $(1.1, 1.2)$.

b) Show that the equation in part a) can be written as $x = p \cos x$, stating the value of p.

c) Using an iterative formula based on part b) and a starting value of $x_0 = 1.1$, find the values up to and including x_8, giving your answers to 4 decimal places. Comment on your findings.

E Q10 The function $f(x) = \cos x + 2x$ has a single root.

a) Show that $f(x) = 0$ can be written as $x = -\dfrac{1}{2} \cos x$.   *[1 mark]*

b) Using the iteration formula $x_{n+1} = -\dfrac{1}{2} \cos x_n$ with $x_0 = -1$, find x_5 as an approximation for the root of $f(x)$. Give your answer to 2 decimal places.   *[2 marks]*

Q11 Each diagram below shows the graph of a function $y = f(x)$ and the graph of $y = x$. A value of x_0 is indicated on the x-axis.

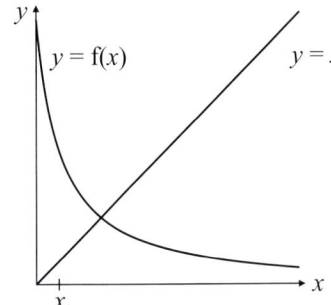
a)

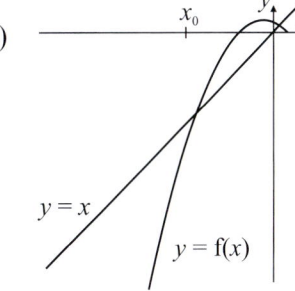
b)

For each diagram, use the position of x_0 given to draw a diagram showing the convergence or divergence of the iterative sequence. Say whether it is a convergent or divergent staircase or cobweb diagram and label x_1 and x_2 on each diagram.

E Q12 The equation $2 \cos x + 1 = x$ has a root α in the interval $[1.3, 1.4]$.

a) On grid paper, draw the graphs of $y = 2 \cos x + 1$ and $y = x$ on the same axes for $-\pi < x < \pi$.   *[1 mark]*

b) On your diagram, show the use of the iteration formula $x_{n+1} = 2 \cos (x_n) + 1$, starting with $x_0 = 1.25$. Label the positions of x_0, x_1, x_2, x_3 and x_4. You do not need to give their values.   *[2 marks]*

Q13 Using the Newton-Raphson method with $x_0 = 1$, find x_6 as an approximation for a root of each of the following functions. Give your answers to 3 decimal places.

a) $f(x) = x^2 + 9x - 4$ b) $f(x) = x^3 + 3x^2 + 5x + 7$

c) $f(x) = e^x - x^4$ d) $f(x) = \sin x \cos x$

E Q14 $f(x) = 5 - 2x - x^2$ has a positive root α and a negative root β.

a) Show that α lies in the interval $1.4 < x < 1.5$.   *[2 marks]*

b) Determine whether the iteration formula $x_{n+1} = \sqrt{5 - 2x_n}$ can be used to estimate α, justifying your answer.   *[2 marks]*

c) Using $x_0 = -4$ as a first approximation to β, apply the Newton-Raphson method three times to obtain a better approximation, giving your answer to 3 s.f.   *[3 marks]*

Q15 Use the trapezium rule with n intervals to estimate the following to 3 s.f.:

a) $\int_0^3 (9 - x^2)^{\frac{1}{2}} \, dx$, $n = 3$ b) $\int_{0.2}^{1.2} x^{x^2} \, dx$, $n = 5$ c) $\int_1^3 2^{x^2} \, dx$, $n = 5$

P Q16 Use the trapezium rule to estimate the value of $\int_0^6 (6x - 12)(x^2 - 4x + 3)^2 \, dx$, first using 4 strips and then again with 6 strips. Calculate the percentage error for each answer.

E Q17
P

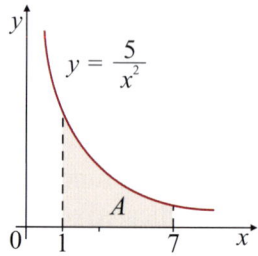

The diagram above shows part of the curve $y = \dfrac{5}{x^2}$.

a) Use the trapezium rule with $n = 4$ to find an estimate for the area A to 2 d.p.   *[3 marks]*

b) Without further calculation, explain whether or not your answer to part a) is an overestimate or underestimate of the true area.   *[1 mark]*

c) Use integration to find the exact value of $\int_1^7 \dfrac{5}{x^2} \, dx$.   *[3 marks]*

d) Hence find the percentage error in your estimate.   *[1 mark]*

Challenge

E Q18 Some experimental data is modelled as a function f(x), where f(x) = $\ln (x + 3) - x + 2$, $x > -3$. f(x) has a single root at $x = m$.

a) Show that m lies between 3 and 4.   *[2 marks]*

b) Find the value of m correct to 2 decimal places using the iteration formula $x_{n+1} = \ln (x_n + 3) + 2$, with $x_0 = 3$.   *[2 marks]*

The diagram below shows the graphs of $y = x$ and $y = \ln (x + 3) + 2$.

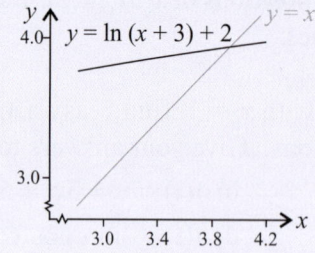

c) Copy and annotate the diagram above to show the convergence of the first two iterations found in part b).   *[2 marks]*

d) Apply the Newton-Raphson method three times, with $x_0 = 3$, to obtain a further approximation to m. Give your answer to 5 decimal places.   *[3 marks]*

e) Explain why the Newton-Raphson method fails with $x_0 = -2$.   *[1 mark]*

E **P** Q19 The diagram below shows the graph of the curve f(x) = $\sqrt{x}$ ln x for 0 < x ≤ 10.
The curve crosses the x-axis at the point (1, 0).

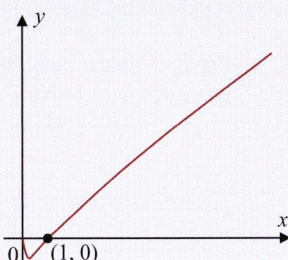

a) Complete the table below, giving the missing values of f(x) to 4 decimal places.

x	2	2.2	2.4	2.6	2.8	3
y	0.9803	1.1695	1.3563			

[1 mark]

b) Use the trapezium rule with all the values of f(x) in the completed table to find
an approximate value for I, where I = $\int_2^3 \sqrt{x} \ln x \, dx$, to 3 decimal places. *[2 marks]*

c) Use integration to show that $\int \sqrt{x} \ln x \, dx = \dfrac{2x^{\frac{3}{2}}}{9}(3 \ln x - 2) + c$. *[5 marks]*

d) Hence find the value of I to 3 decimal places. *[2 marks]*

e) With reference to the graph, explain the relative sizes of your answers to parts b) and d). *[1 mark]*

E **P** **M** Q20 The design for the tail fin of a submarine is shown on the graph below.
The units are in metres.

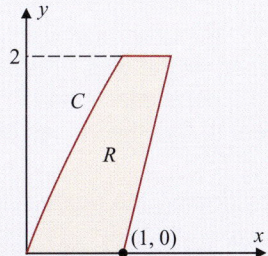

The region R is bounded by the curve C, the x-axis and the lines y = 4x − 4 and y = 2.

C has the equation y = $\dfrac{6x}{\sqrt{x+2}}$ and is defined for 0 ≤ x ≤ 1.

a) Use the trapezium rule with four intervals to find an
approximate value for I = $\int_0^1 \dfrac{6x}{\sqrt{x+2}} dx$, to 3 decimal places. *[3 marks]*

b) Use your answer to part a) to determine an
approximate value for the area of region R. *[2 marks]*

c) Use the substitution u = $\sqrt{x}$ + 2 to find the exact value of I, giving your answer
in the form 12(a + 8 ln b) where a and b are positive rational numbers. *[7 marks]*

d) Hence calculate the error of the answer to part a) compared to the actual value.
Give your answer to the nearest square centimetre. *[2 marks]*

9 Chapter Summary

1   A root of f(x) lies between two values, a and b, if there is a change of sign between f(a) and f(b). This is only true if f(x) is continuous in the interval $a < x < b$. If the function only touches the x-axis there will be no change of sign, so the method fails.

2   To show an approximation of a root is correct to a certain accuracy, substitute the upper and lower bounds of the value into f(x) and work out if there's a sign change between them.

3   Two equations are equal at the points where their graphs intersect. The number of roots of the 'combined' equation (which you get when you set the two equations equal to each other) is the same as the number of points of intersection of the original two graphs.

4   An iteration formula is used to approximate roots to an appropriate level of accuracy when an equation is too difficult to solve algebraically. The formula can be found by rearranging the equation to get a single 'x' on one side.

5   Sometimes iteration formulas will fail to find a root — they may diverge from the root (the numbers get further and further away) or the formula might stop working (e.g. if you have to take the square root of a negative number).

6   Iteration diagrams can be convergent or divergent. Cobweb diagrams spiral towards (or away from) a root, and staircase diagrams look like steps leading towards (or away from) a root.

7   To draw an iteration diagram, first sketch the graphs of $y = $ f(x) (i.e. the iteration formula) and $y = x$ — the point where they intersect is the root. Then draw a vertical line from the start value (x_0) to the curve $y = $ f(x) and a horizontal line to $y = x$. At this point the x-value is the value of the first iteration, x_1. Repeat to find the values of further iterations.

8   The Newton-Raphson method works by finding the tangent to a function at x_0 and using its x-intercept as the next iteration, x_1. It uses the iteration formula $x_{n+1} = x_n - \dfrac{f(x_n)}{f'(x_n)}$.

9   The Newton-Raphson method can't be used to find a root if the function cannot be differentiated or the tangent to $y = $ f(x) is horizontal at x_n (i.e. at a stationary point). If the start point is too far away from the root or the function has a shallow gradient at x_n, the method may also fail as this could cause the iteration sequence to diverge.

10   The trapezium rule is used to approximate the area under a curve when a function is too difficult to integrate. The area under the curve is split into a number of strips (i.e. trapeziums), the areas of which are then calculated and added together.

11   The trapezium rule will give an overestimate or underestimate depending on whether the trapeziums lie above or below the curve. Splitting the area into more strips gives a more accurate approximation.

Vectors

Learning Objectives

Once you've completed this chapter, you should be able to:

- Understand how to represent vectors in three dimensions, using both unit vectors and column vectors.
- Add and subtract 3D vectors and multiply them by scalars.
- Show whether two 3D vectors are parallel, and whether three points are collinear.
- Find the magnitude of any vector in three dimensions.
- Find the unit vector in the direction of any vector in three dimensions.
- Find the distance between any two three-dimensional points using vectors.
- Calculate the angle between any two vectors in three dimensions using trigonometry.

Prior Knowledge Check

1 The vectors **a**, **b** and **c** are defined as $\mathbf{a} = \begin{pmatrix} 2 \\ 4 \end{pmatrix}$, $\mathbf{b} = \begin{pmatrix} 1 \\ -5 \end{pmatrix}$, and $\mathbf{c} = \begin{pmatrix} -3 \\ -1 \end{pmatrix}$.

   Find: a) **a** + **b**　　　　b) 3**c** + 2**a**　　　　c) 2**b** − **c** + 3**a**

   see Year 1 Pure Maths

2 Show that the vectors (8**i** − 12**j**) and (−20**i** + 30**j**) are parallel.

   see Year 1 Pure Maths

3 Determine whether the points A (−8, 0), B (1, 6) and C (6, 9) are collinear.

   see Year 1 Pure Maths

4 Use the cosine rule to find the angle $\theta°$ in the triangle below to 1 d.p.

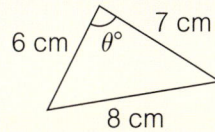

   see Year 1 Pure Maths

5 The vectors **f** and **g** are defined as **f** = (3**i** − 6**j**) and **g** = (−**i** + 4**j**). Find:
   a) the magnitude of **f**
   b) the direction of **g** (to 1 d.p.)
   c) the smallest angle between **f** and **g**

   see Year 1 Pure Maths

10.1 Vectors in Three Dimensions

- Three-dimensional vectors have components in the direction of the x-, y- and z-axes. Imagine that the x- and y-axes lie flat on the page. Then imagine a third axis sticking straight through the page at right angles to it — this is the z-axis.

- The points in three dimensions are given (x, y, z) coordinates.

- The unit vector in the direction of the z-axis is **k**, so three-dimensional vectors can be written like this: ⟶ $x\mathbf{i} + y\mathbf{j} + z\mathbf{k}$ or $\begin{pmatrix} x \\ y \\ z \end{pmatrix}$

- Three-dimensional vectors are used to describe things in three-dimensional space, e.g. an aeroplane moving through the sky.

- Calculating with 3D vectors is just the same as with 2D vectors, as the next example shows.

Example 1

The diagram on the right shows the position of the points P and Q.

a) Write the position vectors $\overrightarrow{OP}$ and $\overrightarrow{OQ}$ as column vectors.

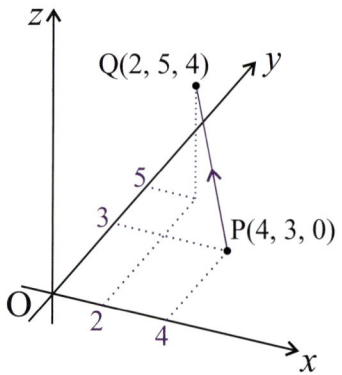

$$\overrightarrow{OP} = 4\mathbf{i} + 3\mathbf{j} + 0\mathbf{k} = \begin{pmatrix} 4 \\ 3 \\ 0 \end{pmatrix} \qquad \overrightarrow{OQ} = 2\mathbf{i} + 5\mathbf{j} + 4\mathbf{k} = \begin{pmatrix} 2 \\ 5 \\ 4 \end{pmatrix}$$

b) Hence find $\overrightarrow{PQ}$ as a column vector.

$$\overrightarrow{PQ} = -\overrightarrow{OP} + \overrightarrow{OQ}$$
$$= -\begin{pmatrix} 4 \\ 3 \\ 0 \end{pmatrix} + \begin{pmatrix} 2 \\ 5 \\ 4 \end{pmatrix} = \begin{pmatrix} 2-4 \\ 5-3 \\ 4-0 \end{pmatrix} = \begin{pmatrix} -2 \\ 2 \\ 4 \end{pmatrix}$$

> Don't forget — you've multiplied $\overrightarrow{OP}$ by the scalar '–1' so all its entries will change signs.

> Add along each row of the column, just like with 2D column vectors.

The point R lies on the line PQ such that $PR : RQ = 3 : 1$.

c) Find the position vector of point R, in unit vector form.

$$\overrightarrow{PR} = \frac{3}{4}\overrightarrow{PQ}$$

> The ratio tells you that point R is $\frac{3}{4}$ of the way along the line PQ from point P.

$$= \frac{3}{4} \times \begin{pmatrix} -2 \\ 2 \\ 4 \end{pmatrix} = \begin{pmatrix} -1.5 \\ 1.5 \\ 3 \end{pmatrix}$$

continued on the next page…

Position vector $\overrightarrow{OR} = \overrightarrow{OP} + \overrightarrow{PR}$:

$$\overrightarrow{OR} = \begin{pmatrix} 4 \\ 3 \\ 0 \end{pmatrix} + \begin{pmatrix} -1.5 \\ 1.5 \\ 3 \end{pmatrix} = \begin{pmatrix} 4-1.5 \\ 3+1.5 \\ 0+3 \end{pmatrix} = \begin{pmatrix} 2.5 \\ 4.5 \\ 3 \end{pmatrix}$$

$$= 2.5\mathbf{i} + 4.5\mathbf{j} + 3\mathbf{k}$$

Again, add along each row of the column.

Tip
You could draw a vector triangle if you needed to.

Don't get confused between a point and its position vector. Points are given by (x, y, z) coordinates and position vectors are the movement needed to get to the point from the origin, given in the form $x\mathbf{i} + y\mathbf{j} + z\mathbf{k}$. You'll need to be able to swap between the forms.

Parallel lines and collinear points

As with two-dimensional vectors, **parallel** vectors in three dimensions are **scalar multiples** of each other.

To determine whether two vectors are parallel, check whether one vector can be produced by multiplying the other by a scalar.

If two vectors are **parallel**, and also have a **point in common**, then they must lie on the same straight line — their points are **collinear**.

Example 2

Show that points A, B and C are collinear, if:

$\overrightarrow{OA} = 2\mathbf{i} - \mathbf{j} + \mathbf{k}$, $\overrightarrow{OB} = \mathbf{i} + 2\mathbf{j} + 3\mathbf{k}$ and $\overrightarrow{OC} = -\mathbf{i} + 8\mathbf{j} + 7\mathbf{k}$.

$\overrightarrow{AB} = -\overrightarrow{OA} + \overrightarrow{OB}$
$= (-2 + 1)\mathbf{i} + (1 + 2)\mathbf{j} + (-1 + 3)\mathbf{k}$
$= -\mathbf{i} + 3\mathbf{j} + 2\mathbf{k}$

$\overrightarrow{BC} = -\overrightarrow{OB} + \overrightarrow{OC}$
$= (-1 - 1)\mathbf{i} + (-2 + 8)\mathbf{j} + (-3 + 7)\mathbf{k}$
$= -2\mathbf{i} + 6\mathbf{j} + 4\mathbf{k}$

Start by using the position vectors given in the question to find vectors $\overrightarrow{AB}$ and $\overrightarrow{BC}$.

$\overrightarrow{BC} = -2\mathbf{i} + 6\mathbf{j} + 4\mathbf{k}$
$= 2(-\mathbf{i} + 3\mathbf{j} + 2\mathbf{k})$
$= 2\overrightarrow{AB}$

$\overrightarrow{BC}$ is a scalar multiple of $\overrightarrow{AB}$, so the vectors are parallel. As point B is part of both vectors, points A, B and C must lie on a straight line.

So A, B and C must all lie on the same line — they are collinear.

Make sure you're comfortable working with vectors in all forms — as column vectors, using $\mathbf{i}$, $\mathbf{j}$, $\mathbf{k}$ notation, and as the movement between points given as Cartesian coordinates.

Exercise 10.1

Q1 R is the point (4, –5, 1) and S is the point (–3, 0, –1).
    Write down the position vectors of R and S, giving your answers:
    a) as column vectors   b) in unit vector form.

Q2 Give $\overrightarrow{GH}$ and $\overrightarrow{HG}$ as column vectors, where $\overrightarrow{OG} = \begin{pmatrix} 2 \\ -3 \\ 4 \end{pmatrix}$ and $\overrightarrow{OH} = \begin{pmatrix} -1 \\ 4 \\ 9 \end{pmatrix}$.

Q3 Triangle JKL has vertices at the points J (4, 0, –3), K (–1, 3, 0) and L (2, 2, 7). Find the vectors $\overrightarrow{JK}$, $\overrightarrow{KL}$ and $\overrightarrow{LJ}$.

> **Q3 Hint** The question doesn't mention **i** and **j** components or column vectors so you can answer it using either.

P Q4 M is a point on the line CD, where C has coordinates (–1, 3, –5),
    M has coordinates (1, 1, –2) and $\overrightarrow{CD} = \begin{pmatrix} 4 \\ -4 \\ 6 \end{pmatrix}$.

    Show that M is the midpoint of CD.

P **M** Q5 A 3D printer is being used to make the plastic toy sketched below.

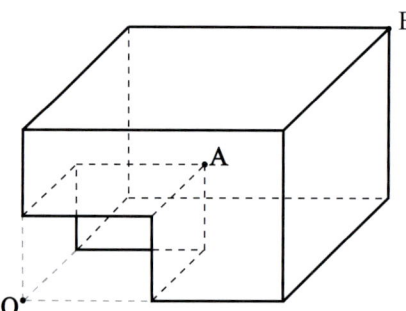

The toy can be modelled as a large cuboid with a smaller cuboid section missing. The large cuboid section is an enlargement of the smaller, 'missing' cuboid, by a scale factor of 2, centred at point O.
    a) Prove that $\overrightarrow{AB} = \overrightarrow{OA}$.
    b) Find the position vector of point B, if $\overrightarrow{OA} = 3\mathbf{i} + \mathbf{j} + 2\mathbf{k}$.

E Q6 Points A and B have coordinates (3, –2, 0) and (4, 1, –1) respectively.
    Show that $\overrightarrow{AB}$ is parallel to the vector $\mathbf{c} = \begin{pmatrix} -4 \\ -12 \\ 4 \end{pmatrix}$.   *[3 marks]*

P Q7 Show that vectors $\mathbf{a} = \frac{3}{4}\mathbf{i} + \frac{1}{3}\mathbf{j} - 2\mathbf{k}$ and $\mathbf{b} = \frac{1}{4}\mathbf{i} + \mathbf{j} - \frac{2}{3}\mathbf{k}$ are <u>not</u> parallel.

E **Q8**   Find the exact values of p and q as surds in their simplest form for which
the vectors $\sqrt{3}\,\mathbf{i} + 18\mathbf{j} + p\mathbf{k}$ and $6\mathbf{i} + q\mathbf{j} + 54\mathbf{k}$ are parallel.   *[4 marks]*

M **Q9**   In a 3D board game, players take turns to position counters at points
inside a cuboid grid. Players score by forming a straight line with
three of their counters, unblocked by their opponent.

Show that counters placed at coordinates (1, 0, 3), (3, 1, 2) and (7, 3, 0) lie on a straight line.

P **Q10**   The points with position vectors $\overrightarrow{OP} = \begin{pmatrix} -2 \\ a \\ -8 \end{pmatrix}$, $\overrightarrow{OQ} = \begin{pmatrix} 1 \\ b \\ -4 \end{pmatrix}$ and $\overrightarrow{OR} = \begin{pmatrix} -5 \\ 6 \\ 3b \end{pmatrix}$ are collinear.

Find the values of a and b.

E **Q11**   The points A, B and C are collinear and have coordinates
P   (3, 4, –1), (2, 1, 2p) and (–1, –8, 19) respectively.
   a) Find the value of p.   *[5 marks]*
   b) State the ratio AB : BC.   *[1 mark]*

E **Q12**
P

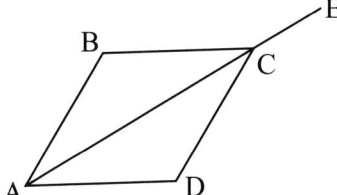

The vertices A, B and C of a rhombus ABCD, shown in the above diagram,
have coordinates (11, 5, –3), (7, –1, 1) and (3, 5, 5) respectively.
The line segment AC is extended to a point E such that CE : AE = 1 : 5.
Find:
   a) the coordinates of vertex D,   *[3 marks]*
   b) the position vector of E.   *[3 marks]*

Challenge

E **Q13**   The base of a cuboid ABCDEFGH is the region bounded by the lines $x = 0$, $x = 2$, $y = 1$
P   and $y = 4$ which all lie within the plane $z = 0$. The point A is the closest vertex to O.
   a) Write down the coordinates of A.   *[1 mark]*

$z \geq 0$ for all vertices of cuboid ABCDEFGH. H is the vertex which is
furthest away from O. P is the point on the line segment AH such that
AP = 3PH. The volume of the cuboid is 30 units³.
   b) Find the position vector of P, giving your answer in the form $\frac{1}{4}\begin{pmatrix} q \\ r \\ s \end{pmatrix}$,
       where q, r and s are integers.   *[5 marks]*

10.2 Calculating with 3D Vectors

Magnitude of a 3D vector

- The **magnitude** (sometimes called **modulus**) of a vector **v** is written as |**v**|, and for a vector $\overrightarrow{AB}$ it is written $|\overrightarrow{AB}|$.

- Magnitude is always a **positive**, **scalar** quantity.

You will have used **Pythagoras' theorem** to find the length of a 2D vector, and it works in **three dimensions** too:

> The distance of point (a, b, c) from the origin is: $\sqrt{a^2 + b^2 + c^2}$

You can **derive** this formula from the **two-dimensional** Pythagoras' theorem:

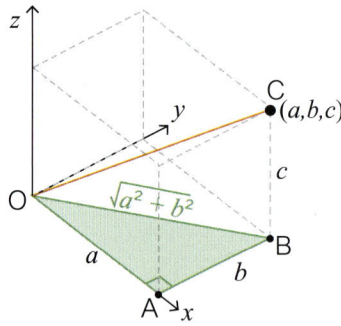

The **orange** line shows the distance of the point (a, b, c) from the origin.

First use Pythagoras on the **green triangle**.

OB is $\sqrt{a^2 + b^2}$.

Tip Remember the position vector of the point (a, b, c) is $a\mathbf{i} + b\mathbf{j} + c\mathbf{k}$, or as a column vector: $\begin{pmatrix} a \\ b \\ c \end{pmatrix}$

Now use Pythagoras on the **blue triangle**.

$OC = \sqrt{\left(\sqrt{a^2 + b^2}\right)^2 + c^2} = \sqrt{a^2 + b^2 + c^2}$

So the distance from the origin to (a, b, c) is $\sqrt{a^2 + b^2 + c^2}$.

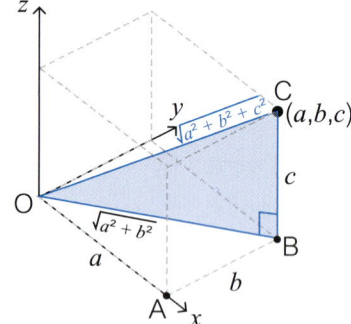

So the **magnitude** of any vector $\mathbf{v} = a\mathbf{i} + b\mathbf{j} + c\mathbf{k}$ is the same as the distance of point (a, b, c) from the origin:

$$|\mathbf{v}| = |a\mathbf{i} + b\mathbf{j} + c\mathbf{k}| = \sqrt{a^2 + b^2 + c^2}$$

Problem Solving A unit vector has a magnitude of 1. Magnitude is always a positive scalar, so the unit vector is always parallel to the vector, and has the same direction.

The **unit vector** in the direction of **v** is $\frac{\mathbf{v}}{|\mathbf{v}|}$, as it is for 2D vectors.

Example 1

The diagram on the right shows the position of point Q.

a) Find $|\overrightarrow{OQ}|$, giving your answer in reduced surd form.

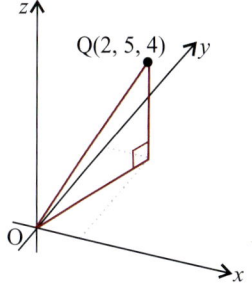
z
$Q(2, 5, 4)$ y
O
x

$\overrightarrow{OQ} = 2\mathbf{i} + 5\mathbf{j} + 4\mathbf{k}$

$|\overrightarrow{OQ}| = \sqrt{x^2 + y^2 + z^2}$

$= \sqrt{2^2 + 5^2 + 4^2}$ ← Put the coordinates, which are the **i, j** and **k** coefficients in $\overrightarrow{OQ}$, into the formula.

$= \sqrt{45} = 3\sqrt{5}$

b) Find the unit vector in the direction of $\overrightarrow{OQ}$.

$\dfrac{\overrightarrow{OQ}}{|\overrightarrow{OQ}|} = \dfrac{1}{3\sqrt{5}}(2\mathbf{i} + 5\mathbf{j} + 4\mathbf{k})$ ← To find a unit vector, divide the vector by its magnitude.

$= \dfrac{2}{3\sqrt{5}}\mathbf{i} + \dfrac{5}{3\sqrt{5}}\mathbf{j} + \dfrac{4}{3\sqrt{5}}\mathbf{k}$

$= \dfrac{2\sqrt{5}}{15}\mathbf{i} + \dfrac{\sqrt{5}}{3}\mathbf{j} + \dfrac{4\sqrt{5}}{15}\mathbf{k}$

Distance between two points in three dimensions

The **distance** between any two points $P(x_1, y_1, z_1)$ and $Q(x_2, y_2, z_2)$ is the **magnitude** of the vector $\overrightarrow{PQ}$:

$\overrightarrow{PQ} = \overrightarrow{OQ} - \overrightarrow{OP} = (x_2\mathbf{i} + y_2\mathbf{j} + z_2\mathbf{k}) - (x_1\mathbf{i} + y_1\mathbf{j} + z_1\mathbf{k}) = (x_2 - x_1)\mathbf{i} + (y_2 - y_1)\mathbf{j} + (z_2 - z_1)\mathbf{k}$

So the magnitude of the vector $\overrightarrow{PQ} = (x_2 - x_1)\mathbf{i} + (y_2 - y_1)\mathbf{j} + (z_2 - z_1)\mathbf{k}$

is $\sqrt{(x_2 - x_1)^2 + (y_2 - y_1)^2 + (z_2 - z_1)^2}$, by Pythagoras' theorem.

Example 2

The position vector of point A is $3\mathbf{i} + 2\mathbf{j} + 4\mathbf{k}$, and the position vector of point B is $2\mathbf{i} + 6\mathbf{j} - 5\mathbf{k}$.

Find $|\overrightarrow{AB}|$ to 1 decimal place.

$|\overrightarrow{AB}| = \sqrt{(x_2 - x_1)^2 + (y_2 - y_1)^2 + (z_2 - z_1)^2}$

$= \sqrt{(2 - 3)^2 + (6 - 2)^2 + (-5 - 4)^2}$ ← A has the coordinates (3, 2, 4), B has the coordinates (2, 6, –5) — sub these coordinates into the formula for magnitude to find $|\overrightarrow{AB}|$.

$= \sqrt{1 + 16 + 81} = \sqrt{98} = 9.9$ (1 d.p.)

Example 3

$A = (-3, -6, 4)$, $B = (2t, 1, -t)$. $|\overrightarrow{AB}| = 3\sqrt{11}$. Find the possible values of t.

$$|\overrightarrow{AB}| = \sqrt{(2t+3)^2 + (1+6)^2 + (-t-4)^2}$$

Put the coordinates in the formula to find $|\overrightarrow{AB}|$ and set the expression equal to $3\sqrt{11}$.

$$\Rightarrow \sqrt{(2t+3)^2 + (1+6)^2 + (-t-4)^2} = 3\sqrt{11}$$
$$\Rightarrow 4t^2 + 12t + 9 + 49 + t^2 + 8t + 16 = 99$$
$$\Rightarrow 5t^2 + 20t + 74 = 99$$
$$\Rightarrow 5t^2 + 20t - 25 = 0$$

Solve the resulting quadratic for t.

$$\Rightarrow t^2 + 4t - 5 = 0$$
$$\Rightarrow (t+5)(t-1) = 0$$
$$\Rightarrow t = -5 \text{ or } t = 1$$

The angle between two vectors

To find the angle between two vectors:

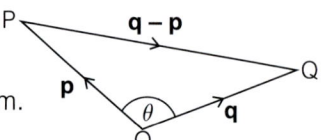

- Create a **triangle** with the vectors as two sides, and angle θ between them.
- Find the **magnitude** (i.e. the length) of each side of the triangle.
- Use the **cosine rule** to find the angle θ from these lengths:

$$\cos A = \frac{b^2 + c^2 - a^2}{2bc}$$

Example 4

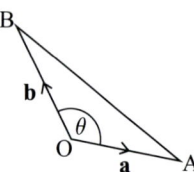

The flight paths of two different aeroplanes taking off from a runway are modelled as position vectors $\mathbf{a} = 5\mathbf{i} - 2\mathbf{j} + \mathbf{k}$, and $\mathbf{b} = -3\mathbf{i} + 3\mathbf{j} + \mathbf{k}$. Find the angle between the two flight paths, in degrees, to 1 d.p.

$$|\mathbf{a}| = \sqrt{5^2 + (-2)^2 + 1^2} = \sqrt{30}$$
$$|\mathbf{b}| = \sqrt{(-3)^2 + 3^2 + 1^2} = \sqrt{19}$$

$\mathbf{a}$ and $\mathbf{b}$ form two sides of the triangle AOB, so start off by finding the lengths of $\mathbf{a}$ and $\mathbf{b}$.

$$\overrightarrow{AB} = \mathbf{b} - \mathbf{a} = (-3\mathbf{i} + 3\mathbf{j} + \mathbf{k}) - (5\mathbf{i} - 2\mathbf{j} + \mathbf{k})$$
$$= -8\mathbf{i} + 5\mathbf{j}$$

Tip

$|\overrightarrow{AB}|$ is the length of side c of the cosine formula.

$$|\overrightarrow{AB}| = \sqrt{(-8)^2 + 5^2 + 0^2} = \sqrt{89}$$

$$\cos\theta = \frac{(\sqrt{30})^2 + (\sqrt{19})^2 - (\sqrt{89})^2}{2 \times \sqrt{30} \times \sqrt{19}} = \frac{-40}{2\sqrt{570}}$$

Use the cosine rule, $\cos A = \dfrac{b^2 + c^2 - a^2}{2bc}$.

$$\Rightarrow \theta = \cos^{-1}\left(\frac{-40}{2\sqrt{570}}\right) = 146.9° \text{ (1 d.p.)}$$

Exercise 10.2

Unless specified, give each answer in this exercise as an integer or as a simplified surd.

Q1 Find the magnitude of each of the following vectors:

a) $\mathbf{i} + 4\mathbf{j} + 8\mathbf{k}$

b) $\begin{pmatrix} 4 \\ 2 \\ 4 \end{pmatrix}$

c) $\begin{pmatrix} -4 \\ -5 \\ 20 \end{pmatrix}$

d) $7\mathbf{i} + \mathbf{j} - 7\mathbf{k}$

Q2 Find the magnitude of the resultant of each pair of vectors.

a) $\mathbf{i} + \mathbf{j} + 2\mathbf{k}$ and $\mathbf{i} + 2\mathbf{j} + 4\mathbf{k}$

b) $2\mathbf{i} + 11\mathbf{j} + 25\mathbf{k}$ and $3\mathbf{j} - 2\mathbf{k}$

c) $\begin{pmatrix} 4 \\ 2 \\ 8 \end{pmatrix}$ and $\begin{pmatrix} -2 \\ 4 \\ 1 \end{pmatrix}$

d) $\begin{pmatrix} 3 \\ 0 \\ 10 \end{pmatrix}$ and $\begin{pmatrix} -1 \\ 5 \\ 4 \end{pmatrix}$

Q3 Two vectors, **a** and **b**, are given by the column vectors $\mathbf{a} = \begin{pmatrix} 8 \\ 4 \\ 10 \end{pmatrix}$ and $\mathbf{b} = \begin{pmatrix} 2 \\ -2 \\ 4 \end{pmatrix}$.

Find $|\mathbf{a} + \mathbf{b}|$.

E **Q4** The vector $\mathbf{a} = \begin{pmatrix} 3p - 2 \\ 2pq \\ p^2q + 5p \end{pmatrix} = \begin{pmatrix} 10 \\ -1 \\ r \end{pmatrix}$. Show that $|\mathbf{a}| = 5\sqrt{s}$ where s is an integer.   *[5 marks]*

Q5 Find the unit vector in the direction of the following vectors:

a) $\mathbf{t} = 4\mathbf{i} - 4\mathbf{j} - 7\mathbf{k}$

b) $\mathbf{u} = -\mathbf{i} + 2\mathbf{j} - 2\mathbf{k}$

c) $\mathbf{v} = 2\mathbf{i} + 3\mathbf{j} - \mathbf{k}$

Q6 Find the distances between each of the following pairs of points:

a) $(3, 4, 5)$, $(5, 6, 6)$

b) $(7, 2, 9)$, $(-11, 1, 15)$

c) $(10, -2, -1)$, $(6, 10, -4)$

d) $(0, -4, 10)$, $(7, 0, 14)$

e) $(-4, 7, 10)$, $(2, 4, -12)$

f) $(7, -1, 4)$, $(30, 9, -6)$

M **Q7** The flight path of a toy aeroplane is modelled by the vector $2\mathbf{m} - \mathbf{n}$,

where $\mathbf{m} = \begin{pmatrix} -5 \\ -2 \\ 6 \end{pmatrix}$ metres, and $\mathbf{n} = \begin{pmatrix} -4 \\ 1 \\ 2 \end{pmatrix}$ metres.

Find the distance of the plane's destination from its starting position.
Give your answer in metres to 1 decimal place.

P **Q8** $\overrightarrow{OA} = \mathbf{i} - 4\mathbf{j} + 3\mathbf{k}$ and $\overrightarrow{OB} = -\mathbf{i} - 3\mathbf{j} + 5\mathbf{k}$. Find $|\overrightarrow{AO}|$, $|\overrightarrow{BO}|$ and $|\overrightarrow{BA}|$.
Show that triangle AOB is right-angled.

Q9 P is the point $(2, -1, 4)$ and Q is the point $(q - 2, 5, 2q + 1)$. Given that the length of the line PQ is 11, find the possible coordinates of the point Q.

Q10 Find the angle, in degrees to 1 d.p., that the vector $\begin{pmatrix} 1 \\ 3 \\ 2 \end{pmatrix}$ makes with:

a) vector $\begin{pmatrix} 3 \\ 2 \\ 1 \end{pmatrix}$ 　　b) the unit vector **j** 　　c) vector $\begin{pmatrix} -3 \\ 1 \\ -2 \end{pmatrix}$ 　　d) vector $\begin{pmatrix} 2 \\ 2 \\ 2 \end{pmatrix}$

P **M** **Q11** A toy rocket is launched at an angle of 60° to the unit vector in the **i** direction, with a velocity **v** = (**i** + a**j** + **k**) ms⁻¹. Find the possible values of a, in surd form.

> **Q11 Problem Solving** If you can form a suitable right-angled triangle, you can use $\cos = \dfrac{\text{adj}}{\text{hyp}}$ instead of the cosine rule.

E **P** **M** **Q12** The courses of two submarines leaving a port P are modelled by the position vectors

$\overrightarrow{PA}$ and $\overrightarrow{PB}$ where $\overrightarrow{PA} = \begin{pmatrix} 3 \\ 4 \\ -1 \end{pmatrix}$ and $\overrightarrow{PB} = \begin{pmatrix} -4 \\ 7 \\ 2 \end{pmatrix}$.

Show that the angle between their courses, APB = 70.7° to one decimal place. *[5 marks]*

E **P** **Q13** The angle θ is made by the vector **a** = x**i** + y**j** + z**k** and the positive x-axis.

a) Prove that angle θ satisfies $\cos\theta = \dfrac{x}{|\mathbf{a}|}$. *[2 marks]*

b) Hence find, to one decimal place, the angle that the vector 3**i** − 2**j** + **k** makes with the positive x-axis. *[3 marks]*

P **M** **Q14** The position of a bee during its flight can be modelled using the vector $\begin{pmatrix} 15t\cos 36° \\ 15t\sin 36° \\ 8t \end{pmatrix}$, where t is the time in seconds, $0 < t \le 5$.

Find an expression, in terms of t, for the bee's displacement from its starting point.

Challenge

E **Q15** Two forces **F₁** and **F₂** have a resultant force **R** = 4**i** + 4**j** − 7**k**, as shown in the diagram.

a) Find the magnitude of **R**. *[1 mark]*

F₁ = 2**i** − 3**j** + 6**k** and the angle between **F₂** and **R** is 42°.

b) Find, in degrees to one decimal place, the angle θ between **F₁** and **F₂**. *[4 marks]*

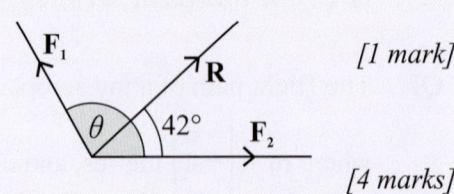

E **P** **M** **Q16** The coordinates of a rocket are given in kilometres by $(15t\tan\dfrac{t}{4}, 12t, 9t)$, where t is the time after launch in seconds, with $0 \le t < 5$, and angles are measured in radians. The distance that the rocket moves from its launch position, d, is in kilometres.

a) Show that $d = 15t\sec\dfrac{t}{4}$ km after t seconds. *[4 marks]*

b) Hence find the exact value of d when $t = \dfrac{2\pi}{3}$, giving your answer in the form $k\sqrt{3}\,\pi$ where k is a rational number to be found. *[3 marks]*

Q1   Give, in unit vector form, the position vector of point P, which has the coordinates (2, –4, 5).

Q2   X is the point (6, –1, 0) and Y is the point (4, –4, 7).
Write the vectors $\overrightarrow{XO}$ and $\overrightarrow{YO}$ in unit vector form and in column vector form.

E Q3   Points A and C have position vectors $-2\mathbf{i} + 5\mathbf{j} + 7\mathbf{k}$ and $-\mathbf{i} + 5\mathbf{j} + 3\mathbf{k}$ respectively.
Vector $\overrightarrow{AB} = -2\mathbf{i} - \mathbf{j} + 6\mathbf{k}$ and M is the midpoint of the line segment AB.
Find the vector $\overrightarrow{MC}$.   *[4 marks]*

Q4   Give two vectors that are parallel to each of the following:
a) $2\mathbf{a}$ b) $3\mathbf{i} + 4\mathbf{j} - 2\mathbf{k}$ c) $\begin{pmatrix} 1 \\ 2 \\ -1 \end{pmatrix}$

Q5   Given that $\mathbf{a} = 2\mathbf{i} - 3\mathbf{j} + \mathbf{k}$, $\mathbf{b} = -\mathbf{i} + 4\mathbf{j} - 7\mathbf{k}$ and $\mathbf{c} = -6\mathbf{i} + 4\mathbf{j} + 10\mathbf{k}$,
show that $2\mathbf{a} + \mathbf{b}$ is parallel to $\mathbf{c}$.

Q6   Show that points A, B and C are collinear, given that $\overrightarrow{OA} = \begin{pmatrix} 2 \\ 1 \\ 3 \end{pmatrix}$, $\overrightarrow{OB} = \begin{pmatrix} 1 \\ 5 \\ 7 \end{pmatrix}$ and $\overrightarrow{OC} = \begin{pmatrix} -2 \\ 17 \\ 19 \end{pmatrix}$.

E Q7   L, M and N are three collinear points.
The point L has position vector $\mathbf{l} = 3\mathbf{i} + \mathbf{j} + 8\mathbf{k}$, and $\overrightarrow{LM} = 2\mathbf{i} - \mathbf{j} - 2\mathbf{k}$.
Given that M divides the line LN in the ratio 2:3, find the position vector for N.   *[3 marks]*

Q8   Find the exact magnitudes of these vectors: a) $3\mathbf{i} + 4\mathbf{j} - 2\mathbf{k}$ b) $\begin{pmatrix} 1 \\ 2 \\ -1 \end{pmatrix}$

Q9   If A is (1, 2, 3) and B is (3, –1, –2), find: a) $|\overrightarrow{OA}|$ b) $|\overrightarrow{OB}|$ c) $|\overrightarrow{AB}|$

Q10   Find the unit vectors in the direction of each of these vectors:

a) $\mathbf{i} - 3\mathbf{k}$ b) $-2\mathbf{i} + 2\mathbf{j} + 5\mathbf{k}$ c) $\begin{pmatrix} -1 \\ -3 \\ 3 \end{pmatrix}$ d) $\begin{pmatrix} 7 \\ -1 \\ 12 \end{pmatrix}$

E Q11   The position vectors of the points A and B are $\mathbf{i} + 4\mathbf{j} - 5\mathbf{k}$ and $4\mathbf{i} - 2\mathbf{j} + \mathbf{k}$ respectively.
Find the unit vector in the direction $\overrightarrow{AB}$, giving your answer in the form
$m(x\mathbf{i} + y\mathbf{j} + z\mathbf{k})$, where x, y and z are integers and m is a real number.   *[3 marks]*

Q12   Find the exact distance between points P and Q given by
position vectors $\overrightarrow{OP} = -\mathbf{i} + 2\mathbf{j} + 3\mathbf{k}$ and $\overrightarrow{OQ} = 2\mathbf{i} - 2\mathbf{j} + 4\mathbf{k}$.

P **Q13** X is the point $(-2, 1, 0)$.

The distance between X and Y is 6. The unit vector in the direction $\overrightarrow{XY}$ is $\left(\frac{2}{3}, \frac{2}{3}, -\frac{1}{3}\right)$.

Find the coordinates of Y.

E **Q14** Two forces, $\mathbf{F_1}$ and $\mathbf{F_2}$, are defined by the vectors $-6\mathbf{j} + 2p\mathbf{k}$ and $\mathbf{i} + \left(\sqrt{p}\right)\mathbf{j} - 2\mathbf{k}$ respectively. Given that the magnitude of $\mathbf{F_1}$ is three times the magnitude of $\mathbf{F_2}$, find the possible values of p. *[5 marks]*

E **Q15** The position vector $\overrightarrow{OA} = 3p\mathbf{i} + (4p - 7)\mathbf{j} + 4r^2\mathbf{k}$
P is equal to the vector $(2q - 1)\mathbf{i} + 5\mathbf{j} + (7r + 15)\mathbf{k}$.

a) Find the possible values of p, q and r. *[4 marks]*
b) Hence find the distance between the two possible positions of point A. *[2 marks]*

E **Q16** The diagram on the right shows a sketch of the parallelogram $ABCD$.

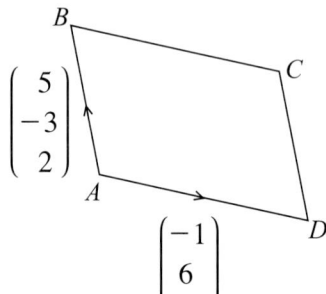

a) Write $\overrightarrow{AC}$ as a single column vector. *[2 marks]*
b) Find the length of the vector $\overrightarrow{AC}$. *[2 marks]*

E **Q17** A particle is acted upon by two forces $\mathbf{F_1}$ and $\mathbf{F_2}$.
P
Initially, $\mathbf{F_1} = \begin{pmatrix} 5 \\ 0 \\ -2 \end{pmatrix}$ and the resultant force is $\mathbf{R_a} = \begin{pmatrix} 2 \\ 1 \\ 6 \end{pmatrix}$.

a) Find the force $\mathbf{F_2}$. *[2 marks]*

Both forces are later changed so that $\mathbf{F_1}$
is doubled and the new resultant force is $\mathbf{R_b} = \begin{pmatrix} 19 \\ -3 \\ -28 \end{pmatrix}$.

b) Find the new force $\mathbf{F_2}$. *[2 marks]*
c) Describe how the new force $\mathbf{F_2}$ acts on the particle
compared with how it acted on the particle initially. *[2 marks]*

E **Q18** The first ten minutes of the flight of a helicopter is modelled such that its position is given
P by the vector $x\mathbf{i} + y\mathbf{j} + z\mathbf{k}$ where $x = 4 + 0.2t^2$, $y = -3t$ and $z = 0.1t$ for $0 \le t \le 10$.
M Time t is measured in minutes and x, y and z are measured in miles.

a) State the coordinates of the starting position of the helicopter. *[1 mark]*
b) If the helicopter is at point A when $t = 5$ and point B when $t = 10$,
find the distance AB. *[4 marks]*
c) Explain clearly why the length of AB is not equal to the distance travelled
by the helicopter between $t = 5$ and $t = 10$. *[2 marks]*

Q19 Two particles have position vectors $2t\mathbf{i} + 4\mathbf{j} - t\mathbf{k}$ and $7\mathbf{i} + t\mathbf{j} - \mathbf{k}$, where the vector components are measured in metres and t is the time in minutes. The particles are exactly $3\sqrt{6}$ metres apart on two occasions. Find, in seconds, the length of the interval between these times. *[5 marks]*

Q20 Given that $\overrightarrow{OA} = \begin{pmatrix} 4 \\ 3 \\ -3 \end{pmatrix}$, $\overrightarrow{OB} = \begin{pmatrix} -1 \\ 2 \\ -4 \end{pmatrix}$,

find the angle between $\overrightarrow{AB}$ and $\overrightarrow{OB}$ to one decimal place.

Q21 A simple mathematical model of the motion of two asteroids following a collision is given in terms of vectors. The velocity of asteroid R is given by $\mathbf{r} = 5\mathbf{i} + 3\mathbf{j} - \mathbf{k}$ ms⁻¹. The velocity of asteroid S is given by $\mathbf{s} = -2\mathbf{i} - 2\mathbf{j} + 7\mathbf{k}$ ms⁻¹.

a) Calculate the exact speed of each asteroid.

b) After 5 seconds, how much further has S travelled than R? Give your answer to the nearest metre.

c) Calculate the angle between the two asteroids' paths to the nearest degree.

d) Comment on the suitability of this model.

Q22 As shown in the diagram on the right, PQR is a triangle and $\overrightarrow{PQ} = \begin{pmatrix} 4 \\ -12 \\ 6 \end{pmatrix}$ and $\overrightarrow{PR} = \begin{pmatrix} 8 \\ 2 \\ -16 \end{pmatrix}$.

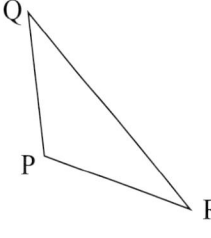

Find the area of the triangle PQR to two decimal places. *[6 marks]*

Q23 Points A, B and C have coordinates $(1, 4, -2)$, $(3, -5, 0)$ and $(-2, 6, 7)$.

a) Show that the triangle ABC is scalene. *[4 marks]*

b) Find the area of triangle ABC. *[4 marks]*

Q24 The points P, Q, R and S have coordinates $(24, -15, 7)$, $(20, -5, 5)$, $(-2, 2, 6)$ and $(0, -3, 7)$ respectively.

a) Show that $\overrightarrow{PQ} = k\overrightarrow{RS}$ where k is a constant to be found. *[2 marks]*

b) Hence explain why the quadrilateral PQRS is a trapezium. *[2 marks]*

The height of the trapezium is equal to $\dfrac{6\sqrt{230}}{5}$.

c) Find the exact area of PQRS, giving your answer in the form $\sqrt{k}$ where k is an integer. *[4 marks]*

E **Q25** The diagram on the right shows a sketch of the triangle XYZ.

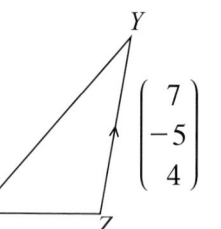

The unit vector in the direction $\overrightarrow{XZ}$ is $\begin{pmatrix} \frac{2}{3} \\ \frac{1}{2} \\ -\frac{\sqrt{11}}{6} \end{pmatrix}$ and $|\overrightarrow{XZ}| = \frac{\sqrt{10}}{5}|\overrightarrow{YZ}|$.

a) Show that $\overrightarrow{XZ} = \begin{pmatrix} 4 \\ 3 \\ -\sqrt{11} \end{pmatrix}$. *[4 marks]*

b) Hence or otherwise find the angle XZY. *[5 marks]*

Challenge

E **Q26** $\overrightarrow{AB} = \mathbf{p}$ and $\overrightarrow{BC} = \mathbf{q}$, and $|\mathbf{p}| + |\mathbf{q}| = |\mathbf{p} + \mathbf{q}|$.

a) Use this information to explain clearly why the points A, B and C are collinear. *[2 marks]*

b) Given that $|\mathbf{p}| = 2$ and $|\mathbf{p} + \mathbf{q}| = 29$ and $\mathbf{q} = r\mathbf{i} + 4r\mathbf{j} + (5r + 9)\mathbf{k}$,
find the possible values of r. *[5 marks]*

E **Q27** Find, to the nearest degree, the acute angle that the vector $\mathbf{a} = \begin{pmatrix} 3 \\ -5 \\ 4 \end{pmatrix}$
makes with the xy-plane. *[4 marks]*

E **Q28** The position vectors of points A and B are given by
P $\overrightarrow{OA} = -\mathbf{i} - 4\mathbf{j} + 8\mathbf{k}$ and $\overrightarrow{OB} = -4\mathbf{i} + 4\mathbf{j} + 7\mathbf{k}$.

a) Show that the triangle OAB is isosceles. *[3 marks]*

X is a point on the side AB such that OX is the shortest distance from O to AB.

b) State the ratio $AX:AB$. *[1 mark]*

c) Hence show that $\tan(XOA) = \frac{\sqrt{185}}{25}$. *[4 marks]*

d) Given that θ is the angle AOB, use a trigonometric identity
to directly find the exact value of $\tan\theta$. *[3 marks]*

E **Q29** Points A, B and C have the following position vectors:
P $\overrightarrow{OA} = \mathbf{i} - 4\mathbf{j} + 2\mathbf{k}$ $\overrightarrow{OB} = 7\mathbf{i} - \mathbf{j} + 4\mathbf{k}$ $\overrightarrow{OC} = 9\mathbf{i} + 2\mathbf{j} + 10\mathbf{k}$

a) Verify that $AB = BC$. *[2 marks]*

Point D is the image of point B when reflected in the line AC.

b) Find the position vector of D. *[2 marks]*

E is a point on line segment BD such that $DE:EB = m:n$ where $m < n$.

c) Describe the quadrilateral AECB. *[1 mark]*

| **10** | Chapter Summary |

1   Three-dimensional vectors have components in the direction of the x-, y- and z-axes.

2   Three-dimensional vectors can either be written in terms of the unit vectors **i**, **j** and **k** or as column vectors.

3   Parallel three-dimensional vectors are scalar multiples of one another.

4   If two 3D vectors are parallel, and also have a point in common, then they must lie on the same straight line — their points are collinear.

5   You can adapt Pythagoras' theorem to find the magnitude of a 3D vector, so that vector $\mathbf{v} = a\mathbf{i} + b\mathbf{j} + c\mathbf{k}$ has magnitude $|\mathbf{v}| = \sqrt{a^2 + b^2 + c^2}$.

6   To find the 3D unit vector in the direction of a 3D vector, divide the vector by its magnitude.

7   The distance between two points P and Q is the magnitude of the vector $\overrightarrow{PQ}$.

8   The cosine rule can be used to find the angle between two 3D vectors.

Edexcel A-Level Mathematics

Paper 1: Pure Mathematics

Time allowed: 2 hours

Centre name					
Centre number					
Candidate number					

Surname
Other names
Candidate signature

In addition to this paper you should have:
- An Edexcel mathematical formula booklet
- A calculator

Instructions to candidates
- Use black ink or ball-point pen.
- A pencil may be used for diagrams, sketches and graphs.
- Write your name and other details in the spaces provided above.
- Show clearly how you worked out your answers.
- Round answers to 3 significant figures unless otherwise stated.

Information for candidates
- There are 14 questions in this paper.
- There are 100 marks available for this paper.
- The marks available are given in brackets at the end of each question.
- You may get marks for method, even if your answer is incorrect.

Advice to candidates
- Work steadily through the paper and try to answer every question.
- Don't spend too long on one question.
- If you have time at the end, go back and check your answers.

For examiner's use			
Q	Mark	Q	Mark
1		8	
2		9	
3		10	
4		11	
5		12	
6		13	
7		14	
Total			

Answer ALL the questions.

Q1　Consider the function $g(x) = 5^x + 3$.

　　a) Sketch the graph of $y = g(x)$ for $x \in \mathbb{R}$.
　　　Label any asymptotes and where the graph meets the axes.

(2)

　　b) Find the inverse function $g^{-1}(x)$.

(3)

Q2　Figure 1 shows circle C_1 with centre $(-2, 6)$.

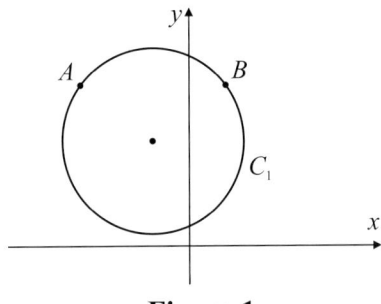

Figure 1

Points A and B lie on the circumference of C_1
and have coordinates $(-6, 9)$ and $(2, 9)$ respectively.

　　a) Find the equation of C_1.

(3)

　　b) (i)　Find the equation of the normal to C_1 at point B.

　　　(ii)　Find the coordinates of the point where the line that is tangent to C_1 at point A
　　　　　meets the line that is the normal to C_1 at point B.

(6)

The line that is the normal to C_1 at point A passes through
the centre of the circle C_2 defined by the equation $(x - 10)^2 + (y + 3)^2 = k$.

　　c) Determine the range of values of k. Explain your answer.

(1)

Q3 a) π is an irrational number. Prove by contradiction that $\pi + 1$ is also irrational.

(3)

  b) Disprove the following statement:

  "If a straight-line graph intercepts the axes at rational coordinates,
  then every point on the line has rational coordinates."

(1)

Q4 The trajectory of a particle is given by the parametric equations $x = t^3 + t^2$, $y = \frac{1}{2}t^2 - 6t$.

  a) Find the gradient $\frac{dy}{dx}$ of the trajectory in terms of t.

(2)

  b) Hence find the turning point of the trajectory.

(2)

  c) Find the equation of the tangent to the trajectory when $t = 12$.

(3)

Q5 An engineer is inspecting weather buoys at sea. The positions of three buoys,
  F, G and H, relative to the engineer's current position, P, are modelled using vectors.
  Buoy F has position vector $(3\mathbf{i} + 4\mathbf{j})$ km,
  Buoy G has position vector $(-2\mathbf{i} + 5\mathbf{j})$ km
  and the position vector of buoy H is equal to $\overrightarrow{GF}$.

  a) Which buoy is closest to the engineer? Show your working.

(4)

  b) Starting and finishing at their current position, the engineer visits all three buoys.
  What is the shortest distance, to 3 significant figures, they could have travelled?

(4)

Q6 An arithmetic sequence begins with the terms -3, 1, 5, 9, ...
  The nth term is given by the expression $kn + m$, where k and m are constants.

  a) Find the values of k and m.

(2)

  b) Using your answer to part a), find the value of the 20th term.

(1)

  c) Hence find the sum of the first 20 terms of the sequence.

(2)

Q7 The curve C has the equation $x^4 - 3x^2y^2 = y^3$.
Find the equation of the tangent to C at $(4, 4)$.

(4)

Q8 $f(x) = \cos x$

a) On the same set of axes, sketch the following in the interval $0 \le x \le 2\pi$, and give the range of each:

(i) $|f(x)|$ (ii) $f(2x) + 1$

(4)

b) Hence or otherwise, find all solutions to the equation $|f(x)| = f(2x) + 1$ in the interval $0 \le x \le 2\pi$.

(5)

Q9 The curve C has parametric equations:

$$x = \frac{1 - \cos\theta}{3} \quad \text{and} \quad y = \frac{\cos\theta \sin 2\theta}{2\sin\theta}, \quad 0 < \theta < 2\pi$$

a) Find the Cartesian equation of the curve in the form $y = f(x)$.

(3)

b) Hence determine whether the curve intersects the line $y + 3 = 2x$.

(2)

Q10 a) Show that $\dfrac{\cos x}{(\cos x + 1)(\cos x - 1)} \equiv -\text{cosec } x \cot x$.

(3)

b) Hence find $\displaystyle\int \frac{\cos x \cdot e^{\frac{1}{\sin x}}}{(\cos x + 1)(\cos x - 1)} \, dx$.

(4)

Q11 The volume of a pond (V m^3) is modelled by the equation $V = \sqrt{h^4 + 2}$, where h is the depth of the pond in metres.

a) Find the value of $\dfrac{dV}{dh}$ to 3 s.f. when $h = 0.4$.

(3)

b) Given that the volume of the pond increases at a constant rate of 0.15 m^3 per hour, find the rate at which the depth of the water is increasing with time when $h = 0.4$. Give your answer to 3 s.f.

(2)

Q12 The table shows values of $y = \dfrac{3x^2}{x^3 + 4}$ to 3 decimal places where appropriate.

x	1	1.1	1.2	1.3	1.4	1.5
y	0.6	0.681	0.754		0.872	

a) Find the missing values for $x = 1.3$ and $x = 1.5$.

(1)

b) Use the trapezium rule and the values from the table
and part a) to estimate $\displaystyle\int_1^{1.5} \dfrac{3x^2}{x^3 + 4}\, dx$ to 2 d.p.

(3)

c) (i) Find the exact value of $\displaystyle\int_1^{1.5} \dfrac{3x^2}{x^3 + 4}\, dx$ in the form $\ln k$, where k is a rational number.
(ii) Hence find the percentage error of your estimate from part b).

(4)

Q13 $f(x) = x^3 + 5x^2 - 8x - 48$ and $g(x) = \dfrac{3x + 40}{x^3 + 5x^2 - 8x - 48}$

a) Given that $(x + 4)$ is a factor of $f(x)$, fully factorise $f(x)$.

(3)

b) Express $g(x)$ in partial fractions.

(4)

c) Hence obtain the series expansion for $g(x)$ giving all the terms
up to and including the term in x^2.

(6)

Q14 Figure 2 shows a sketch of curve C, which can be given by the differential equation $y = 2\dfrac{dy}{dx}$.

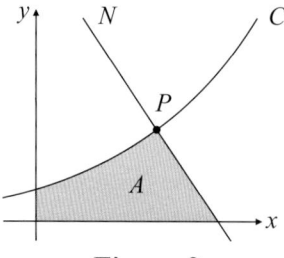

Figure 2

The point $P(1, e)$ lies on the curve, and the line N is normal to C at point P.

a) Find an equation in the form $y = f(x)$ for:

   (i) the line N,   (ii) the curve C.

(5)

b) Hence find the area A, bounded by the curve C, the line N and both axes.

(5)

Edexcel A-Level Mathematics

Paper 2: Pure Mathematics

Time allowed: 2 hours

Centre name					
Centre number					
Candidate number					

Surname	
Other names	
Candidate signature	

In addition to this paper you should have:
- An Edexcel mathematical formula booklet
- A calculator

Instructions to candidates
- Use black ink or ball-point pen.
- A pencil may be used for diagrams, sketches and graphs.
- Write your name and other details in the spaces provided above.
- Show clearly how you worked out your answers.
- Round answers to 3 significant figures unless otherwise stated.

Information for candidates
- There are 14 questions in this paper.
- There are 100 marks available for this paper.
- The marks available are given in brackets at the end of each question.
- You may get marks for method, even if your answer is incorrect.

Advice to candidates
- Work steadily through the paper and try to answer every question.
- Don't spend too long on one question.
- If you have time at the end, go back and check your answers.

For examiner's use			
Q	Mark	Q	Mark
1		8	
2		9	
3		10	
4		11	
5		12	
6		13	
7		14	
Total			

Answer ALL the questions.

Q1   The function f is defined by $f(x) = x^2 + 8k\sqrt{x} - 4$, where $x > 0$, for some constant k.

a) Giving your answers in terms of x and k, find:
(i) $f'(x)$ (ii) $f''(x)$

(3)

b) If $k > 0$, explain how you know that $f(x)$ is an increasing function.

(1)

c) If $k = -4$, show that the curve $y = f(x)$ has exactly one stationary point and describe the nature of that point.

(3)

Q2   Figure 1 shows a right-angled triangle OAC inside a sector AOB with centre O.

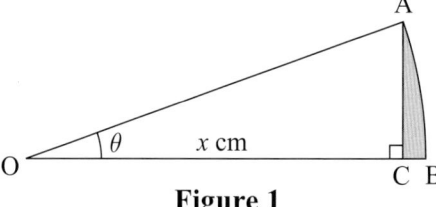

Figure 1

The length of side OC is x cm. The angle AOB is θ radians.

a) Find an expression for the area of the sector AOB.

(2)

b) Given that θ is small, use the small angle approximations to find an expression for:

(i) the area of sector AOB.   (ii) the area of triangle OAC.

(4)

c) Hence or otherwise show that the approximate value
for the shaded area is given by the expression $x^2 \dfrac{\theta^3(4 - \theta^2)}{2(\theta^2 - 2)^2}$.

(3)

Q3   A charity received £20 000 of public donations in its first year, Y_1.
The charity predicts that public donations will increase by 8% each year.

a) Write down an expression for the predicted donations in Y_n.

(1)

b) Assuming the prediction is correct, calculate how much money
the charity will receive in total from Y_6 to Y_{10}, including Y_6 and Y_{10}.

(3)

Q4 Prove the following statement: "$3x^3 + 9x - 1$ is odd for all positive integers x."

(4)

Q5 a) Use a binomial expansion to find the values of constants a, b and c in
the following approximation: $f(x) = (2 - \frac{1}{4}x)^7 \approx a + bx + cx^2$

(2)

     b) i)   Hence, write down the binomial expansion of $f(3x)$ in
ascending powers of x, up to and including the term in x^2.

        ii)   State the range of values for which this expansion is valid.

(2)

     c) Find the percentage error in the gradient of $y = f(3x)$ at $x = 0.05$ if the approximation
in part a) is used. Give your answer to 3 significant figures.

(5)

Q6 a) Sketch the graph of $y = |5 - 2x| - 3$ for $-5 \leq x \leq 10$.

(3)

$|5 - 2x| - 3 = kx$ for some constant k.

     b) Find the range of values that k could take such that,
for $x \in \mathbb{R}$, the equation $|5 - 2x| - 3 = kx$

       (i)   has exactly one solution   (ii)   has exactly two solutions.

(3)

Q7 Figure 2 shows the triangle ABC.

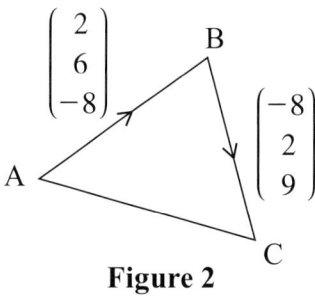

Figure 2

     a) Find $\overrightarrow{AC}$.

(1)

D is a point such that A, B and D are collinear and $AB : AD = 2 : 5$.

     b) Find the angle ACD in degrees to 1 d.p.

(5)

Q8 A function g is defined by $g(x) = \dfrac{x^2}{\sqrt{5x+4}}$.

a) State the maximum possible domain and range of $g(x)$.

(3)

b) (i) Use the quotient rule to find $g'(x)$.

(ii) Hence explain why $y = g(x)$ has exactly one turning point and determine its coordinates.

(6)

c) Sketch the graph of $y = g(x)$.

(3)

Q9 a) Express $\dfrac{t-5}{2t^2 + 7t + 6}$ in partial fractions.

(3)

b) Hence, or otherwise, find the following integral $\displaystyle\int \dfrac{t-5}{2t^2 + 7t + 6}$.

(3)

Q10 For the function, $f(x) = x^3 - 4x^2 - 5x + 6$:

a) Show $f(x) = 0$ has a root, α, in the interval $4.7 < x < 4.8$.

(2)

b) Rearrange $f(x) = 0$ to give an equation for x in the form $x = \sqrt[3]{g(x)}$.

(1)

c) $\alpha = 4.77$ to 3 significant figures.
Using your answer to part b) as an iterative formula with $x_0 = 4.7$, find the smallest n such that x_n is an approximation to α correct to 3 significant figures.

(2)

d) $f(x) = 0$ has a single negative root, β. Using $x_0 = -1.5$ as a first approximation to β, apply the Newton-Raphson method twice to obtain a second approximation. Give your answer to 3 significant figures.

(3)

Q11 a) Prove that $\cot^2 \theta + \sin^2 \theta \equiv \operatorname{cosec}^2 \theta - \cos^2 \theta$.

(2)

b) Hence, or otherwise, solve $\dfrac{\cot^2 \theta + \sin^2 \theta}{\cos^2 \theta} = 3$ for $0 < \theta < \pi$.

(4)

Q12 In a science experiment two different chemicals are added to two samples of bacteria. The population of bacteria cells in sample A (P_A) after t hours is given by $P_A = 1000e^{-0.2t}$.

   a) Estimate how long, to the nearest minute, it will be until the population of sample A is under 600.

(2)

The population of bacteria cells in sample B also decays exponentially. It had an initial population of 1200 and after 30 minutes the population had decreased to 900.

   b) (i) Write an equation in terms of P_B and t to model the population of cells in sample B after t hours. Give any values to 2 significant figures.

      (ii) The experiment was started at 13:00, at what rate will the combined population of the samples be decreasing at 14:30? Give your answer to 2 significant figures.

(6)

Q13 The average daily temperature, $T\,°C$, for a Welsh town is modelled by the equation

$$T = -\frac{5}{2}\sin\left(\frac{t}{60}\right) - \frac{5\sqrt{3}}{2}\cos\left(\frac{t}{60}\right) + 9,$$ where t is time in days since the start of the year.

   a) Write the equation in the form $T = A\sin\left(\frac{t}{60} - B\right) + C$ where $A > 0$, $C > 0$ and $0 \leq B \leq \pi$.

(5)

   b) State the maximum average daily temperature predicted by the model.

(1)

   c) Show that the model predicts that the average daily temperature will first go above 12 °C on the 165th day of the year.

(2)

Q14 The graphs of $y = x^2\ln x$ and $y = -x^2 + x$ are shown in Figure 3. The shaded area between the curves is used to model the shape of the table top for a small desk where units are in metres.

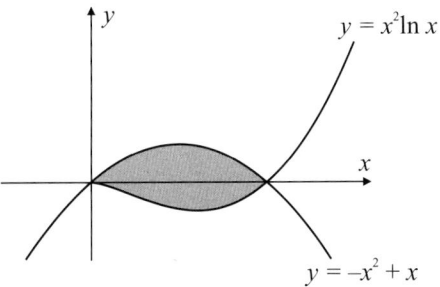

Figure 3

Find the total surface area of table top.

(7)

Answers

Chapter 1: Proof

Prior Knowledge Check

Q1 a) $\frac{a}{b}$ **b)** $2a$ **c)** $2a + 1$

Q2 Take two consecutive numbers n and $n + 1$, where n is an integer. Their sum is $n + (n + 1) = 2n + 1$, which is odd.

Q3 Take an even number, $2a$, and an odd number, $2b + 1$, where a and b are integers. The subtraction gives $2b + 1 - 2a = (2b - 2a) + 1 = 2(b - a) + 1$, which is odd.

Q4 E.g. If $x = 1$, $x^2 = 1^2 = 1$, so $x^2 = x$.
So the statement is not true for all real numbers x.

Chapter 1 Review Exercise

Q1 E.g. If $f(x) = x^2 + 1$, then $f(x) > 0$ for all real values of x, but $f'(x) = 2x < 0$ for $x < 0$.

Q2 E.g. If $a = e$, then $\ln a = \ln e = 1$, which is rational.

Q3 a) Suppose that there is a number x that is the largest odd integer, so it can be written as $x = 2k + 1$ for some integer k. Then $x + 2 = 2k + 3 = 2(k + 1) + 1$ is also an odd integer and is larger than x, which contradicts the initial assumption. So there cannot be a largest odd integer.

b) Suppose that there is a number x that is the largest multiple of 3, so it can be written as $x = 3k$ for some integer k. Then $x + 3 = 3k + 3 = 3(k + 1)$ is also a multiple of 3 and is larger than x, which contradicts the initial assumption. So there cannot be a largest multiple of 3.

Q4 a) Suppose that there is some even number x for which x^2 is odd. Since x is even, it can be written as $x = 2n$ for some integer n. Then $x^2 = (2n)^2 = 4n^2 = 2(2n^2)$ which is even, which contradicts the assumption that x^2 is odd. So if x^2 is odd, then x must also be odd.

b) Suppose there is some even number x for which x^3 is odd. Since x is even, it can be written as $x = 2n$ for some integer n. Then $x^3 = (2n)^3 = 8n^3 = 2(4n^3)$ which is even, which contradicts the assumption that x^3 is odd. So if x^3 is odd, then x must also be odd.

c) Suppose there is some odd number x for which $x^2 + 3$ is odd. Since x is odd, it can be written as $x = 2n + 1$ for some integer n. Then $x^2 + 3 = (2n + 1)^2 + 3 = 4n^2 + 4n + 1 + 3 = 4n^2 + 4n + 4 = 2(2n^2 + 2n + 2)$ which is even, which contradicts the assumption that $x^2 + 3$ is odd. So if $x^2 + 3$ is odd, then x must be even.

Q5 a) $\left(e^{\frac{1}{2}x} - e^{-\frac{1}{2}x}\right)^2 = e^x + e^{-x} - 2$
[2 marks available — 1 mark for two terms correct, 1 mark for fully correct answer]

b) $\left(e^{\frac{1}{2}x} - e^{-\frac{1}{2}x}\right)^2 \geq 0 \implies e^x + e^{-x} - 2 \geq 0$
$\implies e^x + e^{-x} \geq 2$ for all real values of x.
[2 marks available — 1 mark for squared expression ≥ 0, 1 mark for complete proof]

Q6 Suppose there is an integer N that is both even and odd. Since N is even, $N = 2k$, for some integer k. Since N is odd, $N = 2m + 1$, for some integer m. Hence $2k = 2m + 1 \implies 2(k - m) = 1$. 2 is a factor of the LHS but not the RHS, which is a contradiction. Hence there is no integer that is both even and odd.

Q7 Assume that there are integers a and b such that $3a + 5 = 12b$. Then $12b - 3a = 5$, so $3(4b - a) = 5$. 3 is a factor of the LHS but not the RHS, which is a contradiction. Therefore, there are no integer solutions.
There are other ways to prove this — for example, you could show that $4b - a = \frac{5}{3}$, which contradicts the assumption that a and b are integers.

Q8 Assume that b is odd and $b^2 - 4ac = 0$. Then $b^2 = 4ac = 2(2ac)$, which is even, since a and c are integers. But we have assumed b is odd, so $b^2 = $ odd $\times$ odd, which is odd. This contradicts that b^2 is even. Therefore, if b is odd, the discriminant cannot be zero.
[4 marks available — 1 mark for the correct assumption, 1 mark for identifying that b^2 is even, 1 mark for obtaining contradiction, 1 mark for complete proof with conclusion]

Q9 Suppose that the equation does have integer solutions. Then $2x^2 - 6xy - 12y^2 = 1$, so $2(x^2 - 3xy - 6y^2) = 1$, which implies 2 is a factor of 1. This is a contradiction. Therefore, there are no integer solutions.
[3 marks available — 1 mark for the correct assumption, 1 mark for $2(x^2 - 3xy - 6y^2) = 1$, 1 mark for a complete proof with conclusion]

Q10 Rearranging the inequality gives $x^2 + 2x + 3 + \cos x > 0$. Now, $x^2 + 2x + 3 + \cos x = (x + 1)^2 + 2 + \cos x$. Since $(x + 1)^2 \geq 0$ and $\cos x \geq -1$ for all x, then $(x + 1)^2 + 2 + \cos x \geq 0 + 2 - 1 = 1 > 0$. Therefore $x^2 + x + 4 + \cos x > 1 - x$.

Q11 a) Suppose there are positive integers a and b such that $3^a + b^2 = 6$. Since $a, b > 0$, we must have $a = 1$ since $3^2 > 6$. Then $b^2 = 3$, so $b = \pm\sqrt{3}$, which are not integers. This is a contradiction, hence there are no positive integer solutions.

b) $3^a = 6 - b^2 = (\sqrt{6} - b)(\sqrt{6} + b)$
$\implies a = \log_3(\sqrt{6} - b)(\sqrt{6} + b) = \log_3(\sqrt{6} - b) + \log_3(\sqrt{6} + b)$
Hence by part a), there are no positive integer solutions.

Q12 a) Suppose that there is a rational number $x \neq 0$ and an irrational number y such that xy is rational. This means that x can be written as $x = \frac{a}{b}$ and xy can be written as $xy = \frac{c}{d}$, where a, b, c and d are all non-zero integers.
So $xy = \frac{a}{b}y = \frac{c}{d} \implies y = \frac{bc}{ad}$
Since bc and ad are both integers, this means that y is a rational number, which contradicts the assumption that y is irrational. So the product of a non-zero rational number and an irrational number is always irrational.

b) To disprove the statement, find a counter-example. $\sqrt{2}$ is an irrational number, but $\sqrt{2} \times \sqrt{2} = 2$ which is rational, so the statement is false.

Q13 Suppose that there is a smallest positive rational number, and call it x. Since x is rational, it can be written as $x = \frac{a}{b}$, and since it is positive, a and b are both positive integers (or both negative, in which case you can simplify the fraction by dividing top and bottom by -1 to get a and b positive). Then $\frac{a}{b + 1}$ is also a positive rational number, and is smaller than x, which contradicts the assumption that x is the smallest positive rational number. So there cannot be a smallest positive rational number.

Q14 Assume that $1 + \sqrt{2}$ is rational, so it can be written as $\frac{a}{b}$ where a and b are non-zero integers.
So $1 + \sqrt{2} = \frac{a}{b} \implies \sqrt{2} = \frac{a}{b} - 1 \implies \sqrt{2} = \frac{a - b}{b}$
Since a and b are integers, $(a - b)$ is also an integer, which means that $\sqrt{2}$ is rational, which is not true. So $1 + \sqrt{2}$ must be irrational.

Q15 a) Steps 3 and 5. It is not true that if a is an integer and 12 is a factor of a^2, then 12 is a factor of a. For example, consider $a^2 = 36$.
[2 marks available — 1 mark for the correct steps, 1 mark for an explanation]

b) Assume that x is irrational and $2x$ is rational.

Then $2x = \dfrac{p}{q}$ where p and q are non-zero integers.

Then $x = \dfrac{p}{2q}$, so x is rational. This contradicts that x is irrational. Therefore, if x is irrational, so is $2x$.

[2 marks available — 1 mark for correct assumption and writing $2x$ as $\dfrac{p}{q}$, 1 mark for complete proof with conclusion]

c) $\sqrt{12} = 2\sqrt{3}$, so since he knows $\sqrt{3}$ is irrational, $2\sqrt{3}$ must also be irrational from the proof in part b).

[2 marks available — 1 mark for writing $\sqrt{12} = 2\sqrt{3}$, 1 mark for correct explanation]

Q16 a) Suppose that there is an integer x such that x^2 is a multiple of 3 but x is not. If x is not a multiple of 3, then there are two cases to consider: $x = 3k + 1$ and $x = 3k + 2$ for some integer k.

If $x = 3k + 1$, then $x^2 = (3k + 1)^2 = 9k^2 + 6k + 1$
$$= 3(3k^2 + 2k) + 1$$
So x^2 is not a multiple of 3.

If $x = 3k + 2$, then $x^2 = (3k + 2)^2 = 9k^2 + 12k + 4$
$$= 3(3k^2 + 4k + 1) + 1$$
So x^2 is not a multiple of 3.

Therefore, by exhaustion, x^2 cannot be a multiple of 3, which contradicts the initial assumption. So if x^2 is a multiple of 3, then x must also be a multiple of 3.

[3 marks available — 1 mark for each correct expression for x^2, 1 mark for correct interpretation]

b) Suppose that $\sqrt{3}$ is rational, so $\sqrt{3} = \dfrac{a}{b}$ for some non-zero integers a and b that share no common factors.

So $b\sqrt{3} = a \Rightarrow 3b^2 = a^2 \Rightarrow a^2$ is a multiple of 3.
From part a), this means that a is also a multiple of 3, so write $a = 3k$ for some integer k.
$3b^2 = (3k)^2 \Rightarrow 3b^2 = 9k^2 \Rightarrow b^2 = 3k^2$
$\Rightarrow b^2$ is a multiple of 3.
Again, from part a), this means b is a multiple of 3. But it was assumed at the start that a and b had no common factors. So $\sqrt{3}$ cannot be written as an integer fraction, so it is irrational.

[3 marks available — 1 mark for setting $\sqrt{3}$ equal to $\dfrac{a}{b}$, 1 mark for correct deduction that a and b must both be multiples of 3, 1 mark for correct interpretation]

Q17 a) Suppose that the sum of a rational number x and an irrational number y is a rational number.

Then $x + y$ can be written as $\dfrac{a}{b} + y = \dfrac{c}{d}$ for some non-zero integers a and b that share no common factors and some non-zero integers c and d that share no common factors.

So $y = \dfrac{c}{d} - \dfrac{a}{b} = \dfrac{bc - ad}{bd}$. As a, b, c and d are all integers, $(bc - ad)$ and bd are also integers and so y is a rational number, which contradicts the initial assumption. So the sum of a rational number and an irrational number must be an irrational number.

[3 marks available — 1 mark for correctly expressing the sum using fractions, 1 mark for a correct expression for y, 1 mark for correct interpretation]

b) E.g. Suppose there is a number z that is the largest irrational number. Then as proven above, $z + 1$ is also irrational, since 1 is a rational number. $z + 1$ is larger than z, which contradicts the initial assumption. So there cannot be a largest irrational number.

[2 marks available — 1 mark for adding a positive rational number to z, 1 mark for correct interpretation]
You could have added any positive rational number here.

Q18 a) Suppose that the difference between a rational number x and an irrational number y is a rational number.

$x - y$ can be written as $\dfrac{a}{b} - y = \dfrac{c}{d}$ for some non-zero integers a and b that share no common factors and some non-zero integers c and d that share no common factors.

So $y = \dfrac{a}{b} - \dfrac{c}{d} = \dfrac{ad - bc}{bd}$. As a, b, c and d are all integers, $(ad - bc)$ and bd are also integers and so y is a rational number, which contradicts the initial assumption.
So $y - x$ is irrational.
Since $x - y = -(y - x)$, and the negative of an irrational number will also be irrational, the difference $x - y$ is also irrational. So the difference between a rational number and an irrational number must be an irrational number.

[4 marks available — 1 mark for correctly expressing $x - y$ or $y - x$ using fractions, 1 mark for considering the other case, 1 mark for showing a contradiction, 1 mark for correct interpretation]

b) E.g. $1 - \sqrt{2}$ is irrational (from part a)). $\sqrt{2}$ is also irrational.
$(1 - \sqrt{2}) + \sqrt{2} = 1$, which is rational.
So the sum of an irrational number and an irrational number cannot always be irrational.

[2 marks available — 1 mark for choosing two appropriate irrational numbers, 1 mark for correct interpretation]

Q19 Suppose that $\sqrt[3]{5}$ is rational, so $\sqrt[3]{5} = \dfrac{a}{b}$ for some non-zero integers a and b that share no common factors.
So $b\sqrt[3]{5} = a \Rightarrow 5b^3 = a^3 \Rightarrow a^3$ is a multiple of 5.
Using the assumption given, this means that a is also a multiple of 5, so write $a = 5k$ for some integer k.
$5b^3 = (5k)^3 \Rightarrow 5b^3 = 125k^3 \Rightarrow b^3 = 25k^3 = 5(5k^3)$
$\Rightarrow b^3$ is a multiple of 5.
Again, from the given assumption, this means b is a multiple of 5.
But it was assumed at the start that a and b had no common factors. So $\sqrt[3]{5}$ cannot be written as an integer fraction, so it is irrational.

[3 marks available — 1 mark for setting $\sqrt[3]{5}$ equal to $\dfrac{a}{b}$, 1 mark for correct deduction that a and b must both be multiples of 5, 1 mark for correct interpretation]

Q20 Assume that there are two distinct points on L with rational coordinates (p, q) and (r, s), where p, q, r, s are rational.

Then the gradient of L is $\dfrac{s - q}{r - p}$, which is also rational.

This contradicts that m is irrational. Hence there is at most one point on L with rational coordinates.

Q21 a) Assume that $7^{\frac{1}{3}}$ is rational, i.e. $7^{\frac{1}{3}} = \dfrac{a}{b}$, where a and b are non-zero integers that share no common factors. Then $7b^3 = a^3$ so 7 is a factor of a^3 and, since 7 is prime, it must therefore also be a factor of a. We may then write $a = 7k$, where k is an integer, so $7b^3 = (7k)^3 \Rightarrow b^3 = 49k^3$, so 7 is a factor of b^3 and hence of b. This contradicts the assumption that $\dfrac{a}{b}$ is in its simplest form. Therefore, $7^{\frac{1}{3}}$ is irrational.

[4 marks available — 1 mark for the correct assumption, 1 mark for $7b^3 = a^3$, 1 mark for $a = 7k$, 1 mark for complete proof with conclusion]

b) Suppose that $x^{\frac{1}{3}}$ is rational, i.e. $x^{\frac{1}{3}} = \dfrac{a}{b}$ where a, b are integers, $b \neq 0$. Then $x = \dfrac{a^3}{b^3}$, so x is rational. This is a contradiction, hence the cube root of x is irrational.

[3 marks available — 1 mark for the correct assumption, 1 mark for showing that x is rational, 1 mark for complete proof with conclusion]

c) By part a, we have that $7^{\frac{1}{3}}$ is irrational, and by part b, its cube root, $7^{\frac{1}{9}}$ is irrational. By part b, the cube root of $7^{\frac{1}{9}}$, which is $7^{\frac{1}{27}}$, is also irrational.

[2 marks available — 1 mark for spotting the cube root of the cube root of $7^{\frac{1}{3}}$ is $7^{\frac{1}{27}}$, 1 mark for complete explanation using parts a and b]

Q22 Assume that $\ln 5 = \dfrac{p}{q}$ where p and q are non-zero integers that share no common factors. Then $q\ln 5 = p \Rightarrow \ln 5^q = p$ $\Rightarrow 5^q = e^p$. But 5^q is an integer and e^p is irrational. This is a contradiction since integers are rational numbers. Therefore, $\ln 5$ is irrational.

[4 marks available — 1 mark for the correct assumption, 1 mark for using laws of logs correctly, 1 mark for $5^q = e^p$, 1 mark for complete proof with conclusion]

Q23 a) Any odd number can be written as $2n + 1$, where n is an integer. Then $(n + 1)^2 - n^2 = 2n + 1$, so it is the difference of two squares.

b) $197 = 2 \times 98 + 1$ so $197 = 99^2 - 98^2$.

Q24 $2x^2 + 2y^2 + (x - y)^2 \geq 0$
$\Rightarrow 3x^2 + 3y^2 - 2xy \geq 0$
$\Rightarrow 4x^2 + 4y^2 \geq x^2 + 2xy + y^2 = (x + y)^2$
$\Rightarrow \sqrt{4x^2 + 4y^2} = 2\sqrt{x^2 + y^2} \geq \sqrt{(x + y)^2}$
$\Rightarrow 2\sqrt{x^2 + y^2} \geq x + y \Rightarrow \dfrac{x + y}{\sqrt{x^2 + y^2}} \leq 2$

[4 marks available — 1 mark for sum of squares ≥ 0, 1 mark for expanding correctly, 1 mark for rearranging to $4x^2 + 4y^2 \geq x^2 + 2xy + y^2$, 1 mark for complete proof]

Q25 a) Suppose for a contradiction that $a + b \leq c$. Then since $a + b > 0$ and $c > 0$, we have $(a + b)^2 \leq c^2 \Rightarrow a^2 + 2ab + b^2 \leq c^2$. As $a^2 + b^2 = c^2$, we have $c^2 + 2ab \leq c^2$, so $2ab \leq 0$. This is a contradiction since $a, b > 0$. Hence $a + b > c$.

[4 marks available — 1 mark for the correct assumption, 1 mark for squaring, 1 mark for obtaining $2ab \leq 0$, 1 mark for complete proof with conclusion]

b) In a right-angled triangle, the sum of the lengths of the two shorter sides exceeds the length of the hypotenuse.

[1 mark available for a correct explanation]

Chapter 2: Algebra and Functions

Prior Knowledge Check

Q1 $f(x - 3) = 3(x - 3)^2 - 12 = 3(x^2 - 6x + 9) - 3(4) = 3(x^2 - 6x + 5)$
$= 3(x - 5)(x - 1)$

Q2 a) $3^{2x - 1} = 81 \Rightarrow 3^{2x - 1} = 3^4 \Rightarrow 2x - 1 = 4 \Rightarrow x = \dfrac{5}{2}$

b) $\log_4 x + \log_4 8 = 4 \Rightarrow \log_4 8x = 4 \Rightarrow 8x = 4^4 = 256 \Rightarrow x = 32$

Q3

Q4 $f(\tfrac{1}{3}) = 0 \Rightarrow (x - \tfrac{1}{3})$ is a factor $\Rightarrow (3x - 1)$ is a factor.
$3x^3 + 2x^2 - 7x + 2 = (Ax^2 + Bx + C)(3x - 1)$
$= 3Ax^3 - Ax^2 + 3Bx^2 - Bx + 3Cx - C$

Equating coefficients: $3x^3 = 3Ax^3 \Rightarrow 3 = 3A \Rightarrow A = 1$
$2x^2 = -Ax^2 + 3Bx^2 \Rightarrow 2 = -A + 3B = -1 + 3B \Rightarrow 3 = 3B \Rightarrow B = 1$
$2 = -C \Rightarrow C = -2$
So $f(x) = (x^2 + x - 2)(3x - 1)$

Exercise 2.1 — Simplifying Algebraic Fractions

Q1 a) $\dfrac{4}{2x + 10} = \dfrac{4}{2(x + 5)} = \dfrac{2}{x + 5}$

b) $\dfrac{5x}{x^2 + 2x} = \dfrac{5x}{x(x + 2)} = \dfrac{5}{x + 2}$

c) $\dfrac{6x^2 - 3x}{3x^2} = \dfrac{3x(2x - 1)}{3x^2} = \dfrac{2x - 1}{x}$

d) $\dfrac{4x^3}{x^3 + 3x^2} = \dfrac{4x^3}{x^2(x + 3)} = \dfrac{4x}{x + 3}$

Q2 a) $\dfrac{3x + 6}{x^2 + 3x + 2} = \dfrac{3(x + 2)}{(x + 1)(x + 2)} = \dfrac{3}{x + 1}$

b) $\dfrac{x^2 + 3x}{x^2 + x - 6} = \dfrac{x(x + 3)}{(x - 2)(x + 3)} = \dfrac{x}{x - 2}$

c) $\dfrac{2x - 6}{x^2 - 9} = \dfrac{2(x - 3)}{(x - 3)(x + 3)} = \dfrac{2}{x + 3}$

d) $\dfrac{5x^2 - 20x}{2x^2 - 5x - 12} = \dfrac{5x(x - 4)}{(2x + 3)(x - 4)} = \dfrac{5x}{2x + 3}$

e) $\dfrac{3x^2 - 7x - 6}{2x^2 - x - 15} = \dfrac{(3x + 2)(x - 3)}{(2x + 5)(x - 3)} = \dfrac{3x + 2}{2x + 5}$

f) $\dfrac{x^3 - 2x^2}{x^3 - 4x} = \dfrac{x^2(x - 2)}{x(x^2 - 4)} = \dfrac{x^2(x - 2)}{x(x - 2)(x + 2)} = \dfrac{x}{x + 2}$

Q3 a) $\dfrac{1 + \frac{1}{x}}{x + 1} = \dfrac{\left(1 + \frac{1}{x}\right)x}{(x + 1)x} = \dfrac{x + 1}{x(x + 1)} = \dfrac{1}{x}$

b) $\dfrac{3 + \frac{1}{x}}{2 + \frac{1}{x}} = \dfrac{\left(3 + \frac{1}{x}\right)x}{\left(2 + \frac{1}{x}\right)x} = \dfrac{3x + 1}{2x + 1}$

c) $\dfrac{1 + \frac{1}{2x}}{2 + \frac{1}{x}} = \dfrac{\left(1 + \frac{1}{2x}\right)2x}{\left(2 + \frac{1}{x}\right)2x} = \dfrac{2x + 1}{4x + 2} = \dfrac{2x + 1}{2(2x + 1)} = \dfrac{1}{2}$

d) $\dfrac{\frac{1}{3x} - 1}{3x^2 - x} = \dfrac{\left(\frac{1}{3x} - 1\right)3x}{(3x^2 - x)3x} = \dfrac{-(3x - 1)}{3x^2(3x - 1)} = -\dfrac{1}{3x^2}$

e) $\dfrac{2 + \frac{1}{x}}{6x^2 + 3x} = \dfrac{\left(2 + \frac{1}{x}\right)x}{(6x^2 + 3x)x} = \dfrac{2x + 1}{3x^2(2x + 1)} = \dfrac{1}{3x^2}$

f) $\dfrac{x + 4}{x - \frac{16}{x}} = \dfrac{(x + 4)x}{\left(x - \frac{16}{x}\right)x} = \dfrac{(x + 4)x}{x^2 - 16} = \dfrac{(x + 4)x}{(x + 4)(x - 4)} = \dfrac{x}{x - 4}$

Q4 a) $\dfrac{\frac{3x}{x + 2}}{\frac{x}{x + 2} + \frac{1}{x + 2}} = \dfrac{\left(\frac{3x}{x + 2}\right)(x + 2)}{\left(\frac{x}{x + 2} + \frac{1}{x + 2}\right)(x + 2)} = \dfrac{3x}{x + 1}$

b) $\dfrac{2 + \frac{1}{x + 1}}{3 + \frac{1}{x + 1}} = \dfrac{\left(2 + \frac{1}{x + 1}\right)(x + 1)}{\left(3 + \frac{1}{x + 1}\right)(x + 1)} = \dfrac{2(x + 1) + 1}{3(x + 1) + 1} = \dfrac{2x + 3}{3x + 4}$

c) $\dfrac{1 - \frac{2}{x + 3}}{x + 2} = \dfrac{\left(1 - \frac{2}{x + 3}\right)(x + 3)}{(x + 2)(x + 3)} = \dfrac{x + 3 - 2}{(x + 2)(x + 3)}$

$= \dfrac{x + 1}{(x + 2)(x + 3)}$

d) $\dfrac{4 - \frac{1}{x^2}}{2 - \frac{1}{x} - \frac{1}{x^2}} = \dfrac{\left(4 - \frac{1}{x^2}\right)x^2}{\left(2 - \frac{1}{x} - \frac{1}{x^2}\right)x^2} = \dfrac{4x^2 - 1}{2x^2 - x - 1}$

$= \dfrac{(2x + 1)(2x - 1)}{(x - 1)(2x + 1)} = \dfrac{2x - 1}{x - 1}$

e) $\dfrac{\dfrac{4}{x}+\dfrac{x}{4}+2}{\dfrac{4}{x}+1}\times\dfrac{4x}{4x}=\dfrac{\dfrac{16x}{x}+\dfrac{4x^2}{4}+8x}{\dfrac{16x}{x}+4x}$

$=\dfrac{16+x^2+8x}{16+4x}=\dfrac{(4+x)^2}{4(4+x)}=\dfrac{4+x}{4}$

f) $\dfrac{x+\dfrac{13}{2}+\dfrac{3}{x}}{3x-\dfrac{5}{2}-\dfrac{2}{x}}\times\dfrac{2x}{2x}=\dfrac{2x^2+13x+6}{6x^2-5x-4}$

$=\dfrac{(2x+1)(x+6)}{(2x+1)(3x-4)}=\dfrac{x+6}{3x-4}$

Q5 $\dfrac{x^4-16}{x^2+4x+4}=\dfrac{(x^2-4)(x^2+4)}{(x+2)^2}$

$=\dfrac{(x-2)(x+2)(x^2+4)}{(x+2)^2}=\dfrac{(x-2)(x^2+4)}{x+2}$

[3 marks available — 1 mark for factorising the denominator, 1 mark for factorising the numerator as the difference of two squares, 1 mark for the correct answer]

Q6 **a)** Factorise the numerator $f(x)=x^3-4x^2-19x-14$:
$f(-1)=-1-4+19-14=0$, so $(x+1)$ is a factor.
By equating coefficients, you can see that:
$x^3-4x^2-19x-14=(x+1)(x^2-5x-14)$
$=(x+1)(x+2)(x-7)$
So $\dfrac{x^3-4x^2-19x-14}{x^2-6x-7}=\dfrac{(x+1)(x+2)(x-7)}{(x+1)(x-7)}=x+2$

b) Factorise the numerator $f(x)=x^3+6x^2-x-6$:
$f(1)=1+6-1-6=0$, so $(x-1)$ is a factor.
$f(-1)=-1+6+1-6=0$, so $(x+1)$ is a factor.
By equating coefficients, you can see that:
$x^3+6x^2-x-6=(x^2-1)(x+6)=(x-1)(x+1)(x+6)$
And for the denominator $g(x)=x^3+7x^2+4x-12$:
$g(1)=1+7+4-12=0$, so $(x-1)$ is a factor.
Divide $g(x)$ by $(x-1)$ to fully factorise:
$x^3+7x^2+4x-12=(x-1)(Ax^2+Bx+C)$
By equating coefficients, you can see that $A=1$ and $C=12$. Also, $B-A=7\Rightarrow B=7+A=8$. So:
$x^3+7x^2+4x-12=(x-1)(x^2+8x+12)$
$=(x-1)(x+2)(x+6)$
So $\dfrac{x^3+6x^2-x-6}{x^3+7x^2+4x-12}=\dfrac{f(x)}{g(x)}$
$=\dfrac{(x-1)(x+1)(x+6)}{(x-1)(x+2)(x+6)}=\dfrac{x+1}{x+2}$

Q7 $\dfrac{2^{2x}-5\times2^x+6}{2^{2x}-2^{x+3}+15}=\dfrac{(2^x)^2-5\times2^x+6}{(2^x)^2-8\times2^x+15}$

$=\dfrac{(2^x-2)(2^x-3)}{(2^x-5)(2^x-3)}=\dfrac{2^x-2}{2^x-5}$

[3 marks available — 1 mark for correct factorised numerator, 1 mark for correct factorised denominator, 1 mark for correct answer]

Exercise 2.2 — Adding and Subtracting Algebraic Fractions

Q1 **a)** $\dfrac{2x}{3}+\dfrac{x}{5}=\dfrac{10x}{15}+\dfrac{3x}{15}=\dfrac{13x}{15}$

b) $\dfrac{2}{3x}-\dfrac{1}{5x}=\dfrac{10}{15x}-\dfrac{3}{15x}=\dfrac{7}{15x}$

c) $\dfrac{3}{x^2}+\dfrac{2}{x}=\dfrac{3}{x^2}+\dfrac{2x}{x^2}=\dfrac{3+2x}{x^2}$

Q2 **a)** $\dfrac{x+1}{3}+\dfrac{x+2}{4}=\dfrac{4(x+1)}{12}+\dfrac{3(x+2)}{12}$

$=\dfrac{4x+4+3x+6}{12}=\dfrac{7x+10}{12}$

b) $\dfrac{2x}{3}+\dfrac{x-1}{7x}=\dfrac{14x^2}{21x}+\dfrac{3(x-1)}{21x}=\dfrac{14x^2+3x-3}{21x}$

c) $\dfrac{3x}{4}-\dfrac{2x-1}{5x}=\dfrac{15x^2}{20x}-\dfrac{4(2x-1)}{20x}=\dfrac{15x^2-8x+4}{20x}$

d) $\dfrac{2}{x-1}+\dfrac{3}{x}=\dfrac{2x}{x(x-1)}+\dfrac{3(x-1)}{x(x-1)}=\dfrac{2x+3x-3}{x(x-1)}=\dfrac{5x-3}{x(x-1)}$

Q3 **a)** $\dfrac{3}{x+1}+\dfrac{2}{x+2}=\dfrac{3(x+2)}{(x+1)(x+2)}+\dfrac{2(x+1)}{(x+1)(x+2)}$

$=\dfrac{3x+6+2x+2}{(x+1)(x+2)}=\dfrac{5x+8}{(x+1)(x+2)}$

b) $\dfrac{4}{x-3}-\dfrac{1}{x+4}=\dfrac{4(x+4)}{(x-3)(x+4)}-\dfrac{x-3}{(x-3)(x+4)}$

$=\dfrac{4x+16-x+3}{(x-3)(x+4)}=\dfrac{3x+19}{(x-3)(x+4)}$

c) $\dfrac{6}{x+2}+\dfrac{6}{x-2}=\dfrac{6(x-2)}{(x+2)(x-2)}+\dfrac{6(x+2)}{(x+2)(x-2)}$

$=\dfrac{6x-12+6x+12}{(x+2)(x-2)}=\dfrac{12x}{(x+2)(x-2)}$

d) $\dfrac{3}{x-2}-\dfrac{5}{2x+3}=\dfrac{3(2x+3)}{(x-2)(2x+3)}-\dfrac{5(x-2)}{(x-2)(2x+3)}$

$=\dfrac{6x+9-5x+10}{(x-2)(2x+3)}=\dfrac{x+19}{(x-2)(2x+3)}$

e) $\dfrac{3}{x+2}+\dfrac{x}{x+1}=\dfrac{3(x+1)}{(x+2)(x+1)}+\dfrac{x(x+2)}{(x+2)(x+1)}$

$=\dfrac{3x+3+x^2+2x}{(x+2)(x+1)}=\dfrac{x^2+5x+3}{(x+2)(x+1)}$

f) $\dfrac{3}{4x-5}-\dfrac{2}{3x-2}=\dfrac{3(3x-2)}{(4x-5)(3x-2)}-\dfrac{2(4x-5)}{(4x-5)(3x-2)}$

$=\dfrac{9x-6-8x+10}{(4x-5)(3x-2)}=\dfrac{x+4}{(4x-5)(3x-2)}$

Q4 **a)** $\dfrac{3}{x-1}-\dfrac{2}{x-2}=\dfrac{3(x-2)}{(x-1)(x-2)}-\dfrac{2(x-1)}{(x-1)(x-2)}$

$=\dfrac{3x-6-2x+2}{(x-1)(x-2)}=\dfrac{x-4}{(x-1)(x-2)}$ or $\dfrac{x-4}{x^2-3x+2}$

[3 marks available — 1 mark for correct denominator, 1 mark for complete method to combine as single fraction, 1 mark for correct answer]

b) For $x>2$, the denominator $x^2-3x+2=(x-1)(x-2)$ is positive and $x-4<x$, so $\dfrac{x-4}{x^2-3x+2}<\dfrac{x}{x^2-3x+2}$.

[1 mark for a correct explanation]

Q5 **a)** $\dfrac{5x}{(x+1)^2}-\dfrac{3}{x+1}=\dfrac{5x}{(x+1)^2}-\dfrac{3(x+1)}{(x+1)^2}$

$=\dfrac{5x-3x-3}{(x+1)^2}=\dfrac{2x-3}{(x+1)^2}$

b) $\dfrac{5}{x(x+3)}+\dfrac{3}{x+2}=\dfrac{5(x+2)}{x(x+3)(x+2)}+\dfrac{3x(x+3)}{x(x+3)(x+2)}$

$=\dfrac{5x+10+3x^2+9x}{x(x+3)(x+2)}=\dfrac{3x^2+14x+10}{x(x+3)(x+2)}$

c) $\dfrac{x}{x^2-4}-\dfrac{1}{x+2}=\dfrac{x}{(x+2)(x-2)}-\dfrac{1}{x+2}$

$=\dfrac{x}{(x+2)(x-2)}-\dfrac{x-2}{(x+2)(x-2)}$

$=\dfrac{x-x+2}{(x+2)(x-2)}=\dfrac{2}{(x+2)(x-2)}$

d) $\dfrac{3}{x+1}+\dfrac{6}{2x^2+x-1}=\dfrac{3}{x+1}+\dfrac{6}{(x+1)(2x-1)}$

$=\dfrac{3(2x-1)}{(x+1)(2x-1)}+\dfrac{6}{(x+1)(2x-1)}$

$=\dfrac{6x-3+6}{(x+1)(2x-1)}=\dfrac{3(2x+1)}{(x+1)(2x-1)}$

Q6 $\dfrac{x+2}{x-3}+\dfrac{1}{2(x^2-9)}=\dfrac{2(x+3)(x+2)}{2(x+3)(x-3)}+\dfrac{1}{2(x^2-9)}$

$=\dfrac{2(x+3)(x+2)}{2(x^2-9)}+\dfrac{1}{2(x^2-9)}=\dfrac{2x^2+10x+13}{2x^2-18}$

[3 marks available — 1 mark for correct common denominator $2(x^2-9)$ or $2x^2-18$, 1 mark for correct method to add fractions, 1 mark for correct answer]

Answers

Q7 a) $\dfrac{2}{x} + \dfrac{3}{x+1} + \dfrac{4}{x+2}$

$= \dfrac{2(x+1)(x+2)}{x(x+1)(x+2)} + \dfrac{3x(x+2)}{x(x+1)(x+2)} + \dfrac{4x(x+1)}{x(x+1)(x+2)}$

$= \dfrac{2x^2 + 6x + 4 + 3x^2 + 6x + 4x^2 + 4x}{x(x+1)(x+2)} = \dfrac{9x^2 + 16x + 4}{x(x+1)(x+2)}$

b) $\dfrac{3}{x+4} - \dfrac{2}{x+1} + \dfrac{1}{x-2}$

$= \dfrac{3(x+1)(x-2)}{(x+4)(x+1)(x-2)} - \dfrac{2(x+4)(x-2)}{(x+4)(x+1)(x-2)}$

$\qquad\qquad + \dfrac{(x+4)(x+1)}{(x+4)(x+1)(x-2)}$

$= \dfrac{3x^2 - 3x - 6 - 2x^2 - 4x + 16 + x^2 + 5x + 4}{(x+4)(x+1)(x-2)}$

$= \dfrac{2(x^2 - x + 7)}{(x+4)(x+1)(x-2)}$

c) $2 - \dfrac{3}{x+1} + \dfrac{4}{(x+1)^2} = \dfrac{2(x+1)^2}{(x+1)^2} - \dfrac{3(x+1)}{(x+1)^2} + \dfrac{4}{(x+1)^2}$

$= \dfrac{2x^2 + 4x + 2 - 3x - 3 + 4}{(x+1)^2} = \dfrac{2x^2 + x + 3}{(x+1)^2}$

d) $\dfrac{2x^2 - x - 3}{x^2 - 1} + \dfrac{1}{x(x-1)} = \dfrac{(x+1)(2x-3)}{(x+1)(x-1)} + \dfrac{1}{x(x-1)}$

$= \dfrac{2x-3}{x-1} + \dfrac{1}{x(x-1)} = \dfrac{x(2x-3)}{x(x-1)} + \dfrac{1}{x(x-1)}$

$= \dfrac{2x^2 - 3x + 1}{x(x-1)} = \dfrac{(x-1)(2x-1)}{x(x-1)} = \dfrac{2x-1}{x}$

Q8 a) $g(x) = \dfrac{1}{x} + \dfrac{2}{x+1} - \dfrac{1}{x+2}$

$= \dfrac{(x+1)(x+2)}{x(x+1)(x+2)} + \dfrac{2x(x+2)}{x(x+1)(x+2)} - \dfrac{x(x+1)}{x(x+1)(x+2)}$

$= \dfrac{(x+1)(x+2) + 2x(x+2) - x(x+1)}{x(x+1)(x+2)}$

$= \dfrac{x^2 + 3x + 2 + 2x^2 + 4x - x^2 - x}{x(x+1)(x+2)}$

$= \dfrac{2x^2 + 6x + 2}{x(x+1)(x+2)}$ or $\dfrac{2(x^2 + 3x + 1)}{x(x+1)(x+2)}$ or $\dfrac{2x^2 + 6x + 2}{x^3 + 3x^2 + 2x}$

[4 marks available — 1 mark for correct common denominator, 1 mark for all correct numerators, 1 mark for expanding brackets correctly, 1 mark for correct answer with simplified numerator]

b) If $g(x) = 0$, then $2x^2 + 6x + 2 = 0$. This quadratic has discriminant $6^2 - 4 \times 2 \times 2 = 20 > 0$, so g has two real roots.
[3 marks available — 1 mark for identifying numerator of g(x) must be zero, 1 mark for a correct method to find the number of roots, 1 mark for a correct conclusion]
The roots of g are the same as the roots of the numerator because you've written g in its simplest form. If it could be simplified further, this might not be true.

Q9 a) $\dfrac{2x^2 + 5x - 2}{x^2 + 4x + 3} + \dfrac{x-1}{x^2 - x - 2}$

$= \dfrac{2x^2 + 5x - 2}{(x+3)(x+1)} + \dfrac{x-1}{(x-2)(x+1)}$

$= \dfrac{(2x^2 + 5x - 2)(x-2) + (x-1)(x+3)}{(x-2)(x+1)(x+3)}$

$= \dfrac{2x^3 + x^2 - 12x + 4 + x^2 + 2x - 3}{(x-2)(x+1)(x+3)} = \dfrac{2x^3 + 2x^2 - 10x + 1}{(x-2)(x+1)(x+3)}$

[5 marks available — 1 mark for factorising one denominator, 1 mark for factorising the other denominator, 1 mark for method to add fractions, 1 mark for correct expansion of numerator, 1 mark for correct answer]

b) $\dfrac{2x^6 + 5x^3 - 2}{x^6 + 4x^3 + 3} + \dfrac{x^3 - 1}{x^6 - x^3 - 2} = \dfrac{2x^9 + 2x^6 - 10x^3 + 1}{(x^3 - 2)(x^3 + 1)(x^3 + 3)}$
[1 mark]

Exercise 2.3 — Multiplying and Dividing Algebraic Fractions

Q1 a) $\dfrac{2x}{3} \times \dfrac{5x}{4} = \dfrac{x}{3} \times \dfrac{5x}{2} = \dfrac{x \times 5x}{3 \times 2} = \dfrac{5x^2}{6}$

b) $\dfrac{6x^3}{7} \times \dfrac{2}{x^2} = \dfrac{6x}{7} \times \dfrac{2}{1} = \dfrac{6x \times 2}{7 \times 1} = \dfrac{12x}{7}$

c) $\dfrac{8x^2}{3y^2} \times \dfrac{x^3}{4y} = \dfrac{2x^2}{3y^2} \times \dfrac{x^3}{y} = \dfrac{2x^2 \times x^3}{3y^2 \times y} = \dfrac{2x^5}{3y^3}$

d) $\dfrac{8x^4}{3y} \times \dfrac{6y^2}{5x} = \dfrac{8x^3}{1} \times \dfrac{2y}{5} = \dfrac{8x^3 \times 2y}{1 \times 5} = \dfrac{16x^3 y}{5}$

Q2 a) $\dfrac{x}{3} \div \dfrac{3}{x} = \dfrac{x}{3} \times \dfrac{x}{3} = \dfrac{x \times x}{3 \times 3} = \dfrac{x^2}{9}$

b) $\dfrac{4x^3}{3} \div \dfrac{x}{2} = \dfrac{4x^3}{3} \times \dfrac{2}{x} = \dfrac{4x^2}{3} \times \dfrac{2}{1} = \dfrac{4x^2 \times 2}{3 \times 1} = \dfrac{8x^2}{3}$

c) $\dfrac{3}{2x} \div \dfrac{6}{x^3} = \dfrac{3}{2x} \times \dfrac{x^3}{6} = \dfrac{1}{2} \times \dfrac{x^2}{2} = \dfrac{1 \times x^2}{2 \times 2} = \dfrac{x^2}{4}$

d) $\dfrac{2x^3}{3y} \div \dfrac{4x}{y^2} = \dfrac{2x^3}{3y} \times \dfrac{y^2}{4x} = \dfrac{x^2}{3} \times \dfrac{y}{2} = \dfrac{x^2 \times y}{3 \times 2} = \dfrac{x^2 y}{6}$

Q3 a) $\dfrac{x+2}{4} \times \dfrac{x}{3x+6} = \dfrac{x+2}{4} \times \dfrac{x}{3(x+2)} = \dfrac{x}{12}$

b) $\dfrac{4x}{5} \div \dfrac{4x^2 + 8x}{15} = \dfrac{4x}{5} \times \dfrac{15}{4x(x+2)} = \dfrac{3}{(x+2)}$

c) $\dfrac{2}{x^2 + 4x} \div \dfrac{1}{x+4} = \dfrac{2}{x(x+4)} \times \dfrac{x+4}{1} = \dfrac{2}{x} \times \dfrac{1}{1} = \dfrac{2}{x}$

d) $\dfrac{2x^2 + 8x}{x^2 - 2x} \times \dfrac{x-1}{x+4} = \dfrac{2x(x+4)}{x(x-2)} \times \dfrac{x-1}{x+4}$

$= \dfrac{2}{x-2} \times \dfrac{x-1}{1} = \dfrac{2(x-1)}{x-2}$

Q4 a) $\dfrac{x^2 - 4}{9} \div \dfrac{x-2}{3} = \dfrac{(x+2)(x-2)}{9} \times \dfrac{3}{x-2}$

$= \dfrac{x+2}{3} \times \dfrac{1}{1} = \dfrac{x+2}{3}$

b) $\dfrac{2x^2 - 2}{x} \times \dfrac{5x}{3x-3} = \dfrac{2(x-1)(x+1)}{x} \times \dfrac{5x}{3(x-1)}$

$= \dfrac{2(x+1)}{1} \times \dfrac{5}{3} = \dfrac{2(x+1) \times 5}{1 \times 3} = \dfrac{10(x+1)}{3}$

c) $\dfrac{x^2 - 1}{3} \div \dfrac{x^2 + x}{6} = \dfrac{(x+1)(x-1)}{3} \times \dfrac{6}{x(x+1)} = \dfrac{2(x-1)}{x}$

d) $\dfrac{2x^2 - 2x - 24}{x^2 + 7x + 12} \div \dfrac{2}{x+4} = \dfrac{2(x^2 - x - 12)}{x^2 + 7x + 12} \times \dfrac{x+4}{2}$

$= \dfrac{2(x-4)(x+3)}{(x+4)(x+3)} \times \dfrac{x+4}{2} = x - 4$

e) $\dfrac{x^2 + 4x + 3}{x^2 + 5x + 6} \times \dfrac{x^2 + 2x}{x+1} = \dfrac{(x+1)(x+3)}{(x+2)(x+3)} \times \dfrac{x(x+2)}{x+1}$

$= \dfrac{1}{1} \times \dfrac{x}{1} = x$

f) $\dfrac{x^2 + 5x + 6}{x^2 - 2x - 3} \times \dfrac{3x+3}{x^2 + 2x} = \dfrac{(x+2)(x+3)}{(x-3)(x+1)} \times \dfrac{3(x+1)}{x(x+2)}$

$= \dfrac{x+3}{x-3} \times \dfrac{3}{x} = \dfrac{3(x+3)}{x(x-3)}$

g) $\dfrac{x^2 - 4}{6x - 3} \times \dfrac{2x^2 + 5x - 3}{x^2 + 2x} = \dfrac{(x+2)(x-2)}{3(2x-1)} \times \dfrac{(2x-1)(x+3)}{x(x+2)}$

$= \dfrac{x-2}{3} \times \dfrac{x+3}{x} = \dfrac{(x-2)(x+3)}{3x}$

h) $\dfrac{x^2 + 7x + 6}{4x - 4} \div \dfrac{x^2 + 8x + 12}{x^2 - x}$

$= \dfrac{(x+1)(x+6)}{4(x-1)} \times \dfrac{x(x-1)}{(x+2)(x+6)} = \dfrac{x(x+1)}{4(x+2)}$

Q5 a) $\dfrac{3x}{x^2 - 4x - 21} \times \dfrac{9 - x^2}{3x^2 + 21x}$

$= \dfrac{3x}{(x-7)(x+3)} \times \dfrac{(3+x)(3-x)}{3x(x+7)}$

$= \dfrac{3-x}{(x+7)(x-7)}$ or $\dfrac{3-x}{x^2 - 49}$

[4 marks available for the correct answer — 1 mark for factorising denominators, 1 mark for factorising $9 - x^2$, 1 mark for method to multiply, 1 mark for correct answer]

b) $\dfrac{3-(x-3)}{(x-3+7)(x-3-7)} = -\dfrac{x}{(x+4)(x-10)}$

[2 marks available — 1 mark for using a suitable method, 1 mark for the correct answer]

Q6 a) $\dfrac{x^2+4x+4}{x^2-4x+3} \times \dfrac{x^2-2x-3}{2x^2-2x} \times \dfrac{4x-4}{x^2+2x}$

$= \dfrac{(x+2)^2}{(x-3)(x-1)} \times \dfrac{(x-3)(x+1)}{2x(x-1)} \times \dfrac{4(x-1)}{x(x+2)}$

$= \dfrac{x+2}{x-1} \times \dfrac{x+1}{x} \times \dfrac{2}{x} = \dfrac{2(x+2)(x+1)}{x^2(x-1)}$

b) $\dfrac{x}{6x+12} \div \dfrac{x^2-x}{x+2} \times \dfrac{3x-3}{x+1}$

$= \dfrac{x}{6(x+2)} \times \dfrac{x+2}{x(x-1)} \times \dfrac{3(x-1)}{x+1} = \dfrac{1}{2(x+1)}$

c) $\dfrac{x^2+5x}{2x^2+7x+3} \times \dfrac{2x+1}{x^3-x^2} \div \dfrac{x+5}{x^2+x-6}$

$= \dfrac{x(x+5)}{(2x+1)(x+3)} \times \dfrac{2x+1}{x^2(x-1)} \times \dfrac{(x+3)(x-2)}{x+5}$

$= \dfrac{1}{1} \times \dfrac{1}{x(x-1)} \times \dfrac{x-2}{1} = \dfrac{x-2}{x(x-1)}$

d) $\dfrac{3x}{x+2} \div \dfrac{x-2}{x-3} \div \dfrac{x^2-3x}{x^2-4} = \left(\dfrac{3x}{x+2} \times \dfrac{x-3}{x-2}\right) \div \dfrac{x^2-3x}{x^2-4}$

$= \dfrac{3x(x-3)}{(x+2)(x-2)} \times \dfrac{(x+2)(x-2)}{x(x-3)} = 3$

Q7 a) $t^4 - 16 = (t^2-4)(t^2+4) = (t-2)(t+2)(t^2+4)$

[2 marks available — 1 mark for difference of two squares once, 1 mark for the correct answer]

b) $\dfrac{t^2+8t+12}{t^4-16} \div \dfrac{3t+18}{t^2+4}$

$= \dfrac{(t+2)(t+6)}{(t-2)(t+2)(t^2+4)} \times \dfrac{t^2+4}{3(t+6)} = \dfrac{1}{3(t-2)}$

[3 marks available — 1 mark for both numerator and denominator in first fraction factorised, 1 mark for numerator and denominator in second fraction factorised, 1 mark for correct answer]

Q8 $\dfrac{\ln(x)-2}{4x} \div \dfrac{\ln(x^2)-4}{8x^2} = \dfrac{\ln(x)-2}{4x} \times \dfrac{8x^2}{\ln(x^2)-4}$

$= \dfrac{\ln(x)-2}{4x} \times \dfrac{8x^2}{2\ln(x)-4} = \dfrac{\ln(x)-2}{4x} \times \dfrac{8x^2}{2(\ln(x)-2)} = \dfrac{8x^2}{8x} = x$

[3 marks available — 1 mark for writing ln (x²) as 2 ln x, 1 mark for cancelling either (ln x − 2) or 4x, 1 mark for correct answer]

Exercise 2.4 — Algebraic Division

Q1 a) $x^3 - 14x^2 + 6x + 11 \equiv (Ax^2 + Bx + C)(x+1) + D$

The degree of the quotient is the difference between the degrees of the polynomial and the divisor, in this case 3 − 1 = 2. The degree of the remainder must be less than the degree of the divisor (1) so it must be 0.

Set $x = -1$: $-1 - 14 - 6 + 11 = D$, so $D = -10$.

Set $x = 0$: $11 = C + D$, so $C = 21$.

Equating the coefficients of x^3 gives $1 = A$.

Equating the coefficients of x^2 gives:

$-14 = A + B$, so $B = -15$.

So $x^3 - 14x^2 + 6x + 11 \equiv (x^2 - 15x + 21)(x+1) - 10$.

Quotient: $x^2 - 15x + 21$

Remainder: -10

b) $2x^3 + 5x^2 - 8x - 17 \equiv (Ax^2 + Bx + C)(x-2) + D$

Set $x = 2$: $16 + 20 - 16 - 17 = D$, so $D = 3$.

Set $x = 0$: $-17 = -2C + D$, so $C = 10$.

Equating the coefficients of x^3 gives $2 = A$.

Equating the coefficients of x^2 gives:

$5 = -2A + B$, so $B = 9$.

So $2x^3 + 5x^2 - 8x - 17 \equiv (2x^2 + 9x + 10)(x-2) + 3$.

Quotient: $2x^2 + 9x + 10$

Remainder: 3

c) $6x^3 + x^2 - 11x - 5 \equiv (Ax^2 + Bx + C)(2x+1) + D$

Set $x = -\dfrac{1}{2}$: $-\dfrac{6}{8} + \dfrac{1}{4} + \dfrac{11}{2} - 5 = D$, so $D = 0$.

Set $x = 0$: $-5 = C + D$, so $C = -5$.

Equating the coefficients of x^3 gives $6 = 2A$, so $A = 3$.

Equating the coefficients of x^2 gives:

$1 = A + 2B$, so $B = -1$.

So $6x^3 + x^2 - 11x - 5 \equiv (3x^2 - x - 5)(2x+1)$.

Quotient: $3x^2 - x - 5$

Remainder: 0

Q2 $3x^4 - 8x^3 - 6x - 4 \equiv (Ax^3 + Bx^2 + Cx + D)(x-3) + E$

Set $x = 3$: $243 - 216 - 18 - 4 = E$, so $E = 5$.

Set $x = 0$: $-4 = -3D + E$, so $D = 3$.

Equating the coefficients:

x^4 terms $\Rightarrow$ $3 = A$.

x^3 terms $\Rightarrow$ $-8 = -3A + B$, so $B = 1$.

x terms $\Rightarrow$ $-6 = -3C + D$, so $C = 3$.

So $3x^4 - 8x^3 - 6x - 4 \equiv (3x^3 + x^2 + 3x + 3)(x-3) + 5$

or $(3x^4 - 8x^3 - 6x - 4) \div (x-3) = 3x^3 + x^2 + 3x + 3$ remainder 5

[3 marks available — 1 mark for a correct method, 1 mark for any three correct coefficients, 1 mark for the correct answer]

Q3 a)

$$\begin{array}{r} x^2 - 15x + 21 \ \text{r} -10 \\ x+1\overline{)\ x^3 - 14x^2 + 6x + 11} \\ -\ \underline{(x^3 + x^2)} \\ -15x^2 + 6x \\ -\ \underline{(-15x^2 - 15x)} \\ 21x + 11 \\ -\ \underline{(21x + 21)} \\ -10 \end{array}$$

Quotient: $x^2 - 15x + 21$, remainder: -10

b)

$$\begin{array}{r} x^2 + 7x - 6 \ \text{r} 5 \\ x+3\overline{)\ x^3 + 10x^2 + 15x - 13} \\ -\ \underline{(x^3 + 3x^2)} \\ 7x^2 + 15x \\ -\ \underline{(7x^2 + 21x)} \\ -6x - 13 \\ -\ \underline{(-6x - 18)} \\ 5 \end{array}$$

Quotient: $x^2 + 7x - 6$, remainder: 5

c)

$$\begin{array}{r} 2x^2 + 9x + 10 \ \text{r} 3 \\ x-2\overline{)\ 2x^3 + 5x^2 - 8x - 17} \\ -\ \underline{(2x^3 - 4x^2)} \\ 9x^2 - 8x \\ -\ \underline{(9x^2 - 18x)} \\ 10x - 17 \\ -\ \underline{(10x - 20)} \\ 3 \end{array}$$

Quotient: $2x^2 + 9x + 10$, remainder: 3

d)

$$\begin{array}{r} 3x^2 - 15x - 3 \ \text{r} 24 \\ x+5\overline{)\ 3x^3 + 0x^2 - 78x + 9} \\ -\ \underline{(3x^3 + 15x^2)} \\ -15x^2 - 78x \\ -\ \underline{(-15x^2 - 75x)} \\ -3x + 9 \\ -\ \underline{(-3x - 15)} \\ 24 \end{array}$$

Quotient: $3x^2 - 15x - 3$, remainder: 24

e)

$$\begin{array}{r} x^3 + x^2 + x + 1 \\ x-1\overline{)\ x^4 + 0x^3 + 0x^2 + 0x - 1} \\ -\ \underline{(x^4 - x^3)} \\ x^3 + 0x^2 \\ -\ \underline{(x^3 - x^2)} \\ x^2 + 0x \\ -\ \underline{(x^2 - x)} \\ x - 1 \\ -\ \underline{(x - 1)} \\ 0 \end{array}$$

Quotient: $x^3 + x^2 + x + 1$, remainder: 0

Answers

f)

$$2x - 3 \overline{)\,8x^3 - 6x^2 + x - 10\,}$$
$$\quad\ \ 4x^2 + 3x + 5\ \text{r } 25$$
$$-\ (8x^3 - 12x^2)$$
$$\qquad\qquad 6x^2 + x$$
$$\qquad\ -\ (6x^2 - 9x)$$
$$\qquad\qquad\qquad 10x + 10$$
$$\qquad\qquad\ -\ (10x - 15)$$
$$\qquad\qquad\qquad\qquad 25$$

Quotient: $4x^2 + 3x + 5$, remainder: 25

For Questions 4-9, we've given the algebraic long division method, but you could also have used the formula method — you should get the same quotient and remainder.

Q4

$$2x + 1 \overline{)\,10x^3 + 7x^2 - 5x + 21\,}$$
$$\quad\ \ 5x^2 + x - 3\ \text{r } 24$$
$$-\ (10x^3 + 5x^2)$$
$$\qquad\qquad 2x^2 - 5x$$
$$\qquad\ -\ (2x^2 + x)$$
$$\qquad\qquad\qquad -6x + 21$$
$$\qquad\qquad\ -\ (-6x - 3)$$
$$\qquad\qquad\qquad\qquad 24$$

Quotient: $5x^2 + x - 3$, remainder: 24

Q5

$$x - 2 \overline{)\,3x^3 - 8x^2 + 15x - 12\,}$$
$$\quad\ \ 3x^2 - 2x + 11\ \text{r } 10$$
$$-\ (3x^3 - 6x^2)$$
$$\qquad\qquad -2x^2 + 15x$$
$$\qquad\ -\ (-2x^2 + 4x)$$
$$\qquad\qquad\qquad 11x - 12$$
$$\qquad\qquad\ -\ (11x - 22)$$
$$\qquad\qquad\qquad\qquad 10$$

Quotient: $3x^2 - 2x + 11$, remainder: 10

Q6

$$2x - 3 \overline{)\,16x^4 + 0x^3 + 0x^2 + 0x + 0\,}$$
$$\quad\ \ 8x^3 + 12x^2 + 18x + 27\ \text{r } 81$$
$$-\ (16x^4 - 24x^3)$$
$$\qquad\qquad 24x^3 + 0x^2$$
$$\qquad\ -\ (24x^3 - 36x^2)$$
$$\qquad\qquad\qquad 36x^2 + 0x$$
$$\qquad\qquad\ -\ (36x^2 - 54x)$$
$$\qquad\qquad\qquad\qquad 54x + 0$$
$$\qquad\qquad\qquad\ -\ (54x - 81)$$
$$\qquad\qquad\qquad\qquad\qquad 81$$

Quotient: $8x^3 + 12x^2 + 18x + 27$, remainder: 81

Q7

$$2x - 3 \overline{)\,2x^3 - 5x^2 - 21x + 36\,}$$
$$\quad\ \ x^2 - x - 12\ \text{r } 0$$
$$-\ (2x^3 - 3x^2)$$
$$\qquad\qquad -2x^2 - 21x$$
$$\qquad\ -\ (-2x^2 + 3x)$$
$$\qquad\qquad\qquad -24x + 36$$
$$\qquad\qquad\ -\ (-24x + 36)$$
$$\qquad\qquad\qquad\qquad 0$$

$2x^3 - 5x^2 - 21x + 36 = (2x - 3)(x^2 - x - 12)$
$\qquad\qquad\qquad\qquad\quad = (2x - 3)(x - 4)(x + 3)$

So $2x^3 - 5x^2 - 21x + 36 = 0$ has solutions $x = \dfrac{3}{2}$, $x = 4$ and $x = -3$

[3 marks available — 1 mark for a correct method to factorise the cubic, 1 mark for the correct quadratic factor, 1 mark for all three correct solutions]

Q8

$$x + 1 \overline{)\,x^4 + 3x^3 + x^2 + 0x + 1\,}$$
$$\quad\ \ x^3 + 2x^2 - x + 1\ \text{r } 0$$
$$-\ (x^4 + x^3)$$
$$\qquad\qquad 2x^3 + x^2$$
$$\qquad\ -\ (2x^3 + 2x^2)$$
$$\qquad\qquad\qquad -x^2 + 0x$$
$$\qquad\qquad\ -\ (-x^2 - x)$$
$$\qquad\qquad\qquad\qquad x + 1$$
$$\qquad\qquad\ -\ (x + 1)$$
$$\qquad\qquad\qquad\qquad 0$$

$(x + 1)$ is a factor of $x^4 + 3x^3 + x^2 + 1$.
$\Rightarrow x = -1$ is a solution to $x^4 + 3x^3 + x^2 + 1 = 0$

Q9

$$3x - 2 \overline{)\,3x^4 + x^3 - 5x^2 - 4x + 4\,}$$
$$\quad\ \ x^3 + x^2 - x - 2\ \text{r } 0$$
$$-\ (3x^4 - 2x^3)$$
$$\qquad\qquad 3x^3 - 5x^2$$
$$\qquad\ -\ (3x^3 - 2x^2)$$
$$\qquad\qquad\qquad -3x^2 - 4x$$
$$\qquad\qquad\ -\ (-3x^2 + 2x)$$
$$\qquad\qquad\qquad\qquad -6x + 4$$
$$\qquad\qquad\ -\ (-6x + 4)$$
$$\qquad\qquad\qquad\qquad 0$$

So $(3x^4 + x^3 - 5x^2 - 4x + 4) \div (3x - 2) = x^3 + x^2 - x - 2$

Q10 $f(-1) = 1 - 2 + 1 = 0$, so $(x + 1)$ is a factor. So divide by this:

$$x + 1 \overline{)\,x^4 + 2x^3 + 0x^2 + 0x + 1\,}$$
$$\quad\ \ x^3 + x^2 - x + 1$$
$$-\ (x^4 + x^3)$$
$$\qquad\qquad x^3 + 0x^2$$
$$\qquad\ -\ (x^3 + x^2)$$
$$\qquad\qquad\qquad -x^2 + 0x$$
$$\qquad\qquad\ -\ (-x^2 - x)$$
$$\qquad\qquad\qquad\qquad x + 1$$
$$\qquad\qquad\ -\ (x + 1)$$
$$\qquad\qquad\qquad\qquad 0$$

So $f(x) = (x + 1)(x^3 + x^2 - x + 1)$
[3 marks available — 1 mark for correct method, 1 mark for two terms in quotient correct, 1 mark for fully correct quotient]

Q11

$$2y - 5 \overline{)\,2y^4 - 5y^3 + 6y^2 - 11y - 7\,}$$
$$\quad\ \ y^3 + 0y^2 + 3y + 2\ \text{r } 3$$
$$-\ (2y^4 - 5y^3)$$
$$\qquad\qquad 0 + 6y^2 - 11y$$
$$\qquad\ -\ (6y^2 - 15y)$$
$$\qquad\qquad\qquad 4y - 7$$
$$\qquad\qquad\ -\ (4y - 10)$$
$$\qquad\qquad\qquad\qquad 3$$

So the remainder is 3.
[3 marks available — 1 mark for correct method, 1 mark for two terms in quotient correct, 1 mark for the correct remainder]

Q12

$$x - 2 \overline{)\,3x^4 + 7x^3 - 22x^2 - 8x\,}$$
$$\quad\ \ 3x^3 + 13x^2 + 4x\ \text{r } 0$$
$$-\ (3x^4 - 6x^3)$$
$$\qquad\qquad 13x^3 - 22x^2$$
$$\qquad\ -\ (13x^3 - 26x^2)$$
$$\qquad\qquad\qquad 4x^2 - 8x$$
$$\qquad\qquad\ -\ (4x^2 - 8x)$$
$$\qquad\qquad\qquad\qquad 0$$

$3x^4 + 7x^3 - 22x^2 - 8x = (x - 2)(3x^3 + 13x^2 + 4x)$
$= x(x - 2)(3x^2 + 13x + 4) = x(x - 2)(3x + 1)(x + 4)$
So $3x^4 + 7x^3 - 22x^2 - 8x = 0$ has solutions
$x = 0$, $x = 2$, $x = -\dfrac{1}{3}$, $x = -4$

Q13 For $f(x) = 2x^4 - 5x^3 - 50x^2 - 85x - 42$, $f(-1) = 0$, so $(x + 1)$ is a factor. Use this as the divisor:

$$x + 1 \overline{)\,2x^4 - 5x^3 - 50x^2 - 85x - 42\,}$$
$$\quad\ \ 2x^3 - 7x^2 - 43x - 42\ \text{r } 0$$
$$-\ (2x^4 + 2x^3)$$
$$\qquad\qquad -7x^3 - 50x^2$$
$$\qquad\ -\ (-7x^3 - 7x^2)$$
$$\qquad\qquad\qquad -43x^2 - 85x$$
$$\qquad\qquad\ -\ (-43x^2 - 43x)$$
$$\qquad\qquad\qquad\qquad -42x - 42$$
$$\qquad\qquad\ -\ (-42x - 42)$$
$$\qquad\qquad\qquad\qquad 0$$

You need to use algebraic long division again.
For $g(x) = 2x^3 - 7x^2 - 43x - 42$, $g(-2) = 0$, so $(x + 2)$ is a factor.

Use this as the divisor for g(x):

$$\begin{array}{r} 2x^2 - 11x - 21 \ \ r\ 0 \\ x+2\overline{)\ 2x^3 - 7x^2 - 43x - 42} \\ -\underline{(2x^3 + 4x^2)} \\ -11x^2 - 43x \\ -\underline{(-11x^2 - 22x)} \\ -21x - 42 \\ -\underline{(-21x - 42)} \\ 0 \end{array}$$

You're left with a quadratic, which can be easily factorised:

$2x^2 - 11x - 21 = (2x+3)(x-7)$
$\Rightarrow 2x^4 - 5x^3 - 50x^2 - 85x - 42 = (x+1)(x+2)(2x+3)(x-7)$
$\Rightarrow 2x^4 - 5x^3 - 50x^2 - 85x - 42 = 0$ has solutions

$x = -1,\ x = -2,\ x = -\dfrac{3}{2},\ x = 7$

[5 marks available — 1 mark for finding a factor, 1 mark for dividing by factor to give cubic, 1 mark for using algebraic division again, 1 mark for fully factorising, 1 mark for correct answers]

Q14 a) $x^2 + 5 = (ax+b)(x+6) + c$ so substituting $x = -6$, we obtain $41 = c$. Substituting $x = 0$, we obtain $5 = 6b + 41$, so $b = -6$. Equating x^2 coefficients gives $a = 1$.
So $\dfrac{x^2 + 5}{x+6} = x - 6 + \dfrac{41}{x+6}$.

[3 marks available — 1 mark for suitable method to find the quotient, 1 mark for the correct quotient, 1 mark for a fully correct answer]

b) There is a vertical asymptote at $x = -6$. *[1 mark]*

Q15 a)

$$\begin{array}{r} 8x^3 - 4x^2 + 2x - 1 \ \ r\ 1 \\ 2x+1\overline{)\ 16x^4 + 0x^3 + 0x^2 + 0x + 0} \\ -\underline{(16x^4 + 8x^3)} \\ -8x^3 + 0x^2 \\ -\underline{(-8x^3 - 4x^2)} \\ 4x^2 + 0x \\ -\underline{(4x^2 + 2x)} \\ -2x + 0 \\ -\underline{(-2x - 1)} \\ 1 \end{array}$$

So $\dfrac{16x^4}{2x+1} = 8x^3 - 4x^2 + 2x - 1 + \dfrac{1}{2x+1}$

b) $\dfrac{32x^5}{2x+1} = \dfrac{2x \times 16x^4}{2x+1} = 2x\left(8x^3 - 4x^2 + 2x - 1 + \dfrac{1}{2x+1}\right)$

$= 16x^4 - 8x^3 + 4x^2 - 2x + 1 - \dfrac{1}{2x+1}$

Q16 a)

$$\begin{array}{r} x + (3+b) \\ x-b\overline{)\ x^2 + 3x + a} \\ -\underline{(x^2 - bx)} \\ (3+b)x + a \\ -\underline{((3+b)x - b(3+b))} \\ a + b(3+b) \end{array}$$

The quotient is $x + (3+b) = x + b + 3$
and the remainder is $a + b(3+b) = a + b^2 + 3b$.

[3 marks available — 1 mark for a correct method, 1 mark for correct quotient, 1 mark for correct remainder]

b) a and b are positive integers and $13 = a + b(3+b)$
$\Rightarrow b = 1$ and $a = 9$ or $b = 2$ and $a = 3$
When $b = 1$, $Q(x) = x + (3+1) = x + 4$
When $b = 2$, $Q(x) = x + (3+2) = x + 5$

[2 marks available — 1 mark for finding a and b, 1 mark for both correct expressions for Q(x)]

Exercise 2.5 — Mappings and Functions

Q1

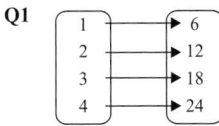

Q2

Q3 a)   **b)**

Q4 $g(0) = \dfrac{1}{2(0)+1} = 1,\ g(2) = \dfrac{1}{2(2)+1} = \dfrac{1}{5}$

Q5 $f(1) = \dfrac{1}{2 + \log_{10} 1} = \dfrac{1}{2+0} = \dfrac{1}{2}$

$f(100) = \dfrac{1}{2 + \log_{10} 100} = \dfrac{1}{2+2} = \dfrac{1}{4}$

[2 marks available — 1 mark for evaluating f(1), 1 mark for the correct answers]

Q6 a) The minimum value of $h(x)$ in the domain is $\sin 0° = 0$, and the maximum is $\sin 90° = 1$. So the range is $0 \le h(x) \le 1$

b) The minimum value of $j(x)$ in this domain is $\cos 180° = -1$, and the maximum is $\cos 0° = 1$. So the range is $-1 \le j(x) \le 1$

Q7 a) For $x > 2$, $r(x) > 18$. For $x \le 2$, $r(x) \ge 1$. So $r(x) \ge 1$ *[1 mark]*

b) $x^2 + 1 = 10 \Rightarrow x^2 = 9 \Rightarrow x = -3$ since $x < 2$.
$2x^3 + 2 = 10 \Rightarrow x^3 = 4 \Rightarrow x = \sqrt[3]{4} = 1.587...$,
which is outside the domain of x, so no solutions from this branch of graph.

[3 marks available — 1 mark for a correct method to find the solution, 1 mark for x = −3, 1 mark for this solution only]

Q8 a) Domain: $x \ge 0$ (as you can't use less than 0 GB).
Range: $P(x) \ge 57$

[2 marks available — 1 mark for the correct domain, 1 mark for the correct range]

b) The cost (in £) per GB of mobile data beyond 5 GB.
[1 mark for the correct answer]

c) $57 + 9.41(x-5) = 87.58 \Rightarrow x - 5 = \dfrac{87.58 - 57}{9.41} = 3.249...$
$\Rightarrow x = 8.25$ GB (2 d.p.)

[2 marks available — 1 mark for a suitable method to find x, 1 mark for the correct answer]

d) For $m > 1000$, the charge for m minutes of calls
$= 0.17(m - 1000)$
$P(6) = 66.41$ so the cost of mobile data (excluding the charge for additional minutes) must be between £57 and £66.41. So:
$66.58 - 66.41 \le 0.17(m - 1000) \le 66.58 - 57$
$\Rightarrow 0.17 \le 0.17(m - 1000) \le 9.58$
$\Rightarrow 1 \le m - 1000 \le 56.352...$
$\Rightarrow 1001 \le m \le 1056$ (to the nearest minute)

[4 marks available — 1 mark for working out P(6), 1 mark for a correct method to find the limits, 1 mark for attempting to simplify, 1 mark for the correct answer]

Q9 a) 3^x is defined for all $x \in \mathbb{R}$, so the domain of $f(x)$ is $\mathbb{R}$.
The value of 3^x gets closer to 0 as x becomes more negative, so its range is $3^x > 0$. So the range of f is $f(x) > -1$

b) $\ln x$ is not defined for $x \le 0$, so the largest possible domain of $g(x)$ is $x > 0$. The range of $\ln x$ is $\mathbb{R}$, but squaring the result means that $g(x)$ will always be positive or zero. So the range is $g(x) \ge 0$.

Q10 a) Yes, it is a function.

Answers

b) No, because the map is not defined for elements 4 and 5 of the domain.

c) No, because a value in the domain can map to more than one value in the range.

Q11 a) $f(\ln 6) = \sqrt{e^{2\ln 6} - e^{\ln 6} - 2} = \sqrt{36 - 6 - 2} = 2\sqrt{7}$
[2 marks available — 1 mark for a correct method, 1 mark for the correct answer]

b) Need $e^{2x} - e^x - 2 \geq 0 \Rightarrow (e^x - 2)(e^x + 1) \geq 0 \Rightarrow e^x \geq 2$ or $e^x \leq -1$. The latter is not possible since $e^x > 0$ for all real values of x. Hence $e^x \geq 2$, which implies $x \geq \ln 2$.
[4 marks available — 1 mark for setting $e^{2x} - e^x - 2 \geq 0$, 1 mark for method to solve, 1 mark for solving quadratic inequality in e^x, 1 mark for correct answer]

c) $e^{2x} - e^x - 2 \geq 0$ for all real values of x, so $f(x) \geq 0$ *[1 mark]*

Q12 a) $ae^{-\ln 2} + be^{\ln 2} = 1 \Rightarrow \frac{1}{2}a + 2b = 1 \Rightarrow a = 2 - 4b$
and $ae^{-\ln 3} + be^{\ln 3} = \frac{5}{4} \Rightarrow \frac{1}{3}a + 3b = \frac{5}{4} \Rightarrow a = \frac{15}{4} - 9b$
Solving simultaneously: $2 - 4b = \frac{15}{4} - 9b \Rightarrow 5b = \frac{7}{4}$
$\Rightarrow b = \frac{7}{20}$, and $a = 2 - 4 \times \frac{7}{20} = \frac{3}{5}$.
[3 marks available — 1 mark for substituting in a pair of values, 1 mark for correctly setting up simultaneous equations, 1 mark for correct answers]

b) e^{-x} and e^x are always positive, and $q(x) = \frac{3}{5}e^{-x} + \frac{7}{20}e^x$.
When $x \geq 0$, $e^x \geq 1$ so $q(x) = \frac{3}{5}e^{-x} + \frac{7}{20}e^x > \frac{7}{20}$.
When $x \leq 0$, $e^{-x} \geq 1$ so $q(x) = \frac{3}{5}e^{-x} + \frac{7}{20}e^x > \frac{3}{5}$.
So the range is not $q(x) > 0$ because it never takes values between 0 and $\frac{7}{20}$.
[2 marks available — 1 mark for some correct working related to the range, 1 mark for any valid justification]
The actual range isn't important here — you're only asked to show why it is not $q(x) > 0$.

Exercise 2.6 — Graphs of Functions

Q1 a) Yes, it is a function.

b) No, because a value of x can map to more than one value of $f(x)$.

Q2 a) 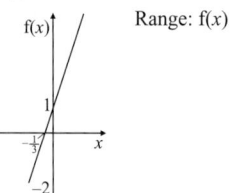 Range: $f(x) \geq -2$

b) 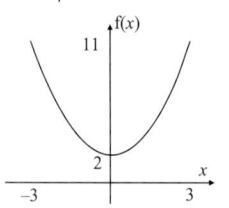 Range: $2 \leq f(x) \leq 11$

c) 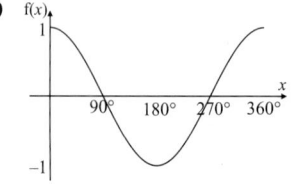 Range: $-1 \leq f(x) \leq 1$

d) 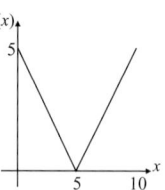 Range: $0 \leq f(x) \leq 5$

Q3 a) The range shown is $3 \leq f(x) \leq 11$, so the domain is:
$\frac{3-1}{2} \leq x \leq \frac{11-1}{2}$, which is $1 \leq x \leq 5$.
In set notation, this is $\{x : 1 \leq x \leq 5\}$.

b) The domain shown is $-1 \leq x \leq 4$. The range is between $f(2)$ and $f(-1)$: $((2)^2 - 4(2) + 5) \leq f(x) \leq ((-1)^2 - 4(-1) + 5)$, which is $1 \leq f(x) \leq 10$, i.e. $\{f(x) : 1 \leq f(x) \leq 10\}$.

Q4 When $x = 0$, $f(x) = 2$. For large x, $x + 2 \approx x + 1$ so as $x \to \infty$, $\frac{x+2}{x+1} \to 1$. So $f(x) = 1$ is an asymptote.
So the range is $1 < f(x) \leq 2$.
[2 marks available — 1 mark for the upper limit of 2, 1 mark for the lower limit of 1]

Q5 The function $f(x) = \frac{1}{x - 2}$ is undefined when $x = 2$, so $a = 2$.

Q6 The function $f(x) = +\sqrt{9 - x^2}$ is only defined when $x^2 \leq 9$, i.e. when $-3 \leq x \leq 3$. So $a = -3$ and $b = 3$.

Q7 The graph of $h(x) = +\sqrt{x + 1}$ is shown below.

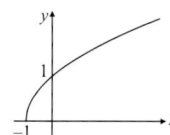

$h(x)$ is undefined when x is less than -1.
Restricting the domain to $\{x : x \geq -1\}$ would make $h(x)$ a function.

Q8 The graph of $k : x \to \tan x$ is shown below.

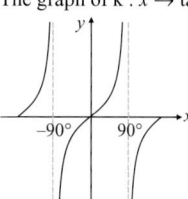

$\tan x$ is undefined at e.g. $-90°$ and $90°$
(it is also undefined periodically either side).
Restricting the domain to e.g. $-90° < x < 90°$ would make this a function.

Q9 a) $f : x \to \frac{2x + 6}{x}$ is undefined at $x = 0$ so the domain can't be $\mathbb{R}$.
[1 mark]

b) $y = f(x) = 2 + \frac{6}{x}$

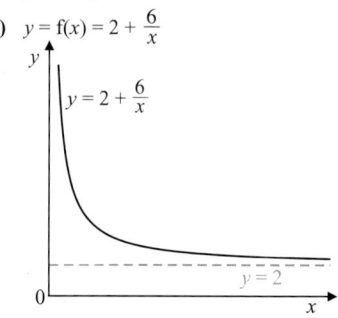

[2 marks available — 1 mark for the correct shape, 1 mark for a horizontal asymptote $y = 2$]

c) f(x) never meets the asymptote f(x) = 2, so f(x) > 2 *[1 mark]*

Q10 The function m(x) = $\dfrac{1}{x^2 - 4}$ is undefined when $x = 2$
and when $x = -2$.
So the largest continuous domain that makes it a function would
be either $x > 2$ or $x < -2$.

Q11 a) It is not a function because it is not defined for all values of x
in the domain (it's not defined at $x = 0$ or $x = 4$).

b) $x \in \mathbb{R}, x \neq 0, x \neq 4$

Q12 a) (i) Let $y = \sqrt{x}$. Then $y^2 - 3y + 2 = 0 \Rightarrow (y - 2)(y - 1) = 0$
$\Rightarrow y = 1$ and $y = 2$ so $x = 1^2 = 1$ and $x = 2^2 = 4$.
*[3 marks available — 1 mark for substituting to form a
quadratic, 1 mark for a correct method to solve, 1 mark
for both correct answers]*

(ii) $a = 1, b = 4$ *[1 mark]*

b) $\sqrt{x}$ is not defined for $x < 0$. *[1 mark]*

c) f(x) ≥ 0, f(x) ≤ -8
[2 marks available — 1 mark for each correct inequality]

Exercise 2.7 — Types of Function
It helps to sketch the graph of each function to identify its type.

Q1 a) One-to-one:

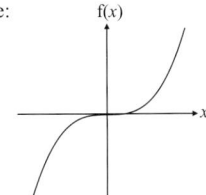

b) Many-to-one:

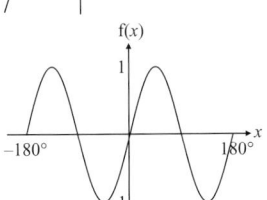

c) One-to-one:

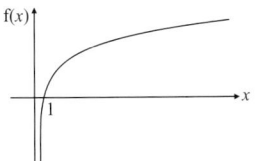

Q2 a) Many-to-one:

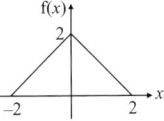

b) Many-to-one:

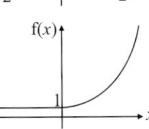

Exercise 2.8 — Composite Functions

Q1 a) Do g(3) = 2(3) + 1 = 7 then f(7) = 7^2 = 49, so fg(3) = 49.

b) gf(3) = g(3^2) = g(9) = 2(9) + 1 = 19

c) f^2(5) = f(5^2) = f(25) = 25^2 = 625

d) g^2(2) = g(2(2) + 1) = g(5) = 2(5) + 1 = 11

Q2 fg(90°) = f(2(90°)) = f(180°) = sin 180° = 0
gf(90°) = g(sin 90°) = g(1) = 2(1) = 2

Q3 a) gf(1) = g$\left(\dfrac{3}{1+2}\right)$ = g(1) = 2(1) = 2.

fg(1) = f(2(1)) = f(2) = $\dfrac{3}{2+2}$ = $\dfrac{3}{4}$.

f^2(4) = f$\left(\dfrac{3}{4+2}\right)$ = f$\left(\dfrac{1}{2}\right)$ = $\dfrac{3}{\frac{1}{2}+2}$ = $\dfrac{6}{1+4}$ = $\dfrac{6}{5}$

b) g(−1) = −2, and f(−2) has a denominator of 0
(which is undefined).

Q4 a) fg(x) = f(2x) = cos 2x *[1 mark]*

b) gf(x) = g(cos x) = 2 cos x *[1 mark]*

Q5 a) fg(x) = f(2^x) = 2(2^x) − 1 (= 2^{x+1} − 1)

b) gf(x) = g(2x − 1) = $2^{(2x-1)}$

c) f^2(x) = f(2x − 1) = 2(2x − 1) − 1 = 4x − 3

Q6 fg(x) = f(x + 4) = $\dfrac{2}{x+4-1}$ = $\dfrac{2}{x+3}$

gf(x) = g$\left(\dfrac{2}{x-1}\right)$ = $\dfrac{2}{x-1}$ + 4

= $\dfrac{2}{x-1}$ + $\dfrac{4x-4}{x-1}$ = $\dfrac{4x-2}{x-1}$ = $\dfrac{2(2x-1)}{x-1}$

Q7 f^2(x) = f$\left(\dfrac{x}{1-x}\right)$ = $\dfrac{\frac{x}{1-x}}{1-\frac{x}{1-x}}$ = $\dfrac{x}{(1-x)-x}$ = $\dfrac{x}{1-2x}$

gfg(x) = gf(x^2) = g$\left(\dfrac{x^2}{1-x^2}\right)$ = $\left(\dfrac{x^2}{1-x^2}\right)^2$ = $\dfrac{x^4}{(1-x^2)^2}$

Q8 a) fg(x) = f(2x − 3) = (2x − 3)², range: fg(x) ≥ 0.

b) gf(x) = g(x^2) = 2x^2 − 3, range: gf(x) ≥ -3.

Q9 a) gf(x) = g$\left(\dfrac{1}{x}\right)$ = ln $\left(\dfrac{1}{x} + 1\right)$
Domain: $\{x : x > 0\}$ (domain of g = range of f = $\{x : x > 0\}$)
Range: $\{$gf(x) : gf(x) > 0$\}$.

b) fg(x) = f(ln (x + 1)) = $\dfrac{1}{\ln (x+1)}$
Domain: $\{x : x > 0\}$ (domain of f = range of g = $\{x : x > 0\}$)
Range: $\{$fg(x) : fg(x) > 0$\}$

Q10 fgh(x) = fg(x^2 + 1) = f(5(x^2 + 1) − 1)
= f(5x^2 + 4) = 3(5x^2 + 4) + 2 = 15x^2 + 14

Q11 a) E.g. f(x) = x + 1. Then ff(x) = x + 2,
so ff(−2) = 0 but f(−2) = −1 ≠ 0.
*[2 marks available — 1 mark for a suitable function,
1 mark for a correct counterexample with value of a]*

b) E.g. If f(x) = x then ff(x) = f(x) = x.
If g(x) = −x then gg(x) = g(−x) = x.
So ff(x) = gg(x) but f(x) ≠ g(x).
*[2 marks available — 1 mark for suitable functions,
1 mark for correct working leading to a counterexample]*

Q12 a) gh(x) = g(x^2) = (x^2)³ − 1 = x^6 − 1
*[2 marks available — 1 mark for a correct method
to find gh(x), 1 mark for the correct answer]*

b) hg(x) = h(x^3 − 1) = (x^3 − 1)² = x^6 − 2x^3 + 1
*[2 marks available — 1 mark for a correct method
to find hg(x), 1 mark for the correct answer]*

c) x^6 ≥ 0 so x^6 − 1 ≥ -1, so gh(x) ≥ -1.
(x^3 − 1)² ≥ 0, so hg(x) ≥ 0.
So the ranges are different.
*[2 marks available — 1 mark for finding the range of either
function, 1 mark for a fully correct explanation]*

Exercise 2.9 — Solving Composite Function Equations

Q1 a) fg(x) = f(3x − 4) = 2(3x − 4) + 1 = 6x − 7
6x − 7 = 23 $\Rightarrow$ x = 5

b) gf(x) = g$\left(\dfrac{1}{x}\right)$ = $\dfrac{2}{x}$ + 5 $\Rightarrow$ $\dfrac{2}{x}$ + 5 = 6 $\Rightarrow$ x = 2

Answers

c) $gf(x) = g(x^2) = \dfrac{x^2}{x^2 - 3}$

$\dfrac{x^2}{x^2 - 3} = 4 \Rightarrow x^2 = 4x^2 - 12 \Rightarrow x^2 = 4 \Rightarrow x = 2$ or $x = -2$

d) $fg(x) = f(3x - 2) = (3x - 2)^2 + 1$

$(3x - 2)^2 + 1 = 50 \Rightarrow x = \dfrac{\pm\sqrt{49} + 2}{3} \Rightarrow x = 3$ or $x = -\dfrac{5}{3}$

e) $fg(x) = f(\sqrt{x}) = 2(\sqrt{x}) + 1$

$2(\sqrt{x}) + 1 = 17 \Rightarrow \sqrt{x} = 8 \Rightarrow x = 64$

f) $fg(x) = f(3 - x) = \log_{10}(3 - x)$

$\log_{10}(3 - x) = 0 \Rightarrow 3 - x = 1 \Rightarrow x = 2$

g) $fg(x) = f(x^2 + 2x) = 2^{(x^2 + 2x)} \Rightarrow 2^{(x^2 + 2x)} = 8 \Rightarrow x^2 + 2x = 3$
$\Rightarrow (x - 1)(x + 3) = 0 \Rightarrow x = 1$ or $x = -3$

Q2 $fg(x) = f(2x - 1) = \dfrac{2x - 1}{2x - 1 + 1} = \dfrac{2x - 1}{2x} = 1 - \dfrac{1}{2x}$

$gf(x) = g\left(\dfrac{x}{x + 1}\right) = 2\left(\dfrac{x}{x + 1}\right) - 1 = \dfrac{x - 1}{x + 1}$

$1 - \dfrac{1}{2x} = \dfrac{x - 1}{x + 1} \Rightarrow 2x(x + 1) - (x + 1) = 2x(x - 1)$

$2x^2 + 2x - x - 1 - 2x^2 + 2x = 0 \Rightarrow 3x = 1 \Rightarrow x = \dfrac{1}{3}$

[4 marks available — 1 mark for finding fg(x), 1 mark for finding gf(x), 1 mark for a correct method to solve equation, 1 mark for the correct answer]

Q3 a) $fg(x) = f(b - 3x) = (b - 3x)^2 + b$
Range: $(b - 3x)^2 \geq 0$, so $fg(x) \geq b$

$gf(x) = g(x^2 + b) = b - 3(x^2 + b) = -3x^2 - 2b$
Range: $-3x^2 \leq 0$, so $gf(x) \leq -2b$

b) $gf(2) = -3(2)^2 - 2b = -8 \Rightarrow 12 + 2b = 8 \Rightarrow b = -2$
$fg(2) = (-2 - 3(2))^2 - 2 = (-8)^2 - 2 = 64 - 2 = 62$

Q4 a) $pq(x) = p(2\ln(x + 1)) = e^{6\ln(x + 1)} - 3 = (x + 1)^6 - 3$.
[2 marks available — 1 mark for a correct unsimplified expression for pq(x), 1 mark for the correct answer]

b) $(x + 1)^6 - 3 = 61 \Rightarrow (x + 1)^6 = 64 \Rightarrow x + 1 = \pm\sqrt[6]{64} = \pm 2$,
so $x = 1$ or $x = -3$. But $x > -1$ so $x = 1$.
[3 marks available — 1 mark for correct method to solve equation, 1 mark for $x + 1 = \pm 2$, 1 mark for the correct answer]

c) $(x + 1)^6 - 3 = 0 \Rightarrow (x + 1)^6 = 3 \Rightarrow x + 1 = \pm\sqrt[6]{3}$
$\Rightarrow x = -1 \pm \sqrt[6]{3}$. But since $x > -1$, we have $x = -1 + \sqrt[6]{3}$.
[2 marks available — 1 mark for a correct method, 1 mark for the correct answer]

d) (i) $qq(x) = q(2\ln(x + 1)) = 2\ln(2\ln(x + 1) + 1)$
$= 2\ln(\ln((x + 1)^2) + 1)$
[1 mark for clear working giving the correct solution]

(ii) $\ln x$ defined only for $x > 0$ so need
$\ln((x + 1)^2) + 1 > 0 \Rightarrow 2\ln(x + 1) + 1 > 0$
$\Rightarrow \ln(x + 1) > -\dfrac{1}{2} \Rightarrow x + 1 > e^{-\frac{1}{2}}$
$\Rightarrow x > e^{-\frac{1}{2}} - 1$, so $a = e^{-\frac{1}{2}} - 1$.
[3 marks available — 1 mark for domain of ln x is x > 0, 1 mark for setting up a correct inequality, 1 mark for the correct answer in terms of e]

Q5 $fg(x) = f(a + \ln x) = \sqrt[3]{2a + 2\ln x + 4}$
$fg(e^2) = 2 \Rightarrow \sqrt[3]{2a + 2\ln e^2 + 4} = 2$
$\Rightarrow 2a + 4 + 4 = 2^3 \Rightarrow 2a = 0 \Rightarrow a = 0$
$g(x) = \ln(x)$ so $gg(x) = \ln(\ln x)$ so
$gg(e^x) = 5 \Rightarrow \ln(\ln e^x) = 5 \Rightarrow \ln(x) = 5 \Rightarrow x = e^5$.
[5 marks available — 1 mark for fg(x), 1 mark for a correct method to find a, 1 mark for a = 0, 1 mark for gg(x), 1 mark for the correct answer]

Exercise 2.10 — Inverse Functions and Their Graphs

Q1 a) Yes, as the graph shows a one-to-one function.

b) No, as it is a many-to-one map, and many-to-one functions do not have inverse functions.

Q2 a) No, as $\sin x$ is a many-to-one function over the domain $x \in \mathbb{R}$.

b) No, as it is a many-to-one function over the domain $x \in \mathbb{R}$.

c) Yes, as it is a one-to-one function over the domain $\{x : x \geq 4\}$.

Q3 a) $f(x) = 3x + 4$ with domain $x \in \mathbb{R}$ has a range $f(x) \in \mathbb{R}$.
Replace $f(x)$ with y: $y = 3x + 4$

Rearrange: $x = \dfrac{y - 4}{3}$

Replace with $f^{-1}(x)$ and x: $f^{-1}(x) = \dfrac{x - 4}{3}$.
The domain of $f^{-1}(x)$ is $x \in \mathbb{R}$ and the range is $f^{-1}(x) \in \mathbb{R}$.

b) $f(x) = 5(x - 2)$ with domain $x \in \mathbb{R}$ has a range $f(x) \in \mathbb{R}$.
Replace $f(x)$ with y: $y = 5(x - 2)$

Rearrange: $x = \dfrac{y}{5} + 2$

Replace with $f^{-1}(x)$ and x: $f^{-1}(x) = \dfrac{x}{5} + 2$.
The domain of $f^{-1}(x)$ is $x \in \mathbb{R}$ and the range is $f^{-1}(x) \in \mathbb{R}$.

c) $f(x) = \dfrac{1}{x + 2}$ with domain $x > -2$ has a range $f(x) > 0$.
Replace $f(x)$ with y: $y = \dfrac{1}{x + 2}$

Rearrange: $x = \dfrac{1}{y} - 2$

Replace with $f^{-1}(x)$ and x: $f^{-1}(x) = \dfrac{1}{x} - 2$.
The domain of $f^{-1}(x)$ is the range of $f(x)$: $x > 0$.
The range of $f^{-1}(x)$ is the domain of $f(x)$: $f^{-1}(x) > -2$.

d) $f(x) = x^2 + 3$ with domain $\{x : x > 0\}$
has a range $\{f(x) : f(x) > 3\}$.
Replace $f(x)$ with y: $y = x^2 + 3$
Rearrange: $x = \sqrt{y - 3}$
Replace with $f^{-1}(x)$ and x: $f^{-1}(x) = \sqrt{x - 3}$.

The domain of $f^{-1}(x)$ is the range of $f(x)$: $\{x : x > 3\}$.
The range of $f^{-1}(x)$ is the domain of $f(x)$: $\{f^{-1}(x) : f^{-1}(x) > 0\}$.

Q4 a) $f(x) = \dfrac{3x}{x + 1}$ with domain $x > -1$ has a range
$f(x) < 3$ — you can work this out by sketching the graph.
If you consider what happens as $x \to \infty$ you'll see that $f(x)$ approaches 3.

Replace $f(x)$ with y: $y = \dfrac{3x}{x + 1}$
Rearrange: $y(x + 1) = 3x \Rightarrow yx + y = 3x$
$\Rightarrow (3 - y)x = y \Rightarrow x = \dfrac{y}{3 - y}$
Replace with $f^{-1}(x)$ and x: $f^{-1}(x) = \dfrac{x}{3 - x}$.
The domain of $f^{-1}(x)$ is the range of $f(x)$: $x < 3$.
The range of $f^{-1}(x)$ is the domain of $f(x)$: $f^{-1}(x) > -1$.

b) $f^{-1}(2) = \dfrac{2}{3 - 2} = 2$

c) $f^{-1}\left(\dfrac{1}{2}\right) = \dfrac{\dfrac{1}{2}}{3 - \dfrac{1}{2}} = \dfrac{1}{6 - 1} = \dfrac{1}{5}$

Q5 **a)** $f(x) = \dfrac{x-4}{x+3}$ with domain $\{x : x > -3\}$ has a range $\{f(x) : f(x) < 1\}$ — again, sketching the graph will help. If you consider what happens as $x \to \infty$ you'll see that $f(x)$ approaches 1.

Replace $f(x)$ with y: $y = \dfrac{x-4}{x+3}$

Rearrange: $y(x+3) = x-4 \Rightarrow yx + 3y = x - 4$

$x(1-y) = 3y + 4 \Rightarrow x = \dfrac{3y+4}{1-y}$

Replace with $f^{-1}(x)$ and x: $f^{-1}(x) = \dfrac{3x+4}{1-x}$.

The domain of $f^{-1}(x)$ is the range of $f(x)$: $\{x : x < 1\}$.

The range of $f^{-1}(x)$ is the domain of $f(x)$: $\{f^{-1}(x) : f^{-1}(x) > -3\}$.

b) $f^{-1}(0) = \dfrac{3(0)+4}{1-0} = 4$

c) $f^{-1}\left(-\dfrac{2}{5}\right) = \dfrac{3\left(-\dfrac{2}{5}\right)+4}{1+\dfrac{2}{5}} = \dfrac{-6+20}{5+2} = \dfrac{14}{7} = 2$

Q6 **a)** $f(x)$ has domain $x > 3$, and range $f(x) \in \mathbb{R}$.
So $f^{-1}(x)$ has domain $x \in \mathbb{R}$ and range $f^{-1}(x) > 3$.

b) $f(x)$ has domain of $1 \le x \le 7$, which will give a range of $(4(1)-2) \le f(x) \le (4(7)-2)$, $2 \le f(x) \le 26$.
So $f^{-1}(x)$ has domain $2 \le x \le 26$ and range $1 \le f^{-1}(x) \le 7$.

c) $f(x)$ has domain $\{x : x < 2\}$ and range $\{f(x) : f(x) < 1\}$.
So $f^{-1}(x)$ has domain $\{x : x < 1\}$ and range $\{f^{-1}(x) : f^{-1}(x) < 2\}$.

d) $f(x)$ has domain $\{x : x \ge 2\}$ and range $\{f(x) : f(x) \ge 3\}$.
So $f^{-1}(x)$ has domain $\{x : x \ge 3\}$ and range $\{f^{-1}(x) : f^{-1}(x) \ge 2\}$.

e) $f(x)$ has domain $0° \le x < 90°$ and range $f(x) \ge 0$.
So $f^{-1}(x)$ has domain $x \ge 0$ and range $0° \le f^{-1}(x) < 90°$.

f) $f(x)$ has domain $\{x : 3 \le x \le 4\}$, which will give a range of $\{f(x) : \ln 9 \le f(x) \le \ln 16\}$.
So $f^{-1}(x)$ has domain $\{x : \ln 9 \le x \le \ln 16\}$ and range $\{f^{-1}(x) : 3 \le f^{-1}(x) \le 4\}$.

Q7 **a)** $y = e^{x+1} \Rightarrow \ln y = x + 1 \Rightarrow (\ln y) - 1 = x$
So $f^{-1}(x) = (\ln x) - 1$
$f(x)$ has domain $x \in \mathbb{R}$ and range $f(x) > 0$,
so $f^{-1}(x)$ has domain $x > 0$ and range $f^{-1}(x) \in \mathbb{R}$.

b) $y = x^3 \Rightarrow \sqrt[3]{y} = x$
So $f^{-1}(x) = \sqrt[3]{x}$
$f(x)$ has both domain $x < 0$ and range $f(x) < 0$,
so $f^{-1}(x)$ also has domain $x < 0$ and range $f^{-1}(x) < 0$.

c) $y = 2 - \log_2(x) \Rightarrow \log_2(x) = 2 - y \Rightarrow x = 2^{2-y}$
So $f^{-1}(x) = 2^{2-x}$
$f(x)$ has domain $x \ge 1$, which means that $\log_2(x) \ge 0$.
So the range of f is $f(x) \le 2$.
So $f^{-1}(x)$ has domain $x \le 2$ and range $f^{-1}(x) \ge 1$.

d) $y = \dfrac{1}{x-2} \Rightarrow \dfrac{1}{y} = x - 2 \Rightarrow \dfrac{1}{y} + 2 = x$

So $f^{-1}(x) = \dfrac{1}{x} + 2$

$f(x)$ has domain $\{x : x \ne 2\}$ and range $\{f(x) : f(x) \ne 0\}$,
so $f^{-1}(x)$ has domain $\{x : x \ne 0\}$ and range $\{f^{-1}(x) : f^{-1}(x) \ne 2\}$.
You can see the range of f(x) more easily from the graph of y = f(x) — there is a horizontal asymptote at y = 0.

e) $y = \dfrac{1}{e^x} = e^{-x} \Rightarrow \ln y = -x \Rightarrow x = -\ln y$

So $f^{-1}(x) = -\ln x$
$f(x)$ has domain $x \in \mathbb{R}$ and range $f(x) > 0$.
So $f^{-1}(x)$ has domain $x > 0$ and range $f^{-1}(x) \in \mathbb{R}$.

f) $y = \log_{10} e^x \Rightarrow 10^y = e^x \Rightarrow \ln 10^y = x$
So $f^{-1}(x) = \ln 10^x = x \ln 10$
$f(x)$ has domain $x \in \mathbb{R}$ and range $f(x) \in \mathbb{R}$
So $f^{-1}(x)$ has domain $x \in \mathbb{R}$ and range $f^{-1}(x) \in \mathbb{R}$.
Even though the functions f(x) and f⁻¹(x) look complicated, they're both just straight lines through the origin.

Q8 **a)** $g(x) = x^2$
Replace $g(x)$ with y: $y = x^2$
Rearrange: $x = \pm\sqrt{y}$ but $x < 0$ so $x = -\sqrt{y}$
Replace with $g^{-1}(x)$ and x: $g^{-1}(x) = -\sqrt{x}$
[2 marks available — 1 mark for a correct method, 1 mark for the correct answer]

b) Range of $g^{-1}(x) = $ domain of $g(x) = \{g^{-1}(x) : g^{-1}(x) < 0\}$.
Domain of $g^{-1}(x) = $ range of $g(x) = \{x : x > 0\}$.
[2 marks available — 1 mark for the correct range, 1 mark for the correct domain]

Q9

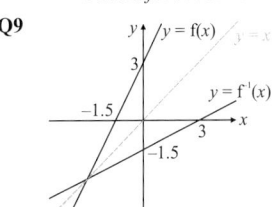

[3 marks available — 1 mark for a correct sketch of y = f(x), 1 mark for a correct sketch of y = f⁻¹(x), 1 mark for all four correctly labelled points where the graphs cross the axes]

Q10 **a)**

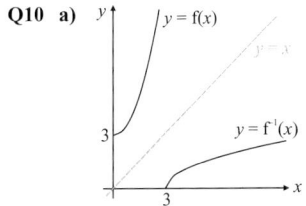

b) $f(x)$ has a domain of $x > 0$, giving a range of $f(x) > 3$.
So $f^{-1}(x)$ has a domain $x > 3$ and range $f^{-1}(x) > 0$.

Q11 **a)** $f(x) = x^2 - 5$
Replace $f(x)$ with y: $y = x^2 - 5$
Rearrange: $x^2 = y + 5 \Rightarrow x = \pm\sqrt{y+5}$
but $x \ge 0$ so $x = \sqrt{y+5}$.
Replace with $f^{-1}(x)$ and x: $f^{-1}(x) = \sqrt{x+5}$
Domain of $f^{-1}(x) = $ Range of $f(x) = \{x : x \ge -5\}$.
[3 marks available — 1 mark for a suitable method to find the inverse function, 1 mark for the correct inverse function, 1 mark for the correct domain]

b)

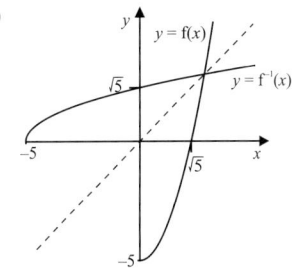

[2 marks available — 1 mark for correct sketch of f(x) for $x \ge 0$, 1 mark for correct sketch of y = f⁻¹(x) for $x \ge -5$]

Answers

c) The point of intersection lies on the line $y = x$,
so $x^2 - 5 = x \Rightarrow x^2 - x - 5 = 0$
$$\Rightarrow x = \frac{-(-1) \pm \sqrt{(-1)^2 - 4(1)(-5)}}{2(1)}$$
$$\Rightarrow x = \frac{1 + \sqrt{21}}{2} \text{ since } x \geq 0$$
[3 marks available — 1 mark for setting f(x) = x,
1 mark for a correct method to solve the quadratic,
1 mark for the correct answer]

Q12 a)

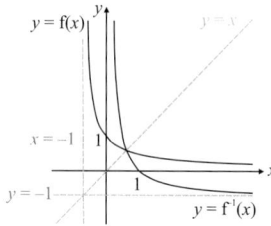

b) There is one point where the graphs intersect.

Q13 a) $f(x) = \frac{1}{x - 3}$ with domain $x > 3$ has a range $f(x) > 0$.
Replace $f(x)$ with y: $y = \frac{1}{x - 3}$
Rearrange: $x = \frac{1}{y} + 3$
Replace with $f^{-1}(x)$ and x: $f^{-1}(x) = \frac{1}{x} + 3$.
The domain of $f^{-1}(x)$ is the range of $f(x)$: $x > 0$.
The range of $f^{-1}(x)$ is the domain of $f(x)$: $f^{-1}(x) > 3$.

b)

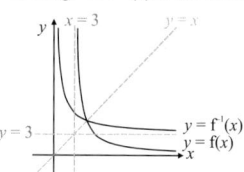

c) There is one solution as the graphs intersect once.

d) $\frac{1}{x - 3} = \frac{1}{x} + 3 \Rightarrow x = (x - 3) + 3x(x - 3)$
$$\Rightarrow x^2 - 3x - 1 = 0$$
So using the quadratic formula gives $x = \frac{3 + \sqrt{13}}{2}$
You can ignore the negative solution to the quadratic equation
because you're only considering the domain x > 3.

Q14 a) $e^{-x} > 0$ so $f(x) = 5e^{-x} + k \Rightarrow f(x) > k$
[1 mark for correct answer]

b) $f(x) = 5e^{-x} + k$
Replace $f(x)$ with y: $y = 5e^{-x} + k$
Rearrange: $y - k = 5e^{-x} \Rightarrow \frac{y - k}{5} = e^{-x}$
$$\Rightarrow -x = \ln\left(\frac{y - k}{5}\right) \Rightarrow x = -\ln\left(\frac{y - k}{5}\right) = \ln\left(\frac{5}{y - k}\right).$$
Replace with $f^{-1}(x)$ and x: $f^{-1}(x) = \ln\left(\frac{5}{x - k}\right)$
[3 marks available — 1 mark for a correct method
to find the inverse, 1 mark for taking logs correctly,
1 mark for the correct answer]

c) Domain of $f^{-1}(x)$ = Range of $f(x) \Rightarrow x > k$
Range of $f^{-1}(x)$ = Domain of $f(x) \Rightarrow f^{-1}(x) \in \mathbb{R}$
[2 marks available — 1 mark for the correct domain,
1 mark for the correct range]

d) $f^{-1}(8) = 0$ so
$\ln\left(\frac{5}{8 - k}\right) = 0 \Rightarrow \frac{5}{8 - k} = 1 \Rightarrow 5 = 8 - k \Rightarrow k = 3$
[2 marks available — 1 mark for a correct method,
1 mark for the correct answer]

Q15 a) Solving $f(x) = 2$, we have
$\sqrt[3]{1 - 4x} = 2 \Rightarrow 1 - 4x = 2^3 = 8 \Rightarrow 4x = -7 \Rightarrow x = -\frac{7}{4}$
so $f^{-1}(2) = -\frac{7}{4}$.
[2 marks available — 1 mark for a correct method,
1 mark for the correct answer]

b) The domain and range of $f(x)$ are both $\mathbb{R}$, so
Domain of $f^{-1}(x)$ = Range of $f^{-1}(x) = \mathbb{R}$
[1 mark for the correct answer]

c)

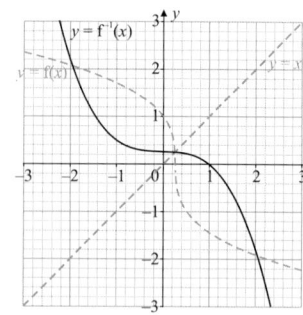

[2 marks available — 1 mark for attempt to reflect in the line
y = x, 1 mark for fully correct graph]

d) $f(x) = f^{-1}(x)$ will have 3 solutions since the graphs of
$y = f(x)$ and $y = f^{-1}(x)$ intersect at three points.
[1 mark for the correct answer with a valid reason]

Q16 a) $y = \frac{3x + 6}{x^2 - 2x - 8} = \frac{3(x + 2)}{(x - 4)(x + 2)} = \frac{3}{x - 4}$
$$\Rightarrow x - 4 = \frac{3}{y} \Rightarrow x = \frac{3}{y} + 4. \text{ So } f^{-1}(x) = \frac{3}{x} + 4$$
$f(x)$ has domain $x < 4$ and range $f(x) < 0$
So $f^{-1}(x)$ has domain $x < 0$ and range $f^{-1}(x) < 4$.

b) $f^2(x) = f(f(x)) = f\left(\frac{3}{x - 4}\right) = \frac{3}{\frac{3}{x - 4} - 4}$
$$= = \frac{3(x - 4)}{3 - 4(x - 4)} = \frac{3(x - 4)}{19 - 4x}$$
$f^2(x) = f^{-1}(x) \Rightarrow \frac{3(x - 4)}{19 - 4x} = \frac{3}{x} + 4$
$$\Rightarrow 3x(x - 4) = 3(19 - 4x) + 4x(19 - 4x)$$
$$\Rightarrow 3x^2 - 12x = 57 - 12x + 76x - 16x^2$$
$$\Rightarrow 19x^2 - 76x - 57 = 0 \Rightarrow x^2 - 4x - 3 = 0$$
$$\Rightarrow x = \frac{4 \pm \sqrt{(-4)^2 - 4 \times 1 \times -3}}{2} = 2 \pm \sqrt{7}$$
But $f^{-1}(x)$ has domain $x < 0$, so only
$x = 2 - \sqrt{7}$ is a valid solution.

Q17 a) As $x \to 0$, $3^x \to 1$ so $\frac{3^x + 1}{3^x - 1} \to \infty$ and $h(x) \to \infty$
As $x \to \infty$, $\frac{3^x + 1}{3^x - 1} \to 1$ and $h(x) \to 0$.
So the range of $h(x)$ is $h(x) > 0$.
[2 marks available — 1 mark for any correct
method, 1 mark for the correct answer]

b) $h(x) = \log_3\left(\frac{3^x + 1}{3^x - 1}\right)$
Replace $h(x)$ with y: $y = \log_3\left(\frac{3^x + 1}{3^x - 1}\right)$
Rearrange: $3^y = \frac{3^x + 1}{3^x - 1} \Rightarrow 3^x 3^y - 3^y = 3^x + 1$
$$\Rightarrow 3^x 3^y - 3^x = 3^y + 1 \Rightarrow 3^x(3^y - 1) = 3^y + 1$$
$$\Rightarrow 3^x = \frac{3^y + 1}{3^y - 1} \Rightarrow x = \log_3\left(\frac{3^y + 1}{3^y - 1}\right)$$
Replace with $h^{-1}(x)$ and x: $h^{-1}(x) = \log_3\left(\frac{3^x + 1}{3^x - 1}\right)$
Domain of $h^{-1}(x)$ = range of $h(x) = \{x : x > 0\}$.
[4 marks available — 1 mark for method to find h^{-1}, 1 mark
for $3^x 3^y - 3^y = 3^x + 1$ or equivalent, 1 mark for the correct
inverse function, 1 mark for correct domain]

Exercise 2.11 — The Modulus Function

Q1 **a)** 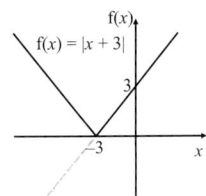 Range: f(x) ≥ 0

f(x) = |x + 3|

b) Range: f(x) ≥ 0

f(x) = |5 − x|

c) f(x) = |3x − 1| Range: f(x) ≥ 0

d) y = |x| − 9 Range: f(x) ≥ −9

e) Range: f(x) ≥ 5

y = 2|x| + 5

f) Range: f(x) ≥ −11

y = 3|x| − 11

Q2 **a)** 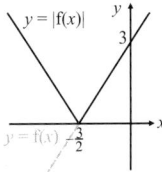 y = |f(x)|

b) 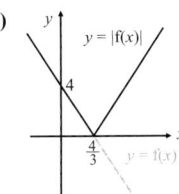 y = |f(x)|

c) 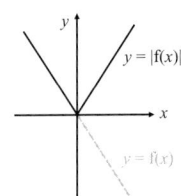 y = |f(x)|

d) 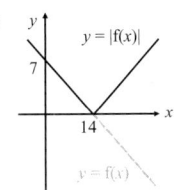 y = |f(x)|

e) 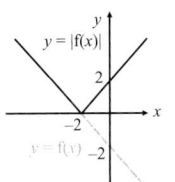 y = |f(x)|

f) 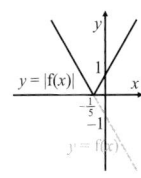 y = |f(x)|

Q3 **a)** 2 **b)** 1 **c)** 4 **d)** 3

Q4 **a)** **(i)** 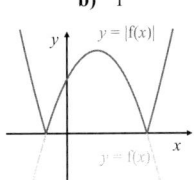 y = |f(x)| **(ii)** 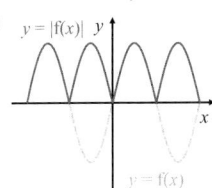 y = |f(x)|

(iii) 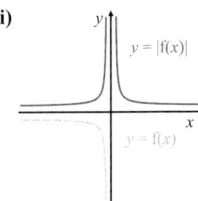 y = |f(x)|

b) 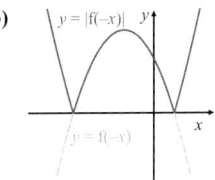 y = |f(−x)|

Q5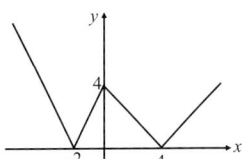

[4 marks available — 1 mark for correct shape for the graph where x < 0, 1 mark for labelling x = −2, 1 mark for correct shape for the graph where x ≥ 0, 1 mark for labelling x = 4 and y = 4]

Q6 **a)** 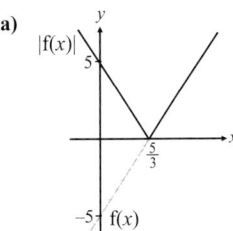 |f(x)|

b) 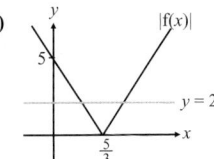 |f(x)|, y = 2

The line y = 2 intersects with the line y = |3x − 5| in two places so there are 2 solutions to |3x − 5| = 2.

Answers

Q7 a)

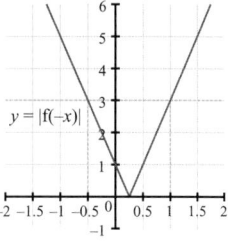

$y = |f(-x)|$

b) There are two solutions to $|-4x + 1| = 3$.
Reading off the graph, these are: $x = -0.5$ and $x = 1$.

Q8 a)

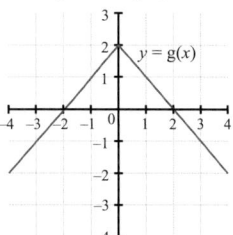

$y = g(x)$

[1 mark]

b)

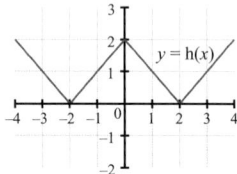

$y = h(x)$

[1 mark]

Q9 a)

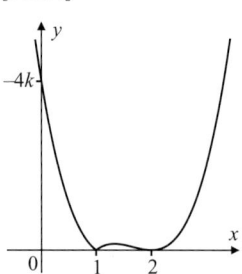

$-4k$

[3 marks available — 1 mark for correct shape, 1 mark for correct x-intercepts, 1 mark for correct y-intercept]

b)

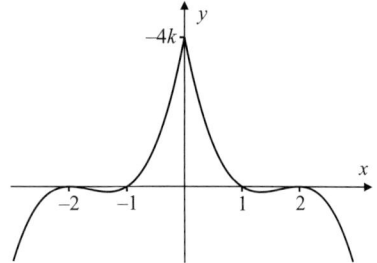

$-4k$

[3 marks available — 1 mark for correct shape, 1 mark for correct x-intercepts, 1 mark for correct y-intercept]

c) $N = 6$ since the horizontal line $y = a$ can intersect the graph in part b) at most six times (just below the x-axis).
[2 marks available — 1 mark for the correct answer, 1 mark for the correct reason]

Q10 a)

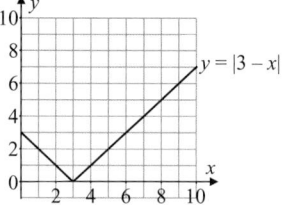

$y = |3 - x|$

[2 marks available — 1 mark for drawing the $0 \leq x \leq 3$ segment of the graph correctly, 1 mark for drawing the $3 \leq x \leq 10$ segment of the graph correctly]

b) $y = \frac{1}{2}x + k$ has a shallower gradient than $y = |3 - x|$ so the only way for one real solution is if it passes through $(3, 0)$.
So $0 = \frac{1}{2}(3) + k \Rightarrow k = -\frac{3}{2}$.
[2 marks available — 1 mark for identifying (3, 0) lies on the line, 1 mark for the correct answer]

c) Try sketching $y = \frac{1}{2}x + k$ for a few different values of k:

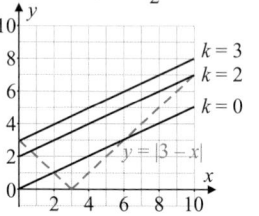

$k = 3$
$k = 2$
$k = 0$
$y = |3 - x|$

It's clear to see that there are two solutions when $k \leq 2$, and there's only one for all values $2 < k \leq 3$.
[2 marks available — 1 mark for a correct method, 1 mark for the correct answer]

Exercise 2.12 — Solving Modulus Equations and Inequalities

Q1 a)

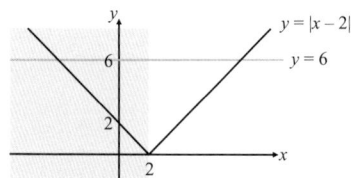

$y = |x - 2|$
$y = 6$

The graph shows that there are two solutions to $|x - 2| = 6$.
$x - 2 \geq 0$ for $x \geq 2$
$x - 2 < 0$ for $x < 2$ (shaded)
So there are two equations to solve:
① $x - 2 = 6 \Rightarrow x = 8$ (this is valid as it's in the range $x \geq 2$)
② (shaded) $-(x - 2) = 6 \Rightarrow x = -4$
(this is also valid as it's in the range $x < 2$)
So the two solutions are $x = 8$ and $x = -4$.

b)

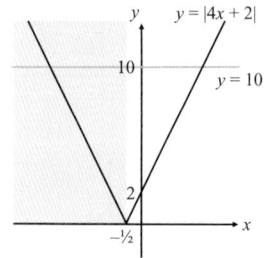

$y = |4x + 2|$
$y = 10$

The graph shows that there are two solutions to $|4x + 2| = 10$.
$4x + 2 \geq 0$ for $x \geq -\frac{1}{2}$

$4x + 2 < 0$ for $x < -\frac{1}{2}$ (shaded)

So there are two equations to solve:
① $4x + 2 = 10 \Rightarrow x = 2$
(this is valid as it's in the range $x \geq -\frac{1}{2}$)
② (shaded) $-(4x + 2) = 10 \Rightarrow x = -3$
(this is also valid as it's in the range $x < -\frac{1}{2}$)
So the two solutions are $x = 2$ and $x = -3$.

c) Rearranging the equation gives $|3x - 4| = 1$.

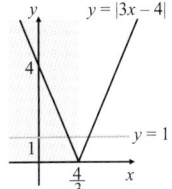

The graph shows that there are two solutions to $|3x - 4| = 1$.
$3x - 4 \geq 0$ for $x \geq \frac{4}{3}$
$3x - 4 < 0$ for $x < \frac{4}{3}$ (shaded)
So there are two equations to solve:
① $3x - 4 = 1 \Rightarrow x = \frac{5}{3}$
(this is valid as it's in the range $x \geq \frac{4}{3}$)
② (shaded) $-(3x - 4) = 1 \Rightarrow x = 1$
(this is also valid as it's in the range $x < \frac{4}{3}$)
So the two solutions are $x = \frac{5}{3}$ and $x = 1$.

d) Rearranging the equation gives $|x + 3| = 9$.

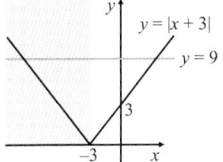

The graph shows that there are two solutions to $|x + 3| = 9$.
$x + 3 \geq 0$ for $x \geq -3$
$x + 3 < 0$ for $x < -3$ (shaded)
So there are two equations to solve:
① $x + 3 = 9 \Rightarrow x = 6$
(this is valid as it's in the range $x \geq -3$)
② (shaded) $-(x + 3) = 9 \Rightarrow x = -12$
(this is also valid as it's in the range $x < -3$)
So the two solutions are $x = 6$ and $x = -12$.

e)

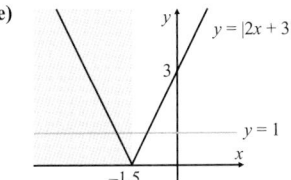

The graph shows that there are two solutions to $|2x + 3| = 1$.
$2x + 3 \geq 0$ for $x \geq -1.5$
$2x + 3 < 0$ for $x < -1.5$ (shaded)
So there are two equations to solve:
① $2x + 3 = 1 \Rightarrow x = -1$
(this is valid as it's in the range $x \geq -1.5$)
② (shaded) $-(2x + 3) = 1 \Rightarrow x = -2$
(this is also valid as it's in the range $x < -1.5$)
So the two solutions are $x = -1$ and $x = -2$.

f)

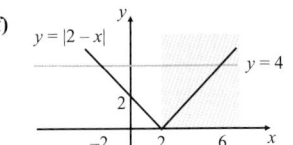

The graph shows that there are two solutions to $|2 - x| = 4$.
$2 - x \geq 0$ for $x < 2$
$2 - x < 0$ for $x \geq 2$ (shaded)

So there are two equations to solve:
① $2 - x = 4 \Rightarrow x = -2$
(this is valid as it's in the range $x < 2$)
② (shaded) $-(2 - x) = 4 \Rightarrow x = 6$
(this is also valid as it's in the range $x \geq 2$)
So the two solutions are $x = -2$ and $x = 6$.

Q2 $|x| = 5 \Rightarrow x = 5$ or $x = -5$
If $x = 5$, $|3x + 2| = |15 + 2| = 17$
If $x = -5$, $|3x + 2| = |-15 + 2| = |-13| = 13$

Q3 $|x| - 2 = -1 \Rightarrow |x| = 1 \Rightarrow x = 1$ or $x = -1$
If $x = 1$, $|7x - 1| = |6| = 6$
If $x = -1$, $|7x - 1| = |-8| = 8$

Q4 $|x| = 3 \Rightarrow x = 3$ or $x = -3$
If $x = 3$, $|-2x + 1| = |-6 + 1| = |-5| = 5$
If $x = -3$, $|-2x + 1| = |6 + 1| = |7| = 7$

Q5 a)

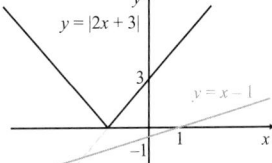

b) You can see from the graph that the lines do not intersect, so there are no solutions to $|f(x)| = g(x)$.

Q6 a)

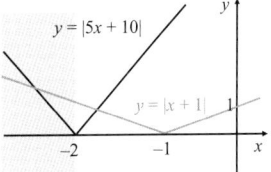

b) There are two solutions to $|f(x)| = |g(x)|$, one where $-2 < x < -1$ and one where $x < -2$ (shaded).
So there are two equations to solve:
① $5x + 10 = -(x + 1) \Rightarrow 6x = -11 \Rightarrow x = -\frac{11}{6}$
(this is valid as it's in the range $-2 < x < -1$)
② (shaded) $-(5x + 10) = -(x + 1) \Rightarrow -4x = 9 \Rightarrow x = -\frac{9}{4}$
(this is also valid as it's in the range $x < -2$)
So the two solutions to $|f(x)| = |g(x)|$ are
$x = -\frac{11}{6}$ and $x = -\frac{9}{4}$.

Q7 a)

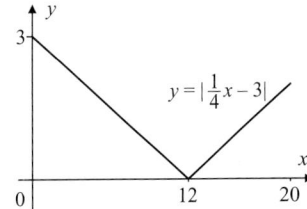

[2 marks available — 1 mark for correct shape in the first quadrant, 1 mark for fully correct graph through (0, 3) and (12, 0)]

Answers

b) $C = 4.5$, so $4.5 = \frac{1}{2}t - 1 + \left|\frac{1}{4}t - 3\right| \Rightarrow \left|\frac{1}{4}t - 3\right| = \frac{11}{2} - \frac{1}{2}t$

So there are two equations to solve:

For $t > 12$: $\frac{1}{4}t - 3 = \frac{11}{2} - \frac{1}{2}t \Rightarrow t - 12 = 22 - 2t \Rightarrow 3t = 34$

$\Rightarrow t = \frac{34}{3}$. But $\frac{34}{3} < 12$ so this solution is not valid.

For $t \leq 12$: $-\left(\frac{1}{4}t - 3\right) = \frac{11}{2} - \frac{1}{2}t \Rightarrow -t + 12 = 22 - 2t$

$\Rightarrow t = 10$. And $10 \leq 12$, so this solution is valid.
So a call costing £4.50 lasts 10 minutes.
[3 marks available — 1 mark for correct rearrangement, 1 mark for suitable method to find t, 1 mark for correct value of t]

c) $C = 36$, so $36 = \frac{1}{2}t - 1 + \left|\frac{1}{4}t - 3\right| \Rightarrow \left|\frac{1}{4}t - 3\right| = 37 - \frac{1}{2}t$

So there are two equations to solve:

For $t > 12$: $\frac{1}{4}t - 3 = 37 - \frac{1}{2}t \Rightarrow \frac{3}{4}t = 40 \Rightarrow t = \frac{160}{3}$

$\frac{160}{3} > 12$, so this solution is valid.

For $t \leq 12$: $-\left(\frac{1}{4}t - 3\right) = 37 - \frac{1}{2}t \Rightarrow \frac{1}{4}t = 34 \Rightarrow t = 136$

But $136 > 12$, so this solution is not valid.

So $t = \frac{160}{3} = 53.33... = 53$ mins 20 seconds.
[3 marks available — 1 mark for correct rearrangement, 1 mark for suitable method to find t, 1 mark for correct value of t]

Q8 a)

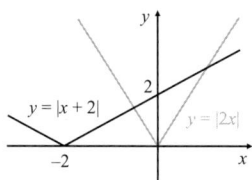

The graph shows that there are two solutions, one where both $x + 2$ and $2x$ are positive, and one where $x + 2$ is positive but $2x$ is negative.
So there are two equations to solve:
① $x + 2 = 2x \Rightarrow x = 2$
② $x + 2 = -2x \Rightarrow x = -\frac{2}{3}$
So the two solutions are $x = 2$ and $x = -\frac{2}{3}$.
Or using the algebraic method, solve:
$(x + 2)^2 = (2x)^2 \Rightarrow x^2 + 4x + 4 = 4x^2$
$\Rightarrow 3x^2 - 4x - 4 = 0 \Rightarrow (3x + 2)(x - 2) = 0$
So $x = 2$ and $x = -\frac{2}{3}$.

b)

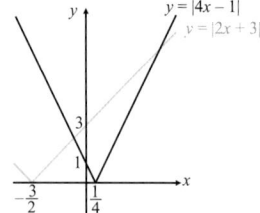

The graph shows that there are two solutions, one where both $2x + 3$ and $4x - 1$ are positive, and one where $2x + 3$ is positive but $4x - 1$ is negative.
So there are two equations to solve:
① $4x - 1 = 2x + 3 \Rightarrow x = 2$
② $-(4x - 1) = 2x + 3 \Rightarrow x = -\frac{1}{3}$
So the two solutions are $x = 2$ and $x = -\frac{1}{3}$.

Or using the algebraic method, solve:
$(2x + 3)^2 = (4x - 1)^2 \Rightarrow 4x^2 + 12x + 9 = 16x^2 - 8x + 1$
$\Rightarrow 12x^2 - 20x - 8 = 0 \Rightarrow 3x^2 - 5x - 2 = 0$
$\Rightarrow (3x + 1)(x - 2) = 0$
So $x = 2$ and $x = -\frac{1}{3}$.

c) Solving algebraically:
$|3x - 6| = |10 - 5x| \Rightarrow (3x - 6)^2 = (10 - 5x)^2$
$9x^2 - 36x + 36 = 100 - 100x + 25x^2$
$16x^2 - 64x + 64 = 0 \Rightarrow x^2 - 4x + 4 = 0 \Rightarrow (x - 2)^2 = 0$
So there is only one solution at $x = 2$.
If you wanted to solve this graphically, you would see that the graphs look like this:

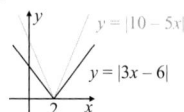

Q9 $|4x + 1| = 3 \Rightarrow 4x + 1 = 3$ or $-(4x + 1) = 3$
$\Rightarrow x = \frac{1}{2}$ or $x = -1$

If $x = \frac{1}{2}$, $2|x - 1| + 3 = 2\left|-\frac{1}{2}\right| + 3 = 1 + 3 = 4$

If $x = -1$, $2|x - 1| + 3 = 2|-2| + 3 = 4 + 3 = 7$

Q10 a) $|x| < 8 \Rightarrow -8 < x < 8$

b) $|x| \geq 5 \Rightarrow x \leq -5$ and $x \geq 5$

c) $|2x| > 12 \Rightarrow 2x < -12$ and $2x > 12 \Rightarrow x < -6$ and $x > 6$

d) $|4x + 2| \leq 6 \Rightarrow -6 \leq 4x + 2 \leq 6 \Rightarrow -8 \leq 4x \leq 4 \Rightarrow -2 \leq x \leq 1$

e) $3 \geq |3x - 3| \Rightarrow -3 \leq 3x - 3 \leq 3 \Rightarrow 0 \leq 3x \leq 6 \Rightarrow 0 \leq x \leq 2$

f) $6 - 2|x + 4| < 0 \Rightarrow 6 < 2|x + 4| \Rightarrow 3 < |x + 4|$
$x + 4 < -3$ and $x + 4 > 3 \Rightarrow x < -7$ and $x > -1$

g)

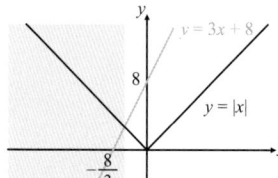

The shaded region shows the values of x that satisfy the inequality. The graph shows that there is one solution to $3x + 8 = |x|$, where $x < 0$.
So the equation to solve is:
$3x + 8 = -x \Rightarrow 4x = -8 \Rightarrow x = -2$ (valid since $-2 < 0$)
So the region that satisfies the inequality is $x < -2$.

h)

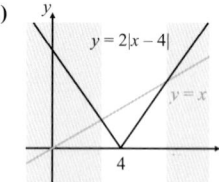

The shaded regions shows the values of x that satisfy the inequality. The graph shows that there are two solutions to $2|x - 4| = x$, one where $x > 4$ and one where $x < 4$.
So there are two equations to solve:
① $2(x - 4) = x \Rightarrow x = 8$ (valid since $8 > 4$)
② $-2(x - 4) = x \Rightarrow x = \frac{8}{3}$ (valid since $\frac{8}{3} < 4$)
So the regions that satisfy the inequality are
$x \leq \frac{8}{3}$ and $x \geq 8$.

i)

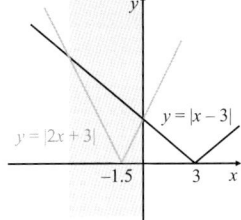

The shaded region shows the values of *x* that satisfy the inequality. The graph shows that there are two solutions to $|x - 3| = |2x + 3|$, one where both $(x - 3)$ and $(2x + 3)$ are negative (i.e. $x < -1.5$) and one where $(x - 3)$ is negative and $(2x + 3)$ is positive (i.e. $-1.5 < x < 3$).

So there are two equations to solve:

① $-(x - 3) = -(2x + 3) \Rightarrow -x + 3 = -2x - 3$
$\Rightarrow x = -6$ (valid since $-6 < -1.5$)

② $-(x - 3) = 2x + 3 \Rightarrow -x + 3 = 2x + 3$
$\Rightarrow x = 0$ (valid since $-1.5 < 0 < 3$)

So the region that satisfies the inequality is $-6 \leq x \leq 0$.

Q11

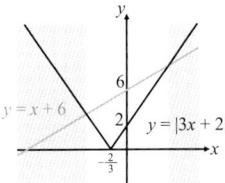

The shaded regions shows the values of *x* that satisfy the inequality. The graph shows that there are two solutions to $x + 6 = |3x + 2|$, one where $(3x + 2) > 0 \left(x > -\frac{2}{3}\right)$ and one where $(3x + 2) < 0 \left(x < -\frac{2}{3}\right)$.

So there are two equations to solve:

① $x + 6 = 3x + 2 \Rightarrow 4 = 2x \Rightarrow x = 2$ (valid since $2 > -\frac{2}{3}$)

② $x + 6 = -3x - 2 \Rightarrow 4x = -8 \Rightarrow x = -2$ (valid since $-2 < -\frac{2}{3}$)

So the regions that satisfy the inequality are $x \geq 2$ and $x \leq -2$.
In set notation, this is $\{x : x \geq 2\} \cup \{x : x \leq -2\}$

[4 marks available — 1 mark for showing that there are two solutions to x + 6 = |3x + 2|, 1 mark for finding the solution to this equation for $x \geq -\frac{2}{3}$, 1 mark for finding the solution for $x \leq -\frac{2}{3}$, 1 mark for the correct answer given in set notation]

Q12 a)

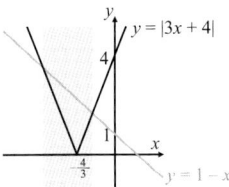

$3x + 4 = 1 - x \Rightarrow 4x = -3 \Rightarrow x = -\frac{3}{4}$ or

$-3x - 4 = 1 - x \Rightarrow 2x = -5 \Rightarrow x = -\frac{5}{2}$ so $x = -\frac{3}{4}, -\frac{5}{2}$.

[3 marks available — 1 mark for $x = -\frac{3}{4}$, 1 mark for a correct method to find the other solution, 1 mark for $x = -\frac{5}{2}$]

b) $\left\{x : -\frac{5}{2} \leq x \leq -\frac{3}{4}\right\}$ *[1 mark]*

Q13 $|1 + 2x| \leq 3 \Rightarrow -3 \leq 1 + 2x \leq 3 \Rightarrow -4 \leq 2x \leq 2 \Rightarrow -2 \leq x \leq 1$
Draw the graph of $y = |5x + 4|$ for these values:

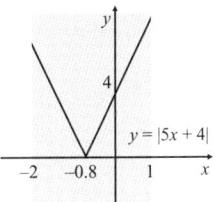

You can see from the graph that the minimum possible value of $|5x + 4|$ is 0, and the maximum is when $x = 1$,
i.e. $|5x + 4| = |5 + 4| = |9| = 9$.
So the possible values are $0 \leq |5x + 4| \leq 9$

Exercise 2.13 — Transformations of Graphs

Q1

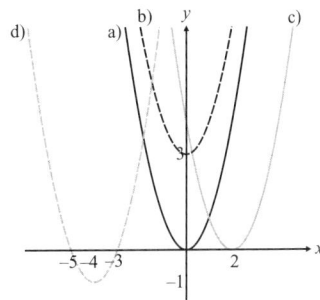

a) Turning point at $(0, 0)$

b) Turning point at $(0, 3)$, translation vector $\begin{pmatrix} 0 \\ 3 \end{pmatrix}$

c) Turning point at $(2, 0)$, translation vector $\begin{pmatrix} 2 \\ 0 \end{pmatrix}$

d) Turning point at $(-4, -1)$, translation vector $\begin{pmatrix} -4 \\ -1 \end{pmatrix}$

Q2 a)

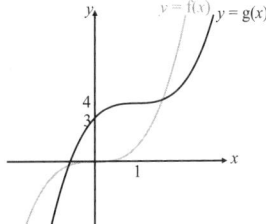

[2 marks available — 1 mark for a correct sketch of f(x), 1 mark for a correct sketch of g(x) with the y-intercept at y = 3 labelled]

b) $\begin{pmatrix} 1 \\ 4 \end{pmatrix}$ *[1 mark]*

c) $y = (x - 1)^3 + 4 \ (= x^3 - 3x^2 + 3x + 3)$ *[1 mark]*

Q3

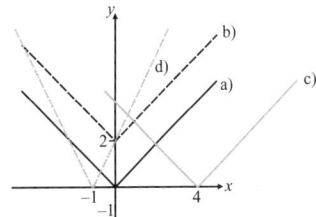

The transformations can be described as follows:

b) A translation of 2 up / by the column vector $\begin{pmatrix} 0 \\ 2 \end{pmatrix}$.

c) A translation of 4 right / by the column vector $\begin{pmatrix} 4 \\ 0 \end{pmatrix}$.

Answers

d) A translation of 1 left / by the column vector $\begin{pmatrix} -1 \\ 0 \end{pmatrix}$

and a stretch vertically by a scale factor of 2.

Q4

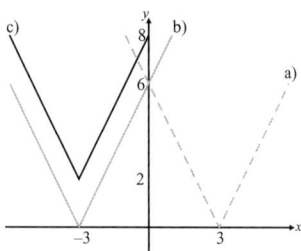

Q5

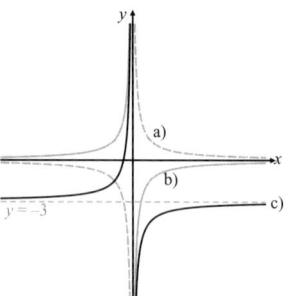

Q6

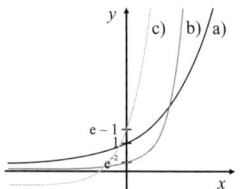

Q7

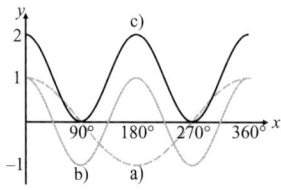

d) The minimum points on c) are $(90°, 0)$ and $(270°, 0)$.

Q8 The maximum value of $\sin x$ is 1, and the minimum is -1:

Transformed Function	New equation	Max value	Min value
$f(x) + 2$	$\sin x + 2$	3	1
$f(x - 90°)$	$\sin(x - 90°)$	1	-1
$f(3x)$	$\sin 3x$	1	-1
$4f(x)$	$4 \sin x$	4	-4

Q9 The point of inflection of x^3 is at $(0, 0)$, so:

Transformed Function	New equation	Coordinates of point of inflection
$f(x) + 1$	$x^3 + 1$	$(0, 1)$
$f(x - 2)$	$(x - 2)^3$	$(2, 0)$
$-f(x) - 3$	$-x^3 - 3$	$(0, -3)$
$f(-x) + 4$	$-x^3 + 4$	$(0, 4)$

Q10 a)

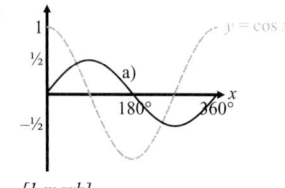

[1 mark]

b) $y = \frac{1}{2}\cos(x - 90°) \ (= \frac{1}{2}\sin x)$ *[1 mark]*

Q11

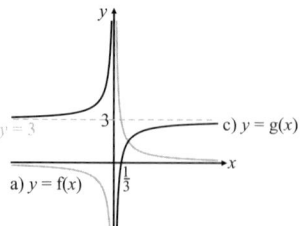

b) Reflect in the y-axis (or x-axis) and

translate by $\begin{pmatrix} 0 \\ 3 \end{pmatrix}$ (3 upwards).

Q12

Original graph	New graph	Sequence of transformations
$y = x^3$	$y = (x - 4)^3 + 5$	Translate by $\begin{pmatrix} 4 \\ 5 \end{pmatrix}$ i.e. 4 right and 5 up.
$y = 4^x$	$y = 4^{3x} - 1$	Stretch horizontally by a factor of $\frac{1}{3}$ and translate by $\begin{pmatrix} 0 \\ -1 \end{pmatrix}$ i.e. 1 down.
$y = \lvert x + 1 \rvert$	$y = 1 - \lvert 2x + 1 \rvert$	Stretch horizontally by a factor of $\frac{1}{2}$, reflect in the x-axis and translate by $\begin{pmatrix} 0 \\ 1 \end{pmatrix}$ i.e. 1 up.
$y = \sin x$	$y = -3 \sin 2x + 1$	Stretch horizontally by a factor of $\frac{1}{2}$, stretch vertically by a factor of 3, reflect in the x-axis and translate by $\begin{pmatrix} 0 \\ 1 \end{pmatrix}$ i.e. 1 up.

Q13 a) $y = 2x^2 - 4x + 6 = 2[x^2 - 2x + 3] = 2[(x - 1)^2 + 2]$

b) Translate by $\begin{pmatrix} 1 \\ 2 \end{pmatrix}$ i.e. 1 right, then 2 up,

then stretch vertically by a factor of 2.

c)

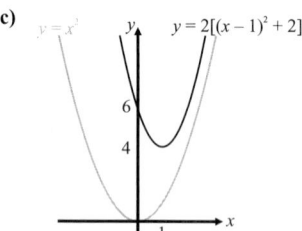

d) The minimum point is at $(1, 4)$.

You can work this out by doing the transformations on the minimum point of the graph $y = x^2$ (which is $(0, 0)$).

Q14 a) Stretch horizontally, scale factor $\frac{1}{3}$, then stretch vertically, scale factor 4.

b) Stretch horizontally, scale factor $\frac{1}{2}$, then reflect in the x-axis, then translate by $\begin{pmatrix} 0 \\ 4 \end{pmatrix}$ i.e. 4 up.

c) Translate by $\begin{pmatrix} 60 \\ 0 \end{pmatrix}$ i.e. 60° right, then stretch vertically, scale factor 2.

Q15 a)

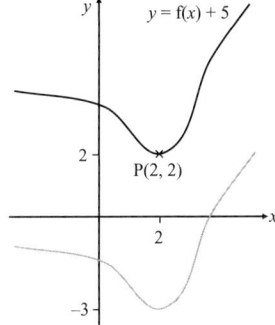

$y = f(x) + 5$
$P(2, 2)$

b)

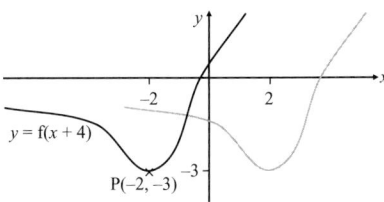

$y = f(x + 4)$
$P(-2, -3)$

c)

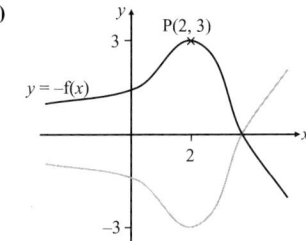

$P(2, 3)$
$y = -f(x)$

Q16 Minimum value is –2. Given that $y = \sin(x)$ is not translated in the vertical direction, and it is stretched vertically by a factor of a, then $a = \pm 2$. But a is positive, so $a = 2$.
$y = \sin(x)$ crosses the x-axis at $x = 0°$ and $x = 180°$ so if $y = 2f(bx)$ crosses at $x = 0°$ and $x = 540°$ then it is a stretch parallel to the x-axis by scale factor 3, so $b = \pm\frac{1}{3}$, but b is positive so $b = \frac{1}{3}$.
[2 marks available — 1 mark for the correct value of a, 1 mark for the correct value of b]

Q17 a)

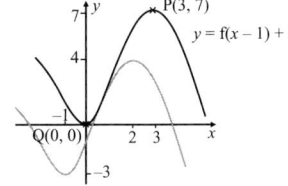

$P(3, 7)$
$y = f(x - 1) + 3$
$Q(0, 0)$

b)

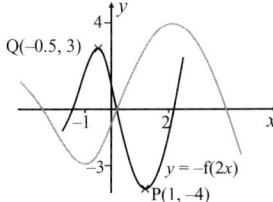

$Q(-0.5, 3)$
$y = -f(2x)$
$P(1, -4)$

c)

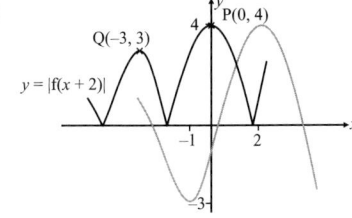

$P(0, 4)$
$Q(-3, 3)$
$y = |f(x + 2)|$

Q18 a)

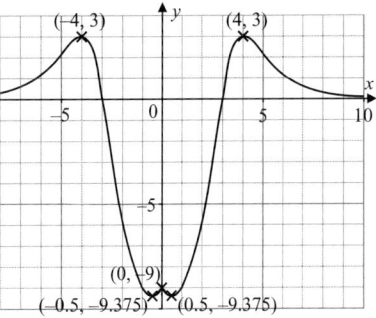

$(-4, 3)$ $(4, 3)$
$(0, -9)$
$(-0.5, -9.375)$ $(0.5, -9.375)$

[2 marks available — 1 mark for correct shape, 1 mark for correct labelling of axis intercepts, maximums and minimums]

b)

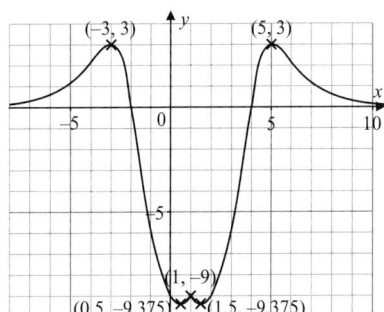

$(-3, 3)$ $(5, 3)$
$(1, -9)$
$(0.5, -9.375)$ $(1.5, -9.375)$

[2 marks available — 1 mark for translation 1 unit to the right, 1 mark for correct maximums and minimums]

c) $f(|x - k|)$ is a translation k units to the right so from the graph in part a), $k = \pm 3$. *[1 mark]*

d) $f(x) + k$ is a translation k units up, so from the graph given in the question, $k = 9$. *[1 mark]*

Q19 $f(x)$ translated 4 units to the right is $f(x - 4)$.
$f(x - 4)$ reflected in the x-axis is $-f(x - 4)$.
$-f(x - 4)$ stretched parallel to x-axis with s.f. $\frac{1}{2}$ is $-f(2x - 4)$.
$-f(2x - 4)$ translated 2 units down is $-f(2x - 4) - 2$.
The new equation of the curve is
$y = -f(2x - 4) - 2 = -[3(2x - 4)^2 + 1] - 2$
$= -[3(4x^2 - 16x + 16) + 1] - 2 = -[12x^2 - 48x + 49] - 2$
$= -12x^2 + 48x - 51$.
[3 marks available — 1 mark for a correct manipulation of f(x) for any two transformations, 1 mark for a correct manipulation of f(x) for all four transformations, 1 mark for correct answer]

Q20 Let $f(x) = |x + 3|$

a) (i) $|x - 5| = f(x - 8)$ so it's a translation by $\binom{8}{0}$. *[1 mark]*

(ii) $|2x + 3| = f(2x)$ so it's a stretch parallel to the x-axis with scale factor $\frac{1}{2}$. *[1 mark]*

b) Translation $\binom{6}{0}$, followed by a stretch scale factor 2 parallel to the y-axis.

Translation $\binom{9}{0}$, followed by a stretch scale factor $\frac{1}{2}$ parallel to the x-axis.
[2 marks available — 1 mark for each correct combination]

Answers

Q21 $f^{-1}(x+1) = \ln x$. f^{-1} can be treated just as any other function, so using rules of transformation, you know that $f^{-1}(x) = \ln(x-1)$.
Find the inverse: let $y = \ln(x-1) \Rightarrow e^y = x-1 \Rightarrow x = e^y - 1$
So $f(x) = e^x - 1$.
The transformation can be described as follows:
starting with the graph of $y = f(x)$, shift the curve up by 1, and then reflect the curve in the line $y = x$.

Exercise 2.14 — Partial Fractions — No Repeated Factors

Q1 $\dfrac{3x+3}{(x-1)(x-4)} \equiv \dfrac{A}{(x-1)} + \dfrac{B}{(x-4)}$

$\Rightarrow \dfrac{3x+3}{(x-1)(x-4)} = \dfrac{A(x-4) + B(x-1)}{(x-1)(x-4)}$

$\Rightarrow \qquad 3x + 3 \equiv A(x-4) + B(x-1)$

Substitution: $\quad x = 4 \Rightarrow 15 = 3B \Rightarrow B = 5$
$\qquad\qquad\qquad x = 1 \Rightarrow 6 = -3A \Rightarrow A = -2$

This gives: $\dfrac{3x+3}{(x-1)(x-4)} \equiv -\dfrac{2}{(x-1)} + \dfrac{5}{(x-4)}$

Q2 $\dfrac{5x-1}{x(2x+1)} \equiv \dfrac{A}{x} + \dfrac{B}{(2x+1)} \Rightarrow 5x - 1 \equiv A(2x+1) + Bx$

Equating coefficients: $\quad x$ terms: $\quad 5 = 2A + B$
$\qquad\qquad\qquad\qquad$ constants: $\quad -1 = A$
$A = -1$, putting this into the first equation gives:
$5 = -2 + B \Rightarrow B = 7$

This gives: $\dfrac{5x-1}{x(2x+1)} \equiv -\dfrac{1}{x} + \dfrac{7}{(2x+1)}$

Q3 $\dfrac{3x-2}{x^2+x-12} \equiv \dfrac{3x-2}{(x+4)(x-3)} \equiv \dfrac{A}{(x+4)} + \dfrac{B}{(x-3)}$

$\Rightarrow \dfrac{3x-2}{(x+4)(x-3)} \equiv \dfrac{A(x-3) + B(x+4)}{(x+4)(x-3)}$

$\Rightarrow \qquad 3x - 2 \equiv A(x-3) + B(x+4)$

Equating coefficients: $\quad x$ terms: $\quad 3 = A + B$
$\qquad\qquad\qquad\qquad$ constants: $\quad -2 = -3A + 4B$
Solving simultaneously gives: $A = 2$, $B = 1$

This gives: $\dfrac{3x-2}{x^2+x-12} \equiv \dfrac{2}{(x+4)} + \dfrac{1}{(x-3)}$

Q4 $\dfrac{2}{x^2-16} \equiv \dfrac{2}{(x+4)(x-4)} \equiv \dfrac{A}{(x+4)} + \dfrac{B}{(x-4)}$

$\Rightarrow 2 \equiv A(x-4) + B(x+4)$

Substitution: $\quad x = 4 \Rightarrow 2 = 8B \Rightarrow B = \dfrac{1}{4}$
$\qquad\qquad\qquad x = -4 \Rightarrow 2 = -8A \Rightarrow A = -\dfrac{1}{4}$

This gives: $\dfrac{2}{x^2-16} \equiv -\dfrac{1}{4(x+4)} + \dfrac{1}{4(x-4)}$

Don't worry if you get fractions for your coefficients — just put the numerator on the top of your partial fraction and the denominator on the bottom.

[3 marks available — 1 mark for correctly factorising the denominator, 1 mark for suitable method to find A or B, 1 mark for correct answer]

Q5 $x^2 - x - 6 = (x-3)(x+2)$

$\dfrac{5}{(x-3)(x+2)} \equiv \dfrac{A}{(x-3)} + \dfrac{B}{(x+2)} \Rightarrow 5 \equiv A(x+2) + B(x-3)$

Equating coefficients: $\quad x$ terms: $\quad 0 = A + B$
$\qquad\qquad\qquad\qquad$ constants: $\quad 5 = 2A - 3B$
Solving simultaneously gives: $A = 1$, $B = -1$

This gives: $\dfrac{5}{x^2-x-6} \equiv \dfrac{1}{(x-3)} - \dfrac{1}{(x+2)}$

Q6 $\dfrac{11x}{2x^2+5x-12} \equiv \dfrac{11x}{(2x-3)(x+4)} \equiv \dfrac{A}{(2x-3)} + \dfrac{B}{(x+4)}$

$\Rightarrow 11x \equiv A(x+4) + B(2x-3)$

Equating coefficients: $\quad x$ terms: $\quad 11 = A + 2B$
$\qquad\qquad\qquad\qquad$ constants: $\quad 0 = 4A - 3B$
Solving simultaneously gives: $A = 3$, $B = 4$

This gives: $\dfrac{11x}{2x^2+5x-12} \equiv \dfrac{3}{(2x-3)} + \dfrac{4}{(x+4)}$

Q7 a) $\dfrac{x+13}{x^2+2x-15} \equiv \dfrac{x+13}{(x+5)(x-3)} \equiv \dfrac{A}{x+5} + \dfrac{B}{x-3}$

$\Rightarrow A(x-3) + B(x+5) \equiv x + 13$
Let $x = 3$: $8B = 16 \Rightarrow B = 2$
Let $x = -5$: $-8A = 8 \Rightarrow A = -1$.
Therefore, $f(x) = \dfrac{2}{x-3} - \dfrac{1}{x+5}$.

[3 marks available — 1 mark for partial fractions in correct form, 1 mark for suitable method to find A or B, 1 mark for correct answer]

b) As $x \to \infty$, the denominators of both fractions get larger, making both fractions smaller, so $f(x) \to 0$.
[2 marks available — 1 mark for correct answer, 1 mark for correct reason.]

Q8 a) $x^3 - 9x = x(x^2 - 9) = x(x+3)(x-3)$

b) $\dfrac{12x+18}{x(x+3)(x-3)} \equiv \dfrac{A}{x} + \dfrac{B}{(x+3)} + \dfrac{C}{(x-3)}$

$\Rightarrow 12x + 18 \equiv A(x+3)(x-3) + Bx(x-3) + Cx(x+3)$
Substitution: $\quad x = 0 \Rightarrow 18 = -9A \Rightarrow A = -2$
$\qquad\qquad\qquad x = -3 \Rightarrow -18 = 18B \Rightarrow B = -1$
$\qquad\qquad\qquad x = 3 \Rightarrow 54 = 18C \Rightarrow C = 3$

This gives: $\dfrac{12x+18}{x^3-9x} \equiv -\dfrac{2}{x} - \dfrac{1}{(x+3)} + \dfrac{3}{(x-3)}$

Q9 $\dfrac{3x+9}{x^3-36x} \equiv \dfrac{3x+9}{x(x^2-36)} \equiv \dfrac{3x+9}{x(x+6)(x-6)}$

$\dfrac{3x+9}{x(x+6)(x-6)} \equiv \dfrac{A}{x} + \dfrac{B}{(x+6)} + \dfrac{C}{(x-6)}$

$\Rightarrow 3x + 9 \equiv A(x+6)(x-6) + Bx(x-6) + Cx(x+6)$
Substitution:

$x = 0 \Rightarrow 9 = -36A \Rightarrow A = -\dfrac{9}{36} = -\dfrac{1}{4}$

$x = -6 \Rightarrow -9 = 72B \Rightarrow B = -\dfrac{9}{72} = -\dfrac{1}{8}$

$x = 6 \Rightarrow 27 = 72C \Rightarrow C = \dfrac{27}{72} = \dfrac{3}{8}$

This gives: $\dfrac{3x+9}{x^3-36x} \equiv -\dfrac{1}{4x} - \dfrac{1}{8(x+6)} + \dfrac{3}{8(x-6)}$

[3 marks available — 1 mark for suitable method to find one of A, B or C, 1 mark for finding another of A, B or C, 1 mark for correct answer]

Q10 a) $f(x) = x^3 - 7x - 6$
$f(-1) = -1 + 7 - 6 = 0 \Rightarrow (x+1)$ is a factor

Once you've found one factor using the factor theorem you can use e.g. algebraic long division to get:
$x^3 - 7x - 6 = (x+1)(x^2 - x - 6)$
Then you can factorise the quadratic:
$x^3 - 7x - 6 = (x+1)(x-3)(x+2)$

b) $\dfrac{6x+2}{x^3-7x-6} \equiv \dfrac{6x+2}{(x+1)(x-3)(x+2)}$

$\dfrac{6x+2}{(x+1)(x-3)(x+2)} \equiv \dfrac{A}{(x+1)} + \dfrac{B}{(x-3)} + \dfrac{C}{(x+2)}$

$\Rightarrow 6x + 2 \equiv A(x-3)(x+2) + B(x+1)(x+2) + C(x+1)(x-3)$
Substitution: $\quad x = -1 \Rightarrow -4 = -4A \Rightarrow A = 1$
$\qquad\qquad\qquad x = 3 \Rightarrow 20 = 20B \Rightarrow B = 1$
$\qquad\qquad\qquad x = -2 \Rightarrow -10 = 5C \Rightarrow C = -2$

This gives: $\dfrac{6x+2}{x^3-7x-6} \equiv \dfrac{1}{(x+1)} + \dfrac{1}{(x-3)} - \dfrac{2}{(x+2)}$

Q11 a) $\dfrac{6x+4}{(x+4)(x-1)(x+1)} \equiv \dfrac{A}{(x+4)} + \dfrac{B}{(x-1)} + \dfrac{C}{(x+1)}$

$\Rightarrow 6x + 4 \equiv A(x-1)(x+1) + B(x+4)(x+1) + C(x+4)(x-1)$

Substitution: $x = -4 \Rightarrow -20 = 15A \Rightarrow A = -\dfrac{4}{3}$
$\qquad\qquad\qquad x = 1 \Rightarrow 10 = 10B \Rightarrow B = 1$
$\qquad\qquad\qquad x = -1 \Rightarrow -2 = -6C \Rightarrow C = \dfrac{1}{3}$

This gives:
$\dfrac{6x+4}{(x+4)(x-1)(x+1)} \equiv -\dfrac{4}{3(x+4)} + \dfrac{1}{(x-1)} + \dfrac{1}{3(x+1)}$

Answers

b) $\dfrac{15x - 27}{x^3 - 6x^2 + 3x + 10} \equiv \dfrac{15x - 27}{(x + 1)(x - 2)(x - 5)}$

You get this by using the factor theorem, as in Q9.

$\dfrac{15x - 27}{(x + 1)(x - 2)(x - 5)} \equiv \dfrac{A}{(x + 1)} + \dfrac{B}{(x - 2)} + \dfrac{C}{(x - 5)}$

$\Rightarrow 15x - 27 \equiv A(x - 2)(x - 5) + B(x + 1)(x - 5) + C(x + 1)(x - 2)$

Substitution: $x = -1 \Rightarrow -42 = 18A \Rightarrow A = -\dfrac{7}{3}$

$x = 2 \Rightarrow 3 = -9B \Rightarrow B = -\dfrac{1}{3}$

$x = 5 \Rightarrow 48 = 18C \Rightarrow C = \dfrac{8}{3}$

This gives:

$\dfrac{15x - 27}{x^3 - 6x^2 + 3x + 10} \equiv -\dfrac{7}{3(x + 1)} - \dfrac{1}{3(x - 2)} + \dfrac{8}{3(x - 5)}$

c) $\dfrac{2x + 7}{x^3 - 2x^2 - 5x + 6} \equiv \dfrac{2x + 7}{(x - 1)(x + 2)(x - 3)}$

$\equiv \dfrac{A}{x - 1} + \dfrac{B}{x + 2} + \dfrac{C}{x - 3}$

$2x + 7 \equiv A(x + 2)(x - 3) + B(x - 1)(x - 3) + C(x - 1)(x + 2)$

Substitution: $x = 1$: $9 = -6A \Rightarrow A = -\dfrac{3}{2}$

$x = -2$: $3 = 15B \Rightarrow B = \dfrac{1}{5}$

$x = 3$: $13 = 10C \Rightarrow C = \dfrac{13}{10}$

$\dfrac{2x + 7}{x^3 - 2x^2 - 5x + 6} \equiv \dfrac{1}{5(x + 2)} + \dfrac{13}{10(x - 3)} - \dfrac{3}{2(x - 1)}$

d) $\dfrac{162}{x^3 - 81x} \equiv \dfrac{162}{x(x + 9)(x - 9)} \equiv \dfrac{A}{x} + \dfrac{B}{x + 9} + \dfrac{C}{x - 9}$

$162 \equiv A(x + 9)(x - 9) + Bx(x - 9) + Cx(x + 9)$

Substitution: $x = 0$: $162 = -81A \Rightarrow A = -2$

$x = -9$: $162 = 162B \Rightarrow B = 1$

$x = 9$: $162 = 162C \Rightarrow C = 1$

$\dfrac{162}{x^3 - 9x} \equiv \dfrac{1}{x - 9} + \dfrac{1}{x + 9} - \dfrac{2}{x}$

e) The coefficients of the cubic in the denominator add to 0, so $(x - 1)$ is a factor. Then using e.g. algebraic long division you can fully factorise the denominator:

$\dfrac{6 - x}{2x^3 - 7x^2 + 7x - 2} \equiv \dfrac{6 - x}{(2x - 1)(x - 2)(x - 1)}$

$\equiv \dfrac{A}{2x - 1} + \dfrac{B}{x - 2} + \dfrac{C}{x - 1}$

$6 - x \equiv A(x - 2)(x - 1) + B(2x - 1)(x - 1) + C(2x - 1)(x - 2)$

Substitution: $x = 2$: $4 = 3B \Rightarrow B = \dfrac{4}{3}$

$x = 1$: $5 = -C \Rightarrow C = -5$

$x = \dfrac{1}{2}$: $\dfrac{11}{2} = \dfrac{3}{4}A \Rightarrow A = \dfrac{22}{3}$

$\dfrac{6 - x}{2x^3 - 7x^2 + 7x - 2} \equiv \dfrac{22}{3(2x - 1)} + \dfrac{4}{3(x - 2)} - \dfrac{5}{x - 1}$

f) $\dfrac{x + 4}{15x^3 - x^2 - 2x} \equiv \dfrac{x + 4}{x(3x + 1)(5x - 2)}$

$\equiv \dfrac{A}{x} + \dfrac{B}{3x + 1} + \dfrac{C}{5x - 2}$

$x + 4 \equiv A(3x + 1)(5x - 2) + Bx(5x - 2) + Cx(3x + 1)$

Substitution: $x = 0$: $4 = -2A \Rightarrow A = -2$

$x = -\dfrac{1}{3}$: $\dfrac{11}{3} = \dfrac{11}{9}B \Rightarrow B = 3$

$x = \dfrac{2}{5}$: $\dfrac{22}{5} = \dfrac{22}{25}C \Rightarrow C = 5$

$\dfrac{x + 4}{15x^3 - x^2 - 2x} \equiv \dfrac{3}{3x + 1} + \dfrac{5}{5x - 2} - \dfrac{2}{x}$

Q12 Start by using algebraic division:

$x^4 = (Jx^2 + Kx + L)(x - 1)(x - 2) + Mx + N$

$= (Jx^2 + Kx + L)(x^2 - 3x + 2) + Mx + N$

$\Rightarrow x^4 = Jx^4 + Kx^3 + Lx^2 - 3Jx^3 - 3Kx^2$
$\qquad - 3Lx + 2Jx^2 + 2Kx + 2L + Mx + N$

$\Rightarrow x^4 = Jx^4 + (K - 3J)x^3 + (L - 3K + 2J)x^2$
$\qquad + (-3L + 2K + M)x + (2L + N)$

Equating coefficients: $x^4 = Jx^4 \Rightarrow J = 1$

$0x^3 = (K - 3J)x^3 \Rightarrow 0 = K - 3(1) \Rightarrow K = 3$

$0x^2 = (L - 3K + 2J)x^2 \Rightarrow 0 = L - 3(3) + 2(1) \Rightarrow L = 7$

$0x = (-3L + 2K + M)x \Rightarrow 0 = -3(7) + 2(3) + M \Rightarrow M = 15$

$0 = 2L + N = 2(7) + N \Rightarrow N = -14$

So $x^4 = (x^2 + 3x + 7)(x - 1)(x - 2) + 15x - 14$

$\Rightarrow \dfrac{x^4}{(x - 1)(x - 2)} = x^2 + 3x + 7 + \dfrac{15x - 14}{(x - 1)(x - 2)}$

Now need to find partial fractions of the final fraction term:

$\dfrac{15x - 14}{(x - 1)(x - 2)} = \dfrac{A}{x - 1} + \dfrac{B}{x - 2}$

$\Rightarrow 15x - 14 \equiv A(x - 2) + B(x - 1)$

Substitution: $x = 2$: $16 = B$.

$x = 1$: $1 = -A \Rightarrow A = -1$.

So $\dfrac{x^4}{(x - 1)(x - 2)} = x^2 + 3x + 7 + \dfrac{16}{x - 2} - \dfrac{1}{x - 1}$

Exercise 2.15 — Partial Fractions — Repeated Factors

Q1 $\dfrac{3x}{(x + 5)^2} \equiv \dfrac{A}{(x + 5)} + \dfrac{B}{(x + 5)^2} \Rightarrow 3x \equiv A(x + 5) + B$

Equating coefficients: x terms: $3 = A$

constants: $0 = 5A + B \Rightarrow B = -15$

This gives: $\dfrac{3x}{(x + 5)^2} \equiv \dfrac{3}{(x + 5)} - \dfrac{15}{(x + 5)^2}$

Q2 $\dfrac{5x + 2}{x^2(x + 1)} \equiv \dfrac{A}{x} + \dfrac{B}{x^2} + \dfrac{C}{(x + 1)}$

$\Rightarrow 5x + 2 \equiv Ax(x + 1) + B(x + 1) + Cx^2$

Equating coefficients:

   constants: $2 = B$

   x terms: $5 = A + B \Rightarrow A = 3$

   x^2 terms: $0 = A + C \Rightarrow C = -3$

This gives: $\dfrac{5x + 2}{x^2(x + 1)} \equiv \dfrac{3}{x} + \dfrac{2}{x^2} - \dfrac{3}{(x + 1)}$

Q3 a) $\dfrac{2x - 7}{(x - 3)^2} \equiv \dfrac{A}{(x - 3)} + \dfrac{B}{(x - 3)^2} \Rightarrow 2x - 7 \equiv A(x - 3) + B$

Substitution: $x = 3 \Rightarrow -1 = B$

Equating coefficients of the x terms: $2 = A$

This gives: $\dfrac{2x - 7}{(x - 3)^2} \equiv \dfrac{2}{(x - 3)} - \dfrac{1}{(x - 3)^2}$

b) $\dfrac{6x + 7}{(2x + 3)^2} \equiv \dfrac{A}{(2x + 3)} + \dfrac{B}{(2x + 3)^2} \Rightarrow 6x + 7 \equiv A(2x + 3) + B$

Substitution: $x = -\dfrac{3}{2} \Rightarrow -2 = B$

Equating coefficients of x: $A = 3$

This gives: $\dfrac{3x + 7}{(2x + 3)^2} \equiv \dfrac{3}{(2x + 3)} - \dfrac{2}{(2x + 3)^2}$

c) $\dfrac{7x}{(x + 4)^2(x - 3)} \equiv \dfrac{A}{(x + 4)} + \dfrac{B}{(x + 4)^2} + \dfrac{C}{(x - 3)}$

$\Rightarrow 7x \equiv A(x + 4)(x - 3) + B(x - 3) + C(x + 4)^2$

Substitution: $x = -4 \Rightarrow -28 = -7B \Rightarrow B = 4$

$\qquad\qquad x = 3 \Rightarrow 21 = 49C \Rightarrow C = \dfrac{3}{7}$

Equating coefficients of the x^2 terms:

$\qquad 0 = A + C \Rightarrow A = -\dfrac{3}{7}$

This gives:

$\dfrac{7x}{(x + 4)^2(x - 3)} \equiv -\dfrac{3}{7(x + 4)} + \dfrac{4}{(x + 4)^2} + \dfrac{3}{7(x - 3)}$

d) $\dfrac{11x - 10}{x(x - 5)^2} \equiv \dfrac{A}{x} + \dfrac{B}{(x - 5)} + \dfrac{C}{(x - 5)^2}$

$\Rightarrow 11x - 10 \equiv A(x - 5)^2 + Bx(x - 5) + Cx$

Substitution: $x = 5 \Rightarrow 45 = 5C \Rightarrow C = 9$

$\qquad\qquad x = 0 \Rightarrow -10 = 25A \Rightarrow A = -\dfrac{2}{5}$

Equating the coefficients of x^2:

$\qquad 0 = A + B \Rightarrow B = \dfrac{2}{5}$

This gives: $\dfrac{11x - 10}{x(x - 5)^2} \equiv -\dfrac{2}{5x} + \dfrac{2}{5(x - 5)} + \dfrac{9}{(x - 5)^2}$

Answers

Q4 $x^3 - 10x^2 + 25x = x(x^2 - 10x + 25) = x(x - 5)(x - 5)$

So $\dfrac{5x + 10}{x^3 - 10x^2 + 25x} \equiv \dfrac{5x + 10}{x(x - 5)^2} \equiv \dfrac{A}{x} + \dfrac{B}{(x - 5)} + \dfrac{C}{(x - 5)^2}$

$\Rightarrow 5x + 10 \equiv A(x - 5)^2 + Bx(x - 5) + Cx$

Substitution: $x = 0 \Rightarrow 10 = 25A \Rightarrow A = \dfrac{2}{5}$

$x = 5 \Rightarrow 35 = 5C \Rightarrow C = 7$

Equating coefficients of x^2 terms:

$0 = A + B \Rightarrow B = -\dfrac{2}{5}$

This gives: $\dfrac{5x + 10}{x^3 - 10x^2 + 25x} \equiv \dfrac{2}{5x} - \dfrac{2}{5(x - 5)} + \dfrac{7}{(x - 5)^2}$

[4 marks available — 1 mark for correctly factorising denominator, 1 mark for method to find A, B or C, 1 mark for correct values of at least 2 of A, B, C, 1 mark for the correct answer]

Q5 $(x - 2)(x^2 - 4) = (x - 2)(x + 2)(x - 2) = (x + 2)(x - 2)^2$

$\dfrac{3x + 2}{(x - 2)(x^2 - 4)} \equiv \dfrac{3x + 2}{(x + 2)(x - 2)^2} \equiv \dfrac{A}{(x + 2)} + \dfrac{B}{(x - 2)} + \dfrac{C}{(x - 2)^2}$

$\Rightarrow 3x + 2 \equiv A(x - 2)^2 + B(x + 2)(x - 2) + C(x + 2)$

Substitution: $x = 2 \Rightarrow 8 = 4C \Rightarrow C = 2$

$x = -2 \Rightarrow -4 = 16A \Rightarrow A = -\dfrac{1}{4}$

Equating coefficients of x^2 terms:

$0 = A + B \Rightarrow B = \dfrac{1}{4}$

This gives: $\dfrac{3x + 2}{(x - 2)(x^2 - 4)} \equiv -\dfrac{1}{4(x + 2)} + \dfrac{1}{4(x - 2)} + \dfrac{2}{(x - 2)^2}$

Q6 $\dfrac{7x + 3}{x^2(3 - 2x)} \equiv \dfrac{A}{x} + \dfrac{B}{x^2} + \dfrac{C}{3 - 2x}$

$\Rightarrow 7x + 3 \equiv Ax(3 - 2x) + B(3 - 2x) + Cx^2$

Substitution: $x = 0 \Rightarrow 3 = 3B \Rightarrow B = 1$

$x = \dfrac{3}{2} \Rightarrow \dfrac{27}{2} = \dfrac{9}{4}C \Rightarrow C = 6$

Equating coefficients of x^2 terms:

$0 = -2A + C \Rightarrow A = 3$

This gives: $\dfrac{7x + 3}{x^2(3 - 2x)} \equiv \dfrac{3}{x} + \dfrac{1}{x^2} + \dfrac{6}{3 - 2x}$

Q7 $\dfrac{6x^2 + 3x + 6}{(x - 2)(x + 1)^2} = \dfrac{A}{x - 2} + \dfrac{B}{x + 1} + \dfrac{C}{(x + 1)^2}$

$\Rightarrow 6x^2 + 3x + 6 = A(x + 1)^2 + B(x + 1)(x - 2) + C(x - 2)$

Substitution: $x = -1 \Rightarrow -3C = 9 \Rightarrow C = -3$

$x = 2 \Rightarrow 9A = 36 \Rightarrow A = 4$

Equating coefficients of x^2 terms: $A + B = 6 \Rightarrow B = 2$

This gives: $\dfrac{6x^2 + 3x + 6}{(x - 2)(x + 1)^2} = \dfrac{4}{x - 2} + \dfrac{2}{x + 1} - \dfrac{3}{(x + 1)^2}$.

[4 marks available — 1 mark for multiplying through by denominator, 1 mark for method to find A, B or C, 1 mark for correct values of at least 2 of A, B, C, 1 mark for the correct answer]

Q8 **a)** $\dfrac{(3x - 1)(2x - 3)}{(x - 1)^2(x - 3)} = \dfrac{6x^2 - 11x + 3}{(x - 1)^2(x - 3)}$

$\dfrac{6x^2 - 11x + 3}{(x - 1)^2(x - 3)} \equiv \dfrac{A}{x - 1} + \dfrac{B}{(x - 1)^2} + \dfrac{C}{x - 3}$

$\Rightarrow A(x - 1)(x - 3) + B(x - 3) + C(x - 1)^2 \equiv 6x^2 - 11x + 3$

Substitution: $x = 1 \Rightarrow -2B = -2 \Rightarrow B = 1$

$x = 3 \Rightarrow 4C = 24 \Rightarrow C = 6$

Equating coefficients of x^2 terms: $A + C = 6 \Rightarrow A = 0$

This gives: $f(x) = \dfrac{1}{(x - 1)^2} + \dfrac{6}{x - 3}$.

[4 marks available — 1 mark for correct form for partial fractions, 1 mark for correct method to find A, B or C, 1 mark for two values of A, B, C correct, 1 mark for fully correct answer]

b) For $x > 3$, both fractions are positive and tend to 0 as $x \to \infty$, so $f(x) > 0 + 0 = 0$. Hence $k = 0$.

[1 mark]

c) On the domain $1 < x < 3$, as $x \to 1$, $\dfrac{1}{(x - 1)^2} \to \infty$.

On the domain $1 < x < 3$, as $x \to 3$, $\dfrac{1}{x - 3} \to -\infty$.

So the range of f(x) is all real numbers $\Rightarrow$ f$(x) \in \mathbb{R}$.

[2 marks available — 1 mark for the explanation, 1 mark for the correct answer]

Q9 $\dfrac{x + 17}{(x + 1)(x + c)^2} \equiv \dfrac{1}{(x + 1)} - \dfrac{1}{(x + c)} + \dfrac{5}{(x + c)^2}$

$\Rightarrow x + 17 \equiv (x + c)^2 - (x + c)(x + 1) + 5(x + 1)$

$x + 17 \equiv x^2 + 2cx + c^2 - x^2 - cx - x - c + 5x + 5$

$x + 17 \equiv (2c - c - 1 + 5)x + (c^2 - c + 5)$

$x + 17 \equiv (c + 4)x + (c^2 - c + 5)$

Equating coefficients of x:

$1 = c + 4 \Rightarrow c = -3$

You can check this by equating constant terms:

$(-3)^2 - (-3) + 5 = 9 + 3 + 5 = 17$

Q10 **a)** You know that $(x + 2)$ is a factor, so e.g. using algebraic long division, you can factorise the expression as:

$2x^3 + 7x^2 + 4x - 4 \equiv (x + 2)(x + 2)(2x - 1)$

b) $\dfrac{x - 13}{(x + 2)^2(2x - 1)} \equiv \dfrac{A}{x + 2} + \dfrac{B}{(x + 2)^2} + \dfrac{C}{2x - 1}$

$\Rightarrow x - 13 \equiv A(x + 2)(2x - 1) + B(2x - 1) + C(x + 2)^2$

Substitution: $x = -2 \Rightarrow -15 = -5B \Rightarrow B = 3$

$x = \dfrac{1}{2} \Rightarrow -\dfrac{25}{2} = \dfrac{25}{4}C \Rightarrow C = -2$

Equating coefficients of x^2 terms:

$0 = 2A + C \Rightarrow A = 1$

This gives: $\dfrac{x - 13}{(x + 2)^2(2x - 1)} \equiv \dfrac{1}{x + 2} + \dfrac{3}{(x + 2)^2} - \dfrac{2}{2x - 1}$

Q11 **a)** $\dfrac{2x^2}{(x + 1)(2x + 1)^2} \equiv \dfrac{A}{x + 1} + \dfrac{B}{2x + 1} + \dfrac{C}{(2x + 1)^2}$

$2x^2 \equiv A(2x + 1)^2 + B(x + 1)(2x + 1) + C(x + 1)$

Substitution: $x = -1 \Rightarrow 2 = A$

$x = -\dfrac{1}{2} \Rightarrow \dfrac{1}{2} = \dfrac{1}{2}C \Rightarrow C = 1$

Equating coefficients of x^2 terms:

$2 = 4A + 2B \Rightarrow 2 = 8 + 2B \Rightarrow 2B = -6 \Rightarrow B = -3$

This gives: $\dfrac{2x^2}{(x + 1)(2x + 1)^2} = \dfrac{2}{x + 1} - \dfrac{3}{2x + 1} + \dfrac{1}{(2x + 1)^2}$

b) $\dfrac{x^2 + 3}{(x + a)(2x + 1)^2} \equiv \dfrac{J}{x + a} + \dfrac{K}{2x + 1} + \dfrac{L}{(2x + 1)^2}$

$x^2 + 3 \equiv J(2x + 1)^2 + K(x + a)(2x + 1) + L(x + a)$

Substitution:

$x = -a, a^2 + 3 = J(-2a + 1)^2 \Rightarrow J = \dfrac{a^2 + 3}{(-2a + 1)^2} = \dfrac{a^2 + 3}{(1 - 2a)^2}$

$x = -\dfrac{1}{2} \Rightarrow \dfrac{1}{4} + 3 = L\left(a - \dfrac{1}{2}\right) \Rightarrow L = \dfrac{\dfrac{1}{4} + 3}{a - \dfrac{1}{2}} = \dfrac{13}{4a - 2}$

Equating coefficients of x^2 terms:

$1 = 4J + 2K \Rightarrow K = \dfrac{1}{2} - 2J = \dfrac{1}{2} - \dfrac{2a^2 + 6}{(1 - 2a)^2}$

$= \dfrac{(1 - 2a)^2 - 4a^2 - 12}{2(1 - 2a)^2}$

$= \dfrac{1 - 4a + 4a^2 - 4a^2 - 12}{2(1 - 2a)^2}$

$= \dfrac{-4a - 11}{2(1 - 2a)^2} = -\dfrac{4a + 11}{2(1 - 2a)^2}$

c) g(x) is undefined when the denominator is zero, i.e. when $(x + a) = 0 \Rightarrow x = -a$

or when $(2x + 1) = 0 \Rightarrow x = -\dfrac{1}{2}$

Since the function of g(x) is restricted to $x \neq \pm b$,

$-\dfrac{1}{2} = -b$ (as $b > 0$) $\Rightarrow b = \dfrac{1}{2}$

and so $-a = \dfrac{1}{2} \Rightarrow a = -\dfrac{1}{2}$

Chapter 2 Review Exercise

Q1 **a)** $\dfrac{4x^2 - 25}{6x - 15} = \dfrac{(2x+5)(2x-5)}{3(2x-5)} = \dfrac{(2x+5)}{3}$

b) $\dfrac{2x+3}{x-2} \times \dfrac{4x-8}{2x^2 - 3x - 9} = \dfrac{(2x+3)\times 4(x-2)}{(x-2)(2x+3)(x-3)} = \dfrac{4}{(x-3)}$

c) $\dfrac{x^2 - 3x}{x+1} \div \dfrac{x}{2} = \dfrac{x(x-3)}{x+1} \times \dfrac{2}{x} = \dfrac{2(x-3)}{x+1}$

Q2 **a)** $f(1) = 1^3 - 1^2 + 1 - 1 = 0$, so $(x-1)$ is a factor of $f(x)$.
[1 mark]

b) Factorise the numerator:
$x^3 - x^2 + x - 1 = (x-1)(Ax^2 + Bx + C)$
$\qquad = Ax^3 + Bx^2 + Cx - Ax^2 - Bx - C$
Equate coefficients: $x^3 = Ax^3 \Rightarrow A = 1$
$-x^2 = Bx^2 - Ax^2 \Rightarrow -1 = B - A = B - 1 \Rightarrow B = 0$
$x = Cx - Bx \Rightarrow 1 = C - B = C - 0 \Rightarrow C = 1$
This gives numerator of $x^3 - x^2 + x - 1 = (x-1)(x^2+1)$
So: $\dfrac{x^3 - x^2 + x - 1}{x^3 + x} = \dfrac{(x-1)(x^2+1)}{x(x^2+1)} = \dfrac{x-1}{x}$ or $1 - \dfrac{1}{x}$
[2 marks available — 1 mark for the correct factorised numerator, 1 mark for the correct answer]

c) $y = \dfrac{x-1}{x} = 1 - \dfrac{1}{x}$
It passes through the x-axis when $y = 0$
$\Rightarrow 1 - \dfrac{1}{x} = 0 \Rightarrow x = 1$

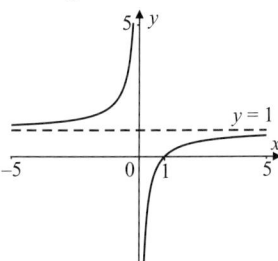

[3 marks available — 1 mark for the correct shape, 1 mark for the correct asymptotes $y = 1$ and $x = 0$, 1 mark for sketch passing through $(1, 0)$]

Q3 **a)** $\dfrac{x^3 - 4x^2 + 3x}{1 - x} = \dfrac{x(x^2 - 4x + 3)}{-(x-1)} = \dfrac{x(x-1)(x-3)}{-(x-1)}$
$= -x(x-3) = 3x - x^2$
[2 marks available — 1 mark for factorising numerator, 1 mark for the correct answer]

b) When $x = 2$, $f(2) = (3 \times 2) - 2^2 = 2$ so the tangent passes through $(2, 2)$.
$f'(x) = 3 - 2x$, so the gradient at $x = 2$ is $f'(2) = -1$.
So the tangent has equation $y - 2 = -1(x-2) \Rightarrow y = 4 - x$
[4 marks available — 1 mark for working out $f(2)$ correctly, 1 mark for differentiating correctly, 1 mark for the gradient, 1 mark for the correct answer]

Q4 **a)** $\dfrac{x}{2x+1} + \dfrac{3}{x^2} + \dfrac{1}{x} = \dfrac{x^3}{x^2(2x+1)} + \dfrac{3(2x+1)}{x^2(2x+1)} + \dfrac{(2x+1)x}{x^2(2x+1)}$
$= \dfrac{x^3 + 2x^2 + 7x + 3}{x^2(2x+1)}$

b) $\dfrac{2}{x^2 - 1} - \dfrac{3x}{x-1} + \dfrac{x}{x+1}$
$= \dfrac{2}{x^2 - 1} - \dfrac{3x(x+1)}{(x-1)(x+1)} + \dfrac{x(x-1)}{(x-1)(x+1)}$
$= \dfrac{2(1 - 2x - x^2)}{(x-1)(x+1)}$

c) $\dfrac{2}{(x+1)^2} - \dfrac{x}{x+1} + \dfrac{1}{3x}$
$= \dfrac{6x}{3x(x+1)^2} - \dfrac{3x^2(x+1)}{3x(x+1)^2} + \dfrac{(x+1)^2}{3x(x+1)^2}$
$= -\dfrac{3x^3 + 2x^2 - 8x - 1}{3x(x+1)^2}$

Q5 **a)** $f(x) = \dfrac{3x-2}{x^2 + 7x + 12} - \dfrac{2}{x+4} = \dfrac{3x-2}{(x+4)(x+3)} - \dfrac{2}{x+4}$
$= \dfrac{3x - 2 - 2(x+3)}{(x+3)(x+4)} = \dfrac{x-8}{(x+3)(x+4)}$
[3 marks available — 1 mark for correct common denominator, 1 mark for correct method to subtract, 1 mark for the correct answer]

b) Undefined for $x = -3$ and $x = -4$.
[1 mark for both correct values]

c) $f(x) = -2 \Rightarrow (x-8) = -2(x+3)(x+4)$
$\Rightarrow x - 8 = -2x^2 - 14x - 24$
$\Rightarrow 2x^2 + 15x + 16 = 0$
$\Rightarrow x = \dfrac{-15 \pm \sqrt{15^2 - 4 \times 2 \times 16}}{2 \times 2} = -1.29$ or -6.21 (to 3 s.f.)
[4 marks available — 1 mark for method to obtain quadratic equation, 1 mark for correct quadratic, 1 mark for method to solve quadratic, 1 mark for the correct answers]

Q6 **a)** $\dfrac{1}{x} - \dfrac{2x}{x+1} = \dfrac{x+1}{x(x+1)} - \dfrac{2x^2}{x(x+1)}$
$= \dfrac{x + 1 - 2x^2}{x(x+1)} = \dfrac{(2x+1)(1-x)}{x(x+1)}$
[3 marks available — 1 mark for attempting to subtract the algebraic fractions, 1 mark for correct factorised denominator, 1 mark for correct factorised numerator]

b) The vertical asymptotes occur where the denominator $h(x) = 0$.
$h(x) = x(x+1) = 0 \Rightarrow x = 0, \ x = -1$.
[2 marks available — 1 mark for each correct asymptote]

Q7 **a)** $\dfrac{x^3 - 2x^2 + x}{2x^2} \times \dfrac{4x}{x-1} = \dfrac{x(x^2 - 2x + 1)}{2x^2} \times \dfrac{4x}{x-1}$
$\dfrac{x(x-1)^2}{2x^2} \times \dfrac{4x}{x-1} = 2(x-1)$
[3 marks available — 1 mark for correct factorised numerator of first fraction, 1 mark for correct method for multiplying, 1 mark for correct answer]

b) Replace x with $\sin x$ in your answer from part a).
$2(\sin x - 1) = \dfrac{1}{4} \Rightarrow \sin x = \dfrac{9}{8}$
But the range of $\sin x$ is $[-1, 1]$ and $\dfrac{9}{8} > 1$ so there are no real solutions.
[2 marks available — 1 mark for obtaining $\sin x = \dfrac{9}{8}$, 1 mark for the correct conclusion]

Q8 **a)** $\dfrac{y-x}{x^2 - 2xy + y^2} \div \dfrac{2}{x^2 + 2xy + y^2}$
$= \dfrac{y-x}{x^2 - 2xy + y^2} \times \dfrac{x^2 + 2xy + y^2}{2}$
$= \dfrac{y-x}{(x-y)^2} \times \dfrac{(x+y)^2}{2} = \dfrac{y-x}{(y-x)^2} \times \dfrac{(x+y)^2}{2} = \dfrac{(x+y)^2}{2(y-x)}$
[3 marks available — 1 mark for $(x-y)^2$ or $(x+y)^2$, 1 mark for correct method for multiplication, 1 mark for correct answer]

b) $(x+y)^2 \geq 0$ for all values of x and y.
So $\dfrac{(x+y)^2}{2(y-x)} < 0 \Rightarrow 2(y-x) < 0 \Rightarrow y < x$
[2 marks available — 1 mark for a correct explanation, 1 mark for $y < x$]

Q9 $2x^3 + 8x^2 + 7x + 8 \equiv (Ax^2 + Bx + C)(x+3) + D$
Set $x = -3$: $-54 + 72 - 21 + 8 = D$, so $D = 5$.
Set $x = 0$: $8 = 3C + D$, so $3C = 3 \Rightarrow C = 1$.
Equating the coefficients of x^3 gives $2 = A$.
Equating the coefficients of x^2 gives
$8 = 3A + B$, so $B = 2$.
So $2x^3 + 8x^2 + 7x + 8 \equiv (2x^2 + 2x + 1)(x+3) + 5$
The result when $2x^3 + 8x^2 + 7x + 8$ is divided by $(x+3)$ is $(2x^2 + 2x + 1)$ remainder 5.

Answers

Q10 $x^4 + x^3 - 5x^2 - 7x - 2 \equiv (Ax^3 + Bx^2 + Cx + D)(x + 1) + E$
Set $x = -1$: $1 - 1 - 5 + 7 - 2 = E$, so $E = 0$
Set $x = 0$: $-2 = D + E$, so $D = -2$
Equating the coefficients of x^4 gives $1 = A$.
Equating the coefficients of x^3 gives $1 = A + B$, so $B = 0$
Equating the coefficients of x^2 gives
$-5 = B + C$, so $C = -5$
So $x^4 + x^3 - 5x^2 - 7x - 2 \equiv (x^3 - 5x - 2)(x + 1)$
$x^4 + x^3 - 5x^2 - 7x = 2 \Rightarrow x^4 + x^3 - 5x^2 - 7x - 2 = 0$
$\Rightarrow (x^3 - 5x - 2)(x + 1) = 0$, so one solution of
$x^4 + x^3 - 5x^2 - 7x = 2$ is $x = -1$

Q11 $2x^3 + 9x^2 - 5x - 39 = x + 1 \Rightarrow 2x^3 + 9x^2 - 6x - 40 = 0$
So call this new function $s(x) = 2x^3 + 9x^2 - 6x - 40$,
and solve $s(x) = 0$.
$s(2) = 2(2)^3 + 9(2)^2 - 6(2) - 40 = 16 + 36 - 12 - 40 = 0$
So by the factor theorem $(x - 2)$ is a factor of $s(x)$.
Dividing $s(x)$ by $(x - 2)$:

$$
\begin{array}{r}
2x^2 + 13x + 20 \ \ \text{r } 0 \\
x - 2 \overline{)\ 2x^3 + 9x^2 - 6x - 40} \\
-\ \underline{(2x^3 - 4x^2)} \\
13x^2 - 6x \\
-\ \underline{(13x^2 - 26x)} \\
20x - 40 \\
-\ \underline{(20x - 40)} \\
0
\end{array}
$$

$s(x) = (x - 2)(2x^2 + 13x + 20) = (x - 2)(2x + 5)(x + 4)$
So the solutions to $r(x) = x + 1$ are $x = 2$, $x = -\frac{5}{2}$, $x = -4$.
[6 marks available — 1 mark for rewriting as one single function, 1 mark for finding one factor of this new function, 1 mark for a suitable method to find a quadratic factor, 1 mark for correct quadratic factor, 1 mark for correctly factorising the quadratic, 1 mark for correct solutions]

Q12 a) Range $f(x) \geq -16$. This is a function, and it's one-to-one (the domain is restricted so every x-value is mapped to a unique value of $f(x)$).

b) Complete the square: $x^2 - 7x + 10 = (x - \frac{7}{2})^2 - \frac{9}{4}$
Sketching this gives:

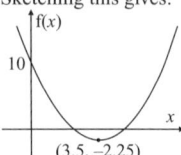

So range $f(x) \geq -2.25$. This is a function, and it's many-to-one.

c) Range $f(x) \geq 0$. This is not a function as $f(x)$ doesn't exist for $x < 0$.
$\sqrt{x}$ means the positive root of x.

Q13 a) $f(0) = 5$, $f\left(\frac{1}{2}\right) = 2\frac{1}{2}$

b)

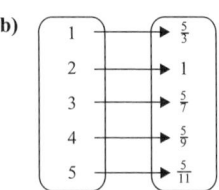

Range $\left\{\frac{5}{3}, 1, \frac{5}{7}, \frac{5}{9}, \frac{5}{11}\right\}$

c) $f(x)$ is not defined when $2x + 1 = 0 \Rightarrow x = -\frac{1}{2}$, but is defined for all other values of x.
$-\frac{1}{2} \notin \mathbb{Z}$, so the mapping is a function for $x \in \mathbb{Z}$

d) No — the mapping is not defined for $x = -\frac{1}{2}$.

Q14 a)

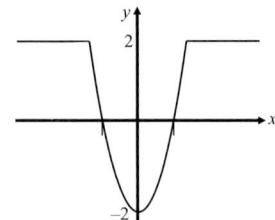

b) $-2 \leq f(x) \leq 2$

Q15 a) The share price, in pounds, on 31st August 2020.
[1 mark]

b)

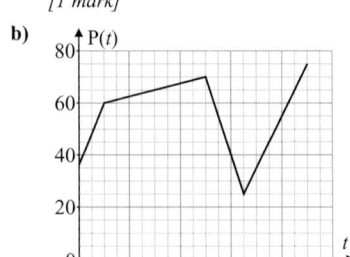

[2 marks available — 1 mark for any two line segments correct, 1 mark for a completely correct graph]

c) Look at the graph for the lowest and highest values of $P(t)$:
$25 \leq P(t) \leq 75$.
[1 mark]

d) $S(t)$ is an increasing function so the range is
$\frac{1}{3}(0) + 30 \leq S(t) \leq \frac{1}{3}(90) + 30 \Rightarrow 30 \leq S(t) \leq 60$
[1 mark]

e) $2.4t + 36 = \frac{1}{3}t + 30$ and $0.25t + 57.5 = \frac{1}{3}t + 30$ have no solutions on the given domain — you can see this algebraically or graphically. Solving $P(t) = S(t)$ for $P(t) = -3t + 220$ and $P(t) = 2t - 105$, we find that
$-3t + 220 = \frac{1}{3}t + 30 \Rightarrow \frac{10}{3}t = 190 \Rightarrow t = 57$ and
$2t - 105 = \frac{1}{3}t + 30 \Rightarrow \frac{5}{3}t = 135 \Rightarrow t = 81$.
These values can also be found graphically by drawing the line onto the graph from part b.
[3 marks available — 1 mark for solving P(t) = S(t), 1 mark for t = 57, 1 mark for t = 81]

f) By drawing on the graph, you can see that the share price of Company B is greater than or equal to that of Company A between the two values of t found in part e). $81 - 57 = 24$, but this only includes one of the days where they're equal, so add one to get 25 days. *[1 mark]*

Q16 a) $fg(x) = f(2x + 3) = \frac{3}{2x + 3}$
$fg(2) = \frac{3}{7}$
$gf(x) = g\left(\frac{3}{x}\right) = \frac{6}{x} + 3$, so $gf(1) = 9$

b) $fg(x) = f(x + 4) = 3(x + 4)^2$
$fg(2) = 3 \times 6^2 = 108$
$gf(x) = g(3x^2) = 3x^2 + 4$, so $gf(1) = 7$

Q17 a) $fg(1) = f(g(1)) = f(100) = 2$
$gf(1) = g(f(1)) = g(0) = 10$
$f^2(10) = f(f(10)) = f(1) = 0$
$g^2(-1) = g(g(-1)) = g(1) = 100$

b) Because $f(1) = \log_{10}1 = 0$, and $f(0) = \log_{10}0$, which is undefined.

Q18 $fg(x) = f(g(x)) = f(x + 7) = 3(x + 7) = 3x + 21$
$gf(x) = g(f(x)) = g(3x) = 3x + 7$
$g^2(x) = g(g(x)) = g(x + 7) = x + 7 + 7 = x + 14$

Q19 a) (i) $fg(x) = f\left(\dfrac{1}{x}\right) = 2\left(\dfrac{1}{x}\right) - 3 = \dfrac{2}{x} - 3$
Domain: $x > 0$, Range: $fg(x) > -3$
[2 marks available — 1 mark for correct (or equivalent) expression for fg(x), 1 mark for both domain and range correct]

(ii) $gf(x) = g(2x - 3) = \dfrac{1}{2x - 3}$
Domain: $x > \dfrac{3}{2}$, Range: $gf(x) > 0$
[2 marks available — 1 mark for correct (or equivalent) expression for fg(x), 1 mark for both domain and range correct]

b) hf(x) is linear, so can be written as $y = mx + c$, where m and c are constants to be found.
hf(x) intercepts the y-axis at –10, so $c = -10$
$\Rightarrow y = mx - 10$.
hf(x) also passes through the point (0.8, 0):
$0 = m(0.8) - 10 \Rightarrow 10 = 0.8m \Rightarrow m = 12.5$
$\Rightarrow hf(x) = 12.5x - 10$
h(x) is also linear, so can be written as $h(x) = ax + b$, where a and b are constants to be found. And you know that
$f(x) = 2x - 3$, so $hf(x) = h(2x - 3) = a(2x - 3) + b = 12.5x - 10$
$\Rightarrow 2ax - 3a + b = 12.5x - 10$
Equating coefficients: $2a = 12.5 \Rightarrow a = 6.25$
$-3a + b = -10 \Rightarrow b - 18.75 = -10 \Rightarrow b = 8.75$
So $h(x) = 6.25x + 8.75$
[4 marks available — 1 mark for finding the expression for hf(x), 1 mark for writing h(x) in terms of unknowns, 1 mark for equating coefficients, 1 mark for correct expression for h(x)]

Q20 f is a one-to-one function so it has an inverse.
Domain: $x \geq 3$, range: $f^{-1}(x) \in \mathbb{R}$.

Q21 Replace f(x) with y: $y = \sqrt{2x - 4} \Rightarrow y^2 = 2x - 4$
$\Rightarrow 2x = y^2 + 4 \Rightarrow x = \dfrac{y^2}{2} + 2$
Replace with $f^{-1}(x)$ and x: $f^{-1}(x) = \dfrac{x^2}{2} + 2$
The domain is $x \geq 0$ and the range is $f^{-1}(x) \geq 2$.
[3 marks available — 1 mark for f⁻¹(x), 1 mark for the correct domain, 1 mark for the correct range]

Q22 In the domain $0 \leq x \leq \dfrac{\pi}{2}$, cos x is a one-to-one function:

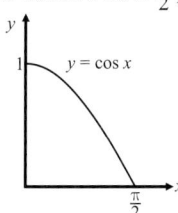

so $f^{-1}(x)$ does exist.

Q23 $y = \dfrac{x}{x - 1} \Rightarrow y(x - 1) = x \Rightarrow yx - y = x$
$\Rightarrow yx - x = y \Rightarrow x(y - 1) = y \Rightarrow x = \dfrac{y}{y - 1}$
Replace with $f^{-1}(x)$ and x: $f^{-1}(x) = \dfrac{x}{x - 1}$
This shows that $f(x) = f^{-1}(x) \Rightarrow f(f(x)) = f(f^{-1}(x)) \Rightarrow f^2(x) = x$

Q24 Replace f(x) with y:
$y = \log_{10}(x + 4) \Rightarrow 10^y = x + 4 \Rightarrow x = 10^y - 4$.
Replace with $f^{-1}(x)$ and x: $f^{-1}(x) = 10^x - 4$

Q25 a)

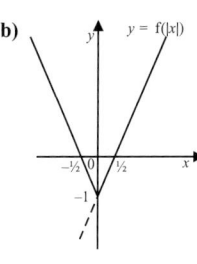

b)

Q26 Solving algebraically:
$|3x - 1| = |4 - x| \Rightarrow (3x - 1)^2 = (4 - x)^2$
$\Rightarrow 9x^2 - 6x + 1 = 16 - 8x + x^2$
$\Rightarrow 8x^2 + 2x - 15 = 0 \Rightarrow (4x - 5)(2x + 3) = 0$
So $x = \dfrac{5}{4}$ and $x = -\dfrac{3}{2}$

Q27 a) The maximum value of y is when $x = 0 \Rightarrow y = 3$ metres
The base of the triangular face is when $y = 0$, so
$y = 3 - 2|x| = 0 \Rightarrow |x| = \dfrac{3}{2} \Rightarrow x = \pm\dfrac{3}{2}$
so the base has length $\dfrac{3}{2} - \left(-\dfrac{3}{2}\right) = 3$ m
Area of the triangular cross-section is $\dfrac{1}{2} \times 3 \times 3 = \dfrac{9}{2}$ m^2.
Therefore the volume is $4 \times \dfrac{9}{2} = 18$ m^3.
[4 marks available — 1 mark for the maximum value of y, 1 mark for a correct method to find x when 3 − 2|x| = 0, 1 mark for the correct base length, 1 mark for correct volume]

b) The area of the base of the tent is $3 \times 4 = 12$ m^2, so a maximum of 6 people.
[1 mark for the correct answer]

c) E.g. Some people are taller/shorter so may require more or less space.
[1 mark for a suitable reason]

Q28 a)

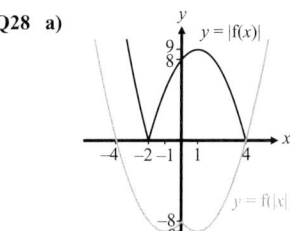

b) There are two solutions to $f(|x|) = -5$ — they correspond to the positive solution to $f(x) = -5$, and the negative solution to $-f(x) = -5$. These are symmetrical about the y-axis.
$f(x) = -5: x^2 - 2x - 8 = -5$
$\Rightarrow x^2 - 2x - 3 = 0 \Rightarrow (x - 3)(x + 1) = 0$.
So the positive solution is $x = 3$.
By symmetry, the other solution is $x = -3$.

Q29 a) (i) and (ii)

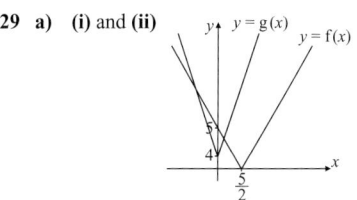

[4 marks available — 1 mark for correct shape and position for y = f(x), 1 mark for correct axis-intercepts labelled for y = f(x), 1 mark for correct shape and position for y = g(x), 1 mark for correct axis-intercept labelled for y = g(x)]

Answers

b) Looking at the sketch, f(x) and g(x) intersect twice.
To the left of the y-axis:
$-(2x - 5) = 3(-x) + 4 \Rightarrow x = -1$
To the right of the y-axis:
$-(2x - 5) = 3x + 4 \Rightarrow x = \frac{1}{5}$
f(x) > g(x) between these two values, i.e. when $-1 < x < \frac{1}{5}$
[3 marks available —1 mark for each correct point
of intersection, 1 mark for correct answer]

c) By looking at the sketch, you can see that if you shift g(x) up
by 1, then f(x) will never be above g(x) + k, and so there will
be no real solutions to the new inequality.
So the smallest value of k is 1 *[1 mark]*

Q30 a)

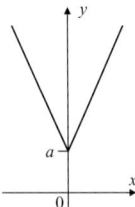

[2 marks available — 1 mark for the correct shape, 1 mark
for the correct graph with the correct minimum point]

b) $3a - x = a + 3x \Rightarrow 4x = 2a \Rightarrow x = \frac{a}{2}$
and $3a - x = a - 3x \Rightarrow 2x = -2a \Rightarrow x = -a$.
[3 marks available — 1 mark for a correct method, 1 mark
for $x = \frac{a}{2}$, 1 mark for $x = -a$]

c) From the graph in part a) and our answers in part b), it is
clear that $a + 3|x| < 3a - x$ when $-a < x < \frac{a}{2}$,
so we must have $-a = -20 \Rightarrow$ so $a = 20$.
[2 marks available — 1 mark for a suitable method
to find a, 1 mark for the correct answer]

Q31 Do the transformation in stages:

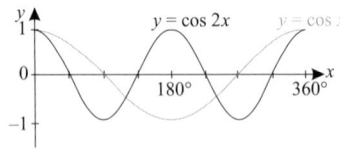

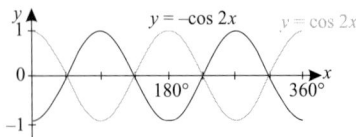

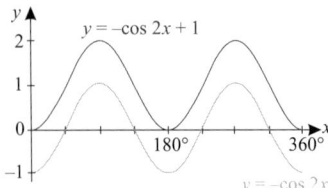

Q32 a)

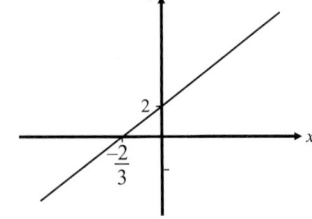

b)

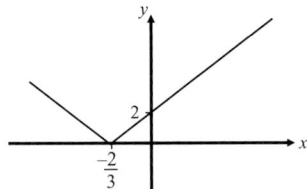

c)

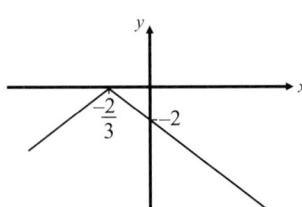

Q33 a)

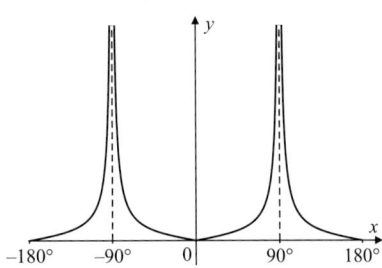

[3 marks available — 1 mark for the correct shape,
1 mark for the correct asymptotes, 1 mark for the
correct graph with the correct axes intercepts]

b) $\tan x = 1$ or $\tan x = -1$, so $x = 45°, 135°, -45°, -135°$.
[2 marks available — 1 mark for 45° and 135°,
1 mark for -45° and -135°]

c)

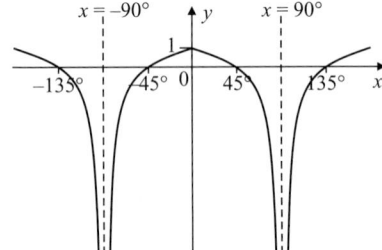

[3 marks available — 1 mark for the correct shape,
1 mark for the correct asymptotes, 1 mark for the correct
graph with the correct axes intercepts]
You can think of this as a reflection of your graph in part a) in
the x-axis followed by a translation of 1 unit up.

Q34 a) Solutions are where the graphs intersect, so sketch
the two curves and find where they meet.

(i) Sketching $y = -f(x)$:

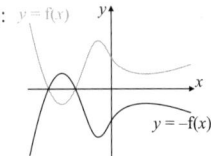

So there are two solutions — at $x = -9$ and $x = -5$.
There are no unknown solutions.
[2 marks available — 1 mark for both solutions,
1 mark for stating that there are no unknown solutions]

(ii) Sketching $y = f(|x|)$ on the interval $-12 \leq x \leq 0$:

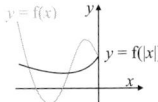

So there is one known solution at $x = 0$.
There are two more unknown solutions.
[2 marks available — 1 mark for the correct solution,
1 mark for stating that there are two unknown solutions]

(iii) Sketching $y = f^{-1}(x)$:

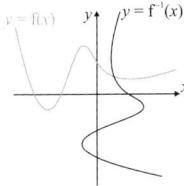

So there is one solution, and it is unknown.
[2 marks available — 1 mark for identifying that there is
one solution, 1 mark for stating that it is unknown]

b)

$y = f(4x) + 2$

The curve has undergone a horizontal stretch by a factor of $\frac{1}{4}$, which means that all x-coordinates (including the limits of the domain) are divided by 4. The curve is then shifted vertically by $+2$, which means the y-coordinates increase by 2. So the curve is valid over the interval $-3 \leq x \leq 3$
[4 marks available — 1 mark for a correct curve that touches the x-axis, 1 mark for each correct axis-intercept, 1 mark for the correct interval]

Q35 $\dfrac{2x}{(x-5)(x+5)} \equiv \dfrac{A}{(x-5)} + \dfrac{B}{(x+5)} \Rightarrow 2x \equiv A(x+5) + B(x-5)$

$x = 5: \ 10 = 10A \Rightarrow A = 1$
$x = -5: \ -10 = -10B \Rightarrow B = 1$
$\Rightarrow \dfrac{2x}{(x-5)(x+5)} \equiv \dfrac{1}{(x-5)} + \dfrac{1}{(x+5)}$

Q36 $\dfrac{2-x}{(3x+2)(x+1)} \equiv \dfrac{A}{(3x+2)} + \dfrac{B}{(x+1)}$

$\Rightarrow 2 - x \equiv A(x+1) + B(3x+2)$
$x = -1: \ 3 = B(-1) \Rightarrow B = -3$
$x = 0: \ 2 = A + 2B = A - 6 \Rightarrow A = 8$

Q37 $\dfrac{x-3}{x^2+3x+2} \equiv \dfrac{A}{(x+1)} + \dfrac{B}{(x+2)}$

$\Rightarrow x - 3 \equiv A(x+2) + B(x+1)$
$x = -1: \ -4 = A(1) \Rightarrow A = -4$
$x = -2: \ -5 = B(-1) \Rightarrow B = 5$
[3 marks available — 1 mark for a suitable method, 1 mark for the correct value of A, 1 mark for the correct value of B]

Q38 a) $\dfrac{4}{x^2+x} \equiv \dfrac{4}{x(x+1)} \equiv \dfrac{A}{x} + \dfrac{B}{x+1}$

$\Rightarrow 4 \equiv A(x+1) + Bx$
$x = 0: \ 4 = A$
$x = -1: \ 4 = -B \Rightarrow B = -4$
$\Rightarrow \dfrac{4}{x^2+x} \equiv \dfrac{4}{x} - \dfrac{4}{(x+1)}$

b) $\dfrac{4x+5}{(x+4)(2x-3)} \equiv \dfrac{A}{x+4} + \dfrac{B}{2x-3}$

$\Rightarrow 4x + 5 \equiv A(2x-3) + B(x+4)$
$x = -4: \ -11 = -11A \Rightarrow A = 1$
$x = 0: \ 5 = -3A + 4B = -3 + 4B \Rightarrow B = 2$
$\Rightarrow \dfrac{4x+5}{(x+4)(2x-3)} \equiv \dfrac{1}{x+4} + \dfrac{2}{2x-3}$

c) $\dfrac{5x}{x^2+x-6} \equiv \dfrac{5x}{(x+3)(x-2)} \equiv \dfrac{A}{x+3} + \dfrac{B}{x-2}$

$\Rightarrow 5x \equiv A(x-2) + B(x+3)$
$x = 2: \ 10 = 5B \Rightarrow B = 2$
$x = -3: \ -15 = -5A \Rightarrow A = 3$
$\Rightarrow \dfrac{5x}{x^2+x-6} \equiv \dfrac{3}{x+3} + \dfrac{2}{x-2}$

d) $\dfrac{10x}{(x+3)(2x+4)} \equiv \dfrac{5x}{(x+3)(x+2)} \equiv \dfrac{A}{x+3} + \dfrac{B}{x+2}$

$\Rightarrow 5x \equiv A(x+2) + B(x+3)$
$x = -3: \ -15 = A(-1) \Rightarrow A = 15$
$x = -2: \ -10 = B$
$\Rightarrow \dfrac{10x}{(x+3)(2x+4)} \equiv \dfrac{15}{x+3} - \dfrac{10}{x+2}$

e) $\dfrac{6x+10}{(2x+1)(2-3x)} \equiv \dfrac{A}{2x+1} + \dfrac{B}{2-3x}$

$\Rightarrow 6x + 10 \equiv A(2-3x) + B(2x+1)$
$x = \frac{2}{3}: \ 14 = \frac{7}{3}B \Rightarrow B = 6$
$x = -\frac{1}{2}: \ 7 = \frac{7}{2}A \Rightarrow A = 2$
$\Rightarrow \dfrac{6x+10}{(2x+1)(2-3x)} \equiv \dfrac{2}{2x+1} + \dfrac{6}{2-3x}$

f) $\dfrac{2x+1}{x^2+3x} \equiv \dfrac{2x+1}{(x+3)x} \equiv \dfrac{A}{x+3} + \dfrac{B}{x}$

$\Rightarrow 2x + 1 \equiv Ax + B(x+3)$
$x = 0: \ 1 = 3B \Rightarrow B = \frac{1}{3}$
$x = -3: \ -5 = -3A \Rightarrow A = \frac{5}{3}$
$\Rightarrow \dfrac{2x+1}{x^2+3x} \equiv \dfrac{5}{3(x+3)} + \dfrac{1}{3x}$

Q39 a) E.g. by using algebraic long division:

$$\begin{array}{r} x^2 + 2x + 1 \text{ r } 0 \\ x-1 \overline{) x^3 + x^2 - x - 1} \\ -\underline{(x^3 - x^2)} \\ 2x^2 - x \\ -\underline{(2x^2 - 2x)} \\ x - 1 \\ -\underline{(x-1)} \\ 0 \end{array}$$

So $g(x) = (x-1)(x^2+2x+1) = (x-1)(x+1)^2$
[3 marks available — 1 mark for a suitable method to find the quadratic factor, 1 mark for the correct quadratic, 1 mark for the correct final answer fully factorised]

b) $\dfrac{x-5}{x^3+x^2-x-1} \equiv \dfrac{x-5}{(x-1)(x+1)^2}$

$\equiv \dfrac{A}{x-1} + \dfrac{B}{x+1} + \dfrac{C}{(x+1)^2}$

$\Rightarrow x - 5 \equiv A(x+1)^2 + B(x-1)(x+1) + C(x-1)$
Substitution: $x = 1: \ -4 = 4A \Rightarrow A = -1$
$x = -1: \ 6 = -2C \Rightarrow C = 3$
$x = 0: \ -5 = A - B - C = -1 - B - 3 \Rightarrow B = 1$
So $\dfrac{x-5}{x^3+x^2-x-1} \equiv \dfrac{3}{(x+1)^2} + \dfrac{1}{x+1} - \dfrac{1}{x-1}$

[4 marks available — 1 mark for writing the expression as the sum of three fractions, 1 mark for finding one unknown, 1 mark for finding the other two unknowns, 1 mark for the correct answer]

Answers

Q40 $\dfrac{2x-5}{(x-5)^2} = \dfrac{A}{(x-5)} + \dfrac{B}{(x-5)^2}$,

so $2x - 5 = A(x-5) + B \Rightarrow 2x - 5 = Ax - 5A + B$
Equating x coefficients gives: $A = 2$
Equating constants gives:
$-5 = -5A + B \Rightarrow -5 = -10 + B \Rightarrow B = 5$.
This gives $\dfrac{2}{(x-5)} + \dfrac{5}{(x-5)^2}$, as required.

Q41 a) $\dfrac{2x+2}{(x+3)^2} \equiv \dfrac{A}{x+3} + \dfrac{B}{(x+3)^2}$

$\Rightarrow 2x + 2 \equiv A(x+3) + B$
$x = -3$: $-4 = B$
$x = 0$: $2 = 3A + B = 3A - 4 \Rightarrow A = 2$
$\Rightarrow \dfrac{2x+2}{(x+3)^2} \equiv \dfrac{2}{x+3} - \dfrac{4}{(x+3)^2}$

b) $\dfrac{-18x+14}{(2x-1)^2(x+2)} \equiv \dfrac{A}{(2x-1)} + \dfrac{B}{(2x-1)^2} + \dfrac{C}{(x+2)}$

$\Rightarrow -18x + 14 \equiv A(2x-1)(x+2) + B(x+2) + C(2x-1)^2$
$x = -2$: $50 = 25C \Rightarrow C = 2$
$x = \dfrac{1}{2}$: $5 = 2\dfrac{1}{2}B \Rightarrow B = 2$
$x = 0$: $14 = -2A + 2B + C \Rightarrow 8 = -2A \Rightarrow A = -4$
$\Rightarrow \dfrac{-18x+14}{(2x-1)^2(x+2)} \equiv \dfrac{-4}{(2x-1)} + \dfrac{2}{(2x-1)^2} + \dfrac{2}{(x+2)}$

c) $\dfrac{x-5}{x^3-x^2} \equiv \dfrac{x-5}{x^2(x-1)} \equiv \dfrac{A}{x^2} + \dfrac{B}{x} + \dfrac{C}{(x-1)}$

$\Rightarrow x - 5 \equiv A(x-1) + Bx(x-1) + Cx^2$
$x = 1$: $-4 = C$
$x = 0$: $-5 = -A \Rightarrow A - 5$
$x = -1$: $-6 = -2A + 2B + C \Rightarrow 8 = 2B \Rightarrow B = 4$
$\Rightarrow \dfrac{x-5}{x^3-x^2} \equiv \dfrac{5}{x^2} + \dfrac{4}{x} - \dfrac{4}{(x-1)}$

Q42 $\dfrac{bx+7}{(x+1)^2(x+2)} = \dfrac{3}{x+1} + \dfrac{2}{(x+1)^2} - \dfrac{3}{x+2}$

$\Rightarrow bx + 7 = 3(x+1)(x+2) + 2(x+2) - 3(x+1)^2$
$= 3(x^2 + 3x + 2) + 2(x+2) - 3(x^2 + 2x + 1)$
$= 3x^2 + 9x + 6 + 2x + 4 - 3x^2 - 6x - 3$
$= 5x + 7 \Rightarrow b = 5$

Q43 a) e.g. This model suggests that for large quantities of fertiliser, the tree will have a negative height, which is impossible.
[1 mark for a suitable answer]

b) $2 - (x-1)^2 = 0 \Rightarrow (x-1)^2 = 2$ so $x = 1 \pm \sqrt{2}$.
But this represents an amount in litres, so can't be negative.
So $n = 2.41$ litres (3 s.f.)
[3 marks available — 1 mark for setting $H(x) = 0$,
1 mark for a correct method to solve the quadratic,
1 mark for the correct answer]

c) e.g. very young lemon trees might not produce any lemons.
[1 mark available for a suitable answer]

d) $MH(x) = \dfrac{9}{2}(2 - (x-1)^2 - 1) + 8 = \dfrac{9}{2}(1 - (x-1)^2) + 8$
$= \dfrac{9}{2}(1 - (x^2 - 2x + 1)) + 8 = -\dfrac{9}{2}x^2 + 9x + 8$
[2 marks available — 1 mark for a suitable method,
1 mark for the correct answer]

e) $(x-1)^2 \geq 0 \Rightarrow 2 - (x-1)^2 \leq 2$, so the maximum height is 2 m, which occurs when the amount of fertiliser is 1 litre.
When $x = 1$, $MH(x) = -\dfrac{9}{2} + 9 + 8 = 12.5$.
Hence $12.5 \times 12 = 150$ kg of lemons produced in one year.
[3 marks available — 1 mark for the correct value of x,
1 mark for a correct method to find the mass of lemons
produced in one month, 1 mark for the correct answer]
You could also just put $y = 2$ into M(y) once you've found the
maximum height.

f) The domain of M(y) is $y > 0.25$, so H(x) > 0.25
$\Rightarrow 2 - (x-1)^2 > 0.25 \Rightarrow 1.75 > (x-1)^2$
$\Rightarrow 1 - \sqrt{1.75} < x < 1 + \sqrt{1.75}$
$\Rightarrow -0.323 < x < 2.323$ (3 d.p.)
But $x \geq 0$ so the domain is $0 \leq x < 2.323$ (3 d.p.)
The upper bound on the range is 12.5, as shown in part e).
The lower bound on the range is the lower bound of the range
of M(y) = $\dfrac{9}{2}(0.25 - 1) + 8 = 4.625$.
So the range of MH(x) is $4.625 < MH(x) \leq 12.5$.
[4 marks available — 1 mark for a correct method to find the
domain, 1 mark for the correct domain, 1 mark for a correct
method to find the range, 1 mark for the correct range]

Q44 a) Find the x-axis intercepts are at:
$5 - (x-1)^2 = 0 \Rightarrow (x-1)^2 = 5 \Rightarrow x = 1 + \sqrt{5}$ (since $x > 1$)
$3x + 2 = 0 \Rightarrow x = -\dfrac{2}{3}$

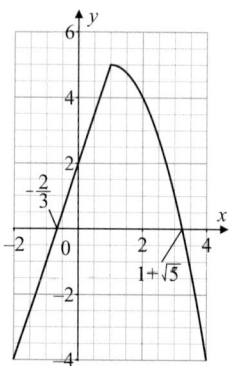

[4 marks available — 1 mark for drawing the straight line
accurately, 1 mark for the correct shape of the quadratic,
1 mark for correctly labelled axis intercepts, 1 mark for a
completely correct graph]

b) Because g is not a one-to-one mapping.
[1 mark for the correct answer]

c) This is the same as asking what is the largest possible domain of g(x) where g(x) > 0 and g(x) is one-to-one.
From part a) you know that g(x) = 0 when
$x = -\dfrac{2}{3}$ and $x = 1 + \sqrt{5}$.
To be one-to-one, either $-\dfrac{2}{3} < x \leq 1$ or $1 \leq x < 1 + \sqrt{5}$.
$1 \leq x < 1 + \sqrt{5}$ is a greater domain. So the largest possible
range for g⁻¹(x) is $1 \leq g^{-1}(x) < 1 + \sqrt{5}$.
[2 marks available — 1 mark for a correct method,
1 mark for the correct answer]
You could also have used the graph to identify where the largest
possible domain of g(x) is.

d) The triangle on the left is $\dfrac{1}{2} \times (1 + \dfrac{2}{3}) \times 5 = \dfrac{25}{6}$.
The area under the curve is
$\displaystyle\int_1^{1+\sqrt{5}} (5 - (x-1)^2)\,dx = \int_1^{1+\sqrt{5}} (-x^2 + 2x + 4)\,dx$
$= \left[-\dfrac{x^3}{3} + x^2 + 4x \right]_1^{1+\sqrt{5}} = 12.1202... - 4.6666... = 7.4535...$
So the total area is $\dfrac{25}{6} + 7.4535... = 11.6$ (3 s.f.)
[5 marks available — 1 mark for the area of the triangle,
1 mark for a correct method to find the area under the curve,
1 mark for integrating correctly, 1 mark for finding the area
under the curve, 1 mark for the correct answer]

Q45 $\left(\dfrac{1}{x} - \dfrac{x}{2x-1}\right) \div \dfrac{x^2 + 2x - 3}{8x - 4} = \left(\dfrac{1}{x} - \dfrac{x}{2x-1}\right) \div \dfrac{(x-1)(x+3)}{8x-4}$

$= \dfrac{(2x-1) - x^2}{x(2x-1)} \div \dfrac{(x-1)(x+3)}{8x-4}$

$= -\dfrac{(x-1)^2}{x(2x-1)} \div \dfrac{(x-1)(x+3)}{4(2x-1)}$

$= -\dfrac{(x-1)^2}{x(2x-1)} \times \dfrac{4(2x-1)}{(x-1)(x+3)} = -\dfrac{4(x-1)}{x(x+3)} = \dfrac{4-4x}{x(x+3)}$

$\dfrac{4-4x}{x(x+3)} \equiv \dfrac{A}{x} + \dfrac{B}{(x+3)} \Rightarrow 4 - 4x \equiv A(x+3) + Bx$

Substitution: $x = -3 \Rightarrow 16 = -3B \Rightarrow B = -\dfrac{16}{3}$

Equating constant terms: $4 = 3A \Rightarrow A = \dfrac{4}{3}$

This gives: $\dfrac{4-4x}{x(x+3)} = \dfrac{4}{3x} - \dfrac{16}{3(x+3)}$

Chapter 3: Trigonometry

Prior Knowledge Check

Q1 **a)**

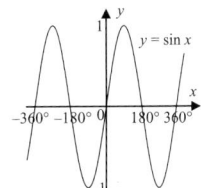

b)

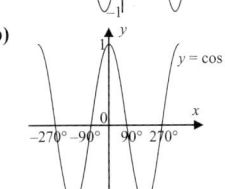

c)

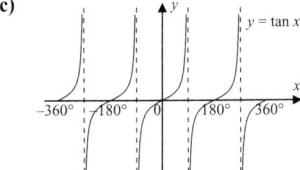

Q2 **a)** $(-2, 5)$ **b)** $(2, -5)$ **c)** $(1, 5)$

Q3 $\cos^2 x \tan x \equiv \cos^2 x \times \dfrac{\sin x}{\cos x} \equiv \sin x \cos x$

Exercise 3.1 — Radians

Q1 **a)** π **b)** $\dfrac{3\pi}{4}$ **c)** $\dfrac{3\pi}{2}$

   d) $\dfrac{7\pi}{18}$ **e)** $\dfrac{5\pi}{6}$ **f)** $\dfrac{5\pi}{12}$

Q2 **a)** $45°$ **b)** $90°$ **c)** $60°$

   d) $450°$ **e)** $135°$ **f)** $420°$

Q3 $\dfrac{2\pi}{5}$ radians $= \dfrac{2\pi}{5} \div \pi \times 180 = 72°$

So projectile B has a steeper launch trajectory.

[1 mark for the correct answer with working]

Exercise 3.2 — Arc Length and Sector Area

Q1 $s = r\theta = 6 \times 2 = 12$ cm

$A = \dfrac{1}{2}r^2\theta = \dfrac{1}{2} \times 6^2 \times 2 = 36$ cm²

Q2 Get the angle in radians:

$46° = \dfrac{46 \times \pi}{180} = 0.802...$ radians

$s = r\theta = 8 \times 0.802... = 6.4$ cm (1 d.p.)

$A = \dfrac{1}{2}r^2\theta = \dfrac{1}{2} \times 8^2 \times 0.802... = 25.7$ cm² (1 d.p.)

Q3 $A = \dfrac{1}{2}r^2\theta \Rightarrow 6\pi = \dfrac{1}{2} \times 4^2 \times \theta \Rightarrow 6\pi = 8\theta \Rightarrow \theta = \dfrac{6\pi}{8} = \dfrac{3\pi}{4}$

Q4 **a)** $s = r\theta = 5 \times 1.2 = 6$ cm

$A = \dfrac{1}{2}r^2\theta = \dfrac{1}{2} \times 5^2 \times 1.2 = 15$ cm²

b) $s = r\theta = 4 \times 0.6 = 2.4$ cm

$A = \dfrac{1}{2}r^2\theta = \dfrac{1}{2} \times 4^2 \times 0.6 = 4.8$ cm²

c) Get the angle in radians:

$80° = \dfrac{80 \times \pi}{180} = \dfrac{4\pi}{9}$ radians

$s = r\theta = 9 \times \dfrac{4\pi}{9} = 4\pi$ cm $= 12.6$ cm (3 s.f.)

$A = \dfrac{1}{2}r^2\theta = \dfrac{1}{2} \times 9^2 \times \dfrac{4\pi}{9} = 18\pi$ cm² $= 56.5$ cm² (3 s.f.)

d) $s = r\theta = 4 \times \dfrac{5\pi}{12} = \dfrac{5\pi}{3}$ cm $= 5.24$ cm (3 s.f.)

$A = \dfrac{1}{2}r^2\theta = \dfrac{1}{2} \times 4^2 \times \dfrac{5\pi}{12} = \dfrac{10\pi}{3}$ cm² $= 10.5$ cm² (3 s.f.)

Q5 Find the radius, r:

$A = \dfrac{1}{2}r^2\theta \Rightarrow 16.2 = \dfrac{1}{2} \times r^2 \times 0.9$

$16.2 = 0.45r^2 \Rightarrow 36 = r^2 \Rightarrow r = 6$ cm

$s = r\theta = 6 \times 0.9 = 5.4$ cm

[2 marks available — 1 mark for correct value of r, 1 mark for the correct answer]

Q6 Get the angle in radians: $20° = \dfrac{20 \times \pi}{180} = \dfrac{\pi}{9}$ radians

$s = r\theta = 3 \times \dfrac{\pi}{9} = \dfrac{\pi}{3}$ cm

$A = \dfrac{1}{2}r^2\theta = \dfrac{1}{2} \times 3^2 \times \dfrac{\pi}{9} = \dfrac{\pi}{2}$ cm²

Q7 Find the radius, r: $s = r\theta \Rightarrow r = \dfrac{s}{\theta} = \dfrac{7}{1.4} = 5$ cm

$A = \dfrac{1}{2}r^2\theta = \dfrac{1}{2} \times 5^2 \times 1.4 = 17.5$ cm²

Q8 Get the angle in radians: $50° = \dfrac{50 \times \pi}{180} = \dfrac{5}{18}\pi$ radians

Find the radius, r:

$A = \dfrac{1}{2}r^2\theta \Rightarrow 20\pi = \dfrac{1}{2} \times r^2 \times \dfrac{5}{18}\pi$

$\Rightarrow r^2 = 144 \Rightarrow r = 12$ cm

Then $s = r\theta = 12 \times \dfrac{5}{18}\pi = \dfrac{10\pi}{3}$ cm

[3 marks available — 1 mark for angle in radians, 1 mark for correct value of r, 1 mark for the correct answer]

Q9 $s = r\theta \Rightarrow r\theta = 16\pi$ ①

$A = \dfrac{1}{2}r^2\theta \Rightarrow \dfrac{1}{2}r^2\theta = 80\pi \Rightarrow r^2\theta = 160\pi$ ②

② $\div$ ①: $\dfrac{r^2\theta}{r\theta} = \dfrac{160\pi}{16\pi} \Rightarrow r = 10$ cm

①: $10\theta = 16\pi \Rightarrow \theta = \dfrac{8\pi}{5} = 5.03$ radians (3 s.f.)

Q10 **a)** Area $= \dfrac{1}{2}r^2\theta = 28$, so $r^2\theta = 56$ and

perimeter $= r\theta + 2r = 22$ so $r = \dfrac{22}{\theta + 2}$.

Substituting into the area equation gives $\dfrac{22^2\theta}{(\theta + 2)^2} = 56$ so

$484\theta = 56(\theta + 2)^2 \Rightarrow 484\theta = 56\theta^2 + 224\theta + 224$

$\Rightarrow 56\theta^2 - 260\theta + 224 = 0$

This has roots $\theta = 3.5$ and $\theta = 1.1428...$, but since $\theta > \pi$, we

have $\theta = 3.5$.

[3 marks available — 1 mark for a correct pair of simultaneous equations, 1 mark for a correct quadratic equation, 1 mark for the correct answer]

b) $r = \dfrac{22}{5.5} = 4$ cm *[1 mark]*

Answers

Q11 Area A $= \frac{1}{2}r^2\theta = \frac{1}{2}(2)^2\theta = 2\theta$

Area B $= \frac{1}{2}(2)^2(\pi - \theta) - \frac{1}{2}(1)^2(\pi - \theta)$

$= 2(\pi - \theta) - \frac{1}{2}(\pi - \theta) = \frac{3}{2}(\pi - \theta)$

A = B, so $2\theta = \frac{3}{2}(\pi - \theta) \Rightarrow 4\theta = 3\pi - 3\theta$

$\Rightarrow 7\theta = 3\pi \Rightarrow \theta = \frac{3\pi}{7}$

The angle of the missing sector would be $\pi - \theta$, since they lie on a straight line.

Q12 a) Using the cosine rule:

$\cos POS = \frac{7^2 + 7^2 - 8^2}{2 \times 7 \times 7} = 0.3469...$

$\Rightarrow POS = \cos^{-1}(0.3469...) = 1.216$ radians (to 3 d.p.)

[3 marks available — 1 mark for using the cosine rule and inputting the numbers correctly, 1 mark for 0.3469, 1 mark for the correct answer]

Alternatively you could have split the isosceles triangle into two identical right-angled triangles that have hypotenuse 7 cm and opposite side 4 cm, then used the sin trig formula.

b) Area $= \frac{1}{2} \times 7^2 \times \sin 1.216... = 23.0$ cm² (to 3 s.f.)

[2 marks available — 1 mark for a correct method, 1 mark for the correct answer]

c) Area not shaded $= \frac{1}{2} \times 3^2 \times 1.216.. = 5.474...$ cm²

So the percentage that is shaded:

$\frac{22.97... - 5.474...}{22.97...} \times 100 = 76\%$ to the nearest per cent.

[3 marks available — 1 mark for the correct sector area, 1 mark for suitable method to find the percentage, 1 mark for the correct answer]

Q13 a) Using completing the square:

$x^2 - 2x + y^2 - 4y = 4$

$\Rightarrow (x-1)^2 - 1^2 + (y-2)^2 - 2^2 = 4$

$\Rightarrow (x-1)^2 + (y-2)^2 = 9,$

so the centre is (1, 2) and the radius is $\sqrt{9} = 3$.

[3 marks available — 1 mark for correct completed square form, 1 mark for the correct centre, 1 mark for the correct radius]

b) A(−2, 2), C(4, 2) *[1 mark for both answers correct]*

c) B has coordinates (−2, 7).

So AB has length 5 and AP has length 3.

$APB = \tan^{-1}\left(\frac{5}{3}\right) = 1.0303... = 1.03$ radians (to 3 s.f.)

[2 marks available — 1 mark for a correct method, 1 mark for the correct answer]

d) Area of circle $= \pi r^2 = \pi \times 3^2 = 9\pi$.

Area of triangle CDP = Area of triangle APB

$= \frac{1}{2} \times 3 \times 5 = \frac{15}{2}$.

Area of sector subtended by angle APB

$= \frac{1}{2} \times 3^2 \times 1.0303... = 4.636...$

So total area $= 9\pi + \frac{15}{2} + \frac{15}{2} - 2 \times 4.636... = 34.0$ (to 3 s.f.)

[4 marks available — 1 mark for the area of the circle, 1 mark for the area of each triangle, 1 mark for area of the sector, 1 mark for the correct answer]

Alternatively you could add together the area of the two triangles, the semicircle and the other sector that isn't part of the triangles.

Exercise 3.3 — Small Angle Approximations

Q1 a) $\sin 0.23 \approx 0.23$

From a calculator, $\sin 0.23 = 0.228$ (3 d.p.)

b) $\cos 0.01 \approx 1 - \frac{1}{2}(0.01)^2 = 1 - 0.00005 = 0.99995$

From a calculator, $\cos 0.01 = 0.999950$ (6 d.p.)

c) $\tan 0.18 \approx 0.18$

From a calculator, $\tan 0.18 = 0.182$ (3 d.p.)

Q2 $f(\theta) = \sin \theta + \cos \theta \approx \theta + 1 - \frac{1}{2}\theta^2$

a) $f(0.3) \approx 0.3 + 1 - \frac{1}{2}(0.3)^2 = 1.255$

$f(0.3) = 1.2509$ (4 d.p.)

b) $f(0.5) \approx 0.5 + 1 - \frac{1}{2}(0.5)^2 = 1.375$

$f(0.5) = 1.3570$ (4 d.p.)

c) $f(0.25) \approx 0.25 + 1 - \frac{1}{2}(0.25)^2 = 1.21875$

$f(0.25) = 1.216316$ (6 d.p.)

d) $f(0.01) \approx 0.01 + 1 - \frac{1}{2}(0.01)^2 = 1.00995$

$f(0.01) = 1.009950$ (6 d.p.)

Q3 a) $\sin \theta \cos \theta \approx \theta\left(1 - \frac{1}{2}\theta^2\right) = \theta - \frac{1}{2}\theta^3$

b) $\theta \tan 5\theta \sin \theta \approx \theta(5\theta)(\theta) = 5\theta^3$

c) $\frac{\sin 4\theta \cos 3\theta}{2\theta} \approx \frac{4\theta\left(1 - \frac{1}{2}(3\theta)^2\right)}{2\theta} = 2\left(1 - \frac{9}{2}\theta^2\right) = 2 - 9\theta^2$

d) $3\tan \theta + \cos 2\theta \approx 3\theta + 1 - \frac{1}{2}(2\theta)^2 = 1 + 3\theta - 2\theta^2$

e) $\sin \frac{1}{2}\theta - \cos \theta \approx \frac{1}{2}\theta - 1 + \frac{1}{2}\theta^2 = \frac{1}{2}(\theta^2 + \theta - 2)$

f) $\frac{\cos \theta - \cos 2\theta}{1 - (\cos 3\theta + 3\sin \theta \tan \theta)} \approx \frac{\left(1 - \frac{1}{2}\theta^2\right) - \left(1 - \frac{1}{2}(2\theta)^2\right)}{1 - \left(1 - \frac{1}{2}(3\theta)^2\right) \quad 3(\theta)(\theta)}$

$= \frac{-\frac{1}{2}\theta^2 + 2\theta^2}{\frac{9}{2}\theta^2 - 3\theta^2} = \frac{\frac{3}{2}\theta^2}{\frac{3}{2}\theta^2} = 1$

Q4 a) $\frac{\sin x}{x} \approx \frac{x}{x} = 1$ *[1 mark]*

b) $\frac{\tan^2 x}{1 - \cos 3x} \approx \frac{x^2}{1 - \left(1 - \frac{(3x)^2}{2}\right)} = \frac{x^2}{\frac{9}{2}x^2} = \frac{2}{9}$

[2 marks available — 1 mark for using the correct small angle approximations, 1 mark for the correct answer]

Q5 $\frac{\sin^3 6x + \tan^4 x}{1 - \cos 2x} \approx \frac{(6x)^3 + x^4}{1 - \left(1 - \frac{(2x)^2}{2}\right)} = \frac{216x^3 + x^4}{2x^2} = \frac{1}{2}x^2 + 108x$

[3 marks available — 1 mark for using the correct small angle approximations, 1 mark for a correct expression without trig functions, 1 mark for working leading to the required result]

Q6 Small angle approximations give

$\frac{1 + \sin \theta}{1 - \cos 2\theta} = 0.62 \Rightarrow \frac{1 + \theta}{1 - \left(1 - \frac{(2\theta)^2}{2}\right)} = 0.62 \Rightarrow \frac{1 + \theta}{2\theta^2} = 0.62$

$\Rightarrow 1.24\theta^2 - \theta - 1 = 0 \Rightarrow \theta = 1.3876...$ and $\theta = -0.5811...$

Reject 1.3876... since θ is negative, so $\theta = -0.581$ (3 s.f.).

[4 marks available — 1 mark for correct small angle approximations, 1 mark for correct quadratic, 1 mark for method to solve quadratic, 1 mark for the correct answer]

$\theta = 1.3876...$ can also be rejected because small angle approximations are not valid for $\theta > 1$.

Q7 $\frac{\tan kx - \cos kx}{\sin 2kx} \approx \frac{kx - \left(1 - \frac{(kx)^2}{2}\right)}{2kx} = \frac{kx - 1 + \frac{1}{2}k^2x^2}{2kx}$

$= \frac{1}{2} - \frac{1}{2kx} + \frac{kx}{4} = \frac{1}{4}kx - \frac{1}{2kx} + \frac{1}{2}$

[3 marks available — 1 mark for using the correct small angle approximations, 1 mark for a correct expression without trig functions, 1 mark for working leading to the required result]

Q8 a) $\cos 2\theta \approx 1 - \frac{(2\theta)^2}{2} = 1 - 2\theta^2$ *[1 mark]*

b) Let $\theta = 5° = \dfrac{\pi}{36}$.

Then $\cos 10° \approx 1 - 2\left(\dfrac{\pi}{36}\right)^2 = 0.9847...$

$\Rightarrow \cos^2 10° \approx 0.9847...^2 = 0.96977... = 0.9698$ (to 4 d.p.).

[2 marks available — 1 mark for correct conversion,
1 mark for the correct answer]

c) $\cos^2 10° = 0.96984...$ so

$\text{error} = \dfrac{0.96984... - 0.96977...}{0.96984...} \times 100 = 0.00784\%$ (to 3 s.f.)

[3 marks available — 1 mark for correct value of $\cos^2 10°$,
1 mark for a correct method to find the percentage error,
1 mark for the correct answer]

Q9 a) $\mathbf{d} = 6\sin\theta\,\mathbf{i} + 6(1 - \cos\theta)\mathbf{j}$

$|\mathbf{d}| = \sqrt{(6\sin\theta)^2 + (6 - 6\cos\theta)^2}$

$= \sqrt{36\sin^2\theta + 36 - 72\cos\theta + 36\cos^2\theta}$

$= \sqrt{36(\sin^2\theta + \cos^2\theta) + 36 - 72\cos\theta}$

$= \sqrt{36 + 36 - 72\cos\theta}$

$= \sqrt{72(1 - \cos\theta)} = 6\sqrt{2(1 - \cos\theta)}$ as required

b) Arc length $s = 6\theta$

$|\mathbf{d}| = 6\sqrt{2(1 - \cos\theta)} \approx 6\sqrt{2\left(1 - \left(1 - \frac{1}{2}\theta^2\right)\right)}$

$= 6\sqrt{2\left(\frac{1}{2}\theta^2\right)} = 6\sqrt{\theta^2} = 6\theta = s$ as required

Q10 a) $\dfrac{AC}{\sin\theta} = \dfrac{11}{\sin\frac{\pi}{6}} \Rightarrow \dfrac{AC}{\sin\theta} = \dfrac{11}{0.5} \Rightarrow AC = 22\sin\theta$ cm

[2 marks available — 1 mark for a correct use of the sine
rule, 1 mark for correct working giving the required result]

b) $AC^2 = AB^2 + BC^2 - 2(AB)(BC)\cos\theta$

$\Rightarrow 484\sin^2\theta = x^2 + 121 - 22x\cos\theta$

Applying small angle approximations gives

$484\theta^2 \approx x^2 + 121 - 22x\left(1 - \dfrac{\theta^2}{2}\right)$, so

$\Rightarrow (484 - 11x)\theta^2 \approx x^2 - 22x + 121 = (x - 11)^2$

$\Rightarrow \theta^2 \approx \dfrac{(x - 11)^2}{484 - 11x}$.

[4 marks available — 1 mark for correct method with cosine
rule, 1 mark for correct equation from cosine rule, 1 mark
for correct application of small angle formulae, 1 mark for
correct rearrangement to give the required result]

Exercise 3.4 — Inverse Trig Functions

Q1 a) If $x = \arccos 1$ then $1 = \cos x$ so $x = 0$.

b) If $x = \arcsin\dfrac{\sqrt{3}}{2}$ then $\dfrac{\sqrt{3}}{2} = \sin x$ so $x = \dfrac{\pi}{3}$.

c) If $x = \arctan\sqrt{3}$ then $\sqrt{3} = \tan x$ so $x = \dfrac{\pi}{3}$.

d) If $x = \arccos\dfrac{1}{\sqrt{2}}$ then $\dfrac{1}{\sqrt{2}} = \cos x$ so $x = \dfrac{\pi}{4}$.

Q2 a) $y = 2\arccos x$

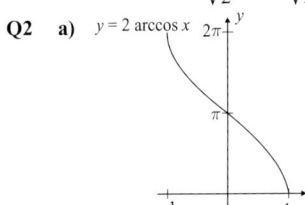

The graph is the same as $y = \arccos x$ but stretched
vertically by a factor of 2, so the y-coordinates of
the endpoints and y-intercept are doubled.

b)

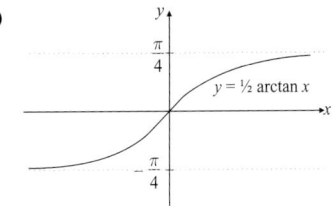

The range is $-\dfrac{\pi}{4} < \dfrac{1}{2}\arctan x < \dfrac{\pi}{4}$.

The graph is the same as $y = \arctan x$ but stretched vertically by
a factor of $\frac{1}{2}$, so the y-coordinates of the asymptotes are halved.

Q3 $y = \cos^{-1} x$

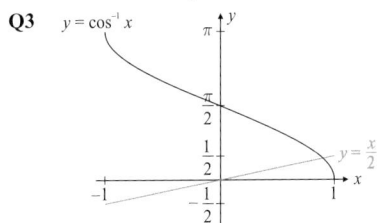

The graphs intersect once, so there is one real root of the equation $\cos^{-1} x = \dfrac{x}{2}$.

Q4 a) $\sin^{-1}(-1) = -\dfrac{\pi}{2}$.

This is one of the endpoints of the arcsin x graph.

b) To find $\cos^{-1}\left(-\dfrac{\sqrt{3}}{2}\right)$, first find the angle a

for which $\cos a = \dfrac{\sqrt{3}}{2}$: 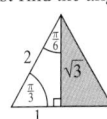 So $\cos\dfrac{\pi}{6} = \dfrac{\sqrt{3}}{2}$.

Now use the CAST diagram to find the negative solutions that lie in the domain $0 \le x \le \pi$:

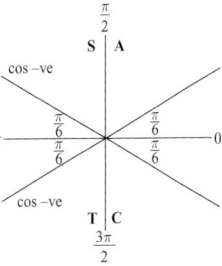

The only negative solution in that domain

is $\pi - \dfrac{\pi}{6} = \dfrac{5\pi}{6}$. So $\cos^{-1}\left(-\dfrac{\sqrt{3}}{2}\right) = \dfrac{5\pi}{6}$.

c) To find $\tan^{-1}\left(-\dfrac{1}{\sqrt{3}}\right)$, first find the angle a

for which $\tan a = \dfrac{1}{\sqrt{3}}$: 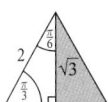 So $\tan\dfrac{\pi}{6} = \dfrac{1}{\sqrt{3}}$.

Now use the CAST diagram to find the negative solutions that lie in the domain $-\frac{\pi}{2} < x < \frac{\pi}{2}$:

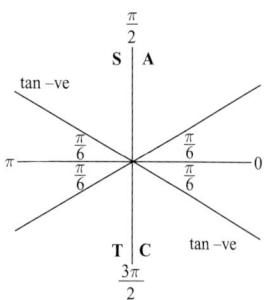

The only negative solution in that domain is $0 - \frac{\pi}{6} = -\frac{\pi}{6}$. So $\tan^{-1}\left(-\frac{1}{\sqrt{3}}\right) = -\frac{\pi}{6}$.

d) To find $\sin^{-1}\left(-\frac{1}{2}\right)$, first find the angle a for which $\sin a = \frac{1}{2}$:

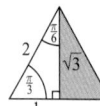

So $\sin\frac{\pi}{6} = \frac{1}{2}$.

Now use the CAST diagram to find the negative solutions that lie in the domain $-\frac{\pi}{2} \leq x \leq \frac{\pi}{2}$:

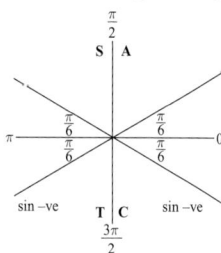

The only negative solution in that domain is $0 - \frac{\pi}{6} = -\frac{\pi}{6}$. So $\sin^{-1}\left(-\frac{1}{2}\right) = -\frac{\pi}{6}$.

Q5 a) $\arcsin\frac{1}{2} = \frac{\pi}{6}$, so $\tan\left(\arcsin\frac{1}{2}\right) = \tan\frac{\pi}{6} = \frac{1}{\sqrt{3}}$.

b) This is just the cos function followed by its inverse function so the answer is $\frac{2\pi}{3}$.

c) $\arcsin\frac{1}{2} = \frac{\pi}{6}$, so $\cos\left(\arcsin\frac{1}{2}\right) = \cos\frac{\pi}{6} = \frac{\sqrt{3}}{2}$.

Q6 To find the inverse of the function, first write as $y = 1 + \sin 2x$, then rearrange to make x the subject:

$\sin 2x = y - 1 \Rightarrow 2x = \sin^{-1}(y-1) \Rightarrow x = \frac{1}{2}\sin^{-1}(y-1)$

Now replace x with $f^{-1}(x)$ and y with x:

$f^{-1}(x) = \frac{1}{2}\sin^{-1}(x-1) = \frac{1}{2}\arcsin(x-1)$

Q7 a) $8y^2 - 6\pi y + \pi^2 = 0 \Rightarrow (4y - \pi)(2y - \pi) = 0 \Rightarrow y = \frac{\pi}{4}$ or $\frac{\pi}{2}$.
[2 marks available — 1 mark for a correct method to solve the quadratic, 1 mark for the correct answers]

b) $\arcsin x = \frac{\pi}{4} \Rightarrow x = \sin\frac{\pi}{4} \Rightarrow x = \frac{\sqrt{2}}{2}$
$\arcsin x = \frac{\pi}{2} \Rightarrow x = \sin\frac{\pi}{2} \Rightarrow x = 1$
[2 marks available — 1 mark for a correct method, 1 mark for both correct answers]

Q8 a) Let $y = \arccos x$.
Then $\cos y = x = \frac{x}{1} = \frac{\text{adj}}{\text{hyp}}$.

So, considering the following right-angle triangle:

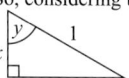

Using Pythagoras' theorem, the missing side is $\sqrt{1^2 - x^2} = \sqrt{1 - x^2}$, so we have $\sin y = \frac{\sqrt{1 - x^2}}{1}$, so $y = \arcsin\sqrt{1 - x^2}$.
Therefore, $\arccos x = \arcsin\sqrt{1 - x^2}$.

b) E.g. $\arccos(-1) = \pi$ whereas $\arcsin\sqrt{1 - (-1)^2} = 0$

Q9 Let $y = \arctan x$. Then $\tan y = x = \frac{x}{1} = \frac{\text{opp}}{\text{adj}}$
So, considering the following right-angle triangle:

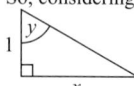

Using Pythagoras' theorem, the hypotenuse is $\sqrt{1^2 + x^2} = \sqrt{1 + x^2}$.
So we have $\sin y = \frac{x}{\sqrt{1 + x^2}}$, so $y = \arcsin\left(\frac{x}{\sqrt{x^2 + 1}}\right)$.
Therefore, $\arcsin\left(\frac{x}{\sqrt{x^2 + 1}}\right) = \arctan x$.

Q10 a) Let $y = \arcsin x$. Then $\sin y = x = \frac{x}{1} = \frac{\text{opp}}{\text{hyp}}$.
So, considering the following right-angle triangle:

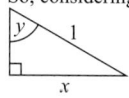

Using Pythagoras' theorem, the missing side is $\sqrt{1^2 - x^2} = \sqrt{1 - x^2}$.
Therefore $\tan(\arcsin x) = \tan y = \frac{x}{\sqrt{1 - x^2}}$.

b) $\arcsin\frac{1}{3} = \arctan k$
$\Rightarrow \tan(\arcsin\frac{1}{3}) = k$
$\Rightarrow k = \frac{\frac{1}{3}}{\sqrt{1 - \left(\frac{1}{3}\right)^2}} = \frac{\frac{1}{3}}{\sqrt{\frac{8}{9}}} = \frac{\frac{1}{3}}{\frac{2\sqrt{2}}{3}} = \frac{1}{2\sqrt{2}} = \frac{\sqrt{2}}{4}$

c) $\sin x \tan x + \frac{\sqrt{2}}{12} = \frac{1}{3}\tan x + \frac{\sqrt{2}}{4}\sin x$
$\Rightarrow \sin x \tan x - \frac{1}{3}\tan x - \frac{\sqrt{2}}{4}\sin x + \frac{\sqrt{2}}{12} = 0$
$\Rightarrow \left(\sin x - \frac{1}{3}\right)\left(\tan x - \frac{\sqrt{2}}{4}\right) = 0$

d) $\sin x = \frac{1}{3}$ or $\tan x = \frac{\sqrt{2}}{4}$ so $x = \arctan\frac{\sqrt{2}}{4}$ or $\arcsin\frac{1}{3}$, but from b) these are equal, so all the solutions are given by $\arcsin\frac{1}{3}$.

Q11 a) It suffices to verify Pythagoras' theorem:
$9^2 + 40^2 = 1681 = 41^2$. *[1 mark for a correct explanation]*

b) Considering the suggested right-angled triangle

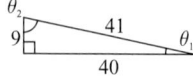

$\cos\theta_1 = \frac{40}{41}$ and $\tan\theta_2 = \frac{40}{9}$
So the angles are $\arccos\left(\frac{40}{41}\right)$, $\arctan\left(\frac{40}{9}\right)$ and $\frac{\pi}{2}$.
The angles must add to π so
$\arccos\left(\frac{40}{41}\right) + \arctan\left(\frac{40}{9}\right) + \frac{\pi}{2} = \pi$
$\Rightarrow \arccos\left(\frac{40}{41}\right) + \arctan\left(\frac{40}{9}\right) = \frac{\pi}{2}$.
[3 marks available — 1 mark for use of triangle, 1 mark for finding the angles in the triangle, 1 mark for completing the proof]

Q12 a) $\arcsin\left(\sin\frac{2\pi}{3}\right) = \arcsin\frac{\sqrt{3}}{2} = \frac{\pi}{3}$ *[1 mark]*

b) Because the range of arcsin is $-\frac{\pi}{2} \le x \le \frac{\pi}{2}$, so if x is outside this range, arcsin(sin x) cannot be equal to x.
[1 mark for a correct explanation]

c)

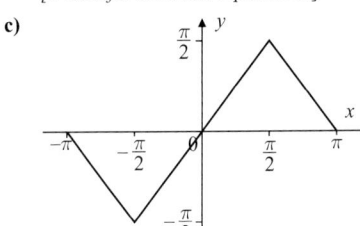

[3 marks available — 1 mark for sketching $y = x$ for $-\frac{\pi}{2} \le x \le \frac{\pi}{2}$, 1 mark for line joining maximum point to $(\pi, 0)$, 1 mark for line joining minimum point to $(-\pi, 0)$]

d) $\sin x = \sin 0.72$ so $x = 0.72$ or
$x = \pi - 0.72 = 2.4215... = 2.42$ (to 2 d.p.).
[2 marks available — 1 mark for each correct answer]

Exercise 3.5 — Graphs of Cosec, Sec and Cot

Q1 a)

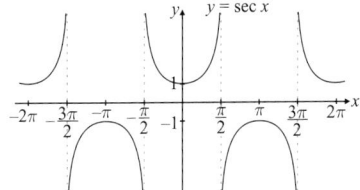

b) The minimum points are at $(-2\pi, 1)$, $(0, 1)$ and $(2\pi, 1)$.

c) The maximum points are at $(-\pi, -1)$ and $(\pi, -1)$.

d) The range is $y \ge 1$ or $y \le -1$.
You could also say that y is undefined for $-1 < y < 1$.

Q2 a)

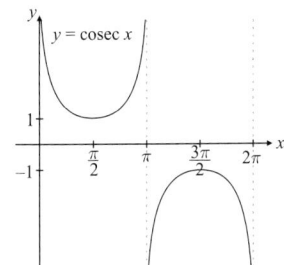

b) There is a maximum at $\left(\frac{3\pi}{2}, -1\right)$ and a minimum at $\left(\frac{\pi}{2}, 1\right)$.

c) The domain is $x \in \mathbb{R}$, $x \ne n\pi$ (where n is an integer).
The range is $y \ge 1$ or $y \le -1$.
The domain is all real numbers except those for which cosec x is undefined (i.e. at the asymptotes).

Q3 A horizontal translation right by $\frac{\pi}{2}$ (or 90°) or
a horizontal translation left by $\frac{3\pi}{2}$ (or 270°).

Q4 a)

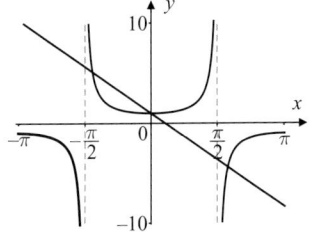

[3 marks available — 1 mark for correct line, 1 mark for correct shape of $y = \sec x$, 1 mark for a fully correct sec graph]

b) $\sec x + 3x = 1 \Rightarrow \sec x = 1 - 3x$
There are 3 points of intersection between
$y = \sec x$ and $y = 1 - 3x$ so 3 real solutions. *[1 mark]*

c) $\sec x + 8x = 4 \Rightarrow \sec x = 4 - 8x$
so the line to be added is $y = 4 - 8x$. *[1 mark]*

Q5 a) If $f(x) = \cot x$, then $y = \cot \frac{x}{4} = f\left(\frac{x}{4}\right)$.
This is a horizontal stretch scale factor 4.

b) The period of $y = \cot x$ is 180°, so the period of
$y = \cot \frac{x}{4}$ is 180° × 4 = 720°.

c)

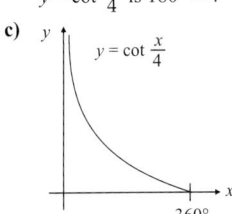

Q6 a) $y = 2 + \sec x$ is the graph of $y = \sec x$ translated
vertically up by 2:

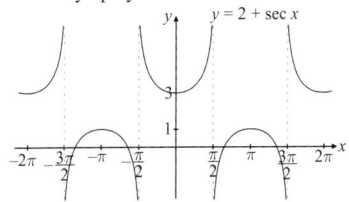

[3 marks available — 1 mark for correct shape, 1 mark for correct number of axes intercepts, 1 mark for correct number of asymptotes]

b) The minimum points are at $(-2\pi, 3)$, $(0, 3)$ and $(2\pi, 3)$.
The maximum points are at $(-\pi, 1)$ and $(\pi, 1)$.
[2 marks available — 1 mark for the correct minimum points, 1 mark for the correct maximum points]
The maximum and minimum points have the same x-coordinates as on the graph of y = sec x, but the y-coordinates have all been increased by 2.

c) The domain is $x \in \mathbb{R}$, $x \ne \left(n\pi + \frac{\pi}{2}\right)$ (where n is an integer).
The range is $y \ge 3$ or $y \le 1$. *[1 mark]*

Q7 a) $y = 2 \csc 2x$ is the graph of $y = \csc x$ stretched
horizontally by a factor of $\frac{1}{2}$ and stretched vertically
by a factor of 2.

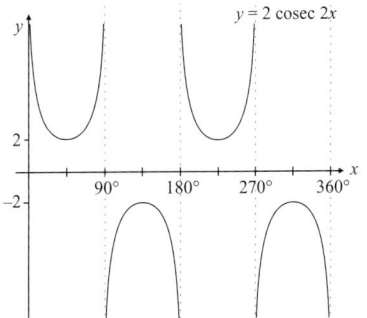

b) The minimum points are at (45°, 2) and (225°, 2).

c) The maximum points are at (135°, –2) and (315°, –2).

d) $y = 2 \csc 2x$ is undefined when
$x = 0°, 90°, 180°, 270°$ and 360°.

Answers

Q8 **a)** If $f(x) = \operatorname{cosec} x$, then $y = 2 + 3 \operatorname{cosec} x = 3f(x) + 2$,
which is a vertical stretch scale factor 3,
followed by a vertical translation of 2 up.
Vertical transformations do not affect the position of the asymptotes, so they are in the same position as for the graph of $y = \operatorname{cosec} x$, i.e. at $n\pi$ or $180n°$, where n is an integer.

b) The period of the graph will be the same as for the graph of $y = \operatorname{cosec} x$, i.e. $360°$.
Vertical transformations will not affect how often the graph repeats itself.

c)

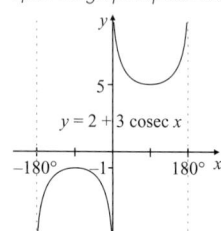

d) The range is $y \geq 5$ or $y \leq -1$.

Q9 **a)**

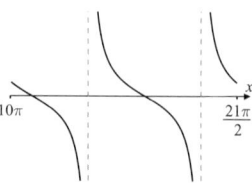

[3 marks available — 1 mark for correct shape, 1 mark for correct number of axes intercepts, 1 mark for correct number of asymptotes]

b) The period of $\cot x$ is π, so the period of $\cot(4x + 1)$ is $\frac{\pi}{4}$.
[1 mark]

c) Divide the interval by the period of $\cot(4x + 1)$
$\dfrac{11\pi}{\frac{\pi}{4}} = 44$ so there are 44 solutions.
[2 marks available — 1 mark for a suitable method, 1 mark for the correct answer]

Exercise 3.6 — Evaluating Cosec, Sec and Cot

Q1 **a)** $\operatorname{cosec} 80° = \dfrac{1}{\sin 80°} = 1.02$

b) $\sec 75° = \dfrac{1}{\cos 75°} = 3.86$

c) $\cot 30° = \dfrac{1}{\tan 30°} = 1.73$

d) $\sec(-70)° = \dfrac{1}{\cos(-70°)} = 2.92$

e) $3 - \cot 250° = 3 - \dfrac{1}{\tan 250°} = 2.64$

f) $2 \operatorname{cosec} 25° = \dfrac{2}{\sin 25°} = 4.73$

Q2 **a)** $\sec 3 = \dfrac{1}{\cos 3} = -1.01$

b) $\cot 0.6 = \dfrac{1}{\tan 0.6} = 1.46$

c) $\operatorname{cosec} 1.8 = \dfrac{1}{\sin 1.8} = 1.03$

d) $\sec(-1) = \dfrac{1}{\cos(-1)} = 1.85$

e) $\operatorname{cosec} \dfrac{\pi}{8} = \dfrac{1}{\sin \frac{\pi}{8}} = 2.61$

f) $8 + \cot \dfrac{\pi}{8} = 8 + \dfrac{1}{\tan \frac{\pi}{8}} = 10.4$

g) $\dfrac{1}{1 + \sec \frac{\pi}{10}} = \dfrac{1}{1 + \dfrac{1}{\cos \frac{\pi}{10}}} = 0.487$

h) $\dfrac{1}{6 + \cot \frac{\pi}{5}} = \dfrac{1}{6 + \dfrac{1}{\tan \frac{\pi}{5}}} = 0.136$

Q3 **a)** $\sec 60° = \dfrac{1}{\cos 60°} = \dfrac{1}{\left(\frac{1}{2}\right)} = 2$

b) $\operatorname{cosec} 30° = \dfrac{1}{\sin 30°} = \dfrac{1}{\left(\frac{1}{2}\right)} = 2$

c) $\cot 45° = \dfrac{1}{\tan 45°} = \dfrac{1}{1} = 1$

d) $\operatorname{cosec} \dfrac{\pi}{3} = \dfrac{1}{\sin \frac{\pi}{3}} = \dfrac{1}{\left(\frac{\sqrt{3}}{2}\right)} = \dfrac{2}{\sqrt{3}} = \dfrac{2\sqrt{3}}{3}$

e) $\sec(-180°) = \dfrac{1}{\cos(-180°)} = \dfrac{1}{\cos 180°} = -1$
The graph of $y = \cos x$ is symmetrical about the y-axis, so $\cos(-x) = \cos x$.

f) $\operatorname{cosec} 135° = \operatorname{cosec}(180° - 45°) = \dfrac{1}{\sin(180° - 45°)}$
The CAST diagram below shows that $\sin 135°$ is the same size as $\sin 45°$, and also lies in a positive quadrant for sin:

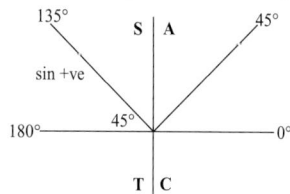

So $\operatorname{cosec} 135° = \dfrac{1}{\sin 45°} = \dfrac{1}{\left(\frac{1}{\sqrt{2}}\right)} = \sqrt{2}$

g) $\cot 330° = \cot(360° - 30°) = \dfrac{1}{\tan(360° - 30°)}$
The CAST diagram below shows that $\tan 330°$ is the same size as $\tan 30°$, but lies in a negative quadrant for tan:

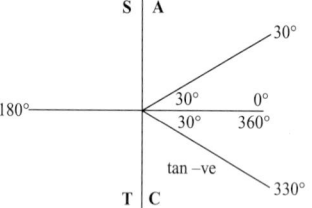

So $\cot 330° = \dfrac{1}{-\tan 30°} = \dfrac{1}{\left(-\frac{1}{\sqrt{3}}\right)} = -\sqrt{3}$

h) $\sec \dfrac{5\pi}{4} = \sec\left(\pi + \dfrac{\pi}{4}\right) = \dfrac{1}{\cos\left(\pi + \frac{\pi}{4}\right)}$
The CAST diagram below shows that $\cos\left(\pi + \dfrac{\pi}{4}\right)$ is the same size as $\cos \dfrac{\pi}{4}$, but lies in a negative quadrant for cos:

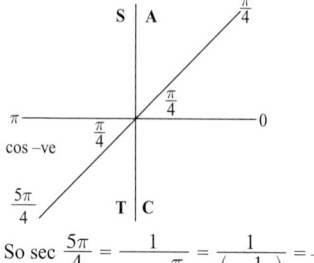

So $\sec \dfrac{5\pi}{4} = \dfrac{1}{-\cos \frac{\pi}{4}} = \dfrac{1}{\left(-\frac{1}{\sqrt{2}}\right)} = -\sqrt{2}$

i) $\csc\dfrac{5\pi}{3} = \csc\left(2\pi - \dfrac{\pi}{3}\right) = \dfrac{1}{\sin\left(2\pi - \frac{\pi}{3}\right)}$

The CAST diagram shows that $\sin\left(2\pi - \dfrac{\pi}{3}\right) = -\sin\dfrac{\pi}{3}$, so:

$\dfrac{1}{\sin\left(2\pi - \frac{\pi}{3}\right)} = \dfrac{1}{-\sin\frac{\pi}{3}} = \dfrac{1}{-\left(\frac{\sqrt{3}}{2}\right)} = -\dfrac{2}{\sqrt{3}} = -\dfrac{2\sqrt{3}}{3}$

j) $\csc\dfrac{2\pi}{3} = \csc\left(\pi - \dfrac{\pi}{3}\right) = \dfrac{1}{\sin\left(\pi - \frac{\pi}{3}\right)}$

The CAST diagram shows that $\sin\left(\pi - \dfrac{\pi}{3}\right) = \sin\dfrac{\pi}{3}$, so:

$\dfrac{1}{\sin\left(\pi - \frac{\pi}{3}\right)} = \dfrac{1}{\sin\frac{\pi}{3}} = \dfrac{1}{\left(\frac{\sqrt{3}}{2}\right)} = \dfrac{2}{\sqrt{3}} = \dfrac{2\sqrt{3}}{3}$

k) $3 - \cot\dfrac{3\pi}{4} = 3 - \cot\left(\pi - \dfrac{\pi}{4}\right) = 3 - \dfrac{1}{\tan\left(\pi - \frac{\pi}{4}\right)}$

$= 3 - \dfrac{1}{-\tan\frac{\pi}{4}} = 3 - \left(\dfrac{1}{-1}\right) = 4$

l) $\dfrac{\sqrt{3}}{\cot\frac{\pi}{6}} = \dfrac{\sqrt{3}}{\left(\frac{1}{\tan\frac{\pi}{6}}\right)} = \sqrt{3}\left(\tan\dfrac{\pi}{6}\right) = \sqrt{3} \times \dfrac{1}{\sqrt{3}} = 1$

Q4 a) $\dfrac{1}{1 + \sec 60°} = \dfrac{1}{1 + \left(\frac{1}{\cos 60°}\right)} = \dfrac{1}{1 + \frac{1}{\left(\frac{1}{2}\right)}} = \dfrac{1}{3}$

b) $\cot 315° = \cot(360° - 45°) = \dfrac{1}{\tan(360° - 45°)}$

$= \dfrac{1}{-\tan 45°} = -1$, so:

$\dfrac{2}{6 + \cot 315°} = \dfrac{2}{6 + (-1)} = \dfrac{2}{5}$

c) $\dfrac{1}{\sqrt{3} - \sec 30°} = \dfrac{1}{\sqrt{3} - \left(\frac{1}{\cos 30°}\right)}$

$= \dfrac{\cos 30°}{\sqrt{3}(\cos 30°) - 1} = \dfrac{\left(\frac{\sqrt{3}}{2}\right)}{\sqrt{3}\left(\frac{\sqrt{3}}{2}\right) - 1} = \dfrac{\frac{\sqrt{3}}{2}}{\frac{3}{2} - \frac{2}{2}} = \dfrac{\sqrt{3}}{3-2} = \sqrt{3}$

d) $1 + \cot 420° = 1 + \cot(360° + 60°) = 1 + \cot 60°$

$= 1 + \dfrac{1}{\tan 60°} = 1 + \dfrac{1}{\sqrt{3}} = \dfrac{3 + \sqrt{3}}{3}$

e) $\cot 150° = \cot(180° - 30°) = \dfrac{1}{\tan(180° - 30°)}$

$= \dfrac{1}{-\tan 30°} = \dfrac{1}{-\left(\frac{1}{\sqrt{3}}\right)} = -\sqrt{3}$, so:

$\dfrac{2}{7 + \sqrt{3}\cot 150°} = \dfrac{2}{7 + \sqrt{3}(-\sqrt{3})} = \dfrac{2}{7 - 3} = \dfrac{1}{2}$

f) $1 - \csc 330° = 1 - \csc(360° - 30°)$

$= 1 - \dfrac{1}{\sin(360° - 30°)} = 1 - \dfrac{1}{-\sin 30°} = 1 - \dfrac{1}{-0.5} = 3$, so:

$\dfrac{2}{1 - \csc 330°} = \dfrac{2}{3}$

Q5 a) $\left[\dfrac{\tan\left(\frac{\pi}{3}\right) - \cot\left(\frac{\pi}{3}\right)}{\sec\left(\frac{\pi}{6}\right)}\right]^2 = \left[\dfrac{\sqrt{3} - \frac{1}{\sqrt{3}}}{\frac{2}{\sqrt{3}}}\right]^2 = \left(\dfrac{3-1}{2}\right)^2 = 1^2 = 1$

[3 marks available — 1 mark for correct values of $\tan\frac{\pi}{3}, \cot\frac{\pi}{3}$ and $\sec\frac{\pi}{6}$, 1 mark for simplifying expression in brackets, 1 mark for correct answer]

b) $\cot^2\left(-\dfrac{\pi}{3}\right) + \cot\left(\dfrac{5\pi}{3}\right) - \sec\left(\dfrac{\pi}{6}\right)$

$= \left(-\dfrac{1}{\sqrt{3}}\right)^2 - \dfrac{1}{\sqrt{3}} - \dfrac{2}{\sqrt{3}} = \dfrac{1}{3} - \dfrac{3}{\sqrt{3}} = \dfrac{1}{3} - \sqrt{3}$.

[3 marks available — 1 mark for correct values of $\cot\left(-\frac{\pi}{3}\right), \cot\left(\frac{5\pi}{3}\right)$ and $\sec\left(\frac{\pi}{6}\right)$, 1 mark for rationalising denominator correctly, 1 mark for the correct answer]

Q6 b is the only variable that shifts the graph left or right.
$y = \csc x°$ has vertical asymptotes every $180°$.
So if the graph has a vertical asymptote at $170°$
and $0 < b < 90$ then $b = 10$ (shifting the graph $10°$ left).
Substitute $b = 10$ and the coordinates given in the question to form two equations and solve simultaneously:
For $(20, 9)$, $9 = a\csc(20 + 10)° + c \Rightarrow 9 = 2a + c$ (1)
For $(260, -3)$, $-3 = a\csc(260 + 10)° + c \Rightarrow -3 = -a + c$ (2)
$(1) - (2)$:
$$9 = 2a + c$$
$$\underline{-(-3 = -a + c)}$$
$$12 = 3a$$
$\Rightarrow a = 4$ and $9 = 2(4) + c \Rightarrow c = 1$
So the graph has equation $y = 4\csc(x + 10)° + 1$.
[3 marks available — 1 mark for correct value of b, 1 mark for method to find a and c, 1 mark for correct values of a and c]

Exercise 3.7 — Solving Equations involving Cosec, Sec and Cot

Q1 a) $\dfrac{1}{\cos x} = \sec x$, so $\sec x + \sec x = 2\sec x$

b) $(\csc^2 x)(\sin^2 x) = \dfrac{1}{\sin^2 x}(\sin^2 x) = 1$

c) $\dfrac{1}{\tan x} = \cot x$, so $2\cot x + \cot x = 3\cot x$

d) $\dfrac{\sec x}{\csc x} = \dfrac{\left(\frac{1}{\cos x}\right)}{\left(\frac{1}{\sin x}\right)} = \dfrac{\sin x}{\cos x} = \tan x$

e) $(\cos x)(\csc x) = \dfrac{\cos x}{\sin x} = \dfrac{1}{\left(\frac{\sin x}{\cos x}\right)} = \dfrac{1}{\tan x} = \cot x$

f) $\dfrac{\csc^2 x}{\cot x} = \dfrac{\left(\frac{1}{\sin^2 x}\right)}{\left(\frac{1}{\tan x}\right)} = \dfrac{\tan x}{\sin^2 x} = \dfrac{\left(\frac{\sin x}{\cos x}\right)}{\sin^2 x}$

$= \dfrac{\sin x}{\cos x \sin^2 x} = \dfrac{1}{\cos x \sin x} = \sec x \csc x$

g) $5\cot x - \dfrac{\csc x}{\sec x} = 5\cot x - \dfrac{\left(\frac{1}{\sin x}\right)}{\left(\frac{1}{\cos x}\right)}$

$= 5\cot x - \cot x = 4\cot x$

h) $\dfrac{1}{\sec^2 x} + \dfrac{1}{\csc^2 x}$

$= \dfrac{1}{\left(\frac{1}{\cos^2 x}\right)} + \dfrac{1}{\left(\frac{1}{\sin^2 x}\right)} = \cos^2 x + \sin^2 x = 1$

Q2 a) $\sin x \cot x = \sin x \left(\dfrac{1}{\tan x}\right) = \sin x \left(\dfrac{\cos x}{\sin x}\right) = \cos x$

b) $\sec x - \cos x = \dfrac{1}{\cos x} - \cos x = \dfrac{1 - \cos^2 x}{\cos x}$

Use the identity $\sin^2 x + \cos^2 x \equiv 1$:

$= \dfrac{\sin^2 x}{\cos x} = \left(\dfrac{\sin x}{\cos x}\right)\sin x = \tan x \sin x$

c) $\dfrac{\sec x}{\cot x} = \dfrac{\frac{1}{\cos x}}{\frac{1}{\tan x}} = \dfrac{\tan x}{\cos x} = \dfrac{\frac{\sin x}{\cos x}}{\cos x} = \dfrac{\sin x}{\cos^2 x} = \sin x \sec^2 x$

d) $\tan x \csc x = \left(\dfrac{\sin x}{\cos x}\right)\left(\dfrac{1}{\sin x}\right) = \dfrac{1}{\cos x} = \sec x$

Answers

e) $\dfrac{(\tan^2 x)(\csc x)}{\sin x} = \dfrac{\left(\dfrac{\sin^2 x}{\cos^2 x}\right)\left(\dfrac{1}{\sin x}\right)}{\sin x}$

$= \dfrac{\left(\dfrac{\sin x}{\cos^2 x}\right)}{\sin x} = \dfrac{1}{\cos^2 x} = \sec^2 x$

f) $\csc x (\sin x + \cos x) = \dfrac{1}{\sin x}(\sin x + \cos x)$

$= \dfrac{\sin x}{\sin x} + \dfrac{\cos x}{\sin x} = 1 + \cot x$

Q3 a) $\sec x = 1.9 \implies \cos x = \dfrac{1}{1.9} = 0.52631...$

$x = \cos^{-1}(0.52631...) = 58.2°$ (1 d.p.)

There is another positive solution in the interval $0° \le x \le 360°$ at $(360° - 58.2°)$ so $x = 301.8°$ (1 d.p.)
Remember, you can use the graphs or the CAST diagram to find other solutions in the interval.

b) $\cot x = 2.4 \implies \tan x = 0.41666...$
$\tan^{-1}(0.41666...) = 22.6°$ (1 d.p.)

There is another positive solution in the interval $0° \le x \le 360°$ at $(180° + 22.6°)$, so $x = 202.6°$ (1 d.p.)

c) $\csc x = -2 \implies \sin x = -0.5$
$\sin^{-1}(-0.5) = -30°$ which is not in the interval $0° \le x \le 360°$.

There are two negative solutions in the interval $0° \le x \le 360°$ at $(180° + 30°)$ and $(360° - 30°)$, so $x = 210°$ and $330°$.

d) $\sec x = -1.3 \implies \cos x = -0.76923...$
$\cos^{-1}(-0.76923...) = 140.3°$ (1 d.p.)

There are two negative solutions in the interval $0° \le x \le 360°$ at $140.3°$ and $(360° - 140.3°)$,
so $x = 140.3°$ and $219.7°$ (1 d.p.)

e) $\cot x = -2.4 \implies \tan x = -0.41666...$
$\tan^{-1}(-0.41666...) = -22.6°$ (1 d.p.)

There are two negative solutions in the interval $0 \le x \le 360°$ at $(-22.6° + 180°)$ and $(-22.6° + 360°)$,
so $x = 157.4°$ and $337.4°$ (1 d.p.)

f) $\dfrac{1}{2}\csc x = 0.7 \implies \csc x = 1.4 \implies \sin x = 0.71428...$
$\sin^{-1}(0.71428...) = 45.6°$ (1 d.p.)

There are two positive solutions in the interval $0° \le x \le 360°$ at $45.6°$ and $(180° - 45.6°)$, so $x = 45.6°$ and $134.4°$ (1 d.p.)

g) $4\sec 2x = -7 \implies \sec 2x = -1.75 \implies \cos 2x = -0.57142...$
$\cos^{-1}(-0.57142...) = 124.84990...$
You need to find all solutions for x in the interval $0° \le x \le 360°$ so $0° \le 2x \le (2 \times 360°)$ so you'll need to look for solutions for 2x in the interval $0° \le 2x \le 720°$.

There are four negative solutions for $2x$ in the interval $0° \le x \le 720°$ at $124.849...$, $(360° - 124.849...°)$, $(360° + 124.849...°)$ and $(720° - 124.849...°)$ and each of these needs to be divided by 2 to give x.
So $x = 62.4°, 117.6°, 242.4°$ and $297.6°$ (1 d.p.)

h) $5\cot 3x = 4 \implies \cot 3x = 0.8 \implies \tan 3x = 1.25$
$\tan^{-1}(1.25) = 51.3°$ (1 d.p.)
You need to find all solutions for x in the interval $0° \le x \le 360°$ so $0° \le 3x \le (3 \times 360°)$ so you'll need to look for solutions for 3x in the interval $0° \le 3x \le 1080°$.

There are six positive solutions for $3x$ in the interval $0° \le x \le 1080°$ at $51.34...°$, $(180° + 51.34...°)$, $(360° + 51.34...°)$, $(540° + 51.34...°)$, $(720° + 51.34...°)$ and $(900° + 51.34...°)$ and each of these needs to be divided by 3 to give x. So $x = 17.1°, 77.1°, 137.1°, 197.1°, 257.1°$ and $317.1°$ (1 d.p.)

Q4 a) $\sec x = 2 \implies \cos x = 0.5 \implies x = \cos^{-1}(0.5) = \dfrac{\pi}{3}$

There is another positive solution in the interval $0 \le x \le 2\pi$ at $\left(2\pi - \dfrac{\pi}{3}\right)$, so $x = \dfrac{5\pi}{3}$.

b) $\csc x = -2 \implies \sin x = -0.5 \implies x = \sin^{-1}(-0.5) = -\dfrac{\pi}{6}$

There are two negative solutions in the interval $0 \le x \le 2\pi$ at $\left(\pi + \dfrac{\pi}{6}\right)$ and $\left(2\pi - \dfrac{\pi}{6}\right)$, so $x = \dfrac{7\pi}{6}$ and $\dfrac{11\pi}{6}$.

c) $\cot 2x = 1 \implies \tan 2x = 1 \implies 2x = \tan^{-1}(1) = \dfrac{\pi}{4}$
Don't forget to double the interval for the next bit — you're looking for solutions for 2x instead of x.

There are 3 other positive solutions for $2x$ in the interval $0 \le 2x \le 4\pi$, at $\left(\pi + \dfrac{\pi}{4}\right)$, $\left(2\pi + \dfrac{\pi}{4}\right)$ and $\left(3\pi + \dfrac{\pi}{4}\right)$, so $x = \dfrac{\pi}{8}, \dfrac{5\pi}{8}, \dfrac{9\pi}{8}$ and $\dfrac{13\pi}{8}$.

d) $\sec 5x = -1 \implies \cos 5x = -1 \implies 5x = \cos^{-1}(-1) = \pi$
In this case you're looking for solutions for 5x — the interval you'll need to look in is $0 \le 5x \le 10\pi$ since $0 \le x \le 2\pi$. Use the fact that the cos graph repeats itself every 2π. $(\pi, -1)$ is a minimum point on the graph, so this will be repeated every 2π.

There are 4 other solutions for $5x$ in the interval $0 \le 5x \le 10\pi$, at $3\pi, 5\pi, 7\pi,$ and 9π, so
$x = \dfrac{\pi}{5}, \dfrac{3\pi}{5}, \pi, \dfrac{7\pi}{5}$ and $\dfrac{9\pi}{5}$.

e) $\csc 3x = -\sqrt{2} \implies \sin 3x = -\dfrac{1}{\sqrt{2}}$

$\implies 3x = \sin^{-1}\left(-\dfrac{1}{\sqrt{2}}\right) = -\dfrac{\pi}{4}$, which isn't in the interval.

Remember to multiply the interval by 3 for the next bit — you're looking for solutions for 3x instead of x.

There are six negative solutions in the interval $0 \le 3x \le 6\pi$ at $\left(\pi + \dfrac{\pi}{4}\right), \left(2\pi - \dfrac{\pi}{4}\right), \left(3\pi + \dfrac{\pi}{4}\right), \left(4\pi - \dfrac{\pi}{4}\right), \left(5\pi + \dfrac{\pi}{4}\right)$ and $\left(6\pi - \dfrac{\pi}{4}\right)$, so $x = \dfrac{5\pi}{12}, \dfrac{7\pi}{12}, \dfrac{13\pi}{12}, \dfrac{5\pi}{4}, \dfrac{7\pi}{4}, \dfrac{23\pi}{12}$.

f) $\cot 4x = \dfrac{1}{\sqrt{3}} \implies \tan 4x = \sqrt{3} \implies 4x = \tan^{-1}(\sqrt{3}) = \dfrac{\pi}{3}$

Remember to multiply the interval by 4 to find the solutions for 4x instead of x.

There are eight positive solutions in the interval $0 \le 4x \le 8\pi$ at $\dfrac{\pi}{3}, \left(\pi + \dfrac{\pi}{3}\right), \left(2\pi + \dfrac{\pi}{3}\right), \left(3\pi + \dfrac{\pi}{3}\right), \left(4\pi + \dfrac{\pi}{3}\right), \left(5\pi + \dfrac{\pi}{3}\right), \left(6\pi + \dfrac{\pi}{3}\right), \left(7\pi + \dfrac{\pi}{3}\right)$

so $x = \dfrac{\pi}{12}, \dfrac{\pi}{3}, \dfrac{7\pi}{12}, \dfrac{5\pi}{6}, \dfrac{13\pi}{12}, \dfrac{4\pi}{3}, \dfrac{19\pi}{12}, \dfrac{11\pi}{6}$.

Q5 a) $\cos x = \sin\left(x + \dfrac{\pi}{2}\right)$ so $\dfrac{1}{\cos x} = \dfrac{1}{\sin\left(x + \dfrac{\pi}{2}\right)}$,

which is the same as $\sec x = \csc\left(x + \dfrac{\pi}{2}\right)$.
[1 mark for a correct explanation]

b) $\sec x = 2 \implies \cos x = \dfrac{1}{2} \implies x = \pm\dfrac{\pi}{3}$
$\csc\left(x + \dfrac{\pi}{2}\right) = 2 \implies \sin\left(x + \dfrac{\pi}{2}\right) = \dfrac{1}{2}$

$\implies x + \dfrac{\pi}{2} = \dfrac{\pi}{6}$ or $x + \dfrac{\pi}{2} = \dfrac{5\pi}{6} \implies x = \pm\dfrac{\pi}{3}$
[3 marks available — 1 mark for both correct solutions for sec x = 2, 1 mark for a correct value of $x + \dfrac{\pi}{2}$, 1 mark for both correct solutions for $\csc\left(x + \dfrac{\pi}{2}\right) = 2$]

Q6 a) $\cot^2 x = 3 \implies \cot x = \pm\sqrt{3} \implies \tan x = \pm\dfrac{1}{\sqrt{3}}$
$\implies x = \pm 30°$
[2 marks available — 1 mark for rewriting in terms of tan x, 1 mark for the correct solutions]

b) $|\text{cosec}\, x| = \dfrac{2\sqrt{3}}{3} \Rightarrow \text{cosec}\, x = \pm\dfrac{2\sqrt{3}}{3} \Rightarrow \sin x = \pm\dfrac{3}{2\sqrt{3}}$

$\Rightarrow \sin x = \pm\dfrac{\sqrt{3}}{2} \Rightarrow x = \pm 60°$

[2 marks available — 1 mark for rewriting in terms of $\sin x$, 1 mark for the correct solutions]

c) $\sec\dfrac{1}{2}x = \dfrac{2\sqrt{3}}{3} \Rightarrow \cos\dfrac{1}{2}x = \dfrac{3}{2\sqrt{3}} \Rightarrow \cos\dfrac{1}{2}x = \dfrac{\sqrt{3}}{2}$

$\Rightarrow \dfrac{1}{2}x = \pm 30° \Rightarrow x = \pm 60°.$

[2 marks available — 1 mark for rewriting in terms of $\cos\dfrac{1}{2}x$, 1 mark for the correct solutions]

Q7 $\cot 2x - 4 = -5 \Rightarrow \tan 2x = -1 \Rightarrow 2x = \tan^{-1}(-1) = -\dfrac{\pi}{4}$

There are 4 solutions for $2x$ in the interval $0 \le 2x \le 4\pi$,

at $\left(\pi - \dfrac{\pi}{4}\right)$, $\left(2\pi - \dfrac{\pi}{4}\right)$, $\left(3\pi - \dfrac{\pi}{4}\right)$ and $\left(4\pi - \dfrac{\pi}{4}\right)$,

so $x = \dfrac{3\pi}{8}, \dfrac{7\pi}{8}, \dfrac{11\pi}{8}$ and $\dfrac{15\pi}{8}$.

[3 marks available — 1 mark for rewriting in terms of $\tan 2x$, 1 mark for the initial solution, 1 mark for the other solutions]

Q8 $2\,\text{cosec}\, 2x = 3 \Rightarrow \sin 2x = \dfrac{2}{3}$

$2x = \sin^{-1}\left(\dfrac{2}{3}\right) = 41.81031...°$

There are three other positive solutions for $2x$ in the interval $0° \le 2x \le 720°$ at $(180° - 41.81...°)$, $(360° + 41.81...°)$ and $(540° - 41.81...°)$, so $x = 20.9°, 69.1°, 200.9°$ and $249.1°$ (1 d.p.)

[3 marks available — 1 mark for rewriting in terms of $\sin 2x$, 1 mark for the initial solution, 1 mark for the other solutions]

Q9 $-2\sec x = 4 \Rightarrow \cos x = -0.5 \Rightarrow x = \cos^{-1}(-0.5) = \dfrac{2\pi}{3}$

There are 2 negative solutions in the interval $0 \le x \le 2\pi$ at $\dfrac{2\pi}{3}$ and $\left(2\pi - \dfrac{2\pi}{3}\right)$, so $x = \dfrac{2\pi}{3}$ and $\dfrac{4\pi}{3}$.

Q10 $\sqrt{3}\,\text{cosec}\, 3x = 2 \Rightarrow \sin 3x = \dfrac{\sqrt{3}}{2} \Rightarrow 3x = \sin^{-1}\left(\dfrac{\sqrt{3}}{2}\right) = \dfrac{\pi}{3}$

There are 5 other positive solutions in the interval $0 \le 3x \le 6\pi$, at $\left(\pi - \dfrac{\pi}{3}\right)$, $\left(2\pi + \dfrac{\pi}{3}\right)$, $\left(3\pi - \dfrac{\pi}{3}\right)$, $\left(4\pi + \dfrac{\pi}{3}\right)$

and $\left(5\pi - \dfrac{\pi}{3}\right)$, so $x = \dfrac{\pi}{9}, \dfrac{2\pi}{9}, \dfrac{7\pi}{9}, \dfrac{8\pi}{9}, \dfrac{13\pi}{9}$ and $\dfrac{14\pi}{9}$.

Q11 a) $\sec^2 x - 2\sqrt{2}\,\sec x + 2$ factorises to $(\sec x - \sqrt{2})^2$.

$(\sec x - \sqrt{2})^2 = 0 \Rightarrow \sec x = \sqrt{2}$

$\Rightarrow \cos x = \dfrac{1}{\sqrt{2}} \Rightarrow x = 45°$

b) $2\cot^2 x + 3\cot x - 2$ factorises to give $(2\cot x - 1)(\cot x + 2)$

So $(2\cot x - 1)(\cot x + 2) = 0$

$\Rightarrow \cot x = \dfrac{1}{2}$ or $\cot x = -2 \Rightarrow \tan x = 2$ or $\tan x = -\dfrac{1}{2}$

The solution for $\tan x = 2$ is $x = 63.4°$.

$\tan^{-1}-\dfrac{1}{2} = -26.6°$, so the other solution is

$x = -26.6° + 180° = 153.4°$. So $x = 63.4°$ and $153.4°$ (1 d.p.)

Q12 $(\text{cosec}\, x - 3)(2\tan x + 1) = 0$ means that either $\text{cosec}\, x = 3$

(and so $\sin x = \dfrac{1}{3}$) or $\tan x = -\dfrac{1}{2}$.

The solutions for $\sin x = \dfrac{1}{3}$ are $x = 19.5°$

or $x = 180° - 19.5° = 160.5°$.

The solutions for $\tan x = -\dfrac{1}{2}$ are

$x = -26.6° + 180° = 153.4°$ or $x = -26.6° + 360° = 333.4°$.

So $x = 19.5°, 160.5°, 153.4°, 333.4°$ (1 d.p.) are all solutions.

Q13 $\sin x + \sin y = 1 \Rightarrow \sin y = 1 - \sin x$

$\text{cosec}\, x + \text{cosec}\, y = 4 \Rightarrow \dfrac{1}{\sin x} + \dfrac{1}{\sin y} = 4$

Substituting $\sin y$ from the first equation into the second gives

$\dfrac{1}{\sin x} + \dfrac{1}{1 - \sin x} = 4$

$\Rightarrow 1 - \sin x + \sin x = 4\sin x(1 - \sin x)$

$\Rightarrow 1 = 4\sin x - 4\sin^2 x$

$\Rightarrow 4\sin^2 x - 4\sin x + 1 = 0$

$\Rightarrow (2\sin x - 1)^2 = 0,$

$\Rightarrow \sin x = \dfrac{1}{2} \Rightarrow x = \dfrac{\pi}{6}, \dfrac{5\pi}{6}.$

Since $\sin x = \dfrac{1}{2}$ and $\sin x + \sin y = 1$ we also have

$\sin y = \dfrac{1}{2} \Rightarrow y = \dfrac{\pi}{6}, \dfrac{5\pi}{6}.$

There are four pairs of solutions:

$x = \dfrac{\pi}{6}, y = \dfrac{\pi}{6} \qquad x = \dfrac{\pi}{6}, y = \dfrac{5\pi}{6}$

$x = \dfrac{5\pi}{6}, y = \dfrac{\pi}{6} \qquad x = \dfrac{5\pi}{6}, y = \dfrac{5\pi}{6}$

[6 marks available — 1 mark for rewriting the second equation in terms of sin, 1 mark for a correct method to solve simultaneously, 1 mark for forming a quadratic in $\sin x$, 1 mark for method to solve the quadratic, 1 mark for at least two correct pairs of solutions, 1 mark for all correct solutions]

Exercise 3.8 — Identities Involving Cosec, Sec and Cot

Q1 $\text{cosec}^2 x + 2\cot^2 x = \text{cosec}^2 x + 2(\text{cosec}^2 x - 1) = 3\,\text{cosec}^2 x - 2$

Q2 $\tan^2 x - \dfrac{1}{\cos^2 x} = \tan^2 x - \sec^2 x = \tan^2 x - (1 + \tan^2 x) = -1$

Q3 a) $\sec y = \sqrt{1 + \tan^2 y} = \sqrt{1 + t^2}$ *[1 mark]*

b) $3\sec^2 y - 4 = 3(1 + \tan^2 y) - 4 = 3(1 + t^2) - 4 = 3t^2 - 1$

[2 marks available — 1 mark for correctly using the trig identity, 1 mark for the correct answer]

c) $\text{cosec}^2 y = 1 + \cot^2 y = 1 + \dfrac{1}{\tan^2 y} = 1 + \dfrac{1}{t^2}$ or $\dfrac{t^2 + 1}{t^2}$

[2 marks available — 1 mark for correctly using the trig identity, 1 mark for the correct answer]

Q4 $x + \dfrac{1}{x} = \sec\theta + \tan\theta + \dfrac{1}{\sec\theta + \tan\theta} = \dfrac{(\sec\theta + \tan\theta)^2 + 1}{\sec\theta + \tan\theta}$

$= \dfrac{\sec^2\theta + 2\sec\theta\tan\theta + \tan^2\theta + 1}{\sec\theta + \tan\theta}$

But since $\sec^2\theta = \tan^2\theta + 1$:

$x + \dfrac{1}{x} = \dfrac{\sec^2\theta + 2\sec\theta\tan\theta + \sec^2\theta}{\sec\theta + \tan\theta}$

$= \dfrac{2\sec\theta(\sec\theta + \tan\theta)}{\sec\theta + \tan\theta} = 2\sec\theta$ as required.

Q5 $\dfrac{36}{25}x^2 - y^2 = \dfrac{36}{25}(5\sec t)^2 - (6\tan t)^2$

$= 36\sec^2 t - 36\tan^2 t = 36(1 + \tan^2 t) - 36\tan^2 t = 36.$

[3 marks available — 1 mark for correct expression for $\dfrac{36}{25}x^2 - y^2$ in terms of t, 1 mark for correctly using the trig identity, 1 mark for fully correct proof]

Q6 a) $\tan^2 x = 2\sec x + 2 \Rightarrow \sec^2 x - 1 = 2\sec x + 2$

$\Rightarrow \sec^2 x - 2\sec x - 3 = 0$

b) Solve $\sec^2 x - 2\sec x - 3 = 0$

This factorises to give: $(\sec x - 3)(\sec x + 1) = 0$

So $\sec x = 3 \Rightarrow \cos x = \dfrac{1}{3}$ and $\sec x = -1 \Rightarrow \cos x = -1$

Solving these over the interval $0° \le x \le 360°$ gives:

$x = 70.5°, 180°$ and $289.5°$ (1 d.p.)

Just use the graph of $\cos x$ or the CAST diagram as usual to find all the solutions in the interval.

Q7 a) $2\,\text{cosec}^2 x = 5 - 5\cot x \Rightarrow 2(1 + \cot^2 x) = 5 - 5\cot x$

$\Rightarrow 2\cot^2 x + 5\cot x - 3 = 0$

Answers

b) Solve $2\cot^2 x + 5\cot x - 3 = 0$
This factorises to give: $(2\cot x - 1)(\cot x + 3) = 0$
So $\cot x = \frac{1}{2} \Rightarrow \tan x = 2$ and $\cot x = -3 \Rightarrow \tan x = -\frac{1}{3}$
Solving these over the interval $-\pi \leq x \leq \pi$ gives:
$x = -2.03, -0.32, 1.11, 2.82$ (2 d.p.)

Q8 a) $2\cot^2 A + 5\csc A = 10$
$\Rightarrow 2(\csc^2 A - 1) + 5\csc A = 10$
$\Rightarrow 2\csc^2 A - 2 + 5\csc A = 10$
$\Rightarrow 2\csc^2 A + 5\csc A - 12 = 0$

b) Solve $2\csc^2 A + 5\csc A - 12 = 0$
This factorises to give:
$(\csc A + 4)(2\csc A - 3) = 0$
So $\csc A = -4 \Rightarrow \sin A = -\frac{1}{4}$
Or $\csc x = \frac{3}{2} \Rightarrow \sin A = \frac{2}{3}$
The solutions (to 1 d.p.) are
$A = 194.5°, 345.5°, 41.8°, 138.2°$.

Q9 $\sec^2 x + \tan x = 1 \Rightarrow 1 + \tan^2 x + \tan x = 1$
$\Rightarrow \tan^2 x + \tan x = 0 \Rightarrow \tan x(\tan x + 1) = 0$
So $\tan x = 0 \Rightarrow x = 0, \pi, 2\pi$.
And $\tan x = -1 \Rightarrow x = \frac{3\pi}{4}, \frac{7\pi}{4}$.

Q10 a) $\csc^2\theta + 2\cot^2\theta = 2 \Rightarrow \csc^2\theta + 2(\csc^2\theta - 1) = 2$
$\Rightarrow 3\csc^2\theta - 2 = 2 \Rightarrow 3\csc^2\theta = 4$
$\Rightarrow \csc^2\theta = \frac{4}{3} \Rightarrow \csc\theta = \pm\frac{2}{\sqrt{3}} \Rightarrow \sin\theta = \pm\frac{\sqrt{3}}{2}$

b) Solving $\sin\theta = \pm\frac{\sqrt{3}}{2}$ over the interval
$0° \leq x \leq 180°$ gives: $\theta = 60°, 120°$.

Q11 $\sec^2 x = 3 + \tan x \Rightarrow (1 + \tan^2 x) = 3 + \tan x$
$\Rightarrow \tan^2 x - \tan x - 2 = 0 \Rightarrow (\tan x - 2)(\tan x + 1) = 0$
So $\tan x = 2$ or $\tan x = -1$.
Solving over the interval $0° \leq x \leq 360°$ gives:
$x = 63.4°, 135°, 243.4°$ and $315°$ (1 d.p.)

Q12 $\cot^2 x + \csc^2 x = 7 \Rightarrow \cot^2 x + (1 + \cot^2 x) = 7$
$\Rightarrow 2\cot^2 x + 1 = 7 \Rightarrow \cot^2 x = 3 \Rightarrow \cot x = \pm\sqrt{3}$
$\Rightarrow \tan x = \pm\frac{1}{\sqrt{3}}$
Solving over the interval $0 \leq x \leq 2\pi$ gives:
$x = \frac{\pi}{6}$ and $\frac{7\pi}{6}$ when $\tan x = +\frac{1}{\sqrt{3}}$
and $x = \frac{5\pi}{6}$ and $\frac{11\pi}{6}$ when $\tan x = -\frac{1}{\sqrt{3}}$.

Q13 $\tan^2 x + 5\sec x + 7 = 0 \Rightarrow (\sec^2 x - 1) + 5\sec x + 7 = 0$
$\Rightarrow \sec^2 x + 5\sec x + 6 = 0 \Rightarrow (\sec x + 2)(\sec x + 3) = 0$
So $\sec x = -2 \Rightarrow \cos x = -\frac{1}{2}$ and $\sec x = -3 \Rightarrow \cos x = -\frac{1}{3}$
Solving over the interval $0 \leq x \leq 2\pi$ gives:
$x = 1.91, 2.09, 4.19$ and 4.37 (2 d.p.)

Q14 Drawing a right-angled triangle will help to solve this question:

$\tan\theta = \frac{60}{11}$

Notice that $180° \leq \theta \leq 270°$ — this puts us in the 3rd quadrant of the CAST diagram so sin will be –ve, cos will be –ve and tan will be +ve.

a) From the triangle, $\sin\theta = -\frac{\text{opp}}{\text{hyp}} = -\frac{60}{61}$.

b) $\cos\theta = -\frac{\text{adj}}{\text{hyp}} = -\frac{11}{61} \Rightarrow \sec\theta = \frac{1}{\left(-\frac{11}{61}\right)} = -\frac{61}{11}$.

c) $\csc\theta = \frac{1}{\sin\theta} = \frac{1}{\left(-\frac{60}{61}\right)} = -\frac{61}{60}$.

Q15 Drawing a right-angled triangle will help to solve this question:

$\csc\theta = -\frac{17}{15}$

$\sin\theta = -\frac{15}{17}$

Notice that $180° \leq \theta \leq 270°$ — this puts us in the 3rd quadrant of the CAST diagram so sin will be –ve, cos will be –ve and tan will be +ve.

a) $\cos\theta = -\frac{\text{adj}}{\text{hyp}} = -\frac{8}{17}$.

b) $\sec\theta = \frac{1}{\cos\theta} = -\frac{17}{8}$.

c) $\tan\theta = \frac{\text{opp}}{\text{adj}} = \frac{15}{8}$ so $\cot\theta = \frac{1}{\tan\theta} = \frac{8}{15}$.

Q16 $\cos x = \frac{1}{6} \Rightarrow \sec x = 6 \Rightarrow \sec^2 x = 36$
So $1 + \tan^2 x = 36 \Rightarrow \tan^2 x = 35 \Rightarrow \tan x = \pm\sqrt{35}$
[2 marks available — 1 mark for use of identity, 1 mark for the correct answer]

Q17 a) $3\tan x + 1 = \sec^2 x \Rightarrow 3\tan x + 1 = 1 + \tan^2 x$
$\Rightarrow \tan^2 x - 3\tan x = 0 \Rightarrow \tan x(\tan x - 3) = 0$
$\Rightarrow \tan x = 0$ or $\tan x = 3$
[3 marks available — 1 mark for use of identity, 1 mark for method to solve quadratic, 1 mark for the correct answers]

b) $\tan x = 0$ has no solutions in the range $(0, \pi)$.
$\tan x = 3 \Rightarrow x = \arctan 3 = 1.2490... = 1.25$ (to 3 s.f.)
[2 marks available — 1 mark for a correct method, 1 mark for the correct answer]

Q18 a) $\sec^2 x - 6 = \sec x \Rightarrow \sec^2 x - \sec x - 6 = 0$
$\Rightarrow (\sec x - 3)(\sec x + 2) = 0 \Rightarrow \sec x = 3$ or $\sec x = -2$
[2 marks available — 1 mark for a correct method to solve the quadratic, 1 mark for the correct answers]

b) $\sec^2 x = (-2)^2 = 4$ and $\sec^2 x = 3^2 = 9$
$\Rightarrow 1 + \tan^2 x = 4$ and $1 + \tan^2 x = 9$
$\Rightarrow \tan^2 x = 3$ and $\tan^2 x = 8$
$\Rightarrow \tan x = \pm\sqrt{3}$ and $\tan x = \pm\sqrt{8} = \pm2\sqrt{2}$
[2 marks available — 1 mark for correct method, 1 mark for correct answers]

Exercise 3.9 — Proving Other Identities

Q1 a) $\sec^2\theta - \csc^2\theta \equiv (1 + \tan^2\theta) - (1 + \cot^2\theta) \equiv \tan^2\theta - \cot^2\theta$

b) $\tan^2\theta - \cot^2\theta$ is the difference of two squares, and so can be written as $(\tan\theta + \cot\theta)(\tan\theta - \cot\theta)$.
So is $\sec^2\theta - \csc^2\theta$, so it can be written $(\sec\theta + \csc\theta)(\sec\theta - \csc\theta)$.
So using the result from part a),
$(\sec\theta + \csc\theta)(\sec\theta - \csc\theta)$
$\equiv (\tan\theta + \cot\theta)(\tan\theta - \cot\theta)$.

Q2 $\csc x - \sin x \equiv \frac{1}{\sin x} - \sin x \equiv \frac{1}{\sin x} - \frac{\sin^2 x}{\sin x}$
$\equiv \frac{1 - \sin^2 x}{\sin x} \equiv \frac{\cos^2 x}{\sin x} \equiv \cos x\,\frac{\cos x}{\sin x} \equiv \cos x\cot x$

Q3 First expand the bracket:
$(\tan x + \cot x)^2 \equiv \tan^2 x + \cot^2 x + 2\tan x\cot x$
$\equiv \tan^2 x + \cot^2 x + \frac{2\tan x}{\tan x}$
$\equiv \tan^2 x + \cot^2 x + 2$
*Split up that '+2' into two lots of '+1'
so it starts to resemble the identities...*
$\equiv (1 + \tan^2 x) + (1 + \cot^2 x)$
$\equiv \sec^2 x + \csc^2 x$

Q4 $\frac{1 - \sec x}{\tan x} \equiv \frac{1 - \frac{1}{\cos x}}{\frac{\sin x}{\cos x}} \equiv \frac{\cos x - 1}{\sin x} \equiv \frac{\cos x}{\sin x} - \frac{1}{\sin x}$
$\equiv \cot x - \csc x$

Q5 a) $(\cos^2 x - 1)(\cot^2 x - \operatorname{cosec}^2 x)$
$\equiv (\cos^2 x - (\sin^2 x + \cos^2 x))((\operatorname{cosec}^2 x - 1) - \operatorname{cosec}^2 x)$
$\equiv (-\sin^2 x)(-1) \equiv \sin^2 x.$
[2 marks available — 1 mark for correct use of known identities, 1 mark for a fully correct proof]

b) $\sqrt{1 + \cot^2 x} \equiv \operatorname{cosec} x \equiv \dfrac{1}{\sin x}$
$\Rightarrow \sin x = -\dfrac{1}{3} \Rightarrow \sin^2 x = \left(-\dfrac{1}{3}\right)^2 = \dfrac{1}{9}$
[2 marks available — 1 mark for correct use of known identities, 1 mark for the correct answer]

Q6 $\tan^2 x + \cos^2 x \equiv (\sec^2 x - 1) + (1 - \sin^2 x) \equiv \sec^2 x - \sin^2 x$
This is the difference of two squares...
$\equiv (\sec x + \sin x)(\sec x - \sin x)$

Q7 $(\sec x + \cot x)(\sec x - \cot x) \equiv \sec^2 x - \cot^2 x$
$\equiv (1 + \tan^2 x) - (\operatorname{cosec}^2 x - 1) \equiv \tan^2 x - \operatorname{cosec}^2 x + 2$

Q8 $\sec^2 \theta \operatorname{cosec}^2 \theta \equiv (1 + \tan^2 \theta)(1 + \cot^2 \theta)$
$\equiv 1 + \cot^2 \theta + \tan^2 \theta + \tan^2 \theta \cot^2 \theta$
$\equiv 1 + \cot^2 \theta + \tan^2 \theta + (\tan^2 \theta)\left(\dfrac{1}{\tan^2 \theta}\right)$
$\equiv 1 + \cot^2 \theta + \tan^2 \theta + 1 \equiv 2 + \cot^2 \theta + \tan^2 \theta$

Q9 $(\sin x - \operatorname{cosec} x)^3 + \cos^3 x \cot^3 x$
$\equiv \left(\sin x - \dfrac{1}{\sin x}\right)^3 + \cos^3 x \cot^3 x \equiv \left(\dfrac{\sin^2 x - 1}{\sin x}\right)^3 + \cos^3 x \cot^3 x$
$\equiv \left(-\dfrac{\cos^2 x}{\sin x}\right)^3 + \cos^3 x \cot^3 x \equiv (-\cos x \cot x)^3 + \cos^3 x \cot^3 x$
$\equiv -\cos^3 x \cot^3 x + \cos^3 x \cot^3 x \equiv 0$

Q10 $\dfrac{\sec^2 x + 1}{\tan^2 x} \equiv \dfrac{(\sec^2 x + 1)(\sec^2 x - 1)}{\tan^2 x(\sec^2 x - 1)}$
The numerator on the right-hand side of the identity is the difference of two squares, so you multiply the top and bottom of the fraction on the left by (sec² x − 1).
$\equiv \dfrac{\sec^4 x - 1}{\tan^2 x(\sec^2 x - 1)} \equiv \dfrac{\sec^4 x - 1}{\tan^2 x(\tan^2 x)} \equiv \dfrac{\sec^4 x - 1}{\tan^4 x}$
You could also start by multiplying top and bottom by tan² x.

Q11 $\dfrac{\cos \theta}{\sin \theta} + \dfrac{\sin \theta}{\cos \theta} \equiv \dfrac{\cos \theta \cos \theta}{\sin \theta \cos \theta} + \dfrac{\sin \theta \sin \theta}{\sin \theta \cos \theta} \equiv \dfrac{\cos^2 \theta + \sin^2 \theta}{\sin \theta \cos \theta}$
$\equiv \dfrac{1}{\sin \theta \cos \theta} \equiv \dfrac{1}{\sin \theta} \times \dfrac{1}{\cos \theta} \equiv \operatorname{cosec} \theta \sec \theta$

Q12 $\dfrac{(\sec x - \tan x)(\tan x + \sec x)}{\operatorname{cosec} x - \cot x} \equiv \dfrac{\sec^2 x - \tan^2 x}{\operatorname{cosec} x - \cot x}$
$\equiv \dfrac{(1 + \tan^2 x) - \tan^2 x}{\operatorname{cosec} x - \cot x} \equiv \dfrac{1}{\operatorname{cosec} x - \cot x}$
Multiply top and bottom by (cosec x + cot x)...
$\dfrac{1}{\operatorname{cosec} x - \cot x} \equiv \dfrac{\operatorname{cosec} x + \cot x}{(\operatorname{cosec} x - \cot x)(\operatorname{cosec} x + \cot x)}$
$\equiv \dfrac{\operatorname{cosec} x + \cot x}{\operatorname{cosec}^2 x - \cot^2 x} \equiv \dfrac{\operatorname{cosec} x + \cot x}{(1 + \cot^2 x) - \cot^2 x} \equiv \cot x + \operatorname{cosec} x$

Q13 $\dfrac{\cot x}{1 + \operatorname{cosec} x} + \dfrac{1 + \operatorname{cosec} x}{\cot x} \equiv \dfrac{\cot^2 x + (1 + \operatorname{cosec} x)^2}{\cot x(1 + \operatorname{cosec} x)}$
$\equiv \dfrac{(\operatorname{cosec}^2 x - 1) + (1 + 2\operatorname{cosec} x + \operatorname{cosec}^2 x)}{\cot x(1 + \operatorname{cosec} x)}$
$\equiv \dfrac{2\operatorname{cosec} x(1 + \operatorname{cosec} x)}{\cot x(1 + \operatorname{cosec} x)} \equiv \dfrac{2\operatorname{cosec} x}{\cot x} \equiv \dfrac{2\tan x}{\sin x}$
$\equiv \dfrac{2 \sin x}{\sin x \cos x} \equiv \dfrac{2}{\cos x} \equiv 2 \sec x$

Q14 $\dfrac{\operatorname{cosec} x + 1}{\operatorname{cosec} x - 1} \equiv \dfrac{(\operatorname{cosec} x + 1)(\operatorname{cosec} x + 1)}{(\operatorname{cosec} x - 1)(\operatorname{cosec} x + 1)}$
$\equiv \dfrac{\operatorname{cosec}^2 x + 2\operatorname{cosec} x + 1}{\operatorname{cosec}^2 x - 1} \equiv \dfrac{\operatorname{cosec}^2 x + 2\operatorname{cosec} x + 1}{(1 + \cot^2 x) - 1}$
$\equiv \dfrac{\operatorname{cosec}^2 x + 2\operatorname{cosec} x + 1}{\cot^2 x} \equiv \dfrac{\operatorname{cosec}^2 x}{\cot^2 x} + \dfrac{2\operatorname{cosec} x}{\cot^2 x} + \dfrac{1}{\cot^2 x}$
$\equiv \dfrac{\tan^2 x}{\sin^2 x} + \dfrac{2\tan^2 x}{\sin x} + \tan^2 x \equiv \dfrac{\sin^2 x}{\cos^2 x \sin^2 x} + \dfrac{2\sin^2 x}{\cos^2 x \sin x} + \tan^2 x$
$\equiv \dfrac{1}{\cos^2 x} + \dfrac{2\sin x}{\cos x \cos x} + \tan^2 x \equiv \dfrac{1}{\cos^2 x} + \dfrac{2\tan x}{\cos x} + \tan^2 x$
$\equiv \sec^2 x + 2\tan x \sec x + (\sec^2 x - 1) \equiv 2\sec^2 x + 2\tan x \sec x - 1$

Q15 a) $\cot x - \operatorname{cosec} x \sec x + \tan x \equiv \dfrac{\cos x}{\sin x} - \dfrac{1}{\sin x \cos x} + \dfrac{\sin x}{\cos x}$
$\equiv \dfrac{\cos^2 x - 1 + \sin^2 x}{\sin x \cos x} \equiv \dfrac{\cos^2 x + \sin^2 x - 1}{\sin x \cos x} \equiv \dfrac{1 - 1}{\sin x \cos x} \equiv 0$
[3 marks available — 1 mark for rewriting in terms of sin and cos, 1 mark for combining into a single fraction, 1 mark for a fully correct proof]

b) $\cos 13° \cot 13° + \cos 13° \tan 13° - \operatorname{cosec} 13°$
$= \cos 13°(\cot 13° + \tan 13° - \operatorname{cosec} 13° \dfrac{1}{\cos 13°})$
$= \cos 13°(\cot 13° - \operatorname{cosec} 13° \sec 13° + \tan 13°)$
$= \cos 13°(0) = 0$
[2 marks available — 1 mark for a correct method, 1 mark for the correct answer]

Q16 $8\sec x + \sec^2 x(4\cot x - \sin x)^2$
$\equiv 8\sec x + \sec^2 x(16\cot^2 x - 8\cot x \sin x + \sin^2 x)$
$\equiv 8\sec x + 16\sec^2 x \cot^2 x - 8\sec^2 x \cot x \sin x + \sec^2 x \sin^2 x$
$\equiv 8\sec x + 16\left(\dfrac{1}{\cos^2 x}\right)\left(\dfrac{\cos^2 x}{\sin^2 x}\right) - 8\left(\dfrac{1}{\cos^2 x}\right)\left(\dfrac{\cos x}{\sin x}\right)\sin x + \dfrac{\sin^2 x}{\cos^2 x}$
$\equiv 8\sec x + 16\dfrac{1}{\sin^2 x} - 8\dfrac{1}{\cos x} + \tan^2 x$
$\equiv 8\sec x + 16\operatorname{cosec}^2 x - 8\sec x + \tan^2 x$
$\equiv \tan^2 x + 16\operatorname{cosec}^2 x$

Q17 a) $\dfrac{\cos 3x}{1 - \cot 3x} + \dfrac{\sin 3x}{1 + \tan 3x}$
$\equiv \dfrac{\cos 3x(1 + \tan 3x) + \sin 3x(1 - \cot 3x)}{(1 - \cot 3x)(1 + \tan 3x)}$
$\equiv \dfrac{\cos 3x + \cos 3x \tan 3x + \sin 3x - \sin 3x \cot 3x}{1 - \cot 3x + \tan 3x - \cot 3x \tan 3x}$
$\equiv \dfrac{\cos 3x + \sin 3x + \sin 3x - \cos 3x}{1 - \cot 3x + \tan 3x - 1}$
$\equiv \dfrac{2\sin 3x}{\tan 3x - \cot 3x} \equiv \dfrac{2\sin 3x}{\dfrac{\sin 3x}{\cos 3x} - \dfrac{\cos 3x}{\sin 3x}}$
$\equiv \dfrac{2\sin 3x}{\dfrac{\sin^2 3x}{\sin 3x \cos 3x} - \dfrac{\cos^2 3x}{\sin 3x \cos 3x}} \equiv \dfrac{2\sin^2 3x \cos 3x}{\sin^2 3x - \cos^2 3x}$
[4 marks available — 1 mark for correct method to add fractions, 1 mark for correct simplified numerator, 1 mark for correct simplified denominator, 1 mark for fully correct proof]

b) $\dfrac{\cos 3x}{1 - \cot 3x} + \dfrac{\sin 3x}{1 + \tan 3x} \equiv \dfrac{2\sin^2 3x \cos 3x}{\sin^2 3x - \cos^2 3x}$
Dividing through by sin²3x, we obtain
$\dfrac{2\sin^2 3x \cos 3x}{\sin^2 3x - \cos^2 3x} \equiv \dfrac{2\cos 3x}{1 - \cot^2 3x}$
$\equiv \dfrac{2\cos 3x}{1 - (\operatorname{cosec}^2 3x - 1)} \equiv \dfrac{2\cos 3x}{2 - \operatorname{cosec}^2 3x}.$
[3 marks available — 1 mark for using appropriate identities, 1 mark for using identities correctly, 1 mark for fully correct proof]

Answers

Q18 a) $\operatorname{cosec} x + \cot x + \dfrac{1}{\operatorname{cosec} x + \cot x}$

$\equiv \dfrac{(\operatorname{cosec} x + \cot x)^2 + 1}{\operatorname{cosec} x + \cot x}$

$\equiv \dfrac{\operatorname{cosec}^2 x + \cot^2 x + 2\operatorname{cosec} x \cot x + 1}{\operatorname{cosec} x + \cot x}$

$\equiv \dfrac{\operatorname{cosec}^2 x + (\operatorname{cosec}^2 x - 1) + 2\operatorname{cosec} x \cot x + 1}{\operatorname{cosec} x + \cot x}$

$\equiv \dfrac{2\operatorname{cosec}^2 x + 2\operatorname{cosec} x \cot x}{\operatorname{cosec} x + \cot x}$

$\equiv \dfrac{2\operatorname{cosec} x(\operatorname{cosec} x + \cot x)}{\operatorname{cosec} x + \cot x} \equiv 2\operatorname{cosec} x$

[3 marks available — 1 mark for writing as a single fraction, 1 mark for use of identity for cosec and cot, 1 mark for fully correct proof]
You could have also left the first cosec x alone and just shown the rest of the expression was also equal to cosec x.

b) Let $y = \operatorname{cosec} x + \cot x$, now square both sides of the identity from part a):

$\left(y + \dfrac{1}{y}\right)^2 \equiv 4\operatorname{cosec}^2 x \Rightarrow y^2 + 1 + 1 + \dfrac{1}{y^2} \equiv 4\operatorname{cosec}^2 x$

$\Rightarrow y^2 + \dfrac{1}{y^2} \equiv 4\operatorname{cosec}^2 x - 2$

So $(\operatorname{cosec} x + \cot x)^2 + \dfrac{1}{(\operatorname{cosec} x + \cot x)^2} \equiv 4\operatorname{cosec}^2 x - 2$

[3 marks available — 1 mark for a correct method, 1 mark for use of identity from part a), 1 mark for a fully correct proof]
You could also have found the result by completing the square and substituting in the identity from part a).

Q19 a) $(x^2 - y^2)(x^4 + x^2 y^2 + y^4) = x^6 + x^4 y^2 + x^2 y^4 - x^4 y^2 - x^2 y^4 - y^6$
$= x^6 - y^6$ *[1 mark for a fully correct proof]*

b) $\operatorname{cosec}^6 x - \cot^6 x$
$\equiv (\operatorname{cosec}^2 x - \cot^2 x)(\operatorname{cosec}^4 x + \operatorname{cosec}^2 x \cot^2 x + \cot^4 x)$
$\equiv (1 + \cot^2 x - \cot^2 x)(\operatorname{cosec}^4 x + \operatorname{cosec}^2 x \cot^2 x + \cot^4 x)$
$\equiv \operatorname{cosec}^4 x + \operatorname{cosec}^2 x \cot^2 x + \cot^4 x$
$\equiv (\operatorname{cosec}^2 x - \cot^2 x)^2 + 3\operatorname{cosec}^2 x \cot^2 x$
$\equiv (1 + \cot^2 x - \cot^2 x)^2 + 3\operatorname{cosec}^2 x \cot^2 x$
$\equiv 1 + 3\operatorname{cosec}^2 x \cot^2 x$

[4 marks available — 1 mark for use of part a, 1 mark for use of identity to remove first bracket, 1 mark for simplifying $(\operatorname{cosec}^2 x - \cot^2 x)^2$, 1 mark for fully correct proof]

c) $\operatorname{cosec}^6 x - \cot^6 x = 1 \Rightarrow 1 + 3\operatorname{cosec}^2 x \cot^2 x = 1$
$\Rightarrow \operatorname{cosec}^2 x \cot^2 x = 0 \Rightarrow \operatorname{cosec} x = 0$ or $\cot x = 0$

$\operatorname{cosec} x = 0$ gives no solutions, so $\cot x = 0 \Rightarrow x = \dfrac{\pi}{2}$ or $\dfrac{3\pi}{2}$.

[3 marks available — 1 mark for correct use of part b, 1 mark for cosec x = 0 or cot x = 0, 1 mark for the correct answers]

Exercise 3.10 — The Addition Formulas — Finding Exact Values

Q1 a) $\cos 72° \cos 12° + \sin 72° \sin 12°$
$= \cos(72° - 12°) = \cos 60° = \dfrac{1}{2}$

b) $\cos 13° \cos 17° - \sin 13° \sin 17°$
$= \cos(13° + 17°) = \cos 30° = \dfrac{\sqrt{3}}{2}$

c) $\dfrac{\tan 12° + \tan 18°}{1 - \tan 12° \tan 18°} = \tan(12° + 18°) = \tan 30° = \dfrac{1}{\sqrt{3}}$

d) $\dfrac{\tan 500° - \tan 140°}{1 + \tan 500° \tan 140°} = \tan(500° - 140°) = \tan 360° = 0$

e) $\sin 35° \cos 10° + \cos 35° \sin 10° = \sin(35° + 10°)$
$= \sin 45° = \dfrac{1}{\sqrt{2}}$

f) $\sin 69° \cos 9° - \cos 69° \sin 9° = \sin(69° - 9°) = \sin 60° = \dfrac{\sqrt{3}}{2}$

Q2 a) $\sin \dfrac{2\pi}{3} \cos \dfrac{\pi}{2} - \cos \dfrac{2\pi}{3} \sin \dfrac{\pi}{2} = \sin\left(\dfrac{2\pi}{3} - \dfrac{\pi}{2}\right)$
$= \sin \dfrac{\pi}{6} = \dfrac{1}{2}$

b) $\cos 4\pi \cos 3\pi + \sin 4\pi \sin 3\pi = \cos(4\pi - 3\pi) = \cos \pi = -1$

c) $\dfrac{\tan \dfrac{5\pi}{12} + \tan \dfrac{5\pi}{4}}{1 - \tan \dfrac{5\pi}{12} \tan \dfrac{5\pi}{4}} = \tan\left(\dfrac{5\pi}{12} + \dfrac{5\pi}{4}\right)$
$= \tan \dfrac{5\pi}{3} = \tan\left(2\pi - \dfrac{\pi}{3}\right) = -\tan \dfrac{\pi}{3} = -\sqrt{3}$

d) $\cos \dfrac{\pi}{9} \cos \dfrac{5\pi}{9} - \sin \dfrac{\pi}{9} \sin \dfrac{5\pi}{9} = \cos\left(\dfrac{\pi}{9} + \dfrac{5\pi}{9}\right)$
$= \cos\left(\dfrac{2\pi}{3}\right) = -0.5$

Q3 a) $\sin(5x - 2x) = \sin 3x$

b) $\cos(4x + 6x) = \cos 10x$

c) $\tan(7x + 3x) = \tan 10x$

d) $5\sin(2x + 3x) = 5\sin 5x$

e) $8\cos(7x - 5x) = 8\cos 2x$

f) $\tan(8x - 5x) = \tan 3x$

Q4 Before answering a)-d), calculate $\cos x$ and $\sin y$:

$\sin x = \dfrac{5}{13} \Rightarrow \sin^2 x = \dfrac{25}{169}$

$\Rightarrow \cos^2 x = 1 - \dfrac{25}{169} = \dfrac{144}{169} \Rightarrow \cos x = \dfrac{12}{13}$

x is acute, meaning cos x must be positive, so take the positive square root.

$\cos y = \dfrac{24}{25} \Rightarrow \cos^2 y = \dfrac{576}{625}$

$\Rightarrow \sin^2 y = 1 - \dfrac{576}{625} = \dfrac{49}{625} \Rightarrow \sin y = \dfrac{7}{25}$

Again, y is acute, so sin y must be positive, so you can take the positive square root. You could have used the triangle method to work out sin y and cos x instead.

a) $\sin(x - y) = \sin x \cos y - \cos x \sin y$
$= \left(\dfrac{5}{13} \times \dfrac{24}{25}\right) - \left(\dfrac{12}{13} \times \dfrac{7}{25}\right) = \left(\dfrac{120}{325}\right) - \left(\dfrac{84}{325}\right) = \dfrac{36}{325}$

b) $\cos(x + y) = \cos x \cos y - \sin x \sin y$
$= \left(\dfrac{12}{13} \times \dfrac{24}{25}\right) - \left(\dfrac{5}{13} \times \dfrac{7}{25}\right) = \left(\dfrac{288}{325}\right) - \left(\dfrac{35}{325}\right) = \dfrac{253}{325}$

c) $\cos(x - y) = \cos x \cos y + \sin x \sin y$
$= \left(\dfrac{12}{13} \times \dfrac{24}{25}\right) + \left(\dfrac{5}{13} \times \dfrac{7}{25}\right) = \left(\dfrac{288}{325}\right) + \left(\dfrac{35}{325}\right) = \dfrac{323}{325}$

d) $\tan(x + y) = \dfrac{\sin(x + y)}{\cos(x + y)}$
$\sin(x + y) = \sin x \cos y + \cos x \sin y$
$= \left(\dfrac{120}{325}\right) + \left(\dfrac{84}{325}\right) = \dfrac{204}{325}$, so:

$\tan(x + y) = \dfrac{\left(\dfrac{204}{325}\right)}{\left(\dfrac{253}{325}\right)} = \dfrac{204}{253}$

Q5 Before answering a)-d), calculate $\cos x$ and $\sin y$:
$\sin x = \dfrac{3}{4} \Rightarrow \sin^2 x = \dfrac{9}{16}$

$\Rightarrow \cos^2 x = 1 - \dfrac{9}{16} = \dfrac{7}{16} \Rightarrow \cos x = \dfrac{\sqrt{7}}{4}$

x is acute, meaning cos x must be positive, so take the positive square root.

$\cos y = \dfrac{3}{\sqrt{10}} \Rightarrow \cos^2 y = \dfrac{9}{10}$

$\Rightarrow \sin^2 y = 1 - \dfrac{9}{10} = \dfrac{1}{10} \Rightarrow \sin y = \dfrac{1}{\sqrt{10}}$

Again, y is acute, so sin y must be positive, so you can take the positive square root. If you don't like using this method, you can use the triangle method to work out sin y and cos x.

a) $\sin(x+y) = \sin x \cos y + \cos x \sin y$

$= \left(\dfrac{3}{4} \times \dfrac{3}{\sqrt{10}}\right) + \left(\dfrac{\sqrt{7}}{4} \times \dfrac{1}{\sqrt{10}}\right) = \dfrac{9+\sqrt{7}}{4\sqrt{10}} = \dfrac{9\sqrt{10}+\sqrt{70}}{40}$

b) $\cos(x-y) = \cos x \cos y + \sin x \sin y$

$= \left(\dfrac{\sqrt{7}}{4} \times \dfrac{3}{\sqrt{10}}\right) + \left(\dfrac{3}{4} \times \dfrac{1}{\sqrt{10}}\right) = \dfrac{3\sqrt{7}+3}{4\sqrt{10}} = \dfrac{3\sqrt{70}+3\sqrt{10}}{40}$

c) $\operatorname{cosec}(x+y) = \dfrac{1}{\sin(x+y)} = \dfrac{40}{9\sqrt{10}+\sqrt{70}} = \dfrac{18\sqrt{10}-2\sqrt{70}}{37}$

d) $\sec(x-y) = \dfrac{1}{\cos(x-y)} = \dfrac{40}{3\sqrt{70}+3\sqrt{10}} = \dfrac{2\sqrt{70}-2\sqrt{10}}{9}$

Q6 $\sin 28° \cos 32° + \sin 30° \cos 30° + \sin 32° \cos 28°$

$= \sin(28°+32°) + \sin 30° \cos 30°$

$= \sin 60° + \sin 30° \cos 30°$

$= \dfrac{\sqrt{3}}{2} + \dfrac{1}{2} \times \dfrac{\sqrt{3}}{2} = \dfrac{3\sqrt{3}}{4}$

[3 marks available — 1 mark for use of addition formula, 1 mark for the correct values of $\sin 60°$, $\sin 30°$ and $\cos 30°$, 1 mark for the correct answer]

Q7 a) $\arcsin(\sin 17° \cos 5° + \sin 5° \cos 17°)$

$= \arcsin(\sin(17°+5°)) = \arcsin(\sin 22°) = 22°$

b) $\arctan\left(\dfrac{2\tan\left(\frac{\pi}{7}\right)}{1-\tan^2\left(\frac{\pi}{7}\right)}\right) = \arctan\left(\dfrac{\tan\left(\frac{\pi}{7}\right)+\tan\left(\frac{\pi}{7}\right)}{1-\tan\left(\frac{\pi}{7}\right)\tan\left(\frac{\pi}{7}\right)}\right)$

$= \arctan\left(\tan\left(\dfrac{\pi}{7}+\dfrac{\pi}{7}\right)\right) = \arctan\left(\tan\left(\dfrac{2\pi}{7}\right)\right) = \dfrac{2\pi}{7}$

Q8 $\cos\dfrac{\pi}{12} = \cos\left(\dfrac{\pi}{4} - \dfrac{\pi}{6}\right) = \cos\dfrac{\pi}{4}\cos\dfrac{\pi}{6} + \sin\dfrac{\pi}{4}\sin\dfrac{\pi}{6}$

$= \left(\dfrac{1}{\sqrt{2}} \times \dfrac{\sqrt{3}}{2}\right) + \left(\dfrac{1}{\sqrt{2}} \times \dfrac{1}{2}\right) = \dfrac{\sqrt{3}+1}{2\sqrt{2}}$

Now rationalise the denominator...

$= \dfrac{(\sqrt{3}+1) \times \sqrt{2}}{(2\sqrt{2}) \times \sqrt{2}} = \dfrac{\sqrt{6}+\sqrt{2}}{4}$

Q9 $\sin 75° = \sin(30°+45°) = \sin 30° \cos 45° + \cos 30° \sin 45°$

$= \left(\dfrac{1}{2} \times \dfrac{1}{\sqrt{2}}\right) + \left(\dfrac{\sqrt{3}}{2} \times \dfrac{1}{\sqrt{2}}\right)$

$= \dfrac{1+\sqrt{3}}{2\sqrt{2}} = \dfrac{(1+\sqrt{3}) \times \sqrt{2}}{(2\sqrt{2}) \times \sqrt{2}} = \dfrac{\sqrt{6}+\sqrt{2}}{4}$

Q10 $\tan 75° = \tan(45°+30°) = \dfrac{\tan 45° + \tan 30°}{1-\tan 45° \tan 30°}$

$= \dfrac{1+\dfrac{1}{\sqrt{3}}}{1-1\times\dfrac{1}{\sqrt{3}}} = \dfrac{\left(\dfrac{\sqrt{3}+1}{\sqrt{3}}\right)}{\left(\dfrac{\sqrt{3}-1}{\sqrt{3}}\right)} = \dfrac{\sqrt{3}+1}{\sqrt{3}-1}$.

Q11 $\sin\dfrac{3\pi}{10} = \sin\left(\dfrac{\pi}{5} + \dfrac{\pi}{10}\right) = \sin\dfrac{\pi}{5}\cos\dfrac{\pi}{10} + \cos\dfrac{\pi}{5}\sin\dfrac{\pi}{10}$

$\Rightarrow \dfrac{1}{4}(\sqrt{5}+1) = \sin\dfrac{\pi}{5}\cos\dfrac{\pi}{10} + \dfrac{1}{4}(\sqrt{5}+1) \times \dfrac{1}{4}(\sqrt{5}-1)$

$\Rightarrow \sin\dfrac{\pi}{5}\cos\dfrac{\pi}{10} = \dfrac{1}{4}(\sqrt{5}+1) - \dfrac{1}{4}(\sqrt{5}+1) \times \dfrac{1}{4}(\sqrt{5}-1)$

$= \dfrac{1}{4}(\sqrt{5}+1) - \dfrac{1}{16}(5-1) = \dfrac{1}{4}(\sqrt{5}+1) - \dfrac{1}{4} = \dfrac{\sqrt{5}}{4}$

[3 marks available — 1 mark for correct use of addition formula, 1 mark for correct method to find $\sin\frac{\pi}{5}\cos\frac{\pi}{10}$, 1 mark for the correct answer]

Q12 a) $\sin x = \dfrac{1}{2} \Rightarrow x = \dfrac{\pi}{6}$ or $x = \dfrac{5\pi}{6} \Rightarrow \cos x = \pm\dfrac{\sqrt{3}}{2}$

[2 marks available — 1 mark for a suitable method, 1 mark for both correct answers]

b) $\sin y = \pm\sqrt{1-\cos^2 y} = \pm\sqrt{1-\left(-\dfrac{1}{4}\right)^2} = \pm\sqrt{\dfrac{15}{16}} = \pm\dfrac{\sqrt{15}}{4}$

[2 marks available — 1 mark for a suitable method, 1 mark for both correct answers]

c) $\cos(x+y) = \cos x \cos y - \sin x \sin y$.

$\cos(x+y) = -\dfrac{1}{4} \times \dfrac{\sqrt{3}}{2} - \dfrac{1}{2} \times \dfrac{\sqrt{15}}{4} = \dfrac{-\sqrt{3}-\sqrt{15}}{8}$

or $\cos(x+y) = -\dfrac{1}{4} \times \left(-\dfrac{\sqrt{3}}{2}\right) - \dfrac{1}{2} \times \dfrac{\sqrt{15}}{4} = \dfrac{\sqrt{3}-\sqrt{15}}{8}$

or $\cos(x+y) = -\dfrac{1}{4} \times \dfrac{\sqrt{3}}{2} - \dfrac{1}{2} \times \left(-\dfrac{\sqrt{15}}{4}\right) = \dfrac{-\sqrt{3}+\sqrt{15}}{8}$

or $\cos(x+y) = -\dfrac{1}{4} \times \left(-\dfrac{\sqrt{3}}{2}\right) - \dfrac{1}{2} \times \left(-\dfrac{\sqrt{15}}{4}\right) = \dfrac{\sqrt{3}+\sqrt{15}}{8}$

[2 marks available — 2 marks for all four correct answers, otherwise 1 mark for at least two correct answers]

Exercise 3.11 — The Addition Formulas — Equations and Identities

Q1 $\tan(A-B) \equiv \dfrac{\sin(A-B)}{\cos(A-B)} \equiv \dfrac{\sin A \cos B - \cos A \sin B}{\cos A \cos B + \sin A \sin B}$

Divide through by $\cos A \cos B$...

$\equiv \dfrac{\left(\dfrac{\sin A \cos B}{\cos A \cos B}\right) - \left(\dfrac{\cos A \sin B}{\cos A \cos B}\right)}{\left(\dfrac{\cos A \cos B}{\cos A \cos B}\right) + \left(\dfrac{\sin A \sin B}{\cos A \cos B}\right)} \equiv \dfrac{\left(\dfrac{\sin A}{\cos A}\right) - \left(\dfrac{\sin B}{\cos B}\right)}{1 + \left(\dfrac{\sin A}{\cos A}\right)\left(\dfrac{\sin B}{\cos B}\right)}$

Now use tan = sin / cos...

$\equiv \dfrac{\tan A - \tan B}{1 + \tan A \tan B}$

Q2 a) $\dfrac{\cos(A-B) - \cos(A+B)}{\cos A \sin B}$

$\equiv \dfrac{(\cos A \cos B + \sin A \sin B) - (\cos A \cos B - \sin A \sin B)}{\cos A \sin B}$

$\equiv \dfrac{2\sin A \sin B}{\cos A \sin B} \equiv \dfrac{2\sin A}{\cos A} \equiv 2\tan A$

b) $\dfrac{1}{2}[\cos(A-B) - \cos(A+B)]$

$\equiv \dfrac{1}{2}[(\cos A \cos B + \sin A \sin B) - (\cos A \cos B - \sin A \sin B)]$

$\equiv \dfrac{1}{2}(2\sin A \sin B) \equiv \sin A \sin B$

c) $\sin(x+90°) \equiv \sin x \cos 90° + \cos x \sin 90°$

$\equiv \sin x (0) + \cos x (1) \equiv \cos x$

d) $\cos(x+180°) \equiv \cos x \cos 180° - \sin x \sin 180°$

$\equiv \cos x (-1) - \sin x (0) \equiv -\cos x$

Q3 $4\sin x \cos\dfrac{\pi}{3} - 4\cos x \sin\dfrac{\pi}{3} = \cos x$

$\Rightarrow 2\sin x - 2\sqrt{3}\cos x = \cos x \Rightarrow 2\sin x = (1+2\sqrt{3})\cos x$

$\Rightarrow \dfrac{\sin x}{\cos x} = \dfrac{1+2\sqrt{3}}{2} = \tan x \Rightarrow x = -1.99$ or 1.15 (2 d.p.)

[3 marks available — 1 mark for correct use of addition formula, 1 mark for correct method to solve for x, 1 mark for the correct answer]

Q4 a) $\tan\left(-\dfrac{\pi}{12}\right) = \tan\left(\dfrac{\pi}{6} - \dfrac{\pi}{4}\right) \equiv \dfrac{\tan\dfrac{\pi}{6} - \tan\dfrac{\pi}{4}}{1 + \tan\dfrac{\pi}{6}\tan\dfrac{\pi}{4}}$

$\equiv \dfrac{\dfrac{1}{\sqrt{3}} - 1}{1 + \dfrac{1}{\sqrt{3}}} \equiv \dfrac{1-\sqrt{3}}{\sqrt{3}+1}$

Now rationalise the denominator...

$\equiv \dfrac{1-\sqrt{3}}{\sqrt{3}+1} \times \dfrac{\sqrt{3}-1}{\sqrt{3}-1} \equiv \dfrac{2\sqrt{3}-4}{2} \equiv \sqrt{3}-2$

b) $\cos x = \cos x \cos\dfrac{\pi}{6} - \sin x \sin\dfrac{\pi}{6}$

$\Rightarrow \cos x = \dfrac{\sqrt{3}}{2}\cos x - \dfrac{1}{2}\sin x \Rightarrow (2-\sqrt{3})\cos x = -\sin x$

$\Rightarrow \dfrac{\sin x}{\cos x} = \tan x = \sqrt{3}-2$

From a), $\tan\left(-\dfrac{\pi}{12}\right) = \sqrt{3}-2$, so one solution for x is $-\dfrac{\pi}{12}$.

To get an answer in the correct interval, add π, since $\tan x$ repeats itself every π radians.

So $x = \dfrac{11\pi}{12}$.

Answers

Q5 $2\sin(x + 30°) \equiv 2\sin x \cos 30° + 2\cos x \sin 30°$

$\equiv 2\sin x \left(\dfrac{\sqrt{3}}{2}\right) + 2\cos x \left(\dfrac{1}{2}\right) \equiv \sqrt{3}\,\sin x + \cos x$

Q6 $\tan\left(x + \dfrac{\pi}{3}\right) = \dfrac{1}{3} \Rightarrow \dfrac{\tan x + \tan\frac{\pi}{3}}{1 - \tan x \tan\frac{\pi}{3}} = \dfrac{1}{3}$

$\Rightarrow \tan x + \sqrt{3} = \dfrac{1}{3}(1 - \sqrt{3}\tan x)$

$\Rightarrow \left(1 + \dfrac{\sqrt{3}}{3}\right)\tan x = \dfrac{1}{3} - \sqrt{3}$

$\Rightarrow \dfrac{3 + \sqrt{3}}{3}\tan x = \dfrac{1 - 3\sqrt{3}}{3}$

$\Rightarrow \tan x = \dfrac{1 - 3\sqrt{3}}{3 + \sqrt{3}} = \dfrac{1 - 3\sqrt{3}}{3 + \sqrt{3}} \times \dfrac{3 - \sqrt{3}}{3 - \sqrt{3}} = \dfrac{6 - 5\sqrt{3}}{3}$

[4 marks available — 1 mark for correct use of addition formula, 1 mark for method to find linear equation in tan x, 1 mark for correct unsimplified expression for tan x, 1 mark for rationalising the denominator]

Q7 a) $\sin(\theta - 45°) \equiv \sin\theta\cos 45° - \cos\theta\sin 45°$

$\equiv \dfrac{1}{\sqrt{2}}\sin\theta - \dfrac{1}{\sqrt{2}}\cos\theta \equiv \dfrac{1}{\sqrt{2}}(\sin\theta - \cos\theta)$

b) $4\sin(\theta - 45°) = \sqrt{2}\cos\theta \Rightarrow \dfrac{4}{\sqrt{2}}(\sin\theta - \cos\theta) = \sqrt{2}\cos\theta$

$\Rightarrow \sin\theta - \cos\theta = \dfrac{1}{2}\cos\theta \Rightarrow \sin\theta = \dfrac{3}{2}\cos\theta$

$\Rightarrow \tan\theta = \dfrac{3}{2} \Rightarrow \theta = \tan^{-1}\left(\dfrac{3}{2}\right) \Rightarrow \theta = 56.30...°$

There is another positive solution at $180° + 56.30...°$
so $\theta = 56.3°$ and $236°$ (3 s.f.)

Q8 $\tan\left(\dfrac{\pi}{3} - x\right) \equiv \dfrac{\tan\frac{\pi}{3} - \tan x}{1 + \tan\frac{\pi}{3}\tan x} \equiv \dfrac{\sqrt{3} - \tan x}{1 + \sqrt{3}\tan x}$

Q9 $\tan(A + B) = \dfrac{\tan A + \tan B}{1 - \tan A \tan B} = \dfrac{1}{4}$

$\Rightarrow \dfrac{\frac{3}{8} + \tan B}{1 - \frac{3}{8}\tan B} = \dfrac{1}{4} \Rightarrow \dfrac{3}{8} + \tan B = \dfrac{1}{4}\left(1 - \dfrac{3}{8}\tan B\right)$

$\Rightarrow \dfrac{3}{8} + \tan B = \dfrac{1}{4} - \dfrac{3}{32}\tan B$

$\Rightarrow \tan B + \dfrac{3}{32}\tan B = \dfrac{1}{4} - \dfrac{3}{8} = -\dfrac{1}{8}$

$\Rightarrow \dfrac{35}{32}\tan B = -\dfrac{1}{8} \Rightarrow \tan B = -\dfrac{1}{8} \times \dfrac{32}{35} = -\dfrac{4}{35}$

Q10 Starting with the addition formulas for sin:
$\sin(x + y) \equiv \sin x \cos y + \cos x \sin y$
$\sin(x - y) \equiv \sin x \cos y - \cos x \sin y$

So $\sin(x + y) + \sin(x - y)$
$\equiv \sin x \cos y + \cos x \sin y + \sin x \cos y - \cos x \sin y \equiv 2\sin x \cos y$

Substitute $A = x + y$ and $B = x - y$.

Then $A + B = 2x$ and $A - B = 2y$,
so $x = \left(\dfrac{A+B}{2}\right)$ and $y = \left(\dfrac{A-B}{2}\right)$.

Putting this back into the identity gives:

$\sin A + \sin B \equiv 2\sin\left(\dfrac{A+B}{2}\right)\cos\left(\dfrac{A-B}{2}\right)$

Q11 a) $\sin x \cos y + \cos x \sin y = 4\cos x \cos y + 4\sin x \sin y$

Dividing through by $\cos x \cos y$ gives:

$\dfrac{\sin x}{\cos x} + \dfrac{\sin y}{\cos y} = 4 + \dfrac{4\sin x \sin y}{\cos x \cos y}$

$\Rightarrow \tan x + \tan y = 4 + 4\tan x \tan y$

$\Rightarrow \tan x - 4\tan x \tan y = 4 - \tan y$

$\Rightarrow \tan x (1 - 4\tan y) = 4 - \tan y \Rightarrow \tan x = \dfrac{4 - \tan y}{1 - 4\tan y}$

b) $\tan x = \dfrac{4 - \tan\frac{\pi}{4}}{1 - 4\tan\frac{\pi}{4}}$

Comparing the equation you have to solve to the one in part a) you can see that $y = \dfrac{\pi}{4}$.

$\Rightarrow \tan x = \dfrac{4 - 1}{1 - 4} = -1 \Rightarrow x = \dfrac{3\pi}{4}$ and $\dfrac{7\pi}{4}$

Q12 a) Use the sin addition formula on $\sin(\theta + 45°)$:
$\sqrt{2}(\sin\theta\cos 45° + \cos\theta\sin 45°) = 3\cos\theta$

$\sqrt{2}\left(\dfrac{1}{\sqrt{2}}\sin\theta + \dfrac{1}{\sqrt{2}}\cos\theta\right) = 3\cos\theta$

$\sin\theta + \cos\theta = 3\cos\theta \Rightarrow \sin\theta = 2\cos\theta$

$\dfrac{\sin\theta}{\cos\theta} = 2 \Rightarrow \tan\theta = 2$

So $\theta = 63.43°$ (2 d.p.) and $63.43° + 180° = 243.43°$ (2 d.p.).

b) Use the cos addition formula:

$2\cos\left(\theta - \dfrac{2\pi}{3}\right) - 5\sin\theta = 0$

$2\left(\cos\theta\cos\dfrac{2\pi}{3} + \sin\theta\sin\dfrac{2\pi}{3}\right) - 5\sin\theta = 0$

$2\left(-\dfrac{1}{2}\cos\theta + \dfrac{\sqrt{3}}{2}\sin\theta\right) - 5\sin\theta = 0$

$-\cos\theta + \sqrt{3}\,\sin\theta - 5\sin\theta = 0$

$-\cos\theta + (\sqrt{3} - 5)\sin\theta = 0$

$\cos\theta = (\sqrt{3} - 5)\sin\theta$

$\dfrac{1}{(\sqrt{3} - 5)} = \dfrac{\sin\theta}{\cos\theta} = \tan\theta$

So $\theta = -0.296... + \pi = 2.84$ (2 d.p.)
and $-0.296... + 2\pi = 5.99$ (2 d.p.).

c) Use the addition formulas:
$\sin(\theta - 30°) - \cos(\theta + 60°) = 0$
$(\sin\theta\cos 30° - \cos\theta\sin 30°)$
$\qquad - (\cos\theta\cos 60° - \sin\theta\sin 60°) = 0$

$\left(\dfrac{\sqrt{3}}{2}\sin\theta - \dfrac{1}{2}\cos\theta\right) - \left(\dfrac{1}{2}\cos\theta - \dfrac{\sqrt{3}}{2}\sin\theta\right) = 0$

$\dfrac{\sqrt{3}}{2}\sin\theta - \dfrac{1}{2}\cos\theta - \dfrac{1}{2}\cos\theta + \dfrac{\sqrt{3}}{2}\sin\theta = 0$

$\sqrt{3}\,\sin\theta - \cos\theta = 0$

$\sqrt{3}\,\sin\theta = \cos\theta$

$\dfrac{\sin\theta}{\cos\theta} = \dfrac{1}{\sqrt{3}} \Rightarrow \tan\theta = \dfrac{1}{\sqrt{3}}$

$\theta = 30°$ and $30° + 180° = 210°$

Q13 Use the sin addition formula for $\sin\left(x + \dfrac{\pi}{6}\right)$:

$\sin\left(x + \dfrac{\pi}{6}\right) = \sin x \cos\dfrac{\pi}{6} + \cos x \sin\dfrac{\pi}{6} = \dfrac{\sqrt{3}}{2}\sin x + \dfrac{1}{2}\cos x$

Use the small angle approximations for sin and cos:

$\approx \dfrac{\sqrt{3}}{2}x + \dfrac{1}{2}\left(1 - \dfrac{1}{2}x^2\right) = \dfrac{\sqrt{3}}{2}x + \dfrac{1}{2} - \dfrac{1}{4}x^2 = \dfrac{1}{2} + \dfrac{\sqrt{3}}{2}x - \dfrac{1}{4}x^2$

Q14 a) $\tan(x + 8°) = \cot(x + 80°)$

$\Rightarrow \dfrac{\sin(x + 8°)}{\cos(x + 8°)} = \dfrac{\cos(x + 80°)}{\sin(x + 80°)}$

$\Rightarrow \sin(x + 8°)\sin(x + 80°) = \cos(x + 8°)\cos(x + 80°)$.

[2 marks available — 1 mark for writing tan and cot in terms of sin and cos, 1 mark for multiplying through by denominators to give the required result]

b) Rearranging, we have
$\cos(x + 8°)\cos(x + 80°) - \sin(x + 8°)\sin(x + 80°) = 0$,
So by the addition formula
$\cos(x + 8° + x + 80°) = 0 \Rightarrow \cos(2x + 88°) = 0$,
$\Rightarrow 2x + 88° = 90°$ or $2x + 88° = 270°$
$\Rightarrow 2x = 2°$ or $2x = 182° \Rightarrow x = 1°$ or $x = 91°$.
[3 marks available — 1 mark for use of the addition formula, 1 mark for values of 2x + 88, 1 mark for correct answers]

Q15 $\cos(x+y) \equiv \cos x \cos y - \sin x \sin y$

So $\sec(x+y) \equiv \dfrac{1}{\cos x \cos y - \sin x \sin y}$.

Hence $\sec(x+y)(1 - \tan x \tan y)$

$\equiv \dfrac{1 - \tan x \tan y}{\cos x \cos y - \sin x \sin y} \equiv \dfrac{1 - \sin x \sin y \sec x \sec y}{\cos x \cos y - \sin x \sin y}$

$\equiv \dfrac{\sec x \sec y(\cos x \cos y - \sin x \sin y)}{\cos x \cos y - \sin x \sin y} \equiv \sec x \sec y$

[4 marks available — 1 mark for correct expression for sec(x + y), 1 mark for writing tan x as sin x sec x, 1 mark for factorising the numerator, 1 mark for a fully correct proof]

Q16 a) $\cot(x+y) \equiv \dfrac{1}{\tan(x+y)} \equiv \dfrac{1 - \tan x \tan y}{\tan x + \tan y}$

$\dfrac{\dfrac{1}{\tan x \tan y} - \dfrac{\tan x \tan y}{\tan x \tan y}}{\dfrac{\tan x}{\tan x \tan y} + \dfrac{\tan y}{\tan x \tan y}} \equiv \dfrac{\cot x \cot y - 1}{\cot x + \cot y}$

b) $\dfrac{\cos(x+y)}{\cos x \cos y} \equiv \dfrac{\cos x \cos y - \sin x \sin y}{\cos x \cos y}$

$\equiv \dfrac{1 - \tan x \tan y}{1} \equiv 1 - \tan x \tan y$

Exercise 3.12 — The Double Angle Formulas — Finding Exact Values

Q1 a) $\sin 2A \equiv 2 \sin A \cos A \Rightarrow 4 \sin A \cos A \equiv 2 \sin 2A$

$\Rightarrow 4 \sin \dfrac{\pi}{12} \cos \dfrac{\pi}{12} = 2 \sin \dfrac{\pi}{6} = 2 \times \dfrac{1}{2} = 1$

b) $\cos 2A \equiv 2 \cos^2 A - 1$

$\Rightarrow \cos \dfrac{2\pi}{3} = 2 \cos^2 \dfrac{\pi}{3} - 1 = 2\left(\dfrac{1}{2}\right)^2 - 1 = -\dfrac{1}{2}$

c) $\sin 2A \equiv 2 \sin A \cos A \Rightarrow \dfrac{\sin 2A}{2} \equiv \sin A \cos A$

$\Rightarrow \dfrac{\sin 120°}{2} = \sin 60° \cos 60° = \dfrac{\sqrt{3}}{2} \times \dfrac{1}{2} = \dfrac{\sqrt{3}}{4}$

d) $\tan 2A \equiv \dfrac{2 \tan A}{1 - \tan^2 A} \Rightarrow \dfrac{\tan A}{2 - 2 \tan^2 A} \equiv \dfrac{\tan 2A}{4}$

$\Rightarrow \dfrac{\tan 15°}{2 - 2 \tan^2 15°} = \dfrac{\tan 30°}{4} = \dfrac{1}{4\sqrt{3}} = \dfrac{\sqrt{3}}{12}$

e) $\cos 2A \equiv 1 - 2 \sin^2 A \Rightarrow 2 \sin^2 A - 1 \equiv -\cos 2A$

$\Rightarrow 2 \sin^2 15° - 1 = -\cos 30° = -\dfrac{\sqrt{3}}{2}$

f) $\tan 2A \equiv \dfrac{2 \tan A}{1 - \tan^2 A} \Rightarrow \tan^2 A = 1 - \dfrac{2 \tan A}{\tan 2A}$

$\Rightarrow \tan^2 \dfrac{\pi}{3} = 1 - \dfrac{2 \tan \dfrac{\pi}{3}}{\tan \dfrac{2\pi}{3}} = 1 - \dfrac{2\sqrt{3}}{-\sqrt{3}} = 1 - (-2) = 3$

Q2 a) $\cos 2A \equiv 1 - 2 \sin^2 A$

$\Rightarrow \cos 2x = 1 - 2 \sin^2 x = 1 - 2\left(\dfrac{1}{6}\right)^2 = \dfrac{17}{18}$

b) First find cos x:

$\cos^2 x = 1 - \sin^2 x = 1 - \left(\dfrac{1}{6}\right)^2 = \dfrac{35}{36} \Rightarrow \cos x = \dfrac{\sqrt{35}}{6}$

x is acute so take the positive root for cos x. Again, if you find it easier you can use the triangle method here.

$\sin 2A \equiv 2 \sin A \cos A \Rightarrow \sin 2x = 2\left(\dfrac{1}{6} \times \dfrac{\sqrt{35}}{6}\right) = \dfrac{\sqrt{35}}{18}$

c) $\tan 2x = \dfrac{\sin 2x}{\cos 2x} = \dfrac{\sqrt{35}}{17}$

d) $\sec 2x = \dfrac{1}{\cos 2x} = \dfrac{18}{17}$

e) $\operatorname{cosec} 2x = \dfrac{1}{\sin 2x} = \dfrac{18}{\sqrt{35}} \left(= \dfrac{18\sqrt{35}}{35}\right)$

f) $\cot 2x = \dfrac{1}{\tan 2x} = \dfrac{17}{\sqrt{35}} \left(= \dfrac{17\sqrt{35}}{35}\right)$

Q3 a) $\cos 2x = 1 - 2 \sin^2 x = 1 - 2\left(-\dfrac{1}{4}\right)^2 = \dfrac{7}{8}$

b) First find cos x:

$\cos^2 x = 1 - \sin^2 x = 1 - \left(-\dfrac{1}{4}\right)^2 = \dfrac{15}{16} \Rightarrow \cos x = -\dfrac{\sqrt{15}}{4}$

x is in the 3rd quadrant of the CAST diagram where cos x is negative, so take the negative root for cos x.

$\sin 2A \equiv 2 \sin A \cos A \Rightarrow \sin 2x = 2\left(-\dfrac{1}{4} \times -\dfrac{\sqrt{15}}{4}\right) = \dfrac{\sqrt{15}}{8}$

c) $\tan 2x = \dfrac{\sin 2x}{\cos 2x} = \dfrac{\sqrt{15}}{7}$

d) $\sec 2x = \dfrac{1}{\cos 2x} = \dfrac{8}{7}$

e) $\operatorname{cosec} 2x = \dfrac{1}{\sin 2x} = \dfrac{8}{\sqrt{15}} \left(= \dfrac{8\sqrt{15}}{15}\right)$

f) $\cot 2x = \dfrac{1}{\tan 2x} = \dfrac{7}{\sqrt{15}} \left(= \dfrac{7\sqrt{15}}{15}\right)$

Q4 a) $\cos 36° = 2 \cos^2 18° - 1 = 2\left(\dfrac{\sqrt{2(5 + \sqrt{5})}}{4}\right)^2 - 1$

$= \dfrac{2(2(5 + \sqrt{5}))}{16} - 1 = \dfrac{5 + \sqrt{5}}{4} - 1 = \dfrac{1 + \sqrt{5}}{4}$

[3 marks available — 1 mark for correct use of the double angle formula, 1 mark for squaring cos 18° correctly, 1 mark for the correct answer]

b) $\cos 72° = 2 \cos^2 36° - 1 = 2\left(\dfrac{1 + \sqrt{5}}{4}\right)^2 - 1$

$= \dfrac{2(1 + 5 + 2\sqrt{5})}{16} - 1 = \dfrac{3 + \sqrt{5}}{4} - 1 = \dfrac{\sqrt{5} - 1}{4}$

So $\sec 72° = \dfrac{4}{\sqrt{5} - 1} = \dfrac{4(\sqrt{5} + 1)}{(\sqrt{5} - 1)(\sqrt{5} + 1)}$

$= \dfrac{4\sqrt{5} + 4}{5 - 1} = \sqrt{5} + 1$

[4 marks available — 1 mark for correct use of the double angle formula, 1 mark for the correct value of cos 72°, 1 mark for the correct method to rationalise the denominator, 1 mark for the correct answer]

Q5 a) Using the sin double angle formula: $\dfrac{\sin 3\theta \cos 3\theta}{3} \equiv \dfrac{\sin 6\theta}{6}$

b) Using the cos double angle formula:

$\sin^2\left(\dfrac{2y}{3}\right) - \cos^2\left(\dfrac{2y}{3}\right) \equiv -\cos\left(\dfrac{4y}{3}\right)$

c) Using the tan double angle formula:

$\dfrac{1 - \tan^2\left(\dfrac{x}{2}\right)}{2 \tan\left(\dfrac{x}{2}\right)} \equiv \dfrac{1}{\tan x} \equiv \cot x$

Q6 a) $\cot x = -\dfrac{3}{5} \Rightarrow \tan x = -\dfrac{5}{3}$

So $\tan 2x = \dfrac{2 \times \left(-\dfrac{5}{3}\right)}{1 - \left(-\dfrac{5}{3}\right)^2} = \dfrac{-\dfrac{10}{3}}{1 - \dfrac{25}{9}} = -\dfrac{30}{-16} = \dfrac{15}{8}$

[3 marks available — 1 mark for correct value of tan x, 1 mark for correct method to find tan(2x), 1 mark for the correct answer]

b) $\tan 3x = \tan(x + 2x) = \dfrac{\tan x + \tan 2x}{1 - \tan x \tan 2x} = \dfrac{-\dfrac{5}{3} + \dfrac{15}{8}}{1 - -\dfrac{5}{3} \times \dfrac{15}{8}} = \dfrac{5}{99}$

[3 marks available — 1 mark for correct use of the addition formula, 1 mark for substituting in values from part a), 1 mark for the correct answer]

c) $\tan 6x = \dfrac{2 \tan 3x}{1 - \tan^2 3x} = \dfrac{2\left(\dfrac{5}{99}\right)}{1 - \left(\dfrac{5}{99}\right)^2} = \dfrac{495}{4888}$

So $\cot 6x = \dfrac{4888}{495}$.

[3 marks available — 1 mark for a suitable method to find tan 6x, 1 mark for correct value of tan 6x, 1 mark for the correct value of cot 6x]

Answers

Q7
$$\frac{3\sin^2(5x) - 3\cos^2(5x)}{\cos^3(5x)\sin(5x) + \sin^3(5x)\cos(5x)}$$

$$= \frac{3\sin^2(5x) - 3\cos^2(5x)}{\cos(5x)\sin(5x)(\cos^2 5x + \sin^2 5x)}$$

$$= \frac{3\sin^2(5x) - 3\cos^2(5x)}{\cos(5x)\sin(5x)} = \frac{-3(\cos^2 5x - \sin^2 5x)}{\cos(5x)\sin(5x)}$$

$$= -\frac{3\cos 10x}{\frac{1}{2}\sin 10x} = -6\cot 10x$$

[4 marks available — 1 mark for use of cos²5x + sin²5x = 1, 1 mark for use of cos double angle formula, 1 mark for use of sin double angle formula, 1 mark for the correct answer]

Exercise 3.13 — The Double Angle Formulas — Equations and Identities

Q1 **a)** Using the cos double angle formula involving sin:
$4(1 - 2\sin^2 x) - 14\sin x = 0 \Rightarrow 4 - 8\sin^2 x - 14\sin x = 0$
$\Rightarrow 8\sin^2 x + 14\sin x - 4 = 0 \Rightarrow 4\sin^2 x + 7\sin x - 2 = 0$
$\Rightarrow (4\sin x - 1)(\sin x + 2) = 0$
So $\sin x = \frac{1}{4}$ or $\sin x = -2$ (not valid)
Solving $\sin x = \frac{1}{4}$ in the interval $0° \le x \le 360°$:
$x = 14.5°$ (1 d.p.) and $(180° - 14.5°) = 165.5°$ (1 d.p.)

b) Using the cos double angle formula involving cos:
$5(2\cos^2 x - 1) + 9\cos x + 7 = 0$
$\Rightarrow 10\cos^2 x + 9\cos x + 2 = 0$
$\Rightarrow (2\cos x + 1)(5\cos x + 2) = 0$
So $\cos x = -\frac{1}{2}$ or $\cos x = -\frac{2}{5}$
$\Rightarrow x = 113.6°, 120°, 240°, 246.4°$ (1 d.p.)

c) Using the tan double angle formula:
$\frac{4(1 - \tan^2 x)}{2\tan x} + \frac{1}{\tan x} = 5 \Rightarrow 2(1 - \tan^2 x) + 1 = 5\tan x$
$\Rightarrow 3 - 2\tan^2 x = 5\tan x \Rightarrow 0 = 2\tan^2 x + 5\tan x - 3$
$\Rightarrow 0 = (2\tan x - 1)(\tan x + 3)$
So $\tan x = \frac{1}{2}$ or $\tan x = -3$
$\Rightarrow x = 26.6°, 108.4°, 206.6°, 288.4°$ (1 d.p.)

d) $\tan x - 5(2\sin x \cos x) = 0 \Rightarrow \frac{\sin x}{\cos x} = 10\sin x \cos x$
$\Rightarrow \sin x = 10\sin x \cos^2 x \Rightarrow \sin x - 10\sin x \cos^2 x = 0$
$\Rightarrow \sin x(1 - 10\cos^2 x) = 0$
So $\sin x = 0$ or $\cos x = \pm\frac{1}{\sqrt{10}}$
Don't forget to find cos⁻¹ of both the positive and negative root...
$x = 0°, 71.6°, 108.4°, 180°, 251.6°, 288.4°, 360°.$

Q2 **a)** $4(2\cos^2 x - 1) - 10\cos x + 1 = 0$
$\Rightarrow 8\cos^2 x - 10\cos x - 3 = 0$
$\Rightarrow (4\cos x + 1)(2\cos x - 3) = 0$
So $\cos x = -\frac{1}{4}$ or $\cos x = \frac{3}{2}$ (not valid)
$\Rightarrow x = 1.82$ and 4.46 (3 s.f.)

b) $\cos 2x - 3 = 6\sin^2 x - 3$
$\Rightarrow (1 - 2\sin^2 x) - 3 - 6\sin^2 x + 3 = 0 \Rightarrow 1 - 8\sin^2 x = 0$
So $\sin x = \pm\frac{1}{2\sqrt{2}} \Rightarrow x = 0.361, 2.78, 3.50, 5.92$ (3 s.f.)

c) $2\sin x \cos x = 4\cos^2 x - 4\sin^2 x$
$\Rightarrow \sin 2x = 4\cos 2x \Rightarrow \tan 2x = 4$
$\Rightarrow 2x = 1.325..., 4.467..., 7.609..., 10.75...$
$\Rightarrow x = 0.662, 2.23, 3.80, 5.38$ (3 s.f.)

d) $2\cos 4x + \sin 2x = -1 \Rightarrow 2(1 - 2\sin^2 2x) + \sin 2x + 1 = 0$
$\Rightarrow 2 - 4\sin^2 2x + \sin 2x + 1 = 0 \Rightarrow 4\sin^2 2x - \sin 2x - 3 = 0$
$\Rightarrow (\sin 2x - 1)(4\sin 2x + 3) = 0$
So $\sin 2x = 1$ or $\sin 2x = -\frac{3}{4}$.
$\sin 2x = 1 \Rightarrow 2x = \frac{\pi}{2}, \frac{5\pi}{2} \Rightarrow x = \frac{\pi}{4}, \frac{5\pi}{4}$
$\Rightarrow x = 0.785, 3.93$ (3 s.f.)
$\sin 2x = -\frac{3}{4} \Rightarrow 2x = -0.848...$ which is outside the interval.
$2x = 3.989..., 5.435..., 10.272..., 11.718...$
There are six negative solutions in total in the interval
$0 \le 2x \le 4\pi$: $x = 0.785, 1.99, 2.72, 3.93, 5.14, 5.86$ (3 s.f)

Q3 **a)** By the double angle formula,
$\cos x = 1 - 2\sin^2\left(\frac{x}{2}\right) \Rightarrow a = 1, b = -2.$
[1 mark for both correct values]

b) Since $\sin\frac{x}{2} \le \frac{x}{2}$,
$1 - 2\sin^2\left(\frac{x}{2}\right) \ge 1 - 2\left(\frac{x}{2}\right)^2 \Rightarrow 1 - 2\sin^2\left(\frac{x}{2}\right) \ge 1 - \frac{2x^2}{4}$
$\Rightarrow 1 - 2\sin^2\left(\frac{x}{2}\right) \ge 1 - \frac{x^2}{2}$
$\Rightarrow \cos x \ge 1 - \frac{1}{2}x^2$
[2 marks available — 1 mark for correct use of $\sin\frac{x}{2} \le \frac{x}{2}$, 1 mark for a fully correct proof]

Q4 **a)** $2\cos^2 x - 1 + 7\cos x = -4 \Rightarrow 2\cos^2 x + 7\cos x + 3 = 0$
$\Rightarrow (2\cos x + 1)(\cos x + 3) = 0$
So $\cos x = -\frac{1}{2}$ or $\cos x = -3$ (not valid)
$\Rightarrow x = \frac{2\pi}{3}$ and $\frac{4\pi}{3}$

b) $2\sin\frac{x}{2}\cos\frac{x}{2} + \cos\frac{x}{2} = 0 \Rightarrow \cos\frac{x}{2}\left(2\sin\frac{x}{2} + 1\right) = 0$
So $\cos\frac{x}{2} = 0$ or $\sin\frac{x}{2} = -\frac{1}{2}$
There are no solutions for $\sin\frac{x}{2} = -\frac{1}{2}$ in the interval $0 \le x \le \pi$ (they both lie in the 3rd and 4th quadrants of the CAST diagram, $\pi \le x \le 2\pi$) so...
$\Rightarrow \frac{x}{2} = \frac{\pi}{2} \Rightarrow x = \pi$

c) $\sin x - \cos 2x = 0 \Rightarrow \sin x - (1 - 2\sin^2 x) = 0$
$\Rightarrow 2\sin 2x + \sin x - 1 = 0 \Rightarrow (\sin x + 1)(2\sin x - 1) = 0$
So $\sin x = -1$ or $\sin x = \frac{1}{2}$.
$x = \sin^{-1}(-1) = \frac{3\pi}{2}$ and $x = \sin^{-1}\left(\frac{1}{2}\right) = \frac{\pi}{6}, \frac{5\pi}{6}$
So $x = \frac{\pi}{6}, \frac{5\pi}{6}, \frac{3\pi}{2}$

d) $\cos x = 7\cos\frac{x}{2} + 3 \Rightarrow 2\cos^2\frac{x}{2} - 1 = 7\cos\frac{x}{2} + 3$
$\Rightarrow 2\cos^2\frac{x}{2} - 7\cos\frac{x}{2} - 4 = 0$
$\Rightarrow \left(\cos\frac{x}{2} - 4\right)\left(2\cos\frac{x}{2} + 1\right) = 0$
$\Rightarrow \cos\frac{x}{2} = 4$ (not valid) or $\cos\frac{x}{2} = -\frac{1}{2}$
$\cos\frac{x}{2} = -\frac{1}{2} \Rightarrow \frac{x}{2} = \frac{2\pi}{3} \Rightarrow x = \frac{4\pi}{3}$

Q5 **a)** $\sin 2x \sec^2 x \equiv (2\sin x \cos x)\left(\frac{1}{\cos^2 x}\right) \equiv \frac{2\sin x}{\cos x} \equiv 2\tan x$

b) $\frac{2}{1 + \cos 2x} \equiv \frac{2}{1 + 2\cos^2 x - 1} \equiv \frac{2}{2\cos^2 x} \equiv \frac{1}{\cos^2 x} \equiv \sec^2 x$

c) $\cot x - 2\cot 2x \equiv \frac{1}{\tan x} - \frac{2}{\tan 2x}$
$\equiv \frac{1}{\tan x} - \frac{2(1 - \tan^2 x)}{2\tan x} \equiv \frac{1 - (1 - \tan^2 x)}{\tan x} \equiv \frac{\tan^2 x}{\tan x} \equiv \tan x$

d) $\tan 2x + \cot 2x \equiv \frac{\sin 2x}{\cos 2x} + \frac{\cos 2x}{\sin 2x} \equiv \frac{\sin^2 2x + \cos^2 2x}{\sin 2x \cos 2x}$
Use $\sin^2 x + \cos^2 x \equiv 1$, and $\sin 4x \equiv 2\sin 2x \cos 2x$...
$\equiv \frac{1}{\frac{1}{2}\sin 4x} \equiv \frac{2}{\sin 4x} \equiv 2\operatorname{cosec} 4x$

Q6 $\dfrac{\sin 2x + \cos 2x + 1}{\sin 2x - \cos 2x + 1} = \dfrac{2\sin x \cos x + 2\cos^2 x - 1 + 1}{2\sin x \cos x + 2\sin^2 x - 1 + 1}$

$= \dfrac{2\cos x(\sin x + \cos x)}{2\sin x(\sin x + \cos x)} = \dfrac{\cos x}{\sin x} = \cot x$

[4 marks available — 1 mark for use of sin 2x double angle

formula, 1 mark for use of cos 2x double angle formulas, 1 mark for factorising, 1 mark for a fully correct proof]

Q7 **a)** $\dfrac{1 + \cos 2x}{\sin 2x} \equiv \dfrac{1 + (2\cos^2 x - 1)}{2\sin x \cos x}$

$\equiv \dfrac{2\cos^2 x}{2\sin x \cos x} \equiv \dfrac{\cos x}{\sin x} \equiv \cot x$

b) Use $4\theta = 2x$, so $x = 2\theta$ and so $\dfrac{1 + \cos 4\theta}{\sin 4\theta} \equiv \cot 2\theta$

Solve $\cot 2\theta = 7 \Rightarrow \tan 2\theta = \dfrac{1}{7}$

$\Rightarrow 2\theta = 8.130°,\ 188.130°,\ 368.130°,\ 548.130°$

$\Rightarrow \theta = 4.1°,\ 94.1°,\ 184.1°,\ 274.1°$ (1 d.p.)

Q8 **a)** $\dfrac{\cot^2 x + 1}{\cot^2 x - 1} \equiv \dfrac{\csc^2 x}{\dfrac{\cos^2 x}{\sin^2 x} - 1} \equiv \dfrac{\dfrac{1}{\sin^2 x}}{\dfrac{\cos^2 x}{\sin^2 x} - 1}$

$\equiv \dfrac{1}{\cos^2 x - \sin^2 x} \equiv \dfrac{1}{\cos 2x} \equiv \sec 2x$

b) $\dfrac{\cot^2 2\alpha + 1}{\cot^2 2\alpha - 1} = \sec 4\alpha$, so solve for $-4\pi \le 4\alpha \le 4\pi$:

$\sec 4\alpha = -3 \Rightarrow \cos 4\alpha = -\dfrac{1}{3} \Rightarrow 4\alpha = 1.910...$

There are 7 other negative solutions at

$(0 - 1.910...),\ (2\pi - 1.910...),\ (2\pi + 1.910...),\ (-2\pi - 1.910...),$
$(-2\pi + 1.910...),\ (4\pi - 1.910...),\ (-4\pi + 1.910...)$

so $4\alpha = -10.656...,\ -8.193...,\ -4.373...,\ -1.910...,$
$1.910...,\ 4.373...,\ 8.193...,\ 10.656...$

$\alpha = -2.66,\ -2.05,\ -1.09,\ -0.478,\ 0.478,\ 1.09,\ 2.05,\ 2.66$ (3 s.f.)

Q9 $\cot x = \tan \dfrac{x}{2} \Rightarrow \dfrac{1}{\tan x} = \tan \dfrac{x}{2} \Rightarrow \dfrac{1 - \tan^2 \dfrac{x}{2}}{2\tan \dfrac{x}{2}} = \tan \dfrac{x}{2}$

$\Rightarrow 1 - \tan^2 \dfrac{x}{2} = 2\tan^2 \dfrac{x}{2} \Rightarrow \tan^2 \dfrac{x}{2} = \dfrac{1}{3}$

So $\tan \dfrac{x}{2} = \pm \dfrac{1}{\sqrt{3}} \Rightarrow \dfrac{x}{2} = \dfrac{\pi}{6}, \dfrac{5\pi}{6} \Rightarrow x = \dfrac{\pi}{3}, \dfrac{5\pi}{3}$

[3 marks available — 1 mark for $\cot\theta = \dfrac{1}{\tan\theta}$ identity, 1 mark for rearranging to $\tan\dfrac{x}{2}$, 1 mark for the correct solutions]

Q10 **a)** $\csc x - \cot \dfrac{x}{2} \equiv \dfrac{1}{\sin x} - \dfrac{\cos \dfrac{x}{2}}{\sin \dfrac{x}{2}}$

$\equiv \dfrac{1}{2\sin \dfrac{x}{2} \cos \dfrac{x}{2}} - \dfrac{\cos \dfrac{x}{2}}{\sin \dfrac{x}{2}} \equiv \dfrac{1 - 2\cos^2 \dfrac{x}{2}}{2\sin \dfrac{x}{2} \cos \dfrac{x}{2}}$

Here we've let $2A = x$ so $x = \dfrac{A}{2}$.

$\equiv \dfrac{-\left(2\cos^2 \dfrac{x}{2} - 1\right)}{2\sin \dfrac{x}{2} \cos \dfrac{x}{2}} \equiv \dfrac{-\cos x}{\sin x}$

$\equiv -\dfrac{1}{\left(\dfrac{\sin x}{\cos x}\right)} \equiv -\dfrac{1}{\tan x} \equiv -\cot x.$

b) Rearranging, $\csc y - \cot \dfrac{y}{2} = -2 = -\cot y$

$\Rightarrow \cot y = 2 \Rightarrow \tan y = \dfrac{1}{2}.$

There are 2 solutions in the interval $-\pi \le y \le \pi$, at $y = 0.464$ (3 s.f.) and $y = 0.464 - \pi = -2.68$ (3 s.f.)

Q11 First find $\cos\theta$. You know $\sin\theta = \dfrac{5}{13}$, so using the triangle method, $\cos\theta = \dfrac{12}{13}$.

a) **(i)** $\cos^2\left(\dfrac{\theta}{2}\right) = \dfrac{1}{2}(1 + \cos\theta) = \dfrac{1}{2}\left(1 + \dfrac{12}{13}\right) = \dfrac{25}{26}$

So $\cos\left(\dfrac{\theta}{2}\right) = \sqrt{\dfrac{25}{26}} = \dfrac{5}{\sqrt{26}}$.

θ is acute, so $\dfrac{\theta}{2}$ is also acute, and $\cos\left(\dfrac{\theta}{2}\right)$ is +ve so we can ignore the negative root.

(ii) $\sin^2\left(\dfrac{\theta}{2}\right) = \dfrac{1}{2}(1 - \cos\theta) = \dfrac{1}{2}\left(1 - \dfrac{12}{13}\right) = \dfrac{1}{26}$

So $\sin\left(\dfrac{\theta}{2}\right) = \sqrt{\dfrac{1}{26}} = \dfrac{1}{\sqrt{26}}$.

Again, θ is acute so $\sin\left(\dfrac{\theta}{2}\right)$ must be positive.

b) $\tan\left(\dfrac{\theta}{2}\right) = \dfrac{\sin\left(\dfrac{\theta}{2}\right)}{\cos\left(\dfrac{\theta}{2}\right)} = \dfrac{\left(\dfrac{1}{\sqrt{26}}\right)}{\left(\dfrac{5}{\sqrt{26}}\right)} = \dfrac{1}{5}$

Q12 **a)** $\tan 4x = \tan(2x + 2x) = \dfrac{2\tan 2x}{1 - \tan^2 2x} = \dfrac{\dfrac{4\tan x}{1 - \tan^2 x}}{1 - \left(\dfrac{2\tan x}{1 - \tan^2 x}\right)^2}$

$= \dfrac{4\tan x(1 - \tan^2 x)}{(1 - \tan^2 x)^2 - 4\tan^2 x}$

$= \dfrac{4\tan x(1 - \tan^2 x)}{\tan^4 x - 2\tan^2 x + 1 - 4\tan^2 x} = \dfrac{4\tan x(1 - \tan^2 x)}{\tan^4 x - 6\tan^2 x + 1}$

[4 marks available — 1 mark for use of double angle formula once, 1 mark for use of double angle formula again, 1 mark for multiplying through by $(1 - \tan^2 x)^2$, 1 mark for simplifying]

b) $\cot x = \dfrac{1}{\sqrt{3}} \Rightarrow \tan x = \sqrt{3}$

$\tan 4x = \dfrac{4\tan x(1 - \tan^2 x)}{\tan^4 x - 6\tan^2 x + 1} = \dfrac{4\sqrt{3}(1 - \sqrt{3}^2)}{\sqrt{3}^4 - 6(\sqrt{3}^2) + 1}$

$= \dfrac{-8\sqrt{3}}{9 - 18 + 1} = \dfrac{-8\sqrt{3}}{-8} = \sqrt{3}$

So $\tan x = \tan 4x$.

[3 marks available — 1 mark for the value of $\tan x$, 1 mark for a correct method to find $\tan 4x$, 1 mark for finding the value of $\tan 4x$]

Q13 **a)** $\tan x \sin 4x = \tan x(\sin(2 \times 2x))$

$= \tan x(2\sin 2x \cos 2x) = \tan x(4\sin x \cos x(1 - 2\sin^2 x))$

$= \dfrac{\sin x}{\cos x}(4\sin x \cos x(1 - 2\sin^2 x)) = \sin x(4\sin x(1 - 2\sin^2 x))$

$= 4\sin^2 x - 8\sin^4 x$

[5 marks available — 1 mark for correct use of double angle formula on sin 4x, 1 mark for correct use of double angle formula on sin 2x, 1 mark for correct use of double angle formula on cos 2x, 1 mark for writing in terms of sin x, 1 mark for simplifying to give required expression]

b) $\tan x \sin 4x = 4\sin^2 x - 8\sin^4 x = 4\left(\dfrac{4}{5}\right)^2 - 8\left(\dfrac{4}{5}\right)^4 = -\dfrac{448}{625}$

[1 mark]

c) $\cot x \csc 4x = 2 \Rightarrow \dfrac{1}{\tan x \sin 4x} = 2$

$\Rightarrow \dfrac{1}{4\sin^2 x - 8\sin^4 x} = 2 \Rightarrow \dfrac{1}{2} = 4\sin^2 x - 8\sin^4 x$

$\Rightarrow 16\sin^4 x - 8\sin^2 x + 1 = 0$

$\Rightarrow (4\sin^2 x - 1)(4\sin^2 x - 1) = 0$

$\Rightarrow 4\sin^2 x - 1 = 0 \Rightarrow \sin x = \pm \dfrac{1}{2}$

$\Rightarrow x = -\dfrac{5\pi}{6},\ x = -\dfrac{\pi}{6},\ x = \dfrac{\pi}{6}$ or $x = \dfrac{5\pi}{6}$

[5 marks available — 1 mark for rewriting in terms of sin x, 1 mark for rearranging to give a quadratic in terms of sin² x, 1 mark for method to solve the quadratic, 1 mark for the values of sin x, 1 mark for all correct final answers]

Answers

Exercise 3.14 — The R Addition Formulas

Q1 $3 \sin x - 2 \cos x \equiv R \sin(x - \alpha)$
$\Rightarrow 3 \sin x - 2 \cos x \equiv R \sin x \cos \alpha - R \cos x \sin \alpha$
$\Rightarrow$ ① $R \cos \alpha = 3$ and ② $R \sin \alpha = 2$

② $\div$ ① gives $\tan \alpha = \dfrac{2}{3} \Rightarrow \alpha = 33.7°$ (1 d.p.)

①² + ②² gives: $R^2 \cos^2 \alpha + R^2 \sin^2 \alpha = 3^2 + 2^2 = 13$
$\Rightarrow R^2 (\cos^2 \alpha + \sin^2 \alpha) = 13$
$\Rightarrow R^2 = 13 \Rightarrow R = \sqrt{13}$

So $3 \sin x - 2 \cos x \equiv \sqrt{13} \sin(x - 33.7°)$.

Q2 $6 \cos x - 5 \sin x \equiv R \cos(x + \alpha)$
$\Rightarrow 6 \cos x - 5 \sin x \equiv R \cos x \cos \alpha - R \sin x \sin \alpha$
$\Rightarrow$ ① $R \cos \alpha = 6$ and ② $R \sin \alpha = 5$

② $\div$ ① gives $\tan \alpha = \dfrac{5}{6} \Rightarrow \alpha = 39.8°$ (1 d.p.)

①² + ②² gives: $R^2 \cos^2 \alpha + R^2 \sin^2 \alpha = 6^2 + 5^2 = 61$
$\Rightarrow R = \sqrt{61}$

So $6 \cos x - 5 \sin x \equiv \sqrt{61} \cos(x + 39.8°)$.

Q3 $\sin x + \sqrt{7} \cos x \equiv R \sin(x + \alpha)$
$\Rightarrow \sin x + \sqrt{7} \cos x \equiv R \sin x \cos \alpha + R \cos x \sin \alpha$
$\Rightarrow$ ① $R \cos \alpha = 1$ and ② $R \sin \alpha = \sqrt{7}$

② $\div$ ① gives $\tan \alpha = \sqrt{7} \Rightarrow \alpha = 1.21$ (3 s.f.)

①² + ②² gives: $R^2 \cos^2 \alpha + R^2 \sin^2 \alpha = 1^2 + (\sqrt{7})^2 = 8$
$\Rightarrow R = \sqrt{8} = \sqrt{4 \times 2} = \sqrt{4} \times \sqrt{2} = 2\sqrt{2}$

So $\sin x + \sqrt{7} \cos x \equiv 2\sqrt{2} \sin(x + 1.21)$.

Q4 $5 \sin \theta - 6 \cos \theta \equiv R \sin(\theta - \alpha)$
$5 \sin \theta - 6 \cos \theta \equiv R \sin \theta \cos \alpha - R \cos \theta \sin \alpha$
$\Rightarrow$ ① $R \cos \alpha = 5$ and ② $R \sin \alpha = 6$

② $\div$ ① gives $\tan \alpha = \dfrac{6}{5} \Rightarrow \alpha = 50.19°$ (2 d.p.)

①² + ②² gives:
$R^2 \cos^2 \alpha + R^2 \sin^2 \alpha = 5^2 + 6^2 = 61 \Rightarrow R = \sqrt{61}$

So $5 \sin \theta - 6 \cos \theta \equiv \sqrt{61} \cos(\theta - 50.19°)$

Q5 $\sqrt{2} \sin x - \cos x \equiv R \sin(x - \alpha)$
$\Rightarrow \sqrt{2} \sin x - \cos x \equiv R \sin x \cos \alpha - R \cos x \sin \alpha$
$\Rightarrow$ ① $R \cos \alpha = \sqrt{2}$ and ② $R \sin \alpha = 1$

② $\div$ ① gives $\tan \alpha = \dfrac{1}{\sqrt{2}}$

①² + ②² gives: $R^2 \cos^2 \alpha + R^2 \sin^2 \alpha = (\sqrt{2})^2 + 1^2 = 3$
$\Rightarrow R^2 (\cos^2 \alpha + \sin^2 \alpha) = 3 \Rightarrow R^2 = 3 \Rightarrow R = \sqrt{3}$

So $\sqrt{2} \sin x - \cos x \equiv \sqrt{3} \sin(x - \alpha)$, where $\tan \alpha = \dfrac{1}{\sqrt{2}}$.

[4 marks available — 1 mark for method to find α, 1 mark for showing $\tan \alpha = \dfrac{1}{\sqrt{2}}$, 1 mark for method to find R, 1 mark for showing $R = \sqrt{3}$]

Q6 $3 \cos 2x + 5 \sin 2x \equiv R \cos(2x - \alpha)$
$3 \cos 2x + 5 \sin 2x \equiv R \cos 2x \cos \alpha + R \sin 2x \sin \alpha$
$\Rightarrow$ ① $R \cos \alpha = 3$ and ② $R \sin \alpha = 5$

② $\div$ ① gives $\tan \alpha = \dfrac{5}{3}$

①² + ②² gives: $R^2 \cos^2 \alpha + R^2 \sin^2 \alpha = 3^2 + 5^2 = 34 \Rightarrow R = \sqrt{34}$

So $3 \cos 2x + 5 \sin 2x \equiv \sqrt{34} \cos(2x - \alpha)$, where $\tan \alpha = \dfrac{5}{3}$.

Q7 **a)** $\sqrt{3} \sin x + \cos x \equiv R \sin(x + \alpha)$
$\Rightarrow \sqrt{3} \sin x + \cos x \equiv R \sin x \cos \alpha + R \cos x \sin \alpha$
$\Rightarrow$ ① $R \cos \alpha = \sqrt{3}$ and ② $R \sin \alpha = 1$

② $\div$ ① gives $\tan \alpha = \dfrac{1}{\sqrt{3}} \Rightarrow \alpha = \dfrac{\pi}{6}$

①² + ②² gives: $R^2 \cos^2 \alpha + R^2 \sin^2 \alpha = (\sqrt{3})^2 + 1^2 = 4$
$\Rightarrow R = \sqrt{4} = 2$

So $\sqrt{3} \sin x + \cos x \equiv 2 \sin\left(x + \dfrac{\pi}{6}\right)$.

b) The graph of $y = 2 \sin\left(x + \dfrac{\pi}{6}\right)$ is the graph of $y = \sin x$ transformed in the following way: a horizontal translation left by $\dfrac{\pi}{6}$, then a vertical stretch by a factor of 2.

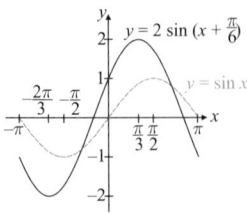

c) The graph of $y = \sin x$ has a minimum at $\left(-\dfrac{\pi}{2}, -1\right)$, a maximum at $\left(\dfrac{\pi}{2}, 1\right)$, and cuts the x-axis at $(-\pi, 0)$, $(0, 0)$ and $(\pi, 0)$. To describe the graph of $y = 2 \sin\left(x + \dfrac{\pi}{6}\right)$, each of these points needs to have $\dfrac{\pi}{6}$ subtracted from the x-coordinates, and the y-coordinates multiplied by 2. So the graph of $y = 2 \sin\left(x + \dfrac{\pi}{6}\right)$ has a minimum at $\left(-\dfrac{2\pi}{3}, -2\right)$, a maximum at $\left(\dfrac{\pi}{3}, 2\right)$, and cuts the x-axis at $\left(-\dfrac{\pi}{6}, 0\right)$ and $\left(\dfrac{5\pi}{6}, 0\right)$.

To find the y-intercept, put $x = 0$ into the equation:
$y = 2 \sin\left(0 + \dfrac{\pi}{6}\right) = 2 \times \dfrac{1}{2} = 1$
So the y-intercept is at $(0, 1)$.

Q8 $\sec x \sin 2x + \dfrac{\cos 2x + 1}{\cos x}$
$= 2 \sec x \sin x \cos x + \dfrac{2 \cos^2 x}{\cos x}$
$\equiv 2 \sin x + 2 \cos x \equiv R \cos(x - \alpha)$
$\Rightarrow 2 \sin x + 2 \cos x \equiv R \cos x \cos \alpha + R \sin x \sin \alpha$
$\Rightarrow$ ① $R \cos \alpha = 2$ and ② $R \sin \alpha = 2$

①² + ②² gives: $R^2 = 2^2 + 2^2 = 8 \Rightarrow R = \sqrt{8} = 2\sqrt{2}$
② $\div$ ① gives: $\tan \alpha = \dfrac{2}{2} = 1 \Rightarrow \alpha = 45°$
Therefore, $\sec x \sin 2x + \dfrac{\cos 2x + 1}{\cos x} = 2\sqrt{2} \cos(x - 45°)$.

[6 marks available — 1 mark for correct use of double angle formulas, 1 mark for fully simplifying original expression, 1 mark for method to find R, 1 mark for the correct value of R, 1 mark for method to find α, 1 mark for the correct value of α]

Q9 **a)** $2 \sin\left(x - \dfrac{\pi}{6}\right) - 2 \cos x$
$= 2 \sin x \cos \dfrac{\pi}{6} - 2 \cos x \sin \dfrac{\pi}{6} - 2 \cos x$
$= \sqrt{3} \sin x - \cos x - 2 \cos x = \sqrt{3} \sin x - 3 \cos x$
[2 marks available — 1 mark for a correct expansion using addition formula, 1 mark for the correct answer]

b) $\sqrt{3} \sin x - 3 \cos x \equiv R \sin(x - \alpha)$
$\Rightarrow \sqrt{3} \sin x - 3 \cos x \equiv R \sin x \cos \alpha - R \cos x \sin \alpha$
$\Rightarrow$ ① $R \cos \alpha = \sqrt{3}$ and ② $R \sin \alpha = 3$
①² + ②² gives: $R^2 = (\sqrt{3})^2 + 3^2 = 12 \Rightarrow R = \sqrt{12} = 2\sqrt{3}$
② $\div$ ① gives: $\tan \alpha = \dfrac{3}{\sqrt{3}} = \sqrt{3} \Rightarrow \alpha = \dfrac{\pi}{3}$.
Hence $2 \sin\left(x - \dfrac{\pi}{6}\right) - 2 \cos x = 2\sqrt{3} \sin\left(x - \dfrac{\pi}{3}\right)$.
[4 marks available — 1 mark for method to find R, 1 mark for correct value of R, 1 mark for method to find α, 1 mark for correct value of α]

c) The period is 2π. *[1 mark]*

Q10 a) $3\cos(x + 45°) + 2\sin(x - 30°)$

$= 3\cos x \cos 45° - 3\sin x \sin 45°$
$\qquad\qquad + 2\sin x \cos 30° - 2\cos x \sin 30°$

$= \dfrac{3}{\sqrt{2}}\cos x - \dfrac{3}{\sqrt{2}}\sin x + \sqrt{3}\sin x - \cos x$

$= \left(\dfrac{3}{\sqrt{2}} - 1\right)\cos x - \left(\dfrac{3}{\sqrt{2}} - \sqrt{3}\right)\sin x.$

$\left(\dfrac{3}{\sqrt{2}} - 1\right)\cos x - \left(\dfrac{3}{\sqrt{2}} - \sqrt{3}\right)\sin x \equiv R\cos(x + \alpha)$

$\Rightarrow \left(\dfrac{3}{\sqrt{2}} - 1\right)\cos x - \left(\dfrac{3}{\sqrt{2}} - \sqrt{3}\right)\sin x$

$\equiv R\cos x \cos \alpha - R\sin x \sin \alpha$

$\Rightarrow$ ① $R\cos \alpha = \left(\dfrac{3}{\sqrt{2}} - 1\right)$ and ② $R\sin \alpha = \left(\dfrac{3}{\sqrt{2}} - \sqrt{3}\right)$

①^2 + ②^2 gives: $R^2 = \left(\dfrac{3}{\sqrt{2}} - 1\right)^2 + \left(\dfrac{3}{\sqrt{2}} - \sqrt{3}\right)^2 = 1.4088...$

$\Rightarrow R = 1.1870$ to 4 decimal places.

$\tan \alpha = \dfrac{\left(\dfrac{3}{\sqrt{2}} - \sqrt{3}\right)}{\left(\dfrac{3}{\sqrt{2}} - 1\right)} \Rightarrow \alpha = 19.1446°$ to 4 decimal places.

Hence $3\cos(x + 45°) + 2\sin(x - 30°)$
$= 1.1870\cos(x + 19.1446°).$

b) $\dfrac{3}{6\cos(2x + 45°) + 4\sin(2x - 30°)}$

$= \dfrac{3}{2 \times 1.1870\cos(2x + 19.1446°)} = 1.3\sec(2x + 19.1446°)$

Exercise 3.15 — Applying the R Addition Formulas

Q1 a) $5\cos \theta - 12\sin \theta \equiv R\cos(\theta + \alpha)$
$\equiv R\cos \theta \cos \alpha - R\sin \theta \sin \alpha$

$\Rightarrow R\cos \alpha = 5, \; R\sin \alpha = 12$

$\tan \alpha = \dfrac{12}{5} \Rightarrow \alpha = 67.4°$ (1 d.p.)

$R = \sqrt{5^2 + 12^2} = \sqrt{169} = 13$

$\Rightarrow 5\cos \theta - 12\sin \theta \equiv 13\cos(\theta + 67.4°)$

b) $13\cos(\theta + 67.4°) = 4$, in the interval
$67.4° \leq (\theta + 67.4°) \leq 427.4°$

$\cos(\theta + 67.4°) = \dfrac{4}{13}$

$\Rightarrow \theta + 67.4° = \cos^{-1}\dfrac{4}{13}$

$= 72.1°$ and $(360° - 72.1°) = 287.9°$

$\Rightarrow \theta = (72.1° - 67.4°)$ and $(287.9° - 67.4°)$
$= 4.7°$ and $220.5°$ (1 d.p.)

c) The maximum and minimum values of $\cos \theta$ are at ± 1. So the maximum and minimum values of $13\cos(\theta + 67.4°)$ are at ± 13.

Q2 a) $2\sin 2\theta + 3\cos 2\theta \equiv R\sin(2\theta + \alpha)$
$\equiv R\sin 2\theta \cos \alpha + R\cos 2\theta \sin \alpha$

$\Rightarrow R\cos \alpha = 2, \; R\sin \alpha = 3$

$\tan \alpha = \dfrac{3}{2} \Rightarrow \alpha = 0.983$ (3 s.f.)

$R = \sqrt{2^2 + 3^2} = \sqrt{13}$

$\Rightarrow 2\sin 2\theta + 3\cos 2\theta \equiv \sqrt{13}\sin(2\theta + 0.983)$

b) $\sqrt{13}\sin(2\theta + 0.983) = 1$ in the interval
$0.983 \leq (2\theta + 0.983) \leq 13.549$.

$\sin(2\theta + 0.983) = \dfrac{1}{\sqrt{13}}$

$\Rightarrow 2\theta + 0.983 = 0.281$ (not in correct interval),
$(\pi - 0.281), (2\pi + 0.281), (3\pi - 0.281), (4\pi + 0.281)$.
There will be 4 solutions for θ between 0 and 2π because you're dealing with $\sin 2\theta$.
$\Rightarrow 2\theta + 0.983 = 2.861, 6.564, 9.144, 12.847$
$\Rightarrow \theta = 0.939, 2.79, 4.08, 5.93$ (3 s.f.)

Q3 a) $3\sin \theta - 2\sqrt{5}\cos \theta \equiv R\sin(\theta - \alpha)$
$\equiv R\sin \theta \cos \alpha - R\cos \theta \sin \alpha$

$\Rightarrow R\cos \alpha = 3, \; R\sin \alpha = 2\sqrt{5}$

$\tan \alpha = \dfrac{2\sqrt{5}}{3} \Rightarrow \alpha = 56.1°$ (1 d.p.)

$R = \sqrt{3^2 + (2\sqrt{5})^2} = \sqrt{29}$

$\Rightarrow 3\sin \theta - 2\sqrt{5}\cos \theta \equiv \sqrt{29}\sin(\theta - 56.1°)$

b) $\sqrt{29}\sin(\theta - 56.1°) = 5$ in the interval
$-56.1° \leq (\theta - 56.1°) \leq 303.9°$.

$\sin(\theta - 56.1°) = \dfrac{5}{\sqrt{29}}$

$\Rightarrow \theta - 56.1° = 68.2°$ and $(180° - 68.2°) = 111.8°$
$\Rightarrow \theta = 124.3°$ and $167.9°$ (1 d.p.)

c) $f(x) = \sqrt{29}\sin(x - 56.1°)$. The maximum of $\sin x$ is 1, so the maximum value of $f(x)$ is $f(x) = \sqrt{29}$.
When $f(x) = \sqrt{29}$, $\sin(x - 56.1°) = 1$
$\Rightarrow x - 56.1° = 90° \Rightarrow x = 146.1°$ (1 d.p.)

Q4 a) $3\sin x + \cos x \equiv R\sin(x + \alpha) \equiv R\sin x \cos \alpha + R\cos x \sin \alpha$
$\Rightarrow R\cos \alpha = 3, \; R\sin \alpha = 1$

$\tan \alpha = \dfrac{1}{3} \Rightarrow \alpha = 18.4°$ (1 d.p.)

$R = \sqrt{3^2 + 1^2} = \sqrt{10}$

$\Rightarrow 3\sin x + \cos x \equiv \sqrt{10}\sin(x + 18.4°)$

b) $\sqrt{10}\sin(x + 18.4°) = 2$ in the interval
$18.4° \leq (x + 18.4°) \leq 378.4°$

$\sin(x + 18.4°) = \dfrac{2}{\sqrt{10}}$

$\Rightarrow x + 18.4° = 39.2°$ and $(180° - 39.2°) = 140.8°$
$\Rightarrow x = 20.8°$ and $122.4°$ (1 d.p.)

c) The maximum and minimum values of $f(x)$ are $\pm \sqrt{10}$.

Q5 a) $4\sin x + \cos x \equiv R\sin(x + \alpha)$
$\equiv R\sin x \cos \alpha + R\cos x \sin \alpha$

$\Rightarrow R\cos \alpha = 4, \; R\sin \alpha = 1$

$\tan \alpha = \dfrac{1}{4} \Rightarrow \alpha = 0.245$ (3 s.f.)

$R = \sqrt{4^2 + 1^2} = \sqrt{17}$

$\Rightarrow 4\sin x + \cos x \equiv \sqrt{17}\sin(x + 0.245)$

b) Maximum value of $4\sin x + \cos x = \sqrt{17}$, so the greatest value of $(4\sin x + \cos x)^4 = (\sqrt{17})^4 = 289$.

c) $\sqrt{17}\sin(x + 0.245) = 1$ in the interval
$0.245 \leq (x + 0.245) \leq 3.387$.

$\sin(x + 0.245) = \dfrac{1}{\sqrt{17}}$

$\Rightarrow x + 0.245 = 0.245$ and $(\pi - 0.245) = 2.897$
$\Rightarrow x = 0$ and 2.65 (3 s.f.)

Q6 a) $8\cos x + 15\sin x \equiv R\cos(x - \alpha)$
$\equiv R\cos x \cos \alpha + R\sin x \sin \alpha$

$\Rightarrow R\cos \alpha = 8, \; R\sin \alpha = 15$

$\tan \alpha = \dfrac{15}{8} \Rightarrow \alpha = 1.08$ (3 s.f.)

$R = \sqrt{8^2 + 15^2} = \sqrt{289} = 17$

$\Rightarrow 8\cos x + 15\sin x \equiv 17\cos(x - 1.08).$

b) So solve for $17\cos(x - 1.08) = 5$ in the interval
$-1.08 \leq (x - 1.08) \leq 5.20$.

$\cos(x - 1.08) = \dfrac{5}{17}$

$\Rightarrow x - 1.08 = 1.27$ and $(2\pi - 1.27) = 5.01$
$\Rightarrow x = 2.35$ and 6.09 (3 s.f.)

Answers

c) $g(x) = (8 \cos x + 15 \sin x)^2 = 17^2 \cos^2 (x - 1.08)$
$$= 289 \cos^2 (x - 1.08)$$
The function $\cos^2 x$ has a minimum value of 0 (since all negative values of $\cos x$ become positive when you square it) so the minimum of $g(x)$ is 0.
This minimum occurs when $\cos^2(x - 1.08) = 0$
so $\cos(x - 1.08) = 0 \Rightarrow x - 1.08 = \frac{\pi}{2} \Rightarrow x = 2.65$ (3 s.f.).

Q7 a) $2 \cos x + \sin x \equiv R \cos (x - \alpha) \equiv R \cos x \cos \alpha + R \sin x \sin \alpha$
$\Rightarrow R \cos \alpha = 2, \; R \sin \alpha = 1$
$\tan \alpha = \frac{1}{2} \Rightarrow \alpha = 26.6°$ (3 s.f.)
$R = \sqrt{2^2 + 1^2} = \sqrt{5}$
$\Rightarrow 2 \cos x + \sin x \equiv \sqrt{5} \cos (x - 26.6°)$.

b) The range of $g(x)$ is between the maximum and minimum values, which are at $\pm \sqrt{5}$.
So $-\sqrt{5} \le g(x) \le \sqrt{5}$.

Q8 $3 \sin \theta - \frac{3}{2} \cos \theta \equiv R \sin (\theta - \alpha) \equiv R \sin \theta \cos \alpha - R \cos \theta \sin \alpha$
$\Rightarrow R \cos \alpha = 3, \; R \sin \alpha = \frac{3}{2}$
$\tan \alpha = \frac{1}{2} \Rightarrow \alpha = 0.464$ (3 s.f.)
$R = \sqrt{3^2 + \left(\frac{3}{2}\right)^2} = \sqrt{\frac{45}{4}} = \frac{3\sqrt{5}}{2}$
$\Rightarrow 3 \sin \theta - \frac{3}{2} \cos \theta \equiv \frac{3\sqrt{5}}{2} \sin (\theta - 0.464)$
So solve $\frac{3\sqrt{5}}{2} \sin (\theta - 0.464) = 3$ in the interval
$-0.464 \le (\theta - 0.464) \le 5.819$.
$\sin (\theta - 0.464) = \frac{2}{\sqrt{5}}$
$\rightarrow \theta - 0.464 = 1.107$ and $(\pi - 1.107) = 2.034$
$\Rightarrow \theta = 1.57$ and 2.50 (3 s.f.)
[6 marks available — 1 mark for method to find α, 1 mark for the correct value of α, 1 mark for method to find R, 1 mark for the correct value of R, 1 mark for the value of $\sin(\theta - 0.464)$, 1 mark for the correct solutions]

Q9 $4 \sin 2\theta + 3 \cos 2\theta \equiv R \sin (2\theta + \alpha)$
$$\equiv R \sin 2\theta \cos \alpha + R \cos 2\theta \sin \alpha$$
$\Rightarrow R \cos \alpha = 4, \; R \sin \alpha = 3$
$\tan \alpha = \frac{3}{4} \Rightarrow \alpha = 0.644$ (3 s.f.)
$R = \sqrt{3^2 + 4^2} = 5$
$\Rightarrow 4 \sin 2\theta + 3 \cos 2\theta \equiv 5 \sin (2\theta + 0.644)$
So solve $5 \sin (2\theta + 0.644) = 2$ in the interval
$0.644 \le (2\theta + 0.644) \le 6.927$.
$\sin (2\theta + 0.644) = \frac{2}{5}$
$\Rightarrow 2\theta + 0.644 = 0.412$ (not in the interval),
$\quad\quad (\pi - 0.412) = 2.730$, and $(2\pi + 0.412) = 6.695$
$\Rightarrow \theta = 1.04$ and 3.03 (3 s.f.)
You could also solve this by writing the function in the form $R \cos (2\theta - \alpha)$ ($R = 5$, $\alpha = 0.927$) — you should get the same solutions for θ either way.

Q10 a) $\sin 2x - \cos 2x \equiv R \sin(2x - \alpha)$
$\Rightarrow \sin 2x - \cos 2x \equiv R \sin 2x \cos \alpha - R \cos 2x \sin \alpha$
$\Rightarrow R \cos \alpha = 1, \; R \sin \alpha = 1$
$\tan \alpha = \frac{1}{1} = 1 \Rightarrow \alpha = \frac{\pi}{4}$
$R = \sqrt{1^2 + 1^2} = \sqrt{2}$
Hence $\sin 2x - \cos 2x = \sqrt{2} \sin \left(2x - \frac{\pi}{4}\right)$.
[4 marks available — 1 mark for method to find α, 1 mark for correct value of α, 1 mark for correct method to find R, 1 mark for correct value of R]

b) Solve $\sqrt{2} \sin \left(2x - \frac{\pi}{4}\right) = \frac{\sqrt{3} - 1}{2}$ in the interval
$-\frac{\pi}{4} \le \left(2x - \frac{\pi}{4}\right) \le \frac{15\pi}{4}$.
$\sin \left(2x - \frac{\pi}{4}\right) = \frac{\sqrt{3} - 1}{2\sqrt{2}}$
$\Rightarrow 2x - \frac{\pi}{4} = \arcsin \frac{\sqrt{3} - 1}{2\sqrt{2}} = \frac{\pi}{12}, \frac{11\pi}{12}, \frac{25\pi}{12}, \frac{35\pi}{12}$
$\Rightarrow 2x = \frac{\pi}{3}, \frac{7\pi}{6}, \frac{7\pi}{3}, \frac{19\pi}{6} \Rightarrow x = \frac{\pi}{6}, \frac{7\pi}{12}, \frac{7\pi}{6}, \frac{19\pi}{12}$.

[4 marks available — 1 mark for rewriting using your result from part a), 1 mark for correct method to solve, 1 mark for the correct values for $2x - \frac{\pi}{4}$, 1 mark for the correct answers]

c) From part a) you know $\sin 2x - \cos 2x = \sqrt{2} \sin \left(2x - \frac{\pi}{4}\right)$.
Start by finding the values of k that have real solutions.
$\sin \left(2x - \frac{\pi}{4}\right)$ has a maximum of 1 and minimum of -1, so:
$-1 \le \sin \left(2x - \frac{\pi}{4}\right) \le 1 \Rightarrow -\sqrt{2} \le \sqrt{2} \sin \left(2x - \frac{\pi}{4}\right) \le \sqrt{2}$
So all values with real solutions are greater than or equal to $-\sqrt{2}$ and less than or equal to $\sqrt{2}$.
So $k < -\sqrt{2}$ or $k > \sqrt{2}$ (or $|k| > \sqrt{2}$)
[2 marks available — 1 mark for a correct method to find the range, 1 mark for the correct answer]

Q11 a) $3 \cos x + \sin x \equiv R \cos(x - \alpha)$
$\Rightarrow 3 \cos x + \sin x \equiv R \cos x \cos \alpha + R \sin x \sin \alpha$
$\Rightarrow R \cos \alpha = 3, \; R \sin \alpha = 1$
$\tan \alpha = \frac{1}{3} \Rightarrow \alpha = 18.4349...° = 18.435°$ (3 d.p.)
$R^2 = 3^2 + 1^2 = 10 \Rightarrow R = \sqrt{10}$
Hence $3 \cos x + \sin x = \sqrt{10} \cos(x - 18.435°)$.
[4 marks available — 1 mark for method to find α, 1 mark for the correct value of α, 1 mark for method to find R, 1 mark for the correct value of R]

b) (i) $f(x) = \frac{1}{(3 \cos x + \sin x)^2} = \frac{1}{(\sqrt{10} \cos(x - 18.435°))^2}$
The least possible value occurs when the denominator is maximal. $-\sqrt{10} \le 3 \cos x + \sin x \le \sqrt{10}$, so the minimum value is $\frac{1}{(\sqrt{10})^2} = \frac{1}{10}$.
[2 marks available — 1 mark for a suitable method, 1 mark for the correct answer]

(ii) Solving $\sqrt{10} \cos(x - 18.435°) = \pm \sqrt{10}$
$\Rightarrow \cos(x - 18.435°) = \pm 1$
$\Rightarrow x - 18.435° = 0°, 180°, -180°, ...$
So the smallest positive value of x will be $18.435°$ (3 d.p.).
[2 marks available — 1 mark for the correct method, 1 mark for the correct answer]

(iii) As the denominator approaches (but never reaches) zero, $f(x)$ becomes greater but with no greatest value.
[1 mark for a correct explanation]

c) As $f(x)$ has no maximum, every turning point of $\sqrt{10} \cos(x - 18.435°)$ corresponds to a minimum point of f. Between $0°$ and $360°$ $\sqrt{10} \cos(x - 18.435°)$ has 2 turning points so f has $2 \times 10 = 20$ turning points in the interval $0° \le x \le 3600°$. Asymptotes occur when $\sqrt{10} \cos(x - 18.435°) = 0$. Again this has two solutions every $360°$, so $2 \times 10 = 20$ solutions in the interval $0° \le x \le 3600°$. Hence $m = 20 = n$.
[3 marks available — 1 mark for correct reasoning, 1 mark for correct value of m, 1 mark for correct value of n]

Exercise 3.16 — Modelling with Trig Functions

Q1 For each sector of the circle, $r = 20$ so

Area $= \frac{1}{2}r^2\theta = (200\theta)$ m², and

Perimeter $= 2r + r\theta = (40 + 20\theta)$ m

Convert each angle to radians then find the area and perimeter:

$120° = \frac{120 \times \pi}{180} = \frac{2\pi}{3}$

Area $= 200 \times \frac{2\pi}{3} = 419$ m² (3 s.f.)

Perimeter $= 40 + 20 \times \frac{2\pi}{3} = 81.9$ m (3 s.f.)

$144° = \frac{144 \times \pi}{180} = \frac{4\pi}{5}$

Area $= 200 \times \frac{4\pi}{5} = 503$ m² (3 s.f.)

Perimeter $= 40 + 20 \times \frac{4\pi}{5} = 90.3$ m (3 s.f.)

$96° = \frac{96 \times \pi}{180} = \frac{8\pi}{15}$

Area $= 200 \times \frac{8\pi}{15} = 335$ m² (3 s.f.)

Perimeter $= 40 + 20 \times \frac{8\pi}{15} = 73.5$ m (3 s.f.)

Q2 a) $\cos\theta$ has a maximum of 1 and a minimum of –1.
You want $A + B\cos\theta$ to have a maximum of 17 and a minimum of 7. So form two equations:

① $A + B(1) = 17$ (at maximum)

② $A + B(-1) = 7$ (at minimum)

①+②: $2A = 24 \Rightarrow A = 12$

Sub back into ①: $12 + B = 17 \Rightarrow B = 5$

So $f(\theta) = 12 + 5\cos\theta$

[3 marks available — 1 mark for a correct method, 1 mark for correct value of A, 1 mark for correct value of B]

You could also think of this as a transformation of the graph of $\cos\theta$ — a vertical stretch with scale factor 5 followed by a translation up by 12.

b) The longest day ($t = 0$) will have 17 hours of daylight, which is the maximum of the function. This will occur when $\cos(Ct + D) = 1$, i.e. when $Ct + D = 0$.

Similarly, the shortest day ($t = 6$) will occur when $\cos(Ct + D) = -1$ i.e. when $Ct + D = \pi$.

So $C(0) + D = 0$ (longest day) $\Rightarrow D = 0$

$C(6) + D = \pi$ (shortest day) $\Rightarrow 6C + 0 = \pi \Rightarrow C = \frac{\pi}{6}$

So $g(t) = 12 + 5\cos\left(\frac{\pi}{6}t\right)$

[3 marks available — 1 mark for correct reasoning, 1 mark for correct value of C, 1 mark for correct value of D]

Because cos is periodic, you might have found $D = 2\pi$.

Also, because cos is symmetrical, $C = -\frac{\pi}{6}$ will give you the same answer — either one is fine.

Q3 a) Ride stops when $v(t) = 0 \Rightarrow \cos\left(\frac{\pi t}{5}\right) = 1$

$\Rightarrow \frac{\pi t}{5}$ is a multiple of 2π, so $t = 10, 20, 30, 40$.

This happens four times during the ride.

[2 marks available — 1 mark for a correct method, 1 mark for the correct answers]

b) (i) The maximum occurs when $\cos\left(\frac{\pi t}{5}\right) = -1$, so 16 ms⁻¹.

[1 mark for the correct answer]

(ii) $\cos\left(\frac{\pi t}{5}\right) = -1 \Rightarrow \frac{\pi t}{5} = \pi \Rightarrow t = 5$ seconds after the start of the ride.

[2 marks available — 1 mark for a correct method, 1 mark for the correct answer]

Q4 a) 1:30 pm is 4.5 hours after 9 am:

$T(4.5) = 19 + 4\sin\left(4.5 - \frac{\pi}{6}\right) = 16.035... = 16.0$ °C (3 s.f.)

[1 mark for the correct answer]

b) (i) $19 + 4\sin\left(t - \frac{\pi}{6}\right) = 17 \Rightarrow \sin\left(t - \frac{\pi}{6}\right) = -\frac{1}{2}$

so $t - \frac{\pi}{6} = -\frac{\pi}{6}, \frac{7\pi}{6} \Rightarrow t = 0, \frac{4\pi}{3}$

Ignore $t = 0$ as the temperature is rising

so $t = \frac{4\pi}{3} = 4.188... \approx 4$ hours and 11 mins

after 9 a.m. so 1:11 pm.

[3 marks available — 1 mark for the correct equation, 1 mark for method to solve, 1 mark for the correct answer]

(ii) The next solution to the equation in part (i) occurs when

$t - \frac{\pi}{6} = \frac{11\pi}{6} \Rightarrow t = 2\pi$ so the radiator is turned on for

$2\pi - \frac{4\pi}{3}$ hours $= \frac{2\pi}{3}$ hours $= 2.094...$ hours

≈ 2 hours 6 minutes.

[3 marks available — 1 mark for a method to find next solution, 1 mark for the correct value of t, 1 mark for the correct answer]

Q5 $h(t) = 14 + 5\sin t + 5\cos t \equiv 14 + R\cos(t - \alpha)$

$\equiv 14 + R\cos t\cos\alpha + R\sin t\sin\alpha$

$\Rightarrow R\cos\alpha = 5, \; R\sin\alpha = 5$

$\tan\alpha = 1 \Rightarrow \alpha = \frac{\pi}{4}$

$R = \sqrt{5^2 + 5^2} = \sqrt{50} = 5\sqrt{2}$

$\Rightarrow h(t) \equiv 14 + 5\sqrt{2}\cos\left(t - \frac{\pi}{4}\right)$

The maximum and minimum of cos are 1 and –1, so the maximum of $h(t) = 14 + 5\sqrt{2} = 21.1$ m (1 d.p.) and the minimum of $h(t) = 14 - 5\sqrt{2} = 6.9$ m (1 d.p.)

[6 marks available — 1 mark for method to find α, 1 mark for the correct value of α, 1 mark for method to find R, 1 mark for the correct value of R, 1 mark for the correct maximum, 1 mark for the correct minimum]

Q6 a) $H = 10 + \frac{7}{2}\sin t - \frac{7\sqrt{3}}{2}\cos t \equiv 10 + R\sin(t - \alpha)$

$\equiv 10 + R\sin t\cos\alpha - R\cos t\sin\alpha$

$\Rightarrow R\cos\alpha = \frac{7}{2}, \; R\sin\alpha = \frac{7\sqrt{3}}{2}$

$\tan\alpha = \sqrt{3} \Rightarrow \alpha = \frac{\pi}{3}$

$R = \sqrt{\left(\frac{7}{2}\right)^2 + \left(\frac{7\sqrt{3}}{2}\right)^2} = \sqrt{\frac{49}{4} + \frac{147}{4}}$

$= \sqrt{\frac{196}{4}} = \sqrt{49} = 7$

$\Rightarrow H \equiv 10 + 7\sin\left(t - \frac{\pi}{3}\right)$

b) Replace $\left(t - \frac{\pi}{3}\right)$ with x to make the working clearer:

$h = 10 + 7\sin x - \cos 2x$

Write $\cos 2x$ in terms of $\sin x$, i.e. $1 - 2\sin^2 x$

$h = 10 + 7\sin x - (1 - 2\sin^2 x)$

$= 10 + 7\sin x - 1 + 2\sin^2 x = 9 + 7\sin x + 2\sin^2 x$

i.e. $A = 9$ and $B = 2$.

c) You need to find $h = 13$, i.e.

$9 + 7\sin x + 2\sin^2 x = 13$

$2\sin^2 x + 7\sin x - 4 = 0$

This is a quadratic in $\sin x$, so factorise:

$(2\sin x - 1)(\sin x + 4) = 0$

$\Rightarrow \sin x = \frac{1}{2}$ or $\sin x = -4$ (not valid since $-1 < \sin x < 1$)

So $\sin x = \frac{1}{2}$ i.e. $\sin\left(t - \frac{\pi}{3}\right) = \frac{1}{2}$

You need to find the smallest t that satisfies this equation.

$\sin\left(t - \frac{\pi}{3}\right) = \frac{1}{2} \Rightarrow t - \frac{\pi}{3} = -\frac{7\pi}{6}, \frac{\pi}{6}, \frac{5\pi}{6}$, etc.

$\Rightarrow t = -\frac{5\pi}{6}, \frac{\pi}{2}, \frac{7\pi}{6}$, etc.

Time cannot be negative, so the smallest solution is $t = \frac{\pi}{2}$

i.e. $t = 1.57$ s (3 s.f.)

Q7 a) $\cos\left(\frac{t}{2}\right) + 5\sin\left(\frac{t}{2}\right) + 6 \equiv A\cos\left(\frac{t}{2} - B\right) + 6$

$\Rightarrow \cos\left(\frac{t}{2}\right) + 5\sin\left(\frac{t}{2}\right) \equiv A\cos\frac{t}{2}\cos B + A\sin\frac{t}{2}\sin B$

$\Rightarrow A\cos B = 1$ and $A\sin B = 5$

$A^2 = 1^2 + 5^2 = 26 \Rightarrow A = \sqrt{26}$

$\tan B = \frac{5}{1} = 5 \Rightarrow B = 1.3734... = 1.373$ (3 d.p.)

So $d(t) = \sqrt{26}\cos\left(\frac{t}{2} - 1.373\right) + 6$.

[4 marks available — 1 mark for correct method to find A, 1 mark for correct method to find B, 1 mark for at least two values correct, 1 mark for all three values correct]

b) Least distance $= 6 - \sqrt{26}$ cm.
Greatest distance $= 6 + \sqrt{26}$ cm.
[2 marks available — 1 mark for each correct answer]

c) $d(t) = 10$ if $\sqrt{26}\cos\left(\frac{t}{2} - 1.3734...\right) = 4$

$\Rightarrow \cos\left(\frac{t}{2} - 1.3734...\right) = \frac{4}{\sqrt{26}}$

$\Rightarrow \frac{t}{2} - 1.373... = -0.668..., 0.668..., 2\pi - 0.668...$

$\Rightarrow t = 2 \times (-0.668... + 1.3734...) = 1.4088...$
and $t = 2 \times (0.668... + 1.3734...) = 4.0847...$

So the total time between the first two times the pendulum is 10 cm away from the object is:

$4.0847... - 1.4088... = 2.6758... = 2.7$ seconds (to 1 d.p.)

[4 marks available — 1 mark for correct equation, 1 mark for correct method to find one solution, 1 mark for correct method to find second solution, 1 mark for the correct answer]

Q8 a) From 8 am to 12 noon there are 240 minutes
When $x = 240$,
$P = 83 + 6\sin(240° - 30°) + 14\cos(240° + 60°) = 87$
[1 mark for the correct answer]

b) $P = 83 + 6\sin(x° - 30°) + 14\cos(x° + 60°)$
$\Rightarrow P = 83 + 6\sin x°\cos 30° - 6\cos x°\sin 30°$
$\qquad\qquad\quad + 14\cos x°\cos 60° - 14\sin x°\sin 60°$
$\Rightarrow P = 83 + 3\sqrt{3}\sin x° - 3\cos x° + 7\cos x° - 7\sqrt{3}\sin x°$
$\Rightarrow P = 83 - 4\sqrt{3}\sin x° + 4\cos x°$

[4 marks available — 1 mark for correct expansion of $\sin(x° - 30°)$, 1 mark for correct expansion of $\cos(x° + 60°)$, 1 mark for the correct value of A, 1 mark for the correct value of B]

c) $83 - 4\sqrt{3}\sin x + 4\cos x \equiv 83 + R\cos(x + \alpha)$
$\Rightarrow 4\cos x - 4\sqrt{3}\sin x \equiv R\cos x\cos\alpha - R\sin x\sin\alpha$
$\Rightarrow R\cos\alpha = 4$, $R\sin\alpha = 4\sqrt{3}$

$\tan\alpha = \frac{4\sqrt{3}}{4} = \sqrt{3} \Rightarrow \alpha = 60°$
$R^2 = 4^2 + (4\sqrt{3})^2 = 64 \Rightarrow R = 8$

So $P = 83 + 8\cos(x + 60°)$.

[4 marks available — 1 mark for correct method to find α, 1 mark for correct value of α, 1 mark for correct method to find R, 1 mark for correct value of R]

d) $P = 90$ when $8\cos(x + 60°) = 7$

$\Rightarrow \cos(x + 60°) = \frac{7}{8} \Rightarrow x + 60° = 28.9550..., 331.0449...$

$\Rightarrow x = -31.0449..., 271.0449...$
So $x \approx 271$ minutes is the smallest positive solution.
That is, Sarah first has high blood pressure 271 minutes after 8 am, which is 12:31 pm.

[4 marks available — 1 mark for correct trig equation, 1 mark for method to solve equation, 1 mark for correct x value, 1 mark for the correct answer]

e) E.g. may not include effects of exercise or stress; blood pressure unlikely to follow a regular cycle.
[1 mark for a suitable answer]

Chapter 3 Review Exercise

Q1 a) (i) $\frac{\pi}{12}$ **(ii)** $\frac{5\pi}{18}$ **(iii)** $\frac{11\pi}{6}$ **(iv)** $\frac{5\pi}{4}$

b) (i) 105° **(ii)** 210° **(iii)** 300° **(iv)** 195°

Q2 Arc length $BC = r\theta = 10 \times 0.7 = 7$ cm
Area $= A = \frac{1}{2}r^2\theta = \frac{1}{2} \times 10^2 \times 0.7 = 35$ cm²

Q3 a) Find the angle in radians: $40° = \frac{40\pi}{180} = \frac{2\pi}{9}$
Area $= \frac{1}{2}r^2\theta = \frac{1}{2} \times 100^2 \times \frac{2\pi}{9} = 3490$ m² (3 s.f.)
[3 marks available — 1 mark for converting the angle into radians, 1 mark for using the formula for the area of a sector, 1 mark for the correct answer]

b) Length of arc $= s = r\theta = 100 \times \frac{2\pi}{9} = 69.81...$ m
Boundary length $= 100 + 100 + s = 270$ m (3 s.f.)
[2 marks available — 1 mark for finding the arc length, 1 mark for the correct answer]

Q4 a) $AC^2 = 50^2 + 50^2 - 2 \times 50 \times 50\cos 1.1 = 2732.0193...$
so $AC = \sqrt{2732.0193...} = 52.27$ m (to 2 d.p.)
[3 marks available — 1 mark for using the cosine rule and inputting the numbers correctly, 1 mark for taking the square root, 1 mark for the correct answer]
Alternatively you could have split the isosceles AOC into two identical right-angled triangles AOE and COE then found the length using trigonometry.

b) Area $= \frac{1}{2} \times 50^2 \times 1.1 = 1375$ m²
[2 marks available — 1 mark for a correct method, 1 mark for the correct answer]

c) $OD = OF = 25$ m.
$DF^2 = 25^2 + 25^2 - 2 \times 25 \times 25\cos 1.1 = 683.0048...$
so $DF = \sqrt{683.0048...}$ m.
$FC = 25$ m. $CB = 50 \times \frac{1}{2}(1.1) = 27.5$ m. $BO = 50$ m
So the total distance $= 25 + \sqrt{683.0048...} + 25 + 27.5 + 50$
$= 153.6343... = 153.63$ m (to 2 d.p.)
[5 marks available — 1 mark for an attempt to use a correct method to find DF, 1 mark for finding DF, 1 mark for the correct method to find CB, 1 mark for finding CB, 1 mark for the correct answer]

Q5 a) (i) $\frac{1}{2}r^2\sin\theta$ *[1 mark]*

(ii) $\frac{1}{2}r^2\theta$ *[1 mark]*

(iii) $\tan\theta = \frac{BC}{r}$ so $BC = r\tan\theta$.
So area of $OBC = \frac{1}{2}r \times r\tan\theta = \frac{1}{2}r^2\tan\theta$.
[3 marks available — 1 mark for a method to find length BC, 1 mark for the correct length BC, 1 mark for the correct answer]

b) (i) Area of triangle OAB < area of sector OAB < area of triangle OBC
$\Rightarrow \frac{1}{2}r^2\sin\theta < \frac{1}{2}r^2\theta < \frac{1}{2}r^2\tan\theta$
$\Rightarrow \sin\theta < \theta < \tan\theta$
[2 marks available — 1 mark for correct inequality with areas, 1 mark for working leading to the correct answer]

(ii) Start by taking the reciprocal of each expression and using $\tan\theta = \frac{\sin\theta}{\cos\theta}$:
$\sin\theta < \theta < \tan\theta \Rightarrow \frac{1}{\sin\theta} > \frac{1}{\theta} > \frac{1}{\tan\theta}$
$\Rightarrow \frac{1}{\sin\theta} > \frac{1}{\theta} > \frac{\cos\theta}{\sin\theta}$
Multiply by $\sin\theta$ (which is positive since θ is acute):
$\cos\theta < \frac{\sin\theta}{\theta} < 1$
[2 marks available — 1 mark for correct reciprocal inequality, 1 mark for working leading to the correct answer]

c) As $\theta \to 0$, $\cos\theta \to 1$ and $1 \to 1$.

So from b)(ii) $\dfrac{\sin\theta}{\theta} \to 1$. Therefore $\sin\theta \approx \theta$ for small θ,

which is the small angle approximation for $\sin\theta$.

[2 marks available — 1 mark for using b)(ii) correctly,
1 mark for a correct explanation leading to the correct result]

Q6 $\tan\theta \approx \theta$, $\cos\theta \approx 1 - \dfrac{1}{2}\theta^2$

so $\tan\theta - \cos\theta \approx \theta - (1 - \dfrac{1}{2}\theta^2) = \theta - 1 + \dfrac{1}{2}\theta^2$.

a) $0.13 - 1 + \dfrac{1}{2}(0.13)^2 = -0.86155$

b) $0.07 - 1 + \dfrac{1}{2}(0.07)^2 = -0.92755$

c) $0.26 - 1 + \dfrac{1}{2}(0.26)^2 = -0.7062$

Q7 a) $\sin 3\theta \tan 4\theta \approx 3\theta \times 4\theta = 12\theta^2$

b) $\cos 4\theta + \cos 8\theta \approx (1 - \dfrac{1}{2}(4\theta)^2) + (1 - \dfrac{1}{2}(8\theta)^2)$

$= 1 - 8\theta^2 + 1 - 32\theta^2 = 2 - 40\theta^2$

c) $\dfrac{2\theta^3}{\sin 2\theta \cos\theta} \approx \dfrac{2\theta^3}{2\theta \times (1 - \dfrac{1}{2}\theta^2)} = \dfrac{2\theta^3}{2\theta - \theta^3} = \dfrac{2\theta^2}{2 - \theta^2}$

Q8 a) Find the angle in radians: $5° = \dfrac{5\pi}{180} = \dfrac{\pi}{36}$

$\tan x \approx x$, so $\tan \dfrac{\pi}{36} \approx \dfrac{\pi}{36} = 0.0873$ (3 s.f.)

[2 marks available — 1 mark for correct conversion,
1 mark for the correct answer]

b) $\tan 5° = 0.0874...$ so the percentage error

is $\dfrac{0.0874... - 0.0872...}{0.0874...} \times 100 = 0.254\%$ (3 s.f.)

[2 marks available — 1 mark for correct method to find
percentage error, 1 mark for correct answer]

Q9 Using common trig angles:

a) If $x = \sin^{-1}\dfrac{1}{\sqrt 2}$, then $\dfrac{1}{\sqrt 2} = \sin x$ so $x = \dfrac{\pi}{4}$.

b) If $x = \cos^{-1}\dfrac{1}{2}$, then $\dfrac{1}{2} = \cos x$ so $x = \dfrac{\pi}{3}$.

c) If $x = \tan^{-1}\dfrac{1}{\sqrt 3}$, then $\dfrac{1}{\sqrt 3} = \tan x$ so $x = \dfrac{\pi}{6}$.

Q10

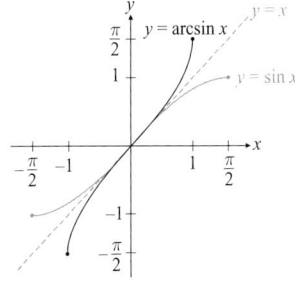

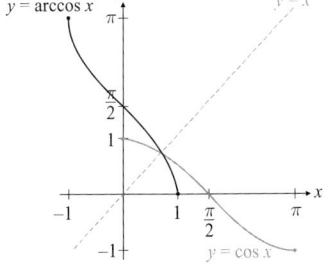

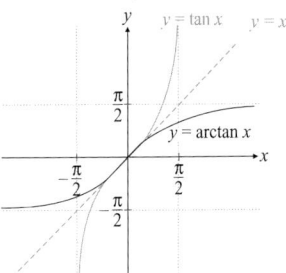

Q11 The domain of $\cos x$ is limited to $0 \le x \le \pi$ in defining $\cos^{-1} x$,

so $y = \cos^{-1} x$ is defined on the domain $-1 \le x \le 1$.

A and C are endpoints, so the x-coordinate of A is -1 and the x-coordinate of C is 1. $\cos x = 1$ when $x = 0$, so $\cos^{-1} 1 = 0$.

Similarly, $\cos x = -1$ when $x = \pi$, so $\cos^{-1} -1 = \pi$.

B is the y-intercept so its x-coordinate is 0.

$\cos x = 0$ when $x = \dfrac{\pi}{2}$, so $\cos^{-1} 0 = \dfrac{\pi}{2}$.

So A $(-1, \pi)$, B $(0, \dfrac{\pi}{2})$ and C $(1, 0)$.

[3 marks available — 1 mark for each correct pair of
coordinates]

Q12

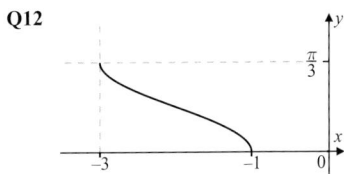

Q13 a) Rearrange to make x the subject:

$1 + \cos x = \dfrac{1}{y} \Rightarrow \cos x = \dfrac{1}{y} - 1 \Rightarrow x = \cos^{-1}(\dfrac{1}{y} - 1)$

Replace x with $f^{-1}(x)$ and y with x:

$f^{-1}(x) = \cos^{-1}(\dfrac{1}{x} - 1) = \arccos(\dfrac{1}{x} - 1)$

b) The range for $f(x)$ is $\dfrac{1}{2} \le f(x) \le 1$,

so the domain for $f^{-1}(x)$ is $\dfrac{1}{2} \le x \le 1$

The domain for $f(x)$ is $0 \le x \le \dfrac{\pi}{2}$,

so the range for $f^{-1}(x)$ is $0 \le f^{-1}(x) \le \dfrac{\pi}{2}$.

c)

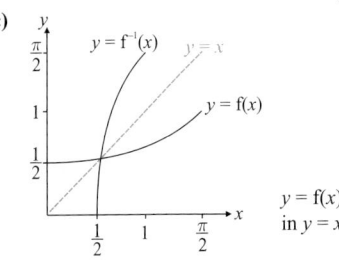

$y = f(x)$ has been reflected in $y = x$.

Q14 a) $f(1) = \sin^{-1}(1) + \cos^{-1}(1) + \tan^{-1}(1) = \dfrac{\pi}{2} + 0 + \dfrac{\pi}{4} = \dfrac{3\pi}{4}$

b) $f(-1) = \sin^{-1}(-1) + \cos^{-1}(-1) + \tan^{-1}(-1)$

$= -\dfrac{\pi}{2} + \pi + -\dfrac{\pi}{4} = \dfrac{\pi}{4}$

Q15

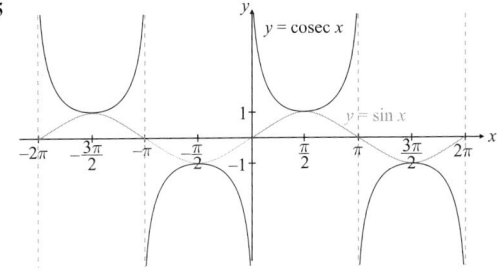

Answers

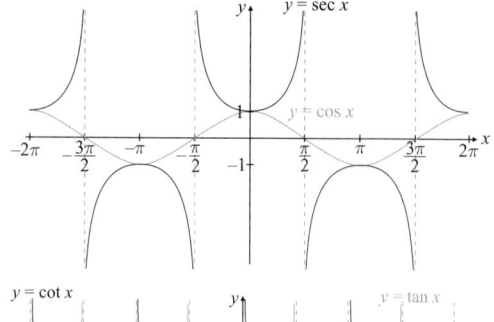

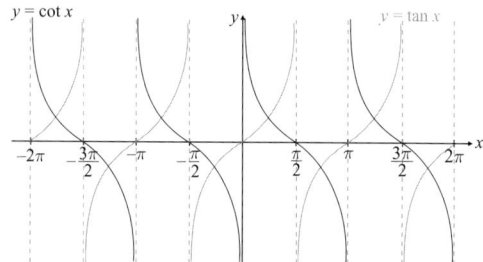

Q16 a) A horizontal stretch with scale factor $\frac{1}{4}$.

b) The period of $y = \sec x$ is 2π, so the period of $y = \sec 4x = 2\pi \div 4 = \frac{\pi}{2}$.

c)

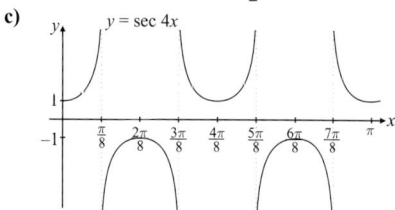

d) $y = \sec 4x$ is undefined when $x = \frac{\pi}{8}, \frac{3\pi}{8}, \frac{5\pi}{8}$ and $\frac{7\pi}{8}$.

Q17 a)

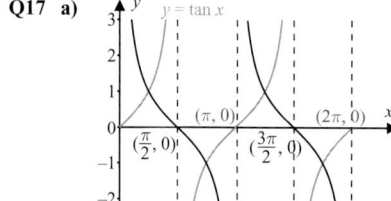

[4 marks available — 1 mark for correct shape of tan graph, 1 mark for correct shape of cot graph, 1 mark for all x-intercepts correct, 1 mark for all asymptotes correct]

b) Four solutions since there are four points of intersection.
[1 mark]

c) $\frac{1}{\tan x} = \tan x \Rightarrow \tan^2 x = 1 \Rightarrow \tan x = \pm 1$
$\Rightarrow x = \arctan 1 = \frac{\pi}{4}$ or $x = \arctan(-1) = -\frac{\pi}{4}$ or $\frac{3\pi}{4}$.
The intersections occur every $\frac{\pi}{2}$,
so $x = \frac{\pi}{4}, x = \frac{3\pi}{4}, x = \frac{5\pi}{4}$ or $x = \frac{7\pi}{4}$.
[4 marks available — 1 mark for $\tan^2 x = 1$, 1 mark for correct values of $\tan x$, 1 mark for a correct solution to $\tan x = \pm 1$, 1 mark for all correct solutions]

Q18 a) Range of $f(x) = \sec x$ is $x \leq -1$ or $x \geq 1$ (or $|f(x)| \geq 1$).
[1 mark for the correct answer]

b) Domain is the range of the original function, so is $|x| \geq 1$.
Range is the domain of the original function, so is $0 \leq f^{-1}(x) \leq \pi$.
[2 marks available — 1 mark for the correct domain, 1 mark for the correct range]

c) Reflecting $y = \sec x$ for $0 \leq x \leq \pi$, gives:

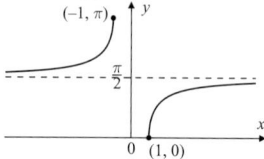

[3 marks available — 1 mark for the correct shape of the graph, 1 mark for correct endpoints, 1 mark for a fully correct graph]

Q19 $\operatorname{cosec}^2 x = 3 \Rightarrow \operatorname{cosec} x = \pm\sqrt{3}$

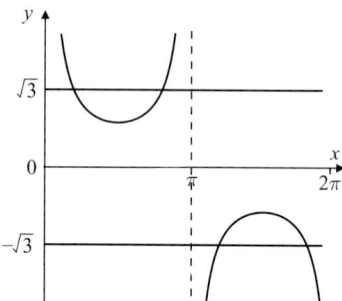

The graph $y = \operatorname{cosec} x$ intersects $y = \pm\sqrt{3}$ four times on the interval $0 < x < 2\pi$, so there are four real solutions.
[3 marks available — 1 mark for $\operatorname{cosec} x = \pm\sqrt{3}$, 1 mark for the correct graphs, 1 mark for the correct answer]

Q20 On the domain $0 < x < \pi$, the range of $\operatorname{cosec} x$ is $1 \leq \operatorname{cosec} x < \infty$.
On the domain $1 \leq \operatorname{cosec} x < \infty$, the range of f(x) is $f(x) \geq 0$.

Q21 a) $\operatorname{cosec} 30° = \frac{1}{\sin 30°} = 1 \div \frac{1}{2} = 2$

b) $\sec 30° = \frac{1}{\cos 30°} = 1 \div \frac{\sqrt{3}}{2} = \frac{2}{\sqrt{3}}$

c) $\cot 30° = \frac{1}{\tan 30°} = 1 \div \frac{1}{\sqrt{3}} = \sqrt{3}$

Q22 a) $\sqrt{1^2 + (\sqrt{3})^2} = 2$ *[1 mark]*

b) Using the top left triangle, $b = \sin 45° = \frac{\sqrt{2}}{2}$.
Using the top right triangle, $a = \sqrt{3}\sin 45° = \frac{\sqrt{6}}{2}$.
[3 marks available — 1 mark for a correct method to find either value, 1 mark for the correct value of b, 1 mark for the correct value of a]

c) (i) $\cos 75° = \frac{a - b}{2} = \frac{\sqrt{6} - \sqrt{2}}{4}$ so
$\sec 75° = \frac{4}{\sqrt{6} - \sqrt{2}} = \frac{4(\sqrt{6} + \sqrt{2})}{(\sqrt{6} - \sqrt{2})(\sqrt{6} + \sqrt{2})}$
$= \frac{4(\sqrt{6} + \sqrt{2})}{6 - 2} = \sqrt{6} + \sqrt{2}$.
[3 marks available — 1 mark for using a correct expression for cos 75°, 1 mark for an unsimplified expression for sec 75°, 1 mark for the correct answer]

(ii) $\tan 15° = \frac{a - b}{a + b} \Rightarrow \cot 15° = \frac{a + b}{a - b} = \frac{\sqrt{6} + \sqrt{2}}{\sqrt{6} - \sqrt{2}}$
$= \frac{(\sqrt{6} + \sqrt{2})(\sqrt{6} + \sqrt{2})}{(\sqrt{6} - \sqrt{2})(\sqrt{6} + \sqrt{2})} = \frac{6 + 2 + 2\sqrt{12}}{6 - 2}$
$= \frac{8 + 4\sqrt{3}}{4} = 2 + \sqrt{3}$

[3 marks available — 1 mark for using a correct expression for tan 15°, 1 mark for an unsimplified expression for cot 15°, 1 mark for the correct answer]

Q23 a) $\tan \dfrac{5\pi}{12} = \tan\left(\dfrac{\pi}{6} + \dfrac{\pi}{4}\right) = \dfrac{\tan\frac{\pi}{6} + \tan\frac{\pi}{4}}{1 - \tan\frac{\pi}{6}\tan\frac{\pi}{4}}$

$= \dfrac{\frac{1}{\sqrt{3}} + 1}{1 - \frac{1}{\sqrt{3}} \times 1} = \dfrac{\frac{1+\sqrt{3}}{\sqrt{3}}}{\frac{\sqrt{3}-1}{\sqrt{3}}} = \dfrac{1+\sqrt{3}}{\sqrt{3}-1}$

$= \dfrac{1+\sqrt{3}}{\sqrt{3}-1} \times \dfrac{\sqrt{3}+1}{\sqrt{3}+1} = \dfrac{4 + 2\sqrt{3}}{2} = 2 + \sqrt{3}$

[3 marks available — 1 mark for splitting $\frac{5\pi}{12}$ into a suitable sum, 1 mark for using the tan addition formula correctly, 1 mark for simplifying to get the correct answer]

b) $\cot \dfrac{5\pi}{12} = \dfrac{1}{\tan\frac{5\pi}{12}} = \dfrac{1}{2+\sqrt{3}} = \dfrac{1}{2+\sqrt{3}} \times \dfrac{2-\sqrt{3}}{2-\sqrt{3}}$

$= \dfrac{2-\sqrt{3}}{1} = 2 - \sqrt{3}$

[1 mark for the correct answer]

Q24 $x^3 = (\cos t + \sec t)^3 = \cos^3 t + 3\cos^2 t\sec t + 3\cos t\sec^2 t + \sec^3 t$
$= y + 3\cos t + 3\sec t = y + 3x$, so $y = x^3 - 3x$.

Q25 a) $\sec x \leq -1$ or $\sec x \geq 1$ so $\sec 3x = \frac{1}{2}$ has no real solutions.
[1 mark for a correct explanation]

b) $\sec x \leq -1$ or $\sec x \geq 1 \Rightarrow \sec^2 3x \geq 1 \; k \Rightarrow k \geq 1$ *[1 mark]*

c) $\sec^2 3x = 4 \Rightarrow \sec 3x = \pm 2 \Rightarrow \cos 3x = \pm\frac{1}{2}$
$\Rightarrow 3x = \pm 60°, \pm 120°, \pm 240°$
$\Rightarrow x = \pm 20°, \pm 40°, \pm 80°$
[3 marks available — 1 mark for correct values of $\cos 3x$, 1 mark for any correct value of x, 1 mark for all correct answers]

Q26 $y = \cot^2 \theta$
By rearranging the identity $\text{cosec}^2 \theta = 1 + \cot^2 \theta$,
you get $\cot^2 \theta = \text{cosec}^2 \theta - 1$:
Since $x = \text{cosec}\,\theta$, $y = x^2 - 1$ as required.

Q27 Square $y = 2\tan\theta$ to get $y^2 = 4\tan^2\theta$.
By rearranging the identity $\sec^2\theta = 1 + \tan^2\theta$,
you get $\tan^2\theta = \sec^2\theta - 1$:
So $y^2 = 4(\sec^2\theta - 1) = 4(x^2 - 1) \Rightarrow y = \pm 2\sqrt{x^2 - 1}$

Q28 a) $\text{cosec}^2 x = \dfrac{3\cot x + 4}{2} \Rightarrow 2\,\text{cosec}^2 x = 3\cot x + 4$
$\Rightarrow 2(1 + \cot^2 x) = 3\cot x + 4$
$\Rightarrow 2\cot^2 x - 3\cot x + 2 - 4 = 0$
$\Rightarrow 2\cot^2 x - 3\cot x - 2 = 0$

[2 marks available — 1 mark for using $\text{cosec}^2\theta \equiv 1 + \cot^2\theta$ identity, 1 mark rearranging into quadratic form]

b) Factorise the quadratic in $\cot x$:
$(\cot x - 2)(2\cot x + 1) = 0$
So $\cot x = 2$ or $\cot x = -\frac{1}{2}$.
$\Rightarrow \tan x = \frac{1}{2}$ or $\tan x = -2$
For $\tan x = \frac{1}{2}$, the first solution is $x = 0.463...$
There is another solution at $0.463... + \pi = 3.605...$
You can find other solutions by sketching a graph or using a CAST diagram.
For $\tan x = -2$, $x = -1.107...$ — this is outside the interval, so add π: $-1.107... + \pi = 2.034...$
There is another solution at $2.034... + \pi = 5.176...$
So $x = 0.46, 3.61, 2.03$ and 5.18 (all to 2 d.p.)
[4 marks available — 1 mark for a correct method to solve the quadratic, 1 mark for correct values of $\cot x$, 1 mark for correct solutions for $\tan x = \frac{1}{2}$, 1 mark for correct solutions for $\tan x = -2$]

Q29 a) $\sec x$ is negative on the interval $\frac{\pi}{2} < x < \frac{3\pi}{2}$.
So $\sec x = -3 \Rightarrow \cos x = -\frac{1}{3} \Rightarrow x = 1.91, 4.37$ (to 2 d.p.)
So $\sec x < -3$ when $\frac{\pi}{2} < x < 1.91$ or $4.37 < x < \frac{3\pi}{2}$.

b) $3 - 4\,\text{cosec}\,x < 1 \Rightarrow \text{cosec}\,x > \frac{1}{2}$.
The range of $\text{cosec}\,x$ is $|f(x)| \geq 1$, so $\text{cosec}\,x > \frac{1}{2}$ is true for all $0 < x < \pi$.

c) $\cot x > 0$ on the interval $0 < x < \frac{\pi}{2}$ and $\pi < x < \frac{3\pi}{2}$.
$\cot x = \frac{3}{2} \Rightarrow \tan x = \frac{2}{3} \Rightarrow x = 0.59, 3.73$ (to 2 d.p.)
So $2\cot x > 3$ when $0 < x < 0.59$ or $\pi < x < 3.73$.

Q30 a) $\sec\theta = \dfrac{1}{\cos\theta} = 1 \div \frac{1}{2} = 2$

b) Draw a right angled triangle to find the third side: $\sqrt{2^2 - 1^2} = \sqrt{3}$.
So $\tan\theta = \dfrac{\sqrt{3}}{1} = \sqrt{3}$.

c) From a), $\sec\theta = 2$, so $\sec^2\theta = 4$
$\Rightarrow 1 + \tan^2\theta = 4 \Rightarrow \tan^2\theta = 3 \Rightarrow \tan\theta = \sqrt{3}$

d) $\cot\theta = \dfrac{1}{\tan\theta} = \dfrac{1}{\sqrt{3}}$

e) From d), $\cot^2\theta = \frac{1}{3}$, so $\text{cosec}^2\theta = 1 + \frac{1}{3} = \frac{4}{3}$
$\Rightarrow \text{cosec}\,\theta = \dfrac{2}{\sqrt{3}} \Rightarrow \sin\theta = \dfrac{1}{\text{cosec}\,\theta} = \dfrac{\sqrt{3}}{2}$
You're told θ is acute, so you know all the trig ratios will be positive.

Q31 $\text{cosec}^2 x = 4 + 2\cot x \Rightarrow 1 + \cot^2 x = 4 + 2\cot x$
$\Rightarrow \cot^2 x - 2\cot x - 3 = 0 \Rightarrow (\cot x + 1)(\cot x - 3) = 0$
$\Rightarrow \cot x = -1$ or $\cot x = 3$
$\cot x = -1 \Rightarrow \tan x = -1 \Rightarrow x = \tan^{-1}(-1) = -\frac{\pi}{4}$
This is outside the interval — there are two solutions in the interval at $\pi - \frac{\pi}{4}$ and $2\pi - \frac{\pi}{4}$ so $x = \frac{3\pi}{4}, \frac{7\pi}{4}$.
$\cot x = 3 \Rightarrow \tan x = \frac{1}{3} \Rightarrow x = \tan^{-1}\left(\frac{1}{3}\right) = 0.3217...$
There is another solution in the interval at $\pi + 0.317...$, so $x = 0.322, 3.46$ (3 s.f.)
So $x = 0.322, 2.36, 3.46, 5.50$ (3 s.f.)
[7 marks available — 1 mark for using the identity $\text{cosec}^2 x \equiv 1 + \cot^2 x$, 1 mark for rearranging to form a quadratic in $\cot x$, 1 mark for factorising the quadratic, 1 mark for finding the two correct values of $\cot x$, 1 mark for using $\cot x = 1/\tan x$, 1 mark for using $\tan^{-1}$ and the periodicity of tan to get two correct values of x, 1 mark for all four values of x correct]

Q32 a) $\sec^2 t = 1 + \tan^2 t$ so $\left(\dfrac{y}{2}\right)^2 = 1 + \left(\dfrac{x}{2}\right)^2$.
Multiplying by 4, we get $y^2 = 4 + x^2 \Rightarrow y = \pm\sqrt{4 + x^2}$.

b) $\text{cosec}^2 t = 1 + \cot^2 t$ so
$(x - 3)^2 = 1 + y^2 \Rightarrow y = \pm\sqrt{x^2 - 6x + 8}$

Q33 a) Using the sin addition formula:
$\sin 2x \cos 9x + \cos 2x \sin 9x = \sin(2x + 9x) = \sin 11x$

b) Using the cos addition formula:
$3\cos 5x \cos 7x - 3\sin 5x \sin 7x = 3\cos(5x + 7x)$
$= 3\cos 12x$

c) Using the tan addition formula:
$\dfrac{\tan 12x - \tan 8x}{1 + \tan 12x \tan 8x} = \tan(12x - 8x) = \tan 4x$

d) Using the sin addition formula:
$12\sin \dfrac{7x}{2}\cos\dfrac{3x}{2} - 12\cos\dfrac{7x}{2}\sin\dfrac{3x}{2}$
$= 12\sin\left(\dfrac{7x}{2} - \dfrac{3x}{2}\right) = 12\sin 2x$

Answers

Q34 $\cos \dfrac{7\pi}{12} = \cos\left(\dfrac{\pi}{3} + \dfrac{\pi}{4}\right) = \cos\dfrac{\pi}{3}\cos\dfrac{\pi}{4} - \sin\dfrac{\pi}{3}\sin\dfrac{\pi}{4}$

$= \dfrac{1}{2} \times \dfrac{1}{\sqrt{2}} - \dfrac{\sqrt{3}}{2} \times \dfrac{1}{\sqrt{2}} = \dfrac{1}{2\sqrt{2}} - \dfrac{\sqrt{3}}{2\sqrt{2}} = \dfrac{1 - \sqrt{3}}{2\sqrt{2}}$

$= \dfrac{1 - \sqrt{3}}{2\sqrt{2}} \times \dfrac{\sqrt{2}}{\sqrt{2}} = \dfrac{\sqrt{2} - \sqrt{2}\sqrt{3}}{4} = \dfrac{\sqrt{2} - \sqrt{6}}{4}$

Q35 $\sin A = \dfrac{4}{5}$, so using the triangle on the right, $\cos A = \dfrac{3}{5}$.

$\sin B = \dfrac{7}{25}$, so from the triangle,

$\cos B = \dfrac{24}{25}$.

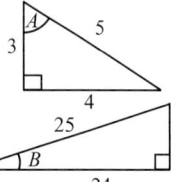

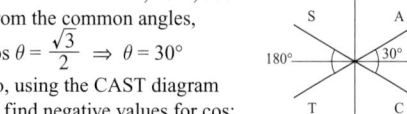

$\sin(A + B) = \sin A \cos B + \cos A \sin B$

$= \dfrac{4}{5} \times \dfrac{24}{25} + \dfrac{3}{5} \times \dfrac{7}{25} = \dfrac{96}{125} + \dfrac{21}{125} = \dfrac{117}{125} = 0.936$

Q36 $\sin 2\theta = 2\sin\theta\cos\theta$, so rewrite the equation as:

$2\sin\theta\cos\theta = -\sqrt{3}\sin\theta \Rightarrow \sin\theta(2\cos\theta + \sqrt{3}) = 0$

$\Rightarrow \sin\theta = 0 \text{ or } \cos\theta = -\dfrac{\sqrt{3}}{2}$

$\sin\theta = 0 \Rightarrow \theta = 0°, 180°, 360°$

From the common angles,

$\cos\theta = \dfrac{\sqrt{3}}{2} \Rightarrow \theta = 30°$

So, using the CAST diagram to find negative values for cos:

$\cos\theta = -\dfrac{\sqrt{3}}{2} \Rightarrow \theta = 150°, 210°$

So $\theta = 0°, 150°, 180°, 210°, 360°$.

Q37 **a)** Using the double angle formula with $A = \dfrac{x}{2}$:

$4\sin x = \sin\dfrac{x}{2} \Rightarrow 4\left(2\sin\dfrac{x}{2}\cos\dfrac{x}{2}\right) = \sin\dfrac{x}{2}$

$\Rightarrow \sin\dfrac{x}{2}\left(8\cos\dfrac{x}{2} - 1\right) = 0$

$\Rightarrow \sin\dfrac{x}{2} = 0 \text{ or } \cos\dfrac{x}{2} = \dfrac{1}{8}$

$\sin\dfrac{x}{2} = 0 \Rightarrow \dfrac{x}{2} = 0, \pi \Rightarrow x = 0, 2\pi$

$\cos\dfrac{x}{2} = \dfrac{1}{8} \Rightarrow \dfrac{x}{2} = 1.445... \Rightarrow x = 2.89 \text{ (3 s.f.)}$

So $x = 0, 2.89 \text{ (3 s.f.)}, 6.28 \text{ (3 s.f.)}$

b) Using the double angle formula with $A = \dfrac{x}{2}$:

$\tan\dfrac{x}{2}\tan x = 2 \Rightarrow \tan\dfrac{x}{2}\dfrac{2\tan\dfrac{x}{2}}{1 - \tan^2\dfrac{x}{2}} = 2$

$\Rightarrow 2\tan^2\dfrac{x}{2} = 2\left(1 - \tan^2\dfrac{x}{2}\right)$

$\Rightarrow 2\tan^2\dfrac{x}{2} = 1 \Rightarrow \tan\dfrac{x}{2} = \pm\dfrac{1}{\sqrt{2}}$

$\tan\dfrac{x}{2} = \dfrac{1}{\sqrt{2}} \Rightarrow \dfrac{x}{2} = 0.6154... \Rightarrow x = 1.23 \text{ (3 s.f.)}$

$\tan\dfrac{x}{2} = -\dfrac{1}{\sqrt{2}} \Rightarrow \dfrac{x}{2} = -0.6154...$

This is outside the interval so add π:

$\dfrac{x}{2} = -0.6154... + \pi = 2.526... \Rightarrow x = 5.05 \text{ (3 s.f.)}$

So $x = 1.23, 5.05 \text{ (to 3 s.f.)}$

Q38 **a)** $2\tan 2x = \tan x \Rightarrow \dfrac{4\tan x}{1 - \tan^2 x} = \tan x$

$\Rightarrow 4\tan x = \tan x(1 - \tan^2 x)$

$\Rightarrow 4\tan x = \tan x - \tan^3 x$

$\Rightarrow \tan x(3 + \tan^2 x) = 0$

$\Rightarrow \tan x = 0 \text{ or } \tan^2 x = -3 \text{ (not valid)}$

$\tan x = 0 \Rightarrow x = 0, \pi, 2\pi$

b) $\sin 6x - \cos 3x = 0$

$\Rightarrow 2\sin 3x\cos 3x - \cos 3x = 0$

$\Rightarrow \cos 3x(2\sin 3x - 1) = 0$

$\Rightarrow \cos 3x = 0 \text{ or } \sin 3x = \dfrac{1}{2}$

$\cos 3x = 0 \Rightarrow 3x = \dfrac{\pi}{2}, \dfrac{3\pi}{2}, \dfrac{5\pi}{2}, \dfrac{7\pi}{2}, \dfrac{9\pi}{2}, \dfrac{11\pi}{2}$

$\Rightarrow x = \dfrac{\pi}{6}, \dfrac{\pi}{2}, \dfrac{5\pi}{6}, \dfrac{7\pi}{6}, \dfrac{3\pi}{2}, \dfrac{11\pi}{6}$

$\sin 3x = \dfrac{1}{2} \Rightarrow 3x = \dfrac{\pi}{6}, \dfrac{5\pi}{6}, \dfrac{13\pi}{6}, \dfrac{17\pi}{6}, \dfrac{25\pi}{6}, \dfrac{29\pi}{6}$

$\Rightarrow x = \dfrac{\pi}{18}, \dfrac{5\pi}{18}, \dfrac{13\pi}{18}, \dfrac{17\pi}{18}, \dfrac{25\pi}{18}, \dfrac{29\pi}{18}$

Q39 $2\sin 2\theta - \cos\theta = 0 \Rightarrow 4\sin\theta\cos\theta - \cos\theta = 0$

$\Rightarrow \cos\theta(4\sin\theta - 1) = 0 \Rightarrow \cos\theta = 0 \text{ or } \sin\theta = \dfrac{1}{4}$

$\cos\theta = 0 \Rightarrow \theta = \dfrac{\pi}{2}, \dfrac{3\pi}{2}$

$\sin\theta = \dfrac{1}{4} \Rightarrow \theta = 0.252..., \pi - 0.252...$

So $\theta = 0.253, 1.57, 2.89 \text{ and } 4.71 \text{ (3 s.f.)}$

[5 marks available — 1 mark for using the identity $\sin 2\theta = 2\sin\theta\cos\theta$, 1 mark for factorising the equation, 1 mark for finding one correct solution of $\cos\theta = 0$, 1 mark for finding one correct solution of $\sin\theta = \dfrac{1}{4}$, 1 mark for all four solutions correctly given to 3 s.f.]

Q40 **a)** $\cot\theta + \tan\theta \equiv \dfrac{\cos\theta}{\sin\theta} + \dfrac{\sin\theta}{\cos\theta} \equiv \dfrac{\cos\theta\cos\theta}{\sin\theta\cos\theta} + \dfrac{\sin\theta\sin\theta}{\sin\theta\cos\theta}$

$\equiv \dfrac{\cos^2\theta + \sin^2\theta}{\sin\theta\cos\theta} \equiv \dfrac{1}{\sin\theta\cos\theta} \equiv \dfrac{1}{\frac{1}{2}\sin 2\theta} \equiv 2\csc 2\theta$

[3 marks available — 1 mark for expressing cot and tan in terms of sin and cos, 1 mark for combining into a single fraction, 1 mark for using the sine double angle formula to show the required result]

b) By replacing θ with 2θ in the result from a):

$\cot 2\theta + \tan 2\theta \equiv 2\csc 4\theta$.

So $2\csc 4\theta = \dfrac{7}{2} \Rightarrow \csc 4\theta = \dfrac{7}{4} \Rightarrow \sin 4\theta = \dfrac{4}{7}$

$\Rightarrow 4\theta = \sin^{-1}\left(\dfrac{4}{7}\right) = 0.6082...$

Look for solutions in the interval $0 \le 4\theta \le 4\pi$. There are three more solutions in this interval at $\pi - 0.6082..., 2\pi + 0.6082... \text{ and } 3\pi - 0.6082...$, so $4\theta = 0.6082..., 2.533..., 6.891... 8.816...$

$\Rightarrow \theta = 0.152, 0.633, 1.72 \text{ and } 2.20 \text{ (all to 3 s.f.)}$

[5 marks available — 1 mark for adjusting the identity from part a), 1 mark for an equation in terms of sin, 1 mark for finding the first solution for 4θ, 1 mark for a method to convert the solutions for 4θ into solutions for θ, 1 mark for all four correct solutions]

Q41 $\tan x(\csc 2x + \cot 2x) \equiv \tan x\left(\dfrac{1}{\sin 2x} + \dfrac{1}{\tan 2x}\right)$

$\equiv \tan x\left(\dfrac{1}{\sin 2x} + \dfrac{\cos 2x}{\sin 2x}\right) \equiv \tan x\left(\dfrac{1 + \cos 2x}{\sin 2x}\right)$

$\equiv \tan x\left(\dfrac{1 + (2\cos^2 x - 1)}{2\sin x\cos x}\right) \equiv \tan x\left(\dfrac{2\cos^2 x}{2\sin x\cos x}\right)$

$\equiv \tan x\left(\dfrac{\cos x}{\sin x}\right) \equiv \tan x\left(\dfrac{1}{\tan x}\right) \equiv 1$

[5 marks available — 1 mark for using the identity $\csc 2x = \dfrac{1}{\sin 2x}$, 1 mark for using $\cot 2x = \dfrac{\cos 2x}{\sin 2x}$, 1 mark for using $\cos 2x = 2\cos^2 x - 1$, 1 mark for $\sin 2x = 2\sin x\cos x$, 1 mark for cancelling to get $\tan x \times \dfrac{1}{\tan x}$ or $\tan x \times \cot x$]

There are often many ways to prove an identity — you'll get the marks for it as long as each step makes sense.

Answers

Q42 a) $3 \cos \theta - 8 \sin \theta \equiv R \cos (\theta + \alpha)$
$\equiv R \cos \theta \cos \alpha - R \sin \theta \sin \alpha$
$\Rightarrow R \cos \alpha = 3, \ R \sin \alpha = 8$
$\tan \alpha = \dfrac{8}{3} \Rightarrow \alpha = 69.44...° = 69.4°$ (1 d.p.)
$R = \sqrt{3^2 + 8^2} = \sqrt{73}$
$\Rightarrow 3 \cos \theta - 8 \sin \theta \equiv \sqrt{73} \ \cos (\theta + 69.4°)$
[4 marks available — 1 mark for correct method to find α, 1 mark for correct value of α, 1 mark for correct method to find R, 1 mark for correct value of R]

b) $\sqrt{73} \ \cos (\theta + 69.4°) = 1.2$, in the interval
$69.4° \leq (\theta + 69.4°) \leq 429.4°$
$\cos (\theta + 69.4°) = \dfrac{1.2}{\sqrt{73}}$
$\Rightarrow \theta + 69.4° = \cos^{-1} \left(\dfrac{1.2}{\sqrt{73}} \right) = 81.9...°$
Using the symmetry of the graph
of cos, there is another solution at
$\theta + 69.4° = (360° - 81.9...°) = 278.0...°$
$\Rightarrow \theta = (81.9...° - 69.4°)$ and $(278.0...° - 69.4°)$
$= 12.5°$ and $208.6°$ (1 d.p.)
[2 marks available — 1 mark for at least one correct solution for θ, 1 mark for both correct solutions for θ]

Q43 $5 \sin t + 3 \cos t \equiv R \sin (t + \alpha) \equiv R \sin t \cos \alpha + R \cos t \sin \alpha$
$\Rightarrow R \cos \alpha = 5$ and $R \sin \alpha = 3$
$\tan \alpha = \dfrac{3}{5} \Rightarrow \alpha = 30.9637...° = 30.964°$ (3 d.p.)
$R = \sqrt{5^2 + 3^2} = \sqrt{34}$
So $h = \sqrt{34} \ \sin (t + 30.964°) + 2$
The maximum height occurs when $\sin (t + 30.964°) = 1$,
so $h_{max} = \sqrt{34} \times 1 + 2 = 5.83... + 2 = 7.83$ m (2 d.p.)

Q44 a) $7 \cos x - 11 \sin x \equiv R \cos (x + \alpha)$
$\equiv R \cos x \cos \alpha - R \sin x \sin \alpha$
$\Rightarrow R \cos \alpha = 7, \ R \sin \alpha = 11$
$\tan \alpha = \dfrac{11}{7} \Rightarrow \alpha = 57.5°$ (3 s.f.)
$R = \sqrt{7^2 + 11^2} = \sqrt{170}$
$\Rightarrow 7 \cos x - 11 \sin x \equiv \sqrt{170} \ \cos (x + 57.5°)$.
[3 marks available — 1 mark for finding $\tan \alpha$, 1 mark for finding α, 1 mark for finding R]

b) $y = \cos x$ has been translated 57.5° to the left and
vertically stretched by a scale factor of $\sqrt{170}$.
[2 marks available — 1 mark for correct translation, 1 mark for correct vertical stretch]

c) The maximum value is $\sqrt{170}$.
$y = 7 \cos x - 11 \sin x$ takes a maximum value when
$\cos (x + 57.5°) = 1$, i.e. when $x + 57.5° = 0 \Rightarrow x = -57.5°$ (3 s.f.)
You could have given x as $-57.5° \pm 360°, \pm 720°$, etc.
[2 marks available — 1 mark for correct maximum value, 1 mark for correct x value]

d) $\sqrt{170} \ \cos (x + 57.5°) = 3$
$\Rightarrow (x + 57.5°) = \cos^{-1} \left(\dfrac{3}{\sqrt{170}} \right) = 76.69...°$
Look for solutions in the interval $57.5° \leq x + 57.5° \leq 417.5°$
There is another solution at $360° - 76.69...°$,
so $x + 57.5° = 76.69...°, \ 283.30...°$.
$\Rightarrow x = 19.2°$ and $225.8°$ (1 d.p.)
[4 marks available — 1 mark for using the form $R \cos (x + \alpha)$, 1 mark for using the inverse cos function, 1 mark for a method seen to convert the solutions of $(x + 57.5°)$ to the solutions of x, 1 mark for both correct values of x rounded to 1 d.p.]

Q45 a) The minimum value of h(t) occurs when
$\sin \left(\dfrac{1}{2}(t - \pi) \right) = -1$, so $9 = a + b(-1) = a - b$.
The maximum value occurs when
$\sin \left(\dfrac{1}{2}(t - \pi) \right) = 1$, so $17 = a + b$.
Solving simultaneously by adding the equations together:
$17 + 9 = 2a \Rightarrow 26 = 2a \Rightarrow a = 13$.
Then $9 = 13 - b \Rightarrow b = 4$.
[3 marks available — 1 mark for using the maximum and minimum properties of sin to generate two equations, 1 mark for solving simultaneously, 1 mark for the correct values of both a and b]

b) Low tide occurs when $\sin \left(\dfrac{1}{2}(t - \pi) \right) = -1$
$\Rightarrow \left(\dfrac{1}{2}(t - \pi) \right) = -\dfrac{\pi}{2} \Rightarrow t = 0$ *[1 mark]*
And when h(t) = 11:
$13 + 4 \sin \left(\dfrac{1}{2}(t - \pi) \right) = 11 \Rightarrow \sin \left(\dfrac{1}{2}(t - \pi) \right) = -\dfrac{1}{2}$
[1 mark]
$\Rightarrow \dfrac{1}{2}(t - \pi) = \sin^{-1} \left(\dfrac{1}{2} \right) = -\dfrac{\pi}{6} \Rightarrow t - \pi = -\dfrac{\pi}{3}$ *[1 mark]*
$\Rightarrow t = \dfrac{2\pi}{3} = 2.09$ (3 s.f.)
The tide reaches 11 feet at
$2.09 - 0 = 2.09$ hours after low tide. *[1 mark]*
[4 marks available — as above]

Q46 $\dfrac{1}{2} \log \left| \dfrac{1 + \sin x}{1 - \sin x} \right| = \dfrac{1}{2} \log \left| \dfrac{\left(\dfrac{1}{\cos x} + \dfrac{\sin x}{\cos x} \right)}{\left(\dfrac{1}{\cos x} - \dfrac{\sin x}{\cos x} \right)} \right|$
$= \dfrac{1}{2} \log \left| \dfrac{\sec x + \tan x}{\sec x - \tan x} \right| = \log \sqrt{\left| \dfrac{\sec x + \tan x}{\sec x - \tan x} \right|}$
$= \log \dfrac{\sqrt{(\sec x + \tan x)(\sec x - \tan x)}}{|\sec x - \tan x|} = \log \dfrac{\sqrt{\sec^2 x - \tan^2 x}}{|\sec x - \tan x|}$
$= \log \left| \dfrac{1}{\sec x - \tan x} \right| = \log \left| \dfrac{\sec x + \tan x}{(\sec x - \tan x)(\sec x + \tan x)} \right|$
$= \log \left| \dfrac{\sec x + \tan x}{\sec^2 x - \tan^2 x} \right| = \log |\sec x + \tan x|$
[6 marks available — 1 mark for attempt to write fraction in terms of sec and tan, 1 mark for correct fraction in terms of sec and tan, 1 mark for using power law of logs correctly, 1 mark for rationalising the denominator, 1 mark for using $\sec^2 x - \tan^2 x = 1$, 1 mark for a fully correct proof]
An alternative method is to multiply the original fraction by $\dfrac{1 + \sin x}{1 + \sin x}$ and then simplify.

Q47 a) $\text{cosec}^2 x - \sec^2 y = -1 \Rightarrow \dfrac{1}{\sin^2 x} - (\sec^2 y - 1) = 0$
$\Rightarrow \dfrac{1}{\sin^2 x} - \tan^2 y = 0 \Rightarrow \dfrac{1}{p^2} - (2p - 3)^2 = 0$
$\Rightarrow 1 - p^2 (2p - 3)^2 = 0 \Rightarrow 1 - p^2 (4p^2 - 12p + 9) = 0$
$\Rightarrow 1 - 4p^4 + 12p^3 - 9p^2 = 0 \Rightarrow 4p^4 - 12p^3 + 9p^2 - 1 = 0$.

b) $(2p^2 - 3p - 1)(Ap^2 + Bp + C) \equiv 4p^4 - 12p^3 + 9p^2 - 1$
Equating coefficients we can see that:
$-1C = -1 \Rightarrow C = 1$
$2Ap^4 = 4p^4 \Rightarrow A = 2$
$-Ap^2 - 3Bp^2 + 2Cp^2 = 9p^2 \Rightarrow -2p^2 - 3Bp^2 + 2p^2 = 9p^2$
$\Rightarrow -3Bp^2 = 9p^2 \Rightarrow B = -3$
Now multiply the brackets to check the other terms:
$(2p^2 - 3p - 1)(2p^2 - 3p + 1) = 4p^4 - 12p^3 + 9p^2 - 1$
Therefore, $2p^2 - 3p - 1$ is a factor of $4p^4 - 12p^3 + 9p^2 - 1$.
You could have used a different method here, e.g. algebraic long division.

c) $(2p^2 - 3p - 1)(2p^2 - 3p + 1) = 0$

$\Rightarrow 2p^2 - 3p - 1 = 0$ or $2p^2 - 3p + 1 = 0$

$2p^2 - 3p - 1 = 0 \Rightarrow p = \dfrac{3 \pm \sqrt{3^2 - 4 \times 2 \times -1}}{2 \times 2} = \dfrac{3 \pm \sqrt{17}}{4}$

$2p^2 - 3p + 1 = 0 \Rightarrow (2p - 1)(p - 1) = 0 \Rightarrow p = 1, \dfrac{1}{2}$

So the solutions are $p = \dfrac{3 - \sqrt{17}}{4}, \dfrac{3 + \sqrt{17}}{4}, 1, \dfrac{1}{2}$.

d) $\sin x = \dfrac{1}{2} \Rightarrow x = 150° \Rightarrow \cos x = -\dfrac{\sqrt{3}}{2} \Rightarrow \sec x = -\dfrac{2\sqrt{3}}{3}$

$\tan y = 2 \times \dfrac{1}{2} - 3 = -2$ so making the following triangle:

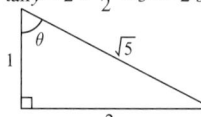

$\tan \theta = 2 \Rightarrow \cos \theta = \dfrac{1}{\sqrt{5}} \Rightarrow \sec \theta = \sqrt{5}$

Since $\tan y < 0$, we know $90° < y < 180°$ so $\sec y < 0$, so we have $\sec y = -\sqrt{5}$.

Therefore, $\sec x + \sec y = -\dfrac{2\sqrt{3}}{3} - \sqrt{5}$.

Q48 a) $\cos(3x) = \cos(2x + x) = \cos 2x \cos x - \sin 2x \sin x$

$= (2\cos^2 x - 1)\cos x - (2\sin x \cos x)\sin x$

$= 2\cos^3 x - \cos x - 2\sin^2 x \cos x$

$= 2\cos^3 x - \cos x - 2(1 - \cos^2 x)\cos x = 4\cos^3 x - 3\cos x$

$\Rightarrow \sec 3x = \dfrac{1}{4\cos^3 x - 3\cos x}$.

[4 marks available — 1 mark for using addition formulae to expand cos(2x + x), 1 mark for correct use of double angle formula to expand cos 2x, 1 mark for correct use of double angle formula to expand sin 2x, 1 mark for a fully correct proof]

b) $\sec 3x = \cos x \Rightarrow \dfrac{1}{4\cos^3 x - 3\cos x} = \cos x$

$\Rightarrow 1 = 4\cos^4 x - 3\cos^2 x \Rightarrow 4\cos^4 x - 3\cos^2 x - 1 = 0$

$\Rightarrow (4\cos^2 x + 1)(\cos^2 x - 1) = 0 \Rightarrow \cos^2 x = -\dfrac{1}{4}$ or $\cos^2 x = 1$

$\cos^2 x = -\dfrac{1}{4}$ has no solutions since $\cos^2 x \geq 0$.

Therefore, $\cos x = \pm 1 \Rightarrow x = 0, \pi, 2\pi$.

[5 marks available — 1 mark for use of part a), 1 mark for correct quartic equation, 1 mark for a correct method to solve, 1 mark for correct values of cos x, 1 mark for the correct solutions]

c) (i) Substituting $y = \dfrac{1}{2}$ into the equation gives

$4\left(\dfrac{1}{2}\right)^3 - 3\left(\dfrac{1}{2}\right) + 1 = \dfrac{4}{8} - \dfrac{12}{8} + 1 = 0$, so $y = \dfrac{1}{2}$ is a root.

[1 mark for a correct explanation]

(ii) $\sec 3x = -1 \Rightarrow \dfrac{1}{4\cos^3 x - 3\cos x} = -1$

$\Rightarrow 4\cos^3 x - 3\cos x = -1 \Rightarrow 4\cos^3 x - 3\cos x + 1 = 0$.

By part (i), $(2\cos x - 1)$ is a factor, so it factorises as

$(2\cos x - 1)(2\cos^2 x + \cos x - 1) = 0$

$\Rightarrow (2\cos x - 1)^2(\cos x + 1) = 0$

$\Rightarrow \cos x = \dfrac{1}{2}$ or $\cos x = -1$.

[4 marks available — 1 mark for correct cubic equation, 1 mark for correct factor from part (i), 1 mark for complete factorisation, 1 mark for the correct answers]

Chapter 4: Coordinate Geometry in the (x, y) Plane

Prior Knowledge Check

Q1 a) $x(x - 1)^2 = 0 \Rightarrow x = 0$ or $x = 1$

b) $2x^2 - 5x = 3 \Rightarrow (2x + 1)(x - 3) = 0 \Rightarrow x = -\dfrac{1}{2}$ or $x = 3$

c) $f(-1) = 0$, so $(x + 1)$ is a factor.

$2x^3 - 7x^2 - 5x + 4 = (x + 1)(2x^2 + nx + 4)$

So $nx^2 + 2x^2 = -7x^2 \Rightarrow n = -9$

Therefore $2x^3 - 7x^2 - 5x + 4 = 0$

$\Rightarrow (x + 1)(2x^2 - 9x + 4) = 0$

$\Rightarrow (x + 1)(2x - 1)(x - 4) = 0$

$\Rightarrow x = -1, x = \dfrac{1}{2}$ or $x = 4$

Q2 a) $\theta = -\dfrac{\pi}{3}, \dfrac{\pi}{3}$

b) $\theta = -\dfrac{3\pi}{4}, \dfrac{\pi}{4}$

c) $\theta = \sin^{-1}(0.4) \approx 0.41, 2.73$ (2 d.p.)

Q3 a) $\cot x = \dfrac{1}{\tan x} = \dfrac{\cos x}{\sin x}$

So $\dfrac{\cot x}{\cos x} = \dfrac{1}{\sin x} = \operatorname{cosec} x$

b) Using the cos double angle formula:

$\sin^2\left(\dfrac{3x}{4}\right) - \cos^2\left(\dfrac{3x}{4}\right) \equiv -\cos\left(\dfrac{3x}{2}\right)$

Exercise 4.1 — Parametric Equations — Finding Coordinates

Q1 a) $x = 3t = 3 \times 5 = 15$

$y = t^2 = 5^2 = 25$, so coordinates are (15, 25).

b) $18 = 3t \Rightarrow t = 6$

c) $36 = t^2 \Rightarrow t = \pm 6 \Rightarrow x = \pm 18$

Q2 a) $x = 2t - 1 = (2 \times 7) - 1 = 13$

$y = 4 - t^2 = 4 - 7^2 = -45$, so coordinates are (13, –45).

b) $2t - 1 = 15 \Rightarrow t = 8$

c) $4 - t^2 = -5 \Rightarrow t = \pm 3 \Rightarrow x = -7$ or 5

Q3 a) $x = 2 + \sin \theta = 2 + \sin \dfrac{\pi}{4} = \dfrac{4 + \sqrt{2}}{2}$

$y = -3 + \cos \theta = -3 + \cos \dfrac{\pi}{4} = \dfrac{\sqrt{2} - 6}{2}$

So coordinates are $\left(\dfrac{4 + \sqrt{2}}{2}, \dfrac{\sqrt{2} - 6}{2}\right)$.

b) $2 + \sin \theta = \dfrac{4 + \sqrt{3}}{2} = 2 + \dfrac{\sqrt{3}}{2} \Rightarrow \sin \theta = \dfrac{\sqrt{3}}{2} \Rightarrow \theta = \dfrac{\pi}{3}$

c) $-3 + \cos \theta = -\dfrac{7}{2} \Rightarrow \cos \theta = -\dfrac{1}{2} \Rightarrow \theta = \dfrac{2\pi}{3}$

The angle in a) was given in radians so make sure you use radians for b) and c). They're both cos values you should know an angle for off by heart. The cos value in part c) is negative so you can use the CAST diagram (have a look at your Year 1 notes) to figure out which angle you need (cos is negative in the 2nd and 3rd quadrants).

Q4 a) $33 = 6 + 3t^2 \Rightarrow t^2 = 9 \Rightarrow t = \pm 3$

b) $6 = t^2 + 5t \Rightarrow t^2 + 5t - 6 = 0 \Rightarrow (t + 6)(t - 1) = 0$

$\Rightarrow t = -6$ or $t = 1$

$\Rightarrow y = 6 + 3(-6)^2 = 114$ or $y = 6 + 3(1)^2 = 9$

So $y = 9$ or 114

Q5

t	–5	–4	–3	–2	–1	1	2	3	4	5
x	–25	–20	–15	–10	–5	5	10	15	20	25
y	$-\dfrac{2}{5}$	$-\dfrac{1}{2}$	$-\dfrac{2}{3}$	–1	–2	2	1	$\dfrac{2}{3}$	$\dfrac{1}{2}$	$\dfrac{2}{5}$

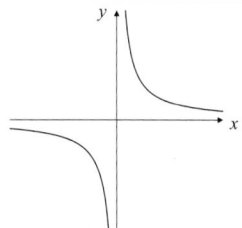

When t = 0, y is undefined, so there must be an asymptote.

Q6

θ	0	$\frac{\pi}{4}$	$\frac{\pi}{3}$	$\frac{\pi}{2}$	$\frac{2\pi}{3}$	$\frac{3\pi}{4}$	π	$\frac{4\pi}{3}$	$\frac{3\pi}{2}$	$\frac{5\pi}{3}$	2π
x	1	1.71	1.87	2	1.87	1.71	1	0.13	0	0.13	1
y	3	2.71	2.5	2	1.5	1.29	1	1.5	2	2.5	3

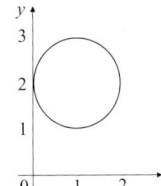

Q7 **a)** When $y = 1$, $x = a$, so:
$1 = 10 - t^2 \Rightarrow t^2 = 9 \Rightarrow t = \pm 3$
When $t = 3$, $x = 2(3)^2 - 7(3) = -3$.
When $t = -3$, $x = 2(-3)^2 - 7(-3) = 39$.
$a > 1$, so $a = 39$.

b) At $(-6, 4)$, $x = -6 \Rightarrow -6 = 2t^2 - 7t$
$\Rightarrow 2t^2 - 7t + 6 = 0 \Rightarrow (2t - 3)(t - 2) = 0$
$\Rightarrow t = \frac{3}{2}, t = 2$
$t = \frac{3}{2} \Rightarrow y = 10 - \left(\frac{3}{2}\right)^2 = \frac{31}{4} \neq 4$
$t = 2 \Rightarrow y = 10 - 2^2 = 6 \neq 4$
So neither of the t-values when $x = -6$ corresponds to $y = 4$,
so $(-6, 4)$ is not on the curve.
You could have started with y = 4 and used a similar method to show that x ≠ −6 at this point.

Q8 **a)** $x \neq 0$ since $\frac{1}{t} \neq 0$, so the graph does not intersect (cross/touch) the y-axis, but approaches it.
[1 mark for any reference to approaching/not crossing etc.]

b) When $x = \frac{1}{2}$, $\frac{1}{t} = \frac{1}{2} \Rightarrow t = 2$.
So $2(2^2 - a) = 0 \Rightarrow a = 4$
[3 marks available — 1 mark for finding t, 1 mark for method to find a, 1 mark for the correct value of a]

c) (i) For very small values of t, $\frac{1}{t}$ becomes very large, so x becomes very large. y has a factor of t so becomes very small (and negative) for very small values of t. *[1 mark]*

(ii) For very large values of t, $\frac{1}{t}$ becomes very small, so x becomes very small. y will increase as t increases so becomes very large when t is very large. *[1 mark]*

d)

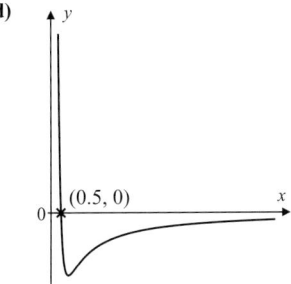

[3 marks available — 1 mark for x-axis intercept, 1 mark for showing a minimum point below the x-axis, 1 mark for general shape/position]

Q9 **a)** When $\theta = 0$: $x = 3 \sin 0 = 3(0) = 0$
$y = 7 + 9 \cos 0 = 7 + 9(1) = 16$
The comet is at $(0, 16)$, so the distance to the Sun at $(0, 0)$ is 16 AU.

b) When $\theta = \frac{\pi}{2}$: $x = 3 \sin \frac{\pi}{2} = 3(1) = 3$
$y = 7 + 9 \cos \frac{\pi}{2} = 7 + 9(0) = 7$
The comet is at $(3, 7)$, so the distance to the Sun at $(0, 0)$ is $\sqrt{3^2 + 7^2} = \sqrt{58} = 7.62$ AU (to 3 s.f.).

Q10 **a)** When $t = 2$, $x = 2^2 + 4(2) = 4 + 8 = 12$
The plane travels 12 m horizontally in the first 2 s.

b) Sonia is standing at the point $x = 21$.
$x = 21 \Rightarrow t^2 + 4t = 21 \Rightarrow t^2 + 4t - 21 = 0$
$\Rightarrow (t + 7)(t - 3) = 0 \Rightarrow t = -7$ or $t = 3$
The parametric equations are only valid for $0 \leq t \leq 5$,
so the plane is at $x = 21$ when $t = 3$.
When $t = 3$, $y = 25 - 3^2 = 25 - 9 = 16$
The plane passes over Sonia's head at a height of 16 m.

Q11 **a)** At $\theta = 0$, $x = 5\cos(\pi - 0) = -5$ and $y = 3 - 3\sin(\pi - 0) = 3$.
At $\theta = \pi$, $x = 5\cos(\pi - \pi) = 5$ and $y = 3 - 3\sin(\pi - \pi) = 3$.
So the coordinates are $(-5, 3)$ and $(5, 3)$.
[3 marks available — 1 mark for a correct method, 1 mark for (−5, 3), 1 mark for (5, 3)]

b) At 4 m horizontally from origin, $x = \pm 4$, so,
$\pm 4 = 5\cos(\pi - \theta) \Rightarrow \cos(\pi - \theta) = \pm 0.8$
$\cos^{-1}(0.8) = 0.6435...$ radians
$\cos^{-1}(-0.8) = 2.4980...$ radians
$\pi - \theta = 0.6435... \Rightarrow \theta = 2.4980... = 2.50$ radians (3 s.f.)
$\pi - \theta = 2.4980... \Rightarrow \theta = 0.6435... = 0.644$ radians (3 s.f.)
[3 marks available — 1 mark for correct value for cos(π − θ), 1 mark for 0.644, 1 mark for 2.50]
These last two lines may seem unnecessary and due to the nature of the graph of y = cos x between 0 and π they are. But technically, in terms of solving the equations as they stand, they are part of the solution.

c) They are at $(-3, 2)$ and the centre of the ellipse is $(0, 3)$.
So they are 3 metres left and 1 metre down from the centre.
$\tan p = \frac{1}{3}$ so $p = \tan^{-1}\left(\frac{1}{3}\right) = 0.3217...$ radians
The 'normal' position on the half-pipe for this angle would be $(5\cos(\pi - 0.3217...), 3 - 3\sin(\pi - 0.3217...))$
$= (-4.7434..., 2.0513...)$
So distance between actual and normal position will be
$d = \sqrt{(-4.74 - -3)^2 + (2.05 - 2)^2} = 1.7$ m (to 1 d.p.)
[6 marks available — 1 mark for method to find p, 1 mark for finding p, 1 mark for method for 'normal' position, 1 mark for coordinates of 'normal' position, 1 mark for method for distance, 1 mark for the correct final answer]

Exercise 4.2 — Parametric Equations — Finding Intersections

Q1 At A: $y = 0 \Rightarrow -2 + t = 0 \Rightarrow t = 2$
$x = 3 + t = 3 + 2 = 5$
At B: $x = 0 \Rightarrow 3 + t = 0 \Rightarrow t = -3$
$y = -2 + (-3) = -5$
So the coordinates are A$(5, 0)$ and B$(0, -5)$.

Q2 **a)** The curve meets the x-axis when $y = 0$, so:
$3t^3 - 24 = 0 \Rightarrow 3(t^3 - 8) = 0 \Rightarrow t^3 = 8 \Rightarrow t = 2$

b) The curve meets the y-axis when $x = 0$, so:
$2t^2 - 50 = 0 \Rightarrow 2(t^2 - 25) = 0 \Rightarrow t^2 = 25 \Rightarrow t = \pm 5$
If you'd have been asked to give the coordinates you would then just put the t values into the parametric equations.

Q3 The curve meets the y-axis when $x = 0$, so:
$64 - t^3 = 0 \Rightarrow t^3 = 64 \Rightarrow t = 4$
$y = \frac{1}{t} = \frac{1}{4}$, so P is $\left(0, \frac{1}{4}\right)$.

Answers

Q4 $y = x - 3 \Rightarrow 4t = (2t + 1) - 3 \Rightarrow 2t = -2 \Rightarrow t = -1$
To find the coordinates when t = −1, just put this value back into the parametric equations...
$x = (2 \times -1) + 1 = -1$, $y = 4 \times -1 = -4$, so P is $(-1, -4)$.

Q5 $y = x^2 + 32 \Rightarrow 6t^2 = (2t)^2 + 32 \Rightarrow 6t^2 = 4t^2 + 32$
$\Rightarrow 2t^2 = 32 \Rightarrow t^2 = 16 \Rightarrow t = \pm 4$
When $t = 4$: $x = 2 \times 4 = 8$, $y = 6 \times 4^2 = 96$
So one point of intersection is $(8, 96)$.
When $t = -4$: $x = 2 \times -4 = -8$, $y = 6 \times (-4)^2 = 96$
So the other point of intersection is $(-8, 96)$.

Q6 $x^2 + y^2 = 32 \Rightarrow (t^2)^2 + (2t)^2 = 32 \Rightarrow t^4 + 4t^2 - 32 = 0$
$\Rightarrow (t^2 - 4)(t^2 + 8) = 0 \Rightarrow t^2 = 4 \Rightarrow t = \pm 2$
(there are no real solutions to $t^2 = -8$).
When $t = 2$: $x = 2^2 = 4$, $y = 2 \times 2 = 4$
So one point of intersection is $(4, 4)$.
When $t = -2$: $x = (-2)^2 = 4$, $y = 2 \times -2 = -4$
So the other point of intersection is $(4, -4)$.

Q7
a) At the point $(0, 4)$:
$x = 0 \Rightarrow a(t - 2) = 0 \Rightarrow t = 2$ (as $a \neq 0$)
$y = 4 \Rightarrow 2at^2 + 3 = 4 \Rightarrow 2a(2^2) + 3 = 4$
$\Rightarrow 8a + 3 = 4 \Rightarrow a = \frac{1}{8}$.

b) The curve would meet the x-axis when $y = 0$.
$y = 2at^2 + 3 = \frac{2t^2}{8} + 3 = \frac{t^2}{4} + 3$
So at the x-axis, $\frac{t^2}{4} + 3 = 0 \Rightarrow t^2 = -12$.
This has no real solutions,
so the curve does not meet the x-axis.

Q8
a) The curve crosses the x-axis when $y = 0$.
$t^2 - 9 = 0 \Rightarrow t = \pm 3$
When $t = 3$, $x = \frac{2}{3}$, so one point is $\left(\frac{2}{3}, 0\right)$.
When $t = -3$, $x = -\frac{2}{3}$, so the other point is $\left(-\frac{2}{3}, 0\right)$.

b) The curve would meet the y-axis when $x = 0$,
i.e. when $\frac{2}{t} = 0$. This has no solutions,
so the curve does not meet the y-axis.

c) $y = \frac{10}{x} - 3 \Rightarrow t^2 - 9 = \frac{10t}{2} - 3 \Rightarrow t^2 - 5t - 6 = 0$
$\Rightarrow (t + 1)(t - 6) = 0 \Rightarrow t = -1$ and $t = 6$
When $t = -1$: $x = -2$, $y = (-1)^2 - 9 = -8$
So one point of intersection is $(-2, -8)$.
When $t = 6$: $x = \frac{2}{6} = \frac{1}{3}$, $y = 6^2 - 9 = 27$
So the other point of intersection is $\left(\frac{1}{3}, 27\right)$.

Q9
a) The curve meets the x-axis when $y = 0$.
$5 \cos t = 0 \Rightarrow \cos t = 0 \Rightarrow t = \frac{\pi}{2}$ and $\frac{3\pi}{2}$.
When $t = \frac{\pi}{2}$, $x = 3 \sin\left(\frac{\pi}{2}\right) = 3$, so $(3, 0)$.
When $t = \frac{3\pi}{2}$, $x = 3 \sin\left(\frac{3\pi}{2}\right) = -3$, so $(-3, 0)$.
So the curve crosses the x-axis twice in the
domain $0 \leq t \leq 2\pi$, at $(-3, 0)$ and $(3, 0)$.
The curve meets the y-axis when $x = 0$.
$3 \sin t = 0 \Rightarrow \sin t = 0 \Rightarrow t = 0$, π and 2π.
When $t = 0$, $y = 5 \cos 0 = 5$, so $(0, 5)$.
When $t = \pi$, $y = 5 \cos \pi = -5$, so $(0, -5)$.
When $t = 2\pi$, $y = 5 \cos 2\pi = 5$, so $(0, 5)$.
So the curve crosses the y-axis twice in the
domain $0 \leq t \leq 2\pi$, at $(0, -5)$ and $(0, 5)$.

b) $y = \left(\frac{5\sqrt{3}}{9}\right)x \Rightarrow 5 \cos t = \left(\frac{5\sqrt{3}}{3}\right)\sin t$
$\Rightarrow \frac{15}{5\sqrt{3}} = \frac{\sin t}{\cos t} = \tan t \Rightarrow \tan t = \frac{3}{\sqrt{3}} = \sqrt{3}$
$\Rightarrow t = \frac{\pi}{3}$ and $\left(\pi + \frac{\pi}{3}\right) \Rightarrow t = \frac{\pi}{3}$ and $\frac{4\pi}{3}$
When $t = \frac{\pi}{3}$: $x = 3 \sin \frac{\pi}{3} = \frac{3\sqrt{3}}{2}$ and $y = 5 \cos \frac{\pi}{3} = \frac{5}{2}$
So one point of intersection is $\left(\frac{3\sqrt{3}}{2}, \frac{5}{2}\right)$.
When $t = \frac{4\pi}{3}$: $x = 3 \sin \frac{4\pi}{3} = -\frac{3\sqrt{3}}{2}$
$y = 5 \cos \frac{4\pi}{3} = -\frac{5}{2}$
So the other point of intersection is $\left(-\frac{3\sqrt{3}}{2}, -\frac{5}{2}\right)$.

Q10 The ships will collide if they have the same x- and y-coordinates at the same time. To find the value of t when they both have the same x-coordinate, set the two equations for x equal to each other:
$x_1 = x_2 \Rightarrow 24 - t = t + 10 \Rightarrow 14 = 2t \Rightarrow t = 7$
Now see if the y-coordinates are the same when $t = 7$:
$t = 7 \Rightarrow y_1 = 10 + 3(7) = 31$,
$\qquad\quad y_2 = 12 + 2(7) - 0.1(7)^2 = 21.1$
$t = 7$ minutes is the only time when $x_1 = x_2$, but $y_1 \neq y_2$ at this time, so they will not collide.
You could have done this the other way around, by setting $y_1 = y_2$ to find the values of t when the y-coordinates are equal. In this case, that's a much more difficult way to solve the question — it gives you a fairly awkward quadratic to solve, and you end up with two values of t to test instead of one.
This question shows how parametric equations can be useful in modelling — if you drew the two Cartesian graphs, you'd find they cross in two places, so you might think the ships would collide. It's only when you think about how the ships' positions depend on time that you can work out what actually happens.

Q11 The alarm should sound the first time when $y = 5$, and stop when $y = 5$ again. So solve $6t - t^2 = 5$.
$t^2 - 6t + 5 = 0 \Rightarrow (t - 5)(t - 1) = 0 \Rightarrow t = 5$ and $t = 1$
So the alarm should sound for $5 - 1 = 4$ seconds.
When $t = 1$, $x = 4 \times 1 + 5 = 9$
When $t = 5$, $x = 4 \times 5 + 5 = 25$
So the ball will travel $25 - 9 = 16$ m horizontally during this time.
[5 marks available — 1 mark for forming a quadratic using $y = 5$, 1 mark for solving quadratic, 1 mark for alarm sounding for 4 seconds, 1 mark for method to find x values at these times, 1 mark for the correct horizontal distance]

Q12
a) At $(3, 8)$ the x and y values (but not necessarily t and u values) of both C_1 and C_2 will be equal.
Substitute $x = 3$ into both x equations to find t and u then substitute those into the y equations to form a pair of simultaneous equations in A and B.
C_1: When $x = 3$, $3 = t - 1 \Rightarrow t = 4$.
So $y = t^2 - At + B \Rightarrow 8 = 16 - 4A + B \Rightarrow 4A - B = 8$
C_2: When $x = 3$, $3 = 2u + 1 \Rightarrow u = 1$.
So $y = (B - 12)u^2 - 6u + A - 1 \Rightarrow 8 = B - 12 - 6 + A - 1$
$\Rightarrow A + B = 27 \Rightarrow B = 27 - A$
Substitute $B = 27 - A$ into $4A - B = 8$
$\Rightarrow 4A - (27 - A) = 8 \Rightarrow 5A = 35 \Rightarrow A = 7$
and $B = 27 - 7 = 20$.
[5 marks available — 1 mark for t = 4, 1 mark for u = 1, 1 mark for a correct method to find A and B, 1 mark for A = 7, 1 mark for B = 20]

b) Substitute x from C_2 into the Cartesian equation from C_1:
$y = (2u + 1)^2 - 5(2u + 1) + 14 \Rightarrow y = 4u^2 - 6u + 10$
Now put this equal to the y equation for C_2 and solve for u.
$4u^2 - 6u + 10 = 8u^2 - 6u + 6 \Rightarrow 4u^2 - 4 = 0 \Rightarrow u = \pm 1$
From part a), $u = 1$ was the value of u for C_2 at (3, 8).
So, the other point of intersection must be when $u = -1$ in C_2.
$y = 4(-1)^2 - 6(-1) + 10 = 20$ and $x = 2(-1) + 1 = -1$.
So the other point of intersection is (–1, 20).
[5 marks available — 1 mark for substituting 2u + 1 into Cartesian equation, 1 mark for setting it equal to y equation for C₂, 1 mark for correct method to solve, 1 mark for u = –1, 1 mark for correct final coordinates]

c) On the x-axis, $y = 0$, so $8u^2 - 6u + 6 = 0$.
Now calculate the discriminant:
$b^2 - 4ac = (-6)^2 - (4 \times 8 \times 6) = 36 - 192 = -156$. Therefore there are no real roots so no intersection with the x-axis.
[2 marks available — 1 mark for a correct method, 1 mark for the complete proof with explanation]
You could have also shown this by completing the square and showing that y is always positive.

Exercise 4.3 — Converting Parametric Equations to Cartesian Equations

Q1 a) $x = t + 3 \Rightarrow t = x - 3$,
so $y = t^2 = (x - 3)^2 = x^2 - 6x + 9$

b) $x = 3t \Rightarrow t = \frac{x}{3}$ so $y = \frac{6}{t} = \frac{18}{x}$

c) $x = 2t^3 \Rightarrow t = \left(\frac{x}{2}\right)^{\frac{1}{3}}$, so $y = t^2 = \left(\frac{x}{2}\right)^{\frac{2}{3}}$

d) $x = t + 7 \Rightarrow t = x - 7$,
so $y = 12 - 2t = 12 - 2(x - 7) = 26 - 2x$

e) $x = t + 4 \Rightarrow t = x - 4$,
so $y = t^2 - 9 = (x - 4)^2 - 9 = x^2 - 8x + 7$

f) $x = \frac{t + 2}{3} \Rightarrow t = 3x - 2$,
so $y = t^2 - t = (3x - 2)^2 - (3x - 2)$
$= 9x^2 - 12x + 4 - 3x + 2 = 9x^2 - 15x + 6$

g) $y = 5 - 8t \Rightarrow t = \frac{5 - y}{8}$,
so $x = t^2 - \frac{t}{2} = \left(\frac{5 - y}{8}\right)^2 - \frac{5 - y}{16} = \frac{y^2 - 6y + 5}{64}$
Here you have to find t in terms of y first and substitute into the equation for x.

h) $x = \sin \theta$, $y = \cos \theta$
Use trig identities here rather than rearranging...
$\sin^2 \theta + \cos^2 \theta \equiv 1 \Rightarrow x^2 + y^2 = 1$

i) $x = \sin \theta$, $y = \cos 2\theta$
The 2θ should make you think of the double angle formulae...
$\cos 2\theta \equiv 1 - 2 \sin^2 \theta \Rightarrow y = 1 - 2x^2$

j) $x = 1 + \sin \theta \Rightarrow \sin \theta = x - 1$
$y = 2 + \cos \theta \Rightarrow \cos \theta = y - 2$
$\sin^2 \theta + \cos^2 \theta \equiv 1 \Rightarrow (x - 1)^2 + (y - 2)^2 = 1$
You can leave this equation in the form it's in — it's the equation of a circle radius 1, centre (1, 2).

k) $x = \cos \theta$, $y = \cos 2\theta$
$\cos 2\theta \equiv 2 \cos^2 \theta - 1 \Rightarrow y = 2x^2 - 1$

l) $x = \cos \theta - 5 \Rightarrow \cos \theta = x + 5$
$y = \cos 2\theta \equiv 2 \cos^2 \theta - 1 \Rightarrow y = 2(x + 5)^2 - 1$
$= 2x^2 + 20x + 49$

Q2 $x = \tan \theta$, $y = \sec \theta$
Using the identity $\sec^2 \theta \equiv 1 + \tan^2 \theta$ gives: $y^2 = 1 + x^2$

Q3 $x = 2 \cot \theta \Rightarrow \cot \theta = \frac{x}{2}$
$y = 3 \csc \theta \Rightarrow \csc \theta = \frac{y}{3}$
Using the identity $\csc^2 \theta \equiv 1 + \cot^2 \theta$ gives:
$\frac{y^2}{9} = 1 + \frac{x^2}{4} \Rightarrow y^2 = 9 + \frac{9x^2}{4}$

Q4 a) From the parametric equations
the centre of the circle is (5, –3), and the radius is 1.

b) $x = 5 + \sin \theta \Rightarrow \sin \theta = x - 5$
$y = -3 + \cos \theta \Rightarrow \cos \theta = y + 3$
Using the identity $\sin^2 \theta + \cos^2 \theta \equiv 1$ gives:
$(x - 5)^2 + (y + 3)^2 = 1$

Q5 a) $x = \frac{1 + 2t}{t} \Rightarrow xt = 1 + 2t \Rightarrow xt - 2t = 1$
$\Rightarrow t(x - 2) = 1 \Rightarrow t = \frac{1}{(x - 2)}$

b) $y = \frac{3 + t}{t^2} = \frac{3 + \frac{1}{(x - 2)}}{\frac{1}{(x - 2)^2}} = 3(x - 2)^2 + (x - 2)$
$= 3(x^2 - 4x + 4) + x - 2$
$= 3x^2 - 11x + 10$
$= (3x - 5)(x - 2)$

c)
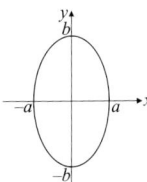

Q6 $x = \frac{2 - 3t}{1 + t} \Rightarrow x(1 + t) = 2 - 3t \Rightarrow x + xt = 2 - 3t$
$\Rightarrow xt + 3t = 2 - x \Rightarrow t(x + 3) = 2 - x \Rightarrow t = \frac{2 - x}{x + 3}$
$y = \frac{5 - t}{4t + 1} = \frac{5 - \left(\frac{2 - x}{x + 3}\right)}{4\left(\frac{2 - x}{x + 3}\right) + 1} = \frac{5(x + 3) - (2 - x)}{4(2 - x) + (x + 3)}$
$\Rightarrow y = \frac{6x + 13}{11 - 3x}$
You could also write this as 6x – 11y + 3xy + 13 = 0

Q7 $x = 5 \sin^2 \theta \Rightarrow \sin^2 \theta = \frac{x}{5}$, $y = \cos \theta$
Using the identity $\sin^2 \theta + \cos^2 \theta \equiv 1$ gives:
$\frac{x}{5} + y^2 = 1 \Rightarrow y^2 = 1 - \frac{x}{5}$

Q8 a) $x = a \sin \theta \Rightarrow \sin \theta = \frac{x}{a}$
$y = b \cos \theta \Rightarrow \cos \theta = \frac{y}{b}$
Using the identity $\sin^2 \theta + \cos^2 \theta \equiv 1$ gives:
$\left(\frac{x}{a}\right)^2 + \left(\frac{y}{b}\right)^2 = 1$

b) To sketch the graph, find the x- and y-intercepts.
When $x = 0$, $\left(\frac{y}{b}\right)^2 = 1 \Rightarrow y = \pm b$.
When $y = 0$, $\left(\frac{x}{a}\right)^2 = 1 \Rightarrow x = \pm a$.
So the curve looks like this:

c) An ellipse.
If a and b were equal it would be a circle.

Q9 a) $y = 2t - 1 \Rightarrow t = \frac{y + 1}{2}$
$x = 3t^2 = 3\left(\frac{y + 1}{2}\right)^2 = \frac{3}{4}(y + 1)^2$

Answers

b) Substitute $y = 4x - 3$ into the equation:

$x = \frac{3}{4}(4x - 3 + 1)^2 \implies 4x = 3(4x - 2)^2$

$4x = 3(16x^2 - 16x + 4) \implies 4x = 48x^2 - 48x + 12$

$12x^2 - 13x + 3 = 0 \implies (3x - 1)(4x - 3) = 0$

$x = \frac{1}{3}$ and $x = \frac{3}{4}$

$y = 4\left(\frac{1}{3}\right) - 3 = -\frac{5}{3}$ and $y = 4\left(\frac{3}{4}\right) - 3 = 0$

So the curve and the line intersect at $\left(\frac{1}{3}, -\frac{5}{3}\right)$ and $\left(\frac{3}{4}, 0\right)$.

Q10 $x = 7t + 2 \implies t = \frac{x - 2}{7}$

$y = \frac{5}{t} = \frac{5}{\left(\frac{x - 2}{7}\right)} = \frac{35}{x - 2}$

This is the graph of $y = \frac{1}{x}$ stretched vertically by a factor of 35 and translated right by 2.

The y-intercept is at $x = 0$, so $y = \frac{35}{-2} = -17.5$.

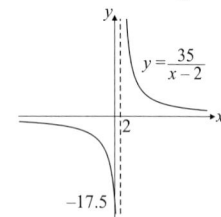

Q11 a) $x = t + 1 \implies t = x - 1$

So $y = f(x) = 3t^2 = 3(x - 1)^2 = 3x^2 - 6x + 3$

[2 marks available — 1 mark for t = x – 1, 1 mark for the correct final answer]

b) $y = f(x) = 3x^2 - 6x + 3 = 3(x^2 - 2x + 1) = 3(x - 1)^2$

When $x = 0$, $y = 3(-1)^2 = 3$.

When $y = 0$, $0 = 3(x - 1)^2 \implies 0 = x - 1 \implies x = 1$

So the graph is u-shaped, intersects the y-axis at $(0, 3)$ and intersects the x-axis at it's minimum point $(1, 0)$.

The graph of $y = f(x - 1) - 3$ is a graph transformation — a translation of $y = f(x)$ by vector $\begin{pmatrix} 1 \\ -3 \end{pmatrix}$.

It has equation $y = 3(x - 2)^2 - 3 = 3x^2 - 12x + 9$

$\implies y = 3(x - 1)(x - 3)$.

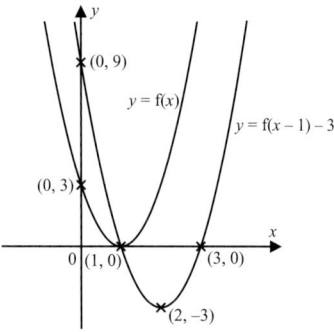

[5 marks available — 1 mark for correct shape of both graphs, 1 mark for turning point of f(x), 1 mark for intercepts of f(x), 1 mark for turning point of f(x – 1) – 3, 1 mark for intercepts of f(x – 1) – 3]

Chapter 4 Review Exercise

Q1 a) $x = \frac{1}{t} \implies t = \frac{1}{x}$. So when $x = \frac{1}{4}$, $t = 4$, and

$y = \frac{2}{t^2} = \frac{2}{4^2} = \frac{1}{8}$.

b) $\frac{2}{t^2} = \frac{1}{50} \implies t^2 = 100 \implies t = \pm 10$

Q2 a)

t	-3	-2	-1	0	1	2	3
x	9	4	1	0	1	4	9
y	$-\frac{27}{2}$	-4	$-\frac{1}{2}$	0	$\frac{1}{2}$	4	$\frac{27}{2}$

[2 marks available — 2 marks for all correct, otherwise 1 mark for a maximum of two errors]

b) $x = t^2 \implies t = \pm\sqrt{x}$

When $x = 25$, $t = \pm 5$ and $y = \pm\frac{125}{2}$

[2 marks available — 2 marks for both t- and y-values, otherwise 1 mark for both t values or for one t-value and one corresponding y-value]

c) The domain of x is $x \geq 0$ because $x = t^2$ and any square value is non-negative. The range of f(x) will be f(x) $\in \mathbb{R}$.

[2 marks available — 1 mark for a valid explanation for the domain of f(x), 1 mark for the correct range]

There are no restrictions on t^3 — it can be positive, negative, or zero.

Q3 a) $A = 2.4$ m. This will be the height, y, at time $t = 0$.

y is a quadratic in t so when $t = 0$ we only need to look for the constant term in the quadratic.

[2 marks available — 1 mark for 2.4, 1 mark for a valid explanation]

b) (i) $y = 3 \implies 2.4 + 1.8t - 0.48t^2 = 3$

$\implies 0.48t^2 - 1.8t + 0.6 = 0$

Using the quadratic formula:

$t = \frac{1.8 \pm \sqrt{(-1.8)^2 - (4 \times 0.48 \times 0.6)}}{2 \times 0.48}$

$\implies t = 3.3801... = 3.38$ (3 s.f.)

and $t = 0.3698... = 0.370$ (3 s.f.)

As the ball is on its way down when it goes in, it will be the later time, so the ball is in the air for 3.38 seconds before a basket is scored.

[2 marks available — 1 mark for a correct method, 1 mark for the correct final answer]

(ii) This will be x when $t = 3.38$.

$x = 4 \times 3.3801... = 13.5207...$

So the basket is 13.5 m (3 s.f.) from where the player took their shot.

[1 mark for 13.5 m or similar sensible rounding]

c) This will be when $y = 0 \implies 2.4 + 1.8t - 0.48t^2 = 0$

Using the quadratic formula:

$t = \frac{-1.8 \pm \sqrt{1.8^2 - (4 \times -0.48 \times 2.4)}}{2 \times -0.48}$

$\implies t = 4.7931... = 4.79$ (3 s.f.)

and $t = -1.0431... = -1.04$ (3 s.f.).

As $t \geq 0$, the ball would be in the air for 4.79 seconds.

[2 marks available — 1 mark for quadratic equal to 0, 1 mark for the correct final answer]

Q4 The curve crosses the y-axis at $x = 0$, so:

$0 = t^3 + t^2 - 6t \implies t(t^2 + t - 6) = 0 \implies t(t + 3)(t - 2) = 0$

So $t = 0$, -3 or 2.

When $t = 0$, $y = 4(0) - 5 = -5$.

When $t = -3$, $y = 4(-3) - 5 = -17$.

When $t = 2$, $y = 4(2) - 5 = 3$.

So the coordinates are $(0, -5)$, $(0, -17)$ and $(0, 3)$.

Q5 Substitute the parametric equations for x and y
in $y = 10x - 8$:

$t^2 - t = 10\left(\dfrac{t+3}{5}\right) - 8 \Rightarrow t^2 - t = 2t + 6 - 8$

$\Rightarrow t^2 - 3t + 2 = 0 \Rightarrow (t - 2)(t - 1) = 0$

So they cross when $t = 1$ and $t = 2$.

When $t = 1$, $x = \dfrac{1+3}{5} = \dfrac{4}{5}$, $y = 1^2 - 1 = 0$.

When $t = 2$, $x = \dfrac{2+3}{5} = 1$, $y = 2^2 - 2 = 2$.

So the coordinates are $\left(\dfrac{4}{5}, 0\right)$ and $(1, 2)$.

Q6 a) The curve crosses the y-axis at $x = 0$, so:

$t^2 - 1 = 0 \Rightarrow t = \pm 1$

When $t = -1$, $y = 4 - 3 = 1$.

When $t = 1$, $y = 4 + 3 = 7$.

So the coordinates are $(0, 1)$ and $(0, 7)$.

b) Substitute the parametric equations for x and y
in $x + 2y = 14$:

$t^2 - 1 + 2\left(4 + \dfrac{3}{t}\right) = 14$

$\Rightarrow t^2 - 1 + 8 + \dfrac{6}{t} - 14 = 0$

$\Rightarrow t^3 - 7t + 6 = 0$

The coefficients add up to 0, so $(t - 1)$ is a factor.
Using e.g. algebraic division:

$(t - 1)(t^2 + t - 6) = 0 \Rightarrow (t - 1)(t - 2)(t + 3) = 0$

So $t = 1$, 2 or -3.

When $t = 1$, $x = 1^2 - 1 = 0$, $y = 4 + \dfrac{3}{1} = 7$.

When $t = 2$, $x = 2^2 - 1 = 3$, $y = 4 + \dfrac{3}{2} = 5.5$.

When $t = -3$, $x = (-3)^2 - 1 = 8$, $y = 4 + \dfrac{3}{-3} = 3$.

So the coordinates are $(0, 7)$, $(3, 5.5)$ and $(8, 3)$.

Q7 a) $e^t \cos 2t = 0$

Since $e^t \neq 0$, $\cos 2t = 0$

$\Rightarrow 2t = \dfrac{\pi}{2}, \dfrac{3\pi}{2}, \dfrac{5\pi}{2}, \dfrac{7\pi}{2} \ldots$

$\Rightarrow t = \dfrac{\pi}{4}, \dfrac{3\pi}{4}, \dfrac{5\pi}{4}, \dfrac{7\pi}{4}$

When $t = \dfrac{\pi}{4}$, $x = -e^{\frac{\pi}{4}} \sin \dfrac{\pi}{2} = -e^{\frac{\pi}{4}}$

When $t = \dfrac{3\pi}{4}$, $x = -e^{\frac{3\pi}{4}} \sin \dfrac{3\pi}{2} = e^{\frac{3\pi}{4}}$

When $t = \dfrac{5\pi}{4}$, $x = -e^{\frac{5\pi}{4}} \sin \dfrac{5\pi}{2} = -e^{\frac{5\pi}{4}}$

When $t = \dfrac{7\pi}{4}$, $x = -e^{\frac{7\pi}{4}} \sin \dfrac{7\pi}{2} = e^{\frac{7\pi}{4}}$

So when $y = 0$, $x = -e^{\frac{\pi}{4}}, e^{\frac{3\pi}{4}}, -e^{\frac{5\pi}{4}}, e^{\frac{7\pi}{4}}$

b) (i) *Use $d = \sqrt{x^2 + y^2}$ to find the distance of the particle from
the origin. Substitute in the equations for x and y...*

$d = \sqrt{(-e^t \sin 2t)^2 + (e^t \cos 2t)^2}$

$\Rightarrow d = \sqrt{e^{2t} \sin^2 2t + e^{2t} \cos^2 2t}$

$\Rightarrow d = \sqrt{e^{2t}(\sin^2 2t + \cos^2 2t)}$

...and use the identity $\sin^2 \theta + \cos^2 \theta \equiv 1$:

$\Rightarrow d = \sqrt{e^{2t}} \Rightarrow d = e^t$

(ii) From b)(i), $d = e^t$, so when $d = e^{\frac{\pi}{8}}$, $t = \dfrac{\pi}{8}$,

and so $x = -e^{\frac{\pi}{8}} \sin \dfrac{\pi}{4} = -\dfrac{\sqrt{2}}{2} e^{\frac{\pi}{8}}$, and

$y = e^{\frac{\pi}{8}} \cos \dfrac{\pi}{4} = \dfrac{\sqrt{2}}{2} e^{\frac{\pi}{8}}$.

So the coordinates are $\left(-\dfrac{\sqrt{2}}{2} e^{\frac{\pi}{8}}, \dfrac{\sqrt{2}}{2} e^{\frac{\pi}{8}}\right)$.

Q8 a) (i) Neither e^t nor e^{2t} can be equal to zero (or negative) and
so neither x nor y can be zero (or negative) either and so
the graph of the curve does not intersect the origin.
[1 mark for a correct explanation]

(ii) Domain: $x > 0$ and Range: $f(x) > 0$.
[1 mark for both domain and range correct]

b) x and y coordinates will be equal when $e^t = 0.1 e^{2t}$

$\dfrac{e^{2t}}{e^t} = \dfrac{1}{0.1} \Rightarrow e^t = 10 \Rightarrow t = \ln 10$

When $t = \ln 10$, $x = e^{\ln 10} = 10$ and $y = 0.1 e^{2\ln 10} = 10$
and so the curve passes through the point $(10, 10)$.
*[4 marks available — 1 mark for forming an equation,
1 mark for method of solving for t, 1 mark for writing t in
terms of ln, 1 mark for (10, 10)]*

c) (i)

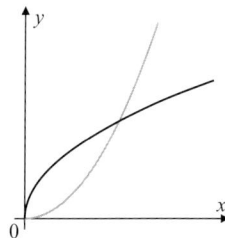

*[2 marks available — 1 mark for a correct method to
reflect the graph in $y = x$, 1 mark for the correct graph]*

(ii) From part b) the second graph will also pass through
$(10, 10)$, so that is the point of intersection. *[1 mark]*
Remember that the origin is not included on either graph.

Q9 a) When $t = 0$, $x = \dfrac{6 - 0}{2} = 3$, $y = 2(0)^2 + 0 + 4 = 4$,

When $t = 1$, $x = \dfrac{6 - 1}{2} = 2.5$, $y = 2(1)^2 + 1 + 4 = 7$

When $t = 2$, $x = \dfrac{6 - 2}{2} = 2$, $y = 2(2)^2 + 2 + 4 = 14$,

When $t = 3$, $x = \dfrac{6 - 3}{2} = 1.5$, $y = 2(3)^2 + 3 + 4 = 25$

b) (i) $-7 = \dfrac{6 - t}{2} \Rightarrow -14 = 6 - t \Rightarrow t = 20$

(ii) $19 = 2t^2 + t + 4 \Rightarrow 2t^2 + t - 15 = 0$
$\Rightarrow (2t - 5)(t + 3) = 0 \Rightarrow t = 2.5, t = -3$

c) $x = \dfrac{6 - t}{2} \Rightarrow t = 6 - 2x$

$y = 2(6 - 2x)^2 + (6 - 2x) + 4$

$y = 2(36 - 24x + 4x^2) + 6 - 2x + 4$

$y = 8x^2 - 50x + 82$

Q10 a) At $x = 0$, $\dfrac{2t - 3}{t} = 0 \Rightarrow t = \dfrac{3}{2}$

$\Rightarrow y = 2\left(\dfrac{3}{2}\right) + 6 = 9$

At $y = 0$, $2t + 6 = 0 \Rightarrow t = -3$

$\Rightarrow x = \dfrac{2(-3) - 3}{-3} = 3$

So the curve crosses the coordinate axes
at points $(0, 9)$ and $(3, 0)$.
*[3 marks available — 1 mark for a correct method,
1 mark for each correct pair of coordinates]*

b) Rearrange the equation for y:

$y = 2t + 6 \Rightarrow t = \dfrac{y - 6}{2}$

Substitute into the equation for x:

$x = \dfrac{2\left(\dfrac{y-6}{2}\right) - 3}{\left(\dfrac{y-6}{2}\right)} = \dfrac{2(y-6) - 6}{y - 6} = \dfrac{2y - 18}{y - 6}$

Rearrange to make y the subject:

$xy - 6x = 2y - 18 \Rightarrow xy - 2y = 6x - 18$

$\Rightarrow y(x - 2) = 6x - 18 \Rightarrow y = \dfrac{6(x - 3)}{x - 2}$

*[3 marks available — 1 mark for equation for t in terms of x
or y, 1 mark for substituting t into other equation, 1 mark for
correct final answer given as y = f(x)]*

c) $1 < t \leq 6$. When $t = 1$, $y = f(x) = 2(1) + 6 = 8$.
When $t = 6$, $y = f(x) = 2(6) + 6 = 18$.
So the range of f(x) is $8 < f(x) \leq 18$.
[2 marks available — 1 mark for substituting the limits of t in an equation for y, 1 mark for correct final answer]

Q11 a) $x = 2t \Rightarrow t = \dfrac{x}{2}$

$\Rightarrow y = \dfrac{x}{2}\left(\dfrac{ax}{2} + 6\right) = \dfrac{a}{4}x^2 + 3x$

If $y = ax - 4$ is a tangent then it touches the curve once, at the point where $\dfrac{a}{4}x^2 + 3x = ax - 4$

$\Rightarrow ax^2 + 12x - 4ax + 16 = 0$
$\Rightarrow ax^2 + (12 - 4a)x + 16 = 0$
There should only be one solution to this equation, so the discriminant should be zero:
$(12 - 4a)^2 - 4 \times a \times 16 = 0$
$\Rightarrow 144 - 96a + 16a^2 - 64a = 0$
$\Rightarrow a^2 - 10a + 9 = 0 \Rightarrow (a - 1)(a - 9) = 0$
So $a = 1$ or $a = 9$
[4 marks available — 1 mark for Cartesian equation, 1 mark for equating the tangent and the curve to form a quadratic, 1 mark for using the discriminant = 0 (or otherwise using the fact that there is only one solution to this quadratic), 1 mark for both correct solutions for a]

b) From a), $a = 1$ or 9, so the Cartesian equation of the curve will be: $y = \dfrac{1}{4}x^2 + 3x = x\left(\dfrac{1}{4}x + 3\right)$

or $y = \dfrac{9}{4}x^2 + 3x = 3x\left(\dfrac{3}{4}x + 1\right)$.

$y = x\left(\dfrac{1}{4}x + 3\right)$ is a u-shaped quadratic (shown in black) through $(0, 0)$ and $(-12, 0)$, with a minimum point where $x = -6$ and $y = \dfrac{1}{4}(-6)^2 + 3(-6) = -9$.

$y = 3x\left(\dfrac{3}{4}x + 1\right)$ is a u-shaped quadratic (shown in grey) through $(0, 0)$ and $\left(-\dfrac{4}{3}, 0\right)$, with a minimum point where

$x = -\dfrac{2}{3}$ and $y = \dfrac{9}{4}\left(-\dfrac{2}{3}\right)^2 + 3\left(-\dfrac{2}{3}\right) = -1$.

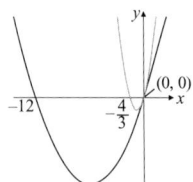

[4 marks available — 1 mark for two graphs with the correct shape, 1 mark for correct intersection at (0, 0), 1 mark for correct axis-intercepts shown on one curve, 1 mark for correct axis-intercepts shown on the other curve]

Q12 a) When $x < 0$, $2t^2 + 3t < 0 \Rightarrow t(2t + 3) < 0 \Rightarrow -\dfrac{3}{2} < t < 0$

When $y < 0$, $2t^2 - 3t < 0 \Rightarrow t(2t - 3) < 0 \Rightarrow 0 < t < \dfrac{3}{2}$

These ranges do not overlap, so there are no values for t for which both x and y are negative.
[3 marks available — 1 mark for finding the values of t for a negative x, 1 mark for finding the values of t for a negative y, 1 mark for showing that they do not overlap]

b) t cannot easily be isolated in either equation. There is an xy term in the final answer, and multiplying the parametric equations together gives the difference of two squares:
$xy = (2t^2 + 3t)(2t^2 - 3t) = 4t^4 - 9t^2$
Adding the parametric equations gives:
$x + y = 2t^2 + 3t + 2t^2 - 3t = 4t^2 \Rightarrow t^2 = \dfrac{x + y}{4}$

Substitute $t^2 = \dfrac{x + y}{4}$ into xy and rearrange:

$xy = 4\left(\dfrac{x + y}{4}\right)^2 - 9\left(\dfrac{x + y}{4}\right)$

$\Rightarrow xy = \dfrac{1}{4}(x^2 + 2xy + y^2) - \dfrac{9}{4}(x + y)$
$\Rightarrow 4xy = x^2 + 2xy + y^2 - 9(x + y)$
$\Rightarrow 0 = x^2 + y^2 - 2xy - 9(x + y)$ — as required,
where $A = -2$ and $B = -9$.
[5 marks available — 1 mark for finding xy, 1 mark for finding x + y, 1 mark for substituting for t², 1 mark for attempt to rearrange in the correct form, 1 mark for final answer with correct values of A and B]

Q13 a) (i) $x = 2 \sin\dfrac{\pi}{4} = \sqrt{2}$

$y = \cos^2\dfrac{\pi}{4} + 4 = \dfrac{9}{2}$

So the coordinates are $\left(\sqrt{2}, \dfrac{9}{2}\right)$.

(ii) $x = 2 \sin\dfrac{\pi}{6} = 1$

$y = \cos^2\dfrac{\pi}{6} + 4 = \dfrac{19}{4}$

So the coordinates are $\left(1, \dfrac{19}{4}\right)$.

b) $x = 2 \sin\theta \Rightarrow \sin\theta = \dfrac{x}{2}$

$y = \cos^2\theta + 4$

$\cos^2\theta \equiv 1 - \sin^2\theta \Rightarrow y = 1 - \left(\dfrac{x}{2}\right)^2 + 4$

$\Rightarrow y = 5 - \dfrac{x^2}{4}$

c) $-1 \leq \sin\theta \leq 1 \Rightarrow -2 \leq 2\sin\theta \leq 2 \Rightarrow -2 \leq x \leq 2$

Q14 $x = \dfrac{\sin\theta}{3} \Rightarrow \sin\theta = 3x$, $y = 3 + 2\cos 2\theta$

Use one of the double angle formulae here:
$\cos 2\theta \equiv 1 - 2\sin^2\theta \Rightarrow y = 3 + 2(1 - 2\sin^2\theta)$
$\Rightarrow y = 3 + 2(1 - 2(3x)^2) \Rightarrow y = 3 + 2 - 4(9x^2) \Rightarrow y = 5 - 36x^2$

Q15 a) $x = 4 - \cos\dfrac{2\pi}{3} = \dfrac{9}{2}$

$y = \sin^4\dfrac{\pi}{3} - \dfrac{1}{2} = \dfrac{1}{16}$

So the coordinates are $\left(\dfrac{9}{2}, \dfrac{1}{16}\right)$

b) $x = 0$ on the y-axis, but $x = 4 - \cos 2\theta$ can never be zero, as $-1 \leq \cos 2\theta \leq 1$, so $3 \leq x \leq 5$.

c) $x = 4 - \cos 2\theta$
Using $\cos 2\theta \equiv 1 - 2\sin^2\theta$:
$x = 4 - (1 - 2\sin^2\theta) \Rightarrow x - 3 = 2\sin^2\theta$

$\Rightarrow \sin^2\theta = \dfrac{x - 3}{2}$

$y = \sin^4\theta - \dfrac{1}{2} = \left(\dfrac{x - 3}{2}\right)^2 - \dfrac{1}{2}$

$\Rightarrow y = \dfrac{x^2 - 6x + 9}{4} - \dfrac{1}{2} \Rightarrow y = \dfrac{x^2 - 6x + 7}{4}$

Q16 a) Use an addition formula to write $y = 5\cos 3t$ in terms of $\cos 2t$, $\cos t$, $\sin 2t$ and $\sin t$:
$y = 5\cos(2t + t) = 5(\cos 2t\cos t - \sin 2t\sin t)$

Use double angle formulae to write everything in terms of $\cos t$ and $\sin t$:
$y = 5[(2\cos^2 t - 1)\cos t - (2\sin t\cos t)\sin t]$
$= 5[2\cos^3 t - \cos t - 2\sin^2 t\cos t]$

Replace $\sin^2 t$ with $1 - \cos^2 t$:
$y = 5[2\cos^3 t - \cos t - 2(1 - \cos^2 t)\cos t]$
$= 5[2\cos^3 t - \cos t - 2\cos t + 2\cos^3 t] = 5\cos t(4\cos^2 t - 3)$

Substitute $\cos t = \frac{x}{2}$:
$y = \frac{5x}{2}\left(\frac{4x^2}{4} - 3\right) = \frac{5}{2}x(x^2 - 3)$ as required,

i.e. $A = \frac{5}{2}$ and $B = 3$.
$-\pi < t < \pi \Rightarrow -1 < \cos t < 1 \Rightarrow -2 < 2\cos t < 2$,
so the domain is $-2 < x < 2$.
[6 marks available — 1 mark for correct use of addition angle formula, 1 mark for correct use of double angle formula for cos 2t, 1 mark for correct use of double angle formula for sin 2t, 1 mark for writing y in terms of cos t only, 1 mark for correct Cartesian equation, 1 mark for correct domain]

b) The range of $y = 5\cos 3t$ is $-5 \le y \le 5$, so $y = 6$ cannot cross the curve *[1 mark]*.

c) A quick sketch will help. C is a positive cubic going through $(0, 0)$ and $(\pm\sqrt{3}, 0)$, with domain $-2 \le x \le 2$, and $y = kx$ is a straight line with a positive gradient through the origin:

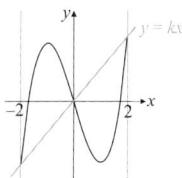

From the graph, any line $y = kx$ whose gradient is greater than the line shown will not cross the curve 3 times. So the gradient of the line shown is the maximum value of k.
When $x = -2$, $y = \frac{5}{2}(-2)(4 - 3) = -5$
When $x = 2$, $y = \frac{5}{2}(2)(4 - 3) = 5$
So the straight line passes through $(-2, -5)$ and $(2, 5)$ when k is at its largest value.
So the gradient of the straight line, $k \le \frac{5 - (-5)}{2 - (-2)}$
$\Rightarrow k \le 2.5$
You're told k is positive, so $0 < k \le 2.5$.
[4 marks available — 1 mark for identifying the limit of k relating to the domain of C, 1 mark for correct end points of the curve, 1 mark for using the end points to find the max gradient of the straight line, 1 mark for a correct range given for k]

Q17 a) Centre is $(a, 1 - 2a)$. Radius is b.
[2 marks available — 1 mark for centre, 1 mark for radius]

b) Substitute the given values of x, y and t into the parametric equations to obtain two simultaneous equations in a and b.
$-4 = a + b\cos\left(\frac{3\pi}{4}\right) \Rightarrow -4 = a - \frac{\sqrt{2}}{2}b$
$2 = 1 - 2a + b\sin\left(\frac{3\pi}{4}\right) \Rightarrow 2 = 1 - 2a + \frac{\sqrt{2}}{2}b$
Adding the equations gives $-4 + 2 = 1 - a \Rightarrow a = 3$
$-4 = 3 - \frac{\sqrt{2}}{2}b \Rightarrow b = -7 \times -\frac{2}{\sqrt{2}} = \frac{14}{\sqrt{2}} = 7\sqrt{2}$
[4 marks available — 1 mark for forming an equation using parametric equation for x, 1 mark for forming an equation using parametric equation for y, 1 mark for eliminating a or b, 1 mark for a and b correct]

c) $(x - 3)^2 + (y + 5)^2 = (7\sqrt{2})^2$ *[1 mark]*
The right-hand side can be written as just 98 too. $(7\sqrt{2})^2 = 98$.

Q18 a) (i) $x = \frac{1}{a}\operatorname{cosec} t \Rightarrow ax = \operatorname{cosec} t \Rightarrow a^2 x^2 = \operatorname{cosec}^2 t$

(ii) $y = \frac{1}{a + 1}\cot t \Rightarrow (a + 1)y = \cot t$
$\Rightarrow (a + 1)^2 y^2 = \cot^2 t$

b) Use the trig identity $\cot^2 A + 1 = \operatorname{cosec}^2 A$:
$(a + 1)^2 y^2 + 1 = a^2 x^2 \Rightarrow (a + 1)^2 y^2 = a^2 x^2 - 1$

c) When $x = \frac{1}{2}$, $y = 0$ (as it intercepts the x-axis).
Substitute these into the Cartesian equation:
$(a + 1)^2(0)^2 = a^2\left(\frac{1}{2}\right)^2 - 1 \Rightarrow a^2 = 4$ and a is a positive integer so $a = 2$.

d) Domain: $x \ge \frac{1}{2}$, Range: $f(x) \in \mathbb{R}$
*The easiest way to see this is by considering the graphs of $y = \operatorname{cosec} x$ and $y = \cot x$ for $0 < x < \pi$.
$\operatorname{cosec} x$ is 1 or greater so $\frac{1}{2}\operatorname{cosec} x$ is $\frac{1}{2}$ or greater.
$\cot x$ takes all values so $\frac{1}{3}\cot x$ takes all values too.*

Q19 a) $x = 10\sin\theta \Rightarrow \sin\theta = \frac{x}{10}$
$y = 6\cos\theta \Rightarrow \cos\theta = \frac{y}{6}$
Now use the identity $\sin^2\theta + \cos^2\theta \equiv 1$:
$\left(\frac{x}{10}\right)^2 + \left(\frac{y}{6}\right)^2 = 1 \Rightarrow \frac{x^2}{100} + \frac{y^2}{36} = 1$
[4 marks available — 1 mark for rearranging parametric equations, 1 mark for using the correct identity, 1 mark for eliminating θ, 1 mark for final answer]

b) At $x = 0$, $y^2 = 36 \Rightarrow y = \pm 6$
At $y = 0$, $x^2 = 100 \Rightarrow x = \pm 10$

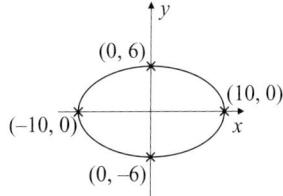

[3 marks available — 1 mark for determining axes intercepts, 1 mark for correct shape of graph, 1 mark for correct intercepts labelled]

c) (i) Either $y = -6$ or $y = 6$ *[1 mark]*
(ii) Either $x = -10$ or $x = 10$ *[1 mark]*

Q20 a) At $t = \pi$, $a\cos\pi = -8 \Rightarrow -a = -8 \Rightarrow a = 8$
Greatest difference in $\sin 2t$ is 2 (max 1 and min −1) so greatest difference in $b\sin 2t$ is $6 \Rightarrow b \times 2 = 6 \Rightarrow b = 3$
[2 marks available — 1 mark for a, 1 mark for b]

b) $3\sin 2t = 0 \Rightarrow \sin 2t = 0$
Solving for $2t$ so rewrite the domain as $0 \le 2t \le 4\pi$.
$2t = 0, \pi, 2\pi, 3\pi, 4\pi \Rightarrow t = 0, \frac{\pi}{2}, \pi, \frac{3\pi}{2}, 2\pi$
[2 marks available — 1 mark for a correct method to solve for 2t, 1 mark for all five correct answers]
A sketch of the graph is always helpful when solving trig equations. Sketching the graph of y = sin 2x will tell you all five answers instantly.

Answers

c) At $\left(4, -\frac{3\sqrt{3}}{2}\right)$, $8\cos t = 4$ and $3\sin 2t = -\frac{3\sqrt{3}}{2}$.

$8\cos t = 4 \implies \cos t = \frac{1}{2} \implies t = \frac{\pi}{3}, \frac{5\pi}{3}, \frac{7\pi}{3}, \frac{11\pi}{3}, \dots$

$3\sin 2t = -\frac{3\sqrt{3}}{2} \implies \sin 2t = -\frac{\sqrt{3}}{2}$

$\implies 2t = \left(-\frac{\pi}{3}\right), \frac{4\pi}{3}, \frac{5\pi}{3}, \frac{10\pi}{3}, \dots \implies t = \frac{2\pi}{3}, \frac{5\pi}{6}, \frac{5\pi}{3}, \dots$

So the obstacle is placed at the position that the speed skater first reaches at $t = \frac{5\pi}{3}$. The speed skater encounters the obstacle at $t = \frac{5\pi}{3}$ and every 2π seconds afterwards.

$\frac{5\pi}{3} = 5.235\dots$, $\frac{5\pi}{3} + 2\pi = \frac{11\pi}{3} = 11.519\dots$,

$\frac{11\pi}{3} + 2\pi = \frac{17\pi}{3} = 17.802\dots$

Clearly this will be the last time within the first 20 seconds. So the speed skater will encounter the obstacle 3 times within the first 20 seconds.

[5 marks available — 1 mark for using coordinates to form equations for t, 1 mark for an attempt to solve the equations, 1 mark for $t = \frac{5\pi}{3}$, 1 mark for attempting to use this value to find the number of obstacle encounters, 1 mark for correct final answer]

Be careful here — it is not enough to rely on just solving $8\cos t = 4$ and using the answer from a calculator. You need to make sure the value of t suits both x and y coordinates.

Q21 a) Firework is at ground level when $y = 0$.

$16t^3 - 84t^2 + 108t = 0 \implies 4t(4t^2 - 21t + 27) = 0$

$\implies 4t(4t - 9)(t - 3) = 0 \implies t = 0, t = \frac{9}{4}, t = 3$

$t = 0$ is a solution but this is clearly before/at launch.

So the firework will hit the ground after $\frac{9}{4} = 2.25$ seconds.

[3 marks available — 1 mark for setting up $y = 0$, 1 mark for solving, 1 mark for the correct answer]

$t = 3$ is greater than $\frac{9}{4}$ so can be rejected. Consider the graph of a positive cubic curve — to obtain $t = 3$ the firework would travel below ground level and then back up.

b) $x = 2t \implies t = \frac{x}{2}$

$y = 16\left(\frac{x}{2}\right)^3 - 84\left(\frac{x}{2}\right)^2 + 108\left(\frac{x}{2}\right)$

$y = 2x^3 - 21x^2 + 54x$

From part a), $0 \le t \le 2.25$ and boundaries of the domain of f(x) will be double the boundaries of t since $x = 2t$.

So domain of f(x) is $0 \le x \le 4.5$.

[3 marks available — 1 mark for eliminating t, 1 mark for correct Cartesian equation, 1 mark for domain of f(x)]

c) Maximum is where $\frac{dy}{dx} = 0$ (and $\frac{d^2y}{dx^2} < 0$)

$\frac{dy}{dx} = 6x^2 - 42x + 54 = 0$

Using the quadratic formula to solve:

$x = \frac{42 \pm \sqrt{(-42)^2 - (4 \times 6 \times 54)}}{2 \times 6}$

$\implies x = 5.3027\dots$ and $x = 1.6972\dots$

$x = 5.3027\dots$ is beyond the domain of f(x) so $x = 1.6972\dots$

Height is the y-coordinate so,

$y = 2(1.6972\dots)^3 - 21(1.6972\dots)^2 + 54(1.6972\dots)$

$= 40.9$ (3 s.f.)

Time at maximum height will be

$t = \frac{x}{2} = \frac{1.6972\dots}{2} = 0.849$ (3 s.f.)

The firework reaches its maximum height of 40.9 m after 0.849 seconds.

[5 marks available — 1 mark for differentiating, 1 mark for $\frac{dy}{dx} = 0$, 1 mark for finding the x-value at maximum height, 1 mark for correct maximum height, 1 mark for correct time]

Chapter 5: Sequences and Series

Prior knowledge Check

Q1

Q2 The difference is 9 and the first term is 5.
$9 - 4 = 5$, so the nth term is $9n - 4$.

Q3 a) 11, 12, 13, 14   **b)** 2, –1, –4, –7
   c) 1, 9, 17, 25   **d)** –8, –10, –12, –14

Q4 5th term: $4(5) - 1 = 19$
25th term: $4(25) - 1 = 99$
100th term: $4(100) - 1 = 399$

Exercise 5.1 — Sequences

Q1 $a_{20} = 3(20) - 5 = 55$

Q2 4th term $= 4(4 + 2) = 24$

Q3 1st term $= (1 - 1)(1 + 1) = 0$
2nd term $= (2 - 1)(2 + 1) = 3$
Using the same method, 3rd, 4th and 5th terms $= 8, 15, 24$

Q4 $29 = 4k - 3 \implies k = 8$

Q5 $a_k = 13 - 6k$ and $a_{k+1} = 13 - 6(k + 1) = 7 - 6k$
The sequence is decreasing if $a_{k+1} < a_k$ for all values of k.
$7 - 6k < 13 - 6k \implies 7 < 13$, which is true
so the sequence is decreasing.

Q6 $3^{-n} = \frac{1}{3^n}$, so $a_k = \frac{1}{3^k}$ and $a_{k+1} = \frac{1}{3^{k+1}}$
The sequence is decreasing if $a_{k+1} < a_k$ for all values of k.
Since k is a positive integer, so is 3^k, so you can multiply through by this without affecting the inequality:
$\frac{1}{3^{k+1}} < \frac{1}{3^k} \implies \frac{3^k}{3^{k+1}} < \frac{3^k}{3^k} \implies \frac{1}{3} < 1$
This is true, so the sequence is decreasing.

Q7 Form equations for the 2nd and 5th terms:
$15 = a(2^2) + b$, $99 = a(5^2) + b$
Solve the equations simultaneously to get $a = 4$, $b = -1$

Q8 Form equations for the first 3 terms:
$9 = (1^2)e + f + g$
$20 = (2^2)e + 2f + g$
$37 = (3^2)e + 3f + g$
Solve the equations simultaneously to get $e = 3$, $f = 2$, $g = 4$.
To solve simultaneous equations with 3 unknowns, you use a similar method to when there are 2 unknowns — it just takes a few more steps.

Q9 $49 = (n - 1)^2$, $n = 8$

Q10 This first 8 terms of the sequence are: 13, 11, 9, 7, 5, 3, 1, –1, ...
The sequence continues to decrease. So 7 terms are positive.
A different way to solve this one would be to use an inequality — set $15 - 2n > 0$ and solve for n (taking the integer value of n).

Q11 $4n(n - 3) + 9 = 4n^2 - 12n + 9 = (2n - 3)^2$
$(2n - 3)^2 > 0$ for all values of $2n - 3$, and so all values in the sequence will be positive.
[2 marks available — 1 mark for showing a perfect square, 1 mark for conclusion/explanation]

Q12 a) **(i)** $u_1 = 0$, $u_2 = -p$, $u_3 = 0$, $u_4 = p$, $u_5 = 0$
   (ii) Assuming p is positive, the maximum would be p and minimum would be $-p$. Vice versa if p is negative.
   (iii) 4 (It takes 4 terms for the sequence to repeat itself)
  b) **(i)** Here, $u_1 = -p$, $u_2 = p$, $u_3 = -p$, so period is 2.

(ii) cos repeats every 2π so the period will be $\dfrac{2\pi}{\left(\dfrac{\pi}{q}\right)} = 2q$

Q13 a) $u_{n+1} - u_n = ((n+1)^2 - 16(n+1) + 55) - (n^2 - 16n + 55)$
$\Rightarrow u_{n+1} - u_n = (n^2 + 2n + 1 - 16n - 16 + 55 - n^2 + 16n - 55)$
$\qquad\qquad\qquad = 2n - 15$
[3 marks available — 1 mark for $(n + 1)^{th}$ term, 1 mark for expanding/simplifying, 1 mark for final answer]

b) Completing the square:
$u_n = n^2 - 16n + 55 = (n - 8)^2 - 64 + 55 = (n - 8)^2 - 9$
So the minimum in the sequence is –9.
This happens when $n = 8$, so it's the 8th term.
[3 marks available — 1 mark for method to find the minimum, 1 mark for the minimum term number, 1 mark for the minimum term value]

c) From part b) you know that the 8th term is the minimum. Given the u-shape of the quadratic graph $y = n^2 - 16n + 55$, you know that it is decreasing from $n = 1$ to the minimum term. So there is a maximum of 8 terms in the sequence.
[1 mark for correct answer with a suitable justification]

Exercise 5.2 — Recurrence Relations

Q1 $u_1 = 10$
$u_2 = 3u_1 = 3(10) = 30$
$u_3 = 3u_2 = 3(30) = 90$
$u_4 = 3u_3 = 3(90) = 270$
$u_5 = 3u_4 = 3(270) = 810$

Q2 $u_1 = 2$
$u_2 = u_1^2 = 2^2 = 4$
$u_3 = u_2^2 = 4^2 = 16$
$u_4 = u_3^2 = 16^2 = 256$

Q3 $u_1 = 4$
$u_2 = \dfrac{-1}{u_1} = -\dfrac{1}{4}$
$u_3 = \dfrac{-1}{u_2} = \dfrac{-1}{\left(\dfrac{-1}{4}\right)} = 4$
$u_4 = \dfrac{-1}{u_3} = -\dfrac{1}{4}$
The sequence is periodic with order 2.

For Q4, 5, 6, you can use any letter in place of u.

Q4 Each term is the previous term doubled and the first term is 3.
$u_{n+1} = 2u_n, \; u_1 = 3$

Q5 a) Each term is 4 more than the previous term.
The first term is 12.
$u_{n+1} = u_n + 4, \; u_1 = 12$

b) Work out the number of 'jumps' of 4 needed to get from 28 to 100:
$100 - 28 = 72, \; 72 \div 4 = 18$
Add on the first 5 terms given in the question:
$18 + 5 = 23$ terms
You could have written an nth term expression and used it to find the position of 100 in the sequence (it's the last term). The nth term expression would be $4n + 8$.

Q6 $u_{n+1} = 11 - u_n$ or $u_{n+1} = 28 \div u_n$, with $u_1 = 7$.
This one is tricky. It's the sort you suddenly go "aah" with. There are other possible answers here, but these are the two simplest.

Q7 $u_1 = 4, \; u_2 = 3(4) - 1 = 11, \; u_3 = 3(11) - 1 = 32$
$u_4 = 3(32) - 1 = 95$, so $k = 4$

Q8 $x_1 = 9, x_2 = (9 + 1) \div 2 = 5$
Keep substituting the result into the formula until...
$x_6 = (\dfrac{3}{2} + 1) \div 2 = \dfrac{5}{4}$, so $r = 6$

Q9 $u_1 = 7, \; u_2 = 7 + 1 = 8, \; u_3 = 8 + 2 = 10$
$u_4 = 10 + 3 = 13, \; u_5 = 13 + 4 = 17$

Q10 Form an equation for getting u_2 from u_1, and an equation for getting u_3 from u_2:
$7 = 6a + b$
$8.5 = 7a + b$
Solve the equations simultaneously to get $a = 1.5, b = -2$

Q11 First 5 terms:
$u_1 = 8, \; u_2 = \dfrac{1}{2}(8) = 4, \; u_3 = \dfrac{1}{2}(4) = 2$
$u_4 = \dfrac{1}{2}(2) = 1, \; u_5 = \dfrac{1}{2}(1) = \dfrac{1}{2}$
The terms are all powers of 2:
$8 = 2^3, 4 = 2^2, 2 = 2^1, 1 = 2^0, \dfrac{1}{2} = 2^{-1}$
so $u_n = 2^{(4-n)}$ or $u_n = 16 \div 2^n$

Q12 a) $u_2 = au_1 + 2 = 3a + 2$
$u_3 = au_2 + 2 = a(3a + 2) + 2 = 3a^2 + 2a + 2$
[2 marks available — 1 mark for each]

b) Solve $3a^2 + 2a + 2 = 87 \Rightarrow 3a^2 + 2a - 85 = 0$
$\Rightarrow (3a + 17)(a - 5) = 0 \Rightarrow a = -\dfrac{17}{3}$ and $a = 5$
[3 marks available — 1 mark for forming the quadratic equation, 1 mark for correctly solving the quadratic equation, 1 mark for both correct answers]

Q13 a) $u_3 = 2, u_4 = 3, u_5 = 5, u_6 = 8$
So there are 5 leaves on branch 5, and 8 leaves on branch 6.

b) Cumulative totals for the Fibonacci sequence are 1, 2, 4, 7, 12, 20, 33, 54. So there are 8 branches on the tree.

Q14 a) $u_3 = 2u_2 - u_1 = 2 \times 6 - 3 = 9$
$u_4 = 2u_3 - u_2 = 2 \times 9 - 6 = 12$
$u_5 = 2u_4 - u_3 = 2 \times 12 - 9 = 15$
[2 marks available — 2 marks for all three correct, otherwise 1 mark for any two correct]

b) (i) $u_3 = 2 \times 2a - a = 3a$
$u_4 = 2 \times 3a - 2a = 4a$
$u_5 = 2 \times 4a - 3a = 5a$
[2 marks available — 2 marks for all three correct, otherwise 1 mark for any two correct]

(ii) $u_n = nu_1$ (or equivalent such as $u_n = na$)
[1 mark for the correct formula]

c) $u_1 = a, u_2 = 3a, u_3 = 6a - a = 5a, u_4 = 10a - 3a = 7a$
This is every "odd" multiple of a.
The nth odd number is given by $2n - 1$. So, $u_n = (2n - 1)u_1$
[2 marks available — 1 mark for a correct method, 1 mark for the correct formula]

Q15 a) $u_2 = p^2 - 1$
$u_3 = (p^2 - 1)^2 - 1 = p^4 - 2p^2$

b) (i) $u_1 = -1, u_2 = 0, u_3 = -1, u_4 = 0$

(ii) All odd terms are equal to –1, all even terms are equal to 0, the sequence repeats every two terms so it is a periodic sequence with period 2.

Answers

c) The property of being periodic with period 2 holds when $u_3 = u_1$ (and not equal to u_2).
$p^4 - 2p^2 = p \Rightarrow p(p^3 - 2p - 1) = 0$
We know $p = -1$ gives the property, so by the factor theorem $(p + 1)$ should be a factor of $(p^3 - 2p - 1)$.
By equating coefficients we get:
$p(p^3 - 2p - 1) = p(p + 1)(p^2 - p - 1) = 0$
$p^2 - p - 1 = 0 \Rightarrow p = \dfrac{1 \pm \sqrt{(-1)^2 - 4 \times 1 \times -1}}{2 \times 1} = \dfrac{1 \pm \sqrt{5}}{2}$
But p must be an integer so you can ignore these solutions.
So $p = 0$ or $p = -1$. When $p = 0$, $u_2 = -1$, $u_3 = 0$, using the result from a), and $u_2 = u_1{}^2 - 1 \Rightarrow u_1 = 0$.
So the sequence has period 2 for $p = 0$. Hence there are only two integer values of p that make this sequence periodic with period 2: $p = 0$ and $p = -1$.

Exercise 5.3 — Arithmetic Sequences

Q1 $a = 7$, $d = 5$
n^{th} term $= a + (n - 1)d = 7 + (n - 1)5 = 5n + 2$
10th term $= 5(10) + 2 = 52$

Q2 **a)** $a = 6$, $d = 3$, so n^{th} term $= 6 + (n - 1)3 = 3n + 3$

b) $a = 4$, $d = 5$, n^{th} term $= 5n - 1$

c) $a = 12$, $d = -4$, n^{th} term $= -4n + 16$

d) $a = 1.5$, $d = 2$, n^{th} term $= 2n - 0.5$

e) $a = 77$, $d = -8$, n^{th} term $= 85 - 8n$

f) $a = -2$, $d = -0.5$, n^{th} term $= -0.5n - 1.5$

Q3 Form equations for 4th and 10th terms:
n^{th} term $= a + (n - 1)d$
$19 = a + (4 - 1)d \Rightarrow 19 = a + 3d$
$43 = a + (10 - 1)d \Rightarrow 43 = a + 9d$
Solving the simultaneous equations: $a = 7$, $d = 4$.

Q4 1st term $= u_1 = a = -5$
Form an equation for the 5th term:
n^{th} term $= a + (n - 1)d$
$19 = a + (5 - 1)d \Rightarrow 19 = -5 + 4d \Rightarrow d = 6$
So $u_{10} = -5 + 9 \times 6 = 49$

Q5 Form equations for 7th and 11th terms:
n^{th} term $= a + (n - 1)d$
$8 = a + (7 - 1)d \Rightarrow 8 = a + 6d$
$10 = a + (11 - 1)d \Rightarrow 10 = a + 10d$
Solving the simultaneous equations: $a = 5$, $d = 0.5$
So $u_3 = 5 + 2(0.5) = 6$

Q6 **a)** $u_n = 6 + (n - 1) \times 4 = 4n + 2$
[2 marks available — 1 mark for using a correct method, 1 mark for the correct final answer]

b) $4n + 2 = 2(n + 1) = 2k$ for some integer k (as n is an integer)
$2k$ is even and so every term of this sequence will be even.
[1 mark for a correct explanation]

c) $4(k + 10) + 2 = 2(4k + 2) + 10$
$\Rightarrow 4k + 42 = 8k + 14 \Rightarrow 4k = 42 - 14 = 28 \Rightarrow k = 7$
[3 marks available — 1 mark for forming the equation, 1 mark for an appropriate method to solve it, 1 mark for the correct answer]

Q7 Form equations for 3rd and 7th terms:
n^{th} term $= a + (n - 1)d$
$15 = a + (3 - 1)d \Rightarrow 15 = a + 2d$
$27 = a + (7 - 1)d \Rightarrow 27 = a + 6d$
Solving the simultaneous equations: $a = 9$, $d = 3$.
Now write an equation for the kth term: $66 = 9 + (k - 1)3 = 6 + 3k$
And solve to find $k = 20$

Q8 **a)** Form equations for 8th and 12th terms:
n^{th} term $= a + (n - 1)d$
$78 = a + (8 - 1)d \Rightarrow 78 = a + 7d$
$102 = a + (12 - 1)d \Rightarrow 102 = a + 11d$
Solving simultaneously: $4d = 24 \Rightarrow d = 6$ and $a = 36$.
[3 marks available — 1 mark for a correct method, 1 mark for a, 1 mark for d]
Alternatively, the 8th and 12th terms are 4 terms apart with a total difference of $102 - 78 = 24$, so the common difference is $24 \div 4 = 6$. The 1st and 8th terms are 7 terms apart, so the 1st term is $78 - 6 \times 7 = 36$.

b) n^{th} term is $u_n = 36 + 6(n - 1) = 6n + 30$
Solve $6n + 30 = 150 \Rightarrow 6n = 120 \Rightarrow n = 20$
[2 marks available — 1 mark for forming the equation, 1 mark for the correct answer]

c) Let the terms be u_k and u_{k+1}
Then, $u_k = 36 + 6(k - 1) = 6k + 30$ and,
$u_{k+1} = 36 + 6(k + 1 - 1) = 6k + 36$
So $6k + 30 + 6k + 36 = 438 \Rightarrow 12k = 372 \Rightarrow k = 31$
Terms and values are $u_{31} = 36 + 30 \times 6 = 216$ and $u_{32} = 222$
[4 marks available — 1 mark for setting up expressions for consecutive terms, 1 mark for forming an equation for the sum, 1 mark for 31 and 32, 1 mark for 216 and 222]

Q9 The difference between all the terms is the same in an arithmetic sequence. Use this fact to set up an equation:
$\ln(x + 8) - \ln x = \ln(x + 48) - \ln(x + 8)$
$\Rightarrow \ln\left(\dfrac{x + 8}{x}\right) = \ln\left(\dfrac{x + 48}{x + 8}\right) \Rightarrow \dfrac{x + 8}{x} = \dfrac{x + 48}{x + 8}$
$\Rightarrow (x + 8)^2 = (x + 48)x \Rightarrow x^2 + 16x + 64 = x^2 + 48x$
$\Rightarrow 64 = 32x \Rightarrow x = 2$
So the common difference is $\ln(2 + 8) - \ln 2 = \ln 5$.
Add $\ln 5$ to the third term to get the next term:
$\ln(x + 48) + \ln 5 = \ln(2 + 48) + \ln 5 = \ln 250$.

Q10 For there to be 20 integer terms above 0 and below 100, there must be 19 'gaps' between the terms. The size of the gap is the common difference, d, where:
$1 + 19d < 100$ or $99 - 19d > 0$
$\Rightarrow d < 99 \div 19 = 5.2105...$ so the progression must have a common difference of $d = \pm 5$.
Using $d = 5$, $n = 11$ and $u_n = 47$:
$47 = a + 10 \times 5 \Rightarrow a = -3$
Using $d = -5$, $n = 11$ and $u_n = 47$:
$47 = a + 10 \times -5 \Rightarrow a = 97$

Exercise 5.4 — Arithmetic Series

Q1 $a = 8$, $d = 3$
n^{th} term $= a + (n - 1)d = 8 + (n - 1)3 = 3n + 5$
10th term $= 3(10) + 5 = 35$
$S_n = \dfrac{n}{2}[2a + (n - 1)d]$
$S_{10} = \dfrac{10}{2}[2(8) + 9(3)] = 215$
Alternatively, you could have used the $S_n = \dfrac{1}{2}n(a + l)$ formula here. You'd worked out the last term earlier in the question.

Q2 Form equations for 2nd and 5th terms:
n^{th} term $= a + (n - 1)d$
$16 = a + d$
$10 = a + 4d$
Solving the simultaneous equations: $a = 18$, $d = -2$.
$S_n = \dfrac{n}{2}[2a + (n - 1)d]$
$S_8 = \dfrac{8}{2}[2(18) + 7(-2)] = 88$

Q3 $a = 12, d = 6$

n^{th} term $= a + (n - 1)d = 12 + (n - 1)6 = 6n + 6$

$u_{100} = 6(100) + 6 = 606$

$S_n = \frac{1}{2}n(a + l)$

$S_{100} = \frac{1}{2} \times 100(12 + 606) = 30\,900$

Q4 1st term $a = 8(1) - 6 = 2$

20th term $l = 8(20) - 6 = 154$

$S_n = \frac{1}{2}n(a + l)$

$S_{20} = \frac{1}{2} \times 20(2 + 154) = 1560$

You could also read off $d = 8$ from the nth term and use the other formula.

Q5 $S_n = \frac{n}{2}[2a + (n - 1)d]$

$S_{10} = \frac{10}{2}[2(-6) + 9d] = 345$

$-60 + 45d = 345$

So $d = 9$

Q6 $u_n = a + (n - 1)d$

For the last term: $79 = 7 + 6(n - 1) \Rightarrow n = 13$

$S_n = \frac{1}{2}n(a + l)$

$S_{13} = \frac{1}{2} \times 13(7 + 79) = 559$

Q7 **a)** $a = 5(1) - 2 = 3$

$l = 5(12) - 2 = 58$

$S_n = \frac{1}{2}n(a + l)$

$S_{12} = \frac{1}{2} \times 12(3 + 58) = 366$

b) $a = 20 - 2(1) = 18, \ l = 20 - 2(9) = 2$

$S_n = \frac{1}{2}n(a + l)$

$S_9 = \frac{1}{2} \times 9(18 + 2) = 90$

Q8 **a)** $u_6 = 80 - 6 \times 6 = 44$

$u_{10} = 80 - 6 \times 10 = 20$

[2 marks available — 1 mark for u_6, 1 mark for u_{10}]

b) i) $\sum_{r=1}^{6} 80 - 6r = S_6$ and $\sum_{r=1}^{10} 80 - 6r = S_{10}$

$u_1 = a = 80 - 6 \times 1 = 74$

From part a) you know the last terms so $S_n = \frac{1}{2}n(a + l)$

$\Rightarrow S_6 = \frac{6}{2}(74 + 44) = 354$

and $S_{10} = \frac{10}{2}(74 + 20) = 470$

[3 marks available — 1 mark for a correct method, 1 mark for S_6, 1 mark for S_{10}]

ii) $\sum_{r=7}^{10} 80 - 6r = S_{10} - S_6 = 470 - 354 = 116$

[1 mark]

Q9 $a = 3, d = 2$

$S_n = \frac{n}{2}[2a + (n - 1)d]$

$960 = \frac{n}{2}[2(3) + 2(n - 1)]$

$960 = n^2 + 2n$

$n^2 + 2n - 960 = 0$

You're expecting a whole number for n, so you should be able to factorise the quadratic — you need two numbers that are 2 apart and multiply to give 960.

$(n + 32)(n - 30) = 0$

Ignore the negative solution since n needs to be positive, so $n = 30$.

Q10 $a = 5(1) + 2 = 7, \ l = 5k + 2$

$S_n = \frac{1}{2}n(a + l)$

$553 = \frac{1}{2}k(7 + 5k + 2)$

$1106 = 5k^2 + 9k$

$5k^2 + 9k - 1106 = 0$

Factorising gives: $(5k + 79)(k - 14) = 0$

Now we can ignore the negative solution, so $k = 14$.

This one looked very tricky to factorise, but you can cheat a little here — you know you're trying to get to $k = 14$, so one of the brackets is going to be $(k - 14)$...

Q11 The first thing to do is use the fact that it's an arithmetic progression to write down some equations — remember, there's a common difference, d, between each term.

$x + 11 + d = 4x + 4 \Rightarrow -3x + d = -7$

$x + 11 + 2d = 9x + 5 \Rightarrow -8x + 2d = -6$

You've now got a pair of simultaneous equations in d and x. Solving these gives $x = -4, d = -19$

So the first term is $a = -4 + 11 = 7$.

Now you can put $a = 7$, $d = -19$ and $n = 11$ into the formula for S_n:

$S_n = \frac{n}{2}[2a + (n - 1)d]$

$S_{11} = \frac{11}{2}[2(7) + 10(-19)]$

$S_{11} = \frac{11}{2} \times -176 = -968$

Q12 $a = 36, d = -4$

$S_n = \frac{n}{2}[2a + (n - 1)d]$

$176 = \frac{n}{2}[2(36) - 4(n - 1)]$

$88 = 19n - n^2$

$n^2 - 19n + 88 = 0$

You know you should be able to factorise the quadratic as n is a whole number:

$(n - 8)(n - 11) = 0$

So $n = 8$ or 11.

Q13 **a)** $u_n = a + (n - 1)d \Rightarrow u_9 = a + 8d = 61 \Rightarrow a = 61 - 8d$

$S_n = \frac{n}{2}[2a + (n - 1)d] \Rightarrow S_{12} = \frac{12}{2}[2a + 11d] = 522$

Solving simultaneously by substitution:

$6[2(61 - 8d) + 11d] = 522 \Rightarrow 122 - 5d = 87$

$\Rightarrow 5d = 35 \Rightarrow d = 7$ and $a = 61 - 8(7) = 5$

[4 marks available — 1 mark for an equation using u_9, 1 mark for an equation using S_{12}, 1 mark for solving simultaneously, 1 mark for both correct answers]

b) $S_n > 1000 \Rightarrow \frac{n}{2}[10 + 7(n - 1)] > 1000$

$\Rightarrow n(7n + 3) > 2000 \Rightarrow 7n^2 + 3n - 2000 > 0$

$7n^2 + 3n - 2000 = 0 \Rightarrow n = \frac{-3 \pm \sqrt{3^2 - 4 \times 7 \times -2000}}{2 \times 7}$

$\Rightarrow n = -17.11...$ or $n = 16.69...$

As n needs to be positive $n > 16.69...$ and as it needs to be an integer, $n = 17$. So the minimum number of terms for the sum to exceed 1000 is 17.

[4 marks available — 1 mark for forming the inequality, 1 mark for rearranging to give a quadratic, 1 mark for a correct method to solve the quadratic, 1 mark for the correct answer with interpretation]

Answers

Q14 $u_n = a + (n-1)d$ and $a = 5 \Rightarrow u_{k-4} = 115 = 5 + (k-4-1)d$

$\Rightarrow d(k-5) = 110 \Rightarrow k = \dfrac{110}{d} + 5$

$S_n = \dfrac{n}{2}[2a + (n-1)d]$

$\Rightarrow S_{k+4} = 2000 = \dfrac{k+4}{2}[10 + (k+4-1)d]$

$\Rightarrow (k+4)[10 + d(k+3)] = 4000$

Solve simultaneously by substitution:

$\left(\dfrac{110}{d} + 5 + 4\right)\left[10 + d\left(\dfrac{110}{d} + 5 + 3\right)\right] = 4000$

$\Rightarrow \left(\dfrac{110}{d} + 9\right)\left[10 + d\left(\dfrac{110}{d} + 8\right)\right] = 4000$

$\Rightarrow \left(\dfrac{110}{d} + 9\right)(120 + 8d) = 4000$

$\Rightarrow (110 + 9d)(120 + 8d) = 4000d$

$\Rightarrow (110 + 9d)(15 + d) = 500d$

$\Rightarrow 9d^2 - 255d + 1650 = 0 \Rightarrow 3d^2 - 85d + 550 = 0$

$\Rightarrow (3d - 55)(d - 10) = 0 \Rightarrow d = \dfrac{55}{3},\ d = 10$

Substituting back in for k: $k = 11$ and $k = 16$

Possible values are $d = 10$ and $k = 16$ or $d = \dfrac{55}{3}$ and $k = 11$

[7 marks available — 1 mark for an equation using u_{k-4}, 1 mark for an equation using S_{k+4}, 1 mark for a correct method of solving simultaneous equations, 1 mark for reducing to a quadratic, 1 mark for a correct method to solve the quadratic, 1 mark for the correct values of k, 1 mark for the correct corresponding values of d]

Exercise 5.5 — Sum of the First n Natural Numbers

Q1 **a)** $S_n = \dfrac{1}{2}n(n+1)$

$S_{10} = \dfrac{1}{2} \times 10 \times 11 = 55$

b) $S_{2000} = \dfrac{1}{2} \times 2000 \times 2001 = 2\,001\,000$

Q2 $S_{32} = \dfrac{1}{2} \times 32 \times 33 = 528$

Q3 $\displaystyle\sum_{n=11}^{20} n = \sum_{n=1}^{20} n - \sum_{n=1}^{10} n = S_{20} - S_{10}$

$S_{10} = \dfrac{1}{2} \times 10 \times 11 = 55$

$S_{20} = \dfrac{1}{2} \times 20 \times 21 = 210$

$\displaystyle\sum_{n=11}^{20} n = 210 - 55 = 155$

Q4 $66 = \dfrac{1}{2}n(n+1)$

$132 = n^2 + n$

$0 = n^2 + n - 132$

$(n+12)(n-11) = 0$

$n = -12$ or 11 — so ignoring the negative answer, the sum of the first 11 terms is 66, so $n = 11$.

Q5 $S_n = \dfrac{1}{2}n(n+1)$

$120 = \dfrac{1}{2}k(k+1)$

$240 = k^2 + k$

$0 = k^2 + k - 240$

$(k+16)(k-15) = 0$

Ignoring the negative solution gives $k = 15$.

Q6 Subtract the sum of the first 15 natural numbers from the sum of the first 35:

$S_{35} = \dfrac{1}{2} \times 35 \times 36 = 630$

$S_{15} = \dfrac{1}{2} \times 15 \times 16 = 120$

So the sum of the series is $630 - 120 = 510$.

Q7 $S_n = \dfrac{1}{2}n(n+1)$

$\dfrac{1}{2}k(k+1) > 1\,000\,000$

$k^2 + k > 2\,000\,000$

$k^2 + k - 2\,000\,000 > 0$

Put the quadratic equal to zero and solve using the quadratic formula to get $k = 1413.7...$ or $-1414.7...$

It's a u-shaped quadratic, so the quadratic is positive when $k > 1413.7$ (ignoring the negative solution).

So the first natural number for which the sum exceeds $1\,000\,000$ is 1414.

Q8 **a)**

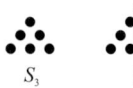

b) This is equivalent to the sum of n natural numbers but starting at 3, so you are missing the first two numbers.

So $150 = S_n - 2 - 1 \Rightarrow S_n = 153$

$\Rightarrow S_n = \dfrac{1}{2}n(n+1) = 153$

$n^2 + n - 306 = 0 \Rightarrow (n+18)(n-17) = 0$

Reject $n = -18$ as a solution as n is a number of rows of balls, so $n = 17$. But you're missing the first 2 rows so there will be $17 - 2 = 15$ rows of balls.

Q9 **a)** $\displaystyle\sum_{n=1}^{80} n = S_{80} = \dfrac{1}{2}(80)(81) = 3240$

b) **(i)** This is the sum of the natural numbers from 2 to 81.

(ii) This is the sum of the first 80 odd numbers.

Q10 $\dfrac{1}{2}(20)(21) - \dfrac{1}{2}k(k+1) = 74$

$\Rightarrow k^2 + k - 272 = 0 \Rightarrow (k-16)(k+17) = 0 \Rightarrow k = 16,\ k = -17$

$\Rightarrow k$ is a number of terms so needs to be positive and so $k = 16$

[4 marks available — 1 mark for setting up equation, 1 mark for simplifying it to give a quadratic, 1 mark for a correct method to solve the quadratic, 1 mark for the correct answer]

Q11 **a)** **(i)** $S_3 + S_4 = \dfrac{1}{2}(3)(4) + \dfrac{1}{2}(4)(5) = 6 + 10 = 16$ *[1 mark]*

(ii) $S_4 + S_5 = 10 + \dfrac{1}{2}(5)(6) = 25$ *[1 mark]*

b) $S_k + S_{k+1} = \dfrac{1}{2}(k)(k+1) + \dfrac{1}{2}(k+1)(k+1+1)$

$= \dfrac{1}{2}(k+1)[k + k + 2]$

$= \dfrac{1}{2}(k+1)[2(k+1)] = (k+1)^2$

Given that k is an integer, $(k+1)$ is also an integer and so $(k+1)^2$ is a square number.

[3 marks available — 1 mark for a correct expression for $S_k + S_{k+1}$, 1 mark for a correct method to simplify the expression, 1 mark for $(k+1)^2$ with justification]

Q12 $S_p = \dfrac{1}{2}p(p+1) = pq \Rightarrow p + 1 = 2q \Rightarrow p = 2q - 1$

You can only do this rearrangement because p is greater than 1 (i.e. it can't be 0, so you can divide by p rather than factorising).

$S_q = \dfrac{1}{2}q(q+1) = 4p - 3q$

Simultaneous equations to solve by substitution:

$q(q+1) = 2(4(2q-1) - 3q) \Rightarrow q^2 + q = 10q - 8$

$\Rightarrow q^2 - 9q + 8 = 0 \Rightarrow (q-8)(q-1) = 0 \Rightarrow q = 8$ or $q = 1$

and $p = 2(8) - 1 = 15$ or $p = 2(1) - 1 = 1$

Since $p > q > 1$, the only solution must be $p = 15$, $q = 8$.

Exercise 5.6 — Geometric Sequences

Q1 E.g. $r = \dfrac{\text{second term}}{\text{first term}} = \dfrac{1875}{3125} = \dfrac{3}{5}$

Q2 Common ratio $r = \dfrac{\text{second term}}{\text{first term}} = \dfrac{3}{2} = 1.5$.
Then:
4th term = 3rd term $\times$ 1.5 = 4.5 $\times$ 1.5 = 6.75
5th term = 4th term $\times$ 1.5 = 6.75 $\times$ 1.5 = 10.125,
6th term = 5th term $\times$ 1.5 = 10.125 $\times$ 1.5 = 15.1875,
7th term = 6th term $\times$ 1.5 = 15.1875 $\times$ 1.5 = 22.78125.
This method isn't as slow as it looks because you can use a scientific calculator to get the terms of the series quickly: press '2 =' then 'x 1.5 =' to get the second term. Pressing '=' repeatedly will give you the following terms. Even so, the method below is quicker, so unless you're asked to find the term after one you've already got, you're better off doing this:
Or: First term $a = 2$, common ratio $r = 1.5$
n^{th} term = $ar^{n-1} = 2 \times (1.5)^{n-1}$
7th term = $2 \times (1.5)^6 = 2 \times 11.390625 = 22.78125$

Q3 Common ratio $r = \dfrac{7^{\text{th}} \text{ term}}{6^{\text{th}} \text{ term}} = \dfrac{6561}{2187} = 3$.
The 6th term is $2187 = ar^5 = a \times 3^5 \Rightarrow a = \dfrac{2187}{3^5} = 9$.
So the 1st term is 9.

Q4 Common ratio: $r = \dfrac{\text{second term}}{\text{first term}} = \dfrac{12}{24} = 0.5$.
First term: $a = 24$.
n^{th} term: $u_n = ar^{n-1} = 24 \times (0.5)^{n-1}$
9th term: $u_9 = ar^8 = 24 \times (0.5)^8 = 0.09375$

Q5 First term: $a = 1.125$
n^{th} term: $u_n = ar^{n-1} = 1.125r^{n-1}$
14th term: $9216 = u_{14} = 1.125r^{13}$
$9216 = 1.125r^{13} \Rightarrow r^{13} = \dfrac{9216}{1.125} = 8192$
$\Rightarrow r = \sqrt[13]{8192} = 2$

Q6 Common ratio: $r = \dfrac{1.1}{1} = 1.1$, first term: $a = 1$.
n^{th} term: $u_n = ar^{n-1} = 1 \times 1.1^{n-1} = 1.1^{n-1}$
So to find the number of terms in the sequence that are less than 4, solve: $u_n = 1.1^{n-1} < 4$
$\Rightarrow \log 1.1^{n-1} < \log 4$
Use the log law, $\log x^n = n(\log x)$:
$\Rightarrow (n-1)\log 1.1 < \log 4$
log 1.1 > 0 so dividing through by log 1.1 doesn't change the direction of the inequality:
$\Rightarrow n - 1 < \dfrac{\log 4}{\log 1.1} \Rightarrow n - 1 < 14.54 \Rightarrow n < 15.54$
so u_n is less than 4 when n is less than 15.54 (2 d.p.). Therefore u_{15} is the last term that's less than 4,
so there are 15 terms that are less than 4.
You could solve this as an equation instead, finding the value of n such that $u_n = 4$ and rounding down.

Q7 $a = 5$, $r = 0.6$, n^{th} term: $u_n = ar^{n-1} = 5 \times (0.6)^{n-1}$
10th term: $u_{10} = 5 \times (0.6)^9 = 0.050388$ (6 d.p.)
15th term: $u_{15} = 5 \times (0.6)^{14} = 0.003918$ (6 d.p.)
Difference: $0.003918 - 0.050388 = -0.04647$ (5 d.p.)
You could also have:
Difference: 0.050388 − 0.003918 = 0.04647 (5 d.p.)

Q8 $a = 25\,000$, $r = 0.8$
n^{th} term: $u_n = ar^{n-1} = 25\,000 \times (0.8)^{n-1}$
to find the first term in the sequence less than 1000, solve:
$u_n = 25\,000 \times (0.8)^{n-1} < 1000$
$25\,000 \times (0.8)^{n-1} < 1000 \Rightarrow (0.8)^{n-1} < \dfrac{1000}{25\,000} = 0.04$
$\Rightarrow \log(0.8)^{n-1} < \log 0.04 \Rightarrow (n-1)\log(0.8) < \log 0.04$
$\Rightarrow n - 1 > \dfrac{\log 0.04}{\log 0.8} \Rightarrow n - 1 > 14.425... \Rightarrow n > 15.425...$
so u_n is less than 1000 when n is greater than 15.425..., therefore u_{16} is the first term that's less than 1000.
0.8 < 1 so log 0.8 < 0 and dividing through by log 0.8 changes the direction of the inequality because log 0.8 is negative. Again, you could solve this as an equation by finding n such that $u_n = 1000$ and rounding up.

Q9 Divide consecutive terms to find the common ratio r:
e.g. $r = \dfrac{\text{second term}}{\text{first term}} = \dfrac{-5}{5} = -1$

Q10 a) Common ratio $r = \dfrac{\text{second term}}{\text{first term}} = \dfrac{\frac{3}{16}}{\frac{1}{4}} = \dfrac{3}{4}$

b) First term: $a = \dfrac{1}{4}$
n^{th} term: $u_n = ar^{n-1} = \dfrac{1}{4} \times \left(\dfrac{3}{4}\right)^{n-1}$
8th term: $u_8 = \dfrac{1}{4} \times \left(\dfrac{3}{4}\right)^7 = \dfrac{1}{4} \times \dfrac{2187}{16\,384} = \dfrac{2187}{65\,536}$

Q11 $r = 0.8$, n^{th} term: $u_n = ar^{n-1} = a(0.8)^{n-1}$
7th term: $196.608 = u_7 = a(0.8)^6$
$196.608 = a(0.8)^6 \Rightarrow a = \dfrac{196.608}{0.8^6} = 750$

Q12 Common ratio: $r = \dfrac{6}{2} = 3$, first term: $a = 2$.
n^{th} term: $u_n = ar^{n-1} = 2 \times 3^{n-1}$
So to find the term which equals 1458, solve:
$u_n = 1458 = 2 \times 3^{n-1} \Rightarrow 3^{n-1} = \dfrac{1458}{2} = 729$
$\Rightarrow \log(3)^{n-1} = \log 729 \Rightarrow (n-1)\log 3 = \log 729$
$\Rightarrow n - 1 = \dfrac{\log 729}{\log 3} = 6 \Rightarrow n = 7$
So the 7th term of the progression is 1458.

Q13 a) Common ratio $r = \dfrac{\text{second term}}{\text{first term}} = \dfrac{-2.4}{3} = -0.8$

b) Continuing the sequence gives -1.536, 1.2288, -0.98304.
So there are 5 terms in the series before a term has modulus less than 1.
You could also answer this part of the question by writing a new series where each term is the modulus of the old series, then using logs to find the first term less than 1. But in this case it's much easier to just find the next few terms.

Q14 $u_8 = ar^7$, $u_6 = ar^5$ So, $ar^7 = 9ar^5 \Rightarrow \dfrac{ar^7}{ar^5} = 9 \Rightarrow r^2 = 9 \Rightarrow r = \pm 3$
[3 marks available — 1 mark for expressions for u_8 and u_6, 1 mark for forming an equation, 1 mark for simplifying to the correct conclusion]

Q15 u_n: $a = 6$, $r = 2$, $u_n = 6(2)^{n-1}$
v_n: $a = 4$, $r = 2$, $v_n = 4(2)^{n-1}$
$u_i - v_i = 6(2)^{i-1} - 4(2)^{i-1} = 2^{i-1}(6-4) = 2(2^{i-1}) = 2^{1+i-1} = 2^i$
[3 marks available — 1 mark for finding both u_n and v_n, 1 mark for finding $u_i - v_i$, 1 mark for rearranging to give the required result]

Q16 a) Using $u_n = ar^{n-1}$, $u_n = 4\left(\dfrac{1}{4}\right)^{n-1} = \left(\dfrac{1}{4}\right)^{-1}\left(\dfrac{1}{4}\right)^{n-1}$
$= \left(\dfrac{1}{4}\right)^{-1+n-1} = \left(\dfrac{1}{4}\right)^{n-2} = (4^{-1})^{n-2} = 4^{2-n}$

b) $4^{2-8} = 4^{-6} = \dfrac{1}{4096} = 0.000\,244$ (3 s.f.)

Answers

c) Find n such that $4^{2-n} < 10^{-9}$ $\Rightarrow$ $(2-n)\ln 4 < \ln 10^{-9}$
$\Rightarrow$ $2 - n < -14.948...$ $\Rightarrow$ $n > 16.948...$
$\Rightarrow$ the 17th term is the first one that is less than one billionth.

Q17 a) $u_n = ar^{n-1}$ $\Rightarrow$ $u_n = x \times 3^{n-1} = x \times 3^n \times 3^{-1} = \frac{1}{3}x(3^n)$
[2 marks available — 1 mark for use of u_n formula, 1 mark for rearranging to give the required result]

b) $\frac{1}{3}x(3^n) > 1\,000\,000x$ $\Rightarrow$ $(3^n) > 3\,000\,000$
$\Rightarrow$ $n\ln 3 > \ln 3\,000\,000$ $\Rightarrow$ $n > \frac{\ln 3\,000\,000}{\ln 3}$
$\Rightarrow$ $n > 13.5754...$
$\Rightarrow$ The 14th term is the first that is greater than $1\,000\,000x$.
[3 marks available — 1 mark for setting up the inequality, 1 mark for a correct method of solving, 1 mark for the correct final answer]

Q18 $\frac{8x+2}{2x+14} = \frac{2x+14}{x-2}$ $\Rightarrow$ $\frac{4x+1}{x+7} = \frac{2(x+7)}{x-2}$
$(4x+1)(x-2) = 2(x+7)^2$ $\Rightarrow$ $4x^2 - 7x - 2 = 2x^2 + 28x + 98$
$2x^2 - 35x - 100 = 0$ $\Rightarrow$ $(2x+5)(x-20) = 0$
$x = -2.5$ would give $r = \frac{2x+14}{x-2} = \frac{9}{-4.5} = -2$
$x = 20$ would give $r = \frac{2x+14}{x-2} = \frac{54}{18} = 3$
[5 marks available — 1 mark for setting up ratios, 1 mark for leading to a quadratic, 1 mark for a correct method to solve the quadratic, 1 mark for both x values, 1 mark for both corresponding r values]

Exercise 5.7 — Geometric Series

Q1 The sum of the first n terms is $S_n = \frac{a(1-r^n)}{(1-r)}$, $a = 8$
and $r = 1.2$, so the sum of the first 15 terms is:
$S_{15} = \frac{a(1-r^{15})}{(1-r)} = \frac{8(1-(1.2)^{15})}{(1-1.2)} = 576.28$ to 2 d.p.

Q2 For a geometric series with first term a and
common ratio r: $\sum_{k=0}^{n-1} ar^k = \frac{a(1-r^n)}{1-r}$,
$\sum_{k=0}^{9} ar^k = \sum_{k=0}^{9} 25(0.7)^k = \frac{25(1-(0.7)^{10})}{1-0.7} = 80.98$ (2 d.p.)

Q3 $a = 3$ and $r = 2$, the sum of the first n terms is:
$\frac{3(1-2^n)}{(1-2)} = -3(1-2^n)$
$196\,605 = S_n = -3(1-2^n)$ $\Rightarrow$ $-65\,535 = 1 - 2^n$
$\Rightarrow$ $65\,536 = 2^n$ $\Rightarrow$ $\log 65\,536 = \log 2^n$
$\Rightarrow$ $\log 65\,536 = n\log 2$ $\Rightarrow$ $n = \frac{\log 65\,536}{\log 2} = 16$

Q4 The first term is $a = 4$,
the common ratio is $r = \frac{\text{second term}}{\text{first term}} = \frac{5}{4} = 1.25$
The sum of the first x terms is
$S_x = \frac{a(1-r^x)}{(1-r)} = \frac{4(1-(1.25)^x)}{(1-1.25)} = 16(1-(1.25)^x)$
So: $103.2 \approx 16(1-(1.25)^x)$ $\Rightarrow$ $-6.45 \approx 1 - (1.25)^x$
$\Rightarrow$ $7.45 \approx 1.25^x$ $\Rightarrow$ $\log 7.45 \approx x\log 1.25$
$\Rightarrow$ $x \approx \frac{\log 7.45}{\log 1.25} = 8.9965...$ $\Rightarrow$ $x = 9$ (as x is an integer)

Q5 a) 3rd term $= ar^2 = 6$, 8th term $= ar^7 = 192$.
Dividing the two equations gives:
$\frac{ar^7}{ar^2} = r^5 = \frac{192}{6} = 32$ $\Rightarrow$ $r = \sqrt[5]{32} = 2$

b) 3rd term $= ar^2 = 6$ and $r = 2$, so $a = \frac{6}{r^2} = \frac{6}{2^2} = 1.5$

c) The sum of the first 15 terms is:
$S_{15} = \frac{a(1-r^{15})}{(1-r)} = \frac{1.5(1-2^{15})}{(1-2)} = 49\,150.5$

Q6 a) $a = 2, r = \frac{1}{2}, S_n = \frac{2(1-(\frac{1}{2})^n)}{1-\frac{1}{2}}$
$= 4(1-2^{-n}) = 4 - 4(2^{-n}) = 4 - (2^2)(2^{-n}) = 4 - 2^{2-n}$
[3 marks available — 1 mark for forming an equation for S_n, 1 mark for partially rearranging/simplifying, 1 mark for a full rearrangement leading to the required result]

b) (i) $u_{10} = ar^9 = 2\left(\frac{1}{2}\right)^9 = \frac{1}{256}$
[2 marks available — 1 mark for a correct method to find u_{10}, 1 mark for u_{10} correct]

(ii) $S_{10} = 4 - 2^{2-n} = 4 - 2^{-8} = \frac{1023}{256}$ *[1 mark]*

Q7 a) Common ratio: $\frac{\text{2nd term}}{\text{1st term}} = \frac{\text{3rd term}}{\text{2nd term}}$, so:
$\frac{m}{m+10} = \frac{2m-21}{m}$ $\Rightarrow$ $m^2 = (m+10)(2m-21)$
$\Rightarrow$ $m^2 = 2m^2 - 21m + 20m - 210$ $\Rightarrow$ $0 = m^2 - m - 210$

b) Factorising $m^2 - m - 210 = 0$ gives:
$(m-15)(m+14) = 0$, so $m = 15$ or $m = -14$,
since $m > 0$, $m = 15$.

c) $m = 15$ gives the first three terms 25, 15, 9.
Common ratio $= \frac{\text{second term}}{\text{first term}} = \frac{15}{25} = 0.6$

d) $a = 25$ and $r = 0.6$, so sum of first 10 terms is:
$S_{10} = \frac{a(1-r^{10})}{(1-r)} = \frac{25(1-0.6^{10})}{(1-0.6)} = 62.12$ to 2 d.p.

Q8 a) $1 + x + x^2 = 3$ $\Rightarrow$ $x^2 + x - 2 = 0$
$\Rightarrow$ $(x-1)(x+2) = 0$ $\Rightarrow$ $x = 1$ or $x = -2$.
Since the terms are all different $x \neq 1$ (as 1 is the first term and $x = 1 \Rightarrow x^2 = 1^2 = 1$), hence $x = -2$.

b) $a = 1$ and $r = \frac{\text{second term}}{\text{first term}} = \frac{-2}{1} = -2$,
so the sum of the first 7 terms is:
$S_7 = \frac{a(1-r^7)}{(1-r)} = \frac{1(1-(-2)^7)}{(1-(-2))} = 43$

Q9 a) $n = 4$: $\sum_{k=1}^{4} 3 \times 4^{k-1} = \frac{3(1-4^4)}{1-4} = 255$
$n = 8$: $\sum_{k=1}^{8} 3 \times 4^{k-1} = \frac{3(1-4^8)}{1-4} = 65\,535$
[2 marks available — 1 mark for each final answer]
For this question the trick is to realise that $\sum_{k=1}^{n} ar^{k-1}$ is exactly the same series as $\sum_{k=0}^{n-1} ar^k$ — they are both S_n which you can see by expanding them to get '$a + ar + ar^2 + ... + ar^{n-1}$'

b) $\sum_{k=5}^{8} 3 \times 4^{k-1} = \sum_{k=1}^{8} 3 \times 4^{k-1} - \sum_{k=1}^{4} 3 \times 4^{k-1}$
$= 65\,535 - 255 = 65\,280$
[2 marks available — 1 mark for a correct method, 1 mark for the correct answer]

c) By copying the method used in parts a) and b):
$\sum_{k=7}^{10} 3 \times 4^{k-1} = \sum_{k=1}^{10} 3 \times 4^{k-1} - \sum_{k=1}^{6} 3 \times 4^{k-1}$
$= \frac{3(1-4^{10})}{1-4} - \frac{3(1-4^6)}{1-4} = 1\,048\,575 - 4095 = 1\,044\,480$
[3 marks available — 1 mark for a correct method, 1 mark for correct working, 1 mark for the correct final answer]

Q10 $a = 7.2$ and $r = 0.38$, so:
$\sum_{k=0}^{9} ar^k = \frac{a(1-r^{10})}{(1-r)} = \frac{7.2(1-0.38^{10})}{(1-0.38)} = 11.61$ to 2 d.p.

Q11 $1.2 = S_8 = \frac{a(1-r^8)}{(1-r)} = \frac{a\left(1-\left(-\frac{1}{3}\right)^8\right)}{\left(1-\left(-\frac{1}{3}\right)\right)} = a(0.749...)$
$\Rightarrow$ $a = 1.60$ to 2 d.p.

Q12 The geometric sequence a, $-2a$, $4a$, ... has first term a and common ratio $r = -2$, so $\sum_{k=0}^{12} a(-2)^k$ is the sum of the first 13 terms of the sequence. Therefore:

$$\sum_{k=0}^{12} a(-2)^k = -5735.1$$

$$\sum_{k=0}^{12} a(-2)^k = \frac{a(1-r^{13})}{(1-r)} = \frac{a(1-(-2)^{13})}{(1-(-2))} = 2731a$$

$$\Rightarrow -5735.1 = 2731a \Rightarrow a = -2.1$$

Q13 a) $u_6 = S_6 - S_5 = 3906 - 781 = 3125$
$u_7 = S_7 - S_6 = 19\,531 - 3906 = 15\,625$
[2 marks available — 1 mark for each correct answer]

b) $u_7 = ar^6 = 15\,625$ and $u_6 = ar^5 = 3125$
Therefore, $r = \frac{15625}{3125} = 5 \Rightarrow a \times 5^6 = 15\,625 \Rightarrow a = 1$
[3 marks available — 1 mark for a correct method, 1 mark for the correct common ratio (r), 1 mark for the correct first term (a)]

Q14 $S_9 = 19\,682$ and $S_{10} = 59\,048$
$u_{10} = S_{10} - S_9 = 59\,048 - 19\,682 = 39\,366$
$u_{10} - u_6 = 38\,880 \Rightarrow u_6 = 39\,366 - 38\,880 = 486$
So now we have $u_6 = ar^5 = 486$ and $u_{10} = ar^9 = 39\,366$
Therefore, $r^4 = \frac{u_{10}}{u_6} = \frac{39366}{486} = 81 \Rightarrow r = -3$ or $r = 3$
Rejecting $r = -3$ since we are told all terms in the series are positive, the common ratio is 3. Substituting to find a,
$a \times 3^5 = 486 \Rightarrow a = 2$, so the first term is 2.

Exercise 5.8 — Convergent Geometric Series

Q1 a) $r = \frac{1.1}{1} = 1.1$, $|r| = |1.1| = 1.1 > 1$,
so the sequence does not converge.

b) $r = \frac{0.8^2}{0.8} = 0.8$, $|r| = |0.8| = 0.8 < 1$,
so the sequence converges.

c) $r = \frac{\frac{1}{4}}{1} = \frac{1}{4}$, $|r| = |\frac{1}{4}| = \frac{1}{4} < 1$, so the sequence converges.

d) $r = \frac{\frac{9}{2}}{3} = \frac{3}{2}$, $|r| = |\frac{3}{2}| = \frac{3}{2} > 1$,
so the sequence does not converge.

e) $r = \frac{-\frac{1}{2}}{1} = -\frac{1}{2}$, $|r| = |-\frac{1}{2}| = \frac{1}{2} < 1$,
so the sequence converges.

f) $r = \frac{5}{5} = 1$, $|r| = |1| = 1$ (and 1 is not less than 1),
so the sequence does not converge.

Q2 Find the common ratio: $r = \frac{2^{nd}\ term}{1^{st}\ term} = \frac{8.1}{9} = 0.9$.
The first term is $a = 9$.
The sum to infinity is:
$$S_\infty = \frac{a}{1-r} = \frac{9}{1-0.9} = \frac{9}{0.1} = 90$$

Q3 $\sum_{k=0}^{\infty} ar^k = S_\infty = \frac{a}{1-r} = \frac{33}{1-0.25} = \frac{33}{0.75} = 44$

Q4 a) $a = 256$, $r = \frac{64}{256} = \frac{1}{4}$, so
$u_{10} = 256\left(\frac{1}{4}\right)^9 = \frac{1}{1024}$ and $u_{12} = 256\left(\frac{1}{4}\right)^{11} = \frac{1}{16\,384}$
[3 marks available — 1 mark finding both a and r, 1 mark for u_{10}, 1 mark for u_{12}]

b) $S_\infty = \frac{256}{1-\frac{1}{4}} = \frac{256}{\frac{3}{4}} = \frac{1024}{3}$
[2 marks available — 1 mark for use of S_∞, 1 mark for the correct answer]

Q5 $S_\infty = \frac{a}{1-r} = 2a \Rightarrow \frac{1}{1-r} = 2 \Rightarrow 1 = 2 - 2r$
$\Rightarrow 2r = 1 \Rightarrow r = 0.5$

Q6 a) Sum to infinity is $13.5 = S_\infty = \frac{a}{1-r}$
Sum of the first three terms is $13 = S_3 = \frac{a(1-r^3)}{(1-r)}$
Divide S_3 by S_∞:
$$\frac{13}{13.5} = \frac{a(1-r^3)}{1-r} \div \frac{a}{1-r} = \frac{a(1-r^3)}{1-r} \times \frac{1-r}{a} = 1 - r^3$$
You can cancel the $1 - r$ because r can't be 1 (as the series converges), so $1 - r \neq 0$.
So $1 - r^3 = \frac{13}{13.5} = \frac{26}{27} \Rightarrow r^3 = \frac{1}{27} \Rightarrow r = \frac{1}{3}$.

b) $13.5 = S_\infty = \frac{a}{1-r} \Rightarrow a = 13.5(1-r) = 13.5 \times \frac{2}{3} = 9$

Q7 $ar = 3$, $12 = S_\infty = \frac{a}{1-r} \Rightarrow 12 - 12r = a$
The first equation gives $a = \frac{3}{r}$,
putting this into the second equation gives:
$12 - 12r = \frac{3}{r} \Rightarrow 12r - 12r^2 = 3$
$\Rightarrow 12r^2 - 12r + 3 = 0$
$\Rightarrow 4r^2 - 4r + 1 = 0$
$4r^2 - 4r + 1 = 0$ factorises to $(2r-1)(2r-1) = 0$
Hence $2r - 1 = 0 \Rightarrow r = 0.5$
Then $a = \frac{3}{r} = \frac{3}{0.5} = 6$.

Q8 a) $a = 6$, $10 = S_\infty = \frac{a}{1-r} = \frac{6}{1-r}$
$\Rightarrow 1 - r = \frac{6}{10} = 0.6 \Rightarrow r = 0.4$

b) 5th term: $u_5 = ar^4 = 6 \times 0.4^4 = 0.1536$.

Q9 a) Second term $= ar = -48$, 5th term $= ar^4 = 0.75$.
Dividing gives: $r^3 = \frac{ar^4}{ar} = \frac{0.75}{-48} = -0.015625$
$\Rightarrow r = -0.25$

b) $ar = -48 \Rightarrow a = \frac{-48}{r} = \frac{-48}{-0.25} = 192$

c) $|r| < 1$ so you can find the sum to infinity:
$$S_\infty = \frac{a}{1-r} = \frac{192}{1-(-0.25)} = \frac{192}{1.25} = 153.6.$$

Q10 a) Each term is 15% less than the previous term so the common ratio is $r = 0.85$. The condition for a convergent geometric series is $|r| < 1$, so this series is convergent.
[2 marks available — 1 mark for stating the common ratio is 0.85, 1 mark for explaining requirement for convergence]

b) L is the sum to infinity which is given by $S_\infty = \frac{a}{1-r}$.
So $30\,000 < \frac{q}{1-0.85} < 35\,000$
$\Rightarrow 30\,000 \times 0.15 < q < 35\,000 \times 0.15 \Rightarrow 4500 < q < 5250$
[2 marks available — 1 mark for setting up the inequality, 1 mark for the correct answer]

Q11 The sum of terms after the 10th is $S_\infty - S_{10}$.
So the question tells you that $S_\infty - S_{10} < \frac{1}{100}S_\infty$
$\Rightarrow \frac{99}{100}S_\infty - S_{10} < 0 \Rightarrow \frac{99}{100}S_\infty < S_{10}$
Then: $0.99(S_\infty) = \frac{0.99a}{1-r} < \frac{a(1-r^{10})}{(1-r)} = S_{10}$
You can cancel and keep the inequality sign because the series is convergent so $|r| < 1 \Rightarrow 1 - r > 0$,
and you know $a > 0$ from the question:
$\Rightarrow 0.99 < 1 - r^{10} \Rightarrow r^{10} < 0.01$
$\Rightarrow |r| < \sqrt[10]{0.01} \Rightarrow |r| < 0.631$ (to 3 s.f.)

Answers

Q12 Using the formulas for S_∞ and S_4

$$\frac{a}{1-r} = \frac{9}{8} \times \frac{a(1-r^4)}{1-r}$$

Cancelling $(1-r)$ and a gives

$$\frac{8}{9} = 1 - r^4 \Rightarrow r^4 = \frac{1}{9} \Rightarrow r = \sqrt[4]{\frac{1}{9}}$$

So, given that r is positive and real, $r = \frac{1}{\sqrt{3}} = \frac{\sqrt{3}}{3}$.

Q13 a) $u_1 = p(2q)^1 = 2pq$
$u_{10} = p(2q)^{10} = 1024pq^{10}$
[2 marks available — 1 mark for each correct answer]

b) The modulus of the common ratio needs to be less than 1. The common ratio is $2q$ and $q > 0$.
So $0 < 2q < 1 \Rightarrow 0 < q < \frac{1}{2}$
[2 marks available — 1 mark for stating that the (modulus of) the common ratio is less than 1, 1 mark for the correct answer]

c) $S_\infty = \frac{2pq}{1-2q} = \frac{2p \times \frac{1}{8}}{1 - 2 \times \frac{1}{8}} = \frac{\frac{1}{4}p}{\frac{3}{4}} = \frac{p}{3}$
[2 marks available — 1 mark for a correct method, 1 mark for the correct answer]

Q14 a) (i) $S_\infty = 100 \Rightarrow \frac{a}{1-r} = 100 \Rightarrow a = 100 - 100r$
$\Rightarrow 100r = 100 - a \Rightarrow r = \frac{100-a}{100} \Rightarrow r = 1 - \frac{a}{100}$
[2 marks available — 1 mark for use of S_∞, 1 mark for rearranging to give the required result]

(ii) For convergence, $-1 < r < 1 \Rightarrow -1 < 1 - \frac{a}{100} < 1$
$\Rightarrow -100 < 100 - a < 100 \Rightarrow -200 < -a < 0$
$\Rightarrow 0 < a < 200$
[2 marks available — 1 mark for a correct method, 1 mark for the correct answer]

b) A periodic sequence with period 2 will repeat itself every 2 values. For this to happen we need $r = -1$:
So $1 - \frac{a}{100} = -1 \Rightarrow \frac{a}{100} = 2 \Rightarrow a = 200$
The series is not convergent — the condition $|r| < 1$ does not hold as $r = -1$.
[3 marks available — 1 mark for a correct method to find a, 1 mark for the correct value of a, 1 mark for stating that the sequence does not converge with a valid explanation]
You could also say that the series is periodic so it jumps between 200 and 0 and therefore never converges.

Exercise 5.9 — Modelling Problems

Q1 $a = 60$, $d = 3$
n^{th} term $= a + (n-1)d = 60 + (n-1)3 = 3n + 57$
12th term $= 3(12) + 57 = 93$
So she earns £93 in her 12th week.
With wordy problems, don't forget to check what the units should be and include them.

Q2 The heights form a geometric sequence (starting from the smallest doll), with $a = 3$, $r = 1.25$.
The nth term is $ar^{n-1} = 3 \times 1.25^{n-1}$,
The first term is the height of the first doll, so the height of the 8th doll is the 8th term. The 8th term is: $3 \times 1.25^7 = 14.3$ cm (1 d.p.)

Q3 $a = 40$, $d = 5$
n^{th} term $= a + (n-1)d = 40 + (n-1)5 = 5n + 35$
$80 = 5n + 35 \Rightarrow n = (80 - 35) \div 5 = 9$
So he'll sell 80 sandwiches on the 9th day.

Q4 $a = 300\,000$, $d = -30\,000$
n^{th} term $= a + (n-1)d = 300\,000 - 30\,000(n-1)$
$= 330\,000 - 30\,000n$
$330\,000 - 30\,000n < 50\,000$
$n > 280\,000 \div 30\,000 = 9.33...$
You want the smallest integer value of n that satisfies the inequality, so round up. Sales will have fallen below £50 000 after 10 months.
$n = 9$ doesn't satisfy the inequality.

Q5 The value decreases by 15% each year so multiply by $1 - 0.15 = 0.85$ to get from one term to the next, so the common ratio $r = 0.85$.
So the nth term $u_n = ar^{n-1} = a(0.85)^{n-1}$
The price when new is the first term in the series and the value after 10 years is the 11th term.
11th term: $2362 = u_{11} = a(0.85)^{10}$
$2362 = a(0.85)^{10} \Rightarrow a = \frac{2362}{0.85^{10}} = 11\,997.496...$
$= 11\,997.50$ to 2 d.p.
When new, the car cost £11 997.50

Q6 a) The cost increases by 3% each year so multiply by $1 + 0.03 = 1.03$ to get from one term to the next, so 2006 cost $= 1.03 \times (2011$ cost$) = 1.03 \times £120 = £123.60$

b) The costs each year form a geometric sequence with common ratio $r = 1.03$ and first term $a = 120$, so the total cost between 2011 and 2016 (including 2011 and 2016) is the sum of the first 6 terms:
$S_6 = \frac{a(1-r^6)}{(1-r)} = \frac{120(1-1.03^6)}{(1-1.03)} = 776.21$ to 2 d.p.
Nigel paid £776.21 (the cost to the nearest penny).

Q7 $a = 6000$, $d = 2000$, $n = 12$ (12 months in the year)
$S_n = \frac{n}{2}[2a + (n-1)d]$
$S_{12} = \frac{12}{2}[2(6000) + 11(2000)] = 204\,000$
204 000 copies will be sold.

Q8 4 weeks is 28 days. The claim is that the leeks' height increases by 15% every 2 days.
$28 \div 2 = 14$, so there are 14 lots of 2 days in 28 days

In 28 days, the height should increase 14 times by 15%. So if the claim is correct, the height of the leeks after 28 days will be the 15th term of a geometric progression with first term 5 and common ratio 1.15.

The first term is the initial height, the second term is the height after one 15% increase and so on, so the 15th term is the height after fourteen 15% increases.

The common ratio is 1.15 because multiplying something by 1.15 is the same as increasing it by 15%.

So $a = 5$ and $r = 1.15$.

The nth term is: $u_n = ar^{n-1} = 5(1.15)^{n-1}$

So the 15th term is: $5 \times 1.15^{14} = 35.3785...$ cm.

So if the claim were true, the leeks would be 35.4 cm tall (to 1 d.p.). Since the leeks only reach a height of 25 cm, the claim on the compost is not justified.

Q9 a) The gnome value goes up by 2% each year, so the price after 1 year is 102% of £80 000:
$80\,000 \times 1.02 = 81\,600$
The value after 1 year is £81 600.

b) To get from one term to the next you multiply by 1.02 (to increase by 2% each time), so the common ratio $r = 1.02$.

c) Price at start = $a = 80\,000$, $r = 1.02$

n^{th} term = $u_n = ar^{n-1} = 80\,000 \times (1.02)^{n-1}$

The value at the start is the 1st term,

the value after 1 year is the 2nd term and so on,

so the value after 10 years is the 11th term:

11th term = $u_{11} = 80\,000 \times (1.02)^{10}$

$= 97\,519.55$ (2 d.p.)

The value after 10 years is £97 519.55 (to the nearest penny)

d) The value after k years is the $(k + 1)^{th}$ term:

$(k + 1)^{th}$ term = $u_{k+1} = ar^k = 80\,000 \times (1.02)^k$

After k years the value is more than 120 000, so:

$u_{k+1} = 80\,000 \times (1.02)^k > 120\,000$

$\Rightarrow (1.02)^k > \dfrac{120\,000}{80\,000} = 1.5$

$\Rightarrow \log(1.02)^k > \log 1.5$

$\Rightarrow k\log(1.02) > \log 1.5$

$\Rightarrow k > \dfrac{\log 1.5}{\log 1.02} = 20.475...$

So u_{k+1} (the value after k years) is more than 120 000 when $k > 20.475...$, therefore the value exceeds 120 000 after 21 years.

Q10 Frazer's series is the natural numbers up to 31.

$S_n = \dfrac{1}{2}n(n+1)$

$S_{31} = \dfrac{1}{2} \times 31 \times 32 = 496$

So Frazer draws 496 dots.

Q11 The thickness of the paper doubles every time you fold it in half. So the paper thickness forms a geometric progression:

After 1 fold, thickness = 0.01×2 cm

After 2 folds, thickness = 0.01×2^2 cm

After n folds, thickness = 0.01×2^n cm

Distance to the moon:

$384\,000$ km = 3.84×10^5 km

$= (3.84 \times 10^5 \times 1000)$ m

$= (3.84 \times 10^5 \times 1000 \times 100)$ cm

$= 3.84 \times 10^{10}$ cm

Therefore the paper reaches the moon when:

$0.01 \times 2^n = 3.84 \times 10^{10}$

$2^n = 3.84 \times 10^{12}$

Taking logs: $n\log 2 = \log(3.84 \times 10^{12})$

$\Rightarrow n = \log(3.84 \times 10^{12}) \div \log 2 = 41.80...$

So after approximately 42 folds the paper would reach the moon.

Q12 a) The distances she runs form a geometric sequence with $a = 12$, $r = 1.03$.

The nth term is $ar^{n-1} = 12 \times 1.03^{n-1}$, so the 10th term is:

$12 \times 1.03^9 = 15.7$ miles (to 1 d.p.).

b) The total distance she runs in 20 days is the sum of the first 20 terms of the sequence.

Using: $S_n = \dfrac{a(1-r^n)}{(1-r)}$, where $a = 12$, $r = 1.03$, $n = 20$,

$S_{20} = \dfrac{12(1 - 1.03^{20})}{(1 - 1.03)} = 322.44...$

In 20 days she runs a total of 322 miles (to the nearest mile).

Q13 Laura's series is the natural numbers. You need to find how many are needed to exceed 1000 (£10 in pence).

$S_n = \dfrac{1}{2}n(n+1)$

$\dfrac{1}{2}(n^2 + n) > 1000$

$n^2 + n > 2000 \Rightarrow n^2 + n - 2000 > 0$

Putting the quadratic equal to zero and solving using the quadratic formula gives $n = 44.2$ (ignoring the negative solution), which by looking at the shape of the quadratic graph gives $n > 44.2$ as the solution to the inequality.

So on the 45th day she'll have over £10.

Q14 After 0 years, $u_1 = a$ and after 1 year, $u_2 = ar$.

So after 10 years, $u_{11} = ar^{10}$. She wants her investment to double so $u_{11} = ar^{10} = 2a \Rightarrow r^{10} = 2$

$\Rightarrow |r| = \sqrt[10]{2} = 1.071773$.

So the interest rate needed is 7.17 % (3 s.f.).

Q15 Arithmetic series with $a = 2.3$ and $d = 3.2$.

The day that daily leakage exceeds 80 litres:

$u_n = 2.3 + 3.2(n-1) = 3.2n - 0.9 > 80 \Rightarrow n > 25.28125...$

So the daily leakage first exceeds 80 litres on day 26.

The day that total leakage exceeds 1000 litres:

$S_n = \dfrac{n}{2}(2 \times 2.3 + (n-1) \times 3.2)$

$= \dfrac{n}{2}(3.2n + 1.4) = 1.6n^2 + 0.7n$

$\Rightarrow 1.6n^2 + 0.7n > 1000 \Rightarrow 16n^2 + 7n - 10\,000 > 0$

Solving $16n^2 + 7n - 10\,000 = 0$

$\Rightarrow n = \dfrac{-7 \pm \sqrt{7^2 - 4 \times 16 \times -10\,000}}{2 \times 16}$

$\Rightarrow n = -25.219...$ or $n = 24.782...$

n will be positive, so $n > 24.782...$

So the total leakage first exceeds 1000 litres on day 25.

So the urgent repair will be deemed necessary on day 25.

As 4th January is day 1, so day 25 is 28th January.

[7 marks available — 1 mark for forming an inequality for the nth term, 1 mark for solving the nth term inequality, 1 mark for forming an inequality for the sum to n terms, 1 mark for a correct method to solve the sum to n terms inequality, 1 mark for the correct solution to the sum to n terms inequality, 1 mark for concluding that the repair is deemed necessary on day 25, 1 mark for the correct date]

Chapter 5 Review Exercise

Q1 $x_8 = 8^2 - 3 = 61$

It's increasing — each term is greater than the previous one (e.g. it continues $x_9 = 78$, $x_{10} = 97...$)

Q2 a) $u_{n+1} - u_n = a(n+1) + b - (an + b)$

$= an + a + b - an - b = a$

[2 marks available — 1 mark for a correct method, 1 mark for the correct answer]

If you recognise this as a linear sequence you could write the answer straight down.

b) (i) For increasing, we require $u_{n+1} - u_n > 0$.

From part a), this means $a > 0$. So a needs to be positive, but there are no restrictions on b.

[2 marks available — 1 mark for the correct restrictions, 1 mark for justification]

(ii) $a < 0$ *[1 mark]*

Q3 $k^2 + 3k + 4 = 44 \Rightarrow k^2 + 3k - 40 = 0 \Rightarrow (k + 8)(k - 5) = 0$

k must be a positive integer, so $k = 5$.

Q4 $u_3 = a(3)^2 + b(3) = 18$

$\Rightarrow 9a + 3b = 18 \Rightarrow 3a + b = 6$ — eqn 1

$u_7 = a(7)^2 + b(7) = 70$

$\Rightarrow 49a + 7b = 70 \Rightarrow 7a + b = 10$ — eqn 2

Solving simultaneously:

Subtract eqn 1 from eqn 2: $4a = 4 \Rightarrow a = 1$

In eqn 1: $3 + b = 6 \Rightarrow b = 3$

Q5 a) $x_1 = 7$

$x_2 = 7 + 3 = 10$

$x_3 = 10 + 3 = 13$

$x_4 = 13 + 3 = 16$

$x_5 = 16 + 3 = 19$

Answers

b) $u_1 = 2$
$u_2 = 6 \div 2 = 3$
$u_3 = 6 \div 3 = 2$
$u_4 = 6 \div 2 = 3$
$u_5 = 6 \div 3 = 2$

Q6 a) Each term is the square root of the previous term,
so $u_{k+1} = \sqrt{u_k}$, $u_1 = 65\,536$

b) The difference between the terms is:
$-2, -4, -6, -8...$ i.e. $-2k$, so $u_{k+1} = u_k - 2k$, $u_1 = 40$

c) Each term from u_3 is the sum of the previous two terms
(it's a Fibonacci sequence), so $u_{k+2} = u_k + u_{k+1}$, $u_1 = 1$, $u_2 = 1$

Q7 $u_{n+1} = ku_n + 3 \Rightarrow 11 = 4k + 3 \Rightarrow k = 2$
$u_3 = 2(11) + 3 = 25$
$u_4 = 2(25) + 3 = 53$

Q8 $u_1 = 8$
$u_2 = 18 - 8 = 10$
$u_3 = 18 - 10 = 8$
$u_4 = 18 - 8 = 10$
$u_5 = 18 - 10 = 8$
The terms are alternating between 8 and 10,
i.e. each term is 9 ± 1, starting with $9 - 1$ for $n = 1$, so:
$u_n = 9 + (-1)^n$. The sequence is periodic.

Q9 Let the three terms be a, b and $(a + b)$.
Then we know, $b - a = 10$ and $a + b = 3a \Rightarrow b = 2a$
Substituting $b = 2a$ into $b - a = 10$ gives,
$2a - a = 10 \Rightarrow a = 10$ and so $b = 2 \times 10 = 20$.
The three terms are 10, 20 and $10 + 20 = 30$.
[4 marks available — 1 mark for setting up expressions for the three terms, 1 mark for creating equations, 1 mark for a correct method to solve the equations, 1 mark for all three terms correct]

Q10 a) $u_{n+2} = \dfrac{u_{n+1} + 1}{u_n}$ and $u_1 = 3$ and $u_2 = 5$ so:

$u_3 = \dfrac{5 + 1}{3} = 2$

$u_4 = \dfrac{2 + 1}{5} = \dfrac{3}{5}$

$u_5 = \dfrac{\frac{3}{5} + 1}{2} = \dfrac{4}{5}$

$u_6 = \dfrac{\frac{4}{5} + 1}{\frac{3}{5}} = 3$

$u_7 = \dfrac{3 + 1}{\frac{4}{5}} = 5$

[2 marks available — 1 mark for substituting u_1 and u_2 into the recurrence relation, 1 mark for correct values of u_6 and u_7]

b) As $u_6 = u_1$ and $u_7 = u_2$ this series is periodic,
so each block of 5 terms will have the same
sum i.e. $3 + 5 + 2 + \dfrac{3}{5} + \dfrac{4}{5} = \dfrac{57}{5}$.
So the sum of the first 25 terms is $5 \times \dfrac{57}{5} = 57$.
[3 marks available — 1 mark for using fact that the series is periodic, 1 mark for correct working, 1 mark for correct final answer]

Q11 There are 28 lots of d between the first term (-2) and the 29th
term (19), so $d = [19 - (-2)] \div 28 = 0.75$.
You could also plug the values for a (-2), n (29) and u_n (19) into the formula for the nth term in an arithmetic sequence.

Q12 a) $u_9 = 20 \times 9 - 4 = 176$ *[1 mark]*

b) Find n such that $u_n > 850 \Rightarrow 20n - 4 > 850 \Rightarrow 20n > 854$
$\Rightarrow n > 42.7 \Rightarrow n = 43$. So month 43 is when the business is
predicted to become profitable.
[3 marks available — 1 mark for setting up the inequality, 1 mark for solving it, 1 mark for the correct conclusion]

c) E.g. The model predicts that the number of customers will
continue to rise indefinitely, but in reality it's likely that the
growth will tail off at some point.
[1 mark for any valid problem]

Q13 $u_n = a + (n - 1)d$
$u_7 = a + 6d = 8$ — eqn 1
$u_{11} = a + 10d = 10$ — eqn 2
eqn 2 – eqn 1:
$4d = 2 \Rightarrow d = 0.5$
In eqn 1: $a + (6 \times 0.5) = 8 \Rightarrow a = 5$
So $u_3 = 5 + 2 \times 0.5 = 6$

Q14 $u_n = a + (n - 1)d$
$u_7 = a + 6d = 36$ — eqn 1
$u_{10} = a + 9d = 30$ — eqn 2
eqn 2 – eqn 1:
$3d = -6 \Rightarrow d = -2$
In eqn 1: $a - 12 = 36 \Rightarrow a = 48$
So n^{th} term $= a + (n - 1)d = 48 + (-2)(n - 1) = -2n + 50$

$S_n = \dfrac{n}{2}[2a + (n - 1)d]$

$S_5 = \dfrac{5}{2}[2 \times 48 + (5 - 1) \times (-2)] = 220$

Q15 $a = 3(1) - 1 = 2$
$l = 3(20) - 1 = 59$
$S_n = \dfrac{1}{2}n(a + l)$
$S_{20} = \dfrac{1}{2} \times 20 \times (2 + 59) = 610$

Q16 a) For an arithmetic sequence: $u_n = a + (n - 1)d$.
$u_6 = a + 5d = p$ — eqn 1
$u_{12} = a + 11d = 4p$ — eqn 2
Subtract eqn 1 from eqn 2:
$6d = 3p \Rightarrow d = \dfrac{p}{2}$
[3 marks available — 1 mark for forming two simultaneous equations in a, d and p, 1 mark for eliminating a, 1 mark for correct final answer]

b) $S_n = \dfrac{n}{2}[2a + (n - 1)d]$
$S_6 = 3[2a + 5(\tfrac{p}{2})] = 6a + \dfrac{15}{2}p = 72$ — eqn 1
$S_{12} = 6[2a + 11(\tfrac{p}{2})] = 12a + 33p = 288$ — eqn 2
Subtract $2 \times$ eqn 1 from eqn 2:
$18p = 144 \Rightarrow p = 8$, so $d = 4$
Substitute $p = 8$ into eqn 2:
$12a + (33 \times 8) = 288 \Rightarrow a = 2$
So the series is:
$u_1 = a = 2$
$u_2 = a + d = 2 + 4 = 6$
$u_3 = a + 2d = 2 + 8 = 10$
[4 marks available — 1 mark for forming two simultaneous equations in a and p, 1 mark for solving for p, 1 mark for solving for a, 1 mark for finding u_2 and u_3]

Q17 a) $u_n = a + (n - 1)d$
$\Rightarrow u_{14} = a + (14 - 1)(-b) = -4$
$\Rightarrow a + -13b = -4 \Rightarrow a + 4 = 13b$
[2 marks available — 1 mark for correct use of u_n formula, 1 mark for rearranging to get the required result]

b) $S_5 = \frac{5}{2}(2a - (5-1)b) = \frac{5}{2}(2a - 4b) = 5a - 10b$

$u_6 = a - (6-1)b = a - 5b$

$S_5 = 7u_6 \Rightarrow 5a - 10b = 7(a - 5b)$

$\Rightarrow 5a - 10b = 7a - 35b \Rightarrow 2a = 25b \Rightarrow a = \frac{25}{2}b$

Solving simultaneously by substituting into part a):

$\frac{25}{2}b + 4 = 13b \Rightarrow \frac{1}{2}b = 4 \Rightarrow b = 8$

So $a = \frac{25}{2} \times 8 = 100$

[5 marks available — 1 mark for S_5, 1 mark for u_6, 1 mark for using $S_5 = 7u_6$ to make an equation, 1 mark for a correct method to solve simultaneously with answer to part a), 1 mark for both a and b]

c) $S_n = 0 \Rightarrow \frac{n}{2}[2 \times 100 - 8(n-1)] = 0$

$\Rightarrow n[100 - 4(n-1)] = 0$

$\Rightarrow n(104 - 4n) = 0 \Rightarrow n = 0$ or $n = 26$

$n = 0$ is the trivial case with no terms, so 26 terms are needed for the sum of the series to be equal to zero.

[3 marks available — 1 mark for $S_n = 0$, 1 mark for a correct method of solving, 1 mark for the correct answer]

Q18 a) $S_n = \frac{1}{2}n(n+1)$

$S_{24} = \frac{1}{2} \times 24 \times 25 = 300$

b) $\sum_{n=13}^{24} n = S_{24} - S_{12}$

$S_{12} = \frac{1}{2} \times 12 \times 13 = 78$

So $S_{24} - S_{12} = 300 - 78 = 222$

c) $S_k = \frac{1}{2}k(k+1) = 630$

$\Rightarrow k^2 + k = 1260 \Rightarrow k^2 + k - 1260 = 0$

$\Rightarrow (k + 36)(k - 35) = 0$

k must be a positive integer, so $k = 35$

Q19 First term $a = 3$, common ratio $r = \frac{-9}{3} = -3$

n^{th} term $= ar^{n-1} = 3(-3)^{n-1} = -(-3)^n$

Q20 a) First term $a = 2$, common ratio $r = \frac{-6}{2} = -3$

$u_n = ar^{n-1} = 2 \times (-3)^{n-1}$

$u_{10} = 2 \times (-3)^9 = -39\,366$

b) $S_n = \frac{a(1 - r^n)}{(1 - r)}$

$\Rightarrow S_{10} = \frac{2(1 - (-3)^{10})}{(1 - (-3))} = \frac{-118\,096}{4} = -29\,524$

Q21 $\sum_{k=0}^{5} 7(0.6)^k = \sum_{k=1}^{6} 7(0.6)^{k-1} = S_6$

$S_n = \frac{a(1 - r^n)}{(1 - r)} \Rightarrow S_6 = \frac{7(1 - 0.6^6)}{(1 - 0.6)} = 16.68$ (2 d.p.)

Q22 The sum of the first n terms of a geometric series is S_n:

$S_n = a + ar + ar^2 + ar^3 + \dots + ar^{n-1}$

Then: $rS_n = ar + ar^2 + ar^3 + ar^4 \dots + ar^n$

Subtracting rS_n from S_n gives:

$(1 - r)S_n = a - ar^n = a(1 - r^n) \Rightarrow S_n = \frac{a(1 - r^n)}{1 - r}$

Q23 a) $r = \frac{\text{second term}}{\text{first term}} = \frac{12}{24} = \frac{1}{2}$

b) First term $a = 24$, $r = \frac{1}{2}$

$u_n = ar^{n-1} = 24 \times \left(\frac{1}{2}\right)^{n-1}$

$u_7 = 24 \times \left(\frac{1}{2}\right)^6 = \frac{3}{8} = 0.375$

c) $S_n = \frac{a(1 - r^n)}{(1 - r)}$

$\Rightarrow S_{10} = \frac{24(1 - 0.5^{10})}{(1 - 0.5)} = 47.953$ (3 d.p.)

d) $S_\infty = \frac{a}{1 - r} = \frac{24}{1 - 0.5} = 48$

Q24 a) $S_\infty = \frac{a}{1 - r} = 8k \Rightarrow a = 8k - 8kr$ — eqn 1

$u_2 = ar = 6$ — eqn 2

Substitute a from eqn 1 into eqn 2:

$(8k - 8kr)r = 6 \Rightarrow 8kr - 8kr^2 = 6$

$\Rightarrow 8kr^2 - 8kr + 6 = 0 \Rightarrow 4kr^2 - 4kr + 3 = 0$ — as required

[3 marks available — 1 mark for forming two simultaneous equations, 1 mark for correct working to solve, 1 mark for rearranging into the required final answer]

b) $4kr^2 - 4kr + 3 = 0$ has one real root, so the discriminant must be zero:

$(-4k)^2 - 4 \times 4k \times 3 = 0 \Rightarrow 16k^2 - 48k = 0$

$\Rightarrow k(k - 3) = 0$, so $k = 0$ or 3, but $k \neq 0$ so $k = 3$.

Substitute into $4kr^2 - 4kr + 3 = 0$ to find r:

$12r^2 - 12r + 3 = 0 \Rightarrow 4r^2 - 4r + 1 = 0$

$\Rightarrow (2r - 1)^2 = 0 \Rightarrow r = \frac{1}{2}$

Using eqn 2 from part a):

$ar = 6 \Rightarrow a \times \frac{1}{2} = 6 \Rightarrow a = 12$.

So the first three terms are 12, 6 and $6 \times \frac{1}{2} = 3$.

[5 marks available — 1 mark for stating that the discriminant is zero, 1 mark for a correct value of k, 1 mark for a correct value of r, 1 mark for a correct value of a, 1 mark for correct third term]

Q25 a) $u_3 = ar^2 = \frac{1}{72}$ and $u_6 = ar^5 = \frac{1}{1944}$

So $\frac{u_6}{u_3} = r^3 = \frac{1}{27} \Rightarrow r = \frac{1}{3}$

[3 marks available — 1 mark for both u_3 and u_6, 1 mark for a correct method to find r, 1 mark for the correct answer]

b) Difference between terms to be less than 10^{-6}.
As terms are decreasing this is $u_n - u_{n+1} < 10^{-6}$.

$\Rightarrow \frac{1}{8}\left(\frac{1}{3}\right)^{n-1} - \frac{1}{8}\left(\frac{1}{3}\right)^n < 10^{-6} \Rightarrow \frac{1}{8}\left(\frac{1}{3}\right)^{n-1}\left(1 - \frac{1}{3}\right) < 10^{-6}$

$\Rightarrow \left(\frac{1}{3}\right)^{n-1} < 12 \times 10^{-6} \Rightarrow 3 \times 3^{-n} < 12 \times 10^{-6}$

$\Rightarrow -n \ln 3 < \ln(4 \times 10^{-6}) \Rightarrow n > \frac{\ln(4 \times 10^{-6})}{-\ln 3}$

$\Rightarrow n > 11.313\dots \Rightarrow n = 12$

[5 marks available — 1 mark for setting up expressions for u_n and u_{n+1}, 1 mark for setting up the inequality, 1 mark for a correct method to simplify, 1 mark for taking logs of both sides, 1 mark for the correct answer]

Q26 The takings can be modelled as an arithmetic sequence with $a = 300$ and $d = 15$.
$u_n = a + (n-1)d$ so find n where:
$a + (n-1)d > 500$
$\Rightarrow 300 + 15(n-1) > 500$
$\Rightarrow 15n > 500 - 285$
$\Rightarrow n > 14.3333\dots$
So the shop will take over £500 on the 15th day.

Q27 The sequence of stones is the sequence of natural numbers $(1, 2, 3\dots)$.
13 weeks $\times$ 5 days $= 65$
So the total number of stones is $S_n = \frac{1}{2}n(n+1)$ with $n = 65$:

$S_{65} = \frac{1}{2} \times 65 \times 66 = 2145$ stones

Answers

Q28 a) The run times can be modelled by an arithmetic sequence with $a = 30$ and $d = 5$, where u_{R_n} is the nth term.
So $u_{R3} = 30 + (3 - 1) \times 5 = 40$ minutes
The cycling times can be modelled by a geometric sequence with $a = 15$ and $r = 1.12$, where u_{C_n} is the nth term.
So $u_{C3} = 15 \times 1.12^{(3-1)} = 18.816$ minutes
The swim times are always 60 minutes per day, so the total exercise time is $40 + 18.816 + 60$
$= 119$ minutes (to the nearest minute)
[3 marks available — 1 mark for correct run time, 1 mark for correct cycle time, 1 mark for correct final answer]

b) The run is longer than the swim when:
$30 + (n - 1) \times 5 > 60 \implies n > 7$, i.e. from day 8.
The cycle is longer than the swim when:
$15 \times 1.12^{(n-1)} > 60 \implies 1.12^{(n-1)} > 4$
$\implies (n - 1)\log 1.12 > \log 4$
$\implies n - 1 > \log 4 \div \log 1.12$
$\implies n > 13.232...$, i.e. from day 14.
So the first day when both the cycle and the run are longer than the swim is day 14.
[4 marks available — 1 mark for solving inequality for the run, 1 mark for forming inequality for the cycle, 1 mark for correctly using logs to solve the inequality for the cycle, 1 mark for correct final answer]

c) Find the sum of the geometric series from day 8 to day 14, i.e. $S_{C14} - S_{C7}$. $S_n = \dfrac{a(1 - r^n)}{1 - r}$, so:
$S_{C14} - S_{C7} = \dfrac{15(1 - 1.12^{14})}{1 - 1.12} - \dfrac{15(1 - 1.12^7)}{1 - 1.12}$
$= 485.889... - 151.335... = 334.553...$
$= 5$ hours 35 minutes (to the nearest minute)
[3 marks available — 1 mark for stating that the answer would be the difference between the sum to 14 terms and the sum to 7 terms, 1 mark for correct use of sum formulas, 1 mark for correct final answer in hours and minutes]

d) E.g. neither the arithmetic or geometric series models allow for the fact that there are a finite number of hours in a day that Samira could use for training.
[1 mark for correct explanation]

Q29 $u_2 = ar = 2$ and $S_\infty = \dfrac{a}{1 - r} = 9 \implies a = 9 - 9r$
So $(9 - 9r)r = 2 \implies 9r - 9r^2 - 2 = 0 \implies 9r^2 - 9r + 2 = 0$
$\implies (3r - 1)(3r - 2) = 0 \implies r = \frac{1}{3}$ or $r = \frac{2}{3}$
$r = \frac{1}{3} \implies \frac{1}{3}a = 2 \implies a = 6$ and $r = \frac{2}{3} \implies \frac{2}{3}a = 2 \implies a = 3$
So either the first term is 6 and the common ratio is $\frac{1}{3}$
or the first term is 3 and the common ratio is $\frac{2}{3}$.
[6 marks available — 1 mark for an equation for u_2, 1 mark for an equation for S_∞, 1 mark for solving simultaneously, 1 mark for factorising the quadratic, 1 mark for both values of a, 1 mark for both corresponding values of r]

Q30 a) $u_{10} = a + 9d$ and $u_4 = a + 3d$
$a + 9d = 2(a + 3d) \implies a - 3d = 0 \implies a = 3d$
[3 marks available — 1 mark for u_{10} and u_4, 1 mark for forming the equation, 1 mark for simplifying to the required result]

b) $u_n = a + (n - 1)d = a + \frac{1}{3}a(n - 1)$
$= a + \frac{1}{3}an - \frac{1}{3}a = \frac{1}{3}an + \frac{2}{3}a = \frac{1}{3}a(n + 2)$
[2 marks available — 1 mark for substituting $d = \frac{1}{3}a$ into the nth term formula, 1 mark for simplifying to the required result]

c) $u_n = \frac{1}{3}(0)(n + 2) = 0$, so all terms of the sequence would be zero, but we are told that neither u_4 nor u_{10} are zero.
[1 mark for valid reason]

d) $u_m = 2u_k$ would be $\frac{1}{3}a(m + 2) = \frac{2}{3}a(k + 2)$
$\implies m + 2 = 2k + 4 \implies m = 2k + 2$
k will be the smaller value so when $k = 1$, $m = 4$.
[2 marks available — 1 mark for a correct method, 1 mark for the correct answer]

Q31 a) $u_1 = (-1)^1 \times 1 = -1 \qquad u_2 = (-1)^2 \times 2 = 2$
$u_3 = (-1)^3 \times 3 = -3 \qquad u_4 = (-1)^4 \times 4 = 4$
This sequence is none of the options given.
[2 marks available — 1 mark for all four terms correct, 1 mark for the correct conclusion]

b) $v_1 = S_1 = -1$,
$v_2 = S_2 = -1 + 2 = 1$,
$v_3 = S_3 = -1 + 2 - 3 = -2$,
$v_4 = S_4 = -1 + 2 - 3 + 4 = 2$
[1 mark for all correct]

c) (i) Even terms are positive and half of n.
So, $v_{32} = \frac{1}{2}(32) = 16$
[2 marks available — 1 mark for the correct answer, 1 mark for suitable justification]

(ii) Odd terms are negative and half of $(n + 1)$.
For odd n, $v_n = -\frac{1}{2}(n + 1)$.
[1 mark for the correct formula]

Q32 a) Radii are a geometric sequence with $a = 2$, $r = 1.5$.
$u_{10} = 2 \times (1.5)^9 = 76.88...$ cm. This is the radius, so diameter $= 76.88... \times 2 = 153.77... = 154$ cm (to the nearest cm).

b) (i) $r = \sqrt{\dfrac{725}{\pi}} = 15.1912...$ cm
Find the value of n where $2(1.5)^{n-1} = 15.1912...$
$\implies (n - 1)\ln 1.5 = \ln 7.595... \implies n - 1 = 5.0006...$
$\implies n$ needs to be an integer, so $n = 6$.
This circle is in the 6th position.

(ii) It is important to know the area has been rounded as the calculations to find n will not result in an integer.

c) If the common ratio of the radii is 1.5 then the common ratio of the areas will be $1.5^2 = 2.25$.

Q33 a) The first few terms are $u_1 = 6(1) - 2 = 4$, $u_2 = 6(2) - 2 = 10$, $u_3 = 6(3) - 2 = 16$, $u_4 = 6(4) - 2 = 22$ and $u_5 = 6(5) - 2 = 28$. These correspond to 4, 10, 16, 22 and 28 hours after midnight on the first day, which are 4 am, 10 am, 4 pm, 10 pm on the first day, and 4 am on the second day. The sequence continues every six hours after that and so will continue the pattern of 4am, 10am, 4pm, 10pm every day during construction.

b) 5 years $= 5 \times 365 = 1825$ days
$1825 \times 4 = 7300$ photos
$u_n = 2n + 6 \implies a = 2(1) + 6 = 8$ and $d = 2$
Now using the series formula $S_n = \frac{n}{2}[2a + (n - 1)d]$
$\implies \frac{n}{2}[2(8) + 2(n - 1)] = 7300$
$\implies n(n + 7) = 7300 \implies n^2 + 7n - 7300 = 0$
$\implies n = \dfrac{-7 \pm \sqrt{7^2 - 4 \times 1 \times -7300}}{2 \times 1}$
$\implies n = -89.011...$ and $n = 82.011...$
Ignore the negative solution, so it will be 82 seconds (to the nearest second).
If you allowed for leap years the values involved differ slightly but still lead to a final answer of 82 seconds.

Q34 a) $a = 10$ and 10% increase $\Rightarrow r = 1.1$
So the formula would be $u_n = 10 \times 1.1^{n-1}$
*[2 marks available — 1 mark for finding a and r,
1 mark for the correct formula]*

b) Find n such that $u_n < 50 \Rightarrow 10 \times 1.1^{n-1} < 50$
$\Rightarrow 1.1^{n-1} < 5 \Rightarrow (n-1)\ln 1.1 < \ln 5 \Rightarrow n < 17.88...$
So, there will be a maximum of 17 stepping stones allowed
in the playground.
*[3 marks available — 1 mark for forming an inequality,
1 mark for solving the inequality, 1 mark for the
correct answer]*
Be careful about the mixture of units given in this question.

c) Total volume is determined by the total of the heights,
as the radius of each is fixed.
Heights form a geometric series with sum:
$S_8 = \dfrac{10(1.1^8 - 1)}{1.1 - 1} = 114.358...$ cm $= 1.143...$ m
Accounting for the 20 cm of each cylinder under the ground,
8×0.2 m $= 1.6$ m, there'll be a total height of
$1.143... + 1.6 = 2.743...$ m across all cylinders.
The company wants to know the total volume required:
$V = \pi(0.2)^2(2.743...) = 0.3447...$ m^3
So the volume of concrete required is 0.345 m^3 (to 3 s.f.)
*[3 marks available — 1 mark for a correct method to
calculate total height, 1 mark for the correct total height,
1 mark for the correct total volume]*

Chapter 6: The Binomial Expansion

Prior Knowledge Check

Q1 $(1 + x)^4 = 1 + 4x + \dfrac{4(4-1)}{1 \times 2}x^2 + \dfrac{4(4-1)(4-2)}{1 \times 2 \times 3}x^3$
$\qquad + \dfrac{4(4-1)(4-2)(4-3)}{1 \times 2 \times 3 \times 4}x^4$
$\qquad = 1 + 4x + 6x^2 + 4x^3 + x^4$

Q2 a) $\dfrac{10}{(x-2)(x+3)} \equiv \dfrac{A}{(x-2)} + \dfrac{B}{(x+3)}$
$\Rightarrow 10 \equiv A(x+3) + B(x-2)$
$x = 2 \Rightarrow 10 = 5A \Rightarrow A = 2$
$x = -3 \Rightarrow 10 = -5B \Rightarrow B = -2$
So $\dfrac{10}{(x-2)(x+3)} \equiv \dfrac{2}{(x-2)} - \dfrac{2}{(x+3)}$

b) $\dfrac{11x + 29}{(x+1)(x+2)(x-5)} \equiv \dfrac{A}{(x+1)} + \dfrac{B}{(x+2)} + \dfrac{C}{(x-5)}$
$\Rightarrow 11x + 29 \equiv A(x+2)(x-5) + B(x+1)(x-5)$
$\qquad + C(x+1)(x+2)$
$x = -1 \Rightarrow 18 = -6A \Rightarrow A = -3$
$x = -2 \Rightarrow 7 = 7B \Rightarrow B = 1$
$x = 5 \Rightarrow 84 = 42C \Rightarrow C = 2$
So $\dfrac{11x + 29}{(x+1)(x+2)(x-5)} \equiv -\dfrac{3}{(x+1)} + \dfrac{1}{(x+2)} + \dfrac{2}{(x-5)}$

c) $\dfrac{4}{(x+1)^2(x+3)} \equiv \dfrac{A}{(x+1)} + \dfrac{B}{(x+1)^2} + \dfrac{C}{(x+3)}$
$\Rightarrow 4 \equiv A(x+1)(x+3) + B(x+3) + C(x+1)^2$
$x = -1 \Rightarrow 4 = 2B \Rightarrow B = 2$
$x = -3 \Rightarrow 4 = 4C \Rightarrow C = 1$
Equating coefficients of x^2 terms:
$0 = A + C \Rightarrow A = -1$
So $\dfrac{4}{(x+1)^2(x+3)} \equiv -\dfrac{1}{(x+1)} + \dfrac{2}{(x+1)^2} + \dfrac{1}{(x+3)}$

Exercise 6.1 — Binomial Expansions where n is a Positive Integer

Q1 $(1+x)^3 = 1 + 3x + \dfrac{3(3-1)}{1 \times 2}x^2 + \dfrac{3(3-1)(3-2)}{1 \times 2 \times 3}x^3$
$\qquad = 1 + 3x + 3x^2 + x^3$

Q2 $(1+x)^7 = 1 + 7x + \dfrac{7(7-1)}{1 \times 2}x^2 + \dfrac{7(7-1)(7-2)}{1 \times 2 \times 3}x^3 + ...$
$\qquad = 1 + 7x + 21x^2 + 35x^3 + ...$
*This isn't the full expansion, so keep the dots
at the end to show it carries on.*

Q3 $(1-x)^4 = 1 + 4(-x) + \dfrac{4(4-1)}{1 \times 2}(-x)^2 + \dfrac{4(4-1)(4-2)}{1 \times 2 \times 3}(-x)^3$
$\qquad + \dfrac{4(4-1)(4-2)(4-3)}{1 \times 2 \times 3 \times 4}(-x)^4$
$\qquad = 1 - 4x + 6x^2 - 4x^3 + x^4$

Q4 $(1+3x)^6 = 1 + 6(3x) + \dfrac{6(6-1)}{1 \times 2}(3x)^2 + ...$
$\qquad = 1 + 6(3x) + 15(9x^2) + ... = 1 + 18x + 135x^2 + ...$

Q5 $(1+2x)^8 = 1 + 8(2x) + \dfrac{8(8-1)}{1 \times 2}(2x)^2 + \dfrac{8(8-1)(8-2)}{1 \times 2 \times 3}(2x)^3 + ...$
$\qquad = 1 + 8(2x) + 28(4x^2) + 56(8x^3) + ...$
$\qquad = 1 + 16x + 112x^2 + 448x^3 + ...$

Q6 $(1-5x)^5 = 1 + 5(-5x) + \dfrac{5(5-1)}{1 \times 2}(-5x)^2 + ...$
$\qquad = 1 + 5(-5x) + 10(25x^2) + ... = 1 - 25x + 250x^2 - ...$

Q7 $(1-4x)^3 = 1 + 3(-4x) + \dfrac{3(3-1)}{1 \times 2}(-4x)^2 + \dfrac{3(3-1)(3-2)}{1 \times 2 \times 3}(-4x)^3$
$\qquad = 1 + 3(-4x) + 3(16x^2) + (-64x^3) = 1 - 12x + 48x^2 - 64x^3$

Q8 $(1+6x)^6 = 1 + 6(6x) + \dfrac{6(6-1)}{1 \times 2}(6x)^2 + \dfrac{6(6-1)(6-2)}{1 \times 2 \times 3}(6x)^3 + ...$
$\qquad = 1 + 6(6x) + 15(36x^2) + 20(216x^3) + ...$
$\qquad = 1 + 36x + 540x^2 + 4320x^3 + ...$

Q9 $\left(1 + \dfrac{2}{3}x\right)^4 = 1 + 4\left(\dfrac{2}{3}x\right) + \dfrac{4(4-1)}{1 \times 2}\left(\dfrac{2}{3}x\right)^2$
$\qquad + \dfrac{4(4-1)(4-2)}{1 \times 2 \times 3}\left(\dfrac{2}{3}x\right)^3$
$\qquad + \dfrac{4(4-1)(4-2)(4-3)}{1 \times 2 \times 3 \times 4}\left(\dfrac{2}{3}x\right)^4$
$\qquad = 1 + \dfrac{8}{3}x + \dfrac{8}{3}x^2 + \dfrac{32}{27}x^3 + \dfrac{16}{81}x^4$
*[2 marks available — 1 mark for a correct expansion,
1 mark for the correct and simplified answer]*

Q10 $\dfrac{5 \times 4 \times 3}{1 \times 2 \times 3}(2k)^3 = \dfrac{8 \times 7 \times 6 \times 5 \times 4}{1 \times 2 \times 3 \times 4 \times 5}(k)^5$
$\Rightarrow 80k^3 = 56k^5 \Rightarrow k^2 = \dfrac{10}{7}$
$\Rightarrow k = \pm\sqrt{\dfrac{10}{7}}$ (or $\pm\dfrac{\sqrt{70}}{7}$ or $\pm 1.19522...$)
*[4 marks available — 1 mark for each correct coefficient,
1 mark for a correct equation, 1 mark for correct values of k]*

Q11 a) $(1 + kx)^n = 1 + n(kx) + \dfrac{n(n-1)}{1 \times 2}(kx)^2 + ...$
$\qquad = 1 + knx + \dfrac{1}{2}k^2 n(n-1)x^2 + ...$
*[2 marks available — 1 mark for a correct expansion,
1 mark for the correct and simplified answer in any form]*

b) $12kn = \dfrac{1}{2}k^2 n(n-1) \Rightarrow 24k(3k) = k^2(3k)(3k-1)$
$\Rightarrow 24 = k(3k-1)$ ('$3k^2$' cancels as $k \neq 0$)
$\Rightarrow 3k^2 - k - 24 = 0$
$\Rightarrow (3k+8)(k-3) = 0$
$\Rightarrow k = -\dfrac{8}{3}$ or $k = 3$
$n = 3k$, so $n = -8$ or $n = 9$.
*[3 marks available — 1 mark for a correct equation,
1 mark for a correct method to solve, 1 mark both correct
values of n]*

Answers

Q12 a) $(1 + ax)^n = 1 + n(ax) + \dfrac{n(n-1)}{1 \times 2}(ax)^2$

$\qquad\qquad + \dfrac{n(n-1)(n-2)}{1 \times 2 \times 3}(ax)^3$

$\qquad\qquad + \dfrac{n(n-1)(n-2)(n-3)}{1 \times 2 \times 3 \times 4}(ax)^4 + \ldots$

$(1 - ax)^n = 1 + n(-ax) + \dfrac{n(n-1)}{1 \times 2}(-ax)^2$

$\qquad\qquad + \dfrac{n(n-1)(n-2)}{1 \times 2 \times 3}(-ax)^3$

$\qquad\qquad + \dfrac{n(n-1)(n-2)(n-3)}{1 \times 2 \times 3 \times 4}(-ax)^4 + \ldots$

Subtracting the two expansions gives:

$(1 + ax)^n - (1 - ax)^n = 2anx + 2a^3\dfrac{n(n-1)(n-2)}{1 \times 2 \times 3}x^3 + \ldots$

All the terms with even powers cancel,
so only terms with odd powers remain.
*Only the terms with odd powers retain the minus sign from
$(-ax)$, so in the subtraction these are added while the terms
with even powers are subtracted.*
*[2 marks available — 1 mark for a correct expansion up to
at least the x^2 term, 1 mark for correctly subtracting the
two expansions]*

b) $(1 + ax)^n - (1 - ax)^n = 2anx + 2a^3\dfrac{n(n-1)(n-2)}{1 \times 2 \times 3}x^3 + \ldots$

$\qquad = 2\left(anx + a^3\dfrac{n(n-1)(n-2)}{1 \times 2 \times 3}x^3 + \ldots\right)$

As a and n are integers, all the coefficients in the expansion
are integers too. As x is an integer, the bracket is a sum of
integer terms and so is also an integer. Hence the expansion
is even.
*[2 mark available — 1 mark for showing a factor of 2,
1 mark for a correct explanation]*

Exercise 6.2 — Binomial Expansions where n is Negative or a Fraction

Q1 $(1 + x)^{-4}$

$= 1 + (-4)x + \dfrac{-4(-4-1)}{1 \times 2}x^2 + \dfrac{-4(-4-1)(-4-2)}{1 \times 2 \times 3}x^3 + \ldots$

$= 1 + (-4)x + \dfrac{-4 \times -5}{2}x^2 + \dfrac{-4 \times -5 \times -6}{6}x^3 + \ldots$

$= 1 - 4x + 10x^2 - 20x^3 + \ldots$

Q2 $\left(1 - \dfrac{3}{4}x\right)^{-2} = 1 + (-2)\left(-\dfrac{3}{4}x\right) + \dfrac{-2 \times (-2-1)}{1 \times 2}\left(-\dfrac{3}{4}x\right)^2 + \ldots$

$\qquad = 1 + \dfrac{3}{2}x + \dfrac{27}{16}x^2 + \ldots$

*[2 marks available — 1 mark for an expansion with at least two
terms correct, 1 mark for the correct and simplified answer]*

Q3 a) $(1 - 6x)^{-3} = 1 + (-3)(-6x) + \dfrac{-3(-3-1)}{1 \times 2}(-6x)^2$

$\qquad\qquad + \dfrac{-3(-3-1)(-3-2)}{1 \times 2 \times 3}(-6x)^3 + \ldots$

$\qquad = 1 + 18x + \dfrac{-3 \times -4}{2}(36x^2)$

$\qquad\qquad + \dfrac{-3 \times -4 \times -5}{6}(-216x^3) + \ldots$

$\qquad = 1 + 18x + 6(36x^2) + (-10)(-216x^3) + \ldots$

$\qquad = 1 + 18x + 216x^2 + 2160x^3 + \ldots$

b) The expansion is valid for $\left|\dfrac{-6x}{1}\right| < 1$, so $|x| < \dfrac{1}{6}$.

Q4 a) $(1 + 4x)^{\frac{1}{3}} = 1 + \dfrac{1}{3}(4x) + \dfrac{\frac{1}{3}\left(\frac{1}{3}-1\right)}{1 \times 2}(4x)^2 + \ldots$

$\qquad = 1 + \dfrac{1}{3}(4x) + \dfrac{\frac{1}{3} \times -\frac{2}{3}}{2}(16x^2) + \ldots$

$\qquad = 1 + \dfrac{1}{3}(4x) + \left(-\dfrac{1}{9}\right)(16x^2) + \ldots$

$\qquad = 1 + \dfrac{4x}{3} - \dfrac{16x^2}{9} + \ldots$

b) $(1 + 4x)^{-\frac{1}{2}} = 1 + \left(-\dfrac{1}{2}\right)(4x) + \dfrac{-\frac{1}{2}\left(-\frac{1}{2}-1\right)}{1 \times 2}(4x)^2 + \ldots$

$\qquad = 1 - \dfrac{1}{2}(4x) + \dfrac{-\frac{1}{2} \times -\frac{3}{2}}{2}(16x^2) + \ldots$

$\qquad = 1 - \dfrac{1}{2}(4x) + \left(\dfrac{3}{8}\right)(16x^2) + \ldots = 1 - 2x + 6x^2 - \ldots$

c) Both a) and b) are valid for $|x| < \dfrac{1}{4}$.

Q5 a) $\left(1 + \dfrac{1}{2}x\right)^{\frac{2}{3}}$

$= 1 + \dfrac{2}{3}\left(\dfrac{1}{2}x\right) + \dfrac{\frac{2}{3} \times -\frac{1}{3}}{1 \times 2}\left(\dfrac{1}{2}x\right)^2 + \dfrac{\frac{2}{3} \times -\frac{1}{3} \times -\frac{4}{3}}{1 \times 2 \times 3}\left(\dfrac{1}{2}x\right)^3 + \ldots$

$= 1 + \dfrac{1}{3}x - \dfrac{1}{36}x^2 + \dfrac{1}{162}x^3 + \ldots$

*[2 marks available — 1 mark for an expansion with at
least two terms correct, 1 mark for the correct and
simplified answer]*

b) The expansion is valid for $\left|\dfrac{1}{2}x\right| < 1 \Rightarrow |x| < 2.$
[1 mark for the correct validity]

Q6 a) $\dfrac{1}{(1 - 4x)^2} = (1 - 4x)^{-2}$

$= 1 + (-2)(-4x) + \dfrac{-2(-2-1)}{1 \times 2}(-4x)^2$

$\qquad + \dfrac{-2(-2-1)(-2-2)}{1 \times 2 \times 3}(-4x)^3 + \ldots$

$= 1 + (-2)(-4x) + \dfrac{-2 \times -3}{2}(16x^2)$

$\qquad + \dfrac{-2 \times -3 \times -4}{6}(-64x^3) + \ldots$

$= 1 + 8x + 3(16x^2) + (-4)(-64x^3) + \ldots$

$= 1 + 8x + 48x^2 + 256x^3 + \ldots$

b) $\sqrt{1 + 6x} = (1 + 6x)^{\frac{1}{2}}$

$= 1 + \dfrac{1}{2}(6x) + \dfrac{\frac{1}{2}\left(\frac{1}{2}-1\right)}{1 \times 2}(6x)^2$

$\qquad + \dfrac{\frac{1}{2}\left(\frac{1}{2}-1\right)\left(\frac{1}{2}-2\right)}{1 \times 2 \times 3}(6x)^3 + \ldots$

$= 1 + \dfrac{1}{2}(6x) + \dfrac{\frac{1}{2} \times -\frac{1}{2}}{2}(36x^2)$

$\qquad + \dfrac{\frac{1}{2} \times -\frac{1}{2} \times -\frac{3}{2}}{6}(216x^3) + \ldots$

$= 1 + \dfrac{1}{2}(6x) + \left(-\dfrac{1}{8}\right)(36x^2) + \dfrac{1}{16}(216x^3) + \ldots$

$= 1 + 3x - \dfrac{9x^2}{2} + \dfrac{27x^3}{2} - \ldots$

c) $\dfrac{1}{\sqrt{1 - 3x}} = (1 - 3x)^{-\frac{1}{2}}$

$= 1 + \left(-\dfrac{1}{2}\right)(-3x) + \dfrac{-\frac{1}{2}\left(-\frac{1}{2}-1\right)}{1 \times 2}(-3x)^2$

$\qquad + \dfrac{-\frac{1}{2}\left(-\frac{1}{2}-1\right)\left(-\frac{1}{2}-2\right)}{1 \times 2 \times 3}(-3x)^3 + \ldots$

$= 1 + \left(-\dfrac{1}{2}\right)(-3x) + \dfrac{-\frac{1}{2} \times -\frac{3}{2}}{2}(9x^2)$

$\qquad + \dfrac{-\frac{1}{2} \times -\frac{3}{2} \times -\frac{5}{2}}{6}(-27x^3) + \ldots$

$= 1 + \dfrac{3x}{2} + \dfrac{3}{8}(9x^2) + \left(-\dfrac{5}{16}\right)(-27x^3) + \ldots$

$= 1 + \dfrac{3x}{2} + \dfrac{27x^2}{8} + \dfrac{135x^3}{16} + \ldots$

d) $\sqrt[3]{1 + \frac{x}{2}} = \left(1 + \frac{1}{2}x\right)^{\frac{1}{3}}$

$= 1 + \frac{1}{3}\left(\frac{x}{2}\right) + \frac{\frac{1}{3}\left(\frac{1}{3} - 1\right)}{1 \times 2}\left(\frac{x}{2}\right)^2$

$\quad + \frac{\frac{1}{3}\left(\frac{1}{3} - 1\right)\left(\frac{1}{3} - 2\right)}{1 \times 2 \times 3}\left(\frac{x}{2}\right)^3 \ldots$

$= 1 + \frac{1}{3}\left(\frac{x}{2}\right) + \frac{\frac{1}{3} \times -\frac{2}{3}}{2}\left(\frac{x}{2}\right)^2$

$\quad + \frac{\frac{1}{3} \times -\frac{2}{3} \times -\frac{5}{3}}{6}\left(\frac{x}{2}\right)^3 \ldots$

$= 1 + \frac{1}{3}\left(\frac{x}{2}\right) + \left(-\frac{1}{9}\right)\left(\frac{x}{2}\right)^2 + \frac{5}{81}\left(\frac{x}{2}\right)^3 \ldots$

$= 1 + \frac{x}{6} - \frac{x^2}{36} + \frac{5x^3}{648} \cdots$

Q7 **a)** $\frac{1}{(1 + 7x)^4} = (1 + 7x)^{-4}$

The x^3 term is $\frac{-4(-4 - 1)(-4 - 2)}{1 \times 2 \times 3}(7x)^3$

$= \frac{-4 \times -5 \times -6}{6}(343x^3) = -20(343x^3) = -6860x^3$

So the coefficient of the x^3 term is -6860.

b) The expansion is valid for $|x| < \frac{1}{7}$.

Q8 **a)** $\sqrt[4]{1 - 4x} = (1 - 4x)^{\frac{1}{4}}$

The x^5 term is:

$\frac{\frac{1}{4}\left(\frac{1}{4} - 1\right)\left(\frac{1}{4} - 2\right)\left(\frac{1}{4} - 3\right)\left(\frac{1}{4} - 4\right)}{1 \times 2 \times 3 \times 4 \times 5}(-4x)^5$

$= \frac{\frac{1}{4} \times -\frac{3}{4} \times -\frac{7}{4} \times -\frac{11}{4} \times -\frac{15}{4}}{120}(-4)^5 x^5$

$= \frac{1 \times -3 \times -7 \times -11 \times -15}{120} \times \left(\frac{-4}{4}\right)^5 \times x^5$

$= \frac{3465}{120} \times -1 \times x^5 = -\frac{231x^5}{8}$

So the coefficient of the x^5 term is $-\frac{231}{8}$.

b) The expansion is valid for $|x| < \frac{1}{4}$.

Q9 **a)** Coefficient of x^2 term is $\frac{n \times (n - 1)}{1 \times 2}\left(-\frac{2}{5}\right)^2 = \frac{1}{40}$

$\Rightarrow \frac{2}{25}n(n - 1) = \frac{1}{40} \Rightarrow n(n - 1) = \frac{5}{16}$

$\Rightarrow 16n^2 - 16n - 5 = 0 \Rightarrow (4n + 1)(4n - 5) = 0$

$\Rightarrow n = -\frac{1}{4}, n = \frac{5}{4}$

[3 marks available — 1 mark for setting up a correct equation in terms of n, 1 mark for a correct method to solve, 1 mark for both correct answers]

b) The expansion is valid for $\left|-\frac{2}{5}x\right| < 1 \Rightarrow |x| < \frac{5}{2}$.

[1 mark for the correct validity]

Q10 **a)** Expansion will be valid when $\left|\frac{ax}{1}\right| < 1 \Rightarrow |x| < \frac{1}{|a|}$.

Therefore $\frac{1}{|a|} = 5 \Rightarrow |a| = \frac{1}{5} \Rightarrow a = \pm\frac{1}{5}$.

[1 mark for both correct answers]

b) Coefficient of the x^2 term for $(1 + ax)^n$ is: $\frac{n(n - 1)}{1 \times 2}a^2$

Coefficient of the x^2 term for $(1 + ax)^{2n}$ is: $\frac{2n(2n - 1)}{1 \times 2}a^2$

So $\frac{1}{2}a^2 n(n - 1) = \frac{1}{2}a^2(2n)(2n - 1) \Rightarrow n^2 - n = 4n^2 - 2n$

$\Rightarrow 3n^2 - n = 0 \Rightarrow n(3n - 1) = 0 \Rightarrow n = \frac{1}{3}$ (as $n \neq 0$)

[4 marks available — 1 mark for each correct coefficient of the x^2 terms, 1 mark for equating coefficients, 1 mark for the correct answer]

Q11 **a)** $(1 - 5x)^{\frac{1}{6}} = 1 + \frac{1}{6}(-5x) + \frac{\frac{1}{6}\left(\frac{1}{6} - 1\right)}{1 \times 2}(-5x)^2 + \ldots$

$= 1 + \frac{1}{6}(-5x) + \frac{\frac{1}{6} \times -\frac{5}{6}}{2}(25x^2) + \ldots$

$= 1 - \frac{5x}{6} + \left(-\frac{5}{72}\right)(25x^2) + \ldots = 1 - \frac{5x}{6} - \frac{125x^2}{72} - \ldots$

b) $(1 + 4x)^4 = 1 + 16x + 96x^2 + \ldots$

So $(1 + 4x)^4(1 - 5x)^{\frac{1}{6}}$

$= (1 + 16x + 96x^2 + \ldots)(1 - \frac{5x}{6} - \frac{125x^2}{72} + \ldots)$

$= 1(1 + 16x + 96x^2) - \frac{5x}{6}(1 + 16x) - \frac{125x^2}{72}(1) + \ldots$

$= 1 + 16x + 96x^2 - \frac{5x}{6} - \frac{40x^2}{3} - \frac{125x^2}{72} + \ldots$

$= 1 + \frac{91x}{6} + \frac{5827x^2}{72} + \ldots$

You're only asked to expand up to the x^2 term, so you don't need to write down any x^3 or higher power terms here.

c) The expansion of $(1 - 5x)^{\frac{1}{6}}$ is valid for $|x| < \frac{1}{5}$.

The expansion of $(1 + 4x)^4$ is valid for all values of x, since n is a positive integer.

So overall, the expansion of $(1 + 4x)^4(1 - 5x)^{\frac{1}{6}}$ is valid for the narrower of these ranges, i.e. $|x| < \frac{1}{5}$.

Q12 **a)** $(1 + x)^2 = 1 + 2x + x^2$

$(1 - 2x)^{-2} = 1 + (-2)(-2x) + \frac{-2(-2 - 1)}{1 \times 2}(-2x)^2 + \ldots$

$= 1 + 4x + 12x^2 + \ldots$

So $(1 + x)^2(1 - 2x)^{-2}$

$= (1 + 2x + x^2)(1 + 4x + 12x^2 + \ldots)$

$= 1(1 + 4x + 12x^2) + 2x(1 + 4x) + x^2(1) + \ldots$

$= 1 + 4x + 12x^2 + 2x + 8x^2 + x^2 + \ldots$

$= 1 + 6x + 21x^2 + \ldots$

The expansion of $(1 + x)^2$ is valid for all values of x, since n is a positive integer.

The expansion of $(1 - 2x)^{-2}$ is valid for $|x| < \frac{1}{2}$.

So overall, the expansion of $(1 + x)^2(1 - 2x)^{-2}$ is valid for the narrower of these ranges, i.e. $|x| < \frac{1}{2}$.

b) $(1 + 2x)^2 = 1 + 4x + 4x^2$

$(1 + 3x)^{-3} = 1 + (-3)(3x) + \frac{-3(-3 - 1)}{1 \times 2}(3x)^2 + \ldots$

$= 1 - 9x + 54x^2 + \ldots$

So $(1 + 2x)^2(1 + 3x)^{-3} = (1 + 4x + 4x^2)(1 - 9x + 54x^2 + \ldots)$

$= 1(1 - 9x + 54x^2) + 4x(1 - 9x) + 4x^2(1) + \ldots$

$= 1 - 9x + 54x^2 + 4x - 36x^2 + 4x^2 + \ldots$

$= 1 - 5x + 22x^2 + \ldots$

The expansion of $(1 + 2x)^2$ is valid for all values of x, since n is a positive integer.

The expansion of $(1 + 3x)^{-3}$ is valid for $|x| < \frac{1}{3}$.

So overall, the expansion of $(1 + 2x)^2(1 + 3x)^{-3}$ is valid for the narrower of these ranges, i.e. $|x| < \frac{1}{3}$.

Answers

c) $(1 + 2x)^{\frac{1}{7}} = 1 + \frac{1}{7}(2x) + \frac{\frac{1}{7}\left(\frac{1}{7} - 1\right)}{1 \times 2}(2x)^2 + ...$

$= 1 + \frac{2}{7}x + \frac{\frac{1}{7} \times -\frac{6}{7}}{2}(4x^2) + ...$

$= 1 + \frac{2}{7}x - \frac{12}{49}x^2 + ...$

So $(1 - 7x)(1 + 2x)^{\frac{1}{7}} = (1 - 7x)(1 + \frac{2}{7}x - \frac{12}{49}x^2 + ...)$

$= 1(1 + \frac{2}{7}x - \frac{12}{49}x^2) - 7x(1 + \frac{2}{7}x) + ...$

$= 1 + \frac{2}{7}x - \frac{12}{49}x^2 - 7x - 2x^2 + ...$

$= 1 - \frac{47}{7}x - \frac{110}{49}x^2 + ...$

$(1 + 7x)$ is valid for all values of x.

The expansion of $(1 + 2x)^{\frac{1}{7}}$ is valid for $|x| < \frac{1}{2}$.

So the expansion of $(1 - 7x)(1 + 2x)^{\frac{1}{7}}$ is valid for the narrower of these ranges, i.e. $|x| < \frac{1}{2}$.

d) $(1 - x)^2 = 1 - 2x + x^2$

$(1 + 4x)^{-\frac{1}{3}} = 1 + \left(-\frac{1}{3}\right)(4x) + \frac{\left(-\frac{1}{3}\right)\left(-\frac{1}{3} - 1\right)}{1 \times 2}(4x)^2 + ...$

$= 1 - \frac{4}{3}x + \frac{-\frac{1}{3} \times -\frac{4}{3}}{2}(16x^2) + ...$

$= 1 - \frac{4}{3}x + \frac{32}{9}x^2 + ...$

So $(1 - x)^2(1 + 4x)^{-\frac{1}{3}} = (1 - 2x + x^2)(1 - \frac{4}{3}x + \frac{32}{9}x^2 + ...)$

$= 1(1 - \frac{4}{3}x + \frac{32}{9}x^2) - 2x(1 - \frac{4}{3}x) + x^2(1) + ...$

$= 1 - \frac{4}{3}x + \frac{32}{9}x^2 - 2x + \frac{8}{3}x^2 + x^2 + ...$

$= 1 - \frac{10}{3}x + \frac{65}{9}x^2 + ...$

The expansion of $(1 - x)^2$ is valid for all values of x, since n is a positive integer.

The expansion of $(1 + 4x)^{-\frac{1}{3}}$ is valid for $|x| < \frac{1}{4}$.

So the expansion of $(1 - x)^2(1 + 4x)^{-\frac{1}{3}}$ is valid for the narrower of these ranges, i.e. $|x| < \frac{1}{4}$.

Q13 a) $\frac{(1 + 3x)^4}{(1 + x)^3} = (1 + 3x)^4(1 + x)^{-3}$

$(1 + 3x)^4 = 1 + 12x + 54x^2 + ...$

$(1 + x)^{-3} = 1 - 3x + 6x^2 - ...$

So $(1 + 3x)^4(1 + x)^{-3} = 1 - 3x + 6x^2 + [12x(1 - 3x)] + 54x^2 + ...$

$= 1 - 3x + 6x^2 + 12x - 36x^2 + 54x^2 + ... = 1 + 9x + 24x^2 + ...$

b) The expansion of $(1 + 3x)^4$ is valid for all values of x, since n is a positive integer.

The expansion of $(1 + x)^{-3}$ is valid for $|x| < 1$.

So overall, the expansion of $\frac{(1 + 3x)^4}{(1 + x)^3}$ is valid for the narrower of these ranges, i.e. $|x| < 1$.

Q14 a) $(1 + 2x)^2(1 + kx)^{-\frac{1}{3}}$

$= (1 + 4x + 4x^2)\left(1 + \left(-\frac{1}{3}\right)(kx) + \frac{\left(-\frac{1}{3}\right)\left(-\frac{4}{3}\right)}{1 \times 2}(kx)^2 + ...\right)$

$= (1 + 4x + 4x^2)\left(1 - \frac{1}{3}kx + \frac{2}{9}k^2x^2 - ...\right)$

$= 1\left(1 - \frac{1}{3}kx + \frac{2}{9}k^2x^2 - ...\right) + 4x\left(1 - \frac{1}{3}kx + ...\right)$

$\qquad + 4x^2(1 - ...) + ...$

$= 1 - \frac{1}{3}kx + \frac{2}{9}k^2x^2 + 4x - \frac{4}{3}kx^2 + 4x^2 + ...$

Extracting the coefficients of terms involving x^2 gives:

$\frac{2}{9}k^2 - \frac{4}{3}k + 4 = 2 \Rightarrow 2k^2 - 12k + 18 = 0$

$\Rightarrow k^2 - 6k + 9 = 0 \Rightarrow (k - 3)^2 = 0 \Rightarrow k = 3$

[5 marks available — 1 mark for a correct expansion of $(1 + kx)^{-\frac{1}{3}}$, 1 mark for a correct expansion up to x^2 terms, 1 mark for setting up a correct equation, 1 mark for a correct method to solve, 1 mark for the correct answer]

b) The first expansion is valid for all x.

The second expansion is valid for $|3x| < 1 \Rightarrow |x| < \frac{1}{3}$.

Therefore, the combined expansion is valid for $|x| < \frac{1}{3}$.

[1 mark for the correct validity of the full expansion]

Q15 a) $\frac{\left(1 - \frac{1}{3}x\right)^{\frac{1}{3}}}{\left(1 + \frac{2}{3}x\right)^{\frac{2}{3}}} = \left(1 - \frac{1}{3}x\right)^{\frac{1}{3}}\left(1 + \frac{2}{3}x\right)^{-\frac{2}{3}}$

$\left(1 - \frac{1}{3}x\right)^{\frac{1}{3}} = 1 + \left(\frac{1}{3}\right)\left(-\frac{1}{3}x\right) + \frac{\left(\frac{1}{3}\right)\left(-\frac{2}{3}\right)}{1 \times 2}\left(-\frac{1}{3}x\right)^2 + ...$

$= 1 - \frac{1}{9}x - \frac{1}{81}x^2 - ...$

$\left(1 + \frac{2}{3}x\right)^{-\frac{2}{3}} = 1 + \left(-\frac{2}{3}\right)\left(\frac{2}{3}x\right) + \frac{\left(-\frac{2}{3}\right)\left(-\frac{5}{3}\right)}{1 \times 2}\left(\frac{2}{3}x\right)^2 + ...$

$= 1 - \frac{4}{9}x + \frac{20}{81}x^2 - ...$

$\frac{\left(1 - \frac{1}{3}x\right)^{\frac{1}{3}}}{\left(1 + \frac{2}{3}x\right)^{\frac{2}{3}}} = \left(1 - \frac{1}{9}x - \frac{1}{81}x^2 - ...\right)\left(1 - \frac{4}{9}x + \frac{20}{81}x^2 - ...\right)$

$= 1\left(1 - \frac{4}{9}x + \frac{20}{81}x^2 - ...\right) - \frac{1}{9}x\left(1 - \frac{4}{9}x + ...\right)$

$\qquad - \frac{1}{81}x^2(1 - ...) + ...$

$\approx 1 - \frac{5}{9}x + \frac{23}{81}x^2$

[4 marks available — 1 mark for rewriting the denominator, 1 mark for each correct expansion, 1 mark for the correct and simplified answer]

b) The numerator expansion is valid for $\left|-\frac{1}{3}x\right| < 1 \Rightarrow |x| < 3$.

The denominator expansion is valid for $\left|\frac{2}{3}x\right| < 1 \Rightarrow |x| < \frac{3}{2}$.

Therefore, the expansion is valid for $|x| < \frac{3}{2}$.

[1 mark for the correct validity of the full expansion]

Q16 a) (i) $(1 + ax)^4 = 1 + 4ax + \frac{4 \times 3}{1 \times 2}(ax)^2 + \frac{4 \times 3 \times 2}{1 \times 2 \times 3}(ax)^3$

$\qquad + \frac{4 \times 3 \times 2 \times 1}{1 \times 2 \times 3 \times 4}(ax)^4$

$= 1 + 4ax + 6a^2x^2 + 4a^3x^3 + a^4x^4$

(ii) $(1 - bx)^{-3} = 1 + (-3)(-bx) + \frac{-3 \times -4}{1 \times 2}(-bx)^2 + ...$

$= 1 + 3bx + 6b^2x^2 + ...$

b) $\frac{(1 + ax)^4}{(1 - bx)^3} = (1 + ax)^4(1 - bx)^{-3}$

$= (1 + 4ax + 6a^2x^2 + ...)(1 + 3bx + 6b^2x^2 + ...)$

$= 1 + 3bx + 6b^2x^2 + 4ax + 12abx^2 + 6a^2x^2 + ...$

$= 1 + (4a + 3b)x + (6a^2 + 12ab + 6b^2)x^2 + ...$

c) Equating coefficients — x^2 terms:

$6a^2 + 12ab + 6b^2 = 24 \Rightarrow 6(a + b)^2 = 24$

$\Rightarrow (a + b)^2 = 4 \Rightarrow a + b = \pm 2$

Equating coefficients — x terms:

$4a + 3b = 1 \Rightarrow a + 3(a + b) = 1$

So $a + b = 2 \Rightarrow a + 3 \times 2 = 1 \Rightarrow a = -5$

$\Rightarrow -5 + b = 2 \Rightarrow b = 7$

$a + b = -2 \Rightarrow a + 3 \times -2 = 1 \Rightarrow a = 7$

$\Rightarrow 7 + b = -2 \Rightarrow b = -9$

So the two pairs of values are: $a = -5$, $b = 7$ and $a = 7$, $b = -9$

You could also have rearranged to get either a or b in terms of the other, and then substituted that into the other equation. You would get the same answer, but the working is a bit longer.

Exercise 6.3 — Expanding $(p + qx)^n$

Q1 **a)** $(2 + 4x)^3 = 2^3(1 + 2x)^3 = 8(1 + 2x)^3$

$(1 + 2x)^3 = 1 + 3(2x) + \dfrac{3 \times 2}{1 \times 2}(2x)^2 + \dfrac{3 \times 2 \times 1}{1 \times 2 \times 3}(2x)^3$

$= 1 + 6x + 3(4x^2) + 8x^3 = 1 + 6x + 12x^2 + 8x^3$

So $(2 + 4x)^3 = 8(1 + 6x + 12x^2 + 8x^3) = 8 + 48x + 96x^2 + 64x^3$

b) $(3 + 4x)^5 = 3^5\left(1 + \dfrac{4}{3}x\right)^5 = 243\left(1 + \dfrac{4}{3}x\right)^5$

$\left(1 + \dfrac{4}{3}x\right)^5 = 1 + 5\left(\dfrac{4}{3}x\right) + \dfrac{5 \times 4}{1 \times 2}\left(\dfrac{4}{3}x\right)^2$

$+ \dfrac{5 \times 4 \times 3}{1 \times 2 \times 3}\left(\dfrac{4}{3}x\right)^3 + \ldots$

$= 1 + \dfrac{20x}{3} + \dfrac{160x^2}{9} + \dfrac{640x^3}{27} + \ldots$

$(3 + 4x)^5 = 243(1 + \dfrac{20x}{3} + \dfrac{160x^2}{9} + \dfrac{640x^3}{27} + \ldots)$

$= 243 + 1620x + 4320x^2 + 5760x^3 + \ldots$

c) $(4 + x)^{\frac{1}{2}} = 4^{\frac{1}{2}}\left(1 + \dfrac{x}{4}\right)^{\frac{1}{2}} = 2\left(1 + \dfrac{x}{4}\right)^{\frac{1}{2}}$

$\left(1 + \dfrac{x}{4}\right)^{\frac{1}{2}} = 1 + \dfrac{1}{2}\left(\dfrac{x}{4}\right) + \dfrac{\frac{1}{2} \times -\frac{1}{2}}{1 \times 2}\left(\dfrac{x}{4}\right)^2$

$+ \dfrac{\frac{1}{2} \times -\frac{1}{2} \times -\frac{3}{2}}{1 \times 2 \times 3}\left(\dfrac{x}{4}\right)^3 + \ldots$

$= 1 + \dfrac{x}{8} - \dfrac{x^2}{128} + \dfrac{x^3}{1024} - \ldots$

$(4 + x)^{\frac{1}{2}} = 2\left(1 + \dfrac{x}{4}\right)^{\frac{1}{2}}$

$= 2(1 + \dfrac{x}{8} - \dfrac{x^2}{128} + \dfrac{x^3}{1024} - \ldots)$

$= 2 + \dfrac{x}{4} - \dfrac{x^2}{64} + \dfrac{x^3}{512} - \ldots$

d) $(8 + 2x)^{-\frac{1}{3}} = 8^{-\frac{1}{3}}\left(1 + \dfrac{2x}{8}\right)^{-\frac{1}{3}} = \dfrac{1}{2}\left(1 + \dfrac{x}{4}\right)^{-\frac{1}{3}}$

$\left(1 + \dfrac{x}{4}\right)^{-\frac{1}{3}} = 1 + \left(-\dfrac{1}{3}\right)\left(\dfrac{x}{4}\right) + \dfrac{-\frac{1}{3} \times -\frac{4}{3}}{1 \times 2}\left(\dfrac{x}{4}\right)^2$

$+ \dfrac{-\frac{1}{3} \times -\frac{4}{3} \times -\frac{7}{3}}{1 \times 2 \times 3}\left(\dfrac{x}{4}\right)^3 + \ldots$

$= 1 - \dfrac{x}{12} + \dfrac{x^2}{72} - \dfrac{7x^3}{2592} + \ldots$

$(8 + 2x)^{-\frac{1}{3}} = \dfrac{1}{2}\left(1 + \dfrac{x}{4}\right)^{-\frac{1}{3}}$

$= \dfrac{1}{2}(1 - \dfrac{x}{12} + \dfrac{x^2}{72} - \dfrac{7x^3}{2592} + \ldots)$

$= \dfrac{1}{2} - \dfrac{x}{24} + \dfrac{x^2}{144} - \dfrac{7x^3}{5184} + \ldots$

Q2 $(a + 5x)^5 = a^5\left(1 + \dfrac{5}{a}x\right)^5$

So the x^2 term is:

$a^5\left(\dfrac{5 \times 4}{1 \times 2}\right)\left(\dfrac{5}{a}x\right)^2 = 10a^5\left(\dfrac{25}{a^2}x^2\right) = \left(\dfrac{250a^5}{a^2}\right)x^2 = 250a^3x^2$

So $250a^3 = 2000 \Rightarrow a^3 = 8 \Rightarrow a = 2$

Q3 **a)** $(2 - 5x)^7 = 2^7\left(1 - \dfrac{5}{2}x\right)^7 = 128\left(1 - \dfrac{5}{2}x\right)^7$

$\left(1 - \dfrac{5}{2}x\right)^7 = 1 + 7\left(-\dfrac{5}{2}x\right) + \dfrac{7 \times 6}{1 \times 2}\left(-\dfrac{5}{2}x\right)^2 + \ldots$

$= 1 - \dfrac{35x}{2} + \dfrac{525x^2}{4} - \ldots$

$(2 - 5x)^7 = 128(1 - \dfrac{35x}{2} + \dfrac{525x^2}{4} - \ldots)$

$= 128 - 2240x + 16\,800x^2 - \ldots$

b) $(1 + 6x)^3 = 1 + 18x + 108x^2 + \ldots$

So $(1 + 6x)^3(2 - 5x)^7 =$

$128 - 2240x + 16\,800x^2 + 18x(128 - 2240x)$

$+ 108x^2(128) + \ldots$

$= 128 + 64x - 9696x^2 + \ldots$

Q4 $(2 + ax)^{-3} = 2^{-3}\left(1 + \dfrac{a}{2}x\right)^{-3}$

The coefficient of the x^2 term:

$(2^{-3})\dfrac{(-3) \times (-4)}{1 \times 2}\left(\dfrac{a}{2}\right)^2 = \dfrac{3}{16}a^2$

The coefficient of the x^3 term:

$(2^{-3})\dfrac{(-3) \times (-4) \times (-5)}{1 \times 2 \times 3}\left(\dfrac{a}{2}\right)^3 = -\dfrac{5}{32}a^3$

So $\dfrac{3}{16}a^2 = 2\left(-\dfrac{5}{32}a^3\right) \Rightarrow a = -\dfrac{3}{5}$

[4 marks available — 1 mark for each correct coefficient, 1 mark for a correct equation, 1 mark for the correct answer]

Q5 **a)** $\left(1 + \dfrac{6}{5}x\right)^{-\frac{1}{2}} = 1 + \left(-\dfrac{1}{2}\right)\left(\dfrac{6}{5}x\right) + \dfrac{\left(-\frac{1}{2}\right) \times \left(-\frac{3}{2}\right)}{1 \times 2}\left(\dfrac{6}{5}x\right)^2$

$+ \dfrac{\left(-\frac{1}{2}\right) \times \left(-\frac{3}{2}\right) \times \left(-\frac{5}{2}\right)}{1 \times 2 \times 3}\left(\dfrac{6}{5}x\right)^3 + \ldots$

$= 1 - \dfrac{3x}{5} + \dfrac{27x^2}{50} - \dfrac{27x^3}{50} + \ldots$

The expansion is valid for $\left|\dfrac{6x}{5}\right| < 1$, so $|x| < \dfrac{5}{6}$.

b) $\sqrt{\dfrac{20}{5 + 6x}} = \dfrac{\sqrt{20}}{\sqrt{5 + 6x}} = 20^{\frac{1}{2}}(5 + 6x)^{-\frac{1}{2}}$

$= (20^{\frac{1}{2}})(5^{-\frac{1}{2}})\left(1 + \dfrac{6}{5}x\right)^{-\frac{1}{2}}$

$= \left(\dfrac{20^{\frac{1}{2}}}{5^{\frac{1}{2}}}\right)\left(1 + \dfrac{6}{5}x\right)^{-\frac{1}{2}} = \left(\dfrac{20}{5}\right)^{\frac{1}{2}}\left(1 + \dfrac{6}{5}x\right)^{-\frac{1}{2}}$

$= 4^{\frac{1}{2}}\left(1 + \dfrac{6}{5}x\right)^{-\frac{1}{2}} = 2\left(1 + \dfrac{6}{5}x\right)^{-\frac{1}{2}}$

So, using the expansion in part a):

$\sqrt{\dfrac{20}{5 + 6x}} = 2(1 - \dfrac{3x}{5} + \dfrac{27x^2}{50} - \dfrac{27x^3}{50} + \ldots)$

$= 2 - \dfrac{6x}{5} + \dfrac{27x^2}{25} - \dfrac{27x^3}{25} + \ldots$

Q6 **a)** $\dfrac{1}{\sqrt{5 - 2x}} = (5 - 2x)^{-\frac{1}{2}} = 5^{-\frac{1}{2}}\left(1 - \dfrac{2}{5}x\right)^{-\frac{1}{2}}$

$\left(1 - \dfrac{2}{5}x\right)^{-\frac{1}{2}} = 1 + \left(-\dfrac{1}{2}\right)\left(-\dfrac{2}{5}x\right) + \dfrac{\left(-\frac{1}{2}\right) \times \left(-\frac{3}{2}\right)}{1 \times 2}\left(-\dfrac{2}{5}x\right)^2$

$+ \ldots$

$= 1 + \dfrac{x}{5} + \left(\dfrac{3}{8}\right)\left(\dfrac{4}{25}x^2\right) + \ldots = 1 + \dfrac{x}{5} + \dfrac{3x^2}{50} + \ldots$

So $\dfrac{1}{\sqrt{5 - 2x}} = 5^{-\frac{1}{2}}\left(1 - \dfrac{2}{5}x\right)^{-\frac{1}{2}} = \dfrac{1}{\sqrt{5}}(1 + \dfrac{x}{5} + \dfrac{3x^2}{50} \ldots)$

$= \dfrac{1}{\sqrt{5}} + \dfrac{x}{5\sqrt{5}} + \dfrac{3x^2}{50\sqrt{5}} + \ldots$

Answers

b) $\dfrac{3+x}{\sqrt{5-2x}} \approx (3+x)\left(\dfrac{1}{\sqrt{5}} + \dfrac{x}{5\sqrt{5}} + \dfrac{3x^2}{50\sqrt{5}}\right)$

$\approx 3\left(\dfrac{1}{\sqrt{5}} + \dfrac{x}{5\sqrt{5}} + \dfrac{3x^2}{50\sqrt{5}}\right) + x\left(\dfrac{1}{\sqrt{5}} + \dfrac{x}{5\sqrt{5}}\right)$

$= \dfrac{3}{\sqrt{5}} + \dfrac{3x}{5\sqrt{5}} + \dfrac{9x^2}{50\sqrt{5}} + \dfrac{x}{\sqrt{5}} + \dfrac{x^2}{5\sqrt{5}}$

$= \dfrac{3}{\sqrt{5}} + \dfrac{8x}{5\sqrt{5}} + \dfrac{19x^2}{50\sqrt{5}}$

Q7 a) $(9+4x)^{-\frac{1}{2}} = 9^{-\frac{1}{2}}\left(1 + \dfrac{4}{9}x\right)^{-\frac{1}{2}}$

$\left(1 + \dfrac{4}{9}x\right)^{-\frac{1}{2}} = 1 + \left(-\dfrac{1}{2}\right)\left(\dfrac{4}{9}x\right) + \dfrac{\left(-\frac{1}{2}\right)\times\left(-\frac{3}{2}\right)}{1\times 2}\left(\dfrac{4}{9}x\right)^2 + \dots$

$= 1 - \dfrac{2x}{9} + \dfrac{2x^2}{27} - \dots$

So $(9+4x)^{-\frac{1}{2}} = \dfrac{1}{\sqrt{9}}\left(1 - \dfrac{2x}{9} + \dfrac{2x^2}{27} - \dots\right)$

$= \dfrac{1}{3} - \dfrac{2x}{27} + \dfrac{2x^2}{81} - \dots$

b) $(1+6x)^4 = 1 + 24x + 216x^2 + \dots$

So $\dfrac{(1+6x)^4}{\sqrt{9+4x}} = (1+6x)^4(9+4x)^{-\frac{1}{2}}$

$\approx (1+24x+216x^2)\left(\dfrac{1}{3} - \dfrac{2x}{27} + \dfrac{2x^2}{81}\right)$

$\approx \dfrac{1}{3} - \dfrac{2x}{27} + \dfrac{2x^2}{81} + 24x\left(\dfrac{1}{3} - \dfrac{2x}{27}\right) + 216x^2\left(\dfrac{1}{3}\right)$

$= \dfrac{1}{3} - \dfrac{2x}{27} + \dfrac{2x^2}{81} + 8x - \dfrac{16x^2}{9} + 72x^2$

$= \dfrac{1}{3} + \dfrac{214x}{27} + \dfrac{5690x^2}{81}$

Q8 a) The 1st error is in the second line — when taking out the factor of 4, the student should also raise this to the power of $-\dfrac{3}{2}$.

The 2nd error is in the third line — the expansion for the x^2 term should be $n(n-1) = \left(-\dfrac{3}{2}\right)\times\left(-\dfrac{5}{2}\right)$, the student has used $n(n+1)$ instead.

[4 marks available — 1 mark for each error identified, 1 mark for each correct explanation]

b) $4^{-\frac{3}{2}}\left(1 + \dfrac{3}{4}x\right)^{-\frac{3}{2}}$

$\approx \dfrac{1}{8}\left[1 + \left(-\dfrac{3}{2}\right)\left(\dfrac{3}{4}x\right) + \dfrac{\left(-\frac{3}{2}\right)\times\left(-\frac{5}{2}\right)}{1\times 2}\left(\dfrac{3}{4}x\right)^2\right]$

$= \dfrac{1}{8} - \dfrac{9}{64}x + \dfrac{135}{1024}x^2$

This is valid for $\left|\dfrac{3}{4}x\right| < 1 \Rightarrow |x| < \dfrac{4}{3}$

[3 marks available — 1 mark for a correct use of the expansion formula, 1 mark for the correct simplified expansion, 1 mark for the correct validity]

Q9 a) $\left(\sqrt[4]{16-80x}\right)^3 = 16^{\frac{3}{4}}(1-5x)^{\frac{3}{4}} = 8(1-5x)^{\frac{3}{4}}$

$(1-5x)^{\frac{3}{4}} = 1 + \left(\dfrac{3}{4}\right)(-5x) + \dfrac{\left(\frac{3}{4}\right)\times\left(-\frac{1}{4}\right)}{1\times 2}(-5x)^2$

$+ \dfrac{\left(\frac{3}{4}\right)\times\left(-\frac{1}{4}\right)\times\left(-\frac{5}{4}\right)}{1\times 2\times 3}(-5x)^3 + \dots$

$= 1 - \dfrac{15}{4}x - \dfrac{75}{32}x^2 - \dfrac{625}{128}x^3 - \dots$

So $\left(\sqrt[4]{16-80x}\right)^3 \approx 8\left(1 - \dfrac{15}{4}x - \dfrac{75}{32}x^2 - \dfrac{625}{128}x^3\right)$

$= 8 - 30x - \dfrac{75}{4}x^2 - \dfrac{625}{16}x^3$

[4 marks available — 1 mark for factorising, 1 mark for an expansion of $(1-5x)^{\frac{3}{4}}$ with at least two terms correct, 1 mark for a simplified expansion of $(1-5x)^{\frac{3}{4}}$, 1 mark for the correct final answer]

b) As x is an integer, $8 - 30x$ will be an integer, so we require $-\dfrac{75}{4}x^2 - \dfrac{625}{16}x^3$ to be an integer.

$-\dfrac{75}{4}x^2 - \dfrac{625}{16}x^3 = -\dfrac{25}{16}x^2(12 - 25x)$

$12 - 25x$ will be an integer (as x is an integer), so we require $\dfrac{-25}{16}x^2$ to be an integer. The smallest integer x can be for this to be true must be 4 to give -25, so $x = 4$.

[3 marks available for suitable working, e.g. 1 mark for '8 – 30x is an integer', 1 mark for stating that $\dfrac{-25}{16}x^2$ needs to be an integer, 1 mark for showing that 4 is the smallest possible integer value of x that makes $\dfrac{-25}{16}x^2$ an integer]

Q10 a) $f(x) = (4+x)^{\frac{3}{2}} = 4^{\frac{3}{2}}\left(1 + \dfrac{1}{4}x\right)^{\frac{3}{2}} = 8\left(1 + \dfrac{1}{4}x\right)^{\frac{3}{2}}$

$\approx 8\left[1 + \left(\dfrac{3}{2}\right)\left(\dfrac{1}{4}x\right) + \dfrac{\left(\frac{3}{2}\right)\times\left(\frac{1}{2}\right)}{1\times 2}\left(\dfrac{1}{4}x\right)^2\right]$

$= 8 + 3x + \dfrac{3}{16}x^2$

$g(x) = (4+9x)^{\frac{1}{2}} = 4^{\frac{1}{2}}\left(1 + \dfrac{9}{4}x\right)^{\frac{1}{2}}$

$\approx 2\left[1 + \left(\dfrac{1}{2}\right)\left(\dfrac{9}{4}x\right) + \dfrac{\left(\frac{1}{2}\right)\times\left(-\frac{1}{2}\right)}{1\times 2}\left(\dfrac{9}{4}x\right)^2\right]$

$= 2 + \dfrac{9}{4}x - \dfrac{81}{64}x^2$

$\dfrac{1}{h(x)} = (3-x)^{-3} = 3^{-3}\left(1 - \dfrac{1}{3}x\right)^{-3}$

$\approx \dfrac{1}{27}\left[1 + (-3)\left(-\dfrac{1}{3}x\right) + \dfrac{(-3)\times(-4)}{1\times 2}\left(-\dfrac{1}{3}x\right)^2\right]$

$= \dfrac{1}{27} + \dfrac{1}{27}x + \dfrac{2}{81}x^2$

$p(x) = f(x)\times g(x)\times h(x)^{-1}$

$\approx \left(8 + 3x + \dfrac{3}{16}x^2\right)\times\left(2 + \dfrac{9}{4}x - \dfrac{81}{64}x^2\right)\times h(x)^{-1}$

$\approx \left(16 + 18x - \dfrac{81}{8}x^2 + 6x + \dfrac{27}{4}x^2 + \dfrac{3}{8}x^2\right)\times h(x)^{-1}$

$\approx \left(16 + 24x - 3x^2\right)\left(\dfrac{1}{27} + \dfrac{1}{27}x + \dfrac{2}{81}x^2\right)$

$\approx \dfrac{16}{27} + \dfrac{16}{27}x + \dfrac{32}{81}x^2 + \dfrac{24}{27}x + \dfrac{24}{27}x^2 - \dfrac{1}{9}x^2$

$= \dfrac{16}{27} + \dfrac{40}{27}x + \dfrac{95}{81}x^2$

b) $f(x)$ is valid when $\left|\dfrac{1}{4}x\right| < 1 \Rightarrow |x| < 4$

$g(x)$ is valid when $\left|\dfrac{9}{4}x\right| < 1 \Rightarrow |x| < \dfrac{4}{9}$

$h(x)$ is valid when $\left|-\dfrac{1}{3}x\right| < 1 \Rightarrow |x| < 3$

So $p(x)$ is valid when $|x| < \dfrac{4}{9}$

Q11 $(3+2x)^n = 3^n\left(1 + \dfrac{2}{3}x\right)^n$

Coefficient of x^5 term: $3^n\dfrac{n(n-1)(n-2)(n-3)(n-4)}{1\times 2\times 3\times 4\times 5}\left(\dfrac{2}{3}\right)^5$

Coefficient of x^3 term: $3^n\dfrac{n(n-1)(n-2)}{1\times 2\times 3}\left(\dfrac{2}{3}\right)^3$

x^5 term $= 2\times(x^3$ term), so

$3^n\dfrac{n(n-1)(n-2)(n-3)(n-4)}{1\times 2\times 3\times 4\times 5}\left(\dfrac{2}{3}\right)^5$

$= 2\times 3^n\dfrac{n(n-1)(n-2)}{1\times 2\times 3}\left(\dfrac{2}{3}\right)^3$

Cancelling terms on both sides gives:

$$\frac{(n-3)(n-4)}{4 \times 5}\left(\frac{2}{3}\right)^2 = 2$$

$$\Rightarrow n^2 - 7n + 12 = 2 \times \frac{180}{4}$$

$$\Rightarrow n^2 - 7n - 78 = 0$$

$$\Rightarrow (n - 13)(n + 6) = 0$$

$$\Rightarrow n = 13 \text{ or } n = -6$$

[6 marks available — 1 mark for factorising, 1 mark for each correct coefficient, 1 mark for a correct equation, 1 mark for a correct method to solve, 1 mark both correct answers]
This question can lead to some awkward algebra, so the key is to cancel as much as possible.

Exercise 6.4 — Using the Binomial Expansion as an Approximation

Q1 **a)** $(1 + 6x)^{-1} = 1 + (-1)(6x) + \frac{-1 \times -2}{1 \times 2}(6x)^2 + ...$

$$= 1 - 6x + 36x^2 - ...$$

b) The expansion is valid if $|x| < \frac{1}{6}$.

c) $\frac{100}{106} = \frac{1}{1.06} = \frac{1}{1 + 0.06} = (1 + 0.06)^{-1}$

This is the same as $(1 + 6x)^{-1}$ with $x = 0.01$.

$0.01 < \frac{1}{6}$ so the approximation is valid.

$(1 + 6(0.01))^{-1} \approx 1 - 6(0.01) + 36(0.01^2)$

$$= 1 - 0.06 + 0.0036 = 0.9436$$

d) $\left|\frac{\left(\frac{100}{106}\right) - 0.9436}{\left(\frac{100}{106}\right)}\right| \times 100 = 0.02\%$ (1 s.f.)

In some of the answers that follow, the expansions will just be stated. Look back at the previous sections of this chapter to check how to set out the working if you need to.

Q2 **a)** $(1 + 3x)^{\frac{1}{4}} = 1 + \frac{3x}{4} - \frac{27x^2}{32} + \frac{189x^3}{128} - ...$

b) The expansion is valid if $|x| < \frac{1}{3}$.

c) $\sqrt[4]{1.9} = 1.9^{\frac{1}{4}} = (1 + 0.9)^{\frac{1}{4}}$

This is the same as $(1 + 3x)^{\frac{1}{4}}$ with $x = 0.3$.

$0.3 < \frac{1}{3}$ so the approximation is valid.

$(1 + 3(0.3))^{\frac{1}{4}} \approx 1 + \frac{3(0.3)}{4} - \frac{27(0.3^2)}{32} + \frac{189(0.3^3)}{128}$

$$= 1.1889 \text{ (4 d.p.)}$$

d) $\left|\frac{(\sqrt[4]{1.9}) - 1.1889}{(\sqrt[4]{1.9})}\right| \times 100 = 1.26\%$ (3 s.f.)

Q3 **a)** $(1 - 2x)^{-\frac{1}{2}} = 1 + x + \frac{3x^2}{2} + \frac{5x^3}{2} + ...$

b) The expansion is valid if $|x| < \frac{1}{2}$.

c) Using $x = \frac{1}{10}$ gives:

$$\left(1 - \frac{2}{10}\right)^{-\frac{1}{2}} \approx 1 + \frac{1}{10} + \frac{3}{200} + \frac{5}{2000}$$

$$\left(\frac{4}{5}\right)^{-\frac{1}{2}} \approx 1 + 0.1 + 0.015 + 0.0025$$

$$\left(\frac{5}{4}\right)^{\frac{1}{2}} \approx 1.1175$$

$$\left(\frac{5}{4}\right)^{\frac{1}{2}} = \sqrt{\frac{5}{4}} = \frac{\sqrt{5}}{\sqrt{4}} = \frac{\sqrt{5}}{2}$$

So $\sqrt{5} \approx 2 \times 1.1175 = 2.235$

d) $\left|\frac{\sqrt{5} - 2.235}{\sqrt{5}}\right| \times 100 = 0.048\%$ (2 s.f.)

Q4 **a)** $(2 - 5x)^6 = 2^6\left(1 - \frac{5}{2}x\right)^6 = 64\left(1 - 15x + \frac{375x^2}{4} - ...\right)$

$$= 64 - 960x + 6000x^2 - ...$$

b) $1.95^6 = (2 - 0.05)^6$

This is the same as $(2 - 5x)^6$ with $x = 0.01$.

$(2 - 5(0.01))^6 \approx 64 - 960(0.01) + 6000(0.01)^2$

$$= 64 - 9.6 + 0.6 = 55$$

c) $\left|\frac{1.95^6 - 55}{1.95^6}\right| \times 100 = 0.036\%$ (2 s.f.)

Q5 **a)** $\sqrt{3 - 4x} = (3 - 4x)^{\frac{1}{2}} = \sqrt{3}\left(1 - \frac{4}{3}x\right)^{\frac{1}{2}}$

$$= \sqrt{3}\left(1 - \frac{2x}{3} - \frac{2x^2}{9} - ...\right) = \sqrt{3} - \frac{2\sqrt{3}x}{3} - \frac{2\sqrt{3}x^2}{9} - ...$$

Remember to factorise the original expression to get it in the form $(1 + ax)^n$.

b) The expansion is valid if $|x| < \frac{3}{4}$.

c) $\sqrt{3 - 4\left(\frac{3}{40}\right)} \approx \sqrt{3} - \frac{2\sqrt{3} \times 3}{3 \times 40} - \frac{2\sqrt{3} \times 9}{9 \times 1600}$

$$\sqrt{3 - \frac{3}{10}} \approx \sqrt{3} - \frac{\sqrt{3}}{20} - \frac{\sqrt{3}}{800}$$

$$\sqrt{\frac{27}{10}} \approx \frac{759\sqrt{3}}{800}$$

$$\frac{3\sqrt{3}}{\sqrt{10}} \approx \frac{759\sqrt{3}}{800} \Rightarrow \frac{3}{\sqrt{10}} \approx \frac{759}{800}$$

d) $\left|\frac{\left(\frac{3}{\sqrt{10}}\right) - \left(\frac{759}{800}\right)}{\left(\frac{3}{\sqrt{10}}\right)}\right| \times 100 = 0.007\%$ (1 s.f.)

Q6 **a)** $\left(1 + \frac{1}{4}x\right)^{\frac{2}{3}} = 1 + \left(\frac{2}{3}\right)\left(\frac{1}{4}x\right) + \frac{\left(\frac{2}{3}\right)\left(-\frac{1}{3}\right)}{1 \times 2}\left(\frac{1}{4}x\right)^2 + ...$

$$= 1 + \frac{1}{6}x - \frac{1}{144}x^2 + ...$$

[2 marks available — 1 mark for a correct expansion, 1 mark for simplifying and writing correct coefficients of constant, x and x^2]

b) $1 + \frac{1}{4}x = \pm\sqrt{1.44}$, so $x = 4(1.2 - 1) = 0.8$ *[1 mark]*

c) $x = 0.8$ gives $\left(1 + \frac{1}{4}(0.8)\right)^{\frac{2}{3}} = \sqrt[3]{(1 + 0.2)^2} = \sqrt[3]{1.44}$, so substitute $x = 0.8$ into expansion from part a) to find $\sqrt[3]{1.44}$.

$\sqrt[3]{1.44} \approx 1 + \frac{1}{6}(0.8) - \frac{1}{144}(0.8)^2 = 1.1289$ to 4 d.p.

[2 marks available — 1 mark for a correct substitution, 1 mark for the correct answer]

Q7 **a)** $(2 - x)^{\frac{1}{2}} = 2^{\frac{1}{2}}\left(1 - \frac{1}{2}x\right)^{\frac{1}{2}}$

$$= \sqrt{2}\left[1 + \left(\frac{1}{2}\right)\left(-\frac{1}{2}x\right) + \frac{\left(\frac{1}{2}\right) \times \left(-\frac{1}{2}\right)}{1 \times 2}\left(-\frac{1}{2}x\right)^2 + ...\right]$$

$$\approx \sqrt{2}\left(1 - \frac{1}{4}x - \frac{1}{32}x^2\right) = \sqrt{2} - \frac{\sqrt{2}}{4}x - \frac{\sqrt{2}}{32}x^2$$

[3 marks available — 1 mark for factorising, 1 mark for expansion, 1 mark for simplifying and writing correct coefficients of constant, x and x^2]

b) $(2 + x)^{\frac{1}{2}} \approx \sqrt{2} + \frac{\sqrt{2}}{4}x - \frac{\sqrt{2}}{32}x^2$ *[1 mark]*

c) $(2 - x)^{\frac{1}{2}}(2 + x)^{\frac{1}{2}}$

$$\approx \left(\sqrt{2} - \frac{\sqrt{2}}{4}x - \frac{\sqrt{2}}{32}x^2\right)\left(\sqrt{2} + \frac{\sqrt{2}}{4}x - \frac{\sqrt{2}}{32}x^2\right)$$

$$= 2 + \frac{1}{2}x - \frac{1}{16}x^2 - \frac{1}{2}x - \frac{1}{8}x^2 - \frac{1}{16}x^2 + ... \approx 2 - \frac{1}{4}x^2$$

[2 marks available — 1 mark for a correct method of multiplying the expansions, 1 mark for the correct answer]

d) $\sqrt{1.98 \times 2.02} = (2 - 0.02)^{\frac{1}{2}}(2 + 0.02)^{\frac{1}{2}}$

So use $x = 0.02$ in the expansion from part c),

$\sqrt{1.98 \times 2.02} \approx 2 - \frac{1}{4}(0.02)^2 = 1.9999$

[2 marks available — 1 mark for substituting $x = 0.02$ into the correct expression, 1 mark for the correct answer]

Answers

Q8

a) $\dfrac{(1-5x)}{(1+3x)^{\frac{1}{3}}} = (1-5x)(1+3x)^{-\frac{1}{3}}$

$= (1-5x)(1-x+2x^2+...) = 1-6x+7x^2+...$

b) The expansion is valid if $|x| < \dfrac{1}{3}$.

c) $\dfrac{(1-5(0.1))}{(1+3(0.1))^{\frac{1}{3}}} = \dfrac{0.5}{1.3^{\frac{1}{3}}} = \dfrac{1}{2^3\sqrt{1.3}}$

$\Rightarrow \dfrac{1}{2^3\sqrt{1.3}} \approx 1-6(0.1)+7(0.01) = 0.47$

d) $\left| \dfrac{\left(\dfrac{1}{2^3\sqrt{1.3}}\right) - 0.47}{\left(\dfrac{1}{2^3\sqrt{1.3}}\right)} \right| \times 100 = 2.6\%$ (2 s.f.)

Q9

a) $\sqrt{5x+4} = (4+5x)^{\frac{1}{2}} = 4^{\frac{1}{2}}\left(1+\dfrac{5}{4}x\right)^{\frac{1}{2}}$

$(1+\dfrac{5}{4}x)^{\frac{1}{2}} = 1+\left(\dfrac{1}{2}\right)\left(\dfrac{5}{4}x\right)+\dfrac{\left(\dfrac{1}{2}\right)\times\left(-\dfrac{1}{2}\right)}{1\times 2}\left(\dfrac{5}{4}x\right)^2$

$+\dfrac{\left(\dfrac{1}{2}\right)\times\left(-\dfrac{1}{2}\right)\times\left(-\dfrac{3}{2}\right)}{1\times 2\times 3}\left(\dfrac{5}{4}x\right)^3 +...$

$\sqrt{5x+4} \approx 2\left(1+\dfrac{5}{8}x-\dfrac{25}{128}x^2+\dfrac{125}{1024}x^3\right)$

$= 2+\dfrac{5}{4}x-\dfrac{25}{64}x^2+\dfrac{125}{512}x^3$

[3 marks available — 1 mark for factorising, 1 mark for expansion, 1 mark for the correct and simplified answer]

b) Expansion is valid when $\left|\dfrac{5}{4}x\right| < 1 \Rightarrow |x| < \dfrac{4}{5}$

[1 mark for the correct validity]

c) $\sqrt{5x+4} = \sqrt{\dfrac{15}{2}} \Rightarrow x = \dfrac{7}{10}$

Since $\dfrac{7}{10} < \dfrac{4}{5}$, this is a valid value of x to use in the expansion and so can be used to estimate the value of $\sqrt{\dfrac{15}{2}}$.

[2 marks available — 1 mark for finding $x = \dfrac{7}{10}$ (= 0.7), 1 mark for a correct conclusion and explanation]

Q10

a) $\text{fg}(x) = \text{f}(\text{g}(x)) = \text{f}(4-3x)$

$= \dfrac{1}{\sqrt{2+(4-3x)}} = \dfrac{1}{\sqrt{6-3x}}$ *[1 mark]*

b) $\text{fg}(x) = (6-3x)^{-\frac{1}{2}} = 6^{-\frac{1}{2}}\left(1-\dfrac{1}{2}x\right)^{-\frac{1}{2}}$

$\approx \dfrac{\sqrt{6}}{6}\left[1+\left(-\dfrac{1}{2}\right)\left(-\dfrac{1}{2}x\right)+\dfrac{\left(-\dfrac{1}{2}\right)\left(-\dfrac{3}{2}\right)}{1\times 2}\left(-\dfrac{1}{2}x\right)^2\right]$

$= \dfrac{\sqrt{6}}{6}\left(1+\dfrac{1}{4}x+\dfrac{3}{32}x^2\right) = \dfrac{\sqrt{6}}{6}+\dfrac{\sqrt{6}}{24}x+\dfrac{\sqrt{6}}{64}x^2$

[3 marks available — 1 mark for factorising, 1 mark for expansion, 1 mark for the correct and simplified answer]

c) The expansion is valid for $\left|-\dfrac{1}{2}x\right| < 1 \Rightarrow |x| < 2$.

$\dfrac{1}{\sqrt{6-3x}} = \dfrac{\sqrt{5}}{5} \Rightarrow \dfrac{5}{\sqrt{5}\sqrt{6-3x}} = 1$

$\Rightarrow 6-3x = 5 \Rightarrow x = \dfrac{1}{3}$

Since $\dfrac{1}{3} < 2$, this is a valid value of x to use in the expansion and so can be used to estimate the value of $\dfrac{\sqrt{5}}{5}$.

[2 marks available — 1 mark for finding $x = \dfrac{1}{3}$, 1 mark for a correct conclusion and explanation]

Q11

a) $\sec^6\theta = \dfrac{1}{\cos^6\theta} = (\cos\theta)^{-6}$

θ is small and measured in radians, so using small angle approximation: $\cos\theta \approx 1-\dfrac{1}{2}\theta^2$

So $\sec^6\theta = (\cos\theta)^{-6} \approx \left(1-\dfrac{1}{2}\theta^2\right)^{-6}$

[2 marks available — 1 mark for using the identity $\sec\theta = \dfrac{1}{\cos\theta}$, 1 mark for using small angle approximation to give the correct answer]

b) $\left(1-\dfrac{1}{2}\theta^2\right)^{-6} \approx 1+(-6)\left(-\dfrac{1}{2}\theta^2\right)+\dfrac{(-6)\times(-7)}{1\times 2}\left(-\dfrac{1}{2}\theta^2\right)^2$

$= 1+3\theta^2+\dfrac{21}{4}\theta^2$

[2 marks available — 1 mark for a correct use of the expansion formula, 1 mark for the correct and simplified answer]

c) Using $\theta = 0.2$: $\sec^6 0.2 \approx 1+3(0.2)^2+\dfrac{21}{4}(0.2)^4 = 1.1284$

[2 marks available — 1 mark for substituting $\theta = 0.2$ into the correct expression, 1 mark for the correct answer]

Q12 $(1+ax)^{-\frac{1}{2}} \approx 1+\left(-\dfrac{1}{2}\right)(ax)+\dfrac{\left(-\dfrac{1}{2}\right)\left(-\dfrac{3}{2}\right)}{1\times 2}(ax)^2$

$= 1-\dfrac{1}{2}ax+\dfrac{3}{8}a^2x^2$

$(1+ax)^{-\frac{1}{2}} = \dfrac{1}{\sqrt{1+\dfrac{a}{2}}}$ gives $x = 0.5$

When $x = 0.5$: $1-\dfrac{1}{4}a+\dfrac{3}{32}a^2 = 0.875 \Rightarrow 32-8a+3a^2 = 28$

$\Rightarrow 3a^2-8a+4 = 0 \Rightarrow (3a-2)(a-2) = 0 \Rightarrow a = \dfrac{2}{3}, \ a = 2$

The expansion is valid when $|ax| < 1 \Rightarrow |x| < \dfrac{1}{|a|}$.

If $a = \dfrac{2}{3}$ this is $|x| < \dfrac{3}{2}$ and so the approximation is valid for $x = 0.5$. If $a = 2$ this is $|x| < \dfrac{1}{2}$ and so the approximation is not valid for $x = 0.5$.

So, $a = \dfrac{2}{3}$ and the value estimated is $\dfrac{1}{\sqrt{1+\dfrac{\frac{2}{3}}{2}}} = \dfrac{1}{\sqrt{\dfrac{4}{3}}} = \dfrac{\sqrt{3}}{2}$.

[6 marks available — 1 mark for correct initial expansion, 1 mark for finding $x = 0.5$, 1 mark for a correct quadratic equation, 1 mark for finding $a = \dfrac{2}{3}$ or $a = 2$, 1 mark for stating $a = \dfrac{2}{3}$ with justification, 1 mark for the correct value being estimated $\left(\dfrac{\sqrt{3}}{2}\right)$]

Exercise 6.5 — Binomial Expansion and Partial Fractions

Q1

a) $5-12x \equiv A(4+3x)+B(1+6x)$

Using the substitution method:

When $x = -\dfrac{4}{3}$, $5+16 = -7B \Rightarrow B = -3$.

When $x = -\dfrac{1}{6}$, $5+2 = \dfrac{7A}{2} \Rightarrow A = 2$.

You could also have used the 'equating coefficients' method.

b) **(i)** $(1+6x)^{-1} = 1-6x+36x^2-...$

(ii) $(4+3x)^{-1} = 4^{-1}\left(1+\dfrac{3}{4}x\right)^{-1}$

$= \dfrac{1}{4}\left(1-\dfrac{3x}{4}+\dfrac{9x^2}{16}-...\right) = \dfrac{1}{4}-\dfrac{3x}{16}+\dfrac{9x^2}{64}-...$

c) From a): $\dfrac{5-12x}{(1+6x)(4+3x)} \equiv \dfrac{2}{(1+6x)}-\dfrac{3}{(4+3x)}$

$\equiv 2(1+6x)^{-1}-3(4+3x)^{-1}$

$2(1+6x)^{-1}-3(4+3x)^{-1}$

$\approx 2(1-6x+36x^2)-3\left(\dfrac{1}{4}-\dfrac{3x}{16}+\dfrac{9x^2}{64}\right)$

$= 2-12x+72x^2-\dfrac{3}{4}+\dfrac{9x}{16}-\dfrac{27x^2}{64}$

$= \dfrac{5}{4}-\dfrac{183x}{16}+\dfrac{4581x^2}{64}$

d) $(1+6x)^{-1}$ is valid if $|x| < \dfrac{1}{6}$.

$(4+3x)^{-1}$ is valid if $|x| < \dfrac{4}{3}$.

So the full expansion is valid if $|x| < \dfrac{1}{6}$.

Q2 a) $\dfrac{6}{(1-x)(1+x)(1+2x)} \equiv \dfrac{A}{(1-x)} + \dfrac{B}{(1+x)} + \dfrac{C}{(1+2x)}$

$6 \equiv A(1+x)(1+2x) + B(1-x)(1+2x) + C(1-x)(1+x)$

Using the substitution method:

When $x = 1$, $6 = 6A \Rightarrow A = 1$

When $x = -1$, $6 = -2B \Rightarrow B = -3$

When $x = -\frac{1}{2}$, $6 = \frac{3}{4}C \Rightarrow C = 8$

Putting these values back into the expression gives:

$\dfrac{6}{(1-x)(1+x)(1+2x)} \equiv \dfrac{1}{(1-x)} - \dfrac{3}{(1+x)} + \dfrac{8}{(1+2x)}$

b) f(x) can also be expressed as:

$(1-x)^{-1} - 3(1+x)^{-1} + 8(1+2x)^{-1}$

Expanding these three parts separately:

$(1-x)^{-1} = 1 + x + x^2 + ...$

$(1+x)^{-1} = 1 - x + x^2 + ...$

$(1+2x)^{-1} = 1 - 2x + 4x^2 + ...$

So f(x) $\approx 1 + x + x^2 - 3(1 - x + x^2) + 8(1 - 2x + 4x^2)$

$= 1 + x + x^2 - 3 + 3x - 3x^2 + 8 - 16x + 32x^2$

$= 6 - 12x + 30x^2$

c) $f(0.01) = \dfrac{6}{(1-0.01)(1+0.01)(1+2(0.01))} = 5.8829...$

From the expansion:

$f(0.01) \approx 6 - 12(0.01) + 30(0.01)^2$

$= 6 - 0.12 + 0.003 = 5.883$

So the % error is:

$\left| \dfrac{5.8829... - 5.883}{5.8829...} \right| \times 100 = 0.0010\%$ (2 s.f.)

Q3 a) $2x^3 + 5x^2 - 3x = x(2x - 1)(x + 3)$

b) $\dfrac{5x - 6}{2x^3 + 5x^2 - 3x} \equiv \dfrac{5x - 6}{x(2x - 1)(x + 3)}$

$\equiv \dfrac{A}{x} + \dfrac{B}{(2x - 1)} + \dfrac{C}{(x + 3)}$

$5x - 6 \equiv A(2x - 1)(x + 3) + Bx(x + 3) + Cx(2x - 1)$

Using the substitution method:

When $x = 0$, $-6 = -3A \Rightarrow A = 2$

When $x = \frac{1}{2}$, $\frac{5}{2} - 6 = \frac{7}{4}B \Rightarrow B = -2$

When $x = -3$, $-15 - 6 = 21C \Rightarrow C = -1$

So $\dfrac{5x - 6}{2x^3 + 5x^2 - 3x} \equiv \dfrac{2}{x} - \dfrac{2}{(2x - 1)} - \dfrac{1}{(x + 3)}$

c) $(2x - 1)^{-1} = -(1 - 2x)^{-1} = -(1 + 2x + 4x^2 + ...)$

$(x + 3)^{-1} = \frac{1}{3}\left(1 + \frac{1}{3}x\right)^{-1} = \frac{1}{3}\left(1 - \frac{x}{3} + \frac{x^2}{9} + ...\right)$

$= \dfrac{1}{3} - \dfrac{x}{9} + \dfrac{x^2}{27} + ...$

$\dfrac{5x - 6}{2x^3 + 5x^2 - 3x} \equiv \dfrac{2}{x} - 2(2x - 1)^{-1} - (x + 3)^{-1}$

$\approx \dfrac{2}{x} - 2[-(1 + 2x + 4x^2)] - \left(\dfrac{1}{3} - \dfrac{x}{9} + \dfrac{x^2}{27}\right)$

$= \dfrac{2}{x} + 2 + 4x + 8x^2 - \dfrac{1}{3} + \dfrac{x}{9} - \dfrac{x^2}{27}$

$= \dfrac{2}{x} + \dfrac{5}{3} + \dfrac{37x}{9} + \dfrac{215x^2}{27}$

d) $(2x - 1)^{-1}$ is valid for $|x| < \frac{1}{2}$.

$(x + 3)^{-1}$ is valid for $|x| < 3$.

$\dfrac{2}{x}$ is valid for $x \neq 0$.

So the full expansion is valid for $|x| < \frac{1}{2}$, $x \neq 0$.

Make sure you don't get caught out here
— the brackets are in the form $(qx + p)^n$, not $(p + qx)^n$.

Q4 a) $\dfrac{55x + 7}{(2x - 5)(3x + 1)^2} \equiv \dfrac{A}{(2x - 5)} + \dfrac{B}{(3x + 1)} + \dfrac{C}{(3x + 1)^2}$

$55x + 7 \equiv A(3x + 1)^2 + B(2x - 5)(3x + 1) + C(2x - 5)$

Using the substitution method:

When $x = \frac{5}{2}$, $\dfrac{275}{2} + 7 = A\left(\dfrac{17}{2}\right)^2$

$\Rightarrow \dfrac{289}{2} = \dfrac{289}{4}A \Rightarrow A = 2$

When $x = -\frac{1}{3}$, $-\dfrac{55}{3} + 7 = C\left(-\dfrac{2}{3} - 5\right)$

$-\dfrac{34}{3} = -\dfrac{17}{3}C \Rightarrow C = 2$

Equating coefficients of x^2:

$0 = 9A + 6B \Rightarrow 6B = -18 \Rightarrow B = -3$.

So: $\dfrac{55x + 7}{(2x - 5)(3x + 1)^2} \equiv \dfrac{2}{(2x - 5)} - \dfrac{3}{(3x + 1)} + \dfrac{2}{(3x + 1)^2}$

b) $(2x - 5)^{-1} = -\dfrac{1}{5}\left(1 - \dfrac{2}{5}x\right)^{-1} = -\dfrac{1}{5}\left(1 + \dfrac{2x}{5} + \dfrac{4x^2}{25} + ...\right)$

$= -\dfrac{1}{5} - \dfrac{2x}{25} - \dfrac{4x^2}{125} - ...$

$(3x + 1)^{-1} = 1 - 3x + 9x^2 + ...$

$(3x + 1)^{-2} = 1 - 6x + 27x^2 + ...$

$f(x) = 2(2x - 5)^{-1} - 3(3x + 1)^{-1} + 2(3x + 1)^{-2}$

$\approx 2\left(-\dfrac{1}{5} - \dfrac{2x}{25} - \dfrac{4x^2}{125}\right) - 3(1 - 3x + 9x^2) + 2(1 - 6x + 27x^2)$

$= -\dfrac{2}{5} - \dfrac{4x}{25} - \dfrac{8x^2}{125} - 3 + 9x - 27x^2 + 2 - 12x + 54x^2$

$= -\dfrac{7}{5} - \dfrac{79x}{25} - \dfrac{3367x^2}{125}$

Q5 a) $\dfrac{q}{(1 - 4x)(1 - 2x)} \equiv \dfrac{A}{1 - 4x} + \dfrac{B}{1 - 2x}$

$\Rightarrow q = A(1 - 2x) + B(1 - 4x)$

When $x = \frac{1}{4}$, $q = \frac{1}{2}A \Rightarrow A = 2q$

When $x = \frac{1}{2}$, $q = -B \Rightarrow B = -q$

So $\dfrac{q}{(1 - 4x)(1 - 2x)} = \dfrac{2q}{1 - 4x} - \dfrac{q}{1 - 2x}$

[3 marks available — 1 mark for setting up partial fractions,
1 mark for finding A and B, 1 mark for the correct answer]

b) (i) $\dfrac{(-1) \times (-2) \times (-3)}{1 \times 2 \times 3}(-4)^3 = 64$ *[1 mark]*

(ii) $\dfrac{(-1) \times (-2) \times (-3)}{1 \times 2 \times 3}(-2)^3 = 8$ *[1 mark]*

c) (i) $(64 \times 2q) - (8 \times q) = 120q$

[2 marks available — 1 mark for a correct method,
1 mark for the correct answer]

(ii) $120q = 60 \Rightarrow q = \dfrac{1}{2}$ *[1 mark]*

Q6 a) (i) $(-1)^3 + 6(-1)^2 + 11(-1) + 6 = -1 + 6 - 11 + 6 = 0$

Therefore, $(1 + x)$ is a factor of $x^3 + 6x^2 + 11x + 6$.

[1 mark for a correct evaluation at $x = -1$]

(ii) $x^3 + 6x^2 + 11x + 6 = (1 + x)(6 + Ax + x^2)$

Equating coefficients of x^2: $6 = 1 + A \Rightarrow A = 5$

So $x^3 + 6x^2 + 11x + 6 = (1 + x)(6 + 5x + x^2)$

$= (1 + x)(2 + x)(3 + x)$

[3 marks available — 1 mark for a correct method to
find a quadratic factor, 1 mark for the correct quadratic
factor, 1 mark for the correct answer]
You could also use long division here.

b) $\dfrac{2}{x^3 + 6x^2 + 11x + 6} \equiv \dfrac{A}{1+x} + \dfrac{B}{2+x} + \dfrac{C}{3+x}$

$2 \equiv A(2+x)(3+x) + B(1+x)(3+x) + C(1+x)(2+x)$

When $x = -1$, $2 = 2A \Rightarrow A = 1$

When $x = -2$, $2 = -B \Rightarrow B = -2$

When $x = -3$, $2 = 2C \Rightarrow C = 1$

So $\dfrac{2}{x^3 + 6x^2 + 11x + 6} \equiv \dfrac{1}{1+x} - \dfrac{2}{2+x} + \dfrac{1}{3+x}$

[3 marks available — 1 mark for using a correct method to find the unknowns (e.g. substitution or equating coefficients), 1 mark for at least one correct value of A, B or C, 1 mark for the correct final answer]

c) $(1+x)^{-1} \approx 1 + (-1)(x) + \dfrac{(-1)\times(-2)}{1\times 2}(x)^2 = 1 - x + x^2$

$(2+x)^{-1} = 2^{-1}\left(1 + \dfrac{1}{2}x\right)^{-1}$

$\approx \dfrac{1}{2}\left[1 + (-1)\left(\dfrac{1}{2}x\right) + \dfrac{(-1)\times(-2)}{1\times 2}\left(\dfrac{1}{2}x\right)^2\right]$

$= \dfrac{1}{2} - \dfrac{1}{4}x + \dfrac{1}{8}x^2$

$(3+x)^{-1} = 3^{-1}\left(1 + \dfrac{1}{3}x\right)^{-1}$

$\approx \dfrac{1}{3}\left[1 + (-1)\left(\dfrac{1}{3}x\right) + \dfrac{(-1)\times(-2)}{1\times 2}\left(\dfrac{1}{3}x\right)^2\right]$

$= \dfrac{1}{3} - \dfrac{1}{9}x + \dfrac{1}{27}x^2$

So binomial expansion (up to x^2) is:

$(1 - x + x^2) - 2\left(\dfrac{1}{2} - \dfrac{1}{4}x + \dfrac{1}{8}x^2\right) + \left(\dfrac{1}{3} - \dfrac{1}{9}x + \dfrac{1}{27}x^2\right)$

$= \dfrac{1}{3} - \dfrac{11}{18}x + \dfrac{85}{108}x^2$

[4 marks available – 1 mark for each correct expansion, 1 mark for the correct final answer]

d) $(1+x)^{-1}$ is valid for $|x| < 1$

$\left(1 + \dfrac{1}{2}x\right)^{-1}$ is valid for $\left|\dfrac{1}{2}x\right| < 1 \Rightarrow |x| < 2$

$\left(1 + \dfrac{1}{3}x\right)^{-1}$ is valid for $\left|\dfrac{1}{3}x\right| < 1 \Rightarrow |x| < 3$

So the expansion in part c) is valid for $|x| < 1$.

[1 mark for the correct validity of the full expansion]

Q7 a) $35x - 33x^2 - 3$

$\equiv A(1+2x)(1-3x) + B(1-x)(1-3x) + C(1-x)(1+2x)$

When $x = 1$, $-1 = -6A \Rightarrow A = \dfrac{1}{6}$

When $x = -\dfrac{1}{2}$, $-\dfrac{115}{4} = \dfrac{15B}{4} \Rightarrow B = -\dfrac{23}{3}$

When $x = \dfrac{1}{3}$, $5 = \dfrac{10C}{9} \Rightarrow C = \dfrac{9}{2}$

[3 marks available — 1 mark for using a correct method to find the unknowns (e.g. substitution or equating coefficients), 2 marks for all three correct values, otherwise 1 mark for at least two correct values]

b) $(1-x)^{-1} \approx 1 + (-1)(-x) + \dfrac{-1\times -2}{1\times 2}(-x)^2 = 1 + x + x^2$

$(1+2x)^{-1} \approx 1 + (-1)(2x) + \dfrac{-1\times -2}{1\times 2}(2x)^2 = 1 - 2x + 4x^2$

$(1-3x)^{-1} \approx 1 + (-1)(-3x) + \dfrac{-1\times -2}{1\times 2}(-3x)^2 = 1 + 3x + 9x^2$

[3 marks available — 3 marks for all three fully correct, otherwise 2 marks for at least one fully correct, or 1 mark for a correct use of the expansion formula]

c) $f(x) = \dfrac{1}{6}(1-x)^{-1} - \dfrac{23}{3}(1+2x)^{-1} + \dfrac{9}{2}(1-3x)^{-1}$

$\approx \dfrac{1}{6}(1 + x + x^2) - \dfrac{23}{3}(1 - 2x + 4x^2) + \dfrac{9}{2}(1 + 3x + 9x^2)$

$= -3 + 29x + 10x^2$

[2 marks available — 1 mark for correctly using the values of A, B and C, 1 mark for the correct answer]

d) $(1-x)^{-1}$ is valid for $|-x| < 1 \Rightarrow |x| < 1$

$(1+2x)^{-1}$ is valid for $|2x| < 1 \Rightarrow |x| < \dfrac{1}{2}$

$(1-3x)^{-1}$ is valid for $|-3x| < 1 \Rightarrow |x| < \dfrac{1}{3}$

So the expansion of f(x) is therefore valid for $|x| < \dfrac{1}{3}$.

[1 mark for the correct validity of the full expansion]

e) $0.1 < \dfrac{1}{3}$, so the approximation $f(x) \approx 10x^2 + 29x - 3$ is valid for $x = 0.1$. As $f(0.1) \approx 10(0.1)^2 + 29(0.1) - 3 = 0$, f(x) must have a root near 0.1.

[2 marks available — 1 mark for stating the approximation is valid at x = 0.1, 1 mark for using the expansion to show x = 0.1 is an approximate root (by substitution or factorising)]

Q8 a) Using the Factor Theorem, if f(0.5) = 0 then $(x - 0.5)$ is a factor $\Rightarrow (2x - 1)$ is a factor.

Using e.g. algebraic long division:

$f(x) = (2x - 1)(6x^2 - x - 1) \Rightarrow f(x) = (2x - 1)(2x - 1)(3x + 1)$

$\Rightarrow f(x) = (3x + 1)(2x - 1)^2$

b) $\dfrac{8 - x}{12x^3 - 8x^2 - x + 1} \equiv \dfrac{8 - x}{(3x + 1)(2x - 1)^2}$

$\equiv \dfrac{A}{(3x + 1)} + \dfrac{B}{(2x - 1)} + \dfrac{C}{(2x - 1)^2}$

$8 - x \equiv A(2x - 1)^2 + B(3x + 1)(2x - 1) + C(3x + 1)$

Using the substitution method:

When $x = \dfrac{1}{2}$, $8 - \dfrac{1}{2} = \dfrac{5}{2}C \Rightarrow C = 3$

When $x = -\dfrac{1}{3}$, $8 + \dfrac{1}{3} = \dfrac{25}{9}A \Rightarrow A = 3$

Equating coefficients of x^2:

$0 = 4A + 6B \Rightarrow 0 = 12 + 6B \Rightarrow B = -2$

So $\dfrac{8 - x}{(3x + 1)(2x - 1)^2} \equiv \dfrac{3}{(3x + 1)} - \dfrac{2}{(2x - 1)} + \dfrac{3}{(2x - 1)^2}$

c) $(3x + 1)^{-1} = (1 + 3x)^{-1} = (1 - 3x + 9x^2 + ...)$

$(2x - 1)^{-1} = -(1 - 2x)^{-1} = -(1 + 2x + 4x^2 + ...)$

$(2x - 1)^{-2} = (-1)^{-2}(1 - 2x)^{-2} = (1 + 4x + 12x^2 + ...)$

$\dfrac{8 - x}{(3x + 1)(2x - 1)^2} \equiv 3(3x + 1)^{-1} - 2(2x - 1)^{-1} + 3(2x - 1)^{-2}$

$\approx 3[1 - 3x + 9x^2] - 2[-1 - 2x - 4x^2] + 3[1 + 4x + 12x^2]$

$= 3 - 9x + 27x^2 + 2 + 4x + 8x^2 + 3 + 12x + 36x^2$

$= 8 + 7x + 71x^2$

d) $(3x + 1)^{-1}$ is valid for $|x| < \dfrac{1}{3}$.

$(2x - 1)^{-1}$ and $(2x - 1)^{-2}$ are valid for $|x| < \dfrac{1}{2}$.

So the full expansion is valid for $|x| < \dfrac{1}{3}$.

e) $g(0.001) = \dfrac{8 - 0.001}{(0.003 + 1)(0.002 - 1)^2}$

$= 8.007071032... = 8.007071 \text{ (6 d.p.)}$

From the expansion:

$g(0.001) \approx 8 + 7(0.001) + 71(0.001)^2$

$= 8 + 0.007 + 0.000071 = 8.007071$

So the estimate is correct to 6 decimal places.

Q9 Using partial fractions:

$\dfrac{p}{(2 - x)(5 - 2x)} \equiv \dfrac{A}{2 - x} + \dfrac{B}{5 - 2x} \Rightarrow p \equiv A(5 - 2x) + B(2 - x)$

When $x = 2$, $p = A(1) \Rightarrow A = p$

When $x = \dfrac{5}{2}$, $p = B\left(-\dfrac{1}{2}\right) \Rightarrow B = -2p$

So $\dfrac{p}{(2 - x)(5 - 2x)} = \dfrac{p}{2 - x} - \dfrac{2p}{5 - 2x}$

We need the coefficients of the x^4 terms in the expansion of each part.

$\dfrac{p}{2 - x} = p(2 - x)^{-1} = p\left[2^{-1}\left(1 - \dfrac{1}{2}x\right)^{-1}\right] = \dfrac{p}{2}\left(1 - \dfrac{1}{2}x\right)^{-1}$

So the coefficient of the x^4 term is:

$$\frac{p}{2}\left[\frac{(-1)(-2)(-3)(-4)}{1\times2\times3\times4}\left(-\frac{1}{2}\right)^4\right]=\frac{p}{32}$$

$$\frac{2p}{5-2x}=2p(5-2x)^{-1}=2p\left[5^{-1}\left(1-\frac{2}{5}x\right)^{-1}\right]=\frac{2p}{5}\left(1-\frac{2}{5}x\right)^{-1}$$

So the coefficient of the x^4 term is:

$$\frac{2p}{5}\left[\frac{(-1)(-2)(-3)(-4)}{1\times2\times3\times4}\left(-\frac{2}{5}\right)^4\right]=\frac{32p}{3125}$$

The coefficient of the x^4 term in expansion of the combined expression is $\frac{p}{32}-\frac{32p}{3125}=0.02101p$,

so $0.02101p=0.06303\Rightarrow p=3$.

[7 marks available — 1 mark for setting up partial fractions, 1 mark for finding A and B, 1 mark for rewriting any of the two expansions, 1 mark for each correct coefficient of the x^4 terms, 1 mark for a correct equation in p, 1 mark for correct answer]

Q10 a) $\frac{2kx}{(1-kx)(1+kx)}\equiv\frac{A}{1-kx}+\frac{B}{1+kx}$

$2kx\equiv A(1+kx)+B(1-kx)$

When $x=\frac{1}{k}$, $2=2A\Rightarrow A=1$

When $x=-\frac{1}{k}$, $-2=2B\Rightarrow B=-1$

So $\frac{2kx}{(1-kx)(1+kx)}\equiv\frac{1}{1-kx}-\frac{1}{1+kx}$

[2 marks available — 1 mark for setting up partial fractions, 1 mark for showing that A = 1 and B = -1]

b) $(1-kx)^{-1}\approx1+(-1)(-kx)+\frac{(-1)\times(-2)}{1\times2}(-kx)^2$

$=1+kx+k^2x^2$

$(1+kx)^{-1}\approx1+(-1)(kx)+\frac{(-1)\times(-2)}{1\times2}(kx)^2$

$=1-kx+k^2x^2$

$\frac{2kx}{(1-kx)(1+kx)}\approx(1+kx+k^2x^2)-(1-kx+k^2x^2)=2kx$

So $2kx=4x\Rightarrow k=2$

[3 marks available — 1 mark for a correct use of the expansion formula, 1 mark for showing the full expansion simplifies to 2kx, 1 mark for the correct answer]

c) The expansion of $(1-2x)^{-1}$ is valid for $|-2x|<1\Rightarrow|x|<\frac{1}{2}$

The expansion of $(1+2x)^{-1}$ is valid for $|2x|<1\Rightarrow|x|<\frac{1}{2}$

So, the combined expansion is valid for $|x|<\frac{1}{2}$.

[1 mark for the correct validity of the full expansion]

Chapter 6 Review Exercise

Q1 a) $(1+2x)^3$

$=1+3(2x)+\frac{3(3-1)}{1\times2}(2x)^2+\frac{3(3-1)(3-2)}{1\times2\times3}(2x)^3$

$=1+6x+12x^2+8x^3$

b) $(1-x)^5=1+5(-x)+\frac{5(5-1)}{1\times2}(-x)^2$

$+\frac{5(5-1)(5-2)}{1\times2\times3}(-x)^3$

$+\frac{5(5-1)(5-2)(5-3)}{1\times2\times3\times4}(-x)^4$

$+\frac{5(5-1)(5-2)(5-3)(5-4)}{1\times2\times3\times4\times5}(-x)^5$

$=1-5x+10x^2-10x^3+5x^4-x^5$

c) $(1-4x)^4=1+4(-4x)+\frac{4(4-1)}{1\times2}(-4x)^2$

$+\frac{4(4-1)(4-2)}{1\times2\times3}(-4x)^3$

$+\frac{4(4-1)(4-2)(4-3)}{1\times2\times3\times4}(-4x)^4$

$=1-16x+96x^2-256x^3+256x^4$

d) $\left(1-\frac{2}{3}x\right)^4=1+4\left(-\frac{2x}{3}\right)+\frac{4(4-1)}{1\times2}\left(-\frac{2x}{3}\right)^2$

$+\frac{4(4-1)(4-2)}{1\times2\times3}\left(-\frac{2x}{3}\right)^3$

$+\frac{4(4-1)(4-2)(4-3)}{1\times2\times3\times4}\left(-\frac{2x}{3}\right)^4$

$=1-\frac{8x}{3}+\frac{8x^2}{3}-\frac{32x^3}{27}+\frac{16x^4}{81}$

Q2 For positive integer values (and zero).

Q3 The x^3 term coefficient is: $\frac{8\times7\times6}{1\times2\times3}k^3=56k^3$

$56k^3=3584\Rightarrow k^3=64\Rightarrow k=4$

[2 marks available — 1 mark for the correct coefficient ($56k^3$), 1 mark for the correct answer]

Q4 $3(1-8x)^{-3}\approx3\left[1+(-3)(-8x)+\frac{(-3)\times(-4)}{1\times2}(-8x)^2\right]$

$=3(1+24x+384x^2)=3+72x+1152x^2$

[2 marks available — 1 mark for expansion, 1 mark for simplifying and writing correct coefficients of constant, x and x^2]

Q5 $\sqrt[3]{(1+2x)^2}=(1+2x)^{\frac{2}{3}}=1+\frac{2}{3}(2x)+\frac{\frac{2}{3}\left(\frac{2}{3}-1\right)}{1\times2}(2x)^2+...$

$=1+\frac{4}{3}x-\frac{4}{9}x^2+...$

[2 marks available — 1 mark for expansion, 1 mark for simplifying and writing correct coefficients of constant, x and x^2]

Q6 The coefficient of x^3 in the expansion of $(1+ax)^{-4}$ is:

$\frac{-4(-4-1)(-4-2)}{1\times2\times3}a^3=-20a^3$

So: $-20a^3=-160\Rightarrow a^3=8\Rightarrow a=2$

[3 marks available — 1 mark for correct coefficient of x^3 in the expansion, 1 mark for equating to -160, 1 mark for correct value of a]

Q7 a) $\frac{-2(-2-1)}{1\times2}a^2=48\Rightarrow3a^2=48\Rightarrow a=4$

You're told that a is a positive integer, so ignore the negative root.

b) $\frac{\left(\frac{1}{3}\right)\left(\frac{1}{3}-1\right)\left(\frac{1}{3}-2\right)\left(\frac{1}{3}-3\right)}{1\times2\times3\times4}(-a)^4=-\frac{10}{3}$

$\Rightarrow-\frac{10a^4}{243}=-\frac{10}{3}\Rightarrow a^4=81\Rightarrow a=3$

Again, you need to take the positive root.

Q8 Coefficient of the term in x^4 is

$\frac{\left(-\frac{1}{2}\right)\times\left(-\frac{3}{2}\right)\times\left(-\frac{5}{2}\right)\times\left(-\frac{7}{2}\right)}{1\times2\times3\times4}(k)^4=\frac{35}{10\,368}$

$\Rightarrow\frac{35}{128}k^4=\frac{35}{10\,368}\Rightarrow k^4=\frac{1}{81}\Rightarrow k=\frac{1}{3}$

(Since k is positive, $k=-\frac{1}{3}$ can be ignored.)

[2 marks available — 1 mark for using the expansion formula, 1 mark for the correct answer]

Q9 $(1+ax)^n=1+nax+\frac{n(n-1)}{1\times2}(ax)^2+...$

Compare coefficients of x: $na=-6\Rightarrow a=-\frac{6}{n}$

Compare coefficients of x^2:

$\frac{n(n-1)a^2}{2}=\frac{45}{2}\Rightarrow(n^2-n)\left(-\frac{6}{n}\right)^2=45$

$\Rightarrow36(n^2-n)=45n^2\Rightarrow0=9n^2+36n$

$\Rightarrow0=9n(n+4)\Rightarrow n=0$ or -4, but $n\neq0$, so $n=-4$

$a=\frac{-6}{-4}=\frac{3}{2}$

[6 marks available — 1 mark for correct expansion, 1 mark for equating the coefficients of x, 1 mark for equating the coefficients of x^2, 1 mark for attempt to solve simultaneously, 1 mark for correct value of n, 1 mark for correct value of a]

Answers

Q10 The expansion is valid for $\left|\dfrac{dx}{c}\right| < 1$ or $|x| < \left|\dfrac{c}{d}\right|$.

Q11 a) $\dfrac{1}{(1+x)^5} = (1+x)^{-5}$

$= 1 + (-5)x + \dfrac{-5(-5-1)}{1\times 2}x^2 + \dfrac{-5(-5-1)(-5-2)}{1\times 2\times 3}x^3 + \ldots$

$= 1 - 5x + 15x^2 - 35x^3 + \ldots$

The expansion is valid for $|x| < 1$.

b) $\dfrac{1}{(1-3x)^3} = (1-3x)^{-3}$

$= 1 + (-3)(-3x) + \dfrac{-3(-3-1)}{1\times 2}(-3x)^2$

$+ \dfrac{-3(-3-1)(-3-2)}{1\times 2\times 3}(-3x)^3 + \ldots$

$= 1 + 9x + 54x^2 + 270x^3 + \ldots$

The expansion is valid for $|x| < \dfrac{1}{3}$.

c) $\sqrt{1-5x} = (1-5x)^{\frac{1}{2}}$

$= 1 + \dfrac{1}{2}(-5x) + \dfrac{\left(\frac{1}{2}\right)\left(\frac{1}{2}-1\right)}{1\times 2}(-5x)^2$

$+ \dfrac{\left(\frac{1}{2}\right)\left(\frac{1}{2}-1\right)\left(\frac{1}{2}-2\right)}{1\times 2\times 3}(-5x)^3 + \ldots$

$= 1 - \dfrac{5x}{2} - \dfrac{25x^2}{8} - \dfrac{125x^3}{16} + \ldots$

The expansion is valid for $|x| < \dfrac{1}{5}$.

d) $\dfrac{1}{\sqrt[3]{1+2x}} = (1+2x)^{-\frac{1}{3}}$

$= 1 + \left(-\dfrac{1}{3}\right)(2x) + \dfrac{\left(-\frac{1}{3}\right)\left(-\frac{1}{3}-1\right)}{1\times 2}(2x)^2$

$+ \dfrac{\left(-\frac{1}{3}\right)\left(-\frac{1}{3}-1\right)\left(-\frac{1}{3}-2\right)}{1\times 2\times 3}(2x)^3 + \ldots$

$= 1 - \dfrac{2x}{3} + \dfrac{8x^2}{9} - \dfrac{112x^3}{81} + \ldots$

The expansion is valid for $|x| < \dfrac{1}{2}$.

Q12 a) (i) $(1-2x)^{\frac{1}{2}} \approx 1 + \left(\dfrac{1}{2}\right)(-2x) + \dfrac{\left(\frac{1}{2}\right)\times\left(-\frac{1}{2}\right)}{1\times 2}(-2x)^2$

$= 1 - x - \dfrac{1}{2}x^2$

(ii) $(1-3x)^{\frac{1}{3}} \approx 1 + \left(\dfrac{1}{3}\right)(-3x) + \dfrac{\left(\frac{1}{3}\right)\times\left(-\frac{2}{3}\right)}{1\times 2}(-3x)^2$

$= 1 - x - x^2$

[3 marks available — 3 marks for both fully correct, otherwise 2 marks for at least one fully correct, or 1 mark for a correct use of the expansion formula]

b) (i) $(1-2x)^{\frac{1}{2}}(1-3x)^{\frac{1}{3}} \approx \left(1-x-\dfrac{1}{2}x^2\right)(1-x-x^2)$

$= 1 - x - x^2 - x + x^2 + x^3 - \dfrac{1}{2}x^2 + \dfrac{1}{2}x^3 + \dfrac{1}{2}x^4$

$\approx 1 - 2x - \dfrac{1}{2}x^2$

[2 marks available — 1 mark for attempting to multiply the expansions together, 1 mark for the correct answer]

(ii) The separate expansions are valid for

$|-2x| \le 1 \Rightarrow |x| < \dfrac{1}{2}$ and $|-3x| < 1 \Rightarrow |x| < \dfrac{1}{3}$.

So, the combined expansion is valid when $|x| < \dfrac{1}{3}$.

[1 mark for the correct validity of the full expansion]

Q13 a) $\dfrac{4}{(6+2x)^2} = \dfrac{4}{6^2\left(1+\frac{1}{3}x\right)^2} = \dfrac{1}{9}\left(1+\dfrac{1}{3}x\right)^{-2}$

$\approx \dfrac{1}{9}\left[1 + (-2)\left(\dfrac{1}{3}x\right) + \dfrac{(-2)\times(-3)}{1\times 2}\left(\dfrac{1}{3}x\right)^2 \right.$

$\left. + \dfrac{(-2)\times(-3)\times(-4)}{1\times 2\times 3}\left(\dfrac{1}{3}x\right)^3\right]$

$= \dfrac{1}{9} - \dfrac{2}{27}x + \dfrac{1}{27}x^2 - \dfrac{4}{243}x^3$

[3 marks available — 1 mark for rewriting the original expression, 1 mark for a correct use of the expansion formula, 1 mark for the correct and simplified expansion]

b) $\dfrac{12x}{(6+2x)^2} = 3x \times \dfrac{4}{(6+2x)^2}$

$\approx 3x\left(\dfrac{1}{9} - \dfrac{2}{27}x + \dfrac{1}{27}x^2 - \dfrac{4}{243}x^3\right) \approx \dfrac{1}{3}x - \dfrac{2}{9}x^2 + \dfrac{1}{9}x^3$

[2 marks available — 1 mark for multiplying by 3x and expanding, 1 mark for the correct, simplified answer]

c) Both expansions are valid when $\left|\dfrac{1}{3}x\right| < 1 \Rightarrow |x| < 3$

[1 mark for the correct validity]

Q14 a) $\left(1-\dfrac{1}{2}x\right)^{-3} \approx 1 + (-3)\left(-\dfrac{1}{2}x\right) + \dfrac{(-3)\times(-4)}{1\times 2}\left(-\dfrac{1}{2}x\right)^2$

$= 1 + \dfrac{3}{2}x + \dfrac{3}{2}x^2$

[2 marks available — 1 mark for expansion, 1 mark for simplifying and writing correct coefficients of constant, x and x^2]

b) $1 - \dfrac{1}{2}x = 0.95 \Rightarrow x = 0.1$

Using this in the expansion from part a) gives:

$(0.95)^{-3} \approx 1 + \dfrac{3}{2}(0.1) + \dfrac{3}{2}(0.1)^2 = 1.165$

[2 marks available — 1 mark for correct x value, 1 mark for substitution leading to correct answer]

Q15 a) $(1-3x)^{\frac{1}{2}} = 1 + \dfrac{1}{2}(-3x) + \dfrac{\frac{1}{2}\left(\frac{1}{2}-1\right)}{1\times 2}(-3x)^2 + \ldots$

$= 1 - \dfrac{3}{2}x - \dfrac{9}{8}x^2 + \ldots$

The expansion is valid for $|3x| < 1$ or $|x| < \dfrac{1}{3}$.

[3 marks available — 1 mark for expansion, 1 mark for simplifying and writing correct coefficients of constant, x and x^2, 1 mark for the correct validity]

b) $1 - 3x = 0.97 \Rightarrow x = 0.01$

Using the expansion:

$\sqrt{0.97} \approx 1 - \dfrac{3}{2}(0.01) - \dfrac{9}{8}(0.01)^2 = 0.9848875$

[2 marks available — 1 mark for correct x value, 1 mark for substitution leading to correct answer]

Q16 a) $(1-4x^2)^{\frac{1}{3}} = 1 + \left(\dfrac{1}{3}\right)(-4x^2) + \dfrac{\left(\frac{1}{3}\right)\times\left(-\frac{2}{3}\right)}{1\times 2}(-4x^2)^2 + \ldots$

$= 1 - \dfrac{4}{3}x^2 - \dfrac{16}{9}x^4 - \ldots$

[2 marks available — 1 mark for an expansion with at least two terms correct, 1 mark for the correct, simplified answer]

b) $|-4x^2| < 1 \Rightarrow |x^2| < \dfrac{1}{4} \Rightarrow |x| < \dfrac{1}{2}$

[1 mark for the correct validity]

c) $\left(1-4\left(\dfrac{1}{6}\right)^2\right)^{\frac{1}{3}} \approx 1 - \dfrac{4}{3}\left(\dfrac{1}{6}\right)^2 - \dfrac{16}{9}\left(\dfrac{1}{6}\right)^4$

$= 1 - \dfrac{1}{27} - \dfrac{1}{729} = \dfrac{701}{729}$

$\left(1-4\left(\dfrac{1}{6}\right)^2\right)^{\frac{1}{3}} = \sqrt[3]{\dfrac{8}{9}} \Rightarrow \sqrt[3]{\dfrac{8}{9}} \approx \dfrac{701}{729}$

$\Rightarrow \dfrac{2}{\sqrt[3]{9}} \approx \dfrac{701}{729} \Rightarrow \sqrt[3]{9} \approx \dfrac{2\times 729}{701} = \dfrac{1458}{701} = 2.079\ldots$

[3 marks available — 1 mark for substituting $x = \dfrac{1}{6}$ into both sides of the expansion, 1 mark for rearranging for $\sqrt[3]{9}$, 1 mark for the correct final answer]

Answers

Q17 a) $(2 - ax)^{-4} = 2^{-4}\left(1 - \frac{a}{2}x\right)^{-4} = \frac{1}{16}\left(1 - \frac{a}{2}x\right)^{-4}$

$\approx \frac{1}{16}\left[1 + (-4)\left(-\frac{a}{2}x\right) + \frac{(-4)(-5)}{1 \times 2}\left(-\frac{a}{2}x\right)^2\right]$

$= \frac{1}{16}\left[1 + 2ax + \frac{5}{2}a^2x^2\right] = \frac{1}{16} + \frac{1}{8}ax + \frac{5}{32}a^2x^2$

[3 marks available — 1 mark for factorising,
1 mark for expansion, 1 mark for simplifying and
writing correct coefficients of constant, x and x^2]

b) When $x = 0.1$, $\frac{1}{16} + \frac{1}{80}a + \frac{5}{3200}a^2 = 0.0765625$

$\Rightarrow 200 + 40a + 5a^2 = 245 \Rightarrow a^2 + 8a - 9 = 0$

$\Rightarrow (a - 1)(a + 9) = 0 \Rightarrow a = 1$ and $a = -9$.

Since a is a positive integer, $a = 1$.

[2 marks available — 1 mark for a correct method to solve,
1 mark for the correct value of a]

Q18 a) (i) $\frac{1}{(3 + 2x)^2} = 3^{-2}\left(1 + \frac{2}{3}x\right)^{-2} = \frac{1}{9}\left(1 + \frac{2}{3}x\right)^{-2}$

$\left(1 + \frac{2}{3}x\right)^{-2} = 1 + (-2)\left(\frac{2x}{3}\right) + \frac{-2(-2-1)}{1 \times 2}\left(\frac{2x}{3}\right)^2 + \dots$

$= 1 - \frac{4x}{3} + \frac{4x^2}{3} - \dots$

$\frac{1}{(3 + 2x)^2} = \frac{1}{9}\left[1 - \frac{4x}{3} + \frac{4x^2}{3} - \dots\right]$

$= \frac{1}{9} - \frac{4}{27}x + \frac{4}{27}x^2 - \dots$

This is valid for $|x| < \frac{3}{2}$.

(ii) $\sqrt[3]{8 - x} = 8^{\frac{1}{3}}\left(1 - \frac{1}{8}x\right)^{\frac{1}{3}} = 2\left(1 - \frac{1}{8}x\right)^{\frac{1}{3}}$

$\left(1 - \frac{1}{8}x\right)^{\frac{1}{3}} = 1 + \frac{1}{3}\left(-\frac{x}{8}\right) + \frac{\frac{1}{3}\left(\frac{1}{3} - 1\right)}{1 \times 2}\left(-\frac{x}{8}\right)^2 + \dots$

$= 1 - \frac{x}{24} - \frac{x^2}{576} - \dots$

$\sqrt[3]{8 - x} = 2\left[1 - \frac{x}{24} - \frac{x^2}{576} - \dots\right]$

$= 2 - \frac{1}{12}x - \frac{1}{288}x^2 - \dots$

This is valid for $|x| < 8$.

b) $\frac{\sqrt[3]{8 - x}}{(3 + 2x)^2} = \left[2 - \frac{1}{12}x - \frac{1}{288}x^2 - \dots\right]$

$\times \left[\frac{1}{9} - \frac{4}{27}x + \frac{4}{27}x^2 - \dots\right]$

$= 2\left[\frac{1}{9} - \frac{4}{27}x + \frac{4}{27}x^2\right] - \frac{x}{12}\left[\frac{1}{9} - \frac{4}{27}x\right] - \frac{x^2}{288}\left[\frac{1}{9}\right] + \dots$

$= \frac{2}{9} - \frac{11x}{36} + \frac{799x^2}{2592} - \dots$

This is valid for the narrower of the two ranges, i.e. $|x| < \frac{3}{2}$.

c) (i) $\sqrt[3]{8 - x} = \sqrt[3]{7}$ when $x = 1$.

$\sqrt[3]{7} \approx 2 - \frac{1}{12}(1) - \frac{1}{288}(1)^2 = \frac{551}{288}$

(ii) The % error is:

$\left|\frac{\sqrt[3]{7} - \left(\frac{551}{228}\right)}{\sqrt[3]{7}}\right| \times 100 = 0.014\%$ (to 2 s.f.)

Q19 a) $(1 - 4x)^{-\frac{1}{4}} \approx 1 + \left(-\frac{1}{4}\right)(-4x) + \frac{\left(-\frac{1}{4}\right) \times \left(-\frac{5}{4}\right)}{1 \times 2}(-4x)^2$

$= 1 + x + \frac{5}{2}x^2$

[2 marks available — 1 mark for expansion, 1 mark for
simplifying and writing correct coefficients of constant,
x and x^2]

b) (i) $\frac{1}{\sqrt[4]{0.92}} = (1 - 4(0.02))^{-\frac{1}{4}}$, so use $x = 0.02$

$\Rightarrow \frac{1}{\sqrt[4]{0.92}} \approx 1 + 0.02 + \frac{5}{2}(0.02)^2 = 1.021$

[2 marks available — 1 mark for using x = 0.02,
1 mark for the correct answer]

(ii) Percentage error

$= \left|\frac{(0.92)^{-\frac{1}{4}} - 1.021}{(0.92)^{-\frac{1}{4}}}\right| \times 100 = 0.0063\%$ (to 2 s.f.)

[2 marks available — 1 mark for a correct method,
1 mark for the correct answer]

Q20 a) $\sqrt{\frac{1 + x}{1 - x}} = (1 + x)^{\frac{1}{2}}(1 - x)^{-\frac{1}{2}}$

$(1 + x)^{\frac{1}{2}} = 1 + \frac{1}{2}x + \frac{\frac{1}{2}\left(\frac{1}{2} - 1\right)}{1 \times 2}x^2 + \frac{\frac{1}{2}\left(\frac{1}{2} - 1\right)\left(\frac{1}{2} - 2\right)}{1 \times 2 \times 3}x^3$

$+ \dots$

$= 1 + \frac{1}{2}x - \frac{1}{8}x^2 + \frac{1}{16}x^3 + \dots$

$(1 - x)^{-\frac{1}{2}} = 1 + \left(-\frac{1}{2}\right)(-x) + \frac{-\frac{1}{2}\left(-\frac{1}{2} - 1\right)}{1 \times 2}(-x)^2$

$+ \frac{-\frac{1}{2}\left(-\frac{1}{2} - 1\right)\left(-\frac{1}{2} - 2\right)}{1 \times 2 \times 3}(-x)^3 + \dots$

$= 1 + \frac{1}{2}x + \frac{3}{8}x^2 + \frac{5}{16}x^3 + \dots$

So $\sqrt{\frac{1 + x}{1 - x}} = [1 + \frac{1}{2}x - \frac{1}{8}x^2 + \frac{1}{16}x^3 + \dots]$

$\times [1 + \frac{1}{2}x + \frac{3}{8}x^2 + \frac{5}{16}x^3 + \dots]$

$= 1[1 + \frac{1}{2}x + \frac{3}{8}x^2 + \frac{5}{16}x^3]$

$+ \frac{1}{2}x[1 + \frac{1}{2}x + \frac{3}{8}x^2]$

$- \frac{1}{8}x^2[1 + \frac{1}{2}x] + \frac{1}{16}x^3 + \dots$

$= 1 + \frac{1}{2}x + \frac{3}{8}x^2 + \frac{5}{16}x^3 + \frac{1}{2}x + \frac{1}{4}x^2$

$+ \frac{3}{16}x^3 - \frac{1}{8}x^2 - \frac{1}{16}x^3 + \frac{1}{16}x^3 + \dots$

$= 1 + x + \frac{1}{2}x^2 + \frac{1}{2}x^3 + \dots$

Both $(1 + x)^{\frac{1}{2}}$ and $(1 - x)^{-\frac{1}{2}}$ are valid for $|x| < 1$,
so the expansion is valid for $|x| < 1$.

[7 marks available — 1 mark for writing as the product of
two expressions, 1 mark for correct use of formula for at
least one expansion, 1 mark for each correct and simplified
expansion, 1 mark for correct working to multiply the
expansions, 1 mark for correct simplified final answer,
1 mark for correct limit]

b) $\frac{1 + x}{1 - x} = 3 \Rightarrow 1 + x = 3 - 3x \Rightarrow 4x = 2 \Rightarrow x = \frac{1}{2}$

Using the expansion:

$\sqrt{3} \approx 1 + \frac{1}{2} + \frac{1}{2}\left(\frac{1}{2}\right)^2 + \frac{1}{2}\left(\frac{1}{2}\right)^3$

$= 1 + \frac{1}{2} + \frac{1}{8} + \frac{1}{16} = \frac{27}{16}$

[2 marks available — 1 mark for correct value of x,
1 mark for correct final answer]

Q21 a) $3x^2 + x + 2 \equiv$
$A(x - 1)(2x + 1) + B(x + 1)(2x + 1) + C(x + 1)(x - 1)$

Using the substitution method:

When $x = -1$, $3 - 1 + 2 = 2A \Rightarrow A = 2$

When $x = 1$, $3 + 1 + 2 = 6B \Rightarrow B = 1$

When $x = -\frac{1}{2}$, $\frac{3}{4} - \frac{1}{2} + 2 = -\frac{3}{4}C \Rightarrow C = -3$

[3 marks available — 1 mark for using a correct method to
find the unknowns (e.g. substitution or equating coefficients),
1 mark for at least two correct values, 1 mark for all three
correct values]

Answers

b) $\dfrac{3x^2 + x + 2}{(x+1)(x-1)(2x+1)} \equiv \dfrac{2}{(x+1)} + \dfrac{1}{(x-1)} - \dfrac{3}{(2x+1)}$

$\equiv 2(x+1)^{-1} + (x-1)^{-1} - 3(2x+1)^{-1}$

$(x+1)^{-1} = (1+x)^{-1} = (1 - x + x^2 + ...)$

$(x-1)^{-1} = -(1-x)^{-1} = -(1 + x + x^2 + ...)$

$(2x+1)^{-1} = (1+2x)^{-1} = (1 - 2x + 4x^2 + ...)$

So $\dfrac{3x^2 + x + 2}{(x+1)(x-1)(2x+1)}$

$\approx 2[1 - x + x^2] - [1 + x + x^2] - 3[1 - 2x + 4x^2]$

$= 2 - 2x + 2x^2 - 1 - x - x^2 - 3 + 6x - 12x^2 = -2 + 3x - 11x^2$

[6 marks available — 1 mark for each correct expansion of the denominators, 1 mark for multiplying the denominator expansions by the correct values, 1 mark for attempting to add the expansions, 1 mark for correct simplified final answer]

c) $(x+1)^{-1}$ and $(x-1)^{-1}$ are valid for $|x| < 1$.

$(2x+1)^{-1}$ is valid for $|x| < \dfrac{1}{2}$.

So the whole expansion is valid for $|x| < \dfrac{1}{2}$.

[1 mark for the correct validity of the full expansion]

Q22 a) Write as an identity:

$\dfrac{5 - 10x}{(1+2x)(2-x)} \equiv \dfrac{A}{(1+2x)} + \dfrac{B}{(2-x)}$

So $5 - 10x \equiv A(2-x) + B(1+2x)$

Using the substitution method:

$x = 2 \Rightarrow 5 - 20 = 5B \Rightarrow B = -3$

$x = -0.5 \Rightarrow 5 + 5 = 2.5A \Rightarrow A = 4$.

So: $\dfrac{5 - 10x}{(1+2x)(2-x)} \equiv \dfrac{4}{(1+2x)} - \dfrac{3}{(2-x)}$ as required

b) $\dfrac{4}{(1+2x)} = 4(1+2x)^{-1}$

$\dfrac{3}{(2-x)} = 3(2-x)^{-1} = 3 \times 2^{-1}\left(1 - \dfrac{1}{2}x\right)^{-1} = \dfrac{3}{2}\left(1 - \dfrac{1}{2}x\right)^{-1}$

$(1+2x)^{-1} = 1 - 2x + 4x^2 - ...$

$\left(1 - \dfrac{1}{2}x\right)^{-1} = 1 + \dfrac{x}{2} + \dfrac{x^2}{4} + ...$

So $\dfrac{5 - 10x}{(1+2x)(2-x)} = 4(1+2x)^{-1} - \dfrac{3}{2}\left(1 - \dfrac{1}{2}x\right)^{-1}$

$\approx 4[1 - 2x + 4x^2] - \dfrac{3}{2}\left[1 + \dfrac{x}{2} + \dfrac{x^2}{4}\right] + ...$

$\approx \dfrac{5}{2} - \dfrac{35x}{4} + \dfrac{125x^2}{8}$

c) The expansion is valid for $|x| < \dfrac{1}{2}$, so using $x = 0.1$ is valid.

$\dfrac{4}{1.2 \times 1.9} = \dfrac{100}{57}$

Expansion gives an approximation of:

$\dfrac{5}{2} - \dfrac{35(0.1)}{4} + \dfrac{125(0.01)}{8} = \dfrac{57}{32}$

The % error is:

$\left| \dfrac{\left(\dfrac{100}{57}\right) - \left(\dfrac{52}{32}\right)}{\left(\dfrac{100}{57}\right)} \right| \times 100 = 1.5\%$ (to 2 s.f.)

Q23 a) $f(x) = x(3 - x - 2x^2) = x(1-x)(3+2x)$

[2 marks available — 1 mark for finding a factor of x, 1 mark for correct answer fully factorised]

b) $\dfrac{15}{f(x)} = \dfrac{15}{x(1-x)(3+2x)} \equiv \dfrac{A}{x} + \dfrac{B}{1-x} + \dfrac{C}{3+2x}$

$15 \equiv A(1-x)(3+2x) + Bx(3+2x) + Cx(1-x)$

When $x = 1$, $15 = 5B \Rightarrow B = 3$

When $x = 0$, $15 = 3A \Rightarrow A = 5$

When $x = 2$, $15 = -7A + 14B - 2C$

$\Rightarrow 15 = -35 + 42 - 2C \Rightarrow C = -4$

So $\dfrac{15}{f(x)} = \dfrac{5}{x} + \dfrac{3}{1-x} - \dfrac{4}{3+2x}$

[4 marks available — 1 mark for splitting appropriately, 1 mark for using a correct method to find the unknowns (e.g. substitution or equating coefficients), 1 mark for at least one of A, B or C, 1 mark for the correct final answer]

c) $(1-x)^{-1} \approx 1 + (-1)(-x) + \dfrac{(-1) \times (-2)}{1 \times 2}(-x)^2$

$+ \dfrac{(-1) \times (-2) \times (-3)}{1 \times 2 \times 3}(-x)^3$

$= 1 + x + x^2 + x^3$

$(3+2x)^{-1} = 3^{-1}\left(1 + \dfrac{2}{3}x\right)^{-1}$

$\approx \dfrac{1}{3}\left[1 + (-1)\left(\dfrac{2}{3}x\right) + \dfrac{(-1) \times (-2)}{1 \times 2}\left(\dfrac{2}{3}x\right)^2 \right.$

$\left. + \dfrac{(-1) \times (-2) \times (-3)}{1 \times 2 \times 3}\left(\dfrac{2}{3}x\right)^3\right]$

$= \dfrac{1}{3} - \dfrac{2}{9}x + \dfrac{4}{27}x^2 - \dfrac{8}{81}x^3$

So $\dfrac{15}{f(x)} = \dfrac{5}{x} + 3(1 + x + x^2 + x^3)$

$- 4\left(\dfrac{1}{3} - \dfrac{2}{9}x + \dfrac{4}{27}x^2 - \dfrac{8}{81}x^3\right)$

$= 5x^{-1} + \dfrac{5}{3} + \dfrac{35}{9}x + \dfrac{65}{27}x^2 + \dfrac{275}{81}x^3$

[5 marks available — 1 mark for expansion of $(1-x)^{-1}$ (or $3(1-x)^{-1}$), 1 mark for rewriting $(3+2x)^{-1}$, 1 mark for expansion of $(3+2x)^{-1}$ (or $4(3+2x)^{-1}$), 1 mark for putting all together correctly, 1 mark for the correct final answer]

d) Expansion of $(1-x)^{-1}$ is valid for $|-x| < 1 \Rightarrow |x| < 1$

Expansion of $(3+2x)^{-1}$ is valid for $\left|\dfrac{2}{3}x\right| < 1 \Rightarrow |x| < \dfrac{3}{2}$

Therefore, the expansion of $\dfrac{15}{f(x)}$ is valid for $|x| < 1$.

[1 mark for the correct validity of the full expansion]

Q24 a) $f(x) = \dfrac{4x^2 - 8x + 9}{x(2x-3)^2} \equiv \dfrac{A}{x} + \dfrac{B}{2x-3} + \dfrac{C}{(2x-3)^2}$

$\Rightarrow 4x^2 - 8x + 9 = A(2x-3)^2 + Bx(2x-3) + Cx$

When $x = \dfrac{3}{2}$, $6 = \dfrac{3}{2}C \Rightarrow C = 4$

When $x = 0$, $9 = 9A \Rightarrow A = 1$

When $x = 1$, $5 = A - B + C \Rightarrow 5 = 1 - B + 4 \Rightarrow B = 0$

So $f(x) = \dfrac{1}{x} + \dfrac{4}{(2x-3)^2}$

[4 marks available — 1 mark for splitting appropriately, 1 mark for using a correct method to find the unknowns (e.g. substitution or equating coefficients), 1 mark for at least one of A, B or C, 1 mark for the correct final answer]

b) $(2x-3)^{-2} = (-3+2x)^{-2} = \left[-3\left(1 - \dfrac{2}{3}x\right)\right]^{-2}$

$= (-3)^{-2}\left(1 - \dfrac{2}{3}x\right)^{-2}$

$\approx \dfrac{1}{9}\left[1 + (-2)\left(-\dfrac{2}{3}x\right) + \dfrac{(-2) \times (-3)}{1 \times 2}\left(-\dfrac{2}{3}x\right)^2\right]$

$= \dfrac{1}{9}\left[1 + \dfrac{4}{3}x + \dfrac{4}{3}x^2\right] = \dfrac{1}{9} + \dfrac{4}{27}x + \dfrac{4}{27}x^2$

[3 marks available — 1 mark for rewriting the expression, 1 mark for a correct use of the expansion formula, 1 mark for the correct and simplified expansion]

c) $f(x) = \dfrac{1}{x} + \dfrac{4}{(2x-3)^2}$

$\approx \dfrac{1}{x} + 4\left[\dfrac{1}{9} + \dfrac{4}{27}x + \dfrac{4}{27}x^2\right]$

$= x^{-1} + \dfrac{4}{9} + \dfrac{16}{27}x + \dfrac{16}{27}x^2$

[2 marks available — 1 mark for substituting the expansion from b), 1 mark for rearranging into the required form]

d) The result is valid when $\left|-\dfrac{2}{3}x\right| < 1 \Rightarrow |x| < \dfrac{3}{2}$

[1 mark for the correct validity]

Answers

Q25 $3(p + 4x)^{\frac{1}{2}} = 3\left[p^{\frac{1}{2}}\left(1 + \frac{4}{p}x\right)^{\frac{1}{2}}\right] = 3p^{\frac{1}{2}}\left(1 + \frac{4}{p}x\right)^{\frac{1}{2}}$

The x^2 coefficient will be: $3p^{\frac{1}{2}}\left[\frac{\left(\frac{1}{2}\right) \times \left(-\frac{1}{2}\right)}{1 \times 2}\left(\frac{4}{p}\right)^2\right] = -6p^{-\frac{3}{2}}$

$\frac{4}{\sqrt{4 - px}} = 4(4 - px)^{-\frac{1}{2}} = 4\left[4^{-\frac{1}{2}}\left(1 - \frac{p}{4}x\right)^{-\frac{1}{2}}\right] = 2\left(1 - \frac{p}{4}x\right)^{-\frac{1}{2}}$

The x^2 coefficient will be: $2\left[\frac{\left(-\frac{1}{2}\right)\left(-\frac{3}{2}\right)}{1 \times 2}\left(-\frac{p}{4}\right)^2\right] = \frac{3}{64}p^2$

So $-6p^{-\frac{3}{2}} = \frac{3}{64}p^2 \Rightarrow p^{\frac{7}{2}} = -128 \Rightarrow p = (-128)^{\frac{2}{7}} \Rightarrow p = 4$

[5 marks available — 1 mark for each rewritten expression, 1 mark for each coefficient of x^2, 1 mark for the correct answer]

Q26 The x^2 term is $\frac{5 \times 4}{1 \times 2}\left(-\frac{k}{k+1}x\right)^2 = \frac{10k^2}{(k+1)^2}x^2$

$\Rightarrow \frac{10k^2}{k^2 + 2k + 1} = \frac{45}{8} \Rightarrow 80k^2 = 45k^2 + 90k + 45$

$\Rightarrow 35k^2 - 90k - 45 = 0 \Rightarrow 7k^2 - 18k - 9 = 0$

$\Rightarrow (7k + 3)(k - 3) = 0 \Rightarrow k = -\frac{3}{7}, \; k = 3$

Since k is an integer, $k = 3$.

[3 marks available — 1 mark for the correct x^2 term coefficient, 1 mark for a correct method to solve for k, 1 mark for the correct answer]

Q27 a) $(1 + x)^6 - (1 - x)^6 = [(1 + x)^3 + (1 - x)^3][(1 + x)^3 - (1 - x)^3]$
$= [(1 + 3x + 3x^2 + x^3)$
$\qquad + (1 - 3x + 3x^2 - x^3)][(1 + 3x + 3x^2 + x^3)$
$\qquad\qquad\qquad - (1 - 3x + 3x^2 - x^3)]$
$= (2 + 6x^2)(6x + 2x^3) = 4x(1 + 3x^2)(3 + x^2)$

[3 marks available — 1 mark for one correct expansion, 1 mark for method (expansion or difference of two squares), 1 mark for simplifying to the required result]

b) On the x-axis, $y = 0 \Rightarrow 4x(1 + 3x^2)(3 + x^2) = 0$.
Since $x^2 \geq 0$, $(1 + 3x^2) \neq 0$ and $(3 + x^2) \neq 0$.
The function only equals zero at $x = 0$,
so the origin is the only x-axis intercept.

[2 marks available — 1 mark for explaining $x^2 \geq 0$, 1 mark for a correct explanation]

Q28 The x^3 term coefficient is:
$\frac{n(n-1)(n-2)}{1 \times 2 \times 3}(-3)^3 = -\frac{9}{2}n(n-1)(n-2)$

$-\frac{9}{2}n(n-1)(n-2) = -108$

$\Rightarrow n(n-1)(n-2) = 24 \Rightarrow n^3 - 3n^2 + 2n - 24 = 0$

Let $f(n) = n^3 - 3n^2 + 2n - 24$, then $f(4) = 64 - 48 + 8 - 24 = 0$.
Factorising gives $f(n) = (n - 4)(n^2 + n + 6)$.
$(n^2 + n + 6)$ has no real solutions, so $n = 4$.

Q29 a) $(6x + a)^{-3} = (a + 6x)^{-3} = a^{-3}\left(1 + \frac{6}{a}x\right)^{-3}$

$\approx a^{-3}\left[1 + (-3)\left(\frac{6}{a}x\right) + \frac{(-3) \times (-4)}{1 \times 2}\left(\frac{6}{a}x\right)^2\right]$

$= \frac{1}{a^3} - \frac{18}{a^4}x + \frac{216}{a^5}x^2$

[3 marks available — 1 mark for factorising, 1 mark for expansion, 1 mark for simplifying and writing correct coefficients of constant, x and x^2]

b) $\frac{(1 + 2x)^2}{(6x + a)^3} = (1 + 2x)^2(6x + a)^{-3}$

$\approx (1 + 4x + 4x^2)\left(\frac{1}{a^3} - \frac{18}{a^4}x + \frac{216}{a^5}x^2\right)$

Extracting the coefficients from x^2 terms:

$\frac{216}{a^5} - \frac{72}{a^4} + \frac{4}{a^3} = -\frac{108}{a^5} \Rightarrow \frac{216 - 72a + 4a^2}{a^5} = -\frac{108}{a^5}$

$\Rightarrow 4a^2 - 72a + 324 = 0 \Rightarrow a^2 - 18a + 81 = 0$

$\Rightarrow (a - 9)^2 = 0 \Rightarrow a = 9$

[4 marks available — 1 mark for showing expansions multiplied, 1 mark for equating x^2 coefficients, 1 mark for reducing to a quadratic, 1 mark for the correct answer]

c) Expansion is valid for $\left|\frac{6}{9}x\right| < 1 \Rightarrow |x| < \frac{3}{2}$

[1 mark for the correct validity]

Q30 $\sqrt{\frac{a + x}{a - x}} = \sqrt{\frac{a\left(1 + \frac{1}{a}x\right)}{a\left(1 - \frac{1}{a}x\right)}} = \frac{\left(1 + \frac{1}{a}x\right)^{\frac{1}{2}}}{\left(1 - \frac{1}{a}x\right)^{\frac{1}{2}}} = \left(1 + \frac{1}{a}x\right)^{\frac{1}{2}}\left(1 - \frac{1}{a}x\right)^{-\frac{1}{2}}$

$= \left[1 + \frac{1}{2}\left(\frac{1}{a}x\right) + \frac{\left(\frac{1}{2}\right) \times \left(-\frac{1}{2}\right)}{1 \times 2}\left(\frac{1}{a}x\right)^2 + \ldots\right]$

$\quad \times \left[1 + \left(-\frac{1}{2}\right)\left(-\frac{1}{a}x\right) + \frac{\left(-\frac{1}{2}\right) \times \left(-\frac{3}{2}\right)}{1 \times 2}\left(-\frac{1}{a}x\right)^2 + \ldots\right]$

$= \left(1 + \frac{1}{2a}x - \frac{1}{8a^2}x^2 + \ldots\right)\left(1 + \frac{1}{2a}x + \frac{3}{8a^2}x^2 + \ldots\right)$

$\approx 1 + \frac{1}{a}x + \frac{1}{2a^2}x^2$

When $x = 0.3$, $1 + \frac{1}{a}(0.3) + \frac{1}{2a^2}(0.3)^2 = 0.745$

$\Rightarrow 1 + \frac{0.3}{a} + \frac{0.045}{a^2} = 0.745 \Rightarrow 1000a^2 + 300a + 45 = 745a^2$

$\Rightarrow 17a^2 + 20a + 3 = 0 \Rightarrow (a + 1)(17a + 3) \Rightarrow a = -1, \; a = -\frac{3}{17}$

[6 marks available — 1 mark for rewriting the original expression, 1 mark for at least one correct expansion, 1 mark for final expansion, 1 mark for setting up equation using $x = 0.3$, 1 mark for reducing to a quadratic, 1 mark for both correct answers]

Q31 a) $f(x) = \frac{1}{(a - 2x)^2} = (a - 2x)^{-2} = a^{-2}\left(1 - \frac{2}{a}x\right)^{-2}$

$= \frac{1}{a^2}\left(1 - \frac{2}{a}x\right)^{-2} = \frac{1}{a^2}\left[1 + \frac{4x}{a} + \frac{12x^2}{a^2} + \ldots\right]$

$= \frac{1}{a^2} + \frac{4x}{a^3} + \frac{12x^2}{a^4} + \ldots$

b) The expansion is valid for $|x| < \left|\frac{a}{2}\right|$.

c) $g(x) = \frac{1}{\sqrt{4 - ax}} = (4 - ax)^{-\frac{1}{2}} = 4^{-\frac{1}{2}}\left(1 - \frac{a}{4}x\right)^{-\frac{1}{2}}$

$= \frac{1}{2}\left(1 - \frac{a}{4}x\right)^{-\frac{1}{2}} = \frac{1}{2}\left[1 + \frac{ax}{8} + \frac{3a^2x^2}{128} + \ldots\right]$

$= \frac{1}{2} + \frac{ax}{16} + \frac{3a^2x^2}{256} + \ldots$

d) The expansion is valid for $|x| < \left|\frac{4}{a}\right|$.

e) $\left[\frac{1}{a^2} + \frac{4x}{a^3} + \frac{12x^2}{a^4} + \ldots\right] \times \left[\frac{1}{2} + \frac{ax}{16} + \frac{3a^2x^2}{256} + \ldots\right]$

$= \frac{1}{a^2}\left[\frac{1}{2} + \frac{ax}{16}\right] + \frac{4x}{a^3}\left[\frac{1}{2}\right] + \ldots = \frac{1}{2a^2} + \frac{ax}{16a^2} + \frac{4x}{2a^3} + \ldots$

$= \frac{1}{2a^2} + \frac{(a^2 + 32)x}{16a^3} + \ldots$

f) When $a = 2$, $\frac{(a^2 + 32)}{16a^3} = \frac{(2^2 + 32)}{16(2)^3} = \frac{9}{32}$

Q32 a) $f(x) = \frac{4x^2 - 6x + 3}{2x^2 - 3x + 1} = \frac{2(2x^2 - 3x + 1) + 1}{2x^2 - 3x + 1}$

$= 2 + \frac{1}{2x^2 - 3x + 1} = 2 + \frac{1}{(1 - x)(1 - 2x)}$

Now use partial fractions:

$\frac{1}{(1 - x)(1 - 2x)} \equiv \frac{B}{1 - x} + \frac{C}{1 - 2x}$

$1 \equiv B(1 - 2x) + C(1 - x)$

When $x = 1$, $1 = -B \Rightarrow B = -1$

When $x = \frac{1}{2}$, $1 = \frac{1}{2}C \Rightarrow C = 2$

So $f(x) = 2 - \frac{1}{1 - x} + \frac{2}{1 - 2x}$

[4 marks available — 1 mark for simplifying to $2 + \frac{1}{2x^2 - 3x + 1}$, 1 mark for a correct attempt at partial fractions, 1 mark for each correct value of B and C]

479

b) (i) $(1-x)^{-1} \approx 1 + (-1)(-x) + \dfrac{(-1)\times(-2)}{1\times 2}(-x)^2$

$= 1 + x + x^2$

(ii) $(1-2x)^{-1} \approx 1 + (-1)(-2x) + \dfrac{(-1)\times(-2)}{1\times 2}(-2x)^2$

$= 1 + 2x + 4x^2$

[3 marks available — 3 marks for both fully correct, otherwise 2 marks for at least one fully correct, or 1 mark for a correct use of the expansion formula]

c) $f(x) \approx 2 - (1 + x + x^2) + 2(1 + 2x + 4x^2) = 3 + 3x + 7x^2$

[2 marks available — 1 mark for a correct method, 1 mark for the correct answer]

d) Expansion for $(1-x)^{-1}$ is valid for $|-x| < 1 \Rightarrow |x| < 1$

Expansion for $(1-2x)^{-1}$ is valid for $|-2x| < 1 \Rightarrow |x| < \dfrac{1}{2}$

So, the combined expansion for $f(x)$ is valid for $|x| < \dfrac{1}{2}$.

[1 mark for the correct validity of the full expansion]

Chapter 7: Differentiation

Prior Knowledge Check

Q1 $\dfrac{dy}{dx} = \lim_{h \to 0} \left[\dfrac{3(x+h)^2 + 4 - (3x^2 + 4)}{(x+h) - x} \right]$

$= \lim_{h \to 0} \left[\dfrac{3x^2 + 6hx + 3h^2 + 4 - (3x^2 + 4)}{(x+h) - x} \right]$

$= \lim_{h \to 0} \left[\dfrac{6hx + 3h^2}{h} \right] = \lim_{h \to 0} [6x + 3h] = 6x$

Q2 $\dfrac{dy}{dx} = 3x^2 + 4x + 1$

$\dfrac{d^2y}{dx^2} = 6x + 4$

So when $x = -1$, $\dfrac{d^2y}{dx^2} = -2$

$\dfrac{d^2y}{dx^2} < 0$ at $(-1, -6)$, so it's a maximum.

Q3 $\dfrac{dy}{dx} = x - 6$, so gradient of curve at $(2, -10)$ is $2 - 6 = -4$.

Gradient of normal at $(2, -10) = \dfrac{1}{4}$.

$y = mx + c \Rightarrow -10 = \left(\dfrac{1}{4} \times 2\right) + c \Rightarrow c = -\dfrac{21}{2}$

So equation of normal is $y = \dfrac{1}{4}x - \dfrac{21}{2}$.

Q4 $\ln 6x = 5 \Rightarrow 6x = e^5 \Rightarrow x = \dfrac{e^5}{6}$

Q5 $\cos \pi = -1$, so $\arccos(-1) = \pi$ radians.

Q6 a) $\sin 5x \cos 5x = \dfrac{1}{2}\sin 10x$

b) $\sin 4x \cos 3x - \cos 4x \sin 3x = \sin(4x - 3x) = \sin x$

Q7 $x = t - 2 \Rightarrow t = x + 2$,

so $y = 2(x+2)^2 - 7$ or $y = 2x^2 + 8x + 1$.

Exercise 7.1-7.3 — Convex/Concave Curves, and Points of Inflection

Q1 a) $y = \dfrac{1}{6}x^3 - \dfrac{5}{2}x^2 + \dfrac{1}{4}x + \dfrac{1}{9}$

$\Rightarrow \dfrac{dy}{dx} = \dfrac{1}{2}x^2 - 5x + \dfrac{1}{4} \Rightarrow \dfrac{d^2y}{dx^2} = x - 5$

The graph is concave when $\dfrac{d^2y}{dx^2} < 0 \Rightarrow x - 5 < 0 \Rightarrow x < 5$

b) $y = 4x^2 - x^4 \Rightarrow \dfrac{dy}{dx} = 8x - 4x^3 \Rightarrow \dfrac{d^2y}{dx^2} = 8 - 12x^2$

The graph is concave when $\dfrac{d^2y}{dx^2} < 0 \Rightarrow 8 - 12x^2 < 0$

$\Rightarrow 8 < 12x^2 \Rightarrow \dfrac{2}{3} < x^2 \Rightarrow x < -\sqrt{\dfrac{2}{3}}$ or $x > \sqrt{\dfrac{2}{3}}$

Q2 $y = \dfrac{3}{2}x^4 - x^2 - 3x \Rightarrow \dfrac{dy}{dx} = 6x^3 - 2x - 3 \Rightarrow \dfrac{d^2y}{dx^2} = 18x^2 - 2$

At the points of inflection, $\dfrac{d^2y}{dx^2} = 0$

$\Rightarrow 18x^2 - 2 = 0 \Rightarrow x^2 = \dfrac{1}{9} \Rightarrow x = \pm\dfrac{1}{3}$

Q3 $f(x) = \dfrac{1}{16}x^4 + \dfrac{3}{4}x^3 - \dfrac{21}{8}x^2 - 6x + 20$

$\Rightarrow f'(x) = \dfrac{1}{4}x^3 + \dfrac{9}{4}x^2 - \dfrac{21}{4}x - 6 \Rightarrow f''(x) = \dfrac{3}{4}x^2 + \dfrac{9}{2}x - \dfrac{21}{4}$

So $f(x)$ is convex when $f''(x) > 0 \Rightarrow \dfrac{3}{4}x^2 + \dfrac{9}{2}x - \dfrac{21}{4} > 0$

$\Rightarrow 3x^2 + 18x - 21 > 0 \Rightarrow x^2 + 6x - 7 > 0$

$\Rightarrow (x+7)(x-1) > 0 \Rightarrow x < -7$ or $x > 1$

So $f(x)$ is convex for $x < -7$ and $x > 1$, and concave for $-7 < x < 1$.

Sketch the graph of f''(x) if you're not sure which way the inequalities should go.

Q4 $f'(x) = \dfrac{2}{3}x^{-\frac{1}{3}}$ and $f''(x) = -\dfrac{2}{9}x^{-\frac{4}{3}} = -\dfrac{2}{9}\left(x^{-\frac{2}{3}}\right)^2$

For all real values of x, $x \neq 0$, $\left(x^{-\frac{2}{3}}\right)^2 > 0 \Rightarrow -\dfrac{2}{9}\left(x^{-\frac{2}{3}}\right)^2 < 0$,

so the graph of $y = f(x)$ is concave.

[3 marks available — 1 mark for the correct first derivative, 1 mark for the correct second derivative, 1 mark for a correct explanation of why the graph is concave]

Q5 $\left(x + \dfrac{1}{x}\right)^2 = x^2 + 2 + \dfrac{1}{x^2}$,

$f'(x) = 2x - \dfrac{2}{x^3}$ so $f''(x) = 2 + \dfrac{6}{x^4}$.

For $x > 0$, $x^4 > 0$, so $f''(x) = 2 + \dfrac{6}{x^4} > 0$

and the graph of $y = f(x)$ is convex.

[4 marks available — 1 mark for correct expansion of brackets, 1 mark for the correct first derivative, 1 mark for the correct second derivative, 1 mark for a correct explanation of why the graph is convex]

You could also differentiate using the chain rule.

Q6 $y = x^2 - \dfrac{1}{x} \Rightarrow \dfrac{dy}{dx} = 2x + \dfrac{1}{x^2} \Rightarrow \dfrac{d^2y}{dx^2} = 2 - \dfrac{2}{x^3}$

At a point of inflection, $\dfrac{d^2y}{dx^2} = 0 \Rightarrow 2 - \dfrac{2}{x^3} = 0$

$\Rightarrow 2 = \dfrac{2}{x^3} \Rightarrow x^3 = 1 \Rightarrow x = 1$

If $x < 1$, $x^3 < 1$, so $\dfrac{2}{x^3} > 2$, so $\dfrac{d^2y}{dx^2}$ is negative.

If $x > 1$, $x^3 > 1$, so $\dfrac{2}{x^3} < 2$, so $\dfrac{d^2y}{dx^2}$ is positive.

So at $x = 1$, $\dfrac{d^2y}{dx^2} = 0$ and the sign of $\dfrac{d^2y}{dx^2}$ changes, so this is a point of inflection.

When $x = 1$, $y = 1^2 - \dfrac{1}{1} = 0$,

so $(1, 0)$ is a point of inflection of $y = x^2 - \dfrac{1}{x}$.

Q7 a) $f(x) = x^3 + 2x^2 + 3x + 3 \Rightarrow f'(x) = 3x^2 + 4x + 3$

$\Rightarrow f''(x) = 6x + 4$

At a point of inflection, $f''(x) = 0 \Rightarrow 6x + 4 = 0 \Rightarrow x = -\dfrac{2}{3}$

When $x > -\dfrac{2}{3}$, $6x + 4 > 0$ and when $x < -\dfrac{2}{3}$, $6x + 4 < 0$

So the graph of $y = f(x)$ has one point of inflection,

at $x = -\dfrac{2}{3}$.

[4 marks available — 1 mark for the correct first derivative, 1 mark for the correct second derivative, 1 mark for the correct x-coordinate, 1 mark for the correct method to show a point of inflection at $x = -\dfrac{2}{3}$]

Answers

b) $f'(x) = 3x^2 + 4x + 3$, so at $x = -\frac{2}{3}$,

$f'(x) = 3\left(-\frac{2}{3}\right)^2 + 4\left(-\frac{2}{3}\right) + 3 = \frac{4}{3} - \frac{8}{3} + 3 = \frac{5}{3}$

$f'(x) \neq 0$ at the point of inflection, so it is not a stationary point.

[2 marks available — 1 mark for substituting into the first derivative, 1 mark for a correct explanation]

Q8 a) $y = 5x^3(x-1)^2 = 5x^3(x^2 - 2x + 1) = 5x^5 - 10x^4 + 5x^3$ so

$\frac{dy}{dx} = 25x^4 - 40x^3 + 15x^2$

$= 5x^2(5x^2 - 8x + 3) = 5x^2(5x - 3)(x - 1)$

so $\frac{dy}{dx} = 0 \Rightarrow 5x^2(5x-3)(x-1) = 0 \Rightarrow x = 0, \frac{3}{5}, 1$

[4 marks available — 1 mark for correct expansion of brackets, 1 mark for the correct derivative, 1 mark for a correct method to solve derivative equal to zero, 1 mark for all three correct answers]

b) $\frac{d^2y}{dx^2} = 100x^3 - 120x^2 + 30x$

At $x = 0$, $\frac{d^2y}{dx^2} = 0$. Check $\frac{d^2y}{dx^2}$ either side of $x = 0$,

e.g. when $x = 0.1$, $\frac{d^2y}{dx^2} = 1.9$ and $x = -0.1$, $\frac{d^2y}{dx^2} = -4.3$, there is a change of sign so there is a point of inflection at $x = 0$.

[2 marks available — 1 mark for the correct second derivative, 1 mark for a correct method to show stationary point of inflection at $x = 0$]

Q9 a) $y = -2x^3 + 5x^2 - 4x$ $\frac{dy}{dx} = -6x^2 + 10x - 4$

$\Rightarrow \frac{d^2y}{dx^2} = -12x + 10$

[2 marks available — 1 mark for the correct first derivative, 1 mark for the correct second derivative]

b) $\frac{d^2y}{dx^2} = 0 \Rightarrow 12x = 10 \Rightarrow x = \frac{5}{6}$

So $y = -2\left(\frac{5}{6}\right)^3 + 5\left(\frac{5}{6}\right)^2 - 4\left(\frac{5}{6}\right) = -\frac{55}{54}$.

When $x = 1$, $\frac{d^2y}{dx^2} = -2 \Rightarrow \frac{d^2y}{dx^2} < 0$,

when $x - 0$, $\frac{d^2y}{dx^2} = 10 \Rightarrow \frac{d^2y}{dx^2} > 0$.

There is a change of sign, so the curve has a point of inflection at $\left(\frac{5}{6}, -\frac{55}{54}\right)$.

[4 marks available — 1 mark for setting the second derivative equal to 0, 1 mark for the correct x-coordinate, 1 mark for the correct y-coordinate, 1 mark for a correct method to show it is a point of inflection]

c) When $x = \frac{5}{6}$, $\frac{dy}{dx} = \frac{1}{6} \neq 0$, so it is not a stationary point of inflection.

[1 mark for the correct answer with explanation]

Q10 $y = \frac{1}{10}x^5 - \frac{1}{3}x^3 + \frac{1}{2}x + 4$

$\Rightarrow \frac{dy}{dx} = \frac{1}{2}x^4 - x^2 + \frac{1}{2} \Rightarrow \frac{d^2y}{dx^2} = 2x^3 - 2x$

At a stationary point, $\frac{dy}{dx} = 0$

$\Rightarrow \frac{1}{2}x^4 - x^2 + \frac{1}{2} = 0 \Rightarrow x^4 - 2x^2 + 1 = 0$

$\Rightarrow (x^2 - 1)^2 = 0 \Rightarrow [(x+1)(x-1)]^2 = 0 \Rightarrow x = \pm 1$

When $x = 1$, $y = \frac{1}{10} - \frac{1}{3} + \frac{1}{2} + 4 = \frac{64}{15}$ and $\frac{d^2y}{dx^2} = 2 - 2 = 0$.

When $x = -1$, $y = -\frac{1}{10} + \frac{1}{3} - \frac{1}{2} + 4 = \frac{56}{15}$

and $\frac{d^2y}{dx^2} = -2 + 2 = 0$.

$\frac{d^2y}{dx^2} = 0$ at both stationary points,

so check what happens to $\frac{d^2y}{dx^2}$ around $x = 1$ and $x = -1$.

$\frac{d^2y}{dx^2} = 2x^3 - 2x = 2x(x^2 - 1) = 2x(x+1)(x-1)$

When $x < -1$, $2x$ is negative, $x + 1$ is negative and $x - 1$ is negative, so $\frac{d^2y}{dx^2}$ is negative for $x < -1$.

When $-1 < x < 0$, $2x$ is negative, $x + 1$ is positive and $x - 1$ is negative, so $\frac{d^2y}{dx^2}$ is positive for $-1 < x < 0$.

So $\frac{d^2y}{dx^2}$ changes sign at $x = -1$.

When $0 < x < 1$, $2x$ is positive, $x + 1$ is positive and $x - 1$ is negative, so $\frac{d^2y}{dx^2}$ is negative for $0 < x < 1$.

When $1 < x$, $2x$ is positive, $x + 1$ is positive and $x - 1$ is positive, so $\frac{d^2y}{dx^2}$ is positive for $1 < x$. So $\frac{d^2y}{dx^2}$ changes sign at $x = 1$.

The graph of $y = \frac{1}{10}x^5 - \frac{1}{3}x^3 + \frac{1}{2}x + 4$ has stationary points at $\left(-1, \frac{56}{15}\right)$ and $\left(1, \frac{64}{15}\right)$, and both stationary points are points of inflection.

Again, it might help to sketch the graph of $\frac{d^2y}{dx^2}$. You could also use the fact that $\frac{dy}{dx} = \frac{1}{2}(x^2 - 1)^2$ to show that the gradient is always ≥ 0, which means that any stationary points must be points of inflection.

Q11 E.g. $f(x) = \sqrt{x}$ is only defined where $x \geq 0$.

$f'(x) = \frac{1}{2}x^{-\frac{1}{2}} = \frac{1}{2\sqrt{x}} > 0$ for all $x > 0$ so $f(x)$ is increasing.

$f''(x) = -\frac{1}{4}x^{-\frac{3}{2}} = -\frac{1}{4\sqrt{x^3}} < 0$ for all $x > 0$,

so the graph of $y = f(x)$ is concave for all positive values of x.

Exercise 7.4 — The Chain Rule

Q1 a) $y = (x + 7)^2$, so let $y = u^2$ where $u = x + 7$

$\Rightarrow \frac{dy}{du} = 2u = 2(x + 7)$, $\frac{du}{dx} = 1$

$\frac{dy}{dx} = \frac{dy}{du} \times \frac{du}{dx} = 2(x + 7) \times 1 = 2(x + 7)$

b) $y = (2x - 1)^5$, so let $y = u^5$ where $u = 2x - 1$

$\Rightarrow \frac{dy}{du} = 5u^4 = 5(2x - 1)^4$, $\frac{du}{dx} = 2$

$\frac{dy}{dx} = \frac{dy}{du} \times \frac{du}{dx} = 5(2x - 1)^4 \times 2 = 10(2x - 1)^4$

c) $y = 3(4 - x)^8$, so let $y = 3u^8$ where $u = 4 - x$

$\Rightarrow \frac{dy}{du} = 24u^7 = 24(4 - x)^7$, $\frac{du}{dx} = -1$

$\frac{dy}{dx} = \frac{dy}{du} \times \frac{du}{dx} = 24(4 - x)^7 \times (-1) = -24(4 - x)^7$

d) $y = (3 - 2x)^7$, so let $y = u^7$ where $u = 3 - 2x$

$\Rightarrow \frac{dy}{du} = 7u^6 = 7(3 - 2x)^6$, $\frac{du}{dx} = -2$

$\frac{dy}{dx} = \frac{dy}{du} \times \frac{du}{dx} = 7(3 - 2x)^6 \times (-2) = -14(3 - 2x)^6$

e) $y = (x^2 + 3)^5$, so let $y = u^5$ where $u = x^2 + 3$

$\Rightarrow \frac{dy}{du} = 5u^4 = 5(x^2 + 3)^4$, $\frac{du}{dx} = 2x$

$\frac{dy}{dx} = \frac{dy}{du} \times \frac{du}{dx} = 5(x^2 + 3)^4 \times 2x = 10x(x^2 + 3)^4$

f) $y = (5x^2 + 3)^2$, so let $y = u^2$ where $u = 5x^2 + 3$

$\Rightarrow \frac{dy}{du} = 2u = 2(5x^2 + 3)$, $\frac{du}{dx} = 10x$

$\frac{dy}{dx} = \frac{dy}{du} \times \frac{du}{dx} = 2(5x^2 + 3) \times 10x = 20x(5x^2 + 3)$

Answers

Q2 **a)** $f(x) = (4x^3 - 9)^8$, so let $y = u^8$ where $u = 4x^3 - 9$

$\Rightarrow \dfrac{dy}{du} = 8u^7 = 8(4x^3 - 9)^7$, $\dfrac{du}{dx} = 12x^2$

$f'(x) = \dfrac{dy}{du} \times \dfrac{du}{dx} = 8(4x^3 - 9)^7 \times 12x^2 = 96x^2(4x^3 - 9)^7$

b) $f(x) = (6 - 7x^2)^4$, so let $y = u^4$ where $u = 6 - 7x^2$

$\Rightarrow \dfrac{dy}{du} = 4u^3 = 4(6 - 7x^2)^3$, $\dfrac{du}{dx} = -14x$

$f'(x) = \dfrac{dy}{du} \times \dfrac{du}{dx} = 4(6 - 7x^2)^3 \times (-14x) = -56x(6 - 7x^2)^3$

c) $f(x) = (x^2 + 5x + 7)^6$, so let $y = u^6$
where $u = x^2 + 5x + 7$

$\Rightarrow \dfrac{dy}{du} = 6u^5 = 6(x^2 + 5x + 7)^5$, $\dfrac{du}{dx} = 2x + 5$

$f'(x) = \dfrac{dy}{du} \times \dfrac{du}{dx} = 6(x^2 + 5x + 7)^5 \times (2x + 5)$
$\qquad = (x^2 + 5x + 7)^5(12x + 30)$

d) $f(x) = (x + 4)^{-3}$, so let $y = u^{-3}$ where $u = x + 4$

$\Rightarrow \dfrac{dy}{du} = -3u^{-4} = -3(x + 4)^{-4}$, $\dfrac{du}{dx} = 1$

$f'(x) = \dfrac{dy}{du} \times \dfrac{du}{dx} = -3(x + 4)^{-4} \times 1 = -3(x + 4)^{-4}$

e) $f(x) = (5 - 3x)^{-2}$, so let $y = u^{-2}$ where $u = 5 - 3x$

$\Rightarrow \dfrac{dy}{du} = -2u^{-3} = -2(5 - 3x)^{-3}$, $\dfrac{du}{dx} = -3$

$f'(x) = \dfrac{dy}{du} \times \dfrac{du}{dx} = -2(5 - 3x)^{-3} \times (-3) = 6(5 - 3x)^{-3}$

f) $f(x) = \dfrac{1}{(5 - 3x)^4} = (5 - 3x)^{-4}$, so let $y = u^{-4}$ where $u = 5 - 3x$

$\Rightarrow \dfrac{dy}{du} = -4u^{-5} = -4(5 - 3x)^{-5}$, $\dfrac{du}{dx} = -3$

$f'(x) = \dfrac{dy}{du} \times \dfrac{du}{dx} = -4(5 - 3x)^{-5} \times (-3) = \dfrac{12}{(5 - 3x)^5}$

g) $f(x) = (3x^2 + 4)^{\frac{3}{2}}$, so let $y = u^{\frac{3}{2}}$ where $u = 3x^2 + 4$

$\Rightarrow \dfrac{dy}{du} = \dfrac{3}{2}u^{\frac{1}{2}} = \dfrac{3}{2}(3x^2 + 4)^{\frac{1}{2}}$, $\dfrac{du}{dx} = 6x$

$f'(x) = \dfrac{dy}{du} \times \dfrac{du}{dx} = \dfrac{3}{2}(3x^2 + 4)^{\frac{1}{2}} \times 6x = 9x(3x^2 + 4)^{\frac{1}{2}}$

h) $f(x) = \dfrac{1}{\sqrt{5 - 3x}} = (5 - 3x)^{-\frac{1}{2}}$,

so let $y = u^{-\frac{1}{2}}$ where $u = 5 - 3x$

$\Rightarrow \dfrac{dy}{du} = -\dfrac{1}{2}u^{-\frac{3}{2}} = -\dfrac{1}{2}(5 - 3x)^{-\frac{3}{2}}$, $\dfrac{du}{dx} = -3$

$f'(x) = \dfrac{dy}{du} \times \dfrac{du}{dx} = -\dfrac{1}{2}(5 - 3x)^{-\frac{3}{2}} \times (-3) = \dfrac{3}{2(\sqrt{5 - 3x})^3}$

i) $f(x) = \dfrac{1}{\sqrt{x^3 + 2x^2}} = (x^3 + 2x^2)^{-\frac{1}{2}}$,

so let $y = u^{-\frac{1}{2}}$ where $u = x^3 + 2x^2$

$\Rightarrow \dfrac{dy}{du} = -\dfrac{1}{2}u^{-\frac{3}{2}} = -\dfrac{1}{2}(x^3 + 2x^2)^{-\frac{3}{2}}$, $\dfrac{du}{dx} = 3x^2 + 4x$

$f'(x) = \dfrac{dy}{du} \times \dfrac{du}{dx} = -\dfrac{1}{2}(x^3 + 2x^2)^{-\frac{3}{2}} \times (3x^2 + 4x)$

$\qquad = -\dfrac{3x^2 + 4x}{2(\sqrt{x^3 + 2x^2})^3}$

Q3 **a)** $y = (5x - 3x^2)^{-\frac{1}{2}}$, so let $y = u^{-\frac{1}{2}}$ where $u = 5x - 3x^2$

$\Rightarrow \dfrac{dy}{du} = -\dfrac{1}{2}u^{-\frac{3}{2}} = -\dfrac{1}{2}(5x - 3x^2)^{-\frac{3}{2}}$, $\dfrac{du}{dx} = 5 - 6x$

$\dfrac{dy}{dx} = \dfrac{dy}{du} \times \dfrac{du}{dx} = -\dfrac{1}{2}(5x - 3x^2)^{-\frac{3}{2}} \times (5 - 6x)$

$\qquad = -\dfrac{5 - 6x}{2(\sqrt{5x - 3x^2})^3}$

When $x = 1$, $\dfrac{dy}{dx} = -\dfrac{5 - (6 \times 1)}{2(\sqrt{(5 \times 1) - (3 \times 1^2)})^3} = \dfrac{1}{4\sqrt{2}}$

b) $y = \dfrac{12}{\sqrt[3]{x + 6}} = 12(x + 6)^{-\frac{1}{3}}$,

so let $y = 12u^{-\frac{1}{3}}$ where $u = x + 6$

$\Rightarrow \dfrac{dy}{du} = -4u^{-\frac{4}{3}} = -4(x + 6)^{-\frac{4}{3}}$, $\dfrac{du}{dx} = 1$

$\dfrac{dy}{dx} = \dfrac{dy}{du} \times \dfrac{du}{dx} = -4(x + 6)^{-\frac{4}{3}} \times 1 = -\dfrac{4}{\sqrt[3]{(x + 6)^4}}$

When $x = 1$, $\dfrac{dy}{dx} = -\dfrac{4}{\sqrt[3]{(1 + 6)^4}} = -\dfrac{4}{7(\sqrt[3]{7})}$

Q4 **a)** $\left(\sqrt{x} + \dfrac{1}{\sqrt{x}}\right)^2 = \sqrt{x}\sqrt{x} + 2\sqrt{x}\dfrac{1}{\sqrt{x}} + \dfrac{1}{\sqrt{x}}\dfrac{1}{\sqrt{x}} = x + \dfrac{1}{x} + 2$

$\dfrac{d}{dx}\left(x + \dfrac{1}{x} + 2\right) = 1 - \dfrac{1}{x^2}$

Remember $\dfrac{1}{x} = x^{-1}$.

b) $y = \left(\sqrt{x} + \dfrac{1}{\sqrt{x}}\right)^2$, so let $y = u^2$ where $u = \sqrt{x} + \dfrac{1}{\sqrt{x}}$

$\Rightarrow \dfrac{dy}{du} = 2u = 2\left(\sqrt{x} + \dfrac{1}{\sqrt{x}}\right)$,

$\dfrac{du}{dx} = \dfrac{1}{2}x^{-\frac{1}{2}} - \dfrac{1}{2}x^{-\frac{3}{2}} = \dfrac{1}{2\sqrt{x}} - \dfrac{1}{2(\sqrt{x})^3}$

$\dfrac{dy}{dx} = \dfrac{dy}{du} \times \dfrac{du}{dx} = 2\left(\sqrt{x} + \dfrac{1}{\sqrt{x}}\right) \times \left(\dfrac{1}{2\sqrt{x}} - \dfrac{1}{2(\sqrt{x})^3}\right)$

$\qquad = 2\left(\dfrac{1}{2} + \dfrac{1}{2x} - \dfrac{1}{2x} - \dfrac{1}{2x^2}\right) = 1 - \dfrac{1}{x^2}$

Using powers notation makes this question easier to handle.

Q5 $y = (x - 3)^5$, so let $y = u^5$ where $u = x - 3$

$\Rightarrow \dfrac{dy}{du} = 5u^4 = 5(x - 3)^4$, $\dfrac{du}{dx} = 1$

$\dfrac{dy}{dx} = \dfrac{dy}{du} \times \dfrac{du}{dx} = 5(x - 3)^4 \times 1 = 5(x - 3)^4$

At the point $(1, -32)$, gradient $= 5(1 - 3)^4 = 80$
The equation of a straight line is $y = mx + c$
$\Rightarrow -32 = (80 \times 1) + c \Rightarrow c = -112$
So the equation of the tangent is $y = 80x - 112$.

Q6 $y = \dfrac{1}{4}(x - 7)^4$, so let $y = \dfrac{1}{4}u^4$ where $u = x - 7$

$\Rightarrow \dfrac{dy}{du} = u^3 = (x - 7)^3$, $\dfrac{du}{dx} = 1$

$\dfrac{dy}{dx} = \dfrac{dy}{du} \times \dfrac{du}{dx} = (x - 7)^3 \times 1 = (x - 7)^3$

When $x = 6$, $y = \dfrac{1}{4}(6 - 7)^4 = \dfrac{1}{4}$ and $\dfrac{dy}{dx} = (6 - 7)^3 = -1$

The gradient of the normal is $-1 \div -1 = 1$.
The equation of a straight line is $y = mx + c$
$\Rightarrow \dfrac{1}{4} = (1 \times 6) + c \Rightarrow c = -\dfrac{23}{4}$

So the equation of the normal is $y = x - \dfrac{23}{4}$

Q7 $y = (7x^2 - 3)^{-4}$, so let $y = u^{-4}$ where $u = 7x^2 - 3$

$\Rightarrow \dfrac{dy}{du} = -4u^{-5} = -4(7x^2 - 3)^{-5}$, $\dfrac{du}{dx} = 14x$

$\dfrac{dy}{dx} = \dfrac{dy}{du} \times \dfrac{du}{dx} = -4(7x^2 - 3)^{-5} \times 14x = -56x(7x^2 - 3)^{-5}$

When $x = 1$, $\dfrac{dy}{dx} = -56(1)(7(1)^2 - 3)^{-5} = -56(4^{-5}) = -\dfrac{7}{128}$

Q8 $y = \dfrac{7}{\sqrt[3]{3 - 2x}}$, so let $y = 7u^{-\frac{1}{3}}$ where $u = 3 - 2x$

$\Rightarrow \dfrac{dy}{du} = -\dfrac{7}{3}u^{-\frac{4}{3}} = -\dfrac{7}{3(\sqrt[3]{3 - 2x})^4}$, $\dfrac{du}{dx} = -2$

$f'(x) = \dfrac{dy}{du} \times \dfrac{du}{dx} = -\dfrac{7}{3(\sqrt[3]{3 - 2x})^4} \times -2 = \dfrac{14}{3(\sqrt[3]{3 - 2x})^4}$

$f'(x)$ could also be written as $\dfrac{14}{3}(3 - 2x)^{-\frac{4}{3}}$.

Q9 $y = \sqrt{5x-1}$, so let $y = u^{\frac{1}{2}}$ where $u = 5x - 1$

$\Rightarrow \dfrac{dy}{du} = \dfrac{1}{2}u^{-\frac{1}{2}} = \dfrac{1}{2}\dfrac{1}{\sqrt{5x-1}}, \dfrac{du}{dx} = 5$

$\dfrac{dy}{dx} = \dfrac{dy}{du} \times \dfrac{du}{dx} = \dfrac{1}{2}\dfrac{1}{\sqrt{5x-1}} \times 5 = \dfrac{5}{2\sqrt{5x-1}}$

when $x = 2$, $\dfrac{dy}{dx} = \dfrac{5}{2\sqrt{10-1}} = \dfrac{5}{6}$

and $y = \sqrt{10-1} = 3$

$y = mx + c \Rightarrow 3 = (\dfrac{5}{6} \times 2) + c \Rightarrow c = \dfrac{4}{3}$

So the equation of the tangent is $y = \dfrac{5}{6}x + \dfrac{4}{3}$.

In the form $ax + by + c = 0$, $5x - 6y + 8 = 0$.

Q10 $y = \sqrt[3]{3x-7}$, so let $y = u^{\frac{1}{3}}$ where $u = 3x - 7$

$\Rightarrow \dfrac{dy}{du} = \dfrac{1}{3}u^{-\frac{2}{3}} = \dfrac{1}{3(\sqrt[3]{3x-7})^2}, \dfrac{du}{dx} = 3$

$\dfrac{dy}{dx} = \dfrac{dy}{du} \times \dfrac{du}{dx} = \dfrac{1}{3(\sqrt[3]{3x-7})^2} \times 3 = \dfrac{1}{(\sqrt[3]{3x-7})^2}$

When $x = 5$, $\dfrac{dy}{dx} = \dfrac{1}{(\sqrt[3]{(3 \times 5)-7})^2} = \dfrac{1}{4}$,

$y = \sqrt[3]{15-7} = 2$

Gradient of normal $= \dfrac{-1}{\frac{1}{4}} = -4$

$y = mx + c \Rightarrow 2 = (-4 \times 5) + c \Rightarrow c = 22$

So the equation of the normal is $y = 22 - 4x$.

Q11 a) $y = \left(\dfrac{1}{x} + 3\right)^4$ so let $y = u^4$ where $u = \dfrac{1}{x} + 3$

$\dfrac{dy}{du} = 4u^3 = 4\left(\dfrac{1}{x} + 3\right)^3, \dfrac{du}{dx} = -x^{-2} = -\dfrac{1}{x^2}$

$\dfrac{dy}{dx} = \dfrac{dy}{du} \times \dfrac{du}{dx} = 4\left(\dfrac{1}{x} + 3\right)^3 \times -\dfrac{1}{x^2} = -\dfrac{4}{x^2}\left(\dfrac{1}{x} + 3\right)^3$

[2 marks available — 1 mark for a correct method, 1 mark for the correct answer]

b) At $x = -1$,

gradient $= -\dfrac{4}{(-1)^2}(-1 + 3)^3 = -32$ and $y = (-1 + 3)^4 = 16$

$y = mx + c \Rightarrow 16 = (-32 \times -1) + c \Rightarrow c = -16$

So the equation of T is $y = -32x - 16$.

$x = 0 \Rightarrow y = (-32 \times 0) - 16 = -16$, so B is $(0, -16)$.

$y = 0 \Rightarrow 0 = -32x - 16 \Rightarrow 16 = -32x$

$\Rightarrow x = -\dfrac{1}{2}$, so A is $\left(-\dfrac{1}{2}, 0\right)$.

So the area of the triangle is $\dfrac{1}{2} \times \dfrac{1}{2} \times 16 = 4$.

[4 marks available — 1 mark for the correct gradient, 1 mark for the correct equation for line T, 1 mark for the correct coordinates of A and B, 1 mark for the correct answer]

c) $\dfrac{dy}{dx} = 0 \Rightarrow -\dfrac{4}{x^2}\left(\dfrac{1}{x} + 3\right)^3 = 0 \Rightarrow \left(\dfrac{1}{x} + 3\right)^3 = 0$

$\Rightarrow \dfrac{1}{x} + 3 = 0 \Rightarrow 3x = -1 \Rightarrow x = -\dfrac{1}{3}$

$y = \left(\dfrac{1}{-\frac{1}{3}} + 3\right)^4 = 0$, so the minimum point is $\left(-\dfrac{1}{3}, 0\right)$.

[2 marks available — 1 mark for a correct method to solve the derivative equal to zero, 1 mark for the correct coordinates]

You can divide by $-\dfrac{4}{x^2}$ to simplify because it can never equal 0.

Q12 $y = (x^4 + x^3 + x^2)^2$, so let $y = u^2$ where $u = x^4 + x^3 + x^2$

$\Rightarrow \dfrac{dy}{du} = 2u = 2(x^4 + x^3 + x^2), \dfrac{du}{dx} = 4x^3 + 3x^2 + 2x$

$\dfrac{dy}{dx} = \dfrac{dy}{du} \times \dfrac{du}{dx} = 2(x^4 + x^3 + x^2) \times (4x^3 + 3x^2 + 2x)$

At $x = -1$, $\dfrac{dy}{dx} = 2(1 - 1 + 1) \times (-4 + 3 + -2) = -6$

$y = ((-1)^4 + (-1)^3 + (-1)^2)^2 = 1$

$y = mx + c \Rightarrow 1 = (-6 \times -1) + c \Rightarrow c = -5$

So the equation of the tangent is $y = -6x - 5$.

Q13 $y = (2x - 3)^7$, so let $y = u^7$ where $u = 2x - 3$

$\Rightarrow \dfrac{dy}{du} = 7u^6 = 7(2x - 3)^6, \dfrac{du}{dx} = 2$

$\dfrac{dy}{dx} = \dfrac{dy}{du} \times \dfrac{du}{dx} = 7(2x - 3)^6 \times 2 = 14(2x - 3)^6$

Use the chain rule again to find $\dfrac{d^2y}{dx^2}$:

$\dfrac{dy}{dx} = 14(2x - 3)^6$, so let $\dfrac{dy}{dx} = 14u^6$ where $u = 2x - 3$

$\Rightarrow \dfrac{d}{du}\left(\dfrac{dy}{dx}\right) = 84u^5 = 84(2x - 3)^5, \dfrac{du}{dx} = 2$

$\dfrac{d^2y}{dx^2} = \dfrac{d}{dx}\left(\dfrac{dy}{dx}\right) = \dfrac{d}{du}\left(\dfrac{dy}{dx}\right) \times \dfrac{du}{dx} = 84(2x - 3)^5 \times 2$

$= 168(2x - 3)^5$

At a point of inflection, $\dfrac{d^2y}{dx^2} = 0 \Rightarrow 2x - 3 = 0 \Rightarrow x = \dfrac{3}{2}$

When $x < \dfrac{3}{2}$, $(2x - 3) < 0 \Rightarrow \dfrac{d^2y}{dx^2} < 0$

When $x > \dfrac{3}{2}$, $(2x - 3) > 0 \Rightarrow \dfrac{d^2y}{dx^2} > 0$

So the sign of $\dfrac{d^2y}{dx^2}$ changes at $x = \dfrac{3}{2}$, so it's a point of inflection.

$x = \dfrac{3}{2} \Rightarrow y = (2 \times \dfrac{3}{2} - 3)^7 = 0$

So the coordinates of the point of inflection are $\left(\dfrac{3}{2}, 0\right)$.

Q14 $y = (\dfrac{x}{4} - 2)^3$, so let $y = u^3$ where $u = \dfrac{x}{4} - 2$

$\Rightarrow \dfrac{dy}{du} = 3u^2 = 3(\dfrac{x}{4} - 2)^2, \dfrac{du}{dx} = \dfrac{1}{4}$

$\dfrac{dy}{dx} = \dfrac{dy}{du} \times \dfrac{du}{dx} = 3(\dfrac{x}{4} - 2)^2 \times \dfrac{1}{4} = \dfrac{3}{4}(\dfrac{x}{4} - 2)^2$

Use the chain rule again to find $\dfrac{d^2y}{dx^2}$:

$\dfrac{dy}{dx} = \dfrac{3}{4}(\dfrac{x}{4} - 2)^2$, so let $\dfrac{dy}{dx} = \dfrac{3}{4}u^2$ where $u = \dfrac{x}{4} - 2$

$\Rightarrow \dfrac{d}{du}\left(\dfrac{dy}{dx}\right) = \dfrac{3}{2}u = \dfrac{3}{2}(\dfrac{x}{4} - 2) = \dfrac{3}{8}x - 3, \dfrac{du}{dx} = \dfrac{1}{4}$

$\dfrac{d^2y}{dx^2} = \dfrac{d}{dx}\left(\dfrac{dy}{dx}\right) = \dfrac{d}{du}\left(\dfrac{dy}{dx}\right) \times \dfrac{du}{dx} = \left(\dfrac{3}{8}x - 3\right) \times \dfrac{1}{4} = \dfrac{3}{32}x - \dfrac{3}{4}$

The curve is convex when $\dfrac{d^2y}{dx^2} > 0 \Rightarrow \dfrac{3}{32}x - \dfrac{3}{4} > 0$

$\Rightarrow \dfrac{3}{32}x > \dfrac{3}{4} \Rightarrow x > 8$

The curve is concave when $\dfrac{d^2y}{dx^2} < 0 \Rightarrow \dfrac{3}{32}x - \dfrac{3}{4} < 0$

$\Rightarrow \dfrac{3}{32}x < \dfrac{3}{4} \Rightarrow x < 8$

So the curve is convex for $x > 8$ and concave for $x < 8$.

Q15 a) $y = (ax^2 + bx + c)^{2020}$, so let $y = u^{2020}$ where $u = ax^2 + bx + c$

$\dfrac{dy}{du} = 2020u^{2019} = 2020(ax^2 + bx + c)^{2019}, \dfrac{du}{dx} = 2ax + b$

$\dfrac{dy}{dx} = \dfrac{dy}{du} \times \dfrac{du}{dx} = 2020(2ax + b)(ax^2 + bx + c)^{2019}$

$= (4040ax + 2020b)(ax^2 + bx + c)^{2019}$

By comparing coefficients,

$4040a = 404$ and $2020b = -1010$.

So $a = \dfrac{1}{10}$ and $b = -\dfrac{1}{2}$.

[3 marks available — 1 mark for both $\dfrac{dy}{du}$ and $\dfrac{du}{dx}$, 1 mark for $\dfrac{dy}{dx}$, 1 mark for both correct answers]

b) When $x = 0$,

$\dfrac{dy}{dx} = (404(0) - 1010)\left(\dfrac{1}{10}(0)^2 + \left(-\dfrac{1}{2}\right)(0) + c\right)^{2019}$

$= -1010c^{2019}$. The equation of the tangent to the curve at $x = 0$ is $y = 1 - 1010x$, so the gradient is -1010.

So $-1010c^{2019} = -1010 \Rightarrow c^{2019} = 1 \Rightarrow c = 1$.

[2 marks available — 1 mark for finding the gradient at $x = 0$ in terms of c, 1 mark for the correct answer]

Q16 a) Differentiate both sides of $f(x) = f(-x)$ with respect to x:

$f'(x) = -f'(-x)$, so f' satisfies the definition of an odd function.

Answers

b) Differentiate both sides of $f(x) = -f(-x)$ with respect to x:
$f'(x) = -\,-f'(-x) = f'(-x)$, so f' satisfies the definition of an even function.

Exercise 7.5 — Finding $\dfrac{dy}{dx}$ when $x = f(y)$

Q1 a) $\dfrac{dx}{dy} = 6y + 5 \Rightarrow \dfrac{dy}{dx} = \dfrac{1}{6y + 5}$

At $(5, -1)$, $y = -1$ so $\dfrac{dy}{dx} = \dfrac{1}{-1} = -1$

b) $\dfrac{dx}{dy} = 3y^2 - 2 \Rightarrow \dfrac{dy}{dx} = \dfrac{1}{3y^2 - 2}$

At $(-4, -2)$, $y = -2$ so $\dfrac{dy}{dx} = \dfrac{1}{10} = 0.1$

c) $x = (2y + 1)(y - 2) = 2y^2 - 3y - 2$

$\dfrac{dx}{dy} = 4y - 3 \Rightarrow \dfrac{dy}{dx} = \dfrac{1}{4y - 3}$

At $(3, -1)$ $y = -1$ so $\dfrac{dy}{dx} = -\dfrac{1}{7}$.

d) $x = \dfrac{4 + y^2}{y} = 4y^{-1} + y$

$\dfrac{dx}{dy} = -4y^{-2} + 1 \Rightarrow \dfrac{dy}{dx} = \dfrac{1}{1 - \dfrac{4}{y^2}} = \dfrac{y^2}{y^2 - 4}$

At $(5, 4)$, $y = 4$ so $\dfrac{dy}{dx} = \dfrac{4}{3}$.

Q2 $x = (2y^3 - 5)^3$, so let $x = u^3$ where $u = 2y^3 - 5$

$\Rightarrow \dfrac{dx}{du} = 3u^2 = 3(2y^3 - 5)^2$, $\dfrac{du}{dy} = 6y^2$

$\dfrac{dx}{dy} = \dfrac{dx}{du} \times \dfrac{du}{dy} = 3(2y^3 - 5)^2 \times 6y^2 = 18y^2(2y^3 - 5)^2$

$\dfrac{dy}{dx} = \dfrac{1}{18y^2(2y^3 - 5)^2}$

Q3 a) $x = \sqrt{4 + y} \Rightarrow x = u^{\frac{1}{2}}$, $u = 4 + y$

$\Rightarrow \dfrac{dx}{du} = \dfrac{1}{2}u^{-\frac{1}{2}} = \dfrac{1}{2}\dfrac{1}{\sqrt{4 + y}}$, $\dfrac{du}{dy} = 1$

$\dfrac{dx}{dy} = \dfrac{dx}{du} \times \dfrac{du}{dy} = \dfrac{1}{2}\dfrac{1}{\sqrt{4 + y}} \times 1 = \dfrac{1}{2\sqrt{4 + y}} = \dfrac{1}{2x}$

$\Rightarrow \dfrac{dy}{dx} = 2x$

Use $x = \sqrt{4 + y}$ from the question to get dx/dy in terms of x.

b) $x = \sqrt{4 + y} \Rightarrow x^2 = 4 + y \Rightarrow y = x^2 - 4 \Rightarrow \dfrac{dy}{dx} = 2x$

Q4 a) $x = \dfrac{1}{\sqrt{2y - 3}} = (2y - 3)^{-\frac{1}{2}} \Rightarrow x = u^{-\frac{1}{2}}$, $u = 2y - 3$

$\Rightarrow \dfrac{dx}{du} = -\dfrac{1}{2}u^{-\frac{3}{2}} = -\dfrac{1}{2(\sqrt{2y - 3})^3}$, $\dfrac{du}{dy} = 2$

$\dfrac{dx}{dy} = \dfrac{dx}{du} \times \dfrac{du}{dy} = -\dfrac{1}{2(\sqrt{2y - 3})^3} \times 2 = -\dfrac{1}{(\sqrt{2y - 3})^3} = -x^3$

$\Rightarrow \dfrac{dy}{dx} = -\dfrac{1}{x^3}$

b) $x = \dfrac{1}{\sqrt{2y - 3}} \Rightarrow x^2 = \dfrac{1}{2y - 3} \Rightarrow 2y - 3 = x^{-2}$

$\Rightarrow y = \dfrac{1}{2}x^{-2} + \dfrac{3}{2} \Rightarrow \dfrac{dy}{dx} = -2 \times \dfrac{1}{2}x^{-3} = -\dfrac{1}{x^3}$

Q5 a) $x = 3y^2 - 2y + 1 \Rightarrow \dfrac{dx}{dy} = 6y - 2$ so $\dfrac{dy}{dx} = \dfrac{1}{6y - 2}$

[2 marks available — 1 mark for correct $\dfrac{dx}{dy}$, 1 mark for the correct answer]

b) Tangent is vertical when $\dfrac{dx}{dy} = 0$,

$6y - 2 = 0 \Rightarrow y = \dfrac{1}{3}$ and $x = 3\left(\dfrac{1}{3}\right)^2 - 2\left(\dfrac{1}{3}\right) + 1 = \dfrac{2}{3}$,

so the point has coordinates $\left(\dfrac{2}{3}, \dfrac{1}{3}\right)$.

[2 marks available — 1 mark for the correct y-value, 1 mark for the correct x-value]

c) At a stationary point $\dfrac{dy}{dx} = 0$. $\dfrac{1}{6y - 2} = 0$ has no solutions since you can never divide 1 by anything to give zero, so there are no stationary points.

[2 marks available — 1 mark for using the fact that $\dfrac{dy}{dx} = 0$ at stationary points, 1 mark for a correct reason why there are no solutions]

Q6 a) $x = 2y^{\frac{1}{2}} - 2y^{-\frac{1}{2}}$ so $\dfrac{dx}{dy} = y^{-\frac{1}{2}} + y^{-\frac{3}{2}}$, so $\dfrac{dy}{dx} = \dfrac{1}{y^{-\frac{1}{2}} + y^{-\frac{3}{2}}}$.

When $y = 4$, $x = 2\sqrt{4} - \dfrac{2}{\sqrt{4}} = 4 - 1 = 3$ and

$\dfrac{dy}{dx} = \dfrac{1}{\left(\dfrac{1}{\sqrt{4}} + \dfrac{1}{\sqrt{4^3}}\right)} = \dfrac{1}{\left(\dfrac{1}{2} + \dfrac{1}{8}\right)} = \dfrac{1}{\left(\dfrac{5}{8}\right)} = \dfrac{8}{5}$.

$y = mx + c \Rightarrow 4 = \left(\dfrac{8}{5} \times 3\right) + c \Rightarrow c = -\dfrac{4}{5}$

So the tangent has equation $y = \dfrac{8}{5}x - \dfrac{4}{5}$.

[4 marks available — 1 mark for correct $\dfrac{dx}{dy}$, 1 mark for a correct method to find $\dfrac{dy}{dx}$ at $y = 4$, 1 mark for correct gradient, 1 mark for the correct answer]

b) When $y = 1$, $\dfrac{dy}{dx} = \dfrac{1}{1 + 1} = \dfrac{1}{2}$

$y = mx + c \Rightarrow 1 = \left(\dfrac{1}{2} \times 0\right) + c \Rightarrow c = 1$

So the tangent has equation $y = \dfrac{1}{2}x + 1$.

The tangents meet when $\dfrac{1}{2}x + 1 = \dfrac{8}{5}x - \dfrac{4}{5}$

$\Rightarrow \dfrac{11}{10}x = \dfrac{9}{5} \Rightarrow x = \dfrac{18}{11}$ and $y = \dfrac{1}{2}\left(\dfrac{18}{11}\right) + 1 = \dfrac{20}{11}$

so P has coordinates $\left(\dfrac{18}{11}, \dfrac{20}{11}\right)$.

[4 marks available — 1 mark for the correct equation of the tangent at (0, 1), 1 mark for a correct method to solve simultaneously to find the x-coordinate, 1 mark for the correct x-value, 1 mark for the correct y-value]

Q7 a) $x = (3y^3 - y - 1)^5 \Rightarrow x = u^5$ where $u = 3y^3 - y - 1$

$\dfrac{dx}{du} = 5u^4 = 5(3y^3 - y - 1)^4$ and $\dfrac{du}{dy} = 9y^2 - 1$

$\dfrac{dx}{dy} = \dfrac{dx}{du} \times \dfrac{du}{dy} = 5(9y^2 - 1)(3y^3 - y - 1)^4$

so $\dfrac{dy}{dx} = \dfrac{1}{5(9y^2 - 1)(3y^3 - y - 1)^4}$

[3 marks available — 1 mark for both $\dfrac{dx}{du}$ and $\dfrac{du}{dy}$, 1 mark for correct $\dfrac{dx}{dy}$, 1 mark for the correct answer]

b) When $y = 1$, $\dfrac{dy}{dx} = \dfrac{1}{5 \times 8 \times 1^4} = \dfrac{1}{40}$, so the gradient of the normal is -40.

$y = mx + c \Rightarrow 1 = (-40 \times 1) + c \Rightarrow c = 41$

So the equation of the normal is $y = 41 - 40x$.

[3 marks available — 1 mark for a correct method to find the gradient of the normal, 1 mark for correct gradient of the normal at (1, 1), 1 mark for the correct answer]

c) The normal to the curve is horizontal when $x = 0$, so the tangent is vertical, meaning the tangent is $x = 0$.

So the lines intersect when $y = 41 - 40(0) = 41$.

So they intersect at $(0, 41)$.

[2 marks available — 1 mark for a correct method, 1 mark for the correct answer]

Exercise 7.6 — Differentiating e^x and $e^{f(x)}$

Q1 a) $y = e^{f(x)}$, where $f(x) = 3x$, so $f'(x) = 3$

$\dfrac{dy}{dx} = f'(x)e^{f(x)} = 3 \times e^{3x} = 3e^{3x}$

b) $\dfrac{dy}{dx} = f'(x)e^{f(x)} = 2 \times e^{2x - 5} = 2e^{2x - 5}$

c) $\dfrac{dy}{dx} = f'(x)e^{f(x)} = 1 \times e^{x+7} = e^{x+7}$

d) $\dfrac{dy}{dx} = f'(x)e^{f(x)} = 3 \times e^{3x+9} = 3e^{3x+9}$

e) $\dfrac{dy}{dx} = f'(x)e^{f(x)} = (-2) \times e^{7-2x} = -2e^{7-2x}$

f) $\dfrac{dy}{dx} = f'(x)e^{f(x)} = 3x^2 \times e^{x^3} = 3x^2 e^{x^3}$

Q2 a) $f'(x) = g'(x)e^{g(x)} = (3x^2 + 3)e^{x^3+3x}$

b) $f'(x) = g'(x)e^{g(x)} = (3x^2 - 3)e^{x^3-3x-5}$

c) $f(x) = e^{2x^2+x}$, so:
$f'(x) = g'(x)e^{g(x)} = (4x + 1) \times e^{2x^2+x} = (4x+1)e^{x(2x+1)}$

Q3 a) e^x differentiates to e^x and e^{-x} differentiates to $-e^{-x}$
so $f'(x) = \dfrac{1}{2} \times (e^x - (-e^{-x})) = \dfrac{1}{2}(e^x + e^{-x})$

b) $f(x) = e^{x^2+7x+12}$
$f'(x) = g'(x)e^{g(x)} = (2x+7)e^{x^2+7x+12}$

c) $\dfrac{d}{dx}(e^{x^4+3x^2}) = (4x^3 + 6x)e^{x^4+3x^2}$ and $\dfrac{d}{dx}(2e^{2x}) = 4e^{2x}$
So $f'(x) = (4x^3 + 6x)e^{x^4+3x^2} + 4e^{2x}$

Q4 $\dfrac{dy}{dx} = f'(x)e^{f(x)} = 2 \times e^{2x} = 2e^{2x}$
At $x = 0$, $\dfrac{dy}{dx} = 2 \times e^{2 \times 0} = 2 \times 1 = 2$
$y = mx + c \Rightarrow 1 = (2 \times 0) + c \Rightarrow c = 1$
So the equation of the tangent is $y = 2x + 1$.
There's no real need to use the $y = mx + c$ formula here as we already know the place it crosses the y-axis is $(0, 1)$ (it's given in the question), but it's good to be safe.

Q5 $\dfrac{dy}{dx} = x - f'(x)e^{f(x)} = x - 2e^{2x-6}$
$\dfrac{d^2y}{dx^2} = 1 - 2 \times f'(x)e^{f(x)} = 1 - 4e^{2x-6}$
At the point of inflection, $\dfrac{d^2y}{dx^2} = 0$
$\Rightarrow 1 - 4e^{2x-6} = 0 \Rightarrow 1 = 4e^{2x-6} \Rightarrow \ln(1) = \ln(4e^{2x-6})$
$\Rightarrow 0 = \ln 4 + \ln(e^{2x-6}) = \ln 4 + 2x - 6 \Rightarrow 2x = 6 - \ln 4$
$\Rightarrow x = 3 - \dfrac{1}{2}\ln 4 = 3 - \ln 2$
[4 marks available — 1 mark for the correct first derivative, 1 mark for the correct second derivative, 1 mark for setting the second derivative equal to zero, 1 mark for the correct answer]
Remember from laws of logs that $\frac{1}{2} \ln 4 = \ln 4^{\frac{1}{2}} = \ln 2$. There are other ways you could have used the log laws in this question — you should get the same answer.

Q6 $\dfrac{dy}{dx} = f'(x)e^{f(x)} = 4xe^{2x^2}$
At $x = 1$, $\dfrac{dy}{dx} = 4 \times 1 \times e^2 = 4e^2$ and $y = e^2$
$y = mx + c \Rightarrow e^2 = (4e^2 \times 1) + c \Rightarrow c = -3e^2$
So the equation of the tangent is $y = 4e^2x - 3e^2$.
[4 marks available — 1 mark for the correct first derivative, 1 mark for correctly finding the gradient, 1 mark for the correct value of y at $x = 1$, 1 mark for the correct equation]

Q7 $f(x) = \sqrt{e^x + e^{2x}}$ so let $y = \sqrt{u} = u^{\frac{1}{2}}$ where $u = e^x + e^{2x}$
$\Rightarrow \dfrac{dy}{du} = \dfrac{1}{2}u^{-\frac{1}{2}} = \dfrac{1}{2\sqrt{u}} = \dfrac{1}{2\sqrt{e^x + e^{2x}}}$,
$\dfrac{du}{dx} = e^x + e^{2x} = e^x + 2e^{2x}$
$\Rightarrow \dfrac{dy}{dx} = \dfrac{dy}{du} \times \dfrac{du}{dx} = \dfrac{1}{2\sqrt{e^x + e^{2x}}} \times (e^x + 2e^{2x}) = \dfrac{e^x + 2e^{2x}}{2\sqrt{e^x + e^{2x}}}$
At $x = 0$, $\dfrac{dy}{dx} = \dfrac{e^0 + 2e^{2(0)}}{2\sqrt{e^0 + e^{2(0)}}} = \dfrac{3}{2\sqrt{2}} = \dfrac{3\sqrt{2}}{4}$
[4 marks available — 1 mark for using the chain rule, 1 mark for $\dfrac{dy}{dx}$, 1 mark for substituting $x = 0$ into the derivative, 1 mark for the correct answer]

Q8 a) $\dfrac{d}{dx}e^x = e^x$ and $\dfrac{d}{dx}e^{-x} = -e^{-x}$
So $f'(x) = \dfrac{e^x - e^{-x}}{2}$ and $f''(x) = \dfrac{e^x + e^{-x}}{2}$.
[3 marks available — 1 mark for correct differentiation of e^{-x}, 1 mark for the correct first derivative, 1 mark for the correct second derivative]

b) $f'(x) = 0 \Rightarrow e^x = e^{-x} \Rightarrow e^{2x} = 1 \Rightarrow 2x = 0 \Rightarrow x = 0$ and
$f(0) = \dfrac{e^0 + e^{-0}}{2} = \dfrac{1+1}{2} = 1$, so the stationary point is $(0, 1)$.
[3 marks available — 1 mark for attempting to solve $f'(x) = 0$, 1 mark for the correct x-value, 1 mark for the correct y-value]

c) At a minimum point $f''(x) > 0$.
When $x = 0$, $f''(0) = \dfrac{e^0 + e^{-0}}{2} = \dfrac{1+1}{2} = 1 > 0$,
so it is a minimum.
[1 mark for correct working leading to the correct conclusion]

d) $f(x) \geq 1$ *[1 mark]*

Q9 $\dfrac{dy}{dx} = f'(x)e^{f(x)} - 1 = 2e^{2x-4} - 1$,
$\dfrac{d^2y}{dx^2} = 2 \times f'(x)e^{f(x)} = 4e^{2x-4}$
$4e^{2x-4} > 0$ for all values of x, so $\dfrac{d^2y}{dx^2}$ is always positive,
which means the graph of $y = e^{2x-4} - x$ is always convex.

Q10 $\dfrac{dy}{dx} = f'(x)e^{f(x)} = (3 \times e^{3x}) = 3e^{3x}$
When it crosses the y-axis, $x = 0$, so $y = e^{3 \times 0} + 3 = 4$
$\dfrac{dy}{dx} = (3 \times e^{3 \times 0}) = (3 \times 1) = 3 \Rightarrow$ Gradient of the normal $= -\dfrac{1}{3}$
$y = mx + c \Rightarrow 4 = (-\dfrac{1}{3} \times 0) + c \Rightarrow c = 4$
So the equation of the normal is $y = -\dfrac{1}{3}x + 4$.

Q11 Let $f(x) = 2x$ and $g(x) = 3 - 4x$, then
$\dfrac{dy}{dx} = 2 \times f'(x)e^{f(x)} - \dfrac{1}{2} \times g'(x)e^{g(x)} = 2 \times 2e^{2x} - \dfrac{1}{2} \times -4e^{3-4x}$
$= 4e^{2x} + 2e^{3-4x}$
$\dfrac{d^2y}{dx^2} = 4 \times f'(x)e^{f(x)} + 2 \times g'(x)e^{g(x)} = 4 \times 2e^{2x} + 2 \times -4e^{3-4x}$
$= 8e^{2x} - 8e^{3-4x}$
At a point of inflection, $\dfrac{d^2y}{dx^2} = 0$
$\Rightarrow 8e^{2x} - 8e^{3-4x} = 0 \Rightarrow e^{2x} = e^{3-4x}$
$\Rightarrow 2x = 3 - 4x \Rightarrow 6x = 3 \Rightarrow x = \dfrac{1}{2}$
$\dfrac{d^2y}{dx^2} > 0$ if $x > \dfrac{1}{2}$ and $\dfrac{d^2y}{dx^2} < 0$ if $x < \dfrac{1}{2}$
$\dfrac{d^2y}{dx^2}$ changes sign from negative to positive at $x = \dfrac{1}{2}$.
So $x = \dfrac{1}{2}$ is a point of inflection.
At $x = \dfrac{1}{2}$, $y = 2e^{2x} - \dfrac{1}{2}e^{3-4x} = 2e^1 - \dfrac{1}{2}e^1 = \dfrac{3}{2}e$.
The graph has one point of inflection, at $\left(\dfrac{1}{2}, \dfrac{3}{2}e\right)$.

Q12 a) $f'(x) = g'(x)e^{g(x)} = -2x \times e^{-x^2} = -2xe^{-x^2}$
Since $e^{-x^2} > 0$ for all real values of x, $f'(x) < 0$ when $x > 0$,
so $f(x)$ is decreasing for $x > 0$.
[3 marks available — 1 mark for a correct method to differentiate, 1 mark for the correct derivative, 1 mark for the correct answer]

b) $f'(x) = 0 \Rightarrow -2xe^{-x^2} = 0 \Rightarrow x = 0$ and $y = e^{-0^2} = 1$
so the maximum point is $(0, 1)$. *[1 mark]*

Answers

c) For all real values of x, $-\infty < -x^2 \le 0$
so the range of f is $0 < e^{-x^2} \le 1$.
[2 marks available — 1 mark for the correct range for $-x^2$, 1 mark for the correct range for f]

d) $y = \sqrt{3 - e^{-x^2}} \Rightarrow y = u^{\frac{1}{2}}$ where $u = 3 - e^{-x^2}$
$\frac{dy}{du} = \frac{1}{2}u^{-\frac{1}{2}} = \frac{1}{2\sqrt{3 - e^{-x^2}}}$ and $\frac{du}{dx} = 2xe^{-x^2}$
$\frac{dy}{dx} = \frac{dy}{du} \times \frac{du}{dx} = \frac{1}{2\sqrt{3 - e^{-x^2}}} \times 2xe^{-x^2} = \frac{xe^{-x^2}}{\sqrt{3 - e^{-x^2}}}$
$\frac{dy}{dx} = 0 \Rightarrow \frac{xe^{-x^2}}{\sqrt{3 - e^{-x^2}}} = 0$
$\Rightarrow x = 0$ and $y = \sqrt{3 - e^{-0^2}} = \sqrt{2}$
so the stationary point has coordinates $(0, \sqrt{2})$.
[4 marks available — 1 mark for using the chain rule, 1 mark for $\frac{dy}{dx}$, 1 mark for setting the derivative equal to 0, 1 mark for the correct answer]

Q13 If y has a stationary point, the gradient $\frac{dy}{dx}$ will be 0.
$\frac{dy}{dx} = f'(x)e^{f(x)} = (3x^2 - 3)e^{x^3 - 3x - 5}$
So if $\frac{dy}{dx} = 0$, either $3x^2 - 3 = 0$ or $e^{x^3 - 3x - 5} = 0$.
If $3x^2 - 3 = 0 \Rightarrow 3(x^2 - 1) = 0 \Rightarrow x^2 = 1 \Rightarrow x = \pm 1$
and if $e^{x^3 - 3x - 5} = 0$, there are no solutions.
So the gradient is 0 when $x = \pm 1$
$\Rightarrow$ the curve has stationary points at $x = \pm 1$.

Q14 Stationary points occur when the gradient is 0.
$\frac{dy}{dx} = f'(x)e^{f(x)} - 6 = 3e^{3x} - 6$
$3e^{3x} - 6 = 0 \Rightarrow e^{3x} = 2 \Rightarrow 3x = \ln 2 \Rightarrow x = \frac{1}{3}\ln 2$
Take ln of both sides to get rid of the exponential.
To find the nature of the stationary point, calculate $\frac{d^2 y}{dx^2}$
$\frac{d^2 y}{dx^2} = 3f'(x)e^{f(x)} = 9e^{3x}$
When $x = \frac{1}{3}\ln 2$, $\frac{d^2 y}{dx^2} = 9e^{\ln 2} = 9 \times 2 = 18$
$\frac{d^2 y}{dx^2}$ is positive, so it's a minimum point.
You could also use the fact that $9e^{3x}$ is always positive to show that it's a minimum without needing to substitute.

Exercise 7.7 — Differentiating ln x and ln (f(x))

Q1 a) $y = \ln (3x) = \ln 3 + \ln x \Rightarrow \frac{dy}{dx} = \frac{1}{x}$
You could use that the derivative of ln (f(x)) is $\frac{f'(x)}{f(x)}$ — it'd give the same answer.

b) $\frac{dy}{dx} = \frac{f'(x)}{f(x)} = \frac{1}{1 + x}$
The coefficient of x is 1 so it's a 1 on top of the fraction.

c) $\frac{dy}{dx} = \frac{f'(x)}{f(x)} = \frac{2}{3 + 2x}$

d) $\frac{dy}{dx} = \frac{f'(x)}{f(x)} = \frac{5}{1 + 5x}$

e) $\frac{dy}{dx} = 4 \times \frac{f'(x)}{f(x)} = 4 \times \frac{4}{4x - 2} = \frac{16}{4x - 2} = \frac{8}{2x - 1}$
Don't forget to simplify your answers if you can.

f) $\frac{dy}{dx} = 9 \times \frac{f'(x)}{f(x)} = 9 \times \frac{3}{3x - 3} = \frac{9}{x - 1}$

Q2 a) $\frac{dy}{dx} = \frac{f'(x)}{f(x)} = \frac{2x}{1 + x^2}$

b) $y = \ln (2 + x)^2 = 2 \ln (2 + x)$
$\frac{dy}{dx} = 2 \times \frac{f'(x)}{f(x)} = 2 \times \frac{1}{2 + x} = \frac{2}{2 + x}$

c) $y = \ln (2x - 8)^3 = 3 \ln (2x - 8)$
$\frac{dy}{dx} = 3 \times \frac{f'(x)}{f(x)} = 3 \times \frac{2}{2x - 8} = \frac{3}{x - 4}$

d) $3 \ln x^3 = 9 \ln x \Rightarrow \frac{dy}{dx} = \frac{9}{x}$

e) $\frac{dy}{dx} = \frac{f'(x)}{f(x)} = \frac{3x^2 + 2x}{x^3 + x^2} = \frac{3x + 2}{x^2 + x}$

f) $y = \ln \sqrt{2x^2 - 4} = \ln (2x^2 - 4)^{\frac{1}{2}} = \frac{1}{2}\ln (2x^2 - 4)$
$\frac{dy}{dx} = \frac{1}{2} \times \frac{f'(x)}{f(x)} = \frac{1}{2} \times \frac{4x}{2x^2 - 4} = \frac{4x}{4x^2 - 8} = \frac{x}{x^2 - 2}$

Q3 a) $f(x) = \ln \frac{1}{x} = \ln x^{-1} = -\ln x \Rightarrow f'(x) = -\frac{1}{x}$

b) $f(x) = \ln \sqrt{x} = \ln x^{\frac{1}{2}} = \frac{1}{2}\ln x \Rightarrow f'(x) = \frac{1}{2x}$

Q4 a) $f'(x) = \frac{1}{x + 2} + 2x = (x + 2)^{-1} + 2x$
so $f''(x) = -(x + 2)^{-2} + 2 = -\frac{1}{(x + 2)^2} + 2$
[2 marks available — 1 mark for the correct first derivative, 1 mark for the correct second derivative]

b) $f''(x) > 0 \Rightarrow 2 - \frac{1}{(x + 2)^2} > 0$
$\Rightarrow 2(x + 2)^2 - 1 > 0 \Rightarrow 2x^2 + 8x + 7 > 0$.
[2 marks available — 1 mark for setting up a correct inequality, 1 mark for showing the correct result]

c) $2x^2 + 8x + 7 = 0 \Rightarrow x = \frac{-8 \pm \sqrt{8^2 - 4 \times 2 \times 7}}{2 \times 2} = \frac{-4 \pm \sqrt{2}}{2}$
so f is convex when $x < \frac{-4 - \sqrt{2}}{2}$ or when $x > \frac{-4 + \sqrt{2}}{2}$.
[2 marks available — 1 mark for the correct roots, 1 mark for the correct answer]

Q5 $\ln ((2x + 1)^2 \sqrt{x - 4}) = \ln (2x + 1)^2 + \ln \sqrt{x - 4}$
$= 2 \ln (2x + 1) + \frac{1}{2} \ln (x - 4)$
First part: $f'(x) = 2 \times \frac{2}{2x + 1} = \frac{4}{2x + 1}$
Second part: $f'(x) = \frac{1}{2} \times \frac{1}{x - 4} = \frac{1}{2(x - 4)}$
Putting it all together:
$f'(x) = \frac{4}{2x + 1} + \frac{1}{2(x - 4)} = \frac{8(x - 4) + 2x + 1}{2(2x + 1)(x - 4)}$
$= \frac{10x - 31}{2(2x + 1)(x - 4)}$

Q6 $g(x) = x - \sqrt{x - 4} \Rightarrow g'(x) = 1 - \left(\frac{dy}{du} \times \frac{du}{dx}\right) = 1 - \frac{1}{2\sqrt{x - 4}}$
$f'(x) = \frac{g'(x)}{g(x)} = \frac{1 - \frac{1}{2\sqrt{x - 4}}}{x - \sqrt{x - 4}} = \frac{2\sqrt{x - 4} - 1}{2(x\sqrt{x - 4} - x + 4)}$

Q7 $\ln \left(\frac{(3x + 1)^2}{\sqrt{2x + 1}}\right) = \ln (3x + 1)^2 - \ln \sqrt{2x + 1}$
$= 2\ln (3x + 1) - \frac{1}{2} \ln (2x + 1)$
First part: $f'(x) = 2 \times \frac{3}{3x + 1} = \frac{6}{3x + 1}$
Second part: $f'(x) = \frac{1}{2} \times \frac{2}{2x + 1} = \frac{1}{2x + 1}$
Putting it all together:
$f'(x) = \frac{6}{3x + 1} - \frac{1}{2x + 1}$
$= \frac{6(2x + 1) - 3x - 1}{(3x + 1)(2x + 1)} = \frac{9x + 5}{(3x + 1)(2x + 1)}$
You could have left your answer as 2 fractions.

Q8 a) $y = \ln (3x)^2 = 2 \ln (3x) = 2 \ln 3 + 2 \ln x \Rightarrow \frac{dy}{dx} = \frac{2}{x}$
When $x = -2$, $\frac{dy}{dx} = -1$ and $y = \ln 36$
$y = mx + c \Rightarrow \ln 36 = [(-1) \times (-2)] + c \Rightarrow c = \ln 36 - 2$
So the equation of the tangent is $y = -x + \ln 36 - 2$

b) When $x = 2$, $\frac{dy}{dx} = 1$ and $y = \ln 36$

$y = mx + c \implies \ln 36 = [1 \times 2] + c \implies c = \ln 36 - 2$
So the equation of the tangent is $y = x + \ln 36 - 2$

Q9 a) $y = \ln (x + 6)^2 = 2 \ln (x + 6)$

$\implies \frac{dy}{dx} = 2 \frac{f'(x)}{f(x)} = \frac{2}{x + 6}$

When $x = -3$, $\frac{dy}{dx} = \frac{2}{3}$ and $y = \ln 9$

So the gradient of the normal $= -\frac{1}{\left(\frac{2}{3}\right)} = -\frac{3}{2}$

$y = mx + c \implies \ln 9 = (-\frac{3}{2} \times -3) + c \implies c = \ln 9 - \frac{9}{2}$

so the equation of the normal is $y = -\frac{3}{2}x + \ln 9 - \frac{9}{2}$.

b) When $x = 0$, $\frac{dy}{dx} = \frac{2}{6} = \frac{1}{3}$ and $y = \ln 36$

So the gradient of the normal $= -\frac{1}{\left(\frac{1}{3}\right)} = -3$

$y = mx + c \implies \ln 36 = (-3 \times 0) + c \implies c = \ln 36$
So the equation of the normal is $y = -3x + \ln 36$

Q10 Stationary points occur when the gradient is 0.

$\frac{dy}{dx} = \frac{f'(x)}{f(x)} = \frac{3x^2 - 6x + 3}{x^3 - 3x^2 + 3x}$

so $3x^2 - 6x + 3 = 0 \implies 3(x - 1)(x - 1) = 0$

You can ignore the denominator here, as it's only the top part that affects when it's equal to 0.
So the gradient is 0 when $x = 1$.
When $x = 1$, $y = 0$ so the stationary point is at $(1, 0)$.

Q11 a) $\frac{dy}{dx} = \frac{f'(x)}{f(x)} = \frac{4x^3 + 4x}{x^4 + 2x^2} = \frac{4x^2 + 4}{x^3 + 2x}$

[2 marks available — 1 mark for a correct method, 1 mark for the correct answer]

b) When $x = 1$, $\frac{dy}{dx} = \frac{4(1^2) + 4}{1^3 + 2(1)} = \frac{8}{3}$;

so the gradient of the normal is $-\frac{3}{8}$.

The y-coordinate is $\ln (1^4 + 2(1^2)) = \ln 3$.

$y = mx + c \implies \ln 3 = \left(-\frac{3}{8} \times 1\right) + c \implies c = \frac{3}{8} + \ln 3$

So the equation of L is $y = -\frac{3}{8}x + \frac{3}{8} + \ln 3$.

[3 marks available — 1 mark for the correct gradient of the normal, 1 mark for a correct method to find equation of normal, 1 mark for the correct answer]

c) When $y = 0$, $-\frac{3}{8}x + \frac{3}{8} + \ln 3 = 0$

so $x = \frac{\frac{3}{8} + \ln 3}{\frac{3}{8}} = 1 + \frac{8}{3} \ln 3$,

so the coordinates are $(1 + \frac{8}{3} \ln 3, 0)$.

[2 marks available — 1 mark for a correct method, 1 mark for the correct answer]

Exercise 7.8 — Differentiating a^x and $a^{f(x)}$

Q1 a) $\frac{dy}{dx} = 5^x \ln 5$

b) Let $u = 2x$ and $y = 3^u$, then

$\frac{dy}{du} = 3^u \ln 3 = 3^{2x} \ln 3$ and $\frac{du}{dx} = 2$

So $\frac{dy}{dx} = \frac{dy}{du} \times \frac{du}{dx} = 3^{2x} \ln 3 \times 2 = (2 \ln 3) 3^{2x}$

c) Let $u = -x$ and $y = 10^u$, then

$\frac{dy}{du} = 10^u \ln 10 = 10^{-x} \ln 10$ and $\frac{du}{dx} = -1$

So $\frac{dy}{dx} = 10^{-x} \ln 10 \times (-1) = -(10^{-x} \ln 10)$

d) Let $u = qx$ and $y = p^u$, then

$\frac{dy}{du} = p^u \ln p = p^{qx} \ln p$ and $\frac{du}{dx} = q$

So $\frac{dy}{dx} = p^{qx} \ln p \times q = (q \ln p) p^{qx}$

Q2 a) Let $u = 4x$, then $y = 2^u$ and

$\frac{dy}{dx} = \frac{d}{du}(2^u) \times \frac{d}{dx}(4x) = 4(2^{4x} \ln 2)$

b) When $x = 2$, $\frac{dy}{dx} = 4(2^8 \ln 2) = 1024 \ln 2$

$y = 2^8 = 256$

Putting this into $y = mx + c$ gives:

$256 = 2048 \ln 2 + c \implies c = 256 - 2048 \ln 2$.

So the equation of the tangent is:
$y = (1024 \ln 2) x + 256 - 2048 \ln 2$
or $y = (1024 \ln 2) x + 256(1 - 8 \ln 2)$

Q3 a) When $x = 1$, $2^p = 32$, so $p = 5$.

You should just know this result, but if not you can take logs ($p \ln 2 = \ln 32$, so $p = \ln 32 \div \ln 2 = 5$).

b) $\frac{dy}{dx} = 5(2^{5x} \ln 2)$

When $x = 1$, $\frac{dy}{dx} = 5(32 \ln 2) = 160 \ln 2$

Q4 a) Let $u = x^3$, then $y = p^u$, and so

$\frac{dy}{dx} = \frac{d}{du}(p^u) \times \frac{d}{dx}(x^3) = 3x^2(p^{x^3} \ln p)$

b) When $x = 2$, $y = p^8 = 6561$.

So $p = \sqrt[8]{6561} = 3$.

You can also work this out by taking logs.
$8 \ln p = \ln 6561$, so $p = \exp\left(\frac{\ln 6561}{8}\right) = 3$.

c) When $x = 1$, $y = 3^1 = 3$.
The gradient at $x = 1$ is $3(3 \ln 3) = 9 \ln 3$.
Putting this into $y = mx + c$ gives:
$3 = 9 \ln 3 + c \implies c = 3 - 9 \ln 3$
So the equation of the tangent at $(1, 3)$ is
$y = (9 \ln 3)x + (3 - 9 \ln 3)$

Q5 When $x = 25$, $y = 4^5 = 1024$, so $a = 1024$.

Let $u - x^{\frac{1}{2}}$ and $y = 4^u$, so

$\frac{dy}{dx} = \frac{d}{du}(4^u)\frac{d}{dx}\left(x^{\frac{1}{2}}\right) = \frac{1}{2}x^{-\frac{1}{2}}(4^{\sqrt{x}} \ln 4)$

The gradient of the tangent when $x = 25$ is

$\frac{1}{10}(1024 \ln 4) = 102.4 \ln 4$

Putting this into $y = mx + c$ gives:
$1024 = 2560 \ln 4 + c \implies c = 1024 - 2560 \ln 4$.
So the equation of the tangent is
$y = (102.4 \ln 4)x + (1024 - 2560 \ln 4)$
$\implies y = 142x - 2520$ to 3 s.f.

Q6 a) Let $y = 3^u$, where $u = 2 - x$

$\frac{dy}{du} = 3^u \ln 3 = 3^{2-x} \ln 3$ and $\frac{du}{dx} = -1$

$f'(x) = 3^{2-x} \ln 3 \times (-1) = -3^{2-x} \ln 3$

$3^{2-x} > 0$ for all real values of x

$\implies -3^{2-x} \ln 3 < 0$, so f is decreasing for all real values of x.

[3 marks available — 1 mark for a correct method to differentiate, 1 mark for the correct derivative, 1 mark for a correct explanation of why f is decreasing]

Answers

b) Use the chain rule again to find the second derivative:

Let $y = -\ln(3) \times 3^u$, where $u = 2 - x$

$\frac{dy}{du} = -\ln(3) \times 3^u \ln 3 = -3^{2-x}(\ln 3)^2$ and $\frac{du}{dx} = -1$

$f''(x) = -3^{2-x}(\ln 3)^2 \times (-1) = 3^{2-x}(\ln 3)^2$

$3^{2-x} > 0$ for all real values of $x \Rightarrow 3^{2-x}(\ln 3)^2 > 0$,

so the curve $y = f(x)$ is convex for all real values of x.

[2 marks available — 1 mark for the correct derivative, 1 mark for a correct explanation of why f is convex]

Q7 a) Let $u = -3x$, then $y = 2^u$, so

$\frac{du}{dx} = -3$ and $\frac{dy}{du} = 2^u \ln 2 = 2^{-3x} \ln 2 \Rightarrow \frac{dy}{dx} = -3(2^{-3x} \ln 2)$

[2 marks available — 1 mark for using the chain rule, 1 mark for correct answer]

b) When $x = 2$, $y = b = 2^{-6} = \frac{1}{64}$ and $\frac{dy}{dx} = -\frac{3}{64} \ln 2$

[2 marks available — 1 mark for b, 1 mark for the gradient of the curve at (2, b)]

c) At $\left(2, \frac{1}{64}\right)$, the gradient of the tangent is $-\frac{3}{64} \ln 2$.

Putting this into $y = mx + c$ gives:

$\frac{1}{64} = -\frac{6}{64} \ln 2 + c \Rightarrow c = \frac{1}{64} + \frac{6}{64} \ln 2$.

So the equation of the tangent is

$y = \frac{1 + 6 \ln 2}{64} - \left(\frac{3 \ln 2}{64}\right)x \Rightarrow 64y = 1 + 6 \ln 2 - (3 \ln 2)x$.

[2 marks available — 1 mark for correctly finding c, 1 mark for rearranging equation to show the correct result]

Alternatively you could rearrange the given equation to show it has the right gradient and that (2, b) lies on the line.

Q8 a) Let $u = x^2$, then $y = 3^u$, so

$\frac{du}{dx} = 2x$ and $\frac{dy}{du} = 3^u \ln 3 = 3^{x^2} \ln 3 \Rightarrow \frac{dy}{dx} = 2x(3^{x^2} \ln 3)$

[2 marks available — 1 mark for using the chain rule, 1 mark for the correct answer]

b) When $x = 2$, $y = p = 3^4 = 81$ and $\frac{dy}{dx} = 4(81 \ln 3) = 324 \ln 3$

[2 marks available — 1 mark for p, 1 mark for the gradient of the curve at (2, p)]

c) At $(2, 81)$, the gradient of the tangent is $324 \ln 3$, so the gradient of the normal is $-\frac{1}{324 \ln 3}$.

Putting this into $y = mx + c$ gives:

$81 = -\frac{1}{324 \ln 3} \times 2 + c \Rightarrow c = 81 + \frac{2}{324 \ln 3}$.

So the equation of the normal is

$y = 81 + \frac{2}{324 \ln 3} - \frac{1}{324 \ln 3}x = 81 + \frac{2 - x}{324 \ln 3}$

[3 marks available — 1 mark for finding the gradient of the normal, 1 mark for correctly finding c, 1 mark for rearranging the equation to show the correct result]

Alternatively you could rearrange the given equation to show it has the right gradient and that (2, p) lies on the line.

Q9 a) $\frac{dy}{dx} = 3^x \ln 3 - 2^x \ln 2$

[2 marks available — 1 mark for a correct method, 1 mark for the correct answer]

b) If $\frac{dy}{dx} = 0$, then $3^x \ln 3 = 2^x \ln 2$,

$\Rightarrow \left(\frac{3}{2}\right)^x = \frac{\ln 2}{\ln 3} \Rightarrow x = \log_{\frac{3}{2}} \frac{\ln 2}{\ln 3}$.

[3 marks available — 1 mark for setting the derivative equal to 0, 1 mark for a correct partial rearrangement, 1 mark for showing the correct result]

c) $\frac{d^2y}{dx^2} = 3^x(\ln 3)^2 - 2^x(\ln 2)^2$

When $x = \log_{\frac{3}{2}} \frac{\ln 2}{\ln 3} = -1.1358...$,

$\frac{d^2y}{dx^2} = 3^{-1.1358...}(\ln 3)^2 - 2^{-1.1358...}(\ln 2)^2 = 0.127... > 0$,

so it is a minimum point.

[3 marks available — 1 mark for a correct method to differentiate, 1 mark for the correct second derivative, 1 mark for substituting in the x-value and a correct conclusion]

d) When $x = 0$, $y = 3^0 - 2^0 = 0$.

When $y = 0$, $3^x - 2^x = 0 \Rightarrow 3^x = 2^x \Rightarrow \left(\frac{3}{2}\right)^x = 1$

$\Rightarrow x = \log_{\frac{3}{2}} 1 = 0$.

Hence the only point of intersection is at $(0, 0)$.

[3 marks available — 1 mark for showing (0, 0) lies on curve, 1 mark for an attempt to find any other x-intercepts, 1 mark for showing (0, 0) is the only x-intercept]

e) The minimum occurs at $x = \log_{\frac{3}{2}} \frac{\ln 2}{\ln 3} = -1.1358...$

So $y = 3^{-1.1358...} - 2^{-1.1358...} = -0.1679...$

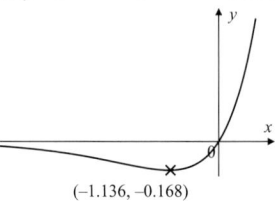

$(-1.136, -0.168)$

[2 marks available — 1 mark for a correct curve shape, 1 mark for y-intercept and stationary point marked correctly]

Exercise 7.9 — Differentiating Sin, Cos and Tan

Q1 a) $y = \sin(3x)$, so let $y = \sin u$ where $u = 3x$

$\Rightarrow \frac{dy}{du} = \cos u = \cos(3x)$, $\frac{du}{dx} = 3$

$\frac{dy}{dx} = \frac{dy}{du} \times \frac{du}{dx} = 3 \cos(3x)$

b) $y = \cos(-2x)$, so let $y = \cos u$ where $u = -2x$

$\Rightarrow \frac{dy}{du} = -\sin(u) = -\sin(-2x)$, $\frac{du}{dx} = -2$

$\frac{dy}{dx} = \frac{dy}{du} \times \frac{du}{dx} = (-\sin(-2x)) \times (-2) = 2 \sin(-2x)$

As you can see, the number you multiply by is always just the coefficient of x inside the trig function — but check whether it should be positive or negative.

c) $\frac{dy}{dx} = \frac{dy}{du} \times \frac{du}{dx} = \frac{1}{2} \times -\sin \frac{x}{2} = -\frac{1}{2} \sin \frac{x}{2}$

d) $\frac{dy}{dx} = \frac{dy}{du} \times \frac{du}{dx} = \cos\left(x + \frac{\pi}{4}\right) \times 1 = \cos\left(x + \frac{\pi}{4}\right)$

e) $\frac{dy}{dx} = \frac{dy}{du} \times \frac{du}{dx} = 6 \times \sec^2 \frac{x}{2} \times \frac{1}{2} = 3 \sec^2 \frac{x}{2}$

f) $\frac{dy}{dx} = \frac{dy}{du} \times \frac{du}{dx} = 3 \times \sec^2(5x) \times 5 = 15 \sec^2(5x)$

Q2 $f'(x) = \frac{dy}{du} \times \frac{du}{dx} = 3 \times \sec^2(2x - 1) \times 2 = 6 \sec^2(2x - 1)$

Q3 First part: $\frac{dy}{dx} = 3 \sec^2 x$

Second part: $\frac{dy}{dx} = \frac{dy}{du} \times \frac{du}{dx} = 3 \sec^2(3x)$

Putting it all together: $f'(x) = 3(\sec^2 x + \sec^2(3x))$

Q4 $f'(x) = \frac{dy}{du} \times \frac{du}{dx} = 2x \cos\left(x^2 + \frac{\pi}{3}\right)$

Q5 $f(x) = \sin^2 x$, so let $y = u^2$ where $u = \sin x$

$f'(x) = \frac{dy}{du} \times \frac{du}{dx} = (2 \sin x) \times \cos x = 2 \sin x \cos x$

Q6 $f(x) = 2 \sin^3 x$, so let $y = 2u^3$ where $u = \sin x$

$f'(x) = \frac{dy}{du} \times \frac{du}{dx} = 6 \sin^2 x \cos x$

Q7 a) $f'(x) = 3\cos x - 2\sin x$

b) $f'(x) = 0 \Rightarrow 3\cos x - 2\sin x = 0 \Rightarrow 3\cos x = 2\sin x$
$\Rightarrow \dfrac{3}{2} = \tan x \Rightarrow x = \tan^{-1}\dfrac{3}{2} = 0.983$ (3 s.f.)

Remember that $\tan x = \dfrac{\sin x}{\cos x}$.

Q8 $y = \dfrac{1}{\cos x} = (\cos x)^{-1}$, so let $y = u^{-1}$ where $u = \cos x$

$\Rightarrow \dfrac{dy}{du} = -u^{-2} = -\dfrac{1}{\cos^2 x}$, $\dfrac{du}{dx} = -\sin x$

$\dfrac{dy}{dx} = \dfrac{dy}{du} \times \dfrac{du}{dx} = -\dfrac{1}{\cos^2 x} \times -\sin x = \dfrac{\sin x}{\cos^2 x} = \sec x \tan x$

Remember that $\dfrac{1}{\cos x} = \sec x$.

Q9 Let $y = f(x)$, where $f(x) = \cos x$.

Then $\dfrac{dy}{dx} = \lim_{h \to 0}\left[\dfrac{f(x+h) - f(x)}{(x+h) - x}\right] = \lim_{h \to 0}\left[\dfrac{\cos(x+h) - \cos x}{(x+h) - x}\right]$

Using the cos addition formula:

$\dfrac{dy}{dx} = \lim_{h \to 0}\left[\dfrac{\cos x \cos h - \sin x \sin h - \cos x}{(x+h) - x}\right]$

$= \lim_{h \to 0}\left[\dfrac{\cos x(\cos h - 1) - \sin x \sin h}{h}\right]$

$= \lim_{h \to 0}\left[\dfrac{\cos x(\cos h - 1)}{h} - \dfrac{\sin x \sin h}{h}\right]$

Using small angle approximations,
$\cos h \approx 1 - \dfrac{1}{2}h^2$ and $\sin h \approx h$, so:

$\dfrac{dy}{dx} = \lim_{h \to 0}\left[\dfrac{\cos x \times \left(-\frac{1}{2}h^2\right)}{h} - \dfrac{\sin x \times h}{h}\right]$

$= \lim_{h \to 0}\left[-\dfrac{h\cos x}{2} - \sin x\right] = -\sin x$

Q10 a) $y = \cos^2 x$, so let $y = u^2$ where $u = \cos x$.

$\dfrac{dy}{dx} = \dfrac{dy}{du} \times \dfrac{du}{dx} = (2\cos x) \times (-\sin x) = -2\sin x \cos x$

b) Using the double angle formula:

$\cos(2x) \equiv 2\cos^2 x - 1 \Rightarrow \cos^2 x = \dfrac{1}{2}(\cos(2x) + 1)$
$\dfrac{dy}{dx} = \dfrac{1}{2} \times 2 \times (-\sin(2x)) = -\sin(2x)$

As the original function was in terms of cos x rather than cos (2x), it would be better to rearrange this.

From the double angle formula for sin:
$-\sin(2x) \equiv -2\sin x \cos x$

Q11 First part:
$y = 6\cos^2 x = 6(\cos x)^2$, so let $y = 6u^2$ where $u = \cos x$
$\Rightarrow \dfrac{dy}{du} = 12u = 12\cos x$, $\dfrac{du}{dx} = -\sin x$

$\dfrac{dy}{dx} = \dfrac{dy}{du} \times \dfrac{du}{dx} = -12\sin x \cos x$

Second part:
$y = 2\sin(2x)$, so let $y = 2\sin u$ where $u = 2x$
$\Rightarrow \dfrac{dy}{du} = 2\cos u = 2\cos(2x)$, $\dfrac{du}{dx} = 2$

$\dfrac{dy}{dx} = \dfrac{dy}{du} \times \dfrac{du}{dx} = 4\cos(2x)$

Putting it all together:
$\dfrac{dy}{dx} = -12\sin x \cos x - 4\cos(2x)$

Double angle formula: $2\sin x \cos x \equiv \sin(2x)$
$\Rightarrow -12\sin x \cos x - 4\cos(2x) = -6\sin(2x) - 4\cos(2x)$

Q12 $\dfrac{dy}{dx} = \cos x$. When $x = \dfrac{\pi}{4}$, $\dfrac{dy}{dx} = \dfrac{1}{\sqrt{2}}$

Q13 $\dfrac{dy}{dx} = -2\sin(2x)$

When $x = \dfrac{\pi}{4}$, $y = 0$ and $\dfrac{dy}{dx} = -2$

So the gradient of the normal is $\dfrac{-1}{-2} = \dfrac{1}{2}$.

$y = mx + c \Rightarrow 0 = \left(\dfrac{1}{2} \times \dfrac{\pi}{4}\right) + c \Rightarrow c = -\dfrac{\pi}{8}$

So the equation of the normal is $y = \dfrac{1}{2}x - \dfrac{\pi}{8}$ (or $4x - 8y - \pi = 0$)

Q14 a) Let $y = \sin u$ where $u = x^{-1}$.

$\Rightarrow \dfrac{dy}{du} = \cos u$, $\dfrac{du}{dx} = -x^{-2}$

$\dfrac{dy}{dx} = \dfrac{dy}{du} \times \dfrac{du}{dx} = \cos u \times -x^{-2} = -\dfrac{1}{x^2}\cos\left(\dfrac{1}{x}\right)$

[2 marks available — 1 mark for a correct method, 1 mark for the correct answer]

b) $\dfrac{dy}{dx} = 0 \Rightarrow \cos\left(\dfrac{1}{x}\right) = 0 \Rightarrow \dfrac{1}{x} = \dfrac{\pi}{2}, \dfrac{3\pi}{2}, \dfrac{5\pi}{2}, \ldots$

$\Rightarrow x = \dfrac{2}{\pi}, \dfrac{2}{3\pi}, \dfrac{2}{5\pi}, \ldots$

Since $x = \dfrac{2}{\pi + 2k\pi}$ is a solution for every integer k, the

curve $y = \sin\left(\dfrac{1}{x}\right)$ has infinitely many stationary points.

[2 marks available — 1 mark for a correct method to find stationary points, 1 mark for showing that $\cos\left(\dfrac{1}{x}\right) = 0$ has infinitely many solutions]

Q15 a) $\dfrac{dx}{dy} = 2\cos(2y)$, $\dfrac{dy}{dx} = \dfrac{1}{2\cos(2y)} = \dfrac{1}{2}\sec(2y)$

At the point $\left(\dfrac{\sqrt{3}}{2}, \dfrac{\pi}{6}\right)$, $\dfrac{dy}{dx} = \dfrac{1}{2\cos\frac{\pi}{3}} = 1$

$y = mx + c \Rightarrow \dfrac{\pi}{6} = \dfrac{\sqrt{3}}{2} + c \Rightarrow c = \dfrac{\pi}{6} - \dfrac{\sqrt{3}}{2}$

So the equation of the tangent is $y = x + \dfrac{\pi}{6} - \dfrac{\sqrt{3}}{2}$.

b) From part a), $\dfrac{dy}{dx} = 1$, so the normal gradient is -1.

$y = mx + c \Rightarrow \dfrac{\pi}{6} = -\dfrac{\sqrt{3}}{2} + c \Rightarrow c = \dfrac{\pi}{6} + \dfrac{\sqrt{3}}{2}$

So the equation of the normal is $y = -x + \dfrac{\pi}{6} + \dfrac{\sqrt{3}}{2}$.

Q16 a) $y = 2\sin(2x)\cos x \Rightarrow y = 4\sin x \cos^2 x$
(from double angle formula sin (2x) ≡ 2 sin x cos x)
$\Rightarrow y = 4\sin x(1 - \sin^2 x)$ *(from sin² x + cos² x ≡ 1)*
$\Rightarrow y = 4\sin x - 4\sin^3 x$

b) First part: $\dfrac{dy}{dx} = 4\cos x$

Second part: $y = 4\sin^3 x = 4(\sin x)^3$,
so let $y = 4u^3$ where $u = \sin x$
$\dfrac{dy}{dx} = \dfrac{dy}{du} \times \dfrac{du}{dx} = 12\sin^2 x \cos x$

Putting it all together: $\dfrac{dy}{dx} = 4\cos x - 12\sin^2 x \cos x$

Q17 a) $y = \sin(\sin x)$, let $y = \sin u$ where $u = \sin x$

$\Rightarrow \dfrac{dy}{du} = \cos u$, $\dfrac{du}{dx} = \cos x$

$\dfrac{dy}{dx} = \dfrac{dy}{du} \times \dfrac{du}{dx} = \cos u \times \cos x = \cos x \cos(\sin x)$

[2 marks available — 1 mark for a correct method, 1 mark for the correct answer]

Answers

b) Stationary points occur when either
$\cos x = 0$ or $\cos(\sin x) = 0$.
$\cos x = 0 \Rightarrow x = \frac{\pi}{2}$ or $x = \frac{3\pi}{2}$.
$\cos(\sin x) = 0$ has no solutions since $|\sin x| \leq 1$.
The corresponding y-values are
$\sin\left(\sin\frac{\pi}{2}\right) = \sin 1$ and $\sin\left(\sin\frac{3\pi}{2}\right) = \sin(-1) = -\sin 1$.
The stationary points are $\left(\frac{\pi}{2}, \sin 1\right)$ and $\left(\frac{3\pi}{2}, -\sin 1\right)$.
When $x = \frac{\pi}{2} - 0.01$, $\frac{dy}{dx} = 0.005... > 0$.
When $x = \frac{\pi}{2} + 0.01$, $\frac{dy}{dx} = -0.005... < 0$.
Therefore, $\left(\frac{\pi}{2}, \sin 1\right)$ is a maximum point.
When $x = \frac{3\pi}{2} - 0.01$, $\frac{dy}{dx} = -0.005... < 0$.
When $x = \frac{3\pi}{2} + 0.01$, $\frac{dy}{dx} = 0.005... > 0$.
Therefore, $\left(\frac{3\pi}{2}, -\sin 1\right)$ is a minimum point.
[6 marks available — 1 mark for method to solve $\frac{dy}{dx} = 0$,
1 mark for the correct x-values of the stationary points,
1 mark for both correct coordinates, 1 mark for a correct
method to determine the nature of the stationary points,
1 mark for identifying the maximum, 1 mark for identifying
the minimum]

c) There must be a point of inflection between the maximum
and the minimum because the curve must change from
concave to convex at some point between those points.
[1 mark for a correct answer with a valid reason]

Exercise 7.10 — Differentiating by Using the Chain Rule Twice

Q1 a) $y = \sin^2(x + 2)$, so let $y = u^2$ where $u = \sin(x + 2)$
$\frac{dy}{du} = 2u = 2\sin(x + 2)$
$u = \sin(x + 2)$, so let $u = \sin v$ where $v = x + 2$
$\Rightarrow \frac{du}{dv} = \cos v = \cos(x + 2)$ and $\frac{dv}{dx} = 1$
$\Rightarrow \frac{du}{dx} = \frac{du}{dv} \times \frac{dv}{dx} = \cos(x + 2)$
So $\frac{dy}{dx} = \frac{dy}{du} \times \frac{du}{dx} = 2\sin(x + 2) \times \cos(x + 2)$
$= 2\sin(x + 2)\cos(x + 2)$

b) $y = \cos^2(x^2)$, so let $y = u^2$ where $u = \cos(x^2)$
$\frac{dy}{du} = 2u = 2\cos(x^2)$
$u = \cos(x^2)$, so let $u = \cos v$ where $v = x^2$
$\Rightarrow \frac{du}{dv} = -\sin v = -\sin(x^2)$ and $\frac{dv}{dx} = 2x$
$\Rightarrow \frac{du}{dx} = \frac{du}{dv} \times \frac{dv}{dx} = -\sin(x^2) \times 2x = -2x\sin(x^2)$
So $\frac{dy}{dx} = \frac{dy}{du} \times \frac{du}{dx} = 2\cos(x^2) \times -2x\sin(x^2)$
$= -4x\sin(x^2)\cos(x^2)$

c) $y = \sqrt{\tan(4x)}$, so let $y = \sqrt{u} = u^{\frac{1}{2}}$ where $u = \tan(4x)$
$\frac{dy}{du} = \frac{u^{-\frac{1}{2}}}{2} = \frac{1}{2\sqrt{u}} = \frac{1}{2\sqrt{\tan(4x)}}$
$u = \tan(4x)$, so let $u = \tan v$ where $v = 4x$
$\Rightarrow \frac{du}{dv} = \sec^2 v = \sec^2(4x)$ and $\frac{dv}{dx} = 4$
$\Rightarrow \frac{du}{dx} = \frac{du}{dv} \times \frac{dv}{dx} = 4\sec^2(4x)$
So $\frac{dy}{dx} = \frac{dy}{du} \times \frac{du}{dx}$
$= \frac{1}{2\sqrt{\tan(4x)}} \times 4\sec^2(4x) = \frac{2\sec^2(4x)}{\sqrt{\tan(4x)}}$

Q2 a) $y = \sin(\cos(2x))$, so let $y = \sin u$ where $u = \cos(2x)$
$\frac{dy}{du} = \cos u = \cos(\cos(2x))$
Using the chain rule again on $\frac{du}{dx}$ gives:
$u = \cos(2x) \Rightarrow \frac{du}{dx} = -2\sin(2x)$
Putting it all together:
$\frac{dy}{dx} = \frac{dy}{du} \times \frac{du}{dx} = -2\sin(2x)\cos(\cos(2x))$

b) $y = 2\ln f(x) \Rightarrow \frac{dy}{dx} = 2\frac{f'(x)}{f(x)}$
$f(x) = \cos(3x) \Rightarrow f'(x) = -3\sin(3x)$
$\frac{dy}{dx} = 2\frac{-3\sin(3x)}{\cos(3x)} = -6\tan(3x)$

c) $y = \ln(\tan^2 x) = \ln(f(x)) \Rightarrow \frac{dy}{dx} = \frac{f'(x)}{f(x)}$
$f(x) = (\tan x)^2$, so let $f(x) = v = u^2$ where $u = \tan x$
$\frac{dv}{du} = 2u = 2\tan x$, $\frac{du}{dx} = \sec^2 x$
$f'(x) = \frac{dv}{dx} = \frac{dv}{du} \times \frac{du}{dx} = 2\tan x\sec^2 x$
Putting it all together:
$\frac{dy}{dx} = \frac{2\tan x\sec^2 x}{\tan^2 x} = 2\sec x\,\text{cosec}\,x$
You could've written ln (tan² x) as 2 ln (tan x)
using the laws of logs and then differentiated
— you'd end up with the same answer.

d) $y = e^{f(x)} \Rightarrow \frac{dy}{dx} = f'(x)e^{f(x)}$
$f(x) = \tan(2x)$, so let $f(x) = v = \tan u$ where $u = 2x$
$\frac{dv}{du} = \sec^2 u = \sec^2(2x)$, $\frac{du}{dx} = 2$
$\Rightarrow f'(x) = \frac{dv}{dx} = \frac{dv}{du} \times \frac{du}{dx} = 2\sec^2(2x)$
$\Rightarrow \frac{dy}{dx} = 2\sec^2(2x)e^{\tan(2x)}$

Q3 a) $\frac{dy}{dx} = \sin^4 x^2 = (\sin x^2)^4$, so let $y = u^4$ where $u = \sin x^2$
$\frac{dy}{du} = 4u^3 = 4\sin^3 x^2$
For $\frac{du}{dx}$, set up another chain rule:
$u = \sin x^2$ so let $u = \sin v$, $v = x^2$
$\frac{du}{dv} = \cos v = \cos x^2$, $\frac{dv}{dx} = 2x$
$\frac{du}{dx} = \frac{du}{dv} \times \frac{dv}{dx} = 2x\cos x^2$
Putting it all together:
$\frac{dy}{dx} = \frac{dy}{du} \times \frac{du}{dx} = 8x\sin^3 x^2\cos x^2$

b) $\frac{dy}{dx} = f'(x)e^{f(x)}$
$f(x) = \sin^2 x = (\sin x)^2$, so let $y = u^2$ where $u = \sin x$
$f'(x) = \frac{dy}{du} \times \frac{du}{dx} = (2\sin x) \times (\cos x) = 2\sin x\cos x$
$\Rightarrow \frac{dy}{dx} = 2e^{\sin^2 x}\sin x\cos x$

c) First part:
$y = \tan^2(3x) = (\tan(3x))^2$, so let $y = u^2$ where $u = \tan(3x)$
$\frac{dy}{du} = 2u = 2\tan(3x)$
For $\frac{du}{dx}$ set up the chain rule again:
$u = \tan(3x)$, so let $u = \tan v$ where $v = 3x$
$\frac{du}{dv} = \sec^2 v = \sec^2(3x)$, $\frac{dv}{dx} = 3$
$\Rightarrow \frac{du}{dx} = \frac{du}{dv} \times \frac{dv}{dx} = 3\sec^2(3x)$
$\Rightarrow \frac{dy}{dx} = \frac{dy}{du} \times \frac{du}{dx} = 6\tan(3x)\sec^2(3x)$

Second part:
$$\frac{dy}{dx} = \cos x$$

Putting it all together:
$$\frac{dy}{dx} = 6\tan(3x)\sec^2(3x) + \cos x$$

With practice, you should be able to do some of the simpler chain rule calculations in your head, e.g. $\frac{d}{dx}\tan^2(3x) = 6\tan(3x)\sec^2(3x)$, which will make these questions much quicker.

d) First part: $y = e^{f(x)}$, $f(x) = 2\cos(2x)$ $\Rightarrow \frac{dy}{dx} = f'(x)e^{f(x)}$
$\quad\quad$ $f(x) = 2\cos(2x) \Rightarrow f'(x) = -4\sin(2x)$
$\quad\quad$ So $\frac{dy}{dx} = -4\sin(2x)e^{2\cos(2x)}$

Second part: $y = \cos^2(2x) = (\cos(2x))^2$,
$\quad\quad$ so let $y = u^2$ where $u = \cos(2x)$
$$\frac{dy}{du} = 2u = 2\cos(2x)$$

For $\frac{du}{dx}$, set up the chain rule again:
$$u = \cos(2x) \Rightarrow \frac{du}{dx} = -2\sin(2x)$$
$$\Rightarrow \frac{dy}{dx} = \frac{dy}{du} \times \frac{du}{dx} = -4\sin(2x)\cos(2x)$$

Putting it all together:
$$\frac{dy}{dx} = -4\sin(2x)\,e^{2\cos(2x)} - 4\sin(2x)\cos(2x)$$

Q4 a) $f(x) = e^{1+\sin 2x}$, so let $y = e^u$, where $u = 1 + \sin 2x$
$\quad\quad \Rightarrow \frac{dy}{du} = e^u$, $\frac{du}{dx} = 2\cos 2x$
$\quad\quad \frac{dy}{dx} = \frac{dy}{du} \times \frac{du}{dx} = e^u \times 2\cos 2x \Rightarrow 2e^u\cos 2x$
$\quad\quad \Rightarrow 2e^{1+\sin 2x}\cos 2x$
$\quad\quad$ When $x = \pi$, $\frac{dy}{dx} = 2e^{1+\sin 2\pi}\cos 2\pi = 2e^{1+0} \times 1 = 2e$
$\quad\quad$ *[4 marks available — 1 mark for $\frac{dy}{du}$, 1 mark for $\frac{du}{dx}$, 1 mark for the correct derivative, 1 mark for the correct value at $x = \pi$]*

b) $-1 \le \sin 2x \le 1 \Rightarrow 0 \le 1 + \sin 2x \le 2$
$\quad\quad \Rightarrow e^0 \le e^{1+\sin 2x} \le e^2$ so the range of f is $1 \le f(x) \le e^2$.
$\quad\quad$ *[2 marks available — 1 mark for a correct method, 1 mark for the correct answer]*

c) Stationary points occur when $f'(x) = 0$.
$\quad\quad 2e^{1+\sin 2x}\cos 2x = 0 \Rightarrow \cos 2x = 0$ since $e^{1+\sin 2x} \ge 1$.
$\quad\quad \cos 2x = 0 \Rightarrow x = \frac{\pi}{4}$ and $x = \frac{3\pi}{4}$
$\quad\quad$ When $x = \frac{\pi}{4}$, $y = e^{1+\sin\frac{2\pi}{4}} = e^2$.
$\quad\quad$ When $x = \frac{3\pi}{4}$, $y = e^{1+\sin\frac{6\pi}{4}} = 1$.
$\quad\quad$ As the range is $1 \le e^{1+\sin 2x} \le e^2$, $x = \frac{\pi}{4}$ must be a maximum and $x = \frac{3\pi}{4}$ must be a minimum.
$\quad\quad$ *[4 marks available — 1 mark for setting f'(x) equal to 0, 1 mark for both x-values, 1 mark for both y-values, 1 mark for identifying the nature of both turning points]*

d) When $x = 0$, $y = e^{1+\sin 2(0)} = e$.

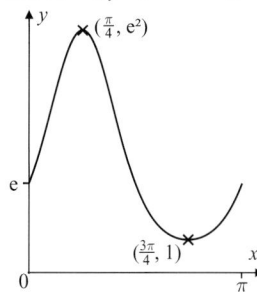

[3 marks available — 1 mark for the correct curve, 1 mark for both turning points marked correctly, 1 mark for the y-intercept marked correctly]

Q5 a) Using the double angle formula:
$$g(x) = (\sin(x^2)\cos(x^2))^2 = \left(\frac{1}{2}\sin(2x^2)\right)^2 = \frac{1}{4}\sin^2(2x^2)$$
$\quad\quad$ Let $y = \frac{1}{4}u^2$ where $u = \sin(2x^2)$
$$\frac{dy}{du} = \frac{1}{2}u = \frac{1}{2}\sin(2x^2)$$
$\quad\quad$ For $u = \sin(2x^2)$, let $u = \sin v$ where $v = 2x^2$
$$\frac{du}{dv} = \cos v = \cos(2x^2) \text{ and } \frac{dv}{dx} = 4x$$
$$\frac{du}{dx} = \frac{du}{dv} \times \frac{dv}{dx} = \cos(2x^2) \times 4x = 4x\cos(2x^2)$$
$$\frac{dy}{dx} = \frac{dy}{du} \times \frac{du}{dx} = \frac{1}{2}\sin(2x^2) \times 4x\cos(2x^2)$$
$$= 2x\sin(2x^2)\cos(2x^2) \text{ or } x\sin(4x^2).$$
$\quad\quad$ *[4 marks available — 1 mark for rewriting using the correct identity, 1 mark for $\frac{du}{dv}$, 1 mark for $\frac{du}{dx}$, 1 mark for the correct answer]*

b) Stationary points occur when $g'(x) = 0$
$\quad\quad \Rightarrow \sin(4x^2) = 0$ or $x = 0$
$\quad\quad x = 0$ is not in the domain of g so is not a solution.
$\quad\quad \sin(4x^2) = 0 \Rightarrow 4x^2 = \pi, 2\pi, 3\pi, ...$
$\quad\quad \Rightarrow x = \sqrt{\frac{\pi}{4}}, \sqrt{\frac{\pi}{2}}, \sqrt{\frac{3\pi}{4}}$
$\quad\quad$ *All other solutions to $\sin(4x^2) = 0$ fall outside the domain of g.*
$\quad\quad$ *[3 marks available — 1 mark for setting the derivative equal to 0, 2 marks for all three x-values correct, otherwise 1 mark for at least one x-value correct]*

c) $g(x) = \frac{1}{4}\sin^2(2x^2)$ so $0 \le g(x) \le \frac{1}{4}$.
$\quad\quad$ The maximum value is $\frac{1}{4}$ and the minimum value is 0.
$\quad\quad$ *[2 marks available — 1 mark for each correct answer]*

Exercise 7.11 — Product Rule

Q1 a) $y = x(x + 2) = x^2 + 2x$
$$\frac{dy}{dx} = 2x + 2$$

b) $u = x$, $v = x + 2 \Rightarrow \frac{du}{dx} = 1$, $\frac{dv}{dx} = 1$
$$\frac{dy}{dx} = u\frac{dv}{dx} + v\frac{du}{dx} = x + (x + 2) = 2x + 2$$

Q2 a) $u = x^2$, $v = (x + 6)^3 \Rightarrow \frac{du}{dx} = 2x$, $\frac{dv}{dx} = 3(x + 6)^2$
$$\frac{dy}{dx} = u\frac{dv}{dx} + v\frac{du}{dx} = [x^2 \times 3(x + 6)^2] + [(x + 6)^3 \times 2x]$$
$$= 3x^2(x + 6)^2 + 2x(x + 6)^3$$
$$= x(x + 6)^2[3x + 2(x + 6)] = x(x + 6)^2(5x + 12)$$

Here the chain rule was used to find $\frac{dv}{dx}$ — write out all the steps if you're struggling.

Answers

b) $u = x^3$, $v = (5x + 2)^4$ $\Rightarrow$ $\dfrac{du}{dx} = 3x^2$, $\dfrac{dv}{dx} = 20(5x + 2)^3$

$\dfrac{dy}{dx} = u\dfrac{dv}{dx} + v\dfrac{du}{dx} = [x^3 \times 20(5x + 2)^3] + [(5x + 2)^4 \times 3x^2]$

$\quad = 20x^3(5x + 2)^3 + 3x^2(5x + 2)^4$

$\quad = x^2(5x + 2)^3[20x + 3(5x + 2)]$

$\quad = x^2(5x + 2)^3(35x + 6)$

c) $u = x^3$, $v = e^x$ $\Rightarrow$ $\dfrac{du}{dx} = 3x^2$, $\dfrac{dv}{dx} = e^x$

$\dfrac{dy}{dx} = u\dfrac{dv}{dx} + v\dfrac{du}{dx} = x^3 e^x + e^x 3x^2 = x^2 e^x(x + 3)$

d) $u = x$, $v = e^{4x}$ $\Rightarrow$ $\dfrac{du}{dx} = 1$, $\dfrac{dv}{dx} = 4e^{4x}$

$\dfrac{dy}{dx} = u\dfrac{dv}{dx} + v\dfrac{du}{dx} = 4xe^{4x} + e^{4x} = e^{4x}(4x + 1)$

e) $u = x$, $v = e^{x^2}$ $\Rightarrow$ $\dfrac{du}{dx} = 1$, $\dfrac{dv}{dx} = 2xe^{x^2}$

$\dfrac{dy}{dx} = u\dfrac{dv}{dx} + v\dfrac{du}{dx} = x \times 2xe^{x^2} + e^{x^2} = e^{x^2}(2x^2 + 1)$

f) $u = e^{2x}$, $v = \sin x$ $\Rightarrow$ $\dfrac{du}{dx} = 2e^{2x}$, $\dfrac{dv}{dx} = \cos x$

$\dfrac{dy}{dx} = u\dfrac{dv}{dx} + v\dfrac{du}{dx} = e^{2x} \times \cos x + \sin x \times 2e^{2x}$

$\quad = e^{2x}(\cos x + 2 \sin x)$

Q3 a) $u = x^3$, $v = (x + 3)^{\frac{1}{2}}$ $\Rightarrow$ $\dfrac{du}{dx} = 3x^2$, $\dfrac{dv}{dx} = \dfrac{1}{2}(x + 3)^{-\frac{1}{2}}$

$f'(x) = u\dfrac{dv}{dx} + v\dfrac{du}{dx} = \left[x^3 \times \dfrac{1}{2}(x + 3)^{-\frac{1}{2}}\right] + \left[(x + 3)^{\frac{1}{2}} \times 3x^2\right]$

$\quad = \dfrac{x^3}{2(x + 3)^{\frac{1}{2}}} + 3x^2(x + 3)^{\frac{1}{2}}$

b) $u = x^2$, $v = (x - 7)^{-\frac{1}{2}}$ $\Rightarrow$ $\dfrac{du}{dx} = 2x$, $\dfrac{dv}{dx} = -\dfrac{1}{2}(x - 7)^{-\frac{3}{2}}$

$f'(x) = u\dfrac{dv}{dx} + v\dfrac{du}{dx}$

$\quad = \left[x^2 \times \left(-\dfrac{1}{2}\right) \times (x - 7)^{-\frac{3}{2}}\right] + \left[(x - 7)^{-\frac{1}{2}} \times 2x\right]$

$\quad = -\dfrac{x^2}{2(\sqrt{x - 7})^3} + \dfrac{2x}{\sqrt{x - 7}}$

c) $u = x^4$, $v = \ln x$ $\Rightarrow$ $\dfrac{du}{dx} = 4x^3$, $\dfrac{dv}{dx} = \dfrac{1}{x}$

$f'(x) = u\dfrac{dv}{dx} + v\dfrac{du}{dx} = \left[x^4 \times \dfrac{1}{x}\right] + [\ln x \times 4x^3]$

$\quad = x^3 + 4x^3 \ln x = x^3(1 + 4 \ln x)$

d) $u = 4x$, $v = \ln x^2 = 2\ln x$ $\Rightarrow$ $\dfrac{du}{dx} = 4$, $\dfrac{dv}{dx} = \dfrac{2}{x}$

$f'(x) = u\dfrac{dv}{dx} + v\dfrac{du}{dx} = \left[4x \times \dfrac{2}{x}\right] + [\ln x^2 \times 4] = 8 + 4 \ln x^2$

e) $u = 2x^3$, $v = \cos x$ $\Rightarrow$ $\dfrac{du}{dx} = 6x^2$, $\dfrac{dv}{dx} = -\sin x$

$f'(x) = u\dfrac{dv}{dx} + v\dfrac{du}{dx} = (-2x^3 \times \sin x) + (\cos x \times 6x^2)$

$\quad = 2x^2(3 \cos x - x \sin x)$

f) $u = x^2$, $v = \cos(2x)$ $\Rightarrow$ $\dfrac{du}{dx} = 2x$, $\dfrac{dv}{dx} = -2\sin(2x)$

$f'(x) = u\dfrac{dv}{dx} + v\dfrac{du}{dx} = [x^2 \times -2\sin(2x)] + [\cos(2x) \times 2x]$

$\quad = 2x \cos(2x) - 2x^2 \sin(2x) = 2x(\cos(2x) - x \sin(2x))$

Q4 a) $u = (x + 1)^2$, $v = x^2 - 1$ $\Rightarrow$ $\dfrac{du}{dx} = 2(x + 1)$, $\dfrac{dv}{dx} = 2x$

$\dfrac{dy}{dx} = u\dfrac{dv}{dx} + v\dfrac{du}{dx} = [(x + 1)^2 \times 2x] + [(x^2 - 1) \times 2(x + 1)]$

$\quad = 2x(x + 1)^2 + 2(x^2 - 1)(x + 1)$

$\quad = 2x^3 + 4x^2 + 2x + 2x^3 + 2x^2 - 2x - 2 = 4x^3 + 6x^2 - 2$

b) $u = (x + 1)^3$, $v = x - 1$ $\Rightarrow$ $\dfrac{du}{dx} = 3(x + 1)^2$, $\dfrac{dv}{dx} = 1$

$\dfrac{dy}{dx} = u\dfrac{dv}{dx} + v\dfrac{du}{dx} = [(x + 1)^3 \times 1] + [(x - 1) \times 3(x + 1)^2]$

$\quad = (x + 1)^3 + 3(x - 1)(x + 1)^2$

$\quad = (x^3 + 3x^2 + 3x + 1) + (3x^3 + 6x^2 + 3x - 3x^2 - 6x - 3)$

$\quad = 4x^3 + 6x^2 - 2$

Use the binomial formula to expand $(x + 1)^3$
— you'll get the coefficients 1, 3, 3, 1.

c) $y = (x + 1)^2(x^2 - 1) = (x + 1)^2(x + 1)(x - 1) = (x + 1)^3(x - 1)$

$x^2 - 1$ is the difference of two squares.

Q5 When $y = xe^x$ is concave, $\dfrac{d^2y}{dx^2}$ is negative.

$u = x$, $v = e^x$ $\Rightarrow$ $\dfrac{du}{dx} = 1$, $\dfrac{dv}{dx} = e^x$

$\dfrac{dy}{dx} = u\dfrac{dv}{dx} + v\dfrac{du}{dx} = xe^x + e^x$

$\dfrac{d^2y}{dx^2} = \dfrac{d}{dx}(xe^x) + \dfrac{d}{dx}(e^x) = [xe^x + e^x] + [e^x] = xe^x + 2e^x$

$\dfrac{d^2y}{dx^2} < 0 \Rightarrow xe^x + 2e^x < 0 \Rightarrow e^x(x + 2) < 0$

Since $e^x > 0$ for all x, this means that $(x + 2) < 0 \Rightarrow x < -2$
So the curve $y = xe^x$ is concave when $x < -2$.

Q6 $u = \sqrt{x + 2}$, $v = \sqrt{x + 7}$ $\Rightarrow$ $\dfrac{du}{dx} = \dfrac{1}{2\sqrt{x + 2}}$, $\dfrac{dv}{dx} = \dfrac{1}{2\sqrt{x + 7}}$

$\dfrac{dy}{dx} = u\dfrac{dv}{dx} + v\dfrac{du}{dx} = \dfrac{\sqrt{x + 2}}{2\sqrt{x + 7}} + \dfrac{\sqrt{x + 7}}{2\sqrt{x + 2}}$

At the point $(2, 6)$, $\dfrac{dy}{dx} = \dfrac{\sqrt{4}}{2\sqrt{9}} + \dfrac{\sqrt{9}}{2\sqrt{4}} = \dfrac{13}{12}$

$y = mx + c \Rightarrow 6 = \left(2 \times \dfrac{13}{12}\right) + c \Rightarrow c = \dfrac{23}{6}$

So the equation of the tangent is $y = \dfrac{13}{12}x + \dfrac{23}{6}$.
To write this in the form $ax + by + c = 0$ where a,
b and c are integers, multiply by 12 and rearrange.
$y = \dfrac{13}{12}x + \dfrac{23}{6} \Rightarrow 12y = 13x + 46 \Rightarrow 13x - 12y + 46 = 0$

Q7 a) $u = (x - 1)^{\frac{1}{2}}$, $v = (x + 4)^{-\frac{1}{2}}$

$\Rightarrow \dfrac{du}{dx} = \dfrac{1}{2\sqrt{x - 1}}$, $\dfrac{dv}{dx} = -\dfrac{1}{2(\sqrt{x + 4})^3}$

$\dfrac{dy}{dx} = u\dfrac{dv}{dx} + v\dfrac{du}{dx} = \dfrac{1}{2\sqrt{x - 1}\sqrt{x + 4}} - \dfrac{\sqrt{x - 1}}{2(\sqrt{x + 4})^3}$

When $x = 5$, $\dfrac{dy}{dx} = \dfrac{1}{2\sqrt{4}\sqrt{9}} - \dfrac{\sqrt{4}}{2(\sqrt{9})^3} = \dfrac{5}{108}$

and $y = \dfrac{\sqrt{4}}{\sqrt{9}} = \dfrac{2}{3}$

$y = mx + c \Rightarrow \dfrac{2}{3} = \left(5 \times \dfrac{5}{108}\right) + c \Rightarrow c = \dfrac{47}{108}$

So the equation of the tangent is $y = \dfrac{5}{108}x + \dfrac{47}{108}$

To write this in the form $ax + by + c = 0$
where a, b and c are integers, multiply by 108 and rearrange.

$y = \dfrac{5}{108}x + \dfrac{47}{108} \Rightarrow 108y = 5x + 47 \Rightarrow 5x - 108y + 47 = 0$

b) Gradient of the normal $= -\dfrac{1}{\left(\dfrac{5}{108}\right)} = -\dfrac{108}{5}$

$y = mx + c \Rightarrow \dfrac{2}{3} = \left(5 \times \left(-\dfrac{108}{5}\right)\right) + c \Rightarrow c = \dfrac{326}{3}$

So the equation of the normal is $y = -\dfrac{108}{5}x + \dfrac{326}{3}$.

To write this in the form $ax + by + c = 0$ where a,
b and c are integers, multiply by 15 and rearrange.

$y = -\dfrac{108}{5}x + \dfrac{326}{3} \Rightarrow 15y = -324x + 1630$

$\Rightarrow 324x + 15y - 1630 = 0$

Answers

Q8 First use the chain rule:

$\frac{dy}{dx} = f'(x)e^{f(x)}$ where $f(x) = x^2\sqrt{x+3}$.

Then use the product rule to find $f'(x)$:

$u = x^2, v = \sqrt{x+3} \Rightarrow \frac{du}{dx} = 2x, \frac{dv}{dx} = \frac{1}{2\sqrt{x+3}}$

$\frac{dy}{dx} = u\frac{dv}{dx} + v\frac{du}{dx} = \left[x^2 \times \frac{1}{2\sqrt{x+3}}\right] + \left[\sqrt{x+3} \times 2x\right]$

$= \frac{x^2 + 4x(x+3)}{2\sqrt{x+3}} = \frac{5x^2 + 12x}{2\sqrt{x+3}}$

Now putting it all together:

$\frac{dy}{dx} = \frac{5x^2 + 12x}{2\sqrt{x+3}} e^{x^2\sqrt{x+3}}$

Q9 Stationary points occur when the gradient is 0.

Differentiate with the product rule:

$u = x, v = e^{x-x^2} \Rightarrow \frac{du}{dx} = 1, \frac{dv}{dx} = (1-2x)e^{x-x^2}$

$\frac{dy}{dx} = u\frac{dv}{dx} + v\frac{du}{dx} = \left[x(1-2x)e^{x-x^2}\right] + \left[e^{x-x^2} \times 1\right]$

$= e^{x-x^2}(x - 2x^2 + 1)$

e^{x-x^2} cannot be 0, so the stationary points occur when $-2x^2 + x + 1 = 0$

$\Rightarrow (2x+1)(-x+1) = 0 \Rightarrow x = 1$ or $x = -\frac{1}{2}$

When $x = 1, y = 1 \times e^0 = 1$.

When $x = -\frac{1}{2}, y = -\frac{1}{2} \times e^{-\frac{3}{4}} = -\frac{e^{-\frac{3}{4}}}{2}$.

So the stationary points are $(1, 1)$ and $\left(-\frac{1}{2}, -\frac{e^{-\frac{3}{4}}}{2}\right)$.

Q10 a) Stationary points occur when the gradient is 0.

$u = (x-2)^2, v = (x+4)^3 \Rightarrow \frac{du}{dx} = 2(x-2), \frac{dv}{dx} = 3(x+4)^2$

$\frac{dy}{dx} = u\frac{dv}{dx} + v\frac{du}{dx}$

$= \left[(x-2)^2 \times 3(x+4)^2\right] + \left[(x+4)^3 \times 2(x-2)\right]$

$= 3(x-2)^2(x+4)^2 + 2(x-2)(x+4)^3$

$= (x-2)(x+4)^2[3x - 6 + 2x + 8] = (x-2)(x+4)^2(5x+2)$

So the stationary points occur when:

$x - 2 = 0 \Rightarrow x = 2$

and $x + 4 = 0 \Rightarrow x = -4$

and $5x + 2 = 0 \Rightarrow x = -\frac{2}{5} = -0.4$

When $x = 2, y = 0 \times 6^3 = 0$

When $x = -4, y = (-6)^2 \times 0 = 0$

When $x = -0.4, y = (-2.4)^2(3.6)^3 = 268.74$ (2 d.p.)

So the stationary points are $(2, 0), (-4, 0)$ and $(-0.4, 268.74)$.

b) To write $\frac{dy}{dx}$ in the form $(Ax^2 + Bx + C)(x + D)^n$,

multiply out the brackets $(x-2)$ and $(5x+2)$, which will give you the quadratic bracket:

$\frac{dy}{dx} = (x-2)(x+4)^2(5x+2) = [5x^2 - 10x + 2x - 4](x+4)^2$

$= (5x^2 - 8x - 4)(x+4)^2$

Now you can differentiate this using the product rule:

$u = (5x^2 - 8x - 4), v = (x+4)^2$

$\Rightarrow \frac{du}{dx} = (10x - 8), \frac{dv}{dx} = 2(x+4)$

$\frac{d^2y}{dx^2} = u\frac{dv}{dx} + v\frac{du}{dx}$

$= [(5x^2 - 8x - 4) \times 2(x+4)] + [(x+4)^2 \times (10x - 8)]$

$= (x+4)(10x^2 - 16x - 8) + (x+4)(10x^2 + 32x - 32)$

$= (x+4)(20x^2 + 16x - 40) = 4(x+4)(5x^2 + 4x - 10)$

When $x = 2, \frac{d^2y}{dx^2} = 4(2+4)(5(2)^2 + 4(2) - 10)$

$= 4(6)(20 + 8 - 10) = 432 (> 0)$

So $(2, 0)$ is a minimum point.

When $x = -4, \frac{d^2y}{dx^2} = 4(-4+4)(5(-4)^2 + 4(-4) - 10)$

$= 4(0)(80 - 16 - 10) = 0$

Check $\frac{d^2y}{dx^2}$ either side of $x = -4$:

When $x > -4, (x+4) > 0$ and $(5x^2 + 4x - 10) > 0$

$\Rightarrow \frac{d^2y}{dx^2} > 0$

When $x < -4, (x+4) < 0$ and $(5x^2 + 4x - 10) > 0$

$\Rightarrow \frac{d^2y}{dx^2} < 0$

So $(-4, 0)$ is a point of inflection.

When $x = -0.4, \frac{d^2y}{dx^2} = 4(-0.4 + 4)(5(-0.4)^2 + 4(-0.4) - 10)$

$= 4(3.6)(0.8 - 1.6 - 10) = -155.52 (< 0)$

So $(-0.4, 268.74)$ is a maximum point.

Q11 a) $f(x) = e^x \sin x$, so let $u = e^x, v = \sin x$

$\Rightarrow \frac{du}{dx} = e^x, \frac{dv}{dx} = \cos x$

so $f'(x) = u\frac{dv}{dx} + v\frac{du}{dx} = e^x \cos x + e^x \sin x$

Now differentiate each part separately to find $f''(x)$:

$e^x \sin x$ is same as $f(x)$, so $f'(x) = e^x \cos x + f(x)$.

Let $g(x) = e^x \cos x$, so for $g'(x)$:

let $u = e^x, v = \cos x \Rightarrow \frac{du}{dx} = e^x, \frac{dv}{dx} = -\sin x$

$g'(x) = u\frac{dv}{dx} + v\frac{du}{dx} = -e^x \sin x + e^x \cos x$

So $f''(x) = g'(x) + f'(x) = -e^x \sin x + e^x \cos x + e^x \cos x + e^x \sin x$

$= 2e^x \cos x$

[4 marks available — 1 mark for a correct method to differentiate, 1 mark for the correct first derivative, 1 mark for a correct method to find the second derivative, 1 mark for the correct answer]

b) Using working from part a),

$\frac{d}{dx}(e^x \sin x) = e^x \cos x + e^x \sin x$,

$\frac{d}{dx}(e^x \cos x) = -e^x \sin x + e^x \cos x$,

and $\frac{d^2f}{dx^2} = 2e^x \cos x$.

So $\frac{d^3f}{dx^3} = 2(-e^x \sin x + e^x \cos x) = -2e^x \sin x + 2e^x \cos x$

$\Rightarrow \frac{d^4f}{dx^4} = -2(e^x \cos x + e^x \sin x) + 2(-e^x \sin x + e^x \cos x)$

$= -4e^x \sin x = -4f(x)$

[2 marks available — 1 mark for the correct third derivative, 1 mark for the correct fourth derivative giving the required result]

c) Differentiating 100 times is the same as finding the fourth derivative 25 times:

$f^{(4)}(x) = -4f(x) \Rightarrow f^{(100)}(x) = (-4)^{25} f(x) = -4^{25} e^x \sin x$.

[2 marks available — 1 mark for a correct method to find 100th derivative, 1 mark for the correct answer]

Q12 a) $y = \sin^3 x$, let $y = u^3$ where $u = \sin x$

$\frac{dy}{du} = 3u^2 = 3\sin^2 x$ and $\frac{du}{dx} = \cos x$

$\frac{dy}{dx} = \frac{dy}{du} \times \frac{du}{dx} = 3\sin^2 x \cos x$

Let $u = 3\sin^2 x, v = \cos x \Rightarrow \frac{dv}{dx} = -\sin x$

Use the chain rule to find $\frac{du}{dx}$.

Let $u = 3w^2$ where $w = \sin x$

$\frac{du}{dw} = 6w = 6\sin x$ and $\frac{dw}{dx} = \cos x$

$\frac{du}{dx} = \frac{du}{dw} \times \frac{dw}{dx} = 6\sin x \cos x$

$\frac{d^2y}{dx^2} = u\frac{dv}{dx} + v\frac{du}{dx}$

$= (3\sin^2 x \times -\sin x) + (\cos x \times 6\sin x \cos x)$

$= -3\sin^3 x + 6\sin x \cos^2 x = -3\sin^3 x + 6\sin x(1 - \sin^2 x)$

$= 6\sin x - 9\sin^3 x$

[5 marks available — 1 mark for using the chain rule to differentiate $\sin^3 x$, 1 mark for the correct first derivative, 1 mark for using the product rule to differentiate again, 1 mark for the correct second derivative, 1 mark for rewriting the second derivative in terms of sin]

Answers

b) The curve M is convex when $f''(x) > 0$,

so $6\sin x - 9\sin^3 x > 0 \Rightarrow 3\sin x(2 - 3\sin^2 x) > 0$.

Now $\sin x > 0$ for $0 < x < \pi$, so

$3\sin x(2 - 3\sin^2 x) > 0 \Rightarrow 2 - 3\sin^2 x > 0$

$2 - 3\sin^2 x > 0 \Rightarrow \sin^2 x < \dfrac{2}{3}$

$\Rightarrow \sin x < \sqrt{\dfrac{2}{3}}$ and $\sin x > -\sqrt{\dfrac{2}{3}}$

$\Rightarrow x < 0.9553...$ and $x > 2.1862...$

So $0 < x < \pi$, $x < 0.9553...$ and $x > 2.1862...$

$\Rightarrow M$ is convex for $0 < x < 0.955$ and $2.186 < x < \pi$.

[4 marks available — 1 mark for using $f''(x) > 0$, 2 marks for all boundary values or 1 mark for at least two correct boundary values, 1 mark for both correct regions]

c) $\dfrac{d^2y}{dx^2} = 6\sin x - 9\sin^3 x = 3\sin x(2 - 3\sin^2 x)$

At $x = n\pi$, $\dfrac{d^2y}{dx^2} = 3\sin(n\pi)(2 - 3\sin^2(n\pi)) = 0$.

$\sin(n\pi + 0.1) = \sin(n\pi)\cos(0.1) + \sin(0.1)\cos(n\pi)$
$= \sin(0.1)\cos(n\pi)$

$\sin(n\pi - 0.1) = \sin(n\pi)\cos(0.1) - \sin(0.1)\cos(n\pi)$
$= -\sin(0.1)\cos(n\pi)$

$\sin x$ does change sign either side of $x = n\pi$.

$2 - 3\sin^2 x$ doesn't change sign either side of $x = n\pi$.

$3\sin x(2 - 3\sin^2 x)$ changes sign either side of $x = n\pi$, so there is a point of inflection at $x = n\pi$ for all integer values of n.

[3 marks available — 1 mark for showing the second derivative is equal to 0 at $x = n\pi$, 1 mark for the correct method to check for points of inflection, 1 mark for a correct explanation leading to the correct conclusion]

d) Since $x = n\pi$ is a point of inflection for each integer n, there are infinitely many points of inflection.

[1 mark for the correct answer]

Exercise 7.12 — Quotient Rule

Q1 a) $u = x + 5$, $v = x - 3 \Rightarrow \dfrac{du}{dx} = 1, \dfrac{dv}{dx} = 1$

$\dfrac{dy}{dx} = \dfrac{v\dfrac{du}{dx} - u\dfrac{dv}{dx}}{v^2}$

$= \dfrac{((x-3) \times 1) - ((x+5) \times 1)}{(x-3)^2} = -\dfrac{8}{(x-3)^2}$

b) $u = (x-7)^4$, $v = (5-x)^3 \Rightarrow \dfrac{du}{dx} = 4(x-7)^3, \dfrac{dv}{dx} = -3(5-x)^2$

$\dfrac{dy}{dx} = \dfrac{v\dfrac{du}{dx} - u\dfrac{dv}{dx}}{v^2}$

$= \dfrac{[(5-x)^3 \times 4(x-7)^3] - [(x-7)^4 \times (-3)(5-x)^2]}{(5-x)^6}$

$= \dfrac{(5-x)^2(x-7)^3[4(5-x) + 3(x-7)]}{(5-x)^6}$

$= \dfrac{(x-7)^3(-x-1)}{(5-x)^4}$

c) $u = e^x$, $v = x^2 \Rightarrow \dfrac{du}{dx} = e^x, \dfrac{dv}{dx} = 2x$

$\dfrac{dy}{dx} = \dfrac{v\dfrac{du}{dx} - u\dfrac{dv}{dx}}{v^2} = \dfrac{x^2e^x - e^x 2x}{x^4} = \dfrac{xe^x(x-2)}{x^4} = \dfrac{e^x(x-2)}{x^3}$

d) $u = 3x$, $v = (x-1)^2 \Rightarrow \dfrac{du}{dx} = 3, \dfrac{dv}{dx} = 2(x-1)$

$\dfrac{dy}{dx} = \dfrac{v\dfrac{du}{dx} - u\dfrac{dv}{dx}}{v^2} = \dfrac{[(x-1)^2 \times 3] - [3x \times 2(x-1)]}{(x-1)^4}$

$= \dfrac{(x-1)[3(x-1) - 6x]}{(x-1)^4} = \dfrac{-3x-3}{(x-1)^3}$

e) $u = \ln x^2 = 2\ln x$, $v = 5x \Rightarrow \dfrac{du}{dx} = \dfrac{2}{x}, \dfrac{dv}{dx} = 5$

$\dfrac{dy}{dx} = \dfrac{v\dfrac{du}{dx} - u\dfrac{dv}{dx}}{v^2} = \dfrac{\left[5x \times \dfrac{2}{x}\right] - [\ln x^2 \times 5]}{25x^2}$

$= \dfrac{10 - 5\ln x^2}{25x^2} = \dfrac{2 - \ln x^2}{5x^2}$

f) $u = 4x^3$, $v = e^x \Rightarrow \dfrac{du}{dx} = 12x^2, \dfrac{dv}{dx} = e^x$

$\dfrac{dy}{dx} = \dfrac{v\dfrac{du}{dx} - u\dfrac{dv}{dx}}{v^2} = \dfrac{12x^2e^x - 4x^3e^x}{(e^x)^2} = \dfrac{12x^2 - 4x^3}{e^x}$

Q2 a) $u = x^3$, $v = (x+3)^3 \Rightarrow \dfrac{du}{dx} = 3x^2, \dfrac{dv}{dx} = 3(x+3)^2$

$f'(x) = \dfrac{v\dfrac{du}{dx} - u\dfrac{dv}{dx}}{v^2} = \dfrac{[(x+3)^3 \times 3x^2] - [x^3 \times 3(x+3)^2]}{(x+3)^6}$

$= \dfrac{3x^2(x+3) - 3x^3}{(x+3)^4} = \dfrac{9x^2}{(x+3)^4}$

b) $u = x^2$, $v = \sqrt{x-7} \Rightarrow \dfrac{du}{dx} = 2x, \dfrac{dv}{dx} = \dfrac{1}{2\sqrt{x-7}}$

$f'(x) = \dfrac{v\dfrac{du}{dx} - u\dfrac{dv}{dx}}{v^2} = \dfrac{[\sqrt{x-7} \times 2x] - \left[x^2\dfrac{1}{2\sqrt{x-7}}\right]}{x-7}$

$= \dfrac{4x(x-7) - x^2}{2(\sqrt{x-7})^3} = \dfrac{3x^2 - 28x}{2(\sqrt{x-7})^3}$

c) $u = e^{2x}$, $v = e^{2x} + e^{-2x} \Rightarrow \dfrac{du}{dx} = 2e^{2x}, \dfrac{dv}{dx} = 2e^{2x} - 2e^{-2x}$

$f'(x) = \dfrac{v\dfrac{du}{dx} - u\dfrac{dv}{dx}}{v^2} = \dfrac{[(e^{2x} + e^{-2x})2e^{2x}] - [e^{2x}(2e^{2x} - 2e^{-2x})]}{(e^{2x} + e^{-2x})^2}$

$= \dfrac{2e^{4x} + 2 - 2e^{4x} + 2}{e^{4x} + e^{-4x} + 2} = \dfrac{4}{e^{4x} + e^{-4x} + 2}$

d) $u = x$, $v = \sin x \Rightarrow \dfrac{du}{dx} = 1, \dfrac{dv}{dx} = \cos x$

$f'(x) = \dfrac{v\dfrac{du}{dx} - u\dfrac{dv}{dx}}{v^2} = \dfrac{\sin x - x\cos x}{\sin^2 x}$

e) $u = \sin x$, $v = x \Rightarrow \dfrac{du}{dx} = \cos x, \dfrac{dv}{dx} = 1$

$f'(x) = \dfrac{v\dfrac{du}{dx} - u\dfrac{dv}{dx}}{v^2} = \dfrac{x\cos x - \sin x}{x^2}$

f) $u = \cos x$, $v = 3x \Rightarrow \dfrac{du}{dx} = -\sin x, \dfrac{dv}{dx} = 3$

$f'(x) = \dfrac{v\dfrac{du}{dx} - u\dfrac{dv}{dx}}{v^2}$

$= \dfrac{-3x\sin x - 3\cos x}{9x^2} = \dfrac{-x\sin x - \cos x}{3x^2}$

Q3 $u = x^2$, $v = \tan x \Rightarrow \dfrac{du}{dx} = 2x, \dfrac{dv}{dx} = \sec^2 x$

$f'(x) = \dfrac{v\dfrac{du}{dx} - u\dfrac{dv}{dx}}{v^2}$

$= \dfrac{[\tan x \times 2x] - [x^2 \times \sec^2 x]}{\tan^2 x} = \dfrac{2x\tan x - x^2\sec^2 x}{\tan^2 x}$

$= \dfrac{2x}{\tan x} - x^2\dfrac{\dfrac{1}{\cos^2 x}}{\dfrac{\sin^2 x}{\cos^2 x}} = 2x\cot x - x^2\cosec^2 x$

Q4 $u = 5x - 4$, $v = 2x^2 \Rightarrow \dfrac{du}{dx} = 5, \dfrac{dv}{dx} = 4x$

$\dfrac{dy}{dx} = \dfrac{v\dfrac{du}{dx} - u\dfrac{dv}{dx}}{v^2} = \dfrac{[2x^2 \times 5] - [(5x-4) \times 4x]}{(2x^2)^2}$

$= \dfrac{10x^2 - (20x^2 - 16x)}{4x^4} = \dfrac{16x - 10x^2}{4x^4} = \dfrac{8 - 5x}{2x^3}$

There is a stationary point when $\dfrac{dy}{dx} = 0$:

$\dfrac{dy}{dx} = 0 \Rightarrow 8 - 5x = 0 \Rightarrow 5x = 8 \Rightarrow x = \dfrac{8}{5}$

When $x = \dfrac{8}{5}$, $y = \dfrac{5\left(\dfrac{8}{5}\right) - 4}{2\left(\dfrac{8}{5}\right)^2} = \dfrac{8-4}{2\left(\dfrac{64}{25}\right)} = \dfrac{25}{32}$

So the coordinates of the stationary point are $\left(\dfrac{8}{5}, \dfrac{25}{32}\right)$.

To determine the nature of the stationary point, differentiate again to find $\dfrac{d^2y}{dx^2}$: $\dfrac{dy}{dx} = \dfrac{8 - 5x}{2x^3}$

$u = 8 - 5x$, $v = 2x^3 \Rightarrow \dfrac{du}{dx} = -5, \dfrac{dv}{dx} = 6x^2$

$$\frac{d^2y}{dx^2} = \frac{v\frac{du}{dx} - u\frac{dv}{dx}}{v^2} = \frac{[2x^3 \times (-5)] - [(8 - 5x) \times 6x^2]}{(2x^3)^2}$$

$$= \frac{-10x^3 - (48x^2 - 30x^3)}{4x^6} = \frac{20x^3 - 48x^2}{4x^6} = \frac{5x - 12}{x^4}$$

When $x = \frac{8}{5}$, $\frac{d^2y}{dx^2} = \frac{5\left(\frac{8}{5}\right) - 12}{\left(\frac{8}{5}\right)^4} = \frac{8 - 12}{\left(\frac{8}{5}\right)^4} = \frac{-4}{\left(\frac{8}{5}\right)^4}$

Since the denominator is positive, $\frac{d^2y}{dx^2}$ is negative, and so the stationary point must be a maximum.

Q5 **a)** Use the quotient rule to find the first derivative:

$u = 4x, v = 1 + x^2 \Rightarrow \frac{du}{dx} = 4, \frac{dv}{dx} = 2x$

$$\frac{dy}{dx} = \frac{v\frac{du}{dx} - u\frac{dv}{dx}}{v^2} = \frac{4(1 + x^2) - (2x)(4x)}{(1 + x^2)^2} = \frac{4 - 4x^2}{(1 + x^2)^2}$$

Use the quotient rule again to find the second derivative:

$u = 4 - 4x^2, v = (1 + x^2)^2 \Rightarrow \frac{du}{dx} = -8x, \frac{dv}{dx} = 4x(1 + x^2)$

$$\frac{d^2y}{dx^2} = \frac{v\frac{du}{dx} - u\frac{dv}{dx}}{v^2} = \frac{-8x(1 + x^2)^2 - (4 - 4x^2) \times 4x(1 + x^2)}{(1 + x^2)^4}$$

$$= \frac{-8x(1 + x^2) - 4x(4 - 4x^2)}{(1 + x^2)^3} = \frac{8x(x^2 - 3)}{(1 + x^2)^3}$$

[5 marks available — 1 mark for a correct method to differentiate once, 1 mark for the correct first derivative, 1 mark for a correct method to differentiate again, 1 mark for the correct second derivative, 1 mark for the correct simplified answer]

b) $\frac{d^2y}{dx^2} = 0 \Rightarrow 8x(x^2 - 3) = 0 \Rightarrow x = 0$, or $x = \pm\sqrt{3}$.

When $x = 0, y = 0$.

When $x = \sqrt{3}, y = \frac{4\sqrt{3}}{1 + 3} = \sqrt{3}$

and when $x = -\sqrt{3}, y = \frac{-4\sqrt{3}}{1 + 3} = -\sqrt{3}$,

so the points of inflection are $(0, 0), (\sqrt{3}, \sqrt{3}), (-\sqrt{3}, -\sqrt{3})$.

[3 marks available — 1 mark for a method to solve second derivative equal to zero, 1 mark for the correct x-values, 1 mark for the correct answers]

You know the curve has three points of inflection, so they must be these points.

Q6 $u = x, v = e^x \Rightarrow \frac{du}{dx} = 1, \frac{dv}{dx} = e^x$

$$\frac{dy}{dx} = \frac{v\frac{du}{dx} - u\frac{dv}{dx}}{v^2} = \frac{[e^x \times 1] - [x \times e^x]}{(e^x)^2} = \frac{e^x - xe^x}{(e^x)^2} = \frac{1 - x}{e^x}$$

Use the quotient rule again to find $\frac{d^2y}{dx^2}$:

$u = 1 - x, v = e^x \Rightarrow \frac{du}{dx} = -1, \frac{dv}{dx} = e^x$

$$\frac{d^2y}{dx^2} = \frac{v\frac{du}{dx} - u\frac{dv}{dx}}{v^2} = \frac{[e^x \times -1] - [(1 - x) \times e^x]}{(e^x)^2}$$

$$= \frac{-e^x - e^x + xe^x}{(e^x)^2} = \frac{-1 - 1 + x}{e^x} = \frac{x - 2}{e^x}$$

$\frac{d^2y}{dx^2} = 0$ when $x - 2 = 0 \Rightarrow x = 2$

Check $\frac{d^2y}{dx^2}$ either side of $x = 2$:

When $x > 2, (x - 2) > 0 \Rightarrow \frac{d^2y}{dx^2} > 0$

When $x < 2, (x - 2) < 0 \Rightarrow \frac{d^2y}{dx^2} < 0$

$\frac{d^2y}{dx^2}$ changes sign at $x = 2$, so it's a point of inflection.

When $x = 2, y = \frac{2}{e^2}$, so the only point of inflection of the graph $y = \frac{x}{e^x}$ occurs at the point $\left(2, \frac{2}{e^2}\right)$.

Q7 **a)** $u = x, v = \cos(2x) \Rightarrow \frac{du}{dx} = 1, \frac{dv}{dx} = -2\sin(2x)$

$$\frac{dy}{dx} = \frac{v\frac{du}{dx} - u\frac{dv}{dx}}{v^2} = \frac{\cos(2x) - [x \times -2\sin(2x)]}{\cos^2(2x)}$$

$$= \frac{\cos(2x) + 2x\sin(2x)}{\cos^2(2x)}$$

b) $\frac{dy}{dx} = 0$ if $\cos(2x) + 2x\sin(2x) = 0$

$\Rightarrow -\cos(2x) = 2x\sin(2x)$

$\Rightarrow 2x = -\frac{\cos(2x)}{\sin(2x)} = -\cot(2x) \Rightarrow x = -\frac{1}{2}\cot(2x)$

Remember cos x/sin x = 1/tan x = cot x.

Q8 **a)** $u = 1, v = 1 + 4\cos x \Rightarrow \frac{du}{dx} = 0, \frac{dv}{dx} = -4\sin x$

$$\frac{dy}{dx} = \frac{v\frac{du}{dx} - u\frac{dv}{dx}}{v^2} = \frac{[(1 + 4\cos x) \times 0] - [1 \times (-4\sin x)]}{(1 + 4\cos x)^2}$$

$$= \frac{4\sin x}{(1 + 4\cos x)^2}$$

When $x = \frac{\pi}{2}, \frac{dy}{dx} = \frac{4}{(1)^2} = 4$ and $y = \frac{1}{1} = 1$

$y = mx + c \Rightarrow 1 = 4\frac{\pi}{2} + c \Rightarrow c = 1 - 2\pi$

So the equation of the tangent is $y = 4x + 1 - 2\pi$.

b) From part a), the gradient of the normal must be $-\frac{1}{4}$. Equation of a straight line:

$y = mx + c \Rightarrow 1 = -\frac{1}{4}\frac{\pi}{2} + c \Rightarrow c = 1 + \frac{\pi}{8}$

So the equation of the normal is $y = -\frac{1}{4}x + 1 + \frac{\pi}{8}$

You could also do the differentiation here by writing y as (1 + 4 cos x)⁻¹, and then using the chain rule. In general, you don't have to use the quotient rule when the numerator is just a number.

Q9 $u = 2x, v = \cos x \Rightarrow \frac{du}{dx} = 2, \frac{dv}{dx} = -\sin x$

$$\frac{dy}{dx} = \frac{v\frac{du}{dx} - u\frac{dv}{dx}}{v^2} = \frac{[\cos x \times 2] - [2x \times (-\sin x)]}{\cos^2 x}$$

$$= \frac{2\cos x + 2x\sin x}{\cos^2 x}$$

When $x = \frac{\pi}{3}, \frac{dy}{dx} = \frac{1 + \frac{\pi\sqrt{3}}{3}}{\left(\frac{1}{2}\right)^2} = 4 + \frac{4\pi\sqrt{3}}{3}$

Q10 $u = x - \sin x, v = 1 + \cos x$

$\Rightarrow \frac{du}{dx} = 1 - \cos x, \frac{dv}{dx} = -\sin x$

$$\frac{dy}{dx} = \frac{v\frac{du}{dx} - u\frac{dv}{dx}}{v^2}$$

$$= \frac{[(1 + \cos x)(1 - \cos x)] - [(x - \sin x)(-\sin x)]}{(1 + \cos x)^2}$$

$$= \frac{1 - \cos^2 x - \sin^2 x + x\sin x}{(1 + \cos x)^2}$$

$$= \frac{1 - 1 + x\sin x}{(1 + \cos x)^2} = \frac{x\sin x}{(1 + \cos x)^2}$$

[3 marks available — 1 mark for attempted use of the quotient rule, 1 mark for the correct expression for the derivative, 1 mark for rearranging to give the correct result]

Use the identity sin² x + cos² x ≡ 1 to simplify the expression.

Answers

Q11 Stationary points occur when the gradient is 0.

$u = \cos x$, $v = 4 - 3\cos x$, $\Rightarrow \dfrac{du}{dx} = -\sin x$, $\dfrac{dv}{dx} = 3\sin x$

$\dfrac{dy}{dx} = \dfrac{v\dfrac{du}{dx} - u\dfrac{dv}{dx}}{v^2} = \dfrac{[(4 - 3\cos x)(-\sin x)] - [\cos x(3\sin x)]}{(4 - 3\cos x)^2}$

$= \dfrac{-4\sin x}{(4 - 3\cos x)^2}$

$\dfrac{dy}{dx} = 0 \Rightarrow -4\sin x = 0 \Rightarrow x = \sin^{-1} 0 = 0$, π and 2π.

When $x = 0$, $y = \dfrac{1}{4 - 3} = 1$

When $x = \pi$, $y = \dfrac{-1}{4 - (-3)} = -\dfrac{1}{7}$

When $x = 2\pi$, $y = \dfrac{1}{4 - 3} = 1$

So the stationary points are $(0, 1)$, $\left(\pi, -\dfrac{1}{7}\right)$ and $(2\pi, 1)$

Q12 First use the chain rule: $y = e^{f(x)} \Rightarrow \dfrac{dy}{dx} = f'(x)e^{f(x)}$

Then use the quotient rule to find $f'(x)$:

$u = 1 + x$, $v = 1 - x \Rightarrow \dfrac{du}{dx} = 1$, $\dfrac{dv}{dx} = -1$

$\dfrac{dy}{dx} = \dfrac{v\dfrac{du}{dx} - u\dfrac{dv}{dx}}{v^2} = \dfrac{[(1 - x)(1)] - [(1 + x)(-1)]}{(1 - x)^2} = \dfrac{2}{(1 - x)^2}$

So $\dfrac{dy}{dx} = f'(x)e^{f(x)} = \dfrac{2e^{\frac{1+x}{1-x}}}{(1 - x)^2}$

Q13 $y = \dfrac{2 + 3x^2}{3x - 1}$ is increasing when $\dfrac{dy}{dx} > 0$.

$u = 2 + 3x^2$, $v = 3x - 1 \Rightarrow \dfrac{du}{dx} = 6x$, $\dfrac{dv}{dx} = 3$

$\dfrac{dy}{dx} = \dfrac{v\dfrac{du}{dx} - u\dfrac{dv}{dx}}{v^2} = \dfrac{[(3x - 1) \times 6x] - [(2 + 3x^2) \times 3]}{(3x - 1)^2}$

$= \dfrac{(18x^2 - 6x) - (6 + 9x^2)}{(3x - 1)^2} = \dfrac{9x^2 - 6x - 6}{(3x - 1)^2}$

Since the denominator is squared, it is always positive.

So $\dfrac{dy}{dx} > 0$ when $9x^2 - 6x - 6 > 0 \Rightarrow 3x^2 - 2x - 2 > 0$

Use the quadratic formula to solve $\dfrac{dy}{dx} = 0$:

$x = \dfrac{2 \pm \sqrt{4 - (4 \times 3 \times -2)}}{6} = \dfrac{2 \pm \sqrt{28}}{6} = \dfrac{1 \pm \sqrt{7}}{3}$

Since the coefficient of x^2 in $9x^2 - 6x - 6$ is positive,

the graph of $\dfrac{dy}{dx}$ is u-shaped. So the values of x where $\dfrac{dy}{dx} > 0$

and where $\dfrac{2 + 3x^2}{3x - 1}$ is increasing are $x < \dfrac{1 - \sqrt{7}}{3}$ and $x > \dfrac{1 + \sqrt{7}}{3}$.

The function y is undefined when $3x - 1 = 0$, i.e. $x = \dfrac{1}{3}$
(the graph has an asymptote), but this isn't in the range
where the function is increasing, so it doesn't affect the answer.

Q14 a) $u = \ln x$, $v = x \Rightarrow \dfrac{du}{dx} = \dfrac{1}{x}$, $\dfrac{dv}{dx} = 1$

$\dfrac{dy}{dx} = \dfrac{v\dfrac{du}{dx} - u\dfrac{dv}{dx}}{v^2} = \dfrac{x \times \left(\dfrac{1}{x}\right) - \ln x \times 1}{x^2} = \dfrac{1 - \ln x}{x^2}$

[3 marks available — 1 mark for attempted use of the
quotient rule, 1 mark for the correct expression for the
derivative, 1 mark for the correct simplified answer]

b) $\dfrac{dy}{dx} = 0 \Rightarrow 1 - \ln x = 0 \Rightarrow \ln x = 1 \Rightarrow x = e$

so $y = \dfrac{\ln e}{e} = \dfrac{1}{e}$ so the stationary point is $\left(e, \dfrac{1}{e}\right)$.

[3 marks available — 1 mark for setting $\dfrac{dy}{dx} = 0$, 1 mark for
the correct x-value, 1 mark for the correct y-value]

c) (i) $u = 1 - \ln x$, $v = x^2 \Rightarrow \dfrac{du}{dx} = -\dfrac{1}{x}$, $\dfrac{dv}{dx} = 2x$

$\dfrac{d^2y}{dx^2} = \dfrac{v\dfrac{du}{dx} - u\dfrac{dv}{dx}}{v^2}$

$= \dfrac{\left(x^2 \times \left(-\dfrac{1}{x}\right)\right) - (1 - \ln x) \times 2x}{x^4} = \dfrac{2x\ln x - 3x}{x^4}$

so when $x = e$, $\dfrac{d^2y}{dx^2} = \dfrac{2e\ln e - 3e}{e^4} = -\dfrac{e}{e^4} = -\dfrac{1}{e^3} < 0$

so there is a maximum at $x = e$.

[3 marks available — 1 mark for a correct method to
find the second derivative, 1 mark for the correct second
derivative, 1 mark for substituting $x = e$ with the correct
conclusion]

(ii) Since $\left(e, \dfrac{1}{e}\right)$ is the maximum point, $\dfrac{\ln 2}{2} < \dfrac{1}{e}$

$\Rightarrow e\ln 2 < 2 \Rightarrow \ln 2^e < 2 \Rightarrow 2^e < e^2$.

[2 marks available — 1 mark for $\dfrac{\ln 2}{2} < \dfrac{1}{e}$, 1 mark for
rearranging to give the correct result]

Exercise 7.13 — Differentiating Cosec, Sec and Cot

Q1 a) $\dfrac{dy}{dx} = -2\text{cosec}\,(2x)\cot(2x)$

b) $\dfrac{dy}{dx} = \dfrac{dy}{du} \times \dfrac{du}{dx} = (2\text{cosec}\,x)(-\text{cosec}\,x\cot x)$
$= -2\text{cosec}^2\,x\cot x$

c) $\dfrac{dy}{dx} = \dfrac{dy}{du} \times \dfrac{du}{dx} = -7\text{cosec}^2\,(7x)$

d) $\dfrac{dy}{dx} = \dfrac{dy}{du} \times \dfrac{du}{dx} = (7\cot^6 x)(-\text{cosec}^2 x) = -7\cot^6 x\,\text{cosec}^2 x$

e) $\dfrac{dy}{dx} = u\dfrac{dv}{dx} + v\dfrac{du}{dx} = [x^4 \times (-\text{cosec}^2 x)] + [\cot x \times 4x^3]$
$= 4x^3\cot x - x^4\text{cosec}^2 x$
$= x^3(4\cot x - x\text{cosec}^2 x)$

f) $\dfrac{dy}{dx} = \dfrac{dy}{du} \times \dfrac{du}{dx} = 2(x + \sec x)(1 + \sec x\tan x)$

g) $\dfrac{dy}{dx} = \dfrac{dy}{du} \times \dfrac{du}{dx} = [-\text{cosec}\,(x^2 + 5)\cot(x^2 + 5)] \times 2x$
$= -2x\,\text{cosec}\,(x^2 + 5)\cot(x^2 + 5)$

h) $\dfrac{dy}{dx} = u\dfrac{dv}{dx} + v\dfrac{du}{dx} = [e^{3x} \times \sec x\tan x] + [\sec x \times 3e^{3x}]$
$= e^{3x}\sec x\,(\tan x + 3)$

i) $\dfrac{dy}{dx} = \dfrac{dy}{du} \times \dfrac{du}{dx} = 3(2x + \cot x)^2(2 - \text{cosec}^2 x)$

Q2 a) $f'(x) = \dfrac{v\dfrac{du}{dx} - u\dfrac{dv}{dx}}{v^2} = \dfrac{(x + 3)\sec x\tan x - \sec x}{(x + 3)^2}$

b) $f'(x) = \dfrac{dy}{du} \times \dfrac{du}{dx} = \left(\sec \dfrac{1}{x}\tan \dfrac{1}{x}\right)\left(-\dfrac{1}{x^2}\right) = -\dfrac{\sec \dfrac{1}{x}\tan \dfrac{1}{x}}{x^2}$

c) $f'(x) = \dfrac{dy}{du} \times \dfrac{du}{dx} = (\sec\sqrt{x}\,\tan\sqrt{x})\left(\dfrac{1}{2} \times \dfrac{1}{\sqrt{x}}\right)$
$= \dfrac{\sec\sqrt{x}\tan\sqrt{x}}{2\sqrt{x}} = \dfrac{\tan\sqrt{x}}{2\sqrt{x}\cos\sqrt{x}}$

Q3 a) $\dfrac{dy}{dx} = u\dfrac{dv}{dx} + v\dfrac{du}{dx} = (x \times -\text{cosec}^2 x) + (\cot x \times 1)$
$= \cot x - x\,\text{cosec}^2 x$

[2 marks available — 1 mark for a correct method,
1 mark for the correct answer]

b) When $x = \dfrac{\pi}{2}$, $\dfrac{dy}{dx} = \cot\left(\dfrac{\pi}{2}\right) - \dfrac{\pi}{2}\text{cosec}^2\left(\dfrac{\pi}{2}\right) = 0 - \dfrac{\pi}{2} = -\dfrac{\pi}{2}$

and $y = 0$ so the tangent has equation

$y = -\dfrac{\pi}{2}\left(x - \dfrac{\pi}{2}\right) \Rightarrow y = -\dfrac{\pi}{2}x + \dfrac{\pi^2}{4} \Rightarrow 4y = -2\pi x + \pi^2$

$\Rightarrow 2\pi x + 4y - \pi^2 = 0$

[3 marks available — 1 mark for the correct gradient, 1 mark
for a correct method to find the equation of the line, 1 mark
for rearranging to give the required result]

Alternatively you could rearrange the given equation to show it
has the right gradient and that $(\frac{\pi}{2}, 0)$ lies on the line.

Q4 $f'(x) = \dfrac{dy}{du} \times \dfrac{du}{dx} = 2(\sec x + \text{cosec } x)(\sec x \tan x - \text{cosec } x \cot x)$

Q5 $f(x) = \dfrac{1}{x \cot x} = \dfrac{\tan x}{x}$

$f'(x) = \dfrac{v\frac{du}{dx} - u\frac{dv}{dx}}{v^2} = \dfrac{x\sec^2 x - \tan x}{x^2}$

Q6 $f'(x) = u\dfrac{dv}{dx} + v\dfrac{du}{dx}$

$= [e^x \times (-\text{cosec } x \cot x)] + [\text{cosec } x \times e^x] = e^x \text{cosec } x(1 - \cot x)$

Q7 $f'(x) = u\dfrac{dv}{dx} + v\dfrac{du}{dx} = [e^{3x} \times (-4\text{ cosec}^2 4x)] + [\cot 4x \times 3e^{3x}]$

$= e^{3x}(3 \cot 4x - 4 \text{ cosec}^2 4x)$

The chain rule was used here to differentiate cot 4x
— use y = cot u and u = 4x.

Q8 $f'(x) = u\dfrac{dv}{dx} + v\dfrac{du}{dx}$

$= [e^{-2x} \times (-4 \text{ cosec } 4x \cot 4x)] + [\text{cosec } 4x \times (-2)e^{-2x}]$

$= -2e^{-2x} \text{cosec } 4x [2 \cot 4x + 1]$

Q9 $f'(x) = u\dfrac{dv}{dx} + v\dfrac{du}{dx} = [\ln x \times (-\text{cosec } x \cot x)] + \left[\text{cosec } x \times \dfrac{1}{x}\right]$

$= \text{cosec } x\left(\dfrac{1}{x} - \ln x \cot x\right)$

Q10 $f'(x) = \dfrac{dy}{du} \times \dfrac{du}{dx} = \left(\dfrac{1}{2} \times \dfrac{1}{\sqrt{\sec x}}\right) \times \sec x \tan x$

$= \dfrac{\sec x \tan x}{2\sqrt{\sec x}} = \dfrac{1}{2}\tan x\sqrt{\sec x}$

Q11 $f'(x) = g'(x)e^{g(x)} = e^{\sec x} \sec x \tan x$

Q12 a) $f'(x) = \dfrac{g'(x)}{g(x)} = \dfrac{-\text{cosec } x \cot x}{\text{cosec } x} = -\cot x$

b) $\ln(\text{cosec } x) = \ln\left(\dfrac{1}{\sin x}\right) = \ln 1 - \ln(\sin x)$

$= -\ln(\sin x) \text{ (as } \ln 1 = 0)$

Here you could also rearrange by saying
ln (1/sin x) = ln (sin x)$^{-1}$ = −ln (sin x).

$f'(x) = \dfrac{g'(x)}{g(x)} = -\dfrac{\cos x}{\sin x} = -\dfrac{1}{\tan x} = -\cot x$

Q13 $f'(x) = \dfrac{g'(x)}{g(x)} = \dfrac{1 + \sec x \tan x}{x + \sec x}$

Q14 $\dfrac{dy}{dx} = \dfrac{dy}{du} \times \dfrac{du}{dx} = \sec\sqrt{x^2 + 5}\, \tan\sqrt{x^2 + 5} \times \dfrac{2x}{2\sqrt{x^2 + 5}}$

$= \dfrac{x\sec\sqrt{x^2 + 5}\, \tan\sqrt{x^2 + 5}}{\sqrt{x^2 + 5}}$

Q15 a) $u = 2\cot 2x,\ v = x \Rightarrow \dfrac{du}{dx} = -4\text{ cosec}^2 2x$ and $\dfrac{dv}{dx} = 1$

$\dfrac{dy}{dx} = \dfrac{v\frac{du}{dx} - u\frac{dv}{dx}}{v^2} = \dfrac{-4x\text{cosec}^2 2x - 2\cot 2x}{x^2}$

When $x = \dfrac{\pi}{3}$, $\dfrac{dy}{dx} = \dfrac{-\frac{4\pi}{3}\text{cosec}^2\frac{2\pi}{3} - 2\cot\frac{2\pi}{3}}{\frac{\pi^2}{9}}$

$= \dfrac{-\frac{4}{3}\pi\left(\frac{4}{3}\right) - 2\left(-\frac{\sqrt{3}}{3}\right)}{\frac{\pi^2}{9}} = \dfrac{-12\pi\left(\frac{4}{3}\right) - 18\left(-\frac{\sqrt{3}}{3}\right)}{\pi^2}$

$= \dfrac{-16\pi + 6\sqrt{3}}{\pi^2} = \dfrac{6\sqrt{3} - 16\pi}{\pi^2}$

[5 marks available — 1 mark for correct $\frac{du}{dx}$, 1 mark for
correct use of the quotient or product rule, 1 mark for correct
$\frac{dy}{dx}$, 1 mark for evaluating the trig functions correctly,
1 mark for simplifying to give the correct answer]

b) When $x = \dfrac{\pi}{3}$, $y = \dfrac{2\left(-\frac{\sqrt{3}}{3}\right)}{\frac{\pi}{3}} = -\dfrac{2\sqrt{3}}{\pi}$, so the tangent has

equation $y + \dfrac{2\sqrt{3}}{\pi} = \dfrac{-16\pi + 6\sqrt{3}}{\pi^2}\left(x - \dfrac{\pi}{3}\right)$

Multiply each side by $3\pi^2$, then rearrange:

$3\pi^2 y + 6\sqrt{3}\,\pi = (-16\pi + 6\sqrt{3})(3x - \pi)$

$\Rightarrow 3\pi^2 y + 6\sqrt{3}\,\pi = 3x(-16\pi + 6\sqrt{3}) + 16\pi^2 - 6\sqrt{3}\,\pi$

$\Rightarrow 3\pi^2 y + 6x(8\pi - 3\sqrt{3}) = 16\pi^2 - 12\pi\sqrt{3}$

$\Rightarrow 3\pi^2 y + 6(8\pi - 3\sqrt{3})x = 4\pi(4\pi - 3\sqrt{3})$

[3 marks available — 1 mark for the correct y-value, 1 mark
for a correct method to find the equation of the tangent,
1 mark for rearranging to give the required result]

Exercise 7.14 — Connected Rates of Change

Q1 $\dfrac{dx}{dt} = -0.1$ (it's negative as the cube is shrinking),

and $V = x^3$, so $\dfrac{dV}{dx} = 3x^2$

$\dfrac{dV}{dt} = \dfrac{dV}{dx} \times \dfrac{dx}{dt} = -0.3x^2 \text{ cm}^3 \text{ min}^{-1}$

Q2 $V = 30x^3$ so $\dfrac{dV}{dx} = 90x^2$, and $\dfrac{dx}{d\theta} = 0.15$.

$\dfrac{dV}{d\theta} = \dfrac{dV}{dx} \times \dfrac{dx}{d\theta} = 13.5x^2 \text{ cm}^3 \text{ °C}^{-1}$

When $x = 3$, $\dfrac{dV}{d\theta} = 121.5 \text{ cm}^3 \text{ °C}^{-1}$

Q3 $A = 4\pi r^2$ so $\dfrac{dA}{dr} = 8\pi r$, and $\dfrac{dr}{dt} = -1.6$.

$\dfrac{dA}{dt} = \dfrac{dA}{dr} \times \dfrac{dr}{dt} = -12.8\pi r \text{ cm}^2 \text{ h}^{-1}$

When $r = 5.5$ cm, $\dfrac{dA}{dt} = -221.17 \text{ cm}^2 \text{ h}^{-1}$ (2 d.p.)

Q4 $\dfrac{dr}{d\theta} = 2 \times 10^{-2} \text{ mm °C}^{-1} = 2 \times 10^{-5} \text{ m °C}^{-1}$,

and $V = \dfrac{4}{3}\pi r^3 \text{ m}^3$, so $\dfrac{dV}{dr} = 4\pi r^2 \text{ m}^3 \text{ m}^{-1}$.

$\dfrac{dV}{d\theta} = \dfrac{dV}{dr} \times \dfrac{dr}{d\theta} = 8 \times 10^{-5}\pi r^2 \text{ m}^3 \text{ °C}^{-1}$

You could have converted to mm instead,
and given your answer in mm^3 °C^{-1}.

Q5 a) (i) Let $y = \sqrt{u} = u^{\frac{1}{2}}$, where $u = 16x + 25$

$\dfrac{dy}{du} = \dfrac{1}{2}u^{-\frac{1}{2}} = \dfrac{1}{2\sqrt{u}} = \dfrac{1}{2\sqrt{16x + 25}}$ and $\dfrac{du}{dx} = 16$

$\dfrac{dy}{dx} = \dfrac{dy}{du} \times \dfrac{du}{dx} = \dfrac{1}{2\sqrt{16x + 25}} \times 16 = \dfrac{8}{\sqrt{16x + 25}}$

[2 marks available — 1 mark for a correct method,
1 mark for the correct answer]

(ii) Let $z = 2\ln(v)$, where $v = y^2 + 1$

$\dfrac{dz}{dv} = \dfrac{2}{v} = \dfrac{2}{y^2 + 1}$ and $\dfrac{dv}{dy} = 2y$

$\dfrac{dz}{dv} = \dfrac{dz}{dv} \times \dfrac{dv}{dy} = \dfrac{2}{y^2 + 1} \times 2y = \dfrac{4y}{y^2 + 1}$

[2 marks available — 1 mark for a correct method,
1 mark for the correct answer]

b) $\dfrac{dz}{dx} = \dfrac{dz}{dy} \times \dfrac{dy}{dx} = \dfrac{4y}{y^2 + 1} \times \dfrac{8}{\sqrt{16x + 25}}$

$= \dfrac{4\sqrt{16x + 25}}{\sqrt{16x + 25}^2 + 1} \times \dfrac{8}{\sqrt{16x + 25}} = \dfrac{32}{\sqrt{16x + 25}^2 + 1}$

$= \dfrac{32}{16x + 25 + 1} = \dfrac{32}{16x + 26} = \dfrac{16}{8x + 13}$

When $x = 6$, $\dfrac{dz}{dx} = \dfrac{16}{8(6) + 13} = \dfrac{16}{61}$

[2 marks available — 1 mark for a correct method,
1 mark for the correct answer]

Answers

Q6 The surface area of the tank, $A = 2(\pi r^2) + 3r(2\pi r) = 8\pi r^2$,
so $\frac{dA}{dr} = 16\pi r$, and $\frac{dH}{dA} = -2$ (it's negative as heat is lost).
$\frac{dH}{dr} = \frac{dH}{dA} \times \frac{dA}{dr} = -32\pi r$ J cm^{-1}
When $r = 12.3$, $\frac{dH}{dr} = -1236.53$ J cm^{-1} (2 d.p.)

Q7 $\frac{dH}{dt} = -0.5$ mm h^{-1} = -0.05 cm h^{-1} (length is decreasing).
$V = \pi r^2 H$, so $\frac{dV}{dH} = \pi r^2$.
$\frac{dV}{dt} = \frac{dV}{dH} \times \frac{dH}{dt} = -0.05\pi r^2$ cm^3 h^{-1}

Q8 **a)** Using Pythagoras, the height of the triangle is
$\sqrt{x^2 - \left(\frac{x}{2}\right)^2} = \frac{\sqrt{3}x}{2}$, so the area of the end is
$A = \frac{1}{2}\left(\frac{\sqrt{3}x}{2}\right)x = \frac{\sqrt{3}x^2}{4}$.
[2 marks available — 1 mark for a correct method, 1 mark for the correct answer]

b) $V = A \times 20 = 5\sqrt{3}\ x^2$, so $\frac{dV}{dx} = 10\sqrt{3}\ x$.
$\frac{dx}{dt} = 0.6$.
$\frac{dV}{dt} = \frac{dV}{dx} \times \frac{dx}{dt} = 6\sqrt{3}\ x$ mm^3 per day
[3 marks available — 1 mark for using the expression from part a) correctly, 1 mark for $\frac{dV}{dx}$, 1 mark for the correct answer]

c) $x - 0.5 \Rightarrow \frac{dV}{dt} = 3\sqrt{3} = 5.20$ mm^3 per day (2 d.p.)
[2 marks available — 1 mark for substituting x = 0.5 into the equation, 1 mark for the correct answer]

Q9 **a)** n is directly proportional to D, so you can write this as
$n = kD$, where k is a constant.
Use the condition that when $D = 2$, $n = 208$:
$208 = 2k \Rightarrow k = 104$
$n = 104D \Rightarrow \frac{dn}{dD} = 104$
$D = 1 + 2^{\lambda t}$, so $\frac{dD}{dt} = \lambda 2^{\lambda t} \ln 2$.
Use the rule for differentiating a^x from earlier in the chapter.
So $\frac{dn}{dt} = \frac{dn}{dD} \times \frac{dD}{dt} = 104\lambda 2^{\lambda t} \ln 2$ per day.

b) If $t = 1$ day and $\lambda = 5$,
$\frac{dn}{dt} = 104 \times 5 \times 2^5 \times \ln 2 = 16640 \ln 2$
$= 1.15 \times 10^4$ per day (3 s.f.)

Q10 $\frac{dV}{dt} = 3$ and $V = \frac{4}{3}\pi r^3$ so $\frac{dV}{dr} = 4\pi r^2$
$\frac{dr}{dt} = \frac{dr}{dV} \times \frac{dV}{dt} = \frac{1}{4\pi r^2} \times 3 = \frac{3}{4\pi r^2}$
When $V = 100$, $\frac{4}{3}\pi r^3 = 100$
$\Rightarrow r^3 = \frac{300}{4\pi} = \frac{75}{\pi} \Rightarrow r = \sqrt[3]{\frac{75}{\pi}}$.
So $\frac{dr}{dt} = \frac{3}{4\pi\left(\sqrt[3]{\frac{75}{\pi}}\right)^2} = 0.0288$ cms^{-1} to 3 s.f.
[4 marks available — 1 mark for $\frac{dV}{dr}$, 1 mark for using the chain rule, 1 mark for finding r when V = 100, 1 mark for the correct answer]

Q11 **a)** Volume of water $V = \pi r^2 h$, so $\frac{dV}{dh} = \pi r^2$.
$\frac{dV}{dt} = -0.3$.
$\frac{dh}{dt} = \frac{dh}{dV} \times \frac{dV}{dt}$
$= \frac{1}{\frac{dV}{dh}} \times \frac{dV}{dt} = \frac{1}{\pi r^2} \times -0.3 = -\frac{3}{10\pi r^2}$ cm s^{-1}
[3 marks available — 1 mark for both $\frac{dh}{dV}$ and $\frac{dV}{dt}$, 1 mark for using the chain rule, 1 mark for the correct answer]

b) $r = 6$, so $\frac{dh}{dt} = -\frac{1}{120\pi}$ cm s^{-1}.
So the water level falls at a constant rate
of $\frac{1}{120\pi}$ cm s^{-1} and $\frac{1}{2\pi} = 0.159$ cm min^{-1} (3 s.f.)
You don't actually need to use h in the calculation.
[2 marks available — 1 mark for using the derivative, 1 mark for the correct answer]

Q12 **a)** $\frac{dV}{d\theta} = k$, $V = \frac{2}{3}\pi r^3$, so $\frac{dV}{dr} = 2\pi r^2$.
$\frac{dr}{d\theta} = \frac{dr}{dV} \times \frac{dV}{d\theta} = \frac{1}{\frac{dV}{dr}} \times \frac{dV}{d\theta} = \frac{1}{2\pi r^2} \times k = \frac{k}{2\pi r^2}$ cm °C^{-1}
[3 marks available — 1 mark for both $\frac{dr}{dV}$ and $\frac{dV}{d\theta}$, 1 mark for using the chain rule, 1 mark for the correct answer]

b) $V = 4$, so $r = \sqrt[3]{\frac{6}{\pi}} = 1.240...$, and $k = 1.5$.
So $\frac{dr}{d\theta} = \frac{1.5}{2\pi(1.240...)^2} = 0.155$ cm °C^{-1} (3 s.f.)
[2 marks available — 1 mark for using the derivative, 1 mark for the correct answer]

Q13 **a)** $V = 3h^2(h + 1) \Rightarrow V = 3h^3 + 3h^2$
$\frac{dV}{dh} = 9h^2 + 6h$ and $\frac{dV}{dt} = 10$ so
$\frac{dh}{dt} = \frac{dh}{dV} \times \frac{dV}{dt} = \frac{1}{\frac{dV}{dh}} \times \frac{dV}{dt} = \frac{1}{9h^2 + 6h} \times 10 = \frac{10}{9h^2 + 6h}$
[3 marks available — 1 mark for $\frac{dV}{dh}$, 1 mark for using the chain rule, 1 mark for $\frac{dh}{dt}$]

b) When $h = 8$, $\frac{dh}{dt} = \frac{10}{9 \times 8^2 + 6 \times 8} = 0.0160$ cm s^{-1} to 3 s.f.
[2 marks available — 1 mark for a correct method, 1 mark for the correct answer]

c) A rate of 2 cm every 3 seconds means $\frac{2}{3}$ cm every second,
so $\frac{dh}{dt} = \frac{2}{3} \Rightarrow \frac{10}{9h^2 + 6h} = \frac{2}{3} \Rightarrow 10 = 6h^2 + 4h$
$\Rightarrow 6h^2 + 4h - 10 = 0 \Rightarrow 3h^2 + 2h - 5 = 0$
$\Rightarrow (3h + 5)(h - 1) = 0 \Rightarrow h = -\frac{5}{3}$ and $h = 1$,
but h must be positive so ignore the negative value.
Then $V = 3(1^2)(1 + 1) = 6$ cm^3.
[3 marks available — 1 mark for a correct method to find h, 1 mark for the correct value of h, 1 mark for the correct volume]

Q14 $K = \frac{1}{2}mv^2 \Rightarrow \frac{dK}{dv} = mv$ and $\frac{dv}{dt} = a$ (as the rate of change of velocity with respect to time is acceleration)
$\Rightarrow \frac{dK}{dt} = \frac{dK}{dv} \times \frac{dv}{dt} = mva$
$\frac{dK}{dt} = 5 \times 1 \times 6 = 30$ joules/sec
[3 marks available — 1 mark for $\frac{dK}{dv}$, 1 mark for $\frac{dK}{dt}$, 1 mark for the correct answer]

Q15 $\frac{1}{r} = \frac{1}{s_1} + \frac{1}{s_2}$ $\Rightarrow r^{-1} = s_1^{-1} + s_2^{-1}$

Finding the derivative of r^{-1} with respect to t:

Let $y = r^{-1} \Rightarrow \frac{dy}{dr} = -r^{-2} = -\frac{1}{r^2}$

$\Rightarrow \frac{d}{dt} r^{-1} = \frac{dy}{dr} \times \frac{dr}{dt} = -\frac{1}{r^2} \times \frac{dr}{dt}$

Similarly for the resistors s_1 and s_2:

$\frac{d}{dt} s_1^{-1} = -\frac{1}{(s_1)^2} \times \frac{ds_1}{dt}$ and $\frac{d}{dt} s_2^{-1} = -\frac{1}{(s_2)^2} \times \frac{ds_2}{dt}$.

So $-\frac{1}{r^2} \times \frac{dr}{dt} = -\frac{1}{(s_1)^2} \times \frac{ds_1}{dt} - \frac{1}{(s_2)^2} \times \frac{ds_2}{dt}$

When $s_1 = 20$, $s_2 = 30$, so $\frac{1}{r} = \frac{1}{20} + \frac{1}{30} = \frac{1}{12}$ $\Rightarrow r = 12$

Hence we have:

$-\frac{1}{12^2} \times \frac{dr}{dt} = -\frac{1}{20^2} \times 0.3 - \frac{1}{30^2} \times (-0.1) = -\frac{23}{36000}$

$\Rightarrow \frac{dr}{dt} = -\frac{23}{36000} \times (-144) = \frac{23}{250} = 0.092$ ohms/min

Exercise 7.15 — Differentiating Parametric Equations

Q1 **a)** $\frac{dx}{dt} = 2t$, $\frac{dy}{dt} = 3t^2 - 1$.

Using the chain rule, $\frac{dy}{dx} = \frac{dy}{dt} \div \frac{dx}{dt}$ so $\frac{dy}{dx} = \frac{3t^2 - 1}{2t}$

b) $\frac{dx}{dt} = 3t^2 + 1$, $\frac{dy}{dt} = 4t$, so $\frac{dy}{dx} = \frac{4t}{3t^2 + 1}$

c) $\frac{dx}{dt} = 4t^3$, $\frac{dy}{dt} = 3t^2 - 2t$, so $\frac{dy}{dx} = \frac{3t^2 - 2t}{4t^3} = \frac{3t - 2}{4t^2}$

d) $\frac{dx}{dt} = -\sin t$, $\frac{dy}{dt} = 4 - 2t$, so $\frac{dy}{dx} = \frac{2t - 4}{\sin t}$

Q2 **a)** $\frac{dx}{dt} = 2t$, $\frac{dy}{dt} = 2e^{2t}$, so $\frac{dy}{dx} = \frac{e^{2t}}{t}$

b) When $t = 1$, $\frac{dy}{dx} = e^2$.

Q3 **a)** $\frac{dy}{dt} = 12t^2 - 4t$, $\frac{dx}{dt} = 3e^{3t}$, so $\frac{dy}{dx} = \frac{12t^2 - 4t}{3e^{3t}}$

b) When $t = 0$, $\frac{dy}{dx} = 0$.

Q4 **a)** $\frac{dx}{dt} = 3t^2$, $\frac{dy}{dt} = 2t \cos t - t^2 \sin t$,

so $\frac{dy}{dx} = \frac{2t \cos t - t^2 \sin t}{3t^2} = \frac{2 \cos t - t \sin t}{3t}$

b) When $t = \pi$, $\frac{dy}{dx} = -\frac{2}{3\pi}$

Q5 **a)** $\frac{dx}{dt} = 2t \sin t + t^2 \cos t$,

$\frac{dy}{dt} = t^3 \cos t + 3t^2 \sin t - \sin t$,

so $\frac{dy}{dx} = \frac{t^3 \cos t + (3t^2 - 1) \sin t}{2t \sin t + t^2 \cos t}$

b) When $t = \pi$, $\frac{dy}{dx} = \pi$.

Q6 **a)** $\frac{dx}{dt} = -6 \sin 2t$, $\frac{dy}{dt} = 2 \sin t \cos t = \sin 2t$

so $\frac{dy}{dx} = \frac{\sin 2t}{-6 \sin 2t} = -\frac{1}{6}$

[3 marks available — 1 mark for $\frac{dx}{dt}$, 1 mark for $\frac{dy}{dt}$, 1 mark for the correct answer]

b) $\cos 2t \equiv 1 - 2 \sin^2 t \Rightarrow x = 3(1 - 2 \sin^2 t) = 3(1 - 2y)$

[2 marks available — 1 mark for correct use of double angle identity, 1 mark for the correct answer]

Alternatively, as the gradient is constant, you know it's a straight-line equation. Put any value of t into the parametric equations in the question (e.g. when t = 0, x = 3 and y = 0) then you have all the information you need to derive the Cartesian equation.

c) $\frac{x}{3} = 1 - 2y \Rightarrow y = \frac{1}{2} - \frac{1}{6}x$. This is a straight line with y-intercept $(0, \frac{1}{2})$, x-intercept $(3, 0)$ and gradient $-\frac{1}{6}$.

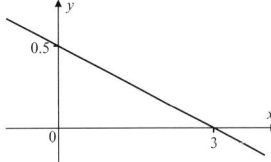

[2 marks available — 2 marks for a complete correct graph, otherwise 1 mark for a straight line with one intercept marked correctly]

Q7 **a)** $\frac{dx}{dt} = \frac{1}{t}$, $\frac{dy}{dt} = 6t - 3t^2$, so $\frac{dy}{dx} = 6t^2 - 3t^3$

b) When $t = -1$, $\frac{dy}{dx} = 9$.

c) At the stationary points $6t^2 - 3t^3 = 3t^2(2 - t) = 0$, so the stationary points occur at $t = 0$ and $t = 2$.

At $t = 0$, x is not defined.

At $t = 2$, the coordinates are $(\ln 2, 4)$.

Q8 **a)** $\frac{dx}{dt} = 2t$, $\frac{dy}{dt} = 9t^2 - 4$, so $\frac{dy}{dx} = \frac{9t^2 - 4}{2t}$

[2 marks available — 1 mark for $\frac{dx}{dt}$ and $\frac{dy}{dt}$, 1 mark for $\frac{dy}{dx}$]

b) At the stationary points $\frac{9t^2 - 4}{2t} = 0$

$\Rightarrow 9t^2 - 4 = 0 \Rightarrow t^2 = \frac{4}{9} \Rightarrow t = -\frac{2}{3}$ or $\frac{2}{3}$

$x = t^2 = \frac{4}{9}$

At $t = -\frac{2}{3}$, $y = 3\left(-\frac{2}{3}\right)^3 - 4\left(-\frac{2}{3}\right) = \frac{16}{9}$

At $t = \frac{2}{3}$, $y = 3\left(\frac{2}{3}\right)^3 - 4\left(\frac{2}{3}\right) = -\frac{16}{9}$

The stationary points are $\left(\frac{4}{9}, -\frac{16}{9}\right)$ and $\left(\frac{4}{9}, \frac{16}{9}\right)$.

[3 marks available — 1 mark for a method to solve the first derivative equal to zero, 1 mark for the x-coordinate of both points, 1 mark for both pairs of coordinates]

Q9 **a)** Using the quotient rule with $u = \cos t$ and $v = t$

$\Rightarrow \frac{du}{dt} = -\sin t, \frac{dv}{dt} = 1 \Rightarrow \frac{dx}{dt} = \frac{-t \sin t - \cos t}{t^2}$

Using the quotient rule with $u = \sin t$ and $v = t$

$\Rightarrow \frac{du}{dt} = \cos t, \frac{dv}{dt} = 1 \Rightarrow \frac{dy}{dt} = \frac{t \cos t - \sin t}{t^2}$

so $\frac{dy}{dx} = \frac{dy}{dt} \times \frac{dt}{dx} = \frac{t \cos t - \sin t}{t^2} \times \frac{t^2}{-t \sin t - \cos t}$

$= \frac{t \cos t - \sin t}{-t \sin t - \cos t} = \frac{\sin t - t \cos t}{\cos t + t \sin t}$

[4 marks available — 1 mark for a correct method to differentiate, 1 mark for $\frac{dx}{dt}$, 1 mark for $\frac{dy}{dt}$, 1 mark for the correct answer]

b) When $t = \pi$, $\frac{dy}{dx} = \frac{\sin \pi - \pi \cos \pi}{\cos \pi - \pi \sin \pi} = -\pi$.

[1 mark for the correct answer]

c) When $y = 0$, $\sin t = 0$ so $t = \pi, 2\pi, 3\pi$.

So $x = \frac{\cos \pi}{\pi}, \frac{\cos 2\pi}{2\pi}, \frac{\cos 3\pi}{3\pi} \Rightarrow x = -\frac{1}{\pi}, \frac{1}{2\pi}, -\frac{1}{3\pi}$

[2 marks available — 1 mark for the correct values of t, 1 mark for all three correct values of x]

Q10 **a)** $\frac{dx}{dt} = 1 + \sin t$, $\frac{dy}{dt} = \cos t$ so $\frac{dy}{dx} = \frac{\cos t}{1 + \sin t}$

[2 marks available — 1 mark for $\frac{dx}{dt}$ and $\frac{dy}{dt}$, 1 mark for $\frac{dy}{dx}$]

Answers

b) $\dfrac{d^2y}{dx^2} = \dfrac{d}{dx}\left(\dfrac{\cos t}{1+\sin t}\right) = \dfrac{d}{dt}\left(\dfrac{\cos t}{1+\sin t}\right) \times \dfrac{dt}{dx}$

Firstly, use the quotient rule to differentiate with respect to t:

Let $u = \cos t$ and $v = 1 + \sin t$

$\dfrac{du}{dt} = -\sin t$ and $\dfrac{dv}{dt} = \cos t$

So $\dfrac{d}{dt}\left(\dfrac{\cos t}{1+\sin t}\right) = \dfrac{-\sin t(1+\sin t) - \cos^2 t}{(1+\sin t)^2}$.

Then find $\dfrac{dt}{dx}$:

$\dfrac{dt}{dx} = \dfrac{1}{\dfrac{dx}{dt}} = \dfrac{1}{1+\sin t}$

$\dfrac{d^2y}{dx^2} = \dfrac{d}{dt}\left(\dfrac{\cos t}{1+\sin t}\right) \times \dfrac{dt}{dx}$

$= \dfrac{-\sin t(1+\sin t) - \cos^2 t}{(1+\sin t)^2} \times \dfrac{1}{1+\sin t}$

$= \dfrac{-\sin t - \sin^2 t - \cos^2 t}{(1+\sin t)^3} = \dfrac{-(\sin t + \sin^2 t + \cos^2 t)}{(1+\sin t)^3}$

$= -\dfrac{1+\sin t}{(1+\sin t)^3} = -\dfrac{1}{(1+\sin t)^2}$

[4 marks available — 1 mark for using the quotient rule to differentiate with respect to t, 1 mark for $\dfrac{dt}{dx}$, 1 mark for the correct method to find $\dfrac{d^2y}{dx^2}$, 1 mark for the correct answer]

c) $\dfrac{d^2y}{dx^2} \neq 0$ for any value of t since the numerator is -1, so there are no points of inflection.

[1 mark for a correct explanation]

Exercise 7.16 — Finding Tangents and Normals of Parametric Curves

Q1 $\dfrac{dx}{dt} = 2t$, $\dfrac{dy}{dt} = 3t^2 - 6$.

Using the chain rule, $\dfrac{dy}{dx} = \dfrac{dy}{dt} \div \dfrac{dx}{dt}$, so $\dfrac{dy}{dx} = \dfrac{3t^2 - 6}{2t}$.

When $t = 3$: $\dfrac{dy}{dx} = \dfrac{21}{6} = \dfrac{7}{2}$, $x = 3^2 = 9$ and $y = 3^3 - 6(3) = 9$.

Putting this into $y = mx + c$ gives: $9 = \dfrac{7}{2}(9) + c \Rightarrow c = -\dfrac{45}{2}$

So the equation of the tangent is:

$y = \dfrac{7}{2}x - \dfrac{45}{2} \Rightarrow 7x - 2y - 45 = 0$.

Q2 $\dfrac{dy}{dx} = \dfrac{3t^2 - 2t + 5}{3t^2 - 4t}$

When $t = -1$: $\dfrac{dy}{dx} = \dfrac{10}{7}$, $x = -3$ and $y = -7$.

Putting this into $y = mx + c$ gives: $-7 = \dfrac{10}{7}(-3) + c \Rightarrow c = -\dfrac{19}{7}$

So the equation of the tangent is $10x - 7y - 19 = 0$.

Q3 $\dfrac{dy}{dx} = \dfrac{3\cos t - t\sin t}{2\cos 2t}$

When $t = \pi$: $\dfrac{dy}{dx} = -\dfrac{3}{2}$, $x = 0$ and $y = -\pi$.

The gradient of the normal is $\dfrac{2}{3}$.

Putting this into $y = mx + c$ gives: $-\pi = \dfrac{2}{3}(0) + c \Rightarrow c = -\pi$

So the equation of the normal is $y = \dfrac{2}{3}x - \pi$.

Q4 $\dfrac{dy}{dx} = \dfrac{3t^2 - 2t}{1 + \ln t}$

When $t = 1$: $\dfrac{dy}{dx} = \dfrac{3-2}{1+0} = 1$, $x = 0$ and $y = 3$.

Putting this into $y = mx + c$ gives: $3 = 0 + c \Rightarrow c = 3$

So the equation of the tangent is: $y = x + 3$

Q5 $\dfrac{dy}{dx} = \dfrac{2\theta + \cos\theta - \theta\sin\theta}{\sin 2\theta + 2\theta\cos 2\theta}$

When $\theta = \dfrac{\pi}{2}$: $\dfrac{dy}{dx} = -\dfrac{1}{2}$, $x = 0$ and $y = \dfrac{\pi^2}{4}$.

The gradient of the normal is 2.

Putting this into $y = mx + c$ gives: $\dfrac{\pi^2}{4} = 2(0) + c \Rightarrow c = \dfrac{\pi^2}{4}$

So the equation of the normal is $y = 2x + \dfrac{\pi^2}{4}$.

Q6 a) $\dfrac{dy}{dx} = \dfrac{3 - 3t^2}{2t - 1}$

When $t = 2$, $\dfrac{dy}{dx} = -3$, $x = 2$ and $y = -2$.

Putting this into $y = mx + c$ gives: $-2 = -3(2) + c \Rightarrow c = 4$

So the equation of the tangent is $y = 4 - 3x$.

b) The gradient of the normal at $t = 2$ is $\dfrac{1}{3}$.

Putting this into $y = mx + c$ gives: $-2 = \dfrac{1}{3}(2) + c \Rightarrow c = -\dfrac{8}{3}$

So the equation of the normal is $3y = x - 8$.

This crosses the x-axis at $y = 0$, so $0 = x - 8$
$\Rightarrow x = 8$ and so the coordinates are $(8, 0)$.

Q7 a) $\dfrac{dx}{dt} = 8t$, $\dfrac{dy}{dt} = 8$ so $\dfrac{dy}{dx} = \dfrac{8}{8t} = \dfrac{1}{t}$.

At $(1, 4)$, $4 = 8t \Rightarrow t = \dfrac{1}{2}$ so the gradient of the curve is 2.

Therefore, the gradient of the normal is $-\dfrac{1}{2}$.

The equation of the normal is $y - 4 = -\dfrac{1}{2}(x - 1)$
$\Rightarrow 2y - 8 = -x + 1 \Rightarrow x + 2y - 9 = 0$.

[4 marks available — 1 mark for $\dfrac{dx}{dt}$ and $\dfrac{dy}{dt}$, 1 mark for $\dfrac{dy}{dx}$, 1 mark for the correct gradient of the normal, 1 mark for showing the correct result]

b) $t = \dfrac{y}{8}$ so $x = 4\left(\dfrac{y}{8}\right)^2 = \dfrac{y^2}{16} \Rightarrow y^2 = 16x$

[2 marks available — 1 mark for rearranging to make t the subject, 1 mark for the correct answer]

c) The branch of the curve for $y > 0$ has equation $y = \sqrt{16x} = 4\sqrt{x}$, so the area between $x = 0$ and $x = 1$ is given by $\displaystyle\int_0^1 4\sqrt{x}\ dx = \int_0^1 4x^{\frac{1}{2}}\ dx = \left[\dfrac{8}{3}x^{\frac{3}{2}}\right]_0^1 = \dfrac{8}{3}$.

The normal crosses the x-axis when $x = 9$.

The area between $x = 1$ and $x = 9$ is a triangle with area $\dfrac{1}{2} \times 8 \times 4 = 16$. So the total area is $\dfrac{8}{3} + 16 = \dfrac{56}{3}$.

[5 marks available — 1 mark for rewriting the equation with y as the subject, 1 mark for a correct method of integration, 1 mark for the correct area under the curve, 1 mark for the correct x-intercept of the normal line, 1 mark for the correct answer]

Q8 a) $\dfrac{dy}{dx} = \dfrac{\theta\cos\theta + \sin\theta}{2\cos 2\theta - 2\sin\theta}$

b) When $\theta = \dfrac{\pi}{2}$: $\dfrac{dy}{dx} = -\dfrac{1}{4}$, $x = 0$ and $y = \dfrac{\pi}{2}$.

Putting this into $y = mx + c$ gives: $\dfrac{\pi}{2} = -\dfrac{1}{4}(0) + c \Rightarrow c = \dfrac{\pi}{2}$

So the equation of the tangent is $y = \dfrac{\pi}{2} - \dfrac{1}{4}x$.

The gradient of the normal is 4. The normal also goes through the point $\left(0, \dfrac{\pi}{2}\right)$, so again $c = \dfrac{\pi}{2}$.

So the equation of the normal is $y = 4x + \dfrac{\pi}{2}$.

Q9 a) The path cuts the y-axis when $x = 0$, so $s^3 \ln s = 0$.

Since $\ln 0$ is undefined, $s = 0$ cannot be a solution, so the only solution is $s = 1$ (i.e. when $\ln s = 0$).

[2 marks available — 1 mark for rejecting s = 0 with justification, 1 mark for stating s = 1 is a solution]

b) $\dfrac{dx}{ds} = \dfrac{s^3}{s} + 3s^2 \ln s = s^2 + 3s^2 \ln s$,

$\dfrac{dy}{ds} = 3s^2 - \left(\dfrac{s^2}{s} + 2s \ln s\right) = 3s^2 - s - 2s \ln s$

so $\dfrac{dy}{dx} = \dfrac{3s^2 - s - 2s \ln s}{s^2 + 3s^2 \ln s} = \dfrac{3s - 1 - 2 \ln s}{s + 3s \ln s}$.

From a), when $x = 0$, $s = 1$.

When $s = 1$: $\dfrac{dy}{dx} = 2$, $x = 0$ and $y = 1$.

Putting this into $y = mx + c$ gives: $1 = 2(0) + c \Rightarrow c = 1$.

So the equation of the tangent is $y = 2x + 1$.

[4 marks available — 1 mark for $\dfrac{dx}{ds}$ and $\dfrac{dy}{ds}$, 1 mark for the expression for $\dfrac{dy}{dx}$, 1 mark for the value of $\dfrac{dy}{dx}$ when $s = 1$, 1 mark for correct working leading to the equation of the tangent]

Q10 a) $\dfrac{dx}{dt} = 2 \sin 2t$, $\dfrac{dy}{dt} = \cos t - \sin t$ so

$\dfrac{dy}{dx} = \dfrac{\cos t - \sin t}{2 \sin 2t} = \dfrac{\cos t - \sin t}{4 \sin t \cos t}$

$= \dfrac{1}{4}\left(\dfrac{1}{\sin t} - \dfrac{1}{\cos t}\right) = \dfrac{1}{4}(\operatorname{cosec} t - \sec t)$

[3 marks available — 1 mark for $\dfrac{dx}{dt}$ and $\dfrac{dy}{dt}$, 1 mark for method to find $\dfrac{dy}{dx}$, 1 mark for correctly rearranging to the required form]

b) When $t = \dfrac{\pi}{6}$,

$\dfrac{dy}{dx} = \dfrac{1}{4}\left(2 - \dfrac{2}{3}\sqrt{3}\right) = \dfrac{3 - \sqrt{3}}{6}$ and $x = 1 - \cos 2\left(\dfrac{\pi}{6}\right) = \dfrac{1}{2}$,

$y = \sin\left(\dfrac{\pi}{6}\right) + \cos\left(\dfrac{\pi}{6}\right) = \dfrac{1 + \sqrt{3}}{2}$

Therefore, the equation of the tangent is

$y - \dfrac{1 + \sqrt{3}}{2} = \dfrac{3 - \sqrt{3}}{6}\left(x - \dfrac{1}{2}\right) \Rightarrow y = \dfrac{3 - \sqrt{3}}{6}x + \dfrac{2\sqrt{3}}{3}$

[4 marks available — 1 mark for the correct gradient, 1 mark for the correct x- and y-values, 1 mark for a correct method to find equation of the tangent, 1 mark for the answer in the correct form]

c) Stationary points are found when $\dfrac{dy}{dx} = 0$

$\Rightarrow \dfrac{1}{4}(\operatorname{cosec} t - \sec t) = 0 \Rightarrow \operatorname{cosec} t = \sec t$

$\Rightarrow \sin t = \cos t \Rightarrow \tan t = 1 \Rightarrow t = \dfrac{\pi}{4}, \dfrac{5\pi}{4}$, so there are two stationary points where $0 \le t \le 2\pi$.

[3 marks available — 1 mark for a correct method to solve derivative = 0, 1 mark for tan t = 1, 1 mark for both correct answers]

d) The tangents will be horizontal as $\dfrac{dy}{dx} = 0$ at the stationary points. When $t = \dfrac{\pi}{4}$, $y = \sqrt{2}$ and when $t = \dfrac{5\pi}{4}$, $y = -\sqrt{2}$ so the tangents have equations $y = \sqrt{2}$ and $y = -\sqrt{2}$.

[2 marks available — 1 mark for a correct method to find tangents, 1 mark for both correct tangents]

Q11 a) Using the quotient rule, $\dfrac{dy}{d\theta} = \dfrac{-\theta^3 \sin \theta - 3\theta^2 \cos \theta}{\theta^6}$

$= \dfrac{-\theta \sin \theta - 3 \cos \theta}{\theta^4}$

Using the product rule, $\dfrac{dx}{d\theta} = \theta^2 \cos \theta + 2\theta \sin \theta$

Hence $\dfrac{dy}{dx} = \dfrac{-\theta \sin \theta - 3 \cos \theta}{\theta^6 \cos \theta + 2\theta^5 \sin \theta}$

This looks a bit complicated but leave it as it is — you'll find the cos and sin terms usually disappear when you substitute.

When $\theta = \pi$, $\dfrac{dy}{dx} = \dfrac{-\pi(0) - 3(-1)}{\pi^6(-1) + 2\pi^5(0)} = -\dfrac{3}{\pi^6}$.

[4 marks available — 1 mark for $\dfrac{dy}{d\theta}$, 1 mark for $\dfrac{dx}{d\theta}$, 1 mark for $\dfrac{dy}{dx}$, 1 mark for substituting $\theta = \pi$ to show the correct result]

b) The gradient of the normal when $\theta = \pi$ is $\dfrac{\pi^6}{3}$,

and $y = -\dfrac{1}{\pi^3}$, $x = 0$.

Putting this into $y = mx + c$ gives:

$-\dfrac{1}{\pi^3} = \dfrac{\pi^6}{3}(0) + c \Rightarrow c = -\dfrac{1}{\pi^3}$

So the equation of the normal is $y = \dfrac{\pi^6}{3}x - \dfrac{1}{\pi^3}$.

[2 marks available — 1 mark for the correct gradient, 1 mark for the correct equation]

Q12 a) Using the quotient rule,

$\dfrac{dx}{d\theta} = \dfrac{\theta^2 \cos \theta - 2\theta \sin \theta}{\theta^4} = \dfrac{\theta \cos \theta - 2 \sin \theta}{\theta^3}$

Using the product rule, $\dfrac{dy}{d\theta} = -2\theta \sin 2\theta + \cos 2\theta$

So $\dfrac{dy}{dx} = \dfrac{dy}{d\theta} \times \dfrac{1}{\left(\dfrac{dx}{d\theta}\right)}$

$= (-2\theta \sin 2\theta + \cos 2\theta) \times \dfrac{\theta^3}{\theta \cos \theta - 2 \sin \theta}$

$= \dfrac{\theta^3 (-2\theta \sin 2\theta + \cos 2\theta)}{\theta \cos \theta - 2 \sin \theta}$

When $\theta = \dfrac{\pi}{2}$, $\dfrac{dy}{dx} = \dfrac{\left(\dfrac{\pi}{2}\right)^3\left(-2\left(\dfrac{\pi}{2}\right)(0) - 1\right)}{\left(\dfrac{\pi}{2}\right)(0) - 2(1)} = \dfrac{\pi^3}{16}$

[4 marks available — 1 mark for $\dfrac{dx}{d\theta}$, 1 mark for $\dfrac{dy}{d\theta}$, 1 mark for $\dfrac{dy}{dx}$, 1 mark for the correct answer]

b) When $\theta = \dfrac{\pi}{2}$, $x = \dfrac{4}{\pi^2}$ and $y = -\dfrac{\pi}{2}$

The tangent to the curve has a gradient of $\dfrac{\pi^3}{16}$ and an equation in the form $y = mx + c$.

So substitute in the values of x, y and m to find c:

$-\dfrac{\pi}{2} = \dfrac{\pi^3}{16} \times \dfrac{4}{\pi^2} + c = \dfrac{\pi}{4} + c \Rightarrow c = -\dfrac{3\pi}{4}$

So the tangent has the equation

$y = \dfrac{\pi^3}{16}x - \dfrac{3\pi}{4}$.

[2 marks available — 1 mark for a correct method, 1 mark for the correct answer]

c) At the x-axis, $y = 0$, so $0 = \theta \cos 2\theta$

$\Rightarrow \theta = 0$ (ignore this result, since $\theta > 0$), or

$\cos 2\theta = 0 \Rightarrow \theta = \dfrac{\pi}{4}, \dfrac{3\pi}{4}, \dfrac{5\pi}{4}$ ... etc.

The solution with $0 < \theta \le \dfrac{\pi}{2}$ is $\theta = \dfrac{\pi}{4}$.

When $\theta = \dfrac{\pi}{4}$, $x = \dfrac{\sqrt{2}}{2} \div \dfrac{\pi^2}{16} = \dfrac{8\sqrt{2}}{\pi^2}$, so the coordinates of the first point where the curve cuts the x-axis are $\left(\dfrac{8\sqrt{2}}{\pi^2}, 0\right)$.

[3 marks available — 1 mark for a correct method to find θ, 1 mark for the correct value of θ, 1 mark for the correct answer]

Exercise 7.17 — Implicit Differentiation

Q1 a) $\dfrac{d}{dx}(y) + \dfrac{d}{dx}(y^3) = \dfrac{d}{dx}(x^2) + \dfrac{d}{dx}(4)$

$\dfrac{dy}{dx} + 3y^2 \dfrac{dy}{dx} = 2x + 0$

$(1 + 3y^2)\dfrac{dy}{dx} = 2x \Rightarrow \dfrac{dy}{dx} = \dfrac{2x}{1 + 3y^2}$

b) $2x + 2y\dfrac{dy}{dx} = 2 + 2\dfrac{dy}{dx}$

$(2y - 2)\dfrac{dy}{dx} = 2 - 2x \Rightarrow \dfrac{dy}{dx} = \dfrac{2 - 2x}{2y - 2} = \dfrac{1 - x}{y - 1}$

c) $9x^2 - 4\dfrac{dy}{dx} = 2y\dfrac{dy}{dx} + 1$

$9x^2 - 1 = (2y + 4)\dfrac{dy}{dx} \Rightarrow \dfrac{dy}{dx} = \dfrac{9x^2 - 1}{2y + 4}$

Answers

d) $5 - 2y\dfrac{dy}{dx} = 5x^4 - 6\dfrac{dy}{dx}$

$5 - 5x^4 = (2y - 6)\dfrac{dy}{dx} \implies \dfrac{dy}{dx} = \dfrac{5 - 5x^4}{2y - 6}$

e) $-\sin x + \cos y\dfrac{dy}{dx} = 2x + 3y^2\dfrac{dy}{dx}$

$(\cos y - 3y^2)\dfrac{dy}{dx} = 2x + \sin x \implies \dfrac{dy}{dx} = \dfrac{2x + \sin x}{\cos y - 3y^2}$

f) $3x^2y^2 + 2x^3y\dfrac{dy}{dx} - \sin x = 4y + 4x\dfrac{dy}{dx}$

$(2x^3y - 4x)\dfrac{dy}{dx} = 4y - 3x^2y^2 + \sin x$

$\implies \dfrac{dy}{dx} = \dfrac{4y - 3x^2y^2 + \sin x}{2x^3y - 4x}$

g) $e^x + e^y\dfrac{dy}{dx} = 3x^2 - \dfrac{dy}{dx}$

$(e^y + 1)\dfrac{dy}{dx} = 3x^2 - e^x \implies \dfrac{dy}{dx} = \dfrac{3x^2 - e^x}{e^y + 1}$

h) $3y^2 + 6xy\dfrac{dy}{dx} + 4xy + 2x^2\dfrac{dy}{dx} = 3x^2 + 4$

$(6xy + 2x^2)\dfrac{dy}{dx} = 3x^2 + 4 - 3y^2 - 4xy$

$\implies \dfrac{dy}{dx} = \dfrac{3x^2 + 4 - 3y^2 - 4xy}{6xy + 2x^2}$

i) $12x^2y^2 + 8x^3y\dfrac{dy}{dx} + 6xy + 3x^2\dfrac{dy}{dx} = 2\cos x - 4x^3$

$\dfrac{dy}{dx}(8x^3y + 3x^2) = 2\cos x - 4x^3 - 12x^2y^2 - 6xy$

$\implies \dfrac{dy}{dx} = \dfrac{2\cos x - 4x^3 - 12x^2y^2 - 6xy}{8x^3y + 3x^2}$

Q2 a) $3x^2 + 2y + 2x\dfrac{dy}{dx} = 4y^3\dfrac{dy}{dx}$

$(4y^3 - 2x)\dfrac{dy}{dx} = 3x^2 + 2y \implies \dfrac{dy}{dx} = \dfrac{3x^2 + 2y}{4y^3 - 2x}$

b) $2xy + x^2\dfrac{dy}{dx} + 2y\dfrac{dy}{dx} = 3x^2$

$(x^2 + 2y)\dfrac{dy}{dx} = 3x^2 - 2xy \implies \dfrac{dy}{dx} = \dfrac{3x^2 - 2xy}{x^2 + 2y}$

c) $y^3 + 3xy^2\dfrac{dy}{dx} + \dfrac{dy}{dx} = \cos x$

$(3xy^2 + 1)\dfrac{dy}{dx} = \cos x - y^3 \implies \dfrac{dy}{dx} = \dfrac{\cos x - y^3}{3xy^2 + 1}$

d) $-y\sin x + \cos x\dfrac{dy}{dx} + \sin y + x\cos y\dfrac{dy}{dx} = y + x\dfrac{dy}{dx}$

$(\cos x + x\cos y - x)\dfrac{dy}{dx} = y + y\sin x - \sin y$

$\implies \dfrac{dy}{dx} = \dfrac{y + y\sin x - \sin y}{\cos x + x\cos y - x}$

e) $e^x + e^y\dfrac{dy}{dx} = y + x\dfrac{dy}{dx}$

$(e^y - x)\dfrac{dy}{dx} = y - e^x \implies \dfrac{dy}{dx} = \dfrac{y - e^x}{e^y - x}$

f) $\dfrac{1}{x} + 2x = 3y^2\dfrac{dy}{dx} + \dfrac{dy}{dx}$

$(3y^2 + 1)\dfrac{dy}{dx} = \dfrac{1}{x} + 2x \implies \dfrac{dy}{dx} = \dfrac{\frac{1}{x} + 2x}{3y^2 + 1} = \dfrac{1 + 2x^2}{3xy^2 + x}$

g) $2e^{2x} + 3e^{3y}\dfrac{dy}{dx} = 6xy^2 + 6x^2y\dfrac{dy}{dx}$

$(3e^{3y} - 6x^2y)\dfrac{dy}{dx} = 6xy^2 - 2e^{2x} \implies \dfrac{dy}{dx} = \dfrac{6xy^2 - 2e^{2x}}{3e^{3y} - 6x^2y}$

h) $\ln x + 1 + \dfrac{y}{x} + \ln x\dfrac{dy}{dx} = 5x^4 + 3y^2\dfrac{dy}{dx}$

$(\ln x - 3y^2)\dfrac{dy}{dx} = 5x^4 - \ln x - 1 - \dfrac{y}{x}$

$\implies \dfrac{dy}{dx} = \dfrac{5x^4 - \ln x - 1 - \frac{y}{x}}{\ln x - 3y^2} = \dfrac{5x^5 - x\ln x - x - y}{x\ln x - 3xy^2}$

Q3 a) At $(0, 1)$:   LHS: $e^0 + 2\ln 1 = 1$

RHS: $1^3 = 1$

So $(0, 1)$ is a point on the curve.

b) $e^x + \dfrac{2}{y}\dfrac{dy}{dx} = 3y^2\dfrac{dy}{dx} \implies \dfrac{dy}{dx} = \dfrac{e^x}{3y^2 - \frac{2}{y}} = \dfrac{ye^x}{3y^3 - 2}$.

At $(0, 1)$ the gradient is $\dfrac{1e^0}{3(1^3) - 2} = 1$

Q4 a) $3x^2 + 2y\dfrac{dy}{dx} - 2y - 2x\dfrac{dy}{dx} = 0 \implies \dfrac{dy}{dx} = \dfrac{2y - 3x^2}{2y - 2x}$

b) Putting $x = -2$ into the equation gives: $-8 + y^2 + 4y = 0$

Complete the square to solve...

$(y + 2)^2 - 4 - 8 = 0 \implies y + 2 = \pm\sqrt{12} = \pm 2\sqrt{3}$

so $y = -2 \pm 2\sqrt{3}$

c) At $(-2, -2 + 2\sqrt{3})$:

$\dfrac{dy}{dx} = \dfrac{(-4 + 4\sqrt{3}) - 3(-2)^2}{(-4 + 4\sqrt{3}) - 2(-2)} = \dfrac{-16 + 4\sqrt{3}}{4\sqrt{3}} = \dfrac{\sqrt{3} - 4}{\sqrt{3}}$

Rationalise the denominator: $= \dfrac{3 - 4\sqrt{3}}{3} = 1 - \dfrac{4}{3}\sqrt{3}$

Q5 a) Putting $x = 1$ into the equation gives:

$1 - y = 2y^2 \implies 2y^2 + y - 1 = 0 \implies (2y - 1)(y + 1) = 0$

So $y = -1$ (given as the other point) and $y = \dfrac{1}{2}$. So $a = \dfrac{1}{2}$.

b) $3x^2 - y - x\dfrac{dy}{dx} = 4y\dfrac{dy}{dx}$

$\dfrac{dy}{dx} = \dfrac{3x^2 - y}{4y + x}$

At $(1, -1)$, $\dfrac{dy}{dx} = \dfrac{3(1^2) - (-1)}{4(-1) + 1} = -\dfrac{4}{3}$

At $\left(1, \dfrac{1}{2}\right)$, $\dfrac{dy}{dx} = \dfrac{3(1^2) - \left(\frac{1}{2}\right)}{4\left(\frac{1}{2}\right) + 1} = \dfrac{5}{6}$

Q6 a) Putting $x = 1$ into the equation gives:

$y + y^2 - y - 4 = 0 \implies y^2 - 4 = 0$

so it cuts the curve at $y = 2$ and $y = -2$.

b) $2xy + x^2\dfrac{dy}{dx} + y^2 + 2xy\dfrac{dy}{dx} = y + x\dfrac{dy}{dx} + 0$

$(x^2 + 2xy - x)\dfrac{dy}{dx} = y - 2xy - y^2$

$\dfrac{dy}{dx} = \dfrac{y - 2xy - y^2}{x^2 + 2xy - x}$

At $(1, 2)$, $\dfrac{dy}{dx} = \dfrac{2 - 2(1)(2) - 2^2}{1^2 + 2(1)(2) - 1} = -\dfrac{3}{2}$

At $(1, -2)$, $\dfrac{dy}{dx} = \dfrac{(-2) - 2(1)(-2) - (-2)^2}{1^2 + 2(1)(-2) - 1} = \dfrac{1}{2}$

Q7 a) Differentiating with respect to x:

$2y\dfrac{dy}{dx} - 2\dfrac{dy}{dx} - \dfrac{1}{x} + 2\pi\sin(2\pi x) = 0$

$\implies 2(y - 1)\dfrac{dy}{dx} = \dfrac{1}{x} - 2\pi\sin(2\pi x) = \dfrac{1 - 2\pi x\sin(2\pi x)}{x}$

$\implies \dfrac{dy}{dx} = \dfrac{1 - 2\pi x\sin(2\pi x)}{2x(y - 1)}$

[3 marks available — 1 mark for two terms correctly differentiated, 1 mark for the other two terms correctly differentiated, 1 mark for rearranging the terms to give the expression for $\dfrac{dy}{dx}$]

b) The gradient is undefined when the denominator of $\dfrac{dy}{dx}$ is zero. So when $2x(y - 1) = 0$ and hence $y = 1$. *[1 mark]*

Q8　**a)**　Differentiating with respect to x:

$$2(x+y)\left(1+\frac{dy}{dx}\right) = y + x\frac{dy}{dx} + 3y^2\frac{dy}{dx}$$

$$\Rightarrow (2x+2y) + (2x+2y)\frac{dy}{dx} = y + x\frac{dy}{dx} + 3y^2\frac{dy}{dx}$$

$$\Rightarrow 2x + 2y - y = x\frac{dy}{dx} + 3y^2\frac{dy}{dx} - 2x\frac{dy}{dx} - 2y\frac{dy}{dx}$$

$$\Rightarrow 2x + y = (3y^2 - x - 2y)\frac{dy}{dx}$$

$$\Rightarrow \frac{dy}{dx} = \frac{2x+y}{3y^2 - x - 2y}$$

[5 marks available — 1 mark for a correct method to differentiate the LHS, 1 mark for the correct differentiation of the LHS, 1 mark for correctly differentiating xy, 1 mark for correctly differentiating y^3, 1 mark for rearranging to give the required answer]

b)　When $y = 1$, $(x + 1)^2 = x + 1 \Rightarrow (x + 1)^2 - (x + 1) = 0$
$\Rightarrow x^2 + 2x + 1 - x - 1 = 0 \Rightarrow x^2 + x = 0 \Rightarrow x(x + 1) = 0$
so $x = 0$, $x = -1$.
[2 marks available — 1 mark for a correct method to solve quadratic equation, 1 mark for the correct answers]

c)　At $(0, 1)$, the gradient is $\frac{1}{3-2} = 1$.
At $(-1, 1)$, the gradient is $\frac{-2+1}{3+1-2} = -\frac{1}{2}$.
[2 marks available — 1 mark for each correct answer]

d)　The gradient changes sign between $x = -1$ and $x = 0$ so there must be a stationary point between these two x-values.
[1 mark]

Q9　**a)**　$y = x^{\tan x} \Rightarrow \ln y = \tan x \ln x$
Differentiating with respect to x (and using the product rule):
$$\frac{1}{y}\frac{dy}{dx} = \sec^2 x \ln x + \frac{\tan x}{x}$$
$$\Rightarrow \frac{dy}{dx} = y\left(\sec^2 x \ln x + \frac{\tan x}{x}\right) = x^{\tan x}\left(\sec^2 x \ln x + \frac{\tan x}{x}\right).$$
[4 marks available — 1 mark for correctly taking logarithms, 1 mark for correctly differentiating ln y, 1 mark for a correct method to differentiate tan x ln x, 1 mark for the correct answer]

b)　When $x = \frac{\pi}{4}$, $\frac{dy}{dx} = \frac{\pi}{4}^{\tan\frac{\pi}{4}}\left(\sec^2\frac{\pi}{4}\ln\frac{\pi}{4} + \frac{\tan\frac{\pi}{4}}{\frac{\pi}{4}}\right)$
$= \frac{\pi}{4}\left(2\ln\frac{\pi}{4} + \frac{4}{\pi}\right) = \frac{\pi}{2}\ln\frac{\pi}{4} + 1$
[2 marks available — 1 mark for a correct method, 1 mark for the correct answer]

Exercise 7.18 — Applications of Implicit Differentiation

Q1　**a)**　Differentiating:
$$2x + 2 + 3\frac{dy}{dx} - 2y\frac{dy}{dx} = 0 \Rightarrow \frac{dy}{dx} = \frac{2x+2}{2y-3}.$$
At the stationary points, $\frac{dy}{dx} = 0$, so:
$2x + 2 = 0 \Rightarrow x = -1$
When $x = -1$, $y^2 - 3y + 1 = 0$
$\Rightarrow y = \frac{3 \pm \sqrt{5}}{2} = 2.62$ or 0.38 (2 d.p.)
So there are 2 stationary points with coordinates $(-1, 2.62)$ and $(-1, 0.38)$.

b)　Putting $x = 0$ into the equation gives:
$3y - y^2 = 0 \Rightarrow y(3 - y) = 0 \Rightarrow y = 0$ and $y = 3$
At $(0, 0)$, $\frac{dy}{dx} = \frac{2(0)+2}{2(0)-3} = -\frac{2}{3}$
The y-intercept is 0 (as it goes through $(0, 0)$).
So the equation of the tangent is $y = -\frac{2}{3}x$ or $3y = -2x$.

At $(0, 3)$, $\frac{dy}{dx} = \frac{2(0)+2}{2(3)-3} = \frac{2}{3}$
The y-intercept is 3 (as it goes through $(0, 3)$).
Putting this into $y = mx + c$ gives the equation of the tangent as $y = \frac{2}{3}x + 3$.

Q2　**a)**　Differentiating:
$$3x^2 + 2x + \frac{dy}{dx} = 2y\frac{dy}{dx} \Rightarrow \frac{dy}{dx} = \frac{3x^2 + 2x}{2y - 1}.$$
At the stationary points, $\frac{dy}{dx} = 0$, so:
$3x^2 + 2x = 0 \Rightarrow x(3x + 2) = 0 \Rightarrow x = 0$ and $x = -\frac{2}{3}$
When $x = 0$, $y = y^2 \Rightarrow y(y - 1) = 0 \Rightarrow y = 0$ and $y = 1$.
When $x = -\frac{2}{3}$, $-\frac{8}{27} + \frac{4}{9} + y = y^2$
$\Rightarrow \frac{4}{27} + y = y^2 \Rightarrow 27y^2 - 27y - 4 = 0$
This has solutions $y = 1.13$ and -0.13 (2 d.p.)
So there are 4 stationary points with coordinates $(0, 0)$, $(0, 1)$, $\left(-\frac{2}{3}, 1.13\right)$ and $\left(-\frac{2}{3}, -0.13\right)$.

b)　Putting $x = 2$ into the equation gives:
$8 + 4 + y = y^2 \Rightarrow y^2 - y - 12 = 0$
$\Rightarrow (y - 4)(y + 3) = 0 \Rightarrow y = 4$ and $y = -3$
At $(2, 4)$, $\frac{dy}{dx} = \frac{3(2^2) + 2(2)}{2(4) - 1} = \frac{16}{7}$
Putting this into $y = mx + c$ gives:
$4 = \frac{16}{7}(2) + c \Rightarrow c = -\frac{4}{7}$
So the equation of the tangent is
$y = \frac{16}{7}x - \frac{4}{7}$ or $16x - 7y - 4 = 0$.
At $(2, -3)$, $\frac{dy}{dx} = \frac{3(2^2) + 2(2)}{2(-3) - 1} = -\frac{16}{7}$
Putting this into $y = mx + c$ gives:
$-3 = -\frac{16}{7}(2) + c \Rightarrow c = \frac{11}{7}$
So the equation of the tangent is $y = \frac{11}{7} - \frac{16}{7}x$
or $16x + 7y - 11 = 0$.

Q3　**a)**　Putting $y = 1$ into the equation gives:
$x^2 + 1 = x + 7 \Rightarrow x^2 - x - 6 = 0$
$\Rightarrow (x - 3)(x + 2) = 0 \Rightarrow x = 3$ and $x = -2$.
Differentiating:
$2xy + x^2\frac{dy}{dx} + 3y^2\frac{dy}{dx} = 1 \Rightarrow \frac{dy}{dx} = \frac{1 - 2xy}{x^2 + 3y^2}$
At $(-2, 1)$, $\frac{dy}{dx} = \frac{1 - 2(-2)(1)}{(-2)^2 + 3(1^2)} = \frac{5}{7}$
so the gradient of the normal is $-\frac{7}{5}$.
Putting this into $y = mx + c$ gives:
$1 = -\frac{7}{5}(-2) + c \Rightarrow c = -\frac{9}{5}$
and so the equation of the normal is $y = -\frac{7}{5}x - \frac{9}{5}$
or $7x + 5y + 9 = 0$.
At $(3, 1)$, $\frac{dy}{dx} = \frac{1 - 2(3)(1)}{(3)^2 + 3(1^2)} = -\frac{5}{12}$
so the gradient of the normal is $\frac{12}{5}$.
Putting this into $y = mx + c$ gives:
$1 = \frac{12}{5}(3) + c \Rightarrow c = -\frac{31}{5}$
so the equation of the normal is $y = \frac{12}{5}x - \frac{31}{5}$
or $12x - 5y - 31 = 0$.
[5 marks available — 1 mark for a correct method to solve the quadratic equation, 1 mark for differentiating implicitly, 1 mark for at least one correct gradient of a normal, 1 mark for each equation of a normal]

Answers

b) The normals intersect when:
$-7x - 9 = 12x - 31 \Rightarrow 22 = 19x \Rightarrow x = \frac{22}{19}$
And so $5y = \frac{264}{19} - 31 = -\frac{325}{19} \Rightarrow y = -\frac{65}{19}$
So they intersect at $\left(\frac{22}{19}, -\frac{65}{19}\right)$.
[2 marks available — 1 mark for a correct method to find the x-coordinate, 1 mark for the correct answer]

Q4 a) Differentiating with respect to x:
$3x^2 - 3y^2 \frac{dy}{dx} = 0 \Rightarrow \frac{dy}{dx} = \frac{x^2}{y^2}$.
At $(0, -1)$, the gradient is $\frac{0^2}{(-1)^2} = 0$.
[3 marks available — 1 mark for differentiating implicitly, 1 mark for substituting in x = 0 and y = – 1, 1 mark for the correct answer]

b) The gradient at $(0, -1)$ is 0, so the tangent is $y = -1$.
At $(1, 0)$, $\frac{dy}{dx}$ is undefined so the tangent is vertical.
Therefore, its equation is $x = 1$. These lines meet at $(1, -1)$.
[3 marks available — 1 mark for the correct tangent at (0, –1), 1 mark for the correct tangent at (1, 0), 1 mark for the correct answer]

Q5 a) Putting $x = 0$ into the equation gives:
$1 + y^2 = 5 - 3y \Rightarrow y^2 + 3y - 4 = 0$
$\Rightarrow (y + 4)(y - 1) = 0 \Rightarrow y = -4$ and $y = 1$.
So $a = -4$ and $b = 1$ ($a < b$).
Differentiating:
$e^x + 2y \frac{dy}{dx} - y - x \frac{dy}{dx} = -3 \frac{dy}{dx} \Rightarrow \frac{dy}{dx} = \frac{y - e^x}{2y - x + 3}$
At $(0, 1)$, $\frac{dy}{dx} = \frac{1 - e^0}{2(1) - 0 + 3} = 0$ so it's a stationary point.
[5 marks available — 1 mark for a correct method to solve quadratic equation, 1 mark for finding both a and b, 1 mark for differentiating implicitly, 1 mark for $\frac{dy}{dx}$, 1 mark for showing the required result]

b) At $(0, -4)$, $\frac{dy}{dx} = \frac{(-4) - e^0}{2(-4) - 0 + 3} = 1$.
So the gradient of the tangent is 1 and the gradient of the normal is -1.
$-4 = 0 + c$ (so $c = -4$) for both, since $x = 0$,
so the equation of the tangent is $y = x - 4$
and the equation of the normal is $y = -x - 4$.
[4 marks available — 1 mark for the gradient of the tangent, 1 mark for the equation for the tangent, 1 mark for the gradient of the normal, 1 mark for the equation for the normal]

Q6 $y = \arccos x \Rightarrow \cos y = x$
Differentiating:
$-\sin y \frac{dy}{dx} = 1 \Rightarrow \frac{dy}{dx} = -\frac{1}{\sin y} = -\frac{1}{\sqrt{\sin^2 y}} = -\frac{1}{\sqrt{1 - \cos^2 y}}$
Using $\cos y = x$: $\frac{dy}{dx} = -\frac{1}{\sqrt{1 - x^2}}$

Q7 $y = \arctan x \Rightarrow \tan y = x$
Differentiating:
$\sec^2 y \frac{dy}{dx} = 1 \Rightarrow \frac{dy}{dx} = \frac{1}{\sec^2 y}$
Using the identity $\sec^2 y \equiv 1 + \tan^2 y$:
$\frac{dy}{dx} = \frac{1}{1 + \tan^2 y}$
Using $\tan y = x$: $\frac{dy}{dx} = \frac{1}{1 + x^2}$ as required.

Q8 a) When $x = 1$:
$0 + y^2 = y + 6 \Rightarrow y^2 - y - 6 = 0$
$\Rightarrow (y - 3)(y + 2) = 0 \Rightarrow y = 3$ and $y = -2$.
So the curve passes through $(1, 3)$ and $(1, -2)$.

b) Differentiating:
$\frac{1}{x} + 2y \frac{dy}{dx} = 2xy + x^2 \frac{dy}{dx}$
$\frac{dy}{dx} = \frac{2xy - \frac{1}{x}}{2y - x^2} = \frac{2x^2 y - 1}{2xy - x^3}$
At $(1, 3)$, $\frac{dy}{dx} = \frac{2(1^2)(3) - 1}{2(1)(3) - (1)^3} = 1$
so the gradient of the normal is -1.
Putting this into $y = mx + c$ gives:
$3 = -1(1) + c \Rightarrow c = 4$
so the equation of the normal is $y = 4 - x$.
At $(1, -2)$, $\frac{dy}{dx} = \frac{2(1^2)(-2) - 1}{2(1)(-2) - (1)^3} = 1$,
so the gradient of the normal is also -1.
$-2 = -1(1) + c \Rightarrow c = -1$
so the equation of the normal is $y = -x - 1$.
Because the gradients are the same,
these lines are parallel and can never intersect.

Q9 a) Differentiating with respect to x:
$2(x + 3) + 4(y - 1)\frac{dy}{dx} = 0$
$\Rightarrow \frac{dy}{dx} = -\frac{2(x + 3)}{4(y - 1)} = \frac{x + 3}{2(1 - y)}$
[3 marks available — 1 mark for the correct differentiation of $(x + 3)^2$, 1 mark for the correct differentiation of $2(y – 1)^2$, 1 mark for the correct answer]

b) $\frac{dy}{dx} = 0 \Rightarrow \frac{x + 3}{2(1 - y)} = 0 \Rightarrow x = -3$
When $x = -3$, $(-3 + 3)^2 + 2(y - 1)^2 = 2 \Rightarrow 2(y - 1)^2 = 2$
$\Rightarrow (y - 1)^2 = 1 \Rightarrow y - 1 = \pm 1 \Rightarrow y = 0$ and $y = 2$
Hence the stationary points are $(-3, 0)$ and $(-3, 2)$.
[3 marks available — 1 mark for the correct x-value, 1 mark for substituting back into the equation of the ellipse, 1 mark for the correct answers]

Q10 When $y = 0$:
$1 + x^2 = 0 + 4x \Rightarrow x^2 - 4x + 1 = 0$
Complete the square to solve...
$(x - 2)^2 - 4 + 1 = 0 \Rightarrow x - 2 = \pm\sqrt{3} \Rightarrow x = 2 \pm \sqrt{3}$
So $a = 2 + \sqrt{3}$ and $b = 2 - \sqrt{3}$.
Differentiating:
$e^y \frac{dy}{dx} + 2x = 3y^2 \frac{dy}{dx} + 4 \Rightarrow \frac{dy}{dx} = \frac{2x - 4}{3y^2 - e^y}$
At $(2 + \sqrt{3}, 0)$, $\frac{dy}{dx} = \frac{4 + 2\sqrt{3} - 4}{3(0) - e^0} = -2\sqrt{3}$
$0 = -2\sqrt{3}(2 + \sqrt{3}) + c \Rightarrow c = 4\sqrt{3} + 6$
So the tangent at this point is $y = 4\sqrt{3} + 6 - 2\sqrt{3}x$.
At $(2 - \sqrt{3}, 0)$, $\frac{dy}{dx} = \frac{4 - 2\sqrt{3} - 4}{3(0) - e^0} = 2\sqrt{3}$
$0 = 2\sqrt{3}(2 - \sqrt{3}) + c \Rightarrow c = 6 - 4\sqrt{3}$
So the tangent at this point is $y = 2\sqrt{3}x + 6 - 4\sqrt{3}$.

Q11 Differentiating:
$\ln x \frac{dy}{dx} + \frac{y}{x} + 2x = 2y \frac{dy}{dx} - \frac{dy}{dx}$
$\frac{dy}{dx} = \frac{\frac{y}{x} + 2x}{2y - \ln x - 1} = \frac{y + 2x^2}{2xy - x \ln x - x}$
$y + 2x^2 = 0 \Rightarrow \frac{dy}{dx} = \frac{y + 2x^2}{2xy - x \ln x - x} = 0$
So if a point on the curve satisfies $y + 2x^2 = 0$,
then it's a stationary point.

Answers

Q12 Use the chain rule to find f′(x):

$y = \arccos u,\; u = x^2$

$\dfrac{dy}{dx} = \dfrac{dy}{du} \times \dfrac{du}{dx} = -\dfrac{1}{\sqrt{1-u^2}} \times 2x = -\dfrac{2x}{\sqrt{1-x^4}}$

When $x = \dfrac{1}{\sqrt{2}}$, $y = \arccos x^2 = \arccos \dfrac{1}{2} = \dfrac{\pi}{3}$

This is the only solution for y in the given interval.

$\dfrac{dy}{dx} = -\dfrac{2\left(\frac{1}{\sqrt{2}}\right)}{\sqrt{1-\left(\frac{1}{\sqrt{2}}\right)^4}} = -\dfrac{\frac{2}{\sqrt{2}}}{\sqrt{1-\frac{1}{(\sqrt{2})^4}}} = -\dfrac{\sqrt{2}}{\sqrt{1-\frac{1}{4}}}$

$= -\dfrac{\sqrt{2}}{\sqrt{\frac{3}{4}}} = -\dfrac{\sqrt{2}}{\frac{\sqrt{3}}{2}} = -\dfrac{2\sqrt{2}}{\sqrt{3}}$

Rationalising the denominator:

$= -\dfrac{2\sqrt{2}\times\sqrt{3}}{\sqrt{3}\times\sqrt{3}} = -\dfrac{2\sqrt{6}}{3}$

So the tangent has a gradient of $-\dfrac{2\sqrt{6}}{3}$ and passes through the point $\left(\dfrac{1}{\sqrt{2}}, \dfrac{\pi}{3}\right)$.

$y = mx + c \;\Rightarrow\; \dfrac{\pi}{3} = -\dfrac{2\sqrt{6}}{3} \times \dfrac{1}{\sqrt{2}} + c$

$c = \dfrac{\pi}{3} + \dfrac{2\sqrt{3}}{3} = \dfrac{\pi + 2\sqrt{3}}{3}$

So the equation of the tangent is $y = -\dfrac{2\sqrt{6}}{3}x + \dfrac{\pi + 2\sqrt{3}}{3}$.

Q13 a) Differentiating:

$2e^{2y}\dfrac{dy}{dx} + e^x = 2x\dfrac{dy}{dx} + 2y$

$\dfrac{dy}{dx} = \dfrac{2y - e^x}{2e^{2y} - 2x}$

When $y = 0$:

$1 + e^x - e^4 = 1 \;\Rightarrow\; e^x = e^4 \;\Rightarrow\; x = 4$

$\dfrac{dy}{dx} = \dfrac{2(0) - e^4}{2e^0 - 2(4)} = \dfrac{e^4}{6}$ (gradient of the tangent)

$0 = \dfrac{e^4}{6}(4) + c \;\Rightarrow\; c = -\dfrac{2e^4}{3}$.

So the equation of the tangent is $y = \dfrac{e^4}{6}(x - 4)$.

b) The gradient of the normal is $-6e^{-4}$.

$0 = -6e^{-4}(4) + c \;\Rightarrow\; c = 24e^{-4}$

So the normal is $y = 6e^{-4}(4 - x)$.

c) The lines intersect when

$\dfrac{e^4}{6}(x - 4) = 6e^{-4}(4 - x) \;\Rightarrow\; e^8(x - 4) = 36(4 - x)$

$\Rightarrow\; e^8 x + 36x = 4e^8 + 144 \;\Rightarrow\; x(e^8 + 36) = 4e^8 + 144$

$\Rightarrow\; x = \dfrac{4e^8 + 144}{e^8 + 36}$

Q14 When $x = 2$:

$2y^2 + 4y - 24 = 4 + 2 \;\Rightarrow\; 2y^2 + 4y - 30 = 0$

$\Rightarrow\; y^2 + 2y - 15 = 0 \;\Rightarrow\; (y + 5)(y - 3) = 0$

$\Rightarrow\; y = -5$ and $y = 3$.

Differentiating:

$y^2 + 2yx\dfrac{dy}{dx} + 2y + 2x\dfrac{dy}{dx} - 9x^2 = 2x$

$\dfrac{dy}{dx} = \dfrac{2x - y^2 - 2y + 9x^2}{2yx + 2x}$

At $(2, -5)$, $\dfrac{dy}{dx} = \dfrac{4 - 25 + 10 + 36}{-20 + 4} = -\dfrac{25}{16}$

This is the gradient of the tangent at $(2, -5)$, so:

$-5 = -\dfrac{25}{16}(2) + c \;\Rightarrow\; c = -\dfrac{15}{8}$

So the equation of the tangent at $(2, -5)$ is:

$y = -\dfrac{25}{16}x - \dfrac{15}{8}$ or $16y = -25x - 30$

At $(2, 3)$, $\dfrac{dy}{dx} = \dfrac{4 - 9 - 6 + 36}{12 + 4} = \dfrac{25}{16}$

This is the gradient of the tangent at $(2, 3)$, so:

$3 = \dfrac{25}{16}(2) + c \;\Rightarrow\; c = -\dfrac{1}{8}$

So the equation of the tangent at $(2, 3)$ is:

$y = \dfrac{25}{16}x - \dfrac{1}{8}$ or $16y = 25x - 2$

The two tangents intersect when:

$-25x - 30 = 25x - 2 \;\Rightarrow\; 50x = -28$

$\Rightarrow\; x = -\dfrac{14}{25}$

And $16y = 25\left(-\dfrac{14}{25}\right) - 2 \;\Rightarrow\; 16y = -16 \;\Rightarrow\; y = -1$.

So they intersect at $\left(-\dfrac{14}{25}, -1\right)$.

Q15 a) $\sin\dfrac{\pi}{6} = \dfrac{1}{2}$ and $\cos^2\left(\dfrac{\pi}{4}\right) = \left(\dfrac{1}{\sqrt{2}}\right)^2 = \dfrac{1}{2}$

so $\left(\dfrac{\pi}{6}, \dfrac{\pi}{4}\right)$ lies on the curve B. *[1 mark]*

b) Differentiating both sides with respect to x:

$\cos x = -2\sin y \cos y \dfrac{dy}{dx}$.

Substituting in the x and y values at P gives:

$\cos\left(\dfrac{\pi}{6}\right) = -2\sin\left(\dfrac{\pi}{4}\right)\cos\left(\dfrac{\pi}{4}\right)\dfrac{dy}{dx} \;\Rightarrow\; \dfrac{\sqrt{3}}{2} = -\dfrac{dy}{dx}$

$\Rightarrow\; \dfrac{dy}{dx} = -\dfrac{\sqrt{3}}{2}$.

[3 marks available — 1 mark for the correct differentiation of sin x, 1 mark for the correct differentiation of cos² y, 1 mark for the correct answer]

c) Gradient of normal is $\dfrac{2}{\sqrt{3}} = \dfrac{2\sqrt{3}}{3}$

$y = mx + c \;\Rightarrow\; \dfrac{\pi}{4} = \left(\dfrac{2\sqrt{3}}{3} \times \dfrac{\pi}{6}\right) + c \;\Rightarrow\; c = -\dfrac{\pi\sqrt{3}}{9} + \dfrac{\pi}{4}$

So the normal has equation $y = \dfrac{2\sqrt{3}}{3}x - \dfrac{\pi\sqrt{3}}{9} + \dfrac{\pi}{4}$

[2 marks available — 1 mark for the correct gradient of the normal, 1 mark for the correct answer]

Q16 a) When $x = \dfrac{\pi}{2}$:

$\cos y \cos\dfrac{\pi}{2} + \cos y \sin\dfrac{\pi}{2} = \dfrac{1}{2}$,

$0 + \cos y = \dfrac{1}{2} \;\Rightarrow\; y = \dfrac{\pi}{3}$

This is the only solution for y in the given interval.

When $x = \pi$:

$\cos y \cos\pi + \cos y \sin\pi = \dfrac{1}{2}$,

$-\cos y + 0 = \dfrac{1}{2} \;\Rightarrow\; y = \dfrac{2\pi}{3}$

You can use the CAST diagram or the graph of cos x to find this solution.

[2 marks available — 1 mark for each correct y-value]

Answers

b) Differentiating:
$$-\cos y \sin x - \sin y \cos x \frac{dy}{dx} + \cos y \cos x - \sin y \sin x \frac{dy}{dx} = 0$$

$$\frac{dy}{dx} = \frac{\cos y \cos x - \cos y \sin x}{\sin y \cos x + \sin y \sin x}$$

At $\left(\frac{\pi}{2}, \frac{\pi}{3}\right)$,

$$\frac{dy}{dx} = \frac{\cos\frac{\pi}{3}\cos\frac{\pi}{2} - \cos\frac{\pi}{3}\sin\frac{\pi}{2}}{\sin\frac{\pi}{3}\cos\frac{\pi}{2} + \sin\frac{\pi}{3}\sin\frac{\pi}{2}} = \frac{0 - \frac{1}{2}}{0 + \frac{\sqrt{3}}{2}} = -\frac{1}{\sqrt{3}}$$

This is the gradient of the tangent, so:
$$\frac{\pi}{3} = -\frac{1}{\sqrt{3}}\left(\frac{\pi}{2}\right) + c \implies c = \frac{(2+\sqrt{3})\pi}{6}$$

So the equation of the tangent at $\left(\frac{\pi}{2}, \frac{\pi}{3}\right)$ is
$$y = -\frac{1}{\sqrt{3}}x + \frac{(2+\sqrt{3})\pi}{6}.$$

At $\left(\pi, \frac{2\pi}{3}\right)$,

$$\frac{dy}{dx} = \frac{\cos\frac{2\pi}{3}\cos\pi - \cos\frac{2\pi}{3}\sin\pi}{\sin\frac{2\pi}{3}\cos\pi + \sin\frac{2\pi}{3}\sin\pi} = \frac{\frac{1}{2} - 0}{-\frac{\sqrt{3}}{2} + 0} = -\frac{1}{\sqrt{3}}$$

This is the gradient of the tangent, so:
$$\frac{2\pi}{3} = -\frac{1}{\sqrt{3}}(\pi) + c \implies c = \frac{(2+\sqrt{3})\pi}{3}$$

So the equation of the tangent at $\left(\pi, \frac{2\pi}{3}\right)$ is
$$y = -\frac{1}{\sqrt{3}}x + \frac{(2+\sqrt{3})\pi}{3}.$$

[4 marks available — 1 mark for using the correct method to differentiate, 1 mark for finding $\frac{dy}{dx}$, 1 mark for the correct equation for the tangent at $\left(\frac{\pi}{2}, \frac{\pi}{3}\right)$, 1 mark for the correct equation for the tangent at $\left(\pi, \frac{2\pi}{3}\right)$]

Q17 Differentiating: $\left(\frac{2}{3}y\right)\frac{dy}{dx} = 18x^2 - \left(2y + 2x\frac{dy}{dx}\right)$

$$\left(\frac{2}{3}y + 2x\right)\frac{dy}{dx} = 18x^2 - 2y$$

$$\frac{dy}{dx} = \frac{18x^2 - 2y}{\frac{2}{3}y + 2x} = \frac{9x^2 - y}{\frac{1}{3}y + x} = \frac{27x^2 - 3y}{y + 3x}$$

When $\frac{dy}{dx} = 0$, $27x^2 - 3y = 0 \implies 3y = 27x^2 \implies y = 9x^2$

Substitute $y = 9x^2$ back into the original equation:
$$\frac{1}{3}(9x^2)^2 = 6x^3 - 2x(9x^2) \implies 27x^4 = 6x^3 - 18x^3$$
$$\implies 27x^4 + 12x^3 = 0 \implies 3x^3(9x + 4) = 0$$

So the stationary points occur when $x = 0$ and $x = -\frac{4}{9}$.
When $x = 0$, $y = 9(0)^2 = 0$
When $x = -\frac{4}{9}$, $y = 9\left(-\frac{4}{9}\right)^2 = \frac{16}{9}$
So the stationary points are at $(0, 0)$ and $\left(-\frac{4}{9}, \frac{16}{9}\right)$.

Chapter 7 Review Exercise

Q1 a) (i) $\frac{dy}{dx} = 6x^2 - 24x + 18 \implies \frac{d^2y}{dx^2} = 12x - 24$

The graph is concave when $\frac{d^2y}{dx^2} < 0$
$$\implies 12x - 24 < 0 \implies x < 2$$

(ii) The graph is convex when $\frac{d^2y}{dx^2} > 0$
$$\implies 12x - 24 > 0 \implies x > 2$$

b) (i) Point of inflection occurs when $\frac{d^2y}{dx^2} = 0$:
$$\implies 12x - 24 = 0 \implies x = 2$$
$$y = 2(2)^3 - 12(2)^2 + 18(2) + 2 = 16 - 48 + 36 + 2 = 6$$
So the point of inflection is at $(2, 6)$.

(ii) Substitute $x = 2$ into $\frac{dy}{dx}$:
$$\frac{dy}{dx} = 6(2)^2 - 24(2) + 18 = 24 - 48 + 18 = -6 \neq 0,$$
So this is not a stationary point of inflection.

Q2 a) $f(x) = (3x + 1)^5$, so let $f(x) = y = u^5$, where $u = 3x + 1$
$$\implies \frac{dy}{du} = 5u^4 = 5(3x + 1)^4, \frac{du}{dx} = 3$$
$$f'(x) = \frac{dy}{dx} = \frac{dy}{du} \times \frac{du}{dx} = 15(3x + 1)^4$$
[2 marks available — 1 mark for a correct method, 1 mark for the correct answer]

b) Use the chain rule on $f'(x)$ to find $f''(x)$:
$f'(x) = 15(3x + 1)^4$, so let $f'(x) = z = 15u^4$, where $u = 3x + 1$
$$\implies \frac{dz}{du} = 15 \times 4u^3 = 60(3x + 1)^3, \frac{du}{dx} = 3$$
$$f''(x) = \frac{dz}{dx} = \frac{dz}{du} \times \frac{du}{dx} = 180(3x + 1)^3$$
At a point of inflection, $f''(x) = 0$
$$\implies 180(3x + 1)^3 = 0 \implies 3x + 1 = 0 \implies x = -\frac{1}{3}$$
When $x < -\frac{1}{3}$, $3x + 1 < 0 \implies f''(x) < 0$
When $x > -\frac{1}{3}$, $3x + 1 > 0 \implies f''(x) > 0$
So the sign of $f''(x)$ changes at $x = -\frac{1}{3}$,
so this is the point of inflection.
And $x = -\frac{1}{3}$ is the only real solution to $f''(x) = 0$,
so it's the only point of inflection.
$$x = -\frac{1}{3} \implies y = f(x) = \left(3\left(-\frac{1}{3}\right) + 1\right)^5 = 0$$
So the point of inflection of $f(x)$ has the coordinates $\left(-\frac{1}{3}, 0\right)$.
[3 marks available — 1 mark for the correct differentiation of answer to part a) (allow errors carried forward), 1 mark for clearly showing f(x) has one point of inflection, 1 mark for the correct coordinates]

c) $f(x)$ is concave where $f''(x) < 0$, so is concave where $x < -\frac{1}{3}$.
[1 mark for the correct answer]

Q3 a) $y = (x^2 - 1)^3$, so let $y = u^3$ where $u = x^2 - 1$
$$\implies \frac{dy}{du} = 3u^2 = 3(x^2 - 1)^2, \frac{du}{dx} = 2x$$
$$\frac{dy}{dx} = \frac{dy}{du} \times \frac{du}{dx} = 3(x^2 - 1)^2 \times 2x = 6x(x^2 - 1)^2$$

b) At $x = 2$, $y = (2^2 - 1)^3 = 27$
Gradient of C at $x = 2$, $\frac{dy}{dx} = 6 \times 2(2^2 - 1)^2 = 108$
So the normal at $x = 2$ has a gradient of $-\frac{1}{108}$ and an equation in the form $y = mx + c$.
Substitute in x, y and m to find c:
$$27 = -\frac{1}{108} \times 2 + c \implies c = \frac{1459}{54}$$
So $y = -\frac{1}{108}x + \frac{1459}{54}$,
which rearranges to $x + 108y - 2918 = 0$.

Q4 a) $y = (x^3 + 2x^2)^{\frac{1}{2}}$, so let $y = u^{\frac{1}{2}}$ where $u = x^3 + 2x^2$
$$\implies \frac{dy}{du} = \frac{1}{2}u^{-\frac{1}{2}} = \frac{1}{2\sqrt{x^3 + 2x^2}}, \frac{du}{dx} = 3x^2 + 4x$$
$$\frac{dy}{dx} = \frac{dy}{du} \times \frac{du}{dx} = \frac{3x^2 + 4x}{2\sqrt{x^3 + 2x^2}}$$

b) $y = e^{f(x)}$, where $f(x) = 5x^2$, so $f'(x) = 10x$
$$\frac{dy}{dx} = f'(x)e^{f(x)} = 10xe^{5x^2}$$

c) $y = \ln f(x)$, where $f(x) = 6 - x^2$, so $f'(x) = -2x$
$$\frac{dy}{dx} = \frac{f'(x)}{f(x)} = \frac{-2x}{6 - x^2}$$

Q5 Using the change of base formula, $\log_a x = \dfrac{\ln x}{\ln a}$, so:

$$\frac{d}{dx}\log_a x = \frac{d}{dx}\left(\frac{\ln x}{\ln a}\right) = \frac{1}{\ln a}\frac{d}{dx}(\ln x) = \frac{1}{\ln a} \times \frac{1}{x} = \frac{1}{x\ln a}$$

Q6 **a)** $y = \ln(f(x)) \;\Rightarrow\; \dfrac{dy}{dx} = \dfrac{f'(x)}{f(x)} = \dfrac{8(2x-3)^3 + 2}{(2x-3)^4 + 2x - 7}$

[2 marks available — 1 mark for a correct method to differentiate, 1 mark for the correct answer]

b) When $x = 3$, $\dfrac{dy}{dx} = \dfrac{8 \times 3^3 + 2}{3^4 + 6 - 7} = \dfrac{218}{80} = \dfrac{109}{40}$

[1 mark for the correct answer]

Q7 **a)** Differentiate each term separately:
$y = 3e^{2x+1} - \ln(1 - x^2) + 2x^3$
$\dfrac{dy}{dx} = 2(3e^{2x+1}) - \dfrac{(-2)x}{1-x^2} + 6x^2 = 6e^{2x+1} + \dfrac{2x}{1-x^2} + 6x^2$

b) Differentiate each term separately:
$y = 16x + e^{\sqrt{x}} + \ln(\cos x)$
$\dfrac{dy}{dx} = 16 + \dfrac{1}{2\sqrt{x}} \times e^{\sqrt{x}} - \dfrac{\sin x}{\cos x} = 16 + \dfrac{e^{\sqrt{x}}}{2\sqrt{x}} - \tan x$

Q8 **a)** **(i)** $x = 2e^{2y} \Rightarrow \dfrac{dx}{dy} = 4e^{2y} = 2x \Rightarrow \dfrac{dy}{dx} = \dfrac{1}{2x}$

(ii) $x = \ln(2y+3) \Rightarrow \dfrac{dx}{dy} = \dfrac{2}{2y+3} = \dfrac{2}{e^x} \Rightarrow \dfrac{dy}{dx} = \dfrac{e^x}{2}$

b) $x = \tan y \Rightarrow \dfrac{dx}{dy} = \sec^2 y = \dfrac{1}{\cos^2 y} \Rightarrow \dfrac{dy}{dx} = \cos^2 y$

Q9 **a)** Let $y = u^2 + 2$ where $u = 5^x - 1$
$\dfrac{dy}{du} = 2u = 2(5^x - 1)$ and $\dfrac{du}{dx} = 5^x \ln 5$
$\dfrac{dy}{dx} = \dfrac{dy}{du} \times \dfrac{du}{dx} = 2(5^x - 1) \times 5^x \ln 5$
$\dfrac{dy}{dx} = 0 \Rightarrow 5^x - 1 = 0 \Rightarrow 5^x = 1 \Rightarrow x = 0$
When $x = 0$, $y = (5^0 - 1)^2 + 2 = 2$.
The stationary point has coordinates (0, 2).
[4 marks available — 1 mark for a correct method to differentiate, 1 mark for $\dfrac{dy}{dx}$, 1 mark for a correct method to solve $\dfrac{dy}{dx} = 0$, 1 mark for the correct answer]

b) $(5^x - 1)^2 \geq 0$ for all real values of x, so the minimum occurs when $5^x - 1 = 0$, i.e. at (0, 2).
[1 mark for a valid explanation]

c) As $x \to -\infty$, $5^x \to 0 \Rightarrow 5^x \ln 5 \to 0$ and $2(5^x - 1) \to -2$
$\Rightarrow \dfrac{dy}{dx} \to 0 \Rightarrow k = 0$
[2 marks available — 1 mark for $5^x \to 0$, 1 mark for the correct answer]

Q10 **a)** $f(x) = a^x + b^x + c^x$
$\Rightarrow f'(x) = a^x \ln a + b^x \ln b + c^x \ln c$
$f'(x) > 0$ since $a^x, b^x, c^x > 0$ and the log of a number greater than 1 is positive. Therefore, $f(x)$ has no stationary points.

b) E.g. $f(x) = 1^x + 2^x + 0.5^x$
$\Rightarrow f'(x) = 1^x \ln 1 + 2^x \ln 2 + 0.5^x \ln 0.5$
$= 2^x \ln 2 + 0.5^x \ln 0.5 = 2^x \ln 2 - 0.5^x \ln 2$
$f'(x) = 0 \Rightarrow 2^x \ln 2 - 0.5^x \ln 2 = 0 \Rightarrow 2^x - 0.5^x = 0$
This has a solution at $x = 0$, so there is a stationary point on the curve $y = f(x)$ at $x = 0$.

Q11 **a)** $f(x) = 2\cos(3x)$, so let $y = 2\cos u$, where $u = 3x$
$\Rightarrow \dfrac{dy}{du} = -2\sin u = -2\sin(3x)$, $\dfrac{du}{dx} = 3$
$f'(x) = \dfrac{dy}{du} \times \dfrac{du}{dx} = -2\sin(3x) \times 3 = -6\sin(3x)$

b) $f(x) = \sqrt{\tan x} = (\tan x)^{\frac{1}{2}}$, so let $y = u^{\frac{1}{2}}$, where $u = \tan x$
$\Rightarrow \dfrac{dy}{du} = \dfrac{1}{2}u^{-\frac{1}{2}} = \dfrac{1}{2\sqrt{\tan x}}$, $\dfrac{du}{dx} = \sec^2 x$
$f'(x) = \dfrac{dy}{du} \times \dfrac{du}{dx} = \dfrac{1}{2\sqrt{\tan x}} \times \sec^2 x = \dfrac{\sec^2 x}{2\sqrt{\tan x}}$

c) $f(x) = e^{\cos(3x)}$, so let $y = e^u$, where $u = \cos(3x)$
$\Rightarrow \dfrac{dy}{du} = e^u = e^{\cos(3x)}$
$u = \cos(3x)$, so let $u = \cos v$, where $v = 3x$
$\Rightarrow \dfrac{du}{dv} = -\sin v = -\sin(3x)$, $\dfrac{dv}{dx} = 3$
$\dfrac{du}{dx} = \dfrac{du}{dv} \times \dfrac{dv}{dx} = -3\sin(3x)$
$f'(x) = \dfrac{dy}{du} \times \dfrac{du}{dx} = e^{\cos(3x)} \times -3\sin(3x) = -3e^{\cos(3x)}\sin(3x)$
You don't need to use the chain rule twice if you can remember the rules of differentiating $e^{f(x)}$ and trig functions.

d) $f(x) = \sin(4x)\tan(x^3)$,
so let $y = uv$, where $u = \sin(4x)$ and $v = \tan(x^3)$
$\Rightarrow \dfrac{du}{dx} = 4\cos(4x)$, $\dfrac{dv}{dx} = 3x^2 \sec^2(x^3)$
$f'(x) = \dfrac{dy}{dx} = u\dfrac{dv}{dx} + v\dfrac{du}{dx}$
$= \sin(4x) \times 3x^2\sec^2(x^3) + \tan(x^3) \times 4\cos(4x)$
$= 3x^2\sin(4x)\sec^2(x^3) + 4\cos(4x)\tan(x^3)$

Q12 **a)** $\frac{1}{2}(\cos(A - B) - \cos(A + B))$
$= \frac{1}{2}(\cos A\cos B + \sin A\sin B - \cos A\cos B + \sin A\sin B)$
$= \frac{1}{2}(2\sin A\sin B) = \sin A\sin B$
[2 marks available — 1 mark for the correct expansion of $\cos(A - B)$ or $\cos(A + B)$, 1 mark for expanding both and simplifying to give the required result]

b) Let $A = \dfrac{15x}{2}$, $B = \dfrac{5x}{2}$. Then from part a),
$\sin\left(\dfrac{15x}{2}\right)\sin\left(\dfrac{5x}{2}\right) = \frac{1}{2}\left(\cos\left(\dfrac{15x}{2} - \dfrac{5x}{2}\right) - \cos\left(\dfrac{15x}{2} + \dfrac{5x}{2}\right)\right)$
$\Rightarrow 2\sin\left(\dfrac{15x}{2}\right)\sin\left(\dfrac{5x}{2}\right) = \cos 5x - \cos 10x$,
so $\cos 10x - \cos 5x = -2\sin\left(\dfrac{15x}{2}\right)\sin\left(\dfrac{5x}{2}\right)$
[2 marks available — 1 mark for correctly using the result from part a), 1 mark for a complete proof]

c) $f'(x) = \dfrac{d}{dx}(\cos 10x - \cos 5x) = -10\sin 10x + 5\sin 5x$
[2 marks available — 1 mark for correctly using the result from part b), 1 mark for the correct answer]

d) $y = -2\sin\left(\dfrac{15x^2}{2}\right)\sin\left(\dfrac{5x^2}{2}\right) = \cos(10x^2) - \cos(5x^2)$
so $\dfrac{dy}{dx} = -20x\sin(10x^2) + 10x\sin(5x^2)$
When $x = 0$,
$\dfrac{dy}{dx} = -20(0)\sin(10(0)^2) + 10(0)\sin(5(0)^2) = 0$
[4 marks available — 1 mark for correctly rewriting y in terms of cos, 1 mark for a correct method to differentiate, 1 mark for the correct derivative, 1 mark for the correct answer]

Q13 **a)** $y = e^{2x}(x^2 - 3)$, so let $y = uv$, where $u = e^{2x}$ and $v = x^2 - 3$
$\Rightarrow \dfrac{du}{dx} = 2e^{2x}$, $\dfrac{dv}{dx} = 2x$
$\dfrac{dy}{dx} = u\dfrac{dv}{dx} + v\dfrac{du}{dx} = e^{2x} \times 2x + (x^2 - 3)2e^{2x}$
$= 2xe^{2x} + 2x^2e^{2x} - 6e^{2x}$
When $x = 0$, $\dfrac{dy}{dx} = 2(0)e^{2(0)} + 2(0)^2e^{2(0)} - 6e^{2(0)} = -6$

b) $y = (\ln x)(\sin x)$, so let $y = uv$, where $u = \ln x$ and $v = \sin x$
$\Rightarrow \dfrac{du}{dx} = \dfrac{1}{x}$, $\dfrac{dv}{dx} = \cos x$
$\dfrac{dy}{dx} = u\dfrac{dv}{dx} + v\dfrac{du}{dx} = \ln x \times \cos x + \sin x \times \dfrac{1}{x}$
$= \ln x\cos x + \dfrac{\sin x}{x}$
When $x = 1$, $\dfrac{dy}{dx} = \ln(1)\cos(1) + \dfrac{\sin(1)}{(1)} = 0.841$ (3 s.f.)
If you got a different answer, check your calculator is set to radians and not degrees.

Answers

Q14 $y = e^{x^2}\sqrt{x+1}$, so let $y = uv$,
where $u = e^{x^2}$ and $v = (x+1)^{\frac{1}{2}} = \sqrt{x+1}$

$\Rightarrow \dfrac{du}{dx} = 2xe^{x^2}, \dfrac{dv}{dx} = \dfrac{1}{2}(x+1)^{-\frac{1}{2}} = \dfrac{1}{2\sqrt{x+1}}$

$\dfrac{dy}{dx} = u\dfrac{dv}{dx} + v\dfrac{du}{dx} = e^{x^2} \times \dfrac{1}{2\sqrt{x+1}} + \sqrt{x+1} \times 2xe^{x^2}$

$= \dfrac{e^{x^2}}{2\sqrt{x+1}} + 2xe^{x^2}\sqrt{x+1}$

When $x = 1$, $\dfrac{dy}{dx} = \dfrac{e^{(1)^2}}{2\sqrt{(1)+1}} + 2(1)e^{(1)^2}\sqrt{(1)+1}$

$= \dfrac{e}{2\sqrt{2}} + 2\sqrt{2}\,e = \dfrac{9e}{2\sqrt{2}} = \dfrac{9\sqrt{2}\,e}{4}$

Q15 By the product rule:
Let $u = a(x-1)$ and $v = (x-2)(x-3)$. $\dfrac{du}{dx} = a$
For $\dfrac{dv}{dx}$ use the product rule again:
Let $u_1 = (x-2)$ and $v_1 = (x-3)$
$\dfrac{du_1}{dx} = 1$ and $\dfrac{dv_1}{dx} = 1$
$\dfrac{dv}{dx} = u_1\dfrac{dv_1}{dx} + v_1\dfrac{du_1}{dx} = (x-2) + (x-3)$
$f'(x) = u\dfrac{dv}{dx} + v\dfrac{du}{dx} = a(x-1)[(x-2)+(x-3)] + a(x-2)(x-3)$
$\Rightarrow f'(x) = a(x-2)(x-3) + a(x-1)(x-3) + a(x-1)(x-2)$
So $\dfrac{f'(x)}{f(x)} = \dfrac{a(x-2)(x-3) + a(x-1)(x-3) + a(x-1)(x-2)}{a(x-1)(x-2)(x-3)}$

$= \dfrac{1}{x-1} + \dfrac{1}{x-2} + \dfrac{1}{x-3}$

[4 marks available — 1 mark for $\dfrac{du}{dx}$, 1 mark for $\dfrac{dv}{dx}$, 1 mark for the correct expression for f'(x), 1 mark for $\dfrac{f'(x)}{f(x)}$ leading to the required result]

Q16 $y = \dfrac{\sqrt{x^2+3}}{\cos(3x)}$, so let $y = \dfrac{u}{v}$, where $u = \sqrt{x^2+3}$ and $v = \cos(3x)$

$\Rightarrow \dfrac{du}{dx} = \dfrac{2x}{2\sqrt{x^2+3}}, \dfrac{dv}{dx} = -3\sin(3x)$

$\dfrac{dy}{dx} = \dfrac{v\dfrac{du}{dx} - u\dfrac{dv}{dx}}{v^2} = \dfrac{\cos(3x)\times\dfrac{2x}{2\sqrt{x^2+3}} - \sqrt{x^2+3}\times -3\sin(3x)}{\cos^2(3x)}$

$= \dfrac{\dfrac{2x\cos(3x)}{2\sqrt{x^2+3}} + 3\sqrt{x^2+3}\sin(3x)}{\cos^2(3x)} = \dfrac{\dfrac{2x}{2\sqrt{x^2+3}} + 3\sqrt{x^2+3}\tan(3x)}{\cos(3x)}$

$= \dfrac{\dfrac{x}{\sqrt{x^2+3}} + 3\sqrt{x^2+3}\tan(3x)}{\cos(3x)} = \dfrac{x + 3(x^2+3)\tan(3x)}{\sqrt{x^2+3}\cos(3x)}$

Here, you have to use the quotient rule — but you have to use the chain rule to differentiate both u and v before you can put them in the quotient rule formula. Finally, you have to do a bit of rearranging to tidy up the fraction.

Q17 a) $y = \dfrac{\cos x^2}{\ln(2x)}$, so let $y = \dfrac{u}{v}$, where $u = \cos x^2$ and $v = \ln(2x)$

$\Rightarrow \dfrac{du}{dx} = 2x \times -\sin x^2 = -2x\sin x^2, \dfrac{dv}{dx} = 2 \times \dfrac{1}{2x} = \dfrac{1}{x}$

$\dfrac{dy}{dx} = \dfrac{v\dfrac{du}{dx} - u\dfrac{dv}{dx}}{v^2} = \dfrac{\ln(2x)\times -2x\sin x^2 - \cos x^2 \times \dfrac{1}{x}}{(\ln(2x))^2}$

$= -\dfrac{2x\sin x^2}{\ln(2x)} - \dfrac{\cos x^2}{x(\ln(2x))^2}$

[4 marks available — 1 mark for attempting to use the quotient (or product) rule, 1 mark for finding $\dfrac{du}{dx}$, 1 mark for finding $\dfrac{dv}{dx}$, 1 mark for the correct answer in any form]

b) At $x = 2$, $\dfrac{dy}{dx} = -\dfrac{2(2)\sin(2^2)}{\ln(2\times2)} - \dfrac{\cos(2^2)}{2(\ln(2\times2))^2} = 2.35$ (3 s.f.)

[1 mark for the correct answer]

Q18 a) Let $u = px - q$ and $v = px + q \Rightarrow \dfrac{du}{dx} = p, \dfrac{dv}{dx} = p$

$\dfrac{dy}{dx} = \dfrac{v\dfrac{du}{dx} - u\dfrac{dv}{dx}}{v^2} = \dfrac{p(px+q) - p(px-q)}{(px+q)^2} = \dfrac{2pq}{(px+q)^2}$

b) Since $p, q > 0$, $\dfrac{dy}{dx} > 0$ for all real values of x
so C has no stationary points.

Q19 a) $y = \cos x \ln x^2$, so let $y = uv$,
where $u = \cos x$ and $v = \ln x^2 = 2\ln x$

$\Rightarrow \dfrac{du}{dx} = -\sin x, \dfrac{dv}{dx} = \dfrac{2}{x}$

$\dfrac{dy}{dx} = u\dfrac{dv}{dx} + v\dfrac{du}{dx} = \cos x \times \dfrac{2}{x} + \ln x^2 \times -\sin x$

$= \dfrac{2\cos x}{x} - \ln x^2 \sin x$

b) $y = \dfrac{e^{x^2-x}}{(x+2)^4}$, so let $y = \dfrac{u}{v}$,

where $u = e^{x^2-x}$ and $v = (x+2)^4$

$\Rightarrow \dfrac{du}{dx} = (2x-1)e^{x^2-x}, \dfrac{dv}{dx} = 4(x+2)^3$

$\dfrac{dy}{dx} = \dfrac{v\dfrac{du}{dx} - u\dfrac{dv}{dx}}{v^2} = \dfrac{(x+2)^4(2x-1)e^{x^2-x} - 4e^{x^2-x}(x+2)^3}{(x+2)^8}$

$= \dfrac{(x+2)(2x-1)e^{x^2-x} - 4e^{x^2-x}}{(x+2)^5} = \dfrac{(2x^2+3x-6)e^{x^2-x}}{(x+2)^5}$

Q20 $y = \dfrac{e^x}{\sqrt{x}}$, so let $y = \dfrac{u}{v}$, where $u = e^x$ and $v = \sqrt{x}$

$\Rightarrow \dfrac{du}{dx} = e^x, \dfrac{dv}{dx} = \dfrac{1}{2}x^{-\frac{1}{2}} = \dfrac{1}{2\sqrt{x}}$

$\dfrac{dy}{dx} = \dfrac{v\dfrac{du}{dx} - u\dfrac{dv}{dx}}{v^2}$

$= \dfrac{\sqrt{x}e^x - \dfrac{e^x}{2\sqrt{x}}}{x} = \dfrac{2xe^x - e^x}{2\sqrt{x^3}} = \dfrac{(2x-1)e^x}{2\sqrt{x^3}}$

When $\dfrac{dy}{dx} = 0$, $\dfrac{(2x-1)e^x}{2\sqrt{x^3}} = 0$

$e^x \neq 0$ for all real x, so $2x - 1 = 0 \Rightarrow x = \dfrac{1}{2}$

When $x = \dfrac{1}{2}$, $y = \dfrac{e^{\frac{1}{2}}}{\sqrt{\frac{1}{2}}} = \dfrac{e^{\frac{1}{2}}}{\frac{\sqrt{2}}{2}} = \sqrt{2}\,e^{\frac{1}{2}}$

So the coordinates of the stationary point are $\left(\dfrac{1}{2}, \sqrt{2}\,e^{\frac{1}{2}}\right)$.

Q21 $y = \dfrac{6x^2+3}{4x^2-1}$, so let $y = \dfrac{u}{v}$, where $u = 6x^2+3$ and $v = 4x^2-1$

$\Rightarrow \dfrac{du}{dx} = 12x, \dfrac{dv}{dx} = 8x$

$\dfrac{dy}{dx} = \dfrac{v\dfrac{du}{dx} - u\dfrac{dv}{dx}}{v^2} = \dfrac{12x(4x^2-1) - 8x(6x^2+3)}{(4x^2-1)^2}$

$= \dfrac{48x^3 - 12x - 48x^3 - 24x}{(4x^2-1)^2} = -\dfrac{36x}{(4x^2-1)^2}$

At $(1, 3)$, $x = 1$ so $\dfrac{dy}{dx} = -\dfrac{36}{(4-1)^2} = -4$

The tangent to the curve at $(1, 3)$ has a gradient of -4 and an equation in the form $y = mx + c$.
Substitute in the values of x, y and m to find c:
$3 = -4(1) + c \Rightarrow c = 7$
So the tangent to the curve at $(1, 3)$ has the equation $y = -4x + 7$, which can be rearranged to $y = 7 - 4x$.

Q22 a) $\dfrac{d}{dx}[x] = \dfrac{d}{dx}\left[\dfrac{y}{y^2-1}\right]$

Using the quotient rule and chain rule:

Let $u = y \Rightarrow \dfrac{du}{dy} = 1,\ \dfrac{du}{dx} = \dfrac{du}{dy} \times \dfrac{dy}{dx}$,

let $v = y^2 - 1 \Rightarrow \dfrac{dv}{dy} = 2y,\ \dfrac{dv}{dx} = \dfrac{dv}{dy} \times \dfrac{dy}{dx}$

$\dfrac{d}{dx}\left[\dfrac{y}{y^2-1}\right] = \dfrac{v\dfrac{du}{dy}\times\dfrac{dy}{dx} - u\dfrac{dv}{dy}\times\dfrac{dy}{dx}}{v^2}$

$= \dfrac{1(y^2-1)\dfrac{dy}{dx} - y(2y)\dfrac{dy}{dx}}{(y^2-1)^2}$

$= \dfrac{y^2-1-2y^2}{(y^2-1)^2} \times \dfrac{dy}{dx} = -\dfrac{y^2+1}{(y^2-1)^2} \times \dfrac{dy}{dx}$

$\dfrac{d}{dx}[x] = 1$, so $1 = -\dfrac{y^2+1}{(y^2-1)^2} \times \dfrac{dy}{dx} \Rightarrow \dfrac{dy}{dx} = -\dfrac{(y^2-1)^2}{y^2+1}$

[4 marks available — 1 mark for $\dfrac{du}{dy}$ and $\dfrac{dv}{dy}$, 1 mark for a correct method to differentiate, 1 mark for correct working for $\dfrac{d}{dx}\left[\dfrac{y}{y^2-1}\right]$, 1 mark for the correct simplified answer]

b) The gradient of the tangent at $(0, 0)$ is $-\dfrac{(0^2-1)^2}{0^2+1} = -1$, so the normal has equation $y = x$.

[2 marks available — 1 mark for the correct gradient of the tangent, 1 mark for the correct equation of the normal]

c) When the normal meets the curve,

$x = \dfrac{x}{x^2-1} \Rightarrow x^3 - x = x \Rightarrow x^3 - 2x = 0$

$\Rightarrow x(x^2-2) = 0 \Rightarrow x = 0, \pm\sqrt{2}$.

So, A and B have coordinates $(-\sqrt{2}, -\sqrt{2})$ and $(\sqrt{2}, \sqrt{2})$.

The length of the line is $\sqrt{(2\sqrt{2})^2 + (2\sqrt{2})^2} = \sqrt{8+8} = 4$.

[4 marks available — 1 mark for the correct method to find points of intersection, 1 mark for the correct x values, 1 mark for a correct method to find length of line, 1 mark for the correct answer]

Q23 $y = 3 \csc \dfrac{x}{4}$, so let $y = 3 \csc u$, where $u = \dfrac{x}{4}$

$\Rightarrow \dfrac{dy}{du} = -3 \csc u \cot u = -3 \csc \dfrac{x}{4} \cot \dfrac{x}{4}$ and $\dfrac{du}{dx} = \dfrac{1}{4}$

$\dfrac{dy}{dx} = \dfrac{dy}{du} \times \dfrac{du}{dx} = -3 \csc \dfrac{x}{4} \cot \dfrac{x}{4} \times \dfrac{1}{4} = -\dfrac{3}{4} \csc \dfrac{x}{4} \cot \dfrac{x}{4}$

At $x = \pi$, $y = 3 \csc \dfrac{\pi}{4} = 3\sqrt{2}$

$\dfrac{dy}{dx} = -\dfrac{3}{4} \csc \dfrac{\pi}{4} \cot \dfrac{\pi}{4} = -\dfrac{3\sqrt{2}}{4}$

The normal to the curve at $x = \pi$ has a gradient of

$-1 \div -\dfrac{3\sqrt{2}}{4} = \dfrac{2\sqrt{2}}{3}$ and an equation in the form $y = mx + c$.

Substitute in the values of x, y and m to find c:

$3\sqrt{2} = \dfrac{2\sqrt{2}\,\pi}{3} + c \Rightarrow c = 3\sqrt{2} - \dfrac{2\sqrt{2}\,\pi}{3}$

So the normal to the curve at $x = \pi$ has the equation

$y = \dfrac{2\sqrt{2}}{3}x + 3\sqrt{2} - \dfrac{2\sqrt{2}\,\pi}{3}$

Q24 $y = \sec(3x-2)$, so let $y = \sec u$, where $u = 3x - 2$

$\Rightarrow \dfrac{dy}{du} = \tan u \sec u = \tan(3x-2) \sec(3x-2)$ and $\dfrac{du}{dx} = 3$

$\dfrac{dy}{dx} = \dfrac{dy}{du} \times \dfrac{du}{dx} = 3 \tan(3x-2) \sec(3x-2)$

At $x = 0$, $\dfrac{dy}{dx} = 3 \tan(3(0)-2) \sec(3(0)-2) = -15.8$ (3 s.f.)

Q25 a) $y = \sqrt{\csc x}$, so let $y = \sqrt{u}$, where $u = \csc x$

$\Rightarrow \dfrac{dy}{du} = \dfrac{1}{2}u^{-\frac{1}{2}} = \dfrac{1}{2\sqrt{\csc x}},\ \dfrac{du}{dx} = -\csc x \cot x$

$\dfrac{dy}{dx} = \dfrac{dy}{du} \times \dfrac{du}{dx} = \dfrac{-\csc x \cot x}{2\sqrt{\csc x}} = \dfrac{-\cot x \sqrt{\csc x}}{2}$

b) $y = \dfrac{\sec x}{x^2}$, so let $y = \dfrac{u}{v}$, where $u = \sec x$ and $v = x^2$

$\Rightarrow \dfrac{du}{dx} = \sec x \tan x,\ \dfrac{dv}{dx} = 2x$

$\dfrac{dy}{dx} = \dfrac{v\dfrac{du}{dx} - u\dfrac{dv}{dx}}{v^2}$

$= \dfrac{x^2 \sec x \tan x - 2x \sec x}{x^4} = \dfrac{\sec x(x\tan x - 2)}{x^3}$

c) $y = \cot(x^2+5)$, so let $y = \cot u$, where $u = x^2 + 5$

$\Rightarrow \dfrac{dy}{du} = -\csc^2 u = -\csc^2(x^2+5),\ \dfrac{du}{dx} = 2x$

$\dfrac{dy}{dx} = \dfrac{dy}{du} \times \dfrac{du}{dx} = -2x \csc^2(x^2+5)$

d) $y = e^{2x} \csc(5x)$, so let $y = uv$,
where $u = e^{2x}$ and $v = \csc(5x)$

$\Rightarrow \dfrac{du}{dx} = 2e^{2x},\ \dfrac{dv}{dx} = -5\csc(5x)\cot(5x)$

$\dfrac{dy}{dx} = u\dfrac{dv}{dx} + v\dfrac{du}{dx}$

$= -5e^{2x}\csc(5x)\cot(5x) + 2e^{2x}\csc(5x)$

$= e^{2x}\csc(5x)(2 - 5\cot(5x))$

Q26 a) $V = \dfrac{4}{3}\pi r^3 \Rightarrow \dfrac{dV}{dr} = 4\pi r^2$

$\dfrac{dV}{d\theta} = \dfrac{dV}{dr} \times \dfrac{dr}{d\theta} = 4\pi r^2 \times -2500 = -10\,000\,\pi r^2$ km^3 K^{-1}

b) Density $(\rho) = \dfrac{m}{V} = \dfrac{m}{kD^3} = \dfrac{m}{k}D^{-3}$

$\Rightarrow \dfrac{d\rho}{dD} = -3\dfrac{m}{k}D^{-4} = -\dfrac{3m}{kD^4}$

From the question $\dfrac{dD}{d\theta} = -215$ km K^{-1},

so $\dfrac{d\rho}{d\theta} = \dfrac{d\rho}{dD} \times \dfrac{dD}{d\theta}$

$= -\dfrac{3m}{kD^4} \times -215$ km K^{-1} $= \dfrac{645m}{kD^4}$ kg km^{-3} K^{-1}

Q27 a) (i) $A = 2(x)(2x) + 2(x)(3x) + 2(2x)(3x)$
$= 4x^2 + 6x^2 + 12x^2 = 22x^2$

$\Rightarrow \dfrac{dA}{dx} = 44x$ *[1 mark for the correct answer]*

(ii) $V = (x)(2x)(3x) = 6x^3 \Rightarrow \dfrac{dV}{dx} = 18x^2$

[1 mark for the correct answer]

b) $\dfrac{dx}{d\theta} = \dfrac{dx}{dV} \times \dfrac{dV}{d\theta} = \dfrac{1}{\left(\dfrac{dV}{dx}\right)} \times \dfrac{dV}{d\theta} = \dfrac{1}{18x^2} \times 3 = \dfrac{1}{6x^2}$

So $\dfrac{dA}{d\theta} = \dfrac{dA}{dx} \times \dfrac{dx}{d\theta} = 44x \times \dfrac{1}{6x^2} = \dfrac{22}{3x}$

[2 marks available — 1 mark for using connected rates of change, 1 mark for clearly showing $\dfrac{dA}{d\theta} = \dfrac{22}{3x}$]
There are several different ways you could have solved this question.

Q28 a) $\dfrac{dx}{dt} = 2t - 6,\ \dfrac{dy}{dt} = 6t^2 - 12t - 18$.

Using the chain rule, $\dfrac{dy}{dx} = \dfrac{dy}{dt} \div \dfrac{dx}{dt}$

$= \dfrac{6t^2 - 12t - 18}{2t - 6} = \dfrac{(2t-6)(3t+3)}{2t-6} = 3(t+1)$

b) When $\dfrac{dy}{dx} = 0$, $3(t+1) = 0 \Rightarrow t = -1$

When $t = -1$, $x = (-1)^2 - 6(-1) = 7$ and
$y = 2(-1)^3 - 6(-1)^2 - 18(-1) = 10$
So the particle has a stationary point at $(7, 10)$.

Answers

Q29 a) $x = t^3 - t^2 \Rightarrow \dfrac{dx}{dt} = 3t^2 - 2t$

$y = t^3 + 3t^2 - 9t \Rightarrow \dfrac{dy}{dt} = 3t^2 + 6t - 9$

$\Rightarrow \dfrac{dy}{dx} = \dfrac{dy}{dt} \div \dfrac{dx}{dt} = \dfrac{3t^2 + 6t - 9}{3t^2 - 2t}$

At turning points, $\dfrac{dy}{dx} = 0 \Rightarrow \dfrac{3t^2 + 6t - 9}{3t^2 - 2t} = 0$

$\Rightarrow t^2 + 2t - 3 = 0 \Rightarrow (t + 3)(t - 1) = 0 \Rightarrow t = -3$ or $t = 1$
When $t = -3$, $x = (-3)^3 - (-3)^2 = -36$
and $y = (-3)^3 + 3(-3)^2 - 9(-3) = 27$
When $t = 1$, $x = (1)^3 - (1)^2 = 0$ and $y = (1)^3 + 3(1)^2 - 9(1) = -5$
So the coordinates of the turning points
are $(-36, 27)$ and $(0, -5)$.

[4 marks available — 1 mark for using $\dfrac{dy}{dx} = \dfrac{dy}{dt} \div \dfrac{dx}{dt}$,
1 mark for correct $\dfrac{dy}{dx}$, 1 mark for setting $\dfrac{dy}{dx} = 0$ and using
it to find values of t, 1 mark for the correct coordinates using
values of t]

b) $y = 0$ at the x-axis, so $t^3 + 3t^2 - 9t = 0 \Rightarrow t(t^2 + 3t - 9) = 0$

$\Rightarrow t = 0$ or $t = \dfrac{-3 \pm \sqrt{3^2 - 4 \times 1 \times (-9)}}{2 \times 1} = \dfrac{-3 \pm 3\sqrt{5}}{2}$

The quadratic formula was used to find two of the values of t.
When $t = 0$, $x = (0)^3 - (0)^2 = 0$ and $y = (0)^3 + 3(0)^2 - 9(0)$,
so C passes through the origin.
[2 marks available — 1 mark for the values of t,
1 mark for clearly showing C passes through the origin]

c) At $t = 2$, $x = (2)^3 - (2)^2 = 4$, $y = (2)^3 + 3(2)^2 - 9(2) = 2$
$\dfrac{dy}{dx} = \dfrac{3(2)^2 + 6(2) - 9}{3(2)^2 - 2(2)} = \dfrac{15}{8}$

The tangent to C at $t = 2$ has the gradient $\dfrac{15}{8}$
and an equation in the form $y = mx + c$.
Substitute in the values of x, y and m to find c:
$2 = \dfrac{15}{8}(4) + c \Rightarrow c = -\dfrac{11}{2}$

So the tangent to C at $t = 2$ has the equation $y = \dfrac{15}{8}x - \dfrac{11}{2}$,
which can be rearranged as $15x - 8y - 44 = 0$.
[2 marks available — 1 mark for substituting t = 2 to find x,
y and $\dfrac{dy}{dx}$, 1 mark for the correct equation for the tangent in
any form]

Q30 a) $x = t \ln t \Rightarrow \dfrac{dx}{dt} = t \times \dfrac{1}{t} + 1 \times \ln t = 1 + \ln t$

$y = 2t^3 - t^2 \Rightarrow \dfrac{dy}{dt} = 6t^2 - 2t$

$\dfrac{dy}{dx} = \dfrac{dy}{dt} \div \dfrac{dx}{dt} = \dfrac{6t^2 - 2t}{1 + \ln t}$

[3 marks available — 1 mark for $\dfrac{dx}{dt}$,
1 mark for $\dfrac{dy}{dt}$, 1 mark for correct answer]

b) At $(0, 1)$, $x = 0$, so $t \ln t = 0$.
t cannot be 0 because $\ln 0$ is undefined $\Rightarrow \ln t = 0 \Rightarrow t = 1$

So $\dfrac{dy}{dx} = \dfrac{6(1)^2 - 2(1)}{1 + \ln(1)} = 4$

So the tangent to the curve at $(0, 1)$ has a gradient of 4 and an
equation in the form $y = mx + c$. Substitute in the values of
x, y and m to find c: $1 = 4(0) + c \Rightarrow c = 1$
So the tangent to the curve at $(0, 1)$
has the equation $y = 4x + 1$.
[3 marks available — 1 mark for finding t = 1,
1 mark for substituting t = 1 into $\dfrac{dy}{dx}$, 1 mark
for correct equation for tangent]

Q31 a) $x = 3se^s$, so let $x = uv$, where $u = 3s$ and $v = e^s$

$\Rightarrow \dfrac{du}{ds} = 3$, $\dfrac{dv}{ds} = e^s$,

$\dfrac{dx}{ds} = u\dfrac{dv}{ds} + v\dfrac{du}{ds} = 3se^s + 3e^s = 3e^s(s + 1)$

$y = e^{2s} + se^{2s} \Rightarrow \dfrac{dy}{ds} = 2e^{2s} + 2se^{2s} + e^{2s}$
$= 3e^{2s} + 2se^{2s} = e^{2s}(3 + 2s)$

$\dfrac{dy}{dx} = \dfrac{dy}{ds} \div \dfrac{dx}{ds} = \dfrac{e^{2s}(3 + 2s)}{3e^s(s + 1)} = \dfrac{e^s(3 + 2s)}{3(s + 1)}$

When $s = 0$, $x = 3(0)e^{(0)} = 0$, $y = e^{2(0)} + (0)e^{2(0)} = 1$,

$\dfrac{dy}{dx} = \dfrac{e^{(0)}(3 + 2(0))}{3((0) + 1)} = 1$

The tangent to the curve at $s = 0$ has an equation in the form
$y = mx + c$. Substitute in the values of x, y and m to find c:
$1 = 1 \times 0 + c \Rightarrow c = 1$
So the tangent to the curve at $s = 0$ has the equation $y = x + 1$.
When $s = 2$, $x = 3(2)e^{(2)} = 6e^2$,
$y = e^{2(2)} + (2)e^{2(2)} = e^4 + 2e^4 = 3e^4$

$\dfrac{dy}{dx} = \dfrac{e^{(2)}(3 + 2(2))}{3((2) + 1)} = \dfrac{7e^2}{9}$

The tangent to the curve at $s = 2$ has an equation in the form
$y = mx + c$. Substitute in the values of x, y and m to find c:
$3e^4 = \dfrac{7e^2}{9} \times 6e^2 + c = \dfrac{42e^4}{9} + c \Rightarrow c = -\dfrac{5e^4}{3}$
So the tangent to the curve at $s = 0$ has

the equation $y = \left(\dfrac{7e^2}{9}\right)x - \dfrac{5e^4}{3}$.

b) The tangents intersect where:

$x + 1 = \left(\dfrac{7e^2}{9}\right)x - \dfrac{5e^4}{3} \Rightarrow 1 + \dfrac{5e^4}{3} = \left(\dfrac{7e^2}{9}\right)x - x$

$\Rightarrow 9 + 15e^4 = x(7e^2 - 9) \Rightarrow x = \dfrac{9 + 15e^4}{7e^2 - 9}$

$y = x + 1 = \dfrac{9 + 15e^4}{7e^2 - 9} + 1$

So the tangents intersect at the point
$\left(\dfrac{9 + 15e^4}{7e^2 - 9}, \dfrac{9 + 15e^4}{7e^2 - 9} + 1\right)$

Q32 At $x = 2$, $4y + (2)^2y^2 = 4(2) \Rightarrow 4y + 4y^2 = 8$
$\Rightarrow y^2 + y - 2 = 0 \Rightarrow (y + 2)(y - 1) = 0$
$\Rightarrow y = -2$ or $y = 1$, so a = 1 and b = -2
Differentiate $4y + x^2y^2 = 4x$: $\dfrac{d}{dx}4y + \dfrac{d}{dx}x^2y^2 = \dfrac{d}{dx}4x$

$\Rightarrow 4\dfrac{dy}{dx} + 2x^2y\dfrac{dy}{dx} + 2xy^2 = 4 \Rightarrow \dfrac{dy}{dx}(4 + 2x^2y) = 4 - 2xy^2$

$\Rightarrow \dfrac{dy}{dx} = \dfrac{4 - 2xy^2}{4 + 2x^2y} = \dfrac{2 - xy^2}{2 + x^2y}$

At $(2, 1)$, $\dfrac{dy}{dx} = \dfrac{2 - (2)(1)^2}{2 + (2)^2(1)} = 0$, so $(2, 1)$ is a stationary point
and the equation of the tangent is $y = 1$.
At $(2, -2)$, $\dfrac{dy}{dx} = \dfrac{2 - (2)(-2)^2}{2 + (2)^2(-2)} = 1$ and the tangent has an
equation in the form $y = mx + c$. Substitute in the values of x, y
and m to find c: $-2 = 2 + c \Rightarrow c = -4$
So the tangent at $(2, -2)$ has the equation $y = x - 4$.
The tangents intersect where $x - 4 = 1 \Rightarrow x = 5$
So the tangents intersect at the point $(5, 1)$.

Q33 a) Differentiate $x^2y + y^2 = x^2 + 1$:
$\dfrac{d}{dx}x^2y + \dfrac{d}{dx}y^2 = \dfrac{d}{dx}x^2 + \dfrac{d}{dx}1 \Rightarrow x^2\dfrac{dy}{dx} + 2xy + 2y\dfrac{dy}{dx} = 2x$
The product rule has been used to differentiate x^2y.

$\Rightarrow \dfrac{dy}{dx}(x^2 + 2y) = 2x - 2xy \Rightarrow \dfrac{dy}{dx} = \dfrac{2x - 2xy}{x^2 + 2y}$

At $(1, -2)$, $\dfrac{dy}{dx} = \dfrac{2(1) - 2(1)(-2)}{(1)^2 + 2(-2)} = -2$

[4 marks available — 1 mark for two terms correctly
differentiated, 1 mark for other two terms correctly
differentiated, 1 mark for terms rearranged to give expression
for $\dfrac{dy}{dx}$, 1 mark for correct gradient]

b) At $(1, a)$, $x = 1 \Rightarrow (1)^2 y + y^2 = (1)^2 + 1 \Rightarrow y + y^2 = 2$

$\Rightarrow y^2 + y - 2 = 0 \Rightarrow (y + 2)(y - 1) = 0 \Rightarrow y = -2$ or $y = 1$

The point $(1, -2)$ is already given in the question, so $a = 1$.

At $(1, 1)$, $\dfrac{dy}{dx} = \dfrac{2(1) - 2(1)(1)}{(1)^2 + 2(1)} = 0$, so it is a turning point.

[2 marks available — 1 mark for the correct value of a,
1 mark for showing $\dfrac{dy}{dx} = 0$ at $(1, a)$]

Q34 a) $x \ln x + x^2 y = y^2 x - 6x \Rightarrow (1)\ln(1) + (1)^2 y = y^2(1) - 6(1)$

$\Rightarrow 0 + y = y^2 - 6 \Rightarrow 0 = y^2 - y - 6 \Rightarrow (y + 2)(y - 3) = 0$

$\Rightarrow y = -2$ or $y = 3$, so $a = 3$ and $b = -2$

b) Differentiate $x \ln x + x^2 y = y^2 x - 6x$:

$\dfrac{d}{dx} x \ln x + \dfrac{d}{dx} x^2 y = \dfrac{d}{dx} y^2 x - \dfrac{d}{dx} 6x$

$\Rightarrow 1 + \ln x + x^2 \dfrac{dy}{dx} + 2xy = y^2 + 2xy \dfrac{dy}{dx} - 6$

$\Rightarrow \dfrac{dy}{dx}(x^2 - 2xy) = y^2 - 2xy - \ln x - 7$

$\Rightarrow \dfrac{dy}{dx} = \dfrac{y^2 - 2xy - \ln x - 7}{x^2 - 2xy}$

At $(1, 3)$, $\dfrac{dy}{dx} = \dfrac{(3)^2 - 2(1)(3) - \ln(1) - 7}{(1)^2 - 2(1)(3)} = \dfrac{4}{5}$

The normal to the curve at $(1, 3)$ has a gradient of $-1 \div \dfrac{4}{5} = -\dfrac{5}{4}$ and an equation in the form $y = mx + c$.

Substitute in the values of x, y and m to find c:

$3 = -\dfrac{5}{4}(1) + c \Rightarrow c = \dfrac{17}{4}$. So the normal to the curve at $(1, 3)$ has the equation $y = -\dfrac{5}{4}x + \dfrac{17}{4}$.

At $(1, -2)$, $\dfrac{dy}{dx} = \dfrac{(-2)^2 - 2(1)(-2) - \ln(1) - 7}{(1)^2 - 2(1)(-2)} = \dfrac{1}{5}$

The normal to the curve at $(1, -2)$ has a gradient of $-1 \div \dfrac{1}{5} = -5$ and an equation in the form $y = mx + c$.

Substitute in the values of x, y and m to find c:

$-2 = -5(1) + c \Rightarrow c = 3$. So the normal to the curve at $(1, -2)$ has the equation $y = -5x + 3$.

c) The normals intersect where:

$-\dfrac{5}{4}x + \dfrac{17}{4} = -5x + 3 \Rightarrow \dfrac{15}{4}x = -\dfrac{5}{4} \Rightarrow x = -\dfrac{1}{3}$

At $x = -\dfrac{1}{3}$, $y = -5\left(-\dfrac{1}{3}\right) + 3 = \dfrac{14}{3}$

So the normals intersect at the point $\left(-\dfrac{1}{3}, \dfrac{14}{3}\right)$.

Q35 a) Differentiate $4x^2 - 2y^2 = 7x^2 y$:

$\dfrac{d}{dx} 4x^2 - \dfrac{d}{dx} 2y^2 = \dfrac{d}{dx} 7x^2 y \Rightarrow 8x - 4y \dfrac{dy}{dx} = 7x^2 \dfrac{dy}{dx} + 14xy$

$\Rightarrow 8x - 14xy = 7x^2 \dfrac{dy}{dx} + 4y \dfrac{dy}{dx} \Rightarrow \dfrac{dy}{dx} = \dfrac{8x - 14xy}{7x^2 + 4y}$

[3 marks available — 1 mark for one term correctly differentiated, 1 mark for the other two terms correctly differentiated, 1 mark for terms rearranged to give expression for $\dfrac{dy}{dx}$ (allow for incorrectly differentiated terms)]

b) (i) At $(1, -4)$, $\dfrac{dy}{dx} = \dfrac{8(1) - 14(1)(-4)}{7(1)^2 + 4(-4)} = -\dfrac{64}{9}$

So the gradient of the tangent to C at $(1, -4)$ is $-\dfrac{64}{9}$.

[1 mark]

(ii) From part b)(i) you know that the gradient of the tangent at $(1, -4)$ is $-\dfrac{64}{9}$, so the gradient of the normal (m) is $-1 \div -\dfrac{64}{9} = \dfrac{9}{64}$.

Substitute x, y and m into the equation for the normal to find c: $y = mx + c \Rightarrow -4 + \dfrac{9}{64}(1) + c \Rightarrow c = -\dfrac{265}{64}$

So the equation of the normal to C at $(1, -4)$ is $y = \dfrac{9}{64}x - \dfrac{265}{64}$, which can be rearranged as $9x - 64y - 265 = 0$.

[3 marks available — 1 mark for the correct value of m, 1 mark for the correct value of c, 1 mark for the correct equation for the normal in the correct form]

Q36 a) Differentiate $x \cos x + y \sin x = y^3$:

$\dfrac{d}{dx} x \cos x + \dfrac{d}{dx} y \sin x = \dfrac{d}{dx} y^3$

$\Rightarrow -x \sin x + \cos x + y \cos x + \left(\dfrac{dy}{dx}\right) \sin x = 3y^2 \dfrac{dy}{dx}$

Use the product rule to differentiate $x \cos x$ and $y \sin x$.

$\Rightarrow \cos x - x \sin x + y \cos x = 3y^2 \dfrac{dy}{dx} - \left(\dfrac{dy}{dx}\right) \sin x$

$\Rightarrow \dfrac{dy}{dx} = \dfrac{\cos x - x \sin x + y \cos x}{3y^2 - \sin x}$

At stationary points $\dfrac{dy}{dx} = 0$, so $\dfrac{\cos x - x \sin x + y \cos x}{3y^2 - \sin x} = 0$

$\Rightarrow \cos x - x \sin x + y \cos x = 0 \Rightarrow y \cos x = x \sin x - \cos x$

$\Rightarrow y = \dfrac{x \sin x - \cos x}{\cos x} \Rightarrow y = x \tan x - 1$

[5 marks available — 1 mark for one term correctly differentiated, 1 mark for other two terms correctly differentiated, 1 mark for terms rearranged to give expression for $\dfrac{dy}{dx}$, 1 mark for setting $\dfrac{dy}{dx} = 0$, 1 mark for clearly showing $y = x \tan x - 1$ at stationary points]

b) When $x = \dfrac{\pi}{2}$, $\left(\dfrac{\pi}{2}\right) \cos\left(\dfrac{\pi}{2}\right) + y \sin\left(\dfrac{\pi}{2}\right) = y^3$

$\Rightarrow \left(\dfrac{\pi}{2}\right)(0) + y(1) = y^3 \Rightarrow 0 = y^3 - y = y(y + 1)(y - 1)$

$\Rightarrow y = 0$ or $y = -1$ or $y = 1$

So the three points have the coordinates $\left(\dfrac{\pi}{2}, 0\right)$, $\left(\dfrac{\pi}{2}, -1\right)$ and $\left(\dfrac{\pi}{2}, 1\right)$.

[3 marks available — 1 mark for substituting in $x = \dfrac{\pi}{2}$, 1 mark for simplifying and factorising the equation, 1 mark for the correct three y-coordinates]

c) At $\left(\dfrac{\pi}{2}, 0\right)$, $\dfrac{dy}{dx} = \dfrac{\cos x - x \sin x + y \cos x}{3y^2 - \sin x} = \dfrac{0 - \frac{\pi}{2} + 0}{0 - 1} = \dfrac{\pi}{2}$

At $\left(\dfrac{\pi}{2}, -1\right)$, $\dfrac{dy}{dx} = \dfrac{0 - \frac{\pi}{2} + 0}{3 - 1} = -\dfrac{\pi}{4}$

At $\left(\dfrac{\pi}{2}, 1\right)$, $\dfrac{dy}{dx} = \dfrac{0 - \frac{\pi}{2} + 0}{3 - 1} = -\dfrac{\pi}{4}$

The tangents at $\left(\dfrac{\pi}{2}, -1\right)$ and $\left(\dfrac{\pi}{2}, 1\right)$ are both $-\dfrac{\pi}{4}$, so they are parallel and will never intersect.

[2 marks available — 1 mark for at least two correct tangent gradients, 1 mark for clearly showing the correct two tangents will not intersect]

Q37 a) Differentiating with respect to x:

$2x - 2 + x\dfrac{dy}{dx} + y + 2y\dfrac{dy}{dx} = 0$ so $\dfrac{dy}{dx}(x + 2y) = 2 - 2x - y$ so $\dfrac{dy}{dx} = \dfrac{2 - 2x - y}{x + 2y}$

[4 marks available — 1 mark for xy differentiated correctly, 1 mark for y^2 differentiated correctly, 1 mark for differentiating the rest of the equation correctly, 1 mark for the correct answer]

b) $\dfrac{dy}{dx} = 0 \Rightarrow 2 - 2x - y = 0 \Rightarrow y = 2 - 2x$

Substituting into the equation of the curve:

$x^2 - 2x + x(2 - 2x) + (2 - 2x)^2 = 20$

$\Rightarrow x^2 - 2x + 2x - 2x^2 + 4x^2 - 8x + 4 = 20$

$\Rightarrow 3x^2 - 8x - 16 = 0 \Rightarrow (3x + 4)(x - 4) = 0$

$\Rightarrow x = -\dfrac{4}{3}$ and $x = 4$.

When $x = -\dfrac{4}{3}$, $y = 2 - 2 \times -\dfrac{4}{3} = \dfrac{14}{3}$

When $x = 4$, $y = 2 - 2 \times 4 = -6$

Therefore, the minimum point is $(4, -6)$.

[4 marks available — 1 mark for setting $\dfrac{dy}{dx} = 0$ and rearranging to make x or y the subject, 1 mark for solving simultaneously with equation of curve, 1 mark for both correct values of x, 1 mark for the correct answer]

Answers

c) Vertical tangents occur when the gradient is undefined,
i.e. when the denominator of the gradient is 0.
$\Rightarrow x + 2y = 0 \Rightarrow x = -2y$
Substituting into equation of curve:
$(-2y)^2 - 2(-2y) + (-2y)(y) + y^2 = 20$
$\Rightarrow 4y^2 + 4y - 2y^2 + y^2 = 20 \Rightarrow 3y^2 + 4y - 20 = 0$
$\Rightarrow (3y + 10)(y - 2) = 0 \Rightarrow y = -\frac{10}{3}$ and $y = 2$
When $y = 2$, $x = -4$ and when $y = -\frac{10}{3}$, $x = \frac{20}{3}$
so P and Q are $(-4, 2)$ and $\left(\frac{20}{3}, -\frac{10}{3}\right)$.
[4 marks available — 1 mark for setting denominator of $\frac{dy}{dx}$
equal to zero, 1 mark for substituting into the equation
of the curve, 1 mark for finding the x- or y-values, 1 mark
for both correct answers]
You could have formed an equation in x instead.

Q38 a) As $x > 1$, $0 < f(x) < \frac{\pi}{2}$.

b) $\csc y = \csc(\text{arccosec } x) \Rightarrow \csc y = x$
Differentiating both sides with respect to x:
$-\csc y \cot y \frac{dy}{dx} = 1 \Rightarrow \frac{dy}{dx} = -\frac{1}{\csc y \cot y} = -\sin y \tan y$
$\csc y = x \Rightarrow \sin y = \frac{1}{x}$
As $x > 1$ you can use Pythagoras' theorem on the right-angled
triangle with opposite side 1 and hypotenuse x to show the
adjacent side is $\sqrt{x^2 - 1} \Rightarrow \tan y = \frac{1}{\sqrt{x^2 - 1}}$.
So $\frac{dy}{dx} = -\frac{1}{x} \times \frac{1}{\sqrt{x^2 - 1}} = -\frac{1}{x\sqrt{x^2 - 1}}$.

c) $x > 1 \Rightarrow \sqrt{x^2 - 1} > 0 \Rightarrow x\sqrt{x^2 - 1} > 0$
So $f'(x) = -\frac{1}{x\sqrt{x^2 - 1}} < 0$, and so $f(x)$ is decreasing.

d) $f'(x) = 0$ has no solutions since the numerator of $-\frac{1}{x\sqrt{x^2 - 1}}$
can never be zero.

e) Let $u = -\frac{1}{x} = -x^{-1}$ and $v = \frac{1}{\sqrt{x^2 - 1}} = (x^2 - 1)^{-\frac{1}{2}}$
$\frac{du}{dx} = x^{-2}$ and $\frac{dv}{dx} = -\frac{1}{2}(x^2 - 1)^{-\frac{3}{2}} \times 2x = -x(x^2 - 1)^{-\frac{3}{2}}$
$f''(x) = -\frac{1}{x} \times -x(x^2 - 1)^{-\frac{3}{2}} + (x^2 - 1)^{-\frac{1}{2}} \times x^{-2}$
$= (x^2 - 1)^{-\frac{3}{2}} + (x^2 - 1)^{-\frac{1}{2}} \times x^{-2}$
$= \frac{1}{(x^2 - 1)^{\frac{3}{2}}} + \frac{1}{x^2(x^2 - 1)^{\frac{1}{2}}}$
$= \frac{x^2}{x^2(x^2 - 1)^{\frac{3}{2}}} + \frac{x^2 - 1}{x^2(x^2 - 1)^{\frac{3}{2}}} = \frac{2x^2 - 1}{x^2(x^2 - 1)^{\frac{3}{2}}}$
The denominator and numerator are always positive for
$x > 1 \Rightarrow f''(x) > 0$ for the entire domain, so $f(x)$ is convex
on the whole domain.

Q39 a) $y = ax^3 + bx^2 + cx + d$
$\Rightarrow \frac{dy}{dx} = 3ax^2 + 2bx + c \Rightarrow \frac{d^2y}{dx^2} = 6ax + 2b$.
Setting $\frac{d^2y}{dx^2} = 0$, we get $6ax + 2b = 0 \Rightarrow x = -\frac{b}{3a}$.
When $x = -\frac{b}{2a}$, $\frac{d^2y}{dx^2} = -b$.
When $x = -\frac{b}{4a}$, $\frac{d^2y}{dx^2} = \frac{b}{2}$.
As $b \neq 0$, there is a sign change either side of the stationary
point, so $x = -\frac{b}{3a}$ is a point of inflection.
[4 marks available — 1 mark for the correct first derivative,
1 mark for the correct second derivative, 1 mark for the
correct x-value, 1 mark for a correct method to show point
of inflection]

b) E.g. $y = x^4$ does not have a point of inflection.
$\frac{dy}{dx} = 4x^3 \Rightarrow \frac{d^2y}{dx^2} = 12x^2$
When $\frac{d^2y}{dx^2} = 0 \Rightarrow x = 0$
When $x = -1$, $\frac{d^2y}{dx^2} = 12(-1)^2 = 12 > 0$.
When $x = 1$, $\frac{d^2y}{dx^2} = 12(1)^2 = 12 > 0$.
So the stationary point at $x = 0$ is not a point of inflection,
meaning the curve does not have a point of inflection.
[2 marks available — 1 mark for a suitable example,
1 mark for a correct explanation]

Q40 a) When $x = 0$, $y = 4 \Rightarrow 2^{a(0)^2 + b(0) + c} = 4 \Rightarrow 2^c = 4 \Rightarrow c = 2$
[1 mark]

b) Let $y = 2^u$ where $u = ax^2 + bx + c$
$\frac{dy}{du} = 2^u = 2^u \ln 2 = 2^{ax^2 + bx + c} \ln 2$ and $\frac{du}{dx} = 2ax + b$
$\frac{dy}{dx} = \frac{dy}{du} \times \frac{du}{dx} = (2ax + b) 2^{ax^2 + bx + c} \ln 2$
[3 marks available — 1 mark for the correct differentiation of
$ax^2 + bx + c$, 1 mark for the correct differentiation of 2^u,
1 mark for the correct answer]

c) Form two simultaneous equations in a and b.
You know that $\ln 2 > 0$ and $2^{\frac{9}{4}} > 0$,
so when $x = \frac{1}{2}$, $\frac{dy}{dx} = 2a \times \frac{1}{2} + b = 0 \Rightarrow a = -b$
Also at $x = \frac{1}{2}$, $2^{\frac{9}{4}} = 2^{ax^2 + bx + c}$ so by equating the powers,
$a\left(\frac{1}{2}\right)^2 + b\left(\frac{1}{2}\right) + c = \frac{9}{4} \Rightarrow \frac{1}{4}a + \frac{1}{2}b + 2 = \frac{9}{4}$
$\Rightarrow \frac{1}{4}a + \frac{1}{2}b = \frac{1}{4} \Rightarrow a + 2b = 1$
Subtracting the two equations gives $b = 1$ and $a = -1$.
[4 marks available — 1 mark for a correct method to form
either equation, 1 mark for one correct equation, 1 mark
for solving simultaneously, 1 mark for the correct values of
a and b]

Q41 a) $f(x) = \frac{2^x}{2^{2x}} + \frac{2^{3x}}{2^{2x}} = 2^{-x} + 2^x$ so $f'(x) = -2^{-x} \ln 2 + 2^x \ln 2$
[3 marks available — 1 mark for rewriting f in a suitable
form, 1 mark for correctly differentiating each term]

b) At $x = 1$, $y = \frac{5}{2}$ and $\frac{dy}{dx} = -\frac{1}{2} \ln 2 + 2 \ln 2 = \frac{3}{2} \ln 2$
so the normal has gradient $-\frac{2}{3 \ln 2}$.
$y = mx + c \Rightarrow \frac{5}{2} = \left(-\frac{2}{3 \ln 2} \times 1\right) + c$
$\Rightarrow c = \frac{5}{2} + \frac{2}{\ln 2^3} \Rightarrow c = \frac{5}{2} + \frac{2}{\ln 8}$
So the equation of the normal is $y = -\frac{2}{\ln 8}x + \frac{5}{2} + \frac{2}{\ln 8}$.
[3 marks available — 1 mark for the correct gradient of the
normal, 1 mark for a correct method to find the equation of
the normal, 1 mark for rearranging to show the correct result]

Q42 $\tan \theta = \frac{10}{x}$ and $\frac{dx}{dt} = 1.5$, so differentiating
implicitly with respect to t:
$\tan \theta = \frac{10}{x} \Rightarrow \sec^2 \theta \frac{d\theta}{dt} = -\frac{10}{x^2} \frac{dx}{dt}$
When the rope between the pulley and statue is 25 m:

So $25^2 = 10^2 + x^2 \Rightarrow x = \sqrt{25^2 - 10^2} \Rightarrow x = 5\sqrt{21}$

$\Rightarrow \cos\theta = \dfrac{5\sqrt{21}}{25}$

$\Rightarrow \sec^2\theta = \left(\dfrac{25}{5\sqrt{21}}\right)^2 = \dfrac{625}{525} = \dfrac{25}{21}$,

hence $\dfrac{25}{21} \times \dfrac{d\theta}{dt} = -\dfrac{10}{(5\sqrt{21})^2} \times 1.5$

$\Rightarrow \dfrac{d\theta}{dt} = -0.024$ radians per second

[6 marks available — 1 mark for an equation relating θ and x, 1 mark for the correct differentiation of $\tan\theta$, 1 mark for the correct differentiation of $\dfrac{10}{x}$, 1 mark for a correct method to find $\sec\theta$, 1 mark for substituting numbers back into the differential equation, 1 mark for the correct answer]
From the diagram you can see $\cos\theta = \dfrac{x}{25}$, so $\sec\theta = \dfrac{25}{x}$.
You could have substituted that directly into the differential equation $\sec^2\theta\dfrac{d\theta}{dt} = -\dfrac{10}{x^2}\dfrac{dx}{dt}$ and solved it.

Q43 Let $u = \sec x + \csc x$ and $v = \cot x$.
Then $\dfrac{du}{dx} = \sec x \tan x - \csc x \cot x$ and $\dfrac{dv}{dx} = -\csc^2 x$,
so by the quotient rule,

$f'(x) = \dfrac{(\sec x\tan x - \csc x\cot x)\cot x - -\csc^2 x(\sec x + \csc x)}{\cot^2 x}$

$= \dfrac{\sec x\tan x}{\cot x} - \csc x + \sec^2 x(\sec x + \csc x)$

$= \sec x\tan^2 x - \csc x + \sec^3 x + \sec^2 x\csc x$

$= \csc x(\sec^2 x - 1) + \sec x(\sec^2 x + \tan^2 x)$

Q44 a) Let $y = u^4$ where $u = \sec\sqrt{x} = \sec(x^{\frac{1}{2}})$
$\dfrac{dy}{du} = 4u^3 = 4\sec^3\sqrt{x}$
$\dfrac{du}{dx} = \dfrac{1}{2}x^{-\frac{1}{2}} \times \sec\sqrt{x}\tan\sqrt{x} = \dfrac{1}{2\sqrt{x}}\sec\sqrt{x}\tan\sqrt{x}$
So $\dfrac{dy}{dx} = \dfrac{dy}{du} \times \dfrac{du}{dx}$
$= 4\sec^3\sqrt{x} \times \dfrac{1}{2\sqrt{x}}\sec\sqrt{x}\tan\sqrt{x} = \dfrac{2\sec^4\sqrt{x}\tan\sqrt{x}}{\sqrt{x}}$
[4 marks available — 1 mark for a correct attempt to use the chain rule, 1 mark for $\dfrac{dy}{du}$, 1 mark for $\dfrac{du}{dx}$, 1 mark for the correct answer]

b) If $2\sec^4\sqrt{x}\tan\sqrt{x} = 0$ then $\tan\sqrt{x} = 0$ as $\sec x \neq 0$
$\Rightarrow \sqrt{x} = \pi \Rightarrow x = \pi^2$
When $x = \pi^2$, $y = \sec^4\pi = 1$, so $(\pi^2, 1)$ is a stationary point.
As $\sec x \leq -1$ or $\sec x \geq 1$, $\sec^4 x \geq 1$, so the stationary point $(\pi^2, 1)$ must be a minimum.
[3 marks available — 1 mark for $\tan\sqrt{x} = 0$, 1 mark for the correct value of x, 1 mark for the correct value of y with justification of it being a minimum point]

c) The other minimum points have x-coordinates $(2\pi)^2, (3\pi)^2, ..., (n\pi)^2$ for all integer values of n. The distance between two consecutive minimum points, $n\pi$ and $(n+1)\pi$, is $((n+1)\pi)^2 - (n\pi)^2 = (n^2 + 2n + 1)\pi^2 - n^2\pi^2$
$= (n^2 - n^2 + 2n + 1)\pi^2 = (2n+1)\pi^2$,
which gets larger as n increases.
[2 marks available — 1 mark for finding the x-coordinates of the minimum points, 1 mark for a sensible explanation]

Chapter 8: Integration

Prior Knowledge Check

Q1 a) $-21\sin 3x$
b) $-2e^{5-2x}$
c) $\dfrac{2x}{x^2 + 1}$
Q2 a) $\sec^2 6x$
b) $\sin 5x$

Q3

t	-1	-0.5	0	0.5	1
x	0	0.375	0	-0.375	0
y	1	0.25	0	0.25	1

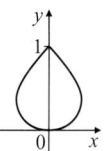

Q4 a) $u = 4x$ so $\dfrac{du}{dx} = 4$. $v = \tan x$ so $\dfrac{dv}{dx} = \sec^2 x$.
$\dfrac{dy}{dx} = 4x\sec^2 x + 4\tan x$

b) $u = e^{3x}$ so $\dfrac{du}{dx} = 3e^{3x}$. $v = \ln(3x^3 + 3)$ so $\dfrac{dv}{dx} = \dfrac{9x^2}{3x^3 + 3}$
$\dfrac{dy}{dx} = \dfrac{9x^2 e^{3x}}{3x^3 + 3} + 3e^{3x}\ln(3x^3 + 3) = \dfrac{3x^2 e^{3x}}{x^3 + 1} + 3e^{3x}\ln(3x^3 + 3)$

Q5 $\dfrac{7}{(2x + 1)(x + 4)} \equiv \dfrac{A}{2x + 1} + \dfrac{B}{x + 4}$
$\Rightarrow 7 \equiv A(x + 4) + B(2x + 1)$
When $x = -4 \Rightarrow 7 = -7B \Rightarrow B = -1$
When $x = -\dfrac{1}{2} \Rightarrow 7 = \dfrac{7}{2}A \Rightarrow A = 2$
So $\dfrac{7}{(2x + 1)(x + 4)} \equiv \dfrac{2}{2x + 1} - \dfrac{1}{x + 4}$

Exercise 8.1 — Integrating $(ax + b)^n$ where $n \neq -1$

Q1 a) $\int (x + 10)^{10}\,dx = \dfrac{1}{1 \times 11}(x + 10)^{11} + C = \dfrac{1}{11}(x + 10)^{11} + C$

b) $\int (5x)^7\,dx = \dfrac{1}{5 \times 8}(5x)^8 + C = \dfrac{1}{40}(5x)^8 + C$
$= \dfrac{1}{40}5^8 x^8 + C = \dfrac{5^8}{40}x^8 + C = \dfrac{78125x^8}{8} + C$

c) $\int (3 - 5x)^{-2}\,dx = \dfrac{1}{-5 \times -1}(3 - 5x)^{-1} + C$
$= \dfrac{1}{5}(3 - 5x)^{-1} + C = \dfrac{1}{5(3 - 5x)} + C$

d) $\int (3x - 4)^{-\frac{4}{3}}\,dx = \dfrac{1}{3 \times \left(-\frac{1}{3}\right)}(3x - 4)^{-\frac{1}{3}} + C$
$= \dfrac{1}{-1}(3x - 4)^{-\frac{1}{3}} + C = -(3x - 4)^{-\frac{1}{3}} + C = \dfrac{-1}{\sqrt[3]{3x - 4}} + C$

Q2 $\int (4 - 2x)^{-7}\,dx = \dfrac{1}{-2 \times -6}(4 - 2x)^{-6} + C = \dfrac{1}{12(4 - 2x)^6} + C$
[2 marks available — 1 mark for a correct method, 1 mark for the correct answer]

Q3 $\int 5(2x + 1)^3 + (4x + 2)^3\,dx$
$= \int 5(2x + 1)^3 + 2^3(2x + 1)^3\,dx = \int 13(2x + 1)^3\,dx$
$= \dfrac{13}{8}(2x + 1)^4 + C$
[3 marks available — 1 mark for a correct method to integrate, 1 mark for simplifying correctly, 1 mark for the correct answer]
You could also have integrated the terms separately and then simplified.

Q4 a) Begin by taking the constant of 8 outside of the integration and then integrate $\int (2x - 4)^4\,dx$ as usual.
$\int 8(2x - 4)^4\,dx = 8\int (2x - 4)^4\,dx$
$= 8 \times \left(\dfrac{1}{2 \times 5}(2x - 4)^5 + c\right)$
$= \dfrac{8}{10}(2x - 4)^5 + C$
$= \dfrac{4}{5}(2x - 4)^5 + C = \dfrac{4(2x - 4)^5}{5} + C$

b) Use your answer to part a). The integral you found will be the same but without the constant of integration — it'll have limits instead.
$\int_{\frac{3}{2}}^{\frac{5}{2}} 8(2x - 4)^4\,dx = \dfrac{4}{5}\left[(2x - 4)^5\right]_{\frac{3}{2}}^{\frac{5}{2}}$
$= \dfrac{4}{5}\left(\left[\left(2\left(\dfrac{5}{2}\right) - 4\right)^5\right] - \left[\left(2\left(\dfrac{3}{2}\right) - 4\right)^5\right]\right)$
$= \dfrac{4}{5}([(1)^5] - [(-1)^5]) = \dfrac{4}{5}([1] - [-1]) = \dfrac{8}{5}$

Answers

Q5 $\int_0^1 (6x+1)^{-3}\,dx = \left[\dfrac{1}{6\times(-2)}(6x+1)^{-2}\right]_0^1$

$= -\dfrac{1}{12}\left[(6x+1)^{-2}\right]_0^1$

$= -\dfrac{1}{12}\left([(7)^{-2}] - [(1)^{-2}]\right)$

$= -\dfrac{1}{12}\left(\dfrac{1}{49} - 1\right) = \dfrac{4}{49}$

Q6 You've been given that $f'(x) = (8-7x)^4$ and you need to find $f(x)$, so integrate with respect to x.

$f(x) = \int f'(x)\,dx = \int (8-7x)^4\,dx$

$= \dfrac{1}{-7\times5}(8-7x)^5 + C = -\dfrac{1}{35}(8-7x)^5 + C$

Substitute in the values of x and y at the point given to find the value of C.

$\dfrac{3}{35} = -\dfrac{1}{35}(8-(7\times1))^5 + C \Rightarrow \dfrac{3}{35} = -\dfrac{1}{35}(1)^5 + C \Rightarrow C = \dfrac{4}{35}$

So $f(x) = -\dfrac{1}{35}(8-7x)^5 + \dfrac{4}{35}$

Q7 $\int_0^b (3x-1)^3\,dx = \left[\dfrac{1}{3\times4}(3x-1)^4\right]_0^b$

$= \left[\dfrac{1}{12}(3x-1)^4\right]_0^b = \dfrac{1}{12}((3b-1)^4 - (-1)^4) = \dfrac{1}{12}((3b-1)^4 - 1)$

So $\dfrac{1}{12}((3b-1)^4 - 1) = 52 \Rightarrow (3b-1)^4 - 1 = 624$

$\Rightarrow (3b-1)^4 = 625 \Rightarrow (3b-1) = \pm5$

Either: $3b-1 = 5 \Rightarrow 3b = 6 \Rightarrow b = 2$

Or: $3b-1 = -5 \Rightarrow 3b = -4 \Rightarrow b = -\dfrac{4}{3}$
but $b > 0$ so ignore this result.

So $b = 2$.

[4 marks available — 1 mark for $\dfrac{1}{12}$, 1 mark for $(3x-1)^4$, 1 mark for a correct method to find b, 1 mark for the correct value of b]

Q8 Start with a quick sketch of the graph:

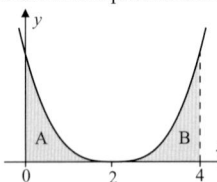

The graph is symmetrical about the line $x = 2$ so area A = area B. Area B is the same graph as if the equation was $y = (x-2)^3$.

Area B $= \int_2^4 (x-2)^3\,dx = \left[\dfrac{1}{1\times4}(x-2)^4\right]_2^4$

$= \left[\dfrac{1}{4}(x-2)^4\right]_2^4 = \dfrac{1}{4}2^4 - \dfrac{1}{4}0^4 = 4$

So the total area is $4 \times 2 = 8$.

If you didn't spot the symmetry of the graph you could have found area A another way. Either by integrating $(x-2)^3$ between 0 and 2 then reversing the sign, or integrating $(-x+2)^3$ between 0 and 2.

Exercise 8.2 — Integrating e^x and e^{ax+b}

Q1 **a)** $\int 2e^x\,dx = 2\int e^x\,dx = 2e^x + C$

b) $\int 4x + 7e^x\,dx = \int 4x\,dx + \int 7e^x\,dx$

$= \int 4x\,dx + 7\int e^x\,dx = 2x^2 + 7e^x + C$

c) $\int e^{10x}\,dx = \dfrac{1}{10}e^{10x} + C$

d) $\int e^{-3x} + x\,dx = \int e^{-3x}\,dx + \int x\,dx$

$= -\dfrac{1}{3}e^{-3x} + \dfrac{1}{2}x^2 + C$

e) $\int e^{\frac{7}{2}x}\,dx = \dfrac{1}{(\frac{7}{2})}e^{\frac{7}{2}x} + C = \dfrac{2}{7}e^{\frac{7}{2}x} + C$

f) $\int e^{4x-2}\,dx = \dfrac{1}{4}e^{4x-2} + C$

g) $\int \dfrac{1}{2}e^{2-\frac{3}{2}x}\,dx = \dfrac{1}{2}\int e^{2-\frac{3}{2}x}\,dx$

$= \dfrac{1}{2}\left(\dfrac{1}{(-\frac{3}{2})}e^{2-\frac{3}{2}x} + c\right) = \left(\dfrac{1}{2} \times -\dfrac{2}{3}e^{2-\frac{3}{2}x}\right) + C$

$= -\dfrac{1}{3}e^{2-\frac{3}{2}x} + C$

h) $\int e^{4(\frac{x}{3}+1)}\,dx = \int e^{\frac{4}{3}x+4}\,dx$

$= \dfrac{1}{(\frac{4}{3})}e^{\frac{4}{3}x+4} + C = \dfrac{3}{4}e^{4(\frac{x}{3}+1)} + C$

Q2 You've been given the derivative of the curve, so integrate it to get the equation of the curve.

$y = \int \dfrac{dy}{dx}\,dx = \int 10e^{-5x-1}\,dx = 10\int e^{-5x-1}\,dx$

$= 10\left(\dfrac{1}{-5}e^{-5x-1} + c\right) = -2e^{-5x-1} + C$

To find C, use the fact that the curve goes through the origin (0, 0). The equation of the curve is $y = -2e^{-5x-1} + C$ and substituting in $x = 0$ and $y = 0$ gives:

$0 = -2e^{-(5\times0)-1} + C = -2e^{-1} + C = -\dfrac{2}{e} + C$ so $C = \dfrac{2}{e}$.

So the curve has equation $y = -2e^{-5x-1} + \dfrac{2}{e}$.

Q3 $\int e^{8y+5}\,dy = \dfrac{1}{8}e^{8y+5} + C$

Q4 $\int e^{\frac{x+10}{5}}\,dx = \int e^{\frac{1}{5}x+2}\,dx = \dfrac{1}{(\frac{1}{5})}e^{\frac{x}{5}+2} + C$

$= 5e^{\frac{x}{5}+2} + C = 5e^2e^{\frac{x}{5}} + C$

So $a = 5e^2$ and $b = \dfrac{1}{5}$ or 0.2.

[3 marks available — 1 mark for a correct method to integrate, 1 mark for the correct value of a, 1 mark for the correct value of b]

Q5 $\int e^{2x+1}\,dx = \dfrac{1}{2}e^{2x+1} + C = f(x)$

$f(-0.5) = \dfrac{1}{2}e^0 + C = \dfrac{1}{2} + C = 0.5 \Rightarrow C = 0$

$f(x) = \dfrac{1}{2}e^{2x+1} \Rightarrow f(x) = \dfrac{1}{2}f'(x)$

So the transformation that maps $f'(x)$ onto $f(x)$ is a stretch with scale factor 0.5 in the y-direction.

[3 marks available — 1 mark for a correct method to integrate, 1 mark for the correct value for C, 1 mark for the correct transformation]

Q6 **a)** $\int_2^3 e^{2x}\,dx = \dfrac{1}{2}[e^{2x}]_2^3 = \dfrac{1}{2}([e^6] - [e^4])$

$= \dfrac{1}{2}(e^6 - e^4)$

b) $\int_{-1}^0 12e^{12x+12}\,dx = \left[12 \times \dfrac{1}{12}e^{12x+12}\right]_{-1}^0$

$= [e^{12x+12}]_{-1}^0 = e^{12} - e^0 = e^{12} - 1$

c) $\int_{-\frac{\pi}{2}}^{\frac{\pi}{2}} e^{\pi-2x}\,dx = \left[\dfrac{1}{-2}e^{\pi-2x}\right]_{-\frac{\pi}{2}}^{\frac{\pi}{2}}$

$= -\dfrac{1}{2}[e^{\pi-2x}]_{-\frac{\pi}{2}}^{\frac{\pi}{2}}$

$= -\dfrac{1}{2}([e^{\pi-\pi}] - [e^{\pi+\pi}])$

$= -\dfrac{1}{2}(e^0 - e^{2\pi}) = \dfrac{1}{2}(e^{2\pi} - 1)$

d) $\int_3^6 \sqrt[6]{e^x} + \dfrac{1}{\sqrt[3]{e^x}}\,dx = \int_3^6 (e^x)^{\frac{1}{6}} + (e^x)^{-\frac{1}{3}}\,dx$

$= \int_3^6 e^{\frac{x}{6}} + e^{-\frac{x}{3}}\,dx = \left[\dfrac{1}{(\frac{1}{6})}e^{\frac{x}{6}} + \dfrac{1}{(-\frac{1}{3})}e^{-\frac{x}{3}}\right]_3^6$

$= [6e^{\frac{x}{6}} - 3e^{-\frac{x}{3}}]_3^6$

$= [6e^1 - 3e^{-2}] - [6e^{\frac{1}{2}} - 3e^{-1}]$

$= 6e - \dfrac{3}{e^2} - 6\sqrt{e} + \dfrac{3}{e}$

Q7 $\int_0^{\ln 2} e^{4x-5} dx = \left[\frac{1}{4}e^{4x-5}\right]_0^{\ln 2} = \frac{1}{4}e^{4\ln 2 - 5} - \frac{1}{4}e^{4(0)-5}$

$= \frac{e^{4\ln 2}}{4e^5} - \frac{1}{4e^5} = \frac{(e^{\ln 2})^4}{4e^5} - \frac{1}{4e^5} = \frac{2^4}{4e^5} - \frac{1}{4e^5} = \frac{15}{4e^5}$

[4 marks available — 1 mark for a correct method to integrate, 1 mark for substituting in the limits correctly, 1 mark for removing e from the numerator, 1 mark for the correct answer]

Q8 **a)** $\int_{-2}^{2} 20e^{5x+10} dx = \left[20 \times \frac{1}{5}e^{5x+10}\right]_{-2}^{2} = \left[4e^{5x+10}\right]_{-2}^{2}$

$= 4e^{5(2)+10} - 4e^{5(-2)+10} = 4e^{20} - 4e^0 = 4e^{20} - 4\ (= 4(e^{20} - 1))$

[3 marks available — 1 mark for a correct method to integrate, 1 mark for substituting the limits, 1 mark for the correct answer]

b) The curve $y = 20e^{5x+10}$ is always above the x-axis so the answer to part a) gives the area between the x-axis and the curve $y = 20e^{5x+10}$ between –2 and 2.
[1 mark for a correct description]

c) $\int_{-k}^{k} 20e^{5x+10} dx = \left[4e^{5x+10}\right]_{-k}^{k}$

$= 4e^{5k+10} - 4e^{-5k+10} = 4e^{10}(e^{5k} - e^{-5k})$

As $k \to \infty$, $e^{-5k} \to 0$ so

$\int_{-k}^{k} 20e^{5x+10} dx \approx 4e^{10}(e^{5k} - 0) = 4e^{10}e^{5k}$

So $A = 4e^{10}$ and $B = 5$.

[3 marks available — 1 mark for substituting the limits correctly, 1 mark for taking out a factor of $4e^{10}$ and showing the negative exponential term tends to 0, 1 mark for the correct answers]

Q9 Find the x-coordinate at A:

$y = e^{3x+2} \Rightarrow \frac{dy}{dx} = 3e^{3x+2}$

$\frac{dy}{dx} = 1 \Rightarrow 3e^{3x+2} = 1 \Rightarrow e^{3x+2} = \frac{1}{3} \Rightarrow 3x + 2 = -\ln 3$

$\Rightarrow x = -\frac{2+\ln 3}{3}$

So area $= \int_{-\frac{2+\ln 3}{3}}^{0} e^{3x+2} dx = \left[\frac{1}{3}e^{3x+2}\right]_{-\frac{2+\ln 3}{3}}^{0} = \frac{1}{3}\left(e^2 - e^{3\left(-\frac{2+\ln 3}{3}\right)+2}\right)$

$= \frac{1}{3}(e^2 - e^{-2-\ln 3 + 2}) = \frac{1}{3}(e^2 - e^{-\ln 3}) = \frac{1}{3}\left(e^2 - \frac{1}{3}\right) = 2.35$ (2 d.p.)

You could also have let the x-coordinate be A, where $e^{3A+2} = \frac{1}{3}$, and then replaced this term directly after substituting in the limits to save a few steps of rearranging.

Exercise 8.3 — Integrating $\frac{1}{x}$ and $\frac{1}{ax+b}$

Q1 **a)** $\int \frac{19}{x} dx = 19 \int \frac{1}{x} dx = 19 \ln|x| + C$

b) $\int \frac{1}{7x} dx = \frac{1}{7} \int \frac{1}{x} dx = \frac{1}{7} \ln|x| + C$

An equivalent answer to b) would be $\frac{1}{7}\ln|7x| + C$ if you used the general formula for integrating $\frac{1}{ax+b}$ instead — both of these answers are valid as the constants of integration would be different.

c) There is no constant term to take out here, so use the general formula:

$\int \frac{1}{7x+2} dx = \frac{1}{7}\ln|7x+2| + C$

d) $\int \frac{4}{1-3x} dx = 4\left(-\frac{1}{3}\ln|1-3x| + c\right) = -\frac{4}{3}\ln|1-3x| + C$

Q2 $\int \frac{1}{8x} - \frac{20}{x} dx = \frac{1}{8}\int \frac{1}{x} dx - 20\int \frac{1}{x} dx$

$= \frac{1}{8}\ln|x| - 20\ln|x| + C = -\frac{159}{8}\ln|x| + C$

You could also notice that $\frac{1}{8x} - \frac{20}{x} = \frac{1-160}{8x} = \frac{-159}{8x}$ and integrate $\frac{-159}{8x}$ using the method for $\frac{1}{x}$.

Q3 She should have taken out a factor of 3^{-1} rather than 3 when factorising (or cancelled the fraction down first). She should have spotted that this integrates to give $9\ln|x+3|$ rather than decreasing the power (assuming working with previous error).

Q4 **a)** $\frac{9x}{3x+6} = \frac{9x+18-18}{3x+6} = \frac{3(3x+6)-18}{3x+6}$

$= 3 - \frac{18}{3x+6} = 3 - \frac{6}{x+2}$

[2 marks available — 1 mark for a correct method, 1 mark for the correct answer]

b) $\int \frac{9x}{3x+6} dx = \int 3 - \frac{6}{x+2} dx = \int 3\, dx - \int \frac{6}{x+2} dx$

$= \int 3\, dx - 6\int \frac{1}{x+2} dx = 3x - 6\ln|x+2| + C$

[2 marks available — 1 mark for a correct method to integrate the fraction, 1 mark for the correct answer]

Q5 **a)** $\int \frac{6}{x} - \frac{3}{x} dx = \int \frac{3}{x} dx = 3\ln|x| + C = \ln|x|^3 + C = \ln|x^3| + C$

b) $\int_4^5 \frac{6}{x} - \frac{3}{x} dx = \left[\ln|x^3|\right]_4^5$

$= \left[\ln 5^3\right] - \left[\ln 4^3\right] = 3\ln 5 - 3\ln 4$

$= 3(\ln 5 - \ln 4) = 3\ln\left(\frac{5}{4}\right)$

You could also have written your answer as $\ln\frac{125}{64}$.

Q6 $\int_b^a 15(5+3x)^{-1} dx = \int_b^a \frac{15}{(5+3x)} dx$

$= \left[15 \times \frac{1}{3}\ln|5+3x|\right]_b^a = 5\left[\ln|5+3x|\right]_b^a$

$= 5(\ln|5+3a| - \ln|5+3b|) = 5\ln\left(\frac{|5+3a|}{|5+3b|}\right)$

$= 5\ln\left|\frac{5+3a}{5+3b}\right| = \ln\left|\frac{5+3a}{5+3b}\right|^5$

Q7 Integrate the derivative to find f(x).

$f(x) = \int f'(x) dx = \int \frac{4}{10-9x} dx$

$= 4 \times \frac{1}{-9}\ln|10-9x| + C = -\frac{4}{9}\ln|10-9x| + C$

The curve passes through the point (1, 2), so substitute these values to find C.

$2 = -\frac{4}{9}\ln|10-(9\times 1)| + C \Rightarrow 2 = -\frac{4}{9}\ln|1| + C \Rightarrow 2 = 0 + C$

So C = 2 and the equation of f(x) is $f(x) = -\frac{4}{9}\ln|10-9x| + 2$.

Q8 **a)** The area required is the shaded area below:

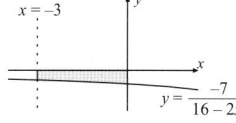

This is found by integrating the curve with respect to x between the limits $x = -3$ and $x = 0$.

So the area is expressed by the integral $\int_{-3}^{0} \frac{-7}{16-2x} dx$

b) $\int_{-3}^{0} \frac{-7}{16-2x} = \left[-7 \times \frac{1}{-2}\ln|16-2x|\right]_{-3}^{0}$

$= \frac{7}{2}[\ln|16-2x|]_{-3}^{0} = \frac{7}{2}(\ln 16 - \ln 22)$

$= \frac{7}{2}\ln\frac{16}{22} = \frac{7}{2}\ln\frac{8}{11} = \ln\left[\left(\frac{8}{11}\right)^{\frac{7}{2}}\right]$

Answers

Q9 Work out the integral and put in the limits to find A.

$\int_1^A \frac{4}{6x-5}\,dx = 10 \Rightarrow \left[4 \times \frac{1}{6}\ln|6x-5|\right]_1^A = 10$

$\Rightarrow \frac{2}{3}[\ln|6x-5|]_1^A = 10 \Rightarrow \frac{2}{3}(\ln|6A-5| - \ln|1|) = 10$

$\Rightarrow \frac{2}{3}(\ln|6A-5| - 0) = 10 \Rightarrow \frac{2}{3}\ln|6A-5| = 10$

$\Rightarrow \ln|6A-5| = \dfrac{10}{\left(\frac{2}{3}\right)} = 15$

Take the exponential of both sides to get rid of the ln.

$|6A-5| = e^{15} \Rightarrow A = \dfrac{e^{15}+5}{6}$

As $A \geq 1$, then $6A - 5$ must be greater than $6 - 5 = 1$, so the modulus can be removed as it'll always be positive.

Q10 $\dfrac{3}{\sqrt{(x+3)}} \times \dfrac{(x+3)^{-0.5}}{6} = \dfrac{3}{\sqrt{(x+3)}} \times \dfrac{1}{6\sqrt{(x+3)}} = \dfrac{1}{2x+6}$

So $\int_{1.5}^5 \dfrac{3}{\sqrt{(x+3)}} \times \dfrac{(x+3)^{-0.5}}{6}\,dx = \int_{1.5}^5 \dfrac{1}{2x+6}\,dx$

$= \left[\frac{1}{2}\ln|2x+6|\right]_{1.5}^5 = \frac{1}{2}\ln|2(5)+6| - \frac{1}{2}\ln|2(1.5)+6|$

$= \frac{1}{2}\ln 16 - \frac{1}{2}\ln 9 = \ln 16^{\frac{1}{2}} - \ln 9^{\frac{1}{2}} = \ln 4 - \ln 3 = \ln\frac{4}{3}$

[5 marks available — 1 mark for simplifying the calculation, 1 mark for integrating correctly, 1 mark for substituting the limits in correctly, 1 mark for using log laws to simplify, 1 mark for the correct answer in the correct form]

Q11 a) $\int \dfrac{x^2+10x+25}{2x^2+13x+15}\,dx = \int \dfrac{(x+5)(x+5)}{(2x+3)(x+5)}\,dx$

$= \int \dfrac{x+5}{2x+3}\,dx = 0.5\int \dfrac{2x+10}{2x+3}\,dx$

$= 0.5\int \dfrac{2x+3+7}{2x+3}\,dx = 0.5\int 1 + \dfrac{7}{2x+3}\,dx$

[3 marks available — 1 mark for factorising and cancelling, 1 mark for taking out a factor of 0.5 correctly, 1 mark for rewriting the fraction in the correct form]

b) $\int_0^a \dfrac{x^2+10x+25}{2x^2+13x+15}\,dx = 0.5\int_0^a 1 + \dfrac{7}{2x+3}\,dx$

$= 0.5[x + 3.5\ln|2x+3|]_0^a$

$= 0.5(a + 3.5\ln|2a+3| - 3.5\ln 3)$

$= 0.5a + 1.75\ln|2a+3| - 1.75\ln 3$

So: $0.5a + 1.75\ln|2a+3| - 1.75\ln 3 = 0.5a + \ln 128$

$\Rightarrow 1.75\ln\dfrac{|2a+3|}{3} = \ln 128 \Rightarrow \left(\dfrac{|2a+3|}{3}\right)^{1.75} = 128$

$\Rightarrow \dfrac{|2a+3|}{3} = 16 \Rightarrow 2a+3 = \pm 48$

$2a+3 = -48 \Rightarrow a = -25.5$ (reject as $a > 0$)

$2a+3 = 48 \Rightarrow a = 22.5$

[4 marks available — 1 mark for integrating correctly, 1 mark for substituting limits in correctly, 1 mark for a correct method to solve using log laws, 1 mark for the correct answer]

Exercise 8.4 — Integrating $\sin x$ and $\cos x$

Q1 a) $\int \frac{1}{7}\cos x\,dx = \frac{1}{7}\int \cos x\,dx = \frac{1}{7}\sin x + C$

b) $\int -3\sin x\,dx = -3\int \sin x\,dx$

$= -3(-\cos x) + C = 3\cos x + C$

c) $\int -3\cos x - 3\sin x\,dx = -3\int \cos x + \sin x\,dx$

$= -3(\sin x - \cos x + c) = -3\sin x + 3\cos x + C$

d) $\int \sin 5x\,dx = -\frac{1}{5}\cos 5x + C$

e) $\int \cos\left(\frac{x}{7}\right)dx = \dfrac{1}{\left(\frac{1}{7}\right)}\sin\left(\frac{x}{7}\right) + C = 7\sin\left(\frac{x}{7}\right) + C$

f) $\int 2\sin(-3x)\,dx = 2\int \sin(-3x)\,dx$

$= 2\left(-\left(-\frac{1}{3}\right)\cos(-3x) + c\right)$

$= \frac{2}{3}\cos(-3x) + C$

An alternative solution would be $\frac{2}{3}\cos(3x) + C$ since $\cos(x) = \cos(-x)$.

g) $\int 5\cos\left(3x + \frac{\pi}{5}\right)dx = 5\left(\frac{1}{3}\sin\left(3x + \frac{\pi}{5}\right) + c\right)$

$= \frac{5}{3}\sin\left(3x + \frac{\pi}{5}\right) + C$

h) $\int -4\sin\left(4x - \frac{\pi}{3}\right)dx = -4\left(-\frac{1}{4}\cos\left(4x - \frac{\pi}{3}\right) + c\right)$

$= \cos\left(4x - \frac{\pi}{3}\right) + C$

i) $\int \cos(4x+3) + \sin(3-4x)\,dx$

$= \frac{1}{4}\sin(4x+3) + \frac{1}{-4}(-\cos(3-4x)) + C$

$= \frac{1}{4}\sin(4x+3) + \frac{1}{4}\cos(3-4x) + C$

Q2 $y = \sin x$ and $\int y\,dx = -\cos x + C$

Q3 $\int \frac{1}{2}\cos 3\theta - \sin\theta\,d\theta = \frac{1}{2}\left(\frac{1}{3}\sin 3\theta\right) - (-\cos\theta) + C$

$= \frac{1}{6}\sin 3\theta + \cos\theta + C$

Q4 a) $\int_0^{\frac{\pi}{2}} \sin x\,dx = [-\cos x]_0^{\frac{\pi}{2}} = -\cos\frac{\pi}{2} + \cos 0 = -0 + 1 = 1$

b) $\int_{\frac{\pi}{6}}^{\frac{\pi}{3}} \sin 3x\,dx = -\frac{1}{3}[\cos 3x]_{\frac{\pi}{6}}^{\frac{\pi}{3}}$

$= -\frac{1}{3}\left(\left[\cos\left(3 \times \frac{\pi}{3}\right)\right] - \left[\cos\left(3 \times \frac{\pi}{6}\right)\right]\right)$

$= -\frac{1}{3}\left(\cos\pi - \cos\frac{\pi}{2}\right) = -\frac{1}{3}(-1 - 0) = \frac{1}{3}$

c) $\int_{-1}^2 3\sin(\pi x + \pi)\,dx = -\frac{3}{\pi}[\cos(\pi x + \pi)]_{-1}^2$

$= -\frac{3}{\pi}(\cos 3\pi - \cos 0)$

$= -\frac{3}{\pi}(-1 - 1) = \frac{6}{\pi}$

Q5 a) $f(\alpha) = \int 4\sin\alpha + \frac{1}{2}\cos 2\alpha + \alpha\,d\alpha$

$= -4\cos\alpha + \frac{1}{4}\sin 2\alpha + \frac{\alpha^2}{2} + C$

$f(0) = 0 \Rightarrow 0 = -4\cos 0 + \frac{1}{4}\sin 0 + \frac{0^2}{2} + C$

$\Rightarrow 0 = -4 + C \Rightarrow C = 4$

So $f(\alpha) = -4\cos\alpha + \frac{1}{4}\sin 2\alpha + \frac{\alpha^2}{2} + 4$

[3 marks available — 1 mark for integrating the cos 2α term correctly, 1 mark for the correct integral, 1 mark for finding the value of C]

b) Try values either side of $\alpha = 0$:

When $\alpha = -0.01$,

$f(\alpha) = -4\cos(-0.01) + \frac{1}{4}\sin(2 \times -0.01) + \dfrac{(-0.01)^2}{2} + 4$

$= -0.00474...$, which is less than 0.

[1 mark for a correct counter-example]

Q6 a) $\int a\sin 2x\,dx = -\frac{a}{2}\cos 2x + C$

[1 mark for the correct answer]

b) From part a), $g'(x) = -\frac{a}{2}\cos 2x + C$

$g(x) = \int -\frac{a}{2}\cos 2x + C\,dx = -\frac{a}{4}\sin 2x + Cx + D$

From the question: $-\frac{a}{4}\sin 2x + Cx + D = -b\sin bx$

$\Rightarrow C = 0, D = 0, b = 2 = \frac{a}{4}$, so $a = 8$.

[3 marks available — 1 mark for the correct integration of the answer to part a), 1 mark for a, 1 mark for b]

Q7 a) Integrate the function with respect to x within the limits 1 and 2:

$$\int_1^2 2\pi \cos\left(\frac{\pi x}{2}\right) dx = \frac{2\pi}{\left(\frac{\pi}{2}\right)}\left[\sin\left(\frac{\pi x}{2}\right)\right]_1^2 = 4\left[\sin\left(\frac{\pi x}{2}\right)\right]_1^2$$

$$= 4\left(\sin(\pi) - \sin\left(\frac{\pi}{2}\right)\right) = 4(0-1) = -4$$

b) Since the function doesn't cross the x-axis for $1 < x < 2$, the described area has to be either entirely above the x-axis or entirely below it. The integral gives a negative value, so all of the area is below the x-axis.

Q8 $\int_{\frac{\pi}{3}}^{\frac{\pi}{2}} \sin(-x) + \cos(-x)\, dx = \int_{\frac{\pi}{3}}^{\frac{\pi}{2}} -\sin x + \cos x\, dx$

$= [-(-\cos x) + \sin x]_{\frac{\pi}{3}}^{\frac{\pi}{2}} = [\cos x + \sin x]_{\frac{\pi}{3}}^{\frac{\pi}{2}}$

$= \left[\cos\left(\frac{\pi}{2}\right) + \sin\left(\frac{\pi}{2}\right)\right] - \left[\cos\left(\frac{\pi}{3}\right) + \sin\left(\frac{\pi}{3}\right)\right]$

$= [0+1] - \left[\frac{1}{2} + \frac{\sqrt{3}}{2}\right] = 1 - \frac{1}{2} - \frac{\sqrt{3}}{2} = \frac{1}{2} - \frac{\sqrt{3}}{2} = \frac{1-\sqrt{3}}{2}$

[4 marks available — 1 mark for a correct method to integrate, 1 mark for substituting the limits in, 1 mark for simplifying using exact trig values, 1 mark for rearranging to give the correct answer]
$\sin(-x) = -\sin x$ and $\cos(-x) = \cos x$ are used in the first step of this solution.

Q9 $\int_{-\frac{\pi}{3}}^{\frac{\pi}{3}} \sin(3x+2) - \cos x\, dx = \left[-\frac{1}{3}\cos(3x+2) - \sin x\right]_{-\frac{\pi}{3}}^{\frac{\pi}{3}}$

$= \left[-\frac{1}{3}\cos\left(3\left(\frac{\pi}{3}\right)+2\right) - \sin\frac{\pi}{3}\right]$

$\qquad -\left[-\frac{1}{3}\cos\left(3\left(-\frac{\pi}{3}\right)+2\right) - \sin\left(-\frac{\pi}{3}\right)\right]$

$= \left[-\frac{1}{3}\cos(\pi+2) - \frac{\sqrt{3}}{2}\right] - \left[-\frac{1}{3}\cos(-\pi+2) - \left(-\frac{\sqrt{3}}{2}\right)\right]$

$= -\sqrt{3}$

Q10 Integrate the function between -2π and π:

$$\int_{-2\pi}^{\pi} 5\cos\frac{x}{6}\, dx = \left[\frac{5}{\left(\frac{1}{6}\right)}\sin\left(\frac{x}{6}\right)\right]_{-2\pi}^{\pi} = 30\left[\sin\left(\frac{x}{6}\right)\right]_{-2\pi}^{\pi}$$

$= 30\left(\sin\left(\frac{\pi}{6}\right) - \sin\left(-\frac{\pi}{3}\right)\right) = 30\left(\frac{1}{2} - \left(-\frac{\sqrt{3}}{2}\right)\right) = 30\left(\frac{1+\sqrt{3}}{2}\right)$

$= 15(1+\sqrt{3})$

Exercise 8.5 — Integrating sec²x

Q1 a) $\int 2\sec^2 x + 1\, dx = 2\tan x + x + C$

b) $\int \sec^2 9x\, dx = \frac{1}{9}\tan 9x + C$

c) $\int 20\sec^2 3y\, dy = 20 \times \frac{1}{3}\tan 3y + C = \frac{20}{3}\tan 3y + C$

d) $\int \sec^2\frac{x}{7}\, dx = \frac{1}{\left(\frac{1}{7}\right)}\tan\left(\frac{x}{7}\right) + C = 7\tan\left(\frac{x}{7}\right) + C$

e) $\int_0^{\frac{\pi}{3}} -\frac{1}{\cos^2\theta}\, d\theta = \int_0^{\frac{\pi}{3}} -\sec^2\theta\, d\theta$

$= [-\tan\theta]_0^{\frac{\pi}{3}} = -\sqrt{3} + 0 = -\sqrt{3}$

f) $\int_0^{\frac{\pi}{4}} 3\sec^2(-3x)\, dx = \left[\frac{3}{-3}\tan(-3x)\right]_0^{\frac{\pi}{4}}$

$= [-\tan(-3x)]_0^{\frac{\pi}{4}} = \left[-\tan\left(-3\times\frac{\pi}{4}\right)\right] - [-\tan(0)]$

$= -1 + 0 = -1$

Q2 Integrate the function between the limits:

$$\int_{\frac{2}{3}\pi}^{\pi} \sec^2 x\, dx = [\tan x]_{\frac{2}{3}\pi}^{\pi} = \tan\pi - \tan\frac{2\pi}{3} = 0 - (-\sqrt{3}) = \sqrt{3}$$

Q3 Between $-\frac{\pi}{4}$ and $\frac{\pi}{4}$, the graph is entirely above the x-axis.

$$\int_{-\frac{\pi}{4}}^{\frac{\pi}{4}} 2\sec^2 x + \cos x\, dx = [2\tan x + \sin x]_{-\frac{\pi}{4}}^{\frac{\pi}{4}}$$

$= \left(2\tan\frac{\pi}{4} + \sin\frac{\pi}{4}\right) - \left(2\tan\left(-\frac{\pi}{4}\right) + \sin\left(-\frac{\pi}{4}\right)\right)$

$= \left(2 + \frac{\sqrt{2}}{2}\right) - \left(-2 - \frac{\sqrt{2}}{2}\right) = 4 + \sqrt{2}$

[3 marks available — 1 mark for integrating correctly, 1 mark for substituting limits, 1 mark for the correct answer]

Q4 The constants α and β do not affect the integration. You only need to worry about the coefficients of x.

$$\int \sec^2(x+\alpha) + \sec^2(3x+\beta)\, dx$$

$$= \tan(x+\alpha) + \frac{1}{3}\tan(3x+\beta) + C$$

Q5 $\int_{\frac{\pi}{12}}^{\frac{\pi}{6}} 5A\sec^2\left(\frac{\pi}{3} - 2\theta\right) d\theta = \left[-\frac{5A}{2}\tan\left(\frac{\pi}{3} - 2\theta\right)\right]_{\frac{\pi}{12}}^{\frac{\pi}{6}}$

$= -\frac{5A}{2}\left[\tan\left(\frac{\pi}{3} - 2\theta\right)\right]_{\frac{\pi}{12}}^{\frac{\pi}{6}} = -\frac{5A}{2}\left(\left[\tan\left(\frac{\pi}{3} - \frac{\pi}{3}\right)\right] - \left[\tan\left(\frac{\pi}{3} - \frac{\pi}{6}\right)\right]\right)$

$= -\frac{5A}{2}\left(\tan(0) - \tan\left(\frac{\pi}{6}\right)\right) = -\frac{5A}{2}\left(0 - \frac{\sqrt{3}}{3}\right) = \frac{5\sqrt{3}A}{6}$

Q6 $\int_1^2 0.5\sec^2(2\theta + \pi)\, d\theta = \left[\frac{1}{2} \times 0.5\tan(2\theta + \pi)\right]_1^2$

$= [0.25\tan(2\theta + \pi)]_1^2 = 0.25\tan(2(2) + \pi) - 0.25\tan(2(1) + \pi)$

$= 0.2894... - (-0.5462...) = 0.8357... = 0.836 \text{ (to 3 s.f.)}$

[3 marks available — 1 mark for integrating correctly, 1 mark for substituting limits, 1 mark for the correct answer]

Q7 a) $\int_{\frac{\pi}{16}}^{\frac{\pi}{12}} (3\sec 4\theta)^2\, d\theta = \int_{\frac{\pi}{16}}^{\frac{\pi}{12}} 9\sec^2 4\theta\, d\theta$

$= \left[\frac{1}{4} \times 9\tan 4\theta\right]_{\frac{\pi}{16}}^{\frac{\pi}{12}} = = \frac{9}{4}[\tan 4\theta]_{\frac{\pi}{16}}^{\frac{\pi}{12}}$

$= \frac{9}{4}\left(\tan\left(4 \times \frac{\pi}{12}\right) - \tan\left(4 \times \frac{\pi}{16}\right)\right)$

$= \frac{9}{4}\left(\tan\left(\frac{\pi}{3}\right) - \tan\left(\frac{\pi}{4}\right)\right) = \frac{9}{4}(\sqrt{3} - 1)\ \left(= \frac{9\sqrt{3} - 9}{4}\right)$

[3 marks available — 1 mark for integrating correctly, 1 mark for substituting limits, 1 mark for the correct answer]

b) $\int_{\frac{\pi}{16}}^{\frac{\pi}{12}} (3\sec 4\theta)^2 + a\sin(4\theta + \pi)\, d\theta$

$= \int_{\frac{\pi}{16}}^{\frac{\pi}{12}} (3\sec 4\theta)^2\, d\theta + a\int_{\frac{\pi}{16}}^{\frac{\pi}{12}} \sin(4\theta + \pi)\, d\theta$

$= \frac{9\sqrt{3} - 9}{4} + a\left[\frac{1}{4} \times -\cos(4\theta + \pi)\right]_{\frac{\pi}{16}}^{\frac{\pi}{12}}$

$= \frac{9\sqrt{3} - 9}{4} - \frac{a}{4}[\cos(4\theta + \pi)]_{\frac{\pi}{16}}^{\frac{\pi}{12}}$

$= \frac{9\sqrt{3} - 9}{4} - \frac{a}{4}\left(\cos\left(\frac{4\pi}{12} + \pi\right) - \cos\left(\frac{4\pi}{16} + \pi\right)\right)$

$= \frac{9\sqrt{3} - 9}{4} - \frac{a}{4}\left(-\frac{1}{2} - \left(-\frac{\sqrt{2}}{2}\right)\right) = \frac{9\sqrt{3}}{4} - \frac{9}{4} + \frac{a}{8} - \frac{a\sqrt{2}}{8}$

So $\frac{9\sqrt{3}}{4} - \frac{9}{4} + \frac{a}{8} - \frac{a\sqrt{2}}{8} = \frac{9(\sqrt{3} - \sqrt{2})}{4}$

$\Rightarrow 18\sqrt{3} - 18 + a - a\sqrt{2} = 18\sqrt{3} - 18\sqrt{2}$

$\Rightarrow a(1 - \sqrt{2}) = 18(1 - \sqrt{2}) \Rightarrow a = 18$

[4 marks available — 1 mark for integrating $a\sin(4\theta + \pi)$ correctly, 1 mark for substituting limits, 1 mark for a fully correct integral in terms of a, 1 mark for finding the correct value of a]

Exercise 8.6 — Integrating Other Trig Functions

Q1 a) $\int \text{cosec}^2 11x\, dx = -\frac{1}{11}\cot 11x + C$

b) $\int 5\sec 10\theta \tan 10\theta\, d\theta = 5 \times \frac{1}{10}\sec 10\theta + C = \frac{1}{2}\sec 10\theta + C$

c) $\int -\text{cosec}(x+17)\cot(x+17)\, dx$

$= -(-\text{cosec}(x+17)) + C = \text{cosec}(x+17) + C$

d) $\int -3\cosec 3x \cot 3x\ dx = -3\left(-\frac{1}{3}\cosec 3x\right) + C$

$= \cosec 3x + C$

e) $\int 13\sec\left(\frac{\pi}{4} - x\right)\tan\left(\frac{\pi}{4} - x\right) dx = 13\left(\frac{1}{-1}\sec\left(\frac{\pi}{4} - x\right)\right) + C$

$= -13\sec\left(\frac{\pi}{4} - x\right) + C$

f) $\int 4\cosec^2(5x + 3)\ dx = 4\left(-\frac{1}{5}\cot(5x + 3)\right) + C$

$= -\frac{4}{5}\cot(5x + 3) + C$

Q2 $\int 10\cosec^2\left(\alpha - \frac{x}{2}\right) - 60\sec(\alpha - 6x)\tan(\alpha - 6x)\ dx$

$= -\frac{10}{\left(-\frac{1}{2}\right)}\cot\left(\alpha - \frac{x}{2}\right) - \frac{60}{-6}\sec(\alpha - 6x) + C$

$= 20\cot\left(\alpha - \frac{x}{2}\right) + 10\sec(\alpha - 6x) + C$

Q3 $\int_{\frac{\pi}{12}}^{\frac{\pi}{8}} 6\sec 2x \tan 2x + 6\cosec 2x \cot 2x\ dx$

$= \left[\frac{6}{2}\sec 2x - \frac{6}{2}\cosec 2x\right]_{\frac{\pi}{12}}^{\frac{\pi}{8}} = 3\left[\sec 2x - \cosec 2x\right]_{\frac{\pi}{12}}^{\frac{\pi}{8}}$

$= 3\left(\left[\sec\frac{\pi}{4} - \cosec\frac{\pi}{4}\right] - \left[\sec\frac{\pi}{6} - \cosec\frac{\pi}{6}\right]\right)$

$= 3\left(\sqrt{2} - \sqrt{2} - \frac{2}{\sqrt{3}} + 2\right) = 3\left(2 - \frac{2}{\sqrt{3}}\right) = 6 - 2\sqrt{3}$

Q4 $\int_{\frac{\pi}{12}}^{\frac{\pi}{6}} \cosec^2(3x)\ dx = \left[-\frac{1}{3}\cot(3x)\right]_{\frac{\pi}{12}}^{\frac{\pi}{6}}$

$= -\frac{1}{3}\left(\cot\left(\frac{\pi}{2}\right) - \cot\left(\frac{\pi}{4}\right)\right) = -\frac{1}{3}(0 - 1) = \frac{1}{3}$

If you know the tan values for the common angles, you can work out the cot values using cot x = 1/tan x.

Q5 $\int_0^{\frac{\pi}{3}} \sec 0.5x \tan 0.5x\ dx = \left[2\sec 0.5x\right]_0^{\frac{\pi}{3}}$

$= 2\sec\left(0.5 \times \frac{\pi}{3}\right) - 2\sec(0.5 \times 0) = \frac{4\sqrt{3}}{3} - 2$ or $\frac{4\sqrt{3} - 6}{3}$

[3 marks available — 1 mark for integrating correctly, 1 mark for correctly substituting the limits, 1 mark for the correct answer]

Q6 a) $\int_{\frac{7\pi}{3}}^{a} \cosec x \cot x\ dx = \left[-\cosec x\right]_{\frac{7\pi}{3}}^{a}$

$= (-\cosec a) - \left(-\cosec\frac{7\pi}{3}\right) = -\cosec a + \frac{2}{\sqrt{3}}$

$-\cosec a + \frac{2}{\sqrt{3}} = 0 \Rightarrow \cosec a = \frac{2}{\sqrt{3}}$

[3 marks available — 1 mark for integrating correctly, 1 mark for correctly substituting the limits, 1 mark for working showing the required result]

b) (i) $\cosec a = \frac{2}{\sqrt{3}} \Rightarrow \sin a = \frac{\sqrt{3}}{2}$

$\Rightarrow a = \frac{\pi}{3}, \frac{2\pi}{3}, \frac{7\pi}{3}, \frac{8\pi}{3}, \frac{13\pi}{3}, \ldots$

As $2\pi < a < 4\pi$ and $a \neq \frac{7\pi}{3} \Rightarrow a = \frac{8\pi}{3}$

[2 marks available — 1 mark for sin a, 1 mark for a]

(ii) As the integral is 0, the area above and below the x-axis must be equal. The graph of $y = \cosec x \cot x$ crosses the x-axis when:

$\cosec x \cot x = 0 \Rightarrow \frac{1}{\sin x} \times \frac{\cos x}{\sin x} = 0$

$\Rightarrow \cos x = 0 \Rightarrow x = \frac{\pi}{2}, \frac{3\pi}{2}, \frac{5\pi}{2}, \ldots$

So $\int_{\frac{7\pi}{3}}^{\frac{5\pi}{2}} \cosec x \cot x\ dx = \int_{\frac{5\pi}{2}}^{\frac{8\pi}{3}} \cosec x \cot x\ dx$.

$\int_{\frac{7\pi}{3}}^{\frac{5\pi}{2}} \cosec x \cot x\ dx = \left[-\cosec x\right]_{\frac{7\pi}{3}}^{\frac{5\pi}{2}}$

$= \left(-\cosec\frac{5\pi}{2}\right) - \left(-\cosec\frac{7\pi}{3}\right) = -1 + \frac{2}{\sqrt{3}}$

Total area $= 2 \times \left(\frac{2}{\sqrt{3}} - 1\right) = \frac{4}{\sqrt{3}} - 2$ or $\frac{4\sqrt{3} - 6}{3}$

[3 marks available — 1 mark for a correct method to split the area into 2 equal sections, 1 mark for substituting the correct limits into the integral, 1 mark for the correct answer]

Exercise 8.7 — Integrating $\frac{f'(x)}{f(x)}$

Q1 a) Differentiating the denominator:

$\frac{d}{dx}(x^4 - 1) = 4x^3 = $ numerator

$\int \frac{4x^3}{x^4 - 1}\ dx = \ln|x^4 - 1| + C$

b) $\frac{d}{dx}(x^2 - x) = 2x - 1 = $ numerator

$\int \frac{2x - 1}{x^2 - x}\ dx = \ln|x^2 - x| + C$

c) $\frac{d}{dx}(3x^5 + 6) = 15x^4$

$\int \frac{x^4}{3x^5 + 6}\ dx = \frac{1}{15}\int \frac{15x^4}{3x^5 + 6}\ dx = \frac{1}{15}\ln|3x^5 + 6| + C$

d) $\frac{d}{dx}(x^4 + 2x^3 - x) = 4x^3 + 6x^2 - 1$

$\int \frac{12x^3 + 18x^2 - 3}{x^4 + 2x^3 - x}\ dx = \int \frac{3(4x^3 + 6x^2 - 1)}{x^4 + 2x^3 - x}\ dx$

$= 3\int \frac{4x^3 + 6x^2 - 1}{x^4 + 2x^3 - x}\ dx = 3\ln|x^4 + 2x^3 - x| + C$

e) $\frac{d}{dx}(x - 2)(2x + 3) = \frac{d}{dx}(2x^2 - x - 6) = 4x - 1$

$\int \frac{8x - 2}{(x - 2)(2x + 3)}\ dx = 2\int \frac{4x - 1}{2x^2 - x - 3}\ dx$

$= 2\ln|(x - 2)(2x + 3)| + C$

f) $\frac{d}{dx}(2x^3 - 9x + 7) = 6x^2 - 9$

$\int \frac{6 - 4x^2}{2x^3 - 9x + 7}\ dx = -\frac{2}{3}\int \frac{6x^2 - 9}{2x^3 - 9x + 7}\ dx$

$= -\frac{2}{3}\ln|2x^3 - 9x + 7| + C$

Q2 a) $\frac{d}{dx}(3x + 2)^3 = 3 \times 3(3x + 2)^2 = 9(3x + 2)^2$

$\int \frac{9(3x + 2)^2}{(3x + 2)^3}\ dx = \ln|(3x + 2)^3| + C$

[2 marks available — 1 mark for differentiating the denominator, 1 mark for the correct answer]

b) $\frac{9(3x + 2)^2}{(3x + 2)^3} = \frac{9}{3x + 2}$

$\frac{d}{dx}(3x + 2) = 3$

$\int \frac{9}{3x + 2}\ dx = 3\int \frac{3}{3x + 2}\ dx = 3\ln|3x + 2| + C$

[2 marks available — 1 mark for a correct method, 1 mark for the correct answer]

You could also just use the standard result for integrating functions of the form $\frac{1}{ax + b}$.

c) Ignoring the constants of integration and using log laws you have: $\ln|(3x + 2)^3| = \ln|3x + 2|^3 = 3\ln|3x + 2|$

[1 mark for correct use of log laws to show equivalence]

Q3 a) Find where the denominator of the fraction is 0:

$3x^2 + 12x + 30 = 0 \Rightarrow x^2 + 4x + 10 = 0$

Discriminant $= b^2 - 4ac = 16 - 40 = -24 < 0$,

so this equation has no real solutions.

Therefore, f(x) is defined for all real numbers,

i.e. the domain is $x \in \mathbb{R}$.

[2 marks available — 1 mark for stating the correct domain, 1 mark for a valid reason]

b) $\frac{d}{dx}(3x^2 + 12x + 30) = 6x + 12 = 3(2x + 4)$

$\int_0^3 \frac{2x + 4}{3x^2 + 12x + 30}\,dx = \frac{1}{3}\int_0^3 \frac{3(2x + 4)}{3x^2 + 12x + 30}\,dx$

$= \left[\frac{1}{3}\ln|3x^2 + 12x + 30|\right]_0^3$

$= \frac{1}{3}\ln|3(3^2) + 12(3) + 30| - \frac{1}{3}\ln|3(0^2) + 12(0) + 30|$

$= \frac{1}{3}\ln 93 - \frac{1}{3}\ln 30 = \frac{1}{3}\ln\left(\frac{93}{30}\right) = \frac{1}{3}\ln\left(\frac{31}{10}\right)$

[4 marks available — 1 mark for differentiating the denominator, 1 mark for the correct integral, 1 mark for substituting the limits, 1 mark for the correct answer]
Modulus signs aren't required since the function exceeds 0 for all x.

c) The domain indicates that the graph is continuous and f(x) ≥ 0 for all x ≥ 0, so yes, the integral will give the area.
[1 mark for a correct answer with a suitable explanation]

Q4 a) $\frac{d}{dx}(e^x + 6) = e^x$

$\int \frac{e^x}{e^x + 6}\,dx = \ln|e^x + 6| + C$

b) $\frac{d}{dx}(e^{2x} + 6e^x) = 2e^{2x} + 6e^x = 2(e^{2x} + 3e^x)$

$\int \frac{2(e^{2x} + 3e^x)}{e^{2x} + 6e^x}\,dx = \ln|e^{2x} + 6e^x| + C$

c) $\frac{d}{dx}(e^x + 3) = e^x$

$\int \frac{e^x}{3(e^x + 3)}\,dx = \frac{1}{3}\int \frac{e^x}{(e^x + 3)}\,dx = \frac{1}{3}\ln|e^x + 3| + C$

d) $\frac{d}{dx}(5e^{4x} + 10x) = 20e^{4x} + 10$

$\int \frac{10e^{4x} + 5}{5e^{4x} + 10x}\,dx = \frac{1}{2}\int \frac{20e^{4x} + 10}{5e^{4x} + 10x}\,dx = \frac{1}{2}\ln|5e^{4x} + 10x| + C$

Q5 a) $\frac{d}{dx}(1 + \sin 2x) = 2\cos 2x$

$\int \frac{2\cos 2x}{1 + \sin 2x}\,dx = \ln|1 + \sin 2x| + C$

b) $\frac{d}{dx}(\cos 3x - 1) = -3\sin 3x$

$\int \frac{\sin 3x}{\cos 3x - 1}\,dx = -\frac{1}{3}\int \frac{-3\sin 3x}{\cos 3x - 1}\,dx$

$= -\frac{1}{3}\ln|\cos 3x - 1| + C$

c) $\frac{d}{dx}(\operatorname{cosec} x - x^2 + 4) = -\operatorname{cosec} x \cot x - 2x$

$\int \frac{3\operatorname{cosec} x \cot x + 6x}{\operatorname{cosec} x - x^2 + 4}\,dx = \int \frac{-3(-\operatorname{cosec} x \cot x - 2x)}{\operatorname{cosec} x - x^2 + 4}\,dx$

$= -3\int \frac{-\operatorname{cosec} x \cot x - 2x}{\operatorname{cosec} x - x^2 + 4}\,dx = -3\ln|\operatorname{cosec} x - x^2 + 4| + C$

d) $\frac{d}{dx}(\tan x) = \sec^2 x$

$\int \frac{\sec^2 x}{\tan x}\,dx = \ln|\tan x| + C$

e) $\frac{d}{dx}(\sec x + 5) = \sec x \tan x$

$\int \frac{\sec x \tan x}{\sec x + 5}\,dx = \ln|\sec x + 5| + C$

f) $\frac{d}{dx}(2\cot x - 1) = -2\operatorname{cosec}^2 x$

$\int \frac{\operatorname{cosec}^2 x}{2\cot x - 1}\,dx = -\frac{1}{2}\int \frac{2\operatorname{cosec}^2 x}{2\cot x - 1}\,dx$

$= -\frac{1}{2}\ln|2\cot x - 1| + C$

Q6 When $y - \ln|\tan 4x| + C$, $\frac{dy}{dx} = \frac{4\sec^2 4x}{\tan 4x}$

So $\int \frac{2f(x)}{3\tan 4x}\,dx \Rightarrow \frac{2}{3}f(x) = 4\sec^2 4x \Rightarrow f(x) = 6\sec^2 4x$

[2 marks available — 1 mark for a correct method, 1 mark for the correct answer]

Q7 $\frac{d}{dx}(\sin(2x + 7)) = 2\cos(2x + 7)$

$\int \frac{4\cos(2x + 7)}{\sin(2x + 7)}\,dx = 2\int \frac{2\cos(2x + 7)}{\sin(2x + 7)}\,dx$

$= 2(\ln|\sin(2x + 7)| + C) = 2(\ln|\sin(2x + 7)| + \ln k)$

$= 2\ln|k\sin(2x + 7)|$

Q8 $\frac{d}{dx}4\theta^2 - 3\sin 2\theta = 8\theta - 6\cos 2\theta$

$\int_\pi^{k\pi} \frac{16\theta - 12\cos 2\theta}{4\theta^2 - 3\sin 2\theta}\,d\theta = 2\int_\pi^{k\pi} \frac{8\theta - 6\cos 2\theta}{4\theta^2 - 3\sin 2\theta}\,d\theta$

$= 2\big[\ln|4\theta^2 - 3\sin 2\theta|\big]_\pi^{k\pi}$

$= 2(\ln|4(k\pi)^2 - 3\sin(2k\pi)| - \ln|4(\pi)^2 - 3\sin(2\pi)|)$

$= 2(\ln|4k^2\pi^2| - \ln|4\pi^2|) = 2\ln\left|\frac{4k^2\pi^2}{4\pi^2}\right|$

$= 2\ln|k^2| = \ln(|k^2|^2) = \ln k^4$

[4 marks available — 1 mark for differentiating the denominator, 1 mark for the correct integral, 1 mark for substituting the limits, 1 mark for rearranging to give the correct answer]

Q9 a) Using the hint, multiply the inside of the integral by a fraction which is the same on the top and bottom (it's equal to 1, so it'll make no difference).

$\int \sec x\,dx = \int \sec x\left(\frac{\sec x + \tan x}{\sec x + \tan x}\right)dx$

$= \int \frac{\sec^2 x + \sec x \tan x}{\sec x + \tan x}\,dx$

Now differentiating the denominator of this integral gives:
$\frac{d}{dx}(\sec x + \tan x) = \sec x \tan x + \sec^2 x$
So the numerator is the derivative of the denominator, so use the result:

$\int \sec x\,dx = \int \frac{\sec^2 x + \sec x \tan x}{\sec x + \tan x}\,dx = \ln|\sec x + \tan x| + C$

b) Use the same method as part a), this time using $\frac{\operatorname{cosec} x + \cot x}{\operatorname{cosec} x + \cot x}$:

$\int \operatorname{cosec} x\,dx = \int \operatorname{cosec} x\left(\frac{\operatorname{cosec} x + \cot x}{\operatorname{cosec} x + \cot x}\right)dx$

$= \int \frac{\operatorname{cosec}^2 x + \operatorname{cosec} x \cot x}{\operatorname{cosec} x + \cot x}\,dx$

Differentiating the denominator of this integral:
$\frac{d}{dx}(\operatorname{cosec} x + \cot x) = -\operatorname{cosec} x \cot x - \operatorname{cosec}^2 x$
$= -(\operatorname{cosec} x \cot x + \operatorname{cosec}^2 x)$
So the numerator is −1 × the derivative of the denominator, so use the result:

$\int \operatorname{cosec} x\,dx = \int \frac{\operatorname{cosec}^2 x + \operatorname{cosec} x \cot x}{\operatorname{cosec} x + \cot x}\,dx$

$= -\int \frac{-\operatorname{cosec}^2 x - \operatorname{cosec} x \cot x}{\operatorname{cosec} x + \cot x}\,dx$

$= -\ln|\operatorname{cosec} x + \cot x| + C$

Q10 a) $\int 2\tan x\,dx = 2\int \frac{\sin x}{\cos x}\,dx = -2\int \frac{-\sin x}{\cos x} = -2\ln|\cos x| + C$

b) $\int \tan 2x\,dx = \int \frac{\sin 2x}{\cos 2x}\,dx = -\frac{1}{2}\int \frac{-2\sin 2x}{\cos 2x}\,dx$

$= -\frac{1}{2}\ln|\cos 2x| + C$

c) $\int 4\operatorname{cosec} x\,dx = 4\int \operatorname{cosec} x\,dx = -4\ln|\operatorname{cosec} x + \cot x| + C$

d) $\int \cot 3x\,dx = \frac{1}{3}\ln|\sin 3x| + C$

e) $\int \frac{1}{2}\sec 2x\,dx = \frac{1}{2}\left(\frac{1}{2}\ln|\sec 2x + \tan 2x| + C\right)$

$= \frac{1}{4}\ln|\sec 2x + \tan 2x| + C$

f) $\int 3\operatorname{cosec} 6x\,dx = 3\left(-\frac{1}{6}\ln|\operatorname{cosec} 6x + \cot 6x| + C\right)$

$= -\frac{1}{2}\ln|\operatorname{cosec} 6x + \cot 6x| + C$

Answers

The next two look pretty complicated, but if you split them into parts and use some standard results they're actually pretty simple.

Q11 $\int \dfrac{4\sin(3-2x)}{3+\cos(3-2x)} + \dfrac{x^2}{6x^3-5}\,dx$

$= 2\int \dfrac{2\sin(3-2x)}{3+\cos(3-2x)}\,dx + \dfrac{1}{18}\int \dfrac{x^2}{6x^3-5}\,dx$

$= 2\ln|3+\cos(3-2x)| + \dfrac{1}{18}\ln|6x^3-5| + C$

Q12 $\int \dfrac{\sec^2 x}{2\tan x} - 4\sec 2x \tan 2x + \dfrac{\cosec 2x \cot 2x - 1}{\cosec 2x + 2x}\,dx$

$= \int \dfrac{\sec^2 x}{2\tan x}\,dx - \int 4\sec 2x \tan 2x\,dx + \int \dfrac{\cosec 2x \cot 2x - 1}{\cosec 2x + 2x}\,dx$

$= \dfrac{1}{2}\int \dfrac{\sec^2 x}{\tan x}\,dx - 2\int 2\sec 2x \tan 2x\,dx$

$\qquad - \dfrac{1}{2}\int \dfrac{-2\cosec 2x \cot 2x + 2}{\cosec 2x + 2x}\,dx$

$= \dfrac{1}{2}\ln|\tan x| - 2\sec 2x - \dfrac{1}{2}\ln|\cosec 2x + 2x| + C$

The first and third integrals were put in the form $\int \dfrac{f'(x)}{f(x)}\,dx$, and the second one you can tackle by reversing the result $\dfrac{d}{dx}(\sec 2x) = 2\sec 2x \tan 2x$.

Q13 a) Intersections between the curve and the *x*-axis are where

$\sin x + \cos x = 0 \Rightarrow \sin x = -\cos x \Rightarrow \tan x = -1$

$\Rightarrow x = -\dfrac{\pi}{4}$ so integrate between $-\dfrac{\pi}{4}$ and 0.

$\dfrac{d}{dx}(\sin x - \cos x) = \cos x - (-\sin x) = \sin x + \cos x$

So the numerator is the derivative of the denominator.

$\int_{-\frac{\pi}{4}}^{0} \dfrac{\sin x + \cos x}{\sin x - \cos x}\,dx = [\ln|\sin x - \cos x|]_{-\frac{\pi}{4}}^{0}$

$= \ln|\sin 0 - \cos 0| - \ln\left|\sin\left(-\dfrac{\pi}{4}\right) - \cos\left(-\dfrac{\pi}{4}\right)\right|$

$= \ln 1 - \ln\sqrt{2} = -\ln\sqrt{2}$ so the area is $\ln\sqrt{2}$.

[5 marks available — 1 mark for correctly finding the limits, 1 mark for differentiating the denominator, 1 mark for the correct integral, 1 mark for substituting the limits, 1 mark for the correct answer]

b) Intersections between the curve and the *x*-axis are where

$\sin kx + \cos kx = 0 \Rightarrow \tan kx = -1 \Rightarrow kx = -\dfrac{\pi}{4}$

$\Rightarrow x = -\dfrac{\pi}{4k}$ so integrate between $-\dfrac{\pi}{4k}$ and 0.

$\dfrac{d}{dx}(\sin kx - \cos kx) = k\sin kx + k\cos kx$

So the numerator in the integral is $\dfrac{1}{k}$ times the derivative of the denominator.

$\int_{-\frac{\pi}{4k}}^{0} \dfrac{\sin kx + \cos kx}{\sin kx - \cos kx}\,dx = \dfrac{1}{k}[\ln|\sin kx - \cos kx|]_{-\frac{\pi}{4k}}^{0}$

$= \dfrac{1}{k}\left(\ln|\sin 0 - \cos 0| - \ln\left|\sin\left(-\dfrac{\pi}{4k}k\right) - \cos\left(-\dfrac{\pi}{4k}k\right)\right|\right)$

$= \dfrac{1}{k}(\ln 1 - \ln\sqrt{2}) = -\dfrac{1}{k}\ln\sqrt{2}$

So the area is $\dfrac{1}{k}\ln\sqrt{2}$, which means it is inversely proportional to *k*.

[4 marks available — 1 mark for correctly finding the limits in terms of k, 1 mark for the correct integral, 1 mark for substituting the limits, 1 mark for the correct answer showing inverse proportionality]

Q14 $\dfrac{d}{dx}(\ln|-5\cos 3x + h(x)| + A) = \dfrac{15\sin 3x + h'(x)}{-5\cos 3x + h(x)}$

Now divide top and bottom by 3 and equate to the given result:

$\dfrac{5\sin 3x + \dfrac{h'(x)}{3}}{-\dfrac{5}{3}\cos 3x + \dfrac{h(x)}{3}} = \dfrac{5\sin 3x + 2\cos x}{g(x)}$

So: $\dfrac{h'(x)}{3} = 2\cos x \Rightarrow h'(x) = 6\cos x \Rightarrow h(x) = 6\sin x + C$

And: $g(x) = -\dfrac{5}{3}\cos 3x + \dfrac{h(x)}{3} = -\dfrac{5}{3}\cos 3x + \dfrac{6\sin x}{3} + D$

$\qquad = -\dfrac{5}{3}\cos 3x + 2\sin x + D$

$g(0) = 0 \Rightarrow -\dfrac{5}{3}\cos 0 + 2\sin 0 + D = 0 \Rightarrow D = \dfrac{5}{3}$

So $g(x) = -\dfrac{5}{3}\cos 3x + 2\sin x + \dfrac{5}{3}$

Exercise 8.8 — Integrating $\dfrac{du}{dx}\mathrm{f}'(u)$

Q1 Let $u = x^2$ so $\dfrac{du}{dx} = 2x$ and $f'(u) = e^u$ so $f(u) = e^u$.

Using the formula: $\int 2xe^{x^2}\,dx = e^{x^2} + C$

Q2 Let $u = 2x^3$ so $\dfrac{du}{dx} = 6x^2$ and $f'(u) = e^u$ so $f(u) = e^u$.

Using the formula: $\int 6x^2 e^{2x^3}\,dx = e^{2x^3} + C$

Q3 Let $u = \sqrt{x}$ so $\dfrac{du}{dx} = \dfrac{1}{2\sqrt{x}}$, and $f'(u) = e^u$ so $f(u) = e^u$.

Using the formula: $\int \dfrac{1}{2\sqrt{x}} e^{\sqrt{x}}\,dx = e^{\sqrt{x}} + C$

Q4 Let $u = x^4$ so $\dfrac{du}{dx} = 4x^3$, and $f'(u) = e^u$ so $f(u) = e^u$.

Use the formula: $\int 4x^3 e^{x^4}\,dx = e^{x^4} + C$

So divide by 4 to get the original integral:

$\int x^3 e^{x^4}\,dx = \dfrac{1}{4}\int 4x^3 e^{x^4}\,dx = \dfrac{1}{4}e^{x^4} + C$

Q5 Let $u = x^2 - \dfrac{1}{2}x$ so $\dfrac{du}{dx} = 2x - \dfrac{1}{2} = \dfrac{1}{2}(4x-1)$,

and $f'(u) = e^u$ so $f(u) = e^u$.

Use the formula: $\int \left(2x - \dfrac{1}{2}\right)e^{\left(x^2 - \frac{1}{2}x\right)}\,dx = e^{\left(x^2 - \frac{1}{2}x\right)} + C$

Multiply by 2 to get the original integral:

$\int (4x-1)e^{\left(x^2 - \frac{1}{2}x\right)}\,dx = \int 2\left(2x - \dfrac{1}{2}\right)e^{\left(x^2 - \frac{1}{2}x\right)}\,dx$

$= 2\int \left(2x - \dfrac{1}{2}\right)e^{\left(x^2 - \frac{1}{2}x\right)}\,dx = 2e^{\left(x^2 - \frac{1}{2}x\right)} + C$

Q6 Let $u = x^2 + 1$ so $\dfrac{du}{dx} = 2x$, and $f'(u) = \sin u$ so $f(u) = -\cos u$.

Use the formula: $\int 2x \sin(x^2 + 1)\,dx = -\cos(x^2 + 1) + C$

Q7 Let $u = x^4$ so $\dfrac{du}{dx} = 4x^3$, and $f'(u) = \cos u$ so $f(u) = \sin u$.

Use the formula: $\int 4x^3 \cos(x^4)\,dx = \sin(x^4) + C$

Now divide by 4 to get the original integral:

$\int x^3 \cos(x^4)\,dx = \dfrac{1}{4}\int 4x^3 \cos(x^4)\,dx = \dfrac{1}{4}\sin(x^4) + C$

Q8 Let $u = x^2$ so $\dfrac{du}{dx} = 2x$, and $f'(u) = \sec^2 u$ so $f(u) = \tan u$.

Use the formula: $\int 2x \sec^2(x^2)\,dx = \tan(x^2) + C$

Now divide by 2 to get the original integral:

$\int x\sec^2(x^2)\,dx = \dfrac{1}{2}\int 2x\sec^2(x^2)\,dx = \dfrac{1}{2}\tan(x^2) + C$

Q9 *It's less obvious which function to choose as u in this one — keep looking out for a function and its derivative. Here we have cos x and sin x. Remember to make u the one which is within another function, i.e. cos x.*

Let $u = \cos x$ so $\dfrac{du}{dx} = -\sin x$, and $f'(u) = e^u$ so $f(u) = e^u$.

Use the formula: $\int -\sin x\, e^{\cos x}\,dx = e^{\cos x} + C$

Multiply by –1 to get the original integral:

$\int \sin x\, e^{\cos x}\,dx = -\int -\sin x\, e^{\cos x}\,dx = -e^{\cos x} + C$

Q10 Let $u = \sin 2x$ so $\dfrac{du}{dx} = 2\cos 2x$, and $f'(u) = e^u$ so $f(u) = e^u$.

Use the formula: $\int 2\cos 2x\, e^{\sin 2x}\,dx = e^{\sin 2x} + C$

Divide by 2 to get the original integral:

$\int \cos 2x\, e^{\sin 2x}\,dx = \dfrac{1}{2}\int 2\cos 2x\, e^{\sin 2x}\,dx = \dfrac{1}{2}e^{\sin 2x} + C$

Q11 Let $u = \tan x$ then $\dfrac{du}{dx} = \sec^2 x$, and $f'(u) = e^u$ so $f(u) = e^u$.

Use the formula: $\displaystyle\int \sec^2 x\, e^{\tan x}\, dx = e^{\tan x} + C$

Q12 Let $u = \sec x$ then $\dfrac{du}{dx} = \sec x \tan x$, and $f'(u) = e^u$ so $f(u) = e^u$.

Use the formula: $\displaystyle\int \sec x \tan x\, e^{\sec x}\, dx = e^{\sec x} + C$

Q13 Let $u = \cot x$ then $\dfrac{du}{dx} = -\text{cosec}^2 x$, and $f'(u) = e^u$, so $f(u) = e^u$.

Use the formula: $\displaystyle\int -\text{cosec}^2 x\, e^{\cot x}\, dx = e^{\cot x} + C$

Now multiply by -2 to get the original integral:
$-2\displaystyle\int -\text{cosec}^2 x\, e^{\cot x}\, dx = -2e^{\cot x} + C$

Q14 Let $u = \text{cosec}\, 3x$ then $\dfrac{du}{dx} = -3\,\text{cosec}\, 3x \cot 3x$, and $f'(u) = e^u$, so $f(u) = e^u$.

Use the formula: $\displaystyle\int -3\,\text{cosec}\, 3x \cot 3x\, e^{\text{cosec}\, 3x}\, dx = e^{\text{cosec}\, 3x} + C$
Divide by -3 to get the original integral:
$-\dfrac{1}{3}\displaystyle\int 3\,\text{cosec}\, 3x \cot 3x\, e^{\text{cosec}\, 3x}\, dx = -\dfrac{1}{3}e^{\text{cosec}\, 3x} + C$

Q15 a) Let $u = 2x^2 + 6x$ so $\dfrac{du}{dx} = 4x + 6$,
and $f'(u) = \sin u$ so $f(u) = -\cos u$.
Use the formula:
$\displaystyle\int (4x + 6) \sin(2x^2 + 6x)\, dx = -\cos(2x^2 + 6x) + C$

[3 marks available — 1 mark for identifying u, 1 mark for differentiating u, 1 mark for the correct answer]

b) $\displaystyle\int 2x \sin(2x^2 + 6x) + 3\sin(2x^2 + 6x)\, dx$

$= \displaystyle\int (2x + 3)\sin(2x^2 + 6x)\, dx = \dfrac{1}{2}\displaystyle\int (4x + 6)\sin(2x^2 + 6x)$

$= -\dfrac{1}{2}\cos(2x^2 + 6x) + C$

[2 marks available — 1 mark for factorising correctly, 1 mark for the correct answer]

Q16 a) Let $u = x^3 + 6x$ so $\dfrac{du}{dx} = 3x^2 + 6$, and $f'(u) = e^u$ so $f(u) = e^u$.

Use the formula: $\displaystyle\int (3x^2 + 6)e^{x^3 + 6x}\, dx = e^{x^3 + 6x} + C$

So $\displaystyle\int (9x^2 + 18)e^{x^3 + 6x}\, dx = 3\displaystyle\int (3x^2 + 6)e^{x^3 + 6x}\, dx$
$= 3e^{x^3 + 6x} + C$

[4 marks available — 1 mark for identifying u, 1 mark for differentiating u, 1 mark for attempting to use the formula, 1 mark for the correct answer]

b) $f(x) = 3e^{x^3 + 6x} + C$
$f(1) = 3e^7 + C = e^7 \Rightarrow C = -2e^7$
So $f(x) = 3e^{x^3 + 6x} - 2e^7$

[2 marks available — 1 mark for a correct method, 1 mark for the correct answer]

Q17 Let $u = 3x^2 + x$ then $\dfrac{du}{dx} = 6x + 1$,
and $f'(u) = \sin u$, so $f(u) = -\cos u$.

Use the formula:
$\displaystyle\int_0^4 (6x + 1)\sin(3x^2 + x)\, dx = \left[-\cos(3x^2 + x)\right]_0^4$

Multiply by 2 to get the original integral:
$\displaystyle\int_0^4 (12x + 2)\sin(3x^2 + x)\, dx = 2\left[-\cos(3x^2 + x)\right]_0^4$
$= 2\left((-\cos(3 \times 4^2 + 4)) - (-\cos(3 \times 0^2 + 0))\right)$
$= 2(0.16299... + 1) = 2.3259... = 2.33$ (to 3 s.f.)

Q18 a) Let $u = \cos(2 - 5x)$ then $\dfrac{du}{dx} = 5\sin(2 - 5x)$,
and $f'(u) = e^u$, so $f(u) = e^u$.

Use the formula: $\displaystyle\int 5\sin(2 - 5x)\, e^{\cos(2 - 5x)}\, dx = e^{\cos(2 - 5x)} + C$

b) $\displaystyle\int_0^1 5\sin(2 - 5x)\, e^{\cos(2 - 5x)}\, dx = \left[e^{\cos(2 - 5x)}\right]_0^1$
$= \left[e^{\cos(2 - 5(1))}\right] - \left[e^{\cos(2 - 5(0))}\right] = \left[e^{\cos(-3)}\right] - \left[e^{\cos(2)}\right]$
$= e^{\cos(-3)} - e^{\cos(2)} = -0.288$ (3 s.f.)
You could also have $e^{\cos 3} - e^{\cos 2}$, since $\cos(-x) = \cos(x)$ for all x.

Q19 a) Let $u = \sin(4x^2 - x)$ then $\dfrac{du}{dx} = (8x - 1)\cos(4x^2 - x)$,
and $f'(u) = e^u$, so $f(u) = e^u$.
Use the formula:
$\displaystyle\int (8x - 1)\cos(4x^2 - x)e^{\sin(4x^2 - x)}\, dx = e^{\sin(4x^2 - x)} + C$

b) $\displaystyle\int_0^\pi (8x - 1)\cos(4x^2 - x)e^{\sin(4x^2 - x)}\, dx$
$= \left[e^{\sin(4x^2 - x)}\right]_0^\pi = \left[e^{\sin(4\pi^2 - \pi)}\right] - \left[e^{\sin(4(0)^2 - 0)}\right] = -0.624$ (3 s.f.)

Q20 a) $u = x^2$, $v = \ln x$, $\dfrac{du}{dx} = 2x$, $\dfrac{dv}{dx} = \dfrac{1}{x}$

Using the quotient rule:
$\dfrac{dy}{dx} = \dfrac{v\dfrac{du}{dx} - u\dfrac{dv}{dx}}{v^2} = \dfrac{2x\ln x - \dfrac{1}{x}x^2}{(\ln x)^2} = \dfrac{2x\ln x - x}{(\ln x)^2}$

[4 marks available — 1 mark for a correct method to differentiate, 1 mark for differentiating u, 1 mark for differentiating v, 1 mark for the correct answer]

b) If $u = \dfrac{x^2}{\ln x}$, then from part a), $\dfrac{du}{dx} = \dfrac{2x\ln x - x}{(\ln x)^2} = \dfrac{\ln x^{2x} - x}{(\ln x)^2}$
and $f'(u) = \sin u$, so $f(u) = -\cos u$.

Using the formula:
$\displaystyle\int \dfrac{\ln x^{2x} - x}{(\ln x)^2}\sin\left(\dfrac{x^2}{\ln x}\right) dx = -\cos\left(\dfrac{x^2}{\ln x}\right) + C$
[2 marks available — 1 mark for attempting to use the formula, 1 mark for the correct answer]

Q21 $\displaystyle\int_0^{\frac{\pi}{6}} \sin x\, e^{3x\sin x} + x\cos x\, e^{3x\sin x}\, dx = \displaystyle\int_0^{\frac{\pi}{6}} (\sin x + x\cos x)e^{3x\sin x}\, dx$

Let $u = 3x\sin x$. Then, using the product rule:
$\dfrac{du}{dx} = 3\sin x + 3x\cos x = 3(\sin x + x\cos x)$
and $f'(u) = e^u$ so $f(u) = e^u$.

Use the formula: $\displaystyle\int_0^{\frac{\pi}{6}} 3(\sin x + x\cos x)e^{3x\sin x}\, dx = \left[e^{3x\sin x}\right]_0^{\frac{\pi}{6}}$
Divide by 3 to get the original integral:
$\displaystyle\int_0^{\frac{\pi}{6}} (\sin x + x\cos x)e^{3x\sin x}\, dx = \dfrac{1}{3}\left[e^{3x\sin x}\right]_0^{\frac{\pi}{6}}$
$= \dfrac{1}{3}\left(e^{3\left(\frac{\pi}{6}\right)\sin\left(\frac{\pi}{6}\right)} - e^{3(0)\sin(0)}\right) = \dfrac{1}{3}\left(e^{\frac{\pi}{4}} - 1\right)$

Q22 $\displaystyle\int_0^k (\sin x\cos x + 2\cos x)e^{\sin^2 x}e^{4\sin x}e^3\, dx$
$= \displaystyle\int_0^k (\sin x\cos x + 2\cos x)e^{\sin^2 x + 4\sin x + 3}\, dx$
Let $u = \sin^2 x + 4\sin x + 3$. Then $\dfrac{du}{dx} = 2\sin x\cos x + 4\cos x$,
and $f'(u) = e^u$, so $f(u) = e^u$.
Use the formula:
$\displaystyle\int_0^k (2\sin x\cos x + 4\cos x)e^{\sin^2 x + 4\sin x + 3}\, dx = \left[e^{\sin^2 x + 4\sin x + 3}\right]_0^k$

Divide by 2 to get the original integral:
$\displaystyle\int_0^k (\sin x\cos x + 2\cos x)e^{\sin^2 x + 4\sin x + 3} = \dfrac{1}{2}\left[e^{\sin^2 x + 4\sin x + 3}\right]_0^k$

$= \dfrac{1}{2}\left[e^{\sin^2 k + 4\sin k + 3} - e^{\sin^2 0 + 4\sin 0 + 3}\right] = \dfrac{1}{2}\left[e^{\sin^2 k + 4\sin k + 3} - e^3\right]$
$= \dfrac{1}{2}\left[e^{\sin^2 k + 4\sin k}e^3 - e^3\right] = \dfrac{e^3}{2}\left[e^{\sin^2 k + 4\sin k} - 1\right]$
From the question: $\dfrac{e^3}{2}\left[e^{\sin^2 k + 4\sin k} - 1\right] = 5e^3$
$\Rightarrow e^{\sin^2 k + 4\sin k} - 1 = 10 \Rightarrow e^{\sin^2 k + 4\sin k} = 11$
$\Rightarrow \sin^2 k + 4\sin k = \ln 11 \Rightarrow \sin^2 k + 4\sin k - \ln 11 = 0$

Answers

Using the quadratic formula gives:

$$\sin k = \frac{-4 \pm \sqrt{4^2 - (4 \times 1 \times -\ln 11)}}{2}$$

$\Rightarrow \sin k = 0.5294...$ and $\sin k = -4.5294...$ (no solutions)

So $k = \sin^{-1}(0.5294...) = 0.5579... = 0.558$ (3 s.f.)

[7 marks available — 1 mark for identifying u, 1 mark for differentiating u, 1 mark for attempting to use the formula, 1 mark for the correct integral, 1 mark for substituting the limits in and equating to 5e³, 1 mark for both correct solutions to the quadratic, 1 mark for the correct answer]

Exercise 8.9 — Integrating f'(x) × [f(x)]ⁿ

Q1 **a)** Let $f(x) = x^2 + 5$ so $f'(x) = 2x$. $n = 2$ so $n + 1 = 3$.

Using the formula: $\int 3 \times 2x(x^2 + 5)^2 \, dx = (x^2 + 5)^3 + C$

So $\int 6x(x^2 + 5)^2 \, dx = (x^2 + 5)^3 + C$

b) Let $f(x) = x^2 + 7x$ so $f'(x) = 2x + 7$. $n = 4$ so $n + 1 = 5$.

Using the formula: $\int 5(2x + 7)(x^2 + 7x)^4 \, dx = (x^2 + 7x)^5 + C$

Divide by 5 to get the original integral:

$\int (2x + 7)(x^2 + 7x)^4 \, dx = \frac{1}{5}(x^2 + 7x)^5 + C$

c) Let $f(x) = x^4 + 4x^2$ so $f'(x) = 4x^3 + 8x$. $n = 3$ so $n + 1 = 4$.

Using the formula: $\int 4(4x^3 + 8x)(x^4 + 4x^2)^3 \, dx = (x^4 + 4x^2)^4 + C$

Divide by 16 to get the original integral:

$\int (x^3 + 2x)(x^4 + 4x^2)^3 \, dx$

$= \frac{1}{16} \int 4(4x^3 + 8x)(x^4 + 4x^2)^3 \, dx = \frac{1}{16}(x^4 + 4x^2)^4 + C$

d) Let $f(x) = x^2 - 1$, so $f'(x) = 2x$. $n = -3$ so $n + 1 = -2$.

Using the formula: $\int -2(2x)(x^2 - 1)^{-3} \, dx = (x^2 - 1)^{-2} + C$

Divide by -2 to get the original integral:

$\int \frac{2x}{(x^2 - 1)^3} \, dx = \int (2x)(x^2 - 1)^{-3} \, dx$

$= -\frac{1}{2}(x^2 - 1)^{-2} + C = -\frac{1}{2(x^2 - 1)^2} + C$

e) Let $f(x) = e^{3x} - 5$, so $f'(x) = 3e^{3x}$. $n = -2$ so $n + 1 = -1$.

Using the formula: $\int -1(3e^{3x})(e^{3x} - 5)^{-2} \, dx = (e^{3x} - 5)^{-1} + C$

Multiply by -2 to get the original integral:

$\int \frac{6e^{3x}}{(e^{3x} - 5)^2} \, dx = \int 6e^{3x}(e^{3x} - 5)^{-2} \, dx$

$= -2 \int -3e^{3x}(e^{3x} - 5)^{-2} \, dx$

$= -2(e^{3x} - 5)^{-1} + C$

f) $\sin x \cos^5 x = \sin x (\cos x)^5$

Let $f(x) = \cos x$ so $f'(x) = -\sin x$. $n = 5$ so $n + 1 = 6$.

Using the formula:

$\int 6(-\sin x)(\cos x)^5 \, dx = (\cos x)^6 + C = \cos^6 x + C$

Divide by -6 to get the original integral:

$\int \sin x \cos^5 x \, dx = \frac{1}{-6} \int 6(-\sin x)(\cos x)^5 \, dx$

$= -\frac{1}{6}\cos^6 x + C$

g) *It's a bit more difficult to tell which is the derivative and which is the function here — both functions are to a power. Remember that the derivative of tan x is sec² x.*

$2 \sec^2 x \tan^3 x = 2 \sec^2 x (\tan x)^3$

Let $f(x) = \tan x$ so $f'(x) = \sec^2 x$. $n = 3$ so $n + 1 = 4$.

Using the formula:

$\int 4 \sec^2 x (\tan x)^3 \, dx = (\tan x)^4 + C = \tan^4 x + C$

Divide by 2 to get the original integral:

$\int 2 \sec^2 x \tan^3 x \, dx = \frac{1}{2} \int 4 \sec^2 x (\tan x)^3 \, dx = \frac{1}{2}\tan^4 x + C$

h) Let $f(x) = e^x + 4$ so $f'(x) = e^x$. $n = 2$ so $n + 1 = 3$.

Using the formula: $\int 3e^x (e^x + 4)^2 \, dx = (e^x + 4)^3 + C$

i) Let $f(x) = e^{4x} - 3x^2$ so $f'(x) = 4e^{4x} - 6x$. $n = 7$ so $n + 1 = 8$.

Using the formula:

$\int 8(4e^{4x} - 6x)(e^{4x} - 3x^2)^7 \, dx = (e^{4x} - 3x^2)^8 + C$

So $\int 16(2e^{4x} - 3x)(e^{4x} - 3x^2)^7 \, dx = (e^{4x} - 3x^2)^8 + C$

Multiply by 2 to get the original integral:

$\int 32(2e^{4x} - 3x)(e^{4x} - 3x^2)^7 \, dx$

$= 2 \int 16(2e^{4x} - 3x)(e^{4x} - 3x^2)^7 \, dx = 2(e^{4x} - 3x^2)^8 + C$

j) Let $f(x) = 2 + \sin x$, so $f'(x) = \cos x$. $n = -4$, so $n + 1 = -3$.

Using the formula:

$\int -3(\cos x)(2 + \sin x)^{-4} \, dx = (2 + \sin x)^{-3} + C$

Divide by -3 to get the original integral:

$\int \frac{\cos x}{(2 + \sin x)^4} \, dx = \int \cos x(2 + \sin x)^{-4} \, dx$

$= -\frac{1}{3}(2 + \sin x)^{-3} + C = -\frac{1}{3(2 + \sin x)^3} + C$

k) *Using the hint, you know the derivative of cosec x is −cosec x cot x.*

Let $f(x) = \operatorname{cosec} x$ so $f'(x) = -\operatorname{cosec} x \cot x$. $n = 4$ so $n + 1 = 5$.

Using the formula:

$\int 5(-\operatorname{cosec} x \cot x)(\operatorname{cosec} x)^4 \, dx = (\operatorname{cosec} x)^5 + C$

Multiply by -1: $\int 5 \operatorname{cosec} x \cot x \operatorname{cosec}^4 x \, dx = -\operatorname{cosec}^5 x + C$

l) *Using the hint, cot x differentiates to −cosec² x.*

Let $f(x) = \cot x$ so $f'(x) = -\operatorname{cosec}^2 x$. $n = 3$ so $n + 1 = 4$.

Using the formula: $\int 4(-\operatorname{cosec}^2 x) \cot^3 x \, dx = \cot^4 x + C$

Divide by -2 to get the original integral:

$\int 2 \operatorname{cosec}^2 x \cot^3 x \, dx = \frac{1}{-2} \int -4 \operatorname{cosec}^2 x \cot^3 x \, dx$

$= -\frac{1}{2}\cot^4 x + C$

Q2 **a)** *sec x differentiates to sec x tan x, so try to write the function as a product of sec x tan x and sec x to a power.*

$6 \tan x \sec^6 x = 6 \tan x \sec x \sec^5 x$

Let $f(x) = \sec x$ so $f'(x) = \sec x \tan x$. $n = 5$ so $n + 1 = 6$.

Using the formula: $\int 6(\tan x \sec x)(\sec^5 x) \, dx = \sec^6 x + C$

So $\int 6 \tan x \sec^6 x \, dx = \sec^6 x + C$

b) *cosec x differentiates to −cot x cosec x so do the same as you did in part a).*

$\cot x \operatorname{cosec}^3 x = \cot x \operatorname{cosec} x \operatorname{cosec}^2 x$

Let $f(x) = \operatorname{cosec} x$ so $f'(x) = -\cot x \operatorname{cosec} x$. $n = 2$ so $n + 1 = 3$.

Using the formula:

$\int 3(-\cot x \operatorname{cosec} x)(\operatorname{cosec}^2 x) \, dx = \operatorname{cosec}^3 x + C$

So $\int -3 \cot x \operatorname{cosec}^3 x \, dx = \operatorname{cosec}^3 x + C$

Divide by -3 to get the original integral.

$\int \cot x \operatorname{cosec}^3 x \, dx = \frac{1}{-3} \int -3 \cot x \operatorname{cosec}^3 x \, dx$

$= -\frac{1}{3}\operatorname{cosec}^3 x + C$

Q3 **a)** Let $f(x) = \sin x$ so $f'(x) = \cos x$ and $n = 3$ so $n + 1 = 4$.

Using the formula: $\int 4 \sin^3 x \cos x \, dx = \sin^4 x + C$

[3 marks available — 1 mark for identifying f(x) and f'(x), 1 mark for using the formula, 1 mark for the correct answer]

b) Let $f(x) = \cos x$ so $f'(x) = -\sin x$ and $n = 3$ so $n + 1 = 4$.

So $\int 4 \cos^3 x \sin x \, dx = -\int 4 \cos^3 x(-\sin x) \, dx$

Using the formula: $= -(\cos^4 x + c) = -\cos^4 x + C$

[3 marks available — 1 mark for identifying f(x) and f'(x), 1 mark for using the formula, 1 mark for the correct answer]

c) $\int 4\sin^3 x\cos x + 4\cos^3 x\sin x\ dx$

$= \int 4\sin^3 x\cos x\ dx\ +\int 4\cos^3 x\sin x\ dx$

$= \sin^4 x - \cos^4 x + C$

$= (\sin^2 x + \cos^2 x)(\sin^2 x - \cos^2 x) + C$

$= \sin^2 x - \cos^2 x + C = -\cos 2x + C$

[2 marks available — 1 mark for factorising correctly, 1 mark for using the correct trig identity to show the result]

Q4 a) *This one looks really complicated, but if you differentiate the bracket ($e^{\sin x} - 5$) using the chain rule, you'll get the function at the front.*

Let $f(x) = e^{\sin x} - 5$ so $f'(x) = \cos x\ e^{\sin x}$. $n = 3$ so $n + 1 = 4$.

Using the formula:

$\int 4(\cos x\ e^{\sin x})(e^{\sin x} - 5)^3\ dx = (e^{\sin x} - 5)^4 + C$

b) Let $f(x) = e^{\cos x} + 4x$ so $f'(x) = -\sin x\ e^{\cos x} + 4$.

$n = 6$ so $n + 1 = 7$. Using the formula:

$\int 7(-\sin x\ e^{\cos x} + 4)(e^{\cos x} + 4x)^6\ dx = (e^{\cos x} + 4x)^7 + C$

So divide by -7 to get the original integral:

$\int (\sin x\ e^{\cos x} - 4)(e^{\cos x} + 4x)^6\ dx$

$= -\frac{1}{7}\int 7(-\sin x\ e^{\cos x} + 4)(e^{\cos x} + 4x)^6\ dx$

$= -\frac{1}{7}(e^{\cos x} + 4x)^7 + C$

Q5 a) $\int \frac{1}{x}(3 + 6x^2)(\ln x + x^2 + 3)^3\ dx$

$= \int 3\left(\frac{1}{x} + 2x\right)(\ln x + x^2 + 3)^3\ dx$

Let $f(x) = \ln x + x^2 + 3$ so $f'(x) = \frac{1}{x} + 2x$ and $n = 3$ so $n + 1 = 4$. Using the formula:

$\int 4\left(\frac{1}{x} + 2x\right)(\ln x + x^2 + 3)^3\ dx = (\ln x + x^2 + 3)^4 + C$

Divide by 4 and multiply by 3 to get the original integral:

$\int \frac{1}{x}(3 + 6x^2)(\ln x + x^2 + 3)^3\ dx$

$= \frac{3}{4}\int 4\left(\frac{1}{x} + 2x\right)(\ln x + x^2 + 3)^3\ dx = \frac{3}{4}(\ln x + x^2 + 3)^4 + C$

[3 marks available — 1 mark for identifying f(x) and f'(x), 1 mark for using the formula, 1 mark for the correct answer]

b) The power on $(\ln x + x^2 + 3)$ is 2 instead of 3, so using your working for part a):

$\int \frac{1}{x}(3 + 6x^2)(\ln x + x^2 + 3)^2\ dx$

$= \frac{3}{3}(\ln x + x^2 + 3)^3 + C = (\ln x + x^2 + 3)^3 + C$

$f(1) = (\ln 1 + 1^2 + 3)^3 + C = 60$

$\Rightarrow 4^3 + C = 60 \Rightarrow C = -4$

So $f(x) = (\ln x + x^2 + 3)^3 - 4$

[2 marks available — 1 mark for using your working for part a) to find the integral with "+ C", 1 mark for the correct answer]

Q6 a) Start by writing the function as $\sec^2 x\ \tan^{-4} x$.

Let $f(x) = \tan x$ so $f'(x) = \sec^2 x$. $n = -4$ so $n + 1 = -3$.

Using the formula: $\int -3\sec^2 x\ \tan^{-4} x\ dx = \tan^{-3} x + C$

Divide by -3 to get the original integral:

$\int \frac{\sec^2 x}{\tan^4 x}\ dx = -\frac{1}{3}\tan^{-3} x + C = -\frac{1}{3\tan^3 x} + C$

b) Start by writing the function as $\cot x\ \mathrm{cosec}\ x\ (\mathrm{cosec}\ x)^{\frac{1}{2}}$.

Let $f(x) = \mathrm{cosec}\ x$ so $f'(x) = -\cot x\ \mathrm{cosec}\ x$.

$n = \frac{1}{2}$ so $n + 1 = \frac{3}{2}$. Using the formula:

$\int \frac{3}{2}(-\cot x\ \mathrm{cosec}\ x)(\mathrm{cosec}\ x)^{\frac{1}{2}}\ dx = (\mathrm{cosec}\ x)^{\frac{3}{2}} + C$

Divide by $-\frac{3}{2}$ to get the original integral:

$\int \cot x\ \mathrm{cosec}\ x\sqrt{\mathrm{cosec}\ x}\ dx = -\frac{2}{3}(\mathrm{cosec}\ x)^{\frac{3}{2}} + C$

$= -\frac{2}{3}(\sqrt{\mathrm{cosec}\ x})^3 + C$

Q7 a) $\int_0^5 (18x + 15)\frac{1}{\sqrt{3x^2 + 5x}}\ dx = \int_0^5 (18x + 15)(3x^2 + 5x)^{-\frac{1}{2}}\ dx$

Let $f(x) = 3x^2 + 5x$ so $f'(x) = 6x + 5$

and $n = -\frac{1}{2}$ so $n + 1 = \frac{1}{2}$.

Using the formula:

$\int_0^5 \frac{1}{2}(6x + 5)(3x^2 + 5x)^{-\frac{1}{2}}\ dx = (3x^2 + 5x)^{\frac{1}{2}} + C$

$= \sqrt{3x^2 + 5x} + C$

Multiply by 6 to get the original integral:

$\int_0^5 (18x + 15)\frac{1}{\sqrt{3x^2 + 5x}}\ dx = 6[\sqrt{3x^2 + 5x}]_0^5$

$= 6[\sqrt{3(5^2) + 5(5)} - \sqrt{3(0^2) + 5(0)}] = 6[\sqrt{100}] = 60$

[5 marks available — 1 mark for identifying f(x) and f'(x), 1 mark for using the formula, 1 mark for the correct integral, 1 mark for substituting limits, 1 mark for the correct answer]

b) Let $f(x) = 3\sin^2 x + 5\sin x$ so

$f'(x) = 6\sin x\cos x + 5\cos x = (6\sin x + 5)\cos x$

So, without doing any more work you can see that this is the same as the first integral substituting x for $\sin x$, and with a different upper limit.

So $\int_0^{\frac{\pi}{2}} (18\sin x + 15)\cos x\frac{1}{\sqrt{3\sin^2 x + 5\sin x}}\ dx$

$= 6[\sqrt{3\sin^2 x + 5\sin x}]_0^{\frac{\pi}{2}}$

$= 6\left[\sqrt{3\sin^2\left(\frac{\pi}{2}\right) + 5\sin\left(\frac{\pi}{2}\right)} - \sqrt{3\sin^2(0) + 5\sin(0)}\right]$

$= 6[\sqrt{3 + 5}] = 6\sqrt{8} = 6\sqrt{4}\sqrt{2} = 12\sqrt{2}$

[3 marks available — 1 mark for explanation of how this relates to part a), 1 mark for substituting in the limits, 1 mark for correct final answer with correct working]

If you didn't spot how this relates to part a), you could also have just performed the whole integration again.

Q8 a) Start by writing the function as $e^{\cot 2x}\ \mathrm{cosec}^2\ 2x\ (e^{\cot 2x})^3$.

Let $f(x) = e^{\cot 2x}$ so $f'(x) = -2\ e^{\cot 2x}\ \mathrm{cosec}^2\ 2x$.

$n = 3$ so $n + 1 = 4$. Using the formula:

$\int 4(-2\ e^{\cot 2x}\ \mathrm{cosec}^2\ 2x)\ e^{\cot 2x} = (e^{\cot 2x})^4 + c$

Divide by -8 to get the original integral:

$\int e^{\cot 2x}\ \mathrm{cosec}^2\ 2x\ e^{\cot 2x} = -\frac{1}{8}e^{4\cot 2x} + C$, as required.

b) $\int_{\frac{1}{2}}^1 e^{\cot 2x}\ \mathrm{cosec}^2\ 2x\ e^{\cot 2x} = -\frac{1}{8}[e^{4\cot 2x}]_{\frac{1}{2}}^1$

$= -\frac{1}{8}[e^{4\cot 2(1)}] - -\frac{1}{8}[e^{4\cot 2(\frac{1}{2})}] = \frac{1}{8}(e^{4\cot 1} - e^{4\cot 2})$

$= 1.61\ (3\ \text{s.f.})$

Exercise 8.10 — Using Trigonometric Identities in Integration

Q1 a) Using the cos double angle formula: $\cos^2 x = \frac{1}{2}(\cos 2x + 1)$

So the integral is:

$\int \cos^2 x\ dx = \int \frac{1}{2}(\cos 2x + 1)\ dx = \frac{1}{2}\left(\frac{1}{2}\sin 2x + x\right) + C$

$= \frac{1}{4}\sin 2x + \frac{1}{2}x + C$

b) $6\sin x\cos x = 3(2\sin x\cos x) = 3\sin 2x$

So the integral is:

$\int 6\sin x\cos x\ dx = \int 3\sin 2x\ dx = -\frac{3}{2}\cos 2x + C$

Answers

c) $\sin^2 6x = \frac{1}{2}(1 - \cos(2 \times 6x)) = \frac{1}{2}(1 - \cos 12x)$

So the integral is:

$\int \sin^2 6x \, dx = \int \frac{1}{2}(1 - \cos 12x) \, dx$

$\qquad = \frac{1}{2}\left(x - \frac{1}{12}\sin 12x\right) + C$

$\qquad = \frac{1}{2}x - \frac{1}{24}\sin 12x + C$

d) Using the tan double angle formula: $\dfrac{2\tan 2x}{1 - \tan^2 2x} = \tan 4x$

So the integral is:

$\int \frac{2\tan 2x}{1 - \tan^2 2x} \, dx = \int \tan 4x \, dx$

$\qquad = -\frac{1}{4}\ln|\cos 4x| + C \left(\text{or} = \frac{1}{4}\ln|\sec 4x| + C\right)$

e) $2\sin 4x \cos 4x = \sin 8x$

So the integral is:

$\int 2\sin 4x \cos 4x \, dx = \int \sin 8x \, dx = -\frac{1}{8}\cos 8x + C$

f) $2\cos^2 4x = 2\left(\frac{1}{2}(\cos 8x + 1)\right) = \cos 8x + 1$

So the integral is:

$\int 2\cos^2 4x \, dx = \int \cos 8x + 1 \, dx = \frac{1}{8}\sin 8x + x + C$

g) $\cos x \sin x = \frac{1}{2}(2\cos x \sin x) = \frac{1}{2}\sin 2x$

So the integral is:

$\int \cos x \sin x \, dx = \int \frac{1}{2}\sin 2x \, dx$

$\qquad = \frac{1}{2}\left(-\frac{1}{2}\cos 2x\right) + C = -\frac{1}{4}\cos 2x + C$

h) $\sin 3x \cos 3x = \frac{1}{2}(2\sin 3x \cos 3x) = \frac{1}{2}\sin 6x$

So the integral is:

$\int \sin 3x \cos 3x \, dx = \int \frac{1}{2}\sin 6x \, dx$

$\qquad = \frac{1}{2}\left(-\frac{1}{6}\cos 6x\right) + C = -\frac{1}{12}\cos 6x + C$

i) $\dfrac{6\tan 3x}{1 - \tan^2 3x} = 3\left(\dfrac{2\tan 3x}{1 - \tan^2 3x}\right) = 3\tan 6x$

So the integral is:

$\int \frac{6\tan 3x}{1 - \tan^2 3x} \, dx = \int 3\tan 6x \, dx = 3\left(-\frac{1}{6}\ln|\cos 6x|\right) + C$

$\qquad = -\frac{1}{2}\ln|\cos 6x| + C \left(\text{or} = \frac{1}{2}\ln|\sec 6x| + C\right)$

j) $5\sin 2x \cos 2x = \frac{5}{2}(2\sin 2x \cos 2x) = \frac{5}{2}\sin 4x$

So the integral is:

$\int 5\sin 2x \cos 2x \, dx = \int \frac{5}{2}\sin 4x \, dx$

$\qquad = \frac{5}{2}\left(-\frac{1}{4}\cos 4x\right) + C = -\frac{5}{8}\cos 4x + C$

k) $(\sin x + \cos x)^2 = \sin^2 x + 2\sin x \cos x + \cos^2 x$
$= \sin^2 x + \cos^2 x + 2\sin x \cos x = 1 + 2\sin x \cos x = 1 + \sin 2x$
$\sin^2 x + \cos^2 x \equiv 1$ has been used to simplify here.

So the integral is:

$\int (\sin x + \cos x)^2 \, dx = \int 1 + \sin 2x \, dx = x - \frac{1}{2}\cos 2x + C$

l) $4\sin x \cos x \cos 2x = 2(2\sin x \cos x)\cos 2x$
$\qquad = 2\sin 2x \cos 2x = \sin 4x$

So the integral is:

$\int 4\sin x \cos x \cos 2x \, dx = \int \sin 4x \, dx = -\frac{1}{4}\cos 4x + C$

m) $(\cos x + \sin x)(\cos x - \sin x)$
$= \cos^2 x - \cos x \sin x + \sin x \cos x - \sin^2 x$
$= \cos^2 x - \sin^2 x = \cos 2x$

So the integral is:

$\int (\cos x + \sin x)(\cos x - \sin x) \, dx$

$= \int \cos 2x \, dx = \frac{1}{2}\sin 2x + C$

n) $\sin^2 x \cot x = \sin^2 x \frac{1}{\tan x} = \sin^2 x \frac{\cos x}{\sin x}$

$\qquad = \sin x \cos x = \frac{1}{2}\sin 2x$

So the integral is:

$\int \sin^2 x \cot x \, dx = \int \frac{1}{2}\sin 2x \, dx$

$\qquad = \frac{1}{2}\left(-\frac{1}{2}\cos 2x\right) + C = -\frac{1}{4}\cos 2x + C$

Q2 a) $\sin^2 x = \frac{1}{2}(1 - \cos 2x)$

So the integral is:

$\int_0^{\frac{\pi}{4}} \sin^2 x \, dx = \int_0^{\frac{\pi}{4}} \frac{1}{2}(1 - \cos 2x) \, dx$

$= \frac{1}{2}\left[\left(x - \frac{1}{2}\sin 2x\right)\right]_0^{\frac{\pi}{4}} = \frac{1}{2}\left(\left(\frac{\pi}{4} - \frac{1}{2}\sin \frac{\pi}{2}\right) - \left(-\frac{1}{2}\sin 0\right)\right)$

$= \frac{1}{2}\left(\left(\frac{\pi}{4} - \left(\frac{1}{2} \times 1\right)\right) - \left(-\frac{1}{2} \times 0\right)\right) = \frac{1}{2}\left(\frac{\pi}{4} - \frac{1}{2}\right) = \frac{\pi}{8} - \frac{1}{4}$

b) $\cos^2 2x = \frac{1}{2}(\cos 4x + 1)$

So the integral is:

$\int_0^{\pi} \frac{1}{2}(\cos 4x + 1) \, dx = \left[\frac{1}{2}\left(\frac{1}{4}\sin 4x + x\right)\right]_0^{\pi}$

$= \left[\left(\frac{1}{8}\sin 4x + \frac{x}{2}\right)\right]_0^{\pi} = \left(\frac{1}{8}\sin 4\pi + \frac{\pi}{2}\right) - \left(\frac{1}{8}\sin 0 + \frac{0}{2}\right)$

$= \left(\frac{1}{8} \times 0 + \frac{\pi}{2}\right) - (0 + 0) = \frac{\pi}{2}$

c) $\sin \frac{x}{2}\cos \frac{x}{2} = \frac{1}{2}\left(2\sin \frac{x}{2}\cos \frac{x}{2}\right) = \frac{1}{2}\sin x$

So the integral is:

$\int_0^{\pi} \sin \frac{x}{2}\cos \frac{x}{2} \, dx = \int_0^{\pi} \frac{1}{2}\sin x \, dx = -\frac{1}{2}[\cos x]_0^{\pi}$

$= -\frac{1}{2}(\cos \pi - \cos 0) = -\frac{1}{2}(-1 - 1) = 1$

d) $\sin^2 2x = \frac{1}{2}(1 - \cos 4x)$

So the integral is:

$\int_{\frac{\pi}{4}}^{\frac{\pi}{2}} \sin^2 2x \, dx = \int_{\frac{\pi}{4}}^{\frac{\pi}{2}} \frac{1}{2}(1 - \cos 4x) \, dx$

$= \frac{1}{2}\left[x - \frac{1}{4}\sin 4x\right]_{\frac{\pi}{4}}^{\frac{\pi}{2}} = \frac{1}{2}\left(\left[\frac{\pi}{2} - \frac{1}{4}\sin 2\pi\right] - \left[\frac{\pi}{4} - \frac{1}{4}\sin \pi\right]\right)$

$= \frac{1}{2}\left(\left[\frac{\pi}{2} - 0\right] - \left[\frac{\pi}{4} - 0\right]\right) = \frac{\pi}{8}$

e) $\cos 2x \sin 2x = \frac{1}{2}(2\sin 2x \cos 2x) = \frac{1}{2}\sin 4x$

So the integral is:

$\int_0^{\frac{\pi}{4}} \cos 2x \sin 2x \, dx = \int_0^{\frac{\pi}{4}} \frac{1}{2}\sin 4x \, dx$

$= \frac{1}{2}\left[\left(-\frac{1}{4}\cos 4x\right)\right]_0^{\frac{\pi}{4}} = -\frac{1}{8}[\cos 4x]_0^{\frac{\pi}{4}}$

$= -\frac{1}{8}(\cos \pi - \cos 0) = -\frac{1}{8}(-1 - 1) = \frac{1}{4}$

f) $\sin^2 x - \cos^2 x = -(\cos^2 x - \sin^2 x) = -\cos 2x$

So the integral is:

$\int_{\frac{\pi}{4}}^{\frac{\pi}{2}} \sin^2 x - \cos^2 x \, dx = \int_{\frac{\pi}{4}}^{\frac{\pi}{2}} -\cos 2x \, dx = -\frac{1}{2}[\sin 2x]_{\frac{\pi}{4}}^{\frac{\pi}{2}}$

$= -\frac{1}{2}\left(\sin \pi - \sin \frac{\pi}{2}\right) = -\frac{1}{2}(0 - 1) = \frac{1}{2}$

Q3 a) Use the cos double angle identities
$\cos 2\theta = 1 - 2\sin^2 \theta$ and $\cos 2\theta = 2\cos^2 \theta - 1$
$\Rightarrow \sin^2 \theta = \frac{1 - \cos 2\theta}{2}$ and $\cos^2 \theta = \frac{1 + \cos 2\theta}{2}$
$\Rightarrow \sin^2 \theta - 2\cos^2 \theta = \frac{1 - \cos 2\theta}{2} - 1 - \cos 2\theta$

$\qquad\qquad\qquad\qquad = -\frac{1}{2} - \frac{3}{2}\cos 2\theta$

[3 marks available — 1 mark for $\sin^2 \theta$ in terms of $\cos 2\theta$, 1 mark for $\cos^2 \theta$ in terms of $\cos 2\theta$, 1 mark for the correct answer]

b) $\int \sin^2\theta - 2\cos^2\theta \ d\theta = \int -\frac{1}{2} - \frac{3}{2}\cos 2\theta \ d\theta$

$\qquad = -\frac{1}{2}\theta - \left(\frac{1}{2} \times \frac{3}{2}\sin 2\theta\right) + C = -\frac{1}{2}\theta - \frac{3}{4}\sin 2\theta + C$

[2 marks available — 1 mark for integrating the cos 2θ term correctly, 1 mark for the correct answer]

Q4 Using the tan addition formula:

$\int_0^{\frac{\pi}{6}} \frac{\tan 7x - \tan 5x}{1 + \tan 7x \tan 5x} \ dx = \int_0^{\frac{\pi}{6}} \tan(7x - 5x) \ dx$

$= \int_0^{\frac{\pi}{6}} \tan 2x \ dx = \left[\frac{1}{2}\ln|\sec 2x|\right]_0^{\frac{\pi}{6}} = \frac{1}{2}\ln 2 - 0 = \frac{\ln 2}{2}$

Q5 a) $\cot^2 x - 4 = (\cosec^2 x - 1) - 4 = \cosec^2 x - 5$

So the integral is:

$\int \cot^2 x - 4 \ dx = \int \cosec^2 x - 5 \ dx = -\cot x - 5x + C$

b) $\tan^2 x = \sec^2 x - 1$

So the integral is:

$\int \tan^2 x \ dx = \int \sec^2 x - 1 \ dx = \tan x - x + C$

c) $3\cot^2 x = 3(\cosec^2 x - 1) = 3\cosec^2 x - 3$

So the integral is:

$\int 3\cot^2 x \ dx = \int 3\cosec^2 x - 3 \ dx = -3\cot x - 3x + C$

d) $\tan^2 4x = \sec^2 4x - 1$

So the integral is:

$\int \tan^2 4x \ dx = \int \sec^2 4x - 1 \ dx = \frac{1}{4}\tan 4x - x + C$

Q6 $\int_{\frac{\pi}{16}}^{\frac{\pi}{12}} \cot^2 4x \ dx = \int_{\frac{\pi}{16}}^{\frac{\pi}{12}} \cosec^2 4x - 1 \ dx = \left[-\frac{1}{4}\cot 4x - x\right]_{\frac{\pi}{16}}^{\frac{\pi}{12}}$

$= \left[-\frac{1}{4}\cot\left(\frac{4\pi}{12}\right) - \frac{\pi}{12}\right] - \left[-\frac{1}{4}\cot\left(\frac{4\pi}{16}\right) - \frac{\pi}{16}\right]$

$= \left[-\frac{1}{4\sqrt{3}} - \frac{\pi}{12}\right] - \left[-\frac{1}{4} - \frac{\pi}{16}\right]$

$= \frac{1}{4} - \frac{\pi}{48} - \frac{1}{4\sqrt{3}} = \frac{1}{4}\left(1 - \frac{\pi}{12} - \frac{1}{\sqrt{3}}\right)$

[4 marks available — 1 mark for giving in terms of cosec, 1 mark for integrating correctly, 1 mark for substituting the limits correctly, 1 mark for rearranging to give the correct answer]

Q7 $\tan^2 x + \cos^2 x - \sin^2 x = (\sec^2 x - 1) + \cos 2x$

So the integral is:

$\int_0^{\frac{\pi}{4}} \tan^2 x + \cos^2 x - \sin^2 x \ dx = \int_0^{\frac{\pi}{4}} \sec^2 x - 1 + \cos 2x \ dx$

$= \left[\tan x - x + \frac{1}{2}\sin 2x\right]_0^{\frac{\pi}{4}}$

$= \left[\tan\frac{\pi}{4} - \frac{\pi}{4} + \frac{1}{2}\sin\frac{2\pi}{4}\right] - \left[\tan 0 - 0 + \frac{1}{2}\sin 0\right]$

$= \left[1 - \frac{\pi}{4} + \frac{1}{2}\right] - [0 - 0 + 0] = \frac{3}{2} - \frac{\pi}{4}$

Q8 $(\sec x + \tan x)^2 = \sec^2 x + 2\tan x \sec x + \tan^2 x$

$= \sec^2 x + 2\tan x \sec x + (\sec^2 x - 1) = 2\sec^2 x + 2\tan x \sec x - 1$

Remember that the derivative of sec x is sec x tan x.

So the integral is:

$\int (\sec x + \tan x)^2 \ dx = \int 2\sec^2 x + 2\tan x \sec x - 1 \ dx$

$\qquad = 2\tan x + 2\sec x - x + C$

Q9 $(\cot x + \cosec x)^2 = \cot^2 x + 2\cot x \cosec x + \cosec^2 x$

$= (\cosec^2 x - 1) + 2\cot x \cosec x + \cosec^2 x$

$= 2\cosec^2 x + 2\cot x \cosec x - 1$

Just keep using the identities that you know until you get to something that you know how to integrate.

So the integral is:

$\int (\cot x + \cosec x)^2 \ dx = \int 2\cosec^2 x + 2\cot x \cosec x - 1 \ dx$

$\qquad = -2\cot x - 2\cosec x - x + C$

Q10 $4 + \cot^2 3x = 4 + (\cosec^2 3x - 1) = 3 + \cosec^2 3x$

So the integral is:

$\int 4 + \cot^2 3x \ dx = \int 3 + \cosec^2 3x \ dx = 3x - \frac{1}{3}\cot 3x + C$

Q11 $\cos^2 4x + \cot^2 4x = \frac{1}{2}(\cos 8x + 1) + (\cosec^2 4x - 1)$

$\qquad = \frac{1}{2}\cos 8x + \cosec^2 4x - \frac{1}{2}$

So the integral is:

$\int \cos^2 4x + \cot^2 4x \ dx = \int \frac{1}{2}\cos 8x + \cosec^2 4x - \frac{1}{2} \ dx$

$\qquad = \frac{1}{2}\left(\frac{1}{8}\sin 8x\right) - \frac{1}{4}\cot 4x - \frac{1}{2}x + C$

$\qquad = \frac{1}{16}\sin 8x - \frac{1}{4}\cot 4x - \frac{1}{2}x + C$

Q12 a) $\frac{\cos 2x}{1 - \cos^2 2x} = \frac{\cos 2x}{\sin^2 2x} = \frac{\cos 2x}{\sin 2x} \times \frac{1}{\sin 2x} = \cot 2x \cosec 2x$

[2 marks available — 1 mark for using trig identity to simplify the denominator, 1 mark for fully correct working]

b) $\int \frac{\cos 2x}{1 - \cos^2 2x} \ dx = \int \cot 2x \cosec 2x \ dx = -\frac{1}{2}\cosec 2x + C$

[2 marks available — 1 mark for converting the integrand to cot and cosec, 1 mark for the correct answer]

Q13 a) $\tan^3 x + \tan^5 x = \tan^3 x(1 + \tan^2 x) = \tan^3 x \sec^2 x = \sec^2 x \tan^3 x$

This is a product containing tan x to a power, and its derivative $\sec^2 x$. Using the formula with f(x) = tan x, $f'(x) = \sec^2 x$, n = 3 and n + 1 = 4 gives:

$\int 4\sec^2 x \tan^3 x \ dx = \tan^4 x + C$

So the integral is:

$\int \tan^3 x + \tan^5 x \ dx = \int \sec^2 x \tan^3 x \ dx$

$\qquad = \frac{1}{4}\int 4\sec^2 x \tan^3 x \ dx = \frac{1}{4}\tan^4 x + C$

b) $\cot^5 x + \cot^3 x = \cot^3 x(\cot^2 x + 1) = \cot^3 x \cosec^2 x$

Again, this is a product of a function to a power and its derivative so use the formula with f(x) = cot x, $f'(x) = -\cosec^2 x$, n = 3 and n + 1 = 4.

$\int -4\cosec^2 x \cot^3 x \ dx = \cot^4 x + C$

So the integral is:

$\int \cot^5 x + \cot^3 x \ dx = \int \cosec^2 x \cot^3 x \ dx$

$\qquad = -\frac{1}{4}\int -4\cosec^2 x \cot^3 x \ dx = -\frac{1}{4}\cot^4 x + C$

c) $\sin^3 x = \sin x \sin^2 x = \sin x(1 - \cos^2 x) = \sin x - \sin x \cos^2 x$

The second term of this function is a product of a function to a power and its derivative. Using the result with f(x) = cos x, f'(x) = -sin x, n = 2 and n + 1 = 3 gives:

$\int -3\sin x \cos^2 x \ dx = \cos^3 x + c$

So the integral is:

$\int \sin^3 x \ dx = \int \sin x - \sin x \cos^2 x \ dx$

$= \int \sin x \ dx + \int -\sin x \cos^2 x \ dx = -\cos x + \frac{1}{3}\cos^3 x + C$

Q14 a) $\frac{2\cos^2 2x}{\sin 4x} = \frac{2\cos^2 2x}{2\sin 2x \cos 2x} = \frac{\cos 2x}{\sin 2x} = \cot 2x$

[2 marks available — 1 mark for using sin double angle identity, 1 mark for simplifying to give the correct answer]

b) $\int_{\frac{\pi}{8}}^{\frac{\pi}{4}} \frac{2\cos^2 2x}{\sin 4x} \ dx = \int_{\frac{\pi}{8}}^{\frac{\pi}{4}} \cot 2x \ dx$

$= \left[\frac{1}{2}\ln|\sin 2x|\right]_{\frac{\pi}{8}}^{\frac{\pi}{4}} = \frac{1}{2}\ln\left|\sin\frac{2\pi}{4}\right| - \frac{1}{2}\ln\left|\sin\frac{2\pi}{8}\right|$

$= \frac{1}{2}\ln 1 - \frac{1}{2}\ln\frac{1}{\sqrt{2}} = -\frac{1}{2}\ln 2^{-\frac{1}{2}} = \ln 2^{-\frac{1}{2}\times-\frac{1}{2}} = \ln 2^{\frac{1}{4}}$

[4 marks available — 1 mark for integrating correctly, 1 mark for substituting the limits, 1 mark for a correct method to simplify, 1 mark for the correct answer in the correct form]

Answers

Q15 You want to find A and B, where $\frac{A+B}{2} = 4x$, and $\frac{A-B}{2} = x$.

Solve simultaneously:

$\frac{A+B}{2} + \frac{A-B}{2} = 4x + x \Rightarrow A = 5x$. So $B = 3x$. Then

$2\sin 4x \cos x \equiv 2\sin\left(\frac{5x+3x}{2}\right)\cos\left(\frac{5x-3x}{2}\right) \equiv \sin 5x + \sin 3x$

So $\int 2\sin 4x \cos x \; dx = \int \sin 5x + \sin 3x \; dx$

$\qquad\qquad = -\frac{1}{5}\cos 5x - \frac{1}{3}\cos 3x + C$

Q16 $\int 3\sin x \cos 2x \; dx = \int 3\sin x (2\cos^2 x - 1)\; dx$

$= \int 6\sin x \cos^2 x - 3\sin x \; dx = \int 6\sin x \cos^2 x \; dx - \int 3\sin x \; dx$

The first integral is of a product of a function to a power and its derivative. Using the formula with $f(x) = \cos x$, $f'(x) = -\sin x$, $n = 2$ and $n + 1 = 3$ gives:

$\int -3\sin x \cos^2 x \; dx \; dx = \cos^3 x + C$

Multiply by -2 to give the original integral:

$\int 6\sin x \cos^2 x \; dx = -2\cos^3 x + C$

So $\int 3\sin x \cos 2x \; dx = \int 6\sin x \cos^2 x \; dx - \int 3\sin x \; dx$

$\qquad\qquad = -2\cos^3 x + 3\cos x + C$

[5 marks available — 1 mark for using cos double angle identity to split up the integral, 1 mark for identifying f(x) and f'(x), 1 mark for a correct method to integrate the first integral, 1 mark for integrating the first integral correctly, 1 mark for the correct answer]

Q17 a) (i) Let $u = \sin^2 x$, then $\frac{du}{dx} = 2\sin x \cos x$,

Let $v = \cos^2 x$, then $\frac{dv}{dx} = -2\sin x \cos x$

Now use the product rule:

$\frac{dy}{dx} = u\frac{dv}{dx} + v\frac{du}{dx} = -2\sin^3 x \cos x + 2\sin x \cos^3 x$

$= 2\sin x \cos x (\cos^2 x - \sin^2 x)$

[3 marks available — 1 mark for attempting to use the product rule, 1 mark for either $\frac{du}{dx}$ or $\frac{dv}{dx}$, 1 mark for the correct answer]

(ii) $y = p$ is the maximum y-value on the curve $y = \sin^2 x \cos^2 x$.

Using part a) the stationary points are

$\sin x = 0 \Rightarrow x = -\pi, 0, \pi, \ldots$

$\cos x = 0 \Rightarrow x = -\frac{\pi}{2}, \frac{\pi}{2}, \ldots$

$\cos^2 x - \sin^2 x = 0 \Rightarrow \cos^2 x - (1 - \cos^2 x) = 0$

$\Rightarrow 2\cos^2 x = 1 \Rightarrow \cos x = \pm\frac{1}{\sqrt{2}} \Rightarrow x = -\frac{\pi}{4}, \frac{\pi}{4}, \ldots$

So $x = \frac{\pi}{4}$ is the lowest positive solution.

[2 marks available — 1 mark for a correct method, 1 mark for the correct answer]

(iii) When $x = \frac{\pi}{4}$, $y = \sin^2\frac{\pi}{4}\cos^2\frac{\pi}{4} = 0.25$. So $p = 0.25$.

[1 mark for the correct answer]

b) (i) $\sin^2 x \cos^2 x = \frac{1 - \cos 2x}{2} \times \frac{\cos 2x + 1}{2}$

$= \frac{1}{4}(1 - \cos^2 2x) = \frac{1}{4}\left(1 - \frac{\cos 4x + 1}{2}\right) = \frac{1 - \cos 4x}{8}$

[3 marks available — 1 mark for correctly using double angle formulas, 1 mark for converting all angles to 2x, 1 mark for correct working leading to the required result]

(ii) From part a) the limits of the integration are $-\frac{\pi}{4}$ and $\frac{\pi}{4}$.

So the required area is:

$\int_{-\frac{\pi}{4}}^{\frac{\pi}{4}} \sin^2 x \cos^2 x \; dx = \int_{-\frac{\pi}{4}}^{\frac{\pi}{4}} \frac{1 - \cos 4x}{8} \; dx$

$= \int_{-\frac{\pi}{4}}^{\frac{\pi}{4}} \frac{1}{8}\; dx - \int_{-\frac{\pi}{4}}^{\frac{\pi}{4}} \frac{\cos 4x}{8} \; dx = \left[\frac{x}{8} - \frac{\sin 4x}{32}\right]_{-\frac{\pi}{4}}^{\frac{\pi}{4}}$

$= \left[\frac{\pi}{32} - \frac{\sin \pi}{32}\right] - \left[\frac{-\pi}{32} - \frac{\sin -\pi}{32}\right] = \frac{\pi}{32} - \left(-\frac{\pi}{32}\right) = \frac{\pi}{16}$

[3 marks available — 1 mark for integrating correctly, 1 mark for substituting the correct limits, 1 mark for the correct answer]

Q18 a) $\frac{\cos x + \sec x}{1 - \cos x} - \sec x \operatorname{cosec}^2 x$

$= \frac{\cos x + \sec x}{1 - \cos x} \times \frac{1 + \cos x}{1 + \cos x} - \sec x \operatorname{cosec}^2 x$

$= \frac{1 + \cos^2 x + \cos x + \sec x}{1 - \cos^2 x} - \sec x \operatorname{cosec}^2 x$

$= \frac{1 + \cos^2 x + \cos x + \sec x}{\sin^2 x} - \sec x \operatorname{cosec}^2 x$

$= \frac{1}{\sin^2 x} + \frac{\cos^2 x}{\sin^2 x} + \frac{\cos x}{\sin^2 x} + \frac{\sec x}{\sin^2 x} - \sec x \operatorname{cosec}^2 x$

$= \operatorname{cosec}^2 x + \cot^2 x + \cot x \operatorname{cosec} x + \sec x \operatorname{cosec}^2 x - \sec x \operatorname{cosec}^2 x$

$= \operatorname{cosec}^2 x + \cot^2 x + \cot x \operatorname{cosec} x$

b) $\int \frac{\cos x + \sec x}{1 - \cos x} - \sec x \operatorname{cosec}^2 x \; dx$

$= \int \operatorname{cosec}^2 x + \cot^2 x + \cot x \operatorname{cosec} x \; dx$

$= \int \operatorname{cosec}^2 x + (\operatorname{cosec}^2 x - 1) + \cot x \operatorname{cosec} x \; dx$

$= \int 2\operatorname{cosec}^2 x - 1 + \cot x \operatorname{cosec} x \; dx$

$= \int 2\operatorname{cosec}^2 x \; dx - \int 1 \; dx + \int \cot x \operatorname{cosec} x \; dx$

$= -2\cot x - x - \operatorname{cosec} x + C$

Exercise 8.11 — Finding Area using Integration

Q1 a) Start by finding the points where the curve and the line intersect. Solve $3x^2 + 4 = 16$:

$\Rightarrow 3x^2 = 12 \Rightarrow x^2 = 4 \Rightarrow x = -2$ or 2.

So they intersect at $x = -2$ and $x = 2$.

(graph: $y = 3x^2 + 4$, $y = 16$, with $x = -2$ and $x = 2$ marked)

So the area is found by subtracting the integral of $3x^2 + 4$ between -2 and 2 from the integral of 16 between -2 and 2. The area under the line is just a rectangle which is 16 by 4 so the area is $16 \times 4 = 64$.

$\int_{-2}^{2} (3x^2 + 4)\; dx = [x^3 + 4x]_{-2}^{2}$

$= (2^3 + 4(2)) - ((-2)^3 + 4(-2))$

$= (8 + 8) - (-8 - 8) = 16 + 16 = 32$

So the area is $64 - 32 = 32$.

b) Solving to find the point of intersection of $x = 2$ and $y = x^3 + 4$: $y = 2^3 + 4 = 12$

Draw a diagram:

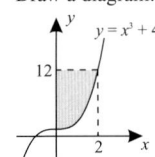

So the required area is the area under $y = 12$ between
$x = 0$ and $x = 2$, minus the area under
$y = x^3 + 4$ between $x = 0$ and $x = 2$.
The first area is a rectangle which is $12 \times 2 = 24$.

$$\int_0^2 x^3 + 4 \, dx = \left[\frac{x^4}{4} + 4x\right]_0^2$$
$$= \left(\frac{2^4}{4} + 8\right) - \left(\frac{0^4}{4} + 0\right) = 12$$

So the shaded area is $24 - 12 = 12$.

c) Solving to find the point of intersection of $y = 4$ and $y = \frac{1}{x^2}$:

$$4 = \frac{1}{x^2} \Rightarrow 4x^2 = 1 \Rightarrow x^2 = \frac{1}{4} \Rightarrow x = \pm\frac{1}{2}$$

so the intersection in the diagram is $x = \frac{1}{2}$:

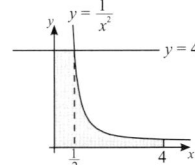

So required area is the area under the line $y = 4$
from $x = 0$ and $x = \frac{1}{2}$ added to the area under
the curve $y = \frac{1}{x^2}$ from $x = \frac{1}{2}$ and $x = 4$.
The first area is just a rectangle which is $\frac{1}{2}$ by 4 so the area
is $\frac{1}{2} \times 4 = 2$.

Area under curve $= \int_{\frac{1}{2}}^4 \frac{1}{x^2} \, dx = \int_{\frac{1}{2}}^4 x^{-2} \, dx = \left[\frac{x^{-1}}{-1}\right]_{\frac{1}{2}}^4 = \left[-\frac{1}{x}\right]_{\frac{1}{2}}^4$

$$= \left(-\frac{1}{4}\right) - \left(-\frac{1}{\left(\frac{1}{2}\right)}\right) = -\frac{1}{4} + 2 = \frac{7}{4}$$

So the area is $2 + \frac{7}{4} = \frac{15}{4}$.

d) Solve $-1 - (x-3)^2 = -5 \Rightarrow 4 = (x-3)^2$
$\Rightarrow x - 3 = \pm 2 \Rightarrow x = 1$ and $x = 5$.
*Be careful — this area is below the x-axis so the integrals will
give negative areas.*
So the area you want is the area above the line
between 1 and 5 minus the area between the curve
and the x-axis between 1 and 5.

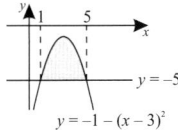

The first area is just a rectangle which is 5 by 4 so the area
is $5 \times 4 = 20$. So the area between the line and the x-axis
between 0 and 5 is 20.
Area between curve and x-axis
$$= \int_1^5 (-1-(x-3)^2) \, dx = \int_1^5 (-1-(x^2-6x+9)) \, dx$$
$$= \int_1^5 (6x - x^2 - 10) \, dx$$
$$= \left[\frac{6x^2}{2} - \frac{x^3}{3} - 10x\right]_1^5 = \left[3x^2 - \frac{x^3}{3} - 10x\right]_1^5$$
$$= \left((3 \times 5^2) - \frac{5^3}{3} - (10 \times 5)\right) - \left((3 \times 1^2) - \frac{1^3}{3} - (10 \times 1)\right)$$
$$= \left(75 - \frac{125}{3} - 50\right) - \left(3 - \frac{1}{3} - 10\right) = -\frac{28}{3}$$

So the area between the curve and the x-axis between
1 and 5 is $\frac{28}{3}$. So the area is $20 - \frac{28}{3} = \frac{32}{3}$.

e) Solve $x^2 = 2x \Rightarrow x^2 - 2x = 0 \Rightarrow x(x-2) = 0$
$\Rightarrow x = 0$ and $x = 2$.

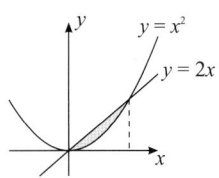

The shaded area is the area under $y = 2x$ from $x = 0$ and
$x = 2$, minus the area under $y = x^2$ from $x = 0$ to $x = 2$.
$$\int_0^2 2x \, dx = [x^2]_0^2 = 2^2 - 0^2 = 4$$
$$\int_0^2 x^2 \, dx = \left[\frac{x^3}{3}\right]_0^2 = \frac{2^3}{3} - \frac{0^3}{3} = \frac{8}{3}$$
So the shaded area is $4 - \frac{8}{3} = \frac{4}{3}$.

f) Solve $x^3 = 4x \Rightarrow x^3 - 4x = 0 \Rightarrow x(x^2-4) = 0$
$\Rightarrow x(x+2)(x-2) = 0 \Rightarrow x = 0$ and $x = \pm 2$.

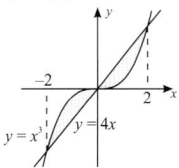

The area above the x-axis is the same as the area below the
x-axis (from the symmetry of the graph).
For the area above the x-axis, find the area
under the line from $x = 0$ to $x = 2$, minus the
area under the curve from $x = 0$ to $x = 2$.
$$\int_0^2 4x \, dx = [2x^2]_0^2 = (2 \times 2^2) - (2 \times 0^2) = 8$$
$$\int_0^2 x^3 \, dx = \left[\frac{x^4}{4}\right]_0^2 = \left(\frac{2^4}{4}\right) - \left(\frac{0^2}{4}\right) = 4$$
So the area above the axis is $8 - 4 = 4$.
So the area below the x-axis is also 4.
The total area is $4 + 4 = 8$.
*For parts e) and f) you could have used the formula for
finding the area of a triangle rather than integrating.*

Q2 a) Solve $x^2 + 4 = x + 4 \Rightarrow x^2 - x = 0 \Rightarrow x(x-1) = 0$
$\Rightarrow x = 0$ and $x = 1$. Draw a diagram:

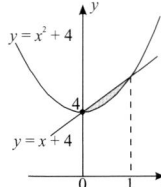

So the area you want is the area under the line minus the area
under the curve, between $x = 0$ and $x = 1$.
$$\int_0^1 (x+4) \, dx = \left[\frac{x^2}{2} + 4x\right]_0^1$$
$$= \left(\frac{1^2}{2} + (4 \times 1)\right) - \left(\frac{0^2}{2} + (4 \times 0)\right)$$
$$= \frac{1}{2} + 4 = \frac{9}{2}$$
$$\int_0^1 (x^2 + 4) \, dx = \left[\frac{x^3}{3} + 4x\right]_0^1$$
$$= \left(\frac{1^3}{3} + (4 \times 1)\right) - \left(\frac{0^3}{3} + (4 \times 0)\right)$$
$$= \frac{1}{3} + 4 = \frac{13}{3}$$
So the area is $\frac{9}{2} - \frac{13}{3} = \frac{1}{6}$.

Answers

b) Solve $x^2 + 2x - 3 = 4x \Rightarrow x^2 - 2x - 3 = 0$
$\Rightarrow (x - 3)(x + 1) = 0 \Rightarrow x = -1$ and $x = 3$.
It'll also help to find where the curve meets the x-axis by
solving $x^2 + 2x - 3 = 0 \Rightarrow (x + 3)(x - 1) \Rightarrow x = 1$ and $x = -3$.
Draw a diagram:

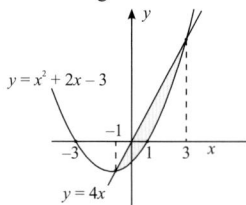

You'll need to find the areas above and below the x-axis
separately.
For the area above the x-axis, integrate $4x$ from $x = 0$ to $x = 3$
and subtract the integral of $x^2 + 2x - 3$ from $x = 1$ to $x = 3$.
$$\int_0^3 4x \, dx = [2x^2]_0^3 = (2(3)^2) - (2(0)^2) = 18$$
$$\int_1^3 (x^2 + 2x - 3) \, dx = \left[\frac{x^3}{3} + x^2 - 3x\right]_1^3$$
$$= \left(\frac{(3)^3}{3} + (3)^2 - 3(3)\right) - \left(\frac{(1)^3}{3} + (1)^2 - 3(1)\right)$$
$$= \left(\frac{27}{3} + 9 - 9\right) - \left(\frac{1}{3} + 1 - 3\right) = \frac{32}{3}$$
So the area above the x-axis is $18 - \frac{32}{3} = \frac{22}{3}$
To find the area below the x-axis you need to find the
positive area between the curve and the axis between -1
and 1 minus the positive area between the line and the axis
between -1 and 0.
$$\int_{-1}^1 (x^2 + 2x - 3) \, dx = \left[\frac{x^3}{3} + x^2 - 3x\right]_{-1}^1$$
$$= \left(\frac{(1)^3}{3} + (1)^2 - 3(1)\right) - \left(\frac{(-1)^3}{3} + (-1)^2 - 3(-1)\right)$$
$$= \left(\frac{1}{3} + 1 - 3\right) - \left(\frac{-1}{3} + 1 + 3\right) = -\frac{16}{3}$$
So the area between the curve and the axis
between -1 and 1 is $\frac{16}{3}$.
$$\int_{-1}^0 4x \, dx = [2x^2]_{-1}^0 = (2(0)^2) - (2(-1)^2) = -2$$
So the area between the line and the
axis between -1 and 1 is 2.
So the area under the x-axis is $\frac{16}{3} - 2 = \frac{10}{3}$
So the total area enclosed by the curve and the line is
$\frac{22}{3} + \frac{10}{3} = \frac{32}{3}$

Q3 Intersection occurs where $x^2 = 4 - x^2$
$\Rightarrow 2x^2 = 4 \Rightarrow x = \pm\sqrt{2}$
The shaded area is entirely above the x-axis, so
integrate both functions between $x = 0$ and $x = \sqrt{2}$.
$$\int_0^{\sqrt{2}} x^2 \, dx = \left[\frac{x^3}{3}\right]_0^{\sqrt{2}} = \frac{(\sqrt{2})^3}{3} - \frac{0^3}{3} = \frac{2}{3}\sqrt{2}$$
$$\int_0^{\sqrt{2}} 4 - x^2 \, dx = \left[4x - \frac{x^3}{3}\right]_0^{\sqrt{2}} = \left(4\sqrt{2} - \frac{(\sqrt{2})^3}{3}\right) - 0 = \frac{10}{3}\sqrt{2}$$
So the shaded area is $\frac{10}{3}\sqrt{2} - \frac{2}{3}\sqrt{2} = \frac{8}{3}\sqrt{2}$
*[5 marks available — 1 mark for finding the x-coordinate of the
intersection point, 1 mark for integrating both functions, 1 mark
for substituting the limits, 1 mark for at least one correct area,
1 mark for the correct answer]*
You could have integrated the difference of the functions
$(4 - x^2 - x^2 = 4 - 2x^2)$ here instead. Using that method
correctly on questions like this will still earn you the marks.

Q4 The intersection points occur when $2 \sin x = \tan x$
Substitute $\tan x = \frac{\sin x}{\cos x}$:
$2 \sin x = \frac{\sin x}{\cos x} \Rightarrow 2 \sin x \cos x = \sin x$
$\Rightarrow (2 \cos x - 1) \sin x = 0$
$\Rightarrow \sin x = 0$ or $\cos x = \frac{1}{2}$
$\Rightarrow x = 0$ or $x = \frac{\pi}{3}$
The shaded area is entirely above the x-axis, so
integrate both functions between $x = 0$ and $x = \frac{\pi}{3}$.
$$\int_0^{\frac{\pi}{3}} 2 \sin x \, dx = [-2 \cos x]_0^{\frac{\pi}{3}} = (-2 \cos \frac{\pi}{3}) - (-2 \cos 0)$$
$$= (-2 \times 0.5) - (-2 \times 1) = 1$$
$$\int_0^{\frac{\pi}{3}} \tan x \, dx = [\ln |\sec x|]_0^{\frac{\pi}{3}} = \ln |\sec \frac{\pi}{3}| - \ln |\sec 0|$$
$$= \ln 2 - \ln 1 = \ln 2$$
So the shaded area is $1 - \ln 2 = 0.3068... = 0.307$ (3 s.f.)
*[6 marks available — 1 mark for using an appropriate trig
identity, 1 mark for finding the x-coordinates of both intersection
points, 1 mark for integrating both functions, 1 mark for
substituting the limits, 1 mark for at least one correct area,
1 mark for the correct answer]*

Q5 Solve $\cos x = 0.5 \Rightarrow x = \cos^{-1} 0.5 = \frac{\pi}{3}$ and $\frac{5\pi}{3}$.

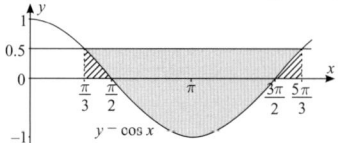

The area above the x-axis is the area of the rectangle
between $\frac{\pi}{3}$ and $\frac{5\pi}{3}$, minus the striped areas.
Rectangle $= (\frac{5\pi}{3} - \frac{\pi}{3}) \times 0.5 = \frac{2\pi}{3}$
One striped area $= \int_{\frac{\pi}{3}}^{\frac{\pi}{2}} \cos x \, dx = [\sin x]_{\frac{\pi}{3}}^{\frac{\pi}{2}}$
$$= \sin\left(\frac{\pi}{2}\right) - \sin\left(\frac{\pi}{3}\right) = 1 - \frac{\sqrt{3}}{2}$$
The two striped areas are the same (because of the
symmetry of the cosine curve). So the area above
the x-axis is: $\frac{2\pi}{3} - 2\left(1 - \frac{\sqrt{3}}{2}\right) = \frac{2\pi}{3} - 2 + \sqrt{3}$
Integrate to find the area below the x-axis:
$$\int_{\frac{\pi}{2}}^{\frac{3\pi}{2}} \cos x \, dx = [\sin x]_{\frac{\pi}{2}}^{\frac{3\pi}{2}} = \sin\left(\frac{3\pi}{2}\right) - \sin\left(\frac{\pi}{2}\right) = -1 - 1 = -2$$
So the area below the x-axis is 2.
The shaded area is $\frac{2\pi}{3} - 2 + \sqrt{3} + 2 = \frac{2\pi}{3} + \sqrt{3}$

Q6 The area of each grey section is the integral of the upper function
minus the integral of the lower function. So the total grey area is:
$$\left(\int_0^\pi 5 \sin x \, dx - \int_0^\pi 4 \sin x \, dx\right)$$
$$+ \left(\int_0^\pi 3 \sin x \, dx - \int_0^\pi 2 \sin x \, dx\right) + \left(\int_0^\pi \sin x \, dx\right)$$
$$= 5 \int_0^\pi \sin x \, dx - 4 \int_0^\pi \sin x \, dx$$
$$+ 3 \int_0^\pi \sin x \, dx - 2 \int_0^\pi \sin x \, dx + \int_0^\pi \sin x \, dx$$
$$= (5 - 4 + 3 - 2 + 1) \int_0^\pi \sin x \, dx = 3 \int_0^\pi \sin x \, dx = 3[-\cos x]_0^\pi$$
$$= 3[(-\cos \pi) - (-\cos 0)] = 3[(-(-1)) - (-1)] = 3(1 + 1) = 6$$

Exercise 8.12 — Parametric Integration

Q1 **a)** $x = 3t^{-1} \Rightarrow \frac{dx}{dt} = -3t^{-2} = -\frac{3}{t^2}$
$y = 4t^2$
$$\int y \, dx = \int y \frac{dx}{dt} \, dt = \int (4t^2) \times \left(-\frac{3}{t^2}\right) dt = \int -12 \, dt$$
You could take the constant outside the integral
— i.e. this could be written as $-12 \int 1 \, dt$.

b) $x = \tan 5\theta \implies \dfrac{dx}{d\theta} = 5\sec^2 5\theta$
$y = \sec^2 5\theta$

$$\int y\,dx = \int y\,\dfrac{dx}{d\theta}\,d\theta = \int (\sec^2 5\theta)(5\sec^2 5\theta)\,d\theta$$
$$= \int 5\sec^4 5\theta\,d\theta$$

Q2 a) $x = (4t - 5)^2 \implies \dfrac{dx}{dt} = 2(4t - 5) \times 4 = 32t - 40$
$y = t^2 - 3t$

$$\int y\,dx = \int y\,\dfrac{dx}{dt}\,dt = \int (t^2 - 3t) \times (32t - 40)\,dt$$
$$= \int (32t^3 - 136t^2 + 120t)\,dt = 8t^4 - \dfrac{136}{3}t^3 + 60t^2 + C$$

b) $x = t^2 + 3 \implies \dfrac{dx}{dt} = 2t$
$y = 4t - 1$

$$\int y\,dx = \int y\,\dfrac{dx}{dt}\,dt = \int (4t - 1) \times (2t)\,dt$$
$$= \int (8t^2 - 2t)\,dt = \dfrac{8}{3}t^3 - t^2 + C$$

Q3 $x = 3t^2 \implies \dfrac{dx}{dt} = 6t$
$y = \dfrac{5}{t}$

So $y\dfrac{dx}{dt} = \dfrac{5}{t} \times 6t = 30$

When $x = 75$, $3t^2 = 75 \implies t = 5$ (since $t > 0$).
When $x = 3$, $3t^2 = 3 \implies t = 1$ (since $t > 0$).

$$\int_3^{75} y\,dx = \int_1^5 y\,\dfrac{dx}{dt}\,dt = \int_1^5 30\,dt = 30\int_1^5 1\,dt$$
$$= 30[t]_1^5 = 30(5 - 1) = 120$$

Q4 Alter the limits from x-values to t-values:
$x = 1 \implies 1 = \dfrac{4}{t} \implies t = 4$, similarly $x = 4 \implies t = 1$
$$\int y\,dx = \int y\,\dfrac{dx}{dt}\,dt \text{ and } \dfrac{dx}{dt} = -\dfrac{4}{t^2}$$
$$\int_1^4 y\,dx = \int_4^1 t^2 \times \left(-\dfrac{4}{t^2}\right)dt = [-4t]_4^1 = (-4 \times 1) - (-4 \times 4) = 12$$

[6 marks available — 1 mark for altering limits, 1 mark for finding $\dfrac{dx}{dt}$, 1 mark for using the formula for integrating parametric equations, 1 mark for integrating correctly, 1 mark for substituting the limits, 1 mark for the correct answer]

Q5 a) When $x = 8$:
$4t(t + 1) = 8 \implies t(t + 1) = 2 \implies t^2 + t - 2 = 0$
$(t - 1)(t + 2) = 0 \implies t = 1$ ($t > 0$).

When $x = 120$:
$4t(t + 1) = 120 \implies t(t + 1) = 30 \implies t^2 + t - 30 = 0$
$(t - 5)(t + 6) = 0 \implies t = 5$ ($t > 0$).

b) $x = 4t^2 + 4t \implies \dfrac{dx}{dt} = 8t + 4$
$y = 3t^3$

So $y\dfrac{dx}{dt} = 24t^4 + 12t^3$

$$\int_8^{120} y\,dx = \int_1^5 y\,\dfrac{dx}{dt}\,dt$$
$$= \int_1^5 24t^4 + 12t^3\,dt = \left[\dfrac{24t^5}{5} + 3t^4\right]_1^5$$
$$= (15\,000 + 1875) - \left(\dfrac{24}{5} + 3\right) = \dfrac{84\,336}{5} = 16\,867.2$$

Q6 Alter the limits from x-values to t-values:
$x = 0 \implies 0 = \cos 2t \implies t = \dfrac{1}{2}\cos^{-1} 0 = \dfrac{\pi}{4}$
$x = \dfrac{1}{2} \implies \dfrac{1}{2} = \cos 2t \implies t = \dfrac{1}{2}\cos^{-1}\dfrac{1}{2} = \dfrac{\pi}{6}$
$$\int y\,dx = \int y\,\dfrac{dx}{dt}\,dt \text{ and } \dfrac{dx}{dt} = -2\sin 2t$$
$$\int_0^{\frac{1}{2}} y\,dx = \int_{\frac{\pi}{4}}^{\frac{\pi}{6}} (2\operatorname{cosec} t)(-2\sin 2t)\,dt$$
$$= \int_{\frac{\pi}{4}}^{\frac{\pi}{6}} -4\dfrac{\sin 2t}{\sin t}\,dt = \int_{\frac{\pi}{4}}^{\frac{\pi}{6}} -4\dfrac{2(\sin t)(\cos t)}{\sin t}\,dt$$
$$= \int_{\frac{\pi}{4}}^{\frac{\pi}{6}} -8\cos t\,dt = [-8\sin t]_{\frac{\pi}{4}}^{\frac{\pi}{6}}$$
$$= (-8\sin\dfrac{\pi}{6}) - (-8\sin\dfrac{\pi}{4})$$
$$= (-8 \times \dfrac{1}{2}) - (-8 \times \dfrac{1}{\sqrt{2}})$$
$$= 4\sqrt{2} - 4\,(= 4(\sqrt{2} - 1))$$

[7 marks available — 1 mark for altering limits, 1 mark for finding $\dfrac{dx}{dt}$, 1 mark for using the formula for integrating parametric equations, 1 mark for using appropriate trig identity, 1 mark for integrating correctly, 1 mark for substituting the limits, 1 mark for the correct answer]

Q7 a) Alter the limits from x-values to t-values:
$x = 0 \implies 0 = 2t - 6 \implies t = 3$
$x = 2 \implies 2 = 2t - 6 \implies t = 4$
$$\int y\,dx = \int y\,\dfrac{dx}{dt}\,dt \text{ and } \dfrac{dx}{dt} = 2$$
$$\int_0^2 y\,dx = \int_3^4 (t^2 - 5t + 6) \times 2\,dt = \left[2\left(\dfrac{1}{3}t^3 - \dfrac{5}{2}t^2 + 6t\right)\right]_3^4$$
$$= 2\left(\dfrac{1}{3}4^3 - \dfrac{5}{2}4^2 + 6(4)\right) - 2\left(\dfrac{1}{3}3^3 - \dfrac{5}{2}3^2 + 6(3)\right)$$
$$= \dfrac{32}{3} - 9 = \dfrac{5}{3}$$

[6 marks available — 1 mark for altering limits, 1 mark for finding $\dfrac{dx}{dt}$, 1 mark for using the formula to integrate parametric equations, 1 mark for integrating, 1 mark for substituting the limits, 1 mark for the correct answer] Alternatively, you could have done some rearranging and substituting to express y in terms of x, then integrated with respect to x. You would still get the marks if you solved it this way as long as the question doesn't specify using parametric integration.

b) The shaded region and the area under the curve form a trapezium. To find the area of the trapezium, start by finding where the normal crosses the y-axis.
$y = t^2 - 5t + 6 \implies \dfrac{dy}{dt} = 2t - 5$
So $\dfrac{dy}{dx} = \dfrac{dy}{dt} \div \dfrac{dx}{dt} = (2t - 5) \div 2 = t - \dfrac{5}{2}$
When $x = 2$, $t = 4$
$\implies y = 4^2 - 5(4) + 6 = 2$ and $\dfrac{dy}{dx} = 4 - \dfrac{5}{2} = \dfrac{3}{2}$

So the normal has gradient $-\dfrac{2}{3}$ and passes through $(2, 2)$.
$y = -\dfrac{2}{3}x + c \implies 2 = -\dfrac{2}{3} \times 2 + c \implies c = \dfrac{10}{3}$
So the y-intercept is $y = \dfrac{10}{3}$.

Area of shaded region
$= $ area of trapezium $-$ area under curve
$= 0.5(2 + \dfrac{10}{3}) \times 2 - \dfrac{5}{3} = \dfrac{11}{3}$

[6 marks available — 1 mark for finding an expression for $\dfrac{dy}{dx}$, 1 mark for finding the y-coordinate of the intersection point, 1 mark for finding the gradient of the normal, 1 mark for finding the y-intercept, 1 mark for finding the area of the trapezium, 1 mark for the correct answer]

Answers

Exercise 8.13 — Integration by Substitution

Q1

a) $u = x + 3 \Rightarrow \frac{du}{dx} = 1 \Rightarrow dx = du$

So $\int 12(x + 3)^5 \, dx = \int 12u^5 \, du = 2u^6 + C = 2(x + 3)^6 + C$

You could also have solved this by using the rule for $(ax + b)^n$.

b) $u = 11 - x \Rightarrow \frac{du}{dx} = -1 \Rightarrow dx = -du$

So $\int (11 - x)^4 \, dx = -\int u^4 \, du = -\frac{1}{5}u^5 + C = -\frac{1}{5}(11 - x)^5 + C$

c) $u = x^2 + 4 \Rightarrow \frac{du}{dx} = 2x \Rightarrow dx = \frac{1}{2x} du$

So $\int 24x(x^2 + 4)^3 \, dx = \int 24x \times u^3 \times \frac{1}{2x} \, du$

$= \int 12u^3 \, du = 3u^4 + C = 3(x^2 + 4)^4 + C$

d) $u = \sin x \Rightarrow \frac{du}{dx} = \cos x \Rightarrow dx = \frac{1}{\cos x} du$

So $\int \sin^5 x \cos x \, dx = \int u^5 \cos x \times \frac{1}{\cos x} \, du$

$= \int u^5 \, du = \frac{1}{6} u^6 + C = \frac{1}{6} \sin^6 x + C$

e) $u = x - 1 \Rightarrow \frac{du}{dx} = 1 \Rightarrow dx = du$

and $u = x - 1 \Rightarrow x = u + 1$

So $\int x(x - 1)^5 \, dx = \int (u + 1)u^5 \, du = \int u^6 + u^5 \, du$

$= \frac{1}{7}u^7 + \frac{1}{6}u^6 + C = \frac{1}{7}(x - 1)^7 + \frac{1}{6}(x - 1)^6 + C$

f) $u = x^2 - 3 \Rightarrow \frac{du}{dx} = 2x \Rightarrow dx = \frac{1}{2x} du$

and $u = x^2 - 3 \Rightarrow x^2 = u + 3$

So $\int 4x^3(x^2 - 3)^6 \, dx = \int 4x^3 u^6 \times \frac{1}{2x} \, du = \int 2x^2 u^6 \, du$

$= \int 2(u + 3)u^6 \, du = \int 2u^7 + 6u^6 \, du = \frac{2}{8}u^8 + \frac{6}{7}u^7 + C$

$= \frac{1}{4}(x^2 - 3)^8 + \frac{6}{7}(x^2 - 3)^7 + C$

Q2

a) Let $u = x + 2 \Rightarrow \frac{du}{dx} = 1 \Rightarrow dx = du$

So $\int 21(x + 2)^6 \, dx = \int 21u^6 \, du = 3u^7 + C = 3(x + 2)^7 + C$

b) Let $u = 5x + 4 \Rightarrow \frac{du}{dx} = 5 \Rightarrow dx = \frac{1}{5} du$

So $\int (5x + 4)^3 \, dx = \int \frac{1}{5} u^3 \, du = \frac{1}{20} u^4 + C = \frac{1}{20}(5x + 4)^4 + C$

c) Let $u = 2x + 3 \Rightarrow \frac{du}{dx} = 2 \Rightarrow dx = \frac{1}{2} du$

and $u = 2x + 3 \Rightarrow x = \frac{u - 3}{2}$

So $\int x(2x + 3)^3 \, dx = \int \frac{u - 3}{2} \times u^3 \times \frac{1}{2} \, du$

$= \frac{1}{4} \int u^4 - 3u^3 \, du = \frac{1}{4}\left(\frac{1}{5}u^5 - \frac{3}{4}u^4\right) + C$

$= \frac{1}{20}u^5 - \frac{3}{16}u^4 + C = \frac{1}{20}(2x + 3)^5 - \frac{3}{16}(2x + 3)^4 + C$

d) Let $u = x^2 - 5 \Rightarrow \frac{du}{dx} = 2x \Rightarrow dx = \frac{1}{2x} du$

So $\int 24x(x^2 - 5)^7 \, dx = \int 24x \times u^7 \times \frac{1}{2x} \, du = \int 12u^7 \, du$

$= \frac{3}{2}u^8 + C = \frac{3}{2}(x^2 - 5)^8 + C$

Q3

a) $u = \sqrt{x + 1} \Rightarrow \frac{du}{dx} = \frac{1}{2\sqrt{x + 1}} = \frac{1}{2u} \Rightarrow dx = 2u \, du$

$u = \sqrt{x + 1} \Rightarrow x = u^2 - 1$

So $\int 6x\sqrt{x + 1} \, dx = \int 6(u^2 - 1) \times u \times 2u \, du$

$= \int 12u^4 - 12u^2 \, du = \frac{12}{5}u^5 - 4u^3 + C$

$= \frac{12}{5}(\sqrt{x + 1})^5 - 4(\sqrt{x + 1})^3 + C$

b) $u = \sqrt{4 - x} \Rightarrow \frac{du}{dx} = -\frac{1}{2\sqrt{4 - x}} = -\frac{1}{2u} \Rightarrow dx = -2u \, du$

$u = \sqrt{4 - x} \Rightarrow x = 4 - u^2$

So $\int \frac{x}{\sqrt{4 - x}} \, dx = \int \frac{4 - u^2}{u} \times -2u \, du$

$= \int -2(4 - u^2) \, du = \int 2u^2 - 8 \, du$

$= \frac{2}{3}u^3 - 8u + C = \frac{2}{3}(\sqrt{4 - x})^3 - 8(\sqrt{4 - x}) + C$

c) $u = \ln x \Rightarrow \frac{du}{dx} = \frac{1}{x} \Rightarrow dx = x \, du$

So $\int \frac{15(\ln x)^4}{x} \, dx = \int \frac{15u^4}{x} \times x \, du = \int 15u^4 \, du$

$= 3u^5 + C = 3(\ln x)^5 + C$

d) $u = \ln(x^2) \Rightarrow \frac{du}{dx} = \frac{2}{x} \Rightarrow dx = \frac{x}{2} du$

So $\int \frac{3}{x(\ln(x^2))^3} \, dx = \int \frac{3}{xu^3} \times \frac{x}{2} \, du$

$= \int \frac{3}{2}u^{-3} \, du = -\frac{3}{4}u^{-2} + C = -\frac{3}{4(\ln(x^2))^2} + C$

Q4 *You might use different substitutions to the ones shown below, but as long as your substitution is valid and you get the same final answer then it's okay.*

a) Let $u = \sqrt{2x - 1} \Rightarrow \frac{du}{dx} = \frac{1}{\sqrt{2x - 1}} = \frac{1}{u} \Rightarrow dx = u \, du$

and $u = \sqrt{2x - 1} \Rightarrow x = \frac{u^2 + 1}{2}$

So $\int \frac{4x}{\sqrt{(2x - 1)}} \, dx = \int 4\left(\frac{u^2 + 1}{2}\right) \times \frac{1}{u} \times u \, du$

$= \int 2u^2 + 2 \, du = \frac{2}{3}u^3 + 2u + C$

$= \frac{2}{3}(\sqrt{2x - 1})^3 + 2(\sqrt{2x - 1}) + C$

b) Let $u = 4 - \sqrt{x} \Rightarrow \frac{du}{dx} = -\frac{1}{2\sqrt{x}} \Rightarrow -2\sqrt{x} \, du = dx$

and $u = 4 - \sqrt{x} \Rightarrow x = (4 - u)^2$

$\Rightarrow dx$ can be written $-2(4 - u) \, du = (2u - 8)du$

So $\int \frac{1}{4 - \sqrt{x}} \, dx = \int \frac{1}{u}(2u - 8) \, du$

$= \int 2 - \frac{8}{u} \, du = 2u - 8\ln|u| + C$

$= 2(4 - \sqrt{x}) - 8\ln|4 - \sqrt{x}| + C = -2\sqrt{x} - 8\ln|4 - \sqrt{x}| + C$

You can leave out any constant terms that appear after you integrate (like the 8 you get out of the term $2(4 - \sqrt{x})$ here) — they just get absorbed into the constant of integration, C.

c) Let $u = 1 + e^x \Rightarrow \frac{du}{dx} = e^x \Rightarrow dx = \frac{1}{e^x} du$

and $u = 1 + e^x \Rightarrow e^x = u - 1$

So $\int \frac{e^{2x}}{1 + e^x} \, dx = \int \frac{e^{2x}}{u} \times \frac{1}{e^x} \, du = \int \frac{e^x}{u} \, du = \int \frac{u - 1}{u} \, du$

$= \int 1 - \frac{1}{u} \, du = u - \ln|u| + c = 1 + e^x - \ln|1 + e^x| + c$

$= e^x - \ln(1 + e^x) + C$

You can remove the modulus as e^x is always positive, so $1 + e^x > 1$.

Q5

a) Compare $3x^2 - 1$ to $3u - 1$, which gives $u = x^2$.

$\Rightarrow \frac{du}{dx} = 2x \Rightarrow dx = \frac{1}{2x} du$

$\int 2x\sqrt{3x^2 - 1} \, dx = \int 2x\sqrt{3u - 1} \frac{1}{2x} \, du = \int (3u - 1)^{0.5} \, du$

[2 marks available — 1 mark for the correct choice of f(x), 1 mark for showing the given result]

b) $\int (3u - 1)^{0.5} \, du = \frac{2}{9}(3u - 1)^{1.5} + C$ *[1 mark]*

c) $\frac{2}{9}(3x^2 - 1)^{1.5} + C$ *[1 mark]*

Q6 Let $u = 3^x + 2 \Rightarrow \dfrac{du}{dx} = 3^x \ln 3 \Rightarrow dx = \dfrac{1}{3^x \ln 3} \, du$

So $\displaystyle\int \dfrac{3^x}{(3^x + 2)^3} \, dx = \int \dfrac{3^x}{u^3} \times \dfrac{1}{3^x \ln 3} \, du = \int \dfrac{1}{u^3 \ln 3} \, du$

$= \dfrac{-u^{-2}}{2 \ln 3} + C = -\dfrac{(3^x + 2)^{-2}}{\ln 9} + C$

[5 marks available — 1 mark for differentiating the expression for u, 1 mark for correctly substituting for an integral in terms of u, 1 mark for correctly integrating with respect to u, 1 mark for substituting back in terms of x, 1 mark for the correct answer]

Q7 Let $u = x^2 - 9 \Rightarrow \dfrac{du}{dx} = 2x \Rightarrow x \, dx = \dfrac{1}{2} \, du$

So $\displaystyle\int x(x^2 - 9)^{0.5} \, dx = \int \dfrac{1}{2} u^{0.5} du = \dfrac{1}{3} u^{1.5} + C = \dfrac{1}{3}(x^2 - 9)^{1.5} + C$

[4 marks available — 1 mark for differentiating the expression for u, 1 mark for correctly substituting for an integral in terms of u, 1 mark for correctly integrating with respect to u, 1 mark for substituting back in terms of x to obtain the required result]

Q8 a) Let $u = \sqrt{x} \Rightarrow \dfrac{du}{dx} = 0.5x^{-0.5} \Rightarrow dx = 2\sqrt{x} \, du = 2u \, du$

$\displaystyle\int \dfrac{4\sqrt{x}}{x(3 - \sqrt{x})} \, dx = \int \dfrac{4u}{u^2(3 - u)} 2u \, du = \int \dfrac{8}{3 - u} \, du$

[2 marks available — 1 mark for differentiating the expression for u, 1 mark for correctly substituting to obtain the given result]

b) $f(x) = \displaystyle\int \dfrac{8}{3 - u} \, du = -8 \ln|3 - u| + C = -8 \ln|3 - \sqrt{x}| + C$

$f(4) = \ln 2 \Rightarrow -8 \ln|3 - \sqrt{4}| + C = \ln 2 \Rightarrow C = \ln 2$

So $f(x) = -8 \ln|3 - \sqrt{x}| + \ln 2$

$= \ln \dfrac{1}{(3 - \sqrt{x})^8} + \ln 2 = \ln \dfrac{2}{(3 - \sqrt{x})^8}$

[4 marks available — 1 mark for integrating with respect to u, 1 mark for finding C, 1 mark for attempting to use log laws to rearrange, 1 mark for correctly rearranging to obtain the given result]

Q9 Let $u = 2 + e^{2x} \Rightarrow \dfrac{du}{dx} = 2e^{2x} \Rightarrow dx = \dfrac{1}{2e^{2x}} \, du$

$\displaystyle\int \dfrac{e^{4x}}{2 + e^{2x}} \, dx = \int \dfrac{(u - 2)^2}{u} \times \dfrac{1}{2(u - 2)} \, du = \int \dfrac{u - 2}{2u} \, du$

$= \displaystyle\int \dfrac{1}{2} - \dfrac{1}{u} \, du = 0.5u - \ln|u| + c$

$= 0.5(2 + e^{2x}) - \ln|2 + e^{2x}| + c = 1 + 0.5e^{2x} - \ln|2 + e^{2x}| + c$

$= 0.5e^{2x} - \ln|2 + e^{2x}| + C$

[7 marks available — 1 mark for choosing a suitable substitution, 1 mark for differentiating the expression for u, 1 mark for correctly substituting for an integral in terms of u, 1 mark for each term correctly integrated with respect to u, 1 mark for substituting back in terms of x, 1 mark for the correct answer]
The '1 + ' from the beginning of the expression has been incorporated into the ' + C' in the final answer, since C is just a number.

Q10 Let $u = f(x) \Rightarrow \dfrac{du}{dx} = f'(x) \Rightarrow dx = \dfrac{1}{f'(x)} \, du$

So $\displaystyle\int (n + 1)f'(x)[f(x)]^n \, dx = \int (n + 1)f'(x)u^n \times \dfrac{1}{f'(x)} \, du$

$= \displaystyle\int (n + 1)u^n \, du = u^{n+1} + C = [f(x)]^{n+1} + C$, as required

Q11 a) $x(x + 1)^3 = x^4 + 3x^3 + 3x^2 + x$
[2 marks available — 1 mark for at least two terms correct, 1 mark for all terms correct]

b) Let $u = x + 1 \Rightarrow \dfrac{du}{dx} = 1 \Rightarrow dx = du$

$\displaystyle\int x(x + 1)^3 dx = \int (u - 1)u^3 du = \dfrac{u^5}{5} - \dfrac{u^4}{4} + C$

$= \dfrac{4u^5}{20} - \dfrac{5u^4}{20} + C = \dfrac{u^4}{20}(4u - 5) + C$

[4 marks available — 1 mark for differentiating the expression for u, 1 mark for correctly substituting for an integral in terms of u, 1 mark for integrating correctly, 1 mark for rearranging to obtain the given result]

c) Substituting $u = x + 1$ into the result from part b):

$\displaystyle\int x(x + 1)^3 dx = \dfrac{u^4}{20}(4u - 5) + C_1$

$= \dfrac{(x + 1)^4}{20}(4x + 4 - 5) + C_1 = \dfrac{(4x - 1)(x + 1)^4}{20} + C_1$

So $\dfrac{(4x - 1)(x + 1)^4}{20} + C_1 = \displaystyle\int x(x + 1)^3 dx$

$= \displaystyle\int x^4 + 3x^3 + 3x^2 + x \, dx = \dfrac{x^5}{5} + \dfrac{3x^4}{4} + x^3 + \dfrac{x^2}{2} + C_2$

Therefore $\dfrac{(4x - 1)(x + 1)^4}{20} = \dfrac{x^5}{5} + \dfrac{3x^4}{4} + x^3 + \dfrac{x^2}{2} + C$

The constants of integration are different for the expanded and non-expanded expressions, so they don't cancel out when you equate the expressions. To find the combined constant, set x equal to zero as that makes all x-terms in the result from integrating the expanded expression equal to zero.

So when $x = 0$, $\dfrac{(-1)(1)^4}{20} = 0 + C \Rightarrow C = -\dfrac{1}{20}$

Therefore $\dfrac{(4x - 1)(x + 1)^4}{20} = \dfrac{x^5}{5} + \dfrac{3x^4}{4} + x^3 + \dfrac{x^2}{2} - \dfrac{1}{20}$

[4 marks available — 1 mark for using the result from b), 1 mark for at least two terms of the binomial expansion integrated correctly, 1 mark for all terms integrated correctly, 1 mark for the correct answer]

Exercise 8.14 — Integration by Substitution — Definite Integrals

Q1 a) $u = 3x - 2 \Rightarrow \dfrac{du}{dx} = 3 \Rightarrow dx = \dfrac{1}{3} \, du$

$x = \dfrac{2}{3} \Rightarrow u = 2 - 2 = 0$

$x = 1 \Rightarrow u = 3 - 2 = 1$

So $\displaystyle\int_{\frac{2}{3}}^{1}(3x - 2)^4 \, dx = \int_0^1 \dfrac{1}{3} u^4 \, du = \left[\dfrac{1}{15} u^5\right]_0^1 = \dfrac{1}{15}$

b) $u = x + 3 \Rightarrow \dfrac{du}{dx} = 1 \Rightarrow dx = 1 \, du$

$u = x + 3 \Rightarrow x = u - 3$

$x = -2 \Rightarrow u = -2 + 3 = 1$

$x = 1 \Rightarrow u = 1 + 3 = 4$

So $\displaystyle\int_{-2}^{1} 2x(x + 3)^4 \, dx = \int_1^4 2(u - 3) \times u^4 \times 1 \, du$

$= \displaystyle\int_1^4 (2u - 6)u^4 \, du = \int_1^4 (2u^5 - 6u^4) \, du = \left[\dfrac{u^6}{3} - \dfrac{6u^5}{5}\right]_1^4$

$= \left(\dfrac{4^6}{3} - \dfrac{6(4)^5}{5}\right) - \left(\dfrac{1^6}{3} - \dfrac{6(1)^5}{5}\right) = \dfrac{687}{5} \; (= 137.4)$

c) $u = \sin x \Rightarrow \dfrac{du}{dx} = \cos x \Rightarrow dx = \dfrac{1}{\cos x} \, du$

$x = 0 \Rightarrow u = \sin(0) = 0$

$x = \dfrac{\pi}{6} \Rightarrow u = \sin \dfrac{\pi}{6} = \dfrac{1}{2}$

So $\displaystyle\int_0^{\frac{\pi}{6}} 8 \sin^3 x \cos x \, dx = \int_0^{\frac{1}{2}} 8u^3 \cos x \times \dfrac{1}{\cos x} \, du$

$= \displaystyle\int_0^{\frac{1}{2}} 8u^3 \, du = [2u^4]_0^{\frac{1}{2}} = 2\left(\dfrac{1}{2}\right)^4 - 0 = \dfrac{1}{8}$

d) $u = \sqrt{x + 1} \Rightarrow \dfrac{du}{dx} = \dfrac{1}{2\sqrt{x + 1}} = \dfrac{1}{2u} \Rightarrow dx = 2u \, du$

$u = \sqrt{x + 1} \Rightarrow x = u^2 - 1$

$x = 0 \Rightarrow u = \sqrt{x + 1} = \sqrt{1} = 1$

$x = 3 \Rightarrow u = \sqrt{4} = 2$

So $\displaystyle\int_0^3 x\sqrt{x + 1} \, dx = \int_1^2 (u^2 - 1) \times u \times 2u \, du$

$= \displaystyle\int_1^2 2u^4 - 2u^2 \, du = \left[\dfrac{2}{5} u^5 - \dfrac{2}{3} u^3\right]_1^2$

$= \left[\left(\dfrac{64}{5} - \dfrac{16}{3}\right) - \left(\dfrac{2}{5} - \dfrac{2}{3}\right)\right] = \dfrac{116}{15}$

Answers

In Q2, 3 and 5 below, there are different substitutions you could use — again, if you get the final answer right and you use a valid substitution, then your method is fine.

Q2 a) Let $u = x^2 - 3 \Rightarrow \dfrac{du}{dx} = 2x \Rightarrow dx = \dfrac{1}{2x}\,du$

$x = 2 \Rightarrow u = 4 - 3 = 1$

$x = \sqrt{5} \Rightarrow u = 5 - 3 = 2$

So $\displaystyle\int_2^{\sqrt{5}} x(x^2-3)^4\,dx = \int_1^2 x \times u^4 \times \dfrac{1}{2x}\,du$

$= \displaystyle\int_1^2 \dfrac{1}{2}u^4\,du = \left[\dfrac{1}{10}u^5\right]_1^2 = \dfrac{32}{10} - \dfrac{1}{10} = \dfrac{31}{10}$ $(= 3.1)$

b) Let $u = 3x - 4 \Rightarrow \dfrac{du}{dx} = 3 \Rightarrow dx = \dfrac{1}{3}\,du$

and $u = 3x - 4 \Rightarrow x = \dfrac{u+4}{3}$

$x = 1 \Rightarrow u = 3 - 4 = -1$

$x = 2 \Rightarrow u = 6 - 4 = 2$

So $\displaystyle\int_1^2 x(3x-4)^3\,dx = \int_{-1}^2 \dfrac{u+4}{3} \times u^3 \times \dfrac{1}{3}\,du$

$= \dfrac{1}{9}\displaystyle\int_{-1}^2 u^4 + 4u^3\,du = \dfrac{1}{9}\left[\dfrac{u^5}{5} + u^4\right]_{-1}^2$

$= \dfrac{1}{9}\left[\left(\dfrac{32}{5} + 16\right) - \left(\dfrac{-1}{5} + 1\right)\right] = \dfrac{1}{9}\left[\dfrac{108}{5}\right] = \dfrac{12}{5}$ $(= 2.4)$

c) Let $u = \sqrt{x-1} \Rightarrow \dfrac{du}{dx} = \dfrac{1}{2\sqrt{x-1}} = \dfrac{1}{2u} \Rightarrow dx = 2u\,du$

and $u = \sqrt{x-1} \Rightarrow x = u^2 + 1$

Using $u = \sqrt{x-1}$, $x = 2 \Rightarrow u = \sqrt{1} = 1$

and $x = 10 \Rightarrow u = \sqrt{9} = 3$

So $\displaystyle\int_2^{10} \dfrac{x}{\sqrt{x-1}}\,dx = \int_1^3 \dfrac{u^2+1}{u} \times 2u\,du = \int_1^3 2u^2 + 2\,du$

$= \left[\dfrac{2}{3}u^3 + 2u\right]_1^3 = \left[(18+6) - \left(\dfrac{2}{3} + 2\right)\right] = \dfrac{64}{3}$

Q3 Let $u = 3 - \sqrt{x} \Rightarrow \dfrac{du}{dx} = -\dfrac{1}{2\sqrt{x}} \Rightarrow dx = -2\sqrt{x}\,du$

$u = 3 - \sqrt{x} \Rightarrow x = (3-u)^2 \Rightarrow dx = (2u-6)\,du$

So $x = 1 \Rightarrow u = 3 - 1 = 2$ and $x = 4 \Rightarrow u = 3 - 2 = 1$

So $\displaystyle\int_1^4 \dfrac{1}{3-\sqrt{x}}\,dx = \int_2^1 \dfrac{1}{u} \times (2u-6)\,du = \int_2^1 2 - \dfrac{6}{u}\,du$

$= \left[2u - 6\ln|u|\right]_2^1 = (2 - 6\ln 1) - (4 - 6\ln 2)$

$= 2 - 0 - 4 + 6\ln 2 = -2 + 6\ln 2$

You could have put 2 as the upper limit and 1 as the lower limit in the integral with respect to u, and put a minus sign in front. Both methods give the right answer, but whichever you use, be careful not to lose any minus signs.

Q4 $u = 1 + e^x \Rightarrow \dfrac{du}{dx} = e^x \Rightarrow dx = \dfrac{1}{e^x}\,du$

$x = 0 \Rightarrow u = 1 + e^0 = 2$

$x = 1 \Rightarrow u = 1 + e$

So $\displaystyle\int_0^1 2e^x(1+e^x)^3\,dx = \int_2^{1+e} 2e^x u^3 \times \dfrac{1}{e^x}\,du$

$= \displaystyle\int_2^{1+e} 2u^3\,du = \left[\dfrac{u^4}{2}\right]_2^{1+e} = \dfrac{(1+e)^4}{2} - 8 = 87.6$ (1 d.p.)

Q5 Let $u = \sqrt{3x+1} \Rightarrow \dfrac{du}{dx} = \dfrac{3}{2\sqrt{3x+1}} = \dfrac{3}{2u} \Rightarrow dx = \dfrac{2}{3}u\,du$

and $u = \sqrt{3x+1} \Rightarrow x = \dfrac{u^2-1}{3}$

$x = 1 \Rightarrow u = \sqrt{4} = 2$

$x = 5 \Rightarrow u = \sqrt{16} = 4$

So $\displaystyle\int_1^5 \dfrac{x}{\sqrt{3x+1}}\,dx = \int_2^4 \dfrac{u^2-1}{3u} \times \dfrac{2}{3}u\,du = \dfrac{2}{9}\int_2^4 u^2 - 1\,du$

$= \dfrac{2}{9}\left[\dfrac{1}{3}u^3 - u\right]_2^4 = \dfrac{2}{9}\left[\left(\dfrac{64}{3} - 4\right) - \left(\dfrac{8}{3} - 2\right)\right] = \dfrac{2}{9}\left[\dfrac{50}{3}\right] = \dfrac{100}{27}$

Q6 Let $u = x^4 + 12 \Rightarrow \dfrac{du}{dx} = 4x^3 \Rightarrow dx = \dfrac{1}{4x^3}\,du$

$x = 0 \Rightarrow u = 0 + 12 = 12$

$x = 1 \Rightarrow u = 1 + 12 = 13$

So $\displaystyle\int_0^1 \dfrac{x^3}{\sqrt{x^4+12}}\,dx = \int_{12}^{13} \dfrac{1}{4u^{\frac{1}{2}}}\,du = \left[\dfrac{1}{2}u^{\frac{1}{2}}\right]_{12}^{13}$

$= \dfrac{1}{2}(\sqrt{13} - \sqrt{12}) = 0.070724... = 0.0707$ (3 s.f.)

[7 marks available — 1 mark for a suitable substitution, 1 mark for finding $\dfrac{du}{dx}$, 1 mark for converting limits to be in terms of u, 1 mark for a correct integral in terms of u, 1 mark for correctly integrating with respect to u, 1 mark for substituting in the u-limits, 1 mark for the correct answer]

Q7 a) Let $u = 1 + \sqrt{x} \Rightarrow \dfrac{du}{dx} = \dfrac{1}{2}x^{-\frac{1}{2}} \Rightarrow dx = 2\sqrt{x}\,du$

$\displaystyle\int \dfrac{5x}{1+\sqrt{x}}\,dx = \int \dfrac{5(u-1)^2}{u} \times 2(u-1)\,du$

$= \displaystyle\int \dfrac{10(u-1)^3}{u}\,du$

$= \displaystyle\int \dfrac{10u^3 - 30u^2 + 30u - 10}{u}\,du$

$= \displaystyle\int 10u^2 - 30u + 30 - \dfrac{10}{u}\,du$

$= \dfrac{10u^3}{3} - 15u^2 + 30u - 10\ln|u| + C$

[5 marks available — 1 mark for finding $\dfrac{du}{dx}$, 1 mark for a correct integral in terms of u, 1 mark for writing the integral in terms of powers of u, 1 mark for all correct u^n terms in the answer, 1 mark for correct final answer]

b) $y = 2.5 \Rightarrow \dfrac{5x}{1+\sqrt{x}} = 2.5 \Rightarrow 5x = 2.5 + 2.5\sqrt{x}$

$\Rightarrow 2x - \sqrt{x} - 1 = 0 \Rightarrow (2\sqrt{x} + 1)(\sqrt{x} - 1)$

$\Rightarrow \sqrt{x} = 1$ or -0.5 (reject as $\sqrt{x} \geq 0$) $\Rightarrow x = 1$

When $x = 1$, $u = 1 + 1 = 2$.

$y = 16 \Rightarrow \dfrac{5x}{1+\sqrt{x}} = 16 \Rightarrow 5x = 16 + 16\sqrt{x}$

$\Rightarrow 5x - 16\sqrt{x} - 16 = 0$ $(5\sqrt{x} + 4)(\sqrt{x} - 4)$

$\Rightarrow \sqrt{x} = 4$ or -0.8 (reject as $\sqrt{x} \geq 0$) $\Rightarrow x = 16$

When $x = 16$, $u = 1 + 4 = 5$.

The required area is shown below, labelled A:

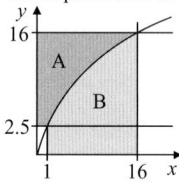

Area B $= \displaystyle\int_1^{16} \dfrac{5x}{1+\sqrt{x}}\,dx = \left[\dfrac{10u^3}{3} - 15u^2 + 30u - 10\ln|u|\right]_2^5$

$\left(\dfrac{1250}{3} - 375 + 150 - 10\ln 5\right) - \left(\dfrac{80}{3} - 60 + 60 - 10\ln 2\right)$

$= 165 - 10(\ln 5 - \ln 2) = 155.837...$

So area A $= (16 \times 16) -$ area B $- (2.5 \times 1)$

$= 256 - 155.837... - 2.5 = 97.7$ (3 s.f.)

[7 marks available — 1 mark for each limit in terms of x, 1 mark for both limits in terms of u, 1 mark for attempting to use the result obtained in part a), 1 mark for the area under the curve, 1 mark for a correct method to find the required area, 1 mark for the correct answer]

Q8 a) Let $u = 2x + 2 \Rightarrow \dfrac{du}{dx} = 2 \Rightarrow dx = \dfrac{1}{2}\,du$

$\displaystyle\int (2x+2)^4\,dx = \int u^4 \times \dfrac{1}{2}\,du = \dfrac{u^5}{10} + C = 0.1(2x+2)^5 + C$

[4 marks available — 1 mark for a suitable substitution, 1 mark for finding $\dfrac{du}{dx}$, 1 mark for attempting to integrate with respect to u, 1 mark for substituting back in terms of x to get the given result]

b) $(2x + 2)^4 \geq 0$, so the graph is entirely above or on the x-axis.
$\int_{-1}^{0} (2x + 2)^4 \, dx = [0.1(2x + 2)^5]_{-1}^{0} = 3.2 - 0 = 3.2$
[2 marks available — 1 mark for a correct method,
1 mark for the correct answer]

c) $y = (2x + 2)^4$ has a line of symmetry, $x = -1$.
Therefore, the lines $x = -2$ and $x = -1$ give the required area.
[2 marks available — 1 mark for a correct method,
1 mark for a correct answer]

d) The graph is translated 2 units to the left, then stretched horizontally by a scale factor of 0.5.
[3 marks available — 1 mark for each transformation,
1 mark for the correct order]
You could also write it as $(2(x + 1))^4$, and do it as a horizontal stretch by scale factor 0.5, followed by a translation 1 unit left — these give the same result as the transformations above.

e) $k = 0.5$, since $y = x^4$ has the same shape over -2 to 2 as $y = (2x + 2)^4$ has over -2 to 0, but it is twice as wide.
[2 marks available — 1 mark for attempting to relate to transformation from part d), 1 mark for the correct answer]

Exercise 8.15 — Integration by Substitution — Using Trig Identities

Q1 $x = \tan\theta \Rightarrow \dfrac{dx}{d\theta} = \sec^2\theta \Rightarrow dx = \sec^2\theta \, d\theta$

$x = 0 \Rightarrow \tan\theta = 0 \Rightarrow \theta = 0$
$x = 1 \Rightarrow \tan\theta = 1 \Rightarrow \theta = \dfrac{\pi}{4}$

So, using the identity $\sec^2\theta \equiv 1 + \tan^2\theta$:

$\int_0^1 \dfrac{1}{1 + x^2} \, dx = \int_0^{\frac{\pi}{4}} \dfrac{1}{1 + \tan^2\theta} \times \sec^2\theta \, d\theta$

$= \int_0^{\frac{\pi}{4}} \dfrac{\sec^2\theta}{\sec^2\theta} \, d\theta = \int_0^{\frac{\pi}{4}} 1 \, d\theta = [\theta]_0^{\frac{\pi}{4}} = \dfrac{\pi}{4}$

Q2 $u = \sin x \Rightarrow \dfrac{du}{dx} = \cos x \Rightarrow dx = \dfrac{1}{\cos x} \, du$

$x = 0 \Rightarrow u = \sin 0 = 0$
$x = \dfrac{\pi}{6} \Rightarrow u = \sin\dfrac{\pi}{6} = \dfrac{1}{2}$

So, using the identity $\sin 2x \equiv 2\sin x \cos x$:

$\int_0^{\frac{\pi}{6}} 3 \sin x \sin 2x \, dx \equiv \int_0^{\frac{\pi}{6}} 6 \sin^2 x \cos x \, dx$

$= \int_0^{\frac{1}{2}} 6u^2 \cos x \times \dfrac{1}{\cos x} \, du = \int_0^{\frac{1}{2}} 6u^2 \, du = [2u^3]_0^{\frac{1}{2}} = \dfrac{1}{4}$

Q3 $x = 2\sin\theta \Rightarrow \dfrac{dx}{d\theta} = 2\cos\theta \Rightarrow dx = 2\cos\theta \, d\theta$

$x = 1 \Rightarrow \sin\theta = \dfrac{1}{2} \Rightarrow \theta = \dfrac{\pi}{6}$

$x = \sqrt{3} \Rightarrow \sin\theta = \dfrac{\sqrt{3}}{2} \Rightarrow \theta = \dfrac{\pi}{3}$

So, using the identity $\sin^2\theta + \cos^2\theta \equiv 1$:

$\int_1^{\sqrt{3}} \dfrac{1}{(4 - x^2)^{\frac{3}{2}}} \, dx = \int_{\frac{\pi}{6}}^{\frac{\pi}{3}} \dfrac{1}{(4 - 4\sin^2\theta)^{\frac{3}{2}}} \times 2\cos\theta \, d\theta$

$= \int_{\frac{\pi}{6}}^{\frac{\pi}{3}} \dfrac{2\cos\theta}{(4 - 4 + 4\cos^2\theta)^{\frac{3}{2}}} \, d\theta = \int_{\frac{\pi}{6}}^{\frac{\pi}{3}} \dfrac{2\cos\theta}{8\cos^3\theta} \, d\theta$

$= \int_{\frac{\pi}{6}}^{\frac{\pi}{3}} \dfrac{1}{4} \sec^2\theta \, d\theta = \dfrac{1}{4}[\tan\theta]_{\frac{\pi}{6}}^{\frac{\pi}{3}} = \dfrac{1}{4}\left(\sqrt{3} - \dfrac{1}{\sqrt{3}}\right) = \dfrac{\sqrt{3}}{6}$

Q4 $x = \cos\theta \Rightarrow \dfrac{dx}{d\theta} = -\sin\theta \Rightarrow dx = -\sin\theta \, d\theta$

$x = \dfrac{1}{2} \Rightarrow \cos\theta = \dfrac{1}{2} \Rightarrow \theta = \dfrac{\pi}{3}$
$x = 1 \Rightarrow \cos\theta = 1 \Rightarrow \theta = 0$

So, using the identity $\sin^2\theta + \cos^2\theta \equiv 1$:

$\int_{\frac{1}{2}}^{1} \dfrac{1}{x^2\sqrt{1 - x^2}} \, dx = \int_{\frac{\pi}{3}}^{0} \dfrac{1}{\cos^2\theta\sqrt{1 - \cos^2\theta}} \times -\sin\theta \, d\theta$

$= \int_{\frac{\pi}{3}}^{0} -\dfrac{\sin\theta}{\cos^2\theta\sin\theta} \, d\theta = \int_0^{\frac{\pi}{3}} \dfrac{\sin\theta}{\cos^2\theta\sin\theta} \, d\theta$

$= \int_0^{\frac{\pi}{3}} \dfrac{1}{\cos^2\theta} \, d\theta = \int_0^{\frac{\pi}{3}} \sec^2\theta \, d\theta = [\tan\theta]_0^{\frac{\pi}{3}} = \sqrt{3} - 0 = \sqrt{3}$

Q5 $u = \sec^2 x \Rightarrow \dfrac{du}{dx} = 2\sec^2 x \tan x$

$\Rightarrow dx = \dfrac{1}{2\sec^2 x \tan x} \, du = \dfrac{1}{2u\tan x} \, du$

And using the identity
$\sec^2 x \equiv 1 + \tan^2 x \Rightarrow \tan^2 x \equiv \sec^2 x - 1 = u - 1$
$\int 2\tan^3 x \, dx = \int 2\tan x(u - 1) \times \dfrac{1}{2u\tan x} \, du = \int \dfrac{u - 1}{u} \, du$

$= \int 1 - \dfrac{1}{u} \, du = u - \ln|u| + C = \sec^2 x - \ln(\sec^2 x) + C$

Q6 a) $u = \cos\theta \Rightarrow du = -\sin\theta \, d\theta \Rightarrow d\theta = -\dfrac{1}{\sin\theta} \, du$

$\theta = \dfrac{\pi}{2} \Rightarrow u = \cos\dfrac{\pi}{2} \Rightarrow u = 0$
$\theta = -\pi \Rightarrow u = \cos -\pi \Rightarrow u = -1$

$\int_{-\pi}^{\frac{\pi}{2}} 3\sin\theta\cos^4\theta \, d\theta = \int_{-1}^{0} 3\sin\theta \, u^4 \times -\dfrac{1}{\sin\theta} \, du$

$= \int_{-1}^{0} -3u^4 \, du = \left[-\dfrac{3u^5}{5}\right]_{-1}^{0} = 0 - \dfrac{3}{5} = -\dfrac{3}{5}$

b) $u = \sec x \Rightarrow du = \sec x \tan x \, dx$

$\Rightarrow dx = \dfrac{1}{\sec x \tan x} \, du = \dfrac{1}{u\tan x} \, du$

$x = \dfrac{\pi}{3} \Rightarrow u = \sec\dfrac{\pi}{3} \Rightarrow u = 2$

$x = \dfrac{\pi}{4} \Rightarrow u = \sec\dfrac{\pi}{4} \Rightarrow u = \sqrt{2}$

$\int_{\frac{\pi}{4}}^{\frac{\pi}{3}} \sec^4 x \tan x \, dx = \int_{\sqrt{2}}^{2} u^4 \tan x \dfrac{1}{u\tan x} \, du$

$= \int_{\sqrt{2}}^{2} u^3 \, du = \left[\dfrac{u^4}{4}\right]_{\sqrt{2}}^{2}$

$= \dfrac{2^4}{4} - \dfrac{\sqrt{2}^4}{4} = 4 - 1 = 3$

Q7 $x = \cot\theta \Rightarrow dx = -\csc^2\theta \, d\theta$

$x = 1 \Rightarrow \tan\theta = 1 \Rightarrow \theta = \dfrac{\pi}{4}$
$x = \sqrt{3} \Rightarrow \tan\theta = \dfrac{1}{\sqrt{3}} \Rightarrow \theta = \dfrac{\pi}{6}$

$\int_1^{\sqrt{3}} \dfrac{4x}{\sqrt{1 + x^2}} \, dx = \int_{\frac{\pi}{4}}^{\frac{\pi}{6}} \dfrac{4\cot\theta}{\sqrt{1 + \cot^2\theta}} \times -\csc^2\theta \, d\theta$

$= \int_{\frac{\pi}{6}}^{\frac{\pi}{4}} 4\cot\theta\csc\theta \, d\theta = [-4\csc\theta]_{\frac{\pi}{6}}^{\frac{\pi}{4}} = -4(\sqrt{2} - 2) = 8 - 4\sqrt{2}$

Q8 a) You can't take the square root of a negative value, so $64 - x^2 \geq 0$. Also, the denominator of the fraction can't be zero, so $64 - x^2 \neq 0$.
So the largest possible domain is $64 - x^2 > 0$
$\Rightarrow x^2 < 64 \Rightarrow -8 < x < 8$
Range: $x^2 \geq 0 \Rightarrow 64 - x^2 \leq 64 \Rightarrow (64 - x^2)^{\frac{1}{2}} \leq 8$
$\Rightarrow \dfrac{4}{(64 - x^2)^{\frac{1}{2}}} \geq 0.5 \Rightarrow y \geq 0.5$
[2 marks available — 1 mark for the correct domain, 1 mark for the correct range]

b)

[4 marks available — 1 mark for the correct shape, 1 mark for each asymptote, 1 mark for the point of intersection]

Answers

c) The area bounded by the graph, the x-axis and the lines $x = 0$ and $x = 4$ is: $\int_0^4 \frac{4}{(64 - x^2)^{\frac{1}{2}}} \, dx$

Let $x = 8 \sin u \Rightarrow \frac{dx}{du} = 8 \cos u \Rightarrow dx = 8 \cos u \, du$

$x = 0 \Rightarrow \sin u = 0 \Rightarrow u = 0$

$x = 4 \Rightarrow \sin u = \frac{1}{2} \Rightarrow u = \frac{\pi}{6}$

So $\int_0^4 \frac{4}{(64 - x^2)^{\frac{1}{2}}} \, dx = \int_0^{\frac{\pi}{6}} \frac{4 \times 8 \cos u}{\sqrt{64 - (8 \sin u)^2}} \, du$

$= \int_0^{\frac{\pi}{6}} \frac{4 \times 8 \cos u}{\sqrt{64(1 - \sin^2 u)}} \, du = \int_0^{\frac{\pi}{6}} \frac{32 \cos u}{8\sqrt{\cos^2 u}} \, du$

$= \int_0^{\frac{\pi}{6}} 4 \, du = [4u]_0^{\frac{\pi}{6}} = \frac{2\pi}{3} - 0 = \frac{2\pi}{3}$

[7 marks available — 1 mark for the required integral stated with the correct limits, 1 mark for a suitable substitution, 1 mark for finding $\frac{dx}{du}$, 1 mark for a correct integral in terms of u only, 1 mark for both limits correct in terms of u, 1 mark for fully simplifying the integral, 1 mark for the correct answer]

Q9 a) $\cos^2 u = \frac{1 + \cos 2u}{2}$ *[1 mark]*

b) Let $x = \sin u \Rightarrow \frac{dx}{du} = \cos u \Rightarrow dx = \cos u \, du$

$\int \sqrt{1 - x^2} \, dx = \int \sqrt{1 - \sin^2 u} \, \cos u \, du = \int \cos^2 u \, du$

$= \int \frac{1 + \cos 2u}{2} \, du = \frac{1}{2} u + \frac{1}{4} \sin 2u + C$

[4 marks available — 1 mark for finding $\frac{dx}{du}$, 1 mark for a correct integral in terms of u only, 1 mark for each correct term in the answer]

c) $y = \sqrt{1 - x^2} \Rightarrow y^2 = 1 - x^2 \Rightarrow x^2 + y^2 = 1$

So the circle has a radius of 1 and centre $(0, 0)$.

Only the half above the y-axis is represented, since y is the positive square root in the original equation.

[2 marks available — 1 mark for correct radius and centre, 1 mark for a correct explanation of which part of the circle is represented]

d) When $x = -1$, $\sin u = -1 \Rightarrow u = -\frac{\pi}{2}$

When $x = 1$, $\sin u = 1 \Rightarrow u = \frac{\pi}{2}$

Area $= \left[\frac{1}{2} u + \frac{1}{4} \sin 2u\right]_{-\frac{\pi}{2}}^{\frac{\pi}{2}}$

$= \left(\frac{\pi}{4} + \frac{1}{4} \sin \pi\right) - \left(-\frac{\pi}{4} + \frac{1}{4} \sin(-\pi)\right)$

$= \frac{\pi}{4} + 0 + \frac{\pi}{4} - 0 = \frac{\pi}{2}$

Using the area of a circle formula gives:

Area $= \frac{1}{2} \pi r^2 = \frac{1}{2} \pi \times 1^2 = \frac{\pi}{2}$, which is the same result.

[3 marks available — 1 mark for converting limits to be in terms of u, 1 mark for the correct area from either method, 1 mark for showing that the two methods give the same result]

Exercise 8.16 — Integration by Parts

Q1 a) Let $u = x$ and $\frac{dv}{dx} = e^x$.

Then $\frac{du}{dx} = 1$ and $v = e^x$.

So $\int x e^x \, dx = x e^x - \int e^x \, dx = x e^x - e^x + C$

b) Let $u = x$ and $\frac{dv}{dx} = e^{-x}$. Then $\frac{du}{dx} = 1$ and $v = -e^{-x}$.

So $\int x e^{-x} \, dx = -x e^{-x} - \int -e^{-x} \, dx = -x e^{-x} - e^{-x} + C$

c) Let $u = x$ and $\frac{dv}{dx} = e^{-\frac{x}{3}}$. Then $\frac{du}{dx} = 1$ and $v = -3 e^{-\frac{x}{3}}$.

So $\int x e^{-\frac{x}{3}} \, dx = -3x e^{-\frac{x}{3}} - \int -3 e^{-\frac{x}{3}} \, dx = -3x e^{-\frac{x}{3}} - 9 e^{-\frac{x}{3}} + C$

d) Let $u = x$ and $\frac{dv}{dx} = e^x + 1$. Then $\frac{du}{dx} = 1$ and $v = e^x + x$.

So $\int x(e^x + 1) \, dx = x(e^x + x) - \int e^x + x \, dx$

$= x e^x + x^2 - e^x - \frac{1}{2} x^2 + C = x e^x - e^x + \frac{1}{2} x^2 + C$

You could also have solved this one by expanding the brackets and integrating $xe^x + x$ separately.

You might have spotted a pattern here — all the parts of this question had $u = x$ and $\frac{dv}{dx}$ as a function involving e.
Your answers might look a bit different if you factorised them.

Q2 a) Let $u = x$ and $\frac{dv}{dx} = \sin x$. Then $\frac{du}{dx} = 1$ and $v = -\cos x$.

So $\int_0^\pi x \sin x \, dx = [-x \cos x]_0^\pi - \int_0^\pi -\cos x \, dx$

$= [-x \cos x]_0^\pi + [\sin x]_0^\pi = (\pi - 0) + (0 - 0) = \pi$

b) Let $u = 2x$ and $\frac{dv}{dx} = \cos x$. Then $\frac{du}{dx} = 2$ and $v = \sin x$.

So $\int 2x \cos x \, dx = 2x \sin x - \int 2 \sin x \, dx$

$= 2x \sin x + 2 \cos x + C$

c) Let $u = 3x$ and $\frac{dv}{dx} = \cos \frac{1}{2} x$.

Then $\frac{du}{dx} = 3$ and $v = 2 \sin \frac{1}{2} x$.

So $\int 3x \cos \frac{1}{2} x \, dx = 6x \sin \frac{1}{2} x - \int 6 \sin \frac{1}{2} x \, dx$

$= 6x \sin \frac{1}{2} x + 12 \cos \frac{1}{2} x + C$

d) Let $u = 2x$ and $\frac{dv}{dx} = 1 - \sin x$. Then $\frac{du}{dx} = 2$ and $v = x + \cos x$.

So $\int_{-\frac{\pi}{2}}^{\frac{\pi}{2}} 2x(1 - \sin x) \, dx$

$= [2x(x + \cos x)]_{-\frac{\pi}{2}}^{\frac{\pi}{2}} - \int_{-\frac{\pi}{2}}^{\frac{\pi}{2}} 2(x + \cos x) \, dx$

$= [2x(x + \cos x)]_{-\frac{\pi}{2}}^{\frac{\pi}{2}} - [x^2 + 2 \sin x]_{-\frac{\pi}{2}}^{\frac{\pi}{2}}$

$= \left[\pi\left(\frac{\pi}{2} + 0\right) - (-\pi)\left(-\frac{\pi}{2} + 0\right)\right] - \left[\left(\frac{\pi^2}{4} + 2\right) - \left(\frac{\pi^2}{4} - 2\right)\right]$

$= -4$

Q3 a) Let $u = \ln x$ and $\frac{dv}{dx} = 2$. Then $\frac{du}{dx} = \frac{1}{x}$ and $v = 2x$.

So $\int 2 \ln x \, dx = 2x \ln x - \int 2 \, dx = 2x \ln x - 2x + C$

b) Let $u = \ln x$ and $\frac{dv}{dx} = x^4$. Then $\frac{du}{dx} = \frac{1}{x}$ and $v = \frac{1}{5} x^5$.

So $\int x^4 \ln x \, dx = \frac{1}{5} x^5 \ln x - \int \frac{1}{5} x^4 \, dx = \frac{1}{5} x^5 \ln x - \frac{1}{25} x^5 + C$

c) Let $u = \ln 4x$ and $\frac{dv}{dx} = 1$. Then $\frac{du}{dx} = \frac{1}{x}$ and $v = x$.

So $\int \ln 4x \, dx = x \ln 4x - \int 1 \, dx = x \ln 4x - x + C$

d) Let $u = \ln x^3$ and $\frac{dv}{dx} = 1$. Then $\frac{du}{dx} = \frac{3}{x}$ and $v = x$.

So $\int \ln x^3 \, dx = x \ln x^3 - \int 3 \, dx = x \ln x^3 - 3x + C$

For parts a), c) and d), if the question hadn't told you to use integration by parts, you could have just used the standard result for integrating ln x (you'd need to rewrite the logs in parts c) and d) as ln 4 + ln x and 3 ln x).

Q4 Let $u = \ln x$ and $\frac{dv}{dx} = x$. Then $\frac{du}{dx} = \frac{1}{x}$ and $v = \frac{x^2}{2}$.

So $\int_1^4 x \ln x \, dx = \left[\frac{x^2 \ln x}{2}\right]_1^4 - \int_1^4 \frac{x}{2} \, dx = \left[\frac{x^2 \ln x}{2}\right]_1^4 - \left[\frac{x^2}{4}\right]_1^4$

$= \left[\frac{16 \ln 4}{2} - \frac{\ln 1}{2}\right] - \left[\frac{16}{4} - \frac{1}{4}\right] = 8 \ln 4 - \frac{15}{4} = 7.34$ (3 s.f.)

[5 marks available — 1 mark for the correct choices of u and $\frac{dv}{dx}$, 1 mark for finding $\frac{du}{dx}$ and v, 1 mark for using the integration by parts formula, 1 mark for substituting in the limits, 1 mark for the correct answer]

Q5 **a)** Let $u = 20x$ and $\frac{dv}{dx} = (x + 1)^3$.

Then $\frac{du}{dx} = 20$ and $v = \frac{1}{4}(x + 1)^4$.

So $\int_{-1}^{1} 20x(x + 1)^3 \, dx = [5x(x + 1)^4]_{-1}^{1} - \int_{-1}^{1} 5(x + 1)^4 \, dx$

$= [5x(x + 1)^4]_{-1}^{1} - [(x + 1)^5]_{-1}^{1} = [5(2^4) - 0] - [(2^5) - 0]$

$= 80 - 32 = 48$

b) Let $u = 30x$ and $\frac{dv}{dx} = (2x + 1)^{\frac{1}{2}}$.

Then $\frac{du}{dx} = 30$ and $v = \frac{1}{3}(2x + 1)^{\frac{3}{2}}$.

So $\int_{0}^{1.5} 30x\sqrt{2x + 1} \, dx$

$= \left[10x(2x + 1)^{\frac{3}{2}}\right]_{0}^{1.5} - \int_{0}^{1.5} 10(2x + 1)^{\frac{3}{2}} \, dx$

$= \left[10x(2x + 1)^{\frac{3}{2}}\right]_{0}^{1.5} - \left[2(2x + 1)^{\frac{5}{2}}\right]_{0}^{1.5}$

$= \left[15(4)^{\frac{3}{2}} - 0\right] - \left[2(4)^{\frac{5}{2}} - 2(1)^{\frac{5}{2}}\right] = 120 - 62 = 58$

Q6 **a)** Let $u = x$ and $\frac{dv}{dx} = 12e^{2x}$. Then $\frac{du}{dx} = 1$ and $v = 6e^{2x}$.

So $\int_{0}^{1} 12xe^{2x} \, dx = [6xe^{2x}]_{0}^{1} - \int_{0}^{1} 6e^{2x} \, dx$

$= 6e^2 - [3e^{2x}]_{0}^{1} = 6e^2 - (3e^2 - 3e^0) = 3e^2 + 3$

b) Let $u = x$ and $\frac{dv}{dx} = 18 \sin 3x$.

Then $\frac{du}{dx} = 1$ and $v = -6 \cos 3x$.

$\int_{0}^{\frac{\pi}{3}} 18x \sin 3x \, dx = [-6x \cos 3x]_{0}^{\frac{\pi}{3}} - \int_{0}^{\frac{\pi}{3}} -6 \cos 3x \, dx$

$= -2\pi \cos \pi + [2 \sin 3x]_{0}^{\frac{\pi}{3}} = 2\pi + [2 \sin \pi - 2 \sin 0] = 2\pi$

c) Let $u = \ln x$ and $\frac{dv}{dx} = \frac{1}{x^2}$. Then $\frac{du}{dx} = \frac{1}{x}$ and $v = -\frac{1}{x}$.

So $\int_{1}^{2} \frac{1}{x^2} \ln x \, dx = \left[-\frac{1}{x} \ln x\right]_{1}^{2} - \int_{1}^{2} -\frac{1}{x^2} \, dx$

$= -\frac{1}{2} \ln 2 + \ln 1 - \left[\frac{1}{x}\right]_{1}^{2}$

$= -\frac{1}{2} \ln 2 - \frac{1}{2} + 1 = \frac{1}{2} - \frac{1}{2} \ln 2$

Q7 **a)** Let $u = x$ and $\frac{dv}{dx} = e^{-2x}$. Then $\frac{du}{dx} = 1$ and $v = -\frac{1}{2}e^{-2x}$.

So $\int \frac{x}{e^{2x}} \, dx = -\frac{x}{2}e^{-2x} - \int -\frac{1}{2}e^{-2x} \, dx = -\frac{x}{2e^{2x}} - \frac{1}{4e^{2x}} + C$

b) Let $u = x + 1$ and $\frac{dv}{dx} = (x + 2)^{\frac{1}{2}}$.

Then $\frac{du}{dx} = 1$ and $v = \frac{2}{3}(x + 2)^{\frac{3}{2}}$.

So $\int (x + 1)\sqrt{x + 2} \, dx$

$= \frac{2}{3}(x + 1)(x + 2)^{\frac{3}{2}} - \int \frac{2}{3}(x + 2)^{\frac{3}{2}} \, dx$

$= \frac{2}{3}(x + 1)(x + 2)^{\frac{3}{2}} - \frac{4}{15}(x + 2)^{\frac{5}{2}} + C$

c) Let $u = \ln(x + 1)$ and $\frac{dv}{dx} = 1$. Then $\frac{du}{dx} = \frac{1}{x + 1}$ and $v = x$.

So $\int \ln(x + 1) \, dx = x \ln(x + 1) - \int \frac{x}{x + 1} \, dx$

$= x \ln(x + 1) - \int \frac{x + 1 - 1}{x + 1} \, dx$

$= x \ln(x + 1) - \int 1 - \frac{1}{x + 1} \, dx$

$= x \ln|x + 1| - x + \ln|x + 1| + C = (x + 1) \ln|x + 1| - x + C$

Q8 **a)** Let $u = 3x$ and $\frac{dv}{dx} = \cos\left(x + \frac{2\pi}{3}\right)$.

Then $\frac{du}{dx} = 3$ and $v = \sin\left(x + \frac{2\pi}{3}\right)$.

So $\int 3x \cos\left(x + \frac{2\pi}{3}\right) dx$

$= 3x \sin\left(x + \frac{2\pi}{3}\right) - \int 3 \sin\left(x + \frac{2\pi}{3}\right) dx$

$= 3x \sin\left(x + \frac{2\pi}{3}\right) + 3 \cos\left(x + \frac{2\pi}{3}\right) + C$

[4 marks available — 1 mark for the correct choices of u and $\frac{dv}{dx}$, 1 mark for finding $\frac{du}{dx}$ and v, 1 mark for substituting into the integration by parts formula, 1 mark for the correct answer]

b) $y = 2\cos x$ is above the x-axis between $x = 0$ and $\frac{\pi}{2}$.

$\int_{0}^{\frac{\pi}{2}} 2 \cos x \, dx = [2 \sin x]_{0}^{\frac{\pi}{2}} = 2(1 - 0) = 2$

The curve $y = 3x \cos\left(x + \frac{2\pi}{3}\right)$ is below the x-axis between $x = 0$ and $\frac{\pi}{2}$.

$\int_{0}^{\frac{\pi}{2}} 3x \cos\left(x + \frac{2\pi}{3}\right) dx$

$= \left[3x \sin\left(x + \frac{2\pi}{3}\right) + 3 \cos\left(x + \frac{2\pi}{3}\right)\right]_{0}^{\frac{\pi}{2}}$

$= \left[\frac{3\pi}{2} \sin \frac{7\pi}{6} + 3 \cos \frac{7\pi}{6}\right] - \left[0 + 3 \cos \frac{2\pi}{3}\right]$

$= \left[-\frac{3\pi}{4} - \frac{3\sqrt{3}}{2}\right] - \left[-\frac{3}{2}\right]$

$= -\frac{3\pi}{4} - \frac{3\sqrt{3}}{2} + \frac{3}{2} = -3.4542...$

So the area of this region is 3.45 (3 s.f.).

So the region bounded by the x-axis from $x = 0$ to $\frac{\pi}{2}$, the line $x = \frac{\pi}{2}$ and the curve $y = 3x \cos\left(x + \frac{2\pi}{3}\right)$ has the greater area.

[5 marks available — 1 mark for integrating 2 cos x, 1 mark for substituting the correct limits into the result from part a), 1 mark for each area correct, 1 mark for the correct conclusion]

Q9 **a)** Let $u = \ln x$ and $v = x^3$. Then $\frac{du}{dx} = \frac{1}{x}$ and $\frac{dv}{dx} = 3x^2$.

So $\frac{dy}{dx} = \frac{x^2 - 3x^2 \ln x}{x^6} = \frac{1 - 3 \ln x}{x^4}$

[4 marks available — 1 mark for attempting to use quotient (or product) rule, 1 mark for finding $\frac{du}{dx}$, 1 mark for finding $\frac{dv}{dx}$, 1 mark for the correct answer in any form]

b) The maximum value of y occurs at the stationary point.

$\frac{dy}{dx} = 0 \Rightarrow 1 - 3 \ln x = 0 \Rightarrow \ln x = \frac{1}{3} \Rightarrow x = e^{\frac{1}{3}}$

When $x = e^{\frac{1}{3}}, y = \frac{\ln\left(e^{\frac{1}{3}}\right)}{\left(e^{\frac{1}{3}}\right)^3} = \frac{1}{3e} = 0.1226...$

So the range is $y \le 0.123$ (3 s.f.).

[2 marks available — 1 mark for a correct method, 1 mark for the correct answer]

c) Let $u = \ln x$ and $\frac{dv}{dx} = x^{-3}$. Then $\frac{du}{dx} = \frac{1}{x}$ and $v = -\frac{1}{2}x^{-2}$.

So $\int \frac{\ln x}{x^3} \, dx = -\frac{1}{2}x^{-2} \ln x + \int \frac{1}{2}x^{-3} \, dx$

$= -\frac{1}{2}x^{-2}\ln x - \frac{1}{4}x^{-2} + C$

[4 marks available — 1 mark for the correct choices of u and $\frac{dv}{dx}$, 1 mark for finding $\frac{du}{dx}$ and v, 1 mark for substituting into the integration by parts formula, 1 mark for the correct answer]

Answers

d) The only place the graph crosses the x-axis is between $x = 0$ and the stationary point at $x = e^{\frac{1}{3}} = 1.395...$, so the graph is above the x-axis for $x > 1.395...$

So the area under the graph for the given limits can be found using the integral from part c):

$$\int_5^{10} \frac{\ln x}{x^3} \, dx = \left[-\frac{1}{2}x^{-2}\ln x - \frac{1}{4}x^{-2}\right]_5^{10}$$

$$= \left[-\frac{1}{200}\ln 10 - \frac{1}{400}\right] - \left[-\frac{1}{50}\ln 5 - \frac{1}{100}\right] = 0.02817...$$

$$\int_{10}^{15} \frac{\ln x}{x^3} \, dx = \left[-\frac{1}{2}x^{-2}\ln x - \frac{1}{4}x^{-2}\right]_{10}^{15}$$

$$= \left[-\frac{1}{450}\ln 15 - \frac{1}{900}\right] - \left[-\frac{1}{200}\ln 10 - \frac{1}{400}\right] = 0.006883...$$

$0.02817... \div 0.006883... = 4.0929...$
So the first area is 409% of the second (to the nearest percentage point).
[5 marks available — 1 mark for justifying that the integrals will give the areas, 1 mark for substituting in the limits correctly, 1 mark for at least one correct area, 1 mark for both correct areas, 1 mark for showing the given result using the correct values]

Exercise 8.17 — Repeated use of Integration by Parts

Q1 a) Let $u = x^2$ and $\frac{dv}{dx} = e^x$. Then $\frac{du}{dx} = 2x$ and $v = e^x$.

So $\int x^2 e^x \, dx = x^2 e^x - \int 2x e^x \, dx$

Integrate by parts again to find $\int 2x e^x \, dx$:

Let $u = 2x$ and $\frac{dv}{dx} = e^x$. Then $\frac{du}{dx} = 2$ and $v = e^x$.
So $\int 2x e^x \, dx = 2x e^x - \int 2e^x \, dx = 2x e^x - 2e^x + c$

So $\int x^2 e^x \, dx = x^2 e^x - \int 2x e^x \, dx = x^2 e^x - (2x e^x - 2e^x + c)$
$= x^2 e^x - 2x e^x + 2e^x + C$

b) Let $u = x^2$ and $\frac{dv}{dx} = \cos x$. Then $\frac{du}{dx} = 2x$ and $v = \sin x$.

So $\int x^2 \cos x \, dx = x^2 \sin x - \int 2x \sin x \, dx$

Integrate by parts again to find $\int 2x \sin x \, dx$:

Let $u = 2x$ and $\frac{dv}{dx} = \sin x$. Then $\frac{du}{dx} = 2$ and $v = -\cos x$.
So $\int 2x \sin x \, dx = -2x \cos x - \int -2 \cos x \, dx$
$= -2x \cos x + 2 \sin x + c$
So $\int x^2 \cos x \, dx = x^2 \sin x - \int 2x \sin x \, dx$
$= x^2 \sin x - (-2x \cos x + 2 \sin x + c)$
$= x^2 \sin x + 2x \cos x - 2 \sin x + C$

c) Let $u = x^2$ and $\frac{dv}{dx} = 4 \sin 2x$.
Then $\frac{du}{dx} = 2x$ and $v = -2 \cos 2x$.

So $\int 4x^2 \sin 2x \, dx = -2x^2 \cos 2x + \int 4x \cos 2x \, dx$

Integrate by parts again to find $\int 4x \cos 2x \, dx$:

Let $u = x$ and $\frac{dv}{dx} = 4 \cos 2x$. Then $\frac{du}{dx} = 1$ and $v = 2 \sin 2x$.

So $\int 4x \cos 2x \, dx = 2x \sin 2x - \int 2 \sin 2x \, dx$
$= 2x \sin 2x + \cos 2x + C$
So $\int 4x^2 \sin 2x \, dx = -2x^2 \cos 2x + \int 4x \cos 2x \, dx$
$= -2x^2 \cos 2x + 2x \sin 2x + \cos 2x + C$

d) Let $u = 40x^2$ and $\frac{dv}{dx} = (2x - 1)^4$.
Then $\frac{du}{dx} = 80x$ and $v = \frac{1}{10}(2x - 1)^5$.

$\int 40x^2(2x - 1)^4 \, dx = 4x^2(2x - 1)^5 - \int 8x(2x - 1)^5 \, dx$

Integrate by parts again to find $\int 8x(2x - 1)^5 \, dx$:
Let $u = 8x$ and $\frac{dv}{dx} = (2x - 1)^5$.
Then $\frac{du}{dx} = 8$ and $v = \frac{1}{12}(2x - 1)^6$

$\int 8x(2x - 1)^5 \, dx = \frac{2x}{3}(2x - 1)^6 - \int \frac{2}{3}(2x - 1)^6 \, dx$
$= \frac{2x}{3}(2x - 1)^6 - \frac{1}{21}(2x - 1)^7 + C$

So $\int 40x^2(2x - 1)^4 \, dx$
$= 4x^2(2x - 1)^5 - \frac{2x}{3}(2x - 1)^6 + \frac{1}{21}(2x - 1)^7 + C$

Q2 Let $u = x^2$ and $\frac{dv}{dx} = (x + 1)^4$. Then $\frac{du}{dx} = 2x$ and $v = \frac{1}{5}(x + 1)^5$.

So $\int_{-1}^0 x^2(x + 1)^4 \, dx = \left[\frac{x^2}{5}(x + 1)^5\right]_{-1}^0 - \int_{-1}^0 \frac{2}{5}x(x + 1)^5 \, dx$

$= 0 - \int_{-1}^0 \frac{2}{5}x(x + 1)^5 \, dx$

Integrate by parts again to find $\int_{-1}^0 \frac{2}{5}x(x + 1)^5 \, dx$:

Let $u = x$ and $\frac{dv}{dx} = \frac{2}{5}(x + 1)^5$. Then $\frac{du}{dx} = 1$ and $v = \frac{1}{15}(x + 1)^6$.

$\int_{-1}^0 \frac{2}{5}x(x + 1)^5 \, dx = \left[\frac{x}{15}(x + 1)^6\right]_{-1}^0 - \int_{-1}^0 \frac{1}{15}(x + 1)^6 \, dx$

$= 0 - \left[\frac{1}{105}(x + 1)^7\right]_{-1}^0 = -\frac{1}{105}$

So $\int_{-1}^0 x^2(x + 1)^4 \, dx = 0 - \int_{-1}^0 \frac{2}{5}x(x + 1)^5 \, dx = -\left(-\frac{1}{105}\right) = \frac{1}{105}$

Q3 Let $u = x^2$ and $\frac{dv}{dx} = e^{-2x}$. Then $\frac{du}{dx} = 2x$ and $v = -\frac{1}{2}e^{-2x}$.

So area $= \int_0^1 x^2 e^{-2x} \, dx = \left[-\frac{x^2}{2}e^{-2x}\right]_0^1 + \int_0^1 x e^{-2x} \, dx$

Integrate by parts again to find $\int_0^1 x e^{-2x} \, dx$:

Let $u = x$ and $\frac{dv}{dx} = e^{-2x}$. Then $\frac{du}{dx} = 1$ and $v = -\frac{1}{2}e^{-2x}$.

So $\int_0^1 x e^{-2x} \, dx = \left[-\frac{x}{2}e^{-2x}\right]_0^1 + \int_0^1 \frac{1}{2}e^{-2x} \, dx$

$= \left[-\frac{x}{2}e^{-2x}\right]_0^1 + \left[-\frac{1}{4}e^{-2x}\right]_0^1$

So area $= \left[-\frac{x^2}{2}e^{-2x}\right]_0^1 + \left[-\frac{x}{2}e^{-2x}\right]_0^1 + \left[-\frac{1}{4}e^{-2x}\right]_0^1$

$= -\frac{1}{2}e^{-2} - \frac{1}{2}e^{-2} - \frac{1}{4}e^{-2} + \frac{1}{4} = \frac{1}{4} - \frac{5}{4}e^{-2}$

Q4 a) $\cos x \geq 0$ between $x = 0$ and $\frac{\pi}{2}$, and $e^x > 0$ for all x. Both graphs are continuous, so the curve $y = e^x \cos x$ lies above the x-axis between $x = 0$ and $\frac{\pi}{2}$, and so the integral will give the area.
[1 mark for a valid explanation]

b) Let $u = e^x$ and $\frac{dv}{dx} = \cos x$. Then $\frac{du}{dx} = e^x$ and $v = \sin x$.

So $\int e^x \cos x \, dx = e^x \sin x - \int e^x \sin x \, dx$

Integrate by parts again to find $\int e^x \sin x \, dx$:

Let $u = e^x$ and $\frac{dv}{dx} = \sin x$, then $\frac{du}{dx} = e^x$ and $v = -\cos x$.

So $\int e^x \sin x \, dx = -e^x \cos x + \int e^x \cos x \, dx$

$\Rightarrow \int e^x \cos x \, dx = e^x \sin x + e^x \cos x - \int e^x \cos x \, dx$

$\Rightarrow 2\int e^x \cos x \, dx = e^x \sin x + e^x \cos x + c$

$\Rightarrow \int e^x \cos x \, dx = \frac{e^x \sin x + e^x \cos x}{2} + C$

[7 marks available — 1 mark for $\frac{du}{dx}$, 1 mark for v, 1 mark for substituting into the integration by parts formula, 1 mark for second $\frac{du}{dx}$, 1 mark for second v, 1 mark for substituting into the integration by parts formula again, 1 mark for the correct answer]

c) $\int_0^{\frac{\pi}{2}} e^x \cos x \, dx = \left[\frac{e^x \sin x + e^x \cos x}{2} \right]_0^{\frac{\pi}{2}}$

$= \left(\frac{e^{\frac{\pi}{2}} \times 1 + e^{\frac{\pi}{2}} \times 0}{2} \right) - \left(\frac{e^0 \times 0 + e^0 \times 1}{2} \right)$

$= \frac{1}{2}(e^{\frac{\pi}{2}} - 1) = 1.9052... = 1.91 \text{ (3 s.f.)}$

[2 marks available — 1 mark for correctly substituting in the limits, 1 mark for the correct answer]

Exercise 8.18 — Integration Using Partial Fractions

Q1 a) First write the function as partial fractions. Factorise the denominator and write as an identity:

$\frac{24(x-1)}{9-4x^2} \equiv \frac{24(x-1)}{(3-2x)(3+2x)} \equiv \frac{A}{(3-2x)} + \frac{B}{(3+2x)}$

Add the fractions and cancel denominators:

$\frac{24(x-1)}{(3-2x)(3+2x)} \equiv \frac{A(3+2x)+B(3-2x)}{(3-2x)(3+2x)}$

$\Rightarrow 24(x-1) \equiv A(3+2x) + B(3-2x)$

Substituting $x = -\frac{3}{2}$ gives: $-60 = 6B \Rightarrow B = -10$

Substituting $x = \frac{3}{2}$ gives: $12 = 6A \Rightarrow A = 2$

So $\frac{24(x-1)}{9-4x^2} \equiv \frac{2}{(3-2x)} - \frac{10}{(3+2x)}$.

So the integral can be expressed:

$\int \frac{24(x-1)}{9-4x^2} \, dx = \int \frac{2}{(3-2x)} - \frac{10}{(3+2x)} \, dx$

$= \frac{2}{-2} \ln|3-2x| - \frac{10}{2} \ln|3+2x| + C$

$= -\ln|3-2x| - 5 \ln|3+2x| + C$

b) $\frac{21x-82}{(x-2)(x-3)(x-4)} \equiv \frac{A}{x-2} + \frac{B}{x-3} + \frac{C}{x-4}$

$\equiv \frac{A(x-3)(x-4)+B(x-2)(x-4)+C(x-2)(x-3)}{(x-2)(x-3)(x-4)}$

$21x-82 \equiv A(x-3)(x-4) + B(x-2)(x-4) + C(x-2)(x-3)$

Substituting $x = 2$ gives: $-40 = 2A \Rightarrow A = -20$

Substituting $x = 3$ gives: $-19 = -B \Rightarrow B = 19$

Substituting $x = 4$ gives: $2 = 2C \Rightarrow C = 1$

So $\frac{21x-82}{(x-2)(x-3)(x-4)} \equiv -\frac{20}{x-2} + \frac{19}{x-3} + \frac{1}{x-4}$

$\equiv \frac{1}{x-4} + \frac{19}{x-3} - \frac{20}{x-2}$

So the integral can be expressed as:

$\int \frac{21x-82}{(x-2)(x-3)(x-4)} \, dx = \int \frac{1}{x-4} + \frac{19}{x-3} - \frac{20}{x-2} \, dx$

$= \ln|x-4| + 19 \ln|x-3| - 20 \ln|x-2| + C$

Q2 First write the function as partial fractions:

$\frac{x}{(x-2)(x-3)} \equiv \frac{A}{x-2} + \frac{B}{x-3} \equiv \frac{A(x-3)+B(x-2)}{(x-2)(x-3)}$

$\Rightarrow x \equiv A(x-3) + B(x-2)$

Substituting $x = 3$: $B = 3$

Substituting $x = 2$: $-A = 2 \Rightarrow A = -2$

So $\frac{x}{(x-2)(x-3)} \equiv \frac{-2}{x-2} + \frac{3}{x-3} \equiv \frac{3}{x-3} - \frac{2}{x-2}$

$\Rightarrow \int_0^1 \frac{x}{(x-2)(x-3)} \, dx = \int_0^1 \frac{3}{x-3} - \frac{2}{x-2} \, dx$

$= [3 \ln|x-3| - 2 \ln|x-2|]_0^1$

$= [3 \ln|1-3| - 2 \ln|1-2|] - [3 \ln|0-3| - 2 \ln|0-2|]$

$= 3 \ln 2 - 2 \ln 1 - 3 \ln 3 + 2 \ln 2 = 0 + 5 \ln 2 - 3 \ln 3$

$= \ln 2^5 - \ln 3^3 = \ln \frac{32}{27}$

Note that the modulus is important in this question, for example $|1 - 2| = |-1| = 1$.

Q3 a) $\frac{5x+5}{(x+2)(x-3)} \equiv \frac{A}{x+2} + \frac{B}{x-3}$

$\Rightarrow 5x+5 \equiv A(x-3) + B(x+2)$

Substituting $x = -2$ gives: $-10 + 5 = -5A \Rightarrow A = 1$

Substituting $x = 3$ gives: $15 + 5 = 5B \Rightarrow B = 4$

So $\frac{5x+5}{(x+2)(x-3)} \equiv \frac{1}{x+2} + \frac{4}{x-3}$

[3 marks available — 1 mark for a correct method, 1 mark for each correct fraction]

b) $\int \frac{5x+5}{(x+2)(x-3)} \, dx = \int \frac{1}{x+2} + \frac{4}{x-3} \, dx$

$= \ln|x+2| + 4\ln|x-3| + C$

[2 marks available — 2 marks for the correct answer, otherwise 1 mark for at least one term integrated correctly]

c) $\int_{-1}^2 \frac{5x+5}{(x+2)(x-3)} \, dx = [\ln|x+2| + 4 \ln|x-3|]_{-1}^2$

$= (\ln 4 + 4 \ln 1) - (\ln 1 + 4 \ln 4) = -3 \ln 4 \text{ (or } \ln \frac{1}{64})$

[2 marks available — 1 mark for substituting in the limits correctly, 1 mark for the correct answer]

Q4 First express the fraction as an identity:

$\frac{f(x)}{g(x)} = \frac{3x+5}{x(x+10)} \equiv \frac{A}{x} + \frac{B}{x+10} \Rightarrow 3x+5 \equiv A(x+10) + Bx$

Now use the equating coefficients method:

Equating constant terms: $10A = 5 \Rightarrow A = \frac{1}{2}$

Equating x coefficients: $A + B = 3 \Rightarrow \frac{1}{2} + B = 3 \Rightarrow B = \frac{5}{2}$

So $\frac{3x+5}{x(x+10)} \equiv \frac{1}{2x} + \frac{5}{2(x+10)}$

Now the integration can be expressed:

$\int_1^2 \frac{f(x)}{g(x)} \, dx = \int_1^2 \frac{3x+5}{x(x+10)} \, dx = \int_1^2 \frac{1}{2x} + \frac{5}{2(x+10)} \, dx$

$= \left[\frac{1}{2} \ln|x| + \frac{5}{2} \ln|x+10| \right]_1^2$

$= \left[\frac{1}{2} \ln|2| + \frac{5}{2} \ln|2+10| \right] - \left[\frac{1}{2} \ln|1| + \frac{5}{2} \ln|1+10| \right]$

$= \left[\frac{1}{2} \ln 2 + \frac{5}{2} \ln 12 \right] - \left[0 + \frac{5}{2} \ln 11 \right]$

$= \frac{1}{2} \ln 2 + \frac{5}{2} \ln 12 - \frac{5}{2} \ln 11 = \frac{1}{2} \left(\ln 2 + 5 \ln \frac{12}{11} \right)$

$= \frac{1}{2} \left(\ln 2 + \ln \left(\frac{12}{11} \right)^5 \right) = 0.564 \text{ (3 d.p.)}$

Q5 a) Factorise the denominator, then write as an identity:

$\frac{6}{2x^2-5x+2} = \frac{6}{(2x-1)(x-2)} \equiv \frac{A}{2x-1} + \frac{B}{x-2}$

$\Rightarrow 6 \equiv A(x-2) + B(2x-1)$

Substituting $x = 2$ gives $6 = 3B$ so $B = 2$.

Substituting $x = \frac{1}{2}$ gives $6 = -\frac{3}{2}A$ so $A = -4$.

So $\frac{6}{2x^2-5x+2} \equiv -\frac{4}{2x-1} + \frac{2}{x-2} = \frac{2}{x-2} - \frac{4}{2x-1}$

b) Using part a)

$\int \frac{6}{2x^2-5x+2} \, dx = \int \frac{2}{x-2} - \frac{4}{2x-1} \, dx$

$= 2 \ln|x-2| - \frac{4}{2} \ln|2x-1| + C = 2 \ln|x-2| - 2 \ln|2x-1| + C$

$x > 2$ so $x - 2 > 0$ and $2x - 1 > 3$ so the modulus signs can be removed. So:

$\int \frac{6}{2x^2-5x+2} \, dx = 2 \ln(x-2) - 2 \ln(2x-1) + C$

$= 2 \ln \left(\frac{x-2}{2x-1} \right) + C = \ln \left[\left(\frac{x-2}{2x-1} \right)^2 \right] + C$

c) Using part b)

$\int_3^5 \frac{6}{2x^2-5x+2} \, dx = \left[\ln \left[\left(\frac{x-2}{2x-1} \right)^2 \right] \right]_3^5$

$= \ln \left[\left(\frac{5-2}{10-1} \right)^2 \right] - \ln \left[\left(\frac{3-2}{6-1} \right)^2 \right]$

$= \ln \frac{1}{9} - \ln \frac{1}{25} = \ln \frac{25}{9}$

Answers

Q6 Begin by writing the function as partial fractions.

$$\frac{-(t+3)}{(3t+2)(t+1)} \equiv \frac{A}{(3t+2)} + \frac{B}{(t+1)}$$

$$\Rightarrow -(t+3) \equiv A(t+1) + B(3t+2)$$

Substituting $t = -1$ gives: $-2 = -B \Rightarrow B = 2$

Equating coefficients of t gives

$A + 3B = -1 \Rightarrow A + 6 = -1 \Rightarrow A = -7$

So $\dfrac{-(t+3)}{(3t+2)(t+1)} \equiv \dfrac{2}{(t+1)} - \dfrac{7}{(3t+2)}$

The integral can be expressed:

$$\int_0^{\frac{2}{3}} \frac{-(t+3)}{(3t+2)(t+1)} \, dt = \int_0^{\frac{2}{3}} \frac{2}{(t+1)} - \frac{7}{(3t+2)} \, dt$$

$$= \left[2\ln|t+1| - \frac{7}{3}\ln|3t+2| \right]_0^{\frac{2}{3}}$$

$$= \left[2\ln\left|\frac{5}{3}\right| - \frac{7}{3}\ln|4| \right] - \left[2\ln|1| - \frac{7}{3}\ln|2| \right]$$

$$= \left[2\ln\left(\frac{5}{3}\right) - \frac{7}{3}\ln(4) \right] - \left[0 - \frac{7}{3}\ln(2) \right]$$

$$= 2\ln\left(\frac{5}{3}\right) - \frac{7}{3}\ln(4) + \frac{7}{3}\ln(2)$$

$$= 2\ln\left(\frac{5}{3}\right) - \frac{7}{3}(\ln(4) - \ln(2)) = 2\ln\left(\frac{5}{3}\right) - \frac{7}{3}\ln\left(\frac{4}{2}\right)$$

$$= 2\ln\frac{5}{3} - \frac{7}{3}\ln 2$$

Q7 a) $\dfrac{5x+7}{(x+3)(x-1)} \equiv \dfrac{A}{x+3} + \dfrac{B}{x-1}$

$\Rightarrow 5x + 7 \equiv A(x-1) + B(x+3)$

Substituting $x = -3$ gives: $-8 = -4A \Rightarrow A = 2$

Substituting $x = 1$ gives: $12 = 4B \Rightarrow B = 3$

$\dfrac{5x+7}{(x+3)(x-1)} = \dfrac{2}{x+3} + \dfrac{3}{x-1}$

[3 marks available — 1 mark for a correct method, 1 mark for each correct fraction]

b) $\displaystyle\int \frac{5x+7}{(x+3)(x-1)} \, dx = \int \frac{2}{x+3} + \frac{3}{x-1} \, dx$

$= 2\ln|x+3| + 3\ln|x-1| + C$

[2 marks available — 2 marks for the correct answer, otherwise 1 mark for at least one term integrated correctly]

c) If $f(x) = \dfrac{5x+7}{(x+3)(x-1)}$, then $f(x-3) = \dfrac{5x-8}{x(x-4)}$

The second function is a translation of the first function by 3 units to the right, so $c = 3$. *[1 mark]*

Q8 a) Asymptotes will be when the denominator of the fraction is 0:

$(x-2)(x+7) = 0 \Rightarrow x = 2$ and $x = -7$ *[1 mark]*

b) $\dfrac{-4x-1}{(x-2)(x+7)} \equiv \dfrac{A}{x-2} + \dfrac{B}{x+7}$

$\Rightarrow -4x - 1 \equiv A(x+7) + B(x-2)$

Substituting $x = 2$ gives: $-8 - 1 = 9A \Rightarrow A = -1$

Substituting $x = -7$ gives: $28 - 1 = -9B \Rightarrow B = -3$

So $\dfrac{-4x-1}{(x-2)(x+7)} \equiv -\dfrac{1}{x-2} - \dfrac{3}{x+7}$

[3 marks available — 1 mark for a correct method, 1 mark for each correct fraction]

c) $\displaystyle\int \frac{-4x-1}{(x-2)(x+7)} \, dx = \int -\frac{1}{x-2} - \frac{3}{x+7} \, dx$

$= -\ln|x-2| - 3\ln|x+7| + C$

[2 marks available — 2 marks for the correct answer, otherwise 1 mark for at least one term integrated correctly]

d) The graph is entirely below the x-axis in this region.

$\displaystyle\int_5^{10} \frac{-4x-1}{(x-2)(x+7)} \, dx = \left[-\ln|x-2| - 3\ln|x+7| \right]_5^{10}$

$= [-\ln 8 - 3\ln 17] - [-\ln 3 - 3\ln 12] = -2.0257...$

So the area is 2.03 (3 s.f.)

[2 marks available — 1 mark for substituting in the limits correctly, 1 mark for the correct answer]

Q9 $u = \sqrt{x} \Rightarrow \dfrac{du}{dx} = \dfrac{1}{2\sqrt{x}} \Rightarrow dx = 2\sqrt{x} \, du = 2u \, du$

and $u = \sqrt{x} \Rightarrow x = u^2$.

Using $u = \sqrt{x}$, $x = 9 \Rightarrow u = \sqrt{9} = 3$

$x = 16 \Rightarrow u = \sqrt{16} = 4$

So $\displaystyle\int_9^{16} \frac{4}{\sqrt{x}(9x-4)} \, dx = \int_3^4 \frac{4}{u(9u^2-4)} 2u \, du = \int_3^4 \frac{8}{(9u^2-4)} \, du$

This function still needs some work before it can be integrated. The next step is to write it as partial fractions.

$$\frac{8}{(9u^2-4)} \equiv \frac{8}{(3u+2)(3u-2)} \equiv \frac{A}{(3u+2)} + \frac{B}{(3u-2)}$$

So $8 = A(3u-2) + B(3u+2)$

Using substitution:

$u = \dfrac{2}{3} \Rightarrow 8 = 4B \Rightarrow 2 = B$

$u = -\dfrac{2}{3} \Rightarrow 8 = -4A \Rightarrow -2 = A$

So $\dfrac{8}{(9u^2-4)} \equiv \dfrac{-2}{(3u+2)} + \dfrac{2}{(3u-2)} \equiv \dfrac{2}{(3u-2)} - \dfrac{2}{(3u+2)}$

So the integral can be expressed:

$$\int_3^4 \frac{8}{(9u^2-4)} \, du = \int_3^4 \frac{2}{(3u-2)} - \frac{2}{(3u+2)} \, du$$

$$= \left[\frac{2}{3}\ln|3u-2| - \frac{2}{3}\ln|3u+2| \right]_3^4$$

$$= \left[\frac{2}{3}\ln\left|\frac{3u-2}{3u+2}\right| \right]_3^4$$

$$= \frac{2}{3}\ln\frac{10}{14} - \frac{2}{3}\ln\frac{7}{11} = \frac{2}{3}\ln\frac{55}{49}$$

Q10 a) $\dfrac{4x+c}{(x+1)(2x-1)} \equiv \dfrac{A}{x+1} + \dfrac{B}{2x-1}$

$\Rightarrow 4x + c \equiv A(2x-1) + B(x+1)$

Substituting $x = -1$ gives: $-4 + c = -3A \Rightarrow A = \dfrac{4-c}{3}$

Substituting $x = 0.5$ gives: $2 + c = 1.5B \Rightarrow B = \dfrac{4+2c}{3}$

[3 marks available — 1 mark for a correct method, 1 mark for A in terms of c, 1 mark for B in terms of c]

b) Find the binomial expansion for each fraction up to the x-term:

$(x+1)^{-1} = 1 - x \dots \Rightarrow \dfrac{A}{x+1} = A - Ax + \dots$

$(2x-1)^{-1} = -(1-2x)^{-1} = -1 - 2x \dots \Rightarrow \dfrac{B}{2x-1} = -B - 2Bx \dots$

So $\dfrac{4x+c}{(x+1)(2x-1)} = (A-B) - (A+2B)x \dots$

Equating coefficients for the constant gives:

$\dfrac{4-c}{3} - \dfrac{4+2c}{3} = -1 \Rightarrow 4 - c - 4 - 2c = -3$

$\Rightarrow -3 = -3c \Rightarrow c = 1$

$A = \dfrac{4-1}{3} = 1$ and $B = \dfrac{4+2}{3} = 2$

[5 marks available — 1 mark for each expansion that has at least one term correct, 1 mark for forming an equation in c, 1 mark for the correct value of c, 1 mark for the correct values of both A and B]

c) $\displaystyle\int \frac{4x+c}{(x+1)(2x-1)} \, dx = \int \frac{1}{x+1} + \frac{2}{2x-1} \, dx$

$= \ln|x+1| + \ln|2x-1| + C$

[2 marks available — 2 marks for the correct answer, otherwise 1 mark for at least one term integrated correctly]

Exercise 8.19 — Differential Equations

Q1 The rate of change of N with respect to t is $\dfrac{dN}{dt}$.

So $\dfrac{dN}{dt} \propto N \Rightarrow \dfrac{dN}{dt} = kN$, for some $k > 0$.

Q2 The rate of change of x with respect to t is $\dfrac{dx}{dt}$.

So $\dfrac{dx}{dt} \propto \dfrac{1}{x^2} \Rightarrow \dfrac{dx}{dt} = \dfrac{k}{x^2}$, for some $k > 0$.

Answers

Q3 Let the variable t represent time.

Then the rate of change of A with respect to t is $\frac{dA}{dt}$.

So $\frac{dA}{dt} \propto \sqrt{A} \Rightarrow \frac{dA}{dt} = -k\sqrt{A}$, for some $k > 0$.

Don't forget to include a minus sign when the situation involves a rate of decrease.

Q4 Let the variable t represent time.

Then the rate of change of y with respect to t is $\frac{dy}{dt}$.

So $\frac{dy}{dt} \propto (y - \lambda) \Rightarrow \frac{dy}{dt} = -k(y - \lambda)$, for some $k > 0$.

Q5 Let the variable t represent time. V is the volume in the container and it is equal to $V_{in} - V_{out}$.

Then the rate of change of V with respect to t

is $\frac{dV}{dt} = \frac{dV_{in}}{dt} - \frac{dV_{out}}{dt}$. $\frac{dV_{in}}{dt}$ is directly proportional to V,

so $\frac{dV_{in}}{dt} = kV$ for some constant k, $k > 0$ and $\frac{dV_{out}}{dt} = 20$.

So the overall rate of change of V is $\frac{dV}{dt} = kV - 20$, for some $k > 0$.

Q6 Let the variable t represent time.

Then the rate of change of p with respect to t is $\frac{dp}{dt}$.

So $\frac{dp}{dt} \propto p^{0.8} \Rightarrow \frac{dp}{dt} = -kp^{0.8}$ for some $k > 0$. *[1 mark]*

Q7 Let the variable t represent time, and θ is the temperature of the jam. The jam is initially colder than 25 °C, so the difference between the temperature of the jam and the room is $(25 - \theta)$.

So $\frac{d\theta}{dt} = k(25 - \theta)$ for some $k > 0$. *[1 mark]*

Q8 a) The candle burns at a constant rate, so the rate of change of height is constant: $\frac{dh}{dt} = -k$ for some $k > 0$. *[1 mark]*

b) The candle has a constant density, so $\frac{m}{V} = d$ for some $d > 0$. The candle is a prism, so $V = h \times A$, where A is the cross-sectional area and is constant.

So $m = V \times d = h \times A \times d = ch$ where $c = A \times d$ is a constant.

Then $\frac{dm}{dh} = c$ for some $c > 0$. *[1 mark]*

c) $\frac{dm}{dh} \times \frac{dh}{dt} = \frac{dm}{dt} \Rightarrow \frac{dm}{dt} = -k \times c = -K$ for some $K > 0$. *[1 mark]*

Exercise 8.20 — Solving Differential Equations

Q1 a) $\frac{dy}{dx} = 8x^3 \Rightarrow dy = 8x^3 \, dx$

$\Rightarrow \int 1 \, dy = \int 8x^3 \, dx \Rightarrow y = 2x^4 + C$

b) $\frac{dy}{dx} = 5y \Rightarrow \frac{1}{y} \, dy = 5 \, dx \Rightarrow \int \frac{1}{y} \, dy = \int 5 \, dx$

$\Rightarrow \ln|y| = 5x + \ln k \Rightarrow y = e^{5x + \ln k} = ke^{5x}$

c) $\frac{dy}{dx} = 6x^2y \Rightarrow \frac{1}{y} \, dy = 6x^2 dx \Rightarrow \int \frac{1}{y} \, dy = \int 6x^2 \, dx$

$\Rightarrow \ln|y| = 2x^3 + \ln k \Rightarrow y = e^{2x^3 + \ln k} = ke^{2x^3}$

d) $\frac{dy}{dx} = \frac{y}{x} \Rightarrow \frac{1}{y} \, dy = \frac{1}{x} \, dx \Rightarrow \int \frac{1}{y} \, dy = \int \frac{1}{x} \, dx$

$\Rightarrow \ln|y| = \ln|x| + \ln k = \ln|kx| \Rightarrow y = kx$

e) $\frac{dy}{dx} = (y + 1)\cos x \Rightarrow \frac{1}{y + 1} \, dy = \cos x \, dx$

$\Rightarrow \int \frac{1}{y + 1} \, dy = \int \cos x \, dx \Rightarrow \ln|y + 1| = \sin x + \ln k$

$\Rightarrow y + 1 = e^{\sin x + \ln k} \Rightarrow y = ke^{\sin x} - 1$

f) $\frac{dy}{dx} = \frac{3xy - 6y}{(x - 4)(2x - 5)} = \frac{(3x - 6)y}{(x - 4)(2x - 5)}$

$\Rightarrow \frac{1}{y} \, dy = \frac{(3x - 6)}{(x - 4)(2x - 5)} \, dx$

$\Rightarrow \int \frac{1}{y} \, dy = \int \frac{(3x - 6)}{(x - 4)(2x - 5)} \, dx$

The integration on the right hand side needs to be split into partial fractions before you can integrate.

$\frac{(3x - 6)}{(x - 4)(2x - 5)} \equiv \frac{A}{(x - 4)} + \frac{B}{(2x - 5)}$

$\Rightarrow (3x - 6) \equiv A(2x - 5) + B(x - 4)$

Substitution: $x = 4$: $6 = 3A \Rightarrow A = 2$

$x = \frac{5}{2}$: $\frac{3}{2} = -\frac{3}{2}B \Rightarrow B = -1$

So $\int \frac{(3x - 6)}{(x - 4)(2x - 5)} \, dx \equiv \int \frac{2}{(x - 4)} - \frac{1}{(2x - 5)} \, dx$

So $\int \frac{1}{y} \, dy = \int \frac{2}{(x - 4)} - \frac{1}{(2x - 5)} \, dx$

$\Rightarrow \ln|y| = 2 \ln|x - 4| - \frac{1}{2} \ln|2x - 5| + \ln k$

$\Rightarrow \ln|y| = \ln|(x - 4)^2| - \ln|\sqrt{2x - 5}| + \ln k$

$\Rightarrow \ln|y| = \ln\left|\frac{k(x - 4)^2}{\sqrt{2x - 5}}\right| \Rightarrow y = \frac{k(x - 4)^2}{\sqrt{2x - 5}}$

Q2 a) $\frac{dy}{dx} = -\frac{x}{y} \Rightarrow y \, dy = -x \, dx \Rightarrow \int y \, dy = \int -x \, dx$

$\Rightarrow \frac{1}{2}y^2 = -\frac{1}{2}x^2 + c \Rightarrow y^2 = -x^2 + C$

So when $x = 0$ and $y = 2$, $C = 4 \Rightarrow y^2 + x^2 = 4$

b) $\frac{dx}{dt} = \frac{2}{\sqrt{x}} \Rightarrow \sqrt{x} \, dx = 2 \, dt \Rightarrow \int x^{\frac{1}{2}} \, dx = \int 2 \, dt$

$\Rightarrow \frac{2}{3}x^{\frac{3}{2}} = 2t + C$

So when $t = 5$ and $x = 9$,

$\frac{2}{3}(27) = 10 + C \Rightarrow 18 = 10 + C \Rightarrow C = 8$

$\Rightarrow \frac{2}{3}x^{\frac{3}{2}} = 2t + 8 \Rightarrow x^{\frac{3}{2}} = 3t + 12 \Rightarrow x^3 = (3t + 12)^2$

You could have left out the last couple of steps here, as the question didn't specify the form of the answer.

c) $\frac{dV}{dt} = 3(V - 1) \Rightarrow \frac{1}{V - 1} \, dV = 3 \, dt$

$\Rightarrow \int \frac{1}{V - 1} \, dV = \int 3 \, dt \Rightarrow \ln|V - 1| = 3t + \ln k$

$\Rightarrow V = ke^{3t} + 1$

So when $t = 0$ and $V = 5$, $5 = k + 1 \Rightarrow k = 4$

$\Rightarrow V = 4e^{3t} + 1$

d) $\frac{dy}{dx} = \frac{\tan y}{x} \Rightarrow \frac{1}{\tan y} \, dy = \frac{1}{x} \, dx \Rightarrow \int \cot y \, dy = \int \frac{1}{x} \, dx$

$\Rightarrow \ln|\sin y| = \ln|x| + \ln k \Rightarrow \sin y = kx$

So when $x = 2$ and $y = \frac{\pi}{2}$,

$1 = 2k \Rightarrow k = \frac{1}{2} \Rightarrow \sin y = \frac{x}{2}$

e) $\frac{dx}{dt} = 10x(x + 1) \Rightarrow \frac{1}{x(x + 1)} \, dx = 10 \, dt$

Using partial fractions, $\frac{1}{x(x + 1)} \equiv \frac{1}{x} - \frac{1}{x + 1}$, so

$\int \frac{1}{x} - \frac{1}{x + 1} \, dx = \int 10 \, dt \Rightarrow \ln|x| - \ln|x + 1| = 10t + \ln k$

$\Rightarrow \ln\left|\frac{x}{x + 1}\right| = 10t + \ln k \Rightarrow \frac{x}{x + 1} = ke^{10t}$

So when $t = 0$ and $x = 1$, $\frac{1}{2} = k \Rightarrow \frac{x}{x + 1} = \frac{1}{2}e^{10t}$

Q3 a) $\frac{dx}{d\theta} = \cos^2 x \cot \theta \Rightarrow \int \frac{1}{\cos^2 x} \, dx = \int \cot \theta \, d\theta$

$\Rightarrow \int \sec^2 x \, dx = \int \cot \theta \, d\theta \Rightarrow \tan x = \ln|\sin \theta| + C$

b) $x = \frac{\pi}{4}$ when $\theta = \frac{\pi}{2}$, so:

$\tan \frac{\pi}{4} = \ln|\sin \frac{\pi}{2}| + C \Rightarrow 1 = \ln 1 + C \Rightarrow C = 1$

So $\tan x = \ln|\sin \theta| + 1$

c) $x = \tan^{-1}(\ln|\sin \theta| + 1)$, so when $\theta = \frac{\pi}{6}$,

$x = \tan^{-1}(\ln|\sin \frac{\pi}{6}| + 1) = 0.298$ (3 s.f.)

Answers

Q4 a) $2y\dfrac{dy}{dx} + x^2 = 5x \Rightarrow 2y\dfrac{dy}{dx} = 5x - x^2$

$\Rightarrow \int 2y\,dy = \int 5x - x^2\,dx \Rightarrow y^2 = \dfrac{5x^2}{2} - \dfrac{x^3}{3} + C$

[3 marks available — 1 mark for separating the variables correctly, 1 mark for at least two terms integrated correctly, 1 mark for the correct answer]

b) $169 = 90 + 72 + C \Rightarrow C = 7$

So $y^2 = \dfrac{5x^2}{2} - \dfrac{x^3}{3} + 7$ *[1 mark]*

Q5 a) $\dfrac{dV}{dt} = a - bV \Rightarrow \dfrac{1}{a - bV}\,dV = dt$

$\Rightarrow \int \dfrac{1}{a - bV}\,dV = \int 1\,dt \Rightarrow -\dfrac{1}{b}\ln|a - bV| = t + C$

$\Rightarrow \ln|a - bV| = -bt - bC$

b and C are both constants, so let $-bC = \ln k$:

$\Rightarrow \ln|a - bV| = -bt + \ln k \Rightarrow a - bV = ke^{-bt}$

$\Rightarrow bV = a - ke^{-bt} \Rightarrow V = \dfrac{a}{b} - Ae^{-bt}$ (letting $A = k \div b$)

b) When $t = 0$ and $V = \dfrac{a}{4b}$,

$\dfrac{a}{4b} = \dfrac{a}{b} - A \Rightarrow A = \dfrac{a}{b} - \dfrac{a}{4b} = \dfrac{4a - a}{4b} = \dfrac{3a}{4b}$

c) As t gets very large, e^{-bt} gets very close to zero, so V approaches $\dfrac{a}{b}$.

Q6 a) $\dfrac{dy}{dx} = xy + x + y + 1 = (x + 1)(y + 1)$

$\Rightarrow \dfrac{1}{y + 1}\,dy = (x + 1)\,dx \Rightarrow \int \dfrac{1}{y + 1}\,dy = \int x + 1\,dx$

$\Rightarrow \ln|y + 1| = \dfrac{x^2}{2} + x + C$

[3 marks available — 1 mark for correctly separating the variables, 1 mark for at least two terms integrated correctly, 1 mark for the correct answer]

b) When $x = 2$, $y = 0 \Rightarrow \ln 1 = \dfrac{4}{2} + 2 + C \Rightarrow C = -4$

$\ln(y + 1) = \dfrac{x^2}{2} + x - 4 \Rightarrow y + 1 = e^{\frac{x^2}{2} + x - 4}$

$\Rightarrow y = \dfrac{(e^x)^{\frac{x}{2} + 1}}{e^4} - 1$

[3 marks available — 1 mark for the correct value of C, 1 mark for attempting to rearrange into the required form, 1 mark for the correct answer]

Q7 $\int \dfrac{1}{y^2}\,dy = \int 3x + 2\,dx \Rightarrow -\dfrac{1}{y} = \dfrac{3}{2}x^2 + 2x + C$

When $y = 0.25$, $x = 2 \Rightarrow -4 = 6 + 4 + C \Rightarrow C = -14$

So $-\dfrac{1}{y} = \dfrac{3}{2}x^2 + 2x - 14$

When $x = 4$, $-\dfrac{1}{y} = 24 + 8 - 14 = 18 \Rightarrow y = -\dfrac{1}{18}$

[5 marks available — 1 mark for correctly separating the variables, 1 mark for at least two terms integrated correctly, 1 mark for the whole function integrated correctly, 1 mark for the correct value of C, 1 mark for the correct answer]

Q8 a) $\int \dfrac{3y^2 + 6y + 11}{y^2 + 2y + 1}\,dy = \int 3x\,dx$ *[1 mark]*

b) $\int \dfrac{3y^2 + 6y + 3 + 8}{y^2 + 2y + 1}\,dy = \int 3x\,dx$

$\Rightarrow \int 3 + \dfrac{8}{(y + 1)^2}\,dy = \int 3x\,dx$

$\Rightarrow 3y - \dfrac{8}{y + 1} = \dfrac{3}{2}x^2 + C$

[3 marks available — 1 mark for simplifying the fraction, 1 mark for at least two terms integrated correctly, 1 mark for the correct answer]

c) When $y = 1$, $x = 1 \Rightarrow 3 - 4 = \dfrac{3}{2} + C \Rightarrow C = -\dfrac{5}{2}$

So $3y - \dfrac{8}{y + 1} = \dfrac{3}{2}x^2 - \dfrac{5}{2}$ *[1 mark]*

Q9 $e^{y + 5 - 2x}\dfrac{dy}{dx} = 1 \Rightarrow \int e^{y + 5}\,dy = \int e^{2x}\,dx \Rightarrow e^{y + 5} = \dfrac{1}{2}e^{2x} + C$

When $x = 3$, $y = 1 \Rightarrow e^6 = \dfrac{1}{2}e^6 + C \Rightarrow C = \dfrac{1}{2}e^6$

So $e^{y + 5} = \dfrac{1}{2}e^{2x} + \dfrac{1}{2}e^6$

When $y = 3$, $e^8 = \dfrac{1}{2}e^{2x} + \dfrac{1}{2}e^6 \Rightarrow 2e^8 - e^6 = e^{2x}$

$\Rightarrow x = \dfrac{1}{2}\ln(2e^8 - e^6)$

[5 marks available — 1 mark for separating the variables, 1 mark for integrating correctly, 1 mark for the correct value of C, 1 mark for substituting y = 3, 1 mark for the correct answer]

Q10 a) $(x + 1)(x - 2)\dfrac{dy}{dx} = y(-x - 7)$

$\Rightarrow \int \dfrac{1}{y}\,dy = \int \dfrac{-x - 7}{(x + 1)(x - 2)}\,dx$

Find the second integral by writing it in partial fractions:

$\dfrac{-x - 7}{(x + 1)(x - 2)} \equiv \dfrac{A}{x + 1} + \dfrac{B}{x - 2}$

$\Rightarrow -x - 7 \equiv A(x - 2) + B(x + 1)$

Substituting $x = 2$ gives: $-2 - 7 = 3B \Rightarrow B = -3$

Substituting $x = -1$ gives: $1 - 7 = -3A \Rightarrow A = 2$

So $\dfrac{-x - 7}{(x + 1)(x - 2)} = \dfrac{2}{x + 1} - \dfrac{3}{x - 2}$

$\Rightarrow \int \dfrac{1}{y}\,dy = \int \dfrac{-x - 7}{(x + 1)(x - 2)}\,dx = \int \dfrac{2}{x + 1} - \dfrac{3}{x - 2}\,dx$

$\Rightarrow \ln|y| = 2\ln|x + 1| - 3\ln|x - 2| + C$

Since $x > 2$, $(x + 1)$ and $(x - 2)$ are both positive, so:

$\Rightarrow \ln|y| = 2\ln(x + 1) - 3\ln(x - 2) + \ln k$

$\Rightarrow \ln|y| = \ln\left(\dfrac{k(x + 1)^2}{(x - 2)^3}\right) \Rightarrow |y| = \dfrac{k(x + 1)^2}{(x - 2)^3}$

[7 marks available — 1 mark for separating the variables, 1 mark for attempting to write the x-integral using partial fractions, 1 mark for each correct partial fraction, 1 mark for at least two terms integrated correctly, 1 mark for all terms integrated correctly, 1 mark for using laws of logs to write in the required form]

b) When $x = 3$, $y = 32$: $32 = \dfrac{k(4)^2}{(1)^3} \Rightarrow 32 = 16k \Rightarrow k = 2$

So $|y| = \dfrac{2(x + 1)^2}{(x - 2)^3}$

[2 marks available — 1 mark for the correct value of k, 1 mark for the correct answer]

Exercise 8.21 — Applying Differential Equations to Real-Life Problems

Q1 a) $\dfrac{dN}{dt} = kN \Rightarrow \dfrac{1}{N}\,dN = k\,dt \Rightarrow \int \dfrac{1}{N}\,dN = \int k\,dt$

$\Rightarrow \ln N = kt + \ln A \Rightarrow N = e^{kt + \ln A} = Ae^{kt}$

Note that you don't need to put modulus signs in ln N here, as N can't be negative — you can't have a negative number of germs in your body. The same principle will apply to a lot of real-life differential equations questions.

b) $t = 0$, $N = 200 \Rightarrow 200 = Ae^0 = A \Rightarrow N = 200e^{kt}$

$t = 8$, $N = 400 \Rightarrow 400 = 200e^{8k} \Rightarrow \ln 2 = 8k$

$\Rightarrow k = \dfrac{1}{8}\ln 2 \Rightarrow N = 200\,e^{\frac{t}{8}\ln 2}$

So $t = 24 \Rightarrow N = 200e^{3\ln 2} = 1600$

c) Some possible answers are:

- The number of germs doubles every 8 hours, so over a long time the model becomes unrealistic because the total number of germs will become very large.

- The number of germs is a discrete variable (you can't have half of a germ) but the model is a continuous function.

- The model does not account for the differences between patients or other conditions such as the presence of other germs or chemicals.

Q2 a) $\dfrac{dV}{dt} \propto V \Rightarrow \dfrac{dV}{dt} = -kV$, for some $k > 0$

$\Rightarrow \dfrac{1}{V}\, dV = -k\, dt \Rightarrow \int \dfrac{1}{V}\, dV = \int -k\, dt$

$\Rightarrow \ln V = -kt + \ln A \Rightarrow V = Ae^{-kt}$

$t = 0,\ V = V_0 \Rightarrow V_0 = Ae^0 = A \Rightarrow V = V_0 e^{-kt}$

b) Using years as the unit of time:

$t = 1,\ V = \tfrac{1}{2}V_0 \Rightarrow \tfrac{1}{2}V_0 = V_0 e^{-k} \Rightarrow \tfrac{1}{2} = e^{-k}$

$\Rightarrow \ln \tfrac{1}{2} = -k \Rightarrow k = \ln 2 \Rightarrow V = V_0 e^{-t \ln 2}$

So $V = 0.05V_0 \Rightarrow 0.05V_0 = V_0 e^{-t \ln 2} \Rightarrow 0.05 = e^{-t \ln 2}$

$\Rightarrow \ln 0.05 = -t \ln 2 \Rightarrow t = \ln 0.05 \div -\ln 2 = 4.322...$

$\Rightarrow t = 4$ years, 4 months (or 52 months)

You could have used months as the units of time instead, and started with $t = 12$. You'd get the same answer.

Q3 a) The rate of change is proportional to S,

so $\dfrac{dS}{dt} \propto S \Rightarrow \dfrac{dS}{dt} = kS$.

b) Initially, $\dfrac{dS}{dt} = 6$ and $S = 30$, so $6 = 30k \Rightarrow k = 0.2$.

Solve the differential equation:

$\dfrac{dS}{dt} = 0.2S \Rightarrow \int \dfrac{1}{S}\, dS = \int 0.2\, dt$

$\Rightarrow \ln S = 0.2t + C \Rightarrow S = e^{0.2t + C} = Ae^{0.2t}$

Initially, $t = 0$ and $S = 30$: $30 = Ae^0 \Rightarrow A = 30$, so $S = 30e^{0.2t}$.

Then $150 = 30e^{0.2t} \Rightarrow e^{0.2t} = 5 \Rightarrow 0.2t = \ln 5$

$\Rightarrow t = 5 \ln 5 = 8.047...$

So it will take 8 weeks (to the nearest week) for the squirrels to take over.

Q4 a) $\dfrac{dN}{dt} \propto N \Rightarrow \dfrac{dN}{dt} = kN$

b) $\dfrac{dN}{dt} = kN \Rightarrow \int \dfrac{1}{N}\, dN = \int k\, dt$

$\Rightarrow \ln N = kt + \ln A \Rightarrow N = e^{kt + \ln A} = Ae^{kt}$

$N = 20$ at $t = 0 \Rightarrow 20 = Ae^0 = A \Rightarrow N = 20e^{kt}$

$N = 30$ at $t = 4 \Rightarrow 30 = 20e^{4k}$

$\Rightarrow k = 0.25 \ln 1.5 \Rightarrow N = 20e^{0.25t \ln 1.5}$

So $N = 1000 \Rightarrow 1000 = 20e^{0.25t \ln 1.5}$

$\Rightarrow \ln 50 = 0.25t \ln 1.5 \Rightarrow t = 4 \ln 50 \div \ln 1.5 = 38.59$

So the field will be overrun in 39 weeks.

c) $\dfrac{dN}{dt} \propto \sqrt{N} \Rightarrow \dfrac{dN}{dt} = k\sqrt{N}$

$\Rightarrow \int \dfrac{1}{\sqrt{N}}\, dN = \int k\, dt \Rightarrow 2\sqrt{N} = kt + C$

$N = 20$ at $t = 0 \Rightarrow 2\sqrt{20} = 4\sqrt{5} = C \Rightarrow 2\sqrt{N} = kt + 4\sqrt{5}$

$N = 30$ at $t = 4 \Rightarrow 2\sqrt{30} = 4k + 4\sqrt{5} \Rightarrow k = \dfrac{\sqrt{30} - 2\sqrt{5}}{2}$

$\Rightarrow 2\sqrt{N} = \dfrac{\sqrt{30} - 2\sqrt{5}}{2}t + 4\sqrt{5}$

So $N = 1000 \Rightarrow 2\sqrt{1000} = \dfrac{\sqrt{30} - 2\sqrt{5}}{2}t + 4\sqrt{5}$

$\Rightarrow t = \dfrac{4\sqrt{1000} - 8\sqrt{5}}{\sqrt{30} - 2\sqrt{5}} = 108.05$

So the field will be overrun in 108 weeks.

Be careful with all these square roots knocking about — it's easy to make a mistake.

d) Some possible answers are:
– The model could be adjusted to use a discrete-valued function, since the number of mice is a discrete variable.
– When there is only 1 mouse (i.e. $N = 1$), the population still increases, which is unrealistic. The model could be adjusted to be more accurate for very low values of N.
– As t increases, N gets larger without any limit. Introducing an upper limit on t or N would show at which point the model stops being accurate.

Q5 a) $\dfrac{dx}{dt} = \dfrac{1}{x^2(t + 1)}$

$V = x^3 \Rightarrow \dfrac{dV}{dx} = 3x^2$

So $\dfrac{dV}{dt} = \dfrac{dV}{dx} \times \dfrac{dx}{dt} = \dfrac{3x^2}{x^2(t+1)} = \dfrac{3}{t+1}$

b) $\dfrac{dV}{dt} = \dfrac{3}{t+1} \Rightarrow \int 1\, dV = \int \dfrac{3}{t+1}\, dt$

$\Rightarrow V = 3 \ln (t+1) + C$

$V = 15$ at $t = 0 \Rightarrow 15 = 3 \ln (1) + C \Rightarrow C = 15$

$\Rightarrow V = 3 \ln (t+1) + 15$

So $V = 18 \Rightarrow 18 = 3 \ln (t+1) + 15$

$\Rightarrow \dfrac{3}{3} = \ln (t+1)$

$\Rightarrow t = e^1 - 1 = 1.72$ seconds (3 s.f.)

Q6 a) $\dfrac{d\theta}{dt} = -k(\theta - 20) \Rightarrow \int \dfrac{1}{\theta - 20}\, d\theta = \int -k\, dt$

$\Rightarrow \ln (\theta - 20) = -kt + C$

When $t = 0,\ \theta = 90 \Rightarrow C = \ln 70$

When $t = 5,\ \theta = 75 \Rightarrow \ln 55 = -5k + \ln 70 \Rightarrow k = -\dfrac{1}{5}\ln \dfrac{55}{70}$

So $\ln (\theta - 20) = \left(\dfrac{1}{5}\ln \dfrac{11}{14}\right)t + \ln 70$

[6 marks available — 1 mark for a correct differential equation linking θ and t, 1 mark for correctly separating the variables, 1 mark for each side of the equation integrated correctly, 1 mark for the correct value of C, 1 mark for the correct value of k]

b) If the temperature in the room had been 30 °C, then the initial equation would have been: $\ln (\theta - 30) = -kt + C$

When $t = 0,\ \theta = 90 \Rightarrow C = \ln 60$

$k = -\dfrac{1}{5}\ln \dfrac{11}{14}$

So $\theta = 75$ °C $\Rightarrow \ln 45 = \left(\dfrac{1}{5}\ln \dfrac{11}{14}\right)t + \ln 60$

$\Rightarrow t = 5.9644... = 6$ minutes (to the nearest minute)

[3 marks available — 1 mark for correctly adapting equation found in part a), 1 mark for the correct value of C, 1 mark for the correct answer]

Q7 a) $\dfrac{dy}{dt} = k(p - y) \Rightarrow \int \dfrac{1}{p - y}\, dy = \int k\, dt$

$\Rightarrow -\ln (p - y) = kt + \ln a \Rightarrow \ln (p - y) = -kt - \ln a$

$\Rightarrow p - y = e^{-kt - \ln a} = e^{-kt}e^{-\ln a} = e^{-kt}e^{\ln \frac{1}{a}} = \dfrac{1}{a}e^{-kt} = Ae^{-kt}$

$\Rightarrow y = p - Ae^{-kt}$

b) If $p = 30\,000$ and $y = 10\,000$ at $t = 0$, then:

$10\,000 = 30\,000 - Ae^0 = 30\,000 - A$

$\Rightarrow A = 20\,000 \Rightarrow y = 30\,000 - 20\,000e^{-kt}$

$t = 5,\ y = 12\,000 \Rightarrow 12\,000 = 30\,000 - 20\,000e^{-5k}$

$\Rightarrow e^{-5k} = 18\,000 \div 20\,000 = 0.9 \Rightarrow -5k = \ln 0.9$

$\Rightarrow k = -0.2 \ln 0.9 \Rightarrow y = 30\,000 - 20\,000e^{0.2t \ln 0.9}$

So $y = 25\,000 \Rightarrow 20\,000e^{0.2t \ln 0.9} = 5000$

$\Rightarrow e^{0.2t \ln 0.9} = 0.25 \Rightarrow 0.2t \ln 0.9 = \ln 0.25$

$\Rightarrow t = 5 \ln 0.25 \div \ln 0.9 = 65.79 = 66$ days

c)

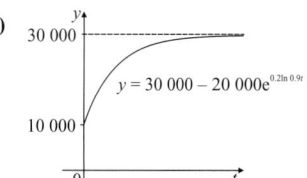

$y = 30\,000 - 20\,000e^{0.2\ln 0.9t}$

Remember that (0.2 ln 0.9) is negative when sketching the graph.

d) $t = 92$

$\Rightarrow y = 30\,000 - 20\,000e^{18.4 \ln 0.9} = 30\,000 - 20\,000e^{-1.939}$

$= 30\,000 - 20\,000(0.1439) = 27\,122$ signatures

So no, the target will not be achieved.

Don't forget, you'll often need to relate your answer back to the question when you've finished calculating.

Answers

e) Some possible answers are:

- The model suggests that the number of signatures eventually reaches the entire population of 30 000 (if you round to the nearest whole number). It may be more accurate to assume that there are some people who will never sign the petition, no matter how much time elapses, and allow for this in the model.
- The number of signatures is a discrete variable, and the model uses a continuous function. Adjusting the function to only allow integer values of y would fix this.
- The model does not account for people entering or leaving the town. Adjusting the function so that p varies with time may make the model more accurate.

Q8 a) $\frac{dV}{dt} \propto r^2 \Rightarrow \frac{dV}{dt} = -kr^2$, where $k > 0$.

Volume of a sphere $V = \frac{4}{3}\pi r^3 \Rightarrow r = \sqrt[3]{\frac{3V}{4\pi}}$

So $\frac{dV}{dt} = -k\left(\frac{3V}{4\pi}\right)^{\frac{2}{3}}$

[3 marks available — 1 mark for forming a differential equation in terms of r, 1 mark for using the formula for the volume of a sphere, 1 mark for the correct answer]

b) $\frac{dV}{dt} = -k\left(\frac{3V}{4\pi}\right)^{\frac{2}{3}} \Rightarrow \left(\frac{4\pi}{3V}\right)^{\frac{2}{3}}dV = -k\ dt$

$\Rightarrow \int\left(\frac{4\pi}{3}\right)^{\frac{2}{3}}V^{-\frac{2}{3}}\ dV = \int -k\ dt \Rightarrow 3\left(\frac{4\pi}{3}\right)^{\frac{2}{3}}V^{\frac{1}{3}} = -kt + C$

When $t = 0$, $V = 8\pi \Rightarrow C = 3\left(\frac{4\pi}{3}\right)^{\frac{2}{3}}(8\pi)^{\frac{1}{3}} = 22.834...$
When $t = 200$, $V = 3.375\pi$

$\Rightarrow 3\left(\frac{4\pi}{3}\right)^{\frac{2}{3}}(3.375\pi)^{\frac{1}{3}} = -200k + 22.834...$

$\Rightarrow k = 0.028543...$
When $V = 0$, $0 = -kt + C \Rightarrow t = 800$ seconds

[6 marks available — 1 mark for separating the variables, 1 mark for each side of the equation integrated correctly, 1 mark for the correct value of C, 1 mark for the correct value of k, 1 mark for the correct answer]

Q9 a) $\frac{dV}{dt} = 100\ln(t + 100)$

Volume of a sphere $V = \frac{4}{3}\pi r^3$

$\Rightarrow \frac{dV}{dr} = 4\pi r^2 \Rightarrow \frac{dr}{dV} = \frac{1}{4\pi r^2}$

Using the chain rule: $\frac{dr}{dt} = \frac{dV}{dt} \times \frac{dr}{dV} = \frac{25\ln(t + 100)}{\pi r^2}$

b) $\int \pi r^2\ dr = \int 25\ln(t + 100)\ dt$

$\int \pi r^2\ dr = \frac{\pi r^3}{3} + C_1$

$\int \ln x\ dx = x\ln x - x + C_2$, so:

$\int 25\ln(t + 100)\ dt = 25((t + 100)\ln(t + 100) - (t + 100)) + c$
$= 25(t + 100)\ln(t + 100) - 25t - 2500 + c$
$= 25(t + 100)\ln(t + 100) - 25t + C_3$

So $\frac{\pi r^3}{3} = 25(t + 100)\ln(t + 100) - 25t + C$ as required.

You could also have used integration by parts to find the integral with respect to t.

c) When $t = 0$, $V = 0$, so $r = 0$:

$0 = 2500\ln 100 + C$ and
$C = -2500\ln 100$

So $\frac{\pi r^3}{3} = 25(t + 100)\ln(t + 100) - 25t - 2500\ln 100$

When $t = 120$ seconds:

$\frac{\pi r^3}{3} = 25(220)\ln 220 - 25(120) - 2500\ln 100$
$= 5500\ln 220 - 3000 - 2500\ln 100$
$= 15152.02...$

$\Rightarrow r = 24.367... = 24.4$ cm (3 s.f.)

Chapter 8 Review Exercise

Q1 a) $\int \frac{1}{\sqrt[3]{(2-11x)}}\ dx = \int (2 - 11x)^{-\frac{1}{3}}\ dx = -\frac{3}{22}(2 - 11x)^{\frac{2}{3}} + C$

b) Integrate the curve between the two limits:

$\int_{-\frac{123}{11}}^{-\frac{62}{11}} \frac{1}{\sqrt[3]{(2-11x)}}\ dx = -\frac{3}{22}\left[(\sqrt[3]{2 - 11x})^2\right]_{-\frac{123}{11}}^{-\frac{62}{11}}$

$= -\frac{3}{22}\left([(\sqrt[3]{2 + 62})^2] - [(\sqrt[3]{2 + 123})^2]\right)$

$= -\frac{3}{22}\left((\sqrt[3]{64})^2 - (\sqrt[3]{125})^2\right) = -\frac{3}{22}(16 - 25) = \frac{27}{22}$

Q2 $y = \int (1 - 7x)^{\frac{1}{2}}\ dx = -\frac{2}{21}(1 - 7x)^{\frac{3}{2}} + C$

y goes through the point (0, 1), so:

$1 = -\frac{2}{21}(1)^{\frac{3}{2}} + C \Rightarrow C = 1 + \frac{2}{21} \Rightarrow C = \frac{23}{21}$

$\Rightarrow y = -\frac{2}{21}(1 - 7x)^{\frac{3}{2}} + \frac{23}{21}$

Q3 a) $\int 4e^{2x}\ dx = 4\int e^{2x}\ dx = 4\left(\frac{1}{2}e^{2x}\right) + C = 2e^{2x} + C$

b) $\int e^{3x-5}\ dx = \frac{1}{3}e^{3x-5} + C$

c) $\int \frac{2}{3x}\ dx = \frac{2}{3}\int \frac{1}{x}\ dx = \frac{2}{3}\ln|x| + C$

d) $\int \frac{2}{2x + 1}\ dx = \ln|2x + 1| + C$

Q4 $\int \frac{8}{2 - x} - \frac{8}{x}\ dx = -8\ln|2 - x| - 8\ln|x| + C$

$= -8(\ln|2 - x| + \ln|x|) + C$
$= -8\ln|2x - x^2| + C = \ln|(2x - x^2)^{-8}| + C$

So $P = (2x - x^2)^{-8}$.

Q5 a) $e^x > 0$ for all x, and $\frac{5}{x} > 0$ for $x > 0$.
So y is always positive over the range $2 \le x \le 5$. *[1 mark]*

b) $\int_2^5 e^x + \frac{5}{x}\ dx = [e^x + 5\ln|x|]_2^5$

$= e^5 + 5\ln 5 - e^2 - 5\ln 2 = e^5 + e^2 + 5\ln \frac{5}{2}$

[3 marks available — 1 mark for integrating correctly, 1 mark for substituting limits, 1 mark for the correct answer]

Q6 $\int_0^A e^{5x}\ dx = \left[\frac{1}{5}e^{5x}\right]_0^A = \frac{1}{5}[e^{5A} - e^0] = \frac{31}{5}$

$\Rightarrow e^{5A} - 1 = 31 \Rightarrow e^{5A} = 32$

$\Rightarrow \ln e^{5A} = \ln 32 = \ln 2^5 = 5\ln 2 \Rightarrow 5A = 5\ln 2 \Rightarrow A = \ln 2$

[4 marks available — 1 mark for integrating the exponential, 1 mark for substituting limits, 1 mark for attempting to solve the equation, 1 mark for the correct value of A]

Q7 a) $\int \cos(x + A)\ dx = \sin(x + A) + C$

b) $\int \text{cosec}^2((A + B)t + A + B)\ dt$

$= -\frac{1}{A + B}\cot((A + B)t + A + B) + C$

Q8 a) $\int_{-\pi}^{\pi} \cos 2x\ dx = \left[\frac{1}{2}\sin 2x\right]_{-\pi}^{\pi} = \frac{1}{2}[\sin(2\pi) - \sin(-2\pi)] = 0$

[3 marks available — 1 mark for integrating correctly, 1 mark for substituting limits, 1 mark for the correct answer]

b) The result suggests that the area above the curve and below the x-axis is equal to the area above the x-axis and below the curve, over the given range. *[1 mark]*

Q9 $\int_{\frac{\pi}{3}}^{\frac{\pi}{2}} \sin(3x - \pi)\ dx = \left[-\frac{1}{3}\cos(3x - \pi)\right]_{\frac{\pi}{3}}^{\frac{\pi}{2}}$

$= -\frac{1}{3}\left[\cos\left(3\frac{\pi}{2} - \pi\right) - \cos\left(3\frac{\pi}{3} - \pi\right)\right] = -\frac{1}{3}\left(\cos\frac{\pi}{2} - \cos 0\right)$

$= -\frac{1}{3}(0 - 1) = \frac{1}{3}$

[3 marks available — 1 mark for integrating sin correctly, 1 mark for substituting limits, 1 mark for correct final value]

Q10 a) $\int \cos 4x - \sec^2 7x \, dx = \frac{1}{4}\sin 4x - \frac{1}{7}\tan 7x + C$

b) $\int 6\sec 3x \tan 3x - \csc^2 \frac{x}{5} \, dx = 2\sec 3x + 5\cot\frac{x}{5} + C$

Q11 Differentiate the denominator: $\frac{d}{dx}(2x^2 + 1) = 4x = $ numerator

So $\int \frac{4x}{2x^2+1}\,dx = \ln|2x^2+1| + C$ *[1 mark]*

The modulus sign isn't required, since $2x^2 + 1 > 0$ for all real x.

Q12 Differentiate the denominator: $\frac{d}{dx}(3x^2 + 18x - 7) = 6x + 18$

Numerator $= \frac{1}{6} \times$ derivative of the denominator:

$\int \frac{x+3}{3x^2+18x-7}\,dx = \frac{1}{6}\int \frac{6x+18}{3x^2+18x-7}\,dx$

$= \frac{1}{6}\ln|3x^2 + 18x - 7| + C$

[2 marks available — 1 mark for differentiating the denominator, 1 mark for integrating the fraction]

Q13 a) Differentiate the denominator: $\frac{d}{dx}(\sin x) = \cos x = $ numerator

So $\int \frac{\cos x}{\sin x}\,dx = \ln|\sin x| + C$

You could also write $\frac{\cos x}{\sin x}$ as cot x and use the standard result.

b) Differentiate the denominator: $\frac{d}{dx}(x^5 + x^3 - 3x) = 5x^4 + 3x^2 - 3$

Numerator $= 4 \times$ derivative of the denominator:

$\int \frac{20x^4 + 12x^2 - 12}{x^5+x^3-3x}\,dx = 4\int \frac{5x^4+3x^2-3}{x^5+x^3-3x}\,dx$

$= 4\ln|x^5 + x^3 - 3x| + C$

Q14 Differentiate the denominator: $\frac{d}{dx}(4x - 3) = 4$

So $\int_2^7 \frac{8}{4x-3}\,dx = 2\int_2^7 \frac{4}{4x-3}\,dx = 2[\ln|4x-3|]_2^7$

$= 2[\ln 25 - \ln 5] = 2\ln 5 = \ln 25$

[4 marks available — 1 mark for differentiating the denominator, 1 mark for integrating the fraction, 1 mark for substituting limits, 1 mark for correct answer]

Q15 a) Let $u = x^3$ so $\frac{du}{dx} = 3x^2$ and $f'(u) = e^u$, so $f(u) = e^u$.

Using the formula: $\int 3x^2 e^{x^3}\,dx = e^{x^3} + C$

b) Let $u = \sin(x^2)$ so $\frac{du}{dx} = 2x\cos(x^2)$ and $f'(u) = e^u$, so $f(u) = e^u$.

Using the formula: $\int 2x\cos(x^2)\,e^{\sin(x^2)}\,dx = e^{\sin(x^2)} + C$

c) Let $u = \sec 4x$ so $\frac{du}{dx} = 4\sec 4x \tan 4x$

and $f'(u) = e^u$, so $f(u) = e^u$.

Using the formula: $\int 4\sec 4x \tan 4x\,e^{\sec 4x}\,dx = e^{\sec 4x} + C$

Divide by 4 to get the original integral:

$\frac{1}{4}\int 4\sec 4x \tan 4x\,e^{\sec 4x}\,dx = \frac{1}{4}e^{\sec 4x} + C$

Q16 Let $u = \cot(x^2)$ so $\frac{du}{dx} = -2x\csc^2(x^2)$ and $f'(u) = e^u$ so $f(u) = e^u$.

Using the formula: $\int -2x\csc^2(x^2)\,e^{\cot(x^2)}\,dx = e^{\cot(x^2)} + C$

Multiply by -2 to get the original integral:

$-2\int -2x\csc^2(x^2)\,e^{\cot(x^2)}\,dx = -2e^{\cot(x^2)} + C$

[4 marks available — 1 mark for identifying u, 1 mark for differentiating u, 1 mark for attempting to use the formula, 1 mark for the correct answer]

Q17 Using the tan double angle formula:

$\int \frac{2\tan 3x}{1 - \tan^2 3x}\,dx = \int \tan 6x\,dx = \frac{1}{6}\ln|\sec 6x| + C$

You could also have $-\frac{1}{6}\ln|\cos 6x| + C$ instead.

Q18 a) Using the cos double angle formula:

$\int 2\sin^2 x\,dx = 2\int \frac{1}{2}(1 - \cos 2x)\,dx$

$= \int 1 - \cos 2x\,dx = x - \frac{1}{2}\sin 2x + C$

b) Using the sin double angle formula:

$\int \sin 2x \cos 2x\,dx = \int \frac{1}{2}\sin 4x\,dx$

$= \frac{1}{2} \times -\frac{1}{4}\cos 4x + C = -\frac{1}{8}\cos 4x + C$

c) Using the identity $\sec^2 x = \tan^2 x + 1$:

$\int \tan^2 x + 1\,dx = \int \sec^2 x\,dx = \tan x + C$

Q19 Using the identity $\sec^2 x = \tan^2 x + 1$:

$\frac{2\tan 2x}{2 - \sec^2 2x} = \frac{2\tan 2x}{1 - \tan^2 2x}$

Using the tan double angle formula:

$\frac{2\tan 2x}{1 - \tan^2 2x} = \tan 4x$

So $\int \frac{2\tan 2x}{2-\sec^2 2x}\,dx = \int \tan 4x\,dx = \frac{1}{4}\ln|\sec 4x| + C$

[3 marks available — 1 mark for converting to be all in terms of tan 2x, 1 mark for converting to tan 4x, 1 mark for the correct answer]

Q20 $\int_0^{\frac{\pi}{4}} 3\sin 2x \cos 2x\,dx = 3\int_0^{\frac{\pi}{4}} \frac{1}{2}\sin 4x\,dx = \frac{3}{2}\left[-\frac{1}{4}\cos 4x\right]_0^{\frac{\pi}{4}}$

$= -\frac{3}{8}(\cos \pi - \cos 0) = -\frac{3}{8}((-1) - (1)) = \frac{3}{4}$

[4 marks available — 1 mark for using the sin double angle formula, 1 mark for integrating sin 4x, 1 mark for substituting limits, 1 mark for correct answer]

Q21 a) Find the points of intersection between the curves:

$6 + 2x - x^2 = \frac{x^2}{2} - x + \frac{3}{2} \Rightarrow 12 + 4x - 2x^2 = x^2 - 2x + 3$

$\Rightarrow 9 + 6x - 3x^2 = 0 \Rightarrow x^2 - 2x - 3 = 0 \Rightarrow (x-3)(x+1) = 0$

The curves intersect at $x = -1$ and $x = 3$.

The shaded area is the area under $y = 6 + 2x - x^2$ between $x = -1$ and $x = 3$ minus the area under $y = \frac{x^2}{2} - x + \frac{3}{2}$ between $x = -1$ and $x = 3$.

$\int_{-1}^3 6 + 2x - x^2\,dx = \left[6x + x^2 - \frac{x^3}{3}\right]_{-1}^3$

$= \left(6\times 3 + 3^2 - \frac{3^3}{3}\right) - \left(6\times -1 + (-1)^2 - \frac{(-1)^3}{3}\right)$

$= (18 + 9 - 9) - \left(-6 + 1 + \frac{1}{3}\right) = 18 - \left(-\frac{14}{3}\right) = \frac{68}{3}$

$\int_{-1}^3 \frac{x^2}{2} - x + \frac{3}{2}\,dx = \left[\frac{x^3}{6} - \frac{x^2}{2} + \frac{3x}{2}\right]_{-1}^3$

$= \left(\frac{3^3}{6} - \frac{3^2}{2} + \frac{3\times 3}{2}\right) - \left(\frac{(-1)^3}{6} - \frac{(-1)^2}{2} + \frac{3\times(-1)}{2}\right)$

$= \left(\frac{9}{2} - \frac{9}{2} + \frac{9}{2}\right) - \left(-\frac{1}{6} - \frac{1}{2} - \frac{3}{2}\right) = \frac{9}{2} - \left(-\frac{13}{6}\right) = \frac{20}{3}$

Total area under the curve $= \frac{68}{3} - \frac{20}{3} = \frac{48}{3} = 16$

Answers

b) Find the points of intersection between the curve and the line:

$\frac{2}{x} = 5 - 2x \Rightarrow 2 = 5x - 2x^2 \Rightarrow 2x^2 - 5x + 2 = 0$
$\Rightarrow (x - 2)(2x - 1) = 0$

The curve and the line intersect at $x = \frac{1}{2}$ and $x = 2$.

The shaded area is the area under $y = 5 - 2x$ between $x = \frac{1}{2}$ and $x = 2$ minus the area under $y = \frac{2}{x}$ between $x = \frac{1}{2}$ and $x = 2$.

$\int_{\frac{1}{2}}^{2} 5 - 2x \, dx = [5x - x^2]_{\frac{1}{2}}^{2}$

$= (5 \times 2 - 2^2) - \left(5 \times \frac{1}{2} - \frac{1}{2}^2\right)$

$= (10 - 4) - \left(\frac{5}{2} - \frac{1}{4}\right) = 6 - \frac{9}{4} = \frac{15}{4}$

$\int_{\frac{1}{2}}^{2} \frac{2}{x} \, dx = [2 \ln x]_{\frac{1}{2}}^{2} = 2 \ln 2 - 2 \ln \frac{1}{2}$

$= \ln\left(\left(2 \div \frac{1}{2}\right)^2\right) = \ln(4^2) = \ln 16$

So the total area is $\frac{15}{4} - \ln 16$.

You could have found the area under $y = 5 - 2x$ using the formula for the area of the trapezium instead of integrating.

Q22 a) $y = \arctan x \Rightarrow x = \tan y$, so integrate x with respect to y:

$\int_{\frac{\pi}{6}}^{\frac{\pi}{4}} \tan y \, dy = [\ln|\sec y|]_{\frac{\pi}{6}}^{\frac{\pi}{4}} = \ln\left|\sec \frac{\pi}{4}\right| - \ln\left|\sec \frac{\pi}{6}\right|$

$= \ln \sqrt{2} - \ln\left(\frac{2}{\sqrt{3}}\right) = \ln\left(\sqrt{2} \div \frac{2}{\sqrt{3}}\right) = \ln\left(\frac{\sqrt{6}}{2}\right)$

b) $y = \frac{1}{x^2} \Rightarrow x = \frac{1}{\sqrt{y}}$, so integrate with respect to x:

$\int_{4}^{16} y^{-\frac{1}{2}} \, dy = \left[2y^{\frac{1}{2}}\right]_{4}^{16} = 2(4 - 2) = 4$

Q23 a) $\int (3 - x)^{\frac{1}{2}} \, dx = -\frac{2}{3}(3 - x)^{\frac{3}{2}} + C$

[2 marks available — 2 marks for the correct answer, otherwise 1 mark for the correct power of $(3 - x)$]

b) The required area can be split as shown:

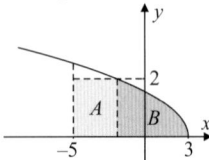

When $y = 2$, $x = -1$, so:

Area $A = 4 \times 2 = 8$

Area $B = \int_{-1}^{3} (3 - x)^{\frac{1}{2}} \, dx = \left[-\frac{2}{3}(3 - x)^{\frac{3}{2}}\right]_{-1}^{3}$

$= [0] - \left[-\frac{2}{3}(4)^{\frac{3}{2}}\right] = \frac{2}{3} \times 8 = \frac{16}{3}$

So total area $= 8 + \frac{16}{3} = \frac{40}{3}$

[3 marks available — 1 mark for substituting the correct limits into the integral from part a), 1 mark for calculating the definite integral correctly, 1 mark for the correct answer]

Q24 a) $4 = t^2 + 3 \Rightarrow t^2 = 1 \Rightarrow t = 1$
$12 = t^2 + 3 \Rightarrow t^2 = 9 \Rightarrow t = 3$

b) $\int y \, dx = \int y \frac{dx}{dt} \, dt$, so

$\int_{1}^{3} (4t - 1) 2t \, dt = \int_{1}^{3} 8t^2 - 2t \, dt = \left[\frac{8t^3}{3} - t^2\right]_{1}^{3}$

$= \left(\frac{8(3)^3}{3} - 3^2\right) - \left(\frac{8(1)^3}{3} - 1^2\right)$

$= (72 - 9) - \left(\frac{8}{3} - 1\right) = \frac{184}{3}$

Q25 Alter the limits from x-values into t-values:
$x = 2 \Rightarrow 2 = 2t^3 \Rightarrow t^3 = 1 \Rightarrow t = 1$
$x = 54 \Rightarrow 54 = 2t^3 \Rightarrow t^3 = 27 \Rightarrow t = 3$

$\int y \, dx = \int y \frac{dx}{dt} \, dt$ and $\frac{dx}{dt} = 6t^2$, so:

$\int_{2}^{54} y \, dx = \int_{1}^{3} \frac{2}{t} \times 6t^2 \, dt = \int_{1}^{3} 12t \, dt = [6t^2]_{1}^{3} = 6(3^2 - 1^2) = 48$

[6 marks available — 1 mark for altering the limits, 1 mark for finding $\frac{dx}{dt}$, 1 mark for using the formula for integrating parametric equations, 1 mark for integrating, 1 mark for substituting the limits, 1 mark for the correct answer]

Q26 a) $u = 5 - x^2 \Rightarrow \frac{du}{dx} = -2x \Rightarrow dx = -\frac{1}{2x} \, du$

So $\int 16x(5 - x^2)^5 \, dx = \int 16x \, u^5 \times -\frac{1}{2x} \, du$

$= \int 8u^5 \, du = \frac{4}{3}u^6 + C = \frac{4}{3}(5 - x^2)^6 + C$

b) $u = e^x - 1 \Rightarrow \frac{du}{dx} = e^x \Rightarrow dx = \frac{1}{e^x} \, du = \frac{1}{u + 1} \, du$

So $\int e^x(e^x + 1)(e^x - 1)^2 \, dx = \int (u + 1)(u + 2)u^2 \times \frac{1}{u + 1} \, du$

$= \int u^3 + 2u^2 \, du = \frac{u^4}{4} - \frac{2u^3}{3} + C$

$= \frac{1}{4}(e^x - 1)^4 + \frac{2}{3}(e^x - 1)^3 + C$

c) $u = x^2 - 4 \Rightarrow \frac{du}{dx} = 2x \Rightarrow dx = \frac{1}{2x} \, du$
$x = 2 \Rightarrow u = 2^2 - 4 = 0$
$x = 4 \Rightarrow u = 4^2 - 4 = 12$

So $\int_{2}^{4} x(x^2 - 4)^3 \, dx = \int_{0}^{12} x \, u^3 \times \frac{1}{2x} \, du = \int_{0}^{12} \frac{1}{2}u^3 \, du$

$= \left[\frac{1}{8}u^4\right]_{0}^{12} = \frac{1}{8}[12^4 - 0^4] = 2592$

d) $u = \sqrt{3x - 8} \Rightarrow \frac{du}{dx} = \frac{3}{2\sqrt{3x - 8}}$

$\Rightarrow dx = \frac{2\sqrt{3x - 8}}{3} \, du = \frac{2u}{3} \, du$

$u = \sqrt{3x - 8} \Rightarrow u^2 = 3x - 8 \Rightarrow x = \frac{u^2 + 8}{3}$

$x = 3 \Rightarrow u = \sqrt{3(3) - 8} = 1$
$x = 11 \Rightarrow u = \sqrt{3(11) - 8} = 5$

So $\int_{3}^{11} \frac{2x}{\sqrt{3x - 8}} \, dx = \int_{1}^{5} \frac{2x}{u} \times \frac{2u}{3} \, du = \int_{1}^{5} \frac{4x}{3} \, du$

$= \int_{1}^{5} \frac{4}{3} \times \frac{u^2 + 8}{3} \, du = \int_{1}^{5} \frac{4}{9}(u^2 + 8) \, du = \frac{4}{9}\left[\frac{u^3}{3} + 8u\right]_{1}^{5}$

$= \frac{4}{9}\left[\left(\frac{125}{3} + 40\right) - \left(\frac{1}{3} + 8\right)\right] = \frac{4}{9}\left(\frac{220}{3}\right) = \frac{880}{27}$

Q27 Let $u = x + 1 \Rightarrow \frac{du}{dx} = 1 \Rightarrow dx = du$ and $u = x + 1 \Rightarrow x = u - 1$

So $\int x(x + 1)^3 \, dx = \int (u - 1)u^3 \, du = \int u^4 - u^3 \, du$

$= \frac{u^5}{5} - \frac{u^4}{4} + C = \frac{(x + 1)^5}{5} - \frac{(x + 1)^4}{4} + C$

[3 marks available — 1 mark for choosing an appropriate substitution, 1 mark for integrating the function in terms of u, 1 mark for the correct integral in terms of x]

Q28 E.g.: He has replaced dx with du, rather than finding dx in terms of du.
He has integrated $\frac{u}{u}$ to $u \ln |u|$, rather than simplifying to 1 and integrating to u.

Answers

Q29 $u = \cos x \Rightarrow \dfrac{du}{dx} = -\sin x \Rightarrow dx = -\dfrac{1}{\sin x}\,du$

$x = 0 \Rightarrow u = \cos 0 = 1$

$x = \dfrac{\pi}{2} \Rightarrow u = \cos \dfrac{\pi}{2} = 0$

Using the sin double angle formula:

$\displaystyle\int_0^{\frac{\pi}{2}} \dfrac{1}{4}\cos x \sin 2x\,dx = \int_0^{\frac{\pi}{2}} \dfrac{1}{2}\cos^2 x \sin x\,dx$

So $\displaystyle\int_0^{\frac{\pi}{2}} \dfrac{1}{4}\cos x \sin 2x\,dx = \int_1^0 \dfrac{1}{2}u^2 \sin x \times -\dfrac{1}{\sin x}\,du$

$\displaystyle = \int_0^1 \dfrac{1}{2}u^2\,du = \left[\dfrac{u^3}{6}\right]_0^1 = \dfrac{1}{6}$

Q30 $x = 3\tan u \Rightarrow \dfrac{dx}{du} = 3\sec^2 u \Rightarrow dx = 3\sec^2 u\,du$

So $\displaystyle\int \dfrac{1}{9 + x^2}\,dx = \int \dfrac{1}{9 + 9\tan^2 u}\,3\sec^2 u\,du$

$\displaystyle = \int \dfrac{1}{9(1 + \tan^2 u)}\,3\sec^2 u\,du = \dfrac{1}{3}\int \dfrac{\sec^2 u}{\sec^2 u}\,du = \dfrac{1}{3}u + C$

But $x = 3\tan u \Rightarrow u = \tan^{-1}\dfrac{x}{3}$, so $\displaystyle\int \dfrac{1}{9 + x^2}\,dx = \dfrac{1}{3}\tan^{-1}\dfrac{x}{3} + C$

You could also write this as $\dfrac{1}{3}\arctan\dfrac{x}{3} + C$.

[6 marks available — 1 mark for $\dfrac{dx}{du}$, 1 mark for a correct integral in terms of u, 1 mark for simplifying the integral, 1 mark for using $\sec^2 u \equiv 1 + \tan^2 u$, 1 mark for integrating, 1 mark for the correct integral in terms of x]

Q31 a) $x = \cos u \Rightarrow \dfrac{dx}{du} = -\sin u \Rightarrow dx = -\sin u\,du$

So $\mathrm{f}(x) = \displaystyle\int x\sqrt{1 - x^2}\,dx = \int \cos u\,\sqrt{1 - \cos^2 u}\,(-\sin u)\,du$

$\displaystyle = \int -\cos u \sin u \sin u\,du = -\dfrac{1}{3}\int 3\cos u \sin^2 u\,du$

This integral is in the form $\displaystyle\int (n + 1)\mathrm{f}'(x)[\mathrm{f}(x)]^n\,dx$,

so it integrates to $[\mathrm{f}(x)]^{n+1} + C$

So $\mathrm{f}(x) = -\dfrac{1}{3}\sin^3 u + C$

[5 marks available — 1 mark for $\dfrac{dx}{du}$, 1 mark for a correct integral in terms of u, 1 mark for simplifying the integral, 1 mark for a correct method to integrate, 1 mark for the correct answer]

b) Max value of $-\dfrac{1}{3}\sin^3 u = \dfrac{1}{3}$, so max value of $\mathrm{f}(x)$ is $\dfrac{1}{3} + C$

This max occurs when $u = \dfrac{3\pi}{2}, \dfrac{7\pi}{2}, \dfrac{11\pi}{2} \ldots$

or in general form, $u = \dfrac{(4n - 1)\pi}{2}$

[3 marks available — 1 mark for the correct maximum value, 1 mark for any correct value of u, 1 mark for the correct general solution]

c) $x = \cos u = 0$ for all of these values of u. *[1 mark]*

Q32 a) Let $u = \ln x$ and $\dfrac{dv}{dx} = 3x^2$. Then $\dfrac{du}{dx} = \dfrac{1}{x}$ and $v = x^3$.

So $\displaystyle\int 3x^2 \ln x\,dx = x^3 \ln x - \int x^3 \times \dfrac{1}{x}\,dx$

$\displaystyle = x^3 \ln x - \int x^2\,dx = x^3 \ln x - \dfrac{x^3}{3} + C$

b) Let $u = 4x$ and $\dfrac{dv}{dx} = \cos 4x$. Then $\dfrac{du}{dx} = 4$ and $v = \dfrac{1}{4}\sin 4x$.

So $\displaystyle\int 4x\cos 4x\,dx = x\sin 4x - \int 4 \times \dfrac{1}{4}\sin 4x\,dx$

$\displaystyle = x\sin 4x - \int \sin 4x\,dx = x\sin 4x + \dfrac{1}{4}\cos 4x + C$

c) Let $u = x^2$ and $\dfrac{dv}{dx} = e^{\frac{x}{2}}$. Then $\dfrac{du}{dx} = 2x$ and $v = 2e^{\frac{x}{2}}$.

So $\displaystyle\int_0^4 e^{\frac{x}{2}}x^2\,dx = \left[2x^2 e^{\frac{x}{2}}\right]_0^4 - \int_0^4 4xe^{\frac{x}{2}}\,dx = 32e^2 - \int_0^4 4xe^{\frac{x}{2}}\,dx$

Integrate by parts again to find $\displaystyle\int_0^4 4xe^{\frac{x}{2}}\,dx$:

Let $u_1 = 4x$ and $\dfrac{dv_1}{dx} = e^{\frac{x}{2}}$. Then $\dfrac{du_1}{dx} = 4$ and $v_1 = 2e^{\frac{x}{2}}$.

So $\displaystyle\int_0^4 4xe^{\frac{x}{2}}\,dx = \left[8xe^{\frac{x}{2}}\right]_0^4 - \int_0^4 8e^{\frac{x}{2}}\,dx = \left[8xe^{\frac{x}{2}}\right]_0^4 - 8\left[2e^{\frac{x}{2}}\right]_0^4$

$= 32e^2 - (16e^2 - 16e^0) = 16e^2 + 16$

So $\displaystyle\int_0^4 e^{\frac{x}{2}}x^2\,dx = 32e^2 - (16e^2 + 16) = 16e^2 - 16$

Q33 a) Let $u = x$ and $\dfrac{dv}{dx} = \sin x$. Then $\dfrac{du}{dx} = 1$ and $v = -\cos x$.

So $\displaystyle\int x\sin x\,dx = -x\cos x + \int \cos x\,dx = -x\cos x + \sin x + C$

[4 marks available — 1 mark for identifying u and $\dfrac{dv}{dx}$, 1 mark for finding $\dfrac{du}{dx}$ and v, 1 mark for using the integration by parts formula, 1 mark for the correct answer]

b) $\displaystyle\int_0^{\frac{\pi}{2}} x\sin x\,dx = [-x\cos x + \sin x]_0^{\frac{\pi}{2}}$

$= \left(-\dfrac{\pi}{2}(0) + 1\right) - (0(1) + 0) = 1$

[2 marks available — 1 mark for substituting the limits, 1 mark for the correct answer]

c) $y = 6x\sin x$ is an enlargement of $y = x\sin x$ by a scale factor 6 in the y-direction. So $\displaystyle\int_0^{\frac{\pi}{2}} 6x\sin x\,dx = 6$ *[1 mark]*

d) The graph is below the axis for some of this domain, so the area for this part would need to be calculated separately. *[1 mark]*

Q34 The graph intersects the x-axis when

$(3 - x)(x + 1)^{\frac{1}{2}} = 0$, i.e. at $x = 3$ and $x = -1$,

so the required area is $\displaystyle\int_{-1}^3 (3 - x)(x + 1)^{\frac{1}{2}}\,dx$

Let $u = 3 - x$ and $\dfrac{dv}{dx} = (x + 1)^{\frac{1}{2}}$.

Then $\dfrac{du}{dx} = -1$ and $v = \dfrac{2}{3}(x + 1)^{\frac{3}{2}}$.

So $\displaystyle\int_{-1}^3 (3 - x)(x + 1)^{\frac{1}{2}}\,dx$

$= \left[\dfrac{2}{3}(x + 1)^{\frac{3}{2}}(3 - x)\right]_{-1}^3 + \int_{-1}^3 \dfrac{2}{3}(x + 1)^{\frac{3}{2}}\,dx$

$= \left[\dfrac{2}{3}(x + 1)^{\frac{3}{2}}(3 - x) + \dfrac{4}{15}(x + 1)^{\frac{5}{2}}\right]_{-1}^3$

$= \left(0 + \dfrac{4}{15} \times 4^{\frac{5}{2}}\right) - (0 + 0) = \dfrac{4}{15} \times 32 = \dfrac{128}{15}$

[7 marks available — 1 mark for the correct limits, 1 mark for identifying u and $\dfrac{dv}{dx}$, 1 mark for finding $\dfrac{du}{dx}$ and v, 1 mark for using the integration by parts formula, 1 mark for the correct integral, 1 mark for substituting the limits, 1 mark for the correct answer]

Q35 Let $u = 10x^2$ and $\dfrac{dv}{dx} = e^{5x}$. Then $\dfrac{du}{dx} = 20x$ and $v = \dfrac{1}{5}e^{5x}$.

So $\displaystyle\int 10x^2 e^{5x}\,dx = 2x^2 e^{5x} - \int 4xe^{5x}\,dx$

Integrate by parts again to find $\displaystyle\int 4xe^{5x}\,dx$:

Let $u_1 = 4x$ and $\dfrac{dv_1}{dx} = e^{5x}$. Then $\dfrac{du_1}{dx} = 4$ and $v_1 = \dfrac{1}{5}e^{5x}$.

So $\displaystyle\int 4xe^{5x}\,dx = \dfrac{4}{5}xe^{5x} - \int \dfrac{4}{5}e^{5x}\,dx = \dfrac{4}{5}xe^{5x} - \dfrac{4}{25}e^{5x} + C$

$\displaystyle\int 10x^2 e^{5x}\,dx = 2x^2 e^{5x} - \int 4xe^{5x}\,dx = 2x^2 e^{5x} - \dfrac{4}{5}xe^{5x} + \dfrac{4}{25}e^{5x} + C$

Answers

Q36 Integrate by parts:

Let $u = 2x^2$ and $\frac{dv}{dx} = e^{3x}$. Then $\frac{du}{dx} = 4x$ and $v = \frac{1}{3}e^{3x}$.

So $\int_0^1 2x^2 e^{3x}\, dx = \left[\frac{2}{3}x^2 e^{3x}\right]_0^1 - \int_0^1 \frac{4}{3}x e^{3x}\, dx = \frac{2}{3}e^3 - \int_0^1 \frac{4}{3}x e^{3x}\, dx$

Integrate by parts again to find $\int_0^1 \frac{4}{3}x e^{3x}\, dx$:

Let $u_1 = \frac{4}{3}x$ and $\frac{dv_1}{dx} = e^{3x}$. Then $\frac{du_1}{dx} = \frac{4}{3}$ and $v_1 = \frac{1}{3}e^{3x}$.

So $\int_0^1 \frac{4}{3}x e^{3x}\, dx = \left[\frac{4}{9}x e^{3x}\right]_0^1 - \int_0^1 \frac{4}{9}e^{3x}\, dx = \left[\frac{4}{9}x e^{3x}\right]_0^1 - \frac{4}{9}\left[\frac{1}{3}e^{3x}\right]_0^1$

$= \frac{4}{9}e^3 - \left(\frac{4}{27}e^3 - \frac{4}{27}\right) = \frac{8}{27}e^3 + \frac{4}{27}$

So $\int_0^1 2x^2 e^{3x}\, dx = \frac{2}{3}e^3 - \left(\frac{8}{27}e^3 + \frac{4}{27}\right)$

$= \frac{10}{27}e^3 - \frac{4}{27} = \frac{2}{27}(5e^3 - 2)$

[7 marks available — 1 mark for identifying u and $\frac{dv}{dx}$, 1 mark for finding v and $\frac{du}{dx}$, 1 mark for using the integration by parts formula, 1 mark for using integration by parts again on the integral, 1 mark for the correct integral, 1 mark for substituting the limits, 1 mark for the correct final answer]

Q37 a) $x^2 \cos 4x = 0 \Rightarrow x = 0$ or $\cos 4x = 0$

$\cos 4x = 0 \Rightarrow 4x = \frac{\pi}{2}, \frac{3\pi}{2}, \frac{5\pi}{2}... \Rightarrow x = \frac{\pi}{8}, \frac{3\pi}{8}, \frac{5\pi}{8}...$

So the first point of intersection for $x > 0$ is at $\left(\frac{\pi}{8}, 0\right)$.
[1 mark]

b) Shaded area $= \int_0^{\frac{\pi}{8}} x^2 \cos 4x\, dx$

Let $u = x^2$ and $\frac{dv}{dx} = \cos 4x$. Then $\frac{du}{dx} = 2x$ and $v = \frac{1}{4}\sin 4x$

So $\int_0^{\frac{\pi}{8}} x^2 \cos 4x\, dx = \left[\frac{1}{4}x^2 \sin 4x\right]_0^{\frac{\pi}{8}} - \int_0^{\frac{\pi}{8}} \frac{1}{2}x \sin 4x\, dx$

Use integration by parts again to find $\int_0^{\frac{\pi}{8}} \frac{1}{2}x \sin 4x\, dx$:

Let $u = \frac{1}{2}x$ and $\frac{dv}{dx} = \sin 4x$.

Then $\frac{du}{dx} = \frac{1}{2}$ and $v = -\frac{1}{4}\cos 4x$

So $\int_0^{\frac{\pi}{8}} \frac{1}{2}x \sin 4x\, dx = \left[-\frac{1}{8}x \cos 4x\right]_0^{\frac{\pi}{8}} + \int_0^{\frac{\pi}{8}} \frac{1}{8}\cos 4x\, dx$

$= \left[-\frac{1}{8}x \cos 4x + \frac{1}{32}\sin 4x\right]_0^{\frac{\pi}{8}}$

So the original integral is:

$\int_0^{\frac{\pi}{8}} x^2 \cos 4x\, dx = \left[\frac{1}{4}x^2 \sin 4x + \frac{1}{8}x \cos 4x - \frac{1}{32}\sin 4x\right]_0^{\frac{\pi}{8}}$

$= \left(\frac{1}{4}\left(\frac{\pi}{8}\right)^2 (1) + 0 - \frac{1}{32}\right) - (0 + 0 - 0)$

$= \frac{\pi^2}{256} - \frac{1}{32} = \frac{\pi^2 - 8}{256}$

[7 marks available — 1 mark for identifying u and $\frac{dv}{dx}$, 1 mark for finding $\frac{du}{dx}$ and v, 1 mark for using the integration by parts formula, 1 mark for finding $\frac{du}{dx}$ and v in the second integration by parts, 1 mark for the correct integral, 1 mark for substituting the limits, 1 mark for the correct answer]

Q38 Write the function as partial fractions:

$\frac{3x + 10}{(2x + 3)(x - 4)} \equiv \frac{A}{2x + 3} + \frac{B}{x - 4}$

$\Rightarrow 3x + 10 \equiv A(x - 4) + B(2x + 3)$

Substituting $x = 4$ gives: $22 = 11B \Rightarrow B = 2$

Substituting $x = -\frac{3}{2}$ gives: $\frac{11}{2} = -\frac{11}{2}A \Rightarrow A = -1$

So $\frac{3x + 10}{(2x + 3)(x - 4)} \equiv \frac{-1}{2x + 3} + \frac{2}{x - 4}$

So the integral can be expressed:

$\int \frac{3x + 10}{(2x + 3)(x - 4)}\, dx = \int \frac{-1}{2x + 3} + \frac{2}{x - 4}\, dx$

$= -\frac{1}{2}\ln|2x + 3| + 2\ln|x - 4| + C$

Q39 Factorise the denominator: $x^2 + 2x - 3 = (x + 3)(x - 1)$

Then express $\frac{5x + 7}{x^2 + 2x - 3}$ as partial fractions:

$\frac{5x + 7}{x^2 + 2x - 3} \equiv \frac{5x + 7}{(x + 3)(x - 1)} \equiv \frac{A}{x + 3} + \frac{B}{x - 1}$

$\Rightarrow 5x + 7 \equiv A(x - 1) + B(x + 3)$

Substituting $x = 1$ gives: $12 = 4B \Rightarrow B = 3$

Substituting $x = -3$ gives: $-8 = -4A \Rightarrow A = 2$

So $\frac{5x + 7}{x^2 + 2x - 3} \equiv \frac{2}{x + 3} + \frac{3}{x - 1}$.

Then $\int \frac{5x + 7}{x^2 + 2x - 3}\, dx = \int \frac{2}{x + 3} + \frac{3}{x - 1}\, dx$

$= 2\ln|x + 3| + 3\ln|x - 1| + C$

[6 marks available — 1 mark for factorising the denominator, 1 mark for expressing as partial fractions, 1 mark for finding A, 1 mark for finding B, 1 mark for integrating one term correctly, 1 mark for integrating the second term correctly]

Q40 $\frac{13x - 18}{(x - 3)^2 (2x + 1)} \equiv \frac{A}{(x - 3)^2} + \frac{B}{(x - 3)} + \frac{C}{(2x + 1)}$

$\Rightarrow 13x - 18 \equiv A(2x + 1) + B(2x + 1)(x - 3) + C(x - 3)^2$

Substituting $x = 3$ gives: $21 = 7A \Rightarrow A = 3$

Substituting $x = -\frac{1}{2}$ gives: $-\frac{49}{2} = \frac{49}{4}C \Rightarrow C = -2$

Substituting $x = 0$ gives:

$-18 = A - 3B + 9C \Rightarrow -18 = 3 - 3B - 18 \Rightarrow 3 = 3B \Rightarrow B = 1$

So $\frac{13x - 18}{(x - 3)^2 (2x + 1)} \equiv \frac{3}{(x - 3)^2} + \frac{1}{(x - 3)} - \frac{2}{(2x + 1)}$

So the integral can be expressed:

$\int_4^9 \frac{3}{(x - 3)^2} + \frac{1}{(x - 3)} - \frac{2}{(2x + 1)}\, dx$

$= \left[-\frac{3}{x - 3} + \ln|x - 3| - \ln|2x + 1|\right]_4^9$

$= \left(-\frac{1}{2} + \ln 6 - \ln 19\right) - (-3 + \ln 1 - \ln 9)$

$= \frac{5}{2} + \ln\left(\frac{6 \times 9}{19}\right) = \frac{5}{2} + \ln\left(\frac{54}{19}\right)$

Q41 a) $\frac{10x + 23}{(x + 2)^2 (x + 5)} \equiv \frac{A}{(x + 2)^2} + \frac{B}{x + 5} + \frac{C}{x + 2}$

$\Rightarrow 10x + 23 \equiv A(x + 5) + B(x + 2)^2 + C(x + 2)(x + 5)$

Substituting $x = -2$ gives: $-20 + 23 = 3A \Rightarrow A = 1$

Substituting $x = -5$ gives: $-50 + 23 = 9B \Rightarrow B = -3$

Comparing coefficients of x^2: $0 = B + C \Rightarrow C = 3$

So $\frac{10x + 23}{(x + 2)^2 (x + 5)} \equiv \frac{1}{(x + 2)^2} - \frac{3}{x + 5} + \frac{3}{x + 2}$

[4 marks available — 1 mark for writing the expression as the sum for three fractions, 1 mark for finding one unknown, 1 mark for finding the other two unknowns, 1 mark for the correct answer]

b) $\int \frac{10x + 23}{(x + 2)^2 (x + 5)}\, dx = \int \frac{1}{(x + 2)^2} - \frac{3}{x + 5} + \frac{3}{x + 2}\, dx$

$= -(x + 2)^{-1} - 3\ln|x + 5| + 3\ln|x + 2| + C$

[3 marks available — 1 mark for each term integrated correctly]

c) $\frac{10x + 23}{(x + 2)^2 (x + 5)} = 0 \Rightarrow 10x + 23 = 0 \Rightarrow x = -2.3$

$\int_{-4}^{-2.3} \frac{10x + 23}{(x + 2)^2 (x + 5)}\, dx$

$= \left[-(x + 2)^{-1} - 3\ln|x + 5| + 3\ln|x + 2|\right]_{-4}^{-2.3}$

$= \left(\frac{1}{0.3} - 3\ln 2.7 + 3\ln 0.3\right) - \left(\frac{1}{2} - 3\ln 1 + 3\ln 2\right)$

$= -5.83778...$

So the required area $= 5.84$ (3 s.f.)

[3 marks available — 1 mark for finding the x-intercept, 1 mark for substituting in the limits, 1 mark for the correct answer]

Q42 $\dfrac{dm}{dt} \propto \dfrac{1}{s} \Rightarrow \dfrac{dm}{dt} = \dfrac{k}{s}$ for some $k > 0$. *[1 mark]*

Q43 $\dfrac{dy}{dx} = \dfrac{1}{y}\cos x \Rightarrow \int y\,dy = \int \cos x\,dx$

$\Rightarrow \dfrac{y^2}{2} = \sin x + C \Rightarrow y^2 = 2\sin x + C$

Q44 $\dfrac{dx}{dt} = kte^t\,dt \Rightarrow x = \int kte^t\,dt$.

Using integration by parts:

Let $u = kt$ and $\dfrac{dv}{dt} = e^t$. Then $\dfrac{du}{dt} = k$ and $v = e^t$.

So $x = \int kte^t\,dt = kte^t - \int ke^t\,dt = kte^t - ke^t + C = ke^t(t-1) + C$

$x = 0$ when $t = 1 \Rightarrow 0 = ke(0) + C \Rightarrow C = 0$

$x = -3$ when $t = 0 \Rightarrow -3 = ke^0(0-1) \Rightarrow k = 3$

So the particular solution is $x = 3e^t(t-1)$.

Q45 a) $\dfrac{dT}{dt} = -k(T - 21) \Rightarrow \int \dfrac{1}{T-21}\,dT = \int -k\,dt$

$\Rightarrow \ln|T - 21| = -kt + C$

$\Rightarrow T - 21 = e^{-kt+C} = Ae^{-kt} \Rightarrow T = Ae^{-kt} + 21$

When $t = 0$, $T = 90$: $90 = Ae^0 + 21 \Rightarrow A = 69$

When $t = 5$, $T = 80$:

$80 = 69e^{-5k} + 21 \Rightarrow 59 = 69e^{-5k} \Rightarrow \ln\dfrac{59}{69} = -5k$

$\Rightarrow k = -\dfrac{1}{5}\ln\dfrac{59}{69} = 0.0313$ (3 s.f.)

So the particular solution is $T = 69e^{-0.0313t} + 21$

b) (i) When $t = 15$, $T = 69e^{-0.0313 \times 15} + 21 = 64.1\ °C$

(ii) $40 = 69e^{-0.0313t} + 21 \Rightarrow \dfrac{19}{69} = e^{-0.0313t}$

$\Rightarrow \ln\dfrac{19}{69} = -0.0313t \Rightarrow t = \ln\dfrac{19}{69} \div -0.0313$

$= 41.2$ minutes

c)

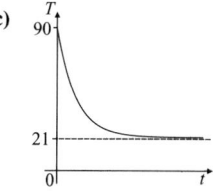

Q46 a) The rate of change of the population is proportional to the population, i.e. $\dfrac{dP}{dt} \propto P$, and the population is decreasing,

so $\dfrac{dP}{dt} = -kP$ where k is a constant, $k > 0$.

Then by separating the variables:

$\int \dfrac{1}{P}\,dP = \int -k\,dt \Rightarrow \ln P = -kt + C \Rightarrow P = e^{-kt+C} = Ae^{-kt}$

When $t = 0$, $P = 75$: $75 = Ae^0 \Rightarrow A = 75$

When $t = 3$, $P = 58$:

$58 = 75e^{-3k} \Rightarrow -3k = \ln\dfrac{58}{75}$

$\Rightarrow k = -\dfrac{1}{3}\ln\dfrac{58}{75} = 0.0857$ (3 s.f.)

So $P = 75e^{-0.0857t}$

[5 marks available — 1 mark for finding an expression for $\dfrac{dP}{dt}$, 1 mark for integrating, 1 mark for finding A, 1 mark for finding k, 1 mark for the correct equation for P]

b) $30 = 75e^{-0.0857t} \Rightarrow -0.0857t = \ln\dfrac{30}{75}$

$\Rightarrow t = \ln\dfrac{30}{75} \div -0.0857 = 10.69...$

So it will take 11 weeks (to the nearest week) for the population to reach 30 birds.

[3 marks available — 1 mark for substituting $P = 30$ into the equation, 1 mark for finding t, 1 mark for the correct answer]

Q47 a) $\dfrac{dn}{dt} \propto n \Rightarrow \dfrac{dn}{dt} = -kn$, where $k > 0$. *[1 mark]*

b) $\dfrac{dn}{dt} = -kn \Rightarrow \int \dfrac{1}{n}\,dn = \int -k\,dt$

$\Rightarrow \ln n = -kt + C \Rightarrow n = e^{-kt+C} = Ae^{-kt}$ for some $A > 0$.

When $t = 0$, $n = 10\,000 \Rightarrow 10\,000 = Ae^0 \Rightarrow A = 10\,000$

So $n = 10\,000e^{-kt}$.

Let $T =$ the time taken for the number of people to reach 9000.

When $t = T$, $n = 9000 \Rightarrow 9000 = 10\,000e^{-kT} \Rightarrow e^{-kT} = 0.9$

When $t = 2T$: $n = 10\,000e^{-k(2T)} = 10\,000(e^{-kT})^2$

$= 10\,000 \times 0.9^2 = 8100$ people

[6 marks available — 1 mark for separating the variables, 1 mark for each side of the equation integrated correctly, 1 mark for using the initial conditions to find the value of one constant, 1 mark for a correct method to find the required value of n, 1 mark for the correct answer]

Q48 First find the points where the line $y = \dfrac{3}{2}x - 1$ intersects the coordinate axes:

$y = \dfrac{3}{2}(0) - 1 \Rightarrow y\text{-intercept} = -1$

and $0 = \dfrac{3}{2}x - 1 \Rightarrow x\text{-intercept} = \dfrac{2}{3}$.

At $x = 2$, $y = \dfrac{3}{2}(2) - 1 = 2$.

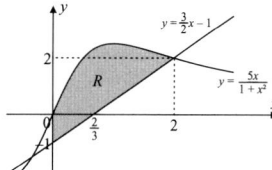

R can be found using the areas:

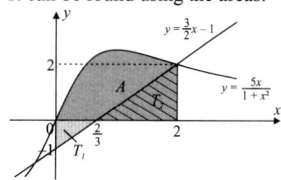

(A is the entire dark grey area, T_1 is the light grey area and T_2 is the hatched area. So $R = A + T_1 - T_2$.)

$A = \displaystyle\int_0^2 \dfrac{5x}{1+x^2}\,dx = \dfrac{5}{2}\big[\ln|1+x^2|\big]_0^2 = \dfrac{5}{2}[\ln 5 - \ln 1] = \dfrac{5}{2}\ln 5$

$T_1 = \dfrac{1}{2} \times \dfrac{2}{3} \times 1 = \dfrac{1}{3}$

$T_2 = \dfrac{1}{2} \times (2 - \dfrac{2}{3}) \times 2 = \dfrac{4}{3}$

So $R = A + T_1 - T_2 = \dfrac{5}{2}\ln 5 + \dfrac{1}{3} - \dfrac{4}{3} = \dfrac{5}{2}\ln 5 - 1$

[7 marks available — 1 mark for finding the axis intercepts, 1 mark for integrating the curve, 1 mark for substituting the limits, 1 mark for the correct value, 1 mark for finding the area of the triangle above the x-axis, 1 mark for finding the area of the triangle below the x-axis, 1 mark for the total area of the shaded region]

Q49 a) $\displaystyle\int_0^a xe^{0.5x^2+x} + e^{0.5x^2+x}\,dx = \int_0^a (x+1)e^{0.5x^2+x}\,dx = \big[e^{0.5x^2+x}\big]_0^a$

$e^{0.5a^2+a} - 1 = 2 \Rightarrow e^{0.5a^2+a} = 3 \Rightarrow \ln 3 = 0.5a^2 + a$

$\Rightarrow a^2 + 2a - 2\ln 3 = 0 \Rightarrow a^2 + 2a - \ln 9 = 0$

[5 marks available — 1 mark for combining the powers of e, 1 mark for integrating correctly, 1 mark for substituting the limits, 1 mark for equating to 2, 1 mark for rearranging correctly to obtain the given result]

Answers

b) Using the quadratic formula or a calculator:

$a = \dfrac{-2 \pm \sqrt{4 + 4\ln 9}}{2} = -1 \pm 1.788...$

$\Rightarrow a = 0.788...$ or $a = -2.788...$ (reject as $a > 0$)

So $a = 0.79$ (2 d.p.)

[2 marks available — 1 mark for a correct method,
1 mark for the correct answer]

Q50 a) $f(x) = \int x^2 e^{0.5x} \, dx$

Let $u = x^2$ and $\dfrac{dv}{dx} = e^{0.5x}$. Then $\dfrac{du}{dx} = 2x$ and $v = 2e^{0.5x}$.

So $\int x^2 e^{0.5x} \, dx = 2x^2 e^{0.5x} - \int 4x e^{0.5x} \, dx$

Integrate by parts again to find $\int 4x e^{0.5x} \, dx$:

Let $u = 4x$ and $\dfrac{dv}{dx} = e^{0.5x}$. Then $\dfrac{du}{dx} = 4$ and $v = 2e^{0.5x}$.

$\int 4x e^{0.5x} \, dx = 8x e^{0.5x} - \int 8 e^{0.5x} \, dx = 8x e^{0.5x} - 16 e^{0.5x} + C$

So $\int x^2 e^{0.5x} \, dx = 2x^2 e^{0.5x} - 8x e^{0.5x} + 16 e^{0.5x} + C$

When $x = 0$, $y = 1 \Rightarrow 1 = 0 - 0 + 16 + C \Rightarrow C = -15$

So $f(x) = 2x^2 e^{0.5x} - 8x e^{0.5x} + 16 e^{0.5x} - 15$

$(= (2x^2 - 8x + 16) e^{0.5x} - 15)$

b) $\int f(x) \, dx = \int 2x^2 e^{0.5x} - 8x e^{0.5x} + 16 e^{0.5x} - 15 \, dx$

From part a):

$\int x^2 e^{0.5x} \, dx = 2x^2 e^{0.5x} - 8x e^{0.5x} + 16 e^{0.5x} + C$

and $\int 4x e^{0.5x} \, dx = 8x e^{0.5x} - 16 e^{0.5x} + C$

So $\int 2x^2 e^{0.5x} - 8x e^{0.5x} + 16 e^{0.5x} - 15 \, dx$

$= 4x^2 e^{0.5x} - 16x e^{0.5x} + 32 e^{0.5x}$
$\qquad - 16x e^{0.5x} + 32 e^{0.5x} + 32 e^{0.5x} - 15x + C$

$= 4x^2 e^{0.5x} - 32x e^{0.5x} + 96 e^{0.5x} - 15x + C$

$(= (4x^2 - 32x + 96) e^{0.5x} - 15x + C)$

c) $\int_0^2 f(x) \, dx = \left[4 e^{0.5x}(x^2 - 8x + 24) - 15x \right]_0^2$

$= (4e(4 - 16 + 24) - 30) - (4(24) - 0)$

$= 48e - 126 = 4.4775... = 4.48$ (3 s.f.)

Q51 Let $u = x^2 + 9$ and $\dfrac{dv}{dx} = 60(x + 2)^3$.

Then $\dfrac{du}{dx} = 2x$ and $v = 15(x + 2)^4$.

So $\int 60(x^2 + 9)(x + 2)^3 \, dx$

$= 15(x^2 + 9)(x + 2)^4 - \int 30x(x + 2)^4 \, dx$

Use integration by parts again to find $\int 30x(x + 2)^4 \, dx$:

Let $u = 30x$ and $\dfrac{dv}{dx} = (x + 2)^4$.

Then $\dfrac{du}{dx} = 30$ and $v = \dfrac{1}{5}(x + 2)^5$.

$\int 30x(x + 2)^4 \, dx = 6x(x + 2)^5 - \int 6(x + 2)^5 \, dx$

$= 6x(x + 2)^5 - (x + 2)^6 + C$

So $\int 60(x^2 + 9)(x + 2)^3 \, dx$

$= 15(x^2 + 9)(x + 2)^4 - 6x(x + 2)^5 + (x + 2)^6 + C$

$f(0) = 2000 \Rightarrow 15(9)(2)^4 - 0 + 64 + C = 2000$

$\Rightarrow C = 2000 - 2160 - 64 = -224$

So $f(x) = 15(x^2 + 9)(x + 2)^4 - 6x(x + 2)^5 + (x + 2)^6 - 224$

$f(1) = 15(10)(3)^4 - 6(3)^5 + (3)^6 - 224$

$= 12\,150 - 1458 + 729 - 224 = 11\,197$

[7 marks available — 1 mark for identifying u and $\dfrac{dv}{dx}$,
1 mark for finding v and $\dfrac{du}{dx}$, 1 mark for using the integration by
parts formula, 1 mark for using integration by parts again on the
integral, 1 mark for the correct integral, 1 mark for finding the
value of C, 1 mark for the correct answer]

Q52 a) $\dfrac{19 - 13x}{(x + 2)(x - 1)^2} \equiv \dfrac{A}{(x + 2)} + \dfrac{B}{(x - 1)} + \dfrac{C}{(x - 1)^2}$

$\Rightarrow 19 - 13x \equiv A(x - 1)^2 + B(x + 2)(x - 1) + C(x + 2)$

Substituting $x = -2$ gives: $19 + 26 = 9A \Rightarrow A = 5$

Substituting $x = 1$ gives: $19 - 13 = 3C \Rightarrow C = 2$

Comparing coefficients of x^2: $0 = A + B \Rightarrow B = -5$

So $\dfrac{19 - 13x}{(x + 2)(x - 1)^2} \equiv \dfrac{5}{(x + 2)} - \dfrac{5}{(x - 1)} + \dfrac{2}{(x - 1)^2}$

[4 marks available — 1 mark for writing the expression as
the sum for three fractions, 1 mark for finding one unknown,
1 mark for finding the other two unknowns, 1 mark for the
correct answer]

b) Required area = area under the line − area under the curve.

Area under the line:

$x = -1 \Rightarrow y = -8 + 16 = 8$

$x = 0.5 \Rightarrow y = 4 + 16 = 20$

So using the area of a trapezium formula:

Area $= \dfrac{1}{2}(8 + 20) \times 1.5 = 14 \times 1.5 = 21$

Area under the curve $= \displaystyle\int_{-1}^{0.5} \dfrac{19 - 13x}{(x + 2)(x - 1)^2} \, dx$

$= \displaystyle\int_{-1}^{0.5} \dfrac{5}{(x + 2)} - \dfrac{5}{(x - 1)} + \dfrac{2}{(x - 1)^2} \, dx$

$= \left[5\ln|x + 2| - 5\ln|x - 1| - \dfrac{2}{(x - 1)} \right]_{-1}^{0.5}$

$= [5\ln 2.5 - 5\ln 0.5 + 4] - [5\ln 1 - 5\ln 2 + 1]$

$= 12.047... + 2.465... = 14.512...$

So required area $= 21 - 14.512... = 6.49$ (3 s.f.)

[7 marks available — 1 mark for a correct method to find the
required area, 1 mark for each fraction integrated correctly
in the integral of the curve, 1 mark for substituting in the
limits, 1 mark for the correct value of the definite integral,
1 mark for the correct answer]

Q53 a) $\dfrac{dS}{dt} \propto \sqrt[3]{S} \Rightarrow \dfrac{dS}{dt} = kS^{\frac{1}{3}}$

When $S = 1\,000\,000$, $\dfrac{dS}{dt} = 60\,000$

$\Rightarrow 60\,000 = 100k \Rightarrow k = 600$

So $\dfrac{dS}{dt} = 600 S^{\frac{1}{3}}$

[2 marks available — 1 mark for a correct differential
equation, 1 mark for finding the value of k]

b) $\dfrac{dS}{dt} = 600 S^{\frac{1}{3}} \Rightarrow \int S^{-\frac{1}{3}} \, ds = \int 600 \, dt \Rightarrow \dfrac{3}{2} S^{\frac{2}{3}} = 600t + C$

When $t = 0$, $S = 64\,000 \Rightarrow \dfrac{3}{2}(64\,000)^{\frac{2}{3}} = 0 + C$

$\Rightarrow C = 2400$

$\dfrac{3}{2} S^{\frac{2}{3}} = 600t + 2400 \Rightarrow S^{\frac{2}{3}} = 400t + 1600$

$\Rightarrow S = (1600 + 400t)^{\frac{3}{2}}$

[3 marks available — 1 mark for separating the variables,
1 mark for finding the value of C, 1 mark for rearranging
correctly to obtain the required result]

c) E.g. The value of $\dfrac{dS}{dt}$ is an 'instantaneous' rate,

while calculating the rate based on the change between
the start of 2011 and 2010 gives the average rate of change
over that year.

[1 mark for a suitable explanation]

Chapter 9: Numerical Methods

Prior Knowledge Check

Q1

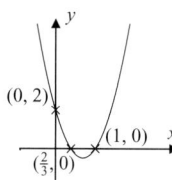

(0, 2)

(1, 0) x

$\left(\frac{2}{3}, 0\right)$

Q2 $\int_{2}^{4} \frac{1}{4}x^2 + 3x - 1 \, dx = \left[\frac{1}{12}x^3 + \frac{3}{2}x^2 - x\right]_{2}^{4}$

$= \left(\frac{16}{3} + 24 - 4\right) - \left(\frac{2}{3} + 6 - 2\right) = \frac{62}{3}$

Q3 **a)** $\frac{dy}{dx} = 12x^3 - 4x$ **b)** $\frac{dy}{dx} = 4e^{2x-1}$

c) $\frac{dy}{dx} = \frac{5}{5x+4}$ **d)** $\frac{dy}{dx} = -3\sin 3x$

Exercise 9.1 — Locating Roots by Changes of Sign

Q1 $f(2) = 2^3 - 5 \times 2 + 1 = -1$
$f(3) = 3^3 - 5 \times 3 + 1 = 13$
There is a sign change (and the function is continuous in this interval) so there is a root in this interval.
They wouldn't ask you if there's a root if the function wasn't continuous in this interval, but it's worth saying anyway just to keep your answer 'strictly true'.

Q2 $f(0.9) = \sin(1.8) - 0.9 = 0.0738...$
$f(1.0) = \sin(2.0) - 1.0 = -0.0907...$
There is a sign change (and the function is continuous in this interval) so there is a root in this interval.

Q3 $f(-5.4) = 2(-5.4)^3 + 5(-5.4)^2 - 26(-5.4) + 21 = -7.728$
$f(-5.3) = 2(-5.3)^3 + 5(-5.3)^2 - 26(-5.3) + 21 = 1.496$
There is a sign change (and the function is continuous in this interval) so there is a root in this interval.

Q4 $f(1.2) = 1.2^3 + \ln 1.2 - 2 = -0.089...$
$f(1.3) = 1.3^3 + \ln 1.3 - 2 = 0.459...$
There is a sign change (and the function is continuous in this interval) so there is a root in this interval.

Q5 **a)** $f(3) = \sin(2 \times 3) = -0.279...$
$f(4) = \sin(2 \times 4) = 0.989...$
There is a sign change (and the function is continuous in this interval) so there is a root in this interval.

b) $f(2.1) = \ln(2.1 - 2) + 2 = -0.302...$
$f(2.2) = \ln(2.2 - 2) + 2 = 0.390...$
There is a sign change (and the function is continuous in this interval) so there is a root in this interval.

c) Rearrange to get $f(x) = x^3 - 4x^2 - 7 = 0$
$f(4.3) = (4.3)^3 - 4(4.3)^2 - 7 = -1.453$
$f(4.5) = (4.5)^3 - 4(4.5)^2 - 7 = 3.125$
There is a sign change (and the function is continuous in this interval) so there is a root in this interval.

d) Rearrange to get $f(x) = e^{2x} + 2e^x - 4 = 0$
$f(0) = e^{2(0)} + 2e^{(0)} - 4 = -1$
$f(0.5) = e^{2(0.5)} + 2e^{(0.5)} - 4 = 2.015...$
There is a sign change (and the function is continuous in this interval) so there is a root in this interval.

Q6 $x^2 + 4x + 4 = 0 \Rightarrow (x + 2)^2 = 0$
$\Rightarrow$ there is a single repeated root $x = -2$, so the graph touches the x-axis at $x = -2$ but doesn't cross it.
$f(x)$ is never negative and there is no change of sign.

Q7 $f(1.2) = (2 \times 1.2^2) - (8 \times 1.2) + 7 = 0.28$
$f(1.3) = (2 \times 1.3^2) - (8 \times 1.3) + 7 = -0.02$
There is a sign change (and the function is continuous in this interval) so there is a root in this interval
— i.e. the bird hits water between 1.2 and 1.3 seconds.

Q8 $f(1.6) = (3 \times 1.6) - 1.6^4 + 3 = 1.24...$
$f(1.7) = (3 \times 1.7) - 1.7^4 + 3 = -0.25...$
There is a sign change (and the function is continuous in this interval) so there is a root in this interval.
$f(-1) = (3 \times (-1)) - (-1)^4 + 3 = -1$
$f(0) = (3 \times 0) - 0^4 + 3 = 3$
There is a sign change (and the function is continuous in this interval) so there is a root in this interval.

Q9 $f(0.01) = e^{0.01-2} - \sqrt{0.01} = 0.0366...$
$f(0.02) = e^{0.02-2} - \sqrt{0.02} = -0.0033...$
There is a sign change (and the function is continuous in this interval) so there is a root in this interval.
$f(2.4) = e^{2.4-2} - \sqrt{2.4} = -0.057...$
$f(2.5) = e^{2.5-2} - \sqrt{2.5} = 0.067...$
There is a sign change (and the function is continuous in this interval) so there is a root in this interval.

Q10 The lower and upper bounds are 2.75 and 2.85.
$f(2.75) = 2.75^3 - (7 \times 2.75) - 2 = -0.45...$
$f(2.85) = 2.85^3 - (7 \times 2.85) - 2 = 1.19...$
There is a sign change between the lower and upper bounds (and the function is continuous in this interval), so a solution to 1 d.p. is $x = 2.8$.

Q11 The lower and upper bounds are 0.35 and 0.45.
$f(0.35) = 3(0.35)^4 + 2(0.35) - 1 = -0.25...$
$f(0.45) = 3(0.45)^4 + 2(0.45) - 1 = 0.02...$
There is a sign change between the lower and upper bounds (and the function is continuous in this interval), so a solution to 1 d.p. is $x = 0.4$.

Q12 Lower and upper bounds are 0.65 and 0.75
$f(0.65) = (2 \times 0.65) - \dfrac{1}{0.65} = -0.23...$
$f(0.75) = (2 \times 0.75) - \dfrac{1}{0.75} = 0.16...$
There is a sign change between the lower and upper bounds (and the function is continuous in this interval), so a solution to 1 d.p. is $x = 0.7$.

Q13 **a)** $f(-1) = 2(-1)^4 - 5(-1)^3 - 1 = 6$
$f(0) = 2(0)^4 - 5(0)^3 - 1 = -1$
There is a sign change (and the function is continuous in this interval) so there is a solution in the interval $-1 < x < 0$.
[2 marks available — 1 mark for correct f(-1) and f(0), 1 mark for identifying sign change]

b) The lower and upper bounds are 2.5305 and 2.5315.
$f(2.5305) = 2(2.5305)^4 - 5(2.5305)^3 - 1 = -0.01...$
$f(2.5315) = 2(2.5315)^4 - 5(2.5315)^3 - 1 = 0.02...$
There is a sign change between the lower and upper bounds (and the function is continuous in this interval), so a solution to 3 d.p. is $x = 2.531$.
[3 marks available — 1 mark for correct lower and upper bounds, 1 mark for correct f(2.5305) and f(2.5315), 1 mark for identifying sign change]

Q14 **a)** $f(-3) = 5(-3)e^{(-3)} + (-3)^2 - 6 = 2.25...$
$f(-2) = 5(-2)e^{(-2)} + (-2)^2 - 6 = -3.35...$
There is a sign change (and the function is continuous in this interval) so $-3 < m < -2$.
[2 marks available — 1 mark for correct f(-3) and f(-2), 1 mark for identifying sign change]

Answers

b) The lower and upper bounds are –2.6355 and –2.6345.
$f(-2.6355) = 5(-2.6355)e^{(-2.6355)} + (-2.6355)^2 - 6 = 0.001...$
$f(-2.6345) = 5(-2.6345)e^{(-2.6345)} + (-2.6345)^2 - 6 = -0.004...$
There is a sign change between the lower and upper bounds (and the function is continuous in this interval), so $m = -2.635$ to 3 d.p.
[3 marks available — 1 mark for correct lower and upper bounds, 1 mark for correct f(–2.6355) and f(–2.6345), 1 mark for identifying sign change]

Q15 Lower and upper bounds are 0.245 and 0.255
$f(0.245) = e^{0.245} - 0.245^3 - (5 \times 0.245) = 0.037...$
$f(0.255) = e^{0.255} - 0.255^3 - (5 \times 0.255) = -0.001...$
There is a sign change between the lower and upper bounds (and the function is continuous in this interval), so a solution to 2 d.p. is $x = 0.25$.

Q16 Rearrange the equation to get $f(x) = 4x - 2x^3 - 15 = 0$
$f(-2.3) = (4 \times -2.3) - (2 \times (-2.3)^3) - 15 = 0.134$
$f(-2.2) = (4 \times -2.2) - (2 \times (-2.2)^3) - 15 = -2.504$
There is a sign change (and the function is continuous in this interval) so there is a root in this interval.

Q17 Rearrange the equation to get $f(x) = 3xe^x - 4 = 0$
$f(0.67) = 3(0.67)e^{(0.67)} - 4 = -0.07...$
$f(0.68) = 3(0.68)e^{(0.68)} - 4 = 0.02...$
There is a sign change (and the function is continuous in this interval) so there is a root in this interval.

Q18 Rearrange the equation to get $f(x) = \ln(x + 3) - 5x = 0$
$f(0.23) = \ln(0.23 + 3) - (5 \times 0.23) = 0.022...$
$f(0.24) = \ln(0.24 + 3) - (5 \times 0.24) = -0.024...$
There is a sign change (and the function is continuous in this interval) so there is a root in this interval.

Q19 Rearrange the equation to get $f(x) = e^{3x}\sin x - 5 = 0$
$f(0) = e^{3 \times 0}\sin 0 - 5 = -5$
$f(1) = e^{3 \times 1}\sin 1 - 5 = 11.9...$
There is a sign change (and the function is continuous in this interval) so there is a root in this interval.

In Q16-19, it's not actually strictly necessary to rearrange to f(x) = 0. It's enough to show that the left-hand side of the original equation is greater than the right-hand side for one of the values of x (x₁ or x₂), and less than the right-hand side for the other.

Exercise 9.2 — Sketching Graphs to Find Approximate Roots

Q1 a)

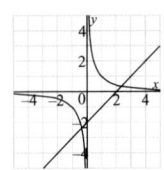

b) The graphs cross twice, so the equation has 2 roots.

c) Rearranging, $f(x) = x - \dfrac{1}{x} - 2$, and the roots are at $f(x) = 0$
$f(2.4) = 2.4 - \dfrac{1}{2.4} - 2 = -0.016...$
$f(2.5) = 2.5 - \dfrac{1}{2.5} - 2 = 0.1$
There is a sign change (and the function is continuous in this interval) so there is a root in this interval.

Q2 a)

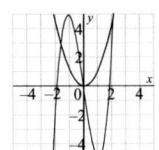

b) The graphs cross 3 times, so the equation has 3 roots.

c) $f(-2) = (2 \times (-2)^3) - (-2)^2 - (7 \times -2) = -6$
$f(-1) = (2 \times (-1)^3) - (-1)^2 - (7 \times -1) = 4$
There is a sign change (and the function is continuous in this interval) so there is a root in this interval.

Q3 a)

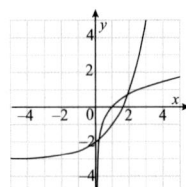

b) The graphs cross twice so the equation has 2 roots.

c) $f(1.8) = \ln(1.8) - 2^{1.8} + 3 = 0.105...$
$f(2.2) = \ln(2.2) - 2^{2.2} + 3 = -0.806...$
There is a sign change (and the function is continuous in this interval), so there is a root in this interval.
$f(2.0) = \ln(2.0) - 2^{2.0} + 3 = -0.306...$
So the root is between 1.8 and 2.0
$f(1.9) = \ln(1.9) - 2^{1.9} + 3 = 0.090...$
So the root is between 1.8 and 1.9
$f(1.85) = \ln(1.85) - 2^{1.85} + 3 = 0.010...$
There is a sign change (and the function is continuous) between 1.85 and 1.9, so the root is at $x = 1.9$ (to 1 d.p.).

Q4 a)

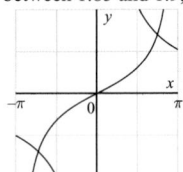

[3 marks available — 1 mark for each correct graph shape, 1 mark for both graphs drawn for the correct range only]

b) The graphs cross twice so the combined equation has two roots.
[1 mark]

c) The roots are at $f(x) = \dfrac{5}{x} - \tan\dfrac{x}{2} = 0$
$f(2.2) = \dfrac{5}{2.2} - \tan\dfrac{2.2}{2} = 0.30...$
$f(2.3) = \dfrac{5}{2.3} - \tan\dfrac{2.3}{2} = -0.06...$
There is a sign change (and the function is continuous in this interval), so α is between 2.2 and 2.3.
[2 marks available — 1 mark for correct f(2.2) and f(2.3), 1 mark for identifying sign change]

d) Using the symmetry of the graph the other root is at $x = -2.284$. *[1 mark]*

Q5 **a)**

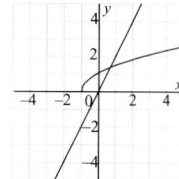

Rearrange the equation to $x = y^2 - 1$ to see that it's a quadratic turned on its side, but only for $x > -1$ because of the square root.

b) The graphs cross once so the equation has 1 root.

c) The roots are at $f(x) = \sqrt{x+1} - 2x = 0$
$f(0.6) = \sqrt{0.6+1} - (2 \times 0.6) = 0.064...$
$f(0.7) = \sqrt{0.7+1} - (2 \times 0.7) = -0.096...$
There is a sign change (and the function is continuous in this interval), so there is a root in this interval.

d) $\sqrt{x+1} = 2x \Rightarrow x + 1 = 4x^2 \Rightarrow 4x^2 - x - 1 = 0$
From quadratic formula root is $x = 0.640$ to 3 s.f.
You can ignore the other solution to this quadratic equation — you want the root between 0.6 and 0.7.

Q6 **a)**

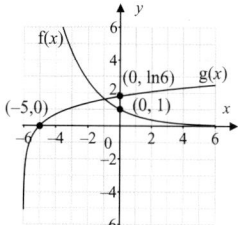

[4 marks available — 1 mark for each correct graph shape in the correct quadrants, 1 mark for the correct coordinates of the y-intercepts, 1 mark for the correct x-intercept of $y = g(x)$]

b) The graphs cross once so the equation has 1 root. *[1 mark]*

c) Let $h(x) = f(x) - g(x)$, then $h(x) = e^{-0.5x} - \ln(x + 6) = 0$.
The lower and upper bounds are −0.965 and −0.955.
$h(-0.965) = e^{-0.5(-0.965)} - \ln(-0.965 + 6) = 0.0037...$
$h(-0.955) = e^{-0.5(-0.955)} - \ln(-0.955 + 6) = -0.0063...$
There is a sign change (and the function is continuous in this interval) so $x = -0.96$ is a solution of $f(x) = g(x)$, correct to 2 s.f.
[2 marks available — 1 mark for correct $f(-0.965)$ and $f(-0.955)$, 1 mark for identifying sign change]
You could have used $h(x) = g(x) - f(x)$ instead of $h(x) = f(x) - g(x)$.

Q7 **a)**

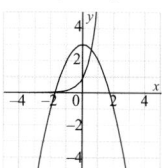

b) If the two functions are set equal to each other they can be rearranged to make $e^{2x} + x^2 = 3$. The graphs cross twice, so the equation has 2 roots.

c) Rearrange to $f(x) = 0$ so $e^{2x} + x^2 - 3 = 0$
$f(-2) = e^{(2 \times -2)} + (-2)^2 - 3 = 1.01...$
$f(-1) = e^{(2 \times -1)} + (-1)^2 - 3 = -1.86...$
There is a sign change (and the function is continuous in this interval) so there is a root in this interval.
$f(-1.5) = e^{(2 \times -1.5)} + (-1.5)^2 - 3 = -0.70...$
So the root is between −1.5 and −2
$f(-1.7) = e^{(2 \times -1.7)} + (-1.7)^2 - 3 = -0.07...$
So the root is between −1.7 and −2
$f(-1.8) = e^{(2 \times -1.8)} + (-1.8)^2 - 3 = 0.26...$
So the root is between −1.7 and −1.8
$f(-1.75) = e^{(2 \times -1.75)} + (-1.75)^2 - 3 = 0.09...$
So the root is between −1.7 and −1.75, so the value to 1 decimal place is $x = -1.7$.

Exercise 9.3 — Using Iteration Formulas

Q1 **a)** $f(1) = 1^3 + 3 \times 1^2 - 7 = -3$
$f(2) = 2^3 + 3 \times 2^2 - 7 = 13$
There is a sign change (and the function is continuous in this interval) so there is a root in this interval.

b) $x_1 = \sqrt{\dfrac{7 - x_0^3}{3}} = \sqrt{\dfrac{7 - 1^3}{3}} = 1.414$,
$x_2 = 1.179$, $x_3 = 1.337$, $x_4 = 1.240$

Q2 **a)** $x_1 = 2 + \ln x_0 = 2 + \ln 3.1 = 3.13140... = 3.1314$ (4 d.p.)
$x_2 = 3.1415$, $x_3 = 3.1447$, $x_4 = 3.1457$,
$x_5 = 3.1460$, $x_6 = 3.1461$ (all to 4 d.p.)

b) The 3rd decimal place stops changing from the 5th iteration, so $\alpha = 3.15$ to 2 d.p.

Q3 **a)** $f(1.4) = 1.4^4 - (5 \times 1.4) + 3 = -0.1584$
$f(1.5) = 1.5^4 - (5 \times 1.5) + 3 = 0.5625$
There is a sign change (and the function is continuous in this interval) so there is a root in this interval.

b) $x_1 = \sqrt[3]{5 - \dfrac{3}{x_0}} = \sqrt[3]{5 - \dfrac{3}{1.4}} = 1.41898... = 1.4190$ (4 d.p.)
$x_2 = 1.4237$, $x_3 = 1.4249$, $x_4 = 1.4251$,
$x_5 = 1.4252$, $x_6 = 1.4252$ (all to 4 d.p.)

c) The 3rd decimal place stops changing from the 4th iteration, so the root is $x = 1.43$ to 2 d.p.

Q4 **a)** $f(5) = 5^2 - (5 \times 5) - 2 = -2$
$f(6) = 6^2 - (5 \times 6) - 2 = 4$
There is a sign change (and the function is continuous in this interval) so there is a root in this interval.

b) $x_1 = \dfrac{2}{x_0} + 5 = \dfrac{2}{5} + 5 = 5.400$,
$x_2 = 5.370$, $x_3 = 5.372$, $x_4 = 5.372$ (all to 4 s.f.)

Q5 **a)** If $x_0 < -1$, $4(1 + x_n) < 0$. So x_1 cannot be obtained since there is no real value for the fourth root of a negative number (any fourth power is the same as squaring twice, so will be ≥ 0).
[1 mark for a correct explanation]

b) $x_1 = \sqrt[4]{4(1 + x_0)} = \sqrt[4]{4(1 + 1.8)} = 1.8293... = 1.829$ (3 d.p.)
$x_2 = 1.8341... = 1.834$ (3 d.p.)
$x_3 = 1.8349... = 1.835$ (3 d.p.)
[2 marks available — 1 mark for correctly substituting 1.8 into the iteration formula to get at least one correct value, 1 mark for three correct values]

c) $x_4 = 1.8350... = 1.835$ (3 d.p.)
$x_5 = 1.8350... = 1.835$ (3 d.p.)
The 3rd decimal place stops changing from the 4th iteration, so $x = 1.84$ is a solution correct to 2 decimal places.
[2 marks available — 1 mark for continued iteration at least as far as x_5, 1 mark for the correct answer]

Answers

Q6 $x_1 = 2 - \ln x_0 = 2 - \ln 1.5 = 1.595$, $x_2 = 1.533$,
$x_3 = 1.573$, $x_4 = 1.547$, $x_5 = 1.563$, $x_6 = 1.553$,
$x_7 = 1.560$, $x_8 = 1.555$, $x_9 = 1.558$ (all to 4 s.f.)
The iterative sequence is bouncing up and down but appears to be converging, so it can be used to estimate a solution to $x = 2 - \ln x$.

Q7 a) $f(3) = e^3 - (10 \times 3) = -9.914...$
$f(4) = e^4 - (10 \times 4) = 14.598...$
There is a sign change (and the function is continuous in this interval) so there is a root in this interval.
[2 marks available — 1 mark for correct f(3) and f(4), 1 mark for identifying sign change]

b) Using starting value $x_0 = 3$:
$x_1 = \ln(10x_0) = \ln(10 \times 3) = 3.4011... = 3.401$ (3 d.p.)
$x_2 = 3.5267... = 3.527$ (3 d.p.), $x_3 = 3.5629... = 3.563$ (3 d.p.)
$x_4 = 3.5731... = 3.573$ (3 d.p.)
[2 marks available — 1 mark for correctly substituting 3 into the iteration formula to get at least one correct value, 1 mark for four correct values]

c) The lower and upper bounds are 3.5765 and 3.5775.
$f(3.5765) = e^{3.5765} - (10 \times 3.5765) = -0.016...$
$f(3.5775) = e^{3.5775} - (10 \times 3.5775) = 0.0089...$
There is a sign change between the lower and upper bounds (and the function is continuous in this interval), so the root to 3 d.p. is $x = 3.577$.
[3 marks available — 1 mark for correct lower and upper bounds, 1 mark for correct f(3.5765) and f(3.5775), 1 mark for identifying change of sign]
You could also show this by continued iteration, but you'd need to go at least as far as x_9 to be correct to 3 decimal places.

d) Using the new iterative formula with $x_0 = 3$:
$x_1 = 2.00855...$, $x_2 = 0.74525...$, $x_3 = 0.21069...$,
$x_4 = 0.12345...$, $x_5 = 0.11313...$, $x_6 = 0.11197...$,
$x_7 = 0.11184...$, $x_8 = 0.11183...$
The iterations are converging, so the formula can be used to estimate another root.
[2 marks available — 1 mark for using the formula correctly at least as far as x_6, 1 mark for the correct conclusion]

Q8 a) $x_1 = \dfrac{x_0^2 - 3x_0}{2} - 5 = \dfrac{(-1)^2 - (3 \times (-1))}{2} - 5 = -3$
$x_2 = 4$, $x_3 = -3$, $x_4 = 4$
The sequence is alternating between -3 and 4.

b) Using the iterative formula given:
$x_1 = 6.32455...$, $x_2 = 6.45157...$, $x_3 = 6.50060...$,
$x_4 = 6.51943...$, $x_5 = 6.52665...$, $x_6 = 6.52941...$,
$x_7 = 6.53047...$, $x_8 = 6.53087...$
The 4th significant figure stops changing from the 7th iteration, so the root is $x = 6.53$ to 3 s.f.

The lower and upper bounds are 6.525 and 6.535.
$f(6.525) = 6.525^2 - (5 \times 6.525) - 10 = -0.049...$
$f(6.535) = 6.535^2 - (5 \times 6.535) - 10 = 0.031...$
There is a sign change between the lower and upper bounds (and the function is continuous in this interval) so this value is correct to 3 s.f.

Exercise 9.4 — Finding Iteration Formulas

Q1 a) $x^4 + 7x - 3 = 0 \Rightarrow x^4 = 3 - 7x \Rightarrow x = \sqrt[4]{3 - 7x}$

b) $x^4 + 7x - 3 = 0 \Rightarrow x^4 + 5x + 2x - 3 = 0$
$\Rightarrow 2x = 3 - 5x - x^4 \Rightarrow x = \dfrac{3 - 5x - x^4}{2}$

c) $x^4 + 7x - 3 = 0 \Rightarrow x^4 = 3 - 7x$
$\Rightarrow x^2 = \sqrt{3 - 7x} \Rightarrow x = \dfrac{\sqrt{3 - 7x}}{x}$

Q2 a) $x^3 - 2x^2 - 5 = 0 \Rightarrow x^3 = 2x^2 + 5 \Rightarrow x = 2 + \dfrac{5}{x^2}$

b) $x_1 = 2 + \dfrac{5}{2^2} = 3.25$
$x_2 = 2.473...$, $x_3 = 2.817...$,
$x_4 = 2.629...$, $x_5 = 2.722...$,
So $x_5 = 2.7$ to 1 d.p.

c) $f(2.65) = 2.65^3 - (2 \times 2.65^2) - 5 = -0.43...$
$f(2.75) = 2.75^3 - (2 \times 2.75^2) - 5 = 0.67...$
There is a sign change between the lower and upper bounds (and the function is continuous in this interval) so this value is correct to 1 d.p.

Q3 a) $x^2 + 3x - 8 = 0 \Rightarrow x^2 = 8 - 3x \Rightarrow x = \dfrac{8}{x} - 3$

b) $f(-5) = (-5)^2 + (3 \times -5) - 8 = 2$
$f(-4) = (-4)^2 + (3 \times -4) - 8 = -4$
There is a sign change (and the function is continuous in this interval) so there is a root in this interval.

c) (i) $x_1 = \dfrac{a}{x_0} + b = \dfrac{8}{-5} - 3 = -4.600$
$x_2 = -4.739$, $x_3 = -4.688$, $x_4 = -4.706$,
$x_5 = -4.700$, $x_6 = -4.702$ (all to 3 d.p.)
(ii) So a root is $x = -4.7$ to 1 d.p.

Q4 a) The graphs intersect when $6 - 1.25x = \sin 0.5x$.
$6 - 1.25x = \sin 0.5x \Rightarrow -1.25x = -6 + \sin 0.5x$
$\Rightarrow x = 4.8 - 0.8\sin 0.5x$,
so $x_{n+1} = 4.8 - 0.8\sin 0.5x_n$ is a possible iteration formula that can be used to find an approximate value of α.
[2 marks available — 1 mark for forming a correct equation and rearranging to make x the subject, 1 mark for the correct iteration formula]

b) $x_1 = 4.8 - 0.8\sin(0.5 \times 4.8) = 4.2596... = 4.260$ (3 d.p.) and
$x_2 = 4.8 - 0.8\sin(0.5 \times 4.2596...) = 4.1217... = 4.122$ (3 d.p.)
[2 marks available — 1 mark for correctly substituting 4.8 into the iteration formula to find x_1, 1 mark for both values correct]

c) $x_3 = 4.094$ (3 d.p.), $x_4 = 4.089$ (3 d.p.), $x_5 = 4.088$ (3 d.p.),
so $n = 5$ is smallest value for which $x_n = 4.088$ to 3 d.p.
[1 mark for the correct answer]

Q5 a) $2^{x-1} = 4\sqrt{x} \Rightarrow 2^{x-1} = 2^2 x^{\frac{1}{2}}$
$\Rightarrow 2^{x-1} \times 2^{-2} = x^{\frac{1}{2}} \Rightarrow 2^{x-3} = x^{\frac{1}{2}}$
$\Rightarrow (2^{x-3})^2 = x \Rightarrow x = 2^{2x-6}$

b) Using the iterative formula given: $x_1 = 0.0625$, $x_2 = 0.0170$,
$x_3 = 0.0160$, $x_4 = 0.0160$ (all 4 d.p.)

c) The lower and upper bounds are 0.01595 and 0.01605.
$f(0.01595) = 2^{0.01595 - 1} - 4\sqrt{0.01595} = 0.00038...$
$f(0.01605) = 2^{0.01605 - 1} - 4\sqrt{0.01605} = -0.00116...$
There is a sign change between the lower and upper bounds (and the function is continuous in this interval) so $x = 0.0160$ is correct to 4 d.p.

Q6 a) $f(0.4) = \ln(2 \times 0.4) + 0.4^3 = -0.159...$
$f(0.5) = \ln(2 \times 0.5) + 0.5^3 = 0.125$
There is a sign change (and the function is continuous in this interval) so there is a root in this interval.

b) $\ln 2x + x^3 = 0 \Rightarrow \ln 2x = -x^3$
$\Rightarrow 2x = e^{-x^3} \Rightarrow x = \dfrac{e^{-x^3}}{2}$

c) Using iterative formula $x_{n+1} = \dfrac{e^{-x_n^3}}{2}$

with starting value $x_0 = 0.4$:

You know the root is between 0.4 and 0.5, so it's a good idea to use one of these as your starting value.

$x_1 = \dfrac{e^{-x_0^3}}{2} = \dfrac{e^{-0.4^3}}{2} = 0.4690...,$

$x_2 = 0.4509..., \ x_3 = 0.4561..., \ x_4 = 0.4547...,$

$x_5 = 0.4551..., \ x_6 = 0.4550...$

So an approximation of the root is $x = 0.46$ to 2 d.p.

Q7 a) $x^2 - 9x - 20 = 0 \Rightarrow x^2 = 9x + 20 \Rightarrow x = \sqrt{9x + 20}$
So an iterative formula is $x_{n+1} = \sqrt{9x_n + 20}$

b) $x_1 = \sqrt{9x_0 + 20} = \sqrt{(9 \times 10) + 20} = 10.488...,$
$x_2 = 10.695..., \ x_3 = 10.782..., \ x_4 = 10.818... = 10.8$ (3 s.f.)

c) $x^2 - 9x - 20 = 0 \Rightarrow x^2 - 5x - 4x - 20 = 0$

$\Rightarrow 5x = x^2 - 4x - 20 \Rightarrow x = \dfrac{x^2 - 4x}{5} - 4$

So an iterative formula is $x_{n+1} = \dfrac{x_n^2 - 4x_n}{5} - 4$

d) $x_1 = \dfrac{x_0^2 - 4x_0}{5} - 4 = \dfrac{1^2 - (4 \times 1)}{5} - 4 = -4.6,$

$x_2 = 3.912, \ x_3 = -4.0688..., \ x_4 = 2.5661...,$
$x_5 = -4.7358..., \ x_6 = 4.2744..., \ x_7 = -3.7653...,$
$x_8 = 1.8479...$

e) The iterations seem to be bouncing up and down without converging to any particular root.

Q8 a) $9 - 0.1t^2 = 3^{0.2t} \Rightarrow 9 - 0.1t^2 - 3^{0.2t} = 0$

$\Rightarrow f(t) = 9 - 0.1t^2 - 3^{0.2t}$ *[1 mark]*

$3^{0.2t} + 0.1t^2 - 9$ *would also be fine here.*

b) $9 - 0.1t^2 - 3^{0.2t} = 0 \Rightarrow 0.1t^2 = 9 - 3^{0.2t}$
$\Rightarrow t^2 = 90 - 10(3^{0.2t})$
$\Rightarrow t = \sqrt{90 - 10(3^{0.2t})}$
So the iteration formula is $t_{n+1} = \sqrt{90 - 10(3^{0.2t_n})}$.
[2 marks available — 1 mark for making t the subject of the formula, 1 mark for the correct iteration formula]

c) $t_1 = \sqrt{90 - 10(3^{(0.2)(6.7)})} = 6.8128..., \ t_2 = 6.7320...,$
$t_3 = 6.7901..., \ t_4 = 6.7485..., \ t_5 = 6.7784...,$
$t_6 = 6.7569... = 6.8$ to 2 s.f.
[2 marks available — 1 mark for correctly substituting 6.7 into the iteration formula to get at least one correct value, 1 mark for the correct final answer]

Q9 a) $V(14) = 4(14)e^{-0.25(14)} + 0.01(14^2) = £3.65$ or 365p
[2 marks available — 1 mark for correctly substituting 14 into the equation, 1 mark for the correct answer to the nearest penny]

b) $e^{-0.25t}(4 - t) + 0.02t = 0 \Rightarrow 4e^{-0.25t} - e^{-0.25t}t + 0.02t = 0$
$\Rightarrow 4e^{-0.25t} = e^{-0.25t}t - 0.02t \Rightarrow 4e^{-0.25t} = (e^{-0.25t} - 0.02)t$
$\Rightarrow t = \dfrac{4e^{-0.25t}}{e^{-0.25t} - 0.02} \Rightarrow t = \dfrac{4}{1 - 0.02e^{0.25t}}$
[2 marks available — 1 mark for a correct attempt to make t the subject of the equation, 1 mark for deriving an equation in the required form]

c) $t_1 = \dfrac{4}{1 - 0.02e^{(0.25)(5)}} = 4.300...,$
$t_2 = 4.249..., \ t_3 = 4.245..., \ t_4 = 4.245...$
So an approximation for t_{max} is 4.25 to 3 s.f.
[3 marks available — 1 mark for correctly substituting 5 into the iteration formula to get at least one correct value, 1 mark for continued iteration at least as far as t_4, 1 mark for the correct answer]

d) $V_{max} = V(4.25) = 4(4.25)e^{-0.25(4.25)} + 0.01(4.25^2)$
$= £6.06$ or 606p

Percentage loss $= \dfrac{6.06 - 3.65}{6.06} \times 100 = 40\%$ to 2 s.f.

[3 marks available — 1 mark for the maximum value of V, 1 mark for a correct method for the percentage loss, 1 mark for the correct answer]

Exercise 9.5 — Cobweb and Staircase Diagrams

Q1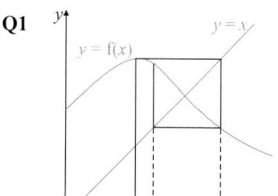

Q2 a)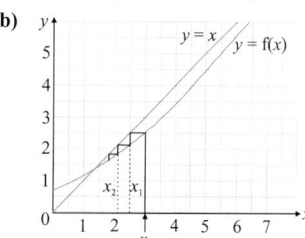

A convergent cobweb diagram.

b)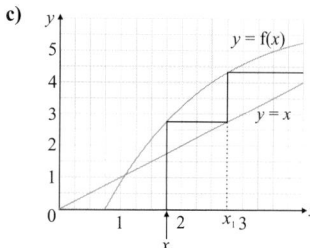

A convergent staircase diagram.

c)
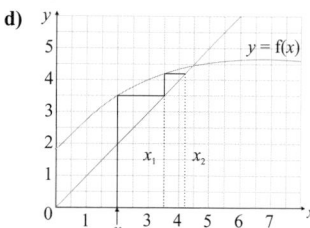

A divergent staircase diagram.
x_2 goes off the scale of the graph so you can't label it.

d)

A convergent staircase diagram.

Answers

e)

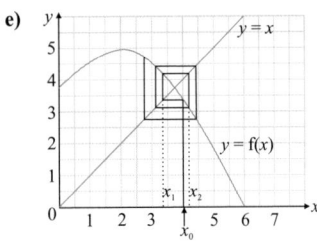

A divergent cobweb diagram.

f)

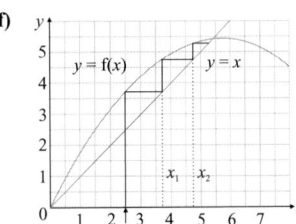

A convergent staircase diagram.

Exercise 9.6 — The Newton-Raphson Method

Q1 **a)** Differentiating f(x) gives f'(x) = 10x.
Putting this into the Newton-Raphson formula gives:
$$x_{n+1} = x_n - \frac{f(x_n)}{f'(x_n)} = x_n - \frac{5x_n^2 - 6}{10x_n}$$

b) Differentiating g(x) gives g'(x) = 3e^{3x} − 8x.
Putting this into the Newton-Raphson formula gives:
$$x_{n+1} = x_n - \frac{g(x_n)}{g'(x_n)} = x_n - \frac{e^{3x_n} - 4x_n^2 - 1}{3e^{3x_n} - 8x_n}$$

c) Differentiating h(x) gives h'(x) = cos x + 3x^2.
Putting this into the Newton-Raphson formula gives:
$$x_{n+1} = x_n - \frac{h(x_n)}{h'(x_n)} = x_n - \frac{\sin(x_n) + x_n^3 - 1}{\cos(x_n) + 3x_n^2}$$

Q2 Differentiating f(x) gives f'(x) = 4x^3 − 6x^2.
Putting this into the Newton-Raphson formula gives:
$$x_{n+1} = x_n - \frac{x_n^4 - 2x_n^3 - 5}{4x_n^3 - 6x_n^2}.$$
Starting with $x_0 = 2.5$,
$$x_1 = 2.5 - \frac{2.5^4 - 2(2.5)^3 - 5}{4(2.5)^3 - 6(2.5)^2} = 2.3875 = 2.39 \text{ (3 s.f.)}$$

Q3 Differentiating f(x) gives f'(x) = 2x − 5.
Putting this into the Newton-Raphson formula gives:
$$x_{n+1} = x_n - \frac{x_n^2 - 5x_n - 12}{2x_n - 5}$$
Starting with $x_0 = -1$,
$$x_1 = -1 - \frac{(-1)^2 - 5(-1) - 12}{2(-1) - 5} = -1.857142...$$
$x_2 = -1.772833...$, $x_3 = -1.772001...$,
$x_4 = -1.772001... = -1.7720$ (5 s.f.)

Q4 **a)** Rearrange the equation to get sin x − x + 1 = 0.
Differentiating the function gives cos x − 1.
Putting this into the Newton-Raphson formula gives:
$$x_{n+1} = x_n - \frac{\sin x_n - x_n + 1}{\cos x_n - 1}$$
Starting with $x_0 = 1$, $x_1 = 1 - \frac{\sin(1) - 1 + 1}{\cos(1) - 1} = 2.830487...$
$x_2 = 2.049555...$, $x_3 = 1.938656...$,
$x_4 = 1.934568... = 1.9346$ (5 s.f.)

b) Rearrange the equation to get x^2 ln x − 5 = 0.
Differentiating using the product rule gives $x(1 + 2 \ln x)$.
Putting this into the Newton-Raphson formula gives:
$$x_{n+1} = x_n - \frac{x_n^2 \ln x_n - 5}{x_n(1 + 2 \ln x_n)}$$
Starting with $x_0 = 2$, $x_1 = 2 - \frac{2^2 \ln 2 - 5}{2(1 + 2 \ln 2)} = 2.466709...$
$x_2 = 2.395369...$, $x_3 = 2.393518...$,
$x_4 = 2.393517... = 2.3935$ (5 s.f.)

c) Rearrange the equation to get $e^{-x} - 2 \cos \frac{1}{2}x = 0$.
Differentiating the function gives $-e^{-x} + \sin \frac{1}{2}x$.
Putting this into the Newton-Raphson formula:
$$x_{n+1} = x_n - \frac{e^{-x_n} - 2 \cos \frac{1}{2}x_n}{-e^{-x_n} + \sin \frac{1}{2}x_n}$$
Starting with $x_0 = 1$, $x_1 = 1 - \frac{e^{-1} - 2 \cos \frac{1}{2}(1)}{-e^{-1} + \sin \frac{1}{2}(1)} = 13.43688...$
$x_2 = 17.73798...$, $x_3 = 14.51784...$,
$x_4 = 15.87165... = 15.872$ (5 s.f.)

Q5 **a)** Differentiating the function gives 12cos 3x + 2e^{2x}.
Putting this into the Newton-Raphson formula:
$$x_{n+1} = x_n - \frac{4 \sin 3x_n + e^{2x_n}}{12 \cos 3x_n + 2e^{2x_n}}$$
Starting with $x_0 = -1$:
$$x_1 = -1 - \frac{4 \sin 3(-1) + e^{2(-1)}}{12 \cos 3(-1) + 2e^{2(-1)}} = -1.03696...$$
$$= -1.0370 \text{ (5 s.f.)}$$
[2 marks available — 1 mark for correct differentiation, 1 mark for the correct value of x_1]

b) $x_2 = -1.03671...$, $x_3 = -1.03671$,
so $\alpha = -1.0367$ (5 s.f.).
[2 marks available — 1 mark for continued iteration at least as far as x_3, 1 mark for the correct answer]

Q6 Differentiating the function gives 6x^2 − 30x.
Putting this into the Newton-Raphson formula:
$$x_{n+1} = x_n - \frac{2x_n^3 - 15x_n^2 + 109}{6x_n^2 - 30x_n}$$
Starting with $x_0 = 1$:
$$x_1 = 1 - \frac{2(1)^3 - 15(1)^2 + 109}{6(1)^2 - 30(1)} = 1 - \frac{96}{-24} = 1 + 4 = 5$$
$$x_2 = 5 - \frac{2(5)^3 - 15(5)^2 + 109}{6(5)^2 - 30(5)} = 5 - \frac{-16}{0}$$
Since this involves dividing by 0, the method fails,
so $x_0 = 1$ cannot be used to find a root.
f'(5) = 0, which means that the tangent line at x_1 is horizontal, and it won't intercept the x-axis.

Q7 **a)** f'(x) = 3x^2 − 10x − 8. At turning points, f'(x) = 0
$\Rightarrow 3x^2 - 10x - 8 = 0 \Rightarrow (3x + 2)(x - 4) = 0$
$\Rightarrow x = -\frac{2}{3}$ or $x = 4$
The x^3 coefficient is positive, so the maximum occurs first
(i.e. where $x = -\frac{2}{3}$). When $x = -\frac{2}{3}$, $y = -\frac{5}{27}$.
f''(x) = 6x − 10. When $x = -\frac{2}{3}$, f''(x) = −14.
The second derivative is negative,
so the turning point $(-\frac{2}{3}, -\frac{5}{27})$ is a maximum.

b)

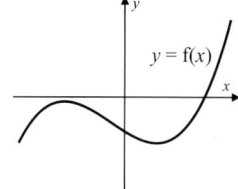

The graph is a positive cubic, so the maximum occurs first, then the minimum. Since the maximum has a negative y-coordinate the minimum must also have a negative y-coordinate, so the graph of $y = f(x)$ will only cross the x-axis once — after both turning points. So there must be exactly one real root of $f(x) = 0$.

c) $x_1 = 6 - \dfrac{f(6)}{f'(6)} = 6 - \dfrac{-15}{40} = 6.375$

d) $x_0 = 4 \Rightarrow x_1 = 4 - \dfrac{f(4)}{f'(4)} = 4 - \dfrac{-51}{0}$
Since this involves dividing by 0, the method fails, so $x_0 = 4$ cannot be used to find the root.

e) The process finds the tangent to $f(x)$ at each x_n value then uses the x-intercept of the tangent as x_{n+1}. Since $x = 4$ is at a stationary point, the tangent is parallel to the x-axis and so does not cross it.

Q8 a) $p = 3(1) - 4\sqrt{1} - \dfrac{6}{(\sqrt{1})^3} = 3 - 4 - 6 = -7$ *[1 mark]*

b) $d(t) = 3t - 4t^{0.5} - 6t^{-1.5}$, so $d'(t) = 3 - 2t^{-0.5} + 9t^{-2.5}$
$t_1 = 1 - \dfrac{d(1)}{d'(1)} = 1 - \dfrac{-7}{10} = 1.7$
[2 marks available — 1 mark for correctly differentiating d(t), 1 mark for using formula to get the correct final answer]

c) The tangent at $t = 1$ will cross the t-axis at a lower value of t than the root since the gradient of the curve is decreasing / the graph is concave. *[1 mark]*

d) $t_2 = 1.7 - \dfrac{d(1.7)}{d'(1.7)} = 2.4322003...$
time taken $= (2.43... - 1) \times 60$
$= 86$ seconds to the nearest second
[3 marks available — 1 mark for correct second iteration, 1 mark for correct method for finding time taken, 1 mark for the correct answer]

Exercise 9.7 — Combining the Iteration Methods

Q1 a) $(x + 2)g(x) = (x + 2)(x^3 - 4x + 1)$
$= x^4 - 4x^2 + x + 2x^3 - 8x + 2 = x^4 + 2x^3 - 4x^2 - 7x + 2 = f(x)$
So $x = -2$ is a root of $f(x)$.

b) $g(1) = 1^3 - 4(1) + 1 = 1 - 4 + 1 = -2$
$g(2) = 2^3 - 4(2) + 1 = 8 - 8 + 1 = 1$
There is a sign change (and the function is continuous in this interval) so there is a root in this interval.

c) $x^3 - 4x + 1 = 0 \Rightarrow x^3 = 4x - 1 \Rightarrow x = \sqrt[3]{4x - 1}$

d) $x_1 = \sqrt[3]{4x_0 - 1} = \sqrt[3]{4(2) - 1} = 1.91293...$
$x_2 = 1.88066..., x_3 = 1.86842..., x_4 = 1.86373...,$
$x_5 = 1.86193... = 1.862$ (4 s.f.)

e) Differentiating $g(x)$ gives $g'(x) = 3x^2 - 4$.
Putting this into the Newton-Raphson formula gives:
$$x_{n+1} = x_n - \dfrac{x_n^3 - 4x_n + 1}{3x_n^2 - 4}$$
$$x_1 = -2 - \dfrac{(-2)^3 - 4(-2) + 1}{3(-2)^2 - 4} = -2.125$$
$x_2 = -2.114975... = -2.115$ (4 s.f.),
$x_3 = -2.114907... = -2.115$ (4 s.f.)
So an approximation for the root is $\beta = -2.115$ to 4 s.f.

f) The Newton-Raphson method fails when $g'(x_n) = 0$ because it involves dividing by zero in the formula. Graphically, it means that the tangent at x_n is horizontal, and so it won't cut the x-axis to find the next iteration.
When $x_n = \dfrac{2\sqrt{3}}{3}$,
$g'\left(\dfrac{2\sqrt{3}}{3}\right) = 3\left(\dfrac{2\sqrt{3}}{3}\right)^2 - 4 = 3\left(\dfrac{12}{9}\right) - 4 = 4 - 4 = 0$

Q2 a) $x^2 - 7x - 12 = 0 \Rightarrow x^2 = 7x + 12 \Rightarrow x = \sqrt{7x + 12}$ *[1 mark]*

b) $x_1 = \sqrt{7(2) + 12} = 5.099... = 5.10$ (2 d.p.) and
$x_2 = \sqrt{7(5.099...) + 12} = 6.906... = 6.91$ (2 d.p.)
[2 marks available — 1 mark for using formula to get at least one correct value, 1 mark for other correct answer]

c) (i) and (ii)

x	0	2	4	6	8	10
y	3.46...	5.10...	6.32...	7.35...	8.25...	9.06...

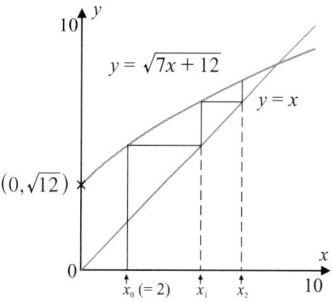

[4 marks available — 1 mark for correctly completing the table, 1 mark for the correct shapes of both graphs, 1 mark for the correct y-intercept of $y = \sqrt{7x + 12}$, 1 mark for correct straight lines added with correct labels]

d) $f(x) = x^2 - 7x - 12$
$f(-1) = (-1)^2 - 7(-1) - 12 = -4$
$f(-2) = (-2)^2 - 7(-2) - 12 = 6$
There is a sign change (and the function is continuous in this interval) so there is a root in this interval.
[2 marks available — 1 mark for correct f(–1) and f(–2), 1 mark for suitable conclusion]

Q3 a) $(x + 4)g(x) = (x + 4)(1 - 2x + 3x^2 - x^3)$
$= -x^4 - 4x^3 + 3x^3 + 12x^2 - 2x^2 - 8x + x + 4$
$= 4 - 7x + 10x^2 - x^3 - x^4 = f(x)$
So $x = -4$ is a root of $f(x)$.

b) Differentiating $g(x)$ gives $g'(x) = -2 + 6x - 3x^2$.
Putting this into the Newton-Raphson formula gives:
$$x_{n+1} = x_n - \dfrac{1 - 2x_n + 3x_n^2 - x_n^3}{-2 + 6x_n - 3x_n^2}$$
$$x_1 = 2 - \dfrac{1 - 2(2) + 3(2)^2 - (2)^3}{-2 + 6(2) - 3(2)^2} = 2 - \dfrac{1}{-2} = 2.5$$
$x_2 = 2.3478..., x_3 = 2.3252..., x_4 = 2.3247...,$
So an approximate solution to $g(x) = 0$ is 2.325 to 4 s.f.

c) $1 - 2x + 3x^2 - x^3 = 0 \Rightarrow 2x + x^3 = 1 + 3x^2$

$\Rightarrow x(2 + x^2) = 1 + 3x^2 \Rightarrow x = \dfrac{1 + 3x^2}{2 + x^2}$, so $x_{n+1} = \dfrac{1 + 3x_n^2}{2 + x_n^2}$

d) $x_1 = \dfrac{1 + 3(2)^2}{2 + 2^2} = 2.1666... = 2.166$ to 4 s.f.

$x_2 = 2.2531... = 2.253$ to 4 s.f.

...

$x_8 = 2.3242... = 2.324$ to 4 s.f.

$x_9 = 2.3245... = 2.325$ to 4 s.f.

So $n = 9$ is the lowest value for which $x_n = 2.325$ to 4 s.f.

e) The Newton-Raphson method converges on the solution more quickly (with fewer iterations) than the other iteration formula, when starting at $x_0 = 2$.
You can't be sure this is true for all starting values.

Q4 a) $K(0) = 6 + e^{-0} - 4 \sin((0.2)(0)) = 7$, so the model predicts 7000 kangaroos will be on the island at the start of the year 2000.
[2 marks available — 1 mark for attempting K(0),
1 mark for the correct answer]

b) Rearrange the equation to get $f(t) = 5 \arccos(-1.25e^{-t}) - t = 0$
$f(7) = 5 \arccos(-1.25e^{-7}) - 7 = 0.859...$
$f(8) = 5 \arccos(-1.25e^{-8}) - 8 = -0.143...$
There is a sign change (and the function is continuous in this interval) so the solution is in this interval.
[2 marks available — 1 mark for correct f(7) and f(8),
1 mark for identifying sign change]

c) $t_1 = 5 \arccos(-1.25e^{-8}) = 7.8561$ to 5 s.f, $t_2 = 7.8564$ to 5 s.f.
[2 marks available — 1 mark for using formula to get at least
one correct value, 1 mark for other correct value]

d) The maximum population in this time interval occurs when $t = 20$, so the difference $= K(20) - K(7.8564)$
$= 9.0272... - 2.0003...$
$= 7.0268...$
This is 7027 kangaroos (to the nearest whole number).
[2 marks available — 1 mark for K(20) − K(7.8564) or
equivalent using answer to part c), 1 mark for 7027]

Exercise 9.8 — The Trapezium Rule

Q1 a) $h = (2 - 0) \div 2 = 1$, so the x-values are 0, 1, 2.

$\displaystyle\int_0^2 \sqrt{x + 2}\, dx \approx \frac{h}{2}[y_0 + 2y_1 + y_2] = \frac{1}{2}[\sqrt{2} + 2\sqrt{3} + \sqrt{4}]$

$= \frac{1}{2}[1.4142 + 3.4641 + 2] = 3.44$ (3 s.f.)

b) $h = (3 - 1) \div 4 = 0.5$,
so the x-values are 1, 1.5, 2, 2.5, 3.

$\displaystyle\int_1^3 2(\ln x)^2\, dx \approx \frac{h}{2}[y_0 + 2(y_1 + y_2 + y_3) + y_4]$

$= \frac{1}{4}[2(\ln 1)^2 + 2(2(\ln 1.5)^2 + 2(\ln 2)^2 + 2(\ln 2.5)^2) + 2(\ln 3)^2]$

$= \frac{1}{4}[0 + 2(0.3288 + 0.9609 + 1.6792) + 2.4139]$

$= 2.09$ (3 s.f.)

c) $h = (0.4 - 0) \div 2 = 0.2$, so the x-values are 0, 0.2, 0.4.

$\displaystyle\int_0^{0.4} e^{x^2}\, dx \approx \frac{h}{2}[y_0 + 2y_1 + y_2] = \frac{0.2}{2}[e^0 + 2e^{0.04} + e^{0.16}]$

$= 0.1[1 + 2.0816 + 1.1735] = 0.426$ (3 s.f.)

d) $h = \left(\dfrac{\pi}{4} - -\dfrac{\pi}{4}\right) \div 4 = \dfrac{\pi}{8}$,

so the x-values are $-\dfrac{\pi}{4}, -\dfrac{\pi}{8}, 0, \dfrac{\pi}{8}, \dfrac{\pi}{4}$.

$\displaystyle\int_{-\frac{\pi}{4}}^{\frac{\pi}{4}} 4x \tan x\, dx \approx \frac{h}{2}[y_0 + 2(y_1 + y_2 + y_3) + y_4]$

$= \dfrac{\pi}{16}\left[-\pi \tan\left(-\dfrac{\pi}{4}\right) + 2\left(-\dfrac{\pi}{2} \tan\left(-\dfrac{\pi}{8}\right) + 0 + \dfrac{\pi}{2} \tan\left(\dfrac{\pi}{8}\right)\right) + \pi \tan\left(\dfrac{\pi}{4}\right)\right]$

$= \dfrac{\pi}{16}[\pi + 2(0.6506 + 0.6506) + \pi] = 1.74$ (3 s.f.)

e) $h = (0.3 - 0) \div 6 = 0.05$, so the x-values are 0, 0.05, 0.1, 0.15, 0.2, 0.25, 0.3.

$\displaystyle\int_0^{0.3} \sqrt{e^x + 1}\, dx \approx \frac{h}{2}[y_0 + 2(y_1 + y_2 + y_3 + y_4 + y_5) + y_6]$

$= \dfrac{0.05}{2}[\sqrt{e^0 + 1} + 2(\sqrt{e^{0.05} + 1} + \sqrt{e^{0.1} + 1}$
$\quad + \sqrt{e^{0.15} + 1} + \sqrt{e^{0.2} + 1} + \sqrt{e^{0.25} + 1}) + \sqrt{e^{0.3} + 1}]$

$= 0.025[\sqrt{2} + 2(\sqrt{2.0513} + \sqrt{2.1052} + \sqrt{2.1618}$
$\quad + \sqrt{2.2214} + \sqrt{2.2840}) + \sqrt{2.3499}]$

$= 0.025[1.4142 + 2(1.4322 + 1.4509 + 1.4703$
$\quad + 1.4904 + 1.5113) + 1.5329]$

$= 0.441$ (3 s.f.)

f) $h = (\pi - 0) \div 6 = \dfrac{\pi}{6}$,
so the x-values are $0, \dfrac{\pi}{6}, \dfrac{\pi}{3}, \dfrac{\pi}{2}, \dfrac{2\pi}{3}, \dfrac{5\pi}{6}, \pi$.

$\displaystyle\int_0^\pi \ln(2 + \sin x)\, dx \approx \frac{h}{2}[y_0 + 2(y_1 + y_2 + y_3 + y_4 + y_5) + y_6]$

$= \dfrac{\pi}{12}[\ln(2 + \sin 0) + 2(\ln(2 + \sin \dfrac{\pi}{6}) + \ln(2 + \sin \dfrac{\pi}{3})$
$\quad + \ln(2 + \sin \dfrac{\pi}{2}) + \ln(2 + \sin \dfrac{2\pi}{3})$
$\quad + \ln(2 + \sin \dfrac{5\pi}{6})) + \ln(2 + \sin \pi)]$

$= \dfrac{\pi}{12}[0.6931 + 2(0.9163 + 1.0529 + 1.0986$
$\quad + 1.0529 + 0.9163) + 0.6931]$

$= 3.00$ (3 s.f.)

Q2 a) $h = 1$, so the x-values are 0, 1, 2 and the y-values are:
$y_0 = \sqrt{2}$, $y_1 = \sqrt{3}$, $y_2 = \sqrt{4}$
Use the simplified versions of the trapezium rule.
The lower bound is found using:
$\displaystyle\int_0^2 \sqrt{x + 2}\, dx \approx h[y_0 + y_1] = 1[\sqrt{2} + \sqrt{3}] = 3.146$ (4 s.f.)
The upper bound is found using:
$\displaystyle\int_0^2 \sqrt{x + 2}\, dx \approx h[y_1 + y_2] = 1[\sqrt{3} + \sqrt{4}] = 3.732$ (4 s.f.)

b) $h = 0.5$, so the x-values are 1, 1.5, 2, 2.5, 3
Substitute these into the function to get y-values:
$y_0 = 0, y_1 = 0.3288..., y_2 = 0.9609...,$
$y_3 = 1.6791..., y_4 = 2.4138...$
Use the simplified versions of the trapezium rule.
The lower bound is found using:
$\displaystyle\int_1^3 2 (\ln x)^2\, dx \approx h[y_0 + y_1 + y_2 + y_3]$
$= 0.5[0 + 0.3288... + 0.9609... + 1.6791...] = 1.484$ (4 s.f.)
The upper bound is found using:
$\displaystyle\int_1^3 2 (\ln x)^2\, dx \approx h[y_1 + y_2 + y_3 + y_4]$
$= 0.5[0.3288... + 0.9609... + 1.6791... + 2.4138...]$
$= 2.691$ (4 s.f.)
The corresponding trapezium rule answers from
Q1 lie inside the lower and upper bounds.

Q3 $h = \dfrac{\pi}{2} \div 3 = \dfrac{\pi}{6}$, so the x-values are $0, \dfrac{\pi}{6}, \dfrac{\pi}{3}, \dfrac{\pi}{2}$.

$\displaystyle\int_0^{\frac{\pi}{2}} \sin^3 \theta\, d\theta \approx \frac{h}{2}[y_0 + 2(y_1 + y_2) + y_3]$

$= \dfrac{\pi}{12}[\sin^3 0 + 2(\sin^3 \dfrac{\pi}{6} + \sin^3 \dfrac{\pi}{3}) + \sin^3 \dfrac{\pi}{2}]$

$= \dfrac{\pi}{12}[0 + 2(0.125 + 0.6495) + 1] = 0.667$ (3 d.p.)

Q4 $h = (7 - 2) \div 5 = 1$, so the x-values are 2, 3, 4, 5, 6, 7.

$\displaystyle\int_2^7 \sqrt{\ln x}\, dx \approx \frac{h}{2}[y_0 + 2(y_1 + y_2 + y_3 + y_4) + y_5]$

$= \dfrac{1}{2}[\sqrt{\ln 2} + 2(\sqrt{\ln 3} + \sqrt{\ln 4} + \sqrt{\ln 5} + \sqrt{\ln 6}) + \sqrt{\ln 7}]$

$= \dfrac{1}{2}[0.8326 + 2(1.0481 + 1.1774 + 1.2686 + 1.3386) + 1.3950]$

$= 5.947$ m² (3 d.p.)

Q5 **a)**

x	0	$\frac{\pi}{8}$	$\frac{\pi}{4}$	$\frac{3\pi}{8}$	$\frac{\pi}{2}$
y	1	1.466	2.028	2.519	2.718

b) **(i)** $\int_0^{\frac{\pi}{2}} e^{\sin x}\, dx \approx \frac{\pi}{8}[1 + 2(2.028) + 2.718] = 3.05$ (2 d.p.)

(ii) $\int_0^{\frac{\pi}{2}} e^{\sin x}\, dx \approx \frac{\pi}{16}[1 + 2(1.466 + 2.028 + 2.519) + 2.718]$
$= 3.09$ (2 d.p.)

c) 3.09 is the better estimate as more intervals have been used in the calculation.

Q6 **a)** $h = (4 - 2) \div 4 = 0.5$, so the x-values are 2, 2.5, 3, 3.5, 4.

$\int_2^4 \frac{3}{\ln x}\, dx \approx \frac{1}{4}\left[\frac{3}{\ln 2} + 2\left(\frac{3}{\ln 2.5} + \frac{3}{\ln 3} + \frac{3}{\ln 3.5}\right) + \frac{3}{\ln 4}\right]$
$= 0.25[4.3281 + 2(3.2741 + 2.7307 + 2.3947) + 2.1640]$
$= 5.82$ (2 d.p.)

b) It is an over-estimate as the top of each trapezium lies above the curve.

Q7 **a)**

x	-2	0	2	4
y	0.125	0.5	2	8

[1 mark for both correct y-values]

b) $\int_{-2}^4 2^{x-1}\, dx \approx \frac{2}{2}[0.125 + 2(0.5 + 2) + 8] = 13.125$
[2 marks available — 1 mark for using the trapezium rule correctly, 1 mark for the correct answer]

c) It is an over-estimate as the top of each trapezium lies above the curve.
[1 mark for the correct answer with explanation]

d) The trapezium rule could be used to give a more accurate estimate by using more intervals.
[1 mark for a suitable suggestion]

e) $\int_{-2}^4 (10 - 2^{x-1})\, dx = \int_{-2}^4 10\, dx - \int_{-2}^4 2^{x-1}\, dx$
$= [10x]_{-2}^4 - 13.125 = (40 - -20) - 13.125$
$= 46.875$
[2 marks available — 1 mark for correctly splitting the integral, 1 mark for the correct answer]

Q8 **a)** There are 6 intervals of $h = \frac{\pi}{6}$ between $-\frac{\pi}{2}$ and $\frac{\pi}{2}$.

x	$-\frac{\pi}{2}$	$-\frac{\pi}{3}$	$-\frac{\pi}{6}$	0	$\frac{\pi}{6}$	$\frac{\pi}{3}$	$\frac{\pi}{2}$
$y = \cos x$	0	0.5	$\frac{\sqrt{3}}{2}$	1	$\frac{\sqrt{3}}{2}$	0.5	0

$\int_{-\frac{\pi}{2}}^{\frac{\pi}{2}} \cos x\, dx \approx \frac{h}{2}[y_0 + 2(y_1 + y_2 + y_3 + y_4 + y_5) + y_6]$

$= \frac{\left(\frac{\pi}{6}\right)}{2}\left[0 + 2\left(0.5 + \frac{\sqrt{3}}{2} + 1 + \frac{\sqrt{3}}{2} + 0.5\right) + 0\right]$

$= \frac{\pi}{12}[0 + 2(2 + \sqrt{3}) + 0] = \pi\left(\frac{2(2 + \sqrt{3})}{12}\right)$

$= \frac{\pi(2 + \sqrt{3})}{6}$ as required

b) It's an under-estimate because between $-\frac{\pi}{2}$ and $\frac{\pi}{2}$ the curve of the graph $y = \cos x$ is concave, and so the top of each trapezium lies below the curve.

Q9 **a)** **(i)** $h = \frac{\pi}{3} \div 2 = \frac{\pi}{6}$, so

x	0	$\frac{\pi}{6}$	$\frac{\pi}{3}$
y	0	$\frac{1}{3}$	3

$\int_0^{\frac{\pi}{3}} \tan^2 x\, dx \approx \left(\frac{1}{2}\right)\left(\frac{\pi}{6}\right)\left[0 + 2\left(\frac{1}{3}\right) + 3\right]$
$= 0.9599$ to 4 d.p.
[3 marks available — 1 mark for correct h, 1 mark for using the trapezium rule correctly, 1 mark for the correct answer]

(ii) $h = \frac{\pi}{3} \div 4 = \frac{\pi}{12}$, so

x	0	$\frac{\pi}{12}$	$\frac{\pi}{6}$	$\frac{\pi}{4}$	$\frac{\pi}{3}$
y	0	$0.071...$	$\frac{1}{3}$	1	3

$\int_0^{\frac{\pi}{3}} \tan^2 x\, dx \approx \left(\frac{1}{2}\right)\left(\frac{\pi}{12}\right)\left[0 + 2\left(0.071... + \frac{1}{3} + 1\right) + 3\right]$
$= 0.7606$ to 4 d.p.
[3 marks available — 1 mark for correct h, 1 mark for using trapezium rule correctly, 1 mark for the correct answer]

b) $I = \int_0^{\frac{\pi}{3}} \tan^2 x\, dx = \int_0^{\frac{\pi}{3}} (\sec^2 x - 1)\, dx = [\tan x - x]_0^{\frac{\pi}{3}}$
$= \tan\left(\frac{\pi}{3}\right) - \frac{\pi}{3} - (\tan(0) - 0) = \sqrt{3} - \frac{\pi}{3}$
[4 marks available — 1 mark for using the identity $\sec^2 x \equiv 1 + \tan^2 x$, 1 mark for correct integration, 1 mark for substitution of the limits shown, 1 mark for correct result]

c) The reduction in percentage error

$= \frac{0.9599 - \left(\sqrt{3} - \frac{\pi}{3}\right)}{\sqrt{3} - \frac{\pi}{3}} \times 100 - \frac{0.7606 - \left(\sqrt{3} - \frac{\pi}{3}\right)}{\sqrt{3} - \frac{\pi}{3}} \times 100$

$= \frac{0.9599 - 0.7606}{\sqrt{3} - \frac{\pi}{3}} \times 100 = 29.1\%$ to 3 s.f.

[2 marks available — 1 mark for a correct method for the difference in percentage error, 1 mark for the correct answer]

Chapter 9 Review Exercise

Q1 **a)** E.g. f(x) may not be continuous, it could have a vertical asymptote so that it does not cross the x-axis between a and b.
f(x) could cross the x-axis several times — a sign change does not guarantee that there is only one root.

b) f(x) is continuous on the interval [2, 4] and f(2) = f(4) = 1, so f(2) and f(4) are both positive, but $x^2 - 6x + 9 = 0$
$\Rightarrow (x - 3)^2 = 0$, so f($x$) has the (repeated) root $x = 3$ in the interval [2, 4]. Therefore Bailey's statement is incorrect.

Q2 Pick two numbers close to and either side of 1.2,
e.g. 1.15 and 1.25
f(1.15) = $(1.15)^3$ + 1.15 − 3 = −0.329...
f(1.25) = $(1.25)^3$ + 1.25 − 3 = 0.203...
There is a sign change in the interval 1.15 < x < 1.25
(where the function is continuous) so x = 1.2 to 1 d.p.

Answers

Q3 a)

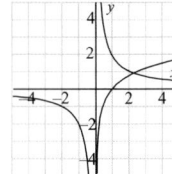

The graphs cross once, so the equation has 1 root.

b) $f(2) = \ln 2 - \frac{2}{2} = -0.306...$

$f(3) = \ln 3 - \frac{2}{3} = 0.431...$

There is a sign change (and the function is continuous in this interval) so there is a root in this interval.

Q4 a)

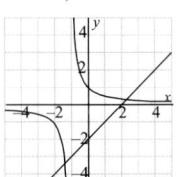

b) $f(-1.4) = \frac{1}{(-1.4)+1} - (-1.4) + 2 = 0.9$

$f(-1.3) = \frac{1}{(-1.3)+1} - (-1.3) + 2 = -0.033...$

There is a sign change (and the function is continuous in this interval) so there is a root in this interval.

c) $\frac{1}{x+1} - x + 2 = 0 \Rightarrow 1 - x(x+1) + 2(x+1) = 0$
$\Rightarrow 1 - x^2 - x + 2x + 2 = 0$
$\Rightarrow -x^2 + x + 3 = 0 \Rightarrow x^2 - x - 3 = 0$

Q5 a)

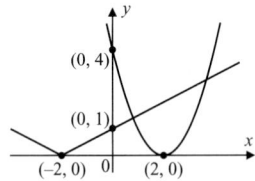

[3 marks available — 1 mark for correct graph shapes, 1 mark for both x-intercepts, 1 mark for both y-intercepts]

b) The graphs cross twice so the equation has 2 roots. *[1 mark]*

c) $f(3) = 3^2 - 4(3) + 4 - |0.5(3) + 1| = -1.5$
$f(4) = 4^2 - 4(4) + 4 - |0.5(4) + 1| = 1$
There is a sign change (and the function is continuous in this interval) so there is a root in the interval [3, 4].
[2 marks available — 1 mark for correct f(3) and f(4), 1 mark for identifying sign change]

Q6 $x_1 = \sqrt{\ln x_0 + 4} = \sqrt{\ln(2) + 4} = 2.1663... = 2.166$ (3 d.p.)
$x_2 = 2.1847... = 2.185$ (3 d.p.), $x_3 = 2.1866... = 2.187$ (3 d.p.)
$x_4 = 2.1868... = 2.187$ (3 d.p.)

Q7 a) Rearrange to get $f(x) = x^x - 3 = 0$
$f(1.5) = (1.5)^{1.5} - 3 = -1.162...$
$f(2) = 2^2 - 3 = 1$
There is a sign change (and the function is continuous in this interval) so there is a root in this interval.

b) Start with $x_0 = 2$:
$x_1 = 3^{\frac{1}{x_0}} = 3^{\frac{1}{2}} = 1.732..., x_2 = 1.885...,$
$x_3 = 1.790..., x_4 = 1.846...$
So an approximation of the root is $x = 1.8$ to 1 d.p.

Q8 a) $x_1 = \sqrt{7(9) + 9} = 8.4852..., x_2 = 8.2702...,$
$x_3 = 8.1787..., x_4 = 8.1394..., x_5 = 8.1225...,$
$x_6 = 8.1152..., x_7 = 8.1121..., x_8 = 8.1108...,$
$x_9 = 8.1102..., x_{10} = 8.1099...$
So an approximation of the root is $x = 8.110$ to 3 d.p.
[2 marks available — 1 mark for using formula to get the first iteration, 1 mark for the correct answer]

b) 8.1095 is the lower bound and 8.1105 is the upper bound, so
$f(8.1095) = (8.1095)^2 - 7(8.1095) - 9 = -0.0025...$
$f(8.1105) = (8.1105)^2 - 7(8.1105) - 9 = 0.0067...$
There is a sign change (and the function is continuous in this interval) so the root is in this interval.
[2 marks available — 1 mark for using a suitable interval, 1 mark for identifying sign change]

c) $x_1 = 10.286, x_2 = 13.828$ and $x_3 = 26.030$ (all to 3 d.p.)
[2 marks available — 1 mark for using formula to get at least one correct value, 1 mark for other correct values]

d) The sequence is divergent so will not find a root. *[1 mark]*

Q9 a) $f(1.1) = 2(1.1) - 5 \cos(1.1) = -0.067...$
$f(1.2) = 2(1.2) - 5 \cos(1.2) = 0.588...$
There is a sign change (and the function is continuous in this interval) so there is a root in this interval.

b) $2x - 5 \cos x = 0 \Rightarrow 2x = 5 \cos x$
$\Rightarrow x = \frac{5}{2} \cos x \Rightarrow p = \frac{5}{2}$

c) Using the iterative formula $x_{n+1} = \frac{5}{2} \cos x_n$:
$x_1 = \frac{5}{2} \cos(1.1) = 1.1340, x_2 = 1.0576,$
$x_3 = 1.2274, x_4 = 0.8418, x_5 = 1.6653,$
$x_6 = -0.2360, x_7 = 2.4307, x_8 = -1.8945$ (all to 4 d.p.)
The sequence at first looks like it might converge to the root in part a) but then it continues to jump up and down and diverges.

Q10 a) $\cos x + 2x = 0 \Rightarrow \cos x = -2x \Rightarrow x = -\frac{1}{2} \cos x$
[1 mark for correctly rearranging formula]

b) In part a) you found an iteration formula, so use it to find the root here:
$x_{n+1} = -\frac{1}{2} \cos x_n \Rightarrow x_1 = -\frac{1}{2} \cos(-1) = -0.270...,$
$x_2 = -0.481..., x_3 = -0.443..., x_4 = -0.451..., x_5 = -0.449...$
So an approximation of the root is $x = -0.45$ to 2 d.p.
[2 marks available — 1 mark for using formula from part a), 1 mark for the correct answer]

Q11 a)

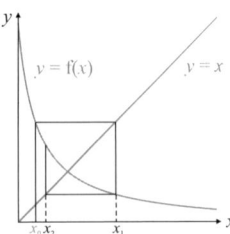

A convergent cobweb diagram.

b)

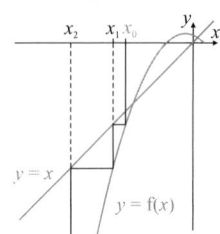

A divergent staircase diagram.

Q12 a) and b)

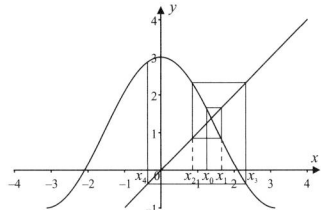

[3 marks available — 1 mark for graphs drawn correctly, 1 mark for the correct iteration diagram shape, 1 mark for correct labels]

Q13 a) $f'(x) = 2x + 9$

Put this into the Newton-Raphson formula:

$$x_{n+1} = x_n - \frac{x_n^2 + 9x_n - 4}{2x_n + 9}$$

Starting with $x_0 = 1$,

$x_1 = 0.4545...$, $x_2 = 0.4245...$, $x_3 = 0.4244...$,
$x_4 = 0.4244...$, $x_5 = 0.4244...$, $x_6 = 0.4244...$,

So an approximation for a root is $x = 0.424$ to 3 d.p.

b) $f'(x) = 3x^2 + 6x + 5$

Put this into the Newton-Raphson formula:

$$x_{n+1} = x_n - \frac{x_n^3 + 3x_n^2 + 5x_n + 7}{3x_n^2 + 6x_n + 5}$$

Starting with $x_0 = 1$,

$x_1 = -0.1428...$, $x_2 = -1.6518...$, $x_3 = -2.3906...$,
$x_4 = -2.2021...$, $x_5 = -2.1797...$, $x_6 = -2.1795...$

So an approximation for a root is $x = 2.180$ to 3 d.p.

c) $f'(x) = e^x - 4x^3$

Put this into the Newton-Raphson formula:

$$x_{n+1} = x_n - \frac{e^{x_n} - x_n^4}{e^{x_n} - 4x_n^3}$$

Starting with $x_0 = 1$,

$x_1 = 2.3406...$, $x_2 = 1.8608...$, $x_3 = 1.5733...$,
$x_4 = 1.4520...$, $x_5 = 1.4302...$, $x_6 = 1.4296...$

So an approximation for a root is $x = 1.430$ to 3 d.p.

d) $f'(x) = -\sin^2 x + \cos^2 x$

Put this into the Newton-Raphson formula:

$$x_{n+1} = x_n - \frac{\sin x_n \cos x_n}{-\sin^2 x_n + \cos^2 x_n}$$

Starting with $x_0 = 1$

$x_1 = 2.0925...$, $x_2 = 1.2339...$, $x_3 = 1.6330...$,
$x_4 = 1.5704...$, $x_5 = 1.5707...$, $x_6 = 1.5707...$

So an approximation for a root is $x = 1.571$ to 3 d.p.

Q14 a) $f(1.4) = 5 - 2(1.4) - (1.4)^2 = 0.24$

$f(1.5) = 5 - 2(1.5) - (1.5)^2 = -0.25$

There is a sign change (and the function is continuous in this interval) so there is a root in this interval.

[2 marks available — 1 mark for correct f(1.4) and f(1.5), 1 mark for identifying sign change]

b) Start with $x_0 = 1.5$:

$x_1 = \sqrt{5 - 2x_0} = \sqrt{5 - 2(1.5)} = 1.414...$,
$x_2 = 1.473...$, $x_3 = 1.432...$, $x_4 = 1.460...$,
$x_5 = 1.441...$, $x_6 = 1.454...$, $x_7 = 1.445...$

The iterations look to be converging, so the formula can be used to estimate the value of α.

[2 marks available — 1 mark for correctly substituting a value between 1.4 and 1.5 into the iteration formula to get at least one correct value, 1 mark for calculating further iterations and stating a correct conclusion]

c) $f'(x) = -2 - 2x$

So $x_{n+1} = x_n - \dfrac{5 - 2x_n - x_n^2}{-2 - 2x_n}$

$x_1 = -4 - \dfrac{5 - 2(-4) - (-4)^2}{-2 - 2(-4)} = -3.5$

$x_2 = -3.45$, $x_3 = -3.449...$,

So a better approximation is $\beta = -3.45$ (3 s.f.)

[3 marks available — 1 mark for correct iteration formula, 1 mark for using formula to get at least one correct value, 1 mark for the correct answer]

Q15 a) $h = (3 - 0) \div 3 = 1$, so the x-values are 0, 1, 2, 3.

$\int_0^3 \sqrt{9 - x^2}\, dx \approx \dfrac{h}{2}[y_0 + 2y_1 + 2y_2 + y_3]$

$= \dfrac{1}{2}[3 + 4\sqrt{2} + 2\sqrt{5} + 0] = 6.56$ (3 s.f.)

b) $h = (1.2 - 0.2) \div 5 = 0.2$,

so the x-values are 0.2, 0.4, 0.6, 0.8, 1.0, 1.2.

$\int_{0.2}^{1.2} x^x\, dx \approx \dfrac{h}{2}[y_0 + 2y_1 + 2y_2 + 2y_3 + 2y_4 + y_5]$

$= \dfrac{1}{10}[0.937... + 1.727... + 1.664... + 1.733... + 2 + 1.300...]$

$= 0.936$ (3 s.f.)

c) $h = (3 - 1) \div 5 = 0.4$,

so the x-values are 1, 1.4, 1.8, 2.2, 2.6, 3.

$\int_1^3 2^{x^2}\, dx \approx \dfrac{h}{2}[y_0 + 2y_1 + 2y_2 + 2y_3 + 2y_4 + y_5]$

$= \dfrac{1}{5}[2 + 7.78... + 18.8... + 57.2... + 216.7... + 512]$

$= 163$ (3 s.f.)

Q16 Using 4 strips:

$h = (6 - 0) \div 4 = 1.5$, so the x-values are 0, 1.5, 3, 4.5, 6

$\int_0^6 (6x - 12)(x^2 - 4x + 3)^2\, dx \approx \dfrac{h}{2}[y_0 + 2y_1 + 2y_2 + 2y_3 + y_4]$

$= \dfrac{3}{4}[-108 - 3.375 + 0 + 826.875 + 5400] = 4586.625$

Using 6 strips:

$h = (6 - 0) \div 6 = 1$, so the x-values are 0, 1, 2, 3, 4, 5, 6

$\int_0^6 (6x - 12)(x^2 - 4x + 3)^2\, dx$

$\approx \dfrac{h}{2}[y_0 + 2y_1 + 2y_2 + 2y_3 + 2y_4 + 2y_5 + y_6]$

$= \dfrac{1}{2}[-108 + 0 + 0 + 0 + 216 + 2304 + 5400] = 3906$

Evaluate the integral:

$\int_0^6 (6x - 12)(x^2 - 4x + 3)^2\, dx$

$= \int_0^6 3(2x - 4)(x^2 - 4x + 3)^2\, dx = [(x^2 - 4x + 3)^3]_0^6$

$= ((6)^2 - 4(6) + 3)^3 - ((0)^2 - 4(0) + 3)^3 = 3375 - 27 = 3348$

The rule $\int (n + 1)f'(x)[f(x)]^n\, dx = [f(x)]^{n+1} + C$ was used here but you could have expanded the brackets instead.

% error using 4 strips $= \dfrac{4586.625 - 3348}{3348} \times 100$

$= 37.0\%$ (3 s.f.)

% error using 6 strips $= \dfrac{3906 - 3348}{3348} \times 100 = 16.7\%$ (3 s.f.)

Q17 a) $h = (7 - 1) \div 4 = 1.5$,

so the x-values are 1, 2.5, 4, 5.5, 7

$\int_1^7 \dfrac{5}{x^2}\, dx \approx \dfrac{h}{2}[y_0 + 2y_1 + 2y_2 + 2y_3 + y_4]$

$= \dfrac{3}{4}[5 + 1.6 + 0.625 + 0.330... + 0.102...] = 5.74$ (2 d.p.)

[3 marks available — 1 mark for using 4 strips, 1 mark for using trapezium rule correctly, 1 mark for the correct answer]

b) E.g. It is an over-estimate as the top of each trapezium lies above the curve.

[1 mark for the correct answer with explanation]

c) $\int_1^7 \frac{5}{x^2}\,dx = \int_1^7 5x^{-2}\,dx = \left[-5x^{-1}\right]_1^7 = \left[-\frac{5}{x}\right]_1^7$

$= -\frac{5}{7} - (-5) = \frac{30}{7} = 4.285...$

[3 marks available — 1 mark for correct integration, 1 mark for substituting in limits, 1 mark for the correct answer]

d) Percentage error $= \frac{5.74 - 4.285...}{4.285...} \times 100 = 34\%$ (2 s.f.)

[1 mark for the correct answer]

Q18 a) $f(3) = \ln(3+3) - 3 + 2 = 0.7917...$
$f(4) = \ln(4+3) - 4 + 2 = -0.0540...$
There is a sign change (and the function is continuous in this interval) so there is a root in this interval.
[2 marks available — 1 mark for correct f(3) and f(4), 1 mark for identifying sign change]

b) $x_{n+1} = \ln(x_n + 3) + 2$, so:
$x_1 = \ln(3+3) + 2 = 3.7917...$
$x_2 = 3.9157..., x_3 = 3.9337..., x_4 = 3.9364..., x_5 = 3.9367...$
So $m = 3.94$ to 2 d.p.
[2 marks available — 1 mark for correctly substituting 3 into the iteration formula to get the first iteration, 1 mark for the correct answer]

c)

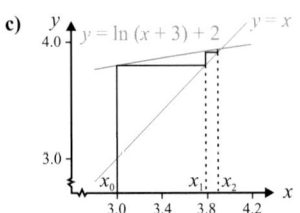

[2 marks available — 1 mark for the correct iteration diagram shape, 1 mark for correct labels]

d) $f'(x) = \frac{1}{x+3} - 1$

So $x_{n+1} = x_n - \frac{\ln(x_n + 3) - x_n + 2}{\frac{1}{x_n + 3} - 1}$

Starting with $x_0 = 3$:

$x_1 = 3.950111..., x_2 = 3.936849..., x_3 = 3.936847...$
So $m = 3.93685$ to 5 d.p.
[3 marks available — 1 mark for correct differentiation, 1 mark for using formula to get the first iteration, 1 mark for the correct answer]

e) $f'(-2) = \frac{1}{-2+3} - 1 = 0$, so the formula will involve division by zero and therefore will fail to give the next iteration / the tangent at this point will be horizontal (as $x = -2$ is a stationary point) so it will not intersect the x-axis to find the next iteration. *[1 mark]*

Q19 a)

x	2	2.2	2.4	2.6	2.8	3
y	0.9803	1.1695	1.3563	1.5407	1.7229	1.9029

[1 mark for all three correct y-values]

b) $\int_2^3 \sqrt{x}\ln x\,dx \approx 0.2 \times \frac{1}{2} \times [0.9803 + 2(1.1695 + 1.3563 + 1.5407 + 1.7229) + 1.9029]$

$= 1.446$ to 3 d.p.

[2 marks available — 1 mark for using the trapezium rule correctly, 1 mark for the correct answer]

c) Using integration by parts, set $u = \ln x$ and $\frac{dv}{dx} = x^{\frac{1}{2}}$.
Then $\frac{du}{dx} = \frac{1}{x}$ and $v = \frac{2}{3}x^{\frac{3}{2}}$.

So $\int \sqrt{x}\ln x\,dx = \frac{2}{3}x^{\frac{3}{2}}\ln x - \int \frac{2}{3}x^{\frac{1}{2}}\,dx$

$= \frac{2}{3}x^{\frac{3}{2}}\ln x - \frac{2}{3}\int x^{\frac{1}{2}}\,dx$

$= \frac{2}{3}x^{\frac{3}{2}}\ln x - \frac{4}{9}x^{\frac{3}{2}} + c$

$= \frac{2x^{\frac{3}{2}}}{9}(3\ln x - 2) + c$

[5 marks available — 1 mark for identifying u and $\frac{dv}{dx}$, 1 mark for finding v and $\frac{du}{dx}$, 1 mark for using integration by parts formula, 1 mark for correct integral, 1 mark for the final answer in the correct form]

d) $I = \left[\frac{2x^{\frac{3}{2}}}{9}(3\ln x - 2)\right]_2^3 = 1.49630... - 0.04993...$
$= 1.446$ to 3 d.p.

[2 marks available — 1 mark for substituting the limits, 1 mark for the correct final answer]

e) The approximate value is the same as the real value to 3 d.p. This is because the curve is almost straight between $x = 2$ and $x = 3$, which makes the trapezium rule very accurate.
[1 mark]

Q20 a) $h = (1 - 0) \div 4 = \frac{1}{4}$, so

x	0	$\frac{1}{4}$	$\frac{1}{2}$	$\frac{3}{4}$	1
y	0	0.6	1.108...	1.570...	2

$\int_0^1 \frac{6x}{\sqrt{x+2}}\,dx \approx \frac{h}{2}[y_0 + 2(y_1 + y_2 + y_3) + y_4]$
$= 0.125[0 + 2(0.6 + 1.108... + 1.570...) + 2]$
$= 1.070$ to 3 d.p.

[3 marks available — 1 mark for correct h, 1 mark for using the trapezium rule correctly, 1 mark for the correct answer]

b) R is made up of the estimated area I and a triangle to the right of the area. The top right vertex of R is where $4x - 4 = 2 \Rightarrow x = 1.5$, so the base length of the triangle is $1.5 - 1 = 0.5$ and its area is $0.5 \times 0.5 \times 2 = 0.5$.
So the area of R is approximately $1.070 + 0.5 = 1.57$ m^2.
[2 marks available — 1 mark for adding a correct triangle area to the answer to part b), 1 mark for 1.57 m^2 with units]

c) Let $u = \sqrt{x} + 2 = x^{\frac{1}{2}} + 2 \Rightarrow \frac{du}{dx} = \frac{1}{2}x^{-\frac{1}{2}} = \frac{1}{2\sqrt{x}}$

$x = 0 \Rightarrow u = 2$ and $x = 1 \Rightarrow u = 3$

$I = \int_0^1 \frac{6x}{\sqrt{x}+2}\,dx = \int_2^3 \frac{6x(2\sqrt{x})}{u}\,du$

$= \int_2^3 \frac{12(\sqrt{x})^3}{u}\,du = 12\int_2^3 \frac{(u-2)^3}{u}\,du$

$= 12\int_2^3 \frac{u^3 - 6u^2 + 12u - 8}{u}\,du$

$= 12\int_2^3 (u^2 - 6u + 12 - \frac{8}{u})\,du$

$= 12\left[\frac{u^3}{3} - 3u^2 + 12u - 8\ln u\right]_2^3$

$= 12[18 - 8\ln 3 - (\frac{44}{3} - 8\ln 2)] = 12(\frac{10}{3} + 8\ln\frac{2}{3})$

[7 marks available — 1 mark for $\frac{du}{dx}$, 1 mark for replacing dx in the integral with $\frac{dx}{du}$ du, 1 mark for obtaining an integral in u only, 1 mark for the correct expansion of $(u-2)^3$, 1 mark for the correct integration, 1 mark for substituting the correct limits, 1 mark for the correct answer]

d) $12(\frac{10}{3} + 8\ln\frac{2}{3}) = 1.07534...$
1.07534... − 1.070 = 0.00534... m²,
which is 53 cm² to the nearest square centimetre.
[2 marks available — 1 mark for the difference between the answers to (a) and (c), 1 mark for 53 cm²]
Remember to multiply by 100 twice as you're converting areas, not lengths.

Chapter 10: Vectors

Prior Knowledge Check

Q1 **a)** $\mathbf{a} + \mathbf{b} = \binom{2}{4} + \binom{1}{-5} = \binom{3}{-1}$

b) $3\mathbf{c} + 2\mathbf{a} = \binom{-9}{-3} + \binom{4}{8} = \binom{-5}{5}$

c) $2\mathbf{b} - \mathbf{c} + 3\mathbf{a} = \binom{2}{-10} - \binom{-3}{-1} + \binom{6}{12} = \binom{11}{3}$

Q2 Find the scale factors for the **i** and **j** components:
$\frac{8}{-20} = -0.4 \qquad \frac{-12}{30} = -0.4$
$8\mathbf{i} - 12\mathbf{j} = -0.4(-20\mathbf{i} + 30\mathbf{j})$, so the vectors are parallel.

Q3 $\overrightarrow{AB} = \overrightarrow{OB} - \overrightarrow{OA} = \binom{1}{6} - \binom{-8}{0} = \binom{9}{6}$

$\overrightarrow{BC} = \overrightarrow{OC} - \overrightarrow{OB} = \binom{6}{9} - \binom{1}{6} = \binom{5}{3}$

$\frac{9}{5} = 1.8 \qquad \frac{6}{3} = 2$
So $\overrightarrow{AB}$ and $\overrightarrow{BC}$ are not parallel, which means A, B and C are not collinear.

Q4 $\cos\theta = \frac{b^2 + c^2 - a^2}{2bc} = \frac{7^2 + 6^2 - 8^2}{2 \times 7 \times 6} = \frac{21}{84} = \frac{1}{4}$
$\Rightarrow \theta = \cos^{-1}(\frac{1}{4}) = 75.522... = 75.5$ (to 1 d.p.)

Q5 **a)** $|\mathbf{f}| = \sqrt{3^2 + (-6)^2} = \sqrt{45} = 3\sqrt{5}$

b) $\mathbf{g} = (-\mathbf{i} + 4\mathbf{j})$
So it looks like this:

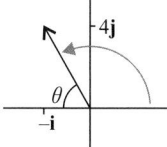

$\tan\theta = \frac{4}{1} \Rightarrow \theta = \tan^{-1}4 = 75.963...$
The direction is anticlockwise from the positive *x*-axis, so 180 − 75.963... = 104.036... = 104.0° (to 1 d.p.)

c) $|\mathbf{f}| = \sqrt{3^2 + (-6^2)} = \sqrt{45}$
$|\mathbf{g}| = \sqrt{(-1)^2 + 4^2} = \sqrt{17}$
If $\mathbf{f} = \overrightarrow{OF}$ and $\mathbf{g} = \overrightarrow{OG}$ then we want to find $|\overrightarrow{FG}|$.
$\overrightarrow{FG} = \overrightarrow{OG} - \overrightarrow{OF} = \mathbf{g} - \mathbf{f} = -4\mathbf{i} + 10\mathbf{j}$
$\Rightarrow |\overrightarrow{FG}| = \sqrt{(-4)^2 + 10^2} = \sqrt{116}$
Using the cosine rule:
$a = \sqrt{116}, b = \sqrt{45}, c = \sqrt{17}$
So $\cos\theta = \frac{b^2 + c^2 - a^2}{2bc} = \frac{45 + 17 - 116}{2 \times 3\sqrt{5} \times \sqrt{17}} = -0.976...$
$\Rightarrow \theta = 167.471...° = 167.5°$ (1 d.p.)

Exercise 10.1 — Vectors in Three Dimensions

Q1 **a)** $\overrightarrow{OR} = \begin{pmatrix} 4 \\ -5 \\ 1 \end{pmatrix} \qquad \overrightarrow{OS} = \begin{pmatrix} -3 \\ 0 \\ -1 \end{pmatrix}$

b) $\overrightarrow{OR} = 4\mathbf{i} - 5\mathbf{j} + \mathbf{k} \qquad \overrightarrow{OS} = -3\mathbf{i} - \mathbf{k}$

*There's no **j** component written for $\overrightarrow{OS}$ because its **j** component is zero (you don't write −3**i** + 0**j** − **k**).*

Q2 $\overrightarrow{GH} = \overrightarrow{OH} - \overrightarrow{OG} = \begin{pmatrix} -1 \\ 4 \\ 9 \end{pmatrix} - \begin{pmatrix} 2 \\ -3 \\ 4 \end{pmatrix} = \begin{pmatrix} -3 \\ 7 \\ 5 \end{pmatrix}$

$\overrightarrow{HG} = -\overrightarrow{GH} = -\begin{pmatrix} -3 \\ 7 \\ 5 \end{pmatrix} = \begin{pmatrix} 3 \\ -7 \\ -5 \end{pmatrix}$

Q3 $\overrightarrow{OJ} = 4\mathbf{i} - 3\mathbf{k}, \overrightarrow{OK} = -\mathbf{i} + 3\mathbf{j}$, and $\overrightarrow{OL} = 2\mathbf{i} + 2\mathbf{j} + 7\mathbf{k}$
$\overrightarrow{JK} = \overrightarrow{OK} - \overrightarrow{OJ} = -\mathbf{i} + 3\mathbf{j} - (4\mathbf{i} - 3\mathbf{k}) = -5\mathbf{i} + 3\mathbf{j} + 3\mathbf{k}$
$\overrightarrow{KL} = \overrightarrow{OL} - \overrightarrow{OK} = 2\mathbf{i} + 2\mathbf{j} + 7\mathbf{k} - (-\mathbf{i} + 3\mathbf{j}) = 3\mathbf{i} - \mathbf{j} + 7\mathbf{k}$
$\overrightarrow{LJ} = \overrightarrow{OJ} - \overrightarrow{OL} = 4\mathbf{i} - 3\mathbf{k} - (2\mathbf{i} + 2\mathbf{j} + 7\mathbf{k}) = 2\mathbf{i} - 2\mathbf{j} - 10\mathbf{k}$

Q4 The position vector of the midpoint of CD is at
$\overrightarrow{OC} + \frac{1}{2}\overrightarrow{CD} = \begin{pmatrix} -1 \\ 3 \\ -5 \end{pmatrix} + \frac{1}{2}\begin{pmatrix} 4 \\ -4 \\ 6 \end{pmatrix} = \begin{pmatrix} -1 \\ 3 \\ -5 \end{pmatrix} + \begin{pmatrix} 2 \\ -2 \\ 3 \end{pmatrix} = \begin{pmatrix} 1 \\ 1 \\ -2 \end{pmatrix} = \overrightarrow{OM}$
So M is the midpoint of CD.

Q5 **a)** Point B is twice the distance of A from the origin in each direction, so $\overrightarrow{OB} = 2\overrightarrow{OA}$.
The position vector of B can also be described as:
$\overrightarrow{OB} = \overrightarrow{AB} + \overrightarrow{OA}$
Substituting for $\overrightarrow{OB}$:
$2\overrightarrow{OA} = \overrightarrow{AB} + \overrightarrow{OA} \Rightarrow 2\overrightarrow{OA} - \overrightarrow{OA} = \overrightarrow{AB}$
So $\overrightarrow{AB} = \overrightarrow{OA}$ — as required.

b) $\overrightarrow{OB} = 2\overrightarrow{OA} = 2(3\mathbf{i} + \mathbf{j} + 2\mathbf{k}) = 6\mathbf{i} + 2\mathbf{j} + 4\mathbf{k}$

Q6 $\overrightarrow{AB} = \begin{pmatrix} 4 \\ 1 \\ -1 \end{pmatrix} - \begin{pmatrix} 3 \\ -2 \\ 0 \end{pmatrix} = \begin{pmatrix} 1 \\ 3 \\ -1 \end{pmatrix}$

$\mathbf{c} = \begin{pmatrix} -4 \\ -12 \\ 4 \end{pmatrix} = -4\begin{pmatrix} 1 \\ 3 \\ -1 \end{pmatrix} = -4\overrightarrow{AB}$, so $\overrightarrow{AB}$ is parallel to **c**.

[3 marks available — 1 mark for attempting to find $\overrightarrow{AB}$, 1 mark for correctly finding $\overrightarrow{AB}$, 1 mark for showing $\mathbf{c} = -4\overrightarrow{AB}$]

Q7 A vector is parallel to another if it is multiplied by the same scalar in each direction.
$\mathbf{a} = (3 \times \frac{1}{4})\mathbf{i} + (\frac{1}{3} \times 1)\mathbf{j} + (3 \times -\frac{2}{3})\mathbf{k}$
$\mathbf{a} = 3\mathbf{b}$ in the **i** and **k** directions, but $\frac{1}{3}\mathbf{b}$ in the **j** direction, so they are not parallel.

Q8 Let $(\sqrt{3}\mathbf{i} + 18\mathbf{j} + p\mathbf{k}) = m(6\mathbf{i} + q\mathbf{j} + 54\mathbf{k})$ then $\sqrt{3} = 6m$
$\Rightarrow m = \frac{\sqrt{3}}{6}$
Then $p = \frac{54\sqrt{3}}{6} = 9\sqrt{3}$ and
$18 = \frac{\sqrt{3}}{6}q \Rightarrow q = 18 \times \frac{6}{\sqrt{3}} = \frac{108}{\sqrt{3}} = \frac{108\sqrt{3}}{3} = 36\sqrt{3}$.
*[4 marks available — 1 mark for comparing the coefficients of **i**, 1 marks for finding the correct ratio between the vectors, 1 mark for the correct value of p, 1 mark for the correct value of q]*

Answers

Q9 Model the counters as points A, B and C, with position vectors
$\overrightarrow{OA} = \mathbf{i} + 3\mathbf{k}$, $\overrightarrow{OB} = 3\mathbf{i} + \mathbf{j} + 2\mathbf{k}$, and $\overrightarrow{OC} = 7\mathbf{i} + 3\mathbf{j}$.
$\overrightarrow{AB} = \overrightarrow{OB} - \overrightarrow{OA} = 2\mathbf{i} + \mathbf{j} - \mathbf{k}$
$\overrightarrow{BC} = \overrightarrow{OC} - \overrightarrow{OB} = 4\mathbf{i} + 2\mathbf{j} - 2\mathbf{k} = 2 \times (2\mathbf{i} + \mathbf{j} - \mathbf{k}) = 2\overrightarrow{AB}$
$\overrightarrow{BC}$ is a scalar multiple of $\overrightarrow{AB}$, therefore the vectors are parallel. They also share a point (B), so they must be collinear. So the counters must lie in a straight line.

Q10 Since the points are collinear, the vectors $\overrightarrow{PQ}$ and $\overrightarrow{QR}$ will be parallel, i.e. scalar multiples of one another.
$$\overrightarrow{PQ} = \begin{pmatrix} 1 \\ b \\ -4 \end{pmatrix} - \begin{pmatrix} -2 \\ a \\ -8 \end{pmatrix} = \begin{pmatrix} 3 \\ b - a \\ 4 \end{pmatrix} \text{ and } \overrightarrow{QR} = \begin{pmatrix} -5 \\ 6 \\ 3b \end{pmatrix} - \begin{pmatrix} 1 \\ b \\ -4 \end{pmatrix} = \begin{pmatrix} -6 \\ 6 - b \\ 3b + 4 \end{pmatrix}$$
Consider the $\mathbf{i}$ components of the two vectors:
$-2 \times 3 = -6$, so $-2\overrightarrow{PQ} = \overrightarrow{QR}$.
For the $\mathbf{k}$ components: $-2(4) = 3b + 4 \Rightarrow -12 = 3b \Rightarrow b = -4$
For the $\mathbf{j}$ components: $-2(b - a) = 6 - b \Rightarrow -2(-4 - a) = 6 - (-4)$
$\Rightarrow 8 + 2a = 10 \Rightarrow a = 1$

Q11 a) $\overrightarrow{AB} = \begin{pmatrix} 2 \\ 1 \\ 2p \end{pmatrix} - \begin{pmatrix} 3 \\ 4 \\ -1 \end{pmatrix} = \begin{pmatrix} -1 \\ -3 \\ 2p + 1 \end{pmatrix}$

$\overrightarrow{BC} = \begin{pmatrix} -1 \\ -8 \\ 19 \end{pmatrix} - \begin{pmatrix} 2 \\ 1 \\ 2p \end{pmatrix} = \begin{pmatrix} -3 \\ -9 \\ 19 - 2p \end{pmatrix}$

Vectors $\overrightarrow{AB}$ and $\overrightarrow{BC}$ are parallel,
so $\overrightarrow{AB} = m\overrightarrow{BC}$, for some number m.
Then use the $\mathbf{i}$ components of $\overrightarrow{AB}$ and $\overrightarrow{BC}$ to find m:
$-3m = -1 \Rightarrow m = \frac{1}{3}$
Now use the $\mathbf{k}$ components of $\overrightarrow{AB}$ and $\overrightarrow{BC}$ to find p:
$2p + 1 = \frac{1}{3}(19 - 2p) \Rightarrow 6p + 3 = 19 - 2p$
$\Rightarrow 8p = 16 \Rightarrow p = 2$
[5 marks available — 1 mark for finding $\overrightarrow{AB}$ and $\overrightarrow{BC}$, 1 mark for a correct method for finding the scale factor, 1 mark for the correct value of the scale factor, 1 mark for forming an equation in p, 1 mark for the correct value of p]

b) Since AB = $\frac{1}{3}$BC, AB : BC is 1 : 3 *[1 mark]*

Q12 a) $\overrightarrow{BC} = \overrightarrow{OC} - \overrightarrow{OB} = (3\mathbf{i} + 5\mathbf{j} + 5\mathbf{k}) - (7\mathbf{i} - \mathbf{j} + \mathbf{k})$
$= -4\mathbf{i} + 6\mathbf{j} + 4\mathbf{k}$
$\overrightarrow{AD} = \overrightarrow{BC}$ as ABCD is a rhombus, so
$\overrightarrow{OD} = \overrightarrow{OA} + \overrightarrow{AD} = \overrightarrow{OA} + \overrightarrow{BC}$
$= (11\mathbf{i} + 5\mathbf{j} - 3\mathbf{k}) + (-4\mathbf{i} + 6\mathbf{j} + 4\mathbf{k})$
$= 7\mathbf{i} + 11\mathbf{j} + \mathbf{k}$, so D is at (7, 11, 1).
[3 marks available — 1 mark for correctly finding vector length of one side, 1 mark for the correct method for $\overrightarrow{OD}$, 1 mark for (7, 11, 1)]

b) $\overrightarrow{AC} = \overrightarrow{OC} - \overrightarrow{OA} = (3\mathbf{i} + 5\mathbf{j} + 5\mathbf{k}) - (11\mathbf{i} + 5\mathbf{j} - 3\mathbf{k})$
$= -8\mathbf{i} + 8\mathbf{k}$
$\overrightarrow{OE} = \overrightarrow{OC} + \overrightarrow{CE} = \overrightarrow{OC} + \frac{1}{4}\overrightarrow{AC}$
$= 3\mathbf{i} + 5\mathbf{j} + 5\mathbf{k} - 2\mathbf{i} + 2\mathbf{k} = \mathbf{i} + 5\mathbf{j} + 7\mathbf{k}$
[3 marks available — 1 mark for the correct $\overrightarrow{AC}$, 1 mark for the correct method for $\overrightarrow{OE}$, 1 mark for $\mathbf{i} + 5\mathbf{j} + 7\mathbf{k}$]

Q13 a) A is at (0, 1, 0) *[1 mark]*

b) The height of the cuboid is $\frac{30}{(2)(3)} = 5$
and $z \geq 0$ so H is at (2, 4, 5).
$$\overrightarrow{AH} = \overrightarrow{OH} - \overrightarrow{OA} = \begin{pmatrix} 2 \\ 4 \\ 5 \end{pmatrix} - \begin{pmatrix} 0 \\ 1 \\ 0 \end{pmatrix} = \begin{pmatrix} 2 \\ 3 \\ 5 \end{pmatrix}$$
$\overrightarrow{OP} = \overrightarrow{OA} + \overrightarrow{AP} = \overrightarrow{OA} + \frac{3}{4}\overrightarrow{AH}$ (since $\overrightarrow{AP} = 3\overrightarrow{PH}$).
So $\overrightarrow{OP} = \begin{pmatrix} 0 \\ 1 \\ 0 \end{pmatrix} + \frac{3}{4}\begin{pmatrix} 2 \\ 3 \\ 5 \end{pmatrix} = \frac{1}{4}\begin{pmatrix} 0 \\ 4 \\ 0 \end{pmatrix} + \frac{1}{4}\begin{pmatrix} 6 \\ 9 \\ 15 \end{pmatrix} = \frac{1}{4}\begin{pmatrix} 6 \\ 13 \\ 15 \end{pmatrix}$.

[5 marks available — 1 mark for the correct height, 1 mark for the correct H, 1 mark for the correct $\overrightarrow{AH}$, 1 mark for the correct method for $\overrightarrow{OP}$, 1 mark for the correct $\overrightarrow{OP}$ in the required form]

Exercise 10.2 — Calculating with 3D Vectors

Q1 a) $\sqrt{1^2 + 4^2 + 8^2} = \sqrt{1 + 16 + 64} = \sqrt{81} = 9$
b) $\sqrt{4^2 + 2^2 + 4^2} = \sqrt{36} = 6$
c) $\sqrt{(-4)^2 + (-5)^2 + 20^2} = \sqrt{441} = 21$
d) $\sqrt{7^2 + 1^2 + (-7)^2} = \sqrt{99} = 3\sqrt{11}$

Q2 a) The resultant is: $(\mathbf{i} + \mathbf{j} + 2\mathbf{k}) + (\mathbf{i} + 2\mathbf{j} + 4\mathbf{k}) = 2\mathbf{i} + 3\mathbf{j} + 6\mathbf{k}$
Its magnitude is: $\sqrt{2^2 + 3^2 + 6^2} = \sqrt{49} = 7$
b) resultant: $2\mathbf{i} + 14\mathbf{j} + 23\mathbf{k}$
magnitude: $\sqrt{2^2 + 14^2 + 23^2} = \sqrt{729} = 27$
c) resultant: $\begin{pmatrix} 2 \\ 6 \\ 9 \end{pmatrix}$, magnitude: $\sqrt{2^2 + 6^2 + 9^2} = 11$
d) resultant: $\begin{pmatrix} 2 \\ 5 \\ 14 \end{pmatrix}$, magnitude: $\sqrt{2^2 + 5^2 + 14^2} = 15$

Q3 $\mathbf{a} + \mathbf{b} = \begin{pmatrix} 10 \\ 2 \\ 14 \end{pmatrix}$. $|\mathbf{a} + \mathbf{b}| = \sqrt{10^2 + 2^2 + 14^2} = 10\sqrt{3}$

Q4 $3p - 2 = 10 \Rightarrow p = 4$
$2pq = -1 \Rightarrow 8q = -1 \Rightarrow q = -\frac{1}{8}$
Now $p^2q + 5p = r \Rightarrow r = 4^2(-\frac{1}{8}) + 5(4) \Rightarrow r = 18$
$\Rightarrow |\mathbf{a}| = \sqrt{10^2 + 1^2 + 18^2} = \sqrt{425} = 5\sqrt{17}$
[5 marks available — 1 mark each for correct values of p, q and r, 1 mark for a correct method for magnitude, 1 mark for $5\sqrt{17}$]

Q5 a) $|\mathbf{t}| = \sqrt{4^2 + (-4)^2 + (-7)^2} = 9$
so the unit vector is $\frac{\mathbf{t}}{|\mathbf{t}|} = \frac{1}{9}\mathbf{t} = \frac{4}{9}\mathbf{i} - \frac{4}{9}\mathbf{j} - \frac{7}{9}\mathbf{k}$
b) $|\mathbf{u}| = \sqrt{(-1)^2 + 2^2 + (-2)^2} = 3$
so the unit vector is $\frac{\mathbf{u}}{|\mathbf{u}|} = \frac{1}{3}\mathbf{u} = -\frac{1}{3}\mathbf{i} + \frac{2}{3}\mathbf{j} - \frac{2}{3}\mathbf{k}$
c) $|\mathbf{v}| = \sqrt{2^2 + 3^2 + (-1)^2} = \sqrt{14}$
so the unit vector is
$\frac{\mathbf{v}}{|\mathbf{v}|} = \frac{1}{\sqrt{14}}\mathbf{v} = \frac{\sqrt{14}}{14}\mathbf{v} = \frac{\sqrt{14}}{7}\mathbf{i} + \frac{3\sqrt{14}}{14}\mathbf{j} - \frac{\sqrt{14}}{14}\mathbf{k}$

Q6 a) $\sqrt{(5-3)^2 + (6-4)^2 + (6-5)^2}$
$= \sqrt{2^2 + 2^2 + 1^2} = \sqrt{4 + 4 + 1} = \sqrt{9} = 3$
b) $\sqrt{(-11-7)^2 + (1-2)^2 + (15-9)^2}$
$= \sqrt{324 + 1 + 36} = \sqrt{361} = 19$

c) $\sqrt{(6-10)^2+(10-(-2))^2+(-4-(-1))^2}$
$= \sqrt{16+144+9} = \sqrt{169} = 13$

d) $\sqrt{(7-0)^2+(0-(-4))^2+(14-10)^2}$
$= \sqrt{49+16+16} = \sqrt{81} = 9$

e) $\sqrt{(2-(-4))^2+(4-7)^2+(-12-10)^2}$
$= \sqrt{36+9+484} = \sqrt{529} = 23$

f) $\sqrt{(30-7)^2+(9-(-1))^2+(-6-4)^2}$
$= \sqrt{529+100+100} = \sqrt{729} = 27$

Q7 $|2\mathbf{m}-\mathbf{n}| = |-6\mathbf{i}-5\mathbf{j}+10\mathbf{k}| = \sqrt{(-6)^2+(-5)^2+10^2}$
$= 12.68857... = 12.7$ m (1 d.p.)

Q8 $|\overrightarrow{AO}| = |\overrightarrow{OA}| = \sqrt{1^2+(-4)^2+3^2} = \sqrt{26}$
$|\overrightarrow{BO}| = |\overrightarrow{OB}| = \sqrt{(-1)^2+(-3)^2+5^2} = \sqrt{35}$
It's pretty clear that the magnitude of $\overrightarrow{AO}$ is going to be the same as $\overrightarrow{OA}$ so there's no need to find $\overrightarrow{AO}$.
$\overrightarrow{BA} = \overrightarrow{OA}-\overrightarrow{OB} = (\mathbf{i}-4\mathbf{j}+3\mathbf{k})-(-\mathbf{i}-3\mathbf{j}+5\mathbf{k}) = 2\mathbf{i}-\mathbf{j}-2\mathbf{k}$
$|\overrightarrow{BA}| = \sqrt{2^2+(-1)^2+(-2)^2} = 3$

Triangle AOB is right-angled because:
$|\overrightarrow{AO}|^2 + |\overrightarrow{BA}|^2 = (\sqrt{26})^2+3^2 = 26+9 = 35 = (\sqrt{35})^2 = |\overrightarrow{BO}|^2$

Q9 $|\overrightarrow{PQ}| = \sqrt{(q-4)^2+6^2+(2q-3)^2}$
So: $\sqrt{(q-4)^2+6^2+(2q-3)^2} = 11$
$\Rightarrow q^2-8q+16+36+4q^2-12q+9 = 121$
$\Rightarrow 5q^2-20q-60 = 0 \Rightarrow q^2-4q-12 = 0 \Rightarrow (q-6)(q+2) = 0$
Either $q = 6$, then Q is (4, 5, 13),
or $q = -2$, then Q is (-4, 5, -3).

Q10 a) Form a triangle between the two vectors (call them **a** and **b**) with angle θ between them. The triangle has side lengths $|\mathbf{a}|$, $|\mathbf{b}|$ and $|\mathbf{b}-\mathbf{a}|$.

$\mathbf{a} = \begin{pmatrix}1\\3\\2\end{pmatrix}$, $|\mathbf{a}| = \sqrt{1^2+3^2+2^2} = \sqrt{14}$

$\mathbf{b} = \begin{pmatrix}3\\2\\1\end{pmatrix}$, $|\mathbf{b}| = \sqrt{3^2+2^2+1^2} = \sqrt{14}$

$\mathbf{b}-\mathbf{a} = \begin{pmatrix}2\\-1\\-1\end{pmatrix}$, $|\mathbf{b}-\mathbf{a}| = \sqrt{2^2+(-1)^2+(-1)^2} = \sqrt{6}$

Using the cosine rule,
$\cos\theta = \dfrac{(\sqrt{14})^2+(\sqrt{14})^2-(\sqrt{6})^2}{2\times\sqrt{14}\times\sqrt{14}} = \dfrac{22}{28} = \dfrac{11}{14}$
$\theta = \cos^{-1}\left(\dfrac{11}{14}\right) = 38.2132... = 38.2°$ (1 d.p.)

b) $|\mathbf{a}| = \sqrt{14}$ as before

$\mathbf{b} = \begin{pmatrix}0\\1\\0\end{pmatrix}$, $|\mathbf{b}| = 1$

$\mathbf{b}-\mathbf{a} = \begin{pmatrix}-1\\-2\\-2\end{pmatrix}$, $|\mathbf{b}-\mathbf{a}| = \sqrt{(-1)^2+(-2)^2+(-2)^2} = 3$

Using the cosine rule,
$\cos\theta = \dfrac{(\sqrt{14})^2+1^2-3^2}{2\times\sqrt{14}\times 1} = \dfrac{6}{2\sqrt{14}} = \dfrac{3}{\sqrt{14}}$
$\theta = \cos^{-1}\left(\dfrac{3}{\sqrt{14}}\right) = 36.6992... = 36.7°$ (1 d.p.)

c) $|\mathbf{a}| = \sqrt{14}$ as before

$\mathbf{b} = \begin{pmatrix}-3\\1\\-2\end{pmatrix}$, $|\mathbf{b}| = \sqrt{(-3)^2+1^2+(-2)^2} = \sqrt{14}$

$\mathbf{b}-\mathbf{a} = \begin{pmatrix}-4\\-2\\-4\end{pmatrix}$, $|\mathbf{b}-\mathbf{a}| = \sqrt{(-4)^2+(-2)^2+(-4)^2} = 6$

Using the cosine rule,
$\cos\theta = \dfrac{(\sqrt{14})^2+(\sqrt{14})^2-6^2}{2\times\sqrt{14}\times\sqrt{14}} = \dfrac{-8}{28} = -\dfrac{2}{7}$
$\theta = \cos^{-1}\left(-\dfrac{2}{7}\right) = 106.6°$ (1 d.p.)

d) $|\mathbf{a}| = \sqrt{14}$ as before

$\mathbf{b} = \begin{pmatrix}2\\2\\2\end{pmatrix}$, $|\mathbf{b}| = \sqrt{2^2+2^2+2^2} = 2\sqrt{3}$

$\mathbf{b}-\mathbf{a} = \begin{pmatrix}1\\-1\\0\end{pmatrix}$, $|\mathbf{b}-\mathbf{a}| = \sqrt{1^2+(-1)^2+0^2} = \sqrt{2}$

Using the cosine rule,
$\cos\theta = \dfrac{(\sqrt{14})^2+(2\sqrt{3})^2-(\sqrt{2})^2}{2\times\sqrt{14}\times 2\sqrt{3}} = \dfrac{24}{4\sqrt{42}} = \dfrac{6}{\sqrt{42}}$
$\theta = \cos^{-1}\left(\dfrac{6}{\sqrt{42}}\right) = 22.2°$ (1 d.p.)

Q11 Vectors **v** and **i** have an angle of 60° between them, so you can use this to make a right-angled triangle, shown on the right: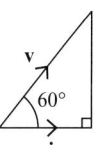

The **i** component of **v** is 1 (the length of **i**), so $|\mathbf{v}| = \sqrt{1^2+a^2+1^2} = \sqrt{2+a^2}$
$\Rightarrow \cos 60° = \dfrac{1}{|\mathbf{v}|} = \dfrac{1}{\sqrt{2+a^2}}$
$\Rightarrow \sqrt{2+a^2} = \dfrac{1}{\cos 60°}$
$\Rightarrow \sqrt{2+a^2} = 2 \Rightarrow 2+a^2 = 4$
$\Rightarrow a^2 = 2 \Rightarrow a = \pm\sqrt{2}$

Q12 $|\overrightarrow{PA}| = \sqrt{3^2+4^2+(-1)^2} = \sqrt{26}$ and
$|\overrightarrow{PB}| = \sqrt{(-4)^2+7^2+2^2} = \sqrt{69}$

$\overrightarrow{AB} = \overrightarrow{PB}-\overrightarrow{AP} = \begin{pmatrix}-7\\3\\3\end{pmatrix}$ so $|\overrightarrow{AB}| = \sqrt{(-7)^2+3^2+3^2} = \sqrt{67}$.

Now $\cos APB = \dfrac{26+69-67}{2\sqrt{26}\sqrt{69}} \Rightarrow APB = 70.698...$
$= 70.7°$ to one decimal place.
[5 marks available — 1 mark for finding $\overrightarrow{AB}$, 1 mark for a correct attempt at a length, 1 mark for correct values for the lengths, 1 mark for correct use of the cosine rule, 1 mark for a fully correct proof]

Q13 a) The **i** component of **a** is x, so the vectors **a** and $x\mathbf{i}$ form a right-angled triangle with the angle of θ between them:

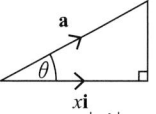

So $\cos\theta = \dfrac{|x\mathbf{i}|}{|\mathbf{a}|} = \dfrac{x}{|\mathbf{a}|}$.
*[2 marks available — 1 mark for forming a right-angled triangle with **a**, x and the angle θ (or using the cosine rule), 1 mark for fully correct proof]*

*To see why this is a right-angled triangle, try picturing **a** as the diagonal of a cuboid with sides x, y and z. If you didn't spot that you could make a right-angled triangle here, you could also answer this question using the cosine rule.*

b) Using $x = 3$, $y = -2$ and $z = 1$:

$$|\mathbf{a}| = \sqrt{(3)^2 + (-2)^2 + (1)^2} = \sqrt{14}$$

From part a), $\cos\theta = \dfrac{x}{|\mathbf{a}|} \Rightarrow \cos\theta = \dfrac{3}{\sqrt{14}}$

$\Rightarrow \theta = \cos^{-1}\dfrac{3}{\sqrt{14}} \Rightarrow \theta = 36.699...° = 36.7°$ (1 d.p.)

[3 marks available — 1 mark for finding the correct value of |a|, 1 mark for the correct numerical expression for cos θ, 1 mark for the correct angle]

Q14 Here, the displacement is just the distance, which is

$$\sqrt{(15t\cos 36°)^2 + (15t\sin 36°)^2 + (8t)^2}$$
$$= \sqrt{(15t)^2\cos^2 36° + (15t)^2\sin^2 36° + 64t^2}$$
$$= \sqrt{225t^2(\cos^2 36° + \sin^2 36°) + 64t^2} = \sqrt{289t^2} = 17t$$

Q15 a) The magnitude of $\mathbf{R} = \sqrt{4^2 + 4^2 + (-7)^2} = 9$ *[1 mark]*

b) The magnitude of $\mathbf{F}_1$ is $\sqrt{2^2 + (-3)^2 + 6^2} = 7$.
Form a triangle with sides $|\mathbf{F}_1|$, $|\mathbf{F}_2|$ and $|\mathbf{R}|$ and
let the angle between sides $|\mathbf{F}_1|$ and $|\mathbf{F}_2|$ of the triangle be α.
Using alternate angles, the angle between $\mathbf{R}$ and $\mathbf{F}_2$ is 42°, so:

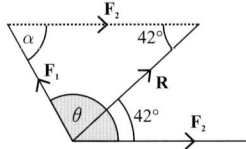

The sine rule gives $\dfrac{\sin\alpha}{9} = \dfrac{\sin 42}{7} \Rightarrow \alpha = 59.35...°$
Using allied angles, $\theta = 180 - \alpha \Rightarrow \theta = 180 - 59.35...$
$= 120.6°$

[4 marks available — 1 mark for the correct magnitude of F_1, 1 mark for a correct application of the sine rule, 1 mark for the correct angle α, 1 mark for the correct angle θ]

Q16 a) $d = \sqrt{(15t\tan\frac{t}{4})^2 + (12t)^2 + (9t)^2}$

$= \sqrt{225\,t^2\tan^2\frac{t}{4} + 225t^2} = \sqrt{225\,t^2(\tan^2\frac{t}{4} + 1)}$

$= \sqrt{225\,t^2(\sec^2\frac{t}{4})} = 15t\sec\frac{t}{4}$

[4 marks available — 1 mark for a correct method for distance, 1 mark for the correct 2-term expression, 1 mark for factorising and using the correct trig identity, 1 mark for a fully correct proof]

b) When $t = \dfrac{2\pi}{3}$,

$$d = \frac{15\left(\frac{2\pi}{3}\right)}{\cos\left(\frac{\left(\frac{2\pi}{3}\right)}{4}\right)} = \frac{10\pi}{\cos\left(\frac{\pi}{6}\right)} = \frac{10\pi}{\frac{\sqrt{3}}{2}} = \frac{20\sqrt{3}}{3}\pi$$

[3 marks available — 1 mark for using sec x = $\frac{1}{\cos x}$, 1 mark for substituting and evaluating, 1 mark for the correct answer in the right form]

Chapter 10 Review Exercise

Q1 $2\mathbf{i} - 4\mathbf{j} + 5\mathbf{k}$

Q2 $\overrightarrow{XO} = -6\mathbf{i} + \mathbf{j} = \begin{pmatrix} -6 \\ 1 \\ 0 \end{pmatrix}$, $\overrightarrow{YO} = -4\mathbf{i} + 4\mathbf{j} - 7\mathbf{k} = \begin{pmatrix} -4 \\ 4 \\ -7 \end{pmatrix}$

Q3 Since M is the midpoint of AB, $\overrightarrow{OM} = \overrightarrow{OA} + \frac{1}{2}\overrightarrow{AB}$
$= (-2\mathbf{i} + 5\mathbf{j} + 7\mathbf{k}) + (-\mathbf{i} - 0.5\mathbf{j} + 3\mathbf{k}) = -3\mathbf{i} + 4.5\mathbf{j} + 10\mathbf{k}$
$\overrightarrow{MC} = \overrightarrow{OC} - \overrightarrow{OM} = (-\mathbf{i} + 5\mathbf{j} + 3\mathbf{k}) - (-3\mathbf{i} + 4.5\mathbf{j} + 10\mathbf{k})$
$= 2\mathbf{i} + 0.5\mathbf{j} - 7\mathbf{k}$

[4 marks available — 1 mark for using a correct method to find an appropriate vector, 1 mark for a correct vector, 1 mark for a correct method to find $\overrightarrow{MC}$ using calculated vector, 1 mark for $\overrightarrow{MC}$ correct]

Q4 a) E.g. **a** and 4**a**

b) E.g. $6\mathbf{i} + 8\mathbf{j} - 4\mathbf{k}$ and $9\mathbf{i} + 12\mathbf{j} - 6\mathbf{k}$

c) E.g. $\begin{pmatrix} 2 \\ 4 \\ -2 \end{pmatrix}$ and $\begin{pmatrix} 4 \\ 8 \\ -4 \end{pmatrix}$

You could give any answer that is a scalar multiple of each vector.

Q5 $2\mathbf{a} + \mathbf{b} = 3\mathbf{i} - 2\mathbf{j} - 5\mathbf{k} = -\frac{1}{2}\mathbf{c}$.
They're scalar multiples, therefore they must be parallel.

Q6 $\overrightarrow{AB} = \begin{pmatrix} -1 \\ 4 \\ 4 \end{pmatrix}$, $\overrightarrow{BC} = \begin{pmatrix} -3 \\ 12 \\ 12 \end{pmatrix} = 3\overrightarrow{AB}$.

This shows $\overrightarrow{AB}$ and $\overrightarrow{BC}$ are parallel, and they share a point (B) so A, B and C must be collinear.

Q7 The vector $\overrightarrow{LN} = \frac{5}{2} \times \overrightarrow{LM}$.
$\frac{5}{2} \times \overrightarrow{LM} = \frac{5}{2}(2\mathbf{i} - \mathbf{j} - 2\mathbf{k}) = 5\mathbf{i} - \frac{5}{2}\mathbf{j} - 5\mathbf{k}$
To find $\overrightarrow{ON}$, add $\overrightarrow{LN}$ to the position vector of L:
$\overrightarrow{ON} = (3\mathbf{i} + \mathbf{j} + 8\mathbf{k}) + (5\mathbf{i} - \frac{5}{2}\mathbf{j} - 5\mathbf{k}) = 8\mathbf{i} - \frac{3}{2}\mathbf{j} + 3\mathbf{k}$

[3 marks available — 1 mark for using the correct method to find $\overrightarrow{LN}$, 1 mark for correct $\overrightarrow{LN}$, 1 mark for correct answer]

Q8 a) $\sqrt{3^2 + 4^2 + (-2)^2} = \sqrt{29}$

b) $\sqrt{1^2 + 2^2 + (-1)^2} = \sqrt{6}$

Q9 a) $|\overrightarrow{OA}| = \sqrt{1^2 + 2^2 + 3^2} = \sqrt{14}$

b) $|\overrightarrow{OB}| = \sqrt{3^2 + (-1)^2 + (-2)^2} = \sqrt{14}$

c) $\overrightarrow{AB} = \overrightarrow{OB} - \overrightarrow{OA} = (2, -3, -5)$
$|\overrightarrow{AB}| = \sqrt{2^2 + (-3)^2 + (-5)^2} = \sqrt{38}$

Q10 a) $|\mathbf{i} - 3\mathbf{k}| = \sqrt{1^2 + (-3)^2} = \sqrt{10} \Rightarrow \frac{1}{\sqrt{10}}\mathbf{i} - \frac{3}{\sqrt{10}}\mathbf{k}$

b) $|-2\mathbf{i} + 2\mathbf{j} + 5\mathbf{k}| = \sqrt{(-2)^2 + 2^2 + 5^2} = \sqrt{33}$
$\Rightarrow -\frac{2}{\sqrt{33}}\mathbf{i} + \frac{2}{\sqrt{33}}\mathbf{j} + \frac{5}{\sqrt{33}}\mathbf{k}$

c) $\left|\begin{pmatrix} -1 \\ -3 \\ 3 \end{pmatrix}\right| = \sqrt{(-1)^2 + (-3)^2 + 3^2} = \sqrt{19} \Rightarrow \begin{pmatrix} -\frac{\sqrt{19}}{19} \\ -\frac{3\sqrt{19}}{19} \\ \frac{3\sqrt{19}}{19} \end{pmatrix}$

d) $\left\|\begin{pmatrix} 7 \\ -1 \\ 12 \end{pmatrix}\right\| = \sqrt{7^2 + (-1)^2 + 12^2} = \sqrt{194} \Rightarrow \begin{pmatrix} \frac{7\sqrt{194}}{194} \\ -\frac{\sqrt{194}}{194} \\ \frac{12\sqrt{194}}{194} \end{pmatrix}$

Q11 $\overrightarrow{AB} = 3\mathbf{i} - 6\mathbf{j} + 6\mathbf{k}$ giving $|\overrightarrow{AB}| = \sqrt{3^2 + (-6)^2 + 6^2} = 9$ so

the required unit vector is $\dfrac{\overrightarrow{AB}}{|\overrightarrow{AB}|} = \dfrac{1}{9}(3\mathbf{i} - 6\mathbf{j} + 6\mathbf{k})$

$= \dfrac{1}{3}(\mathbf{i} - 2\mathbf{j} + 2\mathbf{k})$

[3 marks available — 1 mark for the correct $\overrightarrow{AB}$, 1 mark for a correct unit vector method, 1 mark for the answer in the required form]

Q12 $\overrightarrow{PQ} = \overrightarrow{OQ} - \overrightarrow{OP} = 3\mathbf{i} - 4\mathbf{j} + \mathbf{k}$
$|\overrightarrow{PQ}| = \sqrt{3^2 + (-4)^2 + 1^2} = \sqrt{26}$

Q13 $\overrightarrow{OY} = \overrightarrow{OX} + 6\dfrac{\overrightarrow{XY}}{|XY|} = \begin{pmatrix} -2 \\ 1 \\ 0 \end{pmatrix} + 6\begin{pmatrix} \frac{2}{3} \\ \frac{2}{3} \\ -\frac{1}{3} \end{pmatrix} = \begin{pmatrix} 2 \\ 5 \\ -2 \end{pmatrix}$

So the coordinates of Y are (2, 5, –2).

Q14 $|\mathbf{F_1}| = \sqrt{6^2 + (2p)^2}$ and $|\mathbf{F_2}| = \sqrt{1^2 + (\sqrt{p})^2 + 2^2}$ so
$\sqrt{36 + 4p^2} = 3\sqrt{5 + p} \Rightarrow 36 + 4p^2 = 45 + 9p \Rightarrow 4p^2 - 9p - 9 = 0$
$\Rightarrow (4p + 3)(p - 3) = 0 \Rightarrow p = -\dfrac{3}{4}$ or 3
[5 marks available — 1 mark for a correct method for magnitude, 1 mark for forming the equation, 1 mark for obtaining a three term quadratic, 1 mark for solving it, 1 mark for both correct values]

Q15 a) Equating components gives $4p - 7 = 5 \Rightarrow p = 3$ then
$2q - 1 = 3p \Rightarrow q = 5$. The **k** components give
$4r^2 - 7r - 15 = 0 \Rightarrow (4r + 5)(r - 3) = 0 \Rightarrow r = -\dfrac{5}{4}$ or 3.
[4 marks available — 1 mark for equating components to find p and q, 1 mark for p and q correct, 1 mark for attempting to solve the correct quadratic in r, 1 mark for both correct values for r]

b) The distance required is the difference in the **k** components,
e.g. $4(3)^2 - 4\left(-\dfrac{5}{4}\right)^2 = \dfrac{119}{4}$ or $29\dfrac{3}{4}$ or 29.75.
[2 marks available — 1 mark for attempting to find the difference between the values of $4r^2$ (or $7r - 15$), 1 mark for the correct distance]

Q16 a) $\overrightarrow{AC} = \overrightarrow{AB} + \overrightarrow{BC} = \overrightarrow{AB} + \overrightarrow{AD} = \begin{pmatrix} 5 \\ -3 \\ 2 \end{pmatrix} + \begin{pmatrix} -1 \\ 6 \\ -5 \end{pmatrix} = \begin{pmatrix} 4 \\ 3 \\ -3 \end{pmatrix}$

[2 marks available — 1 mark for correctly identifying that $\overrightarrow{BC} = \overrightarrow{AD}$, 1 mark for correct answer]

b) Length of $\overrightarrow{AC} = \left\|\begin{pmatrix} 4 \\ 3 \\ -3 \end{pmatrix}\right\| = \sqrt{4^2 + 3^2 + (-3)^2} = \sqrt{34}$

[2 marks available — 1 mark for correct attempt at finding length, 1 mark for correct answer]

Q17 a) $\mathbf{F_2} = \mathbf{R}_a - \mathbf{F_1} = \begin{pmatrix} 2 \\ 1 \\ 6 \end{pmatrix} - \begin{pmatrix} 5 \\ 0 \\ -2 \end{pmatrix} = \begin{pmatrix} -3 \\ 1 \\ 8 \end{pmatrix}$

[2 marks available — 1 mark for attempting $R_a - F_1$, 1 mark for the correct F_2]

b) New $\mathbf{F_2} = \mathbf{R}_b - 2\mathbf{F_1} = \begin{pmatrix} 19 \\ -3 \\ -28 \end{pmatrix} - \begin{pmatrix} 10 \\ 0 \\ -4 \end{pmatrix} = \begin{pmatrix} 9 \\ -3 \\ -24 \end{pmatrix}$

[2 marks available — 1 mark for attempting $R_b - 2F_1$, 1 mark for the correct new F_2]

c) The magnitude of F_2 has been trebled *[1 mark]* and it now acts in the opposite direction. *[1 mark]*

Q18 a) $t = 0 \Rightarrow (4, 0, 0)$ miles *[1 mark]*

b) $t = 5 \Rightarrow 9\mathbf{i} - 15\mathbf{j} + 0.5\mathbf{k}$ and $t = 10 \Rightarrow 24\mathbf{i} - 30\mathbf{j} + \mathbf{k}$
$\overrightarrow{AB}$ is then $15\mathbf{i} - 15\mathbf{j} + 0.5\mathbf{k}$ and
Distance AB $= \sqrt{15^2 + 15^2 + 0.5^2} = 21.2$ miles (to 1 d.p.)
[4 marks available — 1 mark for point A or point B correct, 1 mark for $\overrightarrow{AB}$ correct, 1 mark for a correct method for AB, 1 mark for 21.2 miles (or more accurate) with units]

c) The length of AB is linear, but the x component of the position vector is non-linear, so the helicopter does not travel in a straight line. This means the length of the route the helicopter takes between A and B will be longer than the straight line between the points.
[2 marks available — 1 mark for saying the x component is non-linear while AB is linear, 1 mark for stating that the helicopter's path is not straight]

Q19 The vector between points is $(2t - 7)\mathbf{i} + (4 - t)\mathbf{j} + (-t + 1)\mathbf{k}$ and distance is $\sqrt{(2t - 7)^2 + (4 - t)^2 + (-t + 1)^2} = 3\sqrt{6}$
$\Rightarrow (2t - 7)^2 + (4 - t)^2 + (-t + 1)^2 = 54$.
This leads to the quadratic $6t^2 - 38t + 12 = 0$
$\Rightarrow 3t^2 - 19t + 6 = 0 \Rightarrow (3t - 1)(t - 6) = 0 \Rightarrow t = \dfrac{1}{3}$ or $t = 6$.
So interval in seconds is $\left(6 - \dfrac{1}{3}\right)(60) = 340$ s
[5 marks available — 1 mark for finding the difference between the vectors, 1 mark for forming the quadratic equation, 1 mark for the correct quadratic, 1 mark for both solutions, 1 mark for 340 seconds]

Q20 $|\overrightarrow{OA}| = \sqrt{4^2 + 3^2 + (-3)^2} = \sqrt{34}$
$|\overrightarrow{OB}| = \sqrt{(-1)^2 + 2^2 + (-4)^2} = \sqrt{21}$

$\overrightarrow{AB} = \begin{pmatrix} -5 \\ -1 \\ -1 \end{pmatrix} \Rightarrow |\overrightarrow{AB}| = \sqrt{(-5)^2 + (-1)^2 + (-1)^2} = \sqrt{27}$

So you have a triangle:

Use the cosine rule:
$\cos\theta = \dfrac{(\sqrt{27})^2 + (\sqrt{21})^2 - (\sqrt{34})^2}{2 \times \sqrt{27} \times \sqrt{21}} = \dfrac{14}{18\sqrt{7}} = \dfrac{\sqrt{7}}{9}$

$\Rightarrow \theta = \cos^{-1}\left(\dfrac{\sqrt{7}}{9}\right) = 72.9°$ (1 d.p.)

Q21 a) R has speed $\sqrt{5^2 + 3^2 + (-1)^2} = \sqrt{35}$ ms^{-1},
S has speed $\sqrt{(-2)^2 + (-2)^2 + 7^2} = \sqrt{57}$ ms^{-1}.

b) Distance = speed × time = $(\sqrt{57} - \sqrt{35}) \times 5$
$= 8$ m (to the nearest metre)

c) $s - r = -7i - 5j + 8k$

$\Rightarrow |s - r| = \sqrt{(-7)^2 + (-5)^2 + 8^2} = \sqrt{138}$

$\cos\theta = \dfrac{(\sqrt{57})^2 + (\sqrt{35})^2 - (\sqrt{138})^2}{2 \times \sqrt{57} \times \sqrt{35}} = \dfrac{-23}{\sqrt{1995}}$

$\theta = \cos^{-1}\left(\dfrac{-23}{\sqrt{1995}}\right) = 121°$ (to the nearest degree)

d) E.g. This is a suitable model, as there are no resistance forces in space. This model could be improved by including the effect of gravity exerted on the asteroids by other astronomical bodies e.g. planets or moons.

Q22 $|\overrightarrow{PQ}| = \sqrt{4^2 + (-12)^2 + 6^2} = 14$, $|\overrightarrow{PR}| = \sqrt{8^2 + 2^2 + (-16)^2} = 18$

$\overrightarrow{QR} = 4i + 14j - 22k$ so $|\overrightarrow{QR}| = \sqrt{4^2 + 14^2 + (-22)^2} = 2\sqrt{174}$

Cosine rule gives $\cos RPQ = \dfrac{14^2 + 18^2 - (2\sqrt{174})^2}{2(14)(18)} = -\dfrac{22}{63}$

and RPQ = 110.438...

So area = $\dfrac{1}{2}(14)(18)\sin(110.438...) = 118.07$ (2 d.p.)

[6 marks available — 1 mark for one correct length, 1 mark for all three lengths correct, 1 mark for correct use of cosine rule, 1 mark for the correct angle, 1 mark for correct use of area formula, 1 mark for 118.07]

Q23 a) $\overrightarrow{AB} = 2i - 9j + 2k$, $\overrightarrow{BC} = -5i + 11j + 7k$ and $\overrightarrow{AC} = -3i + 2j + 9k$ so $|\overrightarrow{AB}| = \sqrt{2^2 + (-9)^2 + 2^2} = \sqrt{89}$, $|\overrightarrow{BC}| = \sqrt{195}$ and $|\overrightarrow{AC}| = \sqrt{94}$.

Since $|\overrightarrow{AB}| \neq |\overrightarrow{BC}| \neq |\overrightarrow{AC}|$ then triangle ABC is scalene.

[4 marks available — 1 mark for one correct vector, 1 mark for one correct length, 1 mark for all vectors and lengths correct, 1 mark for a fully correct proof with conclusion]

b) Find an angle using the cosine rule – e.g.

$\cos A = \dfrac{89 + 94 - 195}{2\sqrt{89}\sqrt{94}} \Rightarrow A = 93.761...°$

Using the area formula,

area = $\dfrac{1}{2}\sqrt{89}\sqrt{94}\sin(93.761...) = 45.6$ (3 s.f.)

[4 marks available — 1 mark for a correct use of the cosine rule, 1 mark for a correct angle, 1 mark for correct use of the area formula, 1 mark for correct area]

Q24 a) $\overrightarrow{PQ} = -4i + 10j - 2k$ and $\overrightarrow{RS} = 2i - 5j + k$ so $\overrightarrow{PQ} = -2\overrightarrow{RS}$

[2 marks available — 1 mark for one correct vector, 1 mark for finding k = –2]

b) $\overrightarrow{PQ} = -2\overrightarrow{RS}$, so PQRS has a pair of parallel sides which are not equal, so it must be a trapezium.

[2 marks available — 1 mark for "two parallel sides", 1 mark for "not equal"]

c) Area of a trapezium = $\dfrac{1}{2}(a + b)h$.

$a = |\overrightarrow{PQ}| = \sqrt{(-4)^2 + 10^2 + (-2)^2} = \sqrt{120} = 2\sqrt{30}$,

$b = |\overrightarrow{RS}| = \sqrt{2^2 + (-5)^2 + 1^2} = \sqrt{30}$

and $h = \dfrac{6\sqrt{230}}{5}$

$\Rightarrow$ area = $\dfrac{1}{2}(2\sqrt{30} + \sqrt{30})\left(\dfrac{6\sqrt{230}}{5}\right)$

$= \left(\dfrac{3}{2}\right)\left(\dfrac{6}{5}\right)\sqrt{30}\sqrt{230} = \dfrac{9}{5}\sqrt{6900}$

$= \sqrt{\left(\dfrac{81}{25}\right)6900} = \sqrt{22356}$

[4 marks available — 1 mark for a correct length, 1 mark for using the correct area formula, 1 mark for the correct value in any form, 1 mark for $\sqrt{22356}$]

Q25 a) $|\overrightarrow{YZ}| = \left\|\begin{pmatrix} -7 \\ 5 \\ -4 \end{pmatrix}\right\| = \sqrt{(-7)^2 + 5^2 + (-4)^2} = \sqrt{90} = 3\sqrt{10}$

$|\overrightarrow{XZ}| = \dfrac{\sqrt{10}}{5}|\overrightarrow{YZ}| = \dfrac{\sqrt{10}}{5} \times 3\sqrt{10} = \dfrac{30}{5} = 6$

The unit vector in the direction $\overrightarrow{XZ}$ is $\begin{pmatrix} \frac{2}{3} \\ \frac{1}{2} \\ -\frac{\sqrt{11}}{6} \end{pmatrix}$, and $|\overrightarrow{XZ}| = 6$,

so $\overrightarrow{XZ} = \begin{pmatrix} \frac{2}{3} \\ \frac{1}{2} \\ -\frac{\sqrt{11}}{6} \end{pmatrix} \times 6 = \begin{pmatrix} 4 \\ 3 \\ -\sqrt{11} \end{pmatrix}$, as required.

[4 marks available — 1 mark for attempt at finding $|\overrightarrow{YZ}|$, 1 mark for finding $|\overrightarrow{YZ}|$, 1 mark for finding $|\overrightarrow{XZ}|$, 1 mark for multiplying the unit vector by the length to give the required result]

b) $\overrightarrow{XY} = \begin{pmatrix} 7 \\ -5 \\ 4 \end{pmatrix} + \begin{pmatrix} 4 \\ 3 \\ -\sqrt{11} \end{pmatrix} = \begin{pmatrix} 11 \\ -2 \\ 4 - \sqrt{11} \end{pmatrix}$,

so $|\overrightarrow{XY}|^2 = 11^2 + (-2)^2 + (4 - \sqrt{11})^2 = 152 - 8\sqrt{11}$

From part a), $|\overrightarrow{XZ}| = 6$ and $|\overrightarrow{YZ}| = 3\sqrt{10}$

Using the cosine rule:

$\cos Z = \dfrac{(3\sqrt{10})^2 + 6^2 - (152 - 8\sqrt{11})}{2 \times 3\sqrt{10} \times 6} = 0.00468...$

$Z = \cos^{-1}(0.00468...) = 89.7°$ (1 d.p.)

[5 marks available — 1 mark for finding $\overrightarrow{XY}$, 1 mark for attempting to find $|\overrightarrow{XY}|$ or $|\overrightarrow{XY}|^2$, 1 mark for correctly finding $|\overrightarrow{XY}|$ or $|\overrightarrow{XY}|^2$, 1 mark for attempting to use the cosine rule, 1 mark for correct answer]

Q26 a) Points A, B and C must either form a triangle or all lie in a straight line. If they form a triangle, then $|\overrightarrow{AC}| < |\overrightarrow{AB}| + |\overrightarrow{BC}|$ (since each side must be shorter than the other two added together). Otherwise, they must lie in a straight line and $|\overrightarrow{AC}| = |\overrightarrow{AB}| + |\overrightarrow{BC}|$.

$p + q = \overrightarrow{AB} + \overrightarrow{BC} = \overrightarrow{AC}$, so $|p + q| = |\overrightarrow{AC}|$.

So $|p| + |q| = |p + q| \Rightarrow |\overrightarrow{AB}| + |\overrightarrow{BC}| = |\overrightarrow{AC}|$

$\Rightarrow$ A, B and C are collinear.

[2 marks available — 1 mark for $|\overrightarrow{AC}| = |\overrightarrow{AB}| + |\overrightarrow{BC}|$, 1 mark for correct explanation of how this makes A, B and C collinear]

b) $|q| = 27$ so $r^2 + 16r^2 + (5r + 9)^2 = 27^2 \Rightarrow 7r^2 + 15r - 108 = 0$

$\Rightarrow (7r + 36)(r - 3) = 0$ and $r = -\dfrac{36}{7}$ or $r = 3$

[5 marks available — 1 mark for $|q| = 27$, 1 mark for forming the quadratic, 1 mark for the correct quadratic, 1 mark for each correct value of r]

Q27 You can form a triangle by using the vector with the same x and y components as vector **a** but in the xy-plane, i.e., $\begin{pmatrix} 3 \\ -5 \\ 0 \end{pmatrix}$ which has length $\sqrt{3^2 + (-5)^2} = \sqrt{34}$.

$|\mathbf{a}| = \sqrt{3^2 + (-5)^2 + 4^2} = \sqrt{50}$.

Now using right-angled trigonometry, $\cos\theta = \dfrac{\sqrt{34}}{\sqrt{50}}$

$\Rightarrow \theta = 34°$ (to the nearest degree)

[4 marks available — 1 mark each for two correct side lengths, 1 mark for using correct trigonometry, 1 mark for 34°]
You could have used any two sides of the triangle with the appropriate trig formula to find the angle. E.g. the triangle's height is 4, so here you would do $\sin\theta = \dfrac{4}{\sqrt{50}}$.

Q28 a) $\overrightarrow{AB} = -3\mathbf{i} + 8\mathbf{j} - \mathbf{k}$
so $|\overrightarrow{AB}| = \sqrt{(-3)^2 + 8^2 + (-1)^2} = \sqrt{74}$.
$|\overrightarrow{OA}| = \sqrt{(-1)^2 + (-4)^2 + 8^2} = 9$ and
$|\overrightarrow{OB}| = \sqrt{(-4)^2 + 4^2 + 7^2} = 9$
so since OA = OB ≠ AB, triangle OAB is isosceles.
[3 marks available — 1 mark for finding $\overrightarrow{AB}$, 1 mark for one correct distance, 1 mark for a fully correct proof]

b) Since OAB is isosceles with OA = OB, OX will bisect AB at right angles, so AX : AB = 1 : 2 *[1 mark]*

c) $AX = \dfrac{\sqrt{74}}{2}$. By Pythagoras,
$OX = \sqrt{OA^2 - AX^2} = \sqrt{9^2 - \dfrac{37}{2}} = \sqrt{\dfrac{125}{2}} = \dfrac{5\sqrt{10}}{2}$.
Now $\tan XOA = \dfrac{AX}{OX} = \left(\dfrac{\sqrt{74}}{2}\right)\left(\dfrac{2}{5\sqrt{10}}\right) = \dfrac{\sqrt{185}}{25}$
[4 marks available — 1 mark for finding AX, 1 mark for correct use of Pythagoras for OX, 1 mark for using the correct tan ratio, 1 mark for a fully correct proof]

d) $\theta = 2 \times$ angle XOA so using $\tan 2A = \dfrac{2\tan A}{1 - \tan^2 A}$ then
$\tan\theta = \dfrac{2\left(\dfrac{\sqrt{185}}{25}\right)}{1 - \left(\dfrac{\sqrt{185}}{25}\right)^2} = \dfrac{5\sqrt{185}}{44}$
[3 marks available — 1 mark for recognising $\theta = 2 \times$ angle XOA, 1 mark for using the correct trig identity, 1 mark for the correct exact value]

Q29 a) $\overrightarrow{AB} = 6\mathbf{i} + 3\mathbf{j} + 2\mathbf{k}$ and $\overrightarrow{BC} = 2\mathbf{i} + 3\mathbf{j} + 6\mathbf{k}$ so
$AB = BC = \sqrt{2^2 + 3^2 + 6^2}$
[2 marks available — 1 mark for a correct vector, 1 mark for a full verification]

b) All four sides are equal, so ABCD is a rhombus.

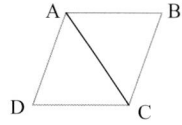

$\overrightarrow{OD} = \overrightarrow{OA} + \overrightarrow{AD} = \overrightarrow{OA} + \overrightarrow{BC}$
$\Rightarrow \overrightarrow{OD} = \mathbf{i} - 4\mathbf{j} + 2\mathbf{k} + 2\mathbf{i} + 3\mathbf{j} + 6\mathbf{k} = 3\mathbf{i} - \mathbf{j} + 8\mathbf{k}$
[2 marks available — 1 mark for a correct method for $\overrightarrow{OD}$, 1 mark for the correct vector]

c) A sketch confirms that AECB is a kite.
E.g.

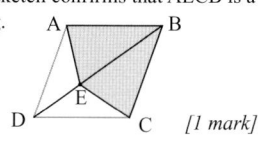

[1 mark]

Practice Paper 1

Q1 a)

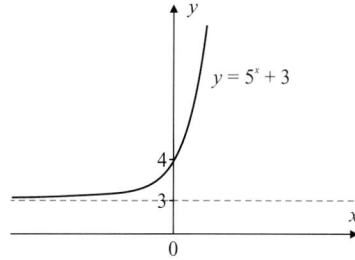

[2 marks available — 1 mark for the correct shape, 1 mark for correctly labelled asymptote and y-intercept]

b) Let $y = 5^x + 3 \Rightarrow y - 3 = 5^x \Rightarrow \log_5(y-3) = \log_5 5^x = x$
Replace x with $g^{-1}(x)$ and y with x: $g^{-1}(x) = \log_5(x-3)$
[3 marks available — 1 mark for correctly rearranging equation, 1 mark for taking logs of both sides, 1 mark for correct answer]
You could also have taken the natural log (ln) of both sides to get the equivalent solution $g^{-1}(x) = \dfrac{\ln(x-3)}{\ln 5}$.

Q2 a) The centre of the circle is $(-2, 6)$, so the equation has the form $(x + 2)^2 + (y - 6)^2 = r^2$.
Now use either point on the circumference to find r^2,
e.g. using $(-6, 9) \Rightarrow (-6 + 2)^2 + (9 - 6)^2 = r^2$
$\Rightarrow (-4)^2 + 3^2 = r^2 \Rightarrow r^2 = 25$
So the equation is $(x + 2)^2 + (y - 6)^2 = 25$.
[3 marks available — 1 mark for using the centre coordinates to form an equation, 1 mark for a correct method to complete the equation, 1 mark for the correct answer]

b) (i) Gradient of the normal at point $B = \dfrac{9 - 6}{2 - (-2)} = \dfrac{3}{4}$
Putting into $y = mx + c$ gives:
$y = \dfrac{3}{4}x + c \Rightarrow 9 = \dfrac{3}{4} \times 2 + c \Rightarrow c = \dfrac{15}{2}$
so the equation is $y = \dfrac{3}{4}x + \dfrac{15}{2}$.
[2 marks available — 1 mark for finding the gradient, 1 mark for a complete correct equation]

(ii) Gradient of the normal at point $A = \dfrac{9 - 6}{(-6) - (-2)} = -\dfrac{3}{4}$
Gradient of tangent of point $A = -1 \div -\dfrac{3}{4} = \dfrac{4}{3}$
Putting into $y = mx + c$ gives:
$y = \dfrac{4}{3}x + c \Rightarrow 9 = \dfrac{4}{3} \times -6 + c \Rightarrow c = 17$
so the equation is $y = \dfrac{4}{3}x + 17$.
This intersects with the line in part (i) when:
$\dfrac{3}{4}x + \dfrac{15}{2} = \dfrac{4}{3}x + 17 \Rightarrow \dfrac{7}{12}x = -\dfrac{19}{2} \Rightarrow x = -\dfrac{114}{7}$
$y = \dfrac{4}{3} \times \left(-\dfrac{114}{7}\right) + 17 = -\dfrac{33}{7}$
So they intersect at $\left(-\dfrac{114}{7}, -\dfrac{33}{7}\right)$.
[4 marks available — 1 mark for finding the gradient of the tangent, 1 mark for the correct equation of the tangent, 1 mark for a correct method to find the intersection, 1 mark for the correct answer]

c) The value of k doesn't change where the centre of the circle is. However the equation is undefined for all negative values of k so $k \geq 0$.
[1 mark for the correct answer with explanation]

Answers

Q3 **a)** Suppose that $\pi + 1$ is a rational number, so $\pi + 1 = \frac{a}{b}$, where a and b are integers and b is non-zero.

So $\pi + 1 = \frac{a}{b} \Rightarrow \pi = \frac{a}{b} - 1 = \frac{a - b}{b}$.

But $a - b$ and b are both integers, so this means that π is rational, in contradiction to the fact that π is irrational as given in the question. The initial assumption must have been wrong, therefore $\pi + 1$ is irrational.

[3 marks available — 1 mark for setting $\pi + 1$ equal to $\frac{a}{b}$, 1 mark for expressing π in terms of a and b, 1 mark for correct interpretation]
You don't have to use a and b — any two letters are fine.

b) E.g. The graph of $y = x + 1$ has an x-intercept at -1 and y-intercept at 1, but passes through $(\pi, \pi + 1)$ which has irrational coordinates by a), so the statement is false.
[1 mark for a correct proof]

Q4 **a)** $x = t^3 + t^2 \Rightarrow \frac{dx}{dt} = 3t^2 + 2t$

$y = \frac{1}{2}t^2 - 6t \Rightarrow \frac{dy}{dt} = t - 6$

$\Rightarrow \frac{dy}{dx} = \frac{dy}{dt} \div \frac{dx}{dt} = \frac{t - 6}{3t^2 + 2t}$

[2 marks available — 1 mark for using $\frac{dy}{dx} = \frac{dy}{dt} \div \frac{dx}{dt}$, 1 mark for correct $\frac{dy}{dx}$]

b) At turning points, $\frac{dy}{dx} = 0 \Rightarrow \frac{t - 6}{3t^2 + 2t} = 0 \Rightarrow t - 6 = 0$
$\Rightarrow t = 6$

When $t = 6$, $x = 6^3 + 6^2 = 252$
and $y = \frac{1}{2} \times 6^2 - 6 \times 6 = -18$

So the coordinates of the turning point are $(252, -18)$.

[2 marks available — 1 mark for setting $\frac{dy}{dx} = 0$ and using it to find value of t, 1 mark for correct coordinates using t]

c) When $t = 12$, $x = 12^3 + 12^2 = 1872$, $y = \frac{1}{2}(12)^2 - 6(12) = 0$
and $\frac{dy}{dx} = \frac{12 - 6}{3(12)^2 + 2(12)} = \frac{1}{76}$

So the tangent to the trajectory when $t = 12$ has a gradient of $\frac{1}{76}$ and an equation in the form $y = mx + c$. Substitute in the values of x, y and m to find c:

$0 = \frac{1}{76} \times 1872 + c \Rightarrow c = -\frac{468}{19}$

So the tangent has the equation $y = \frac{1}{76}x - \frac{468}{19}$.

[3 marks available — 1 mark for finding correct x and y, 1 mark for correct $\frac{dy}{dx}$, 1 mark for the correct equation of the tangent]

Q5 **a)** $|\overrightarrow{PF}| = \sqrt{3^2 + 4^2} = \sqrt{25}$ km
$|\overrightarrow{PG}| = \sqrt{(-2)^2 + 5^2} = \sqrt{29}$ km
$\overrightarrow{PH} = \overrightarrow{GF} = \overrightarrow{GP} + \overrightarrow{PF}$
$= -(-2\mathbf{i} + 5\mathbf{j}) + (3\mathbf{i} + 4\mathbf{j}) = 5\mathbf{i} - \mathbf{j}$
$|\overrightarrow{PH}| = \sqrt{5^2 + (-1)^2} = \sqrt{26}$ km

So buoy F is closest to the engineer.

[4 marks available — 1 mark for a correct method to find any distance, 1 mark for finding any distance correctly, 1 mark for the correct position vector of H, 1 mark for the correct answer with full working]

b) From part a),
$|\overrightarrow{PF}| = \sqrt{25}$ km, $|\overrightarrow{PG}| = \sqrt{29}$ km, $|\overrightarrow{PH}| = \sqrt{26}$ km
First find the distances between each pair of buoys:
$|\overrightarrow{GF}| = |\overrightarrow{PH}| = \sqrt{26}$ km
$\overrightarrow{FH} = \overrightarrow{FP} + \overrightarrow{PH}$
$= -(3\mathbf{i} + 4\mathbf{j}) + (5\mathbf{i} - \mathbf{j}) = 2\mathbf{i} - 5\mathbf{j}$
$|\overrightarrow{FH}| = \sqrt{2^2 + (-5)^2} = \sqrt{29}$ km
$\overrightarrow{GH} = \overrightarrow{GP} + \overrightarrow{PH}$
$= -(-2\mathbf{i} + 5\mathbf{j}) + (5\mathbf{i} - \mathbf{j}) = 7\mathbf{i} - 6\mathbf{j}$
$|\overrightarrow{GH}| = \sqrt{7^2 + (-6)^2} = \sqrt{85}$ km

The shortest path will be avoiding going between G and H.
So $\overrightarrow{PG} + \overrightarrow{GF} + \overrightarrow{FH} + \overrightarrow{HP}$
$= \sqrt{29} + \sqrt{26} + \sqrt{29} + \sqrt{26} = 20.968... = 21.0$ km (3 s.f.)

[4 marks available — 1 mark for a correct method to find distances between buoys, 1 mark for finding all distances between the buoys, 1 mark for choosing the correct path, 1 mark for the correct answer]
If you didn't spot that $\overrightarrow{GH}$ is much longer than the others, you'd have to list each possible route to find the shortest.

Q6 **a)** $a = 1^{st}$ term $= -3$, $d = 4$,
$\Rightarrow n^{th}$ term $= -3 + (n - 1)4 = 4n - 7 \Rightarrow k = 4$ and $m = -7$
[2 marks available — 1 mark for using n^{th} term formula for an arithmetic sequence with correct values of a and d, 1 mark for k and m]

b) $u_n = 4n - 7 \Rightarrow u_{20} = 4 \times 20 - 7 = 73$
[1 mark for correct answer (allow errors carried forward)]

c) From previous parts $a = -3$ and $l = 73$.
$S_n = \frac{1}{2}n(a + l) = \frac{1}{2} \times 20 \times (-3 + 73) = 700$
[2 marks available — 1 mark for correct method, 1 mark for correct answer]

Q7 Differentiate $x^4 - 3x^2y^2 = y^3$:

$\frac{d}{dx}x^4 - \frac{d}{dx}3x^2y^2 = \frac{d}{dx}y^3 \Rightarrow 4x^3 - 6x^2y\frac{dy}{dx} - 6xy^2 = 3y^2\frac{dy}{dx}$

The product rule was used to differentiate $3x^2y^2$.

$\Rightarrow 4x^3 - 6xy^2 = 6x^2y\frac{dy}{dx} + 3y^2\frac{dy}{dx} \Rightarrow \frac{dy}{dx} = \frac{4x^3 - 6xy^2}{6x^2y + 3y^2}$

At $(4, 4)$, $\frac{dy}{dx} = \frac{4(4)^3 - 6(4)(4)^2}{6(4)^2(4) + 3(4)^2} = -\frac{8}{27}$

So the tangent to C at point $(4, 4)$ has a gradient of $-\frac{8}{27}$ and an equation in the form $y = mx + c$.

Substitute in the values of x, y and m to find c:

$4 = -\frac{8}{27}(4) + c \Rightarrow c = \frac{140}{27}$

So the tangent has the equation $y = -\frac{8}{27}x + \frac{140}{27}$.

[4 marks available — 1 mark for two terms correctly differentiated, 1 mark for the other two terms correctly differentiated, 1 mark for correct gradient, 1 mark for the correct answer]

Q8 a)

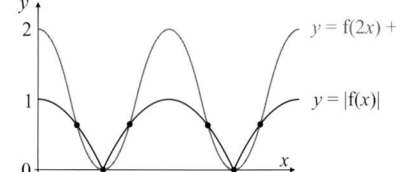

Range of $|f(x)|$ is $0 \leq |f(x)| \leq 1$
Range of $f(2x) + 1$ is $0 \leq f(2x) + 1 \leq 2$
[4 marks available — 1 mark for correct shape and position of $|f(x)|$, 1 mark for correct shape and position of $f(2x) +1$, 1 mark for correct range of $|f(x)|$, 1 mark for correct range of $f(2x) + 1$]

b) $|f(x)| = f(2x) + 1$
On the graph above, you can see there are six solutions.
To find all the solutions, you need to solve $f(x) = f(2x) + 1$ and $-f(x) = f(2x) + 1$.
$f(x) = f(2x) + 1 \Rightarrow \cos x = \cos 2x + 1$
$\Rightarrow \cos x = 2\cos^2 x - 1 + 1 \Rightarrow 2\cos^2 x - \cos x = 0$
$\Rightarrow \cos x(2\cos x - 1) = 0 \Rightarrow \cos x = 0$ or $\cos x = \frac{1}{2}$
If $\cos x = 0$, $x = \frac{\pi}{2}$ and $\frac{3\pi}{2}$, where $0 \leq x \leq 2\pi$.
If $\cos x = \frac{1}{2}$, $x = \frac{\pi}{3}$ and $\frac{5\pi}{3}$, where $0 \leq x \leq 2\pi$.
$-f(x) = f(2x) + 1 \Rightarrow -\cos x = \cos 2x + 1$
$\Rightarrow -\cos x = 2\cos^2 x - 1 + 1 \Rightarrow 2\cos^2 x + \cos x = 0$
$\Rightarrow \cos x(2\cos x + 1) = 0 \Rightarrow \cos x = 0$ or $\cos x = -\frac{1}{2}$
The solutions for $\cos x = 0$ are given above.
If $\cos x = -\frac{1}{2}$, $x = \frac{2\pi}{3}$ and $\frac{4\pi}{3}$, where $0 \leq x \leq 2\pi$.
So the solutions for $|f(x)| = f(2x) + 1$ in the interval $0 \leq x \leq 2\pi$ are $x = \frac{\pi}{3}, \frac{\pi}{2}, \frac{2\pi}{3}, \frac{4\pi}{3}, \frac{3\pi}{2}$ and $\frac{5\pi}{3}$.
[5 marks available — 1 mark for setting $\cos x = \cos 2x + 1$, 1 mark for setting $-\cos x = \cos 2x + 1$, 1 mark for using double angle formula on $\cos 2x$, 1 mark for three correct solutions, 1 mark for all other correct solutions]

Q9 a) Use a double angle formula to write $y = \dfrac{\cos\theta \sin 2\theta}{2\sin\theta}$ in terms of $\cos\theta$:
$y = \dfrac{\cos\theta \sin 2\theta}{2\sin\theta} = \dfrac{\cos\theta(2\sin\theta\cos\theta)}{2\sin\theta} = \cos^2\theta$
Rearrange the equation for x to get $\cos\theta$:
$x = \dfrac{1 - \cos\theta}{3} \Rightarrow \cos\theta = 1 - 3x$
So $y = \cos^2\theta = (1 - 3x)^2 = 9x^2 - 6x + 1$
[3 marks available — 1 mark for correct use of double angle formula, 1 mark for writing y in terms of $\cos\theta$ only, 1 mark for correct Cartesian equation]

b) $y + 3 = 2x \Rightarrow y = 2x - 3$
Substitute this into the Cartesian equation:
$2x - 3 = 9x^2 - 6x + 1 \Rightarrow 0 = 9x^2 - 8x + 4$
Using the discriminant:
$b^2 - 4ac = (-8)^2 - 4 \times 9 \times 4 = -80 < 0$,
so there are no solutions to the quadratic equation and hence the curve does not intersect the line.
[2 marks available — 1 mark for correctly substituting y into Cartesian equation, 1 mark for using discriminant to show no intersection]

Q10 a)
$$\frac{\cos x}{(\cos x + 1)(\cos x - 1)} \equiv \frac{\cos x}{\cos^2 x - 1} \equiv \frac{-\cos x}{1 - \cos^2 x}$$
$$\equiv \frac{-\cos x}{\sin^2 x} \equiv \frac{1}{\sin x} \cdot \frac{-\cos x}{\sin x} \equiv -\text{cosec}\, x \cot x$$
[3 marks available — 1 mark for expanding brackets, 1 mark for using $1 - \cos 2x \equiv \sin 2x$, 1 mark for splitting fraction and simplifying to $-\text{cosec}\,x \cot x$]

b) $\displaystyle\int \frac{\cos x\, e^{\frac{1}{\sin x}}}{(\cos x + 1)(\cos x - 1)}\, dx = \int -\text{cosec}\,x \cot x \times e^{\text{cosec}\,x}\, dx$
Let $u = \text{cosec}\,x$, so $\dfrac{du}{dx} = -\text{cosec}\,x \cot x$ and $f'(u) = e^u$.
Use the formula $\displaystyle\int \frac{du}{dx} f'(u)\, dx = f(u) + C$:
$\displaystyle\int -\text{cosec}\,x \cot x \times e^{\text{cosec}\,x}\, dx = e^{\text{cosec}\,x} + C$
[4 marks available — 1 mark for using part a) to simplify, 1 mark for u, 1 mark for $\frac{du}{dx}$ and $f'(u)$, 1 mark for correct answer]

Q11 a) $V = \sqrt{h^4 + 2} = (h^4 + 2)^{\frac{1}{2}}$, so let $V = u^{\frac{1}{2}}$, where $u = h^4 + 2$
$\Rightarrow \dfrac{dV}{du} = \dfrac{1}{2}u^{-\frac{1}{2}} = \dfrac{1}{2\sqrt{h^4 + 2}}$, $\dfrac{du}{dh} = 4h^3$
$\dfrac{dV}{dh} = \dfrac{dV}{du} \times \dfrac{du}{dh} = \dfrac{1}{2\sqrt{h^4 + 2}} \times 4h^3 = \dfrac{2h^3}{\sqrt{h^4 + 2}}$
When $h = 0.4$, $\dfrac{dV}{dh} = \dfrac{2(0.4)^3}{\sqrt{(0.4)^4 + 2}} = 0.08993...$
$= 0.0899$ (3 s.f.)
[3 marks available — 1 mark for attempting to use the chain rule, 1 mark for correct $\frac{dV}{dh}$, 1 mark for correct final answer]

b) From the question you know that $\dfrac{dV}{dt} = 0.15$ m³ per hour.
You're looking for $\dfrac{dh}{dt}$:
$\dfrac{dV}{dt} = \dfrac{dV}{dh} \times \dfrac{dh}{dt}$, so $\dfrac{dh}{dt} = \dfrac{dV}{dt} \div \dfrac{dV}{dh} = 0.15 \div 0.0899$
$= 1.67$ (3 s.f.)
[2 marks available — 1 mark for correct method, 1 mark for correct answer (allow for errors carried forward from a))]

Q12 a) At $x = 1.3$, $y = \dfrac{3(1.3)^2}{(1.3)^3 + 4} = 0.818$ (3 d.p.)
At $x = 1.5$, $y = \dfrac{3(1.5)^2}{(1.5)^3 + 4} = 0.915$ (3 d.p.)
[1 mark for correct y-values]

b) Using the y-values from the table and part a), use the trapezium with 5 strips and $h = 0.1$.
$\displaystyle\int_1^{1.5} \frac{3x^2}{x^3 + 4}\, dx \approx \frac{h}{2}[y_0 + 2y_1 + 2y_2 + 2y_3 + 2y_4 + y_5]$
$= 0.05[0.6 + 1.362 + 1.508 + 1.636 + 1.744 + 0.915]$
$= 0.39$ (2 d.p.)
[3 marks available — 1 mark for using 5 strips with width of 0.1, 1 mark for using trapezium rule correctly, 1 mark for correct answer]

Answers

c) (i) Differentiating the denominator:

$$\frac{d}{dx} x^3 + 4 = 3x^2 = \text{numerator}$$

So $\int_1^{1.5} \frac{3x^2}{x^3 + 4}\, dx = [\ln|x^3 + 4|]_1^{1.5}$

$$= \ln \frac{59}{8} - \ln 5 = \ln \frac{59}{40}$$

[3 marks available — 1 mark for integrating the fraction, 1 mark for substituting limits, 1 mark for correct final value]

(ii) Percentage error $= \dfrac{0.39 - \ln \frac{59}{40}}{\ln \frac{59}{40}} \times 100 = 0.35\%$ (2 d.p.)

[1 mark for correct answer]

Q13 a) E.g. by using algebraic division:

$$
\begin{array}{r}
x^2 + x - 12 \text{ r } 0 \\
x + 4 \overline{)\, x^3 + 5x^2 - 8x - 48} \\
-(x^3 + 4x^2) \\
\overline{x^2 - 8x} \\
-(x^2 + 4x) \\
\overline{-12x - 48} \\
-(-12x - 48) \\
\overline{0}
\end{array}
$$

So $f(x) = (x + 4)(x^2 + x - 12) = (x + 4)(x + 4)(x - 3)$
$$= (x + 4)^2(x - 3)$$

[3 marks available — 1 mark for a suitable method to find quadratic factor, 1 mark for correct quadratic, 1 mark for correct answer fully factorised]

b) $g(x) \equiv \dfrac{3x + 40}{(x + 4)^2(x - 3)} \equiv \dfrac{A}{x + 4} + \dfrac{B}{(x + 4)^2} + \dfrac{C}{x - 3}$

$$\Rightarrow 3x + 40 \equiv A(x + 4)(x - 3) + B(x - 3) + C(x + 4)^2$$

Substitution: $x = -4$: $28 = -7B \Rightarrow B = -4$
$x = 3$: $49 = 49C \Rightarrow C = 1$

Equating x coefficients: $3 = A + B + 8C = A - 4 + 8$
$$\Rightarrow A = -1$$

So $\dfrac{3x + 40}{(x + 4)^2(x - 3)} \equiv -\dfrac{1}{x + 4} - \dfrac{4}{(x + 4)^2} + \dfrac{1}{x - 3}$

[4 marks available — 1 mark for writing the expression as the sum of three fractions, 1 mark for finding one unknown, 1 mark for finding the other two unknowns, 1 mark for the correct answer]

c) $g(x) \equiv -\dfrac{1}{x + 4} - \dfrac{4}{(x + 4)^2} + \dfrac{1}{x - 3}$

$$\equiv -(x + 4)^{-1} - 4(x + 4)^{-2} + (x - 3)^{-1}$$

$(x + 4)^{-1} = (4 + x)^{-1} = \frac{1}{4}(1 + \frac{x}{4})^{-1}$

$$= \frac{1}{4}(1 - \frac{x}{4} + \frac{x^2}{16} + ...) = (\frac{1}{4} - \frac{x}{16} + \frac{x^2}{64} - ...)$$

$(x + 4)^{-2} = (4 + x)^{-2} = \frac{1}{16}(1 + \frac{x}{4})^{-2} = \frac{1}{16}(1 - \frac{x}{2} + \frac{3x^2}{16} - ...)$

$$= (\frac{1}{16} - \frac{x}{32} + \frac{3x^2}{256} + ...)$$

$(x - 3)^{-1} = -(3 - x)^{-1} = -\frac{1}{3}(1 - \frac{x}{3})^{-1} = -\frac{1}{3}(1 + \frac{x}{3} + \frac{x^2}{9} + ...)$

$$= (-\frac{1}{3} - \frac{x}{9} - \frac{x^2}{27} - ...)$$

So $g(x) \approx -[\frac{1}{4} - \frac{x}{16} + \frac{x^2}{64}] - 4[\frac{1}{16} - \frac{x}{32} + \frac{3x^2}{256}]$

$$+ [-\frac{1}{3} - \frac{x}{9} - \frac{x^2}{27}]$$

$$= -\frac{1}{4} + \frac{x}{16} - \frac{x^2}{64} - \frac{1}{4} + \frac{x}{8} - \frac{3x^2}{64} - \frac{1}{3} - \frac{x}{9} - \frac{x^2}{27}$$

$$= -\frac{5}{6} + \frac{11x}{144} - \frac{43x^2}{432}$$

[6 marks available — 1 mark for each correct expansion of the denominators, 1 mark for multiplying the denominator expansions by the correct values, 1 mark for attempting to add the three expansions, 1 mark for correct simplified final answer]

Q14 a) i) $y = 2\dfrac{dy}{dx}$, so at $(1, e)$, $\dfrac{dy}{dx} = \dfrac{e}{2}$

So the gradient of the tangent at P is $\dfrac{e}{2}$ and the gradient of the normal at P is $-\dfrac{2}{e}$.

The normal has equation in the form $y = mx + c$, so substitute in the values of x, y and m to find c:

$$e = -\frac{2}{e}(1) + c \Rightarrow c = e + \frac{2}{e}$$

So the equation for the normal is $y = -\dfrac{2}{e}x + e + \dfrac{2}{e}$.

[2 marks available — 1 mark for correct gradient of normal at P, 1 mark for equation of normal at P]

ii) Solve differential equation by separating variables:

$$y = 2\frac{dy}{dx} \Rightarrow \int dx = \int \frac{2}{y}\, dy \Rightarrow x = 2\ln|y| + C$$

From the sketch you know y is positive so you don't need the modulus signs.

When $x = 1$, $y = e$, so $1 = 2\ln e + C \Rightarrow C = -1$

$$\Rightarrow x = 2\ln y - 1 \Rightarrow \ln y = \frac{x + 1}{2} \Rightarrow y = e^{\frac{x+1}{2}}$$

[3 marks available — 1 mark for a correct method of solving differential equations, 1 mark for a correct general solution, 1 mark for the correct answer]

b) Split the area into A_1 and A_2 at $x = 1$:

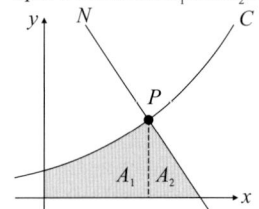

Find where the normal intersects the x-axis:

$$y = 0 \Rightarrow 0 = -\frac{2}{e}x + e + \frac{2}{e} \Rightarrow \frac{2}{e}x = e + \frac{2}{e}$$

$$\Rightarrow 2x = e^2 + 2 \Rightarrow x = \frac{e^2}{2} + 1$$

So $A_2 = \dfrac{1}{2} \times e \times (\dfrac{e^2}{2} + 1 - 1) = \dfrac{1}{2} \times e \times \dfrac{e^2}{2} = \dfrac{e^3}{4}$

The area of a triangle formula was used to find A_2.

For A_1, integrate y with limits $x = 0$ and $x = 1$:

$$\int_0^1 y\, dx = \int_0^1 e^{\frac{x+1}{2}}\, dx = [2e^{\frac{x+1}{2}}]_0^1 = 2e - 2e^{\frac{1}{2}}$$

Add the areas together for A: $A = A_1 + A_2 = 2e - 2e^{\frac{1}{2}} + \dfrac{e^3}{4}$

[5 marks available — 1 mark for finding where normal intersects x-axis, 1 mark for correct A_2, 1 mark for correct method to integrate for A_1, 1 mark for correct A_1, 1 mark for correct total area]

Answers

Practice Paper 2

Q1 a) (i) $f(x) = x^2 + 8k\,x^{\frac{1}{2}} - 4$

$f'(x) = 2x + 4k\,x^{-\frac{1}{2}}$ or $f'(x) = 2x + \dfrac{4k}{\sqrt{x}}$

[2 marks available — 1 mark for a correct method to differentiate all terms, 1 mark for the correct answer]

(ii) $f''(x) = 2 - 2k\,x^{-\frac{3}{2}}$ or $f''(x) = 2 - \dfrac{2k}{(\sqrt{x})^3}$

[1 mark for the correct answer]

b) E.g. As $x > 0$ and $k > 0$, both $2x > 0$ and $\dfrac{4k}{\sqrt{x}} > 0$,

so $f'(x) = 2x + \dfrac{4k}{\sqrt{x}} > 0$ and so $f(x)$ is increasing.

[1 mark for a correct explanation]

c) $k = -4 \Rightarrow f'(x) = 2x - 16\,x^{-\frac{1}{2}}$

The curve $y = f(x)$ has a stationary point when $f'(x) = 0$

$\Rightarrow 2x - 16\,x^{-\frac{1}{2}} = 0 \Rightarrow x^{\frac{3}{2}} = 8 \Rightarrow x = 8^{\frac{2}{3}} = 4$

So there is one stationary point and it's at $x = 4$.

$f''(4) = 2 - 2(-4)(4^{-\frac{3}{2}}) = 3 > 0$, so it's a minimum.

[3 marks available — 1 mark for showing that $f'(x) = 0$ has one solution, 1 mark for showing that $f''(4) > 0$, 1 mark for the correct conclusion]

Q2 a) $\cos\theta = \dfrac{x}{OA} \Rightarrow OA = \dfrac{x}{\cos\theta}$

Area of sector $AOB = \dfrac{1}{2} \times \left(\dfrac{x}{\cos\theta}\right)^2 \times \theta = \dfrac{x^2\theta}{2\cos^2\theta}$

[2 marks available — 1 mark for finding an expression for OA or OB, 1 mark for the correct answer]

b) (i) When θ is small, $\cos\theta \approx 1 - \dfrac{1}{2}\theta^2$

$\Rightarrow$ Area of sector $AOB \approx \dfrac{x^2\theta}{2\left(1 - \frac{1}{2}\theta^2\right)^2}$

[1 mark for the correct answer]

(ii) $\tan\theta = \dfrac{AC}{x} \Rightarrow AC = x\tan\theta$

Area of triangle $AOC = \dfrac{1}{2} \times x\tan\theta \times x = \dfrac{1}{2}x^2\tan\theta$

When θ is small, $\tan\theta \approx \theta$

$\Rightarrow$ Area of triangle $AOC \approx \dfrac{1}{2}x^2\theta$

[3 marks available — 1 mark for a correct method to find the area of the triangle, 1 mark for finding the area in terms of trig functions, 1 mark for the correct answer]

c) Shaded area $\approx \dfrac{x^2\theta}{2\left(1 - \frac{1}{2}\theta^2\right)^2} - \dfrac{1}{2}x^2\theta = x^2\dfrac{\theta}{2}\left(\dfrac{1}{\left(1 - \frac{1}{2}\theta^2\right)^2} - 1\right)$

$= x^2\dfrac{\theta}{2}\left(\dfrac{1}{\frac{1}{4}\theta^4 - \theta^2 + 1} - 1\right) = x^2\dfrac{\theta}{2}\left(\dfrac{4}{\theta^4 - 4\theta^2 + 4} - 1\right)$

$= x^2\dfrac{\theta}{2}\left(\dfrac{-\theta^4 + 4\theta^2}{\theta^4 - 4\theta^2 + 4}\right) = x^2\dfrac{\theta}{2}\left(\dfrac{\theta^2(4 - \theta^2)}{(\theta^2 - 2)^2}\right) = x^2\dfrac{\theta^3(4 - \theta^2)}{2(\theta^2 - 2)^2}$

[3 marks available — 1 mark for an attempt to calculate the difference, 1 mark for expanding and rearranging, 1 mark for factorising correctly leading to the correct answer]

Q3 a) $Y_1 = 20\,000 = a$ and the common ratio (r) is 108%, which is 1.08. So $Y_n = 20\,000(1.08)^{n-1}$

[1 mark for correct expression]

b) Find the sum of the geometric series from Y_1 to Y_5:

$S_5 = \dfrac{a(1 - r^n)}{1 - r} = \dfrac{20\,000(1 - 1.08^5)}{1 - 1.08} = 117\,332.0192$

Find the sum of the geometric series from Y_6 to Y_{10}:

$S_{10} - S_5 = \dfrac{20\,000(1 - 1.08^{10})}{1 - 1.08} - 117\,332.0192$

$= 289\,731.2493 - 117\,332.0192$

$= £172\,399.23$ (to nearest pence)

[3 marks available — 1 mark for understanding answer is difference between sum to 10 terms and sum to 5 terms, 1 mark for correct use of sum of geometric series formula, 1 mark for correct final answer to nearest pence]

Q4 $3x^3 + 9x - 1$ is odd $\Rightarrow 3x^3 + 9x$ is even

Consider the cases when x is even and x is odd.

If x is even then $x = 2n$ for some integer n,

so $3x^3 + 9x = x(3x^2 + 9) = 2n(3(2n)^2 + 9)$.

As $n(3(2n)^2 + 9)$ is an integer, it follows that $2n(3(2n)^2 + 9)$ is a multiple of 2 and therefore even.

If x is odd then $x = 2n + 1$ for some integer n.

So $3x^3 + 9x = 3(2n + 1)^3 + 9(2n + 1)$

$= 3(8n^3 + 12n^2 + 6n + 1) + 18n + 9$

$= 24n^3 + 36n^2 + 18n + 3 + 18n + 9$

$= 24n^3 + 36n^2 + 36n + 12$

$= 2(12n^3 + 18n^2 + 18n + 6)$

$12n^3 + 18n^2 + 18n + 6$ is an integer

$\Rightarrow 2(12n^3 + 18n^2 + 18n + 6)$ is even

So $3x^3 + 9x$ is even in both cases and so $3x^3 + 9x - 1$ is odd.

[4 marks available — 1 mark for considering the two cases, 1 mark for a correct proof when x is even, 1 mark for correct working with x is odd, 1 mark for a correct proof when x is odd leading to the correct conclusion]

Q5 a) $f(x) = 2^7 + 7(2^6)(-\tfrac{1}{4}x) + \dfrac{7(7 - 1)}{1 \times 2}(2^5)\left(-\tfrac{1}{4}x\right)^2 + \ldots$

$= 128 - 112x + 42x^2 + \ldots$

So $a = 128$, $b = -112$ and $c = 42$.

[2 marks available — 1 mark for a correct expansion, 1 mark for simplifying to give the correct answer]

b) (i) $f(3x) = 128 - 112(3x) + 42(3x)^2 + \ldots$

$= 128 - 336x + 378x^2 + \ldots$

[1 mark for the correct answer]

(ii) The expansion is valid for all values of x.

[1 mark for the correct answer]

c) $y = (2 - \tfrac{3}{4}x)^7$

So $\dfrac{dy}{dx} = 7(2 - \tfrac{3}{4}x)^6 \times -\tfrac{3}{4} = -\dfrac{21}{4}(2 - \tfrac{3}{4}x)^6$

When $x = 0.05$, $\dfrac{dy}{dx} = -\dfrac{21}{4}(2 - \tfrac{3}{4}(0.05))^6 = -299.928\ldots$

Using the approximation, $y \approx 128 - 336x + 378x^2$

$\dfrac{dy}{dx} \approx -336 + 756x$

When $x = 0.05$, $\dfrac{dy}{dx} \approx -336 + (756 \times 0.05) = -298.2$

So the percentage error in the approximation is

$\left|\dfrac{-299.928\ldots - (-298.2)}{-299.928\ldots}\right| \times 100 = 0.5762\ldots = 0.576\%$ (3.s.f)

[5 marks available — 1 mark for a correct method to differentiate the exact equation, 1 mark for differentiating the exact equation correctly, 1 mark for differentiating the approximation correctly, 1 mark for finding the values of the gradients at x = 0.05, 1 mark for the correct answer]

Answers

Q6 a) The minimum point is when $x = 2.5$ and $y = -3$.
The y-intercept is $|5 - 2(0)| - 3 = 2$.
The x-intercepts are at $|5 - 2x| - 3 = 0$
$\Rightarrow 5 - 2x = 3$ and $5 - 2x = -3 \Rightarrow x = 1$ and $x = 4$.

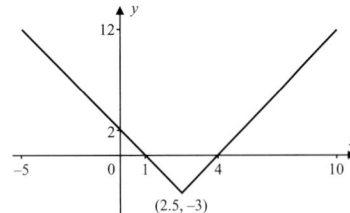

*[3 marks available — 1 mark for a correct shape,
1 mark for intercepts correctly labelled, 1 mark
for minimum point correctly labelled]*

b) (i) The solutions to the equation are given by points where
the graphs of $y = |5 - 2x| - 3$ and $y = kx$ intersect.
The gradient of $y = |5 - 2x| - 3$ is -2 or 2.
If the point of intersection is at $x > 0$, then either
$k \geq 2$ or the graphs intersects at the minimum point:
$y = kx \Rightarrow -3 = k(2.5) \Rightarrow k = -1.2$
If the point of intersection is at $x < 0$, then $k < -2$.
Therefore there will be one solution when
$k \geq 2$, $k = -1.2$, or $k < -2$.
*[2 marks available — 1 mark for at least one region or
point correct, 1 mark for the correct answer]*

(ii) There will be two solutions when $1.2 < k < 2$.
[1 mark for the correct answer]

Q7 a) $\overrightarrow{AC} = \overrightarrow{AB} + \overrightarrow{BC} = \begin{pmatrix} 2 \\ 6 \\ -8 \end{pmatrix} + \begin{pmatrix} -8 \\ 2 \\ 9 \end{pmatrix} = \begin{pmatrix} -6 \\ 8 \\ 1 \end{pmatrix}$

[1 mark for correct answer]

b) $\overrightarrow{AD} = \frac{5}{2}\begin{pmatrix} 2 \\ 6 \\ -8 \end{pmatrix} = \begin{pmatrix} 5 \\ 15 \\ -20 \end{pmatrix}$

$\overrightarrow{CD} = \overrightarrow{AD} - \overrightarrow{AC} = \begin{pmatrix} 5 \\ 15 \\ -20 \end{pmatrix} - \begin{pmatrix} -6 \\ 8 \\ 1 \end{pmatrix} = \begin{pmatrix} 11 \\ 7 \\ -21 \end{pmatrix}$

$|\overrightarrow{AC}| = \sqrt{(-6)^2 + 8^2 + 1^2} = \sqrt{101}$

$|\overrightarrow{AD}| = \sqrt{5^2 + 15^2 + (-20)^2} = \sqrt{650} = 5\sqrt{26}$

$|\overrightarrow{CD}| = \sqrt{11^2 + 7^2 + (-21)^2} = \sqrt{611}$

Using the cosine rule:
$\cos ACD = \frac{(\sqrt{101})^2 + (\sqrt{611})^2 - (5\sqrt{26})^2}{2 \times \sqrt{101} \times \sqrt{611}} = 0.124...$

$ACD = \cos^{-1}(0.124...) = 82.8°$ (1 d.p.)
*[5 marks available — 1 mark for $\overrightarrow{AD}$, 1 mark for $\overrightarrow{CD}$, 1 mark
for correct magnitudes, 1 mark for attempting to use cosine
rule, 1 mark for the correct answer]*

Q8 a) The function is undefined when $5x + 4 \leq 0$
$\Rightarrow 5x \leq -4 \Rightarrow x \leq -0.8$.
So the maximum domain of $g(x)$ is $x > -0.8$.
When $x = 0$, $g(x) = 0$ and for all other values $g(x)$
is positive because x^2 and $\sqrt{5x + 4}$ are both positive.
As $x \to \infty$, x^2 gets a lot bigger than $\sqrt{5x + 4}$ so $g(x) \to \infty$.
so the range of $g(x)$ is $g(x) \geq 0$.
*[3 marks available — 1 mark for a correct method to find
either domain or range, 1 mark for the correct domain,
1 mark for the correct range]*

b) (i) Let $u = x^2$ and $v = \sqrt{5x + 4} = (5x + 4)^{\frac{1}{2}}$.
Then $\frac{du}{dx} = 2x$ and $\frac{dv}{dx} = \frac{5}{2}(5x + 4)^{-\frac{1}{2}}$.
Using the quotient rule:

$g'(x) = \dfrac{v\dfrac{du}{dx} - u\dfrac{dv}{dx}}{v^2} = \dfrac{2x(5x + 4)^{\frac{1}{2}} - \frac{5}{2}x^2(5x + 4)^{-\frac{1}{2}}}{5x + 4}$

$= \dfrac{2x(5x + 4) - \frac{5}{2}x^2}{(5x + 4)(5x + 4)^{\frac{1}{2}}} = \dfrac{10x^2 + 8x - \frac{5}{2}x^2}{(5x + 4)^{\frac{3}{2}}}$

$= \dfrac{20x^2 + 16x - 5x^2}{2(5x + 4)^{\frac{3}{2}}} = \dfrac{x(15x + 16)}{2(5x + 4)^{\frac{3}{2}}}$

*[3 marks available — 1 mark for identifying
and differentiating u, 1 mark for identifying and
differentiating v, 1 mark for the correct answer]*
*You don't need to have fully simplified your answer to be
awarded the marks but it will help for the next part.*

(ii) Turning points occur when $g'(x) = 0 \Rightarrow x(15x + 16) = 0$
$\Rightarrow x = 0$ and $x = -\frac{16}{15}$ but $-\frac{16}{15}$ isn't in the domain of
$g(x)$ so there is only one turning point.
When $x = 0$, $g(x) = 0$ so the turning point is $(0, 0)$.
*[3 marks available — 1 mark for setting g'(x) equal to 0,
1 mark for correctly identifying the x value of the turning
point (excluding the other solution), 1 mark for the
correct answer]*

c)

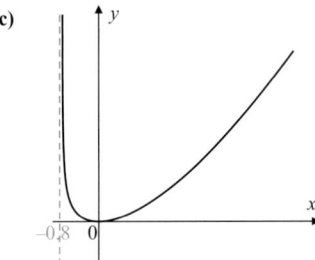

*[3 marks available — 1 mark for the correct shape,
1 mark for the correct turning point,
1 mark for asymptote at x = -0.8]*

Q9 a) $\dfrac{t - 5}{2t^2 + 7t + 6} = \dfrac{t - 5}{(2t + 3)(t + 2)}$

$\dfrac{t - 5}{(2t + 3)(t + 2)} = \dfrac{A}{2t + 3} + \dfrac{B}{t + 2}$

$\Rightarrow t - 5 = A(t + 2) + B(2t + 3)$
$\Rightarrow$ ① $A + 2B = 1$ and ② $2A + 3B = -5$
$\Rightarrow 2 \times$ ① $-$ ②: $B = 7$
So $A = 1 - (2 \times 7) = -13$

So $\dfrac{t - 5}{2t^2 + 7t + 6} = \dfrac{-13}{2t + 3} + \dfrac{7}{t + 2}$

*[3 marks available — 1 mark for a correct method to write as
partial fractions, 1 mark for correct value of A, 1 mark for the
correct value of B]*

b) $\int \frac{t-5}{2t^2+7t+6}\,dt = \int \frac{-13}{2t+3}\,dt + \int \frac{7}{t+2}\,dt$

$= -\frac{13}{2}\int \frac{2}{2t+3}\,dt + 7\int \frac{1}{t+2}\,dt$

$= -\frac{13}{2}\ln(2t+3) + 7\ln(t+2) + C$

[3 marks available — 1 mark for a correct method to integrate, 1 mark for either term integrated correctly, 1 mark for the correct answer]

Q10 a) $f(4.7) = (4.7)^3 - 4(4.7)^2 - 5(4.7) + 6 = -2.037$
$f(4.8) = (4.8)^3 - 4(4.8)^2 - 5(4.8) + 6 = 0.432$
There is a sign change (and the function is continuous in this interval) so there is a root in this interval.
[2 marks available — 1 mark for correct f(4.7) and f(4.8), 1 mark for identifying sign change]

b) $f(x) = 0$, so $x^3 - 4x^2 - 5x + 6 = 0$
$\Rightarrow x^3 = 4x^2 + 5x - 6 \Rightarrow x = \sqrt[3]{4x^2 + 5x - 6}$
[1 mark for correct answer]

c) $x_1 = \sqrt[3]{4x_0^2 + 5x_0 - 6} = 4.730..., \ x_2 = 4.749...,$
$x_3 = 4.762..., \ x_4 = 4.769... = 4.77$ (3 s.f.)
So $n = 4$ is the smallest n such that x_n is correct to 3 s.f.
[2 marks available — 1 mark for correctly substituting 4.7 into the iteration formula to get at least one correct value, 1 mark for correct answer]

d) $f'(x) = 3x^2 - 8x - 5$
So $x_{n+1} = x_n - \frac{x_n^3 - 4x_n^2 - 5x_n + 6}{3x_n^2 - 8x_n - 5}$
$x_1 = -1.5 - \frac{(-1.5)^3 - 4(-1.5)^2 - 5(-1.5) + 6}{3(-1.5)^2 - 8(-1.5) - 5} = -1.581...$
$x_2 = -1.578...,$
So better approximation to the root β is -1.58 (3 s.f.).
[3 marks available — 1 mark for correct iteration formula, 1 mark for using formula to get at least one correct value, 1 mark for correct answer]

Q11 a) $\cot^2\theta + \sin^2\theta$
$\equiv (\csc^2\theta - 1) + (1 - \cos^2\theta) \equiv \csc^2\theta - \cos^2\theta$
[2 marks available — 1 mark for using $\csc^2\theta \equiv 1 + \cot^2\theta$, 1 mark for using $\sin^2\theta + \cos^2\theta \equiv 1$ and simplifying]

b) $\frac{\cot^2\theta + \sin^2\theta}{\cos^2\theta} = 3$
$\Rightarrow \frac{\csc^2\theta - \cos^2\theta}{\cos^2\theta} = 3$ (from part a)
$\Rightarrow \frac{1}{\sin^2\theta\cos^2\theta} - 1 = 3 \Rightarrow 1 = 4\sin^2\theta\cos^2\theta$
$\Rightarrow 1 = (2\sin\theta\cos\theta)^2 \Rightarrow 1 = \sin^2 2\theta$
$\Rightarrow \sin 2\theta = 1$ and $\sin 2\theta = -1$
$\Rightarrow 2\theta = \frac{\pi}{2}$ and $2\theta = \frac{3\pi}{2} \Rightarrow \theta = \frac{\pi}{4}$ and $\theta = \frac{3\pi}{4}$
[4 marks available — 1 mark for using part a) and simplifying, 1 mark for using any correct identity (e.g. double angles), 1 mark for a correct method to solve, 1 mark for both correct answers]

Q12 a) $600 = 1000e^{-0.2t}$
$\Rightarrow e^{-0.2t} = 0.6 \Rightarrow -0.2t = \ln(0.6) \Rightarrow -0.2t = -0.5108...$
$\Rightarrow t = 2.5541...$ hours $= 153$ minutes (to the nearest minute)
[2 marks available — 1 mark for a correct method, 1 mark for the correct answer]

b) (i) $P_B = me^{-nt}$
When $t = 0$, $P_B = 1200 \Rightarrow 1200 = me^{-n(0)} \Rightarrow m = 1200$
When $t = 0.5$, $P_B = 900 \Rightarrow 900 = 1200e^{-0.5n}$
$\Rightarrow 0.75 = e^{-0.5n} \Rightarrow \ln(0.75) = -0.5n$
$\Rightarrow n = 0.5753... = 0.58$ (to 2 s.f.)
So $P_B = 1200e^{-0.58t}$
[3 marks available — 1 mark for the value of m, 1 mark for the correct method to find n, 1 mark for the correct equation]

(ii) $P_A + P_B = 1000e^{-0.2t} + 1200e^{-0.5753...t}$
Differentiating with respect to t gives a rate of:
$-200e^{-0.2t} - 690.4369...e^{-0.5753...t}$
At 14:30, $t = 1.5$ so the combined rate of decay is
$-200e^{-0.2(1.5)} - 690.4369...e^{-0.5753...(1.5)}$
$= -148.1636... - 291.2780... = -439.4417...$
$= -440$ (to 2 s.f)
So the combined population will be decreasing at a rate of 440 cells per hour.
[3 marks available — 1 mark for a correct method to differentiate, 1 mark for the correct differentiation, 1 mark for the correct answer]
You should get the same answer using the rounded value from part b)(i) too.

Q13 a) $T = A\sin\left(\frac{t}{60} - B\right) + C$
$\Rightarrow T = A\sin\left(\frac{t}{60}\right)\cos B - A\cos\left(\frac{t}{60}\right)\sin B + C$
Comparing this to the equation in the question you can see that $C = 9$ and:
①$\ A\cos B = -\frac{5}{2}$ ②$\ A\sin B = \frac{5\sqrt{3}}{2}$
①^2 + ②^2 gives: $A^2 = \left(-\frac{5}{2}\right)^2 + \left(\frac{5\sqrt{3}}{2}\right)^2 = 25 \Rightarrow A = 5$
②$\ \div$ ① gives: $\tan B = \frac{5\sqrt{3}}{2} \div -\frac{5}{2} = -\sqrt{3}$
$\Rightarrow B = -\frac{\pi}{3}, \frac{2\pi}{3}, ...$
$0 < B < \pi$, so $B = \frac{2\pi}{3}$ and $T = 5\sin\left(\frac{t}{60} - \frac{2\pi}{3}\right) + 9$.
[5 marks available — 1 mark for the correct value for C, 1 mark for a correct method to find A, 1 mark for a correct method to find B, 1 mark for finding either A or B, 1 mark for the correct answer]

b) The maximum value of $\sin\left(\frac{t}{60} - \frac{2\pi}{3}\right)$ is 1 so the maximum value of $5\sin\left(\frac{t}{60} - \frac{2\pi}{3}\right) + 9$ is $5(1) + 9 = 14$ °C.
[1 mark for the correct answer]

c) $5\sin\left(\frac{t}{60} - \frac{2\pi}{3}\right) + 9 = 12 \Rightarrow 5\sin\left(\frac{t}{60} - \frac{2\pi}{3}\right) = 3$
$\Rightarrow \sin\left(\frac{t}{60} - \frac{2\pi}{3}\right) = 0.6 \Rightarrow \frac{t}{60} - \frac{2\pi}{3} = 0.6435...$
$\Rightarrow \frac{t}{60} = 2.7378... \Rightarrow t = 164.2738...$
So the temperature goes above 12 °C on the 165th day.
[2 marks available — 1 mark for forming and rearranging the equation to remove sin, 1 mark for the correct working leading to the correct answer]
You only need the principal solution to the trig equation (i.e. the one your calculator gives) since this will give the smallest positive value for t.

Answers

Q14 $y = -x^2 + x = x(1 - x)$ so the roots of this equation are 0 and 1. The roots are the start and end points of the shaded area and therefore are the limits of the integration.

$$\int_0^1 -x^2 + x \, dx = \left[-\frac{1}{3}x^3 + \frac{1}{2}x^2 \right]_0^1$$

$$= \left(-\frac{1}{3}(1^3) + \frac{1}{2}(1^2) \right) - \left(-\frac{1}{3}(0^3) + \frac{1}{2}(0^2) \right) = -\frac{1}{3} + \frac{1}{2} = \frac{1}{6}$$

Use integration by parts to integrate $x^2 \ln x$:

Let $u = \ln x$ and $\frac{dv}{dx} = x^2$.

Then $\frac{du}{dx} = \frac{1}{x}$ and $v = \frac{x^3}{3}$

$$\int_0^1 x^2 \ln x \, dx = \left[\ln x \times \frac{x^3}{3} \right]_0^1 - \int_0^1 \frac{x^3}{3} \times \frac{1}{x} \, dx$$

$$= \left[\frac{x^3 \ln x}{3} \right]_0^1 - \int_0^1 \frac{x^2}{3} \, dx = \left[\frac{x^3 \ln x}{3} \right]_0^1 - \left[\frac{x^3}{9} \right]_0^1$$

$$= \left[\frac{1^3 \ln 1}{3} - \frac{0^3 \ln 0}{3} \right] - \left[\frac{1^3}{9} - \frac{0^3}{9} \right] = -\frac{1}{9}$$

So the total area is $\frac{1}{6} + \frac{1}{9} = \frac{5}{18}$ m².

[7 marks available — 1 mark for finding the limits of the integration, 1 mark for a correct method to integrate $-x^2 + x$, 1 mark for the correct value of that integral, 1 mark for using integration by parts on $x^2 \ln x$, 1 mark for correctly integrating $x^2 \ln x$, 1 mark for the correct value of this integral, 1 mark for the correct answer]

A

Absolute value
Another name for the modulus.

Algebraic division
Dividing one algebraic expression by another.

Algebraic fraction
A fraction made up of algebraic expressions.

Arc
The curved edge of a sector of a circle.

Arccos
The inverse of the cosine function, also written as arccosine or $\cos^{-1}$.

Arcsin
The inverse of the sine function, also written as arcsine or $\sin^{-1}$.

Arctan
The inverse of the tangent function, also written as arctangent or $\tan^{-1}$.

Arithmetic sequence/series
A sequence or series where successive terms have a common difference.

B

Binomial coefficient
The coefficients of terms of a binomial expansion $(1 + x)^r$. The coefficient of x^r is $\dfrac{n!}{r!(n-r)!}$.

Binomial expansion
A method of expanding functions of the form $(q + px)^n$. Can be used to give a finite expansion or to find an approximation of a value.

C

Cartesian equation
An equation relating the perpendicular axes x and y in 2D (or x, y and z in 3D). Cartesian coordinates are given in the form (x, y) in 2D (or (x, y, z) in 3D).

Chain rule
A method for differentiating a function of a function.

Collinear points
Three or more points are collinear if they all lie on the same straight line.

Component
The effect of a vector in a given direction.

Composite function
A combination of two or more functions acting on a value or set of values.

Concave curve
A curve with a negative second derivative — i.e. $f''(x) \leq 0$ for all x.

Constant of integration
A constant term coming from an indefinite integration representing any number.

Convergence
A sequence converges if the terms get closer and closer to a single value.

Convergent sequence/series
A sequence/series that tends towards a limit.

Convex curve
A curve with a positive second derivative — i.e. $f''(x) \geq 0$ for all x.

Cosec
The reciprocal of the sine function, sometimes written as cosecant.

Cot
The reciprocal of the tangent function, sometimes written as cotangent.

D

Definite integral
An integral that is evaluated over an interval given by two limits, representing the area under the curve between those limits.

Degree
The highest power of x in a polynomial.

Derivative
The result after differentiating a function.

Differential equation
An equation connecting variables with their rates of change.

Differentiation
A method for finding the rate of change of a function with respect to a variable — the opposite of integration.

Divergence
A sequence diverges if the terms get further and further apart.

Divergent sequence/series
A sequence/series that does not have a limit.

Divisor
The number or expression that you're dividing by in a division.

Domain
The set of values that can be input into a mapping or function. Usually given as the set of values that x can take.

Glossary

E

e
An irrational number for which the gradient of $y = e^x$ is equal to e^x.

Equating coefficients
Making the coefficients of equivalent terms on each side of an identity equal in order to calculate the value of unknowns in the identity.

Exponential function
A function of the form $y = a^x$. $y = e^x$ is known as 'the' exponential function.

F

Function
A type of mapping which maps every number in the domain to only one number in the range.

G

Geometric sequence/series
A sequence or series in which you multiply by a common ratio to get from one term to the next.

I

i unit vector
The standard horizontal unit vector (i.e. along the x-axis).

Identity
An equation that is true for all values of a variable, usually denoted by the '≡' sign.

Implicit differentiation
A method of differentiating an implicit relation.

Implicit relation
An equation in x and y written in the form $f(x, y) = g(x, y)$, instead of $y = f(x)$.

Indefinite integral
An integral which contains a constant of integration that comes from integrating without limits.

Integer
A whole number, including 0 and negative numbers. The set of integers has the notation $\mathbb{Z}$.

Integral
The result you get when you integrate something.

Integration
Process for finding the equation of a function, given its derivative — the opposite of differentiation.

Integration by parts
A method for integrating a product of two functions. The reverse process of the product rule.

Integration by substitution
A method for integrating a function of a function. The reverse process of the chain rule.

Inverse function
An inverse function, e.g. $f^{-1}(x)$, reverses the effect of the function $f(x)$.

Irrational number
A number that can't be expressed as the quotient (division) of two integers. Examples include surds and π.

Iteration
A numerical method for solving equations that allows you to find the approximate value of a root by repeatedly using an iteration formula.

Iteration sequence
The list of results x_1, x_2... etc. found with an iteration formula.

J

j unit vector
The standard vertical unit vector (i.e. along the y-axis).

K

k unit vector
The standard unit vector used in 3D to represent movement along the z-axis.

L

Limit (sequences and series)
The value that the individual terms in a sequence, or the sum of the terms in a series, tends towards.

Limits (integration)
The numbers between which you integrate to find a definite integral.

Logarithm
The logarithm to the base a of a number x (written $\log_a x$) is the power to which a must be raised to give that number.

Lower bound
The lowest value a number could take and still be rounded up to the correct answer.

M

Magnitude
The size of a vector.

Many-to-one function
A function where some values in the range correspond to more than one value in the domain.

Mapping
An operation that takes one number and transforms it into another.

Model
A mathematical approximation of a real-life situation, in which certain assumptions are made about the situation.

Modulus
The modulus of a number is its positive numerical value. The modulus of a function, $f(x)$, makes every value of $f(x)$ positive by removing any minus signs. The modulus of a vector is the same as its magnitude.

N

Natural logarithm
The inverse function of e^x, written as $\ln x$ or $\log_e x$.

Natural number
A positive integer, not including 0. The set of natural numbers has the notation $\mathbb{N}$.

^nC_r
The binomial coefficient of x^r in the binomial expansion of $(1 + x)^n$.
Also written $\begin{pmatrix} n \\ r \end{pmatrix}$.

Newton-Raphson Method
An iterative method for finding a root of an equation, using the formula:
$$x_{n+1} = x_n - \frac{f(x_n)}{f'(x_n)}$$

O

One-to-one function
A function where each value in the range corresponds to one and only one value in the domain.

P

Parameter
The variable linking a set of parametric equations (usually t or θ).

Parametric equations
A set of equations defining x and y in terms of another variable, called the parameter.

Partial fractions
A way of writing an algebraic fraction with several linear factors in its denominator as a sum of fractions with linear denominators.

Particular solution
A solution to a differential equation where known values have been used to find the constant term.

Percentage error
The difference between a value and its approximation, as a percentage of the real value.

Point of inflection
A stationary point on a graph where the gradient doesn't change sign on either side of the point.

Polynomial
An algebraic expression made up of the sum of constant terms and variables raised to positive integer powers.

Product rule
A method for differentiating a product of two functions.

Progression
Another word for sequence.

Proof
Using mathematical arguments to show that a statement is true or false.

Proof by contradiction
Assuming that a statement is false, then showing that this assumption is impossible, to prove that the statement is true.

Q

Quotient
The result when you divide one thing by another, not including the remainder.

Quotient rule
A method of differentiating one function divided by another.

R

Radian
A unit of measurement for angles. 1 radian is the angle in a sector of a circle with radius r that has an arc of length r.

Range
The set of values output by a mapping or function. Usually given as a set of values that y or $f(x)$ can take.

Rational expression
A function that can be written as a fraction where the numerator and denominator are both polynomials.

Rational number
A number that can be written as the quotient (division) of two integers, where the denominator is non-zero. The set of rational numbers has the notation $\mathbb{Q}$.

Real number
Any positive or negative number (or 0) including all rational and irrational numbers, e.g. fractions, decimals, integers and surds. The set of real numbers has the notation $\mathbb{R}$.

Remainder (algebraic division)
The expression left over following an algebraic division that has a degree lower than the divisor.

Resultant vector
The single vector which has the same effect as two or more vectors added together.

Root
A value of x at which a function is equal to 0.

S

Sec
The reciprocal of the cosine function, sometimes written as secant.

Second order derivative
The result of differentiating a function twice.

Sector
A section of a circle formed by two radii and part of the circumference.

Separating variables
A method for solving differential equations by first rewriting in the form $\frac{1}{g(y)}\, dy = f(x)\, dx$ in order to integrate.

Sequence
An ordered list of numbers (referred to as terms) that follow a set pattern. E.g. 2, 6, 10, ... or –4, 1, –4, 1, ...

Series
An ordered list of numbers, just like a sequence, but where the terms are being added together (to find a sum).

Glossary

Sigma notation
Used for the sum of series.
E.g. $\displaystyle\sum_{n=1}^{15}(2n+3)$ is the sum
of the first 15 terms of the
series with n^{th} term $2n + 3$.

Small angle approximations
Functions that approximate $\sin x$,
$\cos x$ and $\tan x$ for small values of x.

Stationary point
A point on a curve where the
gradient is 0.

Sum to infinity
The sum to infinity of a series is the
value that the sum tends towards as
more and more terms are added.
Also known as the limit of a series.

T

Trapezium rule
A way of estimating the area
under a curve by dividing it up
into trapezium-shaped strips.

Turning point
A stationary point that is a (local)
maximum or minimum point of a
curve.

U

Unit vector
A vector of magnitude one unit.

Upper bound
The upper limit of the values that
a number could take and still be
rounded down to the correct answer.

V

Vector
A quantity which has both a
magnitude and a direction.

Index

Index

These are the formulas you'll be given in the exam, but make sure you know exactly when you need them and how to use them.

Trigonometric Identities

$$\sin(A \pm B) = \sin A \cos B \pm \cos A \sin B$$

$$\sin A + \sin B = 2\sin\frac{A+B}{2}\cos\frac{A-B}{2}$$

$$\cos(A \pm B) = \cos A \cos B \mp \sin A \sin B$$

$$\sin A - \sin B = 2\cos\frac{A+B}{2}\sin\frac{A-B}{2}$$

$$\tan(A \pm B) = \frac{\tan A \pm \tan B}{1 \mp \tan A \tan B} \quad (A \pm B \neq (k + \tfrac{1}{2})\pi)$$

$$\cos A + \cos B = 2\cos\frac{A+B}{2}\cos\frac{A-B}{2}$$

$$\cos A - \cos B = -2\sin\frac{A+B}{2}\sin\frac{A-B}{2}$$

Small Angle Approximations

$$\sin\theta \approx \theta \qquad \cos\theta \approx 1 - \frac{1}{2}\theta^2 \qquad \tan\theta \approx \theta \qquad \text{where } \theta \text{ is measured in radians}$$

Exponentials and Logarithms

$$\log_a x = \frac{\log_b x}{\log_b a}$$

$$e^x \ln a = a^x$$

Mensuration

Surface area of sphere $= 4\pi r^2$

Area of curved surface of cone
$= \pi r \times$ slant height

Binomial Series

$$(a + b)^n = a^n + \binom{n}{1}a^{n-1}b + \binom{n}{2}a^{n-2}b^2 + \dots + \binom{n}{r}a^{n-r}b^r + \dots + b^n \quad (n \in \mathbb{N})$$

$$\text{where } \binom{n}{r} = {}^nC_r = \frac{n!}{r!(n-r)!}$$

$$(1 + x)^n = 1 + nx + \frac{n(n-1)}{2!}x^2 + \dots + \frac{n(n-1)\dots(n-r+1)}{r!}x^r + \dots \quad (|x| < 1, n \in \mathbb{R})$$

Arithmetic Series

$$S_n = \frac{1}{2}n(a + l) = \frac{1}{2}n[2a + (n-1)d]$$

Geometric Series

$$S_n = \frac{a(1 - r^n)}{1 - r}$$

$$S_\infty = \frac{a}{1 - r} \text{ for } |r| < 1$$

Differentiation From First Principles

$$f'(x) = \lim_{h \to 0} \frac{f(x + h) - f(x)}{h}$$

Differentiation

$f(x)$	$f'(x)$
$\tan kx$	$k \sec^2 kx$
$\sec kx$	$k \sec kx \tan kx$
$\cot kx$	$-k \csc^2 kx$
$\csc kx$	$-k \csc kx \cot kx$

Quotient rule:

for $y = \dfrac{f(x)}{g(x)}$,

$$\frac{dy}{dx} = \frac{f'(x)g(x) - f(x)g'(x)}{(g(x))^2}$$

Integration (+ a constant)

$f(x)$	$\int f(x)\,dx$
$\sec^2 kx$	$\frac{1}{k}\tan kx$
$\tan kx$	$\frac{1}{k}\ln|\sec kx|$
$\cot kx$	$\frac{1}{k}\ln|\sin kx|$

$f(x)$	$\int f(x)\,dx$
$\csc kx$	$-\frac{1}{k}\ln|\csc kx + \cot kx|$, $\frac{1}{k}\ln\left|\tan\left(\frac{1}{2}kx\right)\right|$
$\sec kx$	$\frac{1}{k}\ln|\sec kx + \tan kx|$, $\frac{1}{k}\ln\left|\tan\left(\frac{1}{2}kx + \frac{\pi}{4}\right)\right|$

Integration by parts: $\int u\dfrac{dv}{dx}\,dx = uv - \int v\dfrac{du}{dx}\,dx$

Numerical Methods

Trapezium rule: $\int_a^b y\,dx \approx \frac{1}{2}h[y_0 + 2(y_1 + y_2 + \ldots + y_{n-1}) + y_n]$, where $h = \dfrac{b - a}{n}$

The Newton-Raphson formula: $x_{n+1} = x_n - \dfrac{f(x_n)}{f'(x_n)}$

MEPMT62